POTATO BIOLOGY AND BIOTECHNOLOGY

Cover images:

The images are credited to the Scottish Crop Research Institute. We are grateful for the use of the photographs they provided.

POTATO BIOLOGY AND BIOTECHNOLOGY
ADVANCES AND PERSPECTIVES

Edited by

DICK VREUGDENHIL
Laboratory of Plant Physiology
Wageningen University and Research Centre
Wageningen,
The Netherlands

with

JOHN BRADSHAW
CHRISTIANE GEBHARDT
FRANCINE GOVERS
DONALD K.L. MACKERRON
MARK A. TAYLOR
HEATHER A. ROSS

ELSEVIER

Amsterdam – Boston – Heidelberg – London – New York – Oxford
Paris – San Diego – San Francisco – Singapore – Sydney – Tokyo

Elsevier
The Boulevard, Langford Lane, Kidlington, Oxford OX5 1GB, UK
Radarweg 29, PO Box 211, 1000 AE Amsterdam, The Netherlands

First edition 2007

British Library Cataloguing in Publication Data
A catalogue record for this book is available from the British Library

Library of Congress Cataloging-in-Publication Data
A catalog record for this book is available from the Library of Congress

ISBN-13: 978-0-444-51018-1

Transferred to digital print 2008
Printed and bound by CPI Antony Rowe, Eastbourne

Preface

The potato is the fourth most important food crop in the world after wheat, maize and rice with 311 million tonnes produced from 19 million hectares at an average fresh weight yield of 16.4 t/ha in 2003 (FAO statistics), but with a huge range from 2 to 44 t/ha by country. As well as being a staple food the potato is grown as a vegetable for table use, is processed into French fries and chips (crisps) and is used for dried products and starch production. Processing is the fastest growing sector of the world potato economy, and today, processors are building factories in countries where the potato is primarily grown as a staple food. In some countries, the potato is still fed to animals but this use is decreasing. In many countries in Asia, Africa and Central and South America, there is a need for increased and stable potato production to meet increasing demands for food from human population growth during a period of environmental (including climate) change. Potatoes with improved nutritional and health properties are desirable, but the overriding need is for increased and stable yields to eradicate human hunger and poverty. In those countries where food security has been achieved, the potato industries are trying to increase potato usage in an economically and environmentally sustainable way. The emphasis is on more yield of saleable product at less cost of production, reduced use of pesticides and fungicides, better use of water and fertilizers and meeting consumer demands for healthy convenience foods and novel products. These objectives will be met only through new cultivars, better crop management and utilization of resources, better post-harvest storage, better control of pests and diseases and a better understanding of the social, economic and market factors that influence global production and distribution.

Today there is a tremendous opportunity to harness recent advances in potato biology and biotechnology in these endeavours. We therefore considered it timely to ask a number of experts to help us review the current state-of-knowledge in all aspects of the potato crop, from basic science to production, processing and marketing. Therefore, this book includes a wide variety of chapters, describing potato markets, genetics and genetic resources, plant growth and development, response to the environment, tuber quality, pests and diseases, biotechnology and crop management. We gave authors as much freedom as possible over the content and style of their chapters, consistent with the subject matter forming a comprehensive and coherent book without unnecessary duplication. We did our best to help authors make their contributions as readable and free from errors as is possible in a human enterprise. The idea to compile such a comprehensive volume was put forward and initiated by our colleagues Howard Davies and Roberto Viola. We are indebted to them and to many other colleagues for their support, especially Philip Smith for his proof reading, and to the publishers for their advice and encouragement.

We hope that the finished product will be of value not only to potato biologists but also to all those people throughout the world interested in ensuring that the potato continues to make a major contribution to the feeding of humankind.

Dick Vreugdenhil
John Bradshaw
Christiane Gebhardt
Francine Govers
Donald K.L. MacKerron
Mark A. Taylor
Heather A. Ross

Acknowledgement

The editors are grateful for the funding of colour prints received from those listed below.

The British Potato Council
UK

SaKa-Ragis Pflanzenzucht GbR
Hamburg, Germany

Böhm-Nordkartoffel Agrarproduktion OHG
Lüneburg, Germany

Intersnack Knabber-Gebäck GmbH & Co. KG
Köln, Germany

Agrico Research BV
Emmeloord, The Netherlands

HZPC Holland BV
Metslawier, The Netherlands

C. Meijer BV
Rilland, The Netherlands

Contents

List of contributors

Eric J. Allen	Agronomy Centre, Cambridge University Farm, 219b Huntingdon Road, Cambridge CB3 0DL, United Kingdom
Frederik A.J. Börnke	Department of Biochemistry, Friedrich-Alexander-Universität, Erlangen-Nürnberg, Staudtstr. 5, 91058 Erlangen, Germany
John E. Bradshaw	Scottish Crop Research Institute, Invergowrie, Dundee DD2 5DA, United Kingdom
Glen J. Bryan	Scottish Crop Research Institute, Invergowrie, Dundee DD2 5DA, Scotland, United Kingdom
Marcel Bucher	ETH Zurich, Institute of Plant Sciences, Plant Biochemistry & Physiology Group, Experimental Station Eschikon 33, CH-8315 Lindau, Switzerland
A.J. Conner	New Zealand Institute for Crop and Food Research, Private Bag 4704, Christchurch, New Zealand
Solke H. De Boer	Canadian Food Inspection Agency, Charlottetown Laboratory, 93 Mount Edward Road, Charlottetown, C1A 5T1, Canada
Ludwig De Temmerman	Section Agro-ecochemistry, Veterinary and Agrochemical Research Centre, VAR-CODA-CERVA, Leuvensesteenweg 17, B-3080 Tervuren, Belgium
Danielle J. Donnelly	Department of Plant Science, Macdonald Campus of McGill University, 21,111 Lakeshore Rd., Ste Anne de Bellevue, QC, H9X 3V9 Canada
David M. Firman	Agronomy Centre, Cambridge University Farm, 219b Huntingdon Road, Cambridge CB3 0DL, United Kingdom

Tatjana Gavrilenko
Plant Genetics, Cytogenetics and Biotechnology, N.I. Vavilov Institute of Plant Industry (VIR), Bolshaja Morskja Street, 42/44, St.-Petersburg 190 000, Russia

Christiane Gebhardt
Plant Breeding and Genetics, Max Planck Institute for Plant Breeding Research, Carl-von-Linné-Weg 10, 50829 Köln, Germany

Francine Govers
Laboratory of Phytopathology, Wageningen University, Binnenhaven 5, 6709 PD Wageningen, The Netherlands

John P. Hammond
Warwick HRI, University of Warwick, Wellesbourne, Warwick CV35 9EF, United Kingdom

David J. Hannapel
Interdepartmental Plant Physiology Major, 253 Horticulture Hall, Iowa State University, Ames, IA 50011-1100, USA

Anton J. Haverkort
Wageningen University and Research Centre, Plant Research International, P.O. Box 16, 6700 AA Wageningen, The Netherlands

Daniel Hofius
Institute of Molecular Biology and Physiology, University of Copenhagen, Øster Farimagsgade 2A, 1353 Copenhagen K, Denmark

Mirjam M.J. Jacobs
Plant Research International (PRI), Wageningen University, 6700 AA Wageningen, The Netherlands

Shelley H. Jansky
USDA-ARS, Department of Horticulture, University of Wisconsin-Madison, 1575 Linden Drive, Madison, WI 53706, USA

Michael A. Kirkman
3, The Coach House, Main Street, Ravenstone LE67 2AS, United Kingdom

Jens Kossmann
Director, Institute for Plant Biotechnology, Botany and Zoology Department, Stellenbosch University, Private Bag X1, Matieland, South Africa 7602

Abdelaziz Lagnaoui
The World Bank, 1818 H Street NW, Washington, DC 20433, USA

Edward C. Lulai
Sugarbeet & Potato Research Unit, USDA-ARS, Northern Crop Science Laboratory, 1307 18th St. N, Fargo, North Dakota 58105-5677, USA

D.K.L. MacKerron
(formerly) Scottish Crop Research Institute, Invergowrie, Dundee DD5 1QX, United Kingdom

Bruce Marshall
Scottish Crop Research Institute, Invergowrie, Dundee, Scotland DD2 5DA, United Kingdom

Gordon J. McDougall	Quality Health and Nutrition Department, Scottish Crop Research Institute, Invergowrie, Dundee DD2 5DA, Scotland, United Kingdom
Iain McGregor	John Hannah Building, Auchincruive, Ayr KA6 5HW, United Kingdom
Steve Millam	Institute of Molecular Plant Sciences, University of Edinburgh, Daniel Rutherford Building, Kings Buildings, Edinburgh EH9 3JR, United Kingdom
Didier Mugniéry	UMR Bio3P, INRA Domaine de la Motte-au-Vicomte, BP 32327, 35653 Le Rheu, France
R.M. Patel	Department of Plant Science, Macdonald Campus of McGill University, 21,111 Lakeshore Rd., Ste Anne de Bellevue, QC, H9X 3V9 Canada
Mark S. Phillips	Scottish Crop Research Institute, Invergowrie, Dundee, Scotland DD2 5DA, United Kingdom
S.O. Prasher	Bioresource Engineering, Macdonald Campus of McGill University, 21,111 Lakeshore Road, Ste Anne de Bellevue, QC, H9X 3V9, Canada
Edward B. Radcliffe	Department of Entomology, 219 Hodson Hall, 1980 Folwell Ave., St. Paul, MN 55108-6125, USA
Heather A. Ross	Quality Health and Nutrition Department, Scottish Crop Research Institute, Invergowrie, Dundee DD2 5DA, Scotland, United Kingdom
Sanjeev Kumar Sharma	Gene Expression Programme, Scottish Crop Research Institute, Invergowrie, Dundee DD2 5DA, Scotland, United Kingdom
Ivan Simko	USDA-ARS, Crop Improvement and Protection Research Unit, 1636 East Alisal Streets, Salinas, CA 93905 USA
Joe R. Sowokinos	Department of Horticultural Science, University of Minnesota, 311 5th Avenue NE, East Grand Forks, MN 56721, USA
David M. Spooner	USDA-ARS, Department of Horticulture, University of Wisconsin-Madison, 1575 Linden Drive, Madison Wisconsin 53706-1590, USA

Sarah A. Stephenson	Biological Science Aid, USDA-ARS, Department of Horticulture, University of Wisconsin-Madison, 1575 Linden Drive, Madison Wisconsin 53706-1590, USA
Derek Stewart	Quality Health and Nutrition Department, Scottish Crop Research Institute, Invergowrie, Dundee DD2 5DA, Scotland, United Kingdom
Michael Storey	British Potato Council, 4300 Nash Court, Oxford Business Park South, Oxford OX4 2RT, United Kingdom
Paul C. Struik	Crop and Weed Ecology, Plant Sciences Group, Wageningen University, Haarweg 333, 6709 RZ Wageningen, The Netherlands
Jeffrey C. Suttle	Sugarbeet & Potato Research Unit, USDA-ARS Northern Crop Science, Laboratory, 1307 18th St. N, Fargo, North Dakota 58105-5677, USA
Mark A. Taylor	Quality Health and Nutrition Dept., Scottish Crop Research Institute, Invergowrie, Dundee DD2 5DA, Scotland, United Kingdom
Aad J. Termorshuizen	Biological Farming Systems, Wageningen University, Marijkeweg 22, 6709 PG Wageningen, The Netherlands
Jari P.T. Valkonen	Department of Applied Biology, PO Box 27 (Street address: Latokartanonkaari 7), FIN-00014, University of Helsinki, Finland
Ronald G. van den Berg	Wageningen University, 6700 AA Wageningen, The Netherlands
Jan M. van der Wolf	WUR-Plant Research International, Wageningen, The Netherlands
Herman J. van Eck	Laboratory of Plant Breeding, Wageningen University, Droevendaalsesteeg 1, 6708 PB Wageningen, The Netherlands
Cees D. van Loon	Oostrandpark 103, 8212 AT Lelystad, The Netherlands
Karine Vandermeiren	Department Agro-Ecochemistry, Veterinary and Agrochemical Research Centre, VAR-CODA-CERVA, Leuvensesteenweg 17, B-3080 Tervuren, Belgium
Marcel van Oijen	CEH-Edinburgh, Bush Estate, Pinicuik, EH26 0QB UK

Jan Vos Crop and Weed Ecology, Wageningen University and Research Centre, Haarweg 333, 6709 RZ Wageningen, The Netherlands

Dick Vreugdenhil Laboratory of Plant Physiology, Wageningen University, Arboretumlaan 4, 6703 BD Wageningen, The Netherlands

Ron E. Wheatley Environment-Plant Interactions, Scottish Crop Research Institute, Dundee DD2 5DA, United Kingdom

Philip J. White Scottish Crop Research Institute, Invergowrie, Dundee DD2 5DA, Scotland, United Kingdom

Kefeng Zhang Warwick HRI, University of Warwick, Wellesbourne, Warwick CV35 9EF, United Kingdom

Part I

THE MARKETS

Chapter 1

The Fresh Potato Market

Iain McGregor

SAC, Auchincruive, Ayr KA6 5HW, United Kingdom

The potato has arguably revolutionized western society as much as trains, planes and automobiles. It has been and still is a delicacy, a fast food and a hedge against famine. This chapter will review the market for fresh potatoes largely within the UK, a market that is generally accepted as one of the most sophisticated potato markets worldwide.

1.1 INTRODUCTION AND OVERVIEW

The importance of potatoes within the UK's agricultural industry and how the crop compares with other crop enterprises are summarized in Table 1.1. The potato sector contributes £482 million in total to food and agricultural output in the UK (Table 1.2) and represents the second most important crop economically after wheat.

Potato growing in the UK accounts for an area equivalent to approximately 5% of the cereals' area. Yet its contribution per hectare to the agriculture industry's revenue is more than seven times that generated by cereals. Potato growing produces a much higher value crop per hectare than cereal enterprises, but potato growers face much greater fluctuations in the level of prices and revenue as mentioned in the section 1.7.

According to the calendar-year accounts of the Department of the Environment and Rural Affairs (DEFRA) between the mid-1990s and 2005, the value of output from the potato sector fell by 41% albeit with significant fluctuations in output during the intervening years and the cereal sector's receipts fell by a similar amount (43%) over the same period.

1.2 PRODUCTION

Although the registered planted area of potatoes has declined from around 250 000 hectares in the early 1960s to 116 400 hectares in 2005 (Fig. 1.1), gross production over the same period has remained relatively constant at around 6 million tonnes (Fig. 1.2). The fresh supply chain accounted for around half of this (49%) at 2 958 000 tonnes in 2004–05.

In 2005, plantings were 4% down, average yields were 2% down and the associated production figure of 5.65 million tonnes represented the lowest production figure recorded in the UK since 1984. With significant annual variations in value due, among other

Potato Biology and Biotechnology: Advances and Perspectives
D. Vreugdenhil (Editor)

Table 1.1 Agricultural output in the UK at current market prices (£ million).

	2005	2004	2003	2002	2001	1994–95
Cereals	1389	1675	1602	1460	1386	2443
Industrial crops	779	736	765	814	693	905
Forage plants	97	93	104	90	103	97
Vegetables and horticulture	1784	1715	1769	1699	1705	1677
Potatoes	482	633	517	480	656	814
Fruit	369	319	310	251	239	265
Other crops	31	31	32	26	38	37
Total crop output	4931	5202	5099	4820	4820	6238
Total agricultural output	14077	14305	14067	13328	13283	17194

Source: DEFRA, 2006a.
Note: Calendar-year accounts.

Table 1.2 Output from potatoes in the UK, 1999–2005.

	2005	2004	2003	2002	2001	Average 1994–96	Change (%)
Total crop output (£ million)	4931	5202	5099	4820	4820	6238	− 5.2
Potatoes output (£ million)	482	633	517	480	656	814	−23.8
Potatoes as % of total crop output	9.8	12.2	10.1	10.0	13.6	13.0	−19.7
Potatoes as % of total agricultural output	3.4	4.4	3.7	3.6	4.9	4.7	−22.7

Source: DEFRA, 2006a.
'Change' denotes percentage change between 2004 and 2005.

things, to weather patterns, the underlying trend in average yield has been an increase from around 23 tonnes per hectare in 1960 to around 47 tonnes per hectare currently (Fig. 1.2). This information, including a breakdown between early and maincrop varieties, is available in more detail on the web site of the British Potato Council (BPC) at http://www.potato.org.uk/. The increases in yield have only been possible as a result of improved agronomy and crop management, skills which have developed in line with increasing concentration and specialization among growers in the UK.

One may question whether it was as a result of these improved yields that production has remained relatively stable despite the significant reduction in the area planted or was it the rising yields that enabled lower prices and led to the reduction in planted area or, thirdly, was it lower prices that forced the smaller producers from the market.

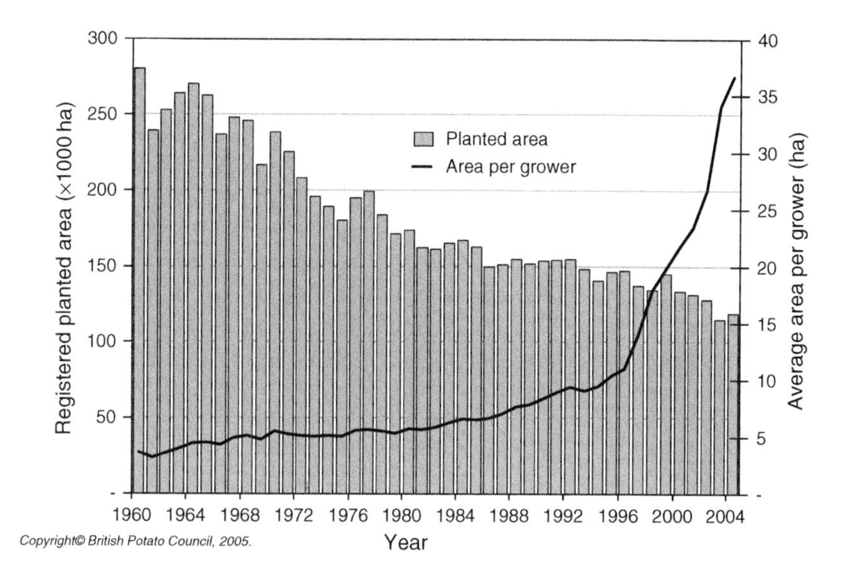

Fig. 1.1. Registered area of potatoes in Great Britain, 1960–2004 [with permission of British Potato Council (BPC, 2006), Market Information and Statistics].

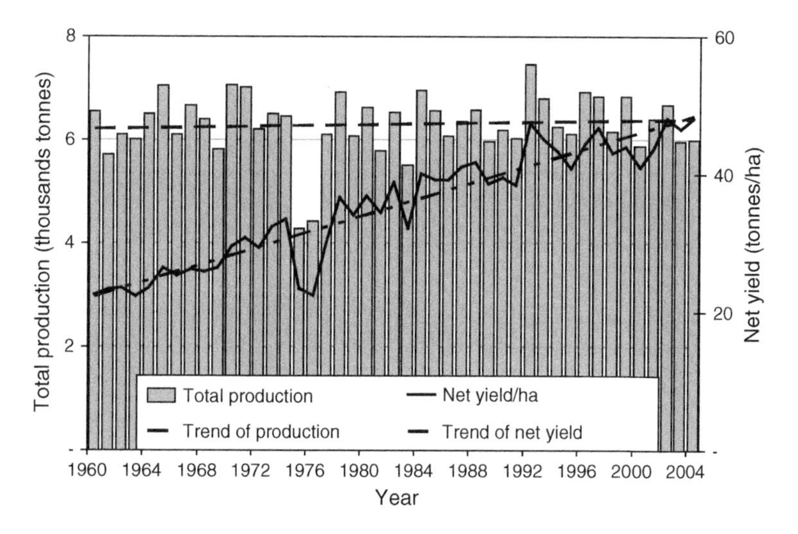

Fig. 1.2. Total production of potatoes in Great Britain, 1960–2004 [with permission of British Potato Council (BPC, 2006), Market Information and Statistics]. Since 1960, total production in Great Britain has remained at about 6 million tonnes. Annual fluctuations are mainly due to weather – see drought of 1975–76. Yield per hectare has increased steadily compensating for decreasing planted areas.

Grower numbers in Great Britain (GB) have declined dramatically over the last 45 years and continue to decline year on year as the trend in specialization and concentration continues. To illustrate the scale and significance of the industry shakeout, the author points out that in 1960 there were 76 825 registered growers in the country whereas in 2005 only 3064 (4%) remain. The average area planted per grower is now around 10 times more than it was back in 1960 (Fig. 1.1). In 2005 the average planting per grower is 38 hectares compared with 3.65 hectares in 1960. The trend in concentration at production level continues to gain momentum illustrated by the fact that in 1996, 28% of registered growers were within the 20-hectare-plus size band. Almost 10 years later in 2005, 50% of growers are within the over 20-hectare size band (Table 1.3).

Variety choice for the ware market as summarized in Table 1.4 has changed very little over the years with the top 10 varieties grown accounting for two-thirds (66%) of the total national ware crop and with Maris Piper and Estima continuing to dominate the market in 2005, the former with 20% of the ware area and the latter with 13%. Maris Peer and the crisping variety Lady Rosetta swapped places in the 2005 rankings at third and fourth places, respectively. Marfona moved up a couple of places to the number eight

Table 1.3 Distribution of British Potato Council (BPC)-registered growers by size band (with permission of BPC, 2006, Market Information and Statistics).

Size band (ha)	2005	2004	2003	2002	2001	2000	1999	1998	1997	1996
1–2	–	–	–	674	783	883	992	1149	1426	1472
3–9	842	880	1074	1287	1451	1641	1881	2061	2327	2619
10–19	678	735	779	897	990	1060	1161	1197	1324	1428
20–49	903	910	914	1028	1084	1160	1316	1262	1319	1419
50–99	412	422	409	465	463	477	524	484	484	527
100+	229	233	209	230	229	213	226	187	175	179
Total	3064	3180	3385	4581	5000	5434	6100	6340	7055	7644

From 1996 to 2002, growers planting 1 ha or more are included. From 2003, only growers with 3 ha or more are included.

Table 1.4 Top 10 varieties of ware potato, 1999–2005 (ha), from British Potato Council (BPC)-registered ware planting (with permission of BPC, 2006, Market Information and Statistics).

	Variety	2005	%	2004	2003	2002	2001	2000	1999
1.	Maris Piper	20 636	20	23 509	24 287	24 129	24 199	25 748	27 355
2.	Estima	13 371	13	12 769	11 371	13 915	15 765	14 179	14 515
3.	Lady Rosetta	5735	6	6364	5654	5220	4927	5175	4654
4.	Maris Peer	5282	5	4891	4325	3433	3307	3591	3265
5.	Pentland Dell	4386	4	4352	4547	4768	4223	4938	5831
6.	Nadine	4305	4	4573	4209	6514	7095	6338	6623
7.	Saturna	4231	4	4142	4098	4319	4244	5781	6271
8.	Marfona	3807	4	3501	3369	4801	5643	4573	4731
9.	King Edward	2937	3	3031	2841	3013	2579	2299	2502
10.	Saxon	2770	3	2256	1765	2288	2330	1868	1403
	Other	35 586	35	36 476	35 715	42 222	43 576	105 980	54 733
Total		103 046	100	105 864	102 181	114 622	117 888	120 501	131 883

slot probably on the back of continued demand from the domestic buyer and from the pubs and high-street cafes for good-quality baking potatoes.

Between 2000 and 2005 there has been a continuous change among the less frequently occurring varieties; for example, Hermes, King Edward (long-standing in the market), Osprey and Harmony all recently achieved higher proportions of the planted area, whereas Désirée, Cara and Maris Bard all declined in popularity. The second early Osprey variety is suited to the pre-pack and general ware market and may end up replacing Cara in the market. Harmony, with 1268 hectares grown for ware in 2005, is an early maincrop variety. Whether any of these changes will become progressive and 'permanent' remains to be seen.

Market research carried out by the BPC clearly demonstrated that many consumers want to know more about potato varieties. They were in agreement that more should be done to tell customers about the differing qualities of varieties – a King Edward compared with a Maris Piper for example. This is an area that will be explored in Section 1.6.

1.3 SUPPLY

Total consumption of potatoes in GB is around 6 million tonnes similar to the total GB supply. However, the figure for supply includes 18% waste and 7% seed supply so that the net GB supply is around 4.5 million tonnes. Processed products are the main import, totalling 1.3 million tonnes raw equivalent with some fresh imports for processing. Fresh ware imports and exports are nearly in balance, with a slight surplus (77 000 tonnes) on imports. The supply chains for fresh and processed potatoes are currently similar in volume. The volume of potatoes entering the fresh supply chain in GB is estimated by the BPC to have been 2.96 million tonnes in 2004–05. This represents almost half (49%) of the total GB consumption figure of 6076 tonnes. Figure 1.3 provides a schematic flowchart representing the several market components. Table 1.5 gives further details on the supply and disposal of potatoes in GB and indicates the annual variation between components. The proportions of the planted area that are intended for processing, pre-pack or other fresh variety have been fairly stable between 2000 and 2006, but there is a trend towards increasing fractions that are planted and grown under contract rather than without contract (free-buy) (Table 1.6).

1.4 DEMAND

Shoppers in UK spent £819.38 million (Table 1.7) on fresh potatoes purchased from retail outlets in the year ending 18 June 2006. This showed a small reduction (2.2%) in the previous year's total consumer outlay, although the quantity of potatoes purchased was up by 1.7% because the average price of potatoes in the fresh market was £0.52 per kilogram, down by 2 pence over the year. That change could be explained by the multiple retailers' continuing quest to drive down prices.

Over 80% of fresh potato sales are now made in supermarkets, 70% in the four biggest ones, and this gives the multiple retailers massive purchasing power and influence over

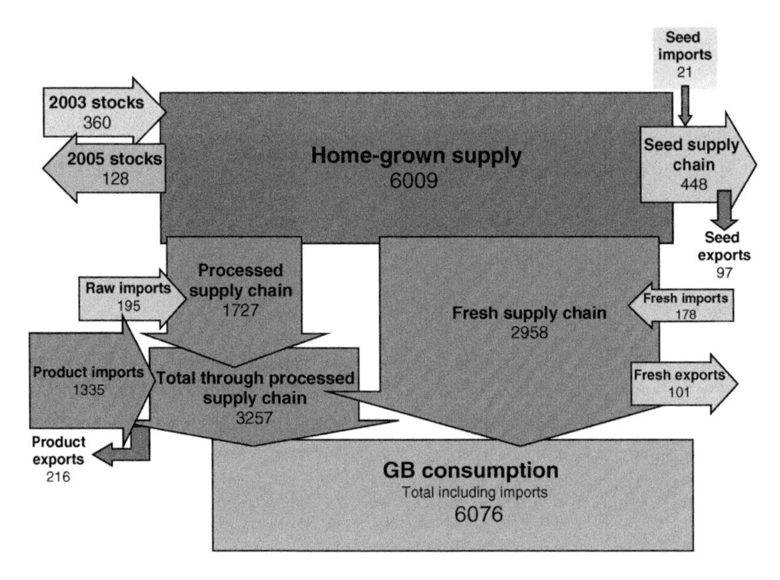

Fig. 1.3. Flowchart for potato supply – figures for the 2004–05 crop year, in thousands of metric tonnes, raw equivalent [with permission of British Potato Council (BPC, 2006), Market Information and Statistics].

Table 1.5 Annual balance sheet of Great Britain potato supplies and disposals, 1999–2004 (1000 tonnes) [with permission of British Potato Council (BPC, 2006), Market Information and Statistics].

	2004	2003	2002	2001	2000	1999	Change
Supplies							
Opening stocks	360	302	372	192	224	347	+19.2
Home grown supply	6009	5819	6694	6417	5998	6854	+3.3
Imports							
New	178	231	193	195	201	214	−22.9
Ware	195	104	164	224	467	59	+87.5
Processed[a]	1335	1368	1418	1274	1084	1121	−2.4
Seed	21	40	28	37	37	19	−47.5
Total	8098	7864	8870	8338	8011	8613	+2.9
Disposals							
Human consumption	6076	5615	6691	6255	6391	6450	+8.2
Exports							
Ware/new	101	81	223	122	91	150	+24.7
Processed[a]	216	206	165	149	123	115	+4.8
Seed	97	62	75	78	71	71	+56.5
Seed for next crop	372	348	348	393	398	409	+6.9
Closing stock	128	360	302	372	192	224	−64.4
Stockfeed and wastage	1108	1192	1067	970	746	1194	−7.0
Total	8098	7864	8870	8338	8011	8613	+2.9

'Change' is percentage change between 2003 and 2004.

[a] Raw equivalent.

Table 1.6 Planted area by market sector, 2004–05 [with permission of British Potato Council (BPC, 2006), Market Information and Statistics].

Market sector	2005			2004		
	ha	%	%	ha	%	%
Fresh						
Bags	10 301	8.8	20.2	10 172	8.5	20.3
Chipping	13 328	11.4		13 997	11.7	
Pre-pack						
Contract	21 493	18.4	41.1	17 500	14.7	40.8
Free-buy	26 474	22.7		31 143	26.1	
Processing						
Contract	28 756	24.6	26.9	29 326	24.6	27.7
Free-buy	2695	2.3		3 726	3.1	
Seed	13 717	11.7		13 431	11.3	
Total	116 764	100.0		119 295	100.0	
Contract % of ware			47.5			44.2

Source: BPC.

Table 1.7 Retail potato sales by type – year to 18 June 2006 (includes all potatoes – home produced and imports) [with permission of British Potato Council (BPC, 2006), Market Information and Statistics].

	Value (£000s)	Change in value (% year on year)	Quantity (tonne)	Change in quantity (% year on year)	Average retail price (£ per kg)
Total fresh (retail)	819 380	−2.2	1 588 459	1.7	0.52
Total frozen	435 300	−3.5	445 267	−0.9	0.98
Reconstituted and convenience	26 877	1.6	12 314	10.1	2.18
Chilled potatoes	16 528	6.3	6 060	−0.4	2.73
Canned potatoes	14 487	1.7	25 072	−4.9	0.58
Potato crisps	566 454	3.6	118 552	3.8	4.78

Source: BPC/Taylor Nelson Sofres.

the supply chain. Of course, this situation is not confined to the potato sector as in 2006, the four big grocery leaders (Tesco, Sainsbury, Asda and Morrison's) control 75% of the £120 UK billion grocery market (Table 1.8).

Altogether 1 588 459 tonnes of fresh potatoes were bought and 468 595 tonnes (29.5%) of these were sold by Tesco. The store's sales showed a 5.0% increase in quantity over the previous year and were the largest among the supermarkets by a wide margin [Source: Taylor Nelson Sofres (TNS)].

Table 1.8 Retailer share of fresh potato market – year to 18 June 2006 (includes all potatoes – produced in Great Britain and imported) [with permission of British Potato Council (BPC, 2006), Market Information and Statistics].

	Value %			Quantity %		
	Total	**Pre-pack**	**Loose**	**Total**	**Pre-pack**	**Loose**
Tesco	27.7	28.9	22.8	29.5	30.8	21.4
Sainsbury	16.5	16.9	15.0	14.0	14.0	13.8
Morrison	12.0	12.5	9.9	12.1	12.5	9.9
Asda	13.3	13.9	10.6	15.0	15.7	10.5
Somerfield	4.8	4.6	5.5	4.2	4.1	4.5
Co-op	4.4	4.6	3.6	4.4	4.7	3.0
Marks & Spencer	4.0	4.3	2.9	1.6	1.6	1.6
Waitrose	4.0	3.6	5.6	2.7	2.4	4.2
Aldi	1.8	2.3	0.1	2.3	2.7	0.1
Lidl	1.6	1.9	0.1	2.5	2.9	0.1
All others	10.9	6.5	23.9	11.7	8.6	30.9
Total	100.00	100.00	100.00	100.00	100.00	100.00

Source: BPC/Taylor Nelson Sofres.

Fresh potato sales are a large market but a relatively stable one. Some retailers are still experiencing growth in this sector, but this is largely at the expense of other retailers rather than overall growth. Increasingly, the supermarkets are turning to their suppliers for assistance in marketing potatoes, developing category management principles that have been seen for many years with fast moving consumer goods. Table 1.8 provides details of the retailer share of the fresh potato market by volume and value in UK.

As the supermarkets have grown their market share, their associated specialist supplier/packers have developed in parallel. Between the years 2000 to 2006 there has been continued concentration through mergers and acquisitions within both the multiple-retail sector and the packing industry. In UK now, four major multiples control three-quarters of the fresh market. As the retail multiples continue to rationalize their supply base, the number of specialist packers declines. Typically, supermarkets now work with one major packer/supplier supplemented by one or two 'secondary' packers. Table 1.9 summarizes this point.

During 2004 and 2005, multiples such as Tesco and Sainsbury's re-evaluated and streamlined their supplier base, dropping some groups to sign more exclusive contracts with others. For example, at the beginning of 2005, Sainsbury's cut its potato suppliers to just three: Greenvale AP, Boston-based Naturally Best Packing (Formerly Hoche International) and QV Foods. According to the supermarket chain, such streamlining means a better position for improving efficiencies.

The suppliers gaining such contracts have the advantage of additional security, although it does mean that they are reliant on a narrower customer base. The need for critical mass to effectively supply the multiples has also driven activity. The grouping together of companies to form strategic alliances has been a feature of the fresh produce sector

Table 1.9 UK Retailer/packer alliances (Potato Newsletter, March 2006).

Tesco	Sainsbury	Morrison	Asda
Branston/QV Foods	Greenvale AP	St Nicholas Court Farms	Taypack
Greenvale AP	QV Foods		Fenmarc
St Nicholas Court Farms	A Bartlett & Sons	Own in-house packing	E Park & Sons Ltd

Note: This is a fluid situation and represents a best estimate at time of writing only.

since the late 1990s. These deals allow smaller producers to gain greater bargaining power with their customers or to provide a wider or less-seasonal product range. An example of such a trading arrangement is the Perthshire-based growers' group Taygrow, who dedicate their total potato crop of approximately 80 000 tonnes from 1620 hectares to Taypack, one of three UK potato suppliers to Asda supermarkets. There is a mutual commitment between Taygrow members, Taypack and Asda. Taygrow potato producers have a guaranteed market, whereas Asda has a guaranteed long-term and reliable potato supply.

Pre-packed fresh potatoes are much preferred, outselling loose fresh potatoes by six to one in quantity (Table 1.10), but loose potatoes have a higher average selling price of 77 pence per kilogram compared with 48 pence per kilogram for pre-packed so that the ratio of values of pre-packed potatoes to loose ones is only 4:1. Pre-packed potatoes are purchased by more households, with only 7 of 10 households buying loose potatoes in the course of a year.

There are progressive changes in both value and quantities of constituent parts of the fresh market. 'Organic' potatoes continue to show the largest increase in market value at 12.6% year on year, with pre-packed baking potatoes showing 6.2% growth in value and pre-packed new potatoes 6.1%. That last figure reflects the high convenience factor of cooking straight from the pack to the pan without the need for scraping or peeling. This is borne out by the increase in the area grown for the main salad varieties Charlotte and Maris Peer by 2 and 5%, respectively.

Organic potatoes also continue to show good volume growth at 12.2%, but this is overtaken by pre-packed baking potatoes at 13.8%. Whereas the growth in quantities of potatoes grown for pre-packing has shown progressive growth – pre-packed baking at 13.8%, pre-packed new potatoes at 7.9% and over-all 4% – the apparent growth in the quantity of organic potatoes (12.2%; Table 1.10) has to be seen against the large increase in planted area between 1998 and 2002 and the subsequent fluctuations (Table 1.11).

The area of land under 'organic' horticultural production increased by 4.5% from 7400 hectares in April 2004 to 7700 hectares in January 2005, according to the Organic Market Report (Soil Association, 2005). Given the fluctuations in area planted to 'organic' potatoes and the fact that the area planted to other 'organic' root vegetables decreased by 28%, concerns have been expressed by the Soil Association over increasing levels of imported organic produce. The contribution of the organic potato crop should be kept in context, however, as the Soil Association's estimated production area of 1886

Table 1.10 Fresh potato sales by pack type – year to 18 June 2006 (includes all potatoes – produced in Great Britain and imported) [with permission of British Potato Council (BPC, 2006), Market Information and Statistics].

	Value (£000s)	Change in value (% year on year)	Quantity (tonne)	Change in quantity (% year on year)	Average retail price (£ per kg)
Pre-packed					
New	215 244	6.1	302 283	7.9	0.71
Maincrop	343 481	−4.8	917 389	1.3	0.37
Baking	95 216	6.2	155 073	13.8	0.61
Total pre-packed	653 941	0.1	1 374 745	4.0	0.48
Loose					
New	69 206	−12.6	97 737	−12.1	0.71
Maincrop	23 700	−2.2	40 001	−5.4	0.59
Baking	72 533	−10.1	75 976	−12.0	0.95
Total loose	165 439	−10.2	213 714	−10.9	0.77
Organic	23 590	12.6	28 959	12.2	0.81
Conventional	795 789	−2.5	1 559 501	1.5	0.51
TOTAL FRESH (retail)	819 380	−2.2	1 588 459	1.7	0.52

Source: BPC/Taylor Nelson Sofres.

Table 1.11 Organic potato production in the UK (hectares) 2002–05.

	2005	2004	2003	2002	1998
Potatoes	1886	1689	1860	3000	911
Change from previous year (%)	11.6	−9.2	−38	57.3 per annum	

Source: Soil Association, 1999.

hectares represents only 1.6% of the BPC's total registered area. There is a further caveat that much of the organic production will be in units smaller than the BPC's 3-hectare registration limit and, therefore, will not be registered. Nevertheless, the organic area provides potential for further development (Fig. 1.4).

1.5 EXPENDITURE AND CONSUMPTION

There are shifting patterns in potato consumption in the home. Since 2003–04, potato consumption has fallen by 5%. This could be attributed partially to the popularity of low-carbohydrate diets such as 'Atkins'.

Fig. 1.4. Organic potatoes from Marks & Spencer's at £1.99 per kg.

'Low-carb' diets such as Atkins have seen a shift in the type of produce eaten; it is a strong proponent of the consumption of plenty of green vegetables but shuns the use of peas and citrus fruit. Unfortunately, one strongly promoted condition is the consumption of potatoes: the BPC reported that there was some impact on the sale of potatoes, which, although low in fat and calories, are high in carbohydrates. However, low-carb dieting appeared to be declining in popularity at the beginning of 2005, with glycaemic index (GI) dieting taking its place.

Potatoes constituted one-third of all carbohydrates consumed in 2005. Practicality is thought to be the main consumption driver, accounting for 51.1%, as potatoes naturally complement many types of food and boiling or steaming is used on 39% of the occasions when potatoes are prepared at home (Fig. 1.5).

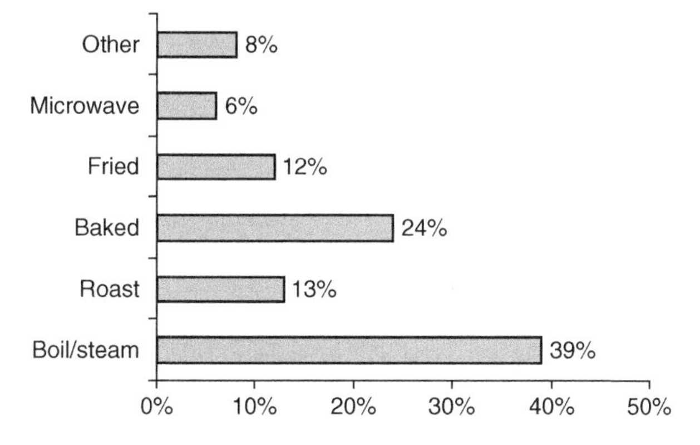

Fig. 1.5. How potatoes are prepared in the home (The Grocer, 2005). *Source*: Taylor Nelson Sofres.

Health is less of a consideration in determining the choice of carbohydrate, with only 12.5% of all potato-eating occasions being cited as 'healthy'. This is much lower than for all other carbohydrates.

According to market researchers TNS, the key age group for potato consumption is the over-45s of both sexes – over-45s account for 43.3% of all potato consumption – although their use of potatoes is also down by 2.7% year on year, in line with the overall trend. Although they are the least significant consumer age group, 17- to 25-year-olds are much more likely to eat potatoes than other carbohydrates. They are most attuned to dieting but choose to eat potatoes on the few occasions when they actually do consume carbohydrates.

Consumption peaks on Sundays when 17.2% of potatoes are eaten. Consumption of the balance is fairly evenly spread through the rest of the week at 12–15% per day. The evening meal is the key occasion for consumption with almost two-thirds of potatoes eaten at that time.

1.6 THE CONSUMERS' VIEWS

The good news is that almost all households in the UK (97%) purchase fresh potatoes at least every 2 weeks. The product is purchased by all socioeconomic groups in relatively large amounts. The frequency with which fresh potatoes are purchased is testimony to their enduring place in the national diet. Research undertaken on behalf of the BPC by The Oxford Partnership into fresh-potato-purchasing habits shows that although consumption and marketing of potatoes are both doing well, there remains room for improvement. Consumers made this clear through discussion groups and interviews with the findings falling into seven key areas – packaging, convenience, usage, marketing, product innovation, health and product quality.

Table 1.12 summarizes the opportunities (good news) and threats (bad news) for the fresh potato sector as it stands at time of writing.

Some of the findings from the research included the fact that 60% of shoppers would look out for 'British' if the packs were labelled more clearly and 42% said packaging could be more attractive. When shown four labelling format options differing only in which components had the largest lettering, 42% opted for the label highlighting the variety compared with 40% for highlighting use (e.g. chipping, boiling, mashing and baking).

Over 80% suggested that more recipes should be available to ensure that potatoes can be cooked in different ways. Conversely, 61% wanted potatoes that are quick and easy to cook, and 46% thought there should be more potatoes that can be transferred straight from the pack to the pan for cooking.

Considering in-store presentation, it tended to be the chilled displays that were rated more highly than fresh or frozen products as they provided opportunities for finding 'something special' and for 'tempting experimentation'. Furthermore, 56% of shoppers like to see fresh, chilled and frozen potatoes displayed together. That might be convenient for the shopper but would not be particularly practical for store layout and logistics.

Table 1.12 The good and bad news for fresh potatoes.

Good news for fresh potatoes	Bad news for fresh potatoes
People whose children have left home and retirees are fans	Per capita consumption drifting lower
Packed with minerals and vitamins	Pasta and rice have better image and are an increasing share of carbohydrates eaten
Chilled value-added, 'fresh' segment is in the early growth stage	Not perceived as convenient
Marketing sophistication behind the product – BPC, packers and retailers	Families with older children are more likely to choose frozen products
	Confused perception on health
	Marketing sophistication – FMCG competitors light years ahead
	Retailers are questioning the shelf space allocated to fresh potatoes given their low retail profitability.

Source: Dr D Hughes, 2005, Imperial College, London.
FMCG, fast moving consumer goods.

Two further points to emerge from the BPC's market research work are that 88% of people agreed that 'much more could be done to tell people how healthy fresh potatoes are' and 41% admitted they would buy more fresh potatoes if the quality was less variable. Clear quality differences were apparent between different retailers and regions of the country. This remains a challenge for all members within the supply chain for fresh potatoes.

1.6.1 When potatoes are consumed

Much work has been carried out by the BPC on potato-purchasing habits. One outcome from this research has been the development of the meal-occasion model for potato purchases. The model identifies seven meal-occasions based on whether the meal is 'everyday' or 'special' in nature and the extent to which it is 'pre-planned'. The model is shown graphically in Fig. 1.6 with 'entertaining friends' understandably characterized as special and requiring pre-planning, whereas a quick and simple kids' meal is the opposite.

The seven meal-occasions can indicate which potato type and variety are suitable for each meal situation, what packaging format is required, how the product needs to be prepared and how the consumer will handle it, which other carbohydrates are perceived as competitors, and where the current potato offer is strong and where the offer is weak.

The type or variety of potato purchased varies according to the meal occasion, and this also influences packaging or presentation expected. Price is not a critical factor in any segment and least of all for a special meal and entertaining friends. Practical examples of how this approach translates into purchasing habits are that jacket potatoes are a popular choice for those looking for a healthy meal and roast potatoes are a firm favourite for

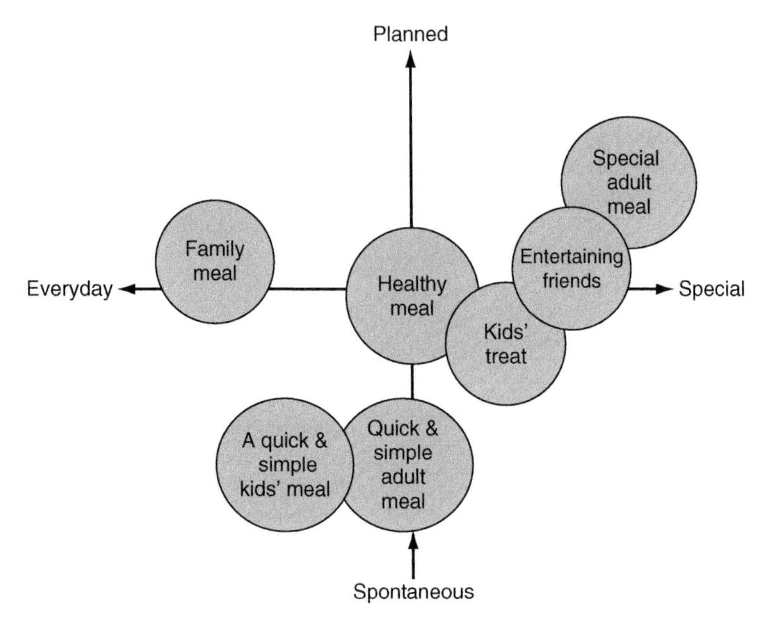

Fig. 1.6. The British Potato Council (BPC) meal-occasion model (with permission of BPC, 2002, Market Information and Statistics).

family meals, whereas chips (French fries) are rated highly for quick and simple kids' meals. Planned special meals offer an opportunity for adding value, but potatoes are not always an inspiring choice, and rice and pasta provide strong competition. This is a challenge for the industry.

It is the four planned meal occasions – family meal, healthy meal, entertaining friends and simple adult meal – that offer development opportunities for the fresh market (Table 1.13). Table 1.13 provides details of market value and growth for the seven

Table 1.13 Using the British Potato Council (BPC) meal-occasion model (with permission of BPC, Market Information and Statistics).

Meal occasion	Market value share %	Growth or decline?	Importance to fresh potatoes
Family meal	30	Decline	High
Adult simple meal	21	Growth	High
Adult special meal	7	Decline	Medium
Entertaining	3	Decline	High
Kid's simple meal	6	Decline	Medium
Kid's treat meal	5	Decline	Low
Healthy meal	13	Growth	High
Snack	15	Growth	Low

meal-occasions together with snacking occasions. The table also offers an indication of the importance of these occasions for the fresh potato market.

As consumption figures for fresh potatoes illustrate, the underlying trend is that of decline. According to DEFRA's Family Food report, the UK currently consumes less that half (43%) of the potatoes that it did back in 1974. The worry, therefore, is that the market will decline even further if nothing is done. There is an opportunity to address this slow decline in consumption by increasing consumer expenditure on added-value products. This may however create volume growth in the chilled sector rather than the fresh market.

Opportunities include addressing the growing demand for products at the premium end of the market. This may include special packaging, exotic varieties and varieties with health benefits, etc. During 2005/06 there have been a number of limited examples of attempts to do exactly this. So, Tesco and their packer supplier Branston introduced the Dutch salad variety Exquisa, a waxy early maincrop variety with yellow skin and flesh, to Tesco's 'Finest' range. There has also been renewed interest in heritage varieties largely driven by Waitrose and their sole packer Solanum. Varieties such as Shetland Black, Ratte, Pink Fir Apple and Kerr's Pink have been offered to consumers who are looking for something exotic or different. In 2006, Greenvale AP has offered the Scottish Crop Research Institute (SCRI)-bred variety Mayan Gold to the market, selected from the species *Solanum phureja*. This is a yellow-fleshed gourmet potato with a unique flavour and requiring 75% less cooking time.

Adding value to the fresh potato sector is always a challenge, but the Waitrose packer Solanum managed just that. In May 2005, Waitrose customers were able to purchase packs of fresh potatoes containing recipes within the bag. Having invested in equipment to insert objects into potato packs before they are sealed, Solanum now has the facility to include inserts such as recipe cards, nutritional information, scratch cards, money-off vouchers or sachets of sauce or herbs. The investment followed market research into consumers' priorities, which suggested that shoppers were looking for 'potato solutions'. This includes help with choice of variety for specific cooking purposes and advice on recipe ideas.

Solanum and Waitrose were to combine once again to offer their customers full traceability for their fresh potatoes from November 2005. The scheme, which was developed in conjunction with Syngenta, is based on customers entering a code number from the pack label onto the Solanum website. Details of the farm in which the potatoes were grown are then provided via the web. That seems to be a neat idea and a useful marketing tool, but it would be interesting to know how many Waitrose customers actually use it.

Another example of brand development is the promotion of the variety Rooster by the packer Albert Bartlett. The variety gained media coverage through awards for food and drink excellence; Rooster potatoes won the 'Retail Fruit and Vegetable' and 'Healthy Eating Marketing Promotion' categories at the Scottish Food and Drink Excellence Awards. More headlines were generated when, in summer 2005, the G8 world leaders were to eat Rooster potatoes while dining at Gleneagles during their stay at the G8 Summit.

Other options for adding value include offering local provenance and building on the associated trust. Research consistently shows consumer preferences for locally produced food. This is difficult logistically with supply chains based on centralized packhouses

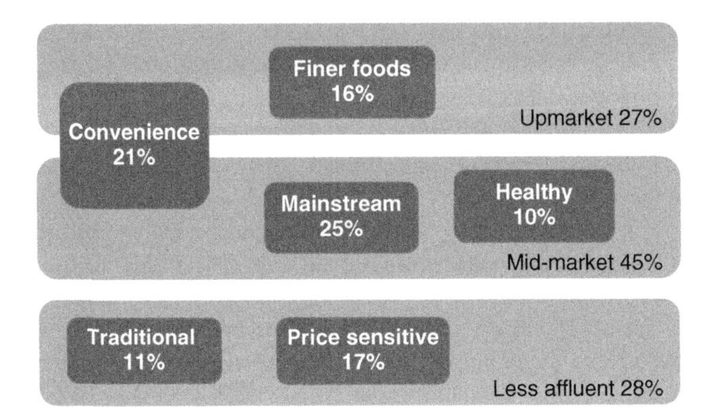

Fig. 1.7. How Tesco segments its customers. *Source*: Tesco. (84% of shopper buy 'Value' lines, 64% buy both 'Value' and 'Finest' lines.)

serving the supermarkets' centralized regional distribution centres. For example, Asda supplier Taypack, based at Inchture between Perth and Dundee, serves all Asda stores in Scotland and all stores in the north of England as far south as North Wales in the west and the River Humber in the east.

Offering consumers more guarantees is another way of adding value and building trust. This is particularly true when it comes to health and environmental guarantees. However, there is a danger of confusing consumers who may be suffering already from information overload concerning farm assurance schemes together with the range of food marques and logos of offer.

There is a clear opportunity for greater market segmentation within the fresh market. While the meal-occasion model is based on meal solutions, fresh potatoes can also be targeted at market segments based on price sensitivity together with consumers' needs and priorities. The method by which Tesco segments its customers is shown graphically in Fig. 1.7, which illustrates these segmentation variables. Customers are categorized according to their affluence and then into six groupings depending on their shopping habits and lifestyle priorities. This allows Tesco to develop a range of food products, we hope including fresh potatoes, to meet the needs of its customer groups.

1.7 PRICES PAID TO PRODUCERS

The relations between area planted, average yield and potato prices are inextricably linked within an unprotected environment where the laws of supply and demand operate with almost textbook accuracy. Prior to 1994, price comparisons were influenced by market regulation through the former Potato Marketing Board. Previously, growers received direct and indirect support under the Potato Marketing Scheme (PMS) in the form of area quota, intervention buying and deficiency payments. There is no support mechanism for potatoes under the Common Agricultural Policy (CAP), thus the level of support for the UK potato sector is determined by the government. The most important element in the

PMS was the area quota. This was 'justified' partly on the grounds that the underlying trend in yields was upwards and consumption was more or less static. It was also 'justified' on the grounds that potato yields tend to fluctuate widely from year to year.

In the 1980s, government policy swung in favour of deregulated markets, and the level of support provided to the potato sector under the PMS began to decline. In 1985 the deficiency payments scheme was abolished, and in 1993 the government announced that the PMS – essentially area quotas – would end in 1997. Given the volatility of yields, many growers feared that the ending of quotas would result in massive instability not only for their businesses but also throughout the supply chain. Fortunately, these fears have not been borne out although years with high production are associated with lower prices and the reverse also applies. Prices are noticeably weaker in years where production exceeds 6 million tonnes. While there is a longer-term trend for declining prices in real terms, there may also be an underlying tendency for price fluctuations to be decreasing. Both these tendencies may be a result of the increasing area of potatoes grown under contract (48% of the ware area) and the increasing tonnage grown on a 'cost plus' basis – a payment system where growers know they will get a guaranteed price above the costs of production. Although there are now fewer speculative growers left in the industry and forward contracts have created more stability in the market, yet there is a positive relation between the average potato price and the area planted in the following year (Fig. 1.8).

Potato prices, as with all other farm gate prices, continue to be a sensitive and contentious issue among farmers. The continuing accusation is that supermarkets are 'putting a lid on prices' and preventing farmers from benefiting from improving free-market prices in times when the market is strong. These are claims that are frequently levelled at the supermarkets by the farming industry via their representative body, the National Farmers Union (NFU). This unfortunately is the corollary of the contract cost-plus system. Farmers are grateful for the cost-plus price and guaranteed market when trade is poor but get agitated when the market is strong, and they are stuck with the lower pre-arranged contract price. What rubs salt into the wound is the fact that supermarkets – particularly UK's largest, Tesco – continue to lodge record profits at a time when farming is being squeezed. The NFU is hoping that all these issues will be thoroughly investigated in the

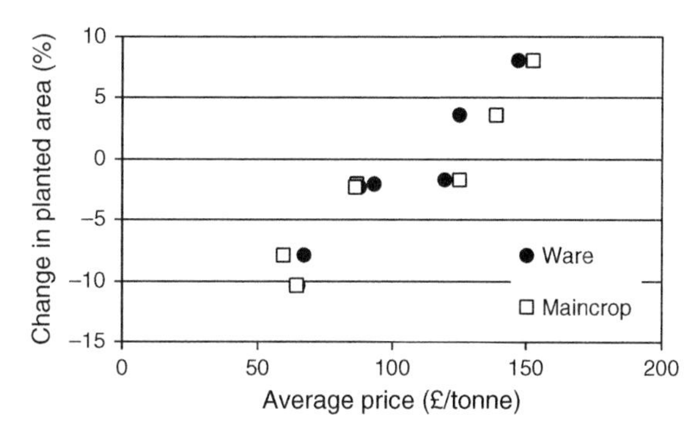

Fig. 1.8. Relation between average ware and maincrop prices and the change in planted area in the following year (1998–2005) [derived from data from British Potato Council (BPC, 2006)].

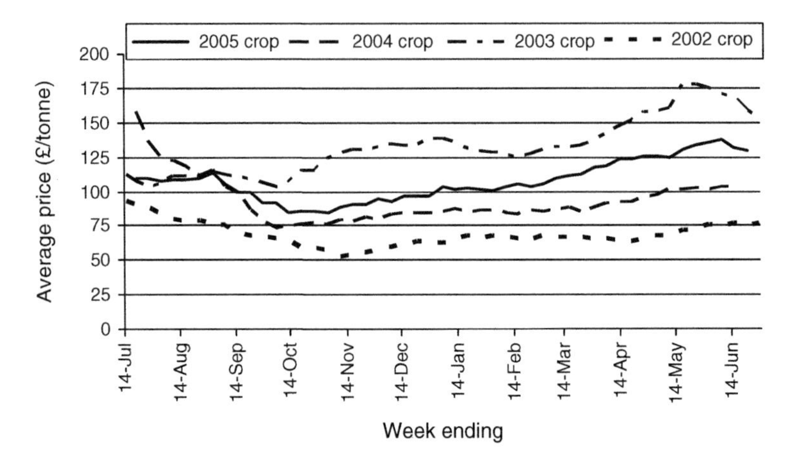

Fig. 1.9. Weekly average price of ware potatoes in successive years 2002–2005 (£ per tonne) (Potato Market weekly, 2006). *Source*: British Potato Council (BPC), Market Information and Statistics, with permission.

Competition Commission enquiry into the way that supermarkets deal with their suppliers that starts during late summer 2006. A concentrated but highly competitive retail sector has created new pressures for food and drink manufacturers, and the fresh potato market is no different.

Years with higher yields and lower prices tend to be balanced by years of lower yields and higher prices. The degree of price volatility was more commonplace in the past when less of the national crop was grown under forward contract and there was less use of cost-plus pricing. The other noticeable market response is for speculative growers to plant a larger area following a year of relatively high prices.

A case in point is 1999–2000 when the average price collapsed to £67 per tonne following the 1998 crop year, which gave an average price of £147 per tonne. The main cause of the fall in price, despite little change in yield from the previous year, was the 8% increase in the area planted.

In most years, there is a strong seasonal cycle of changing prices for ware potatoes, but the amplitude of the cycle and its baseline differ from year to year (Fig. 1.9).

1.8 POTATOES AND THE HEALTH ISSUE

Evolving consumer trends pose both opportunities and challenges for food and drink market suppliers. Along with convenience and pleasure, health is one of the key drivers for consumers when it comes to food shopping (Fig. 1.10). Fortunately, fresh potatoes have much to offer health conscious consumers. As is demonstrated in Table 1.14, the potato is nutritious, relatively low in calories, virtually free of fat and cholesterol, and high in vitamin C and potassium. It is also high in fibre especially when the potatoes are served with their skins.

The BPC run a number of promotional campaigns to heighten demand for British potatoes and to increase the competitiveness of the UK potato industry. Current BPC

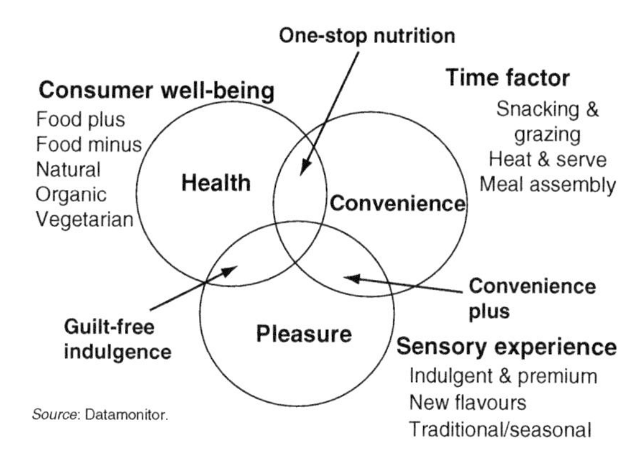

Fig. 1.10. Key consumer influences and trends.

Table 1.14 Nutrition from the potato per 175-g serving – based on boiled new potatoes in skins [with permission of British Potato Council (BPC, 2004b), Market Information and Statistics].

	Per 175 g	% Daily value
Vitamin C	15.75 mg	44
Vitamin B6	0.58 mg	29
Potassium	753 mg	22
Iron	2.8 mg	20
Fibre	2.6 mg	14
Magnesium	31.5 mg	11
Vitamin B1	0.23 mg	16
Folate	33 μg	16
Calories	115.5 kcal	6
Fat	0.5 g	0.7

Percentage of daily allowances based on 2000 kilocalorie controlled diet.

promotional work is built around three 'inter-linked platforms': National Chip Week (February), Health (May–August) and Convenience (September–November). The BPC's 2006 health campaign, 'Fab not Fad', was aimed at the health-conscious market especially – women – promoting potatoes as low fat, low salt and rich in vitamin C.

The 'Help Me To Be Healthy' campaign of summer 2006 followed research that showed that consumers are confused about the health benefits of potatoes – 66% of people surveyed rate pasta and rice as healthier than potatoes and one in seven children think potatoes make you fat.

There is always jockeying for position among our daily carbohydrate requirements between potatoes, pasta and rice. A balanced diet should include 50% of total energy from carbohydrate. This is the equivalent of a minimum of 100 g of carbohydrate per day.

Table 1.15 Nutritional values of potatoes, rice and pasta – based on new potatoes boiled in skins, boiled white rice, and cooked penne pasta [with permission of British Potato Council (BPC, 2004b), Market Information and Statistics].

	Potatoes (175 g)	**Pasta (230 g)**	**Rice (180 g)**
Price per portion	17 p	20 p	21 p
Energy value	126 kcal	198 kcal	248 kcal
Carbohydrate	27 g	43 g	56 g
Fat	0.17 g	1.15 g	2.99 g
Protein	3.15 g	6.9 g	4.68 g
Fibre	2.1 g	2.07 g	0.18 g
Vitamin C	0.5 mg	None	None
Vitamin B6	0.58 mg	0.023 mg	0.13 mg
Folate	45.5 μg	6.9 μg	7.2 μg

Carbohydrates are needed to prevent fatigue and dangerous fluid imbalance. To make sure we get enough, we should eat 6–11 servings each day from the carbohydrate group that includes bread, cereal, rice, pasta and potatoes. One of the health and nutrition messages promoted by the BPC is that potatoes contain less fat and fewer calories than other carbohydrates (Table 1.15).

Between 2003 to 2006 there were a number of recent attempts to endorse and promote potato varieties with health benefits. One example is the variety Adora, which was claimed to have 75% of the calories carried by other potatoes and 66% of the carbohydrates. The first early HZPC variety was re-launched in Florida in early 2005 as the 'Atkins-friendly' potato. In the UK, Naturally Best also played the health card with its variety Vivaldi claiming that it has 26% less carbohydrate and 33% fewer calories than the average variety. These claims represent both imaginative and opportunistic marketing and invite the cynical response that they simply contain more water. Promoting the healthy eating attributes of potatoes is not a new marketing strategy, however, as selenium-rich potatoes have been promoted in Italy for some time.

1.8.1 Glycaemic indices

As a result of fashionable interest in low-carbohydrate diets, there is increasing interest and debate surrounding the GI of foods. This is a measure of how quickly and by how much a food raises blood sugar levels. The GI, introduced by Jenkins et al. (1981), is a tool to rank foods (on an index of 1–100) by whether they raise blood sugar levels dramatically (a high index rating of 70+), moderately (56–69) or a little (55 or less). Potatoes generally have one of the highest GI values of any food. However, GI ratings for foods vary dramatically depending on whether they are eaten alone or in combination with other foods. For example, (Table 1.16) a potato eaten on its own may have a GI rating of 88, but when eaten with cheddar cheese, e.g. in a baked potato, the GI rating of the combination is low at 39.

Table 1.16 Glycaemic index (GI) of potatoes [with permission of British Potato Council (BPC, 2004a), Market Information and Statistics].

	GI value[a]
Potato alone	
Yam	37
Sweet potato	44
New potatoes	57
Potato crisps	57
Canned potatoes	65
White-skinned mashed potatoes	70
French fries	75
Baked potatoes	85
Instant mashed potatoes	86
Red-skinned boiled potatoes	88
With accompaniment	
Potato with cheddar cheese	39
Potato with chilli con carne	83
Potato with baked beans	62
Potato with tuna	80

Source: BPC, University of Dundee.
[a] Low GI < 55, medium GI 56–69 and high GI > 70.

Furthermore, research work carried out by Oxford Brookes University in 2004 on behalf of the BPC found that different potato varieties exhibit a wide range of GI responses from 56 to 94. Potatoes with waxy textures produced medium GI values whereas floury potatoes had high GI values (Table 1.17).

So what is the worth of the GI? Determining the GI of a meal is not as simple as reading a number off a chart. For example, the addition of fat and protein slows down

Table 1.17 Glycaemic index (GI) values and GI classification by variety of potato [with permission of British Potato Council (BPC, 2004a), Market Information and Statistics].

Variety	GI value	Classification
Maris Peer	94	High
Maris Piper	85	High
Desiree	77	High
Estima	66	Medium
Charlotte	66	Medium
Marfona	56	Medium
King Edward	75	High

Source: BPC, Oxford Brookes University.

the absorption of carbohydrate. Therefore, chocolate has a medium GI because of its fat content, and crisps and chips have a lower GI than potatoes cooked without fat.

The GI of food only tells you how quickly or slowly it raises blood glucose when the food is eaten on its own. In practice, we usually eat foods as meals in combination with other foods, so cutting out all high GI foods is not the answer. A further complication with potatoes is that while they score high on the GI scale, they also score high on the satiety index (SI). The higher the SI of a food, the more satisfied a person is between meals and therefore is less likely to snack.

So what are the implications of all this for the potato industry? If people were to confine themselves to low-GI foods, their diet would be unbalanced resulting in an inappropriate amount of energy coming from fat, which could lead to weight gain and increase their risk of heart disease. For this reason, it is important not to focus exclusively on GI and instead to think more in terms of the overall balance of the diet. Describing potatoes simply in terms of their high GI value does not do justice to their nutritional benefits and the fact that they are an important part of a healthy and balanced diet. If a low-GI potato variety can be identified, however, it could be used to lower the overall glycaemic load of the western diet and thus decrease the risk of type 2 diabetes, coronary heart disease and obesity. Confusion among health-conscious consumers about GI and potato consumption is one for organizations like the BPC to address as part of their continuing marketing and educational remit.

Finally, the UK government's 'Five-a-day' campaign urging consumers to eat at least five portions of fruit and vegetables a day is continuing, backed by bodies such as the Food Standards Agency and various supplier groups in the produce sector. Consumers are being encouraged to eat fruit and vegetables at a wider range of eating occasions in order to meet the 'Five-a-day' requirement. This can only be good news for the sector.

1.9 SUMMARY, CONCLUSIONS AND FUTURE PROSPECTS

The potato industry in general can offer a number of lessons to UK agriculture. The way in which the potato sector operates as an unsupported agricultural sector in a totally free market is evidence that farming enterprises can survive without support while at the same time coming to terms with wide fluctuations in producer prices and annual incomes. The potato sector operates with a short three-link supply chain (grower, packer and retailer), and there is a positive willingness among growers to form groups and vertical relationships. For example, there are currently 10 grower groups operating in Scotland alone. These transactional arrangements help to reduce market uncertainty and maintain competitiveness. A continuing problem is the industry's understandable but irrational response to plant a larger area following a year of higher prices. The replacement of 'hard' contracts between growers, packers and retailers with vertical partnerships would help reduce these irrational responses to price fluctuations and help growers and their customers to meet the ever more exacting consumer demands.

The BPC believes that building sales of fresh potatoes should be based on promotional activity that focuses on health, convenience and better value. Research findings from

Somerfield supermarket shoppers concluded that three broad types of promotional activity should be developed:

(1) Promote the health benefits of potatoes.
(2) Create convenient meal solutions with fresh potatoes.
(3) Create better value – for both the retailer and its value-conscious customers.

The research concluded that promotional activity should be focused on those of the multiple's shoppers now buying a lot less potatoes than previously and those of its regular shoppers who buy their potatoes elsewhere. It was reckoned that the biggest impact and best pay-back would be derived from those sections of their customers.

1.9.1 Key points

- The fresh market will continue to decline in value terms relative to the processed market.
- There will be more market segmentation within the fresh market – look at the other staple food product, bread, and how that sector now offers greater specialization and segmentation to its customers.
- The chilled market will grow in line with increasing demand for added-value, convenience food products.
- The processed market will experience increasing pressure from imports.
- Higher transport costs will continue to provide a degree of protection to the fresh potato sector. But higher fuel costs for field operations and storage will reduce growers' margins.
- There will be continued rationalization in production.
- There will be continued concentration throughout the supply chain – this includes growers, packers, processors and retailers.
- Continued yield increases will lead to less land under potatoes.
- Soil sickness and, increasingly, water availability will become more important in determining production.
- Vertical supply chain partnerships will ensure the future viability of the sector.

Getting away from a commodity product and commodity mentality is key to the future of the industry. In too many cases, fresh potatoes are still seen within the trade and by consumers as simply reds or whites with variety, provenance, cookability, taste and texture given little attention. The one thing we do know about consumers is that while they are interested in differentiated products up to a point, they consistently demand both quality and choice. Improvements in both areas are yet another way of pushing up fresh consumption. There is also a strong argument for better labelling and more attractive packaging, something which the processing sector recognized some time ago.

The good educational work being carried out by the BPC particularly at school level must be carried on so that consumers of tomorrow appreciate the value of this valuable staple food product and make use of its qualities to full potential.

REFERENCES

Agribusiness News, January 2006, Potatoes – review and outlook, SAC, U.K. http://www.sac.ac.uk

British Potato Council, 2002, Potatoes – what does the future hold? http://www.potato.org.uk

British Potato Council, 2004a, Glycaemic index (GI) values for potatoes, Ref: R256, http://www.potato.org

British Potato Council, 2004b, Potatoes – a healthy market. http://www.potato.org.uk

British Potato Council, 2006, *Yearbook of Potato Statistics in Great Britain May 2006 edition.* http://www.potato.org

Department for Environment and Rural Affairs, 2006a, *Agriculture in the United Kingdom 2005*, London: TSO, U.K. http://www.defra.gov.uk

Department for Environment and Rural Affairs, 2006b, *Family Food 2004–05*, London: TSO, U.K. http://www.defra.gov.uk

Hughes D., 2005, What do tomorrow's consumers want? Conference Paper British Potato 2005 Harrogate, 30 November 2005.

Jenkins D.J.A., T.M.S. Wolever, R.H. Taylor, H. Barker, H. Fielden, J.M. Baldwin, A.C. Bowling, H.C. Newman, A.L. Jenkins and D.V. Goff, 1981, *Am. J. Clin. Nutr.* 34, 362.

Mintel, 2005, *Fruit and Vegetables – U.K.* May 2005, U.K. http://www.mintel.com

Potato Markets Weekly, 2006, No 1519 July 4 2006, Great Britain, Agra-Informa Ltd, U.K. http://www.agra-net.com

Potato Newsletter, March 2006, Is there a place for potatoes in an overweight population?, SAC, U.K. http://www.sac.ac.uk

Potato Review, November/December 2005, *Fresh Sales Experience – A Renaissance*, Aremi Publishing U.K. http://www.potatoreview.com

Rickard S., 2000, *Challenges and Prospects – Potato Sector Leads the Way*, Lloyds TSB Business, U.K.

Soil Association, 1999, The Organic Food and Farming Report 1999. http://www.soilassociation.org

Soil Association, 2005, Organic Market Report 2005. http://www.soilassociation.org

The Grocer, October 15 2005, 67, Consumer insight. http://www.thegrocer.co.uk

Daphne MacCarthy, *The 2006 Potato Yearbook & Buyer's Guide 10th Edition*, ATC Publishing, U.K.

Chapter 2

Global Markets for Processed Potato Products

Michael A. Kirkman

3 The Coach House, Main Street, Ravenstone, Leicestershire LE67 2AS, United Kingdom

2.1 INTRODUCTION

This chapter is an essay on the global potato-processing industry from the personal perspective of a former Agro R&D manager in a major snack food company. During my employment of 25 years in the industry, significant expansion has occurred not only in the global market for chips (crisps in the UK) but also for French fries (chips in the UK) and other processed potato products.

In the pages that follow, I have tried to marry my understanding of potato science with a new understanding about markets, based on research undertaken during the preparation of this chapter. In this endeavour, considerable help was forthcoming from a number of sources, which I have acknowledged in the final section.

2.2 PROCESSED POTATO PRODUCTS

In developed nations, up to 60% of potato in everyday diet is consumed in processed form. Demand for convenience food in the home, fast food in restaurants and snacking has given rise to a wide variety of products. These include potato chips, French fries and various other frozen products, dehydrated potato products, chilled-peeled potatoes and canned potatoes.

Potato chips are deep-fried, thinly sliced potatoes, with a finished moisture concentration of 1.3–1.5%. They are fried in different types of vegetable oil, with a range of added flavours, e.g. salt, salt and vinegar, cheese and onion, paprika and cool lemon to name a few. They are sold in 25–400 g portions, with a shelf life of 12–14 weeks; freshness is preserved as far as possible in a nitrogen atmosphere within sealed polylaminated bags.

The term 'potato chip' also denotes chips formed from a dough made with dehydrated potato, and either baked or flash-fried. Baked Lays® and Pringles® are examples of this product type. In later discussion of global production (Section 2.4), the term 'chip' will be used inclusively.

French fries are made from potatoes that have been cut into thin strips, washed briefly in cold water, partly dried to remove surface moisture and deep fried in vegetable oil to a light golden colour. Frozen fry manufacturers ship their products raw, par-fried, or partially cooked and drizzled with oil (for baking), to suit the end user. Cutting style

Potato Biology and Biotechnology: Advances and Perspectives
D. Vreugdenhil (Editor)

varies, and the consumer or restaurant selects a product according to the final cooking method. Restaurants generally employ deep fat frying, and preferences in home use include baking and microwaving.

Other frozen potato products include waffles, wedges, hashed brown potatoes, rösti, pre-formed mashed potatoes, patties, potato rounds, diced potatoes, baby roasts and a variety of shaped potato products with child-appeal. Manufacturing and trade statistics are difficult to disaggregate according to different product types. Typically, frozen potato products other than French fries amount to 10% of the total.

Canned potato production constitutes a few percentage of the overall market for processed potatoes. In Europe and subsequently in the UK, production of chilled-peeled potatoes became established during 1995–2005 to supply a growing demand by restaurants, takeaways and the catering business.

2.3 HISTORY OF POTATO PROCESSING

Potato chips have a US origin dating back to 1853 in a hotel kitchen at Saratoga Springs, New York. The first commercial production got underway in 1895 (Gould, 1999). William Tappenden of Cleveland, Ohio, made chips for his restaurant and neighbouring stores. Business thrived, and the first potato chip plant was established in a converted barn.

According to Gould, eleven new chip plants started production during 1895–1928, giving rise to a number of familiar brands that are still in the market in 2005. In the UK, Smith's Crisps began production in 1920. Several innovations mark the industry's subsequent development.

During the 1920s, Herman Lay developed the mechanical potato peeler, which stimulated production for a mass market. About the same time, Laura Scudder pioneered the packaging of potato chips in sealed bags, initially made from waxed paper, later from cellophane. Seasoning technology was developed in the 1950s by the owner of 'Tayto' in Ireland. This allowed controlled amounts of salt and a range of natural and artificial flavours to be added automatically. In recent years, microprocessor-controlled weighing heads (1985), optical sorting to remove defective product (1990), and nitrogen-fill to preserve freshness (1995) have contributed to state-of-the-art potato chip manufacture.

Herman Lay peddled potato chips from Atlanta to Tennessee, selling from the trunk of his car. In due course, Lay's potato chips became the first national brand. In 1961, Lay's company was merged with Frito, a Dallas-based corn chip manufacturer, to become Frito-Lay. Part of PepsiCo since 1965, Frito-Lay dominates the North American snacks market, with its nearest competitor having an 8% share. Frito-Lay is the only global player in the international snacks market.

The industrial production of French fries started in the USA after World War II (WWII), but the product was already familiar to soldiers returning from the war of 1914–18. *Pommes frites* had been a culinary item in France and later in Belgium since 1800. The invention of the industrial process for French fry production is generally attributed to Jack Simplot of the J.R. Simplot Company in Idaho (Guenthner, 2001).

During WWII, Simplot and numbers of other companies engaged in the manufacture of dehydrated potato to supply the US military. French fry manufacture got underway in the early 1950s. In 2005, Simplot produced 1.4 million tonnes of frozen product, though McCain surpasses this figure and operates in more countries. Global players in the frozen fries market also include Lamb Weston.

This steeply increasing trend is reflected in US production figures, moving from zero potato usage for French fries in 1939, 3000 tonnes in 1944 and 2.4 million tonnes in 1972 to approximately 7 million tonnes of frozen product in 2004. This increase parallels the burgeoning popularity of fast food. McDonald's is reputed to utilize 27% of the world's production of frozen fries.

Potatoes have been frozen and dried for 3000 years. The Incas are credited as the first producers of chuño, which is still made in the 21st century in local South American communities. A US patent was granted in 1912, and dehydrated potato was manufactured in the USA and Britain to supply military personnel during WWII. According to the Union of the European Processing Companies (UEITP, 2004), in 2002 the process accounted for 2 million tonnes of potato utilization in the EU. During the same year, 2.2 million tonnes of potatoes were channelled into dehydration in the USA (National Potato Council, 2005).

2.4 CURRENT DIMENSIONS

2.4.1 Global production and consumption

On a country-by-country basis, the annually updated database of the United Nations Food and Agriculture Organization (FAOSTAT) is the only comprehensive source of information for potato production, utilization and trade. FAOSTAT data for 2002 showed that 12.6 million tonnes of potatoes were used in food manufacture, i.e. processing, but does not disaggregate according to product type. Data are missing for North America, so the overall figure that can be derived for the proportion of potatoes in the world's diet consumed in processed form will be significantly underestimated.

US National Potato Council figures (2005) indicate that 7 million tonnes of potatoes were utilized in the USA in 2002 for frozen potato products and 2 million tonnes in Canada. Potato utilization for potato chip manufacture was cited as 2.3 million tonnes in 2002. Dehydration accounted for 2.2 million tonnes. Global consumption of potatoes in processed form for that year thus becomes 26.1 million tonnes, 13% of the overall world consumption of 202 million tonnes of potatoes for food. The estimate is likely to be conservative.

How much of the total is consumed as potato chips is not easy to resolve. Datamonitor (2004) has assessed the 2002 world market volume for chips as 2.3 billion kg, with a value of $14.3 billion. Using a conversion of 3.6 tonnes of potatoes to 1 tonne of finished product yields a figure of 8.3 million tonnes of potatoes converted to chips, representing 32% of the world total of 26.1 million tonnes of processed potatoes.

A combination of FAOSTAT and UEITP (2004) data for the 2002 UK import and production of frozen fries and potato chips yields a comparable figure of 27% potatoes consumed as chips of total processed potatoes in the UK diet. This is confirmed by independent data from the British Potato Council (2004). For the USA and Canada, the figure is approximately 20% (National Potato Council, 2005). Despite a significant and growing export of frozen fries to developing countries reported by Guenthner (2001), it is arguable that potato chips become established sooner than French fries in the diet of developing nations, for reasons that will be explored later in section 2.5 on potato processing companies and locations.

2.4.2 Trends

According to data compiled by J. Guenthner (personal communication) from the American Frozen Food Institute (AFFI), US potato utilization for the manufacture of frozen products amounted to 3.5 million tonnes in 1980 and 7.1 million tonnes in 2002. This represents a 3.3% compound annual growth rate (CAGR) for 2002. During the period 1994–2003, the USA changed from being a net exporter of frozen fries to becoming a net importer (USDA, 2004). On the contrary, Canada increased its net export trade over the same period. The Netherlands continues as the world's largest exporter, primarily to EU countries. The UK is its biggest customer. Belgium, Argentina and Australia also have a share of the international export trade in frozen fries. Japan has a significant share of imports. Mexico, China and Korea account for relatively small but increasing import shares.

New frozen fry production facilities are being established around the globe by McCains, Lamb Weston and the J.R. Simplot Company. It is unclear to what extent this expansion might be supplemented by local manufacture. However, it is reasonable to conclude that global production of frozen potato products is expanding with a CAGR of not less than 3.3%.

For the global potato chip industry, CAGR in mass terms (kg of product) was 3.7% between 1997 and 2003 (Datamonitor, 2004). This figure represents a steady but relatively moderate growth from 2 million tonnes of finished product in 1999 to 2.35 million tonnes in 2003. The corresponding figure for potato utilization in chip manufacture is 8.3 million tonnes. Many potato chip markets around the world have reached maturity. However, in the UK the market appears to have experienced negative growth in the past few years (The Grocer, 2005).

Conclusions by Datamonitor indicate an expected upswing in the global potato chip market from 2.3 to 2.7 million tonnes of product from 2003 to 2008, with a CAGR of 3.2%. In terms of potato utilization, the 2008 figure represents approximately 9.8 million tonnes.

From a combined CAGR for fries and chips of 3.25%, by 2020 the annual increase in global potato consumption in processed form compounds from 13% in 2002 to 17.7% by 2020, equivalent to 46.4 million tonnes of raw potatoes. This calculation is based on the prediction of world potato production in 2020 of 403 million tonnes (Scott et al., 2000) and on the assumption from historical FAOSTAT data that 65% of production enters the human food chain.

2.4.3 Drivers

Patterns of food consumption fall into three categories (Guenthner, 2001). At the most basic level, people eat to survive. Subsistence revolves around a few staples. At the second level, people eat to meet energy needs. The diet is likely to incorporate more variety, including fruit and vegetables, livestock products and some processed foods. At the third level, which applies in most developed nations, people consume food for pleasure. Higher levels of disposable income allow more choices, e.g. to pay for ready-prepared and processed products that appeal on flavour, texture, appearance and convenience.

In developed nations, convenience foods, with their ready availability and ease of preparation, make a substantial contribution to everyday eating. Societies have evolved to place larger demands on time spent in the workplace. Both women and men are affected. Mothers caring for young children are often occupied in full-time jobs. As nations grow richer, more disposable income becomes available to pay for fast food in restaurants, home-delivery, ready-meals, snacks and frozen potato products. It is interesting to note that the trend towards paying higher premiums for the service and pleasure elements of food has been accompanied by a decline in the proportion of income spent on all food. According to the UK Office of National Statistics (2005), the typical UK household spending on food and non-alcoholic beverages has declined from 20% of disposable income in 1980 to 14% in 2004.

Guenthner (2001) reported a 1995 study of factors affecting the demand for potatoes and potato products. For snacks, positive demand drivers included consumer income, meals away from home and advertising. Price and advertising for competing products had a negative effect. For frozen potato products, positive demand drivers included the price of fresh potatoes, consumer income, females in the labour force, meals away from home and advertising. The price of a hamburger and the increasing home use of microwave ovens had a negative effect.

The 2005 situation in developed nations of the EU and North America may not be so clear-cut. Issues around nutrition and health have tended to check market growth. The industry appears to be responding in a positive way, putting its case forward to counteract some of the myths and misunderstandings, at the same time as introducing a variety of healthy-eating alternatives.

It is arguable that the McDonald's Index (MI) provides a snapshot of how the global industry is progressing. On a country-by-country basis, the MI, as defined by Guenthner (2001), is the number of thousands of inhabitants in a population per McDonald's restaurant. For example in the USA, the MI is 23, meaning that there is a McDonald's for every 23 000 people in the population. It has been 23 for quite a few years, indicating that this is the level achievable in a mature market.

Figure 2.1 compares MI and corresponding GDP for selected nations. In 2004, there were 31 600 McDonald's restaurants worldwide for 6.5 billion people, giving an overall MI of 206. With relatively high GDPs, Scandinavian countries lie above the trend line, with fewer McDonald's than might otherwise be anticipated. Italy, Mexico, Russia and China also fall into this category. Other countries, notably the Philippines, Malaysia, New Zealand, Singapore and Taiwan have more McDonald's than expected according to the level of GDP.

It is tempting to predict that annual changes in the MI reflect the potential expansion for all potato products, including snacks and frozen potatoes. This assertion relies on the evidence that CAGR for both fries and chips in world markets are closely similar,

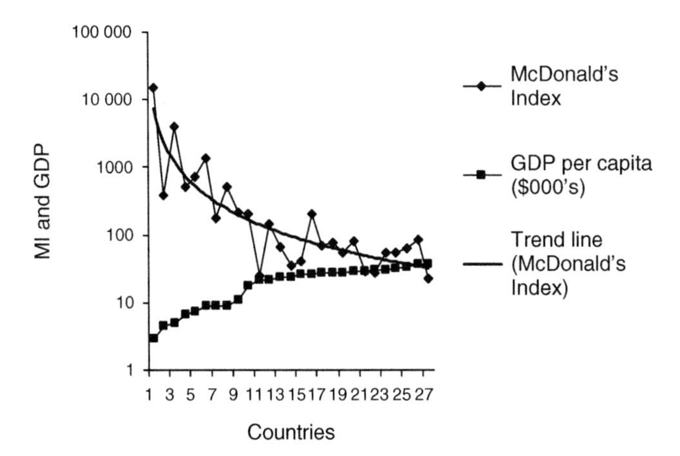

Fig. 2.1. McDonald's Index and GDP in selected countries (Minnesota State University, 2005; Nationmaster, 2005).

as previously demonstrated. A more refined econometric model to predict expansion in world markets would probably include the following factors:

(1) Growth in GDP and disposable income.
(2) Predisposition to eat potatoes in the diet.
(3) The availability of suitable potato varieties and growing skills to support a local or national potato supply.
(4) Population.
(5) Price of potatoes.

In regard to price, evidence cited by Guenthner (2001) indicates that the potato/rice price ratio declined from 1950 to 1980 in various developing countries, including Thailand (3.9 to 0.8) and Bangladesh (2.4 to 0.7).

2.5 POTATO-PROCESSING COMPANIES AND LOCATIONS

Frito-Lay dominates the world market for potato chips. The company operates approximately 67 plants in 27 countries. Utilization of potatoes specifically for chips formed from the raw potato exceeds 3.5 million tonnes. When pre-formed potato products are included, e.g. Baked Lays® and Stax®, Frito-Lay's estimated potato usage for snacks/chips increases to 4 million tonnes, representing approximately half of the world total. Frito-Lay is the only global player in this market.

During the decade 1985–95, Frito-Lay's policy for expansion could be regarded as mainly opportunistic, with a major thrust on acquisition of existing snack/chip companies. During the decade 1995–2005, the plan changed to include building de novo manufacturing facilities in countries where investment in the long term could reap dividends. In the few years prior to 2005, outlay on new plant and infrastructure was probably highest in

China, Russia and India. Plans for a start-up in Pakistan are well advanced. Frito-Lay's potato chip operations include plants in the USA and Canada (28), Europe, Africa and Middle East (15), Asia (11), South and Central America (10) and Oceania (3), a total of 65.

The Canadian family-owned company McCain Foods dominates the world market for frozen potato products, primarily French fries, with a one-third share. According to the company website, McCain Foods operates 55 plants in 13 countries (not all of which may be producing French fries). These include Argentina, Australia, Canada, China, France, India, The Netherlands, New Zealand, Poland, South Africa, Spain, the UK and the USA. McCain Foods markets its products in 110 countries. Coincidentally, this figure corresponds approximately to the distribution of McDonald's restaurants in 114 countries.

Other players in the world market include Lamb Weston and the J.R. Simplot Company. Part of ConAgra Inc., Lamb Weston operates 11 plants worldwide mainly in the USA and Canada. Lamb Weston has two plants in the Netherlands and one in Turkey. Simplot has operations in the USA, Canada, Australia and China. It is a major supplier of frozen fries to McDonald's worldwide, utilizing approximately 3 million tonnes of potatoes for its products.

Two Dutch companies, Farm Frites and Aviko, process frozen potato products from, respectively, 1.2 and 1.4 million tonnes of potatoes annually. Both companies have a manufacturing presence in five countries. One of Aviko's plants is located in the USA.

Whereas international trade in frozen potato products is significant, trade in snacks and potato chips is relatively minor. The contrast is accounted for by product specific gravity (SG). Snacks and chips are characterized by low weight per unit volume resulting in high transport costs. For French fries, with added value and a somewhat higher weight per volume than the potato raw material from which they are manufactured, it becomes more attractive to transport finished goods rather than potatoes. It follows that French fry factories are almost invariably located close to their raw material sources and that the finished product may travel long distances to reach the markets. On the contrary, for potato chip manufacture raw potatoes may be transported long distances, and frequently cross-national boundaries, whereas finished products have a more limited distribution.

The situation described has an important bearing on future volume expansion in the international trade for frozen potato products. Guenthner (2001) noted that the trade value was expanding at the rate of $143 million annually and likely to reach $3 billion in 2007 and $4 billion in 2014. So for the French fry industry, the question of where to build new plants to meet future demand may be just as crucial as for the chip industry.

2.6 POTATO SUPPLY

2.6.1 Supply chain

A continuing potato supply of acceptable quality for processing into chips and French fries is critical to success. In the evolving history of the industry, and new-build enterprises around the globe, a reliable potato supply has always been the first thing to get right.

Increasing involvement by industry technical managers in production chain management has become a feature of the global potato-processing industry. This has been particularly marked during the 1990s, and subsequently. Initially, emphasis was focused on integrating and controlling the sequence of activities and processes in the supply chain to optimize potato quality for processing. Beyond 2000, the combination of supply, cost, quality, product safety and environmental care has taken precedence in modern manufacturing.

Because manufacturers in both sectors require specific varieties to meet demand for quality, the chain starts with the production of nuclear stock seed potatoes. The processing companies generally provide guarantees to the producers of nuclear seed to establish a balanced supply of high-grade seed in the right varieties.

At different stages down the chain, the companies provide technical support. Agronomy, bruise control, chemical maturity monitoring (CMM), storage data monitoring and storage management are activities where company technical managers have led recent innovation. To a greater or lesser extent, expertise is supplemented by technical managers employed by the individual potato suppliers.

Generally, for the potato chip industry approximately 100% of the potato supply is contracted. Manufacturers of frozen potato products tend to contract a lower proportion of their requirement and supplement supplies with free-buy potatoes. The same quality standards apply in the two situations, but price and potato variety may vary in the latter. It also follows that there is less direct participation in production chain management by this sector for some portion of its raw material supplies.

In the UK, potato suppliers to both sectors of the industry must sign up as members of an approved scheme that assures safe and environmentally friendly production of fruit, salads and vegetables. The Assured Produce scheme crop protocol for potatoes extends to 62 pages (Assured Produce Ltd, 2004). Although the main focus is on pesticide usage, it incorporates recommendations on agronomy and a section on producing potatoes for processing, which promotes raw material quality for that purpose. The counterpart to the UK Assured Produce scheme, which is being developed for Europe and the rest of the world by EU member nations, is EurepGAP (2005).

2.6.2 Variety requirements

Figures 2.2–2.5 indicate trends in the production of varieties that are favoured by the processing companies. For chipping varieties, production is approximately equivalent to

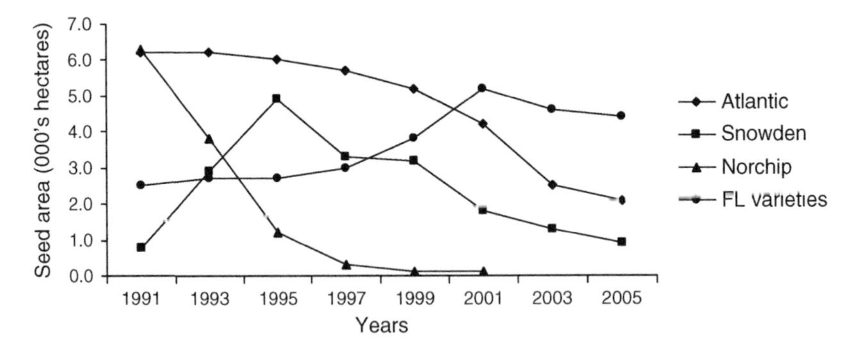

Fig. 2.2. US/Canada trends in chipping variety production (US National Potato Council).

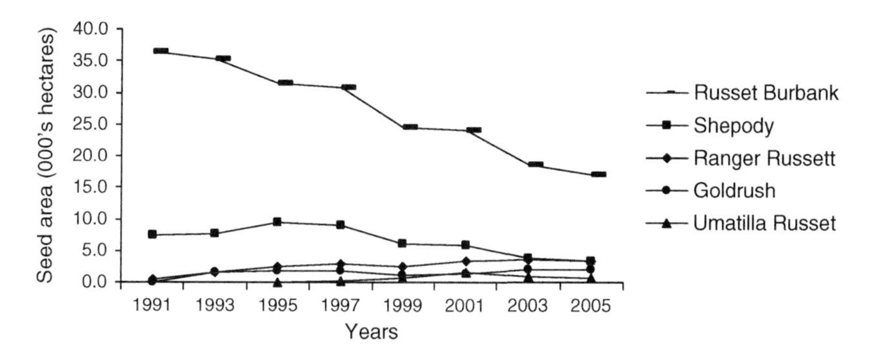

Fig. 2.3. US/Canada trends in the production of French fry varieties (US National Potato Council).

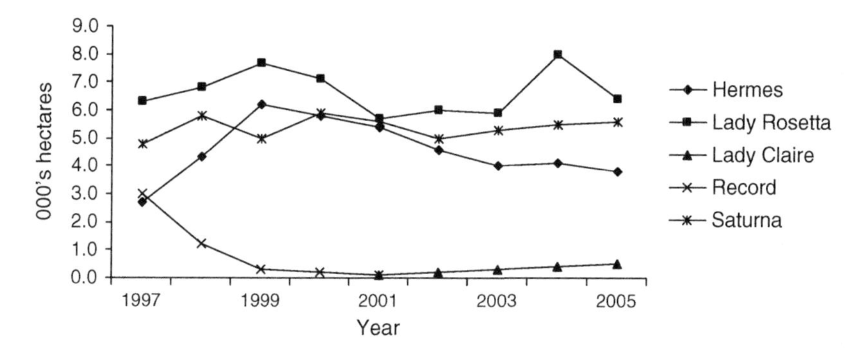

Fig. 2.4. UK trends in chipping varieties (British Potato Council, 2005).

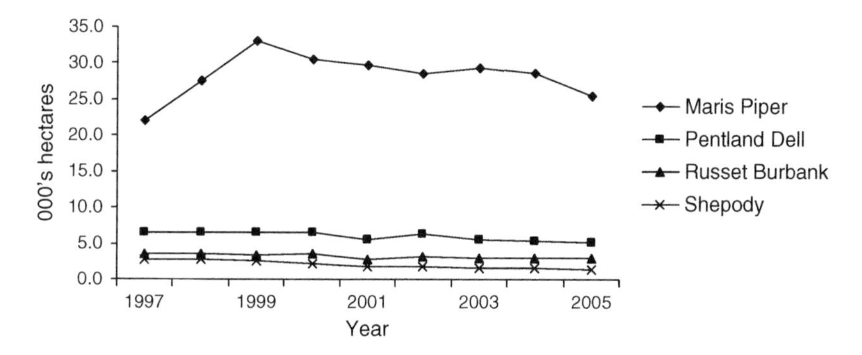

Fig. 2.5. UK trends in French fry varieties (British Potato Council, 2005).

utilization. However, the majority demand for French fry varieties such as Russet Burbank and Maris Piper is for table-stock. UK ware production figures of chipping varieties reflect European usage. Ten per cent of UK seed exports are accounted for by the variety Hermes, mainly to the Mediterranean and Middle East.

Some processing companies have invested significant effort and resource to breed and select improved varieties. Frito-Lay has its own potato-breeding programme, based in Wisconsin. It supplies new varieties for regional evaluation around the globe. McCain's strategy is to collaborate with various breeding partners to meet specific regional requirements for superior raw material to feed its factories.

In both cases, an underlying objective is proprietary ownership, creating the possibility for competitive advantage. With about 80% proprietary usage in the USA and Canada, Frito-Lay has realized its ambition for that region. Frito-Lay varieties are also making significant inroads in other regions, notably Asia, Oceania and Latin America. But in Europe, different disease pressures, consumer preference for yellow chips and competition from established varieties represent significant challenges.

Potato breeding is a hard grind. Usually, it takes 15 years from an initial parental crossing to commercialization, at which point a few thousand tonnes of a successful variety will be making its way into commercial processing. Accelerating the process, for example, by applying progeny selection to identify superior crosses, probably works to cut down the timescale by a few years. Accelerating the process by massively expanding the nuclear seed of a promising variety in the early stages of evaluation (e.g. through state-of-the-art minituber production) is fraught with the risk of serious financial exposure. Even so, a few examples can be cited where this strategy has been successful.

To reiterate the point, breeding successful potato varieties is a big challenge. Using a gross average, each sector can reasonably expect one new variety to enter commercialization every decade. Table 2.1 demonstrates that in a historical context, varieties in current use for processing originated over an extended timescale.

The list of varieties in current use is incomplete. For example, Bintje and Agria are used widely in mainland Europe for French fry manufacture. Asterix, Innovator, Markies,

Table 2.1 Year of introduction and initial commercialization of selected varieties utilized by the potato-processing industry in Europe and North America.

Year	Variety	Origin
1876	Russet Burbank	USA, Luther Burbank
1942	Record	The Netherlands, R.J. de Vroome
1960	Saturna	Holland, Agrico
1960	Pentland Dell	UK/SCRI
1964	Monona	USA/Frito-Lay
1966	Maris Piper	UK, Plant Breeding Institute
1968	Norchip	USA, North Dakota State University
1973	Hermes	Austria, NÖS
1974	Atlantic	USA, United States Department of Agriculture
1978	Snowden	USA, University of Wisconsin
1980	Shepody	Canada, Ag Canada/Alberta
1988	Lady Rosetta	The Netherlands, Meijer
1995	Umatilla Russet	USA, United States Department of Agriculture
1999	Lady Claire	The Netherlands, Meijer

Source: Hils and Pieterse (2005).

Lady Olympia and Victoria are also used to some extent. New US varieties for French fry production include Bannock Russet, Gem Russet and Ranger Russet. Russet Burbank appears to be in decline, most probably reflecting a change in table-stock usage in favour of newer varieties with improved uniformity and disease resistance. Atlantic also appears to be in decline, supplanted mainly by Frito-Lay varieties. However, this variety is still utilized extensively for production of chips in Asia, Latin America and some Middle Eastern countries.

A substantial breeding effort to develop superior processing varieties continues in addition to the proprietary programmes previously mentioned. In North America, the USDA, Ag Canada and several universities and institutions participate. In Europe, privately owned breeding companies, which have devoted significant effort to breeding for chips and French fries, include HZPC, Agrico, Meijer, Solana, Norika, Böhm-Nordkartoffel and Germicopa.

2.7 POTATO COST

2.7.1 Theory and practice

There appear to be two situations that characterize the evolution from small to large volumes of potatoes utilized by different sectors of the processing industry. The first is historical and follows the pattern below:

Stage 1. One or a few buyers, few suppliers.
Stage 2. Few buyers, many suppliers.
Stage 3. Many buyers, many suppliers.
Stage 4. One or a few buyers, few suppliers.

This sequence broadly reflects the development of the potato-processing industry in the USA and the UK. Development of the French fry industry has been concentrated into the last 50 years to 2005, whereas the chip industry developed over a longer timescale. In the early years, for both sectors of the processing industry, price paid for potatoes tended to be higher than in the later stages. Even within the period 1995–2005, price of chipping potatoes in both the USA and the UK has declined in real terms. The same trend probably also applies to the price of potatoes for French fries.

The second situation where potato processing has the potential to evolve from small to large volumes is 'green-field' manufacturing, usually new plant in developing nations where either indigenous production is small and fragmented or non-existent. Again, high prices are paid for potatoes initially, although the starting position is more typically with one buyer and many suppliers.

As time progresses, declining price may be attributed to agronomic improvements leading to enhanced yield and to the consolidation of suppliers to provide economies of scale. In the UK since 1960, yield has increased from an average for all maincrop potatoes of 25 to 45 tonnes per hectare in 2005. In the USA, similar yield improvements have occurred. Also, the supplier base to the processing companies has declined.

For example, by 2005 in both nations, fewer than 20 suppliers to Frito-Lay supported 80% of the company's requirements.

Guenthner (2001) has discussed commoditization. The acid test to determine whether an item of trade is a commodity is whether price in real terms (adjusted by the retail price index) declines in time. Potatoes supplied to the processing industries in mature and maturing markets behave as a commodity. Other potatoes (not including seed potatoes) also behave in this way. Product differentiation is not possible, particularly in the light of the fact that processing companies drive for uniformity and consistency of supply.

In developed nations, the cost of potato production tends to be higher than for other arable crops, not least because the contribution of mechanization is typically around 30% of (fixed + variable) costs. Potato machinery is expensive single-purpose equipment, which requires significant investment in purchasing, insurance, maintenance and repair. Asset fixity locks growers into their potato-growing enterprises, and it is not a realistic option to grow for other markets if the price on offer from the processors does not appeal. Little if any negotiating power on price resides with the growers. The situation stimulates cost efficiency and eliminates numbers of growers who cannot compete. But the beginning and end points in the historical record characterized by one or a few buyers and a few suppliers occur for very different reasons.

Superficially, the price outlook for potato suppliers to the processing industry does not appear encouraging. However, other factors compensate. These revolve around balancing supply and demand through contracts, bonus payments on good performance and an unwritten rule that purchasing managers in the processing companies should reckon on a supplier's profit margin of 15%.

2.7.2 Contracts

The processing industry is vertically integrated through the use of production contracts between growers and processors. As a generalization, the chip industry contracts approximately 100% of its potato supply, whereas manufacturers of frozen potato products contract upwards of 70%. Contracts co-ordinate a changing periodic supply of specific varieties and ensure that quality standards are upheld.

The benefit of contracting to growers is market security and income stability. They are not exposed to price volatility, which characterizes the free market. In addition, potato suppliers to the processing industry benefit from technology transfer of new and sometimes novel scientific and technical applications to different stages of the production chain relevant to supply, cost, quality, product safety and environmental care.

There is no universal model for contracting across the globe by either sector of the industry. Indeed, it is likely that one or another global player will contract in different ways in different locations, depending on market maturity and the implications for supply and quality. However, in general the chip industry contracts at a base-price, which increases according to delivery times from fresh harvest through late storage. Different chip manufacturers differ on whether to pay a quality bonus, and if so on what basis. Most commonly, bonus payments focus on chip colour defects and dry matter concentration. Risk of rejection occurs below 20% dry matter concentration or above 15% colour defects in the finished product.

One UK manufacturer of chips pays a bonus on freedom from colour defects, with different levels applicable during different periods of supply. Fresh supply has attracted a bonus of up to 9% of the base-price below 6% defects. From the middle and late storage periods, up to 18% of the base-price (adjusted for storage interval) could be paid, the amounts increasing to the maximum of 18% as defect incidence declines below 10%.

Potato supplies for French fry manufacture are bonused or penalized (with a price reduction subtracted from the base-price) on the basis of several potato quality characteristics that influence factory yields and quality of the finished product. For example, for the variety Shepody, one UK manufacturer pays a composite bonus based on the following:

(1) Tuber dry matter concentration (too high or too low is penalized, bonus points for the middle range).
(2) Tuber count per 10 kg (an indicator of tuber size distribution).
(3) Percentage of tubers >75 mm in length (low numbers penalized, higher numbers attract bonus points).
(4) Potato defects.
(5) Fry colours.

The possibility of attracting bonus points or penalties is highest for fry colours and potato defects. Potato defects are measured at factory intake and include greening, bruising, growth cracks, secondary growth, pitted scab and damage due to mechanical harvesting, slugs, grubs or frost. The company reserves its right to reject loads of potatoes that do not meet specification on the basis of a low cumulative points score or on the basis of serious deviations in tuber count, fry colour or tuber dry matter concentration. Variety mix is another reason for rejection.

In the above example, differing prices and schedules apply to different varieties, which adds a level of complexity missing from the more straightforward approach by the chip industry. However, it is clear from the examples cited that bonuses, penalties and rejection risk combine together to enhance uniformity, quality and output of the manufacturing process. Growers are rewarded for good performance.

2.8 POTATO QUALITY

2.8.1 Introduction

A great deal has been written about potato quality requirements for the potato-processing industry (e.g. Talburt and Smith, 1987; Gould, 1999). Quality parameters are fixed and have varied little over time or across manufacturing regions.

2.8.2 Tuber shape, size and dry matter composition

In regard to tuber shape, the industry standard for chips is round to short-oval. A long-axis to transverse-axis ratio of 1.25:1 is acceptable; 1.33:1 is borderline, whereas 1.5:1

is unacceptable. The ratios refer to variety characteristics as a mean value, not to natural variations in shape which may occur within a variety. Tubers should be uniform and not excessively disfigured by secondary growth, infolded ends or growth cracks. In the UK and most of Europe, the size range 40–95 mm is acceptable for chip manufacture. Indicative tuber counts per 10 kg should fall within the range 72–112. In the USA, the acceptable size range is 45–105 mm, with no indicative tuber count.

In the UK, premium French fry manufacture is based on long tubers, with incremental bonus points being paid on long varieties such as Shepody and Russet Burbank for the proportion of tubers above 75 mm. Other varieties such as Maris Piper and Morene, whose shape characteristic is round-oval, attract a bonus based on tuber width (transverse axis) greater than 70 mm. Potatoes are acceptable with the transverse axis as little as 50 mm. Penalties are incurred for the proportion of a factory delivery that has the transverse axis less than 50 mm. If this exceeds 12%, outright rejection may ensue. The acceptable tuber count range is 34–67 per 10 kg.

For both sides of the industry, dry matter concentration is a critical component of efficiency in manufacturing. For example, a 0.005 increase in SG yields an extra kilogram of chips per 100 kg finished product (Gould, 1999). Generally, the cut-off below which factory deliveries are not acceptable is 19.5% dry matter (SG = 1.077) for French fries and 20% dry matter (SG = 1.079) for chips. Upper limits do not apply, though penalties may be incurred for >25% dry matter (SG = 1.103) in French fry manufacture.

2.8.3 Blemishing diseases and disorders

Any disease or disorder that leads to undesirable colour, e.g. brown or black spots and marks in the finished product, or a reduction in process efficiency may lead to penalties or rejection of the potatoes for processing. Generally, varieties in current use for processing are resilient, if not wholly resistant to blemishing diseases and disorders. Inevitably, each variety has it own specific susceptibilities, and no variety is perfect.

For example, a consideration of European chipping varieties reveals that Hermes grown in Mediterranean countries can be afflicted by high levels of Potato Virus Y^{NTN}, causing internal necrosis. Also causing internal necrosis, Potato Mop Top Virus (PMTV), vectored by Powdery Scab (*Spongospora subterranea*), has been a feature of Scandinavian Saturna production. In the worst years, up to a third of the crop can be affected. The control of Rhizoctonia in seed potatoes of Lady Rosetta is crucial. In the ware crop of Lady Rosetta, Rhizoctonia leads to deep fissures and infolded ends, which create inefficiencies in manufacturing. Atlantic, the world standard for chipping, can suffer extensive damage from deep-pitted scab and hollow heart. Shape disorders in tubers of Russet Burbank, the world standard for French fry manufacture, constitute the biggest problem for the variety.

Occasionally, a new disease becomes prevalent. Such is the case with 'zebra chip' defect thought to be caused by a phytoplasma, with sporadic outbreaks in Mexico, Texas, Colorado, Kansas and North Dakota (Potato Review, 2005). It has affected some of the most popular Frito-Lay varieties, Atlantic and commonly grown table-stock varieties.

Defects arising from mechanical damage and bruising, incurred during harvesting, are commonplace. The causes are well understood. Advances in technology and management practices have combined in recent years to minimize the problem.

2.8.4 Sugars and fry colours

Arguably, the biggest problem for potato processing revolves around meeting consumer preference on fry colour. For both sectors of the industry, this is 'light-golden'. The industry standard defining colour for chips is based on the measurement of 'L-value', and for French fries, it is based on Agtron measurement of fry colour in strips of fried potato, referenced to a USDA chart.

Fry colours are defined exactly by the reducing sugar concentration in tubers. A great deal of understanding has been developed by the industry from 1980 onwards about most, if not all, of the factors that contribute to fry colour variability.

(1) Variety characteristics.
(2) Agronomic and meteorological conditions prevailing during the growth period.
(3) Maturity at harvest.
(4) Storage temperature regime.
(5) Ethylene in the storage atmosphere.

In practical terms, items (3) and (5) are probably the most interesting, for the reason that trends in current practice have reaped significant rewards.

In the chip industry, CMM is the term given to measuring sucrose and glucose concentration of tubers during a 4-week period in the run-up to harvest (Sowokinos and Preston, 1988). An estimate of preferred storage treatment can then be defined. Immature crops can be pre-conditioned at a higher temperature over a longer period; crops with low and stable sugars can be placed in storage for the long term. It is important to continue monitoring sugar levels frequently during storage. The trend lines for glucose, sucrose, fry colour and the incidence of colour defects can then be used to order and re-order the sequence of store unloading. The aim is to optimize quality, enhance manufacturing efficiency and reward suppliers for good performance. CMM finds application mainly in temperate and continental climates.

It came to light during the British Potato Council's sponsored research on the sprout suppressant CIPC at Glasgow University (Dowd, 2004) that fry colours are deleteriously affected by ethylene, which occurs as a by-product of the combustion process in thermal fogging. The recommendation from this research was to ventilate within a maximum 8-hour interval after fogging. This has become an industry standard and provided significant benefit to the processing companies.

2.9 CURRENT ISSUES AND FUTURE DEVELOPMENT

2.9.1 Acrylamide

In 2002 a surprise discovery was made by Swedish researchers that significant levels of acrylamide occur in a variety of common foodstuffs, including chips and French fries. This led to a food scare, which temporarily impacted sales of processed potato products in a number of European countries. At the time, the currently available toxicological data indicated that acrylamide might be directly or indirectly carcinogenic. In the meantime, numerous studies have failed to confirm the relevance of this claim to the relatively low dietary exposure level to acrylamide, compared for example with occupational exposure. Nor in the latter context have case studies revealed enhanced risk.

Acrylamide is encountered in a range of foods and food products, resulting from the Maillard reaction between asparagine and reducing sugars during frying, baking and roasting. In potatoes, asparagine and glutamine are the main components of the free amino acid pool and occur in several-fold molar excess to the typical concentration of reducing sugars found in processing potato varieties (Duncan and Hardie, 2003). Therefore, the reducing sugar concentration is limiting. Practical measures to minimize reducing sugars can help to lower acrylamide concentration in finished products. Other mitigation steps that can be taken are detailed in 'The CIAA Acrylamide Toolbox' (Confederation of the European Food & Drink Industries, 2005).

2.9.2 Obesity

In the context of obesity, processed potato products continue to suffer bad press due to the perception that trends towards increased snacking and consumption of fast food, coupled with competitive advertising, and appeal to the child market are somehow responsible. The World Health Organisation (2005) cites a figure of 1.2 billion adults overweight (one-sixth of the world's population), 30% of whom are clinically obese. The consequences for health and well-being relate to chronic diseases such as type 2 diabetes, cardiovascular disease, hypertension, stroke and certain forms of cancer. It cannot be denied that processed potato products tend to be energy-dense. But it is interesting to compare obesity rates with the MI to discover whether a correlation exists.

From Fig. 2.6, there appears to be no correlation. Some nations with relatively low obesity levels in the population, e.g. Japan (3.2%) and Sweden (9.7%) with MIs, respectively, of one restaurant per 28 000 and one per 40 000, have relatively high numbers of McDonald's outlets. Conversely, Mexico, the Slovak Republic and Greece have obesity rates above 20%, with MIs of 521, 540 and 223, respectively. It was previously argued in this chapter that MI is indicative of the more general predilection in a population for

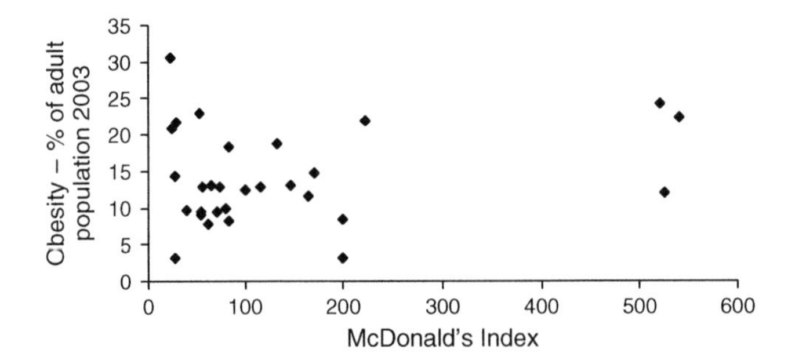

Fig. 2.6. The prevalence of obesity in the adult population of 29 nations (OECD, 2005) plotted against McDonald's Index.

snacks and fast food. World Health Organisation (2005) identified increased consumption of energy-dense foods high in saturated fats as a primary factor in the obesity pandemic. Clearly, a number of factors must be at work. The situation is complex, and potato-processing companies are actively engaged in the current dialogue.

At first sight, it may seem paradoxical that the global chip and fast-food industries are investing to establish new businesses in developing nations, where hunger may prevail in a significant part of the population. J. Guenthner (personal communication) takes the view that from an industry perspective the problems with mature markets and concerns about obesity are somewhat offset by opportunities for market expansion in developing countries. In addition to making new 'better-for-you' products for those concerned about obesity, the industry can also focus on target markets where current concerns are different. Scott et al. (2000) predicted a more rapid expansion of potato consumption in developing countries. A growing predilection for potatoes in the diet is one of the drivers for expansion in the market for processed potatoes. As economies develop, so will the demand for processed potatoes, concomitant with growth in GDP and disposable income.

2.9.3 Nutritional value

Whereas, in general, the nutritional profile of processed potatoes has few deficiencies, future developments are likely to see an increasing diversification of products to provide healthy-eating alternatives. Global manufacturers are engaged in reformulations and new product development to meet the challenge. Trans-fats have been mostly eliminated. A range of low-fat and fat composition changes has recently been launched. For snack manufacturers, frying in olive oil, high-oleic sunflower oil or zero calorie olestra has led to new products. Salt levels have been reduced.

The possibilities envisaged by biotechnologists to engineer components of potatoes, vegetable oils and flavours to provide consumer benefits cannot be disregarded. The composition and content of starch in relation to the glycaemic index, vitamins, antioxidants, proteins, free amino acids, reducing sugars and the reduction of glycoalkaloids could provide legitimate targets aimed at consumers of processed potato products through the first few decades of the 21st century.

ACKNOWLEDGEMENTS

The author gratefully acknowledges help from numerous friends and colleagues. Professor Joe Guenthner of the University of Idaho deserves special mention. Joe Guenthner provided additional information, valuable insights and guidance, and reviewed this chapter before publication. Any mistakes in fact or logic are the author's. Former colleagues Dr Robert Hoopes, Kan Moorthy, Newton Yorinori and Jimmy Mackenzie provided comments, suggestions, information and moral support. Dr Robert Hoopes also provided a valuable review of the draft.

REFERENCES

Assured Produce Ltd, 2004, Crop specific protocol for potatoes. http://www.assuredproduce.co.uk
British Potato Council, 2004, BPC Annual Balance Sheet of Potato Supplies and Disposals 1988–2004. http://www.potato.org.uk
British Potato Council, 2005, Total potato plantings for Great Britain in 2005 (a number of previous editions were consulted). http://www.potato.org.uk
Confederation of the European Food & Drink Industries, 2005, The CIAA Acrylamide "Toolbox", 29 pp. http://www.ciaa.be/documents/positions
Datamonitor, 2004, *Potato Chips: Global Industry Guide*. Datamonitor, New York, U.S.A.
Dowd G., 2004, Investigation of the potentially detrimental effects of CIPC application on the processing quality of stored potatoes. Ph.D. thesis, University of Glasgow.
Duncan H.J. and S. Hardie, 2003, Literature review of the status of amino acids and soluble sugars and their influence on the formation of acrylamide in processed potatoes. A report compiled for the FNK Potato Processors' Research Group. http://www.slv.se/upload/Heatox/Dokument/FNK-rapport.pdf
EurepGAP, 2005. http://www.eurep.org
FAOSTAT, 2005, Food and Agriculture Organization of the United Nations, FAOSTAT Home Page, http://www.faostat.fao.org
Gould W.A., 1999, *Potato Production, Processing and Technology*. CTI Publications, Inc., U.S.A.
Guenthner J., 2001, *The International Potato Industry*. Woodhead Publishing Ltd, Cambridge, U.K.
Hils U. and L. Pieterse, 2005, *World Catalogue of Potato Varieties*. Agrimedia GmbH, Germany.
Minnesota State University, 2005, Population Counter – Country Index. http://www.mnsu.edu
National Potato Council, 2005, *NPC Potato Statistical Yearbook*, Denver, U.S.A. (various annual editions).
Nationmaster, 2005, Map & Graph: Countries by Food: McDonald's restaurants. http://www.nationmaster.com
OECD, 2005, Obesity: percentage of the adult population with a BMI>30. http://www.oecd.org/dataoecd/7/38/35530193.xls
Office of National Statistics, 2005, Family spending – a report on the 2004–05 expenditure and food survey. http://www.statistics.gov.uk
Potato Review, September/October 2005, 'Zebra' chips headache for US growers. Aremi Publishing Ltd, U.K. http://www.potatoreview.com
Scott G.J., M.W. Rosegrant and C. Ringler, 2000, Roots and tubers for the 21st century – trends, projections, and policy options. International Food Policy Research Institute. http://www.ifpri.org
Sowokinos J.R. and D.A. Preston, 1988, Maintenance of potato processing quality by chemical maturity monitoring (CMM). Station Bulletin no. 568–1988. Minnesota Agriculture Experiment Station, St. Paul, MN, U.S.A.
Talburt W.F. and O. Smith, 1987, *Potato Processing* (4th edition). Van Nostrand Reinhold Company, New York, U.S.A.
The Grocer, June 4 2005, 41–54, Snack makers step up their game. http://www.thegrocer.co.uk
UEITP, 2004, Union of the European Potato Processing Companies potato processing statistics, unpublished information from the UK Potato Processors' Association (PPA).
USDA, 2004, The US and world situation: fresh and processed potatoes. http://www.fas.usda.gov
World Health Organisation, 2005, Global strategy on diet, physical activity and health – obesity and overweight. http://www.who.int/dietphysicalactivity/publications/facts/obesity/en/

Chapter 3

The seed potato market

Kees D. van Loon

Oostrandpark 103, 8212 AT Lelystad, The Netherlands

For the production of potatoes, generally, seed tubers are used. If these are larger than about 25 mm in diameter, they are indicated as conventional seed tubers. However in recent years also mini-tubers, that have been derived from in vitro plantlets, are used in the first phase of potato seed multiplication programmes. Furthermore, true potato seed (TPS) is also used, on a limited scale, by direct sowing in the field or by sowing it in beds to raise seedlings that either can be transplanted into the field or grown in nurseries for the production of small seedling tubers that are to be used as seed in the next season.

3.1 SEED TUBERS

Seed potatoes are produced everywhere potatoes are grown. However, in many cases part of the ware crop – often the small tubers – are used for this purpose. Such seed is often of a poor quality. Good quality seed needs to be grown specifically for this purpose. This implies use of healthy (certified) initial seed, land free from soil-borne diseases and special care being taken in the control of diseases and pests during the growing season. The product should then be inspected by a certification agency.

Difficulties in the production of good quality seed potatoes include:

- Unsuitable climatic conditions
- The presence of many aphids during the growing season
- The presence of soil-borne diseases and pests
- Lack of a well-organized seed certification agency

An outstanding problem in potato seed production is the relatively high degeneration rate of initially healthy seed. In many regions, it will only take two or three multiplications to result in poor quality seed. In a seed programme based on clonal selection, 8–10 years of production are required before larger commercial quantities are available. In many potato-growing countries, phytosanitary conditions are such that it is not possible to keep the seed potatoes healthy for such a long period. In such countries, quality seed is produced by making use of rapid multiplication techniques, allowing the production of commercial quantities in 3–4 years, or quality seed is imported.

In countries without an adequate supply of certified seed, there are often 'traditional' seed producing areas, e.g. at higher elevation, where there is slower degeneration because

Potato Biology and Biotechnology: Advances and Perspectives
D. Vreugdenhil (Editor)

of fewer aphids transmitting virus diseases. Seed produced in these areas is of better quality than from other areas (Beukema and van der Zaag, 1990; Dongyu et al., 2004). Ware potato growers in such countries may buy seed from 'traditional' areas to refresh their own stock and use this for the production of ware potatoes or for multiplication as seed for a next crop.

In summary, growers of ware and starch potatoes are making use of seed from four basic seed sources:

- Home-grown certified seed
- Imported certified seed
- Home-grown, but not certified seed
- Self-produced (un)certified seed

In most countries, two or more of these sources are used.

In developing countries, especially sufficient certified seed is often not available. But also in Western Europe, there are potato-producing countries that import large amounts of seed potatoes, e.g. Italy, Spain, France and others. The reason, often, is unfavourable conditions for (basic) seed production. Also the lack of specific cultivars can be a reason for importing seed. Switzerland only imports basic seed. This is multiplied to produce certified seed to be used by ware potato growers. The Swiss deem the conditions for the production of basic seed insufficient in their country. In Greece, until the end of the 1990s only 20–25% of the seed used by ware potato growers was certified (Grigoriadou and Leventakis, 1999).

3.2 SEED MARKET

3.2.1 'Conventional' seed tubers

The most important characteristic of a seed potato is its quality. The most important quality characteristics of potato seed are as follows:

- Freedom from diseases and pests: The seed should be free from quarantine diseases such as wart disease, ring rot and brown rot, and the contamination levels of other, less detrimental diseases such as virus diseases, bacterial diseases (*Erwinia* spp.) and fungal diseases such as Black scurf, common scab, and silver scurf should be low. For basic seed that will be further multiplied, a higher health standard is required than for certified seed, that is used for the immediate production of ware or starch potatoes.
- Growth vigour of the seed: Defined as the potential of a tuber to produce sprouts and plants under conditions favourable for growth (Van der Zaag and van Loon, 1987). If the seed is planted within a few months after harvest, the vigour of growth is suboptimal. The same is true for seed that has been stored for a long period, especially if at high temperatures.
- Seed size: The value of small tubers with a certain health standard is greater than that of big tubers. The reason is that small tubers will produce more stems per unit of weight than large tubers, as is shown in Table 3.1.

Table 3.1 Number of tubers and tuber weight in three seed sizes, required for a stem density of 18 per m^2 and the value of the seed based on the number of stems produced for cultivar Bintje (size 35/45 mm = 100) (*Source*: Van Loon, 1993).

Seed size (mm)	Number of stems per tuber	Number of tubers	Weight in (kg)	Weight in relative figures	Value as seed in relative figures
28/35	3,5	51.500	1450	72	140
35/45	5	36.000	2000	100	100
45/50	6	30.000	2650	133	75

The price of seed potatoes mainly depends on supply and demand. Other factors are the seed grade – basic seed (better health status) has a higher price than certified seed; the appearance of the seed tubers – users generally prefer, and will pay a higher price for, clean tubers with a smooth skin; the provenance of the seed – there are countries and regions within countries with a better quality image for seed potatoes than others; tuber size – in some countries, e.g. The Netherlands, tubers of certain sizes, e.g. size 28–35 mm, are much more expensive per kg than tubers of size 45–50 mm because of a higher seed value (Table 3.1).

3.2.1.1 Domestic market

In developed countries, most farmers buy certified seed potatoes as seed source for the production of ware and starch potatoes. Other farmers buy each year an amount of basic seed on the market and multiply this on their own farm. The harvest is used in the next season as seed, for e.g., ware production. In The Netherlands, farmers of ware potatoes are legally obliged to use only certified seed. They are allowed to multiply basic seed on their own farm but only if this is inspected and certified by the official seed agency.

In the present seed market, cultivars protected by breeder's rights play a dominant role. Only the possessor of the rights and his licensees are allowed to sell seed potatoes of these cultivars. The owner of the breeder's rights determines the acreage of seed potatoes to be grown. For very popular cultivars, this can be a tool to keep prices high. Only for cultivars of which the breeder's rights have expired, free competition among merchants is possible.[1]

Most developing countries do not have a large domestic potato seed market. If there is a seed potato multiplication programme in the country, often only relatively small quantities of certified seed are made available. Moreover, the quality of this seed is not always meeting the requirements of potato farmers, so they prefer, if available, imported

[1] Breeder's rights are acknowledged in countries that have ratified the Convention of the International Union for the Protection of New Varieties of Plants (UPOV) (at the end of 2004, 58 countries). States wanting to accede to the UPOV Convention have laws on plant variety protection in line with the 1991 Act of the Convention. That means that they recognize breeder's rights (duration for potatoes at least 20 years) and accept to pay levies to the breeder when growing his cultivar for seed (Jördens, 2000).

seed. If no better alternatives are available, farmers try to buy seed from 'healthy' regions or from colleagues who are specializing in the production of seed potatoes.

3.2.1.2 International market

Apart from home-grown seed, many developed countries import basic and certified seed (Table 3.2).

Basic seed is mostly used for further multiplication, certified seed for the production of ware and starch potatoes. In Western Europe, Italy, Spain, France and Germany are large importers of potato seed. Countries such as France and Germany both have a rather extensive seed production themselves. One of the reasons for importation of seed in these countries is that they want specific cultivars that are protected by breeder's rights elsewhere. Other reasons can be, the imported seed is of a better quality and less favourable conditions for several multiplications of seed in the field. In such cases, basic seed is imported to be multiplied once or twice.

In developing countries, where potatoes are an important crop, also large amounts of seed potatoes are imported from mainly European countries. In other parts of the world, only Canada is exporting considerable quantities of seed (Table 3.3), mainly to Central and South America.

Algeria, Egypt, Morocco and Tunisia are the main importing countries of seed potatoes in this group. Although these countries are trying to become more self-sufficient for seed potatoes, it appears to be very difficult under their growing conditions to reach a satisfactory level of quality. During the past 20 years, hardly any extension of potato seed production has occurred in these countries.

Table 3.2 Main importing countries for seed potatoes ($\times 1,000$ tonnes).

Country	1995/96	1996/97	2000/01	2002/02	2003/04
Algeria[a]	19.4	62.4	79.0	117.5	109.7
Italy	97.7	104.8	89.8	92.4	93.8
Spain	72.0	60.4	75.1	63.3	78.3
Portugal	62.8	61.8	61.2	41.7	53.0
Egypt[a]	61.1	43.6	69.6	59.2	59.5
France	65.5	62.5	72.3	52.3	51.2
Germany	61.3′	52.7	56.0	67.4	56.1
Morocco[a]	52.6	23.0	36.1	42.8	35.8
United Kingdom	31.5	19.0	28.6	8.6	16.4
Belgium	57.1	43.4	51.1	48.2	54.3
Netherlands	18.9	5.5	17.6	41.4	44.7
Greece	22.8	22.5	12.4	18.1	22.8
Tunisia[a]	13.7	24.0	21.6	46.0	28.5
Israel[a]	11.4	14.6	13.1	22.1	25.0
Russia[a]	6.3	5.4	6.5	21.6	33.2

[a] From EU–15 countries only.

Source: Eurostat.

Table 3.3 Main exporting countries of seed potatoes in the world ($\times 1000$ tons).

Country	1995/96	1996/97	2000/01	2002/03	2003/2004
The Netherlands	617	499	644	669	708
France	89	67	84	96	98
United Kingdom	72	77	69	109	93
Denmark	37	41	42	43	42
Belgium	29	37	42	51	65
Germany	30	24	28	47	39
Canada	182	–	107	126	–
USA	23	–	27	20	21

Source: Eurostat, Statistics Canada, US Potato Board.

In sub-tropical and tropical regions, it is often possible to grow potatoes twice a year or even more frequently. Countries in these regions import seed potatoes for ware potato production in spring and summer. However, seed for the autumn season is produced by multiplying imported basic seed planted in winter or spring.

The main seed potato exporting countries in the world are Canada, France, The Netherlands and the UK, of which The Netherlands are by far the largest exporter (Table 3.3). The Netherlands have relatively favourable climatic and growing conditions for the production of healthy seed; they have a large number of cultivars and a good infrastructure for storage, handling and exporting of seed potatoes.

There is a very large number of cultivars exported by the countries mentioned previously. However, the following cultivars produced in Europe are among the most popular: Désirée, Spunta, Monalisa, Bintje, Agria, Maris Piper and Cara. In North America, cultivars such as Russet Burbank, Shepody and Yukon Gold are among the most exported cultivars.

3.2.2 Mini-tubers

New rapid multiplication techniques for the production of mini-tubers are based on tissue culture, e.g. the use of hydroponics, and allow the relatively cheap mass production of tuberlets. The hydroponic technique consists of the production of mini-tubers by repeated harvesting on liquid cultures in a greenhouse.

Mini-tubers often have a diameter of 5–20 mm and weigh 0.5–5 g. Provided that the initial plant material was free from diseases, including latent ones, these mini-tubers can be used as pre-basic seed. In many countries, healthy mini-tubers are the basis for seed multiplication programmes, as this reduces the number of multiplications and hence the risk of contamination with diseases and pests in the field. More details on the production of mini-tubers are given in Chapter 32 (Millam and Sharma, this volume).

In recent years, an international market has developed for this type of pre-basic seed. An Australian/Indian company, with branches in a number of countries, is market leader at present. This company is mainly producing in and for Asian countries.

3.2.3 True potato seed

Since the beginning of the 1980s, ware potato production based on TPS has been adopted in a number of developing countries such as India, Bangladesh and Vietnam (Almekinders et al., 1996). In 1999 in Vietnam, 10% of the total potato area was planted with TPS-derived seedlings (Fuglie et al., 2001). Use of TPS is cheaper than seed tubers of the same health status. There are no storage costs, and transportation costs are much lower than those of seed tubers, particularly where transportation is to distant, inaccessible regions (Almekinders et al., 1996).

It soon proved that production of seedling tubers on nursery beds gave better results than direct drilling in the field or transplanting into the field from nursery beds (Wiersema, 1986). Although the technology has been adopted by farmers among a number of potato-producing countries mainly in Asia, there has not been a major breakthrough until now. One of the major drawbacks of TPS is its heterogeneity in tuber quality. Moreover, tuber size is often small. Nowadays, hybrid seed is almost always used at a rate of 120–350 g per ha (Fuglie et al., 2001). Chilver et al. (1999) concluded that widespread geographic applicability of TPS is not to be expected, even in India where breeding for TPS families has been applied since the early 1980s. Potential for TPS is greatest when seed costs exceed 20% of the production value.

Most TPS is produced in India and Chile. An indicative price for TPS is US$600–800 per kg (Fuglie et al., 2001).

3.3 BARRIERS TO MARKETS IN SEED POTATOES

On the international market for seed potatoes, there are a number of barriers hampering trade. The most important are the presence of quarantine pests and diseases and also non-quarantine ones in seed potatoes. Other restrictions include required absence of soil, etc. and the breeder's rights.

3.3.1 Quarantine diseases and pests

Many countries do not allow seed potatoes to enter the country if they are contaminated with a quarantine disease or pest. Quarantine organisms have been internationally defined as 'a pest of potential economic importance to the area endangered thereby and not yet present, or present but not widely distributed and being officially controlled' (Duringhof, 2000). The European and Mediterranean Plant Protection Organization (EPPO) of which all countries in these regions are a member has two lists of quarantine diseases and pests, A1 with diseases and pests not present in the EPPO region and A2 containing diseases and pests already introduced into parts of the EPPO region.

Potato diseases and pests figuring on these lists include Potato smut (*Thecaphora solani*), False root knot nematode (*Nacobbus aberrans*) (A1), Tomato spotted wilt virus, Potato tuber spindle viroid, brown rot (*Ralstonia solanacearum*), ring rot (*Clavibacter michiganensis* subsp. *sepedonicus*), wart disease (*Synchitrium endobioticum*), White potato cyst nematode (*Globodera pallida*), Root knot nematode (*Meloidogyne chitwoodi/fallax)* and Colorado beetle (*Leptinotarsa decemlineata*) (A2) (OEPP/EPPO, 2004).

Importing countries normally do not allow consignments of potatoes containing quarantine diseases to enter the country. Even the presence of a quarantine disease in just one region of an exporting country can be a reason for other countries not to accept seed potatoes from there.

3.3.2 Non-quarantine diseases and pests

Most seed importing countries are setting their own phytosanitary standards for diseases such as common scab (*Streptomyces scabies*), black scurf (*Rhizoctonia solani*) and silver scurf (*Helminthosporium solani*). Some importing countries have the seed potatoes inspected by their own inspectors at the port of the exporting country prior to shipping, to avoid refusal at the port of destination, causing extra costs.

3.3.3 Breeder's rights

Breeder's rights may prevent export of protected cultivars to countries that have not ratified the UPOV convention. In this case, the breeder risks multiplication of seed of his cultivar without receiving royalties. As a consequence, he will be reluctant to export seed potatoes to such countries.

REFERENCES

Almekinders C.J.M., A.S. Chilver and H.M. Renia, 1996, *Potato Res.* 39, 289–303.
Beukema H.P. and D.E. van der Zaag, 1990, PUDOC, Wageningen, 208.
Chilver A., T.S. Walker, V.S. Khatana, H. Fano, R. Suherman and A. Rizk, 1999, Working Paper no. 1999–3, CIP, Lima, Peru.
Duringhof H.A., 2000, Proc. 4th World Potato Congress, Amsterdam, 49–50.
Dongyu Q., X. Kaiyun, J. Liping, B. Chunsong and D. Shaoguang, 2004, Proc. 5th World Potato. Congr., Kunming, Yunnan, China.
Fuglie K.O., Do Thi Bich Nga, Dao Huy Chien and Nguyen Thi Hoa, 2001, The International Potato Center, Lima, Peru, 25.
Grigoriadou K. and N. Leventakis, 1999, *Potato Res.* 42, 607–610.
Jördens R., 2000, Breeder's rights, an adequate instrument to the benefit of breeders, farmers and the society, Proc. 4th World Potato Congress, Amsterdam, 267–277; www.upov.int.
OEPP/EPPO, 2004, Standard PM 1/2 (13); www.eppo.org.
Van der Zaag D.E. and C.D. van Loon, 1987, *Potato Res.* 30, 452
Van Loon C.D., 1993, Production of ware potatoes, Growers manual no. 57, PAGV, Lelystad, 34 (in Dutch).
Wiersema S.G., 1986, *Potato Res.* 29, 225–237.

Part II

GENETICS AND GENETIC RESOURCES

Chapter 4

Molecular Taxonomy

Ronald G. van den Berg[1] and Mirjam M.J. Jacobs[2]

[1]*Biosystematics Group, Wageningen University, Gen. Foulkesweg 37, 6703 BL Wageningen, The Netherlands;*
[2]*Plant Research International, PO Box 16, 6700 AA Wageningen, The Netherlands*

4.1 INTRODUCTION

In this chapter, we look at the results of analyses of molecular data sets that have been used to answer questions on the taxonomy of the group of tuber-bearing *Solanum* spp., *Solanum* sect. *Petota*, the cultivated potato and its wild relatives.

4.2 TAXONOMIC BACKGROUND

4.2.1 Wild and cultivated potatoes

The cultivated potato is an unusual crop in that it has an extremely large secondary genepool consisting of related wild species that are tuber-bearing, albeit with small unedible tubers. The taxonomy of the cultivated potato and its wild relatives has been the subject of study for many years. Most of these studies relied on morphological observations and, on a limited scale, experimental methods like cytogenetics and hybridization experiments. More than 200 species have been described and many infraspecific taxa. These taxa have been classified in series, with different authors recognizing different numbers of series, often with different circumscriptions. Two authorative treatments (Correll, 1962; Hawkes, 1990) recognized 26 and 21 series, respectively (Table 4.1). Hawkes (1989) suggested a division of the series into two superseries, *Stellata* and *Rotata*, emphasizing the outline of the corolla as a major distinctive character.

Some of the series contain only one or just a few species, indicating that their relationship to the other species is not clear. On the contrary, series such as *Piurana* and, especially, *Tuberosa*, are large groups of species that may not be closely related to each other.

Hijmans and Spooner (2001) and Hijmans et al. (2002) documented the geographic distribution of wild potato species, with the majority occurring in Argentina, Bolivia, Mexico and Peru, many with only restricted distribution areas.

Potato Biology and Biotechnology: Advances and Perspectives
D. Vreugdenhil (Editor)

Table 4.1 Series according to Hawkes (1990).

Subsection *Estolonifera*
Series *Etuberosa*
Series *Juglandifolia*
Subsection *Potatoe*
Superseries *Stellata*
Series *Morelliforme*
Series *Bulbocastana*
Series *Pinnatisecta*
Series *Polyadenia*
Series *Commersoniana*
Series *Circaeifolia*
Series *Lignicaulia*
Series *Olmosiana*
Series *Yungasensa*
Superseries *Rotata*
Series *Megistacroloba*
Series *Cuneoalata*
Series *Conicibaccata*
Series *Piurana*
Series *Ingifolia*
Series *Maglia*
Series *Tuberosa*
Series *Acaulia*
Series *Longipedicellata*
Series *Demissa*

There is a polyploid series present with diploids, triploids, tetraploids, pentaploids and hexaploids. The polyploids are considered to be allopolyploids derived from hybdridization events involving $2n$ gametes (Chapter 10, Gavrilenko, this volume). The odd-numbered polyploids, while mostly sterile, are able to maintain themselves vegetatively through the tubers.

The cultivated potato, *Solanum tuberosum* L., is accommodated in series *Tuberosa*, a rather large and variable group without clear diagnostic characters. The link between wild and cultivated potatoes, the direct ancestors of the crop, must be looked for in the so-called brevicaule complex, a group of morphologically variable, diploid species within series *Tuberosa*. Within this complex, about 20 species have been distinguished, but Ugent (1966) suggested that these could be drastically reduced to one species (*Solanum brevicaule*) and Van den Berg et al. (1998) by and large confirmed that conclusion. Morphologically, many of the wild species in the brevicaule complex are similar to some of the cultivated potatoes, the main differences being found in leaf dissection, in corolla colour and – obviously – in the tuber. The origin of the cultivated potatoes has been described as the result of successive hybridizations between diploid members of the brevicaule complex, accompanied by chromosome doubling leading to the tetraploid forms. The crop itself has been classified into seven cultivated species (*Solanum ajanhuiri*,

Table 4.2 Alternative classifications of the cultivated potatoes

Cultigen	Groups (Dodds, 1962)	Groups (Huaman and Spooner, 2002)
Solanum tuberosum	Tuberosum group	[a]
Solanum stenotomum	Stenotomum group	Stenotomum Group
Solanum phureja	Phureja group	Phureja Group
Solanum chaucha	Chaucha group	Chaucha Group
Solanum andigena	Andigena group	Andigenum Group
Solanum curtilobum	*S.* × *curtilobum*	Curtilobum Group
Solanum juzepczukii	*S.* × *juzepczukii*	Juzepczukii Group
Solanum ajanhuiri		Ajanhuiri Group
		Chilotanum Group

[a] Modern varieties, cultivar-group name(s) yet to be proposed.

Solanum chaucha, Solanum curtilobum, Solanum juzepczukii, Solanum phureja, Solanum stenotomum and *S. tuberosum* with two subspecies, *tuberosum* and *andigena*), showing several ploidy levels. The discussion about the taxonomic status of cultivated plant material (Hetterscheid and Brandenburg, 1995) suggests that the taxon 'species' (with its connotation of a product resulting from evolutionary processes) is not suitable for the classification of cultivated plants as the influence of humans seriously disturbs the patterns of variation used to classify species. Rather, cultivated material should be treated as artificial entities such as landraces or cultivars and classified into cultivar-groups as advocated in the International Code of Nomenclature of Cultivated Plants (ICNCP, 2004). This was anticipated by Dodds (1962), who, in an appendix to Correll's book, suggested the informal groups Stenotomum, Phureja, Chaucha, Andigena and Tuberosum within the species *S. tuberosum* to accommodate the cultivated potatoes. Huaman and Spooner (2002) suggested a similar solution with eight groups (Ajanhuiri, Juzepczukii, Curtilobum, Chilotanum, Andigenum, Chaucha, Phureja and Stenotomum).

The crop 'potato', making up the total of these groups, can still be assigned to the 'species' *S. tuberosum*, if so desired. This species name should then be considered as a cultigen (a species consisting of cultivated plants only, and as such without wild representatives, without a natural geographic distribution area and without a natural population structure). If the six other species names are used, these too are to be considered as cultigens (Table 4.2).

4.2.2 The evolutionary framework

The place of origin of the group of tuber-bearing potato species has been suggested to be the Mexican/Central American area, where those species are found that are considered to be phylogenetically primitive. These species are diploids, with stellate corollas and an endosperm balance number (EBN) of 1 (EBN refers to a genetic isolating mechanism that allows crosses between species with the same EBN and prevents crosses between different EBN groups; there are five combinations of ploidy level and EBN that determine crossability groups: 2x/EBN1, 2x/EBN2, 4x/EBN2, 4x/EBN4 and 6x/EBN4; Hawkes, 1990).

The further history of the group has been principally determined by two migrations across the landbridge between North and South America. A first migration southward from the Mexican/Central American area introduced the diploid tuber-bearing species to the variety of niches available in the South American continent, especially those in the mountain range of the Andes. This provoked a rapid speciation, producing the numerous species now occurring in Ecuador, Peru, Bolivia and Argentina. This speciation was accompanied by an increase in EBN and chromosome doubling. Morphologically, the corolla shape developed from stellate to rotate. A northward migration led to the establishment of polyploid species of the series *Conicibaccata* in Central America and, finally, to the derived polyploids nowadays found in Mexico, which include the well-known hexaploid species *Solanum demissum*.

The cultivated forms originated in the area around lake Titicaca, on the border of Peru and Bolivia, where several members of the brevicaule complex still occur.

4.2.3 Remaining taxonomic problems

Much of what is known about the taxonomy of potato is due to the work of two formidable taxonomists, Jack Hawkes and Carlos Ochoa. They described numerous species, classified them in series and provided keys based on morphological characters. However, these keys are generally difficult to apply because of the extensive variability in characters such as leaf dissection, pubescence and corolla colour. Many described species are extremely similar to each other, and the group of tuber-bearing *Solanum* spp. seems to be somewhat over-classified, with the application of a rather narrow typological species concept, where all deviations from the 'typical' habit are considered to be due to hybridization. Although hybridization within EBN groups is certainly taking place, another approach would be the recognition of a smaller number of broader circumscribed species, applying a polythetic species concept that allows overlap of character states among species. In certain groups, there is a lack of distinctive characters and species boundaries are difficult to trace. Especially the interaction between wild and cultivated forms and the influence of human selection have obscured species boundaries, and in some cases, described species might be weedy relatives of cultivated plants or escapes from cultivation.

Also, the series classification is problematic, with some series difficult to distinguish from each other and others containing subgroups that could be distinguished as separate series. Consistent with work in other groups within the genus *Solanum* (Knapp, 1991, 2000), it would be advisable to apply the informal concept of 'species groups' instead of the formal taxon 'series'.

The advance of molecular methods has offered the hope to arrive at solutions of the aforementioned problems and improve our understanding of the taxonomy of the potato.

4.3 MOLECULAR DATA

4.3.1 Molecular markers applied to tuber-bearing *Solanum* spp.

The available morphological data on potato species have been supplemented with data from cytology, serology, isozymes and several types of DNA data. The cytological data have helped in acquiring more insight into the origin and distribution of polyploids in the group (Swaminathan and Howard, 1953), the serological data gave indications of interrelationships among groups of species but were difficult to interpret (Lester, 1965) and the isozyme data provided valuable information mainly on the diversity of the cultivated forms (Quiros and McHale, 1985; Douches and Quiros, 1988).

DNA data are basically in one of two types: restriction site and primer-based data [restriction fragment length polymorphism (RFLP), random amplified polymorphic DNA (RAPD), amplified fragment length polymorphism (AFLP), simple sequence repeat (SSR); Chapter 5, Gebhardt, this volume], giving genome-wide, multilocus information and sequence data providing detailed, single locus information on only a small part of the genome. There is a strong relationship between the level of variability of a molecular marker and its suitability at a given taxonomic level.

4.3.2 Methods of analysis of molecular data sets – phenetic versus cladistic approaches

The preferred method to visualize taxonomic interrelationship is to construct bifurcating trees (although scatter plots from ordination techniques have also been found useful). It is important to distinguish the two fundamentally different approaches to tree building, i.e. phenetic versus cladistic approaches, which are distance-based versus character-based, respectively. The distance-based approach calculates the pairwise distances between all combinations of the investigated entities [often called operational taxonomic units (OTUs)], resulting in a triangular distance (or similarity) matrix. Starting from this matrix, different algorithms are used to produce distance-based trees or dendrograms. Consecutive clustering of OTUs with the smallest distances results in an Unweighted Pair Group Method using arithmic averages (UPGMA) tree, whereas for Neighbour Joining trees at each clustering step the effect on the total tree length is taken into account. Both approaches make use of overall similarity based on all the characters simultaneously. Character-based approaches try to construct a tree topology where the character states of each character can be placed in a consistent way (e.g. such that a character state changes into another state just once on the tree). The branches in the tree are considered to be natural groups, called clades (hence cladistic approach). Usually, an analysis generates many possible character-based trees (and just one or very few distance-based trees), making it necessary to adhere to an optimality criterion to choose the 'best' tree. In the character-based approach, this is often the parsimony criterion where the shortest tree (with the minimum number of steps between character states) is considered best.

A comparison of the results of the distance- and character-based approaches showing the topologies of the resulting trees is given in Fig. 4.1.

4.3.3 Application of molecular data to the taxonomy of the tuber-bearing *Solanum* spp.

4.3.3.1 Delimitation of the group

The genus *Solanum* has been subdivided into seven subgenera. The group of wild relatives of the potato is classified within the subgenus *Potatoe* in section *Petota*. Hawkes (1989)

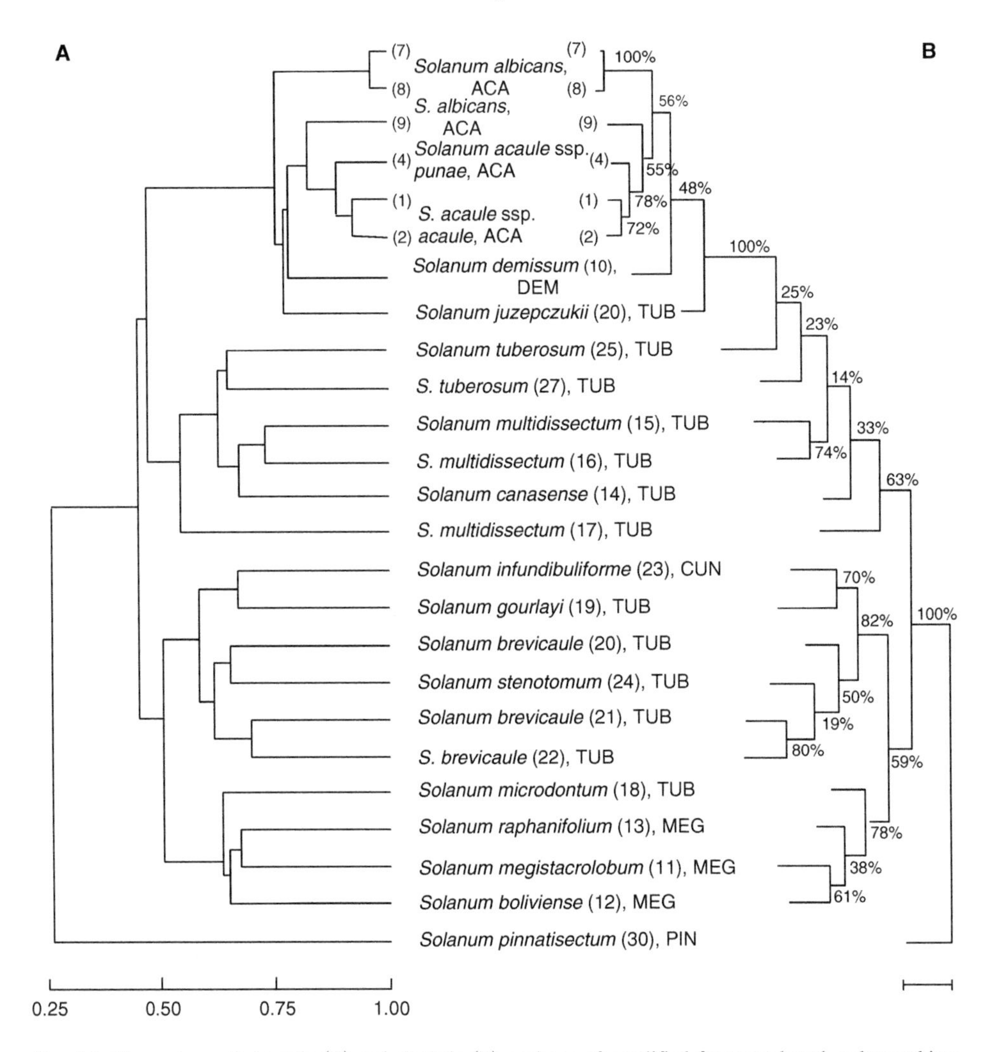

Fig. 4.1. Comparison of phenetic (A) and cladistic (B) analyses of amplified fragment length polymorphism (AFLP) data (Kardolus et al., 1998).

subdivided this section into two subsections, *Potatoe* and *Estolonifera*, accommodating two non-tuber-bearing series (*Etuberosa* and *Juglandifolia*) in the latter. Using chloroplast DNA RFLPs, Spooner et al. (1993) showed that these two non-tuber-bearing series were in fact less closely related to the tuber-bearing series than to the tomato and should be excluded from section *Petota*. This article also presented conclusive evidence for the inclusion of the genus *Lycopersicon* in the genus *Solanum*, as a section closely related to, but separate from, the potatoes. The nomenclatural consequences of this were published by Child (1990).

Kardolus (1998) showed a scatterplot (Fig. 4.2) of the first two multidimensional scaling axes calculated from an AFLP data set, where most of the investigated tuber-bearing

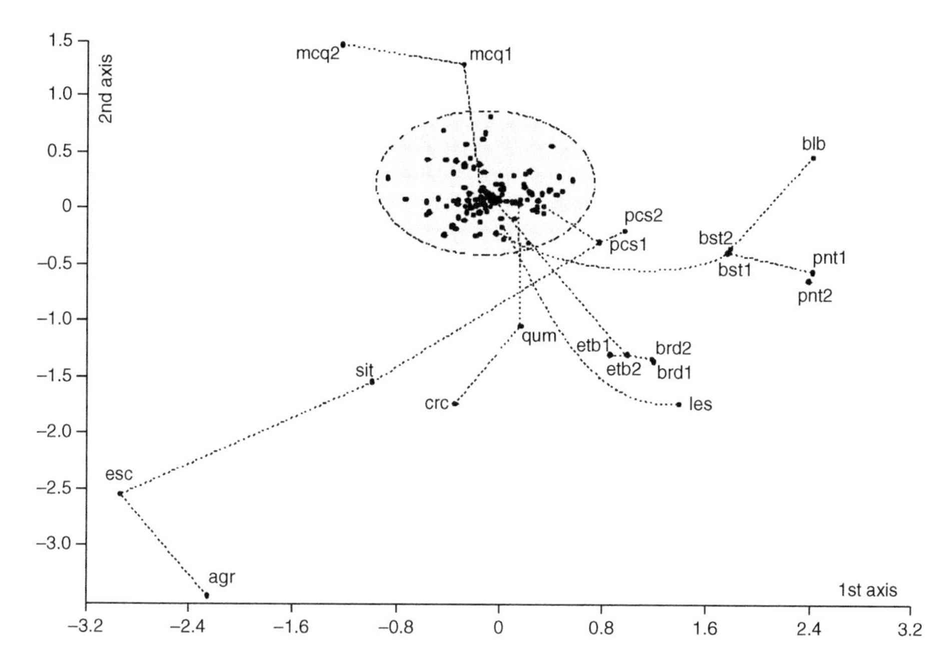

Fig. 4.2. Multidimensional scaling plot of 141 *Solanum* spp., calculated from an amplified fragment length polymorphism (AFLP) data set, with a Minimum Spanning Tree superimposed (Kardolus, 1998).

species form a dense cluster with only a few species outside this core group. Besides three tomato species and members of series *Etuberosa*, also representatives of the Mexican diploid species, series *Circaeifolia* and two accessions of *Solanum mochiquense* were plotted away from the dense central cluster. This would indicate that the Mexican diploid species and the South American species *Solanum circaeifolium* and *S. mochiquense* are relatively distantly related to the South American species and the Mexican polyploids.

4.3.3.2 Overall phylogeny of the group based on molecular markers

Many studies have focused on the elucidation of the structure within the group of tuber-bearing *Solanum* spp., using several molecular markers. Most of these have used RFLPs of the chloroplast (Hosaka et al., 1984; Spooner et al., 1991a; Spooner and Sytsma, 1992; Spooner and Castillo, 1997) or the nuclear genome (Debener et al., 1990). Later, the so-called AFLPs were applied (Kardolus, 1998; Kardolus et al., 1998), and recently, SSR (Bryan et al., 1999) and sequence data (Volkov et al., 2001, 2003) became available.

Hosaka et al. (1984) used cpDNA digest patterns to study the interrelationships of 26 species of section *Petota*, supplemented with four outgroup species of the genus *Lycopersicon* and the series *Juglandifolia* and *Etuberosa*. They found clear differences between the outgroup species and the wild potato species, but within section *Petota* were only able to distinguish the Mexican diploid species from the rest (comprising the Mexican polyploids and the South American species).

Debener et al. (1990) used nuclear RFLPs, studying 14 wild and 2 cultivated potato species, with *Solanum etuberosum* as an outgroup. They could distinguish clearly separated groups, with all the species of series *Tuberosa* in two related groups, one with wild representatives and one with the cultivated potato, *S. tuberosum*, clustering with *S. stenotomum* and *Solanum canasense*. *Solanum acaule* and *S. demissum* together formed a well-separated branch, and *S. etuberosum* was most distant.

Spooner and collaborators (Spooner et al., 1991a; Spooner and Sytsma, 1992; Spooner and Castillo, 1997) used probes rather than directly observed cpDNA digest patterns and greatly extended the number of species studied. They provided evidence for four clades (Fig. 4.3):

(1) The Mexican diploids, but excluding *Solanum bulbocastanum, Solanum cardiophyllum* and *Solanum verrucosum*.
(2) *S. bulbocastanum* and *S. cardiophyllum*.
(3) Members of series *Piurana*, with a number of species from other series.
(4) *Solanum verrucosum*, all remaining South American species and the polyploid species from Mexico and Central America.

Kardolus et al. (1998) and Kardolus (1998) applied AFLPs. In the latter study, 53 species were investigated (Fig. 4.4). The method proved to be highly efficient in producing 997 markers with three primer combinations. Three tomato species and two species from series *Etuberosa* constituted the outgroups. Representatives of the Mexican diploids were placed as the sistergroup of the rest of the tuber-bearing species. The species of series *Tuberosa* were subdivided into geographical groups, with *S. tuberosum* in the Peruvian group associated with species such as *S. canasense, Solanum bukasovii* and *Solanum multidissectum* and other members of the brevicaule complex from Bolivia and Argentina grouping together. *Solanum demissum* was united with species of series *Acaulia*, recalling the results of Debener et al. (1990). Series *Circaeifolia* was placed as the most primitive group of the South American species.

Bryan et al. (1999) used polymorphic SSRs from the chloroplast genome (cpSSRs), studying 24 species and 30 cultivars. This marker system detected high levels of inter-specific cpDNA variation, and the authors suggest its utility in population genetics, germplasm management and phylogenetic studies. The resulting UPGMA tree, however, does not provide much resolution, with cultivated accessions clustering among the wild species (indicating the introgression from wild species into cultivated material) and a tree topology that does not enable the recognition of clear subgroups.

Volkov et al. (2001) used nucleotide sequences of 5S ribosomal DNA genes of 26 wild species, 4 *S. tuberosum*-breeding lines and a tomato accession, with *Solanum dulcamara* as outgroup. This first sequence data set proved difficult to analyse because of the high abundance of indels in comparison with base substitutions, and the dendrograms resulting from different clustering algorithms differed essentially from each other. Because the dendrogram topology was extremely unstable, the authors evaluated the indels 'manually', producing a schematic representation of the molecular evolution. This shows the Mexican diploids (series *Polyadenia, Pinnatisecta* and *Bulbocastana*) as basal and a group with rather conserved 5S rDNA organization comprising *Solanum brevidens, Solanum commersonii, S. circaeifolium* and – surprisingly – *S. bukasovii*, which is far removed

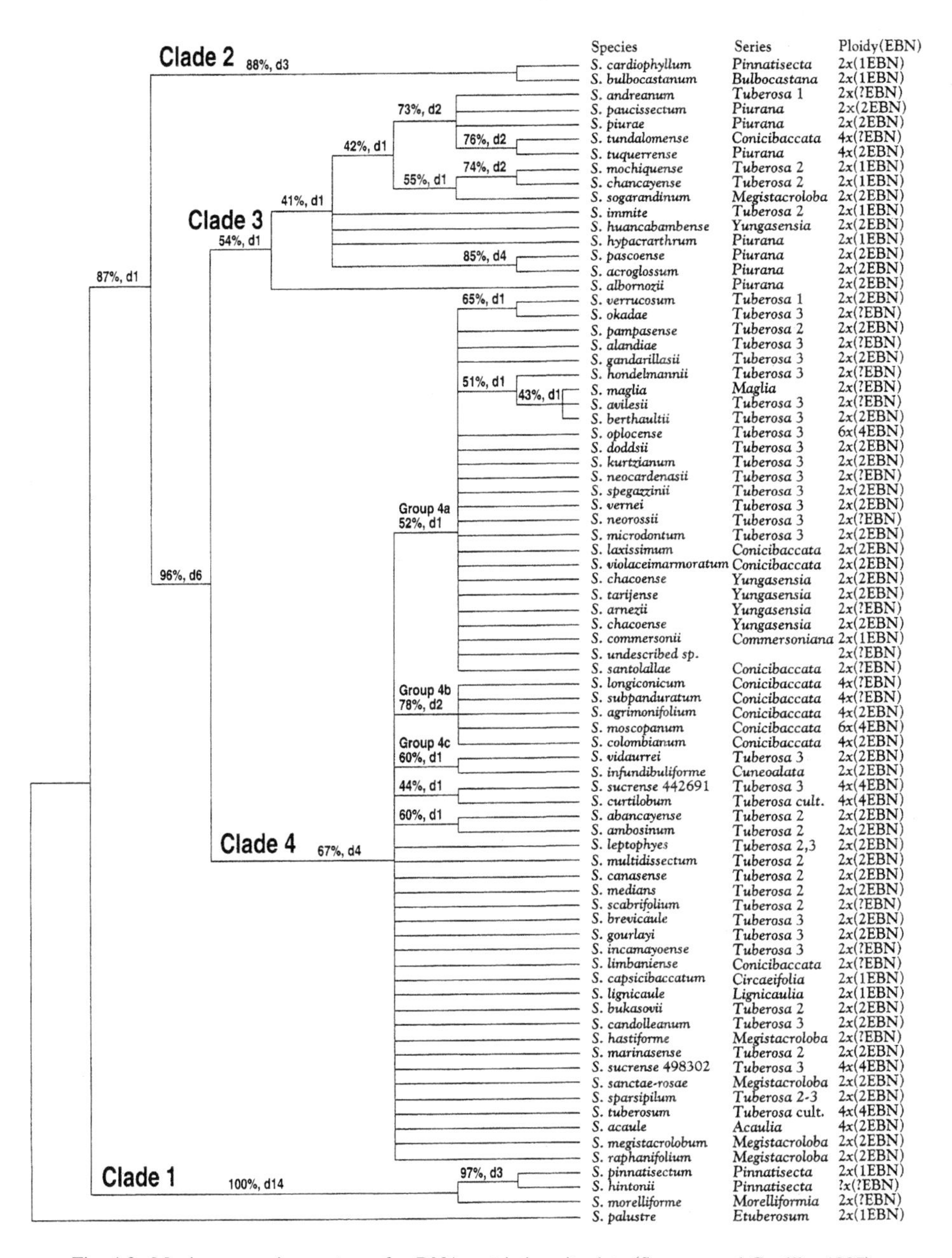

Fig. 4.3. Maximum parsimony tree of cpDNA restriction site data (Spooner and Castillo, 1997).

from the remaining cluster of species belonging to superseries *Rotata*. Within the latter group, the species of series *Tuberosa* are divided into several subgroups, and *S. acaule* and *S. demissum* are grouped together. Although the overall picture conforms with earlier results, the sequenced region does not seem to be optimal for phylogenetic reconstruction.

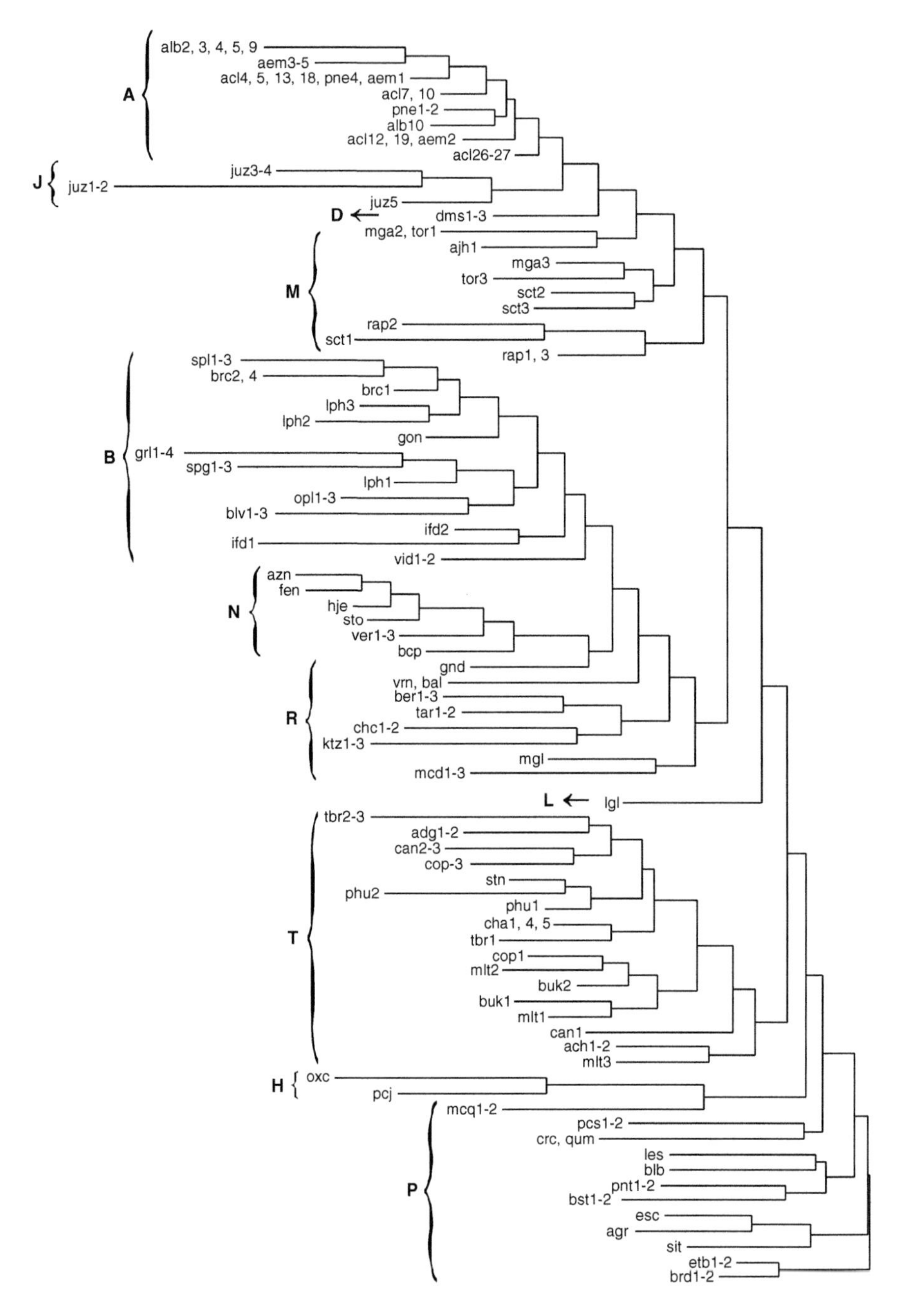

Fig. 4.4. Maximum parsimony tree of amplified fragment length polymorphism (AFLP) data (Kardolus, 1998).

Volkov et al. (2003) turned to the 5′ external transcribed spacer (ETS) region of rDNA, comparing 30 species of *Solanum* sect. *Petota*, with *S. dulcamara* as outgroup. Three structural variants of ETS (variants A–C) could be recognized (Fig. 4.5). Variant A is present in the outgroup *S. dulcamara*, in the non-tuber-bearing species of series *Etuberosa* and in the representatives of the Mexican diploid series *Bulbocastana, Pinnatisecta* and *Polyadenia*. Species of the series *Commersoniana* and *Circaeifolia* possess variant B, and variant C is present in all other investigated species. Variant C can be subdivided into two subgroups, C1 and C2. Group C1 contains species from the series *Megistacroloba, Conicibaccata* and *Acaulia*, whereas group C2 consists of all diploids of series *Tuberosa*. The dendrograms presented show many polytomies (multiple branching instead of dichotomously branching), indicating that resolution within the groups is mostly lacking. Also, representatives of a species like *S. demissum* are widely separated in all trees, indicative of intraspecific variation. According to the authors, the groups are defined by large rearrangements, while base substitutions allow additional discrimination of closely related species, and this broad range of resolving power is taken to suggest the utility of this marker system for phylogeny reconstruction. The authors further suggest – in contrast with the evolutionary scenario in Hawkes (1990) – an origin of primitive *Petota* spp. in South America, followed by a migration of primitive *Stellata* spp. to Mexico, and a development in South America from other primitive *Stellata* towards more advanced *Stellata* and *Rotata* spp.

Summarizing the data from the various studies mentioned above, it seems clear that our insight into the phylogenetic structure of the group of tuber-bearing *Solanum* spp. has been improved by molecular studies. The phylogenetic position of certain species, like e.g. the Mexican diploids, and the series *Circaeifolia*, and the reality of a *S. acaule/S. demissum* assemblage, are supported by several sources. However, the lack of resolution within section *Petota* (4 clades instead of 20 series based on chloroplast RFLP data) seems to be a real phenomenon. Except for the rather distinctive groups in Mexico, the differentiation among the other South American groups is not large, and it remains difficult to subdivide the group into natural units.

The studies discussed above have one serious problem in common: most of them were not able to sample the complete width of the variation of the group of tuber-bearing *Solanum* spp., and undersampling can influence the results of (especially cladistic) analyses. The most complete effort has been the studies by Spooner and collaborators. If one combines these three studies, 86 species from most of the series are considered, but this still is only less than half of the total number of species.

The most promising ways forward could be the extension of a molecular data set to encompass all the relevant taxa of the group (Jacobs et al., 2005) and the search for suitable nuclear sequences as undertaken by Spooner and collaborators [nitrate reductase (NIA), Rodriguez and Spooner, 2004; single-copy waxy gene (GBSSI), Spooner et al., 2004; internal transcribed spacer (ITS), Stephenson et al., 2004].

4.3.3.3 Detailed studies of parts of the group

Molecular data have been utilized to study certain groups of species in detail. For convenience sake, these studies will be discussed according to the series names applied,

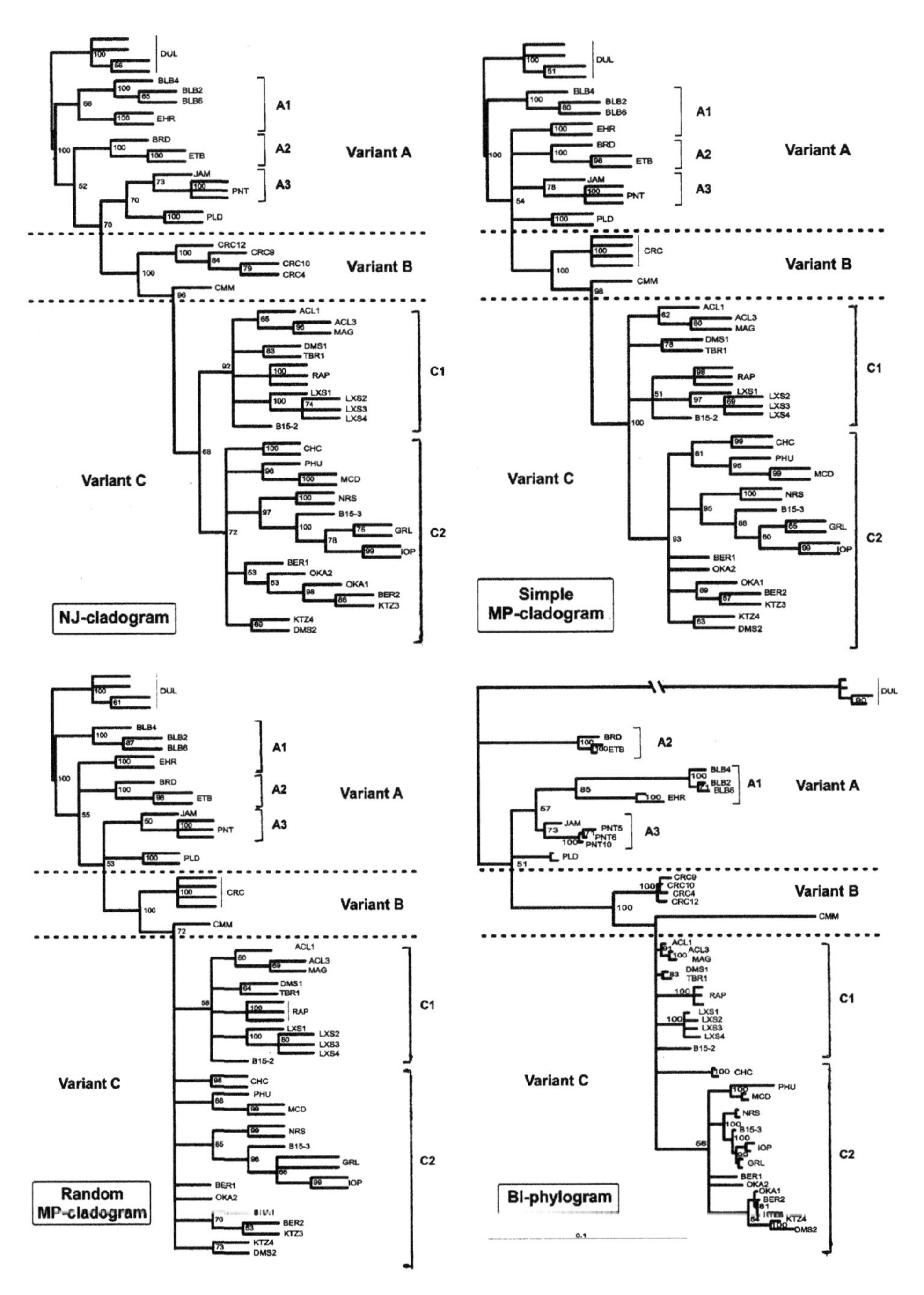

Fig. 4.5. Phenetic and cladistic trees of rDNA EST sequences (Volkov et al., 2003).

even though not all series received support in the phylogenetic studies mentioned in the previous paragraph.

Series *Acaulia*

Series *Acaulia* has attracted many workers, due to the extreme frost tolerance present in the species *S. acaule*. The pentaploid cultivated potato *S. curtilobum*, which is used to produce the freeze-dried 'chuno', resulted from crosses between Andigenum-type potatoes and the triploid cultigen *S. juzepczukii*, itself a cross between *S. acaule* and diploid Stenotomum potatoes. The taxa within the series comprise tetraploids and hexaploids, which have been recognized at different taxonomic levels by different authors.

Hosaka and Spooner (1992), using RFLPs of genomic DNA, studied 105 accessions of *S. acaule* including all four subspecies (*acaule, aemulans, albicans* and *punae*) that were recognized at that time. The results placed subspecies *albicans* as most distant (this hexaploid taxon was later raised to the species level), could not distinguish subspecies *acaule* and *punae* and divided subspecies *aemulans* into two groups, from the provinces La Rioja and Jujuy (Argentina), respectively.

Kardolus (1998), studying this group with AFLPs, also could not consistently distinguish the subspecies *acaule* and *punae*, but recognized a new, hexaploid subspecies, subspecies *palmirense*. The occurrence of this hexaploid cytotype within the species *S. acaule* may indicate the need to re-evaluate the recognition of *Solanum albicans* on the species level.

McGregor et al. (2002) investigated 314 accessions of the Centre for Genetic Resources, The Netherlands (CGN) germplasm collection with AFLPs and concluded that most plants were grouped in an UPGMA tree according to the species and subspecies designations in the passport data. The subspecies *acaule* and *punae* were distinguishable, although only separated by a small genetic distance. The classification of the hexaploid *palmirense* taxon in *S. acaule*, separate from *S. albicans*, was confirmed.

Nakagawa and Hosaka (2002) combined RFLP data from chloroplast and nuclear DNA to study the relationships between *S. acaule, S. albicans* and 27 morphologically closely related species. They found high similarity between *S. acaule, S. albicans* and *S. demissum*, and suggested *Solanum megistacrolobum* and *Solanum sanctae-rosae* as the closest relatives, and possibly involved in the origin of the series *Acaulia* spp.

Series *Circaeifolia*

Van den Berg et al. (2001a), using AFLPs and RAPDs, reinvestigated the taxa within this small series that is endemic to Bolivia. Hawkes and Hjerting (1989) and Ochoa (1990) had provided different classifications of these taxa, with different numbers of species, and using either subspecies or varieties to accommodate minor variants. An earlier morphological study (Van den Berg and Groendijk-Wilders, 1999) doubted the distinction of the recognized taxa, but especially the AFLPs supported four distinct groups, which can either be recognized at the species level or as infraspecific taxa within the species *S. circaeifolium*. These results indicate that certain taxa that can only be recognized with difficulty morphologically might be real entities, and polythetic support for species (where species lack unique diagnostic characters but must be distinguished by combinations of characters that often display overlapping ranges) is frequent in section *Petota*.

Series *Conicibaccata*

Castillo and Spooner (1997) generated cpDNA restriction data of 23 of the possibly 40 species in this group. Their cladistic analysis indicated that a history of hybridization and polyploidization render a reconstruction of the phylogeny extremely difficult. They recognized three groups separating the polyploids of Central Mexico to Southern Ecuador from the diploids of Northern Peru and Bolivia, with a third group suspected to be more related to series *Piurana*, and noted discordances between morphological species as currently defined and the cpDNA clades.

Series *Longipedicellata*

Van den Berg et al. (2001b) applied AFLPs, RAPDs and cpSSRs to study the species boundaries of the six tetraploid species assigned to series *Longipedicellata*. Confirming earlier morphological work (Spooner et al., 2001), not all species were supported by the AFLP and RAPD data (cpSSRs failed to distinguish any of the species), and a reduction of the number of species to be recognized is recommended. The cpDNA microsatellite results were completely discordant with the AFLP and RAPD results, which may have been caused by the lower number of markers generated.

Series *Bulbocastana, Morelliformia, Pinnatisecta* and *Polyadenia*

Lara-Cabrera and Spooner (2004) used AFLPs to study the species boundaries and taxonomic relationships of the diploid species of the United States, Mexico and Central America. They analysed the data set using both cladistic and phenetic approaches. The resulting cladograms and dendrograms showed approximately the same set of species groups, with incongruence occurring mainly within these groups and not among them.

Also, there were many points of agreement between the AFLP results and earlier cpDNA (Spooner and Sytsma, 1992) results. For instance, the separation of *S. bulbocastanum* and *Solanum clarum*, which had been classified in one series *Bulbocastana* by Hawkes (1990), and the combination of *S. clarum* and *Solanum morelliforme* in one group, was confirmed. Also, the subspecies of *S. cardiophyllum* were separated and should most likely be treated as different species, with subspecies *ehrenbergii* more related to the *Solanum brachistotrichium/Solanum stenophyllidium* assemblage. On the contrary, cpDNA data had combined *S. bulbocastanum* with *S. cardiophyllum* (exclusive of subspecies *ehrenbergii*), but these taxa were widely separated in the AFLP results, which could indicate the phenomenon of chloroplast capture. Most clades were species specific, and although many sistergroup relationships between pairs of species were recovered, the traditional series were not apparent.

It is clear that molecular data can be used to solve problems of species or series boundaries, but their utility strongly depends on the amount of variation generated and the application of appropriate methods of analysis. Chloroplast data are generally more suitable for higher taxonomic levels (species groups, series and subsections) and can be profitably complemented by AFLP markers to investigate species boundaries. SSRs do not provide much phylogenetic signal but seem useful at the population level. Lack of support for species in AFLP data may be taken to imply the need for a reduction in species numbers.

4.3.3.4 *Hybridization hypotheses*

Spooner and collaborators investigated many putative hybrid species (Spooner et al., 1991b: *Solanum raphanifolium*; Clausen and Spooner, 1998: *Solanum rechei*; Lara-Cabrera and Spooner, 2004: *Solanum michoacanum* and *Solanum sambucinum*). *Solanum raphanifolium* had been hypothesized to be a hybrid species between *S. canasense* and *S. megistacrolobum*. This hypothesis was tested with cpDNA and nuclear rDNA characters, with the expectation that the hybrid would possess a cpDNA pattern identical to one of the parents and an additive nuclear rDNA pattern from both parents. The results indicated that the putative parents were more closely related to each other than either was to *S. raphanifolium*, and the hybridization hypothesis was refuted, the striking morphological intermediacy notwithstanding. The hybrid origin of *S. rechei* from a cross between *Solanum microdontum* and *Solanum kurtzianum* was verified using nuclear RFLP profiles that showed additivity, although some of the bands present in the putative parents were also present in other species, and *S. rechei* also possessed unique RFLP bands. In the AFLP study by Lara-Cabrera and Spooner (2004) discussed above (Section 4.3.3.3), the two hybrid species *S. michoacanum* and *S. sambucinum* grouped with one of their parents (*Solanum pinnatisectum*) and not with the other. Unfortunately, additive species-specific AFLP markers were not found.

Raimondi et al. (2005) used 4 chloroplast and 11 nuclear SSRs to study the hybrid origin of *Solanum ruiz-lealii* from a cross between *S. kurtzianum* and *Solanum chacoense*. In most of the presented phenograms and principal coordinate analysis plots, the three species are clearly separated from each other, with *S. ruiz-lealii* closer to *S. chacoense*. Supplementing their molecular data with morphological and cytological observations, the authors conclude that *S. ruiz-lealii* is not a recent hybrid.

As the literature on the taxonomy of the potato abounds in hybridization hypotheses that are mainly based on perceived morphological intermediacy, the application of molecular data sets provides a strong tool to establish the true status of these putative hybrids.

4.3.3.5 *The cultivated potato*

Although, as mentioned in the first paragraph of this chapter, a more correct way to refer to cultivated potatoes would be by using cultivar-groups within the cultigen *S. tuberosum*, the species names that are used in almost all the cited literature will be used in this section. Seven cultigenic species *S. stenotomum, S. ajanhuiri, S. chaucha, S. phureja, S. curtilobum, S. juzepczukii* and *S. tuberosum* (with two subspecies: *andigena* and *tuberosum*) are currently recognized as cultivated species (Hawkes, 1990). All tetraploid South-American landraces are classified in *S. tuberosum* ssp. *andigena*. All modern cultivars known to us as the common potato can be accommodated in *S. tuberosum* ssp. *tuberosum*. The transition from subspecies *andigena* to subspecies *tuberosum* apparently resulted from transporting material from the short-day environment of the Peruvian/Bolivian Andes to long-day circumstances. This transport accompanied by adaptation is believed by Hawkes (1990) to have occurred twice: the first event would have taken place in Chile where original subspecies *andigena* material, brought here by migrating Indian tribes from the Andes, underwent adaptation to long-day length and cool climatic conditions, and the second time this development took place was in Europe after the Spaniards introduced the potato there.

Hawkes (1990) regarded the cultigen *S. stenotomum* as being the most primitive of the cultivated material and as the progenitor of the other cultivated 'taxa'. A wild diploid species like *Solanum leptophyes* would have been the progenitor of *S. stenotomum. Solanum tuberosum* ssp. *andigena* originated from a hybridization event between *S. stenotomum* and the wild species *Solanum sparsipilum. Solanum tuberosum* ssp. *tuberosum* later developed from subsp. *andigena.* Grun (1990) described a similar origin of the cultivated potato, with the primitive diploid cultigen *S. stenotomum* arising from a wild progenitor from the brevicaule complex.

The most extensive study using molecular data on the origin of *S. tuberosum*, the relationships among the cultivated species and the relationships between wild and cultivated species, has been conducted by Hosaka and co-workers. In a series of publications ranging from 1986 to 2004, they focused on restriction data of cpDNA of wild and cultivated potatoes.

Hosaka (1986) distinguished seven different chloroplast haplotypes in a selection of wild and cultivated species:

(1) type T was restricted to *S. tuberosum* ssp. *tuberosum*;
(2) type A was characteristic for *S. tuberosum* ssp. *andigena* and *Solanum maglia*;
(3) type S was found in *Solanum goniocalyx, S. phureja, S. stenotomum, S. chaucha* and one accession of subspecies *andigena*;
(4) type C was found in *S. acaule, S. bukasovii, S. canasense, S. multidissectum* and *S. juzepczukii*;
(5) type W was found in wild species and was considered as the most primitive type;
(6) type W′ was found in *S. chacoense* f. *gibberulosum*;
(7) type W″ was found in *Solanum tarijense.*

The author concluded that, indeed, the cultivated potatoes derived from *S. stenotomum*, which itself might have developed from *S. canasense*, and, furthermore, that the chloroplast genome of the European potato derived from Chilean material, which itself was the result of the combination of the nuclear genome of subspecies *andigena* with cytoplasm from an unknown species.

In 1988, a series of three articles on cpDNA data of potato were published by Hosaka and co-workers. Hosaka et al. (1988) showed that the differences between types T and W found with five different restriction enzymes in the earlier study (Hosaka, 1986) were in fact all caused by one physical deletion in the chloroplast genome of the T-type chloroplast. The authors concluded that the T-type chloroplast could easily have evolved from the primitive W type, whereas in the former publication this had not seemed probable.

Hosaka and Hanneman (1988a) found a geographical cline from the Andean region to coastal Chile, supporting the Andean origin of Chilean subspecies *tuberosum.* Material considered as a relic of the first European potato (a hybrid of the cultivar 'Myatt's Ashleaf') showed the A-type chloroplast, confirming Hawkes' opinion that the first

European potatoes were subspecies *andigena*, later replaced by subspecies *tuberosum* from Chile.

Hosaka and Hanneman (1988b) noted extensive cpDNA variation in cultivated potatoes as well as in wild potato species. They hypothesized that the Andean cultivated tetraploid potato, subspecies *andigena*, could have arisen many times from the cultivated diploids.

Hosaka (1995) determined the chloroplast types of 35 accessions of *S. stenotomum* and 97 accessions of putative ancestral wild species, including *S. brevicaule, S. bukasovii, Solanum candolleanum, S. canasense, S. leptophyes* and *S. multidissectum*. Except for *S. brevicaule*, which had only the W type, the wild species proved polymorphic for cpDNA types. Sexual polyploidization formed a wide cpDNA diversity among the Andean tetraploid potatoes and selection caused the limited diversity found in Chilean tetraploid potatoes.

Hosaka (2002) explored the maternal ancestry of the common potato by determining the presence/absence of a 241-bp deletion characteristic for the T-type cpDNA. Sixteen of 80 accessions of *S. tarijense, S. berthaultii* and *S. neorossii* showed the same deletion at the same position. Hosaka (2003) found that all the T-type accessions of cultivated potatoes shared this haplotype only with some accessions of *S. tarijense*. The author concluded that some populations of *S. tarijense* acted as the maternal ancestor of potato. Hosaka (2004), investigating 215 accessions of *S. stenotomum* and 286 accessions of *S. tuberosum* subsp. *andigena*, noted the absence of T-type chloroplast in *S. stenotomum* while this type was present in nine accessions of subsp. *andigena* and concluded that *S. stenotomum* did not play a role in the formation of the tetraploid potatoes.

All the data presented above are based on cpDNA and therefore only show maternal inheritance. Furthermore, the cpDNA types do not seem to be monomorphic within species, which makes it difficult to discuss ancestor/derivative relationships between species. There is a need for suitable nuclear markers (such as AFLPs or a suitable nuclear sequence) to complement the work done on the chloroplast genome. Many studies have taken this approach.

Debener et al. (1991) showed with nuclear RFLPs that *S. andigena, S. stenotomum* and *S. canasense* were very closely related to each other and could in fact not be distinguished with the single locus information. Miller and Spooner (1999) used single to low-copy nuclear RFLPs and RAPDs to investigate the species boundaries and relationships among the members of the brevicaule complex. They confirmed the separation of populations from Peru and immediately adjacent northwestern Boliva, including most cultivated accessions, and of populations from northwestern Boliva and Argentina. This had been found by Van den Berg et al. (1998) using morphological data and Kardolus et al. (1998) using AFLPs, which also showed *S. tuberosum* clustering together with the wild Peruvian species *S. canasense* and *S. multidissectum*. Miller and Spooner (1999) indicated the paraphyletic nature of the brevicaule complex and the need to reduce the number of species names in this group.

Raker and Spooner (2002) tested the genetic differences between accessions of *S. tuberosum* ssp. *andigena* and *S. tuberosum* ssp. *tuberosum* using nuclear DNA microsatellites. The two subspecies could be separated from each other although the separation is not very firm. Other cultivated species (*S. stenotomum* and *S. phureja*) and wild species (*S. bukasovii, S. multidissectum* and *S. canasense*) used in this study were mixed with *S. tuberosum* ssp. *andigena*.

Sukhotu et al. (2004) combined data on the cpDNA types of Hosaka with chloroplast microsatellite markers and nuclear RFLPs. The differences among cpDNA types were highly correlated with the microsatellite markers. The nuclear RFLPs supported the differentiation between the W type versus the C, S and A types, but not the differentiation among the three latter types, suggesting frequent genetic exchange among them. In a UPGMA dendrogram of the nuclear DNA restriction data, three clusters could be identified, with both *S. tuberosum* ssp. *andigena* and *S. tuberosum* ssp. *tuberosum* accessions placed together with most other cultivated *Andean* spp. and members of the brevicaule complex.

In a recent study of the brevicaule complex, Spooner et al. (2005a), using AFLP data, reconfirmed the distinction of the northern and southern subgroups within the complex and argued that cultivated potatoes have had a monophyletic origin in the northern part of the distribution area of the brevicaule complex, as all the landrace populations form a clade in the parsimony cladogram. The progenitor of the cultivated potato should thus be sought in the members of the brevicaule complex occurring in southern Peru. The authors note that these species are poorly defined and may have to be reduced to a single species, the earliest valid name being *S. bukasovii*. The brevicaule complex itself is designated to be polyphyletic.

The origin of our modern cultivated potato varieties and the manner of introduction of the cultivated potato in Europe have been the subject of controversy. According to many authors, the first potato material to be introduced in Europe belonged to *S. tuberosum* ssp. *andigena*. Most of the potato stock derived from this original material was believed to have been wiped out during the late-blight outbreak in Europe in the 1840s. After this, the breeding stock would have been replaced with introductions from Chile of *S. tuberosum* ssp. *tuberosum* material (Grun, 1990). Juzepczuk and Bukasov (1929) had, however, suggested that the early European introductions already consisted of subspecies *tuberosum* germplasm from Chile, because of the similarity in morphology and growing conditions.

Spooner et al. (2005b) published results from a nuclear microsatellite analysis of mainly Indian potato cultivars. The analysis included several accessions that were considered to be derived from *S. tuberosum* ssp. *andigena* to test the idea that the first potato introductions in the old world were actually subspecies *andigena*. Late blight was not recorded in India until 1870, so only after the late blight disaster in Europe. The andigena germplasm in India would therefore not have been eliminated by the epidemic. The microsatellite results showed, however, that all Indian cultivars, including those that were thought to be derived from subspecies *andigena*, clustered together with the subspecies *tuberosum* landraces and European cultivars. All 12 tested subspecies *andigena* landraces from Central and South America clustered together separately from this group. The andigena introduction theory was thus not supported. Spooner et al. (2005b) concluded that no remnant landraces of subspecies *andigena* were involved in the development of the Indian germplasm. Considering this evidence and other historical and cytological information, they suggested that the early introductions of cultivated potatoes of India (and Europe) came from both the Chile and the Andes. The Chilean landraces became the predominant breeding germplasm before the outbreak of late blight, likely because of their pre-adaptation to long-day/cool climate conditions.

Remarkably, five Indian cultivars that, based on the nuclear microsatellite data were linked to subspecies *tuberosum*, lacked the typical 241 bp deletion. This could have been caused by either a subspecies *tuberosum* progenitor lacking the typical deletion or the incorporation of other non-tuberosum accessions as maternal material.

Huaman and Spooner (2002) examined morphological support for the classification of landrace populations of cultivated potatoes. They recognized all landrace populations as a single species, *S. tuberosum*, with eight cultivar groups. Following the philosophy of cultivar-group classification, the remaining cultivated materials, e.g. the modern varieties, were not automatically classified as a ninth 'Tuberosum' group.

Many authors have suggested that molecular markers are appropriate to identify cultivars and reveal infraspecific variation (e.g. Debener et al., 1990; Hosaka et al., 1994; Bryan et al. 1999; Bornet et al., 2002), but these methods have not been used to produce an overall classification of cultivars. Most studies are restricted to the assessment of genetic diversity of cultivars or their discrimination with fingerprinting techniques (Görg et al., 1992).

Provan et al. (1999) used polymorphic chloroplast and nuclear SSRs to study the diversity in most modern potato cultivars grown in the UK. In total, 151 of 178 accessions tested showed the same chloroplast haplotype, named haplotype A, which corresponds with the T type of Hosaka (1986). A much higher diversity was found in the remaining accessions outside the T-type group, which were assigned to 25 different haplotypes. The diversity of the nuclear SSR loci did not show this difference between the T-type group and the rest. The authors suggested that the dominance of the T-type cytoplasm was caused by the use of only a limited number of maternal lineages in breeding programmes.

Bryan et al. (1999) using cpSSRs demonstrated that among a set of 30 tetraploid potato cultivars, a single chloroplast haplotype was prevalent and they attributed this to the widespread use as a female parent of the imported US cultivar 'Purple Chili' in the latter half of the 19th century. The chloroplast diversity that is present has arisen through introgression from wild and primitive cultivated material.

The low level of genetic diversity of European cultivated potatoes was confirmed in an analysis using ISSRs by Bornet et al. (2002). Their results showed that European potatoes are quite homogenous, and the genetic diversity was very low compared with Argentinian cultivars.

Molecular data have been used to address three main issues about the cultivated potato:

(1) the mode of origin of the crop and the relationships with its wild relatives;
(2) the relationship between the Andigena and Tuberosum groups and the introduction of the cultivated potato from South America to Europe and the rest of the world;
(3) the genetic diversity of the crop.

The conclusions about these issues are not unequivocal. Results from the chloroplast and nuclear genome conflict as to the role taxa like *S. tarijense, S. stenotomum* and the brevicaule complex have played in the origin of the crop. Different data sets give rise to different hypotheses on the multiple or single domestication event(s) that occurred, most probably, in southern Peru. The role that Chilean material played in the introduction of the cultivated potato in Europe has been clarified. The genetic diversity of the crop has been shown to have suffered a severe maternal bottleneck during the development of the modern cultivated potato. Finally, the classification of the modern cultivars of potato in subgroups has not really been addressed yet with molecular markers.

4.4 CONCLUSION

Molecular data have been used to establish the phylogeny of the group of tuber-bearing *Solanum* spp., to evaluate hybridization hypotheses, to evaluate infraspecific classifications, to establish the ancestry of the cultivated potato, to trace introgression from wild species and to assess genetic diversity within species and cultivated material. In the context of genebank management, the effect of seed increases on the diversity of genebank accessions (Del Rio and Bamberg, 2003), and the extent of redundancy (McGregor et al., 2002) has been studied. Furthermore, molecular data allow checking for misidentifications and can be utilized in risk-assessment studies.

Although the search for the phylogenetic structure of the group has suffered from a lack of resolution, at the species level, the utility of AFLPs is evident, as long as closely related taxa are compared. There remains a need for a suitable nuclear marker to fill the gap between the high level chloroplast derived data and the fingerprinting data like SSRs, but this will most probably be forthcoming in the near future.

REFERENCES

Bornet B., F. Goraguer, G. Joly and M. Branchard, 2002, *Genome* 45, 481.

Bryan G.J., J. McNicoll, G. Ramsay, R.C. Meyer and W.S. de Jong, 1999, *Theor. Appl. Genet.* 99, 859.

Castillo R.O. and D.M. Spooner, 1997, *Syst. Bot.* 22, 45.

Child A., 1990, *Feddes Repertorium* 101, 209.

Clausen A.M. and D.M. Spooner, 1998, *Crop Sci.* 3, 858.

Correll D.S., 1962, *The Potato and Its Wild Relatives.* Contributions from the Texas Research Foundation, Botanical Studies 4, 1.

Debener T., F. Salamini and C. Gebhardt, 1990, *Theor. Appl. Genet.* 79, 360.

Debener T., F. Salamini and C. Gebhardt, 1991, *Plant Breed.* 106, 173.

Del Rio A.H. and J.B. Bamberg, 2003, *Am. J. Potato Res.* 80, 215.

Dodds K.S., 1962, In: D.S. Correll (ed.), *The Potato and Its Wild Relatives.* Contributions from the Texas Research Foundation, Botanical Studies 4, 517.

Douches D.S. and C.F. Quiros, 1988, *J. Hered.* 79, 377.

Görg R., U. Schachtschabel, E. Ritter, F. Salamini and C. Gebhardt, 1992, *Crop Sci.* 32, 815.

Grun P., 1990, *Economic Bot.* 44 (3 Suppl.), 39.

Hawkes J.G., 1989, *Taxon* 38, 489.

Hawkes J.G., 1990, *The Potato: Evolution, Biodiversity & Genetic Resources.* Belhaven Press, London, U.K.

Hawkes J.G. and J.P. Hjerting, 1989, *The Potatoes of Argentina, Brazil, Paraguay and Uruguay. A Biosystematic Study.* Oxford University Press, Oxford, U.K.

Hetterscheid W.L.A. and W.A. Brandenburg, 1995, *Taxon* 44, 161.

Hijmans R.J. and D.M. Spooner, 2001, *Am. J. Bot.* 88, 2101.

Hijmans R.J., D.M. Spooner, A.R. Salas, L. Guarino and J. de la Cruz, 2002, *Atlas of Wild Potatoes.* IPGRI, Rome, Italy.

Hosaka K., 1986, *Theor. Appl. Genet.* 72, 606.
Hosaka K., 1995, *Theor. Appl. Genet.* 90, 356.
Hosaka K., 2002, *Am. J. Potato Res.* 79, 119.
Hosaka K., 2003, *Am. J. Potato Res.* 80, 1.
Hosaka K., 2004, *Am. J. Potato Res.* 81, 153.
Hosaka K., G.A. de Zoeten and R.E. Hanneman Jr., 1988, *Theor. Appl. Genet.* 75, 741.
Hosaka K. and R.E. Hanneman Jr., 1988a, *Theor. Appl. Genet.* 76, 172.
Hosaka K. and R.E. Hanneman Jr., 1988b, *Theor. Appl. Genet.* 76, 333.
Hosaka K., M. Mori and K. Ogawa, 1994, *Am. Potato J.* 71, 535.
Hosaka K., Y. Ogihara, M. Matsubayashi and K. Tsunewaki, 1984, *Jpn. J. Genet.* 59, 349.
Hosaka K. and D.M. Spooner, 1992, *Theor. Appl. Genet.* 84, 851.
Huaman Z. and D.M. Spooner, 2002, *Am. J. Bot.* 89, 947.
ICNCP, 2004, *Acta Hortic.* 647.
Jacobs M., R.G. van den Berg, R. Hoekstra and B. Vosman, 2005, *XVII International Botanical Congress*, July 17–23 2005, Vienna, 361.
Juzepczuk S.W. and S.M. Bukasov, 1929, *Proc. USSR Congr. Genet. Plant Anim. Breed.* 3, 592.
Kardolus J.P., 1998, *A Biosystematic Analysis of Solanum acaule.* PhD dissertation, Wageningen Agricultural University, Wageningen, The Netherlands.
Kardolus J.P., H.J. van Eck and R.G. van den Berg, 1998, *Plant Syst. Evol.* 210, 87.
Knapp S., 1991, *Bot. J. Linn. Soc.* 105, 179.
Knapp S., 2000, *Bull. Nat. Hist. Mus. London (Bot.)* 30, 13.
Lara-Cabrera S.I. and D.M. Spooner, 2004, *Plant Syst. Evol.* 248, 129.
Lester R.N., 1965, *Ann. Bot.* 29, 609.
McGregor C.E., R. van Treuren, R. Hoekstra and Th.J.L. van Hintum, 2002, *Theor. Appl. Genet.* 104, 146.
Miller J.T. and D.M. Spooner, 1999, *Plant Syst. Evol.* 214, 103.
Nakagawa K. and K. Hosaka, 2002, *Am. J. Potato Res.* 79, 85.
Ochoa C.M. 1990. *The Potatoes of South America: Bolivia.* Cambridge University Press, Cambridge, UK.
Provan J., W. Powell, H. Dewar, G.J. Bryan, C.C. Machray and R. Waugh, 1999, *Proc. R. Soc. Lond. B Biol. Sci.* 266, 633.
Quiros C.F. and N. McHale, 1985, *Genetics* 111, 131.
Raimondi J.P., I.E. Peralta, R.W. Masuelli, S. Feingold and E.L. Camadro, 2005, *Plant Syst. Evol.* 253, 33.
Raker C. and D.M. Spooner, 2002, *Crop Sci.* 42, 1451.
Rodriguez F. and D.M. Spooner, 2004, *Abstract Botany 2005 Meeting*, 13–17 Aug 2005, Austin.
Spooner D.M., G.J. Anderson and R.K. Jansen, 1993, *Am. J. Bot.* 80, 676.
Spooner D.M. and R.T. Castillo, 1997, *Am. J. Bot.* 84, 671.
Spooner D.M., K. McLean, G. Ramsay, R. Waugh and G.J. Bryan, 2005a, *Proc. Natl. Acad. Sci. U.S.A.* 102, 14694.
Spooner D.M., J. Nunez, F. Rodriguez, P.S. Naik and M. Ghislain, 2005b, *Theor. Appl. Genet.* 110, 1020.
Spooner D.M., S.A. Stephenson, H.E. Ballard and Z. Polgar, 2004, *Abstract, Plant and Animal Genome Conference*, 10–14 Jan 2004, San Diego, CA.
Spooner D.M. and K.J. Sytsma, 1992, *Syst. Bot.* 17, 432.
Spooner D.M., K.J. Systma and E. Conti, 1991a, *Am. J. Bot.* 78, 1354.
Spooner D.M., K.J. Systma and J.F. Smith, 1991b, *Evolution* 45, 757.
Spooner D.M., R.G. van den Berg and J.T. Miller, 2001, *Am. J. Bot.* 88, 113.
Stephenson S.A., D.M. Spooner and H.E. Ballard, 2004, *Abstract, Plant and Animal Genome Conference*, 10–14 Jan 2004, San Diego, CA.
Sukhotu T., O. Kamijima and K. Hosaka, 2004, *Genome* 47, 46.
Swaminathan M.S. and H.W. Howard, 1953, *Bibliogr. Genet.* 16, 1.
Ugent D., 1966, *Hybrid Weed Complexes in Solanum, Section Tuberarium.* PhD dissertation, University of Wisconsin, Madison.
Van den Berg R.G., G.J. Bryan, A. del Rio and D.M. Spooner, 2001b, *Theor. Appl. Genet.* 105, 1109.
Van den Berg R.G. and N. Groendijk-Wilders, 1999, In: M. Nee, D.E. Symon, R.N. Lester and J.P. Jessop (eds), *Solanaceae IV.* Royal Botanic Gardens, Kew, UK, 213.

Van den Berg R.G., N. Groendijk-Wilders, M.J. Zevenbergen and D.M. Spooner, 2001a, In: R.G. van den Berg, G.M. van der Weerden, G.W.M. Barendse and C. Mariani (eds), *Solanaceae V*. Nijmegen University Press, The Netherlands, 73.

Van den Berg R.G., J.T. Miller, M.L. Ugarte, J.P. Kardolus, J. Villand, J. Nienhuis and D.M. Spooner, 1998, *Am. J. Bot.* 85, 92.

Volkov R.A., N.Y. Komarova, I.I. Panchuk and V. Hemleben, 2003, *Mol. Phylogenet. Evol.* 29, 187.

Volkov R.A., C. Zanke, I.I. Panchuk and V. Hemleben, 2001, *Theor. Appl. Genet.* 103, 1273.

Chapter 5

Molecular Markers, Maps and Population Genetics

Christiane Gebhardt

Department for Plant Breeding and Genetics, Max-Planck Institute for Plant Breeding Research, Carl von Linné Weg 10, 50829 Köln, Germany

5.1 INTRODUCTION

Potato genetics was initiated around 100 years ago when R. N. Salaman applied the newly discovered Mendelian principles of inheritance to characters such as male sterility, haulm characters, tuber shape, colour and eye depth, and immunity to *Phytophthora infestans* (Salaman, 1910–1911). He stated: 'Although the subject material of this research was my own choice, at the time it was determined on I was quite ignorant of the very special advantages as well as disadvantages which the Potato offers for the Mendelian student' (Salaman, 1910–1911). At that time, the tetraploidy and tetrasomic inheritance of the cultivated potato *Solanum tuberosum* was not recognized, which makes the potato a rather unfortunate choice for Mendelian genetics of simple traits. Recessive alleles are difficult to uncover, alleles occur in four dosages (simplex, duplex, triplex and quadruplex) instead of only two in diploids (homozygous and heterozygous), pure lines are not obtained due to severe inbreeding depression after repeated selfing and multiple allelism generates a multitude of genotypes (Fig. 5.1) that are difficult to distinguish based on comparing expected with observed segregation ratios. Inheritance studies of Mendelian characters in potato were therefore mostly restricted to dominant traits such as anthocyanin pigmentation genes (reviewed by Black, 1933; de Jong, 1991) or major genes for pathogen resistance (Cockerham, 1970). Genetic linkage between genes was rarely detected (Dodds and Long, 1956; Cockerham, 1970), and a linkage map was never constructed based on classical morphological markers.

Two technical developments provided the enabling tools for extensive linkage studies and map construction in potato, the manipulation of the ploidy level and the advent of DNA-based genetic markers. Ploidy reduction from the tetraploid ($2n = 4x = 48$) to the diploid level ($2n = 2x = 24$) became possible either by pollination of tetraploid *S. tuberosum* ssp. *tuberosum* with certain genotypes of *Solanum phureja* ($2n = 2x = 24$) which induces the parthenogenetic development of diploid female gametes into plants (Ivanovskaja, 1939; Hougas and Peloquin, 1958; Hougas et al., 1964; Hermsen and Verdenius, 1973), or by regenerating diploid plants from male gametes of tetraploid plants through anther or microspore culture (Dunwell and Sunderland, 1973; Powell and Uhrig, 1987). Diploid potatoes are highly self-incompatible (gametophytic incompatibility).

Potato Biology and Biotechnology: Advances and Perspectives
D. Vreugdenhil (Editor)

	$A_1 A_2 A_3 A_4 \times A_5 A_6 A_7 A_8$					
2n gametes	$A_5 A_6$	$A_5 A_7$	$A_5 A_8$	$A_6 A_7$	$A_6 A_8$	$A_7 A_8$
$A_1 A_2$	$A_1 A_2 A_5 A_6$	$A_1 A_2 A_5 A_7$	$A_1 A_2 A_5 A_8$	$A_1 A_2 A_6 A_7$	$A_1 A_2 A_6 A_8$	$A_1 A_2 A_7 A_8$
$A_1 A_3$	$A_1 A_3 A_5 A_6$	$A_1 A_3 A_5 A_7$	$A_1 A_3 A_5 A_8$	$A_1 A_3 A_6 A_7$	$A_1 A_3 A_6 A_8$	$A_1 A_3 A_7 A_8$
$A_1 A_4$	$A_1 A_4 A_5 A_6$	$A_1 A_4 A_5 A_7$	$A_1 A_4 A_5 A_8$	$A_1 A_4 A_6 A_7$	$A_1 A_4 A_6 A_8$	$A_1 A_4 A_7 A_8$
$A_2 A_3$	$A_2 A_3 A_5 A_6$	$A_2 A_3 A_5 A_7$	$A_2 A_3 A_5 A_8$	$A_2 A_3 A_6 A_7$	$A_2 A_3 A_6 A_8$	$A_2 A_3 A_7 A_8$
$A_2 A_4$	$A_2 A_4 A_5 A_6$	$A_2 A_4 A_5 A_7$	$A_2 A_4 A_5 A_8$	$A_2 A_4 A_6 A_7$	$A_2 A_4 A_6 A_8$	$A_2 A_4 A_7 A_8$
$A_3 A_4$	$A_3 A_4 A_5 A_6$	$A_3 A_4 A_5 A_7$	$A_3 A_4 A_5 A_8$	$A_3 A_4 A_6 A_7$	$A_3 A_4 A_6 A_8$	$A_3 A_4 A_7 A_8$

Fig. 5.1. Genotypic classes obtained in the F_1 when crossing tetraploid, fully heterozygous potato individuals, each having four alleles at a locus *A*. Based on the assumption of tetrasomic inheritance and random formation of 2n gametes, 36 genotype classes of equal frequency are expected. With reduction of heterozygosity of the parents (less than eight different alleles), the number and frequency of individual genotype classes vary, depending of the allelic state of the parents.

At the diploid level, the complexity of genetic analysis in potato is therefore equivalent to human genetics: partially heterozygous parents generate segregating offspring. Compared to *Homo sapiens*, diploid potato has the advantage that experimental F_1 families with hundreds of sibs can be generated by crossing compatible parental genotypes.

DNA-based genetic markers essentially detect point mutations, insertions, deletions or inversions in allelic DNA fragments, which differentiate the individuals of the same species. These DNA polymorphisms are usually selectively neutral and hence a transient phase of molecular evolution, in which they are maintained in a species by mutational input and random extinction. DNA-based markers show Mendelian inheritance, are available in unlimited numbers and most of them are phenotypically neutral. Based on DNA polymorphisms and their detection methods described below, the construction of molecular linkage maps of the 12 chromosomes of one potato genome complement became feasible since the mid-1980s (Bonierbale et al., 1988; Gebhardt et al., 1989). Molecular maps are the basis for the tremendous advances made over the last 20 years in structural and functional characterization of the potato genome.

5.2 DNA MARKER TYPES USEFUL FOR POTATO GENETICS

5.2.1 Restriction fragment length polymorphism

Restriction enzymes cut DNA at specific recognition sites. A point mutation within this site results in the loss or gain of a recognition site, giving rise to restriction fragments of different lengths. Insertion, deletion or inversion of DNA stretches can also lead to a length variation of restriction fragments. Restriction fragment length polymorphisms (RFLPs) (Botstein et al., 1980) are detected on Southern gel blots (Southern, 1975). In potato, the genomic DNA of partially heterozygous, diploid genotypes is cleaved with an enzyme of choice, the fragments are size separated by electrophoresis in a gel matrix, transferred

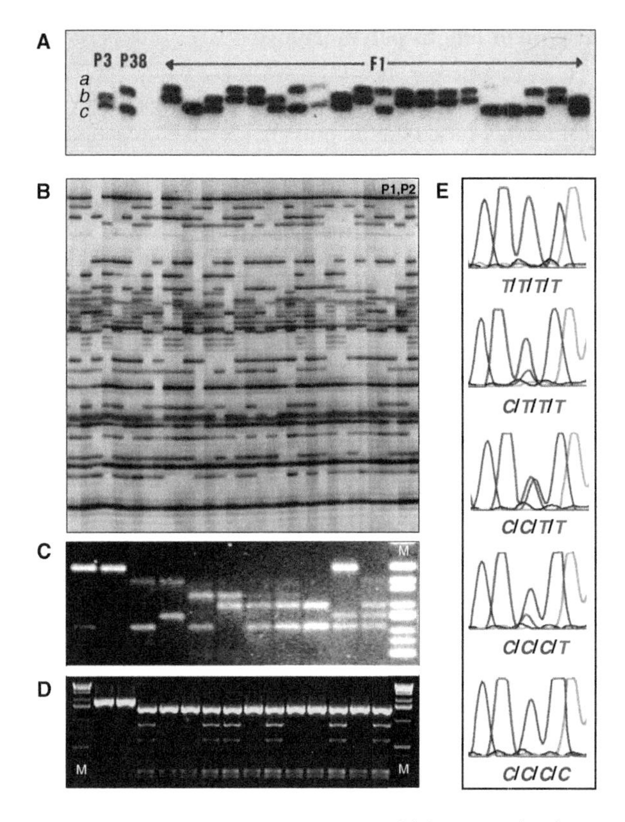

Fig. 5.2. Examples of different types of DNA-based markers useful for genotyping in potato. (A) Co-dominant segregation of a restriction fragment length polymorphism (RFLP) marker with three alleles *a*, *b* and *c* in F_1 progeny of heterozygous diploid parents P3 (*b*/*c*) and P38 (*a*/*c*). (B) Dominant segregation of AFLP markers in F_1 progeny of heterozygous diploid parents P1 and P2. (C) Simple sequence repeat (SSR) alleles scored in 11 tetraploid potato genotypes related by descent. Simple sequence repeat alleles were size separated on the Elchrom gel electrophoresis system (Elchrom Scientific AG, Cham, Switzerland). The number of SSR alleles per genotype varies between one and three, and the allele dosage can be estimated based on number of bands and band intensity. M = size marker. (D) Dominant segregation of a cleaved amplified polymorphic sequence (CAPS) marker in diploid F_1 progeny. M = size marker. (E) Single-nucleotide polymorphism (SNP) marker scored in the DNA sequences of the amplicons of five tetraploid individuals representing the five allelic states (*TTTT*, *CTTT*, *CCTT*, *CCCT* and *CCCC*) of a bi-allelic SNP (*C*/*T*). Allele dosage is estimated based on the height ratio of the ‘T’ peak (red) versus the ‘C’ peak (blue) in electropherograms of automated sequencers (Applied Biosystems).

to a carrier membrane and hybridized against a suitably labelled probe. Fig. 5.2A shows the example of an RFLP marker with three alleles *a*, *b* and *c* of different length, which are present in the heterozygous state in the diploid potato genotypes P3 (*b*, *c*) and P38 (*a*, *c*) (Schäfer-Pregl et al., 1998). In F_1 progeny of the cross P3 × P38, the three alleles segregate as co-dominant Mendelian markers and combine in four genotype classes: *ac*, *bc*, *ab* and *cc*. This type of segregation pattern is most informative for linkage analysis (Section 5.3) but is observed for only a subset of marker loci.

5.2.2 Amplified fragment length polymorphism

Amplified fragment length polymorphism™ (AFLP™) (Zabeau and Vos, 1993; Vos et al., 1995) is a marker technology based on the polymerase chain reaction (PCR) (Saiki et al., 1988). Using cleaved genomic DNA as template, multi-locus DNA markers are generated by PCR without prior sequence information. Briefly, the DNA is cleaved with two different restriction enzymes, one having a four-base pair, the other having a six-base pair recognition site. The fragments are ligated with recognition site-specific adapter oligonucleotides. A subset of approximately 50–100 restriction fragments is then selectively amplified in two consecutive PCRs. The amplified restriction fragments are radioactive or fluorescent labelled and size separated by electrophoresis on high-resolution polyacrylamide gels. Selectivity is achieved by designing PCR primers that anneal specifically to the adaptor and the recognition site and carry one to three arbitrary chosen nucleotides at the 3′ end. Only those restriction fragments will be amplified that have on both ends nucleotides complementary to the adapter and the recognition site plus the arbitrary nucleotide extension. DNA fragment polymorphisms between different genotypes result from point mutations in the enzyme recognition sites or the nucleotide extensions, which either allow or prevent PCR amplification of a specific DNA fragment. Insertions or deletions between recognition sites can also give rise to AFLPs. Figure 5.2B shows a section of an AFLP pattern of two heterozygous diploid potato genotypes P1 and P2 and their F_1 progeny. Each AFLP fragment segregates as a dominant Mendelian marker, with presence of the fragment being dominant over its absence. AFLP markers are very well suited to quickly generate a large number of segregating DNA markers for fingerprinting and linkage analysis (Menendez et al., 2002). AFLPs are the markers of choice for linkage map construction in segregating populations derived from crossing tetraploid parents (Meyer et al., 1998) and when aiming at a high-density map (Isidore et al., 2003) to aid map-based cloning and physical mapping.

5.2.3 Simple sequence repeat or microsatellite

Simple sequence repeats (SSRs) or microsatellites are DNA stretches consisting of short, tandemly repeated di-, tri-, tetra- or penta-nucleotide motifs. Simple sequence repeats have been found in all eukaryotic species that were scrutinized for them (Tautz and Renz, 1984). To amplify SSRs by PCR, information on unique flanking DNA sequences is required for primer design. The amplification products are size separated by electrophoresis and visualized by silver staining or fluorescent dyes. The amplicons from different genotypes frequently show length polymorphisms due to allelic variation of the number of repeat motifs in the microsatellite. Potato genotypes are highly heterozygous for multiple SSR alleles (Provan et al., 1996; Milbourne et al., 1998), which makes SSR markers highly informative and useful for fingerprinting and linkage studies in potato. All four possible genotype classes in diploid F_1 progeny may be identified, similar as shown for the RFLP marker in Fig. 5.1A. Under very good circumstances of amplification, separation and visualization, the dosage of co-dominant SSR alleles can be assessed in heterozygous tetraploid potato cultivars by band intensity (Fig. 5.2C).

5.2.4 Cleaved amplified polymorphic sequence, sequence characterized amplified region and allele-specific amplification

Because DNA sequencing became a universal, high throughput and affordable tool in molecular genetics, it is now possible to develop PCR-based markers for almost any sequence of interest, provided DNA polymorphisms between individuals exist in the amplified region. Sequence-specific primers are used to amplify by PCR allelic fragments from genomic DNA of different potato genotypes, tetraploids as well as diploids. An insertion/deletion polymorphism (Indel) in the amplicon is then detected directly by electrophoretic size separation in appropriate gel matrices [sequence characterized amplified region (SCAR), Marczewski et al., 2001]. Point mutations can be detected by digestion of the amplicons with appropriate restriction enzymes and subsequent fragment separation [cleaved amplified polymorphic sequence (CAPS), Fig. 5.2D]. Comparative allele sequencing is the basis for developing PCR primers that specifically amplify a single allele [allele-specific amplification (ASA), Li et al., 2005]. Locus-specific, user-friendly PCR-based markers such as SSR, SCAR, CAPS or ASA markers are most appropriate for practical breeding applications.

5.2.5 Single-nucleotide polymorphism

The most direct and also most informative method to identify point mutations is the sequencing of amplicons obtained with the same primer pair from DNA templates of different potato genotypes, tetraploids as well as diploids. Automated sequencing detects each of the four deoxynucleotides G, C, T and A with a different fluorescent dye. The deoxynucleotide sequence is recorded as an electropherogram (trace file), with G's represented by black, C's by blue, T's by red and A's by green peaks. Due to the heterozygosity of potato, single-nucleotide polymorphisms (SNPs) are detected as alternative coloured nucleotide peaks at a given position in sequence trace files (Fig. 5.2E). High quality trace files allow the estimation of the allele dosage in tetraploid genotypes based on the height ratio of two co-migrating nucleotide peaks (Fig. 5.2E). Insertion/deletion polymorphisms can also be detected by observing specific positions in trace files, downstream of which different nucleotide sequences overlap. A first assessment of the frequency of SNPs and Indels in the potato genome was performed by comparative sequencing of 78 amplicons in a panel of 17 tetraploid and 11 diploid genotypes (Rickert et al., 2003). This revealed, on average, one SNP every 21 base pairs and one Indel every 243 base pairs, confirming the high degree of natural DNA sequence variation present in potato (Gebhardt et al., 1989). Once identified, SNPs can be scored with a variety of techniques, e.g. pyrosequencing or single-nucleotide primer extension (SNuPE) (Rickert et al., 2002). In the future, array technologies will be become available, which should allow the parallel scoring of many SNPs on a chip. Single-nucleotide polmorphisms are the markers of choice for comprehensive population genetics and linkage disequilibrium studies. For breeding applications, SNPs linked to agronomic traits can be converted into easy-to-use allele-specific PCR assays (Niewöhner et al., 1995; Sattarzadeh et al., 2006).

5.3 PRINCIPLES OF LINKAGE MAP CONSTRUCTION

The basis for constructing molecular linkage maps for the 12 potato chromosomes is an F_1 or backcross (BC) progeny of partially heterozygous parents, which segregates for a sufficient number of DNA polymorphisms (RFLP, AFLP, SSR, CAPS, SCAR, ASA and SNP). Figure 5.3 shows the scheme for generating such a mapping population by crossing diploid parents. The resolution of the genetic map depends on the population size. For example, a map resolution of 1 cM (on average 1% recombination) requires the genotyping of at least 100 individuals. Parents and progeny are scored for DNA polymorphisms, the more the better. The genetic distance between pairs of markers is estimated by counting recombinant and non-recombinant individuals in the progeny, and linked markers are arranged in linear linkage groups. Acceptable map coverage is achieved when most of the markers scored can be arranged in a number of linkage groups that corresponds to the number of chromosomes per genome complement (12 in potato). Unlike in F_2 or BC populations derived from inbred lines, in a BC or F_1 population descending from partially heterozygous, non-inbred parents, the alleles at different marker loci can segregate according to several different genetic models, depending on the number and heterozygosity of the parental alleles at any given marker locus. These models have been described in detail by Ritter et al. (1990). However, when considering only the presence or absence of single DNA fragments in the parents and their progeny, with presence of the fragment being dominant over its absence, all models are reduced to three observable segregation patterns, which are valid for all DNA marker types. Fragments that are present in the heterozygous state either in the seed or in the pollen parent are expected to segregate in F_1 with a 1:1 ratio. Fragments that are present in the heterozygous state in both parents are expected to segregate in F_1 with a 3:1 ratio (presence versus absence, Fig. 5.3). The same segregation patterns apply to tetraploid mapping populations

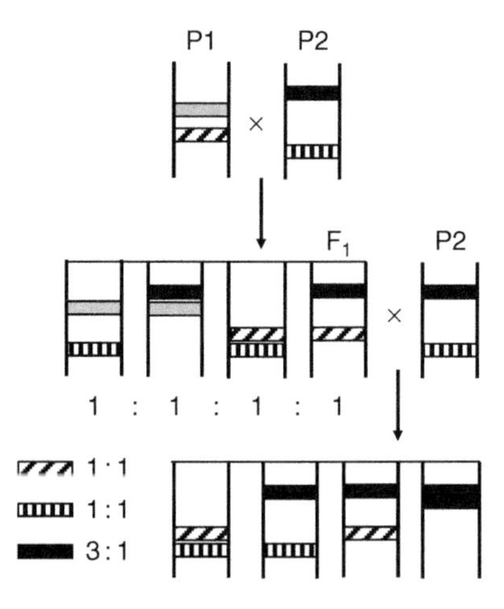

Fig. 5.3. Crossing scheme for generating mapping populations in diploid potato and DNA marker segregation in F_1 and backcross progeny.

when the DNA fragment is present as a single-dose (simplex) allele. In addition, double-dose (duplex) alleles are also required for linkage mapping in tetraploid crosses (Hackett et al., 1998; Meyer et al., 1998). The frequency of recombination is estimated between all fragments descending from the seed parent or the pollen parent, taking into account coupling and repulsion phase linkages (Ritter et al., 1990), and two independent linkage groups are constructed, based on meiotic recombination either in the seed parent or in the pollen parent. Linkage subgroups can also be constructed based on DNA fragments common to both parents, although with lower accuracy. The parental linkage groups are connected and oriented relative to each other based on highly informative RFLP or SSR marker loci with three alleles (Fig. 5.2A), one of which is shared among the parents and therefore forms an 'allelic bridge' between the parental linkage groups (Fig. 5.3, Ritter et al., 1990). Software packages have been developed, which help in constructing molecular linkage maps, taking into consideration the appropriate genetic models for non-inbred potato [MAPRF, available from E. Ritter, NEIKER, Granja Modelo de Arkaute (Alava), Apdo. 46, 01080 Vitoria-Gasteiz, Spain; JoinMap, Stam, 1993]. Luo et al. (2001) have developed the theory for construction of a linkage map in an autotetraploid species using either co-dominant or dominant markers scored on two parents and their full-sib progeny. A 'Windows version' of the original software (Hackett and Luo, 2003) is now available and called 'TetraploidMap for Windows'.

5.4 MOLECULAR MAPS OF POTATO

The first potato genetic map was constructed at Cornell University (Ithaca, USA) based on 65 diploid F_1 individuals derived from an inter-species cross between a *S. phureja* seed parent and a *S. tuberosum* × *Solanum chacoense* hybrid used as pollen parent. The markers used included a few isozymes but were mostly RFLPs, which were identified with tomato markers of known position on the tomato molecular maps ('TG', 'CD' and 'CT' markers, Bonierbale et al., 1988). This was the first comparative mapping experiment in plants and revealed the high degree of co-linearity between the potato and tomato genomes (Section 5.5). A second inter-species map between *S. tuberosum* and *Solanum berthaultii* was constructed using tomato RFLP markers in a population of 155 BC_1 individuals (Tanksley et al., 1992; http://www.sgn.cornell.edu). At the Max-Planck Institute for Plant Breeding Research (Cologne, Germany), two diploid reference potato RFLP maps have been constructed primarily in a *S. tuberosum* background, each consisting of 24 linkage groups, 12 each for the chromosome complement of the seed and pollen parent. The first intra-species mapping population ($BC916^2$) consisted of 67 diploid BC_1 individuals. DNA markers were selected de novo from potato cDNA ('CP' markers) and genomic libraries ('GP' markers) and mapped as RFLPs (Gebhardt et al., 1989, 1991, 2001). This map has been anchored to the tomato molecular maps by mapping a set of chromosome-specific tomato RFLP markers ('TG' markers) in the $BC916^2$ mapping population (Gebhardt et al., 1991) and vice versa, by mapping a set of chromosome-specific potato RFLP markers in the inter-species tomato reference mapping population (Tanksley et al., 1992). By means of linkage group-specific DNA markers and fluorescence in situ hybridization (FISH), the genetic map of potato was linked to the cytogenetic map (Dong et al., 2000). The BC916^2 population has also been instrumental for mapping the first set of potato SSR markers ('STM' markers) developed at the Scottish Crop Research Institute (SCRI) (Invergowrie, Scotland) by Milbourne et al. (1998). The second RFLP map was constructed in a population of 92 F_1 individuals (population F1840), obtained by crossing a *S. tuberosum* seed parent with a *S. tuberosum* × *Solanum spegazzinii* hybrid as pollen parent (Barone et al., 1990; Gebhardt et al., 1991; Leister et al., 1996). The population F1840 was instrumental for comparative mapping with *Arabidopsis thaliana* (Gebhardt et al., 2003). The F1840 and BC916^2 maps are well aligned to each other due to shared RFLP markers, and together they consist of nearly 1000 DNA marker loci including expressed sequence tags (ESTs) and cloned potato genes. Both maps are accessible through internet in the PoMaMo (Potato Maps and More) database at https://gabi.rzpd.de/PoMaMo.html (Meyer et al., 2005). Tomato 'TG', 'CD' or 'CT' and potato 'GP' or 'CP' markers with known position on the reference RFLP maps have been widely used in many different mapping experiments in Solanaceous species, mainly tomato, potato and pepper, which aimed at the localization of genetic factors controlling agronomic traits such as pathogen resistance and tuber quality [Chapter 6 (van Eck) and Chapter 7 (Simko et al.), this volume]. The anchor markers shared between independent mapping experiments allow to some extent the integration of positional information obtained in different Solanaceous species, different genetic materials and different laboratories into 'function maps' for pathogen resistance (Leister et al., 1996; Grube et al., 2000; Gebhardt and Valkonen, 2001; https://gabi.rzpd.de/PoMaMo.html) or tuber quality traits (Chen et al., 2001; Gebhardt, 2004). The majority of potato and tomato markers originally mapped as RFLPs have been completely or partially sequenced, and the sequences are now accessible in databases (http://www.sgn.cornell.edu, https://gabi.rzpd.de/PoMaMo.html). These 'sequence tagged sites' (STS) or 'ESTs' in the potato/tomato genome are a valuable resource for developing locus-specific PCR-based markers and for anchoring the potato genetic maps, including the mapped factors for agronomic traits, to plant genome sequences of Arabidopsis, rice, tomato, potato and other plant species, which may become available in the future.

Within a European collaborative project, an ultra-high-density (UHD) genetic map was constructed at Wageningen University (Wageningen, The Netherlands), based on AFLP fingerprinting of 130 diploid F_1 individuals derived from crossing *S. tuberosum* parents (SH × RH, Isidore et al., 2003; van Os, 2005; http://www.dpw.wageningen-ur.nl/uhd/). New software had to be developed for reliable linkage group assignment and ordering of approximately 10 000 AFLP fragments (van Os et al., 2005a,b). This UHD map is the basis for constructing a potato physical map as prerequisite for sequencing the potato genome. Another reference potato map was constructed at Wageningen University based on 67 *S. tuberosum* BC individuals (C × E, Jacobs et al., 1995; Van Eck et al., 1995), which includes RFLP, AFLP and several classical markers. As this map contains a number of tomato 'TG' markers and some isozyme loci, it can be anchored to other potato and tomato maps.

The only example for a linkage map of tetraploid potato is the AFLP linkage map constructed at the SCRI in a tetraploid mapping population of 94 F_1 plants derived from crossing the cultivar Stirling with an advanced breeding line. Three hundred and forty-six

AFLP markers were assigned to linkage groups, covering approximately 25% of the genome (Meyer et al., 1998).

Multi-allelic, co-dominant SSR markers of known map position are increasingly important for anchoring anonymous AFLP linkage groups to specific chromosome regions (Ghislain et al., 2001; Bradshaw et al., 2004) and for genotype identification (Ghislain et al., 2004). Genomics resources such as potato EST sequence databases can be mined to find microsatellites, and existing RFLP maps are enriched by mapping new SSRs (Feingold et al., 2005).

5.5 COMPARING THE POTATO WITH OTHER PLANT GENOMES

The RFLP assay is based on nucleic acid hybridization between a labelled marker probe and a membrane-bound genomic target sequence. Hence, depending on the experimental conditions used, cross-hybridization is detected between DNA sequences that are not identical but similar and RFLP markers originating from one species can be used for the construction of linkage maps in related species. Most (but not all) sequences of Solanaceous species that were tested, readily cross-hybridize to each other due to high sequence similarity (>80%). Tomato RFLP markers were used to construct potato molecular maps (Bonierbale et al., 1988; Tanksley et al., 1992). These experiments revealed extensive conservation of genetic linkage (synteny) between RFLP loci detected by the same marker probes in tomato and potato. The marker order was inverted in the potato maps relative to tomato maps in parts of five linkage groups, indicative of paracentric inversions of arms of potato chromosomes V, IX, X, XI and XII (Tanksley et al., 1992). Restriction fragment length polymorphism maps using tomato markers were also constructed in inter-species mapping populations of pepper (*Capsicum* ssp., Livingstone et al., 1999), the non-tuber bearing *Solanum* species *Solanum palustre* and *Solanum etuberosum* (Perez et al., 1999) and eggplant (*Solanum melongena*, Doganlar et al., 2002), thereby allowing comparisons of the tomato/potato genome structure with other Solanaceous species. Clear genome co-linearity with tomato/potato was found in all cases, but the syntenic genome segments were more fragmented and rearranged in the other Solanaceous species when compared with tomato and potato. The comparisons of genome structure as well as evidence from molecular taxonomy (Spooner et al., 1993; Olmstead and Palmer, 1997) suggest that the taxonomic placement of tomato and potato in different genera is not justified and that *Lycopersicon esculentum* should be renamed as *Solanum lycopersicum* (Peralta I.E., Knapp S. and Spooner D. M., unpublished monograph, http://www.sgn.cornell.edu).

Synteny between plant species that are members of distantly related taxonomic families is difficult to detect with hybridization-based methods due to increasing sequence divergence. With the genome sequence of *Arabidopsis thaliana* completed (The Arabidopsis Genome Initiative, 2000) and available in public databases, the comparison between the potato molecular map and the Arabidopsis genome became feasible. The sequences of 293 RFLP markers including 31 EST markers of Arabidopsis that were used to construct the F1840 potato map (Section 5.4) were compared to the Arabidopsis genome sequence and mapped in silico to the physical map of Arabidopsis (Gebhardt et al., 2003). Based on conserved linkage between groups of at least three different markers with high sequence similarity on the genetic map of potato and the physical map of Arabidopsis, 94 putative syntenic blocks were identified covering 41% of the potato genetic map and 50% of the Arabidopsis physical map. The existence and distribution of syntenic blocks suggested a higher degree of structural conservation in some parts of the potato genome when compared to others. The syntenic blocks were redundant: most potato syntenic blocks were related to several Arabidopsis genome segments and vice versa. Particularly striking in this respect were the short arms of potato chromosomes I and VI, which were related to each other and to 13 and 11 Arabidopsis genome segments, respectively. These conserved regions of the potato genome may contain the remains of an archaic plant genome. Some duplicated potato syntenic blocks correlated well with ancient segmental duplications in Arabidopsis. Syntenic relationships between different genomic segments of potato and the same segment of the Arabidopsis genome indicated that potato genome evolution included ancient intra- and inter-chromosomal duplications.

5.6 POPULATION GENETICS

Analysis of genetic linkage is based on measuring the frequency of recombination between pairs of loci in a single meiotic generation. Alleles that are physically linked in the parents (parental haplotypes) are transmitted together to the offspring, except when recombination separates the two alleles. With each subsequent generation, the common transmission of physically linked alleles (linkage disequilibrium) is reduced in proportion to the fraction of recombination until linkage equilibrium is reached, that is, recombinant and non-recombinant haplotypes are equally distributed in a population of individuals related by descent. The closer the linkage between two loci, the longer linkage disequilibrium persists over multiple meiotic generations in a population due to low frequency of recombination. When a molecular marker is closely linked to a locus controlling a phenotype of interest, for example, a quantitative trait locus (QTL) for pathogen resistance, linkage disequilibrium may exist between the marker locus and the QTL, and specific haplotypes or marker alleles may be associated with resistance values measured in a population of individuals related by descent. To detect association between a DNA marker and an unknown factor controlling a quantitative character, the marker must be physically tightly linked, whereas detection of linkage is possible over large distances (up to around 40 cM) between pairs of loci (Lander and Schork, 1994). Ideally, the marker is located within the gene that underlies the character of interest. In this case, linkage disequilibrium is complete and marker polymorphisms are directly associated with allelic variation of the phenotypic trait. The advantage of marker-trait associations as compared to marker-trait linkages is that a marker associated with a phenotypic character is diagnostic in wide germplasm pools, whereas a linked marker is diagnostic primarily in progeny descending from a specific carrier of a specific trait allele.

Although the concept had been known for a long time in human population genetics (Schlosstein et al., 1973), association studies were not feasible on a larger scale due to the lack of sufficient numbers of genetic markers that could fulfil the criterion of tight linkage. This situation changed when, on the one hand, DNA-based markers and DNA sequences became available on a large scale and, on the other hand, genome research started to provide candidate genes for complex characters such as quantitative resistance to pathogens in potato (Chapter 7, Simko et al., this volume) or tuber quality traits (Chapter 6, van Eck, this volume). There are two strategies for association studies: the genome-wide approach and the candidate gene approach. In the genome-wide approach, a large number of markers covering the whole genome at regular intervals are genotyped in a population of individuals related by descent. The same individuals are evaluated for the phenotypes of interest. Appropriate statistical methods are then used to test for significant associations between genotype and phenotype (Hirschhorn and Daly, 2005). The candidate gene approach uses physiological, biochemical, genetic and molecular information available for a trait of interest to make learned hypotheses about the function or structure of the gene(s) underlying the trait of interest and focuses on genotyping in genomic regions harbouring such candidate genes.

In potato, collections of varieties, breeding clones, landraces and wild species accessions can be considered as populations of individuals related by descent. As a vegetatively propagated crop, relatively few meiotic generations separate individual genotypes in the contemporary pool of potato cultivars and breeding lines (Gebhardt et al., 2004; Potato Pedigree Database: http://www.dpw.wau.nl/pv/query.asp). The feasibility of detecting marker-trait associations in populations of tetraploid potato varieties and breeding clones has been recently demonstrated. In three experiments reported so far, the candidate gene approach was used. In a first experiment, a collection of 600 tetraploid potato cultivars bred between 1850 and 1990 in different countries and maintained by the IPK germplasm bank at Groß-Lüsewitz (Germany) was genotyped with five ASA, CAPS and SCAR markers linked to previously mapped QTL for resistance to late blight and plant maturity on potato chromosome V. Polymorphic DNA fragments were tested for association with these quantitative traits based on available evaluation data for the cultivars. Significant marker-trait associations were detected with an ASA marker derived from *R1*, a major gene for resistance to late blight (Ballvora et al., 2002), and markers flanking the *R1* locus at 0.2 cM genetic distance (Gebhardt et al., 2004). In the second experiment, an SSR marker closely linked (1.5 cM) to a locus that is orthologous to the tomato *Ve* locus for resistance to *Verticillium dahliae* on tomato/potato chromosome IX was shown to be associated with quantitative resistance to *V. dahliae*, which was evaluated in 137 tetraploid potato cultivars, mostly from North America (Simko et al., 2004). In the third experiment, DNA variation at the *invGE/GF* candidate locus was analysed in 188 tetraploid potato cultivars, which have been assessed for chip quality and tuber starch content. The *invGE/GF* locus on potato chromosome IX encodes two invertase genes *invGE* and *invGF* (Maddison et al., 1999) and co-localizes with cold-sweetening QTL *Sug9* (Menendez et al., 2002). Two closely correlated invertase alleles were associated with better chip quality, and a third allele was associated with lower tuber starch content (Li et al., 2005).

The finding of marker-trait associations in tetraploid, advanced potato breeding materials is exciting as it closes the gap between linkage mapping of quantitative and qualitative traits in experimental, mostly diploid populations and DNA marker applications in breeding programs. It opens the possibility to develop PCR-based markers of general diagnostic value for parental screening and marker-assisted selection (Simko, 2004).

REFERENCES

Ballvora A., M.R. Ercolano, J. Weiß, K. Meksem, C. Bormann, P. Oberhagemann, F. Salamini and C. Gebhardt, 2002, *Plant J.* 30, 361.
Barone A., E. Ritter, U. Schachtschabel, T. Debener, F. Salamini and C. Gebhardt, 1990, *Mol. Gen. Genet.* 224, 177.
Black W., 1933, *J. Genet* 27, 319.
Bonierbale M.W., R.L. Plaisted and S.D. Tanksley, 1988, *Genetics* 120, 1095.
Botstein D., R.L. White, M. Skolnick and R.W. Davis, 1980, *Am. J. Hum. Genet.* 32, 314.
Bradshaw J.E., B. Pande, G.J. Bryan, C.A. Hackett, K. McLean, H.E. Stewart and R. Waugh, 2004, *Genetics* 168, 983.
Chen X., F. Salamini and C. Gebhardt, 2001, *Theor. Appl. Genet.* 102, 284.
Cockerham G., 1970, *Heredity* 25, 309.
de Jong H., 1991, *Am. Potato J.* 68, 585.
Dodds K.S. and D.H. Long, 1956, *J. Genetics* 54, 27.
Doganlar S., A. Frary, M.-C. Daunay, R.N. Lester and S.D. Tanksley, 2002, *Genetics* 161, 1697.
Dong F., J. Song, S.K. Naess, J.P. Helgeson, C. Gebhardt and J. Jiang, 2000, *Theor. Appl. Genet.* 101, 1001.
Dunwell J.M. and N. Sunderland, 1973, *Euphytica* 22, 317.
Feingold S., J. Lloyd, N. Norero, M. Bonierbale and J. Lorenzen, 2005, *Theor. Appl. Genet.*, 111, 456.
Gebhardt C., 2004, In: Lörz, H. and G. Wenzel (eds), *Molecular Marker Systems, Biotechnology in Agriculture and Forestry*, Vol. 55, Springer-Verlag, Berlin, Heidelberg, pp 215–227.
Gebhardt C., A. Ballvora, B. Walkemeier, P. Oberhagemann and K Schüler, (2004), *Mol. Breeding* 13, 93.
Gebhardt C., E. Ritter, A. Barone, T. Debener, B. Walkemeier, U. Schachtschabel, H. Kaufmann, R.D. Thompson, M.W. Bonierbale, M.W. Ganal, S.D. Tanksley and F. Salamini, 1991, *Theor. Appl. Genet.* 83, 49.
Gebhardt C., E._Ritter, T. Debener, U. Schachtschabel, B. Walkemeier, H. Uhrig and F. Salamini, 1989, *Theor. Appl. Genet.* 78, 65.
Gebhardt C., E. Ritter and F. Salamini, 2001, In: Phillips, R.L. and I.K. Vasil (eds), *DNA-based Markers in Plants, 2nd edition. Advances in Cellular and Molecular Biology of Plants*, Vol. 6, Kluwer Academic Publishers, Dordrecht/Boston/London, pp 319–336.
Gebhardt C. and J.P.T. Valkonen, 2001, *Annu. Rev. Phytopathol.* 39, 79.
Gebhardt C., B. Walkemeier, H. Henselewski, A. Barakat, M. Delseny and K. Stüber, 2003, Plant J. 34, 529.
Ghislain M., D.M. Spooner, F. Rodriguez, F. Villamón, J. Nunez, C. Vásquez, R. Waugh and M. Bonierbale, 2004, *Theor. Appl. Genet.* 108, 881.
Ghislain M., B. Trognitz, M. Herrera, J. Solis, G. Casallo, C. Vasquez, O. Hurtado, R. Castillo, L. Portal and M. Orillo, 2001, *Theor. Appl. Genet.* 103, 433.
Grube R.C., E.R. Radwanski and M. Jahn, 2000, *Genetics* 155, 873.
Hackett C.A. and Z.W. Luo, 2003, *J. Hered.* 94, 358.
Hackett C.A., J.E. Bradshaw, R.C. Meyer, J.W. McNicol, D. Milbourne and R. Waugh, 1998, *Genet. Res.* 71, 143.
Hermsen J.G.T. and J. Verdenius, 1973, *Euphytica* 22, 244.
Hirschhorn J.N. and M.J. Daly, 2005, *Nat. Rev. Genet.* 6, 95.
Hougas R.W. and S.J. Peloquin, 1958, *Am. Potato J.* 35, 701.
Hougas R.W., S.J. Peloquin and A.C. Gabert, 1964, *Crop Sci.* 4, 593.
Isidore E., H. van Os, S. Andrzejewski, J. Bakker, I. Barrena, G.J. Bryan, B. Caromel, H.J. van Eck, B. Ghareeb, W. de Jong, P. van Koert, V. Lefebvre, D. Milbourne, E. Ritter, J.N.A.M. Rouppe van der Voort, F. Rousselle-Bourgeois, J. van Vliet and R. Waugh, 2003, *Genetics* 165, 2107.
Ivanovskaja E.V., 1939, *C. R. Acad. Sc. URSSS* 24, 517.
Jacobs J.M.E., H.J. van Eck, P. Arens, B. Verkerk-Bakker, B. te Lintel Hekkert, H.J.M. Bastiaanssen, A. El Kharbotly, A. Pereira, E. Jacobsen and W.J. Stiekema, 1995, *Theor. Appl. Genet.* 91, 289.
Lander E.S. and N.J. Schork, 1994, *Science* 265, 2037.
Leister D., A. Ballvora, F. Salamini and C. Gebhardt, 1996, *Nat. Genet.* 14, 421.
Li L., J. Strahwald, H.-R. Hofferbert, J. Lübeck, E. Tacke, H. Junghans, J. Wunder and C. Gebhardt, 2005, *Genetics* 170, 813.

Livingstone K.D., V.K. Lackney, J.R. Blauth, R. van Wijk and M. Kyle-Jahn, 1999, *Genetics* 152, 1173.
Luo Z.W., C.A. Hackett, J.E. Bradshaw, J.W. McNicol and D. Milbourne, 2001, *Genetics* 157, 1369.
Maddison A.L., P.E. Hedley, R.C. Meyer, N. Aziz, D. Davidson and G.C. Machray, 1999, *Plant. Mol. Biol.* 41, 741.
Marczewski W., B. Flis, J. Syller, R. Schäfer-Pregl and C. Gebhardt, 2001, *Mol. Plant Microbe Interact.* 14, 1420.
Menendez C.M., E. Ritter, R. Schäfer-Pregl, B. Walkemeier, A. Kalde, F. Salamini and C. Gebhardt, 2002, *Genetics* 162, 1423.
Meyer R.C., D. Milbourne, C.A. Hackett, J.E. Bradshaw, J.W. McNicol and R. Waugh, 1998, *Mol. Gen. Genet.* 259, 150.
Meyer S., A. Nagel and C. Gebhardt, 2005, *Nucleic. Acids Res.* 33, Database issue, D666. doi:10.1093/nar/gki018.
Milbourne D., R.C. Meyer, A.J. Collins, L.D. Ramsay, C. Gebhardt and R. Waugh, 1998, *Mol. Gen. Genet.* 259, 233.
Niewöhner J., F. Salamini and C. Gebhardt, 1995, *Mol. Breeding* 1, 65.
Olmstead R.G. and J.D. Palmer, 1997, *Syst. Bot.* 22, 19.
Perez F., A. Menendez, P. Dehal, C.F. Quiros, 1999, *Theor. Appl. Genet.* 98, 1183.
Powell W. and H. Uhrig, 1987, *Plant Cell Tiss. Org. Cult.* 11, 13.
Provan J, W. Powell and R. Waugh, 1996, *Theor. Appl. Genet.* 92, 1078.
Rickert A.M., J.H. Kim, S. Meyer, A. Nagel, A. Ballvora, P.J. Oefner and C. Gebhardt, 2003, *Plant Biotechnol. J.* 1, 399.
Rickert A.M., A. Premstaller, C. Gebhardt and P.J. Oefner, 2002, *Biotechniques* 32, 592, 596, 600.
Ritter E., C. Gebhardt and F. Salamini, 1990, *Genetics* 125, 645.
Saiki R.K., D.H. Gelfand, S. Stoffel, S.J. Scharf, R. Higuchi, G.T. Horu, K.B. Mullis and H.A. Ehrlich, 1988, *Science* 239, 487.
Salaman R.N., 1910–1911, *J. Genet* 1, 7.
Sattarzadeh A., U. Achenbach, J. Lübeck, J. Strahwald, E. Tacke, H.-R. Hofferbert, T. Rothsteyn and C. Gebhardt, 2006, Single nucleotide polymorphism (SNP) genotyping as basis for developing a PCR-based marker highly diagnostic for potato varieties with high resistance to *Globodera Pallida* Pathotype Pa2/3. *Mol. Breed.*, 18, 301–312.
Schäfer-Pregl R., E. Ritter, L. Concilio, J. Hesselbach, L. Lovatti, B. Walkemeier, H. Thelen, F. Salamini and C. Gebhardt, 1998, *Theor. Appl. Genet.* 97, 834.
Schlosstein L., J.I. Terasaki, R. Bluestone and C.M. Pearson, 1973, *N. Engl. J. Med.* 288, 704.
Simko I., 2004, *Trends Plant Sci.* 9, 441.
Simko I., S. Costanzo, K.G. Haynes, B.J. Christ and R.W. Jones, 2004, *Theor. Appl. Genet.* 108, 217.
Southern E.M., 1975, *J. Mol. Biol.* 98, 503.
Spooner D.M., G.J. Anderson and R.K. Jansen, 1993, *Am. J. Bot.* 80, 676.
Stam P., 1993, *Plant J.* 3, 739.
Tanksley S.D, M.W. Ganal, J.P. Prince, M.C. de Vicente, M.W. Bonierbale, P. Broun, T.M. Fulton, J.J. Giovannoni, S. Grandillo, G.B. Martin, R. Messeguer, J.C. Miller, L. Miller, A.H. Paterson, O. Pineda, M.S. Röder, R.A. Wing, W. Wu and N.D. Young, 1992, *Genetics* 132, 1141.
Tautz D. and M. Renz, 1984, *Nucleic Acid Res.* 12, 4127.
The Arabidopsis Genome Initiative, 2000, *Nature* 408, 796.
Van Eck H.J., J. Rouppe van der Voort, J. Draaistra, P. van Zandvoort, E. van Enckevort, B. Segers, J. Peleman, E. Jacobsen, J. Helder and J. Bakker, 1995, *Mol. Breeding* 1, 397.
Van Os H., 2005, Thesis Wageningen University, CIP-DATA Koninkijke Bibliotheek, Den Haag, ISBN 90-8504-221-6.
Van Os H., P. Stam, R.G.F. Visser and H.J. van Eck, 2005a, Theor Appl Genet., 112, 30.
Van Os H., P. Stam, R.G.F. Visser and H.J. van Eck, 2005b, *Theor. Appl. Genet.*, 112, 187.
Vos P., R. Hogers, M. Bleeker, M. Reijans, T. van de Lee, M. Hornes, A. Frijters, J. Pot, J. Peleman, M. Kuiper and M. Zabeau, 1995, *Nucleic Acids Res.* 23, 4407.
Zabeau M. and P. Vos, 1993, European Patent Application, publication no. EP 0534858-A1, No 92402629.7

Chapter 6

Genetics of Morphological and Tuber Traits

Herman J. van Eck

Laboratory of Plant Breeding, Wageningen University, Wageningen, The Netherlands

6.1 INTRODUCTION

In the genomics era, the scope of geneticists tends to diversify into molecular biology, bioinformatics, metabolomics and so on. This chapter will describe most tuber traits from a strict genetics point of view. Genetics in the strict sense simply refers to understanding the heritable basis of a trait. This can be achieved by using an experimental design that allows correlating phenotypic variation with variation at the allele or genotype level. The genotype level is nothing else than the distribution of allelic variation at a locus, which usually follows the Mendelian segregation rules. This allelic variation is usually represented by genetic symbol A/a irrespective of the method to obtain genotypic information. Currently, the genotypic information is obtained with molecular marker techniques (Chapter 5, Gebhardt, this volume). This implies that the geneticist has the ability to tag any locus with a marker. In this way, the geneticist can offer conclusive evidence for the presence of a locus involved in heritable trait variation. The identity of the locus in terms of genes, candidate genes, DNA polymorphisms or the molecular function of this locus in biochemical or signalling functions does not have to be known.

6.1.1 The breeder's perspective

Why is there a demand for genetic information? In most cases, the potato breeder will ask for the hereditary basis of a phenotype, with the aim to rationalize his/her breeding efforts. The answer that the breeder might expect is (1) the number of loci involved, (2) the dominance relationships between alleles within and between loci and (3) the size of the effects of the alleles and the size of the allele interactions. Often the geneticist or breeder will not address the issue of gene function, because this is simply not a prerequisite for breeding work. In this chapter, the issue of gene function is not entirely out of scope, but it will not have emphasis.

6.1.2 What is heritable variation?

However, before any attempt to answer the question on the hereditary basis of a phenotype, it is important to realize that phenotypic variation may not have a heritable basis at all. In many cases, there may not be any genetic variation despite severe phenotypic

Potato Biology and Biotechnology: Advances and Perspectives
D. Vreugdenhil (Editor)
2007 Published by Elsevier B.V.

differences. The phenotypic differences could be simply due to environmental influences on the crop or the pathogen. For example, it is easy to observe remarkable differences in infection of silver scurf (*Helminthosporium solani*) across genotypes. However, the entire cultivated gene pool should be regarded as susceptible, and therefore, the variation in silver scurf infection is not heritable but entirely due to environmental effects. Therefore, the first and most obvious aspect that needs to be investigated is the heritability of a trait H^2. More specifically, we refer to broad-sense heritability in potato, because in non-inbred species, the heritability is always the summation of additive and dominance genetic variance. Heritability is the ratio of the variation caused by genetic factors to the total variation caused by genes, environment and genotype–environment ($G \times E$) interactions. Heritability is by no means a fixed value for a certain trait. Across the literature, the heritability estimates for a certain trait can show dramatic variation. Heritability estimates are positively influenced by minimizing experimental noise or by maximizing the phenotypic contrasts of parental genotypes.

6.1.3 Morphological and tuber traits discussed in this chapter

The inheritance of many tuber traits was first reviewed by Swaminathan and Howard (1953) and later by Howard (1970) and Bradshaw and Mackay (1994). In this chapter, not all classic papers will be discussed once more. Only when more recent publications have shed new light on a classical tuber trait, the added value of the new papers will be reviewed. The topics for this chapter can be subdivided into a number of classes of tuber traits. Some of these classes may overlap, but this overlap is ignored to organize a common sense order in the presentation of tuber traits. In Table 6.1, the tuber traits are listed. Although this table is far from complete, it offers a comprehensive list of traits that could be included when describing the idiotype of potato tubers. A subset of the tuber traits is only presented in the table and not discussed in the text because of lack of relevant hereditary information. For poorly understood traits such as 'internal distribution of dry matter' or watery pith and growing defects such as 'hollow hearts' or 'growth cracks', and 'tuber size distribution' or 'marketable yield', the amount of heritable variation is unclear, or heritable variation is lacking, like the example of silver scurf, which has been described above.

6.2 CLASSICAL POTATO GENETICS WITH MOLECULAR TECHNIQUES

6.2.1 The characteristics of classical genetic analysis

Molecular tools have fundamentally changed the experimental design of genetic studies. Conceptually, we currently think in terms of candidate genes and retrieve DNA sequence, marker and single-nucleotide polymorphism (SNP) information from web-based databases, such as the Solanaceae Genomics Network (http://www.sgn.cornell.edu), the ultra dense (AFLP) map (Van Os et al., 2006; http://potatodbase.dpw.wau.nl), the potato expressed sequence tag (EST) sequencing effort as presented by, for example, The Institute for Genomic Research (TIGR) (Ronning et al., 2003), which includes sequences from

Table 6.1 Overview of morphological and tuber traits.

Morphological traits
Tuber flesh colour
Tuber skin colour and flower colour
Tuber shape
Eye depth
Tuber skin characters
Stolon length
Physiological tuber traits
Tuberization
Dormancy
Tuber quality traits
Starch content
Discolouration
After-cooking blackening
Enzymatic discolouration
Black spot bruising
Maillard reaction, processing quality, cold-sweetening
Chlorophyll discolouration
Cooking type, texture
Glycoalkaloid content
Growing defects (hollow hearts, growth cracks and secondary growth)
Tuber size uniformity

the Danish (Crookshanks et al., 2001) and Canadian (Flinn et al., 2005) EST sequencing effort (http://www.tigr.org/tdb/potato), and the German database Potato Maps and More (Meyer et al., 2005; http://www.gabi.rzpd.de/PoMaMo.html). These resources have fundamentally changed the research strategies, and within some years, the entire genome sequence will become available (http://www.potatogenome.net). In the genomics era, it is even harder to imagine how genetic studies were performed until the eighties of the previous century. Usually, the genetic control of a potato trait was studied using several F1 populations that were obtained from contrasting parental clones. Some, a few, or all of the F1 populations would show phenotypic segregation. To fit the observed segregation ratios of the descendants into Mendelian classes was not always easy. Analysis of F1 populations alone was not acceptable to study the inheritance of a given trait. Also selfings (if possible) and backcrosses (BC1) with either parent were required to obtain 3:1 ratios to provide evidence for single-gene models, or 9:7, 13:3 or other classical Mendelian ratios to prove two-gene models with or without epistatic interactions. The summit of this classical work is for example the three-gene models for russet skin proposed by De Jong (1981) and Pavek and Corsini (1981). It should be regarded as a tour de force working with such complex gene models.

Despite the above-mentioned good examples of potato genetics, there are many examples where genetic inferences were made solely on the basis of a large number of cultivar ×

cultivar crosses (Lunden, 1937, 1974). In view of the complicated Mendelian ratios that can be expected at the tetraploid level and the blurring effect of double reduction and non-disjunction (Catcheside, 1959), the conclusions obtained from such studies should be interpreted with care.

With this remark, I do not intend to diminish the relevance of many inheritance studies on the cultivated potato of the previous century, which often involved workloads we can hardly afford nowadays. Somehow, it is important to be very critical while reading old literature, because alternative genetic models are far from being excluded. Occam's razor is teaching us that when the segregation data are not in disagreement with the simplest genetic model, there is no sense in proposing a more complex genetic model. Consequently, important genetic loci may go unnoticed just because the genetic material was not polymorphic at that locus or because not all parental combinations were crossed or backcrossed.

In the early days of genetics, there were quite a few geneticists who chose the potato as their model organism. But gradually, the model organisms such as *Drosophila* or crops that could be selfed, such as tomato and barley, pulled geneticists away from potato. In species that can be selfed, important genetic stock collections became available to the research community. One of the few geneticists who understood and contributed to the development of such stocks was De Jong (1991). Such genetic stocks would be beneficial for genetic studies, allowing complementation tests. With such crosses, the allelic relations between phenotypes could be understood. The transition of classical genetics to molecular genetic studies was marked by the attempt to generate a classical genetic map of potato comprising morphological and isozyme markers (Jacobs et al., 1995; van Eck, 1995). Currently, the value of a classical genetic map is no longer recognized, because the marker numbers and the efficiency of molecular techniques offer a valid substitute for morphological markers (Chapter 5, Gebhardt, this volume).

6.2.2 The characteristics of molecular genetic analysis

The field of potato genetics has considerably changed with the advent of molecular genetic markers. Co-localization of a phenotype on a genetic linkage map has replaced the complementation test for allelism. Hence, the lack of genetic stock collections is not perceived as a limitation.

Ploidy level is still perceived as a limitation in genetic studies, and this has not been alleviated by molecular markers. Most molecular marker techniques do not allow full classification (all alleles can be identified) at the tetraploid level, not even the most multi-allelic microsatellite loci. On the contrary, most often only two or three alleles per locus are displayed with co-dominant techniques, whereas AFLP displays only one allele per locus.

This lack of full classification is a severe problem for the detection of linkage between marker loci, because in tetrasomic inheritance, linkage in repulsion (*trans* configuration) between marker alleles will be observed as independent segregation. Only marker alleles linked in coupling phase (*cis* configuration) will demonstrate association, where simplex genotypes are most easy to interpret.

Only one research group can be recognized for its contributions to tetraploid genetics in potato. The Scottish Crop Research Institute (SCRI) has developed not only a tetraploid map (Meyer et al., 1998) but also a series of analysis tools to calculate linkage maps at the tetraploid level (Hackett et al., 2001, 2003; Luo et al., 2001). The principle behind the construction of a tetraploid linkage map is the placement of sufficient single-dose markers on the 96 linkage groups (2 parents × 4 homologous chromosomes × 12) in conjunction with duplex markers that can bridge between linkage groups from the same homologous chromosome.

Quantitative trait loci (QTL) analysis will be feasible for simplex QTL alleles, such as quantitatively inherited resistance traits (Meyer et al., 1998; Bryan et al., 2002, 2004; Bradshaw et al., 2004). The genetic model assuming simplex inheritance does not follow from the study per se but is justified from pedigree information on the introgression of resistance from wild *Solanum* species. QTL analysis allowing the modelling of multiple alleles and multiple allele interactions at QTL loci is probably feasible once potato has reached a stage of identification of haplotypes that is currently pursued in human genetics.

At the diploid level, not only the genetics is simpler but also phenotypic assessment and trait selection are less ambiguous. Diploids show their phenotypes much better! A single diploid offspring shows segregation (usually transgressive) for a large number of traits, but the spectrum of phenotypes easily exceeds the entire diversity in trait values of tetraploid cultivars shown in the National List. Some examples of trait loci with a wide spectrum of phenotypic variation are given by Celis Gamboa (2002) and Celis Gamboa et al. (2003). The genetic explanation for this phenomenon (that diploids show their phenotypes better) is the probability of finding two outstanding alleles in one diploid versus the chance of having four of those alleles in a tetraploid. As an illustration, we assume two different alleles (+ and −) in diploid heterozygous crossing parents, resulting in $1/4{+}{+}$, $1/2{+}{-}$, $1/4{-}{-}$ offspring, whereas duplex crossing parents will generate only $1/36{+}{+}{+}{+}$ and $1/36{-}{-}{-}{-}$ offspring at the tetraploid level.

6.2.3 Quantitative and qualitative genetic approaches

Throughout this chapter, listing tuber traits as qualitatively or quantitatively inherited characters has been avoided. This distinction is obvious in the case of textbooks that need to explain the difference between Mendelian genetics and quantitative genetics, but in reality, this distinction is artificial. Many traits can be viewed and studied from both the qualitative and the quantitative perspective. For example, the breeder's eye can classify segregating offspring genotypes. For flesh colour, the classes white or yellow and, for tuber shape, the classes round or long will allow the unambiguous identification of a single locus on the genetic map of potato. On the contrary, flesh colour and tuber shape phenotypes can be recorded in ordinal classes or metric units, respectively. Ordinal classification of flesh colour ranging from 1 to 9 (1 = white. . . 9 = intense yellow) can be applied to show the distribution of phenotypes in a mapping population. The resulting chart may show a bi-modal distribution, which offers a justification for the breeder's eye classification of the offspring in Mendelian classes. The variation observed within the major classes 'white' and 'yellow' still needs an explanation. Besides the locus that explains the presence or absence of flesh colour, there is no data on any different locus that can explain flesh colour intensity differences within the classes 'white' and 'yellow'.

Likewise, tuber shape can be recorded as a metric trait (length/width ratio) for QTL analysis. The chromosomal position of the trait locus obtained with metric data in a QTL analysis approach should coincide with the position obtained with the Mendelian segregation of phenotypic classes (round vs. long) in a normal linkage mapping approach where a morphological marker is mapped along with molecular markers (van Eck et al., 1994). The ability to switch from the discrete classifications to continuously distributed metric observations (from linkage mapping to QTL mapping) is a requirement to achieve accurate results. An integration of both approaches allows exploiting the information offered by flanking markers to distinguish between offspring with and without recombination events near the QTL. This can improve the mapping of the trait locus position considerably (Peleman et al., 2005) both at high and at low offspring numbers.

6.3 THE GENETICS OF MORPHOLOGICAL TRAITS

6.3.1 Tuber flesh colour

A large number of tuber flesh phenotypes are shown in Fig. 6.1. From the natural diversity in pigmentation patterns, a collection of colourful crisps were produced and sold as 'Inca crisps'. These pigmentation phenotypes comprise combinations of different pigments and pigment patterns. The red and blue pigments are caused by anthocyanins, and their localized expression suggests tissue-specific transcription of the underlying genes involved in the synthesis of anthocyanins. The tissues in which anthocyanin pigmentation can be observed are the vascular bundle or one or more of the three distinct parenchyma tissue types. Parenchyma tissues differ by their ontogeny: the cortex, the peri-medullary zone and the central pith. The arms of medullary parenchyma often radiate from the pith into the peri-medullary zone (Fig. 6.2).

The genetic loci involved in the biosynthesis of anthocyanins are described in section 6.3.2 dealing with tuber skin colour. Studies by Lewis et al. (1998a,b) offer a quantification of the major anthocyanins, flavonoids and phenolic acids in tubers of wild *Solanum* species and cultivated potato with coloured skins and/or flesh using analytical high-performance liquid chromatography (HPLC).

The genes involved in tissue-specific regulation of anthocyanin biosynthesis belong to a complex locus, first analysed by Dodds (1955). The complex locus probably represented a series of multiple alleles or a clustered gene family comprising the *B-I-F* linkage group (Dodds and Long, 1956). The *B-I-F* group represents closely linked genes involved in the pigmentation of many tissue types. Alleles of the *B*-locus, *Bd* > *Bc* > *Bb* > *Ba* > *b* in order of dominance, are distinguished. *Bd* produces in the seed a band of pigment on the node of the hypocotyl of the embryo and, in the mature plant, a band of pigment at each node, a coloured floral abscission layer and a coloured eyebrow on the tuber. With *Bc*, a band of pigment at each node of the mature plant is absent, but other effects are like those of *Bd*. *Bb* gives a coloured floral abscission layer and a coloured eyebrow on the tuber. With *Ba*, only the floral abscission layer is coloured (Dodds and Long, 1956). The remaining two loci of this cluster are flower colour (*F*-locus) and tuber skin colour (*I*-locus), which have been mapped close to TG63 on potato chromosome X (van Eck et al., 1993a,b).

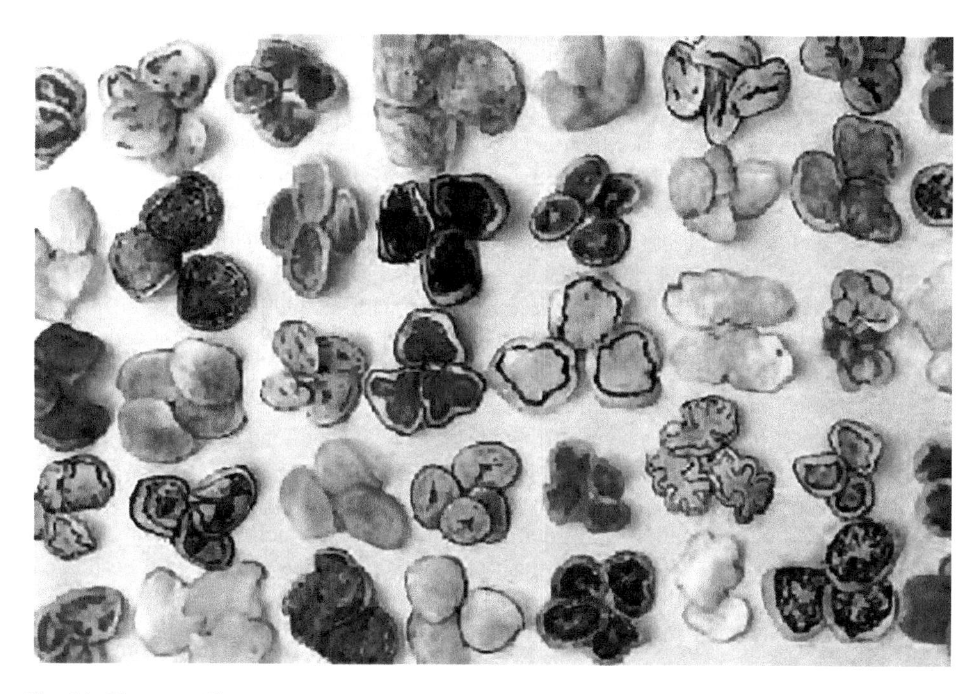

Fig. 6.1. Tissue-specific regulation of anthocyanin pigmentation observed in chips made from native Andean potatoes (© International Potato Center, W. Amorós).

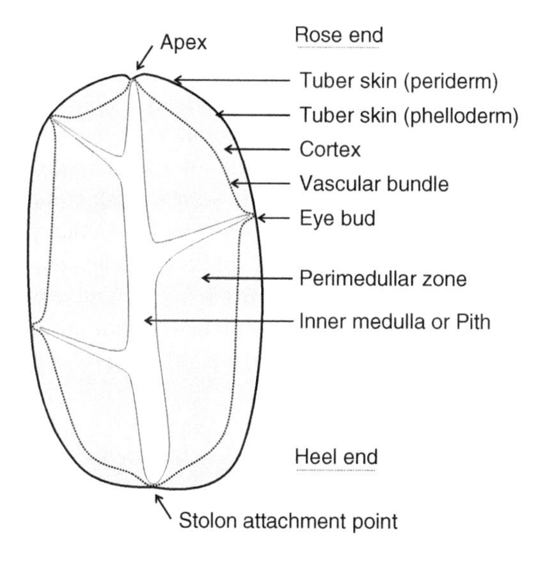

Fig. 6.2. Anatomy of a potato tuber.

After this detour, we need to return to anthocyanin pigmentation of tuber flesh. It could be hypothesized that this trait could also be controlled by the complex *B-I-F* cluster on chromosome X. Evidence for this hypothesis was offered by De Jong (1987), who observed close linkage between a single locus involved in pigmented flesh (*Pf*-locus) and the *I*-locus involved in tuber skin colour. The *I*-locus, which controls the distribution of pigment to the tuber skin, was found to be necessary for the expression of *Pf*.

More general than anthocyanins, tuber flesh can exhibit yellow pigmentation due to carotenoids such as lutein and zeaxanthin (Lu et al., 2001). The regulation of carotenogenesis during the tuber life cycle in potato tubers was recently described by Morris et al. (2004). Genetic studies on tuber flesh colour started with qualitative analysis of phenotypic variation using two classes, yellow and white, disregarding shades of yellowness within these classes. This resulted in the postulation of the *Y*-locus, where yellow (*Y*.) is dominant over white (*yy*) (Frühwirth, 1912). The different degrees of yellowness within these main classes have always been attributed to genetic background or modifiers. Brown et al. (1993) described another flesh colour phenotype: orange. The trait was found in *Solanum stenotomum* and *Solanum phureja* germplasm. The orange phenotype appeared to be controlled by the *Or*-allele, a third allele at the *Y*-locus, which is dominant over the effect of the *Y* or *y* allele. The orange flesh phenotype appeared to be highly correlated with the content of zeaxanthin in the flesh (Brown et al., 1993).

The *Y*-locus has been mapped to chromosome III (Bonierbale et al., 1988; Jacobs et al., 1995), offering a simple morphological marker locus. The gene(s) underlying the *Y*-locus remains to be identified. On the basis of syntenic map positions on tomato and pepper chromosomes, two candidate genes have been implied for the *Y*-locus in potato (Thorup et al., 2000). They proposed a phytoene synthase, the *Psy* gene, and a β-carotene hydroxylase, the *CrtZ-2* gene encoding an enzyme that converts β-carotene to zeaxanthin. Metabolic engineering using a phytoene synthase indeed resulted in high-carotenoid potato tubers containing enhanced levels of β-carotene and lutein (Ducreux et al., 2005). Another example of metabolic engineering is provided by Diretto et al. (2006), who could enhance potato tuber carotenoid levels through tuber-specific silencing of the lycopene epsilon cyclase (*CrtL-e*) gene. At this moment, there is no further information on how allelic variation in these genes relates to allelic variation in the *Y*-locus.

Quantitative genetic analysis of the intensity of yellow pigmentation has not yet resulted in additional QTL positions involved in yellow flesh colour at other chromosomal locations. This leaves the question unanswered as to whether the *Y*-locus on potato chromosome III is only involved in qualitative differences and what is causing the quantitative differences in yellowness. Significant differences among environments and G × E interaction have been observed for yellow flesh intensity, but in view of the high broad-sense heritability of 0.93 (Haynes et al., 1996), the influence of environmental factors on the yellow trait should be of limited concern.

Carotenoid content (neoxanthin, violaxanthin, lutein-5,6-epoxide, lutein, zeaxanthin and an unknown carotenoid) appeared to be highly correlated with the intensity of yellow pigmented tuber flesh colour (Lu et al., 2001). This exponential relationship between total carotenoid content and tuber yellow intensity suggests that selecting for more intense yellow flesh will result in increasingly higher levels of carotenoids (Wu et al., 2001).

Collectively, these tuber flesh pigments, carotenoids, anthocyanins and their phenolic and flavonoid precursors, are regarded as beneficial secondary metabolites for human consumption and pathogen defence. Most secondary metabolites act as potent antioxidants. Concentrations of these components that underline the high nutritional value of potato are given by Brown (2005). Further enrichment of potato with these functional food components seems feasible in view of the superior breeding lines identified by Brown et al. (2005), but it will never turn potato into a red beet or carrot. Furthermore, consumer acceptance of intensely coloured potato flesh, having different taste characteristics, should not be overestimated. Phenolic acids and flavonoids also aid in plant defence (Hahlbrock and Scheel, 1989), and the observation that much higher concentrations of phenolic acids, flavonoids and anthocyanins are found in the skin rather than in the flesh supports this hypothesis, as the skin is the first barrier of defence against pathogens and pests.

6.3.2 Tuber skin and flower colour

The skin of the potato tuber is composed of two layers, the cortex and the epidermis, often referred to as the phelloderm (outermost cell layers of the cortex) and the periderm (the epidermal part of the tuber skin that is susceptible to shear in immature condition), respectively. Pigmentation of the skin can be observed in the cortex (controlled by locus I_{co}) and/or epidermis (controlled by locus I_{ep}) on potato chromosome X (van Eck et al., 1993b). In the previous literature, four different names, *R, E, I* and *PSC*, have been used for the locus on chromosome X involved in tuber skin colour. The symbol *PSC* (Gebhardt et al., 1989) was only used because it was unknown whether the segregation for skin colour was due to segregation of a locus involved in anthocyanin production or a locus for tissue-specific expression. The symbols *R* and *E* are loci proposed in older literature on tetraploid potato but should no longer be used. The phenotype of *R* (Lunden, 1937) differs from R/R^{pw} (Dodds and Long, 1955), but both suggest pigment production rather than tissue-specific gene regulation. For the same reason, symbol *E* (Lunden, 1937) is rejected as it should result in a weak pink pigmentation of the tuber epidermis, even in the absence of *D* or *P*. The symbol *I* (Dodds and Long, 1956) is to be preferred to indicate tissue-specific expression of anthocyanin pigmentation in the tuber skin (van Eck et al., 1993b). To accommodate for the difference between pigmented cortex and epidermis, it was proposed to add a suffix to symbol *I* (I_{co} and I_{ep}; van Eck et al., 1993b). It is not clear whether I_{co} and I_{ep} should be regarded as different alleles (like the series of alleles at the *B*-locus; Dodds and Long, 1956) or two different loci that comprise the *B-I-F* linkage group (Dodds and Long, 1956).

Apart from the two layers of the skin, the pigmentation can be restricted to certain areas within these skin layers. The phenotypes are named ‘splashed’ or ‘spotted’ (pigmentation in a zone surrounding the eyes), ‘spectacled’ (a coloured tuber without pigmentation around the eyes), ‘pigmented eyebrow’, which is nothing else than a nodal band controlled by locus *B*, and ‘pigmented eyes’. Because no recent work has been published on these phenotypes, the earlier reviews by Howard (1970) and Bradshaw and Mackay (1994) are still adequate.

Recent progress in the identification of the genes controlling tuber pigmentation was made by De Jong et al. (2004), who published a series of papers that described loci involved in the biosynthesis of anthocyanins. With primers for all published genes in the pathway (pathway mapping), the candidates for the loci *R*, *P* and *I* were identified (De Jong et al., 2004).

The *R*-locus on chromosome II (van Eck et al., 1993a) is required for the production of red pelargonidin-based anthocyanin pigments in any tissue of the plant, including tuber skin and flower petals. The production of pelargonidins in plants requires the activity of dihydroflavonol 4-reductase (DFR) to catalyse the reduction of dihydrokaempferol into leucopelargonidin. To test the hypothesis that potato *R* encodes DFR, portions of both *dfr* alleles were sequenced from a diploid potato clone known to be heterozygous *Rr*. Sequence comparison revealed a sequence polymorphism, diagnostic for the phenotype of *R*, both in three mapping populations and in a wide range of cultivars. The red allele (*R*) is predicted to encode a 382-amino acid protein that differs at 10 amino acid positions from the gene products of the two alternative alleles (*r*). The observation of a single *R*-allele (haplotype) is also consistent with a half-century-old hypothesis (Dodds and Long, 1955), suggesting that *R* arose just once during the domestication of potato (De Jong et al., 2003a).

Using allele-specific fluorescently labelled primers in a TaqMan assay, the allele dosage could be analysed. This allowed not only the determination of the zygosity at locus *R* in diploids but also the assignment of quadruplex, triplex, duplex, simplex and nulliplex genotypes for the red *dfr* allele at the tetraploid level (De Jong et al., 2003b).

The *P*-locus, mapped to chromosome XI (van Eck et al., 1993a), is required for the production of blue/purple anthocyanin pigments in any tissue of the potato plant such as tubers, flowers or stems. Locus *P* in potato co-localized to a candidate gene coding for the anthocyanin biosynthetic enzyme flavonoid 3′, 5′-hydroxylase ($f3'5'h$), to the same region on chromosome 11 in tomato (De Jong et al., 2003a). Evidence that locus *P* encodes this enzyme was obtained by a complementation test. A potato f3′5′h cDNA clone was introduced into the red-skinned cultivar Desiree, and transformants displayed purple delphinidin-based pigments in tuber and stem tissues that are coloured red in Desiree (Jung et al., 2005).

Locus *I* on chromosome X, required for tissue-specific expression in tuber skin, appeared to correspond to *Petunia* locus *an2*, encoding an *R2R3 Myb* domain transcriptional regulator of the anthocyanin pathway (Borovsky et al., 2004; De Jong et al., 2004).

Locus *F* was mentioned already, as it belongs to the *B-I-F* linkage group. The *F*-locus (Lunden, 1937) controls pink, blue or purple pigmentation of the flower in the presence of *D.pp, ddP.* and *D.P.*, respectively, or white flowers (*D.P.ff* or *ddppF.*). The recessive *ff* genotypes as described by Dodds and Long (1956) have flecked flowers, which is probably a leaky allele. Flower colour patterns with only pigmentation in the tips of the corolla or the reverse – white tips – could be the result of other (leaky) alleles of the *F*-locus.

6.3.3 Tuber shape

Tuber shape is a syndrome of many characters, which have contributed to a rich spectrum of cultivar names such as ‘Kidney’, having a strong degree of flatness, ‘Pink Fir Apple’ with a tuberosed form resembling a pinecone (due to extreme bulking of the tuber area between the eyes), ‘Asparges’, a very long tuber, ‘Banana’, ‘Long Pinkeye’ and ‘muizen’, an ovate form tapering at the apical end. For pictures, I refer to page 277 in the publication by Bradshaw and Mackay (1994). More systematically, the tuber shape characters can be subdivided into the following components: (1) length/width ratio (l) describing the overall shape, (2) tapering of the heel-end and/or the rose-end, where oblong tuber shapes could be viewed as the other extreme of highly tapered tuber ends, and (3) straightness of the length axis. The older literature has been reviewed by De Jong and Burns (1993) and Bradshaw and Mackay (1994).

Only for the overall tuber shape is the inheritance somewhat understood. The basic notion described in various older publications indicates that tuber shape is regulated by a single locus *Ro*, where round (*Ro_*) is dominant over long (*roro*). De Jong and Rowe (1972) confirmed linkage between loci involved in tuber shape and skin colour. Both loci indeed map close to TG63 on potato chromosome X (van Eck et al., 1993b, 1994; Jacobs et al., 1995). Nevertheless, the genetic model described here does not satisfactorily explain the observation of tuber shapes ranging from round ($l < 1.4$) to oval ($1.5 < l < 1.9$) to long ($l > 2.0$). In the paper of 1993, De Jong and Burns could phenotypically identify all three possible genotypes (*RoRo, Roro* and *roro*) that were segregating in their genetic material. This suggests incomplete dominance at the *Ro*-locus. Nevertheless, tuber shape phenotypes are not confined to these three classes but display a continuous distribution. Continuously distributed traits are often interpreted as being controlled by polygenes, but in the case of tuber shape, the single gene hypothesis is maintained. At the *Ro*-locus, a series of multiple alleles can explain all intermediate shapes between round (going to flat) and long (van Eck et al., 1994). At the tetraploid level, multiple alleles can make large numbers of allele combinations and intralocus allele interactions, which may explain the continuous range of tuber shape phenotypes. At the molecular level, the function of the *Ro*-locus is not understood, and once this gene has been cloned, further analysis of alleles and allele effects can be investigated.

Tuber shape has been studied in correlation with tissue composition (Tai and Misener, 1994). They concluded a positive association between the length of the tuber and the narrowness of the pith, suggesting that long potatoes have a narrow pith and were inclined to have a smaller volume of pith. No clear association with other traits, such as specific gravity, was observed.

6.3.4 Eye depth

After decades without research on the inheritance of eye depth, Li et al. (2005b) published a paper that not only offers a good overview of past hypothesis on eye depth but also identifies a monogenic locus *Eyd/eyd* on chromosome X at 4-cM distance from tuber shape locus *Ro/ro*. The correlation between eye depth and tuber shape was noticed before (Maris, 1962, 1966), but correlation may imply two different hypotheses – (1) one locus with a pleiotropic effect on tuber shape and eye depth and (2) two closely linked loci that independently influence eye depth and tuber shape.

The first hypothesis would imply that one gene involved in the elongation or shortening of the longitudinal growth acts not only on the heel–apical bud axis but also on the axis towards the lateral buds. The observation of long tubers with deep eyes and round tubers with shallow eyes argues against this hypothesis or suggests more than one mechanism to regulate eye depth. By choosing other parental clones, these hypotheses could be unravelled. Segregating offspring for eye depth that are not polymorphic for tuber shape would be free of pleiotropic effects caused by tuber shape and preferably in both a long and a round tuber genetic background.

Li et al. (2005b) classified both tuber shape and eye depth in three classes, where the intermediate classes contained only a few descendants. Li et al. (2005b) did not specify a genetic model for tuber shape and eye depth that accommodates the intermediate classes. This is not trivial, because the genetic distance between both traits is largely based on the intermediate phenotypes. Recombinant analysis using the information of flanking marker loci would reinforce their conclusions on the issue of genetic distance between the *Eyd* and *Ro* loci. Flanking markers would also allow clarifying the issue of the intermediate phenotypes for tuber shape and eye depth.

6.3.5 Tuber skin characters

The tuber skin displays many heritable characters such as skin thickness, smoothness and transparency, appearance of lenticels, skin adhesion strength and susceptibility to scuffing damage during harvest. Breeders' preferences for specific skin types are highly divergent across continents. In North America, Russet skin is regarded as a positive character, whereas European breeders prefer a thin, smooth, shiny and transparent skin. In the latter case, the cream to yellow tuber colour is predominantly determined by the underlying flesh colour. This skin type is not without risk, because light-induced greening of tuber flesh will show off immediately. Nevertheless, the selling of washed and pre-packed potatoes, as well as the ability to serve unpeeled potato dishes, is a trend in Europe that demands the smooth skin. The inheritance of most skin-related characters is largely unknown, but most of the above-mentioned characters are relatively stable across environments, suggesting a high value for the heritability.

Bowen et al. (1996) showed clear genotypic differences in skin adhesion strengths and independence of this character from other skin morphological characteristics such as skin thickness, cell size and suberin content. The inheritance of resistance to scuffing damage (skin-set) is not yet clear. Studies by Lulai and Freeman (2001) established a paradigm whereby the thickening and strengthening of tuber phellogen cell walls upon periderm maturation are the determinant for resistance to tuber excoriation. These cellular processes and their interaction with environmental factors will result in a low heritability for this trait. Nevertheless, it appeared to be feasible to determine genotypic differences in skin-set (Lulai and Orr, 1993).

Segregation between brown and colourless skin was observed by Kukimura (1972) in the cross Spartaan × Maris Piper, and they suggested that brown tuber skin could be dominant. The most studied skin character is russeting, which was studied by De Jong (1981) and Pavek and Corsini (1981). Both papers describe the requirement of three complementary dominant genes to develop a russet skin. No correlation between these three loci was observed, suggesting positions on different linkage groups. For the dominant alleles, additivity was concluded, because homozygosity results in a stronger russet phenotype. Early selection for russet skin in seedling generation relies on the heritability of the phenotype and the lack of G × E interactions. Love et al. (1997) concluded that positive selection for russet skin is feasible in early generations. Negative selection should be applied on the incidence and severity of russet patchiness. Early selection for uniformity of skin russeting was not recommended, suggesting a stronger influence of non-heritable effects.

6.4 GENETICS OF TUBER PHYSIOLOGY

6.4.1 Tuberization

In this book, an entire section is dedicated to plant growth and development, including chapters on tuber induction, tuber growth, dormancy and sprouting. Therefore, this paragraph will only briefly touch on genetic aspects of tuberization and dormancy. The identification of QTL involved in tuberization (Van den Berg et al., 1996a) will require a specific set of conditions in the first place that are permissive to the ability to tuberize. Therefore, analysis of tuberization will be strongly confounded by those environmental conditions, referred to as G × E interactions. Jackson (1999) lists six key factors that promote tuber formation: short day length, phytochrome, high light, high sucrose, low temperature and low nitrogen levels. Reproducible results with high heritabilities can be obtained only if these factors are controlled well during genetic experiments. These factors influence multiple signalling pathways for which many candidate genes can be proposed, a list of genes almost too large to be tested for association with tuber formation QTL. Furthermore, epistatic interactions (G × G) should also be expected between these many loci. Recently, Dr. Christiane Gebhardt (Max-Planck Institute for Plant Breeding Research, Cologne, Germany) mapped a large number of candidate genes involved in perception of environmental conditions, hormone synthesis and perception (unpublished work, personal communication), as well as candidate genes involved in carbohydrate metabolism (Chen et al., 2001).

These candidate genes can be mapped to identify co-localization with QTL (see below), but also to perform reverse genetic studies, or to deploy these genes in genetically modified (GM) potato cultivars. Two examples of GM applications of candidate genes are given here. The enhanced expression of phytochrome B promotes tuber formation and yield. The *PHYB* transgenics also showed higher maximum photosynthesis due to increased leaf stomatal conductance (Boccalandro et al., 2003). Constitutive expression of CONSTANS impairs tuberization under short-day inductive conditions (Martínez-García et al., 2002).

Besides these reverse genetic approaches, two forward genetic studies have resulted in the publication of QTL positions involved in tuberization. Van den Berg et al. (1996a) used short-day-induced nodal cuttings to measure the rate of tuber induction. Eleven loci on seven chromosomes were associated with variation in tuberization. Most of the loci had small effects, but a *Solanum tuberosum*-derived QTL allele explained 27% of the variance and mapped close to TG441 on chromosome V. The second study by Simko et al. (1999) showed that similar results could be obtained with greenhouse and in vitro grown plants. The most significant QTL for earliness of tuberization in vitro was located on chromosome VIII, coinciding with QTL for sucrose concentration in leaf exudates. Simko et al. (1999) have also discussed the relation between earliness of tuberization and early crop maturity. In this connection, the QTL *gt5.1* for earliness of tuberization on the whole plant (interval TG441–CD31, on chromosome V) has been consistently detected as the most significant QTL (logarithm of odds (LOD) ranging from 3.3 to 10.4) associated with tuberization and maturity in eight independent experiments (including Van den Berg et al., 1996a,b) and regardless of the way tuberization was measured. Furthermore, Simko et al. (1999) have noted that this QTL maps to the same general location as the one detected for potato maturity (van Eck, 1995; van Eck and Jacobsen, 1996) and foliage blight resistance (Simko, 2002).

6.4.2 Dormancy, sprouting

The first two studies on QTL involved in potato tuber dormancy or sprouting (Freyre et al., 1994; Van den Berg et al., 1996b) are not easily compared because of the different wild diploid crossing parents. Freyre et al. (1994) detected QTL for tuber dormancy in a (*S. tuberosum* × *Solanum chacoense*) × *S. phureja* mapping population. *S. phureja* has much shorter dormancy than *S. tuberosum* and contributed dominant genes for tuber dormancy (Freyre et al., 1994), whereas *Solanum berthaultii* in the second study (Van den Berg et al., 1996b) had a much longer dormancy than *S. tuberosum* and contributed recessive genes for long dormancy (Simko et al., 1997). In second study, eight QTL involved with tuber dormancy were detected (Van den Berg et al., 1996b), and these loci were somewhat congruent with those obtained for tuberization (Van den Berg et al., 1996a). Coinciding QTL for dormancy and tuberization might be expected if genes at these QTL are involved in similar phytohormone signalling pathways. Therefore, Simko et al. (1997) analysed variation in abscisic acid (ABA) in this mapping population. At least three loci (TG234, TG155 and TG499) on three chromosomes (2, 4 and 7, respectively) were associated with variation in ABA content. No correlation was observed between ABA content and dormancy with conventional correlation testing, but relationships between ABA level and dormancy could be demonstrated through the similarity between QTL positions and epistatic interactions (Simko et al., 1997). Another follow-up study resulted in the mapping of more hormone QTL (polyamines, ABA, tuberonic acid, tuberonic acid glucoside, zeatin riboside and gibberellin A1), some of which coincided with the QTL for tuberization or dormancy (Ewing et al., 2004).

Parallel to the association of phytohormones, the carbohydrate metabolism also appears to be involved in dormancy. The expression of transgenes resulted in the observations that a bacterial phosphoglucomutase (PGM) could considerably delay tuber sprouting (Lytovchenko et al., 2005) and a bacterial pyrophosphatase could both delay (Hajirezaei and Sonnewald, 1999) and accelerate sprouting (Farré et al., 2001). A vast majority of loci involved in carbohydrate metabolism have been mapped on the potato function map (Chen et al., 2001), but future research is required to investigate candidate genes that co-localize with QTL associated with tuber formation and dormancy.

6.5 TUBER QUALITY TRAITS

6.5.1 Starch content

Often dry matter content and starch content are used as synonyms. This is only partially true, as protein – for example – is included in dry matter as well. However, it is obvious that dry matter content is largely determined by starch content. Other names for the same trait are 'specific gravity' and 'underwater weight'. In view of the strong correlation with starch content, the genetics of tuber yield or yield is addressed as well.

The trait is most important for the processing industry (starch, crisps and French fries) and moderately important for table use, although fashionable 'low-carb diets' may increase the demand for cultivars with lower starch content. For the processing industry, not only dry matter content is important but also the distribution of dry matter in the tuber. Watery pith results in reduced quality of finished product (fries or crisps). The pith usually has reduced starch content, but the size and shape of the pith and the level of starch reduction strongly depend on genotype. So far, nothing has been published on this trait. As mentioned before, Tai and Misener (1994) could not detect a correlation between pith characters and specific gravity of the entire tuber.

With a growing season of 80 days, early maturing cultivars will not be able to produce the same amounts of starch or tuber yield, as compared with late maturing cultivars, which grow for 120 days. The profound impact of maturity type will overshadow the remaining genetic variation for starch content and tuber yield. This sharp effect is comparable with the overruling effect of a dwarfing (short-straw) gene segregating in cereal mapping studies on grain yield.

QTL analysis of starch content (as many other traits) certainly benefits from the inclusion of maturity type as co-variable in statistical analysis or the mapping of the residuals. Such QTL analysis methods were explored in a study on foliage late blight, also known to be associated with maturity (Visker et al., 2003).

Studies that describe starch content or yield QTL (Freyre and Douches, 1994a,b; Schäfer-Pregl et al., 1998; Gebhardt et al., 2005) allow the conclusion that tuber starch content and tuber yield are truly quantitative and polygenic traits. Genes contributing to the phenotypic effects are located on all chromosomes (Gebhardt et al., 2005). This also applies to reducing sugars, as described in the next section. Furthermore, extensive co-localization of QTL for tuber starch, sugar content and yield has been observed (Chen et al., 2001; Menéndez et al., 2002; Gebhardt et al., 2005).

Besides starch content, starch quality properties (such as the ratio of amylose to amylopectin, degree of phosphorylation, granule size and gelatinization temperature) are of great importance for various applications of starch and starch derivatives in food and non-food industry. Strong genotypic effects have been observed for starch quality traits. However, the genetics of starch quality properties are considered to be outwith the scope of this chapter (e.g. the reviews by Ball et al., 1998; Blennow et al., 2002; Davis et al., 2003; Jobling, 2004). Most candidate genes implied for starch quality properties have been mapped in the context of the QTL studies (Chen et al., 2001) discussed in this section.

6.5.2 Discolouration

6.5.2.1 After-cooking darkening

Discolouration of potato tuber products is caused enzymatically, black-spot bruising and enzymatic oxidative discolouration, or non-enzymatically, after-cooking darkening and the Maillard reaction (Kolbe and Haasse, 1997). In the next sections, these traits will be discussed.

After-cooking darkening is an undesirable tuber trait, which is observed as a bluish-grey haze after cooking when potatoes are exposed to air. Breeding programmes aiming for table potatoes or processing industry (blanched, steam peeled, pre-packed or French fries) have to select cultivars with low susceptibility for this trait. Chemically, the blackening is due to the formation of a colourless reduced ferrous ion–chlorogenic acid complex in the potato. The ferrous complex is then non-enzymatically oxidized to a dark ferric complex following exposure to oxygen in the air (reviewed by Friedman, 1997).

The potential for darkening can be measured in two ways: (1) by a test involving a waiting period of 1–12 h for full colour development after steaming and (2) by a chemical test based on reaction of chlorogenic acid with a mixture of urea, tartaric acid and sodium nitrite (Mann and Lambert, 1989). The latter is a rapid, histological staining method based on the formation of a cherry-red colour of nitrosylated chlorogenic acid. The histological method for visualizing chlorogenic acid in potato tissues agreed with after-cooking darkening in a blanching fry test. The severity of darkening is dependent on the ratio of chlorogenic acid to citric acid concentrations in the potato tubers. Higher ratio normally results in darker tubers (Silva et al., 1991). Furthermore, chelating agents inhibit potato blackening by competing with chlorogenic acid for ferrous ion chelating sites. The concentration of chlorogenic and citric or ascorbic acids is genetically controlled and influenced by environmental conditions (Wang-Pruski et al., 2003; Wang-Pruski and Nowak, 2004). Chlorogenic acid is the major polyphenol in potato and may account for about 80% of the total phenolic acids (Brown, 2005).

A number of candidate genes could be proposed, from the shikimate and the phenylpropanoid pathways, but these will influence many more phenotypes caused by the intermediates and products of this pathway. The final step resulting in chlorogenic acid is catalysed by quinate-hydroxycinnamoyl transferase (HQT), which couples caffeoyl-D-glucose to D-quinic acid (Friedman, 1997). The potato *HQT* gene sequence has been identified in cultivar Kuras via an EST sequencing project (Crookshanks et al., 2001). Overexpression of *HQT* in tomato caused plants to accumulate higher levels of chlorogenic acid, with no side effects on the levels of other soluble phenolics, and to show improved resistance to infection by a bacterial pathogen (Niggeweg et al., 2004).

Besides *HQT*, there are candidate genes that are associated with the level of citric acid. Enhancement of citric acid would not only reduce after-cooking blackening but may also influence acrylamide formation during the production of baked or fried products (Jung et al., 2003). Asparagine, a major free amino acid in potato, is the crucial participant in the production of acrylamide during the Maillard reaction, but lowering the pH limits the formation of acrylamide (Jung et al., 2003).

6.5.2.2 Enzymatic discolouration and black-spot bruising

Enzymatic discolouration is the process in which phenolic compounds are oxidized by the enzyme polyphenol oxidase (PPO) to quinones and the quinones transformed to dark pigments (Friedman, 1997). Prevention of enzymatic browning is currently achieved by the use of sulfiting agents – an undesirable additive to human food – but clear genotypic differences in the level of enzymatic discolouration could be exploited as well. The inheritance of susceptibility to discolouration is determined by the inheritance of the levels of phenolic compounds, including tyrosine (Corsini et al., 1992), and the enzyme PPO. PPO has been proved to be one of the key factors in enzymatic discolouration. Antisense inhibition of PPO clearly blocked discolouration (Bachem et al., 1994). PPO in potato consists of a gene family of at least six genes, namely *POTP1* and *POTP2* (Hunt et al., 1993) and *POT32, POT33, POT41* and *POT72* (Thygesen et al., 1995). The genes are differentially expressed, with *POT32* being the major form expressed in tubers besides *POT33* and *POT72. POTP1* and *POTP2* are mainly expressed in leaves and flowers (Thygesen et al., 1995). The tissue-specific expression can also be deduced from the work of Newman et al. (1993), where they found differential expression of the PPOs in tomato, with these PPOs sharing greater identity with their interspecific homolog (tomato–potato) than with the other intraspecific PPOs (paralogs). Therefore, *POT32* is now strongly implicated as a candidate gene for enzymatic discolouration.

The phenolic compounds oxidized by PPO are chlorogenic acid and tyrosine, but no correlation was observed between the level of chlorogenic acid and discolouration (Friedman, 1997). A high correlation between tyrosine content and the results of the abrasive peel test, comparable with enzymatic discolouration, was detected by Corsini et al. (1992). Another determinant was the partitioning of tyrosine between tuber protein and the free amino acid pool, where high free amino acid levels corresponded with a high level of discolouration. However, Mondy and Munshi (1993) reported that although a correlation between free tyrosine and discolouration is observed, tyrosine does not seem to be the sole determining factor for enzymatic discolouration.

External impact on potato tubers can result in different types of subsurface damage, such as shatter bruising and black-spot bruising (reviewed by Storey and Davies, 1992; Bradshaw and Mackay, 1994; McGarry et al., 1996). Young tubers, at harvest, with high turgor have a lower potential to develop black spot but will develop shatter bruising in return. Black-spot bruising appears as melanin formation surrounding the starch grains and adjacent to the cell wall, resulting in blue/grey to black patches of the flesh. Quick testing for black-spot susceptibility (colour development of blended tuber tissue) allows measurement of discolouration potential, but this tends to overestimate black-spot susceptibility. The blending will disrupt different cellular compartments, bringing enzyme (PPO) and substrate (tyrosine and cysteine) together. More realistic, but time consuming, tests use a device to reproducibly generate a certain impact level and quantify black-spot development. Tuber impact sensitivity determines whether a given impact will cause a bruise and has two components: bruise threshold and bruise resistance. Bruise threshold is the impact at which a tuber will just begin to bruise. Bruise resistance is the ratio of bruising energy to the resulting bruise volume (Baritelle and Hyde, 2003). In contrast to enzymatic discolouration, black-spot bruising is strongly correlated with specific gravity of the tuber, where higher specific gravity tubers are more sensitive to impact. Tuber specific gravity depends on cultivar, quality of the growing conditions and dehydration. Baritelle and Hyde (2003) show that Russet Burbank and Atlantic have lower bruise resistance and lower bruise thresholds with increasing specific gravity. Snowden, however, showed higher bruise threshold but lower bruise resistance with increasing specific gravity. These results demonstrate that variation in resistance to damage adds to the complexity of black-spot bruising. The interpretation of QTL studies on black-spot bruising should take all these factors into account for correct interpretation of the inheritance of this trait. Understanding the inheritance of PPO enzyme activity and substrate level remains the simplest way to obtain cultivars with reduced susceptibility to black-spot bruising.

6.5.2.3 Discolouration due to the Maillard reaction

As a result of the reaction between reducing sugars and free amino acids at high temperatures, a brown discolouration is formed, also known as the Maillard reaction. Excessive discolouration and the development of off-flavours severely reduce the quality of French fries and crisps. The main determinant is the concentration of reducing sugars, resulting from the level of starch mobilization, which depends on genotype and storage conditions. A lower storage temperature (4 °C) especially results in an accumulation of reducing sugars. At more ambient temperatures, tubers can be reconditioned. This is the lowering of the concentration of reducing sugars via increasing the metabolic activity.

The trait is often analysed at three moments in time and hence indicated by different names: (1) processing quality before storage, (2) processing quality after storage at low temperatures [also called cold-sweetening, sugar stability or (cold) chipping quality] and (3) processing quality after reconditioning. These three traits are poorly correlated. That so-called 'cold chippers' can be processed immediately from cold storage suggests the high heritability of sugar stability. However, the ability of cultivars to recover after reconditioning is highly heritable.

The regulation of reducing sugar concentrations in tuber tissue is described in other chapters in this book, describing carbohydrate metabolism, and will not be copied here. From this chapter, a number of enzymes can be proposed as candidate genes for processing quality. The most important enzymes involved in cold-sweetening and/or reconditioning are (1) amylases, because sucrose is the first sugar to accumulate (Sowokinos, 1990), (2) PGM, (3) UDP-glucose pyrophosphorylase (UGPase; Sowokinos, 2001), (4) glucose-6-P/Pi translocator (GPT), (5) sucrose phosphate synthase (SPS), (6) invertases (Li et al., 2005a), (7) ADP-glucose pyrophosphorylase (AGPase) and (8) sucrose synthase (SuSy). These and many more candidate genes involved in carbohydrate metabolism have been mapped (Chen et al., 2001) and compared with the position of QTL involved in cold-sweetening (Menéndez et al., 2002). With a few markers the first six QTL were detected (Douches and Freyre, 1994), suggesting that many loci are involved in cold-sweetening. This is confirmed by Menéndez et al. (2002), who mapped QTL for glucose, fructose and sucrose content on all potato chromosomes. Candidate genes that collocated with QTL (some explained more than 10% of the phenotypic variance or were consistent across environments) encoded invertase, SuSy3, SPS, AGPase, sucrose transporter 1 and a putative sucrose sensor.

Two candidate genes received more attention. First, Sowokinos (2001) confirmed the correlation between UGPase and cold-sweetening. Possibly, this locus is similar to *Sug11a* (Menéndez et al., 2002), but this QTL on chromosome XI is even more closely linked to *Sut1*, a sugar transporter (Menéndez et al., 2002). Secondly, the role of the apoplastic invertase locus *invGE/GF* on chromosome IX has been confirmed by Li et al. (2005a). This locus consisting of duplicated invertase genes is collocated with cold-sweetening QTL *Sug9*.

Linkage disequilibrium studies may further confirm the importance of these two candidate genes and their utility in marker-assisted breeding.

6.5.2.4 Chlorophyll discolouration

Exposure of tubers to light initiates another form of discolouration, also named 'tuber-greening'. Greening is caused by the development of chlorophyll pigments in chloroplasts. The conversion of amyloplasts into chloroplasts begins after one-day exposure to illumination (Muraja-Fras et al., 1994). Tuber-greening has a negative influence on visual quality but not on food safety. Food safety is only at risk because of light-induced increase of glycoalkaloids. Several studies examined the supposed correlation between light-induced chlorophyll and glycoalkaloid accumulation (Dale et al., 1993), but the papers by Edwards and co-workers (1997, 1999) demonstrate that greening and light-enhanced glycoalkaloid accumulation are unrelated processes and, hence, separate breeding goals.

Varietal differences have been observed in the time lag before greening appeared and the amount of light necessary for greening (Brown and Riley, 1976). Greening reactions of 144 potato varieties (16 russet, 29 red and 99 white) were evaluated by Reeves (1988), who concluded that discolouration could be subdivided into three independent components of greening (external colour, internal colour and depth of colour). Significant differences in greening susceptibility between potato cultivars were also observed by Dale et al. (1993). At this moment, little is known on the inheritance of this trait, but the observation of varietal differences with reasonably high heritability suggests that selection for low greening susceptibility is feasible.

6.5.3 Texture

Texture of cooked potatoes is a character that must have a high heritability, because cultivar name alone is often sufficient to know cooking type. Cooking type is often indicated in classes A (solid) to D (mealy), but texture can be described with many more characteristics, such as mealy/non-mealy, firmness/breaking of cooked tissue, crumbly/smooth and brittle/stickiness, which could be discriminated (Van Marle, 1997). A major issue is to translate these sensory descriptors in tuber tissue composition and the effect of cooking on tuber tissue, as a first step, to unravel the genes that are associated with texture.

The composition of the cell wall and middle lamellae, as well as the gelatinization characteristics of starch, has been studied in relation to texture. A mealy cultivar such as Irene will release more pectic materials. The pectic polysaccharides contain more and/or longer side chains for cultivar Irene than for the non-mealy cultivar Nicola, which had more branched and a higher amount of branched pectic polysaccharides (Van Marle et al., 1997).

Texture is probably under polygenic control and should be studied via QTL mapping or association studies. So far, a first genetic locus involved in cooking type was identified with QTL analysis, where a significant QTL mapped to the distal part of the long arm of chromosome IX. This locus co-localized with an eQTL [quantitative variation in gene expression determined with quantitative reverse transcription–polymerase chain reaction (RT–PCR)] of a candidate gene *StTLRP*. This tyrosine- and lysine-rich protein was identified on a cDNA microarray of potato, hybridized with mRNA from bulked segregants that differed in cooking type (Kloosterman, 2006).

6.5.4 Glycoalkaloids

Potato breeders have to select genotypes with low glycoalkaloid contents in view of their potential toxicity. Glycoalkaloids have two basic constituents – a glycosidic grouping and a steroid alkaloid skeleton, such as solanidine and solasodine (also referred to as the aglycone constituent). Total glycoalkaloid content (TGA) in tubers for consumption may not exceed food safety values of 200 mg kg^{-1} fresh weight (FW). Toxicity, however, is highly variable among individual glycoalkaloids, and furthermore, specific combinations of glycoalkaloids may have a synergistic effect on toxicity (Rayburn, 1995). Korpan et al. (2004) recently published a critical review on toxicity issues and the TGA threshold. The TGA threshold is easily surpassed when tubers are exposed to light, although genotype specific differences were observed (De Maine et al., 1988; Dale et al., 1993). Percival (1999) showed that cultivar Maris Piper was light-insensitive, which represents a valuable observation to stimulate further breeding and research efforts. Another light-induced character, tuber-greening cannot be used as an indicator of light-induced elevation of TGA level, because these were shown to be unrelated processes (Edwards and Cobb, 1997; Edwards et al., 1999).

Breeding for pathogen resistance often employs wild *Solanum* species, which may accumulate glycoalkaloids exceeding 2200 mg kg^{-1} FW (Van Gelder et al., 1988). Furthermore, these wild species often produce glycoalkaloids that are different from those found in cultivated potato. This has severe implications for potato breeding, where care should be taken to avoid unacceptably high TGA levels and unintended introgression of such a character in elite germplasm. Not only in wild species but also in cultivated potato, the genotype is the most important factor determining glycoalkaloid content, next to environmental effects such as temperature, light and damage. Further details are found in reviews by Valkonen et al. (1996) and Friedman and McDonald (1997).

Both forward and reverse genetic approaches have been used to study the genetic control of glycoalkaloid content. The forward genetic studies resulted in QTL (see below), which have not been compared with the loci of genes used in reverse genetic studies. Candidate genes implied to control TGA could be derived from the sterol pathway resulting in a reduction of the aglycone content of genes involved in the glucosylation step. An example of the latter is offered by studies on solanidine–UDP-glucose glucosyltransferase (GST; Moehs et al., 1997). Using this GST sequence in an antisense construct, it was possible to reduce the glycoalkaloid contents of potato tubers (McCue et al., 2003). The study by Arnqvist et al. (2003) is an example that aims to interfere with the formation of sterols. Transgenic potato plants overexpressing a sterol methyltransferase were used to study sterol biosynthesis. A decreased glycoalkaloid level could be obtained and cholesterol was implied as a precursor in glycoalkaloid biosynthesis (Arnqvist et al., 2003).

Mutation breeding, which should be included among the reverse genetic approaches, has been examined for reducing the glycoalkaloid level (Love et al., 1996). Indeed, mutants with reduced levels of glycoalkaloids were obtained, but screening for a phenotype remains the rate-limiting step in mutagenesis experiments. With the recent expansion of sequence information on candidate genes, alternative screening methods for locus-to-phenotype reverse genetic strategies have been proposed. Tagging of mutant alleles with the TILLING approach (McCallum et al., 2000) could be exploited to select low glycoalkaloid mutants.

The first forward genetic studies using molecular markers resulted in QTL on many chromosomes, indicating a polygenic inheritance. Yencho et al. (1998) analysed the segregation of the glycoalkaloids solanine, chaconine, solasodine and solamargine to identify QTL for the production of the aglycones solanidine and solasodine. Several QTL for the accumulation of solasodine and solanidine were identified on chromosomes 4, 6 and 12 for the BCB mapping population and on chromosomes 4, 8 and 11 for the BCT population. Two QTL for solanidine were identified in BCT on chromosomes 1 and 4, where the QTL located on chromosome 1 in the TG70–TG71 interval was most significant. The locus explained 17–22% of the phenotypic variation and was also detected in BCB (Yencho et al., 1998). The importance of the short arm of chromosome 1 has been confirmed by Ronning et al. (1999) and by Hutvagner et al. (2001), who both mapped a QTL involved in the production of several glycoalkaloids from *S. chacoense*. The work by Ronning et al. (1999) focussed on the development of Colorado potato beetle-resistant potato varieties via enhanced glycoalkaloid content. This is only feasible if leaf and tuber glycoalkaloid are under independent genetic control.

6.5.5 Growing defects (hollow hearts, growth cracks, second growth, internal heat necrosis)

No recent work has shed new light on the genetic control of susceptibility for growing defects such as hollow hearts and growth cracks. A publication by Claassens and Vreugdenhil (2000) concludes that dormancy breaking and tuber formation are highly related, although complementary processes. Second growth, resulting in glassy-end tubers, is also related to these complementary processes, but apart from these physiological considerations, there are no novel insights into the inheritance of second growth susceptibility. Susceptibility to internal heat necrosis (IHN), resulting in necrotic tissue in the pith of the tubers, was recently studied by Sterrett et al. (2003, 2006). They analysed interspecific 4x–2x *S. tuberosum* × *S. phureja*–*S. stenotomum* hybrids with significantly different IHN resistance and suggested that broadening the genetic basis of cultivated potato could offer some room for breeding for IHN resistance.

6.5.6 Tuber size uniformity

Uniformity of tuber size is of great economic importance, because it directly affects the marketable yield, as well as costs involved in size grading. Most studies focus on the processes of stolon initiation, elongation, swelling, tuber growth and resorption within a given genotype, across a wide range of environmental factors (reviewed by Ewing and Struik, 1992). Genetic variation in size distribution across a wide range of diploid genotypes was studied by Celis Gamboa et al. (2003). Their results suggest that the final marketable yield and size distribution of potato tubers are defined by the degree of stolon branching, the duration of the stolon tip swelling period, the ability of the small growing tubers to reach a marketable size and tuber resorption. Therefore, understanding of the inheritance of tuber size uniformity might not be feasible without first understanding the inheritance of all underlying factors, which are known to be highly genotype-dependent.

REFERENCES

Arnqvist L., P.C. Dutta, L. Jonsson and F. Sitbon, 2003, *Plant Physiol.* 131, 1792.

Bachem C.W.B., G.J. Speckmann, P.C.G. van der Linde, F.T.M. Verheggen, M.D. Hunt, J.C. Steffens and M. Zabeau, 1994, *Bio/Technology* 12, 1101.

Ball S.G., M.H.B.J. van de Wal and R.G.F. Visser, 1998, *Trends Plant Sci.* 3, 462.

Baritelle A.L. and G.M. Hyde, 2003, *Postharvest Biol. Technol.* 29, 279.

Blennow A., T.H. Nielsen, L. Baunsgaard, R. Mikkelsen and S.B. Engelsen, 2002, *Trends Plant Sci.* 7, 445.

Boccalandro H.E., E.L. Ploschuk, M.J. Yanovsky, R.A. Sánchez, C. Gatz and J.J. Casal, 2003, *Plant Physiol.* 133, 1539.

Bonierbale M.W., R.L. Plaisted and S.D. Tanksley, 1988, *Genetics* 120, 1095.

Borovsky Y., M. Oren-Shamir, R. Ovadia, W.D. de Jong and I. Paran, 2004, *Theor. Appl. Genet.* 109, 23.

Bowen S.A., A.Y. Muir and C.T. Dewar, 1996, *Potato Res.* 39, 313.

Bradshaw J.E. and G.R. Mackay, 1994, *Potato Genetics.* CAB International, Wallingford, U.K.

Bradshaw J.E., B. Pande, G.J. Bryan, C.A. Hackett, K. McLean, H.E. Stewart and R. Waugh, 2004, *Genetics* 168, 983.

Brown C.R., 2005, *Am. J. Potato Res.* 82, 163.

Brown C.R., D. Culley, C.P. Yang, R. Durst and R. Wrolstad, 2005, *J. Am. Soc. Hort. Sci.* 130, 174.

Brown C.R., C.G. Edwards, C.P. Yang and B.B. Dean, 1993, *J. Am. Soc. Hort. Sci.* 118, 145.

Brown E. and W. Riley, 1976, *J. Natl. Inst. Agr. Botany* 14, 70.

Bryan G.J., K. McLean, J.E. Bradshaw, W.S. de Jong, M. Phillips, L. Castelli and R. Waugh, 2002, *Theor. Appl. Genet.* 105, 68.

Bryan G.J., K. McLean, B. Pande, A. Purvis, C.A. Hackett, J.E. Bradshaw and R. Waugh, 2004, *Mol. Breed.* 14, 105.

Catcheside D.G., 1959, *Heredity* 13, 403.

Celis Gamboa C., 2002, PhD thesis, Wageningen University, The Netherlands. http://www.library.wur.nl/wda/.

Celis Gamboa C., P.C. Struik, E. Jacobsen and R.G.F. Visser, 2003, *Ann. Appl. Biol.* 143, 175.

Chen X., F. Salamini and C. Gebhardt, 2001, *Theor. Appl. Genet.* 102, 284.

Claassens M.M.J. and D. Vreugdenhil, 2000, *Potato Res.* 43, 347.

Corsini D.L., J.J. Pavek and B. Dean, 1992, *Am. Potato J.* 69, 423.

Crookshanks M., J. Emmersen, K.G. Welinder and K.L. Nielsen, 2001, *FEBS Lett.* 506, 123.

Dale M.F.B., D.W. Griffiths, H. Bain and D. Todd, 1993, *Ann. Appl. Biol.* 123, 411.

Davis J.P., N. Supatcharee, R.L. Khandelwal and R.N. Chibbar, 2003, *Starch/Stärke* 55, 107.

De Jong H., 1981, *Potato Res.* 24, 309.

De Jong H., 1987, *Am. Potato J.* 64, 337.

De Jong H., 1991, *Am. Potato J.* 68, 585.

De Jong H. and V.J. Burns, 1993, *Am. Potato J.* 70, 267.

De Jong W.S., D.M. De Jong, H. de Jong, J. Kalazich and M. Bodis, 2003a, *Theor. Appl. Genet.* 107, 1375.

De Jong W.S., D.M. de Jong and M. Bodis, 2003b, *Theor. Appl. Genet.* 107, 1384.

De Jong W.S., N.T. Eannetta, D.M. De Jong and M. Bodis, 2004, *Theor. Appl. Genet.* 108, 423.

De Jong H. and P.R. Rowe, 1972, *Potato Res.* 15, 200.

De Maine M.J., H. Baine and J.A.L. Joyce, 1988, J. Agric.Sci., UK, 111 (1), 57–58.

Diretto G., R. Tavazza, R. Welsch, D. Pizzichini, F. Mourgues, V. Papacchioli, P. Beyer and G. Giuliano, 2006, *BMC Plant Biol.* 6, 13.
Dodds K.S., 1955, *Nature* 175, 394.
Dodds K.S. and D.H. Long, 1955, *J. Genet.* 53, 136.
Dodds K.S. and D.H. Long, 1956, *J. Genet.* 54, 27.
Douches D.S. and R. Freyre, 1994, *Am. Potato J.* 71, 581.
Ducreux L.J.M., W.L. Morris, P.E. Hedley, T. Shepherd, H.V. Davies, S. Millam and M.A. Taylor, 2005, *J. Exp. Botany* 56, 409, 81.
Edwards E.J. and A.H. Cobb, 1997, *J. Agr. Food Chem.* 45, 1032.
Edwards E.J., R.E. Saint and A.H. Cobb, 1999, *J. Sci. Food Agr.* 76, 327.
Ewing E.E., I. Simko, E.A. Omer and P.J. Davies, 2004, *Am. J. Potato Res.* 81, 281.
Ewing E.E. and P.C. Struik, 1992, *Hortic. Rev.* 14, 89–198.
Farré E.M., A. Bachmann, L. Willmitzer and R.N. Trethewey, 2001, *Nat. Biotechnol.* 19, 268.
Flinn B., C. Rothwell, R. Griffiths, M. Lägue, D. DeKoeyer, R. Sardana, P. Audy, C. Goyer, X.-Q. Li, G. Wang-Pruski and S. Regan, 2005, *Plant Mol. Biol.* 59, 407.
Freyre R. and D.S. Douches, 1994a, *Theor. Appl. Genet.* 87, 764.
Freyre R. and D.S. Douches, 1994b, *Crop Sci.* 34, 1361.
Freyre R., S. Warnke, B. Sosinski and D.S. Douches, 1994, *Theor. Appl. Genet.* 89, 474.
Friedman M., 1997, *J. Agr. Food Chem.* 45, 1523.
Friedman M. and G.M. McDonald, 1997, *Crit. Rev. Plant Sci.* 16, 55.
Frühwirth C., 1912, *Deutsches Landwirtschaftlichte Presse* 39, 551, 565.
Gebhardt C., C. Menéndez, X. Chen, L. Li, R. Schäfer-Pregl and F. Salamini, 2005, *Acta Hortic.* 684, 85.
Gebhardt C., E. Ritter, U. Schachtschabel, B. Walkemeier, H. Uhrig and F. Salamini, 1989, *Theor. Appl. Genet.* 78, 65.
Hackett C.A., J.E. Bradshaw and J.W. McNicol, 2001, *Genetics* 159, 1819.
Hackett C.A., B. Pande and G.J. Bryan, 2003, *Theor. Appl. Genet.* 106, 1107.
Hahlbrock K. and P. Scheel, 1989, *Ann. Rev. Plant Physiol. Plant Mol. Biol.* 40, 347.
Hajirezaei M. and U. Sonnewald, 1999, *Potato Res.* 42(suppl), 353–372.
Haynes K.G., J.B. Sieczka, M.R. Henninger and D.L. Fleck, 1996, *J. Am. Soc. Hort. Sci.* 121, 175.
Howard H.W., 1970, *The Genetics of Potato.* Springer Verlag, New York.
Hunt M.D., N.T. Eannetta, H.F. Yu, S.M. Newman and J.C. Steffens, 1993, *Plant Mol. Biol.* 21, 59.
Hutvagner G., Z. Banfalvi, I. Milankovics, D. Silhavy, Z. Polgar, S. Horvath, P. Wolters and J.P. Nap, 2001, *Theor. Appl. Genet.* 102, 1065.
Jackson S.D., 1999, *Plant Physiol.* 119, 1.
Jacobs J.M.E., H.J. van Eck, P. Arens, B. Verkerk-Bakker, B. te Lintel Hekkert, H.J.M. Bastiaanssen, A. El-Kharbotly, A. Pereira, E. Jacobsen and W.J. Stiekema, 1995, *Theor. Appl. Genet.* 91, 289.
Jobling S., 2004, *Curr. Opin. Plant Biol.* 7, 210.
Jung M.Y., D.S. Choi and J.W. Ju, 2003, *J. Food Sci.* 68, 1287.
Jung C.S., H.M. Griffiths, D.M. de Jong, S.P. Cheng, M. Bodis and W.S. de Jong, 2005, *Theor. Appl. Genet.* 110, 269.
Kloosterman, B., 2006, PhD thesis, Wageningen University, The Netherlands. http://library.wur.nl/wda/abstracts/ab3984.html
Kolbe H. and N.U. Haase, 1997, *Kartoffelbau* 48, 234.
Korpan Y.I., E.A. Nazarenko, I.V. Skryshevskaya, C. Martelet, N. Jaffrezic-Renault and A.V. El'skaya, 2004, *Trends Biotechnol.* 22, 147.
Kukimura H., 1972, *Potato Res.* 15, 106.
Lewis C.E., J.R.L. Walker, J.E. Lancaster and K.H. Sutton, 1998a, *J. Sci. Food Agr.* 77, 45.
Lewis C.E., J.R.L. Walker, J.E. Lancaster and K.H. Sutton, 1998b, *J. Sci. Food Agr.* 77, 58.
Li X.-Q., H. de Jong, D.M. de Jong and W.S. de Jong, 2005b, *Theor. Appl. Genet.* 110, 1068.
Li L., J. Strahwald, H.-R. Hofferbert, J. Lübeck, E. Tacke, H. Junghans, J. Wunder and C. Gebhardt, 2005a, *Genetics* 170, 813.
Love S.L., T.P. Baker, J.A. Thompson and B.K. Werner, 1996, *Plant Breed.* 115, 119.
Love S.L., B.K. Werner and J.J. Pavek, 1997, *Am. Potato J.* 74, 199.
Lu W., K. Haynes, E. Wiley and B. Clevidence, 2001, *J. Am. Soc. Hort. Sci.* 126, 722.

Lulai E.C. and T.P. Freeman, 2001, *Ann. Bot.* 88, 555.
Lulai E.C. and P.H. Orr, 1993, *Am. Potato J.* 70, 599.
Lunden A.P., 1937, *Meld. fra Norges Landbrukshogskole* 30, 1.
Lunden A.P., 1974, *Meld. fra Norges landbrukshogskole* 53, 1.
Luo Z.W., C.A. Hackett, J.E. Bradshaw, J.W. McNicol and D. Milbourne, 2001, *Genetics* 157, 1369.
Lytovchenko A., M. Hajirezaei, I. Eickmeier, V. Mittendorf, U. Sonnewald, L. Willmitzer and A.R. Fernie, 2005, *Planta* 221, 915.
Mann J.D. and L.D. De Lambert, 1989, *N.Z. J. Crop Hort. Sci.* 17, 207.
Maris B., 1962, PhD thesis, Agricultural University, Wageningen, The Netherlands.
Maris B., 1966, *Euphytica* 15, 18.
Martínez-García J.F., A. Virgós-Soler and S. Prat, 2002, *Proc. Natl. Acad. Sci. U.S.A.* 99, 15211.
McCallum C.M., L. Comai, E.A. Greene and S. Henikoff. 2000, *Nat. Biotechnol.* 18, 455.
McCue K.F., P.V. Allen, D.R. Rockhold, M.M. Maccree, W.R. Belknap, L.V.T. Shepherd, H. Davies, P. Joyce, D.L. Corsini and C.P. Moehs, 2003, *Acta Hortic.* 619, 77.
McGarry A., C.C. Hole, R.L.K. Drew and N. Parsons, 1996, *Postharvest Biol. Technol.* 8, 239.
Menéndez C.M., E. Ritter, R. Schäfer-Pregl, B. Walkemeier, A. Kalde, F. Salamini and C. Gebhardt, 2002, *Genetics*, 162, 1423.
Meyer R.C., D. Milbourne, C.A. Hackett, J.E. Bradshaw, J.W. McNicol and R. Waugh, 1998, *Mol. Gen. Genet.* 259, 150.
Meyer S., A. Nagel and C. Gebhardt, 2005, *Nucleic Acids Res.* 33(Database issue), D666.
Moehs C.P., P.V. Allen, M. Friedman and W.R. Belknap, 1997, *Plant J.* 11, 227.
Mondy N.I. and C.B. Munshi, 1993, *J. Agr. Food Chem.* 41, 1868.
Morris W.L., L. Ducreux, D.W. Griffiths, D. Stewart, H.V. Davies and M.A. Taylor, 2004, *J. Exp. Bot.* 55, 975.
Muraja-Fras J., M. Krsnik-Rasol and M. Wrischer, 1994, *J. Plant Physiol.* 144, 58.
Newman S.M., N.T. Eannetta, H.F. Yu, J.P. Prince and M.C. de Vicente, 1993, *Plant Mol. Biol.* 21, 1035.
Niggeweg R., A.J. Michael and C. Martin, 2004, *Nat. Biotechnol.* 22, 746.
Pavek J.J. and D.L. Corsini, 1981, *Am. Potato J.* 58, 515.
Peleman J.D., C. Wye, J. Zethof, A.P. Sørensen, H. Verbakel, J. van Oeveren, T. Gerats and J.N.A.M. Rouppe van der Voort, 2005, *Genetics* 171, 1341.
Percival G., 1999, *J. Sci. Food Agr.* 79, 1310.
Rayburn J.R., 1995, *Food Chem. Toxicol.* 33, 1013.
Reeves A.F., 1988, *Am. Potato J.* 65, 651.
Ronning C.M., S.S. Stegalkina, R.A. Ascenzi, O. Bougri, A.L. Hart, T.R. Utterbach, S.E. Vanaken, S.B. Riedmuller, J.A. White, J. Cho, G.M. Pertea, Y. Lee, S. Karamycheva, R. Sultana, J. Tsai, J. Quackenbush, H.M. Griffiths, S. Restrepo, C.D. Smart, W.E. Fry, R. van der Hoeven, S.D. Tanksley, P. Zhang, H. Jin, M.L. Yamamoto, B.J. Baker and C.R. Buell, 2003, *Plant Physiol.* 131, 419.
Ronning C.M., J.R. Stommel, S.P. Kowalski, L.L. Sanford, R.S. Kobayashi and O. Pineada, 1999, *Theor. Appl. Genet.* 98, 39.
Schäfer-Pregl R., E. Ritter, L. Concilio, J. Hesselbach, L. Lovatti, B. Walkemeier, H. Thelen, F. Salamini and C. Gebhardt, 1998, *Theor. Appl. Genet.* 97, 834.
Silva G.H., R.W. Chase, R. Hammerschmidt and J.N. Cash, 1991, *J. Agr. Food Chem.* 39, 871.
Simko I., 2002, *Am. J. Potato Res.* 79, 125.
Simko I., S. McMurry, H.-M. Yang, A. Manschot, P.J. Davies and E.E. Ewing, 1997, *Plant Physiol.* 115, 1453.
Simko I., D. Vreugdenhil, C.S. Jung and G.D., May 1999, *Mol. Breed.* 5(5), 417.
Sowokinos J.R., 1990, In: M.E. Vayda and W.D. Park (eds), *The Molecular and Cellular Biology of the Potato.* CAB International, Wallingford, Oxon, U.K., p. 137.
Sowokinos J.R., 2001, *Am. J. Potato Res.* 78, 57.
Sterrett S.B., K.G. Haynes, G.C. Yencho, M.R. Henninger and B.T. Vinyard, 2006, *Crop Sci.* 46, 1471.
Sterrett S.B., M.R. Henninger, G.C. Yencho, W. Lu, B.T. Vinyard and K.G. Haynes, 2003, *Crop Sci.* 43, 790.
Storey R.M.J. and H.V. Davies, 1992, Tuber quality, In: P.M. Harris (ed.), *The Potato Crop*, 2nd edn. Chapman and Hall, London, U.K.
Swaminathan M.S. and H.W. Howard, 1953, *Bibliographica Genetica* 16, 1.
Tai G.C.C. and G.C. Misener, 1994, *Potato Res.* 37, 353.

Thorup T.A., B. Tanyolac, K.D. Livingstone, S. Popovsky, I. Paran and M. Jahn, 2000, *Proc. Natl. Acad. Sci. U.S.A.* 97, 11192.
Thygesen P.W., I.B. Dry and S.P. Robinson, 1995, *Plant Physiol.* 109, 525.
Valkonen J.P.T., M. Keskitalo, T. Vasara and L. Pietila, 1996, *Crit. Rev. Plant Sci.* 15, 1.
Van den Berg J.H., E.E. Ewing, R.L. Plaisted, S. McMurry and M.W. Bonierbale, 1996a, *Theor. Appl. Genet.* 93, 307.
Van den Berg J.H., E.E. Ewing, R.L. Plaisted, S. McMurry and M.W. Bonierbale, 1996b, *Theor. Appl. Genet.* 93, 317.
van Eck H.J., 1995, PhD thesis, Wageningen University, The Netherlands. http://www.library.wur.nl/wda/.
van Eck H.J., J.M.E. Jacobs, J. van Dijk, W.J. Stiekema and E. Jacobsen, 1993a, *Theor. Appl. Genet.* 86, 295.
van Eck H.J., J.M.E. Jacobs, P.M.M.M. van den Berg, W.J. Stiekema and E. Jacobsen, 1993b, *Heredity* 73, 410.
van Eck H.J., J.M.E. Jacobs, P. Stam, J. Ton and E. Jacobsen, 1994, *Genetics* 137, 303.
van Eck H.J. and E. Jacobsen, 1996, In: P.C. Struik, J. Hoogendoorn, J.K. Kouwenhoven, L.J. Mastenbroek, L.J. Turkensteen, A. Veerman and J. Vos (eds), *Abstracts of the 13th Triennial Conference EAPR*, Wageningen, p. 130.
Van Gelder W.M.J., J.H. Vinke and J.J.C. Scheffer, 1988, *Euphytica* 37S, 147.
Van Marle J.T., 1997, PhD thesis, Wageningen University, The Netherlands. http://www.library.wur.nl/wda/.
Van Marle J.T., K. Recourt, C. van Dijk, H.A. Schols and A.G.J. Voragen, 1997, *J. Agr. Food Chem.* 45, 1686.
Van Os H., S. Andrzejewski, E. Bakker, I. Barrena, G.J. Bryan, B. Caromel, B. Ghareeb, E. Isidore, W. de Jong, P. van Koert, V. Lefebvre, D. Milbourne, E. Ritter, J.N.A.M. Rouppe van der Voort, F. Rousselle-Bourgeois, J. van Vliet, R. Waugh, R.G.F. Visser, J. Bakker and H.J. van Eck, 2006, *Genetics* 173, 1075.
Visker M.H.P.W., L.C.P. Keizer, H.J. van Eck, E. Jacobsen, L.T. Colon and P.C. Struik, 2003, *Theor. Appl. Genet.* 106, 317.
Wang-Pruski G., H. de Jong, T. Astatkie and Y. Leclerc, 2003, *Acta Hortic.* 619, 45.
Wang-Pruski G. and J. Nowak, 2004, *Am. J. Potato Res.* 81, 7.
Yencho G.C., S.P. Kowalski, R.S. Kobayashi, S.L. Sinden, M.W. Bonierbale and K.L. Deahl, 1998, *Theor. Appl. Genet.* 97, 563.

Chapter 7

Genetics of Resistance to Pests and Disease

Ivan Simko[1], Shelley Jansky[2], Sarah Stephenson[2] and David Spooner[2]

[1] *USDA-ARS, Crop Improvement and Protection Unit, Salinas, CA 93905, USA;*
[2] *USDA, University of Wisconsin-Madison, Department of Horticulture, Madison, WI 53706, USA*

Potato is a host to many pathogens that affect all parts of the plant and cause reductions in the quantity and quality of yield. The development of new cultivars that are more resistant to economically important pests and diseases is therefore one of the top priorities for potato-breeding programs worldwide. Numerous resistance genes have been discovered in *Solanum* species and introgressed into the cultivated potato. Since the 1980s, the introduction of molecular marker techniques has facilitated gene mapping and shifted orientation from phenotype-based resistance genetics to genotype-based approaches. A number of loci conferring quantitative resistance [quantitative resistance loci (QRL)] and around forty single dominant genes (R-genes) conferring qualitative resistance have been positioned on the potato molecular map. As of 2006, eight of the mapped R-genes have been isolated and molecularly characterized. The analysis of mapped and cloned resistance genes shows that they often occur in clusters and that some of them can respond to more than one elicitor. Ongoing research on resistance gene evolution will help us in understanding the dynamic interaction between potato plants and pathogens and opens a way for the development of more resistant cultivars.

7.1 RESISTANCE SCREENING

To carry out breeding and genetic studies, one must correctly identify genotypes based on phenotypes. This may sound like a trivial matter. However, especially in disease and pest assays, the concept of resistance is necessarily linked to the screening method. Screening data must be used to assign resistance phenotypes, and rankings of clones may vary with the screening method. There is often no black and white difference between resistant and susceptible phenotypes, so judgments must be made to assign phenotypes and, consequently, genotypes. Finally, most host–pathogen interactions are strongly influenced by the environment, so it is important to ultimately evaluate plants under the range of field conditions in which the crop will be grown.

7.1.1 Field screening

Field assessments of resistance are especially valuable because they evaluate plants under the conditions in which they will eventually be grown. However, field screening can be

Potato Biology and Biotechnology: Advances and Perspectives
D. Vreugdenhil (Editor)
2007 Published by Elsevier B.V.

expensive in terms of both time and resources, and screening results may be variable because of environmental heterogeneity.

Disease pressure is an important variable in field trials, but it can be difficult to control for two reasons. First, the amount of inoculum that is naturally present in the field cannot be easily manipulated in most cases. Second, disease pressure is typically influenced by environmental factors such as temperature and humidity, which cannot be controlled. If disease pressure is too high, then clones with moderate, but possibly acceptable, levels of resistance will be eliminated. Stewart et al. (1983) found that if young seedlings were inoculated with *Phytophthora infestans* (the causal agent of late-blight disease), then even resistant plants died. Inoculations of older seedlings allowed for the successful identification of resistant plants. Similarly, Hilton et al. (2000) found that if tubers were exposed to *Helminthosporium solani* for too long under ideal conditions for the silver scurf fungus, then no differences were seen between resistant and susceptible cultivars. On the contrary, if disease pressure is too low, then selection for resistance will also be ineffective. For example, Hoyos et al. (1993) assayed vascular colonization by *Verticillium dahliae* in seedling transplants and identified putatively resistant clones. However, few of the putatively resistant clones selected as seedlings were resistant in the first clonal generation, probably because selection pressure was too low in the seedling generation.

Because breeders typically evaluate large populations of segregating plants, they must strive toward a balance between quantity and quality of disease-scoring data. On one hand, it is important to be able to quickly evaluate large numbers of clones in segregating populations. On the other hand, the data are meaningless if they do not effectively identify resistant phenotypes. It is important to identify screening methods that will consistently identify resistant plants in breeding programs with large numbers of genotypes. For example, a common Verticillium wilt-resistance-scoring technique involves the plating of stem tissue from infected plants, followed by the counting of *Verticillium* colonies that grew from that tissue. Because stem-to-stem variation for pathogen populations is very high, it is important to evaluate a large number of stems, which is time-consuming. A compromise offered by Treadwell (1991) substituted a rating scale for colony counts, allowing a larger number of stems to be evaluated. Similarly, Christ (1991) found that estimating leaf area covered by early-blight (*Alternaria solani*) lesions was faster than counting individual lesions.

Resistance assays can be based on disease symptoms or pathogen levels in the plant. Disease symptom expression is typically an easy, fast, and inexpensive assay. For example, Dale and Brown (1989) determined that foliage symptom expression correlated with yield loss due to potato cyst nematode (*Globodera pallida*) infection. Consequently, selection for foliage vigor allows for the identification of tolerant genotypes. Christ (1991) also effectively identified early-blight-resistant clones based on symptom development in the field.

Although symptom expression is commonly used for disease scoring, there are problems associated with this technique. It may be difficult to distinguish between symptoms caused by the pathogen of interest and those caused by other pathogens or by abiotic stresses. In addition, the interaction between the pathogen and both its biotic and abiotic environment may result in a wide range of symptoms, some of which are not typical. Consequently, methods have been developed to quantify pathogen populations in plant tissues. These methods are discussed in Section 7.1.3.

It is often desirable to partition resistance into components, each of which can be selected individually and may be controlled by its own genetic system. For example, resistance to potato leaf roll virus (PLRV) can result from resistance to infection, resistance to virus accumulation, and reduced translocation from leaves to tubers (Wilson and Jones, 1993). Each of these components appears to be controlled by a different genetic system. It is possible to select plants for each of these components, and combining the components may provide even more effective resistance (Solomon-Blackburn et al., 1994). Late-blight resistance can also be partitioned into components such as infection efficiency, latent period, lesion size, and sporulation. However, unlike PLRV resistance components, which are controlled by major genes, these components of late-blight resistance appear to be quantitatively inherited and they interact with each other to determine the resistance phenotype (Birhman and Singh, 1995). The heritability estimates for each component vary, as does the contribution of each to the resistance genotype. Similarly, scab (*Streptomyces scabies*) resistance can be partitioned into surface area infected, lesion type, and proportion of scabby tubers (Goth et al., 1993). Cluster analysis based on lesion type and surface area infected appears to most effectively identify resistant clones.

Host plant maturity is an important variable in most disease and pest resistance evaluations. Typically, plants are more resistant early in their life cycle, becoming more susceptible with age. Immature plant resistance is commonly observed for late blight (Dorrance and Inglis, 1997), early blight (Boiteux et al., 1995), and Verticillium wilt (Busch and Edgington, 1967). Consequently, it is important to evaluate disease resistance as plants senesce in the field. This can be a problem when scoring wild *Solanum* species, which typically do not senesce during the growing season in North temperate regions. In contrast to immature plant resistance, resistance to PLRV appears to increase in mature plants (DiFonzo et al., 1994).

7.1.2 Greenhouse screening

Greenhouse-based resistance assays can provide an attractive compromise between costly, time-consuming field trials and laboratory assays, which do not typically evaluate entire plants. They can identify putatively resistant plants that can then undergo more extensive field evaluations in subsequent studies.

A major advantage of greenhouse screening is that the environment can be controlled to optimize disease pressure while minimizing the effects of confounding biotic and abiotic factors. Greenhouse assays allow control over the timing, dose, and virulence of the pathogen. Stewart and Bradshaw (1993) optimized the greenhouse environment for early-blight resistance screening by controlling humidity, inoculum dose, and plant age at inoculation. Most plants identified as resistant or susceptible remained so in field tests. Similarly, Dorrance and Inglis (1997) found that resistance scores from greenhouse inoculations with *P. infestans* corresponded to those from field evaluations.

Another advantage of greenhouse screening is that it allows for the evaluation of resistance to pathogens that cannot feasibly be introduced to the outside environment. For example, greenhouse assays were used to evaluate resistance to ring rot (*Clavibacter michiganensis* ssp. *sepedonicus*) because field trials would release a serious pathogen to the environment (Kriel et al., 1995b). Similarly, in regions where breeders cannot risk spreading *P. infestans* to growers' fields, they use greenhouse evaluations to screen for late-blight resistance.

Greenhouse screening in potato also allows plants to be grown under short day conditions in winter in order to induce tuberization in wild *Solanum* species. Breeders can then evaluate tuber traits, which cannot be studied in the field because wild species do not tuberize under the long days of summer in North temperate regions. Hosaka et al. (2000) transplanted wild species seedlings into pots containing *S. scabies*. As the plants tuberized in the greenhouse during the winter, they were evaluated for resistance to common scab. Because short day conditions stimulate wild species to senesce, winter greenhouse trials also allow screening for true resistance rather than immature plant resistance, as in Section 7.1.1.

One limitation of greenhouse screening is that the controlled environment may not adequately mimic the complexities of a field trial. Consequently, levels of resistance may not correlate strongly with those in the greenhouse. For example, DiFonzo et al. (1994) reported that the variety 'Cascade' was more resistant to PLRV in the field than in the greenhouse. Apparently, resistance was more complex than the greenhouse screen alone could assess. Similarly, 'Desirée' was more resistant to early blight in field than greenhouse assays (Stewart and Bradshaw, 1993).

7.1.3 Laboratory screening

Laboratory-based assays are based on samples ranging from whole plants to individual cells. A number of biochemical assays have been developed to quantify pathogen levels in host plant tissue. A main advantage of these assays is that they can confirm the presence of the pathogen in both symptomatic (susceptible) and asymptomatic (tolerant) plants. Consequently, they can distinguish between tolerant and resistant plants.

Pathogen populations in inoculated host plant tissues can be quantified in a number of ways. For example, fungal pathogens can be quantified using culture plating, an enzyme-linked immunosorbent assay (ELISA), or the polymerase chain reaction (PCR). Culture-plating assays are commonly used for Verticillium wilt resistance screening. Hoyos et al. (1991) plated sap, whereas Davis et al. (1983) plated dried stem tissue from plants grown in *V. dahliae*-infested fields. Both reported a strong correlation between the numbers of colony-forming units and wilt scores in the field. Davis et al. (2001) suggest that the plating of roots on *V. dahliae* on selective medium provides an even more effective measure of colonization by the pathogen. Culture plating is simple and inexpensive, but it can be time-consuming and does not provide information until the pathogen grows, which may take weeks. In addition, it is sometimes difficult to distinguish between the pathogen and other organisms that grow on the plate. ELISA provides rapid results and does not require specialized equipment other than a plate reader. It has been used to assay for resistance to late blight (Harrison et al., 1991; Beckman et al., 1994), Verticillium wilt (Plasencia et al., 1996), and potato viruses (Valkonen et al., 1992a; Singh et al., 2000). PCR is now commonly used to quantify pathogen levels in disease resistance assays. The limit of detection of this technique is typically lower than that of ELISA tests or culture plating. Quantitative PCR has been used to distinguish between Verticillium wilt-resistant and Verticillium wilt-tolerant clones (Dan et al., 2001). In addition, PCR detection can be species specific. For example, PCR can specifically detect *V. dahliae*, whereas ELISA tests detect both pathogenic forms of *Verticillium* (*Verticillium albo-atrum* and *V. dahliae*) (Plasencia et al., 1996; Dan et al., 2001). However, a limitation of both PCR and ELISA methods is that they cannot distinguish between living and dead pathogens, whereas culture plating detects only living material. Moreover, the pathogen population levels do not always mirror disease symptoms. Jansky and Rouse (2000) noted that Verticillium wilt symptoms did not correlate with levels of stem colonization. Similarly, Harrison et al. (1991) noted that levels of *P. infestans* in leaf tissue based on ELISA assays did not always relate to visual estimates of disease.

As an alternative to pathogen exposure in the field, plants or plant parts may be inoculated in the laboratory. Laboratory inoculations have been used effectively in resistance screening for silver scurf (Rodriguez et al., 1995; Hilton et al., 2000), soft rot (*Erwinia carotovora*) (Łojkowska and Kelman, 1994), powdery scab (*Spongospora subterranea*) (Merz et al., 2004), early blight (Bussey and Stevenson, 1991), and late blight (Dorrance and Inglis, 1998). However, the latter study illustrates a limitation of laboratory-based assays – they cannot simulate a field environment. Tubers of some cultivars that were susceptible in the laboratory assay were resistant in the field, probably because placement of tubers in the hill allowed them to escape infection in the field.

In some cases, it may be possible to screen for resistance using toxins produced by the pathogen rather than the pathogen itself. Christinzio and Testa (1999) screened potato leaves for late-blight resistance by applying fungal culture filtrates and then measuring electrolyte leakage. Similarly, Lawrence et al. (1990) were able to induce scab lesions by inoculating minitubers with thaxtomin, the toxin produced by *S. scabies*. Lynch et al. (1991) found that in vitro assays for early-blight resistance using culture filtrates containing toxic metabolites produced results similar to those using fungal spores.

7.2 RESISTANCE GENETICS IN POTATO

Wild and cultivated relatives of potato are credited with contributing the majority of disease resistance genes to breeding programs. It is interesting that most emphasis has been placed on a few species, such as late-blight resistance genes from *Solanum demissum* and virus resistance genes from *Solanum stoloniferum*. Only 13 wild species have contributed germplasm to European cultivars (Hawkes, 1990). Consequently, wild *Solanum* germplasm represents a largely untapped reservoir of genetic diversity for disease resistance genes. Ruiz de Galarreta et al. (1998) found that over 70% of the 98 wild *Solanum* species accessions they screened expressed resistance to one or more diseases. Much work remains to be done to identify and characterize the resistance factors in these species.

7.2.1 Resistance breeding

7.2.1.1 Exotic sources of disease and pest resistance

Reports of disease resistance in wild and cultivated relatives of potato are abundant (Table 7.1). Some species are especially potent sources of resistance to a number of

Table 7.1 Sources of potato disease and pest resistance.

Pathogen/Pest	Source of resistance	Reference
Virus		
Alfalfa mosaic virus (AMV)	*Solanum palustre*	(Valkonen et al., 1992b)
Andean potato latent virus	*S. palustre*	(Valkonen et al., 1992b)
Cucumber mosaic virus (CMV)	*Solanum fernandezianum, Solanum stoloniferum*	(Horvath, 1994; Valkonen et al., 1995)
Henbane mosaic virus	*S. stoloniferum*	(Horvath and Wolf, 1991)
Potato leaf roll virus (PLRV)	*Solanum chacoense, S. fernandezianum, S. palustre, Solanum sparsipilum, Solanum spegazzinii*	(Helgeson et al., 1986; Valkonen et al., 1992a; Brown and Thomas, 1994; Ruiz de Galarreta et al., 1998)
Potato virus A (PVA)	*Solanum polyadenium*	(Valkonen, 1997)
Potato virus M (PVM)	*S. palustre, S. sparsipilum*	(Valkonen et al., 1992b; Ruiz de Galarreta et al., 1998)
Potato virus S (PVS)	*S. palustre*	(Valkonen et al., 1992b)
Potato virus V (PVV)	*Solanum maglia*	(Valkonen, 1997)
Potato virus X (PVX)	*Solanum acaule, Solanum commersonii, Solanum lesteri, Solanum marinasense, Solanum oplocense, S. palustre, S. sparsipilum*	(Horvath et al., 1988; Tozzini et al., 1991; Valkonen et al., 1992b)
Potato virus Y (PVY)	*S. acaule, Solanum achacachense, Solanum acroscopicum, Solanum ambosinum, Solanum arnezii, S. chacoense, Solanum doddsii, S. fernandezianum, Solanum megistacrolobum, S. palustre, S. polyadenium, Solanum polytrichon, S. sparsipilum, S. stoloniferum, Solanum sucrense, Solanum tarnii, Solanum trifidum*	(Horvath and Wolf, 1991; Horvath et al., 1988; Valkonen et al., 1992a; Singh et al., 1994; Bosze et al., 1996; Valkonen, 1997; Takacs et al., 1999a)
Potato-yellowing virus	*S. palustre*	(Valkonen et al., 1992b)
Tobacco etch virus	*S. commersonii*	(Valkonen, 1997)
Bacteria		
Clavibacter michiganensis	*S. acaule, Solanum phureja, Solanum sanctae-rosae, Solanum stenotomum, Solanum verrucosum*	(Kurowski and Manzer, 1992; Ishimaru et al., 1994; Kriel et al., 1995a)

Erwinia carotovora/ Erwinia chrysanthemi	*Solanum bukasovii, Solanum bulbocastanum, Solanum circaeifolium, Solanum demissum, S. marinasense, S. phureja, S. stoloniferum, S. tarijense*	(Łojkowska and Kelman, 1989; Carputo et al., 1996; Chen et al., 2003)
Streptomyces scabies	*Solanum boliviense, S. bukasovii, Solanum canasense, Solanum multidissectum*	(Hosaka et al., 2000)
Fungi		
Fusarium sambucinum	*S. boliviense, Solanum fendleri, Solanum gandarillasii, Solanum gourlayi, Solanum kurtzianum, Solanum microdontum, S. oplocense, S. sanctae-rosae, Solanum vidaurrei*	(Lynch et al., 2003)
Helminthosporium solani	*Solanum albicans, S. chacoense, S. demissum, S. oplocense, Solanum oxycarpum, S. stoloniferum*	(Rodriguez et al., 1995)
Verticillium albo-atrum	*S. chacoense*	(Concibido et al., 1994; Lynch et al., 1997)
Verticillium dahliae	*Solanum berthaultii, S. chacoense, S. commersonii, S. phureja, Solanum raphanifolium, S. sparsipilum, S. tarijense*	(Corsini et al., 1988; Concibido et al., 1994; Bastia et al., 2000)
Oomycete		
Phytophthora infestans	*S. ambosinum, S. berthaultii, Solanum brachycarpum, S. bulbocastanum, Solanum cardiophyllum, S. chacoense, S. circaeifolium, S. commersonii, S. demissum, S. fendleri, Solanum guerreroense, Solanum iopetalum, S. megistacrolobum, S. microdontum, S. pinnatisectum, S. raphanifolium, S. sparsipilum, S. stoloniferum, S. sucrense, S. verrucosum*	(Holley et al., 1987; Tooley, 1990; Ruiz de Galarreta et al., 1998; Micheletto et al., 1999; Douches et al., 2001; Perez et al., 2001; Chen et al., 2003)

(*Continued*)

Table 7.1 *(Continued)*

Pathogen/Pest	Source of resistance	Reference
Nematodes		
Globodera pallida	*S. acaule, Solanum brevicaule, S. sparsipilum*	(Jackson et al., 1988)
Meloidogyne arenaria	*S. chacoense*	(Di Vito et al., 2003)
Meloidogyne chitwoodi	*S. acaule, Solanum andreanum, S. boliviense, S. bulbocastanum, S. fendleri, Solanum hougasii*	(Brown et al., 1989, 1991, 1999, 2004; Mojtahedi et al., 1995)
Meloidogyne hapla	*S. bulbocastanum, S. chacoense*	(Brown et al., 1995; Di Vito et al., 2003)
Meloidogyne incognita	*S. chacoense*	(Di Vito et al., 2003)
Meloidogyne javanica	*S. chacoense, S. commersonii, S. tarijense*	(Di Vito et al., 2003)
Insects		
Empoasca fabae	*S. berthaultii*	(De Medeiros et al., 2004)
Leptinotarsa decemlineata	*S. berthaultii, S. chacoense, S. circaeifolium, Solanum jamesii, Solanum neocardenasii, Solanum okadae, S. oplocense, S. pinnatisectum, S. polyadenium, S. tarijense, S. trifidum*	(Dimock et al., 1986; Groden and Casagrande, 1986; Sinden et al., 1986; Cantelo et al., 1987; Neal et al., 1989; Pelletier and Smilowitz, 1990, 1991; Neal et al., 1991; Franca and Tingey, 1994; Franca et al., 1994; Bamberg et al., 1996; Sikinyi et al., 1997; Rangarajan et al., 2000; Pelletier and Tai, 2001; Pelletier et al., 2001; Chen et al., 2003)
Myzus persicae	*S. berthaultii, S. bukasovii, S. bulbocastanum, Solanum chiquidenum, Solanum chomatophilum, S. circaeifolium, Solanum etuberosum, S. trifidum*	(Lapointe and Tingey, 1986; Radcliffe et al., 1988)
Phthorimaea operculella	*S. berthaultii*	(Malakar and Tingey, 1999)
Thrips palmi	*S. chacoense*	(Fernandez and Bernardo, 1999)

Potato accessions rated as resistant/highly resistant (or hypersensitive/immune for virus resistance) are included in the table. More information about accessions can be found on the United States Potato Genebank website (http://www.ars-grin.gov/ars/MidWest/NR6/).

diseases and pests. Resistance to ring rot, potato cyst nematode, root knot nematode (*Meloidogyne chitwoodi*), potato virus X (PVX), and potato virus Y (PVY) has been reported in *Solanum acaule*; resistance to Colorado potato beetle (CPB) (*Leptinotarsa decemlineata*), green peach aphid (*Myzus persicae*), potato tuberworm (*Phthorimaea operculella*), late blight, and Verticillium wilt has been reported in *Solanum berthaultii*; resistance to silver scurf, CPB, four species of root knot nematode, late blight, PLRV, PVY, thrips (*Thrips palmi*), and both Verticillium wilt species has been reported in *Solanum chacoense*; resistance to root knot nematode, late blight, PVX, tobacco etch virus, and Verticillium wilt has been reported in *Solanum commersonii*; resistance to potato cyst nematode, late blight, PLRV, Verticillium wilt, and potato viruses M, X, and Y has been reported in *Solanum sparsipilum*; resistance to soft rot, silver scurf, late blight, cucumber mosaic virus (CMV), henbane mosaic virus, and PVY has been reported in *S. stoloniferum*; and resistance to soft rot, CPB, root knot nematode, and Verticillium wilt has been reported in *Solanum tarijense*. The outgroup *Solanum palustre* seems to be an especially rich source of virus resistance genes. It is important to note that there is tremendous diversity within wild species and even within accessions, so fine screening is necessary to identify individual clones with resistance genes. From a breeding standpoint, it is encouraging to note that several of the wild species that are rich in disease resistance genes (e.g. *S. berthaultii, S. chacoense, S. sparsipilum*, and *S. tarijense*) are also easily accessible through the simple ploidy manipulations outlined below.

7.2.1.2 Sexual hybridization

Resistance genes are often derived from wild *Solanum* relatives, most of which are diploid. Consequently, the ploidy of selected wild species clones can be doubled somatically or through sexual polyploidization to make them crossable with cultivars. Alternatively, diploid species clones can be crossed to parthenogenetically derived haploids ($2n = 2x = 24$) of cultivars, followed by polyploidization. The track record for transferring disease resistance to tetraploids through sexual polyploidization is impressive. It has successfully created hybrids with resistance to bacterial wilt (*Pseudomonas solanacearum*) (Watanabe et al., 1992), early blight (Herriott et al., 1990), common scab (Murphy et al., 1995), potato cyst nematode (De Maine et al., 1986; Ortiz et al., 1997), Verticillium wilt (K.E. Frost and S.H. Jansky, unpublished), and soft rot (Carputo et al., 2000; Capo et al., 2002). Iwanaga et al. (1989) crossed root knot nematode-resistant diploids to 'Atzimba' (4x) and to a haploid (2x) of 'Atzimba', both of which are susceptible to the nematode. A significantly higher proportion of resistant offspring (25%) was obtained in the $4x \times 2x$ crosses than in the $2x \times 2x$ crosses (11%). Presumably, alleles at loci between the centromere and the first crossover on each chromosome were transferred to offspring intact in $2n$ gametes, whereas those alleles were randomly reassorted in n gametes. This may explain why $2n$ gametes transmitted resistance to bacterial wilt, root knot nematodes, late blight, and glandular trichomes to a high proportion of $4x \times 2x$ offspring (Watanabe et al., 1999).

Diploid, 1 endosperm balance number (1EBN) species are often good sources of disease resistance genes. Several strategies have been developed to tap into this germplasm resource. Carputo et al. (1997) doubled the chromosome number of *S. commersonii* and crossed the resulting 4x, 2EBN clone with a 2x, 2EBN Tuberosum Group–Phureja Group hybrid, which was introgressed into tetraploid clones, some of which exhibited resistance to bacterial soft rot (Carputo et al., 2002; Iovene et al., 2004). Ramon and Hanneman (2002) used embryo rescue to incorporate late-blight resistance from the 2x, 1EBN species *Solanum pinnatisectum* into *Solanum tuberosum*.

Extremely wide hybridizations may introduce novel resistance genes into the potato gene pool. Colon et al. (1993) used embryo rescue to introduce late-blight resistance genes into the potato from the Solanaceous weed species *Solanum nigrum* and *Solanum villosum*. Valkonen et al. (1995) used embryo rescue to transfer the extreme resistance to PVY found in the non-tuber-bearing 2x, 1EBN species *Solanum brevidens* to the cultivated potato. Chavez et al. (1988) used bridging crosses, ploidy manipulations, and embryo rescue to transfer PLRV resistance from the non-tuber-bearing 2x, 1EBN species *Solanum etuberosum* to tuber-bearing species.

Breeding strategies may be designed to incorporate multiple sources of disease resistance genes. Solomon-Blackburn and Barker (1993) created clones with strong PLRV resistance by combining genes that limit virus multiplication with those for resistance to infection. Spitters and Ward (1988) found that resistance to potato cyst nematodes was more durable in clones with two resistance genes instead of one. Colon et al. (1995a) combined minor genes for late-blight resistance from four wild *Solanum* species with diploid Tuberosum Group clones. Murphy et al. (1999) used conventional hybridization between two tetraploid breeding clones, each with different disease resistance traits, to create a clone with resistance to several diseases. Hybrids containing a large proportion of wild germplasm may express multiple resistances because wild *Solanum* relatives are rich in disease resistance genes. Jansky and Rouse (2003) identified resistance to several diseases in populations of diploid interspecific hybrids. Chen et al. (2003) identified wild species genotypes with multiple resistances to late blight, CPB, and blackleg (*E. carotovora*). Similarly, De Maine et al. (1993) argue that Phureja Group is a valuable source of multiple disease resistance genes. The authors noted that resistance levels decrease as the exotic germplasm is diluted following recurrent backcrosses to *S. tuberosum*.

7.2.1.3 Somatic hybridization

High levels of disease resistance are sometimes found in wild *Solanum* species that are sexually incompatible (or nearly so) with the cultivated potato. Somatic fusion offers an effective strategy to access this germplasm. The 2x non-tuber-bearing species (*Solanum fernandezianum, S. brevidens*, and *S. etuberosum*) seem to be especially rich in disease resistance genes. For example, somatic hybrids between cultivated potatoes and *S. brevidens* express resistance to PLRV (Austin et al., 1985), PVX, PVY, and PLRV (Gibson et al., 1988; Valkonen and Rokka, 1998), and soft rot (Austin et al., 1988; Zimnoch-Guzowska and Łojkowska, 1993; Allefs et al., 1995; McGrath et al., 2002). Somatic fusion hybrids containing the 2x, 1EBN species *S. commersonii* are resistant to bacterial wilt (Leferriere et al., 1999; Kim-Lee et al., 2005), soft rot (Carputo et al., 1997), and Verticillium wilt (Bastia et al., 2000), whereas those with *Solanum bulbocastanum* are resistant to late blight (Helgeson et al., 1998; Song et al., 2003).

Somatic hybrids involving non-tuber-bearing species generally tuberize poorly, and as backcrosses are made to Tuberosum Group to improve tuberization, levels of resistance may decrease (Austin et al., 1988; Allefs et al., 1995). Rokka et al. (1995) suggest that repeated cycles of somatic fusion and anther culture may provide an effective strategy to improve tuberization while retaining virus resistance. In other cases, though, resistance genes are retained and expressed following multiple backcrosses to Tuberosum Group (Novy et al., 2002; Tek et al., 2004).

Another application of somatic fusion is to combine resistance genes from two sexually compatible parents. Thach et al. (1993) fused diploids carrying major genes for resistance to PVX (*Rx*) or PVY (*Ry*). When *Rx*-carrying clones were fused with *Ry*-carrying clones, most of the hybrids were resistant to both PVX and PVY. Rasmussen et al. (1996) fused a Tuberosum Group haploid carrying the gene for resistance to potato cyst nematode pathotype *Pa2* with another haploid carrying the *Pa3* resistance gene. Some of the hybrids exhibited a high level of resistance to both pathotypes of the nematode. Valkonen and Rokka (1998) fused a *S. brevidens* clone with a Tuberosum Group clone, both of which are resistant to PVA, PVX, PVY, and CMV, but through different mechanisms. Surprisingly, the simultaneous expression of two resistance mechanisms in the hybrids resulted in variable responses to the pathogens but not typically complementation for resistance. Carrasco et al. (2000) fused *Solanum verrucosum* with Tuberosum Group to combine PLRV resistance from *S. verrucosum* with adaptation and tuber yield from Tuberosum Group.

It is interesting to note that a dilution effect is sometimes reported for resistance traits in somatic fusion hybrids. Cooper-Bland et al. (1994) fused a potato cyst nematode-susceptible Tuberosum Group haploid with a resistant one and produced hybrids that were more similar to the susceptible parent. Somatic hybrids created by Rasmussen et al. (1998) had lower levels of quantitative resistance to both tuber and foliar late blight than the resistant donors. Similarly, soft rot resistance was reduced in backcross generations compared with *S. brevidens* + Tuberosum Group fusions (McGrath et al., 2002) and *S. commersonii* + Tuberosum Group fusions (Carputo et al., 2002). Gavrilenko et al. (2003) reported that 6x somatic fusion hybrids with four doses of the resistant parent (*S. etuberosum*) genome expressed extreme resistance to PVY, whereas 6x hybrids with four doses of the susceptible parent (Tuberosum Group) genome were susceptible. On the contrary, Zimnoch-Guzowska et al. (2003) reported that somatic hybrids produced from fusions of Tuberosum Group with the wild species *S. nigrum* were often more resistant than the resistant wild parent, perhaps due to complementation of resistance genes.

7.2.2 Resistance genetics based on disease phenotype

7.2.2.1 Vertical resistance

Disease resistance genetic studies are often based on tetraploid families. However, even major genes are difficult to identify at the tetraploid level due to the complexities of tetrasomic segregation. For example, Brown et al. (1997) intercrossed tetraploid Tuberosum Group clones and obtained high heritability estimates for PLRV resistance. The authors suggest that a few major genes are mainly responsible for resistance, but individual genes and their effects could not be elucidated. Based on crosses among tetraploid Tuberosum Group clones, Barker et al. (1994) proposed a two-gene model for PLRV resistance.

Barker and Solomon (1990) observed an approximately 1:1 segregation ratio in a cross between a susceptible tetraploid clone and a clone with resistance to PLRV multiplication. They suggest that a single dominant gene may confer resistance, but it was not possible to determine the genotypes of the parents. In contrast, when Brown and Thomas (1994) carried out inheritance studies at the diploid level, with the wild species *S. chacoense*, a single dominant resistance gene was identified and parental genotypes were determined based on offspring ratios.

Major genes for resistance to other potato viruses have also been identified. Solomon-Blackburn and Barker (2001) listed 28 major genes/alleles responsible for virus resistance in potato. Most originate from wild *Solanum* species, although Tuberosum and Stenotomum Groups have also contributed resistance genes. Some virus resistance genes may be found in closely linked clusters or they may be single genes that confer broad-spectrum resistance. For example, Barker (1997) suggested that a single dominant gene confers resistance to PVA, PVX, and PVY.

Dominant R-genes[1] for late-blight resistance have been used by breeders for decades. Eleven R-gene alleles have been identified in the Mexican wild species *S. demissum* (Black et al., 1953). An additional resistance gene, *Rpi1*, has been identified in the Mexican wild species *S. pinnatisectum* (Kuhl et al., 2001). Characterization of this gene indicates that it might correspond to R9 from *S. demissum*. These resistance genes are presumed to have evolved as a result of long-term interactions between the host wild species and the late-blight pathogen in Mexico. It is interesting that R-genes have been identified in Argentinian wild species even though these species have not evolved in the presence of *P. infestans* (Micheletto et al., 1999). In breeding studies using clones with major gene resistance, an excess of susceptible progeny is sometimes detected in segregating populations. A dominant suppressor of R-genes has been proposed to explain these data (El-Kharbotly et al., 1996b; Ordoñez et al., 1997). A long history of R-gene-resistant cultivar releases has demonstrated that vertical resistance to late blight is not durable because the pathogen is capable of evolving quickly to overcome R-genes.

Surprisingly, a single late-blight resistance gene derived from *S. bulbocastanum, RB*, appears to differ from other R-genes by conferring durable resistance. Clones containing the *RB* gene are even resistant to 'super-races' of the fungus that can overcome all 11 R-genes (Song et al., 2003; van der Vossen et al., 2003).

It is interesting that there are several examples of two-gene disease resistance systems in potato. Vallejo et al. (1995) suggested that PVY resistance in a diploid Phureja–Stenotomum Group population is controlled by complementary action of two dominant genes. Both genes must be present to confer resistance. Similarly, Kriel et al. (1995b)

[1] Classification of the genes involved in plant resistance can sometimes be confusing with different authors using somewhat different terminology. In this chapter, the term 'R-gene' refers to a phenotypically defined single gene that specifically recognizes an avirulence gene product of the pathogen. The R-gene-based type of resistance is often described in literature as monogenic, qualitative, vertical, narrow, specific, or complete. On the contrary, if the resistance phenotype is not controlled by individually recognizable R-genes and the resistance is measured quantitatively rather than qualitatively, the loci associated with the resistance are referred to as 'quantitative resistance loci' (QRL). The QRL-based type of resistance is frequently described as polygenic, quantitative, horizontal, broad, general, partial, or field resistance. However, the distinction between the two types of resistance is often equivocal at the phenotypic level.

found that complementary gene action is responsible for resistance to ring rot in *S. acaule*. In addition, Singh et al. (2000) determined that resistance to PVA in potato cultivars due to two independent genes with complementary gene action. Vallejo et al. (1995) also found that two dominant genes control resistance to PVX in Phureja–Stenotomum Group hybrids. However, this system exhibits duplicate dominant epistasis, as only one of the two genes is necessary for resistance. Jansky et al. (2004) also reported that a similar genetic mechanism exists for resistance to Verticillium wilt in diploid interspecific hybrids. Conversely, Lynch et al. (1997) identified a single dominant gene resistance is sufficient for Verticillium wilt resistance in *S. chacoense*. Perhaps a second resistance gene could not be detected because it was not segregating in that population.

7.2.2.2 Horizontal resistance

There is considerable interest in the development of potato cultivars with durable late-blight resistance due to polygenes. The so-called field resistance is complexly inherited and may involve the production of phytoalexins, phenolics, and glycoalkaloids (Andreu et al., 2001). Breeding progress has been slow because the genetic basis of resistance is not yet understood. Horizontal resistance to late blight, presumably due to minor genes, has been reported in wild *Solanum* species (Rivera-Peña, 1990; Colon et al., 1995b), in cultivated relatives of potato (Cañizares and Forbes, 1995; Haynes and Christ, 1999; Trognitz et al., 2001; Zlesak and Thill, 2004), and in Tuberosum Group cultivars (Colon et al., 1995b; Grünwald et al., 2002; Porter et al., 2004). Late-blight resistance is present in some heirloom cultivars and appears to be durable, because the resistance levels are similar to those when the cultivars were released more than 40 years ago. However, the most resistant cultivars are also late-maturing, and it is not known whether durable resistance can be combined with early maturity.

Resistance to early blight (*A. solani*) appears to be quantitatively inherited. Good sources of resistance are rare, and the combination of resistance and early maturity is even more difficult to find (Boiteux et al., 1995). Christ and Haynes (2001) reported a relatively high heritability estimate (0.61) for early-blight resistance in a Phureja–Stenotomum Group population, indicating that additive genetic variance is important. Similarly, Herriott et al. (1990) and Ortiz et al. (1993) found that additive genetic variance is important for early-blight resistance. According to Brandolini et al. (1992) and Gopal (1988), non-additive gene action contributes to resistance as well. In most populations developed for early-blight resistance, the most resistant clones are late to mature. However, Boiteaux et al. (1995) surveyed a large number of clones (934) and found some with both early maturity and early-blight resistance.

A genetic study using diploid interspecific hybrids determined that broad- and narrow-sense heritability values for soft rot resistance are high (0.92 and 0.89, respectively) (Lebecka and Zimnoch-Guzowska, 2004). Consequently, additive genetic variance is more important than non-additive variance. Although individual resistance genes were not identified in this study, the high heritability estimates may indicate that only a few genes control resistance.

Several studies have identified quantitative resistance for Verticillium wilt. Tsror and Nachmias (1995) suggest that minor genes are responsible for resistance in some cultivars.

Pavek and Corsini (1994) also favor a horizontal resistance model, with additive genetic variance contributing significantly to resistance. Recently, Simko et al. (2004b) identified four quantitative trait loci that contribute to Verticillium wilt resistance in diploid populations. It is interesting to note that different types of Verticillium wilt resistance may be controlled by different genetic systems. Lynch et al. (1997) believe that, in diploid *S. chacoense*, tolerance is a polygenic trait, whereas resistance to infection and colonization is due to a major gene.

7.3 MOLECULAR ANALYSIS OF POTATO RESISTANCE

7.3.1 Experimental strategies for gene mapping and cloning

7.3.1.1 Mapping plant resistance genes

The mapping of plant resistance genes is typically carried out on segregating populations derived from parents with contrasting phenotypes. To localize genes associated with particular resistance on a molecular linkage map, the resistance phenotype has to be assessed for the individuals in the mapping population. Then, linkage between marker loci and the resistance trait is calculated. Unfortunately, the cultivated potato (*S. tuberosum* ssp. *tuberosum*) is a highly heterozygous autotetraploid ($2n = 4x = 48$) species with complex genetic inheritance that complicates gene mapping. To limit the complexity of potato genetics, diploid ($2n = 2x = 24$) individuals are frequently used as parents for molecular map construction and linkage analysis. Diploids can be derived from tetraploid genotypes through anther or pollen culture, or through interspecific hybridization with certain genotypes of *Solanum phureja* ($2n = 2x = 24$). However, the diploid potato genotypes are self-incompatible (or are having large inbreeding depression), the feature that precludes development of pure lines. Therefore, a number of common mapping approaches based on homozygous lines, and often used in plant genetics, cannot be applied in potato. On the contrary, because alleles of heterozygous parents segregate in meiosis, already an F1-hybrid population can be used for potato gene mapping.

Typically, the initial screening is performed on a population consisting of 100–200 individuals and a series of markers ideally spaced at even intervals of about 10 cM on each chromosome. The necessary requirement for gene detection is a phenotype that segregates within the population and can be clearly scored in each individual. Following detection of linkage between the resistance phenotype and a molecular marker, saturation of the genomic region with more markers can be carried out on an expanded population of several hundred or even thousands of individuals. The large population size allows detection of flanking markers that are more closely linked to the resistance gene. Closely linked markers then may be used for either marker-assisted selection (MAS) or the map-based cloning of the resistance gene.

It should be noted, however, that populations originating from two diploid parental genotypes sample only a small proportion of all possible alleles. Moreover, if wild species are used for the development of mapping populations, the observed gene effects are not often representative of those encountered in elite cultivars. Detecting variation in economically important traits within genetic backgrounds that are relevant to plant breeders can be improved by complementary mapping techniques, such as association mapping. The association mapping method is a linkage disequilibrium (LD)-based technique that exploits biodiversity observed in existing cultivars and breeding lines without developing new mapping populations. This method has previously been used in diploid species and was recently successfully applied to map resistance genes in tetraploid potato (Gebhardt et al., 2004; Simko, 2004; Simko et al., 2004a,b). The association mapping approach effectively incorporates the effect of many past generations of recombinants into a single analysis. Because no mapping population needs to be created for the study, the linkage test can be performed relatively quickly and inexpensively. The association mapping technique that provides the means for detecting genes underlying the variation of a trait among existing genotypes is thus complementary to linkage-mapping methods that effectively locate genes segregating in a population originating from two individuals. A significant difference between association mapping in a general population and genetic linkage mapping in a defined segregating population is that association mapping generally identifies the association of common alleles (rare alleles do not reach statistical significance), whereas a population originating from a biparental cross enables the identification of alleles rare in the population at large (Simko, 2004). The resolution of the association-mapping approach depends on the structure of LD within the test population. The extent of LD in potato is not yet known; however, preliminary analysis of 66 loci indicates a relatively fast decay of LD within 1kb, but slow decay afterwards (Simko et al., 2006b).

7.3.1.2 Map-based cloning of resistance genes

Thus far, eight potato resistance genes have been cloned, and all of them were isolated using variants of the map-based strategy often combined with the candidate gene approach. Therefore, we will describe briefly this cloning technique, although other approaches can also be used for gene isolation.

After detecting molecular markers that are flanking a resistance gene locus, the screening of a large-insert genomic library (usually a bacterial artificial chromosome – BAC) with the identified flanking markers is performed. If the flanking markers are separated by a large distance, then the 'chromosome walking' method (Bender et al., 1983) is used to identify cloned DNA between the two markers through a stepwise analysis of successive overlapping clones. Alternatively, the 'chromosome landing' (Tanksley et al., 1995) approach may be used to pinpoint a resistance gene location with more accuracy. The idea behind the chromosome-landing method is that several thousand markers can be screened on the population to search for markers at a genetic distance corresponding to the physical size of large-insert clones. Identified markers are then used for screening genomic libraries. Ideally, the markers will land directly in the clones that contain, among other genes, the target resistance gene.

Once a large-insert clone(s) that encompasses the resistance gene is detected, the resistance gene identity needs to be established. This is usually done through a combination of sequencing, bioinformatic analyses (to reveal relevant candidates), and functional testing of the candidate alleles in transgenic plants.

7.3.2 Resistance factors mapped in potato

Numerous genes conferring resistance to viruses, nematodes, bacteria, fungi, oomycetes, and insects have been mapped in potato, and their location is summarized in Tables 7.2 and 7.3.

Two common types of single gene resistance to viruses in potato are hypersensitive resistance and extreme resistance. The genes for hypersensitive resistance are often virus strain group specific. When plants carrying these genes are inoculated with viruses, they usually develop either local necrotic lesions in the infected tissue or systemic necrosis. Several genes coding hypersensitive resistance to potato viruses A, S, X, and Y have been mapped in potato. On the contrary, very limited (or no) necrosis is observed on plants having genes for extreme resistance. The extreme resistance genes confer comprehensive resistance to several virus strains, and only an extremely low level of virus can be detected in some of the inoculated plants. Genes for extreme resistance to PVX and PVY originating from at least four different potato species have been placed on the potato molecular map. QRL for resistance to PLRV have also been reported in some mapping progenies.

Resistance genes to three economically important species of nematodes have been mapped in potato. Two of the species (*Globodera rostochiensis* and *G. pallida*) are root cyst nematodes, whereas *M. chitwoodi* is a root knot nematode. The first nematode resistance gene (*H1*) was discovered in the 1950s (Toxopeus and Huijsman, 1953), and since then, it has been introgressed into many commercially available cultivars to control *G. rostochiensis* pathotypes. The gene is located on potato chromosome 5. Additional dominant genes for qualitative resistance to *G. rostochiensis* and *G. pallida* have been mapped, together with several major QRL. However, only a single resistance locus against *M. chitwoodi* species has been identified so far in the potato genome. The R_{Mc1} gene from *S. bulbocastanum* was introgressed into cultivated potato by somatic hybridization (Brown et al., 1996). It was demonstrated later that the resistance spectrum of R_{Mc1} includes not only *M. chitwoodi* and the related species *Meloidogyne fallax* but also a genetically distinct population of *Meloidogyne hapla* (Rouppe van der Voort et al., 1999).

Quantitative resistance loci against bacteria *E. carotovora* ssp. *atroseptica*, a causal agent (together with other *Erwinia* species) of potato black leg and tuber soft rot, were detected in a diploid population with complex pedigree that included three *Solanum* species: *Solanum yungasense, S. tuberosum*, and *S. chacoense*. Genetic factors affecting resistance to *E. carotovora* ssp. *atroseptica* were found on all 12 potato chromosomes (Zimnoch-Guzowska et al., 2000).

Potato is affected by a number of fungal pathogens; however, only a limited number generate major loss of a crop. As of 2006, monogenic resistance to *Synchytrium endobioticum* (potato wart), and QRL for resistance to *A. solani* (early blight), *V. dahliae*, and *V. albo-atrum* (Verticillium wilt) have been identified.

The most economically important disease of potato is late blight caused by oomycete *P. infestans*. Twenty R-genes, conferring potato foliage resistance against late blight, have been placed on a molecular map so far. Eleven of the genes (*R1, R2, R3a, R3b, R5–R11*) come from *S. demissum*, four genes (*RB/Rpi-blb1, Rpi-blb2, Rpi-blb3, Rpi-abpt*) from *S. bulbocastanum*, and one each from *S. berthaultii* (R_{ber}/R_{Pi-ber}), *S. pinnatisectum*

Table 7.2 R-genes mapped in potato.

Chromosome	R-gene	Anchor marker	Pathogen resistance	Source of resistance	Reference	Class of resistance proteins	Reference
4	Ny_{tbr}	TG316	PVY	*Solanum tuberosum* ssp. *tuberosum*	(Celebi-Toprak et al., 2002)		
4	*Rpi-abpt*	TG370	*Phytophthora infestans*	*Solanum bulbocastanum* ssp. *bulbocastanum*	(Park et al., 2005c)		
4	*Rpi-blb3*	TG370	*P. infestans*	*S. bulbocastanum* ssp. *dolichophyllum*	(Park et al., 2005a)		
4	*R2*	TG370	*P. infestans*	*Solanum demissum*	(Li et al., 1998)		
4	*R2*-like	TG370	*P. infestans*	–	(Park et al., 2005b)		
5	*R1*	GP21	*P. infestans*	*S. demissum*	(Leonards-Schippers et al., 1992)	CC–NBS–LRR	(Ballvora et al., 2002)
5	*Nb*	GP21	PVX	–	(De Jong et al., 1997)		
5	*Rx2*	GP21	PVX	*Solanum acaule*	(Ritter et al., 1991)	CC–NBS–LRR	(Bendahmane et al., 2000)
5	*Grp1*	GP21	*Globodera rostochiensis, Globodera pallida*	–	(Rouppe van der Voort et al., 1998)		
5	*H1*	CP113	*G. rostochiensis*	*S. tuberosum* ssp. *andigena*	(Gebhardt et al., 1993; Pineda et al., 1993)		
5	*GroV1*	TG69	*G. rostochiensis*	*Solanum vernei*	(Jacobs et al., 1996)		
6	*Rpi-blb2*	CT119	*P. infestans*	*S. bulbocastanum*	(van der Vossen et al., 2005)	CC–NBS–LRR	(van der Vossen et al., 2005)

(*Continued*)

Table 7.2 (*Continued*)

Chromosome	R-gene	Anchor marker	Pathogen resistance	Source of resistance	Reference	Class of resistance proteins	Reference
7	*Rpi1*	TG20a	*P. infestans*	*Solanum pinnatisectum*	(Kuhl et al., 2001)		
7	*Gro1*	CP56	*G. rostochiensis*	*Solanum spegazzinii*	(Barone et al., 1990)	TIR–NBS–LRR	(Paal et al., 2004)
8	*RB/Rpi-blb1*	CP53	*P. infestans*	*S. bulbocastanum*	(Naess et al., 2000)	CC–NBS–LRR	(Song et al., 2003; van der Vossen et al., 2003)
8	*Ns*	CP16	PVS	*S. tuberosum* ssp. *andigena*	(Marczewski et al., 1998, 2002)		
9	Nx_{phu}	TG424	PVX	*Solanum phureja*	(Tommiska et al., 1998)		
9	*Rpi-moc1*	TG328	*P. infestans*	*Solanum mochiquense*	(Smilde et al., 2005)		
9	Ry_{chc}	CT220	PVY	*Solanum chacoense*	(Hosaka et al., 2001; K. Hosaka, personal communication)		
10	R_{ber}/R_{Pi-ber}	TG63	*P. infestans*	*Solanum berthaultii*	(Ewing et al., 2000; Rauscher et al., 2006)		
11	Ry_{adg}	CP58	PVY	*S. tuberosum* ssp. *andigena*	(Hämäläinen et al., 1997)		
11	Ry_{sto}[a]	CP58	PVY	*Solanum stoloniferum*	(Brigneti et al., 1997)		
11	*Sen1*	CP58	*Solanum endobioticum*	–	(Hehl et al., 1999)		
11	R_{Mc1}	TG523	*M. chitwodii*	*S. bulbocastanum*	(Brown et al., 1996)		

11	Na_{adg}	TG523	PVA	*S. tuberosum* ssp. *andigena*	(Hämäläinen et al., 2000)		
11	*R3a*	GP185	*P. infestans*	*S. demissum*	(El-Kharbotly et al., 1994; Huang et al., 2004)	CC–NBS–LRR	(Huang et al., 2005)
11	*R3b*	GP185	*P. infestans*	*S. demissum*	(El-Kharbotly et al., 1994; Huang et al., 2004)		
11	*R5*	GP185	*P. infestans*	*S. demissum*	(Huang, 2005)		
11	*R6*	GP185	*P. infestans*	*S. demissum*	(El-Kharbotly et al., 1996a)		
11	*R7*	GP185	*P. infestans*	*S. demissum*	(El-Kharbotly et al., 1996a)		
11	*R8*	GP185	*P. infestans*	*S. demissum*	(Huang, 2005)		
11	*R9*	GP185	*P. infestans*	*S. demissum*	(Huang, 2005)		
11	*R10*	GP185	*P. infestans*	*S. demissum*	(Huang, 2005; Bradshaw et al., 2006)		
11	*R11*	GP185	*P. infestans*	*S. demissum*	(Huang, 2005; Bradshaw et al., 2006)		
12	$Ry-f_{sto}$ [b]	GP122	PVY	*S. stoloniferum*	(Flis et al., 2005)		
12	Ry_{sto} [b]	STM0003d	PVY	*S. stoloniferum*	(Song et al., 2005)		
12	*Gpa2*	GP34	*G. pallida*	*S. tuberosum* ssp. *andigena*	(Rouppe van der Voort et al., 1997)	CC–NBS–LRR	(van der Vossen et al., 2000)
12	*Rx* [c]	GP34	PVX	*S. tuberosum* ssp. *andigena*	(Bendahmane et al., 1997)	CC–NBS–LRR	(Bendahmane et al., 1999)
12	*Rx1* [c]	GP34	PVX	*S. tuberosum* ssp. *andigena*	(Ritter et al., 1991)		

[a] Song et al. (2005) dispute position of this gene.
[b] $Ry-f_{sto}$ and Ry_{sto} genes might be identical.
[c] *Rx* and *Rx1* genes might be identical.

Table 7.3 Quantitative resistance loci (QRL) mapped in potato.

Chromosome	Pathogen resistance	Tested tissue	Reference
1	*Phytophthora infestans*	Foliage	(Collins et al., 1999; Oberhagemann et al., 1999; Simko, 2002)
1	*Erwinic carotovora* ssp. *atroseptica*	Foliage	(Zimnoch-Guzowska et al., 2000)
1	*E. carotovora* ssp. *atroseptica*	Tubers	(Zimnoch-Guzowska et al., 2000)
1	*Leptinofarsa decemlineata*	Foliage	(Yencho et al., 1996)
2	*P. infestans*	Foliage	(Leonards-Schippers et al., 1994; Collins et al., 1999)
2	*P. infestans*	Tubers	(Collins et al., 1999; Simko et al., 2006a)
2	*E. carotovora* ssp. *atroseptica*	Foliage	(Zimnoch-Guzowska et al., 2000)
2	*E. carotovora* ssp. *atroseptica*	Tubers	(Zimnoch-Guzowska et al., 2000)
2	*Verticillium albo-atrum*	Roots[a]	(Simko et al., 2004b)
3	*P. infestans*	Foliage	(Leonards-Schippers et al., 1994; Collins et al., 1999; Oberhagemann et al., 1999; Ewing et al., 2000; Ghislain et al., 2001; Bink et al., 2002; Visker et al., 2003; Bormann et al., 2004; Costanzo et al., 2005)
3	*E. carotovora* ssp. *atroseptica*	Foliage	(Zimnoch-Guzowska et al., 2000)
3	*E. carotovora* ssp. *atroseptica*	Tubers	(Zimnoch-Guzowska et al., 2000)
3	*Globodera rostochiensis*	Roots	(Kreike et al., 1996)
4	*P. infestans*	Foliage	(Leonards-Schippers et al., 1994; Meyer et al., 1998; Collins et al., 1999, Oberhagemann et al., 1999; Sandbrink et al., 2000; Bormann et al., 2004; Bradshaw et al., 2004a,b)[(b)]
4	*E. carotovora* ssp. *atroseptica*	Foliage	(Zimnoch-Guzowska et al., 2000)
4	*E. carotovora* ssp. *atroseptica*	Tuber	(Zimnoch-Guzowska et al., 2000)
4	*G. pallida*	Roots	(Bradshaw et al., 1998; Bryan et al., 2004)
4	*Alternaria solani*	Foliage	(Zhang, 2004)
5	*P. infestans*	Foliage	(Leonards-Schippers et al., 1994; van Eck and Jacobsen, 1996; Collins et al., 1999; Oberhagemann et al., 1999; Sandbrink et al., 2000; Ghislain et al., 2001; Simko, 2002; Visker et al., 2003; Bormann et al., 2004; Bradshaw et al., 2004a,b; Costanzo et al., 2005; Mayton et al., 2005)

5	*P. infestans*	Tuber	(Collins et al., 1999; Oberhagemann et al., 1999; Bradshaw et al., 2004a,b; Mayton et al., 2005)
5	*E. carotovora* ssp. *atroseptica*	Foliage	(Zimnoch-Guzowska et al., 2000)
5	*E. carotovora* ssp. *atroseptica*	Tuber	(Zimnoch-Guzowska et al., 2000)
5	*G. pallida*	Roots	(Kreike et al., 1994; Rouppe van der Voort et al., 2000; Bryan et al., 2002; Caromel et al., 2003, 2005)
5	*A. solani*	Foliage	(Zhang, 2004)
5	PLRV	Foliage	(Marczewski et al., 2001)
5	*L. decemlineata*	Foliage[c]	(Yencho et al., 1996)
6	*P. infestans*	Foliage	(Leonards-Schippers et al., 1994; Collins et al., 1999; Oberhagemann et al., 1999)
6	*P. infestans*	Tubers	(Simko et al., 2006a)
6	*E. carotovora* ssp. *atroseptica*	Foliage	(Zimnoch-Guzowska et al., 2000)
6	*E. carotovora* ssp. *atroseptica*	Tubers	(Zimnoch-Guzowska et al., 2000)
6	*G. pallida*	Roots	(Caromel et al., 2003)
6	*V. albo-atrum*	Roots[a]	(Simko et al., 2004b)
6	PLRV	Foliage	(Marczewski et al., 2001)
7	*P. infestans*	Foliage	(Leonards-Schippers et al., 1994; Collins et al., 1999; Ghislain et al., 2001; Costanzo et al., 2005)
7	*E. carotovora* ssp. *atroseptica*	Foliage	(Zimnoch-Guzowska et al., 2000)
8	*P. infestans*	Foliage	(Collins et al., 1999; Oberhagemann et al., 1999; Ghislain et al., 2001; Simko, 2002; Bormann et al., 2004)
8	*P. infestans*	Tubers	(Simko et al., 2006a)
8	*E. carotovora* ssp. *atroseptica*	Foliage	(Zimnoch-Guzowska et al., 2000)
8	*L. decemlineata*	Foliage[c]	(Yencho et al., 1996)
9	*P. infestans*	Foliage	(Leonards-Schippers et al., 1994; Collins et al., 1999; Oberhagemann et al., 1999; Simko, 2002; Bormann et al., 2004)

(*Continued*)

Table 7.3 (*Continued*)

Chromosome	Pathogen resistance	Tested tissue	Reference
9	*P. infestans*	Tuber	(Collins et al., 1999)
9	*E. carotovora* ssp. *atroseptica*	Foliage	(Zimnoch-Guzowska et al., 2000)
9	*E. carotovora* ssp. *atroseptica*	Tuber	(Zimnoch-Guzowska et al., 2000)
9	*V. albo-atrum*	Roots[a]	(Simko et al., 2004b)
9	*V. dahliae*	Roots[a]	(Simko et al., 2004a)
9	*G. pallida*	Roots	(Rouppe van der Voort et al., 2000; Bryan et al., 2002)
9	*A. solani*	Foliage	(Zhang, 2004)
10	*P. infestans*	Foliage	(Sandbrink et al., 2000; Mayton et al., 2005)
10	*P. infestans*	Tubers	(Mayton et al., 2005; Simko et al., 2006a)
10	*E. carotovora* ssp. *atroseptica*	Foliage	(Zimnoch-Guzowska et al., 2000)
10	*E. carotovora* ssp. *atroseptica*	Tubers	(Zimnoch-Guzowska et al., 2000)
10	*G. rostochiensis*	Roots	(Kreike et al., 1993)
10	*L. decemlineata*	Foliage[c]	(Yencho et al., 1996)
11	*P. infestans*	Foliage	(Leonards-Schippers et al., 1994; Collins et al., 1999; Oberhagemann et al., 1999; Ghislain et al., 2001; Bormann et al., 2004; Costanzo et al., 2005)
11	*E. carotovora* ssp. *atroseptica*	Foliage	(Zimnoch-Guzowska et al., 2000)
11	*E. carotovora* ssp. *atroseptica*	Tubers	(Zimnoch-Guzowska et al., 2000)
11	*G. rostochiensis*	Roots	(Kreike et al., 1993)
11	*G. pallida*	Roots	(Bryan et al., 2004; Caromel et al., 2005)
11	PLRV	Foliage	(Marczewski et al., 2001; Marczewski et al., 2004)
11	*A. solani*	Foliage	(Zhang, 2004)
12	*P. infestans*	Foliage	(Leonards-Schippers et al., 1994; Collins et al., 1999; Ghislain et al., 2001; Bormann et al., 2004)
12	*E. carotovora* ssp. *atroseptica*	Tubers	(Zimnoch-Guzowska et al., 2000)
12	*G. pallida*	Roots	(Caromel et al., 2003)
12	*V. albo-atrum*	Roots[a]	(Simko et al., 2004b)
12	*A. solani*	Foliage	(Zhang, 2004)

[a] Fungus infects plant through roots, but the disease symptoms were observed on foliage.

[b] Meyer et al. (1998) originally detected the resistance locus on chromosome 8. The locus was later placed on chromosome 4.

[c] Possibly, a trichome-related resistance.

(*Rpi1*), and *Solanum mochiquense* (*Rpi-moc1*). Nine R-genes (*R3a, R3b, R5–R11*) are clustered in the late-blight resistance hotspot on the distal part of chromosome 11. In addition to phenotypically characterized classical R-genes originating from *S. demissum* (Black et al., 1953), several of the fully functional allelic versions or duplications of the resistance genes were detected on chromosomes 4 (*R2-like*) (Park et al., 2005b) and 11 (*SH R3, Ma R3, Sc R3*, and *FS R3*) (Huang et al., 2005).

Polygenic factors affecting foliage resistance to late blight have been identified on all 12 chromosomes. Most of the QRL have a relatively small or moderate effect on the resistance, although about one-tenth of the mapped QRL explain 30–50% of the trait variation. Rarely, the detected QRL explains more than half of the total phenotypic variation. Because QRL are positioned on the molecular map with less precision than single genes and the confidence interval of the QRL position may exceed 40 cM (Simko, 2002), it is problematic to compare the location of resistance QRL from different studies. Nevertheless, when the likelihood of QRL being detected by different studies in the same general genomic region was estimated from a binomial distribution, regions on chromosomes 3, 4, and 5 were identified with a high probability as late-blight resistance hotspots (Simko, 2002; Jones and Simko, 2005). The most obvious and consistent resistance hotspot is located on chromosome 5, near the marker locus GP179. Unfortunately, plants having alleles at this locus, which increases foliage resistance, usually exhibit later maturity. Linkage between resistance and maturity at this chromosomal location was confirmed by association mapping performed on almost 600 cultivars. In this study, the markers tightly linked to the *R1* gene were significantly associated with quantitative resistance to late blight and late maturity (Gebhardt et al., 2004). There is a possibility that a single gene near the GP179 marker locus has a pleiotropic effect on both plant resistance and maturity (Bradshaw et al., 2004b). However, the presence of two tightly linked loci cannot be ruled out with one locus having a pleiotropic effect on both late-blight resistance and foliage maturity, and another having merely an effect on resistance (Visker et al., 2003).

While several genes have been mapped for foliage resistance to late blight, there is relatively little information about the genes affecting resistance in tubers. In many cases, the relationship between late-blight resistance in foliage and tuber can be ambiguous. Oberhagemann et al. (1999) hypothesized that a differential expression in leaves and tubers of multiple alleles and allele combinations results in a differential effect on late-blight resistance in the two tissues.

Very little information has been published concerning natural insect resistance loci in potato. In one study, two reciprocal backcross *S. tuberosum* × *S. berthaultii* potato progenies were screened for resistance to CPB consumption, oviposition, and defoliation (Yencho et al., 1996). Most of the QRL for resistance to CPB were linked to the loci for glandular trichome traits (Bonierbale et al., 1994). However, a relatively strong and consistent QRL for trichome-independent insect resistance was observed in both backcross populations on chromosome 1 (Yencho et al., 1996).

Depicting the location of known R-genes and QRL on the potato linkage map reveals clustering of resistance genes (Fig. 7.1). These hotspots contain multiple gene families conferring resistance to a range of different pathogens. The most prominent R-gene clusters are located on chromosomes 4, 5, 9, 11, and 12. Such clustered resistance gene families

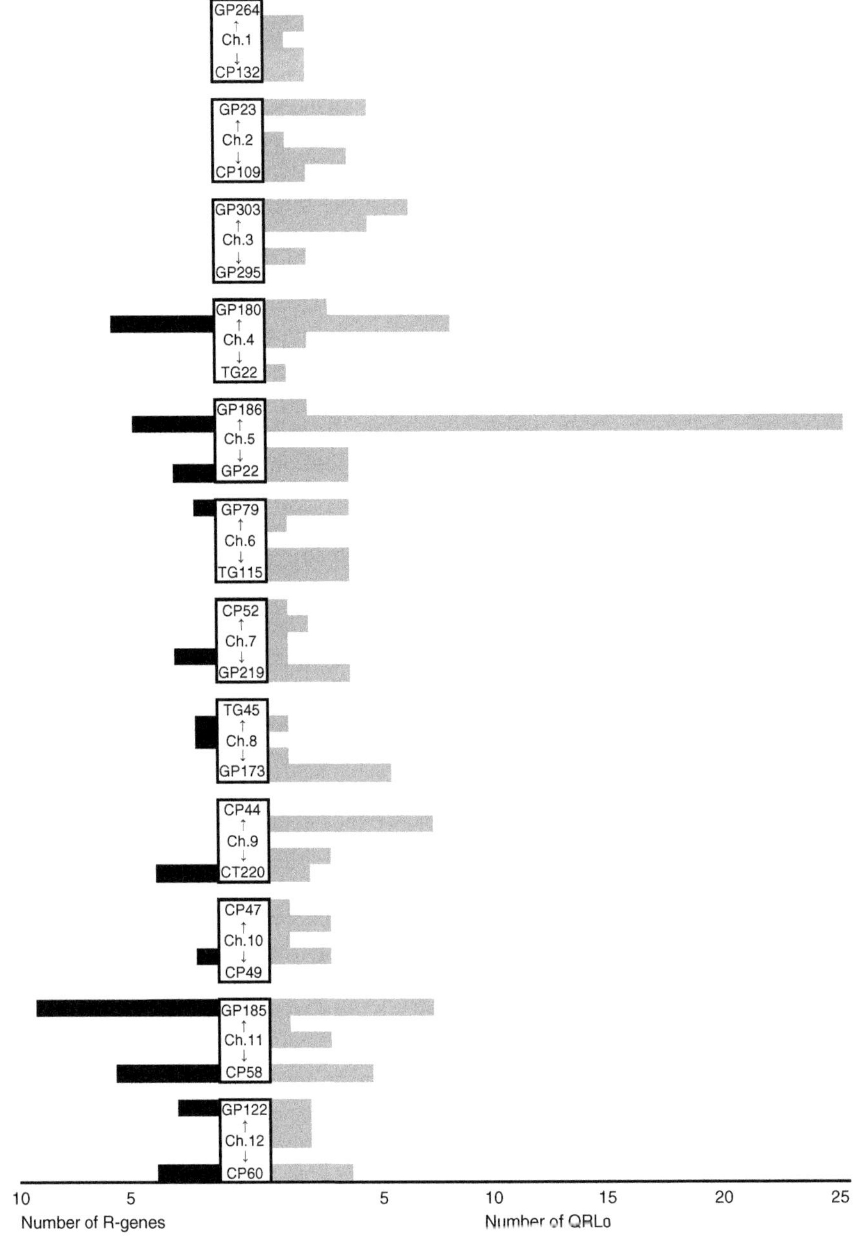

Fig. 7.1. Genomic position of R-genes (black bars) and quantitative resistance loci (QRL) (gray bars) conferring resistance to potato pathogens. Relative location of the loci associated with resistance to viruses, bacteria, fungi, oomycetes, and nematodes (Tables 7.2 and 7.3) was estimated from anchor markers. Loci coding mechanically different (trichome-related) type of resistance to insects are not included. Two restriction fragment length polymorphism (RFLP) markers per chromosome indicate orientation of the molecular linkage map according to Dong et al. (2000). More information about resistance loci is available in Tables 7.2 and 7.3. Ch., chromosome.

likely evolved from common ancestors by means of gene duplication with subsequent structural and functional diversification. Integrating the position of genes for quantitative resistance into the functional map shows a positional linkage between clusters of R-genes and QRL. A major QRL cluster on chromosome 5 comprises loci conferring resistance to virus, fungi, oomycete, bacteria, and nematodes. As new resistance genes are continuously mapped, the reader is referred to the potato functional map of pathogen resistance in the PoMaMo (Meyer et al., 2005) database (https://gabi.rzpd.de/PoMaMo.html).

7.3.3 Resistance genes cloned and characterized

Plant R-genes are presumed to enable the detection of avirulence (*Avr*) gene-specified pathogen molecules, initiate signal transduction to activate defenses, and have the capacity to evolve new specificities rapidly (Hammond-Kosack and Jones, 1997). Despite interactions with a wide range of pathogens, plant R-genes encode only five classes of resistance proteins. The majority of the cloned R-genes encode proteins containing a predicted nucleotide-binding site (NBS) followed by a series of leucine-rich repeats (LRR) at their C termini. NBS–LRR resistance proteins generally contain one of two types of N-terminal domains that have homology with the Toll and interleukin-1 receptor (TIR) proteins or a predicted coiled-coil domain (CC) (Ellis et al., 2000; Pan et al., 2000b; Dangl and Jones, 2001). The other four classes encode LRR, CC, kinase, and LRR plus kinase-conserved domains (Dangl and Jones, 2001) (Fig. 7.2). As of 2006, eight different resistance genes have been cloned from potato, and all of them are members of the NBS–LRR superfamily. The cloned genes are involved in the recognition of avirulence factors of viruses, nematodes, and oomycetes. It has been estimated that the potato genome contains at least 100–200 genes of this class (Gebhardt and Valkonen, 2001).

7.3.3.1 Virus resistance genes Rx *and* Rx2

The first resistance gene that was cloned and sequenced from potato was the *Rx* gene for extreme resistance against PVX (Bendahmane et al., 1999). The *Rx*-mediated extreme resistance in potato does not involve a necrotic hypersensitive response at the site of initial infection; however, the *Rx* protein is structurally similar to products of disease resistance genes conferring the hypersensitive response (Bendahmane et al., 1999). The highest degree of similarity is between *Rx* and a subclass of CC–NBS–LRR resistance proteins. Sequence analysis has revealed that *Rx* encodes a protein of 937 amino acid residues with molecular weight of 107.5 kDa (Bendahmane et al., 1999). The *Rx* locus has been introgressed into a cultivated *S. tuberosum* potato from *Solanum andigena* and maps to chromosome 12 (Bendahmane et al., 1997). In contrast, *Rx2*, the second PVX resistance gene cloned from potato (Bendahmane et al., 2000), was introgressed from *S. acaule* and is located on chromosome 5 (Ritter et al., 1991). The two genes share 95% sequence identity. Based on sequence conservation in *Rx* and *Rx2*, it is clear that there is a direct evolutionary relationship between these proteins, though the proteins are encoded on different chromosomes of different species (Bendahmane et al., 2000).

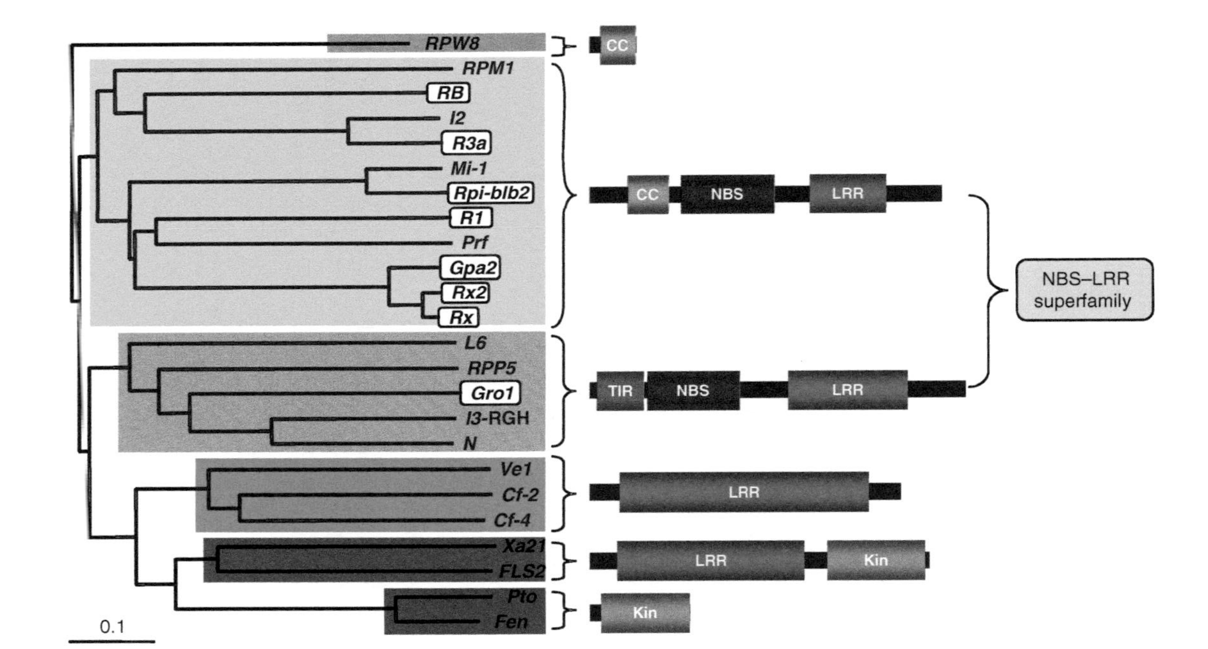

Fig. 7.2. Dendogram and representation of the structure of the five classes of plant disease resistance proteins. Amino acid alignment was carried out on the proteins that are (1) typical for the five main resistance classes (*Cf-2, Cf-4, Fen, FLS2, L6, N, Pto, RPP5, RPM1, RPW8, Ve1, Xa21*) (Hammond-Kosack and Jones, 1997; Dangl and Jones, 2001; Kawchuk et al., 2001) or (2) similar to the potato resistance proteins (*I2, I3*-linked resistance gene homolog, *Mi-1, Prf*) (Ballvora et al., 2002; Paal et al., 2004; Huang, 2005; van der Vossen et al., 2005) or (3) known potato resistance proteins (*Gpa2, Gro1, R1, R3a, RB, Rpi-blb2, Rx, Rx2*) (Bendahmane et al., 1999; Bendahmane et al., 2000; van der Vossen et al., 2000; Ballvora et al., 2002; Song et al., 2003; Paal et al., 2004; Huang et al., 2005; van der Vossen et al., 2003; van der Vossen et al., 2005). Conserved domains in the resistance proteins are CC, coiled-coil; Kin, kinase; LRR, leucine-rich repeats; NBS, nucleotide-binding site; and TIR, Toll and interleukin-1 receptor (Dangl and Jones, 2001). Resistance proteins isolated from potato are boxed.

7.3.3.2 Nematode resistance genes Gpa2 *and* Gro1

The isolation of the nematode resistance gene *Gpa2* (van der Vossen et al., 2000) on chromosome 12 was achieved when the genomic region containing the *Rx* resistance gene was further analyzed. Molecular analysis of this chromosomal region revealed that the *Gpa2* locus is one of four highly homologous genes in a region of approximately 115 kb. At least two of these genes are active. One homolog corresponds to the previously isolated *Rx* gene that confers resistance to PVX, whereas the other corresponds to the *Gpa2* gene that confers resistance to potato cyst nematode *G. pallida*. The deduced open reading frame of the *Gpa2* gene encodes a predicted polypeptide of 912 amino acids with a molecular weight of 104.5 kDa (van der Vossen et al., 2000). The *Gpa2* protein belongs to the CC–NBS–LRR class of plant resistance genes and shares over 88% amino acid identity with *Rx* and *Rx2*. Sequence conservation points to a direct evolutionary relationship of these proteins and to the fact that genes for resistance to distinct pathogens might be structurally similar. Isolation of *Gpa2* from the *Rx*-gene cluster demonstrates that within a single haplotype, members of the same R-gene cluster can evolve to confer resistance to distinct pathogen species (van der Vossen et al., 2000).

Gro1-4, a major, dominant locus conferring resistance to *G. rostochiensis* that is localized on potato chromosome 7 (Barone et al., 1990; Ballvora et al., 1995), has been recently cloned (Paal et al., 2004). *Gro1-4* is a member of the *Gro1* candidate gene family consisting of 15 closely related candidate resistance genes. The gene encodes a protein of 1136 amino acids that contains a Toll–interleukin 1 receptor (TIR) domain (Paal et al., 2004) and belongs to the TIR–NBS–LRR group of resistance genes (Dangl and Jones, 2001). *Gro1-4* is the first potato resistance gene of this type. At the nucleotide level, *Gro1-4* is more than 90% identical with two NBS–LRR-type R-gene homologs of tomato, which map to syntenic positions and are tightly linked to the tomato *I3* gene for resistance to *Fusarium oxysporum* (Pan et al., 2000a). Analysis of the susceptible cultivar Desirée complemented with *Gro1-4* showed that the gene is able to confer resistance to *G. rostochiensis* pathotype Ro1. Reverse transcriptase (RT)–PCR products of the gene family were generated from stem, leaf, flower, tuber, and stolon cDNA, demonstrating that the members of the *Gro1* gene family are expressed in all tested potato tissues. The possibility exists that other members of the *Gro1* gene family are also functional nematode R-genes conferring resistance to pathotypes other than Ro1 (Paal et al., 2004).

7.3.3.3 Oomycete resistance genes R1, R3a, RB/Rpi-blb1, *and* Rpi-blb2

Ballvora et al. (2002) used a combination of positional cloning and a candidate gene approach to clone *R1*, the race-specific gene for resistance to late blight. The *R1* gene encodes a protein of 1293 amino acids with a molecular mass of 149.4 kDa. Based on the deduced protein sequence, *R1* is a member of the CC–NBS–LRR class of plant resistance genes (Hammond-Kosack and Jones, 1997). The most closely related plant resistance gene (36% identity) is the *Prf* gene for resistance to *Pseudomonas syringae* of tomato (Ballvora et al., 2002). Two potato genes for resistance to PVX, *Rx2* and *Nb*, map to similar positions as *R1* (Ritter et al., 1991; De Jong et al., 1997). The *Rx2* gene has been cloned and is, like *R1*, a member of the CC–NBS–LRR class of resistance genes (Bendahmane et al., 2000). The two resistance genes share 32% sequence identity and are therefore different members of the same superfamily of genes. Analysis of the *R1* genomic region indicates that further paralogous members may be located in the region. There is the possibility that the *R1* locus hosts more than one R-gene with the same race specificity, or, in addition to *R1*, other R-genes with so far unknown specificities. The interaction between *R1* and the late-blight pathogen *P. infestans* is in accordance with the gene-for-gene concept (Flor, 1971). Transfer of a single gene to susceptible cultivar Desirée was sufficient to elicit the hypersensitive resistance response upon infection with a *P. infestans* race incompatible with *R1* (Ballvora et al., 2002). The *R1* gene is located within a resistance hotspot on chromosome 5 that includes major QRL for resistance to *P. infestans* (Gebhardt and Valkonen, 2001; Simko, 2002; Jones and Simko, 2005).

Huang et al. (2005) used the comparative genomics approach to isolate the *R3a* resistance gene. Analysis of *R3a* revealed that the gene encodes a predicted polypeptide of 1282 amino acids with a relative molecular mass of 145.9 kDa. The *R3a* gene encodes a putative CC–NBS–LRR protein and shares 88% DNA identity and 83% amino acid similarity to *I2* of tomato that confers resistance to fungus *F. oxysporum*. The *R3a* and *I2* proteins are more related to each other than to other known R-proteins. Thus, the *I2* and *R3a* genes belong to the same R-gene family. The *R3a* protein bears only limited similarity (15 and 30% amino acid identity) to the other two late-blight R-proteins, *R1* and *RB/Rpi-blb1* respectively (Huang et al., 2005). The *R3a*-gene family contains at least four full-length and many truncated paralogues. Constitutive expression was observed for the *R3a* gene, as well as some of its paralogues whose functions remain unknown (Huang et al., 2005). Comparative analyses of the *R3* complex locus with the corresponding *I2* complex locus in tomato suggest that this is an ancient locus involved in plant innate immunity against oomycete and fungal pathogens. However, the *R3* complex locus has evolved after its divergence from tomato. This expansion has resulted in an increase in the number of R-genes and in functional diversification (Huang et al., 2005). The *R3* locus is composed of two genes with distinct specificity. The two genes – *R3a* and *R3b* – have been mapped 0.4 cM apart and both have been introgressed from *S. demissum* (Huang et al., 2004). At least 10 additional *I2* gene analogs have been found in the *R3b* region (Huang et al., 2005).

Potato germplasm derived from *S. bulbocastanum* has shown durable and effective resistance against *P. infestans* in field tests. Two groups independently isolated the resistance gene from *S. bulbocastanum*-derived material. The coding sequence of the *RB* gene (Song et al., 2003) is identical to that of the *Rpi-blb1* gene (van der Vossen et al., 2003), which suggests that the two isolated genes are identical. However, when flanking sequences and the intron of these two genes were compared, single nucleotide polymorphisms (SNPs) were found at a frequency ranging from 0.4 to 0.8%. Van der Vossen et al. (2003) therefore concluded that *RB* and *Rpi-blb1* are allelic, although functionally equivalent. Molecular analysis of the resistance locus identified a cluster of resistance genes belonging to the CC–NBS–LRR class of resistance genes. The *RB/Rpi-blb1* gene family includes one truncated and four complete genes within a 40-kb region. The four complete genes are similar in length (2895–2979 bp) and have conserved intron–exon structures (Song et al., 2003; van der Vossen et al., 2003). The *RB/Rpi-blb1* gene encodes a polypeptide of 970 amino acids and is more closely related to the *I2* protein of tomato (30% identity, 47% similarity over 1070 amino acids) than it is to any other known R protein. *RB/Rpi-blb1* has limited similarity with the protein of *R1* (22% identity, 49% similarity over 902 amino acids), a gene derived from *S. demissum* that confers hypersensitive resistance to potato late blight (Song et al., 2003; van der Vossen et al., 2003). The *S. bulbocastanum* resistance gene is able to complement the susceptible phenotype in a *S. tuberosum* and tomato background, demonstrating the potential of interspecific transfer of broad-spectrum late-blight resistance to cultivated *Solanaceae* from sexually incompatible host species (van der Vossen et al., 2003). Similarly, as with the original *S. bulbocastanum* material, the transgenic plants with the *RB/Rpi-blb1* gene developed limited lesions on the lower leaves when inoculated with *P. infestans* (Song et al., 2003). The resistance gene transcript was detected in unchallenged plants, indicating the gene is constitutively expressed (Song et al., 2003).

Recently, Van der Vossen et al. (2005) cloned another resistance gene originating from *S. bulbocastanum* – *Rpi-blb2*. Similarly to *RB/Rpi-blb1*, the *Rpi-blb2* gene confers resistance to complex isolates of *P. infestans*. The *Rpi-blb2* locus is located on chromosome 6 and encodes a protein that shares 82% sequence identity with the *Mi-1* protein. However, whereas *Rpi-blb2* confers resistance to late blight in potato, the *Mi-1* gene confers resistance to root knot nematodes (*Meloidogyne* spp.), aphids (*Macrosiphum euphorbiae*), and whiteflies (*Bemisia tabaci*) in tomato. Molecular analysis of *S. bulbocastanum*-derived BAC clones spanning the *Rpi-blb2* locus has identified at least 14 additional *Mi-1* homologs (van der Vossen et al., 2005).

7.3.3.4 Resistance genes in potato

Eight resistance genes have been cloned so far from potato. All of these molecularly characterized genes have been found in clusters of various tightly linked R-genes and/or R-gene homologs of unknown function. Most of the homologs appear to encode proteins similar to that of the functional R-genes. Further complementation studies are required to elucidate the function of these homologs in the plant–pathogen interaction. An intriguing question concerning resistance genes is whether the rate of their evolution is directly related to gene specificity; hence, broad-spectrum resistance genes (e.g. *Rpi-blb1, Rpi-blb2*, and *Rpi-abpt*) evolve at a slower rate than race-specific genes (e.g. *R1–R11*). The isolation and molecular analysis of additional resistance genes may help with the identification of molecular motifs that determine resistance protein specificity and the rate of gene evolution. Given the fast progress in genetics and genomics, and the rapidly expanding knowledge about functionality of resistance genes, the number of mapped, isolated, and sequenced R-genes and QRL in potato will increase rapidly in the near future. Conserved domains from known resistance loci (Leister et al., 1996) and plant defense gene families (Trognitz et al., 2002) can facilitate more efficient detection and cloning of resistance genes through the 'candidate gene' approach.

7.3.4 Synteny of resistance loci in *Solanaceae*

In pepper, two major QRL for resistance to *Phytophthora capsici* (Lefebvre and Palloix, 1996) are located in the genomic regions corresponding to the potato late-blight R-gene clusters on chromosomes 4 and 11, respectively. Three R-genes (*Ph-1, Ph-2*, and *Ph-3*) confer resistance to *P. infestans* in tomato. All three loci map to genomic positions corresponding to potato monogenic resistance. The *Ph-1* gene from tomato chromosome 7 (Peirce, 1971) appears to be positioned similarly to the *Rpi1* gene from *S. pinnatisectum* (Kuhl et al., 2001), whereas *Ph-3* (Chunwongse et al., 2002) is located on chromosome 9 in the same genomic region as *Rpi-moc1* from *S. mochiquense* (Smilde et al., 2005). *Ph-2* (Moreau et al., 1998b) from tomato and R_{ber}/R_{Pi-ber} from *S. berthaultii* (Ewing et al., 2000; Rauscher et al., 2006) are both located on chromosome 10. However, the *Ph-2* and R_{ber}/R_{Pi-ber} genes do not appear to be orthologous because they show different race specificity when tested with the US-7 race of *P. infestans* (Ewing et al., 2000). Nonetheless, the similarity in the location of genes for resistance to *Phytophthora* species in potato, tomato, and pepper suggests possible evolutionary conservation of resistance genes. Grube et al. (2000) hypothesized that the general function of resistance alleles (e.g. initiation of the resistance response) may be conserved at homologous loci in related plant genera, although the taxonomic specificity of host resistance genes may be evolving rapidly. In case of potato, tomato, and pepper, it is possible that the resistance gene specificity across genera remains relatively conserved and those homologous genes are still conferring resistance to *Phytophthora* species.

In addition to resistance against *Phytophthora*, the *Solanaceae* family shows a conserved position of genes conferring resistance to some other pathogens. Three potato genes encoding resistance to PVY (Ry_{adg} and Ry_{sto}) and PVA (Na_{adg}) reside in the resistance-gene hotspot on the long arm of chromosome 11 (Brigneti et al., 1997; Hämäläinen et al., 1998; Hämäläinen et al., 2000). The corresponding region in tobacco (*Nicotiana tabacum*) carries the *N* gene conferring resistance to tobacco mosaic virus (TMV) (Leister et al., 1996). The *Ve* locus from tomato confers resistance to *Verticillium* species. The locus is located on the short arm of chromosome 9 (Diwan et al., 1999). Positional cloning of the gene identified not one but two closely linked inverted genes (*Ve1* and *Ve2*) that independently confer resistance to the same pathogen. When the tomato *Ve1* and *Ve2* genes were expressed functionally in potato, resistance was observed in otherwise susceptible plants (Kawchuk et al., 2001). This observation indicates that all necessary components of the resistance response are present and functional in the related host genus. A probe derived from the tomato *Ve1* gene identified homologous sequences on potato chromosome 9 that are significantly associated with resistance to both *V. dahliae* (Simko et al., 2004a) and *V. albo-atrum* (Simko et al., 2004b). It appears that the *Verticillium* resistance genes in potato and tomato retain positional and functional synteny.

However, even within a single species, the positional and structural synteny of loci does not necessarily imply functional synteny. It is known that tightly linked and highly similar genes can confer resistance to different pathogens, for example *Rx* to virus and *Gpa2* to nematodes. Therefore, it is plausible to assume that structurally similar loci from corresponding chromosomal regions of different *Solanaceae* species might confer resistance to distinct pathogens, a fact that can be used for gene isolation. The *I2* locus encodes resistance to the fungi *F. oxysporum* in tomato (Ori et al., 1997; Simons et al., 1998). The resistance gene is located at the distal end of the short arm of chromosome 11, in the same genomic region as the *R3a* gene from potato conferring resistance to the oomycete *P. infestans* (Huang et al., 2004). Information from the *I2* locus was used to develop resistance gene analog (RGA)-specific primers to amplify all candidate RGAs at the target region. This elegant approach eventually led to the isolation of the *R3a* gene with 88% DNA identity to *I2* (Huang et al., 2005). Similarly, the *Rpi-blb2* resistance gene from *S. bulbocastanum* (van der Vossen et al., 2005) is positioned in the same genomic region of chromosome 6 as the *Mi-1* gene from tomato (Milligan et al., 1998). The two genes share 82% protein identity, although they encode resistance to very different pathogens: *Rpi-blb2* to *P. infestans* (van der Vossen et al., 2005) whereas *Mi-1* to nematodes, aphids, and whiteflies (Milligan et al., 1998; Nombela et al., 2003).

These examples demonstrate how synteny within *Solanaceae* can be effectively used for comparative mapping and gene isolation. The completion of the tomato genome-sequencing project (http://www.sgn.cornell.edu/help/about/tomato_sequencing.html) will provide additional extremely valuable information for the efficient cloning of other potato resistance genes.

7.3.5 Marker-assisted resistance breeding

Selection of individuals with desirable traits from a breeding population can be based on phenotype, genotype (assessed with molecular markers), or a combination of the two. Phenotypic selection is more efficient for a trait with high heritability because it uses the sources of variation of all the loci, while markers can use only those loci to which they are linked (Charcosset and Gallais, 2003). MAS will be more effective than phenotypic selection when the proportion of additive variance accounted for by the marker loci is greater than the heritability of the trait (Dudley, 1993). Computer simulation shows that MAS can be more efficient than selection based only on a phenotype if the heritability of the trait is between 0.05 and 0.5 and the markers are close to the loci of interest (Moreau et al., 1998a).

If the evaluation of a target trait is time-consuming, difficult, or costly, MAS has an added advantage over phenotypic selection. For example, in breeding for resistance, MAS allows the breeder to conduct selection without depending on the natural occurrence of the pathogen or pest, or to perform selection during the off-season. Moreover, molecular markers in genomic regions of interest enable the selection of individuals with resistance to several pathogens simultaneously, unlike a phenotype-based selection that requires a number of individual trials. Obviously, when the percentage of genetic variation explained by the marker(s) is high, the efficacy of molecular marker(s) for MAS is also high. Hence, MAS is the simplest and most effective for R-gene-based resistance traits with markers tightly linked with, or residing within, the resistance gene itself. However, as the recombination frequency between the marker and the resistance locus increases, the value of the marker for MAS decreases.

In potato, molecular markers have been developed and successfully tested for a gene conferring extreme resistance to PVY. Kasai et al. (2000) developed a sequence-characterized amplified region (SCAR) marker to the PVY resistance gene Ry_{adg} (Hämäläinen et al., 1998). The marker was generated only in genotypes carrying Ry_{adg}, when tested on 103 breeding lines and cultivars with diverse genetic backgrounds (Kasai et al., 2000). Other known R-genes that are tagged with molecular markers can be conveniently used in MAS as well. For example, markers linked to the *Ns* gene conferring resistance to PVS are currently being used for indirect selection in diploid breeding programs (Marczewski et al., 2002). With allele-specific primers, the presence of the R-gene can be followed even if other R-genes exist in the same plant material. Thus, resistance genes from diverse sources can be incorporated into a single genotype. Whether R-gene polyculture (multilane development) or pyramiding of several resistance genes or the opposite approach of eliminating the race-specific R-genes is an objective, MAS is a valuable tool to achieve the goal more efficiently. Recently, Gebhardt et al. (2006) elegantly demonstrated how MAS could be efficiently used in resistance breeding programs. The authors applied screening with PCR-based molecular markers to develop breeding material that carries combination of four resistance genes: Ry_{adg} for extreme resistance to PVY, *Gro1* for resistance to *G. rostochiensis, Rx1* for extreme resistance to PVX, or *Sen1* for resistance to potato wart. When tested in the presence of pathogen, all selected plants showed expected resistant phenotype. However, an important requirement for molecular markers used in MAS is their universality in a wide gene pool, not just in a specific cross. This problem was well documented when markers linked to the *Gro1* and *H1* resistance genes in the diploid population were tested on 136 unrelated tetraploid cultivars. The *Gro1*-specific marker was not correlated with the resistance phenotype, whereas the *H1*-specific marker was indicative of resistance in only four cultivars (Niewöhner et al., 1995).

Lately, markers tagging QRL conferring resistance to *P. infestans, V. dahliae*, and *V. albo-atrum* have been developed and effectively tested on sizable tetraploid populations. A highly significant association was detected between PCR markers specific for the *R1* gene (or markers flanking the *R1* locus) and the QRL for resistance to foliar and tuber blight. The marker–trait association was tested on 600 cultivars originating from different countries. The marker alleles associated with increased resistance were traced to an introgression from the wild species *S. demissum* (Gebhardt et al., 2004). Simko et al. (2004a) observed a highly significant association between a simple sequence repeat (SSR) marker and quantitative resistance to *V. dahliae* in a collection of 137 North American potato cultivars. When the pedigree of these cultivars was analyzed, the marker traced the origin of the resistance and susceptibility to two breeding lines widely used in the North American breeding program (Simko et al., 2004c). Similarly, a highly significant association was detected between the resistance to *V. albo-atrum* and an SNP marker developed from the *StVe1* locus. The marker–trait association was confirmed on 150 tetraploid cultivars and breeding lines as well as on a population developed from crosses between highly susceptible and resistant cultivars (Simko et al., 2004b). However, as each individual QRL explains only a relatively small portion of the variation in resistance, for a practical application, additional disease resistance loci need to be identified and tagged with molecular markers. After detecting and tagging enough resistance loci, the MAS will facilitate more efficient development of new potato cultivars carrying a desirable resistance gene combination.

REFERENCES

Allefs J.J.H.M., W. Van Dooijeweert, E.R. De Jong, W. Prummel and J. Hoogendoorn, 1995, *Potato Res.* 38, 11.

Andreu A., C. Oliva, S. Distel and G. Daleo, 2001, *Potato Res.* 44, 1.

Austin S., M.A. Baer and J.P. Helgeson, 1985, *Plant Sci.* 39, 75.
Austin S., E. Łojkowska, M.K. Ehlenfeldt, A. Kelman and J.P. Helgeson, 1988, *Phytopathology* 78, 1216.
Ballvora A., M.R. Ercolano, J. Weiss, K. Meksem, C.A. Bormann, P. Oberhagemann, F. Salamini and C. Gebhardt, 2002, *Plant J.* 30, 361.
Ballvora A., J. Hesselbach, J. Niewöhner, D. Leister, F. Salamini and C. Gebhardt, 1995, *Mol. Gen. Genet.* 249, 82.
Bamberg J.B., C.A. Longtine and E.B. Radcliffe, 1996, *Am. Potato J.* 73, 211.
Barker H., 1997, *Theor. Appl. Genet.* 95, 1258.
Barker H. and R.M. Solomon, 1990, *Theor. Appl. Genet.* 80, 188.
Barker H., R.M. Solomon-Blackburn, J.W. McNicol and J.E. Bradshaw, 1994, *Theor. Appl. Genet.* 88, 754.
Barone A., E. Ritter, U. Schachtschabel, T. Debener, F. Salamini and C. Gebhardt, 1990, *Mol. Gen. Genet.* 224, 177.
Bastia T., N. Carotenuto, B. Basile, A. Zoina and T. Cardi, 2000, *Euphytica* 116, 1.
Beckman K.B., J.G. Harrison and D.S. Ingram, 1994, *Physiol. Mol. Plant Pathol.* 44, 19.
Bendahmane A., K. Kanyuka and D.C. Baulcombe, 1997, *Theor. Appl. Genet.* 95, 153.
Bendahmane A., K. Kanyuka and D.C. Baulcombe, 1999, *Plant Cell* 11, 781.
Bendahmane A., M. Querci, K. Kanyuka and D.C. Baulcombe, 2000, *Plant J.* 21, 73.
Bender W., P. Spierer and D.S. Hogness, 1983, *J. Mol. Biol.* 168, 17.
Bink M.C.A.M., P. Uimari, M.J. Sillanpaa, L.L.G. Janss and R.C. Jansen, 2002, *Theor. Appl. Genet.* 104, 751.
Birhman R.K. and B.P. Singh, 1995, *Ann. Appl. Biol.* 127, 353.
Black W., C. Mastenbroek, W.R. Mills and L.C. Peterson, 1953, *Euphytica* 2, 173.
Boiteux L.S., F.J.B. Reifschneider, M.E.N. Fonseca and J.A. Buso, 1995, *Euphytica* 83, 63.
Bonierbale M.W., R.L. Plaisted, O. Pineda and S.D. Tanksley, 1994, *Theor. Appl. Genet.* 87, 973.
Bormann C.A., A.M. Rickert, R.A.C. Ruiz, J. Paal, J. Lubeck, J. Strahwald, K. Buhr and C. Gebhardt, 2004, *Mol. Plant-Microbe Interact.* 17, 1126.
Bosze Z., G. Kazinczi and J. Horvath, 1996, *Acta Phytopathol. Entomol. Hung.* 31, 169.
Bradshaw J.E., G.J. Bryan, C.A. Hackett, K. McLean, B. Pande, H.E. Stewart and R. Waugh, 2004a, *Euphytica* 137, 13.
Bradshaw J.E., G.J. Bryan, A.K. Lees, K. McLean and R.M. Solomon-Blackburn, 2006, *Theor. Appl. Genet.* 112, 744.
Bradshaw J.E., R.C. Meyer, D. Milbourne, J.W. McNicol, M.S. Phillips and R. Waugh, 1998, *Theor. Appl. Genet.* 97, 202.
Bradshaw J.E., B. Pande, G.J. Bryan, C.A. Hackett, K. McLean, H.E. Stewart and R. Waugh, 2004b, *Genetics* 168, 983.
Brandolini A., P.D.S. Caligari and H.A. Mendoza, 1992, *Euphytica* 61, 37.
Brigneti G., J. Garcia-Mas and D.C. Baulcombe, 1997, *Theor. Appl. Genet.* 94, 198.
Brown C.R., D. Corsini, J. Pavek and P.E. Thomas, 1997, *Plant Breed.* 116, 585.
Brown C.R., H. Mojtahedi and J. Bamberg, 2004, *Am. J. Potato Res.* 81, 415.
Brown C.R., H. Mojtahedi and G.S. Santo, 1989, *Plant Dis.* 73, 957.
Brown C.R., H. Mojtahedi and G.S. Santo, 1991, *Am. Potato J.* 68, 445.
Brown C.R., H. Mojtahedi and G.S. Santo, 1995, *Euphytica* 83, 71.
Brown C.R., H. Mojtahedi and G.S. Santo, 1999, *J. Nematol.* 31, 264.
Brown C.R. and P.E. Thomas, 1994, *Euphytica* 74, 51.
Brown C.R., C.P. Yang, H. Mojtahedi, G.S. Santo and R. Masuelli, 1996, *Theor. Appl. Genet.* 92, 572.
Bryan G.J., K. McLean, J.E. Bradshaw, W.S. De Jong, M. Phillips, L. Castelli and R. Waugh, 2002, *Theor. Appl. Genet.* 105, 68.
Bryan G.J., K. McLean, B. Pande, A. Purvis, C.A. Hackett, J.E. Bradshaw and R. Waugh, 2004, *Mol. Breed.* 14, 105.
Busch L.V. and L.V. Edgington, 1967, *Can. J. Bot.* 45, 691.
Bussey M.J. and W.R. Stevenson, 1991, *Plant Dis.* 75, 385.
Cañizares C.A. and G.A. Forbes, 1995, *Potato Res.* 38, 3.
Cantelo W.W., L.W. Douglass, L.L. Sanford, S.L. Sinden and K.L. Deahl, 1987, *J. Entomol. Sci.* 22, 245.
Capo A., M. Cammareri, F. Della Rocca, A. Errico, A. Zoina and C. Conicella, 2002, *Am. J. Potato Res.* 79, 139.

Caromel B., V. Lefebvre, S. Andrzejewski, P. Rousselle, F. Rousselle-Bourgeois, D. Mugniéry, D. Ellissèche and M.C. Kerlan, 2003, *Theor. Appl. Genet.* 106, 1517.
Caromel B., D. Mugniéry, M.C. Kerlan, S. Andrzejewski, A. Palloix, D. Ellissèche, F. Rousselle-Bourgeois and V. Lefebvre, 2005, *Mol. Plant-Microbe Interact.* 18, 1186.
Carputo D., B. Basile, T. Cardi and L. Frusciante, 2000, *Am. J. Potato Res.* 43, 135.
Carputo D., T. Cardi, M. Speggiorin, A. Zoina and L. Frusciante, 1997, *Am. Potato J.* 74, 161.
Carputo D., L. Frusciante, L. Monti, M. Parisi and A. Barone, 2002, *Am. J. Potato Res.* 79, 345.
Carputo D., M. Speggiorin, P. Garreffa, A. Raio and L.M. Monti, 1996, *J. Genet. Breed.* 50, 221.
Carrasco A., J.I. Ruiz De Galarreta, A. Rico and E. Ritter, 2000, *Potato Res.* 43, 31.
Celebi-Toprak F., S.A. Slack and M.M. Jahn, 2002, *Theor. Appl. Genet.* 104, 669.
Charcosset A. and A. Gallais, 2003, In: D. de Vienne (Ed.), *Molecular Markers in Plant Genetics and Biotechnology*. Science Publishers, Enfield, New Hampshire, U.S.A.
Chavez R., C.R. Brown and M. Iwanaga, 1988, *Theor. Appl. Genet.* 76, 497.
Chen Q., L.M. Kawchuk, D.R. Lynch, M.S. Goettel and D.K. Fujimoto, 2003, *Am. J. Potato Res.* 80, 9.
Christ B.J., 1991, *Plant Dis.* 75, 353.
Christ B.J. and K.G. Haynes, 2001, *Plant Breed.* 120, 169.
Christinzio G. and A. Testa, 1999, *Potato Res.* 42, 101.
Chunwongse J., C. Chunwongse, L. Black and P. Hanson, 2002, *J. Hortic. Sci. Biotechnol.* 77, 281.
Collins A., D. Milbourne, L. Ramsay, R. Meyer, C. Chatot-Balandras, P. Oberhagemann, W. De Jong, C. Gebhardt, E. Bonnel and R. Waugh, 1999, *Mol. Breed.* 5, 387.
Colon L.T., R. Eijlander, D.J. Budding, M.T. Van Ijzendoorn, M.M.J. Pieters and J. Hoogendoorn, 1993, *Euphytica* 66, 55.
Colon L.T., R.C. Jansen and D.J. Budding, 1995a, *Theor. Appl. Genet.* 90, 691.
Colon L.T., L.J. Turkensteen, W. Prummel, D.J. Budding and J. Hoogendoorn, 1995b, *Eur. J. Plant Pathol.* 101, 387.
Concibido V.C., G.A. Secor and S.H. Jansky, 1994, *Euphytica* 76, 145.
Cooper-Bland S., M.J. De Maine, M.L.M.H. Fleming, M.S. Phillips, W. Powell and A. Kumar, 1994, *J. Exp. Bot.* 45, 1319.
Corsini D.L., J.J. Pavek and J.R. Davis, 1988, *Plant Dis.* 72, 148.
Costanzo S., I. Simko, B.J. Christ and K.G. Haynes, 2005, *Theor. Appl. Genet.* 111, 609.
Dale M.F.B. and J. Brown, 1989, *Ann. Appl. Biol.* 115, 313.
Dan H., S.T. Ali-Khan and J. Robb, 2001, *Plant Dis.* 85, 700.
Dangl J.L. and J.D.G. Jones, 2001, *Nature* 411, 826.
Davis J.R., O.C. Huisman, D.O. Everson and A.T. Schneider, 2001, *Am. J. Potato Res.* 78, 291.
Davis J.R., J.J. Pavek and D.L. Corsini, 1983, *Phytopathology* 73, 1009.
De Jong W., A. Forsyth, D. Leister, C. Gebhardt and D.C. Baulcombe, 1997, *Theor. Appl. Genet.* 95, 246.
De Maine M.J., C.P. Carroll, H.E. Stewart, R.M. Solomon and R.L. Wastie, 1993, *Potato Res.* 36, 21.
De Maine M.J., L.A. Farrer and M.S. Phillips, 1986, *Euphytica* 35, 1001.
De Medeiros A.H., W.M. Tingey and W.S. De Jong, 2004, *Am. J. Potato Res.* 81, 431.
Di Vito M., N. Greco, D. Carputo and L. Frusciante, 2003, *Nematropica* 33, 65.
DiFonzo C.D., D.W. Ragsdale and E.B. Radcliffe, 1994, *Plant Dis.* 78, 1173.
Dimock M.B., S.L. Lapointe and W.M. Tingey, 1986, *J. Econ. Entomol.* 79, 1269.
Diwan N., R. Fluhr, Y. Eshed, D. Zamir and S.D. Tanksley, 1999, *Theor. Appl. Genet.* 98, 315.
Dong F., J. Song, S.K. Naess, J.P. Helgeson, C. Gebhardt and J. Jiang, 2000, *Theor. Appl. Genet.* 101, 1001.
Dorrance A.E. and D.A. Inglis, 1997, *Plant Dis.* 81, 1206.
Dorrance A.E. and D.A. Inglis, 1998, *Plant Dis.* 82, 442.
Douches D.S., J.B. Bamberg, W. Kirk, K. Jastrzebski, B.A. Niemira, J. Coombs, D.A. Bisognin and K.J. Felcher, 2001, *Am. J. Potato Res.* 78, 159.
Dudley J.W., 1993, *Crop Sci.* 33, 660.
El-Kharbotly A., C. Leonards-Schippers, D.J. Huigen, E. Jacobsen, A. Pereira, W.J. Stiekema, F. Salamini and C. Gebhardt, 1994, *Mol. Gen. Genet.* 242, 749.
El-Kharbotly A., C. Palomino-Sanchez, F. Salamini, E. Jacobsen and C. Gebhardt, 1996a, *Theor. Appl. Genet.* 92, 880.
El-Kharbotly A., A. Pereira, W.J. Stiekema and E. Jacobsen, 1996b, *Euphytica* 90, 331.

Ellis J., P. Dodds and T. Pryor, 2000, *Curr. Opin. Plant Biol.* 3, 278.
Ewing E.E., I. Simko, C.D. Smart, M.W. Bonierbale, E.S.G. Mizubuti, G.D. May and W.E. Fry, 2000, *Mol. Breed.* 6, 25.
Fernandez E.C. and E.N. Bernardo, 1999, *Philippine Entomol.* 13, 21.
Flis B., J. Hermig, D. Strzelczyk-Żyta, C. Gebhardt and W. Marezewski, 2005, *Mol. Breed.* 15, 95.
Flor H.H., 1971, *Annu. Rev. Phytopathol.* 9, 275.
Franca F.H., R.L. Plaisted, R.T. Roush, S. Via and W.M. Tingey, 1994, *Entomol. Exp. Appl.* 73, 101.
Franca F.H. and W.M. Tingey, 1994, *J. Am. Soc. Hortic. Sci.* 119, 915.
Gavrilenko T., R. Thieme, U. Heimbach and T. Thieme, 2003, *Euphytica* 131, 323.
Gebhardt C., A. Ballvora, B. Walkemeier, P. Oberhagemann and K. Schuler, 2004, *Mol. Breed.* 13, 93.
Gebhardt C., D. Bellin, H. Henselewski, W. Lehmann, J. Schwarzfischer and J.P.T. Valkonen, 2006, *Theor. Appl. Genet.* 112, 1458.
Gebhardt C., D. Mugniery, E. Ritter, F. Salamini and E. Bonnel, 1993, *Theor. Appl. Genet.* 85, 541.
Gebhardt C. and J.P.T. Valkonen, 2001, *Annu. Rev. Phytopathol.* 39, 79.
Ghislain M., B. Trognitz, M.D. Herrera, J. Solis, G. Casallo, C. Vasquez, O. Hurtado, R. Castillo, L. Portal and M. Orrillo, 2001, *Theor. Appl. Genet.* 103, 433.
Gibson R.W., M.G.K. Jones and N. Fish, 1988, *Theor. Appl. Genet.* 76, 113.
Gopal J., 1988, *Potato Res.* 41, 311.
Goth R.W., K.G. Haynes and D.R. Wilson, 1993, *Plant Dis.* 77, 911.
Groden E. and R.A. Casagrande, 1986, *J. Econ. Entomol.* 79, 91.
Grube R.C., E.R. Radwanski and M. Jahn, 2000, *Genetics* 155, 873.
Grünwald N.J., M.A. Cadena Hinojosa, O.R. Covarrubias, A.R. Peña, J.S. Niederhauser and W.E. Fry, 2002, *Phytopathology* 92, 688.
Hämäläinen J.H., T. Kekarainen, C. Gebhardt, K.N. Watanabe and J.P.T. Valkonen, 2000, *Mol. Plant-Microbe Interact.* 13, 402.
Hämäläinen J.H., V.A. Sorri, K.N. Watanabe, C. Gebhardt and J.P.T. Valkonen, 1998, *Theor. Appl. Genet.* 96, 1036.
Hämäläinen J.H., K.N. Watanabe, J.P.T. Valkonen, A. Arihara, R.L. Plaisted, E. Pehu, L. Miller and S.A. Slack, 1997, *Theor. Appl. Genet.* 94, 192.
Hammond-Kosack K.E. and J.D.G. Jones, 1997, *Annu. Rev. Plant Physiol. Plant Mol. Biol.* 48, 575.
Harrison J.G., R. Lowe and J.M. Duncan, 1991, *Plant Pathol.* 40, 431.
Hawkes J.G., 1990, The Potato Evolution, Biodiversity & Genetic Resources. Smithsonian Institution Press, Washington, D.C., U.S.A.
Haynes K.G. and B.J. Christ, 1999, *Plant Breed.* 118, 431.
Hehl R., E. Faurie, J. Hesselbach, F. Salamini, S. Whitham, B. Baker and C. Gebhardt, 1999, *Theor. Appl. Genet.* 98, 379.
Helgeson J.P., G.J. Hunt, G.T. Haberlach and S. Austin, 1986, *Plant Cell Rep.* 5, 212.
Helgeson J.P., J.D. Pohlman, S. Austin, G.T. Haberlach, S.M. Wielgus, D. Ronis, L. Zambolim, P. Tooley, J.M. McGrath, R.V. James and W.R. Stevenson, 1998, *Theor. Appl. Genet.* 96, 738.
Herriott A.B., F.L. Haynes, Jr and P.B. Shoemaker, 1990, *HortScience* 25, 224.
Hilton A.J., H.E. Stewart, S.L. Linton, M.J. Nicolson and A.K. Lees, 2000, *Potato Res.* 43, 263.
Holley J.D., R.R. King and R.P. Singh, 1987, *Can. J. Plant Pathol.* 9, 291.
Horvath J., 1994, *Acta Phytopathol. Entomol. Hung.* 29, 105.
Horvath J., M. Kolber and I. Wolf, 1988, *Acta Phytopathol. Entomol. Hung.* 23, 465.
Horvath J. and I. Wolf, 1991, *Indian J. Virol.* 7, 176.
Hosaka K., Y. Hosaka, M. Mori, T. Maida and H. Matsunaga, 2001, *Am. J. Potato Res.* 78, 191.
Hosaka K., H. Matsunaga and K. Senda, 2000, *Am. J. Potato Res.* 77, 41.
Hoyos G.P., F.I. Lauer and N.A. Anderson, 1993, *Am. Potato J.* 70, 535.
Hoyos G.P., P.J. Zambino and N.A. Anderson, 1991, *Am. Potato J.* 68, 727.
Huang S., 2005, The discovery and characterization of the major late blight resistance complex in potato: genomic structure, functional diversity and implications. Ph.D. thesis, Wageningen Agricultural University, Wageningen, the Netherlands.
Huang S., E.A.G. van der Vossen, H. Kuang, V.G.A.A. Vleeshouwers, N. Zhang, T.J.A. Borm, H.J. van Eck, B. Baker, E. Jacobsen and R.G.F. Visser, 2005, *Plant J.* 42, 251.

Huang S.W., V.G.A.A. Vleeshouwers, J.S. Werij, R.C.B. Hutten, H.J. van Eck, R.G.F. Visser and E. Jacobsen, 2004, *Mol. Plant-Microbe Interact.* 17, 428.
Iovene M., A. Barone, L. Fusciante, L. Monti and D. Carputo, 2004, *Theor. Appl. Genet.* 109, 1139.
Ishimaru C.A., N.L.V. Lapitan, A. Vanburen, A. Fenwick and K. Pedas, 1994, *Am. Potato J.* 71, 517.
Iwanaga M., P. Jatala, R. Ortiz and E. Guevara, 1989, *J. Am. Soc. Hortic. Sci.* 114, 1008.
Jackson M.T., J.G. Hawkes, B.S. Malekayiwa and N.W.M. Wanyera, 1988, *Plant Breed.* 101, 261.
Jacobs J.M.E., H.J. van Eck, K. Horsman, P.F.P. Arens, B. Verkerk-Bakker, E. Jacobsen, A. Pereira and W.J. Stiekema, 1996, *Mol. Breed.* 2, 51.
Jansky S.H. and D.I. Rouse, 2000, *Potato Res.* 43, 239.
Jansky S.H. and D.I. Rouse, 2003, *Plant Dis.* 87, 266.
Jansky S., D.I. Rouse and P.J. Kauth, 2004, *Plant Dis.* 88, 1075.
Jones R.W. and I. Simko, 2005, In: M.K. Razdan and A.K. Mattoo (Eds), *Genetic Improvement of Solanaceous Crops*. Science Publishers, Enfield, New Hampshire, U.S.A.
Kasai K., Y. Morikawa, V.A. Sorri, J.P.T. Valkonen, C. Gebhardt and K.N. Watanabe, 2000, *Genome* 43, 1.
Kawchuk L.M., J. Hachey, D.R. Lynch, F. Kulcsar, G. van Rooijen, D.R. Waterer, A. Robertson, E. Kokko, R. Byers, R.J. Howard, R. Fischer and D. Prufer, 2001, *Proc. Natl. Acad. Sci. U.S.A.* 98, 6511.
Kim-Lee H., J.S. Moon, Y.J. Hong, M.S. Kim and H.M. Cho, 2005, *Am. J. Potato Res.* 82, 129.
Kreike C.M., J.R.A. Dekoning, J.H. Vinke, J.W. Vanooijen, C. Gebhardt and W.J. Stiekema, 1993, *Theor. Appl. Genet.* 87, 464.
Kreike C.M., J.R.A. Dekoning, J.H. Vinke, J.W. Vanooijen and W.J. Stiekema, 1994, *Theor. Appl. Genet.* 88, 764.
Kreike C.M., A.A. KokWesteneng, J.H. Vinke and W.J. Stiekema, 1996, *Theor. Appl. Genet.* 92, 463.
Kriel C.J., S.H. Jansky, N.C. Gudmestad and D.H. Ronis, 1995a, *Euphytica* 82, 125.
Kriel C.J., S.H. Jansky, N.C. Gudmestad and D.H. Ronis, 1995b, *Euphytica* 82, 133.
Kuhl J.C., R.E. Hanneman and M.J. Havey, 2001, *Mol. Genet. Genomics* 265, 977.
Kurowski C.J. and F.E. Manzer, 1992, *Am. Potato J.* 69, 289.
Lapointe S.L. and W.M. Tingey, 1986, *J. Econ. Entomol.* 79, 1264.
Lawrence C.H., M.C. Clark and R.R. King, 1990, *Phytopathology* 80, 606.
Lebecka R. and E. Zimnoch-Guzowska, 2004, *Am. J. Potato Res.* 81, 395.
Lefebvre V. and A. Palloix, 1996, *Theor. Appl. Genet.* 93, 503.
Leferriere L.T., J.P. Helgeson and C. Allen, 1999, *Theor. Appl. Genet.* 98, 1272.
Leister D., A. Ballvora, F. Salamini and C. Gebhardt, 1996, *Nat. Genet.* 14, 421.
Leonards-Schippers C., W. Gieffers, F. Salamini and C. Gebhardt, 1992, *Mol. Gen. Genet.* 233, 278.
Leonards-Schippers C., W. Gieffers, R. Schäfer-Pregl, E. Ritter, S.J. Knapp, F. Salamini and C. Gebhardt, 1994, *Genetics* 137, 67.
Li X., H.J. van Eck, J.N.A.M. Rouppe van der Voort, D.J. Huigen, P. Stam and E. Jacobsen, 1998, *Theor. Appl. Genet.* 96, 1121.
Lynch D.R., L.M. Kawchuk, Q. Chen and M. Kokko, 2003, *Am. J. Potato Res.* 80, 353.
Lynch D.R., L.M. Kawchuk and J. Hachey, 1997, *Plant Dis.* 81, 1011.
Lynch D.R., R.L. Wastie, H.E. Stewart, G.R. Mackay, G.D. Lyon and A. Nachmias, 1991, *Potato Res.* 34, 297.
Łojkowska E. and A. Kelman, 1989, *Am. Potato J.* 66, 379.
Łojkowska E. and A. Kelman, 1994, *Am. Potato J.* 71, 99.
Malakar R. and W.M. Tingey, 1999, *J. Econ. Entomol.* 92, 497.
Marczewski W., B. Flis, J. Syller, R. Schäfer-Pregl and C. Gebhardt, 2001, *Mol. Plant-Microbe Interact.* 14, 1420.
Marczewski W., B. Flis, J. Syller, D. Strzelczyk-Żyta, J. Hennig and C. Gebhardt, 2004, *Theor. Appl. Genet.* 109, 1604.
Marczewski W., J. Hennig and C. Gebhardt, 2002, *Theor. Appl. Genet.* 105, 564.
Marczewski W., K. Ostrowska and E. Zimnoch-Guzowska, 1998, *Plant Breed.* 117, 88.
Mayton H., W. De Jong, I. Simko and W.E. Fry, 2005, 9th International Workshop on Plant Disease Epidemiology, Plant Disease Epidemiology: Facing 21st Century Challenges, April 11–15, 2005, Landerneau, France.
McGrath J.M., C.E. Williams, G.T. Haberlach, S.M. Wielgus, T.F. Uchytil and J.P. Helgeson, 2002, *Am. J. Potato Res.* 79, 19.

Merz U., V. Martinez and R. Schwärzel, 2004, *Eur. J. Plant Pathol.* 110, 71.
Meyer R.C., D. Milbourne, C.A. Hackett, J.E. Bradshaw, J.W. McNichol and R. Waugh, 1998, *Mol. Gen. Genet.* 259, 150.
Meyer S., A. Nagel and C. Gebhardt, 2005, *Nucleic Acids Res.* 33, D666.
Micheletto S., M. Andreoni and M.A. Huarte, 1999, *Euphytica* 110, 133.
Milligan S.B., J. Bodeau, J. Yaghoobi, I. Kaloshian, P. Zabel and V.M. Williamson, 1998, *Plant Cell* 10, 1307.
Mojtahedi H., C.R. Brown and G.S. Santo, 1995, *J. Nematol.* 27, 86.
Moreau L., A. Charcosset, F. Hospital and A. Gallais, 1998a, *Genetics* 148, 1353.
Moreau P., P. Thoquet, J. Olivier, H. Laterrot and N. Grimsley, 1998b, *Mol. Plant-Microbe Interact.* 11, 259.
Murphy A.M., H. De Jong and K.G. Proudfoot, 1999, *Can. J. Plant Pathol.* 21, 207.
Murphy A.M., H. De Jong and G.C.C. Tai, 1995, *Euphytica* 82, 227.
Naess S.K., J.M. Bradeen, S.M. Wielgus, G.T. Haberlach, J.M. McGrath and J.P. Helgeson, 2000, *Theor. Appl. Genet.* 101, 697.
Neal J.J., R.L. Plaisted and W.M. Tingey, 1991, *Am. Potato J.* 68, 649.
Neal J.J., J.C. Steffens and W.M. Tingey, 1989, *Entomol. Exp. Appl.* 51, 133.
Niewöhner J., F. Salamini and C. Gebhardt, 1995, *Mol. Breed.* 1, 65.
Nombela G., V.M. Williamson and M. Muniz, 2003, *Mol. Plant-Microbe Interact.* 16, 645.
Novy R.G., A. Nasruddin, D.W. Ragsdale and E.B. Radcliffe, 2002, *Am. J. Potato Res.* 79, 9.
Oberhagemann P., C. Chatot-Balandras, R. Schäfer-Pregl, D. Wegener, C. Palomino, F. Salamini, E. Bonnel and C. Gebhardt, 1999, *Mol. Breed.* 5, 399.
Ordoñez M.E., G.A. Forbes and B.R. Trognitz, 1997, *Euphytica* 95, 167.
Ori N., Y. Eshed, I. Paran, G. Presting, D. Aviv, S. Tanksley, D. Zamir and R. Fluhr, 1997, *Plant Cell* 9, 521.
Ortiz R., J. Franco and M. Iwanaga, 1997, *Euphytica* 96, 339.
Ortiz R., C. Martin, M. Iwanaga and H. Torres, 1993, *Euphytica* 71, 15.
Paal J., H. Henselewski, J. Muth, K. Meksem, C.M. Menendez, F. Salamini, A. Ballvora and C. Gebhardt, 2004, *Plant J.* 38, 285.
Pan Q.L., Y.S. Liu, O. Budai-Hadrian, M. Sela, L. Carmel-Goren, D. Zamir and R. Fluhr, 2000a, *Genetics* 155, 309.
Pan Q.L., J. Wendel and R. Fluhr, 2000b, *J. Mol. Evol.* 50, 203.
Park T.H., J. Gros, A. Sikkema, V.G.A.A. Vleeshouwers, M. Muskens, S. Allefs, E. Jacobsen, R. Visser and E.A.G. van der Vossen, 2005a, *Mol. Plant-Microbe Interact.* 18, 722.
Park T.H., V.G.A.A. Vleeshouwers, D.J. Huigen, E.A.G. van der Vossen, H.J. van Eck and R. Visser, 2005b, *Theor. Appl. Genet.* 111, 591.
Park T.H., V.G.A.A. Vleeshouwers, R.C.B. Hutten, H.J. van Eck, E. van der Vossen, E. Jacobsen and R.G.F. Visser, 2005c, *Mol. Breed.* 16, 33.
Pavek J.J. and D.L. Corsini, 1994, In: J.E. Bradshaw and G.R. Mackay (Eds), *Potato Genetics.* CAB International, Wallingford, U.K.
Peirce, L.C., 1971, *Rep. Tomato Genet. Coop.* 21, 30.
Pelletier Y., C. Clark and G.C. Tai, 2001, *Entomol. Exp. Appl.* 100, 31.
Pelletier Y. and Z. Smilowitz, 1990, *J. Chem. Ecol.* 16, 1547.
Pelletier Y. and Z. Smilowitz, 1991, *Can. J. Zool.* 69, 1280.
Pelletier Y. and G.C.C. Tai, 2001, *J. Econ. Entomol.* 94, 572.
Perez W., A. Salas, R. Raymundo, Z. Huaman, R. Nelson and M. Bonierbale, 2001, Scientist and Farmer. Partners in Research for the 21st Century. CIP Program Report 1999–2000, Lima, Peru.
Pineda O., M.W. Bonierbale, R.L. Plaisted, B.B. Brodie and S.D. Tanksley, 1993, *Genome* 36, 152.
Plasencia J., R. Jemmerson and E.E. Banttari, 1996, *Phytopathology* 86, 170.
Porter L.D., D.A. Inglis and D.A. Johnson, 2004, *Plant Dis.* 88, 965.
Radcliffe E.B., W.M. Tingey, R.W. Gibson, L. Valencia and K.V. Raman, 1988, *J. Econ. Entomol.* 81, 361.
Ramon M. and R.E.J. Hanneman, 2002, *Euphytica* 127, 421.
Rangarajan A., A.R. Miller and R.E. Veilleux, 2000, *J. Am. Soc. Hortic. Sci.* 125, 689.
Rasmussen J.O., J.P. Nepper, H.-G. Krik, K. Tolstrup and O.S. Rasmussen, 1998, *Euphytica* 102, 363.
Rasmussen J.O., J.P. Nepper and O.S. Rasmussen, 1996, *Theor. Appl. Genet.* 92, 403.
Rauscher G.M., C.D. Smart, I. Simko, M. Bonierbale, H. Mayton, A. Greenland and W.E. Fry, 2006, *Theor. Appl. Genet.* 112, 674.

Ritter E., T. Debener, A. Barone, F. Salamini and C. Gebhardt, 1991, *Mol. Gen. Genet.* 227, 81.
Rivera-Peña A., 1990, *Potato Res.* 33, 479.
Rodriguez D.A., G.A. Secor, N.C. Gudmestad and G. Grafton, 1995, *Am. Potato J.* 72, 669.
Rokka V.M., J.P.T. Valkonen and E. Pehu, 1995, *Plant Sci.* 112, 85.
Rouppe van der Voort J., W. Lindeman, R. Folkertsma, R. Hutten, H. Overmars, E. van der Vossen, E. Jacobsen and J. Bakker, 1998, *Theor. Appl. Genet.* 96, 654.
Rouppe van der Voort J., E. van der Vossen, E. Bakker, H. Overmars, P. van Zandroort, R. Hutten, R.K. Lankhorst and J. Bakker, 2000, *Theor. Appl. Genet.* 101, 1122.
Rouppe van der Voort J., P. Wolters, R. Folkertsma, R. Hutten, P. van Zandvoort, H. Vinke, K. Kanyuka, A. Bendahmane, E. Jacobsen, R. Janssen and J. Bakker, 1997, *Theor. Appl. Genet.* 95, 874.
Rouppe van der Voort J.N.A.M., G.J.W. Janssen, H. Overmars, P.M. van Zandvoort, A. van Norel, O.E. Scholten, R. Janssen and J. Bakker, 1999, *Euphytica* 106, 187.
Ruiz de Galarreta J.I., A. Carrasco, A. Salazar, I. Barrena, E. Iturritxa, R. Marquinez, F.J. Legorburu and E. Ritter, 1998, *Potato Res.* 41, 57.
Sandbrink J.M., L.T. Colon, P.J.C.C. Wolters and W.J. Stiekema, 2000, *Mol. Breed.* 6, 215.
Sikinyi E., D.J. Hannapel, P.M. Imerman and H.M. Stahr, 1997, *J. Econ. Entomol.* 90, 689.
Simko I., 2002, *Am. J. Potato Res.* 79, 125.
Simko I., 2004, *Trends Plant Sci.* 9, 441.
Simko I., S. Costanzo, K.G. Haynes, B.J. Christ and R.W. Jones, 2004a, *Theor. Appl. Genet.* 108, 217.
Simko I., S. Costanzo, V. Ramanjulu, B.J. Christ and K.G. Haynes, 2006a, *Plant Breed.* 125, 385.
Simko I., K.G. Haynes, E.E. Ewing, S. Costanzo, B.J. Christ and R.W. Jones, 2004b, *Mol. Genet. Genomics* 271, 522.
Simko I., K.G. Haynes and R.W. Jones, 2004c, *Theor. Appl. Genet.* 108, 225.
Simko I., R.W. Jones and K.G. Haynes, 2006b, *Genetics*, 173, 2237.
Simons G., J. Groenendijk, J. Wijbrandi, M. Reijans, J. Groenen, P. Diergaarde, T. Van der Lee, M. Bleeker, J. Onstenk, M. de Both, M. Haring, J. Mes, B. Cornelissen, M. Zabeau and P. Vos, 1998, *Plant Cell* 10, 1055.
Sinden S.L., L.L. Sanford, W.W. Cantelo and K.L. Deahl, 1986, *Environ. Entomol.* 15, 1057.
Singh M., R.P. Singh and T.H. Somerville, 1994, *Am. Potato J.* 71, 567.
Singh R.P., X. Nie and G.C.C. Tai, 2000, *Theor. Appl. Genet.* 100, 401.
Smilde W.D., G. Brigneti, L. Jagger, S. Perkins and J.D.G. Jones, 2005, *Theor. Appl. Genet.* 110, 252.
Solomon-Blackburn R.M. and H. Barker, 1993, *Ann. Appl. Biol.* 122, 329.
Solomon-Blackburn R.M. and H. Barker, 2001, *Heredity* 86, 8.
Solomon-Blackburn R.M., G.R. Mackay and J. Brown, 1994, *J. Agric. Sci.* 122, 231.
Song J.Q., J.M. Bradeen, S.K. Naess, J.A. Raasch, S.M. Wielgus, G.T. Haberlach, J. Liu, H.H. Kuang, S. Austin-Phillips, C.R. Buell, J.P. Helgeson and J.M. Jiang, 2003, *Proc. Natl. Acad. Sci. U.S.A.* 100, 9128.
Song Y.S., L. Hepting, G. Schweizer, L. Hartl, G. Wenzel and A. Schwarzfischer, 2005, *Theor. Appl. Genet.* 111, 879.
Spitters C.J.T. and S.A. Ward, 1988, *Euphytica* 39, 87.
Stewart H.E. and J.E. Bradshaw, 1993, *Potato Res.* 36, 35.
Stewart H.E., K. Taylor and R.L. Wastie, 1983, *Potato Res.* 26, 363.
Takacs A.P., J. Horvath, G. Kazinczi and D. Pribek, 1999a, 51st International Symposium on Crop Protection. Part II, May 4, 1999, Gent, Belgium.
Takacs A.P., G. Kazinczi, J. Horvath, Z. Bosze and D. Pribek, 1999b, 51st International Symposium on Crop Protection. Part II, May 4, 1999, Gent, Belgium.
Tanksley S.D., M.W. Ganal and G.B. Martin, 1995, *Trends Genet.* 11, 63.
Tek A.L., W.R. Stevenson, J.P. Helgeson and J. Jiang, 2004, *Theor. Appl. Genet.* 109, 249.
Thach N.Q., U. Frei and G. Wenzel, 1993, *Theor. Appl. Genet.* 85, 863.
Tommiska T.J., J.H. Hämäläinen, K.N. Watanabe and J.P.T. Valkonen, 1998, *Theor. Appl. Genet.* 96, 840.
Tooley P.W., 1990, *Am. Potato J.* 67, 491.
Toxopeus H.J. and C.A. Huijsman, 1953, *Euphytica* 2, 180.
Tozzini A.C., M.F. Ceriani, M.V. Saladrigas and H.E. Hopp, 1991, *Potato Res.* 34, 317.
Treadwell F.J., 1991, Breeding for Resistance to Verticillium Wilt in Potato. Ph.D. thesis, University of Minnesota, St. Paul, Minnesota, U.S.A.
Trognitz B.R., M. Orrillo, L. Portal, C. Román, P. Ramón, S. Perez and G. Chacón, 2001, *Plant Pathol.* 50, 281.

Trognitz F., P. Manosalva, R. Gysin, D. Nino-Liu, R. Simon, M.D. Herrera, B. Trognitz, M. Ghislain and R. Nelson, 2002, *Mol. Plant-Microbe Interact.* 15, 587.
Tsror L. and A. Nachmias, 1995, *Isr. J. Plant Sci.* 43, 315.
Valkonen J.P.T., 1997, *Ann. Appl. Biol.* 130, 91.
Valkonen J.P.T., G. Brigneti, F. Salazar, E. Pehu and R.W. Gibson, 1992a, *Ann. Appl. Biol.* 120, 301.
Valkonen J.P.T., A. Contreras, E. Pehu and L.F. Salazar, 1992b, *Potato Res.* 35, 411.
Valkonen J.P.T. and V.M. Rokka, 1998, *Plant Sci.* 131, 85.
Valkonen J.P.T., S.A. Slack and K.N. Watanabe, 1995, *Ann. Appl. Biol.* 126, 143.
Vallejo R.L., W.W. Collins and J.B. Young, 1995, *J. Hered.* 86, 89.
van der Vossen E., A. Sikkema, B.T.L. Hekkert, J. Gros, P. Stevens, M. Muskens, D. Wouters, A. Pereira, W. Stiekema and S. Allefs, 2003, *Plant J.* 36, 867.
van der Vossen E.A.G., J. Gros, A. Sikkema, M. Muskens, D. Wouters, P. Wolters, A. Pereira and S. Allefs, 2005, *Plant J.* 44, 208.
van der Vossen E.A.G., J.N.A.M. Rouppe van der Voort, K. Kanyuka, A. Bendahmane, H. Sandbrink, D.C. Baulcombe, J. Bakker, W.J. Stiekema and R.M. Klein-Lankhorst, 2000, *Plant J.* 23, 567.
van Eck H.J. and E. Jacobsen, 1996, 13th Triennial conference of the European Association for Potato Research, Wageningen, The Netherlands.
Visker M.H.P.W., L.C.P. Keizer, H.J. Van Eck, E. Jacobsen, L.T. Colon and P.C. Struik, 2003, *Theor. Appl. Genet.* 106, 317.
Watanabe J.A., M. Orrillo and K.N. Watanabe, 1999, *Breed. Sci.* 49, 53.
Watanabe K., H.M. El-Nashaar and M. Iwanaga, 1992, *Euphytica* 60, 21.
Wilson C.R. and R.A.C. Jones, 1993, *Aust. J. Exp. Agric.* 33, 83.
Yencho G.C., M.W. Bonierbale, W.M. Tingey, R.L. Plaisted and S.D. Tanksley, 1996, *Entomol. Exp. Appl.* 81, 141.
Zhang R., 2004, Genetic Characterization and Mapping of Partial Resistance to Early Blight in Diploid Potato, Ph.D. thesis, Pennsylvania State University, University Park, Pennsylvania, U.S.A.
Zimnoch-Guzowska E., R. Lebecka, A. Kryszczuk, U. Maciejewska, A. Szczerbakowa and B. Wielgat, 2003, *Theor. Appl. Genet.* 107, 43.
Zimnoch-Guzowska E. and E. Łojkowska, 1993, *Potato Res.* 36, 177.
Zimnoch-Guzowska E., W. Marczewski, R. Lebecka, B. Flis, R. Schäfer-Pregl, F. Salamini and C. Gebhardt, 2000, *Crop Sci.* 40, 1156.
Zlesak D.C. and C.A. Thill, 2004, *Am. J. Potato Res.* 81, 421.

Chapter 8

Potato-Breeding Strategy

John E. Bradshaw

Scottish Crop Research Institute, Invergowrie, Dundee DD2 5DA, United Kingdom

8.1 INTRODUCTION

Potato-breeding strategy can be viewed as the key decisions that a breeder makes concerning the objectives of a breeding programme, what germplasm and breeding methods to use, whether new cultivars will be propagated vegetatively or through true potato seed (TPS), whether or not new cultivars will be genetically modified and how to achieve durable disease and pest resistance. Sound decisions require knowledge of the evolution of the modern crop, target environments and end uses for new cultivars, the reproductive biology of cultivated potatoes and their wild relatives, and the population structure of pathogens and the epidemiology of diseases. Genetic knowledge is also required and has increased dramatically for the potato since the first molecular marker map appeared in 1988 (Bonierbale et al., 1988, 2003). Advances in potato genetics are covered in Chapter 7 (Simko et al., this volume) and Chapter 9 (Bryan, this volume). This chapter concentrates on their application to breeding new cultivars that will be required for sustainable increases in potato production during a period of environmental change and human population growth. The examples chosen for discussion are those most familiar to the author and are not intended to be a comprehensive set.

8.2 EVOLUTION OF THE MODERN POTATO CROP

Wild tuber-bearing *Solanum* species are distributed from the southwestern USA (38 °N) to central Argentina and adjacent Chile (41 °S) and hence cover a great ecogeographical range (Spooner and Hijmans, 2001). Hawkes (1990) recognized 219 species and grouped them into 19 series, but their taxonomy is difficult and is being revised (Chapter 4, Van den Berg and Jacobs, this volume). Simmonds (1995) concluded that just a few closely related species in series Tuberosa (e.g. *Solanum brevicaule, Solanum leptophyes* and *Solanum canasense*) were domesticated in the Andes of southern Peru and northern Bolivia about 7000 years ago. More recently, Spooner et al. (2005a) have provided evidence for a single domestication in Peru from the northern group of members of the *S. brevicaule* complex of species. The result of domestication was a diploid species *Solanum stenotomum*, also referred to as a form of *Solanum tuberosum* (Group Stenotomum), from which other cultivated species were derived, including diploid *Solanum phureja* (or Group Phureja), tetraploid *S. tuberosum* subsp. *andigena* (or Group Andigena) and tetraploid *S. tuberosum*

Potato Biology and Biotechnology: Advances and Perspectives
D. Vreugdenhil (Editor)

subsp. *tuberosum* (or Group Tuberosum). Phureja potatoes were selected by Andean farmers from Stenotomum for lack of tuber dormancy and faster tuber development so that they could grow up to three crops per year in the lower, warmer, eastern valleys of the Andes. Andigena potatoes became the most widely grown form in South America, presumably because farmers found them superior to the diploids for yield and other traits. Tuberosum potatoes were selected from Andigena types for tuber production in long days in coastal Chile and are referred to as Chilean Tuberosum. The involvement of wild as well as cultivated species as progenitors of Andigena and Chilean Tuberosum potatoes is still a matter of debate, but they are genetically distinct groups of potatoes with different cytoplasms (Raker and Spooner, 2002; Hosaka, 2004).

Potatoes (tetraploid *S. tuberosum*) were introduced into Europe in the 1570s, and then starting in the seventeenth century, they were taken from Europe and cultivated in many other parts of the world (Hawkes and Francisco-Ortega, 1993; Pandey and Kaushik, 2003). It is likely that the early introductions came from both the Andes and coastal Chile but were few in number and hence captured only some of the biodiversity present in the cultivated potatoes of South America (Hawkes, 1990; Spooner et al., 2005b). It is therefore remarkable that today potatoes are grown in 149 countries from latitudes 65 °N to 50 °S and at altitudes from sea level to 4000 m (Hijmans, 2001) and that potatoes are the fourth most important food crop after wheat, maize and rice (Lang, 2001). Their distribution reflects the adaptation of *S. tuberosum* first to the short summer days of the highland tropics and subtropics, then to the long summer days of lowland temperate regions and finally to the short winter days of the lowland subtropics and tropics. The distribution also reflects growing conditions that are neither too hot nor too cold and where there is adequate water, whether rain or irrigation. Potatoes have also been adapted to a wide range of end uses. As well as being a staple food, the potato is grown as a vegetable for table use, is processed into French fries and crisps (chips) and is used for dried products and starch production. In some countries it is still fed to animals, but this use is decreasing.

8.3 POTATO BREEDING AND THE NEED FOR NEW CULTIVARS

8.3.1 Potato breeding

The adaptation of the potato to a wide range of environments and end uses was helped by its reproductive biology, which is ideal for creating and maintaining variation. Potatoes flower and set true seed in berries following natural pollination by insects, particularly bumblebees. Outcrossing is enforced in cultivated (and most wild) diploid species by a single S-locus, multiallelic, gametophytic self-incompatibility system (Dodds, 1965). Although this system does not operate in tetraploid *S. tuberosum*, 40% (range 21–71%) natural cross-pollination was estimated to occur in subsp. *andigena* in the Andes (Brown, 1993) and 20% (range 14–30%) in an artificially constructed Andigena population (Glendinning, 1976). This sexual reproduction creates an abundance of diversity by recombining the variants of genes that arose by mutation, and as a consequence, potatoes are highly heterozygous individuals that display inbreeding depression on selfing.

The genetically unique seedlings that grow from true seeds produce tubers that can be replanted as seed tubers, and hence distinct clones can be established and maintained by asexual vegetative reproduction. No doubt many Andean cultivars were produced by farmer selection from naturally occurring variation. Domestication must have involved selection of less bitter and hence less toxic tubers, and Andean farmers certainly retained a much wider variety of tuber shapes and skin and flesh colours than seen in wild species (Simmonds, 1995). Subsequent selection for early maturity, appropriate dormancy and resistance to abiotic and biotic stresses must have occurred in many environments.

Modern potato breeding began in 1807 in England when Knight made the first recorded hybridizations between varieties by artificial pollination (Knight, 1807). It flourished in Britain and elsewhere in Europe and North America during the second half of the nineteenth century when many new cultivars were produced by farmers, hobby breeders and seedsmen. At least two Chilean Tuberosum introductions were used in breeding during the nineteenth century. The cultivar Daber was introduced into Germany in 1830, most probably from Chile (Plaisted and Hoopes, 1989), and likewise the cultivar Rough Purple Chili was introduced into the USA in 1851 (Goodrich, 1863). North America's most popular potato cultivar, Russet Burbank, was derived from it by three generations of open pollination with selection and released in 1914 (Ortiz, 2001). The descendents of Rough Purple Chili were widely employed as female parents in crosses with European Tuberosum at the end of the nineteenth century, and Chilean Tuberosum cytoplasm predominates in modern cultivars. This uniformity of cytoplasm has been a cause for concern since the epidemics of Southern corn leaf blight on maize hybrids with T (Texas) cytoplasm in 1970–71 (Ullstrup, 1972) and still needs to be addressed by potato breeders. Modern potato breeding started later in the 1930s in China and India, but these countries are now two of the leading potato producers in the world (Gaur and Pandey, 2000; Jin et al., 2004). The extent of progress since 1807 can be judged by the latest *World Catalogue of Potato Varieties* (Hils and Pieterse, 2005), which lists over 4000 cultivars from more than 100 countries, a remarkable achievement for a crop which outside of Latin America was derived from a narrow genetic base. Although these present-day cultivars are the foundation for future breeding, breeders should also seek to make full use of the diversity that exists in native Latin American cultivars and their cross-compatible wild relatives.

8.3.2 Need for new cultivars

There is certainly a need for new cultivars despite the large number currently available. At least two contrasting scenarios can be seen. In the EU, the potato industry is trying to increase potato usage in an economically and environmentally sustainable way. New cultivars must give economic benefits through more yield of saleable product at less cost of production, whether the potatoes are for processing or table use. They must have inbuilt resistances to pests and diseases that give environmental benefits through reduced use of pesticides and fungicides. Increased water and mineral use efficiency are also desirable for better use of water and fertilizers, both nitrogen and phosphate. Finally, they must help meet consumer demands for convenience foods, improved nutritional and health benefits, improved flavour and novel products. By contrast, in Asia and Africa there is a need for increased and stable potato production to meet increased demand for food.

New cultivars must deliver higher yields under low inputs, disease and pest attacks, and environmental stresses such as heat, cold, drought and salinity. If possible, they should also have improved nutritional and health properties.

8.3.3 True potato seed

Today breeders normally raise seedlings from crosses between pairs of parents to find the best genotype in the best progeny to clonally (vegetatively) propagate as a new cultivar. However, there are circumstances where cultivars based on TPS propagation are an attractive proposition despite being genetically variable and inferior to the best genotype that exists within the progeny. In the torrid zones of the lowland tropics and subtropics, reduced seed costs (due to much smaller amounts of planting material required), flexibility of planting time (no physiological age of seed tubers) and freedom from tuber borne diseases (particularly viruses, but not viroids) can outweigh the difficulty of establishing the crop, later maturation and less uniformity (Golmirzaie et al., 1994). TPS potatoes are established in Bangladesh, China, Egypt, India, Indonesia, Nicaragua, Peru, Philippines, southern Italy and Vietnam (Almekinders et al., 1996; Ortiz, 1997; Simmonds, 1997). Although TPS can be propagated by direct drilling and transplanting, preference is now given to rapid multiplication by cuttings or shoot-tip culture, which gives rise to give tuberlets (first-generation tubers) or mini-tubers which can be chitted before planting (Simmonds, 1997). Methods involving multiplication allow some selection to be practised within a TPS progeny and hence are a compromise between clonal (vegetative) and true seed multiplication. Whether to breed cultivars for propagation by tubers or true seed is a key strategic decision.

8.4 ADAPTATION TO ENVIRONMENTS AND END USES

8.4.1 Genotype by environment interactions

Another key strategic decision is the number of new cultivars required for a given range of target environments and end uses. The answer requires an assessment of genotype × environment (including end use) interactions because it is unlikely that one of many potential new cultivars will be best in all environments and for all uses. Genotype × environment interactions can be analysed and visualized through a principal component analysis as done by Forbes et al. (2005) to determine the stability of resistance to *Phytophthora infestans*. A related question concerns the environments in which to practise selection. It is known that selection in one type of environment has consequences for performances in different types of environment, and Falconer and Mackay (1996) explained how these can be quantified. The improvement of performance in one environment as a result of selection in a different environment can be viewed as a correlated response and compared with the expected response from direct selection in the target environment. The potato's vegetative means of reproduction does lend itself to selection experiments in contrasting environments, but extensive studies have not been done. In practice, the logistics of seed tuber multiplication mean that potato breeders are likely to select their early generations at local seed and ware sites and then test relatively few potential cultivars in a much wider range of environments. Brown et al. (1996) found a greater correlation for total marketable yield between the Scottish ware site, where clones were selected, and sites in England (0.43–0.70) than with sites in the Mediterranean (0.00–0.67). Although the very best clones from the Scottish ware site performed reasonably well in the Mediterranean, the results supported the idea that selection would be optimized by selecting in environments more similar to those in which the cultivars are to be grown. Potato growing countries should therefore have their own breeding programmes targeted at adaptation to their local environments and end uses, notwithstanding commercial companies wanting to see their new cultivars grown as widely as possible.

8.4.2 Ideotypes

Recently, Haverkort and Grashoff (2004) have developed a decision-support system for potato breeding based on ideotyping that may lead to the use of physiological selection criteria for adaptation to target environments in ways that have not proved feasible in practice in the past.

They define an ideotype as the ideal genotype for a particular environment. The ideotype has a growth cycle with a length that matches that of the available growing season, which is characterized by a temperature window (neither too cold nor too hot) and resources such as solar radiation, water and nutrients. They claim that the optimal set of 'genes' can be calculated before starting a breeding programme for any environment. In practice, it is a question of whether or not physiological selection criteria can be defined, which prove superior to selecting for saleable yield per se in the target environment, and this is a matter for determination by experiment.

The kinds of questions that Haverkort and Grashoff (2004) ask are certainly worthy of consideration and may help breeding to accommodate future environmental changes, including climate change. The aim is to quantify the benefits of frost tolerance for earlier planting, greater sprout growth rate for earlier emergence, lower base temperature for crop growth, higher initial leaf areas at emergence and/or higher leaf area development rates, thinner leaves in temperate climates with less intense solar radiation than in the tropics, changes in leaf area index and patterns of senescence, and deeper rooting and increased water use efficiency. Optimization of nine parameter values in simulation runs resulted in predicted increases of tuber dry matter production of 78% for unirrigated crops grown in dry and sunny years and over 40% for crops with irrigation. The possibility of such optimization should be taken seriously as fresh weight yields of 120 tonnes per hectare have been achieved experimentally in the absence of pests and pathogens, and with adequate inputs of water and fertilizers (Mackay, 1996), yet the average world yield is only 17 tonnes per hectare (Lang, 2001).

8.5 GERMPLASM AVAILABLE

The genetic resources available for potato breeding and their utilization are summarized in Fig. 8.1.

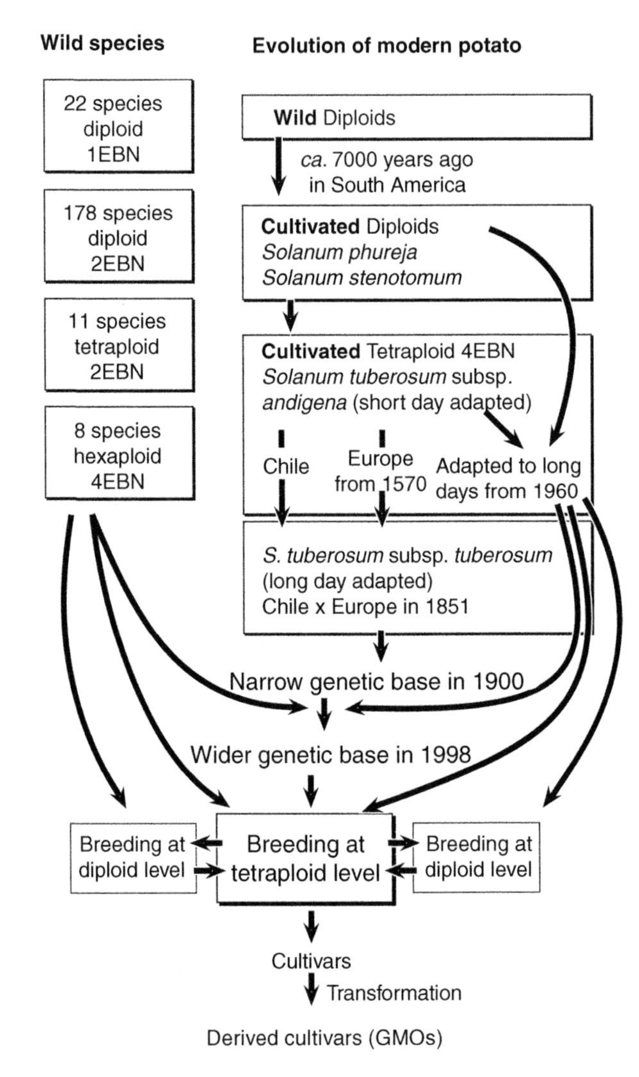

Fig. 8.1. Genetic resources and their utilization in potato breeding (from the Scottish Crop Research Institute Annual Report 1997/98).

8.5.1 Wild species

Recognition of Central and South America as the centres of origin and diversity of tuber-bearing *Solanum* species resulted in numerous collecting expeditions, from those pioneered by the Russians in the 1920s (Hawkes, 1990) to the more recent ones of the 1990s (Spooner and Hijmans, 2001). These in turn led to the establishment of a number of potato germplasm collections worldwide. The main ones are the world collection at the International Potato Centre (CIP, Lima, Peru), the Commonwealth Potato Collection

(CPC, Dundee, Scotland), the Dutch-German Potato Collection (CGN, Wageningen, The Netherlands), the Gro Lüsewitz Potato Collection (GLKS, IPK, Groβ Lüsewitz, Germany), the Potato Collection of the Vavilov Institute (VIR, St Petersburg, Russia), the US Potato Genebank (NRSP-6, Sturgeon Bay, USA) and Potato Collections in Argentina, Bolivia and Peru. Together they comprise the Association for Potato Intergenebank Collaboration and have established an Inter-genebank Potato Database (IPD), which can be accessed through the World Wide Web (http://www.potgenebank.org). The IPD contains 7112 different accessions of 188 taxa (species, subspecies, varieties and forms) out of the 247 tuber-bearing wild potato taxa recognized by Hawkes (Huaman et al., 2000a). Data are available for more than 33 000 evaluations of wild potato accessions covering 55 traits (dry matter, starch, reducing sugar and glycoalkaloid content and resistances to fungi, bacteria, viruses, viroids, insects and environmental stresses such as frost and heat/drought). Clearly, these germplasm collections contain a lot of biodiversity, but it is pertinent to ask to what extent they have been utilized in the past and what are the prospects for their further use in the future.

The introgression of genes from wild species has been significant but fairly limited in number. Resistance to late blight was introgressed into cultivated potato from *Solanum demissum* and *Solanum stoloniferum*, resistance to viruses from these species together with *Solanum chacoense* and *Solanum acaule*, and resistance to potato cyst nematodes from *Solanum vernei* and *Solanum spegazzinii*. By the end of the 1980s, these wild species together with cultivated *S. tuberosum* subsp. *andigena* and *S. phureja* had been used extensively in the breeding of successful cultivars in Europe (Ross, 1986), and of these, *S. demissum*, *S. chacoense* and *S. acaule* dominated in North American cultivars (Plaisted and Hoopes, 1989). Likewise, one of the most successful cultivars to be introduced into China by CIP, CIP-24, had *S. acaule*, *S. demissum* and *S. stoloniferum* in its pedigree (Ortiz, 2001). The genetic base of modern Indian potato selections can be traced to only 49 ancestors, of which 10 from the UK account for 41% of the total genomic constitution (Gopal and Oyama, 2005). The most frequent ancestors were two clones from the UK, 2814a1 and 3069d4, which trace back to a cross between *Solanum rybinii* (a variant of *S. phureja*) and *S. demissum*, which was made by Dr Black in 1937 to introgress late blight resistance from *S. demissum*.

Further improvements in the resistance of cultivated potatoes to abiotic and biotic stresses should come from the greater use of wild species in potato breeding, given the wide range of habitats in which they have evolved. Indeed, breeders should seek to make full use of the diversity that exists in the world's major potato germplasm collections. The screening of these collections for desirable traits will continue and so will genetic studies on their inheritance, followed by introgression of desirable alleles into Tuberosum, as described by Bradshaw and Ramsay (2005) for the CPC. These approaches will be supplemented by an increasing amount of molecular marker and DNA sequence data (Chapter 7, Simko et al., this volume; Chapter 9, Bryan, this volume). McGregor et al. (2002) analysed the entire CGN collection of 314 accessions of series *Acaulia* with two amplified fragment length polymorphism (AFLP) primer pairs and were able to verify taxonomic classification and to identify redundancies. However, there was no consistent relationship between the geographic distances of collection sites and genetic distance. Hence, the challenge is to use the new knowledge being gained to help breeders find desirable novelty amongst the available germplasm. The outcome may depend on whether or not allele frequencies and distributions can be associated with features of the natural habitats of the accessions.

8.5.2 Cultivated species

Following its creation in 1971, CIP in Peru assembled a potato collection of more than 15 000 accessions of native potato cultivars from nine countries in Latin America (Argentina, Bolivia, Chile, Colombia, Ecuador, Guatemala, Mexico, Peru and Venezuela). CIP identified those which were duplicate accessions of the same cultivar and selected 3527 individual cultivars among them, of which 552 were diploids, 128 triploids, 2836 tetraploids (2644 subsp. *andigena*, 144 subsp. *tuberosum* and 48 hybrids) and 11 pentaploids (Huaman et al., 1997). By 1997, researchers at CIP had already conducted 46 124 evaluations on the collection for the reactions of cultivars to biotic and abiotic stresses and for other desirable traits. Then to aid utilization, Huaman et al. (2000b) used cluster analysis of morphological data to establish a core set of 306 accessions from the 2379 cultivars of subsp. *andigena* still held at CIP. The set was chosen to represent the widest morphological diversity and to maximize geographical representation. Evaluation data were taken into account in choosing the representative accession of each cluster. Isozyme analysis of the entire collection confirmed that the core collection had captured a representative sample of the alleles at nine allozyme loci with only the loss of rare alleles (frequency less than 0.05%) (Huaman et al., 2000c). However, a simulation study revealed that a core collection size of 600 would be required to adequately represent allele frequencies and locus heterozygosity (Chandra et al., 2002). The core collection should nevertheless be a valuable resource for detailed evaluation and future breeding both in Latin America and worldwide.

Most of the cultivars from Latin America are adapted to tubering in short days. However, since the first experiment started by Simmonds in 1959 (Simmonds, 1969), a number of programmes worldwide have demonstrated that, through simple mass selection under northern latitude, long-day summer conditions, subsp. *andigena* will adapt and produce parents suitable for direct incorporation into European and North American potato-breeding programmes (Bradshaw and Mackay, 1994). Likewise, during the period 1962–79, Carroll (1982) employed a mass selection method to produce a population of *S. phureja/S. stenotomum* adapted to long-day North European conditions. Diploid cultivars, such as Mayan Gold, have now been produced from this population but are targeted at niche markets because their yield is only two-thirds that of Tuberosum potatoes. Direct hybridization of members of this improved diploid population with tetraploid potato cultivars through unreduced pollen grains ($4x \times 2x$ crosses) resulted in tetraploid hybrids, some of which were superior to standard tetraploid cultivars in both total and marketable yield (Carroll and De,Maine, 1989). Similar work has been done in North America (Plaisted and Hoopes, 1989; Haynes and Christ, 1999; Haynes and Lu, 2005). However, relatively few clones of Neotuberosum (long-day subsp. *andigena*) and long-day *S. phureja/S. stenotomum* have been used to any extent in the breeding of successful cultivars, despite displaying yield heterosis in crosses with Tuberosum (Glendinning,

1969; Cubillos and Plaisted, 1976; Tai and De Jong, 1980; Tarn and Tai, 1983; Maris, 1989; Buso et al., 1999). Part of the reason for this is that whilst adaptation to tubering in long days was quickly achieved, other problems remained. Neotuberosum clones lacked the regularity of tuber shape of intensively selected subsp. *tuberosum* clones, and long-day adapted *S. phureja* clones lacked tuber dormancy. Hence, these populations need, and are receiving, selection for further improvements to achieve the original goal of direct use as parents in breeding finished cultivars (i.e. contributing 50% of their genes to their offspring, or 25% in backcrosses). The improved populations, along with modern cultivars, should be compared with CIP's core collections for molecular marker diversity to see how many common alleles have been captured in the long-day populations and cultivars. Decisions can then be made on whether or not there is a need to create and select new populations that capture more biodiversity to further broaden the genetic base in breeding for long-day temperate climates. Even if future populations of long-day-adapted Andigena and Phureja/Stenotomum fail to live up to expectation, biodiverse populations that can be assessed for tuber characteristics in long days should prove to be valuable sources of desirable alleles for molecular-marker-assisted incorporation into Tuberosum and thence into finished cultivars. Using molecular markers and Tuberosum, Neotuberosum and Phureja germplasm in the Cornell University potato-breeding programme, Bonierbale et al. (1993) found that specific combinations of individual marker fragments were more important than maximum heterozygosity for tuber yield heterosis. The big strategic question that still needs answering is how inferior to modern cultivars do biodiverse populations have to be for targeted gene transfer to be a better strategy than base broadening.

8.6 INTROGRESSION OF GENES FROM WILD SPECIES

8.6.1 Sexual and somatic hybridization of *S. tuberosum* with wild species

Wild tuber-bearing *Solanum* species form a polyploid series from diploid ($2n = 2x = 24$) to hexaploid ($2n = 6x = 72$), in which nearly all of the diploid species are self-incompatible outbreeders and the tetraploids and hexaploids are mostly self-compatible allopolyploids which display disomic inheritance (Hawkes, 1990). Crossability groups occur, each defined by an endosperm balance number (EBN), although interspecific pollen–pistil incompatibility and nuclear–cytoplasmic male sterility can occur (Camadro et al., 2004). Today, by manipulation of ploidy, with due regard to EBN, virtually any potato species can be utilized for the introgression of desirable genes into *S. tuberosum* (Ortiz, 1998, 2001; Jansky, 2006). The schemes used for ploidy level manipulations exploit the facts that unreduced $2n$ gametes are common in *Solanum* species and maternal haploids (dihaploids) of tetraploid *S. tuberosum* can be extracted following crosses with pollinator clones of *S. phureja*, thus allowing breeding at the diploid level before returning to the tetraploid level by sexual hybridization for cultivar production (Jansky et al., 1990; Carputo and Barone, 2005). Schemes have also been devised, in which somatic fusion is used to return to the tetraploid level and maximum heterozygosity (Wenzel et al., 1979). Somatic fusion has also allowed the production of hybrids between tetraploid 4EBN *S. tuberosum* and diploid 1EBN species, e.g. the non-tuber-bearing species *Solanum brevidens*, which has tuber soft rot and early blight resistances (Tek et al., 2004), and *Solanum bulbocastanum*, which has a major gene for broad-spectrum resistance to late blight (Naess et al., 2000).

8.6.2 Molecular-marker-assisted introgression and gene cloning

It took from three to seven backcrosses to transfer a major dominant resistance gene from a wild species into a successful cultivar, but fewer generations from cultivated potatoes (Ross, 1986; Bradshaw and Ramsay, 2005). Molecular-marker-assisted introgression offers the possibility of faster progress because not only can one select phenotypically or genotypically for the desired gene(s) from the wild species, but one can also select genotypically against the wild species genome (Hermsen, 1994; Barone, 2004; Iovene et al., 2004). In fact, with adequate molecular marker coverage of all 12 potato chromosomes, it is possible to estimate the optimal combination of population sizes and number of backcross generations and to select in a very precise way for the desired products of meiosis in each backcross generation (Hospital, 2003).

Where introgression is performed at the tetraploid level, the result may not be a genotype with 48 Tuberosum chromosomes including one or more with the introgressed gene(s). This need not affect the commercial success of any new cultivar produced and propagated vegetatively but could affect the fertility and value of such a cultivar as a parent for further breeding. The result of the introgression from *S. brevidens* referred to above was a high-yielding clone, C-75-5 + 297, with resistances to both tuber soft rot and early blight. Using both molecular and cytogenetic approaches, Tek et al. (2004) showed that C-75-5 + 297 had 47 chromosomes, including four copies of chromosome 8, three from potato and one from *S. brevidens*, which was the only part of the wild species genome present. By contrast, Barone et al. (2001) did obtain 48 chromosomes and evidence of recombination between *Solanum commersonii* (an 1EBN diploid) and *S. tuberosum* chromosomes in their molecular-marker-assisted introgression of tuber soft rot resistance.

As potatoes are heterozygous outbreeders, use of the same recurrent parent during introgression would result in a self of the recurrent parent and hence inbreeding depression. This can be avoided by using different Tuberosum parents for each backcross but would result in an entirely new cultivar, which may or may not be the desired outcome. The only way to introduce a gene into an existing cultivar is by the transgenic route. Hence, the molecular cloning of natural resistance genes and their transfer into well-adapted but susceptible cultivars are being pursued in a number of laboratories worldwide, and examples can be found in Chapter 7 (Simko et al., this volume).

8.6.3 Base broadening versus introgression

Most of the evolutionary, more advanced, tuber-bearing wild species are diploid with an EBN of 2, the same as haploids (dihaploids) of *S. tuberosum*, with which they will readily cross. These wild species are sources of genetic diversity in general, as well as genetic resistances to diseases, pests and abiotic stresses, but they do not tuber in long days. By contrast, Jansky et al. (2004) found that in 125 of 154 families of haploid × wild species crosses, more than 50% of the plants did tuberize in long days and could therefore be selected for tuber characteristics. Hence, if they also produce $2n$ gametes by first division restitution (FDR), the genetic diversity of the wild species can be efficiently transferred to the tetraploid cultivated form in $4x \times 2x$ crosses and result in about 25% of the wild species genes in the final product. This led Peloquin and his coworkers to propose such $4x \times 2x$ crosses as a novel breeding strategy designed to both introgress specific characteristics and broaden the genetic base (Hermundstad and Peloquin, 1987; Jansky et al., 1990), in a way similar to that envisaged by Chase (1963) in his analytic breeding scheme. However, Tai (1994) concluded that haploid × wild species hybrids need to be improved before they are used in $4x \times 2x$ crosses, e.g. through population improvement by recurrent selection (Rousselle-Bourgeois and Rouselle, 1992). In summary, the strategic question that still needs to be answered by experiment is whether or not wild species contain sufficient desirable alleles to warrant base broadening as opposed to the introgression of just a few desirable genes.

8.7 BREEDING CULTIVARS AT THE TETRAPLOID LEVEL FOR CLONAL PROPAGATION

8.7.1 Parents

Potato breeding worldwide has traditionally involved making crosses between pairs of parents with complementary features, and this is still the main route to new cultivars. Increasingly, parents will have genes introgressed from wild species and they may also be from complementary groups of germplasm such as Neotuberosum and Andigena to exploit yield heterosis. The aim is to generate genetical variation on which phenotypic selection can be practised over a number of vegetative generations for clones with as many desirable characteristics as possible for release as new cultivars. The choice of parents is important because breeding can never simply be a number game. If each of the 100 or so countries listed in the *World Catalogue of Potato Varieties* (Hils and Pieterse, 2005) had 10 breeding programmes each raising 100 000 seedlings annually, the total global effort would be 100 000 000 seedlings. However, crossing the 4000 cultivars in the catalogue in all possible combinations would generate 7 998 000 progenies for evaluation, and 500 seedlings of each would give a staggering total of 3 999 000 000. By contrast, a phenotypic assessment of 4000 cultivars is feasible, and so is a genotypic assessment of diversity and content with molecular markers. Hence, breeders can now think in terms of capturing allelic diversity in a smaller core set of parents and of using association (linkage disequilibrium) genetics to choose parents genotypically as well as phenotypically (Simko, 2004). They can also use genetic distance based on molecular markers (Powell et al., 1991) to complement coancestry/pedigree analysis (Tarn et al., 1992; Gopal and Oyama, 2005) to avoid closely related parents, and hence inbreeding depression, and to ensure genetical variation for continued progress. Both analyses are required as clustering based on molecular markers can be different to that based on pedigree (Sun et al., 2003).

As genetic knowledge accumulates, it will be possible to choose parents for use in pair crosses such that one or both parents have desired major genes and quantitative trait loci (QTLs) alleles of large effect. Major genes have been mapped for flesh, skin and flower colour, for tuber shape and eye depth, and for resistances to late blight, nematodes, viruses such as potato virus (PV)X, PVY, PVA, PVM and PVS and wart. QTLs of large effect have been mapped for maturity and resistances to late blight, Verticillium wilt, potato cyst nematodes and potato leaf roll virus (PLRV). By contrast, many economically important traits still appear to be complex polygenic traits and these include dormancy, dry matter and starch content, fry colour, resistance to *Erwinia*, tuberization and yield. For these traits, breeders will still have to rely on phenotypic data and use knowledge of offspring–midparent regressions to determine crossing strategy. A statistically significant regression is evidence of heritable variation, and the slope of the regression line is a measure of heritability. With a highly heritable trait like fry colour, the midparent value is a good predictor of the mean performance of the offspring and a few carefully chosen crosses can be made (Bradshaw et al., 2000). By contrast, with only a moderately heritable trait such as yield, offspring mean is less predictable and more crosses need to be made to ensure that they include the best possible.

It is not clear how quickly information about expressed genes from microarray experiments and gene sequence data will translate into genes for use in breeding programmes. But again, it cannot simply be a number game. If the potato turns out to have 30 000 genes, as few as two alleles per locus (one desirable, the other undesirable) would generate a large enough number of combinations to be infinite for all practical purposes. Hence, in the genomics age, gene discovery will need to be targeted at those loci likely to have the biggest social and economic impact, and understanding the genetic control of key biochemical pathways may result in the fastest progress in the immediate future.

8.7.2 Early generations

The programme at the Scottish Crop Research Institute (SCRI) before 1982 was typical in its handling of the early generations (Bradshaw and Mackay, 1994). Visual selection reduced the number of potential cultivars from 100 000 in the seedling generation (SG) in the glasshouse to 40 000 spaced plants at a high-grade seed site in the first clonal generation (FCG), then to 4000 four-plant plots at the seed site in the second clonal generation (SCG) and finally to 1000 clones in replicated yield trials at a ware site in the third clonal generation (TCG). Several independent reviews concluded that such intense early-generation visual selection was very ineffective (Tai and Young, 1984; Caligari, 1992; Tarn et al., 1992; Bradshaw and Mackay, 1994).

One solution to the problem is the breeding strategy developed at SCRI, which avoids the ineffective practice of intense early-generation visual selection between seedlings in a glasshouse and spaced plants at a seed site (Bradshaw et al., 2003). Once pair crosses have been made, progeny tests are used to discard whole progenies before starting conventional within-progeny selection at the unreplicated small-plot stage. Clones are also visually selected from the best progenies for use as parents in the next cycle of crosses whilst they are multiplied to provide enough tubers for assessment of their yield and quality. Midparent values, as well as progeny tests, are then used to select between the resultant crosses. Material from other breeding programmes can be included in the parental assessments and progeny tests and used in the next cycle of crosses if superior. Finally, in seeking new cultivars, the number of clones on which selection is to be practised can be increased by sowing more true seed of the best progenies, but without selection until the small-plot stage. Currently, seedling progeny tests are used at SCRI for resistance to late blight [*P. infestans* (Mont.) de Bary], resistance to the white potato cyst nematode [*Globodera pallida* (Stone)] and tuber yield and appearance, as visually assessed by breeders. Tuber progeny tests are used for fry colour and a second visual assessment of tuber yield and appearance. However, breeders wishing to use the scheme will need to adjust the use of progeny tests according to the economic importance of traits in their situations. The use of progeny tests for key traits also means that full-sib family selection can be operated on a 3-year cycle for these traits, an improvement on the practice of individual clonal selection after a further six vegetative generations (i.e. not using potential cultivars as new parents until they are entered into official National List Trials).

8.7.3 Intermediate and later generations

SCRI can again be considered typical in being able to handle 1000 clones in the first year of replicated yield trials at a ware site. However, the relatively slow rate of natural vegetative reproduction (approximately eight-fold per year under SCRI conditions), together with the complicated logistics of accurately assessing 1000 or more clones for a very large number of traits, meant that another 6 years elapsed before one or a few potential cultivars could be confidently entered into official statutory (National List) trials. During this period, decreasing numbers of selected clones were grown in increasingly sophisticated trials over as wide a geographical range as economics permitted and breeding objectives demanded. During these intermediate and final stages of selection, the production of seed tubers was separated from the trials that were grown under ware conditions, designed as far as possible to approximate to those of good commercial practice. In addition to yield and agronomic performance, clones undergoing selection were assessed for their cooking and processing characteristics and tested for their resistances to numerous pests and diseases. Different breeders have no doubt given different priorities to different traits according to their situation. However, the selection criteria and testing procedures in their multitrait, multistage schemes will have been largely governed by practical considerations and experience of the reliability of the various tests used, rather than by genetical knowledge, such as heritabilities or genetical correlations between traits. It might therefore still be worthwhile trying to develop a more robust decision-making process based on multitrait, multistage selection theory and estimated genetical parameters (Bradshaw and Mackay, 1994).

8.7.4 Genetic knowledge and molecular-marker-assisted selection

As knowledge increases about the number and chromosomal locations of genes affecting economically important traits, breeders should be able to design better breeding programmes. As well as selecting parents that complement one another genotypically, they will be able to determine the seedling population size required for certainty of finding the desired genotype, or more realistically, the number of cycles of crossing and selection required before this is achievable in practice in the size of population they can handle. A big impact on the efficiency and rate of progress would be the identification of superior clones genotypically as seedlings in the glasshouse and the use of modern methods of rapid multiplication to progress them to commercialization. This will require molecular-marker-assisted selection or preferably direct recognition of the desired allele at a genetic locus, as has recently been achieved for the *RB* gene for late blight resistance from *S. bulbocastanum* (Colton et al., 2006). Progress has been slow but is expected to increase, and some possibilities can be found in the literature. Kasai et al. (2000) have developed sequence characterized amplified region (SCAR) markers to the PVY resistance gene *Ry* (from Andigena, on chromosome 11), which should allow marker-assisted selection as they showed high accuracy for detection of the *Ry* gene, and one marker RYSC3 was generated only in genotypes carrying *Ry*, namely 14 of 103 breeding lines and cultivars with diverse genetic backgrounds. Bakker et al. (2004) have identified an AFLP marker (EM1) which co-segregates with the *H1* gene for resistance to *Globodera rostochiensis* pathotype Ro1. EM1 and *H1* were present in 19 resistant cultivars and absent from 26 susceptible ones. However, Bakker et al. (2004) recommended conversion to a cleaved amplified polymorphic sequence (CAPS) marker for use in marker-assisted selection as such markers are cheaper and easier to handle. A marker linked to the QTL for *G. pallida* resistance on chromosome 5 was converted to a single-locus PCR-based marker and shown to detect the presence of the QTL in diploid and tetraploid potato germplasm (Bryan et al., 2002). As there was good evidence that it was specific to an introgressed segment of DNA from *S. vernei*, it should prove useful in marker-assisted selection for the QTL.

8.8 BREEDING CULTIVARS FOR TPS

Breeding cultivars for TPS was started by CIP in 1972 with the aim of high yields and acceptable uniformity. None of the current breeding methods can deliver genetic uniformity, and hence all of them involve selection for acceptable uniformity. Methods and progress have been reviewed by Golmirzaie et al. (1994) and Ortiz (1997). Breeding strategy has to take account of the reproductive biology of the potato and certain genetical considerations. Clearly, flower production and berry and seed set are important, but as the transport of true seed is relatively cheap, TPS can be produced in a favourable environment which is distant from where it will be evaluated and grown. Current TPS breeding aims to produce tetraploid cultivars, from $4x \times 4x$ crosses in which heterosis is exploited between Tuberosum and Andigena (Simmonds, 1997), or from $4x \times 2x$ crosses in which the $2x$ parent produces a high frequency of $2n$ gametes by FDR so that 83% of its heterozygosity is transmitted to the offspring (Golmirzaie et al., 1994; Clulow et al., 1995), or from $2x \times 2x$ crosses in which both parents produce $2n$ gametes, again with a very high frequency of $2n$ pollen produced by FDR (otherwise the offspring will contain more diploids than tetraploids) (Ortiz and Peloquin, 1991). However, Simmonds (1997) argued that diploid TPS cultivars should not be ruled out for the future, and De,Maine (1996) produced TPS families of long-day-adapted Phureja which appeared as uniform for tuber size and shape as selected clones. Open-pollination of diploids will result in almost 100% outcrossing because of self-incompatibility, whereas open-pollination of tetraploids will normally result in varying amounts of self-pollination and inbreeding depression. The inbreeding depression of the tetraploids may not be outweighed by their apparent intrinsic yield advantage over diploids. Furthermore, the seed fertility (seed yield) of diploids is greater than that of tetraploids. Whether or not diploid inbred lines and true F1 hybrids could be produced in the future for maximum heterosis and uniformity is a matter for further research. The inbreeding depression from selfing also means that hand-pollinated tetraploids are superior to open-pollinated ones, but their production is more expensive because of the cost of emasculation of the flowers of the female parent (Simmonds, 1997). The use of male sterility that does not involve loss of visits by bumblebees is an attractive proposition, e.g. the use of protoplast fusion to transfer cytoplasmic male sterility into a TPS parental line (Golmirzaie et al., 1994). Another possibility is the use of a dominant marker in the male parent so that hybrids can be selected from the seedlings for transplanting and tuber production (Ortiz, 1997).

The genetical basis of heterosis and inbreeding depression is also an issue, particularly for tetraploid TPS cultivars. The theory of inbreeding and crossbreeding in tetraploids is more complicated than in diploids. One generation of random mating does not remove the effects of inbreeding in tetraploids and double-cross hybrids can be superior to single-cross hybrids (Bradshaw, 1994). When selfing for a number of generations is started with an individual with four non-identical alleles A1A2A3A4, individuals with four and three alleles quickly disappear, whereas those with two different alleles initially increase in frequency before declining. The issue for TPS breeding is whether or not interactions between four and three different alleles are important compared with interactions between two different alleles. The evidence is inconclusive, but Golmirzaie et al. (1994) concluded that crossing unrelated genetic material as a means of exploiting heterosis in a TPS programme offers promising perspectives for rapid progress. Hence, tetraploid crosses between Tuberosum and Andigena do make sense, as does the use of diploid hybrids between Tuberosum and wild species as the diploid parents in $4x \times 2x$ crosses.

8.9 GENETICALLY MODIFIED POTATOES

As explained earlier, the molecular cloning of natural resistance genes and their transfer into well-adapted but susceptible cultivars is an attractive and efficient way of improving existing cultivars. Furthermore, where the control of biochemical pathways of interest for improving traits is understood, down-regulation of gene expression using antisense technology has proved useful. For example, bruise-resistant potatoes can be produced by down-regulation of polyphenol oxidase (Bachem et al., 1994), high-carotenoid (pigments with health-promoting attributes) potatoes by down-regulation of zeaxanthin epoxidase (Romer et al., 2002), high-amylopectin starch by down-regulation of granule bound starch synthase (Visser et al., 1991) and high-amylose starch by down-regulating two starch-branching enzymes, A and B (Schwall et al., 2000). The genetic modification of existing potato cultivars also offers the possibility of introducing genes not present in cultivated potatoes and their wild relatives and hence of novel biochemistry and desirable traits. Specific examples can be found in Chapter 30 (Millam, this volume) as well as more general issues in Section 7 on 'Biotechnology', but a number of points are worth making here in the context of breeding strategies.

Potato transformation using *Agrobacterium*-mediated systems (e.g. *Agrobacterium tumefaciens* Ti plasmid-mediated gene transfer) is relatively straightforward, but regeneration can still be a problem with some genotypes (Dale and Hampson, 1995). Briefly, the gene of interest is incorporated into the bacterial plasmid along with a selectable marker such as resistance to an antibiotic, the bacterium is co-cultured with freshly-cut tuber discs or leaf or internode explants of the potato, and regeneration of shoots with the selectable marker takes place in plant tissue culture in the presence of the selectable agent. The products of transformation require screening in order to select the best transformants for commercialization, including demonstration of substantial equivalence to the parent cultivar (Davies, 2002). Commercialization also involves demonstration that it is safe both to grow and eat the genetically modified potatoes. The gradual elimination of antibiotic resistant markers will help to deal with one of the safety issues, but the demonstration of economic, environmental and health benefits will be crucial in convincing sceptical consumers in some countries of the value of genetically modified potatoes. The best strategy will be to modify high-yielding and well-adapted cultivars for specific traits. Without going into the details in this chapter, it is nevertheless worth mentioning the kinds of improvements that can already be made.

Genes that encode proteins from the bacterium *Bacillus thuringiensis* have been shown to confer resistance to the Colorado potato beetle (CPB, *Leptinotarsa decemlineata* Say) and to the potato tuber moth (*Phthorimaea operculella*) (Mohammed et al., 2000; Duncan et al., 2002; Davidson et al., 2004). Genes that encode cysteine proteinase inhibitors (cystatins) have shown promise in conferring resistance to potato cyst nematodes (*G. rostochiensis* and *G. pallida*) (Urwin et al., 2003). Replicase and coat protein genes from PLRV and PVY, respectively, provide resistance to these viruses (Duncan et al., 2002). A gene encoding a chicken lysozyme enzyme enhances resistance to blackleg and soft rot caused by infection with *Erwinia carotovora* subsp. *atroseptica* (Serrano et al., 2000). Transgenes that confer resistance to more than one disease would be particularly useful. A recent example is expression of a gene for a derivative of the antimicrobial peptide dermaseptin B1, from the arboreal frog *Phyllomedusa bicolor*, which has been shown to increase resistance to diseases such as late blight, dry rot and pink rot and to markedly extend the shelf life of tubers (Osusky et al., 2005). Protein content has been increased and amino acid composition improved by expression of a non-allergenic seed albumin gene (*AmA1*) from *Amaranthus hypochondriacus* (Chakraborty et al., 2000), and carbohydrate composition has been improved by inulin production from the expression of fructosyltransferases from globe artichoke (*Cynara scolymus*) (Hellwege et al., 2000). Inulins enrich food with dietary fibre and replace sugar and fat. The conversion of sucrose to glucose and fructose, and hence cold sweetening (the cause of dark and bitter tasting fry products), has been minimized by expressing a putative vacuolar invertase inhibitor protein from tobacco (Greiner et al., 1999).

8.10 ACHIEVING DURABLE DISEASE AND PEST RESISTANCE

Most potato-breeding programmes will have disease and pest resistance as an objective. In any particular programme, the breeder must be realistic in deciding priorities for improving resistance and recognize that the commercial success of new cultivars is usually determined by their yield and quality. For major disease and pest problems, new and durable forms of resistance will be actively sought, together with ways of incorporating them into new cultivars as quickly as possible. For those that are less important, potential cultivars will simply be screened to avoid extreme susceptibility. This can be done in tests, or when natural epidemics of minor diseases occur. The prerequisites for successful breeding are a source of durable resistance, a reliable screen for resistance using the most appropriate isolate(s) of the pathogen and finally an understanding of the inheritance of resistance. Breeding for disease resistance in potato has been reviewed by Jansky (2000).

Strategies are required for achieving durability of resistance because this is not a property of the new cultivar per se. It depends on the population genetics of the pathogen and the epidemiology of the disease, and whether or not isolates of the pathogen that can overcome resistance genes are less fit in the absence of those genes than other isolates so that resistance genes can be appropriately managed in the environment. Durability is still difficult to predict with certainty and there have been mixed fortunes in the past. The major gene resistances to PVY have proved durable and so have those to PVX despite the occurrence of resistance-breaking strains (Jones, 1985). The *H1* gene has remained effective against *G. rostochiensis* in Britain because Ro1 is still the main pathotype, but its widespread deployment has encouraged the spread of *G. pallida*. Transgenic Bt-protein-producing potatoes provide an example of management practices recommended to extend the durability of resistance to CPB. Monsanto advised growers to plant 20% of their field area with a non-Bt-containing potato as a refuge for susceptible beetles and to rotate crops and fields on which transgenic potatoes were grown (Duncan et al., 2002). The *S. demissum*-derived R-genes for resistance to late blight failed to provide durable resistance either singly or in combination due to the evolution of new races of *P. infestans* (Malcolmson, 1969). As a consequence, many breeders started to select for quantitative field resistance either by using races of *P. infestans* compatible with the R-genes present in their material or by creating R-gene-free germplasm so that screening could be done with any race (Toxopeus, 1964; Black, 1970; Wastie, 1991; Ortiz, 2001). A good example is the well-known Group B germplasm from CIP, which is being made available to developing countries through their National Agricultural Research Systems (Trognitz et al., 2001). Forbes et al. (2005) have been able to demonstrate that quantitative resistance in cultivars from the Netherlands, Scotland, Peru, Colombia and Mexico was relatively stable across locations representing a wide range of latitudes and altitudes. Although this is an encouraging result, it does not guarantee durability over time. Nevertheless, it does seem worthwhile trying to combine quantitative resistance with the other traits required for commercial success, such as early maturity, yield and the demanding quality specifications in developed countries. Currently, there is much interest and debate over whether or not the R-genes being found in other wild species will be more durable per se or can be deployed in a more durable way such a multiline produced by map-based cloning and *Agrobacterium*-mediated transformation of a popular but susceptible cultivar (Huang, 2005; Smilde et al., 2005). Finally, there is the question of whether or not transgenic resistance can be durable. The *R3a* gene in cultivar Pentland Ace has been cloned (Huang et al., 2005) and the corresponding avirulence gene, *Avr3a*, in *P. infestans* has been shown to encode a 147 amino acid extracellular protein, which is recognized in the cytoplasm and triggers *R3a*-dependent cell death (Armstrong et al., 2005). A transgenic solution to resistance would be to produce a genetically modified potato containing *R3a* in which a pathogen-inducible promoter was linked to *Avr3a* so that cell death would only occur in those cells infected with *P. infestans*. Not surprisingly, the search for pathogen-inducible promoters is currently an active area of research, as is the expression of pathogenesis-related (PR) genes and peptides or proteins with antimicrobial activity in plants (Gurr and Rushton, 2005).

8.11 CONCLUSIONS

The evolution of the modern potato since domestication can be seen in Fig. 8.2, which compares the tubers of wild species and primitive cultivated species with those of a modern cultivar. Worldwide, new cultivars will continue to come from crosses between pairs of parents with complementary features but adapted to local growing conditions and end uses. However, breeders should seek to make full use of the diversity that exists in the world's major potato germplasm collections that comprise both potato cultivars native to Latin America and their cross-compatible wild species. A major strategic question that needs answering is when to use targeted gene transfer by molecular-marker-assisted introgression or gene cloning and when to use wider base-broadening approaches. Successful cultivars should be genetically modified to achieve further improvements by the introduction of genes not present in cultivated potatoes and their wild relatives. Other strategic decisions that need to be made are when to produce cultivars for propagation by true seed rather than tubers and how to achieve durable resistance to diseases and pests. Increasing genetic knowledge is expected to impact on both the design and execution of breeding programmes.

Fig. 8.2. Evolution of the modern potato.

REFERENCES

Almekinders C.J.M., A.S. Chilver and H.M. Renia, 1996, *Potato Res.* 39, 289.

Armstrong M.R., S.C. Whisson, L. Pritchard, J.I.B. Bos, E. Venter, A.O. Avrova, A.P. Rehmany, U. Bohme, K. Brooks, I. Cherevach, N. Hamlin, B. White, A. Fraser, A. Lord, M.A. Quail, C. Churcher, N. Hall, M. Berriman, S. Huang, S. Kamoun, J.L. Beynon and P.J.R. Birch, 2005, *Proc. Natl. Acad. Sci. U.S.A.* 102, 7766.

Bachem C.W.B., G.J. Speckmann, P.C.G. Van Der Linde, F.T.M. Verheggen, M.D. Hunt, J.C. Steffens and M. Zabeau, 1994, *Biotechnology* 12, 1101.

Bakker E., U. Achenbach, J. Bakker, J. van Vliet, J. Peleman, B. Segers, S. van der Heijden, P. van der Linde, R. Graveland, R. Hutten, H. van Eck, E. Coppoolse, E. van der Vossen, J. Bakker and A. Goverse, 2004, *Theor. Appl. Genet.* 109, 146.

Barone A., 2004, *Am. J. Potato Res.* 81, 111.

Barone A., A. Sebastiano, D. Carputo, F. della Rocca and L. Frusciante, 2001, *Theor. Appl. Genet.* 102, 900.

Black W., 1970, *Am. Potato J.* 47, 279.

Bonierbale M.W., R.L. Plaisted and S.D. Tanksley, 1988, *Genetics* 120, 1095.

Bonierbale M.W., R.L. Plaisted and S.D. Tanksley, 1993, *Theor. Appl. Genet.* 86, 481.

Bonierbale M.W., R. Simon, D.P. Zhang, M. Ghislain, C. Mba and X.-Q. Li, 2003, In: H.J. Newbury (ed.), *Plant Molecular Breeding*, p. 216. Blackwell, Oxford.

Bradshaw J.E., 1994, In: J.E. Bradshaw and G.R. Mackay (eds), *Potato Genetics*, p. 71. CAB International, Wallingford, UK.

Bradshaw J.E., M.F.B. Dale and G.R. Mackay, 2003, *Theor. Appl. Genet.* 107, 36.

Bradshaw J.E. and G.R. Mackay, 1994, In: J.E. Bradshaw and G.R. Mackay (eds), *Potato Genetics*, p. 467. CAB International, Wallingford, UK.

Bradshaw J.E. and G. Ramsay, 2005, *Euphytica* 146, 9.

Bradshaw J.E., D. Todd and R.N. Wilson, 2000, *Theor. Appl. Genet.* 100, 772.

Brown C.R., 1993, *Am. Potato J.* 70, 725.

Brown J., M.F.B. Dale and G.R. Mackay, 1996, *J. Agric. Sci., Camb.* 126, 441.

Bryan G.J., K. McLean, J.E. Bradshaw, W.S. De Jong, M. Phillips, L. Castelli and R. Waugh, 2002, *Theor. Appl. Genet.* 105, 68.

Buso J.A., L.S. Boiteux and S.J. Peloquin, 1999, *Euphytica* 109, 191.

Caligari P.D.S., 1992, In: P. Harris (ed.), *The Potato Crop*, 2nd edition, p. 334. Chapman and Hall, London.

Camadro E.L., D. Carputo and S.J. Peloquin, 2004, *Theor. Appl. Genet.* 109, 1369.

Carputo D. and A. Barone, 2005, *Ann. Appl. Biol.* 146, 71.

Carroll C.P., 1982, *J. Agric. Sci., Camb.* 99, 631.

Carroll C.P. and M.J. De,Maine, 1989, *Potato Res.* 32, 447.

Chakraborty S., N. Chakraborty and A. Datta, 2000, *Proc. Natl. Acad. Sci. U.S.A.* 97, 3724.

Chandra S., Z. Huaman, S. Hari Krishna and R. Ortiz, 2002, *Theor. Appl. Genet.* 104, 1325.

Chase S.S., 1963, *Can. J. Genet. Cytol.* 5, 359.

Clulow S.A., J. McNicoll and J.E. Bradshaw, 1995, *Theor. Appl. Genet.* 90, 519.

Colton L.M., H.I. Groza, S.M. Wielgus and J. Jiang, 2006, *Crop Sci.* 46, 589.

Cubillos A.G. and R.F. Plaisted, 1976, *Am. Potato J.* 53, 143.

Dale P.J. and K.K. Hampson, 1995, *Euphytica* 85, 101.

Davidson M.M., R.C. Butler, S.D. Wratten and A.J. Conner, 2004, *Ann. Appl. Biol.* 145, 271.

Davies H.V., 2002, In: V. Valpuesta (ed.), *Fruit and Vegetable Biotechnology*, p. 222. Woodhead Publishing Limited, Cambridge.

De,Maine M.J., 1996, *Potato Res.* 39, 323.

Dodds K.S., 1965, In: J.B. Hutchinson (ed.), *Essays in Crop Plant Evolution*, p. 123. Cambridge University Press, Cambridge.

Duncan D.R., D. Hammond, J. Zalewski, J. Cudnohufsky, W. Kaniewski, M. Thornton, J.T. Bookout, P. Lavrik, G.J. Rogan and J. Feldman-Riebe, 2002, *HortScience* 37, 275.

Falconer D.S. and T.F.C. Mackay, 1996, *Introduction to Quantitative Genetics*, 4th edition. Longman, Harlow, England.

Forbes G.A., M.G. Chacon, H.G. Kirk, M.A. Huarte, M. Van Damme, S. Distel, G.R. Mackay, H.E. Stewart, R. Lowe, J.M. Duncan, H.S. Mayton, W.E. Fry, D. Andrivon, D. Ellisseche, R. Pelle, H.W. Platt, G. MacKenzie, T.R. Tarn, L.T. Colon, D.J. Budding, H. Lozoya-Saldana, A. Hernandez-Vilchis and S. Capezio, 2005, *Plant Pathol.* 54, 364.
Gaur P.C. and S.K. Pandey, 2000, In: S.M. Paul Khurana, G.S. Shekhawat, B.P. Singh and S.K. Pandey (eds), *Potato, Global Research and Development-Volume 1*, p. 52. Indian Potato Association, Shimla, India.
Glendinning D.R., 1969, *Eur. Potato J.* 12, 13.
Glendinning D.R., 1976, *Potato Res.* 19, 27.
Golmirzaie A.M., P. Malagamba and N. Pallais, 1994, In: J.E. Bradshaw and G.R. Mackay (eds), *Potato Genetics*, p. 499. CAB International, Wallingford, UK.
Goodrich C.E., 1863, *Trans. N. Y. State Agric. Soc.* 23, 89.
Gopal J. and K. Oyama, 2005, *Euphytica* 142, 23.
Greiner S., T. Rausch, U. Sonnewald and K. Herbers, 1999, *Nat. Biotechnol.* 17, 708.
Gurr S.J. and P.J. Rushton, 2005, *Trends Biotechnol.* 23, 283.
Haverkort A.J. and C. Grashoff, 2004, In: D.K.L. MacKerron and A.J. Haverkort (eds), *Decision Support Systems in Potato Production*, p. 199. Wageningen Academic Publishers, Wageningen.
Hawkes J.G., 1990, *The Potato: Evolution, Biodiversity and Genetic Resources.* Belhaven Press, London.
Hawkes J.G. and J. Francisco-Ortega, 1993, *Euphytica* 70, 1.
Haynes K.G. and B.J. Christ, 1999, *Plant Breed.* 118, 431.
Haynes K.G. and W. Lu, 2005, In: M.K. Razdan and A.K. Mattoo (eds), *Genetic Improvement of Solanaceous Crops Volume I: Potato*, p. 101. Science Publishers, Inc., Enfield.
Hellwege E.M., S. Czapla, A. Jahnke, L. Willmitzer and A.G. Heyer, 2000, *Proc. Natl. Acad. Sci. U.S.A.* 97, 8699.
Hermsen J.G.T., 1994, In: J.E. Bradshaw and G.R. Mackay (eds), *Potato Genetics*, p. 515. CAB International, Wallingford, UK.
Hermundstad S.A. and S.J. Peloquin, 1987, In: G.J. Jellis and D.E. Richardson (eds), *The Production of New Potato Varieties*, p. 197. Cambridge University Press, Cambridge.
Hijmans R.J., 2001, *Am. J. Potato Res.* 78, 403.
Hils U. and L. Pieterse, 2005, *World Catalogue of Potato Varieties.* Agrimedia GmbH, Bergen/Dumme, Germany.
Hosaka K., 2004, *Am. J. Potato Res.* 81, 153.
Hospital F., 2003, In: H.J. Newbury (ed.), *Plant Molecular Breeding*, p. 30. Blackwell, Oxford.
Huaman Z., A. Golmirzaie and W. Amoros, 1997, In: D. Fuccillo, L. Sears and P. Stapleton (eds), *Biodiversity in Trust: Conservation and Use of Plant Genetic Resources in CGIAR Centres*, p. 21. Cambridge University Press, Cambridge, UK.
Huaman Z., R. Hoekstra and J.B. Bamberg, 2000a, *Am. J. Potato Res.* 77, 353.
Huaman Z., R. Ortiz and R. Gomez, 2000b, *Am. J. Potato Res.* 77, 183.
Huaman Z., R. Ortiz, D. Zhang and F. Rodriguez, 2000c, *Crop Sci.* 40, 273.
Huang S., 2005, Discovery and characterization of the major late blight resistance complex in potato, Thesis. Wageningen University, The Netherlands.
Huang S., E.A.G. van der Vossen, H. Kuang, V.G.A.A. Vleeshouwers, N. Zhang, T.J.A. Borm, H.J. van Eck, B. Baker, E. Jacobsen and R. Visser, 2005, *Plant J.* 42, 251.
Iovene M., A. Barone, L. Frusciante, L. Monti and D. Carputo, 2004, *Theor. Appl. Genet.* 109, 1139.
Jansky S., 2000, In: J. Janick (ed.), *Plant Breeding Reviews*, Volume 19, p. 69. John Wiley & Sons, New York.
Jansky S., 2006, *Plant Breed.* 125, 1.
Jansky S.H., G.L. Davis and S.J. Peloquin, 2004, *Am. J. Potato Res.* 81, 335.
Jansky S.H., G.L. Yerk and S.J. Peloquin, 1990, *Plant Breed.* 104, 290.
Jin L.P., D.Y. Qu, K.Y. Xie, C.S. Bian and S.G. Duan, 2004, Potato germplasm, breeding studies in China. Proceedings of the Fifth World Potato Congress, p. 175. Kunming, China.
Jones R.A.C., 1985, *Plant Pathol.* 34, 182.
Kasai K., Y. Morikawa, V.A. Sorri, J.P.T. Valkonen, C. Gebhardt and K.N. Watanabe, 2000, *Genome* 43, 1.
Knight T.A., 1807, *Trans. Hort. Soc. Lond.* 1, 57.
Lang J., 2001, *Notes of a Potato Watcher.* Texas A&M University Press, College Station.
Mackay G.R., 1996, *Potato Res.* 39, 387.

Malcolmson J.F., 1969, *Trans. Br. Mycol. Soc.* 53, 417.
Maris B., 1989, *Euphytica* 41, 163.
McGregor C.E., R. van Treuren, R. Hoekstra and T.J.L. van Hintum, 2002, *Theor. Appl. Genet.* 104, 146.
Mohammed A., D.S. Douches, W. Pett, E. Grafius, J. Coombs, W. Liswidowati, W. Li and M.A. Madkour, 2000, *J. Econ. Entomol.* 93, 472.
Naess S.K., J.M. Bradeen, S.M. Wielgus, G.T. Haberlach, J.M. McGrath and J.P. Helgeson, 2000, *Theor. Appl. Genet.* 101, 697.
Ortiz R., 1997, *Plant Breed. Abstr.* 67, 1355.
Ortiz R., 1998, In: J. Janick (ed.), *Plant Breeding Reviews*, Volume 16, p. 15. John Wiley & Sons, New York.
Ortiz R., 2001, In: H.D. Cooper, C. Spillane and T. Hodgkin (eds), *Broadening the Genetic Base of Crop Production*, p. 181. CABI Publishing, Wallingford, Oxon.
Ortiz R. and S.J. Peloquin, 1991, *Euphytica* 57, 103.
Osusky M., L. Osuska, W. Kay and S. Misra, 2005, *Theor. Appl. Genet.* 111, 711.
Pandey S.K. and S.K. Kaushik, 2003, In: S.M.P. Khurana, J.S. Minhas and S.K. Pandey (eds), *The Potato – Production and Utilization in Sub-Tropics*, p. 15. Mehta Publishers, New Delhi.
Plaisted R.L. and R.W. Hoopes, 1989, *Am. Potato J.* 66, 603.
Powell W., M.S. Phillips, J.W. McNicol and R. Waugh, 1991, *Ann. Appl. Biol.* 118, 423.
Raker C.M. and D.M. Spooner, 2002, *Crop Sci.* 42, 1451.
Romer S., J. Lubeck, F. Kauder, S. Steiger, C. Adomat and G. Sandmann, 2002, *Metab. Eng.* 4, 263.
Ross H., 1986, *Potato Breeding – Problems and Perspectives. Advances in Plant Breeding*, Volume 13. Paul Parey, Berlin and Hamburg.
Rousselle-Bourgeois F. and P. Rousselle, 1992, *Agronomie* 12, 59.
Schwall G.P., R. Safford, R.J. Westcott, R. Jeffcoat, A. Tayal, Y.C. Shi, M.J. Gidley and S.A. Jobling, 2000, *Nat. Biotechnol.* 18, 551.
Serrano C., P. Arce-Johnson, H. Torres, M. Gebauer, M. Gutierrez, M. Moreno, X. Jordana, A. Venegas, J. Kalazich and L. Holuigue, 2000, *Am. J. Potato Res.* 77, 191.
Simko I., 2004, *Trends Plant Sci.* 9, 441.
Simmonds N.W., 1969, Scottish Plant Breeding Station, Forty-Eighth Annual Report 1968–69, 18.
Simmonds N.W., 1995, In: J. Smartt and N.W. Simmonds (eds), *Evolution of Crop Plants*, 2nd edition, p. 466. Longman Scientific & Technical, Singapore.
Simmonds N.W., 1997, *Potato Res.* 40, 191.
Smilde W.D., G. Brigneti, L. Jagger, S. Perkins and J.D.G. Jones, 2005, *Theor. Appl. Genet.* 110, 252.
Spooner D.M. and R.J. Hijmans, 2001, *Am. J. Potato Res.* 78, 237.
Spooner D.M., K. McLean, G. Ramsay, R. Waugh and G.J. Bryan, 2005a, *Proc. Natl. Acad. Sci. U.S.A.* 102, 14694.
Spooner D.M., J. Nunez, F. Rodriguez, P.S. Naik and M. Ghislain, 2005b, *Theor. Appl. Genet.* 110, 1020.
Sun G., G. Wang-Pruski, M. Mayich and H. De Jong, 2003, *Theor. Appl. Genet.* 107, 110.
Tai G.C.C., 1994, In: J.E. Bradshaw and G.R. Mackay (eds), *Potato Genetics*, p. 109. CAB International, Wallingford, UK.
Tai G.C.C. and H. De Jong, 1980, *Can. J. Genet. Cytol.* 22, 277.
Tai G.C.C. and D.A. Young, 1984, *Am. Potato J.* 61, 419.
Tarn T.R. and G.C.C. Tai, 1983, *Theor. Appl. Genet.* 66, 87.
Tarn T.R., G.C.C. Tai, H. De Jong, A.M. Murphy and J.E.A. Seabrook, 1992, In: J. Janick (ed.), *Plant Breeding Reviews*, Volume 9, p. 217. John Wiley & Sons, New York.
Tek A.L., W.R. Stevensen, J.P. Helgeson and J. Jiang, 2004, *Theor. Appl. Genet.* 109, 249.
Toxopeus H.J., 1964, *Euphytica* 13, 206.
Trognitz B.R., M. Bonierbale, J.A. Landeo, G. Forbes, J.E. Bradshaw, G.R. Mackay, R. Waugh, M.A. Huarte and L. Colon, 2001, In: H.D. Cooper, C. Spillane and T. Hodgkin (eds), *Broadening the Genetic Base of Crop Production*, p. 385. CABI Publishing, Wallingford, Oxon.
Ullstrup A.J., 1972, *Annu. Rev. Phytopathol.* 10, 37.
Urwin P.E., J. Green and H.J. Atkinson, 2003, *Mol. Breed.* 12, 263.
Visser R.G.F., I. Somhorst, G.J. Kuipers, N.J. Ruys, W.J. Feenstra and E. Jacobsen, 1991, *Mol. Gen. Genet.* 225, 289.
Wastie R.L., 1991, *Adv. Plant Pathol.* 7, 193.
Wenzel G., O. Schieder, T. Przewozny, S.K. Sopory and G. Melchers, 1979, *Theor. Appl. Genet.* 55, 49.

Chapter 9

Genomics

Glenn J. Bryan

Genome Dynamics Programme, Scottish Crop Research Institute, Invergowrie, Dundee DD2 5DA, United Kingdom

9.1 INTRODUCTION

Cultivated potato (*Solanum tuberosum* ssp. *tuberosum*) is a key member of the Solanaceae family and, along with tomato, pepper, eggplant and tobacco, forms a group of important crop species which have been well studied genetically since the first use of DNA-based markers in crop plants (Chapter 5, Gebhardt, this volume). Tomato (*Solanum lycopersicum*) has a similar genome size to potato (Arumuganathan and Earle, 1991) and is seen as a genetic and genomic model for the Solanaceae family, doubtless largely because of it being both diploid and self-fertile. Genetical research in potato is hampered by its high levels of heterozygosity and autotetraploidy and that it suffers acutely from inbreeding depression. To counter this problem, most genetic mapping in potato is performed in diploid populations produced by the crossing of heterozygous diploid parents (Chapter 5, Gebhardt, this volume). Moreover, significant progress has been made in both linkage and quantitative trait locus (QTL) mapping at the tetraploid level although this type of approach has not been broadly adopted. Tomato and potato genetic maps show extremely high levels of colinearity and have been shown to differ by only five major inversions (Tanksley et al., 1992). A great deal of trait genetic analysis has been performed in potato, for tuber traits (Chapter 6, Van Eck, this volume) and pest/disease resistance (Chapter 6, Simko et al., this volume). The early history of potato molecular genetics was characterized by many successful gene cloning efforts that were based on several different approaches and that did not rely on the existence of detailed molecular genetic maps. The earliest genes from potato were those thought to be involved in important tuber biochemical processes and often involved the use of sequence information from other plants, notably *Arabidopsis*. More recent map-based cloning efforts have had the concomitant effects of facilitating the development of genomic resources for potato, such as large-insert genomic libraries and localized physical maps. More latterly, there has been the development of considerable quantities of expressed sequence tag (EST) data, facilitating the availability of potato microarrays for gene expression studies. In summary, potato genomics has really 'taken off' since the start of the new millennium, and potato is now equal to almost any other crop plant in terms of the genome resources available to those that work on the world's fourth most important crop plant. This chapter outlines the history of potato molecular genetics and the notable recent advances in potato genomics and development of genomic resources made to date.

Potato Biology and Biotechnology: Advances and Perspectives
D. Vreugdenhil (Editor)

9.2 CHARACTERISTICS OF THE POTATO GENOME

Studies of the organization and evolution of the potato genome are quite few in number. Cytogenetic studies are hampered by the extremely small size of the chromosomes, although recent advances using meiotic tissue for fluorescence in situ hybridization (FISH) and the use of extended DNA 'fibres' have overcome some of the barriers posed by chromosome size (Chapter 10, Gavrilenko, this volume). Cultivated potato behaves as an autotetraploid and has $2n = 4x = 48$ chromosomes. It is well established that potato has a genome size (850–1000 Mb) that is very similar to that of tomato (Arumuganathan and Earle, 1991) and that their genetic maps show extremely high levels of colinearity, having been shown to differ by only five major inversions (Tanksley et al., 1992). Consequently, information on the structure and organization of the tomato genome is likely to be equally applicable to potato. An early study showed that the tomato genome is essentially composed of low copy number sequences that diverge rapidly in evolutionary time (Zamir and Tanksley, 1988). One of the earliest attempts to characterize the potato genome in terms of the amounts of different classes of repetitive DNA was performed by Schweizer et al. (1993), who isolated some of the major highly repeated sequences from potato and determined the distribution and organization of these in the genome of potato cultivars, diploid breeding lines and wild species. These genomic fractions, in combination with ribosomal DNA, represented only 4–7% of the potato genome, suggesting that the potato genome, as a whole, was relatively devoid of highly repetitive DNA sequences. This observation is supportive of the earlier tomato study and augured well for the future of potato genomics and activities such as map-based cloning and physical mapping. There have been surprisingly few attempts to follow up these studies with more detailed studies of genome organization in potato, although recent gene isolation projects have yielded 'bacterial artificial chromosome (BAC)-length' tracts of DNA sequence that can be used to glean insights into potato genome structure. Tomato EST and BAC sequence data have been used to estimate that the tomato genome encodes approximately 35 000 genes, largely present in euchromatin, which corresponds to only 25% or so of the DNA in the tomato nucleus (Van der Hoeven et al., 2002). A key observation for tomato and potato genome projects is that the majority of tomato heterochromatin is found in centromeric regions with almost all the euchromatic DNA distally in long uninterrupted tracts (Wang et al., 2005). Stupar et al. (2002) have reported the existence of rDNA-related tandem repeats within highly condensed pericentric heterochromatin.

9.3 GENE ISOLATION

9.3.1 Early gene cloning and expression studies

Potato molecular genetics really began in the mid-1980s, with the construction of the first genetic maps (see Chapter 5, Gebhardt, on mapping in potato) and the isolation of the first potato genes by molecular cloning. This followed the development of the technique of restriction fragment length polymorphism (RFLP) for linkage mapping in human genetic studies (Botstein et al., 1980) and the widespread availability of phage and plasmid vectors for library construction. Thus, the first genomic libraries of potato were constructed, and these were used both for probe generation for mapping studies and for screening for clones containing particular potato genes. The subsequent isolation of many of these genes depended in many cases on knowledge of tuber biochemistry and targeted genes for particular known enzymatic pathways, for example carbohydrate metabolism and important tuber biochemical processes affecting potato quality traits. In many cases, genes were identified from genomic or cDNA libraries of potato using heterologous probes/primers from other plants (e.g. *Arabidopsis*, other crops) or antibodies directed against particular proteins but did not involve the use of molecular genetic maps that were not yet available. Many of these studies were accompanied by expression analysis of the cloned genes. The first potato genes to be isolated included light-inducible genes (Eckes et al., 1986), the patatin storage protein gene (Rosahl et al., 1986) and the gene for granule-bound starch synthase (GBSS; Visser et al., 1989) using a maize cDNA probe. Possibly, the first study of a gene family in potato was performed by Wolter et al. (1988), who examined five genes from the *rbcS* family, three of which were found to be tightly clustered. Nap et al. (1992) found a lambda clone that contained two tandemly arranged patatin genes, demonstrating close physical linkage between two members of the patatin gene family. Taylor et al. (1994, 1995) used a potato stolon tip cDNA library to isolate cDNAs for genes encoding a fructokinase and two beta-tubulins. These genes were analysed for their expression patterns, and it was also established that there were several copies of the beta-tubulin genes in the potato genome, with considerable levels of divergence in their 3′ non-coding sequences. Kang and Hannapel (1995, 1996) isolated and characterized a potato MADS box gene (*POTM1*) that showed high levels of homology to flower-specific homeotic genes of tomato and *Arabidopsis*, indicating that *POTM1* gene is a homologue of the APETALA1 (AP1) gene family. Gene expression data suggested that this gene may function as a transcription factor regulating plant developmental processes in a number of tissue types. Many potato genes were isolated using antibody screening, illustrating how knowledge of potato biochemistry has proved useful for this type of approach. For example, des Francs-Small et al. (1993) were one of the first groups to isolate a potato gene using a specific antibody raised against an abundant mitochondrial protein, nicotinamide adenine dinucleotide (NAD)-dependent formate dehydrogenase, to screen a cDNA expression library. The polymerase chain reaction (PCR) and derived methods have been used to screen libraries for the presence of potato genes. Graeve et al. (1994) screened a cDNA library with primers specific to a glucose-6-phosphate dehydrogenase (*G6PDH*) gene and obtained a full-length cDNA showing high levels of homology to animal and yeast *G6PDH* genes. These are but a very few of the gene isolation projects that were performed, but they give a flavour of the types of gene targeted and the different methods used to isolate them. A key point here is that potato tubers, being relatively large and simple organs, permit the isolation of nucleic acids or proteins involved in physiologically important tuber processes in large quantities, thus making it fairly straightforward to produce antibodies and 'expression' libraries.

Owing to the relative ease of performing stable transformation in potato, considerable effort has been expended in the isolation of potato promoter sequences. Some of these efforts were directed towards the goal of finding promoters that would direct tuber-specific expression of transgenes, thus permitting direct insights into the function of such transgenes in tuber biochemical processes. For example, Herbers et al. (1994) isolated a clone encoding a cathepsin D inhibitor from a potato genomic library. The promoter of this gene was found to drive tuber-specific β-glucuronidase (GUS) gene expression in transgenic plants, and it was also observed that the promoter shared regions of sequence homology with promoter regions of tuber-specific class I patatin genes. These promoters and others with similar properties may eventually find use in transgenic cultivars with improved tuber traits.

9.3.2 Map-based gene isolation

The development of genetic and molecular maps for potato has been reviewed in Chapter 5. These mapping efforts have led to the accumulation of knowledge concerning the trait 'architecture' of a large number of characters, including pest and disease resistance, tuber quality traits, dormancy, tuber shape, eye depth and colour. These mapping endeavours have led to several successful attempts to isolate potato genes using a map-based approach, with most of these being aimed at the isolation of major genes for resistance to the more serious pests and pathogens of potato, the late blight pathogen *Phytophthora infestans* (Mont. de Bary), potato cyst nematodes (PCN) and viruses. These activities have necessitated the development of dense genetic maps around the targeted resistance loci, as well as concomitant generation of genomic resources, chiefly large-insert genomic DNA libraries, such as cosmid or BAC libraries and methods for their screening. A BAC library is a genomic library composed of large (typically ~100–120 kb average size) segments of DNA ligated into a low copy number plasmid specially designed for the accommodation of large inserts. Another consequence of gene cloning efforts has been the increase in our understanding of the structure of the potato genome, through the sequencing of a considerable number of BAC and other genomic clones containing disease resistance loci. A notable example is the sequencing of 19 BAC clones from the *Solanum demissum* R1 locus on chromosome V (Kuang et al., 2005). This study, in conjunction with other studies of the region (Ballvora et al., 2002), has provided invaluable insights into the structure of an important location in the potato genome, a hot spot for disease resistance. It is clear that this region harbours many near-identical resistance gene-like sequences and that the locus is extremely variable with respect to gene copy number and physical length. This suggests that physical mapping in regions of the genome harbouring important disease resistances may be somewhat problematic. In February 2007, there are approximately 50 fully sequenced potato BAC clones in the sequence databases. This number will doubtlessly increase dramatically in the next few years through further gene cloning efforts and through potato genome sequencing projects. These impending resource development efforts will facilitate a comprehensive analysis of genome organization and structure in potato.

9.3.3 Use of candidate gene approaches for gene isolation

Candidate gene approaches for the isolation of plant trait genes have been reviewed by Pflieger et al. (2001). The principles of this approach are based on the concept whereby a structural gene (such as an EST fragment or other coding DNA sequence) either maps genetically near a known trait locus or encodes a putative function similar to the target trait, then the said gene is a 'candidate' for the trait locus itself. A gene is a 'candidate' based either on its map location or on its known or putative function. In potato, there have been a few notable successful examples of the use of this approach. The isolation of the *Gro1* nematode resistance gene is a good example (Paal et al., 2004). In this case, the hypothesis was that the target gene was a particular type of NBS-LRR gene, and a sequence-linked amplified length amplified polymorphism (AFLP) marker fragment showed homology to this type of gene. Subsequently, a large number of candidate genes were identified from the region which were then reduced using a combination of genetic and functional approaches.

One of the most clear-cut uses of a candidate gene approach in potato has been the isolation of the *P* gene in potato, required for the production of blue/purple anthocyanin pigments in potato tissues, such as tubers, flowers or stems. Jung et al. (2005) have isolated a copy of the gene encoding the anthocyanin biosynthetic enzyme, flavonoid 3′, 5′-hydroxylase (*f3′5′h*), previously shown to map to the *P* locus. A *Petunia f3′5′h* gene was used to screen a potato cDNA library prepared from purple-coloured flowers and stems. Six positively hybridizing cDNA clones were sequenced, and all appeared to be derived from a single gene that shares 85% sequence identity at the amino acid level with *Petunia f3′5′h*. The potato gene co-segregated with purple-coloured tuber in a diploid population and, moreover, was found to be expressed in tuber skin only in the presence of the anthocyanin regulatory locus I. A potato *f3′5′h* cDNA clone was placed under the control of a doubled CaMV 35S promoter and transformed into the red-skinned cultivar 'Desiree'. Tuber and stem tissues that are red in Desiree were purple in 9 of 17 independently transformed lines, confirming the hypothesis that the transformed gene corresponds to the *P* locus.

Li et al. (2005) have analysed DNA sequence variation at the *invGE/GF* locus on potato chromosome IX that co-localizes with a cold-sweetening QTL. This locus consists of duplicated invertase genes *invGE* and *invGF*. The study focused on 188 tetraploid potato cultivars, which were assessed for chip quality and tuber starch content. Two closely linked invertase alleles, *invGE-f* and *invGF-d*, were associated with better chip quality in three breeding populations. Moreover, allele *invGF-b* was associated with lower tuber starch content. The potato *invGE* gene is orthologous to the tomato invertase gene *Lin5*, causal for a fruit-sugar yield QTL (Fridman et al., 2004). These results suggest that natural variation of sugar yield in tomato fruits and that sugar content of potato tubers is controlled by functional variants of orthologous invertase genes.

These few examples clearly demonstrate the potential of using the candidate gene approach in potato. It is also clear that the extensive knowledge of tuber biochemistry and the large number of potato gene sequences should enable its further application for tuber quality traits. Studies whereby large numbers of genes involved in a process or a trait can be genetically mapped will surely help in this type of activity (Leister et al., 1996; Chen et al., 2001). In addition, knowledge of the location, structure and sequence of resistance gene complexes should also enable this approach to be used for the isolation of more potato resistance genes and QTL.

9.4 STRUCTURAL GENOMIC RESOURCES

9.4.1 Large-insert genomic libraries

Bacterial artificial chromosome libraries have become the predominant method for the generation of large-insert genomic representations of plant genomes. Such libraries allow an entire potato 'genome equivalent' to be represented by approximately 10 000 clones (assuming average insert size of 100–120 kb). These libraries, in combination with dense genetic maps, provide the main genomic resource for performing map-based gene cloning and for the generation of local or genome-wide physical maps. Several BAC libraries are now available for potato, constructed from a range of cultivars, breeding lines and wild species. The BAC libraries known to exist for potato are listed in Table 9.1. These libraries have, in general, been constructed for the purpose of isolating potato genes coding for resistance to one of the major potato pathogens using a map-based cloning approach. A BAC library constructed from the male parent of the potato ultra high density (UHD) genetic map is being used for the construction of a genome-wide physical map of potato (see Section 9.8). The UHD map of potato is a highly dense, but low resolution, genetic map that contains over 10 000 genetic markers, primarily AFLPs (van Os et al., 2006). The potato BAC libraries constructed prior to february 2007 represent a diverse range of potato species: cultivated *S. tuberosum* (tetraploid), *S. demissum* (hexaploid), *Solanum pinnatisectum* (diploid) and *Solanum bulbocastanum* (diploid) and so should represent a useful resource for the study of comparative genome organization and evolution in potato. Other developments arising from the use of these BAC libraries include the use of BAC clones and FISH to develop chromosome-specific cytogenetic DNA markers for chromosome identification in potato (Dong et al., 2000). A potato BAC library was screened with genetically mapped RFLP markers, and BACs thus identified were then labelled as probes for FISH analysis. A panel of 12 chromosome-specific BAC clones was established, and FISH signals derived from these BAC clones serve as convenient and reliable cytological markers for potato chromosome identification. The same group has also shown that large-insert binary vector binary bacterial artificial chromosome (BIBAC) and transformation-competent artificial chromosome (TAC) clones containing genomic fragments larger than 100 kb are not stable in the *Agrobacterium* strains used for potato transformation (Song et al., 2003). This effect was directly related to the size of the insert and was even observed when *recA*-deficient strains were used. These authors have developed a potentially useful transposon-based technique for subcloning BAC inserts into a small number of more stable smaller constructs that overcomes the instability problem.

9.4.2 Expressed sequence tag resources

The generation of large EST collections has become the primary route for large-scale gene discovery in most organisms. Expressed sequence tag sequencing is generally either a substitute for or a prelude to full genome sequencing. Basically, RNA is isolated from a particular tissue (or a mixture of tissues for 'normalized' libraries), converted to cDNA and cloned into a sequencing vector. Random sequencing of many thousands of cDNA clones generates a large collection of 'expressed' sequences that represent

Table 9.1 Bacterial artificial chromosome (BAC) libraries constructed from potato.

Species	Cultivar/clone/ accession	Number of clones (x1000)	Average insert size (kb)	Restriction enzyme	Genome coverage	Citation
Solanum tuberosum ssp. *tuberosum*	cv. Cara self SC-781	160	100	*Hind*III	16	Kanyuka et al., 1999
S. tuberosum ssp. *tuberosum*	SH83-92-488	60	100	*Hind*III	6	Rouppe van der Voort et al., 1999
S. tuberosum ssp. *tuberosum*	P6/210	~ 200	70–80	*Hind*III, *Eco*RI	15	Ballvora et al., 2002 and unpublished
S. tuberosum ssp. *tuberosum*	3704-76(JP)	41	98	*Hind*III	~ 4	Bryan et al., unpublished
S. tuberosum ssp. *tuberosum*	RH89-039-16	35.7	102	*Hind*III	~ 4	Bryan et al., unpublished
S. tuberosum ssp. *tuberosum*	RH89-039-16	70	130	*Hind*III/ *Eco*RI	10	De Boer et al., unpublished
S. tuberosum ssp. *tuberosum*	2X(V-2)7	29	125	*Hind*III	3.3	Zhang et al., 2003
S. demissum	PI161729	397	125	*Hind*III	17	Kuang et al., 2005
S. bulbocastanum	8005	130	100	*Hind*III?	15	Van der Vossen et al., 2003
S. bulbocastanum	PT29	24	155	*Hind*III	3.7	Song et al., 2003
S. pinnatisectum	PI253214	40	125	*Bam*HI	4.1	Chen et al., 2004
S. pinnatisectum	PI275233	17	135	*Eco*RI	1.9	Chen et al., 2004
Solanum. tuberosum x Solanum. phureja	PD59	~ 50	80	*Hind*III	~ 4	Castillo Ruiz et al., 2005

a proportion of the coding fraction of the source genome. Such an approach generates a great deal of redundant information, but for an outbreeder such as potato, this redundancy can yield important and useful information [e.g. single-nucleotide polymorphism (SNP) and simple sequence repeat (SSR) markers]. The first significant potato EST project was reported by Crookshanks et al. (2001), who analysed 6077 ESTs, of which 2254 were full length, from a mature tuber cDNA library made from field-grown potatoes (*S. tuberosum* var. Kuras). These ESTs were assembled into 828 clusters (i.e. an assembly of more than one overlapping fragment from the same gene) and 1533 singletons (individual fragments with no overlapping fragments from the same gene). Clustering requires the use of analytical software that can discriminate between orthologous (same gene) and paralogous (different member of a family of closely related genes) sequences. More than 94% of these ESTs showed homology to genes from other organisms. Genes involved in protein synthesis, protein targeting and cell defence predominated in tuber compared with stolon, shoot and leaf organs, and 1063 clones were unique to tuber. Transcripts of starch metabolizing enzymes showed similar relative levels in tuber and stolon. Recently, the number of potato ESTs has increased very dramatically. Ronning et al. (2003) report the sequencing of 61 940 ESTs from a wide range of diverse potato tissues, both below and above ground, and including pathogen-challenged material. These assembled into 19 892 unique sequences that formed 8741 contigs tentative consensus (TCs) and 11 151 singletons. Putative function was identified for approximately 44% of these sequences, and a small number of sequences was found to be constitutively expressed in all the tissues sampled, with 21% being expressed in only a single tissue. Table 9.2 provides a breakdown of the numbers of library-specific EST sequences from the various libraries generated by Ronning et al. (2003). This EST collection has subsequently been greatly expanded, both by The Institute for Genomic Research (TIGR) and with the addition of more than 85 000 sequences from the Canadian Potato

Table 9.2 Identification of sequences specific to a single cDNA library.

Library	Library-specific sequences (%)
Stolon	2 021 (20)
Leaf	2 273 (22)
Sprouting eye I	176 (20)
Sprouting eye II	2 045 (22)
Dormant tuber	851 (17)
Microtuber	775 (14)
Root	2 356 (23)
Compatible leaf	1 371 (27)
Incompatible leaf	1 200 (22)
Total	**13 068 (21)**

The number of library-specific sequences was determined by adding the TC and the singleton expressed sequence tags (ESTs) that were detected only in a single cDNA library. The numbers in parentheses are the per cent of library-specific sequences within all the sequences determined from that library. [*Source: Ronning et al. (2003).*]

Genome Project (CPGP) (Flinn et al., 2005). The Potato Gene Index maintained at TIGR (http://www.tigr.org/tigr-scripts/tgi/T_index.cgi?species=potato) now contains over 190 000 sequences and has been assembled into 21 063 contigs and 17 077 singletons. A mixed assembly of sequences from TIGR and the CPGP show that approximately 22% of the EST sequences from the CPGP do not assemble with the TIGR sequences, so the additional effort appears to be well justified.

These EST collections comprise a major genomics resource for potato researchers. They contain a wealth of data on a large component of the potato gene repertoire and will be an important source of candidate genes and markers in the foreseeable future. A drawback of this resource is that the map locations of only a relatively small proportion of potato genes or ESTs are known (Gebhardt et al., 2003). Given the possibility of full potato and tomato genome sequences in the near future, the genetic location of most potato genes should soon be established.

9.5 ANALYSIS OF POTATO GENE EXPRESSION

Isolation of cDNA or genomic clones corresponding to particular potato genes has very often been accompanied by the analysis of the expression of these genes in different potato tissues. There are far too many of these studies to discuss in detail here. Global gene expression studies have been facilitated by the development of methods for performing transcriptional profiling. These methodologies can be categorized according to whether they are 'open' or 'closed' systems. Open systems include serial analysis of gene expression (SAGE), subtractive suppressive hybridization (SSH), and cDNA AFLP, require no prior knowledge of transcript sequences and can lead to the discovery of previously unknown transcripts. Closed systems (e.g. most microarrays) require knowledge of transcript sequences for the design of probe sets. A wide range of these technologies is now available, and some technologies have been used by potato researchers. Here now follows a brief summary of the ways in which these types of study have been employed in potato. Expression analysis is a discipline that is still in transition, and it is likely to undergo significant development in the future.

Differential screening of cDNA libraries has been used, for example, to identify genes whose expression is modulated as a result of different types of abiotic stress. Space considerations preclude a full discussion of these studies, but a few representative studies will be outlined. Van Berkel et al. (1994) isolated 16 clones corresponding to cold-inducible transcripts by differential screening of a cDNA library, which were classified into four non–cross-hybridizing groups. Some of these showed homologies to small heat-shock proteins from other plant species, dehydrins, and to cold-induced proteins from *Arabidopsis* and spinach. Rorat et al. (1997) isolated 12 cold-induced cDNA clones by differential screening of a cDNA library prepared with mRNA from a cold-resistant potato species. All the clones clearly respond to cold, and accumulation of the corresponding transcripts was observed after 1 day of cold hardening; however, they varied with regard to the timing of mRNA accumulation. The isolated genes corresponded to known genes in the databases, such as S-adenosyl-L-methionine decarboxylases, chloroplast chaperonins, cell-division cycle proteins, malate dehydrogenase, elongation factors and proteins known to be abscisic acid (ABA) inducible in tomato. The authors postulated that the identified genes may be involved in two different functions referring to cold resistance, one protecting chloroplasts and other cell functions under cold stress and the other involved in metabolic adjustment to cold. Kim et al. (2004) analysed changes of expression in cold-regulated mRNA levels in potato (*S. tuberosum* L. cultivar Superior) by reverse northern blot analysis of 12 000 cDNAs. A total of 245 cDNA clones were sequenced from a cDNA library constructed from the cold treatment. The analysed genes were classified into 12 groups according to their putative functions, where 20.2% of group I was associated with energy metabolism, 13.1% of group XI with cell rescue and defence and 2.6% of group VIII with signal transduction. Most of the cold-treated clones showed overexpression compared with the control, whereas some showed down-regulation. In general, it was found that cold stress-related genes were overexpressed more than two-fold. Kim et al. (2003) have also examined the expression patterns of potato genes under cold stress using small microarrays. Results showed that seven clones of cold stress protein were up-regulated by cold treatment.

The cDNA–AFLP technique (Bachem et al., 1996) is based on an AFLP-derived transcript fingerprinting method and has primarily been used to study gene expression during different aspects of the tuber life cycle (Bachem et al., 2000; Trindade et al., 2004). These authors have used cDNA–AFLP to analyse gene expression from stolon formation to sprouting in a range of different tissues. Approximately 18 000 transcript-derived fragments (TDFs) were observed, and over 200 'process-specific' TDFs were isolated and sequenced throughout the potato tuber life cycle. The sequence similarities of these TDFs to known genes give insights into the kinds of processes occurring during tuberization, dormancy and sprouting. One of the disadvantages of the cDNA method is that it does not provide gene sequence information and requires laborious excision or elution of TDFs from polyacrylamide gels for further characterization. However, this technique is extremely sensitive and can detect differences among gene family members indistinguishable by northern blotting.

A further development of cDNA display technology has been the realization that cDNA–AFLP fragments can show genetic polymorphism in segregating populations and can be mapped as transcriptome genetic polymorphisms (Brugmans et al., 2002). Importantly, these markers show less clustering than AFLP markers derived directly from genomic DNA and appear to be targeted specifically to transcriptionally active regions of the genome. These markers, as they tend to be relatively low copy, may be particularly suited to use in genetic analysis, marker-assisted breeding and map-based cloning.

Other applications of cDNA–AFLP include its use to isolate a stolon-specific promoter from potato (Trindade et al., 2003) and identification of genes involved in cell wall biosynthesis (Oomen et al., 2003). More recently, the method has been used to perform a large-scale survey of genes differentially expressed during the tuber life cycle and the isolation of some of their promoter regions (Trindade et al., 2004). Most genes expressed in the tuber life cycle are involved in defence, stress, storage and signal transduction pathways. Twelve cis-acting elements were identified and are known to be responsive to environmental stimuli known to play an important role during the tuber life cycle (light, sugars, hormones, etc.).

An example of the use of another useful 'open' gene profiling technology has been reported by Faivre-Rampant et al. (2004a), who have used the SSH approach to make a cDNA library enriched for 385 different genes that are up-regulated in the potato tuber apical bud on dormancy release. One of these cDNAs was identified as encoding a member of the auxin response factor family (*ARF6*), and the expression pattern of this gene was determined byISH.

An alternative to microarray technology for gene expression profiling is SAGE, which generates short cDNA sequence tags (Velculescu et al., 1995; Saha et al., 2002) using a concatemerization-based method. Nielsen et al. (2005) have used LongSAGE to examine global gene expression in potato tubers, generating 58 322 sequence tags (of length 19 nucleotides) representing 22 233 different tags. Putative functions were assigned to almost 700 of those tags occurring at least 10 times. Almost 70% of these tags matched a known potato EST sequence. This technology, such as SSH, has the advantage over microarray technology in being an 'open' technology, with the possibility of discovering 'new' transcripts. Rapid amplification of cDNA ends (RACE) cloning was used to verify the reliability of SAGE tag annotation using EST sequences from more than one cultivar. Seventy-two per cent of tags represented genes that participated in a known biological process, with the largest group (43%) consisting of transcripts active in physiological processes, about half of which were involved in metabolism (Fig. 9.1). There were no transcripts found that were involved in photosynthesis. Of the 50 most abundant transcripts from the mature tuber, protease inhibitors were the dominant class, which is in good agreement with previous EST projects (Crookshanks et al., 2001; Ronning et al., 2003).

The methodologies used in this section are in some ways the prelude to microarray studies and share with all such studies the 'problem' that they are only indicative of the function of particular genes or sets of genes in biological processes. These studies require functional analyses whereby the function of the candidate genes is compromised or enhanced in some way. This topic is discussed later in this chapter.

9.6 MICROARRAYS

Existing potato EST resources comprise a large, but as yet unknown, portion of the total gene complement of potato. The ESTs are from a variety of genotypes, tissues, exposure to different environmental influences and developmental changes, and this allows researchers to begin to examine differences in global gene expression patterns under various conditions and to identify key genes active in these conditions. A non-redundant set of 10 000 of these ESTs has been used by TIGR to develop a spotted cDNA potato microarray that is available to the research community at minimal cost. Moreover, the same organization offers a transcription profiling service to allow the evaluation of these arrays by a wide range of users working on different *Solanaceous* plant species and different biological questions. The performance of these experiments under a common experimental system has allowed the amassment of a large amount of microarray data that are publicly available (http://www.tigr.org/tdb/potato/profiling_service2.shtml#AProcedure). However, this platform has the disadvantage of only having approximately 25–30% of the available gene resource. Furthermore, the fact that these arrays are based on spotted

Biological process

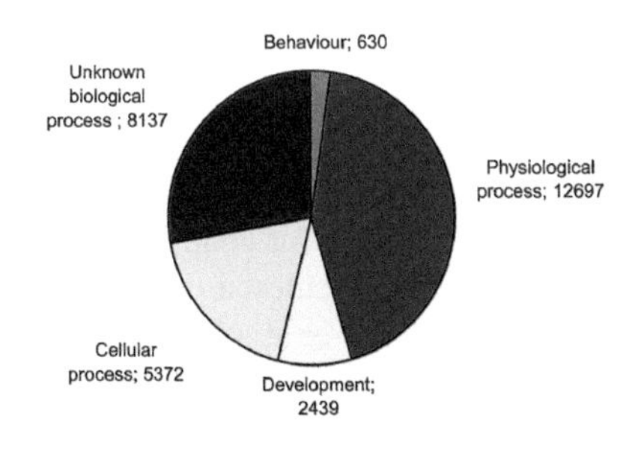

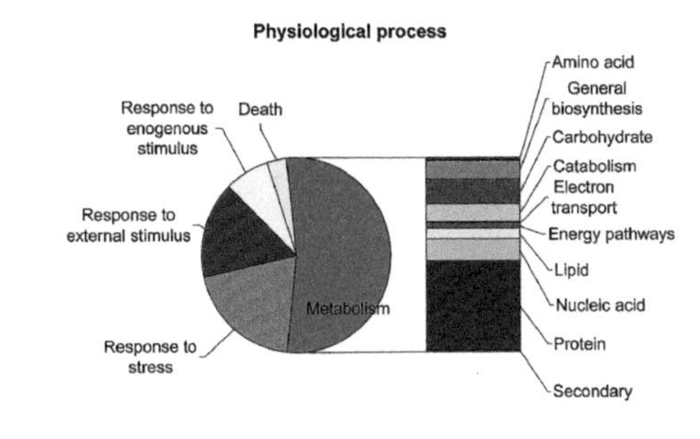

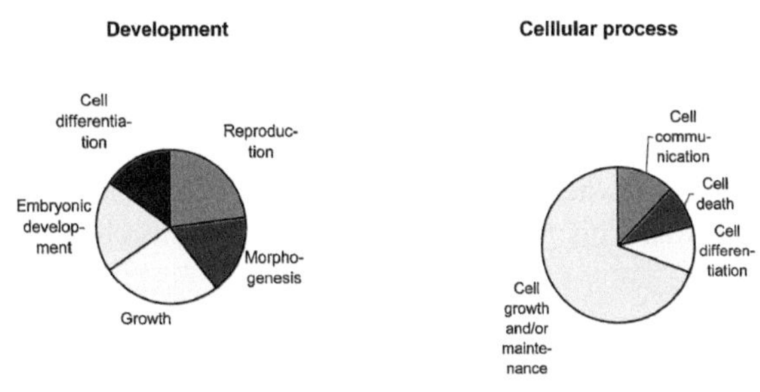

Fig. 9.1. Gene ontology (GO) functional annotation of biological process of potato tuber transcripts. Transcripts were classified as described in 'Experimental procedures'. The upper pie is the highest functional level defined in the GO hierarchy. The lower pies are lower level details of the selected functional classes: physiological process, development and cellular process. [*Source: Nielsen et al. (2005).*]

cDNAs, with which it is inherently difficult to achieve a high level of reproducibility, makes it impossible to compare experiments without common references. Rensink et al. (2005) have used this platform to identify genes involved in abiotic stress responses in potato. Over 3000 genes were found to be significantly up- or down-regulated in response to at least one of the stress conditions used (cold, heat and salt). The genes can be partitioned into stress-specific or shared response genes, suggesting that there are both general and stress-specific response pathways. In a very detailed study, Kloosterman et al. (2005) have followed the expression of 1315 genes during tuber development. Transient changes in gene expression were relatively uncommon, and genes previously not

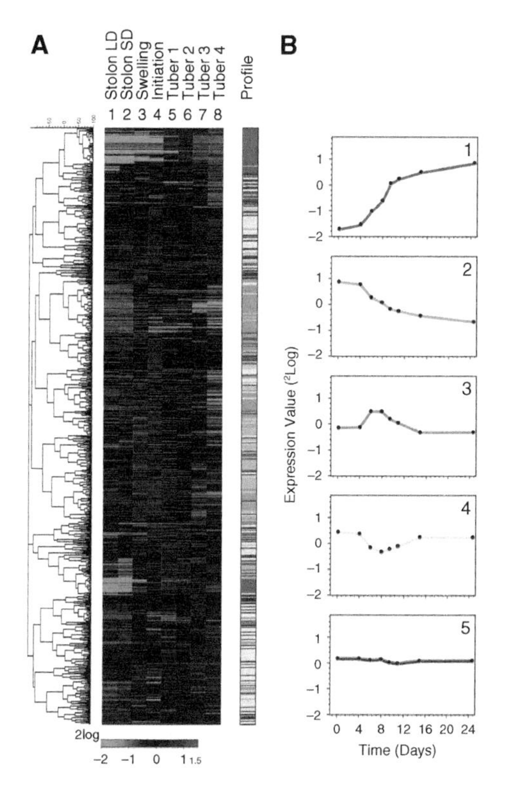

Fig. 9.2. Relative gene expression data of 1315 expressed sequence tags (ESTs) of the eight potato tuber developmental stages (Fig. 9.1), clustered and grouped according to their relative expression profile. (A) Hierarchical clustering of developmental stages. (1–8) with, on the right, a colour bar grouping all genes in six major expression profiles: ■, up-regulated; ■, down regulated; ■, transient up-regulated; ■, transient down-regulated; ■, constitutive expression; ■, rest group. (B) Graphs 1–5; averaged expression patterns of the identified expression profile groups during tuber developmental stages 1–8. [*Source: Kloosterman et al. (2005).*]

known to be differentially expressed during tuber development were identified (Fig. 9.2). These studies, although both useful and informative, highlight the perennial dilemma faced by plant molecular biologists in prioritizing genes for further study from a large number of candidate genes in the absence of genetic information and mutations in target trait genes.

The potato research community has, for the longer term, been looking towards the establishment of an accurate, flexible and reproducible platform for microarray projects that will permit comparability of experimental results between laboratories. Many of these features are well represented in the 'printed' long oligonucleotide arrays manufactured by various technology providers. The capacity to produce long oligonucleotides with great accuracy is a particular advantage with potato, where the high degree of allelic heterozygosity is more likely to result in misinterpretations when using short oligos. The potato oligo chip initiative (POCI) has selected the Agilent '44K feature platform' system for this purpose. This will entail the designing of 60-mer oligonucleotides for up to 42 000 genes and was made available for use in 2006.

9.7 FUNCTIONAL GENOMIC RESOURCES

9.7.1 The phenotype gap

There now exists a wealth of information pertaining to the location of genes and QTL coding for important potato traits, including pest and disease resistance and tuber traits. The volume of gene sequence information for potato, notably EST data, is now substantial. However, it is very clear that the recent developments in genetics and structural genomics have not been matched by concomitant development of functional genomic tools and resources that are available for model systems and some other crop plants. These tools and resources comprise mutant populations, gene tagging resources, high-throughput gene knockout and silencing systems and comprehensive microarrays. This deficiency renders the identification of gene function problematic for most of the unique genes (more than 38 000) currently held in the TIGR potato gene index. Potato breeders and geneticists have always been at a disadvantage in having comparatively little knowledge of the genetics of the morphological and physiological traits with which they work. Exceptions are the knowledge that has accumulated on the inheritance of tuber traits, such as shape, flesh and skin colour (Chapter 6, Van Eck, this volume), and the induction of an amylose-free starch mutant of potato (Jacobsen et al., 1989). As for many other crops (with the exception of rice), the number of genes genetically mapped in potato is quite small, making the use of a 'positional' candidate gene approach, whereby a mapped gene is a trait candidate by virtue of mapping to or near the trait locus, infeasible for most traits. For potato, as with other crop plants, there is a need for relatively high-throughput methods for testing and assessing gene function. The availability of mutant populations of potato would be of tremendous value in this regard. However, such populations are not yet available in potato, largely because of its ploidy and preferred mating system, and there is a dearth of described mutational variation in potato. Despite the extensive knowledge of resistance and tuber traits, there is little information regarding the genetics of plant architecture and biochemical characters. In the longer term, the lack of mutants may be addressed by recourse to the use of diploid inbreeding lines of potato for the development of mutant populations. In the shorter term, we are left with the use of transformation-based techniques or use of viral vector-mediated gene delivery systems for the establishment of information regarding gene function. There are tantalizing current developments in the availability of functional genomic tools and resources for potato. Of course, gene expression profiling or microarray studies have a role to play in the identification of a pool of candidate genes potentially involved in any given biological process. These methods, in combination with other functional genomic tools such as RNA interference (RNAi), virus-induced gene silencing (VIGS) and activation-tagged lines, have the potential to facilitate the identification of the role of thousands of potato genes. Furthermore, combining structural genetic approaches (such as QTL and candidate gene mapping) with functional genomics information (such as microarray-derived gene expression data for candidate genes) has great potential for the dissection of many complex, polygenic potato traits.

9.7.2 Transgenic approaches for the study of gene function

Transformation of potato is a relatively straightforward matter, and so it is not surprising that many attempts to demonstrate or establish gene function of cloned genes in potato have involved the use of transgenic technology. These have included direct complementation by stable transformation, overexpression and antisense technology. The most common and direct use of transgenic complementation has been for the establishment of proof of gene function for several potato resistance genes thus far isolated. This type of approach is only generally useful for 'dominant' phenotypes and points to the lack of recessive mutations in potato. Overexpression has been used in several cases to investigate gene function in potato, with mixed success. Rosin et al. (2003a) have performed overexpression of a knotted-like homeobox gene (*POTH1*). Transcripts of this gene, isolated from an early stage tuber cDNA library, were found in a wide range of tissues, including meristematic cells, leaf primordia and the vascular procambium of the young stem using ISH. Overexpression of *POTH1* produced dwarf plants with altered leaf morphology. Results of biochemical studies suggested that *POTH1* mediates the development of potato by acting as a negative regulator of giberellin (GA) biosynthesis. Oomen et al. (2004a) have shown that overexpression of two different potato UDP-Glc 4-epimerases (UGE) can increase the galactose content of potato tuber cell walls, that the two proteins have different biochemical properties and that they have a different function in plant development.

Suppression has been used by Rosin et al. (2003b) to examine the effects of a vegetative MADS box gene (*POTM1*) in activating axillary meristem development. The RNA of *POTM1* is most abundant in vegetative meristems of potato. Transgenic lines with reduced levels of POTM1 mRNA exhibited decreased apical dominance accompanied by a compact growth habit and a reduction in leaf size. Suppression lines produced truncated shoot clusters from stem buds and, in a model system, exhibited enhanced axillary bud growth instead of producing a tuber. Tuber yields were reduced, and rooting of cuttings was strongly inhibited. Results implied that POTM1 mediates the control of axillary bud development by regulating cell growth in vegetative meristems.

Oomen et al. (2004b) have modified the cellulose content of tuber cell walls by antisense expression of four different members of the cellulose synthase (*CesA*) family. None of the several transformants showed an overt developmental phenotype, but several of them showed decreased levels of cellulose. Their results suggested that the cellulose content in potato tubers can be reduced down to 40% of the wild type (WT) level without affecting normal plant development and that the constructs used are specific and sufficient to down-regulate cellulose biosynthesis. Liu et al. (1999) characterized an anther-expressed protein kinase gene in *Solanum berthaultii* and examined its antisense inhibition in transgenic plants. Transformants fell into two classes, one displaying a typical antisense phenotype and a second type arising from a co-suppression event. Both types of transformants gave rise to increased proportions of defective pollen, although the co-suppression effect was more severe. The results indicate that the anther-expressed protein kinase is essential for pollen development in potato.

These experiments demonstrate that, although some knowledge about gene function can be gleaned, the use of transgenesis to modify gene expression does not always give unequivocal results. Complications can be caused due to unintended silencing effects, presence of gene orthologues and other related phenomena.

9.7.3 Transposon tagging

The problem of a lack of mutational variation in potato has been addressed by harnessing heterologous 'transposon tagging' systems in potato. Transposon tagging is based on the use of mobile genetic elements to disrupt gene function through insertion in or near coding sequences. For the development of an efficient transposon tagging strategy, it is important to generate populations of plants containing large numbers of unique independent transposon insertions that will mutate genes of interest at detectable frequencies. For example, the Ac/Ds maize transposon system has been introduced into potato (Knapp et al., 1988) and, more recently, into a self-compatible diploid potato clone (Pereira et al., 1992). Significant progress has been made in generating independent insertions of Ds transposons in mapped genomic locations that, in principle, can be used as 'launching pads' to mutagenize closely linked genes. Mapping of insertions is achieved by inverse PCR (IPCR) to generate DNA fragments flanking the transposon-bearing T-DNA insertions, which are then mapped using RFLP. The strategy envisaged is to recombine a closely linked transposon 'cis' to a target gene and to rely on the known propensity of Ds to transpose to closely linked sites to mutagenize the target gene. Despite these advances, it is far from clear that these considerable, not to mention ingenious, efforts have yielded an effective gene tagging and knockout system for potato. Problems remain with generating homozygous-tagged lines, and the rate of transposition is not sufficiently high to allow efficient tagging of all loci. A subsequent study by the same group of authors (van Enckevort et al., 2001) has followed the behaviour of Ac/Ds in potato. An excision assay was developed for Ac and was used to monitor excision from GBSS (*Waxy*) gene in somatic starch-forming tissues. Excision was found to be somewhat imprecise. Genotypes useful for tagging strategies were used for crosses, and the frequency of independent germinal transpositions was assessed. High levels of Ds transposition were found in lines with Ac in multiple copies, and the authors concluded that the materials developed are useful in somatic insertional mutagenesis aimed at the isolation of tagged mutations in diploid potato.

9.7.4 Virus-induced gene silencing

Virus-induced gene silencing has become an extremely powerful tool for plant functional genomics. Virus-induced gene silencing can be defined as the down-regulation of host plant gene expression because of infection by a viral vector containing host gene sequences. Virus-induced gene silencing has been used routinely in *Nicotiana benthamiana* to assess functions of candidate genes and as a way to discover new genes required for diverse pathways, especially disease resistance signalling. Virus-induced gene silencing is being increasingly used to generate transient loss-of-function assays and has potential as a powerful reverse-genetics tool in functional genomic programmes as a more rapid alternative to stable transformation. However, in potato, there have been problems in establishing the two most commonly used *Solanaceous* viral vectors, tobacco rattle virus (TRV) and potato virus X (PVX), because of problems due to innate host resistance and a lack of infectivity. Despite these setbacks, there have been significant advances in establishing VIGS in potato. Faivre-Rampant et al. (2004b) have shown that a PVX-based VIGS vector is effective in triggering VIGS of a phytoene desaturase (*PDS*) gene in both diploid and cultivated tetraploid *Solanum* species. Silencing was maintained throughout the foliar tissues of potato plants and was also observed in tubers. They also showed that VIGS can be triggered and sustained on in vitro-micropropagated tetraploid potato for several cycles and on in vitro-generated microtubers. This approach may help to facilitate large-scale functional analysis of potato ESTs and to provide a non-invasive reverse-genetic approach to study mechanisms involved in tuber and micro-tuber development. Brigneti et al. (2004) have also reported the successful silencing of genes in a diverse set of potato species, using a TRV viral vector to achieve silencing of the *PDS* gene in the diploid wild species *S. bulbocastanum* and *Solanum okadae*, in cultivated potato and in the distant hexaploid relative *Solanum nigrum*. In doing so, they assessed the effectiveness of different methods of inoculation (Fig. 9.3A), concluding that agro-infiltration and sap-inoculation were more effective than simple spray inoculation

Fig. 9.3. Comparison of inoculation methods. (a) Spray inoculation. (b) Aggro-infiltration. (c) Sap inoculation. *S. okadae* plants showing phytoene desaturase (PDS) silencing. All plants were inoculated at the same time and pictures were taken at 3 weeks' post-inoculation. The plant in (a) shows the initial stages of PDS silencing, whereas plants in (b and c) show a more developed photo bleaching. [*Source: Brigneti et al. (2004).*]

of virus, with sap-inoculation being more suitable for high-throughput application and producing less undesirable necrotic symptoms. Furthermore, they test the hypothesis that the VIGS system developed may be useful in testing the function of candidate resistance genes in potato and its wild relatives, by silencing some cloned resistance genes, including RB, suggesting that the VIGS system is an effective method of rapidly assessing gene function in potato (Fig. 9.3B). Ryu et al. (2004) have developed a novel and effective agro-inoculation method ('Agrodrench') for VIGS in roots and diverse *Solanaceous* species. This method uses TRV-derived VIGS vectors expressed from binary vectors within *Agrobacterium* to induce RNA silencing. The method relies on a novel and simple method of agro-inoculation, whereby soil around the plant root is 'drenched' with an *Agrobacterium* suspension carrying the virus-derived VIGS vectors. Using this method, these authors successfully silenced the expression of several genes in *N. benthamiana*, as well as in economically important crops including potato. Agrodrench can be used for VIGS in very young seedlings, which is not possible with leaf infiltration-based methods. Another benefit is that VIGS has been shown to silence target genes more efficiently than other methods in plant roots. This methodology has the potential for allowing rapid large-scale functional analysis of cDNA libraries and can also be applied to plants that are not currently amenable to VIGS technology by conventional inoculation methods.

9.7.5 Activation tagging

Activation tagging is an established method for the generation of dominant gain-of-function mutants in plants (Weigel et al., 2000). In contrast to T-DNA insertions that are designed to interrupt or interfere with gene expression, activation tagging is suitable for heterozygous, polyploidy crops such as potato, as gene activation can result in dominant phenotypes, thus offering possibilities for mutagenesis in potato. As cultivated potato is tetraploid and most potato diploids are either self-incompatible or relatively infertile, recessive mutants have been traditionally difficult to generate. In activation tagging, strong promoters, such as the cauliflower mosaic virus 35S promoter, are transformed into the plant genome by *Agrobacterium*-mediated transformation. Insertion of these strong promoters activates genes adjacent to the promoter resulting in overexpression of the genes and phenotypic expression of a gain-of-function mutant for that gene. Based on the phenotypic identification of an activation-tagged line, the corresponding gene can be isolated by standard PCR techniques because of the presence of a known sequence within the activation construct. The CPGP had an original objective of establishing 10 000 activation-tagged lines in potato using the variety cultivar Bintje, and this ambitious goal required these researchers to establish a highly efficient genetic transformation system giving stable integration of the gene construct used (Gustafson et al., 2006). It is not clear whether, at the time of writing, these activities have generated a useful resource that will be taken up by the potato community. The large number of clones generated require clonal propagation and may be subject to prohibitive plant health regulations on import to other countries.

9.8 TOWARDS A GENOME-WIDE PHYSICAL MAP AND A POTATO GENOME SEQUENCE

The UHD genetic map of potato (van Os et al., 2006) forms the framework for the construction of a genome-wide physical map of the potato genome. Physical map construction is being carried out in two phases. First, a BAC library from the diploid breeding clone RH89-039-16 (RH), comprising approximately 73 000 clones, has been fingerprinted using a non-selective AFLP-based method. The fingerprint data have been used to assemble the RH BACs into 7134 BAC contigs, with 7580 'singletons' (i.e. single BAC clones) remaining. The second phase entails anchoring of the contigs and single BACs to the UHD map using a BAC pooling method, which should also reduce the number of contigs and increase the average contig size. Subsequent contiging will use a reduced stringency alignment approach that will reduce the number of contigs still further. At present, it is envisaged that the integrated genetic and physical map that will result from these activities will be the platform used for obtaining the DNA sequence of the potato genome, as proposed by the Potato Genome Sequencing Consortium (http://www.genomics.nl/homepage/research/funding_opportunities/pgsc/). In common with the ongoing tomato sequencing project, this recent development will have huge implications for all those working in the area of potato genetics and genomics, although it will make the need for functional genomic tools even more apparent.

9.9 PROTEOMICS AND METABOLOMICS

The sciences known as proteomics and metabolomics are now widely established in plants. Proteomics is the large-scale study of proteins, particularly their structures and functions. Metabolomics is the systematic study of small-molecule metabolite profiles, without bias towards a specific metabolite or group of metabolites. These activities have led to the coining of the terms 'proteome' and 'metabolome' to define the collections of all proteins and metabolites in a biological organism, which are the end products of gene expression. Metabolic profiling can be used to obtain an instantaneous 'snapshot' of the physiology of a particular tissue or cell type. One of the major challenges facing the emerging science of plant systems biology is to integrate proteomics, transcriptomics and metabolomics information to obtain a more complete picture of the biology of plants.

Space does not permit a detailed discussion of the different analytical methods used for these disciplines. Proteomics is heavily dependent on electrophoretic separation of proteins, in either one or two dimensions. It could be argued that biochemists have been engaged in metabolomics for the latter part of the 20th century, and the new technologies are based on pre-existing biochemical techniques. There are two major issues to be addressed for metabolite analysis: separation of the analytes (usually by chromatography or electrophoresis) and detection of the analytes [usually by mass spectrometry (MS)]. Gas chromatography (GC) allied to MS is one of the most widely used and powerful methods.

In plants, these technologies, notably metabolic profiling, have started to make a significant impact of our understanding of plant cellular processes particularly in the model *Arabidopsis* (Fiehn et al., 2000). These authors were able to detect over 300 metabolites in leaf tissue and showed that the differences among profiles of distinct ecotypes were greater than those between ecotypes and their derived mutant lines. In potato, a lot of effort has been focused on the analysis of tuber metabolites, an obvious target given its economic importance and its well-studied biochemistry.

Roessner et al. (2000) have pioneered the use of GC coupled to MS to quantitatively and qualitatively detect more than 150 compounds within potato tubers. It is important to stress that this type of approach is a relatively unbiased 'open' technology, which permits the detection of unpredicted changes in levels of metabolites. The method was found to be highly reproducible, and biological variability exceeded experimental error by a factor of up to 10. Moreover, the method is suitable for high-throughput analysis. This method has been applied to soil- and in vitro-grown tubers. Owing to the simultaneous analysis of a wide range of metabolites, it was immediately apparent that these systems differ significantly in their metabolism, and it was possible to draw conclusions about the underlying physiological differences between both tuber systems. Transgenic lines modified in sucrose catabolism or starch synthesis were also analysed, and some unexpected changes in disaccharides and sugar alcohols were observed. In a similar vein, Lehesranta et al. (2005) have compared the proteomes of potato cultivars, landraces and transgenic lines. Over 90% of more than 1100 proteins examined show statistically significant variation across cultivars and landraces. However, a very small proportion (9 of 730) showed significant differences between genetically modified (GM) lines and controls. These authors accept the limitations of the 2-DE (dimensional electrophoresis) method used and suggest the need for combined application of transcriptomics, metabolomics and proteomics to resolve biological questions and to monitor food safety.

Metabolite profiling is essentially a very detailed phenotypic analysis performed at the chemical level. A systems biology approach to biological questions is being advocated, whereby studies in which the molecules described by transcriptomic, proteomic and metabolomic approaches are being analysed in parallel. For example, Urbanczyk-Wochniak et al. (2003) compared the discriminatory power of metabolic and RNA profiling to distinguish between different potato tuber systems, and they suggest that metabolic profiling has a higher resolution than expression profiling. Moreover, these authors found that about 2% of the approximately 27 000 possible transcript–metabolite pairs showed strong correlations, and most of these involved nutritionally important metabolites. This exciting observation suggests a route to the rapid identification of candidate genes for modifying the metabolite content of potato and other biological systems.

In another study, Roessner et al. (2001) examined four distinct potato clones modified in various aspects of sucrose metabolism for the presence of hydrophilic metabolites. A combination of clustering and ordination analyses was used to assign clusters to the individual plant systems and to determine relative distances between these clusters. Extraction analysis and correlation analysis allowed identification of the most important metabolic components of the observed clusters and, perhaps more significantly, revealed close linkages between a broad spectrum of metabolites. These authors also compared their data with metabolite profiles from environmentally manipulated and transgenic potato plants, concluding that certain growth conditions can 'phenocopy' genetic modifications. This area has been reviewed extensively (Roessner et al., 2002; Stitt and Fernie, 2003).

These and other authors compare current methodologies for metabolic profiling in plants, highlighting the strategies for biochemical phenotyping of plants by determining the steady-state concentrations of a broad spectrum of metabolites. A primary objective is to assimilate the huge amount of information gained from metabolite profiling into a useful, accessible body of knowledge for the research community. Very recently, the main groups working on the potato metabolome have expressed the need to examine metabolic flux to augment steady-state metabolite analysis and expression profiling (Fernie et al., 2005) in the growing field of plant functional genomics.

9.10 GENOMIC DATABASES

During the early to mid-1990s, crop plant genetics and associated bioinformatic activities really took off. It very quickly became evident that Web-based publicly accessible databases representing repositories of genetic information would be very useful tools for the respective research communities. SolGenes along with GrainGenes and RiceGenes were some of the first plant genome databases based on the ACeDB software. SolGenes was Solanaceae-wide in scope, and its core consisted of genetic maps of the various species, stored and displayed in ways that allowed for ready cross-referencing. The maps included were mostly RFLP maps of potato, tomato and pepper. This database became defunct in 2002 and has largely been replaced by the Solanaceous Genome Network (SGN), which remains heavily biased towards the interests of the tomato research community (http://www.sgn.cornell.edu). A database for potato genetic data (PoMaMo, Potato Maps and More) has recently been established (Meyer et al., 2005). PoMaMo contains information on molecular maps, DNA sequences and putative gene functions, results from BLAST analysis, SNP and InDel information from potato germ plasm and publication references and links to other public databases such as GenBank (http://www.ncbi.nim.nih.gov/) or SGN. PoMaMo has several search and data visualization tools that facilitate easy access to the data via the internet (https://gabi.rzpd.de/PoMaMo.html). Chromosome maps can be displayed interactively by clicking on an element of interest. The 'GreenCards' interface allows a text-based data search by marker, sequence or genotype name and by sequence accession number, gene function, BLAST Hit or publication reference. The PoMaMo database is a comprehensive database for different types of potato genome data, and it is hoped that users will contribute further to this database in the future to allow it to become the key Web resource for potato geneticists and breeders. However, with the availability of vast EST resources, and the imminent generation of a full genome sequence, the potato research community will be required to make joint decisions about future database resources for the curation of this information.

9.11 SUMMARY

Potato has been considered the 'poor relative' in relation to other major crop plants in the availability of genetic and genomic resources. Early genetic and genomic studies in the

Solanaceae were heavily biased towards tomato, rightly viewed as a model species for the Solanaceae, despite a similar genome size to potato. The potato research community has seen the rapid development of extensive genomic resources and tools. Map-based gene isolation projects have led to the development of several large-insert genomic libraries for diploid and tetraploid cultivated potato and a few key wild relatives. The development of large potato EST collections and microarray platforms has mirrored the generation of similar resources for tomato, leading to a wealth of gene sequence data for the Solanaceae research community. A global potato physical mapping strategy has been put into place and will serve as a template for a potato genome sequence. Given the advanced state of tuber biology and biochemistry, some newer technologies, notably expression and metabolic profiling, have started to have a deep impact on our understanding of potato biology.

ACKNOWLEDGEMENTS

I acknowledge the financial support of the Scottish Executive Environmental and Rural Affairs Department.

REFERENCES

Arumuganathan K. and E.D. Earle, 1991, *Plant Mol. Biol. Rep.* 9, 211.

Bachem C.W.B., R.S. van der Hoeven, S.M. de Bruijn, D. Vreugdenhil, M. Zabeau and R.G.F. Visser, 1996, *Plant J.* 9, 745.

Bachem C., R. Van Der Hoeven, J. Lucker, R. Oomen, E. Casarini, E. Jacobsen and R. Visser, 2000, *Potato Res.* 43, 297.

Ballvora A., M.R. Ercolano, J. Weiss, K. Meksem, C.S.A. Bormann, P. Oberhagemann, F. Salamini and C. Gebhardt, 2002, *Plant J.* 30, 361.

Botstein D., R.L. White, M. Skolnick and R.W. Davies, 1980, *Am. J. Hum. Genet.* 32, 314.

Brigneti G., A.M. Martin-Hernandez, H.L. Jin, J. Chen, D.C. Baulcombe, B. Baker and J.D.G. Jones, 2004, *Plant J.* 39, 264.

Brugmans B., A.F. del Carmen, C.W.B. Bachem, H. van Os, H.J. van Eck and R.G.F. Visser, 2002, *Plant J.* 31, 211.

Castillo Ruiz R.A., C. Herrera, M. Ghislain and C. Gebhardt, 2005, *Mol. Gen. Genet.* 274, 168.

Chen X., F. Salamini and C. Gebhardt, 2001, *Theor. Appl. Genet.* 102, 284.

Chen Q., S. Sun, Q. Ye, S. McCuine, E. Huff and H.B. Zhang, 2004, *Theor. Appl. Genet.* 108, 1002.

Crookshanks M., J. Emmersen, K.G. Welinder and K.L. Nielsen, 2001, *FEBS Lett.* 506, 123.

des Francs-Small C.C., F. Ambard-Bretteville, I.D. Small and R. Remy, 1993, *Plant Physiol.* 102, 1171.

Dong F., J. Song, S.K. Naess, J.P. Helgeson, C. Gebhardt and J. Jiang, 2000, *Theor. Appl. Genet.* 101, 1001.

Eckes P., S. Rosahl, J. Schell and L. Willmitzer, 1986, *Mol. Gen. Genet.* 205, 14.

Faivre-Rampant O., L. Cardle, D. Marshall, R. Viola and M.A. Taylor, 2004a, *J. Exp. Bot.* 55, 613.

Faivre-Rampant O., E.M. Gilroy, K. Hrubikova, I. Hein, S. Millam, G.J. Loake, P. Birch, M. Taylor and C. Lacomme, 2004b, *Plant Physiol.* 134, 1308.

Fernie A.R., P. Geigenberger and M. Stitt, 2005, *Curr. Opin. Plant Biol.* 8, 174.

Fiehn O., J. Kopka, P. Dormann, T. Altmann, R.N. Trethewey and L. Willmitzer, 2000, *Nat. Biotechnol.* 18, 1157.

Flinn B., C. Rothwell, R. Griffiths, M. Lague, D. Dekoeyer, R. Sardana, P. Audy, C. Goyer, X.Q. Li, G. Wang-Pruski and S. Regan, 2005, *Plant Mol. Biol.* 59, 407.

Fridman E., F. Carrari, Y.S. Liu, A.R. Fernie and D. Zamir, 2004, *Science* 305, 1786.
Gebhardt C., B. Walkemeier, H. Henselewski, A. Barakat, M. Delseny and K. Stüber, 2003, *Plant J.* 34, 529.
Graeve K., A. Von Schaewen and R. Scheibe, 1994, *Plant J.* 5, 353.
Gustafson V., S. Mallubhotla, J. MacDonnell, M. Sanyal-Bagchi, B. Chakravarty, G. Wang-Pruski, C. Rothwell, P. Audy, D. DeKoeyer, M. Siahbazi, B. Flinn and S. Regan, 2006, *Plant Cell Tissue Organ Cult.* 85, 361.
Herbers K., S. Prat and L. Willmitzer, 1994, *Plant Mol. Biol.* 26, 73.
Jacobsen E., J.H.M. Hovenkamp-Hermelink, H.T. Krijgsheld, H. Nijdam, L.P. Pijnacker, B. Witholt and W.J. Feenstra, 1989, *Euphytica* 44, 43.
Jung C.S., H.M. Griffiths, D.M. De Jong, S.P. Cheng, M. Bodis and W.S. De Jong, 2005, *Theor. Appl. Genet.*, 110, 269.
Kang S.G. and D.J. Hannapel, 1995, *Gene* 166, 329.
Kang S.G. and D.J. Hannapel, 1996, *Plant Mol. Biol.* 31, 379.
Kanyuka K., A. Bendahmane, J. van der Voort, E.A.G. van der Vossen and D.C. Baulcombe, 1999, *Theor. Appl. Genet.* 98, 679.
Kim D.Y., H.B. Kwon, H.E. Lee, N.J. Jeong, S.J. Go and M.O. Byun, 2004, *Korean J. Genet.* 26, 251.
Kim D.Y., H.E. Lee, K.W. Yi, S.E. Han, H.B. Kwon, S.J. Go and M.O. Byun, 2003, *Korean J. Genet.* 25, 345.
Kloosterman B., O. Vorst, R.D. Hall, R.G.F. Visser and C.W. Bachem, 2005, *Plant Biotechnol. J.* 3, 505.
Knapp S., G. Coupland, H. Uhrig, P. Starlinger and F. Salamini, 1988, *Mol. Gen. Genet.* 213, 285.
Kuang H.H., F.S. Wei, M.R. Marano, U. Wirtz, X.X. Wang, J. Liu, W.P. Shum, J. Zaborsky, L.J. Tallon, W. Rensink, S. Lobst, P.F. Zhang, C.E. Tornqvist, A. Tek, J. Bamberg, J. Helgeson, W. Fry, F. You, M.C. Luo, J.M. Jiang, C.R. Buell and B. Baker, 2005, *Plant J.* 44, 37.
Lehesranta S.J., H.V. Davies, L.V.T. Shepherd, N. Nunan, J.W. McNicol, S. Auriola, K.M. Koistinen, S. Suomalainen, H.I. Kokko and S.O. Kärenlampi, 2005, *Plant Physiol.* 138, 1690.
Leister D., A. Ballvora, F. Salamini and C. Gebhardt, 1996, *Nat. Genet.* 14, 421.
Li L., J. Strahwald, H.R. Hofferbert, J. Lubeck, E. Tacke, H. Junghans, J. Wunder and C. Gebhardt, 2005, *Genetics* 170, 813.
Liu J.Q., G. Leggewie, S. Varotto and R.D. Thompson, 1999, *Sex. Plant Reprod.* 11, 336.
Meyer S., A. Nagel and C. Gebhardt, 2005, *Nucleic Acids Res.* 33, D666.
Nap J.P., W.G. Dirkse, J. Louwerse, J. Onstenk, R. Visser, A. Loonen, F. Heidekamp and W.J. Stiekema, 1992, *Plant Mol. Biol.* 20, 683.
Nielsen K.L., K. Gronkjaer, K.G. Welinder and J. Emmersen, 2005, *Plant Biotechnol. J.* 3, 175.
Oomen R., M. Bergervoet, C.W.B. Bachem, R.G.F. Visser and J.P. Vincken, 2003, *Plant Physiol. Biochem.* 41, 965.
Oomen R.J.F., B. Dao-Thi, E.N. Tzitzikas, E.J. Bakx, H.A. Schols, R.G.F. Visser and J.P. Vincken, 2004a, *Plant Sci.* 166, 1097.
Oomen R.J.F., E.N. Tzitzikas, E.J. Bakx, I. Straatman-Engelen, M.S. Bush, M.C. McCann, H.A. Schols, R.G.F. Visser and J.P. Vincken, 2004b, *Phytochemistry* 65, 535.
Paal J., H. Henselewski, J. Muth, K. Meksem, C.M. Menendez, F. Salamini, A. Ballvora and C. Gebhardt, 2004, *Plant J.* 38, 285.
Pereira A., J.M.E. Jacobs, W.T. Lintel-Hekkert, E. Rutgers, E. Jacobsen and W.J. Stiekema, 1992, *Neth. J. Plant Pathol.* 98, 215.
Pflieger S., V. Lefebvre and M. Causse, 2001, *Mol. Breeding* 7, 275.
Rensink W., S. Iobst, A. Hart, S. Stegalkina, J. Liu and C.R. Buell, 2005, *Funct. Integr. Genomics* 5, 201.
Roessner U., A. Luedemann, D. Brust, O. Fiehn, T. Linke, L. Willmitzer and A.R. Fernie, 2001, *Plant Cell* 13, 11.
Roessner U., C. Wagner, J. Kopka, R.N. Trethewey and L. Willmitzer, 2000, *Plant J.* 23, 131.
Roessner U., L. Willmitzer and A.R. Fernie, 2002, *Plant Cell Rep.* 21, 189.
Ronning C.M., S.S. Stegalkina, R.A. Ascenzi, O. Bougri, A.L. Hart, T.R. Utterbach, S.E. Vanaken, S.B. Riedmuller, J.A. White, J. Cho, G.M. Pertea, Y. Lee, S. Karamycheva, R. Sultana, J. Tsai, J. Quackenbush, H.M. Griffiths, S. Restrepo, C.D. Smart, W.E. Fry, R. van der Hoeven, S. Tanksley, P.F. Zhang, H.L. Jin, M.L. Yamamoto, B.J. Baker and C.R. Buell, 2003, *Plant Physiol.* 131, 419.
Rorat T., W. Irzykowski and W.J. Grygorowicz, 1997, *Plant Sci.* 124, 69.
Rosahl S., R. Schmidt, J. Schell and L. Willmitzer, 1986, *Mol. Gen. Genet.* 203, 214.
Rosin F.M., J.K. Hart, H.T. Horner, P.J. Davies and D.J. Hannapel, 2003a, *Plant Physiol.* 132, 106.
Rosin F.M., J.K. Hart, H. van Onckelen and D.J. Hannapel, 2003b, *Plant Physiol.* 131, 1613.

Rouppe van der Voort J.R., K. Kanyuka, E. Van Der Vossen, A. Bendahmane, P. Mooijman, R. Klein-Lankhorst, W. Stiekema, D. Baulcombe and J. Bakker, 1999, *Mol. Plant Microbe Interact.* 12, 197.
Ryu C.M., A. Anand, L. Kang and K.S. Mysore, 2004, *Plant J.* 40, 322.
Saha S., A.B. Sparks, C. Rago, V. Akmaev, C.J. Wang, B. Vogelstein, K.W. Kinzler and V.E. Velculescu, 2002, *Nat. Biotechnol.* 20, 508.
Schweizer G., N. Borisjuk, L. Borisjuk, M. Stadler, T. Stelzer, L. Schilde and V. Hemleben, 1993, *Theor. Appl. Genet.* 85, 801.
Song J., J.M. Bradeen, S.K. Naess, J.P. Helgeson and J. Jiang, 2003, *Theor. Appl. Genet.* 107, 958.
Stitt M. and A.R. Fernie, 2003, *Curr. Opin. Biotech.* 14, 136.
Stupar R.M., J.Q. Song, A.L. Tek, Z.K. Cheng, F.G. Dong and J.M. Jiang, 2002, *Genetics* 162, 1435.
Tanksley S.D., M.W. Ganal, J.P. Prince, M.C. de Vicente, M.W. Bonierbale, P. Broun, T.M. Fulton, J.J. Giovannoni, S. Grandillo, G.B. Martin, R. Messeguer, J.C. Miller, L. Miller, A.H. Paterson, O. Pineda, M.S. Roder, R.A. Wing, W. Wu and N.D. Young, 1992, *Genetics* 132, 1141.
Taylor M.A., H.A. Ross, A. Gardner and H.V. Davies, 1995, *J. Plant Physiol.* 145, 253.
Taylor M.A., F. Wright and H.V. Davies, 1994, *Plant Mol. Biol.* 26, 1013.
Trindade L.M., B. Horvath, C. Bachem, E. Jacobsen and R.G.F. Visser, 2003, *Gene* 303, 77.
Trindade L.M., B.M. Horvath, R. van Berloo and R.G.F. Visser, 2004, *Plant Sci.* 166, 423.
Urbanczyk-Wochniak E., A. Luedemann, J. Kopka, J. Selbig, U. Roessner-Tunali, L. Willmitzer and A.R. Fernie, 2003, *EMBO Rep.* 4, 989.
Van Berkel J., F. Salamini and C. Gebhardt, 1994, *Plant Physiol.* 104, 445.
Van der Hoeven R., C. Ronning, J. Giovannoni, G. Martin and S. Tanksley, 2002, *Plant Cell* 14, 1441.
Van der Vossen E., A. Sikkema, B.T.L. Hekkert, J. Gros, P. Stevens, M. Muskens, D. Wouters, A. Pereira, W. Stiekema and S. Allefs, 2003, *Plant J.* 36, 867.
van Enckevort L.J.G., J. Lasschuit, W.J. Stiekema, E. Jacobsen and A. Pereira, 2001, *Mol. Breeding* 7, 117.
van Os H., S. Andrzejewski, E. Bakker, I. Barrena, G. Bryan, B. Caromel, B. Ghareeb, E. Isidore, W. de Jong, P. van Koert, V. Lefebvre, D. Milbourne, E. Ritter, J.N.A.M. Rouppe van der Voort, F. Rousselle-Bourgeois, J. van Vliet, R. Waugh, R.G.F. Visser, J. Bakker and H.J. van Eck, 2006, *Genetics* 173, 1075.
Velculescu V.E., L. Zhang, B. Vogelstein and K.W. Kinzler, 1995, *Science* 270, 484.
Visser R.G.F., M. Hergersberg, F.R. Vanderleij, E. Jacobsen, B. Witholt and W.J. Feenstra, 1989, *Plant Sci.* 64, 185.
Wang Y., R.S. van der Hoeven, R. Nielsen, L.A. Mueller and S.D. Tanksley, 2005, *Theor. Appl. Genet.* 112, 72.
Weigel D., J.H. Ahn, M.A. Blazquez, J.O. Borevitz, S.K. Christensen, C. Fankhauser, C. Ferrandiz, I. Kardailsky, E.J. Malancharuvil, M.M. Neff, J.T. Nguyen, S. Sato, Z.Y. Wang, Y. Xia, R.A. Dixon, M.J. Harrison, C.J. Lamb, M.F. Yanofsky and J. Chory, 2000, *Plant Physiol.* 122, 1003.
Wolter F.P., C.C. Fritz, L. Willmitzer, J. Schell and P.H. Schreier, 1988, *Proc. Natl. Acad. Sci. U.S.A.* 85, 846.
Zamir D. and S.D. Tanksley, 1988, *Mol. Gen. Genet.* 213, 254.
Zhang H.N., J.P.T. Valkonen and K.N. Watanabe, 2003, *Breeding Sci.* 53, 155.

Chapter 10

Potato Cytogenetics

Tatjana Gavrilenko

N.I. Vavilov Institute of Plant Industry, B. Morskaya Str. 42/44, 190000, St. Petersburg, Russia

10.1 INTRODUCTION

Common potato, *Solanum tuberosum*, belongs to the section *Petota*, which is subdivided into 21 series with 228 wild and 7 cultivated species (Hawkes, 1994). According to the latest view, the section contains 199 wild and 1 cultivated species (Spooner and Hijmans, 2001; Huamán and Spooner, 2002; Chapter 4, van den Berg and Jacobs, this volume). Cytogenetic research helped to create the genome concept of wild and cultivated potato species (reviewed by Matsubayashi, 1991), to study haploid production and to use haploids in genetics and breeding (reviewed by Peloquin et al., 1991), to monitor the chromosome status of hybrid material (reviewed by Hermsen, 1994) and to investigate chromosome instability (reviewed by Wilkinson, 1994). This chapter surveys the application of cytogenetic methods for the investigation of genomic, evolutionary and species relationships, the integration of genetic and cytological maps, the analysis of genome structure and the detection of introgressions of alien chromatin. Besides traditional cytogenetic methods, the potential of new molecular techniques is considered.

10.2 BASIC CHROMOSOME NUMBER AND POLYPLOID COMPLEXES

Determination of chromosome number for *S. tuberosum* was the beginning of cytogenetic studies of potato. The haploid chromosome number ($n = 24$) was established for the first time by Kihara (1924). Later, the somatic chromosome number ($2n = 48$) was provided by Stow (1926) for varieties of the common potato. Approximately at the same time, the first indications of the existence of different ploidy levels in the wild potatoes were provided by investigators studying meiosis in pollen mother cells of *Solanum chacoense, Solanum jamesii, Solanum fendleri, Solanum* × *edinense* and *Solanum demissum* (Salaman, 1926; Smith, 1927; Vilmorin and Simonet, 1927). Rybin (1929, 1933) first described the whole polyploid series in wild potatoes (2x-3x-4x-5x-6x) and established an entire polyploid series in cultivated species (2x-3x-4x-5x). Rybin (1929) proposed to use differences in ploidy level for taxonomic classification of cultivated potatoes. All species of the section *Petota* have the same basic chromosome number (x = 12). Of the potato species with known chromosome number, 73% are classified as diploid ($2n = 2x = 24$), 4% triploid ($2n = 3x = 36$), 15% tetraploid ($2n = 4x = 48$), 2% pentaploid ($2n = 5x = 60$) and 6% hexaploid ($2n = 6x = 72$) (Hawkes, 1990).

Potato Biology and Biotechnology: Advances and Perspectives
D. Vreugdenhil (Editor)

Two major mechanisms have been proposed to explain the origin of polyploidy: chromosome doubling of somatic cells and formation of unreduced gametes (sexual polyploidization). Harlan and De Wet (1975) argued that almost all polyploids in nature have originated through sexual polyploidization. This is particularly true for the species of the section *Petota*, many of which often form both $2n$ pollen and $2n$ eggs (Watanabe and Peloquin, 1991). $2n$ gametes provide opportunities for gene flow between species with different ploidy levels and/or different endosperm balance numbers (EBNs) (Den Nijs and Peloquin, 1977). Thus, in addition to causing polyploidization, the ability to form $2n$ gametes also facilitated interspecific hybridization, which has played an important role in the evolution of wild and cultivated potatoes and in the formation of polyploid complexes in the section *Petota*. There are two major types of polyploids: autopolyploids, which received their homeologous set of chromosomes from one species, and allopolyploids, which received their homeologous set of chromosomes from different species. Determination of the type of polyploidy for species in the section *Petota* has been based mainly on the analysis of chromosome pairing in species and their hybrids. In general, strict allotetraploid and allohexaploid species show regular meiosis with bivalent chromosome pairing and extremely low frequency of multivalents. Triploid, pentaploid and autotetraploid species show high frequency of multivalents at metaphase I (MI), irregular meiosis and sterility or very low level of fertility. These species are maintained mainly by vegetative propagation. Some of the polyploids are classified as segmental allopolyploids; they are characterized by 'intermediate' frequencies of multivalents – lower than in autopolyploids and higher than in strict allopolyploids of corresponding ploidy levels.

10.3 GENOME AND SPECIES RELATIONSHIPS

The genome concept has been developed for potato species based on the crossability rate in interspecific combinations, hybrid viability, pollen fertility and the degree of chromosomal homology (Marks, 1955, 1965; Hawkes, 1958; Irikura, 1976; Ramanna and Hermsen, 1981; Hawkes, 1990; Lopez and Hawkes, 1991; Matsubayashi, 1991). Chromosome-pairing relationships in interspecific hybrids and in polyploid species have been interpreted by genome formulas, although authors gave them different symbols. In the last most of the authors agree on the genome hypothesis of Matsubayashi (1991). According to this hypothesis, five genomes (A, B, C, D and P) are recognized in tuber-bearing species of the section *Petota*. A genome E (Ramanna and Hermsen, 1981) is recognized in non-tuber-bearing species of the closely related section *Etuberosum*.

10.3.1 Genomic designation and relationships of diploid potato species

According to Matsubayashi (1991), all diploid tuber-bearing species growing under extremely diverse climatic conditions and exhibiting a wide range of morphological differences comprise one major genomic group A. No diploid species have ever been identified with B, C, D and P genomes. The basic genome A was proposed for diploid species of the four series, *Tuberosa, Commersoniana, Cuneoalata* and *Megistacroloba*, which all have identical (or very similar) genome(s). As reviewed by Matsubayashi (1991), hybrids between diploid species with the AA genome show 12 bivalents at MI, regular meiosis and fertile pollen. Diploid hybrids between species having the A genome and the other diploid potatoes show more or less reduced pollen fertility, and their amphidiploids are characterized by preferential pairing (reviewed by Matsubayashi, 1991). It was hypothesized that genomic variants of diploid potatoes of the *Bulbocastana, Ingifolia, Conicibaccata, Morelliformia, Pinnatisecta, Piurana* and *Polyadenia* series differ from the basic A genome by cryptic structural differences and that genomic variants of diploid species of the *Olmosiana* series and *Solanum rachialatum (Ingifolia* series) differ from other variants of the A genome by definite structural differences (Matsubayashi, 1991). The genomic variants of diploid species belonging to the above-mentioned eight series were designated by Matsubayashi (1991) as genome formula A with superscripts corresponding to each taxonomical series. Dvořák (1983) gave another explanation of differential affinity between the genomic variants of diploid potato species. He suggested that rapid evolution of non-coding sequences caused the differentiation of genomes of diploid tuber-bearing species.

10.3.2 Genomic nature and relationships in polyploid potato species

Relatively few polyploid members of the section *Petota* have been identified that appear to be autopolyploids. Multiple cytotypes ('cytotype' – any variety of a species whose chromosome complement differs quantitatively or qualitatively from the standard complement of the species; Rieger et al., 1991) of diploid species may be of autopolyploid origin. Triploid and tetraploid cytotypes are known for many typically diploid potato species (Hawkes, 1990). Triploid cytotypes derive from the union of unreduced ($2n$) and normal (n) gametes of the same diploid species, and tetraploid cytotypes can be produced by the fertilization of $2n$ egg cells with $2n$ pollen of a diploid species. Autotriploids should have a high frequency of trivalents at MI. Indeed, Sanuda Palazuelos (1962) observed up to eight trivalents in a triploid cytotype ($2n = 36$) of *Solanum cardiophyllum*, which is similar to the 8.4–10.3 trivalents per cell formed at MI in synthetic autotriploids (Irikura, 1976).

Among even-level polyploid potato species, multivalents occur very rarely. The frequency of multivalents at MI in *S. tuberosum* ($2n = 4x = 48$) ranging from 1.5 to 5.2 (Matsubayashi, 1991) is much higher than in other tetraploid species but lower than in synthetic autotetraploids. Chromosomes of *S. tuberosum* pair, recombine and segregate randomly as common potato displays tetrasomic inheritance ratios (Bradshaw and Mackay, 1994). Thus, *S. tuberosum* is one of the exceptional examples of a polysomic polyploid (autotetraploid – AAAA genome) in the section *Petota*. Both regular bivalent pairing and univalents at MI were quite frequently observed in dihaploids ('dihaploid' – an individual produced from a tetraploid form, which possesses half the tetraploid number of chromosomes; Rieger et al., 1991) of common potato. Unpaired segments in bivalents of some dihaploids have been reported (Matsubayashi, 1991). Therefore, segmental allotetraploidy and the genome formula AAA^tA^t were proposed by Matsubayashi (1991) for common potato. One possible explanation for the disagreements about the polyploid nature of *S. tuberosum* is the introgression of germplasm of wild and cultivated species into Andigena and Chilean landraces and into varieties of common potato.

Hawkes (1990) hypothesized that about 12% of potato species have a hybrid origin. Allopolyploids can originate from spontaneous interploid crosses between species possessing the same EBN or spontaneous crosses between species with functional $2n$ gametes and different EBNs or crosses between diploid species with the same EBN and mitotic polyploidization following the hybridization event or fertilization between $2n$ male and female gametes of two diploid species. For instance, the triploid species *S.× vallis-mexici* is a natural hybrid between *Solanum stoloniferum* ($2n = 48$, EBN $= 2$) and *Solanum verrucosum* ($2n = 24$, EBN $= 2$) (Marks, 1958). The pentaploid species *Solanum curtilobum* derived from the fusion of an unreduced (3x) gamete of *Solanum juzepczukii* and a normal (2x) gamete of *Solanum andigenum* ssp. *andigena* (Hawkes, 1962).

Segmental allopolyploidy has been proposed for polyploids of the series *Tuberosa, S. chaucha* (AAA^t), *S. juzepczukii* (AAA^a), *S. curtilobum* ($AAAA^aA^t$) and *S. sucrense* (AAA^sA^s), and for the wild species *Solanum acaule* of the *Acaulia* series (AAA^aA^a) by comparing the frequency of multivalent formation at MI in the species and their haploids or hybrids (Matsubayashi, 1991). We also suppose segmental polyploidy for the tetraploid species *Solanum tuguerrense* of the *Piurana* series, although Matsubayashi (1991) considered it as a strict allotetraploid (A^pA^pPP). However, the observation of a high frequency of trivalents at MI (4.5 trivalents +7.5 bivalents +7.5 univalents per cell) in triploid hybrids (AA^pP) of *S. tuguerrense* with *S. verrucosum* (AA) (Marks, 1965) indicates partial homology of the A^p and P genomes. For comparison, in triploid hybrids (AAA^a) between the segmental allotetraploid *S. acaule* and several diploid A-genome species, the frequency of trivalents at MI ranged from 3.0 to 6.5 (Propach, 1937; Swaminathan and Howard, 1953; Irikura, 1976).

Wild polyploid species of the series *Longipedicellata, Conicibaccata* and *Demissa* are considered as strict allopolyploids (disomic polyploids) based on the results of meiotic studies that showed regular bivalent pairing (Marks, 1955, 1965; Hawkes, 1958; Irikura, 1976; Lopez and Hawkes, 1991; Matsubayashi, 1991). According to Dvořák (1983), bivalent chromosomal pairing in allopolyploid potato species can be explained by genetically controlled regulatory mechanisms preventing intergenomic pairing. However, no convincing data confirming this hypothesis have ever been obtained.

All authors agree that strict allopolyploids share one common component genome, which is highly homologous to the A genome of diploid potato species (Marks, 1965; Irikura, 1976; Matsubayashi, 1991). Based on the analysis of chromosome pairing in hybrids, the diploid species *S. verrucosum* (AA) was suggested as the putative contributor of the common A genome of natural allopolyploids (Marks, 1965). A common origin of *S. verrucosum* and Mexican polyploid species was supported by the similarity of their cpDNA (Spooner and Sytsma, 1992) and by geographical and morphological data. Amplified fragment-length polymorphism (AFLP) results also support a close relationship between *S. verrucosum* and members of the *Longipedicellata, Demissa* and *Acaulia* series (Kardolus, 1998).

All authors also agree that strict allopolyploids differ from one another by their second component genome (Marks, 1965; Irikura, 1976; Matsubayashi, 1991). According to Irikura (1976), allopolyploid species differ from one another by the genomic variants of a merged B genome. Thus, genome designation AAB^sB^s was proposed for allotetraploid species of the *Longipedicellata* series, $AAB^sB^sB^dB^d$ for allohexaploid species of the

Demissa series and AAB^aB^a for segmental allotetraploid species *S. acaule* (Irikura, 1976). According to the genome hypothesis of Matsubayashi (1991), strict allopolyploid species differ from one another by their second specific distinct component genomes B, C, P and D. The B component genome has been recognized in the allopolyploid species of the *Longipedicellata* series (AABB). Genome C has been recognized in the allotetraploid species of the *Conicibaccata* series ($A^cA^cC^\backslash C^\backslash$), genome P in the allotetraploid species of the *Piurana* series and D genomes in the allohexaploid species of the *Demissa* series ($AADDD^\backslash D^\backslash$) (Matsubayashi, 1991). A more complex genome composition has been proposed for allohexaploid species of the *Conicibaccata* and *Acaulia* series. It was suggested that *Solanum moscopanum* ($2n = 6x$) contains a genome of *Solanum colombianum* ($A^cA^cC^\backslash C^\backslash$) and an additional, distinct MM genome of unknown diploid species origin (Lopez and Hawkes, 1991). *Solanum albicans* contains a genome of *S. acaule* and an additional, distinct XX genome of unknown origin (Hawkes, 1963; Matsubayashi, 1991). Nuclear restriction fragment-length polymorphism (RFLP) data confirm that *S. acaule* (AAA^aA^a) is an ancestor of *S. albicans* (Nakagawa and Hosaka, 2002).

Hawkes (1990) hypothesized that the B genome was a 'primitive' indigenous genome from Mexico. Irikura (1976) considered *S. cardiophyllum* as a possible donor of the second merged B genome in natural allopolyploids (Irikura, 1976). However, no experimental evidence was provided. Most authors agree that the origin of the second component genomes of natural allopolyploids is still unknown. It is unlikely that all diploid progenitors of the A^a, B, C and D genomes disappeared. It is possible that the A^a, B and D genomes were derived from a common ancestor and were then modified during the speciation of allopolyploids. This assumption is supported by molecular data that cluster the A^a, B and D genome-containing species (Kardolus, 1998; Nakagawa and Hosaka, 2002). The meiotic behaviour in hybrids also indicates similarity between the A^a, one of the D genomes and the B genomes. For instance, the high frequency of bivalents (5.3 univalents + 24.4 bivalents + 0.7 trivalents + 0.9 quadrivalents; Bains, 1951) in a pentaploid hybrid (AAA^aDD^d) of *S. demissum* ($AADDD^dD^d$) with *S. acaule* (AAA^aA^a) indicates that parental species share two common genomes. Meiotic configurations (15–17 univalents + 20–21 bivalents + 1 trivalent) in pentaploid hybrids ($AABDD^d$) of *S. demissum* ($AADDD^dD^d$) and *S. stoloniferum* (AABB) mean that bivalents are formed between the two A genomes and that most chromosomes of the B genome and one of the D genomes are paired. To reflect the close relationships between *S. demissum* and members of the *Acaulia* and *Longipedicellata* series, Kardolus (1998) proposed the new genome formula $AAA^aA^aB^dB^d$ for *S. demissum*.

During the evolution of natural allopolyploids, the second component genome could be significantly modified compared with the original ancestral genome donor. The hypothesis of Zohary and Feldman (1962) suggested different rates of parental genome modification in allopolyploid species. According to this hypothesis, one subgenome of natural allopolyploids remains stable and very close to the ancestral genome, whereas the second subgenome is modified relative to its progenitor because of introgressive hybridization. It might be suggested that in potato allopolyploids the A subgenome is stable and the second component genome was significantly modified. For instance, hybrids (genome AAA^aB) between *S. acaule* and species of the *Longipedicellata* series are characterized by a high multivalent frequency (0.8–1.3 quadrivalents + 2.2–3.4 trivalents + 14.2–15.8

bivalents + 6.1–4.5 univalents; Matsubayashi, 1991) that probably could reflect structural chromosomal changes accumulated in the A^a and B subgenomes.

It should be mentioned that some cytogenetic studies lack important information either due to limitations associated with the use of single genotype crosses, a single hybrid clone and a single accession of a polyploid species or due to an insufficient number of meiotic cells analysed. Chromosomal configurations were analysed at MI, whereas a true reflection of pairing has to be observed at the pachytene or zygotene stages. Meiotic studies have been performed by conventional methods with limited power to definitely interpret genome affinity in allopolyploids due to the inability to distinguish intergenomic and intragenomic pairing. Besides, the type of meiotic configurations (bivalents, trivalents or quadrivalents) alone is not a sufficient indicator for determining the nature of polyploidy. Predominantly, bivalent chromosome pairing has been described for several autopolyploid species with tetrasomic inheritance (Crawford and Smith, 1984; Samuel et al., 1990). In such cases, natural pressure for high fertility could select mutations in pairing control genes and result in change from random to preferential pairing in autopolyploids. Studies of inheritance patterns of molecular markers would provide more information about the polysomic or disomic inheritance type of polyploids. Obviously, the existing genome concepts of polyploid species of the section *Petota* need to be developed by further studies.

10.3.3 Genomic designation and relationships of potato and non-tuber-bearing species from closely related sections *Etuberosum, Juglandifolium* and *Lycopersicum*

All species of the section *Petota* and the closest non-tuber-bearing relatives from sections *Etuberosum, Juglandifolium* and *Lycopersicum* (Spooner et al., 1993) have the same basic chromosome number (x = 12) and similar karyotype morphology. Genome symbol E was given to the species of the section *Etuberosum* based on the specificity of meiotic behaviour and sterility of their diploid hybrids with A-genome tuber-bearing potato species (Ramanna and Hermsen, 1979, 1981). The distinct genome symbol S has been postulated for *Solanum sitiens* and *Solanum lycopersicoides* of the section *Juglandifolium* based on the differences detected among genetic maps of these species and tomato (Pertuze et al., 2002). Symbol L was proposed for tomato (section *Lycopersicum*) on the basis of preferential chromosome pairing and clear-cut parental genome discrimination by using genomic in situ hybridization (GISH) in amphidiploids of the LLEE type between tomato and *Solanum etuberosum* (Gavrilenko et al., 2001).

The results of comparative mapping studies revealed a high level of conservation of most linkage groups of the A, L, S and E genomes as well as genetically detected inversions, translocations and transpositions (Tanksley et al., 1992; Perez et al., 1999; Pertuze et al., 2002).

These results indicate that S- and L-genome species are most closely related and characterized by the lowest genome differentiation. Differentiation between L and A genomes is more profound, and the E genome is the most divergent within these taxa indicating distinctiveness of the section *Etuberosum*.

10.4 KARYOTYPING OF POTATO SPECIES

Potato is not an ideal species for cytogenetic research. Small somatic metaphase chromosomes of *S. tuberosum* ranging in length from 1.0 to 3.5 μm (Dong et al., 2000) are critical for identification. Low level of karyotype divergence among potato species as well as of those from the closely related sections complicates the application of traditional cytogenetic approaches to the analysis of introgression. Another disadvantage of cytogenetic research in potato is the absence of aneuploid stocks such as monosomic and nullisomic lines and lack of well-characterized structural chromosome mutants with translocations, inversions or deletions, which are routinely employed in other species for assigning linkage groups to individual chromosomes or for locating genes on specific chromosomes.

The first attempts to identify specific somatic chromosomes of potato stained with DNA-binding dyes such as aceto-carmine were based on the analysis of chromosome length, centromere position and the presence of secondary constrictions (Shepeleva, 1937; Lamm, 1945; Swaminathan, 1954). However, the small size and slight differences in morphology did not allow to distinguish precisely specific metaphase chromosomes. The distribution of highly repetitive DNA sequence on potato chromosomes was studied using Giemsa C-banding techniques with the aim to distinguish specific chromosomes (Mok et al., 1974; Lee and Hanneman, 1976; Pijnacker and Ferwerda, 1984). Even though significant progress has been made in the identification of Giemsa-stained chromosomes, difficulties persisted in the discrimination among chromosomes with similar morphology and similar C-banding patterns.

The pachytene chromosome complement was described for several diploid species and dihaploid clones of common potato (Haynes, 1964; Yeh and Peloquin, 1965; Marks, 1969; Ramanna and Wagenvoort, 1976; Wagenvoort, 1988). Potato chromosomes at pachytene show dark staining heterochromatin in pericentromeric regions and light staining euchromatin in terminal regions. These staining patterns together with other chromosomal landmarks such as position of centromeres, heterochromatin knobs and the size of telomeres allow to distinguish each of the 12 potato chromosomes. However, wide application of pachytene karyotyping was limited in cytogenetic research of potato because this method is elaborate and time consuming, and it can be applied only to diploid clones with excellent quality of chromosomal preparations.

10.4.1 Fluorescent in situ hybridization-based cytogenetic mapping

Development of fluorescent in situ hybridization (FISH) techniques for plant species provided new opportunities for the characterization of the potato genome, including chromosome identification and analysis of genome structure. The use of FISH with genomic DNA cloned in large-insert vectors such as bacterial artificial chromosomes (BACs), called BAC-FISH, has been an effective approach in mapping small probes containing only a few kilobases of DNA to physical chromosomes (Jiang et al., 1995). This approach has been used by Jiang and colleagues for correlating specific chromosomes with molecular linkage groups of potato. BACs with large genomic DNA insertions of the wild diploid species *Solanum bulbocastanum* were screened using mapped RFLP markers (Song et al., 2000). RFLP marker-specific BAC clones were labelled as FISH probes that were successfully applied to identify each of the 12 somatic metaphase chromosomes of potato (Dong et al., 2000; Fig. 10.1A). As a result, a larger set of new, chromosome-specific cytogenetic DNA markers (CSCDMs) was established for potato karyotyping to integrate the genetic and cytological maps of potato. This system has the following methodical advantages: CSCDMs clearly discriminate between different chromosomes with similar morphology, CSCDMs can be applied to polyploids with larger

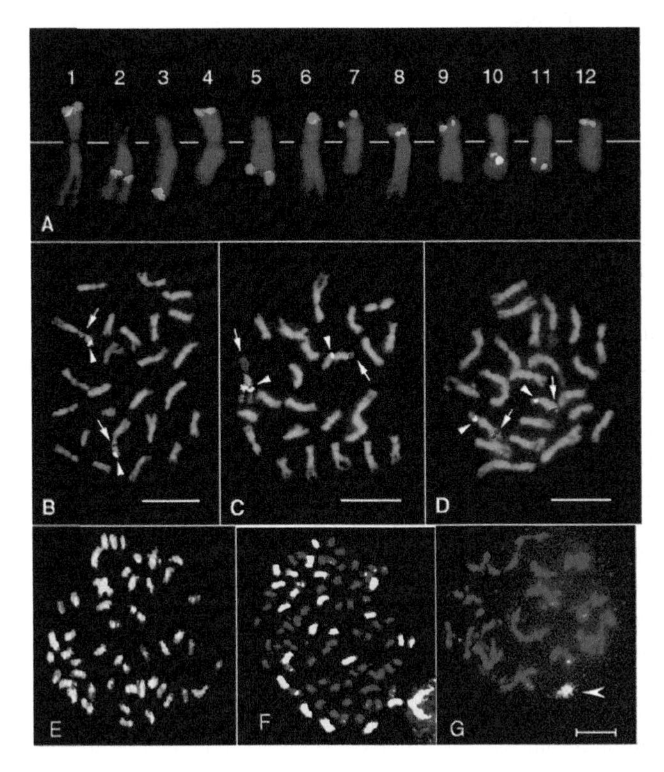

Fig. 10.1. (A) Twelve individual potato chromosomes with fluorescent in situ hybridization (FISH) signals derived from the chromosome-specific cytogenetic DNA markers (CSCDMs). (B) The 5S rRNA genes (red colour and arrows) are located near the centromeres at the same chromosome as chromosome 1-specific DNA marker (yellow colour and arrowheads). (C) The 45S rRNA genes (red colour and arrows) were mapped to the distal region on the short arm of the same chromosome where chromosome 2-specific DNA marker (yellow colour and arrowheads) was located. (D) Bacterial artificial chromosome (BAC) clone, 32A07, which is linked to a potato late blight resistance gene (red colour and arrows), was mapped to the long arm of the same chromosome where the chromosome 8-specific marker (yellow colour and arrowheads) was located. (A–D: from Dong et al., 2000, with kind permission of Springer Science and Business Media.) (E) Genomic in situ hybridization (GISH) of mitotic cells of BC_2 hybrid with 39 chromosomes of potato (yellow colour) and 12 chromosomes of *Solanum etuberosum* (red colour) (Gavrilenko et al., 2003). (F) Somatic cell of hybrid derived from *Solanum nigrum* (+) potato backcross programme with 22 chromosomes of *S. nigrum* (yellow colour) and 36 chromosomes of potato (red colour) (Horsman et al., 2001). (G) Diakinesis stage in the monosomic addition for chromosome 8 of tomato into the potato genome, showing the alien chromosome as a univalent (arrowhead) (Garriga-Calderé et al., 1999). All bars are 10 μm.

chromosome numbers and the quality of chromosome preparations is not so important (Dong et al., 2000).

Visser et al. (1988) were the first to apply in situ hybridization techniques using radioactively labelled repetitive DNA sequences to study genome organization of potato. In further studies, FISH has been used to characterize the distribution of different types of repetitive sequences. Simultaneous hybridization of ribosomal DNA (rDNA) probes with CSCDMs, each labelled with a different fluorochrome, has resulted in mapping two large functionally important families of rDNA sequences of potato (Dong et al., 2000). 5S rDNA genes were located at a single locus near the centromere on the short arm of chromosome 1 (Dong et al., 2000; Fig. 10.1B). A similar location of a single 5S rDNA locus has been detected in tomato using FISH and pachytene analysis (Xu and Earle, 1996a). Only one 5S rDNA locus was found in the S-genome species of the section *Juglandifolium* (Ji et al., 2004). Therefore, no polymorphisms were detected in the number of 5S rDNA loci among the A, L and S genomes.

One major 45S rDNA locus containing 18S-5,8S-26S rRNA genes was found in the nucleolus organizer region (NOR) on the short arm of chromosome 2 in the A, L and S genomes (Fig. 10.1C). Variation in a genome-specific manner was only detected in the number and distribution patterns of minor 45S rDNA loci. Pachytene karyotyping of tomato in combination with FISH revealed four minor 45S rDNA loci that were located in the heterochromatic regions on four chromosomes of the L genome (2L, 6L, 9S and 11S arms) (Xu and Earle, 1996b). In the chromosome complements of the S-genome species, only one minor 45S rDNA locus was detected on chromosomes other than the nucleolar chromosome (Ji et al., 2004). No minor 45S rDNA loci have been reported for the A genome of potato (Dong et al., 2000).

Using FISH, tandemly repeated DNA elements that are highly homologous to the intergenic spacer (IGS) of the 18S–25S rDNA sequence of potato were located at distinct loci in a pericentromeric heterochromatic region on a single (not nucleolar) chromosome of *S. tuberosum* (Stupar et al., 2002). In *S. bulbocastanum*, the same repeated DNA elements were located close to centromeres and distributed on four different chromosomes (Stupar et al., 2002). The other classes of tandem repeats – interstitial telomeric repeats (ITRs) – have been located using FISH in highly condensed centromeric regions of two to seven different chromosomes in several *Solanum* species, and the number of the FISH signals did not correspond to species ploidy level (Tek and Jiang, 2004). The results of FISH on extended DNA fibres revealed that these ITRs are organized in long tandem clusters, suggesting extensive amplification of the ITRs during divergence of potato species (Tek and Jiang, 2004). Both IGS-related repeats and ITRs are highly diverged among a wide range of *Solanum* species indicating their dynamic nature (Stupar et al., 2002; Tek and Jiang, 2004). These results indicate that genome differentiation of the structurally similar, A-genome diploid potatoes might be due to divergence in nucleotide sequences and amplification of different classes of highly repetitive DNA.

Fluorescent in situ hybridization with tandemly repeated, species-specific DNA sequences can be used for comparative karyotyping and for studying introgression. For instance, the pSB1 and pSB7 repeats specific to the E-genome species of the *Etuberosum* section were located mostly in the telomeric and in some centromeric and interstitial areas of the *Solanum brevidens* chromosomes, but not in the *S. tuberosum* chromosomal complement. Whereas the potato clone pST3 showed signals in telomeric regions of a few chromosomes of *S. tuberosum*, this signal was not detected in *S. brevidens* (Rokka et al., 1998a). Moreover, FISH with *S. brevidens*-specific sequences helped to clarify the genomic composition of hybrids between potato and *S. brevidens* (Rokka et al., 1998b).

Genomic in situ hybridization, based on the use of total genomic DNA as probe, has been developed by Schwarzacher et al. (1989) to identify chromosomes and chromosomal segments of different origin. The ability to discriminate chromatin of different genomes depends on the degree of sequence homology and stringency conditions in the GISH experiments. The standard GISH protocol allows to distinguish genomes sharing 80–85% or less sequence homology (Schwarzacher et al., 1989). Using standard GISH protocols, parental chromosomes were discriminated in wide hybrids between distantly related *Solanum* species belonging to different sections, such as *Petota* (potato) and *Lycopersicum* (tomato) (Garriga-Calderé et al., 1997), *Petota* (potato) and *Etuberosum* (*S. etuberosum* and *S. brevidens*) (Dong et al., 1999, 2001; Gavrilenko et al., 2002, 2003), *Petota* (potato) and *Solanum (Solanum nigrum)* (Horsman et al., 2001), *Etuberosum (S. etuberosum)* and *Lycopersicum* (tomato) (Gavrilenko et al., 2001), *Juglandifolium (S. lycopersicoides* and *S. sitiens*) and *Lycopersicum* (tomato) (Ji et al., 2004). Because the A, E, L and S genomes in wide hybrids can be easily discriminated using standard GISH protocols, these genomes are supposed to have a high level of divergence in their dispersed repetitive DNA sequences. Chromosomes of closely related genomes sharing up to 90–95% sequence homology can be discriminated under higher stringency conditions in combination with an excess of unlabelled blocking DNA in the hybridization mixture (Parokonny et al., 1997). Application of such modified GISH protocols allowed to discriminate chromosomes of closely related parental species belonging to the same section – *Lycopersicum* (Parokonny et al., 1997) or *Juglandifolium* (Ji et al., 2004).

Genomic in situ hybridization was successfully used to establish genome composition of wide hybrids and their derivatives (Fig. 10.1E and F), to discriminate between intergenomic and intragenomic pairing in the genomes of wide hybrids (Garriga-Calderé et al., 1999; Gavrilenko et al., 2001; Ji et al., 2004), to study the specificity of genome interactions such as preferential elimination of chromosomes of one parental genome (Garriga-Calderé et al., 1997; Gavrilenko et al., 2001) and to determine intergenomic translocations (Garriga-Calderé et al., 1997; Dong et al., 2001).

Despite the effectiveness of GISH in detecting chromatin of different origin, GISH alone cannot determine genetic identity of alien chromosomes. Sequential GISH and FISH with CSCDMs performed on the same chromosome preparations made it possible to identify precisely specific homeologous chromosomes of the E and A genomes in breeding lines derived from potato (+) *S. brevidens* hybrids (Dong et al., 2001, 2005; Tek et al., 2004). Combination of GISH and FISH with CSCDMs also allowed to determine the specificity of chromosomal re-arrangements (Dong et al., 2001).

10.5 CYTOGENETICS IN POTATO IMPROVEMENT

Wild potato species have been recognized as an important source of useful genes for resistance to pathogens and abiotic stresses (Hawkes, 1994). These gene pools are useful for the improvement of common potato that has a narrow genetic basis as many other crop species (Ross, 1986). Wild germplasm has been actively utilized in potato breeding for at least 70 years (Bukasov, 1937). Following interspecific crosses and backcrossing, all 11 known *R* genes conferring race-specific resistance to late blight have been introduced into potato varieties from *S. demissum* ($AADDD^dD^d$) (Umareus and Umareus, 1994). The virus resistance genes *Ry, Ra, Na* and *Rx2* have been introgressed into potato from *S. stoloniferum* (AABB) and *S. acaule* (AAA^aA^a), respectively (Solomon-Blackburn and Barker, 2001). Methods used for ploidy manipulation (Hougas and Peloquin, 1958) make most of the potato species with different EBNs cross-compatible with *S. tuberosum*. However, some potentially useful species, e.g. $A^{\backslash}A^{\backslash}$ genome-containing diploid Mexican species or $E^{\backslash}E^{\backslash}$ genome-containing species, cannot be hybridized easily because of the crossing barriers (Hermsen, 1994). The range of hybridization has been broadened using biotechnological methods that allowed to bring into breeding programmes new species such as *S. bulbocastanum, Solanum tarnii, S. etuberosum, S. brevidens* and *S. nigrum*. Following protoplast fusion, backcrossing and embryo or ovule rescue, fertile progenies derived from crosses of wide somatic hybrids with common potato have been produced. Some of these derivatives showed high levels of resistance to diseases. The list includes broad-spectrum resistance to late blight from *S. bulbocastanum* (A^bA^b) (Helgeson et al., 1998; Naess et al., 2000), resistance to tuber soft rot from *S. brevidens* (E^bE^b) (Tek et al., 2004) and resistance to viruses and aphids from *S. etuberosum* (E^eE^e) (Novy et al., 2002; Gavrilenko et al., 2003).

The most recent achievements in detecting introgression are discussed here briefly. Molecular markers and in situ hybridization techniques have been essential for detecting genetic material of wild species at the level of whole chromosomes, chromosomal segments and individual genes. These methods were useful for the development and characterization of heteromorphic aneuploid lines derived from crosses between distantly related taxa. For instance, an entire series of monosomic alien addition lines (MAALs) and two disomic addition lines for tomato chromosomes 10 and 11 ($AAAA+L^{10}$ and $AAAA+L^{11}$) into potato have been established using RFLP and GISH (Garriga-Calderé et al., 1998; Haider Ali et al., 2001; Fig. 10.1G). The application of sequential GISH and FISH with CSCDMs allowed to distinguish addition and substitution lines (Dong et al., 2005). Seven of 12 possible MAALs ($AAAA+E^b$) and one monosomic substitution for chromosome 6 of the E^b genome of *S. brevidens* have been extracted from BC_2 to BC_3 progenies derived from potato (+) *S. brevidens* hybrids (Dong et al., 2005). Importantly, the experiments provided the first evidence for the ability of chromosomes of the two distinct genomes (A and E) to substitute for each other. For practical purposes, these cytogenetic stocks can be useful for assigning unmapped gene(s) to chromosomes. Intergenomic translocations have been identified by using in situ hybridization methods in breeding lines originated from fusion hybrids of potato with tomato (Garriga-Calderé et al., 1997, 1999) and potato with *S. brevidens* (Dong et al., 2001). It must be pointed out that alien chromosome(s) or large alien translocated segments may not be stable when transmitted through backcrossing. Stable introgression can be achieved through crossing over. Following crossing of MAALs or substitution lines with common potato, it might be possible to select genotypes carrying chromosomes that originated because of homeologous recombination. However, selection of genotypes with recombinant chromosomes can be very laborious because of extremely low level of chromosome pairing between the parental genomes A and L (Garriga-Calderé et al., 1999) and limited level of crossing over between A and E genomes (McGrath et al., 1996).

Starting from 2000, new approaches based on molecular markers and genomics have been developed to overcome such limitations. Cloned resistance genes of wild species can be transferred through genetic engineering in susceptible varieties by passing the crossing barriers. Already durable and broad-spectrum resistance against all known races of the late blight pathogen *Phytophthora infestans* has been introgressed from *S. bulbocastanum* into potato by somatic hybridization and subsequent backcrossing (Helgeson et al., 1998; Naess et al., 2001). The major late blight resistance gene *RB* of *S. bulbocastanum* was physically mapped by FISH on potato chromosome VIII (Dong et al., 2000; Fig. 10.1D). *RB* was then cloned using a map-based approach and transformed into susceptible potato varieties (Song et al., 2003).

In conclusion, the introduction of in situ hybridization methods has promoted a significant progress in potato cytogenetics, which has led to the integration of genetic and cytological maps, getting new information about genome structure and detecting introgressions with higher precision. Furthermore, the development and use of molecular techniques will be of great help in better understanding genome evolution and polyploid formation, further development of genetic and physical mapping of genes controlling economically important traits in potato and providing new knowledge about their genetic basis.

AKNOWLEDGEMENTS

I thank Drs Munikote Ramanna, Jiming Jiang and Svetlana Temnykh for reading the manuscript and for helpful suggestions; Dr Jiming Jiang for providing Figs 10.1A–D; and Springer Science and Business Media, NRC Research Press and Blackwell Publishing for permission to publish Fig. 10.1A–D, F and G, respectively.

REFERENCES

Bains G.S., 1951, M.Sc.dissertation, University of Cambridge, Cambridge, UK.

Bradshaw J.E. and G.R. Mackay, 1994, In: J.E. Bradshaw and G.R. Mackay (eds) *Potato Genetics*, p. 467. CAB International, Wallingford, UK.

Bukasov S.M., 1937, In: N.I. Vavilov (ed.), *Theoretical Basis of Plant Breeding*, vol. 3, p. 3. Sate-Publisher Moscow-Leningrad.

Crawford D.G. and E.B. Smith, 1984, *Syst. Bot.* 9, 219.

Den Nijs T.P.M. and S.J. Peloquin, 1977, *Euphytica* 26, 585.

Dong F., J.M. McGrath, J.P. Helgeson and J. Jiang, 2001, *Genome* 44, 729.

Dong F., J.P. Novy, J.P. Helgeson and J. Jiang, 1999, *Genome* 42, 987.

Dong F., J. Song, S.K. Naess, J.P. Helgeson, C. Gebhardt and J. Jiang, 2000, *Theor. Appl. Genet.* 101, 1001.

Dong F., A.L. Tek, A.B.L. Frasca, J.M. McGrath, S.M. Wielgus, J.P. Helgeson and J. Jiang, 2005, *Cytogenet. Genome Res.* 109, 368.

Dvořák J., 1983, *Can. J. Genet. Cytol.* 25, 530.

Garriga-Calderé F., D.J. Huigen, A. Angrisano, E. Jacobsen and M.S. Ramanna, 1998, *Theor. Appl. Gen.* 96, 155.
Garriga-Calderé F., D.J. Huigen, F. Filotico, E. Jacobsen and M.S. Ramanna, 1997, *Genome* 40, 666.
Garriga-Calderé F., D.J. Huigen, E. Jacobsen and M.S. Ramanna, 1999, *Genome* 42, 282.
Gavrilenko T., J. Larkka, E. Pehu and V.-M. Rokka, 2002, *Genome* 45, 442.
Gavrilenko T., R. Thieme, U. Heimbach and T. Thieme, 2003, *Euphytica* 131, 323.
Gavrilenko T., R. Thieme and V.-M. Rokka, 2001, *Theor. Appl. Genet.* 103, 231.
Haider Ali S.N., M.S. Ramanna, E. Jacobsen and R.G.F. Visser, 2001, *Theor. Appl. Genet.* 103, 687.
Harlan J.R. and J.M.J. De Wet., 1975, *Bot. Rev.* 41, 361.
Hawkes J.G., 1958, In: H. Kappert and W. Rudorf (eds), *Handbuch Pflanzenzüchtung*, p. 1. Paul Parey, Berlin and Hamburg.
Hawkes J.G., 1962, *Z. Pflanzenzüchtung* 47, 1.
Hawkes J.G., 1963, *Scott. Plant Breed. Station Record*, 76.
Hawkes J.G., 1990, *The Potato: Evolution, Biodiversity and Genetic Resources*. Belhaven Press, Oxford, UK.
Hawkes J.G., 1994, In: J.E. Bradshaw and G.R. Mackay (eds), *Potato Genetics*, p. 3. CAB International, Wallingford, UK.
Haynes F., 1964, *J. Hered.* 55, 168.
Helgeson J., J. Pohlman, S. Austin, G. Haberlach, S. Wielgus, D. Ronnis, L. Zambolim, P. Tooley, J. McGrath, R. James and W. Stevenson, 1998, *Theor. Appl. Genet.* 96, 738.
Hermsen J.G.Th., 1994, In: J.E. Bradshaw and G.R. Mackay (eds), *Potato Genetics*, p. 515. CAB International, Wallingford, UK.
Horsman K., T. Gavrilenko, J. Bergervoet, D.-J. Huigen, A.T.W. Joe and E. Jacobsen, 2001, *Plant Breed.* 120, 201.
Hougas R.W. and S.J. Peloquin, 1958, *Am. Potato J.* 35, 701.
Huamán Z. and D.M. Spooner, 2002, *Am. J. Bot.* 89, 947.
Irikura Y., 1976, *Res. Bull. Hokkaido Nat. Agric. Exper. Station* 115, 1.
Ji Y., R. Pertuze and R.T. Chetelat, 2004, *Chromosome Res.* 12, 107.
Jiang J., B.S. Gill, G.L. Wang, P.C. Ronald and D.C. Ward, 1995, *Proc. Natl. Acad. Sci. U.S.A.* 92, 4487.
Kardolus J.P., 1998, *A Biosystematic Analysis of Solanum acaule*. Thesis, Wageningen, p. 100.
Kihara H., 1924, *Memoirs Coll. Sc. Kyoto Imp. Univ. Ser. B 1* Art. 1.
Lamm R., 1945, *Hereditas* 31, 1.
Lee H.K. and R.E. Hanneman Jr., 1976, *Can. J. Genet. Cytol.* 18, 297.
Lopez L.E. and J.G. Hawkes, 1991, In: J.G. Hawkes, R.N. Lester, M. Nee and R. Estrada (eds), *Solanaceae III: Taxonomy, Chemistry, Evolution*, p. 327. Royal Botanic Gardens, Kew, UK.
Marks G.E., 1955, *J. Genet.* 53, 262.
Marks G.E., 1958, *New Phytol.* 57, 300.
Marks G.E., 1965, *New Phytol.* 64, 293.
Marks G.E., 1969, *Caryologia* 23, 161.
Matsubayashi M., 1991, In: T. Tsuchiya and P.K. Gupta (eds), *Chromosome Engineering in Plants: Genetics, Breeding, Evolution, Part B*, p. 93. Elsevier, Amsterdam.
McGrath J.M., S.M. Wielgus and J.P. Helgeson, 1996, *Genetics* 142, 1335.
Mok D.W.S., K.L. Heiyoung and S.J. Peloquin, 1974, *Am. Potato J.* 51, 337.
Naess S., J. Bradeen, S. Wielgus, G. Haberlach, J. McGrath and J.P. Helgeson, 2000, *Theor. Appl. Genet.* 101, 697.
Naess S., J. Bradeen, S. Wielgus, G. Haberlach, J. McGrath, and J.P. Helgeson, 2001, *Mol. Gen. Genet.* 265, 694.
Nakagawa K. and K. Hosaka, 2002, *Am. J. Potato Res.* 79, 85.
Novy R., A. Nasruddin, D.W. Ragsdale and E.B. Radcliffe, 2002, *Am. J. Potato Res.* 79, 9.
Parokonny A.S., J.A. Marshall, M.D. Bennett, E.C. Cocking, M.R. Davey and J.B. Power, 1997, *Theor. Appl. Genet.* 94, 713.
Peloquin S.J., E.J. Werner and G.L. Yerk, 1991, In: T. Tsuchiya and P.K. Gupta (eds), *Chromosome Engineering in Plants: Genetics, Breeding, Evolution, Part B*, p. 79. Elsevier, Amsterdam.
Perez F., A. Menendez, P. Dehal and C.F. Quiros, 1999, *Theor. Appl. Genet.* 98, 1183.
Pertuze R.A., Y. Ji and R.T. Chetelat, 2002, *Genome* 45, 1003.

Pijnacker L.P. and M.A. Ferwerda, 1984, *Can. J. Genet. Cytol.* 26, 415.
Propach H., 1937, *Z. für Induktive Abstammungs und Vererbungslehre* 73, 143.
Ramanna M.S. and J.G.Th. Hermsen, 1979, *Euphytica* 28, 9.
Ramanna M.S. and J.G.Th. Hermsen, 1981, *Euphytica* 30, 15.
Ramanna M.S. and M. Wagenvoort, 1976, *Euphytica* 25, 233.
Rieger R., A. Michaelis and M.M. Green, 1991, *Glossary of Genetics Classical and Molecular*, Springer-Verlag, Germany.
Rokka V.-M., M.S. Clark, D.L. Knudson, E. Pehu and N.L.V. Lapitan, 1998a, *Genome* 41, 487.
Rokka V.-M., N.L.V. Lapitan, D.L. Knudson and E. Pehu, 1998b, *Agric. Food Sci. Finland* 7, 31.
Ross H., 1986, *Potato Breeding – Problems and Perspectives*. Paul Parey, Berlin.
Rybin V.A., 1929, *Bull. Appl. Bot. Genet. Breed.* 20, 655 (in Russian).
Rybin V.A., 1933, *Bull. Appl. Bot. Genet. Breed. Seria II* 3 (in Russian).
Salaman R.N., 1926, *Potato Varieties*. Cambridge University Press, Cambridge, UK.
Samuel R., W. Pinsker and F. Ehrendorfer, 1990, *Heredity* 65, 369.
Sanuda Palazuelos A., 1962, *Am. Inst. Nac. Agron. Madr.* 11, 191.
Schwarzacher T., A.R. Leitch, M.D. Bennett and J.S. Heslop-Harrison, 1989, *Ann. Bot.* 64, 315.
Shepeleva E.M., 1937, *Doklady Academii Nauk SSSR* 15, 207 (in Russian).
Smith H.B., 1927, *Genetics* 12, 84.
Solomon-Blackburn R. and H. Barker, 2001, *Heredity* 86, 8.
Song J., J.M. Bradeen, S.K. Naess, J.A. Raasch, S.L. Wielgus, G.T. Haberlach, J. Liu, H. Kuang, S. Austin-Philips, C.R. Buell, J.P. Helgeson and J. Jiang, 2003, *Proc. Natl. Acad. Sci. U.S.A.* 100, 9128.
Song J., F. Dong and J. Jiang, 2000, *Genome* 43, 199.
Spooner D.M., G.J. Anderson and R.K. Jansen, 1993, *Am. J. Bot.* 80, 676.
Spooner D.M. and R.J. Hijmans, 2001, *Am. J. Potato Res.* 78, 237.
Spooner D.M. and K.J. Sytsma, 1992, *Syst. Bot.* 17, 432.
Stow I., 1926, *Proc. Imperial Acad. Jpn.* 2, 426.
Stupar R.M., J. Song, A.L. Tek, Z.K. Cheng, F. Dong, and J. Jiang, 2002, *Genetics* 162, 1435.
Swaminathan M.S., 1954, *Genetics* 39, 59.
Swaminathan M.S. and H.W. Howard, 1953, *Bibliogr. Genet.* 16, 1.
Tanksley S.D., M.W. Ganal, J.P. Prince, M.C. Vicente, M.W. Bonierbale, P. Broun, T.M. Fulton, J.J. Giovannoni, S. Grandillo, G.B. Martin, R. Messeguer, J.C. Miller, L. Miller, A.H. Patterson, O. Pineda, M.S. Röder, R.A. Wing, W. Wu and N.D. Young, 1992, *Genetics* 132, 1141.
Tek A.L. and J. Jiang, 2004, *Chromosoma* 113, 77.
Tek A.L., W.R. Stevenson, J.P. Helgeson and J. Jiang, 2004, *Theor. Appl. Genet.* 109, 249.
Umareus V. and M. Umareus, 1994, In: J.E. Bradshaw and G.R. Mackay (eds), Potato Genetics, p. 365. CAB International, Wallingford, UK.
Vilmorin R. De and M. Simonet, 1927, *C.R. Acad. Sci. Paris* 184, 164.
Visser R.G.F., R. Hoekstra, F.R. Leij, L.P. Pijnacker, B. Witholt and W.J. Feenstra, 1988, *Theor. Appl. Genet.* 76, 420.
Wagenvoort M., 1988, *Theor. Appl. Genet.* 82, 621.
Watanabe K. and S.J. Peloquin, 1991, *Theor. Appl. Genet.* 82, 621.
Wilkinson M.J., 1994, In: J.E. Bradshaw and G.R. Mackay (eds) *Potato Genetics*, p. 43. CAB International, Wallingford, UK.
Xu J. and E.D. Earle, 1996a, *Genome* 39, 216.
Xu J. and E.D. Earle, 1996b, *Chromosoma* 104, 545.
Yeh B.P. and S.J. Peloquin, 1965, *Am. J. Bot.* 52, 1014.
Zohary D. and M. Feldman, 1962, *Evolution* 16, 44.

Part III

PLANT GROWTH AND DEVELOPMENT

Chapter 11

Above-Ground and Below-Ground Plant Development

Paul C. Struik

Crop and Weed Ecology, Plant Sciences Group, Wageningen University, Haarweg 333, 6709 RZ Wageningen, The Netherlands

11.1 INTRODUCTION

The potato plant (*Solanum tuberosum* L.) is a perennial herb, but in agriculture it is used as an annual crop. It is usually propagated using seed tubers. Figure 11.1 illustrates the traditional way of multiplying the potato. Seed tubers produce sprouts in their eyes, which develop into shoots, and produce roots from primordia on the sprouts. On these shoots, the stems, foliage, stolons, roots, inflorescences and the next generation of tubers are formed. For definitions of the different plant parts see Sections 11.4–11.9.

One may use very different propagules to propagate the potato plant or even the crop. These may include individual cells, meristems, tissues, sprouts, tuber parts, true (botanical) seeds and leaf or stem cuttings (Struik and Wiersema, 1999). This type of propagule has a strong effect on the above-ground and below-ground development. The size of the main shoot (number of branches, number of nodes per stem part, size of the individual leaves, abundance of flowering), the number and distribution of stolons along the below-ground part of the stem and the position, number and size of tubers are all affected by the type of propagule used to grow the plant. In this chapter, we focus on the development of the plant grown from a seed tuber.

11.2 GENERAL MORPHOLOGY

Figure 11.2 shows the morphology of a normal *S. tuberosum* spp. *tuberosum* plant. Included in this figure are a full-grown plant (A), a flower (B), a berry (C), a seedling (D, for comparison only), the below-ground parts (E), a tuber (F) and a single sprout on a seed tuber (G). Details on the morphology of the stem system, the leaves, the stolon system and the tubers are given in Sections 11.4, 11.5, 11.6 and 11.7, respectively.

The morphology of both the below- and the above-ground plant parts strongly depends on the size and the physiological age of the seed tubers planted. This is illustrated in Fig. 11.3 for the number of stems per seed tuber. However, at the same time, also the growth pattern of each individual stem is influenced through specific effects on vigour, number of leaves formed per stem element, time of stolonization, stolon

Potato Biology and Biotechnology: Advances and Perspectives
D. Vreugdenhil (Editor)

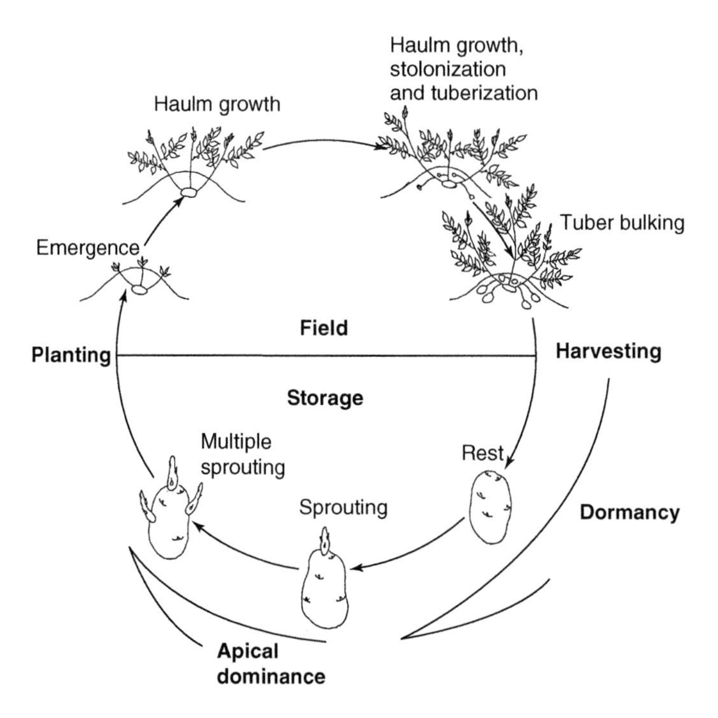

Fig. 11.1. Traditional way of multiplying the potato. Seed tubers may be harvested and stored under proper conditions to form sprouts. Seed tubers may be planted (either before or after sprouting) and will then produce new plants that produce progeny tubers [*Source: Struik and Wiersema (1999)*].

branching, tuberization and flowering, duration of tuber set and tuber growth, the onset of plant senescence and the length of the growth cycle. For the description of the morphology of the plant, we take as a norm a plant grown from a normal-sized seed tuber or seed piece (40–80 g fresh weight) of a 'normal' physiological age, that is an age usually encountered when seed potato tubers are taken from a normal seed potato crop, desiccated before tuber bulking has ended naturally, harvested a few weeks after haulm destruction and stored at 2–4 °C for about 8 months until planting in the next year.

11.3 SPROUT DEVELOPMENT

The sprouts that develop on the seed tuber go through different stages when the seed tuber is ageing. This development of the sprout is partly dependent on the ageing processes taking place in the seed tuber but can – to some extent – also be moderated independently from them by chemical treatment or exposure to certain conditions during storage (light, photoperiod, temperature and relative humidity). Sprout development is crucial for the performance of the seed tuber. An illustration of the morphological development of the sprout and its consequences for the crop is provided in Fig. 11.4.

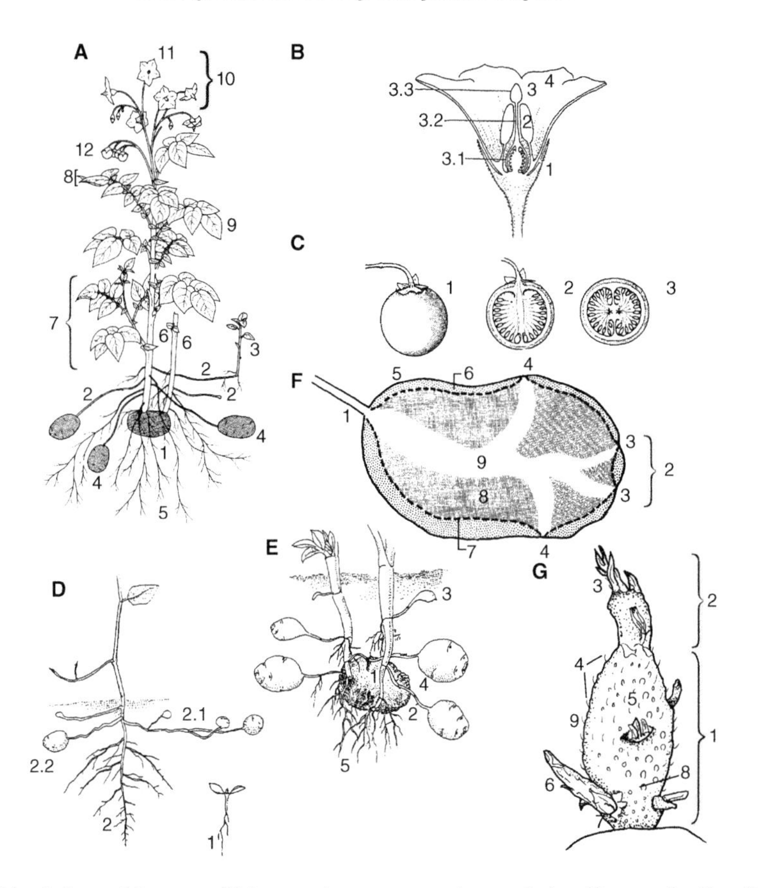

Fig. 11.2. Morphology of the potato (*Solanum tuberosum* spp. *tuberosum*) plant [*Source: Struik and Wiersema (1999)*]. (A) Entire plant: 1, seed tuber; 2, stolon; 3, stolon turning into a below-ground lateral stem; 4, tuber; 5, roots; 6, main stem; 7, above-ground lateral stem; 8, compound leaf; 9, leaflet; 10, inflorescence; 11, flower and 12, berries. (B) Flower: 1, calyx; 2, stamen, consisting of a filament and an anther; 3, female structure, with 3.1, ovary; 3.2, style and 3.3, stigma and 4, corolla, consisting of five petals that are often fused. (C) Berries: 1, general appearance; 2, longitudinal cut showing the position of the true seeds and 3, transversal cut, showing the position of the true seeds. (D) Seedling: 1, seedling shortly after emergence and 2, seedling shortly after tuberization, with 2.1, tuber-bearing stolon and 2.2, small tuber. (E) Below-ground plant parts of a plant from a seed tuber [1, seed tuber; 2, tuber-bearing stolon; 3, incipient tuber; 4, small tuber, with eyes consisting of an eyebrow and buds and with lenticels (not visible) that are important for gas exchange and 5, roots]. (F) Tuber: 1, basal (heel or stolon end); 2, apical (or rose) end; 3, apical eyes; 4, lateral eyes; 5, skin; 6, cortex; 7, vascular system; 8, storage parenchyma and 9, pith. (G) Single sprout on a seed tuber, with 1, basal part; 2, tip; 3, terminal bud; 4, hair; 5, undeveloped bud; 6, developed bud for lateral stem; 7, root tips; 8, lenticel and 9, main sprout developing into main stem.

11.4 THE SHOOT SYSTEM

From a seed tuber, usually several main stems are derived, because several buds produce a viable shoot. As there is more than one bud per eye, the number of stems per eye may even be more than one. An individual potato plant as it commonly manifests itself in

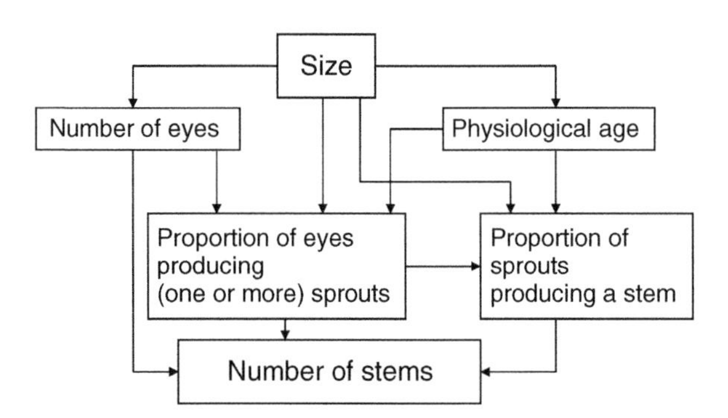

Fig. 11.3. The effect of the size and physiological age of the seed tuber on the number of main stems per seed tuber [*Based on: Struik and Wiersema (1999)*].

agriculture is therefore in fact a cluster of stems, originating from one seed tuber or seed piece. During the first part of their growing period, these stems share resources from the same seed tuber unit, but shortly after planting, these individual stems become independent units albeit that they are in strong competition with each other for light, water and nutrients.

Also each stem consists of various elements, which we will call stem segments of various orders [main (primary) segment, secondary stem segment, tertiary stem segment etc.]. The entire shoot system is illustrated in Fig. 11.5.

Often two types of potato varieties are distinguished: the determinate type and the indeterminate type. Determinate types tend to remain short, as they do not produce many successive orders within one stem. They also tend to have a short life cycle. Indeterminate types, however, tend to grow tall, may produce many different levels and have a long cycle. The yield potential of indeterminate types is higher than that of determinate types, but they need a much longer growing season to realize their potential.

The below-ground parts of the stems are usually round and massive, whereas in the upper part, the internodes can be hollow. Above-ground stem parts are angular, usually with the shape of a triangle when cut in a transverse direction and often winged. Depending on variety, the colour of the stem may be green, red or purple.

11.5 THE LEAVES

The potato plant has one major leaf per node. The early leaves are small, whereas the later leaves are alternate and pinnate compound with three or four pairs of large, ovate to ovate elliptical leaflets with smaller ones in between. The rachis ends in a top leaflet that is often the largest one, with a shape that sometimes deviates from the other large leaflets. The small leaflets are subsessile, ovate to suborbicular.

Firman et al. (1991) showed that 20–40 leaf primordia are produced during storage before the start of flower initiation. The number of leaf primordia thus shows considerable variation. However, the number of nodes of a main stem above ground is usually

Physiological age	Young					Old
Tuber characteristics	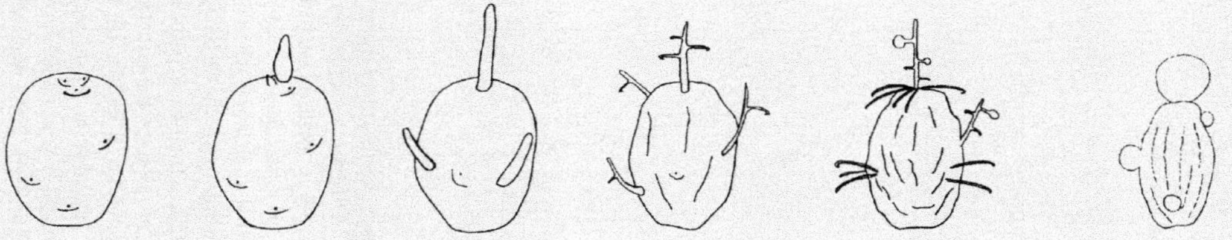					
Physiological stage	Dormancy	Apical dominance	Normal sprouting	Normal sprouting	Senility	Incubation
Sprouting	No sprouts	Apical sprouts only	Few sprouts	Multiple, branched sprouts	-Excessive sprouting -Excessive branching -Hair sprouts -Some little tubers	No sprouts, direct formation of daughter tubers on seed tuber
Crop condition	No or delayed emergence	Single-stemmed plants	Few vigorous stems per plant	Many stems per plant	Weak, multi-stemmed plants	No plants
Yielding ability in short season	None	Low	Moderately high	High	Hardly any	None
Yielding ability in long season	Very low	Relatively low	Very high	Relatively high	Hardly any	None

Fig. 11.4. Effect of physiological age of seed tubers on the morphological aspects of sprouting behaviour and the resulting potential of the crop produced by seed tubers of different ages [*Source: Struik and Wiersema (1999)*].

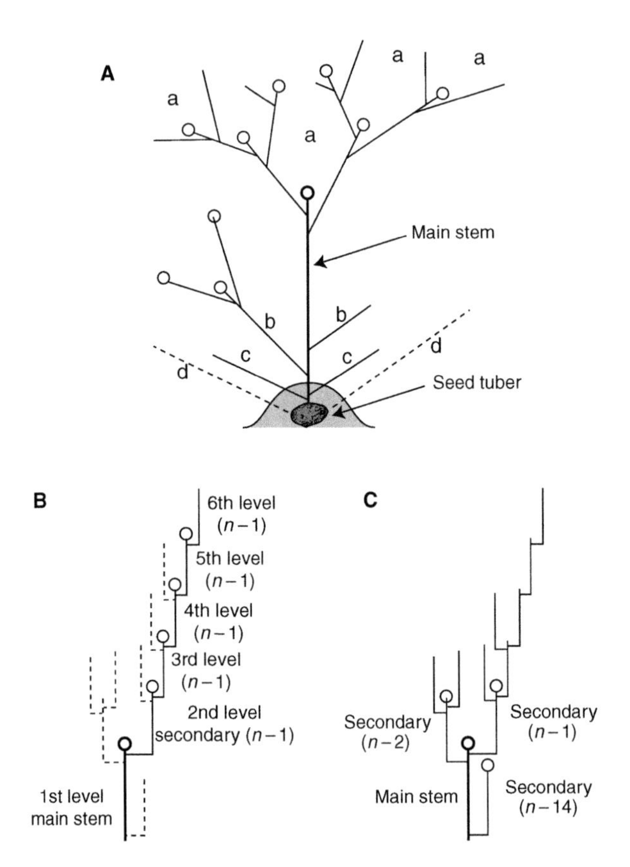

Fig. 11.5. Structure of the stem system of the potato plant. [*Sources: Struik and Ewing (1995), Vos (1995), Almekinders and Struik (1996) and Struik and Wiersema (1999)*]. (A) Different types of stems of a potato plant, consisting of a cluster of main stems. Figure A illustrates schematically a potato plant with three main stems, of which one has been illustrated in detail. This detailed main stem has basal, lateral stems from below-ground nodes which may be difficult to distinguish morphologically from normal main stems when they arise close to the seed tuber. They can produce their own stolons and tubers. The main stem also shows basal, lateral stems from above-ground nodes. These usually do not produce stolons or tubers, unless their basic parts are covered by earthening up or when urged to produce aerial tubers because of diseases or physiological malfunctioning of the below-ground stolon and tuber system. One single main stem will terminate in an inflorescence, but below the apex, several apical lateral stems of successive orders may appear. a, apical, lateral stems of successive orders; b, basal, lateral stems from above-ground node; c, basal, lateral stems from below-ground node; d, other main stems and o, inflorescence.(B) This illustrates a diagram of a complex single main stem. Each main stem may consist of a main primary stem segment, some basal lateral stems and different layers of apical lateral stems (higher-order stem segments). Usually the $n-1$ and the $n-2$ secondary stems develop, with n being the uppermost leaf of the main primary stem. In this way, there can be many inflorescences per individual stem. The structure consisting of the stem segments of increasing order from main primary stem until the sixth level stem segment as illustrated in B is called a sympodium. (C) Simplified version of (B) to illustrate the different types of secondary and higher-order stem segments on the main primary stem. Secondary stem segments occur on the nodes immediately below the inflorescence of the primary main stem and on the nodes close to the soil level. Tertiary and higher-order stem segments are usually only formed on the secondary stems just below the inflorescence of the first level, only when genotype, climatic conditions (long days and relatively high temperatures), weather conditions (rainfall, temperature and light) and crop management (irrigation, nitrogen fertilisation, etc.) allow.

rather constant within a cultivar, showing only little variation caused by variation in environment (Almekinders and Struik, 1996). When the seed tubers are properly pre-sprouted, the number of leaves that will be produced until the first inflorescence appears, is – within a cultivar – rather constant over the years. There is also little variation in time needed to start reproductive development of the first primary main stem. Flowering induction may even be completed before planting, and then the number of above-ground nodes preceding the primary inflorescence is completely determined by processes determining sprout elongation.

For the lateral branches, this is entirely different. There is a strong position effect on the number of leaves produced before the first inflorescence of the above-ground lateral branches. The lower the branch is inserted the higher its number of leaves. The maximum number of leaves in the lower branches can even be higher than that in the main stem. Both long days and higher temperatures increase the number of leaves on the secondary stems. At the higher temperature, the photoperiod effect is stronger. These phenomena are illustrated in Fig. 11.6 (Almekinders and Struik, 1994, 1996).

The development of the number of leaves along the different stem segments of the central sympodium (from primary main stem until higher order) of a complex main stem is shown in Fig. 11.7. The number of leaves per stem segment decreases with an increase in order, although the trend is not always clear cut.

Figure 11.8 gives an indication of the appearance rate of compound leaves at abundant fertiliser application with plants receiving ample light in a pot experiment in the greenhouse. Leaf appearance is linear during the first phase of plant growth. Typically the rate

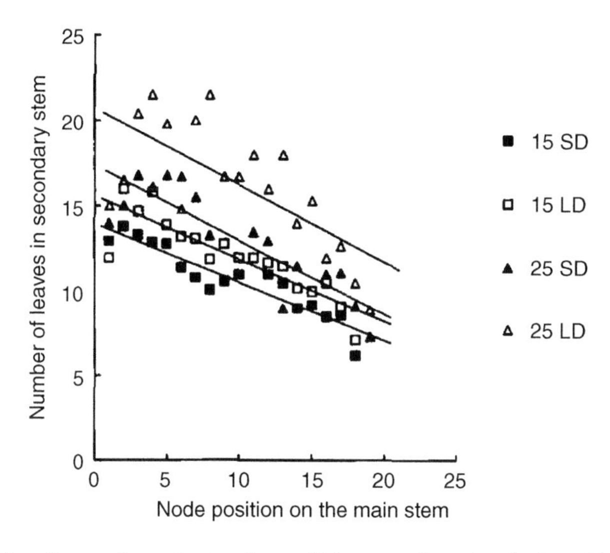

Fig. 11.6. Effect of position on the main stem from which a secondary stem (or lateral branch) develops on the number of leaves until the first inflorescence of that lateral branch (i.e. the secondary inflorescence). Node position 1 is the lowest leaf. Results from cv. Atzimba grown at two temperatures (24 h averages 15 and 25 °C) and two photoperiods (SD = short days = 12 h and LD = long days = 16 h).

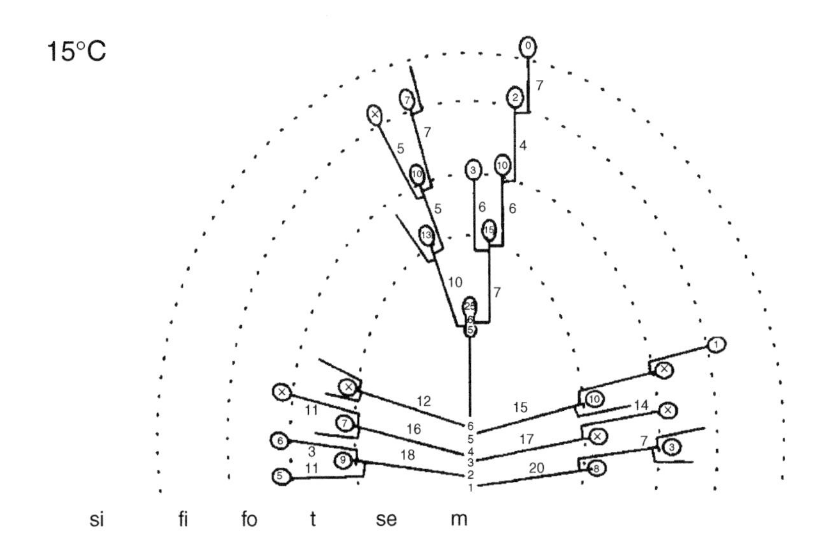

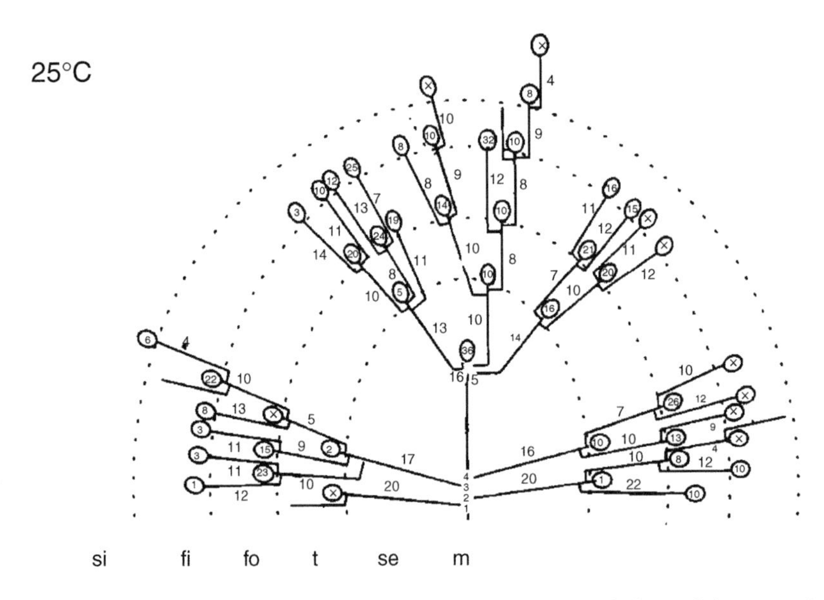

Fig. 11.7. A schematic diagram of the development of the shoot system of plants of the potato Atzimba, grown at 16-h photoperiod and 15 and 25 °C, 24-h average (Almekinders and Struik, 1994, 1996). The shoot systems are presented as a cluster of stems of the first (main, m), second (se), third (t), fourth (fo), fifth (fi) and sixth (si) order. The numbers on the primary part of the main stem indicate the node position from which the secondary stems with inflorescences developed. The numbers near the secondary stem segments and the stem segments from higher levels of growth indicate the number of leaves produced. The circles represent the inflorescences terminating the individual stem segments and numbers in the circles indicate the number of flowers per inflorescence. Aborted inflorescences are indicated by an x in the circle.

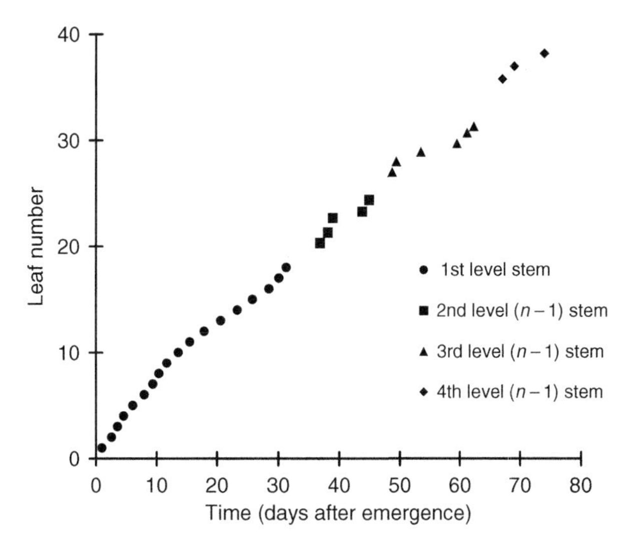

Fig. 11.8. The number of leaves as a function of time for potato plants (cv. Bintje) with abundant light and nutrients [*Based on: Vos and Biemond (1992)*].

of leaf appearance is 0.53 per day or one leaf per 28 °C day (assuming a base temperature of 0 °C) (Vos and Biemond, 1992). Vos (1995) showed a table with a compilation of leaf appearance rates of main stem leaves of potato in experiments with different cultivars and average temperatures of 13.0–17.4 °C. In this data set, the leaf appearance rate ranged from 0.33 to 0.53 per day or 0.0242 to 0.0317 leaf per °C day, that is one leaf per 32–41 °C per day. This shows that the typical leaf appearance rate indicated by Vos and Biemond (1992) is in the upper range of leaf appearance rates. The last leaves appear at a slower rate as at that time there is competition between above-ground growth and below-ground growth.

Figure 11.9 presents different growth characteristics of individual leaves. The rate of leaf area increase is determined by the position of the leaf on the plant. The rate of area increase initially increases with an increase in leaf insertion number until about leaf 13 (depending on cultivar) on the main primary (first order) stem. It then gradually decreases with a further increase in leaf insertion number. The first leaves on the secondary order stem segment differ little in rate of expansion, but with a further increase in leaf insertion number, the expansion rate starts to decline again. The same trends are observed for the following orders. The effects of leaf insertion number on mature leaf size are very similar to the effects on rate of leaf area increase as there is a close, curvilinear relationship (not shown) between rate of area increase and mature area (Vos and Biemond, 1992). The life span of the individual leaves increases with an increase in leaf insertion number until leaves 12–13 of the first order main stem and then slowly declines. The life span varies between 30 and over 100 days. Leaves with the largest area (the middle leaves on the primary order) also have the longest life span.

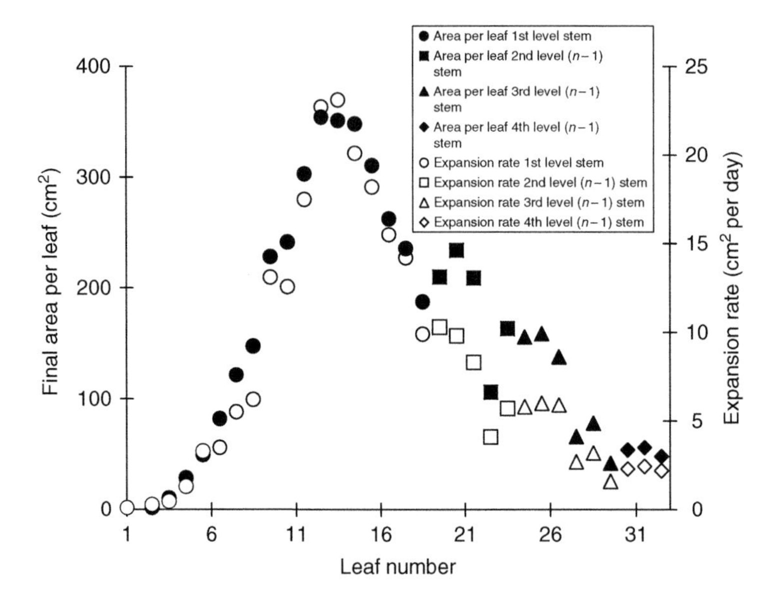

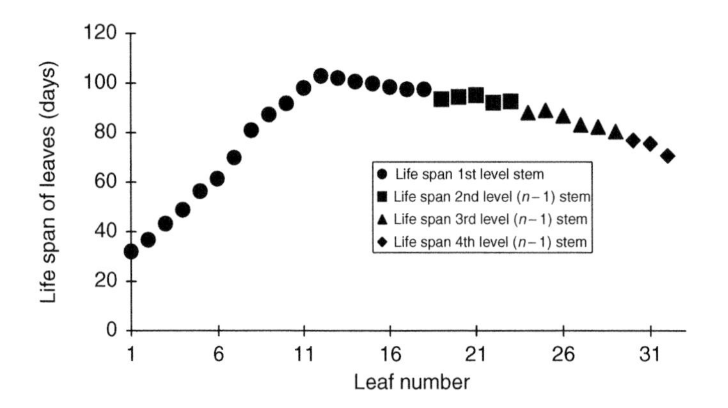

Fig. 11.9. Effects of leaf insertion number on rate of area increase, mature area and life span of individual leaves of cv. Bintje. The first sequence of leaf numbers starting with 1 pertains to the main stem segment (primary order), the second series pertains to the $(n-1)$ secondary order branch, the third series pertains to the $(n-1)$ $(n-1)$ tertiary order branch and the fourth pertains to the $(n-1)$ $(n-1)$ $(n-1)$ quarterly level stem segment. Data for the highest nitrogen level from Vos and Biemond (1992) [*Based on: Vos and Biemond (1992)*].

Specific leaf area of the whole potato plant increases over time until it reaches a peak at about 100 days after emergence (Vos and Biemond, 1992); for an individual leaf, there is a similar trend, but the timing of the peak obviously depends on time of appearance and thus on leaf insertion number.

The proportional distribution of the leaf area over different stem types is relevant as not all stem segments contribute to tuber bulking to the same extent. Table 11.1 summarizes the proportional distribution (in percentage) over main stem segments, apical

Table 11.1 The proportional distribution (in percentage) of the leaf area over main stems and branches and total leaf area over time for cultivars Bintje and Saturna under Dutch field conditions.

Cultivar	Time in days after planting	Main (primary) stem segment	Apical laterals		Basal laterals	Total leaf area index (m^2/m^2)
			2nd order	3rd order		
Bintje	50	100	0	0	0	3.8
	83	55	31	4	10	6.3
	131	21	55	21	3	3.1
Saturna	50	82	0	0	18	4.4
	83	35	26	7	32	8.6
	131	6	18	23	53	3.6

[*Based on*: Struik and Wiersema (1999)].

lateral branches and basal lateral branches. The contribution of the apical lateral branches increases over time. The contribution of the basal laterals strongly depends on the cultivar.

11.6 THE STOLON SYSTEM

The rhizomes of the potato plant are usually called stolons. Stolons are diageotropic (or plagiotropic) shoots or stems, with strongly elongated internodes and rudimentary (scale) leaves. They are usually hooked at the tip and usually originate below ground from the basal stem nodes. Three stolons per node may arise, one main stolon and two from the axillary buds at the same node.

Stolonization is believed to start at the nodes closest to the seed tuber and to progress acropetally (Cutter, 1992). The first stolons grow faster and become longer than later-initiated stolons (Struik and van Voorst, 1986). This is illustrated schematically in Fig. 11.10. However, this pattern is not always clear cut. Other researchers have found a lower rate of elongation growth for the first two or three stolons, a very high rate of growth of stolons four to eight and a lower rate of growth with increasing stolon rank number (A. van der Maarl and P.C. Struik, unpublished data).

Figure 11.11 illustrates the development over time of the number of stolons and tubers. It shows that the stolonization starts very soon after emergence. It may even start before the shoot emerges. Tuberization starts before all stolons have been formed. The tubers that finally grow out to substantial sizes are initiated within a very short period of time.

Before tuberization starts, the longitudinal stolon growth must slow down. Many investigations have tried to link the patterns of longitudinal growth to the onset of tuberization. For example, Helder et al. (1993) analysed the linkage in very much detail for the wild species *Solanum demissum*, but they could not find a clear pattern.

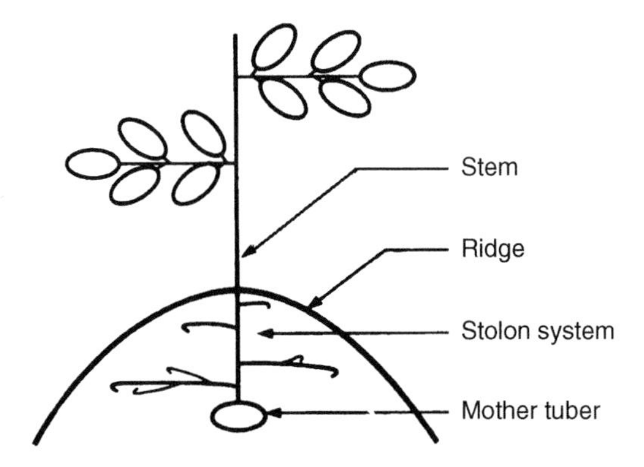

Fig. 11.10. Schematic diagram of a stolon system of a potato plant. Note the effect of stolon position on length of the stolon.

The occurrence of subapical swelling was not linked to the rate of longitudinal growth of the stolons or stolon branches during the period before stolon swelling.

However, there tends to be more delay between stolon initiation and tuber initiation in the earlier-formed stolons than in the later-formed stolons (Vreugdenhil and Struik, 1989). This is consistent with the schematic diagram of the stolon system given in Fig. 11.10.

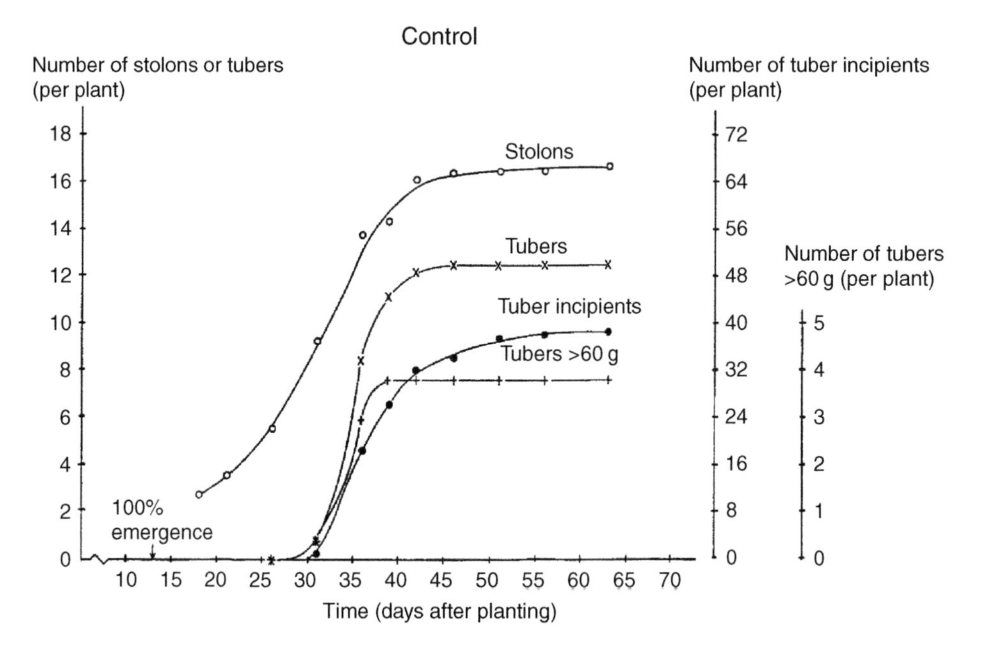

Fig. 11.11. Development over time for the number or primary stolons (open circles), tuber incipients (closed circles), tubers with a diameter >1 cm (x) and tubers that will finally reach a size of at least 60 g fresh weight (+) [*Source: Struik et al. (1988)*].

The first stolons often show extensive branching, thus increasing the number of potential tuber sites (Helder et al., 1993). However, having more than one tuber per stolon causes strong competition for resources and has a large effect on the size distribution of tubers.

11.7 THE TUBERS

Normally, the tubers are the swollen parts of the subterranean rhizomes or stolons and are globose to ellipsoid in shape. The tuber skin has scars of scale leaves (eye brows) with axillary buds (eyes). Tubers vary in size, skin and flesh colour and skin texture. Although tubers are usually formed on stolons, with strong induction to tuberize, they can also be sessile. Tubers may also be formed above ground on all buds and even in the inflorescence (Ewing and Struik, 1992). Originating as the swollen part of stolons, tubers are in fact greatly shortened and thickened stem segments. They contain high levels of starch and storage proteins and even may bear roots.

Tubers start as stolon swellings and can go through different phases of tuber set, tuber growth and tuber maturation. When induction to tuberize is interrupted, they can show secondary growth, especially when the plant is exposed to heat or irregular water supply. At stolon tip swelling, the plane of cell division changes, and there is an increase in radial cell expansion (Xu et al., 1998; Vreugdenhil et al., 1999). One of the first signs of tuber formation is the increase in the dry matter content of the stolon tip and a change in sugar metabolism. The starch content increases, whereas the glucose and fructose contents drop (Helder, 1993). Another significant biochemical change upon tuberization is the accumulation of storage proteins called patatins (glycoproteins) (Park, 1990; Hannapel, 1991). They play a major role during sprouting (Ewing and Struik, 1992).

Figure 11.11 illustrates the development over time of the number of tuber incipients and the number of tubers set. Figure 11.12 presents a schematic representation of the change over time of the total number of tuber incipients and the number of tubers that will grow out to marketable tuber sizes for two cases. Case 1 reflects a situation with abundant tuber initiation; case 2 shows an example of much less abundant tuberization. In the first case, many tuber incipients will not grow out to marketable sizes. They are resorbed during the growing season. In the second case, almost all tubers that are initiated grow out to marketable sizes. Tuber initiation may come in different waves creating an irregular tuber size distribution and severe competition between different categories of tubers.

It is commonly believed that the largest tuber at one point in time is also the largest tuber at the end of the growing season. However, individual tuber growth rate and growth duration may show different and dynamic hierarchies among tubers. Thus, there are crossovers in individual tuber growth rate as illustrated by Schnieders et al. (1988). In general, however, the relative variability in tuber size (the coefficient of variation of the tuber size) is rather constant once tuber set is completed and depends on tuber number, although the relation may be complex (Struik et al., 1990, 1991). Based on the fixed relative variability and the close relation between yield and average tuber size, tuber size distributions can easily be predicted once tuber numbers do not further change drastically.

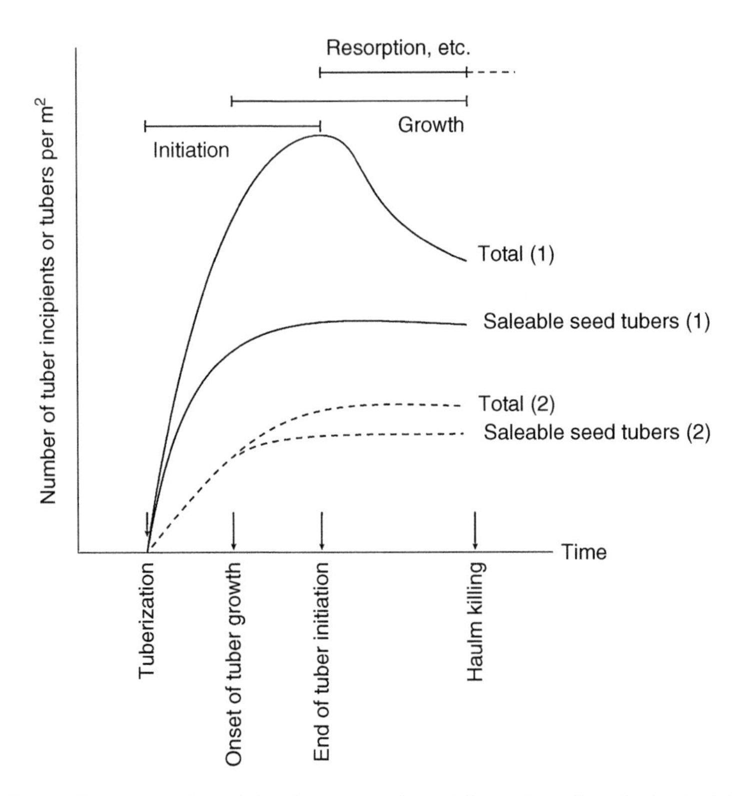

Fig. 11.12. Schematic representation of the change over time of the total number of tuber incipients and the number of tubers that will grow out to marketable sizes for two cases. Case 1 reflects a situation with abundant tuber initiation; case 2 shows an example of much less abundant tuberization.

The position of the large tubers very much depends not only on the environmental conditions to which the entire plant is exposed (Struik et al., 1991) but also to the micro-environment around each individual stolon (Struik and Wiersema, 1999).

11.8 ORGANS OF SEXUAL REPRODUCTION

The inflorescence is a cymose panicle, formed in the axil of the leaves on several stem segments (Figs 11.2 and 11.5). An individual flower consists of a greenish campanulate calyx composed of five sepals and five petals (varying in colour, often white, yellow or violet). It has five stamens. The anthers of the stamens are joined laterally forming a cone-shaped structure, which conceals the ovary.

The number of inflorescences and the number of flowers per inflorescence strongly depend on growing conditions and on the order of the stem segment bearing the inflorescence (Fig. 11.7). Especially, photoperiod, temperature, plant density and nitrogen supply are important. Higher-order stem segments usually have fewer flowers per inflorescence (Fig. 11.7).

The berries are globular, have two carpels and may contain many seeds. Berry size is determined to a great extent by berry position within the inflorescence and the position of the inflorescence. There are specific relations among berry weight, number of seeds per berry and 100-seed weight (Almekinders et al., 1995).

11.9 ROOT SYSTEM

The potato has a fibrous root system (Fig. 11.2). Root initials can already be present on the sprouts during pre-sprouting (Fig. 11.2G). The root system is rather weak, and therefore, water and nutrient use efficiencies in potato are low, and the crop is very sensitive to drought and poor soil structure. Roots occur not only on the stems but also on stolons and sometimes even on tubers. Stolon and tuber roots may be essential in supplying nutrients to the growing tubers, especially because the intensity of root growth declines after the onset of tuber bulking (Steckel and Gray, 1974).

11.10 ASSOCIATION BETWEEN DEVELOPMENT OF ABOVE-GROUND AND BELOW-GROUND PLANT PARTS

There is a delicate balance between the above-ground and below-ground development of the potato plant. Events above ground and below ground are linked because of

1. the initiation of flower primordia on the main stem calling the primary order development to a halt and triggering the formation of apical lateral branches;
2. the mutually interacting initiation of stolons, stolon branches and tuber incipients, and the formation and resorption of tubers;
3. the shift of assimilate partitioning to tubers after tuberization and
4. the resulting cessation of shoot growth which limits the growth cycle of the crop.

The life cycle of the potato crop can be described in terms of the period until induction of flowering, the time to tuber initiation, the time to onset of rapid tuber bulking, the time to cessation of shoot growth and the time to onset of senescence. These events are closely associated, in terms of both crop physiology and time, but the time lapse between the different events depends on the genotype, the environment and the genotype by environment interaction. Not only the timing of these events but also the intensity of several of these changes may differ. The shift in assimilate partitioning towards tubers may be very abrupt and complete but may also take some time and be gradual. The same is true for cessation of shoot growth.

Initiation of flowering and tuber initiation are interlinked. It is commonly accepted that the potato plant is a short day or day neutral plant for tuber initiation. Development scales of the crop often link tuber formation to flowering. Almekinders and Struik (1996) also argued that the potato is a short day or day neutral plant for flowering. Nevertheless, flowering is more abundant when tuber formation is delayed, because a delay of tuber formation will allow more basal and apical lateral branches to be formed, more stem segments to be formed per sympodium and will enhance the development of individual inflorescences and flowers.

For yield formation, however, the link between the number of sprouts, number of stems, number of stolons and number of tubers and the link between onset of tuber formation and the shift in dry matter partitioning and thus the cessation of shoot growth are probably more relevant but rather variable. The first link is illustrated in Fig. 11.13. The second link can be derived by comparing the two temperatures (one allowing relatively strong tuber induction and the other one delaying tuber formation), which is illustrated in Fig. 11.7.

A striking phenomenon in the physiology of the potato crop at higher latitudes is the fact that there is a long overlap between stolonization and tuberization (Vreugdenhil and Struik, 1989). Tuber formation already starts when stolon formation is still in full progress. This illustrates that there is not a sudden and complete switch in the level of induction at plant level but that the trigger to stop longitudinal growth and start radial growth and expansion is local. This is crucial for determining tuber hierarchies, allocation of resources and final size distributions.

How plastic the order of events in the potato crop, is illustrated in Table 11.2.

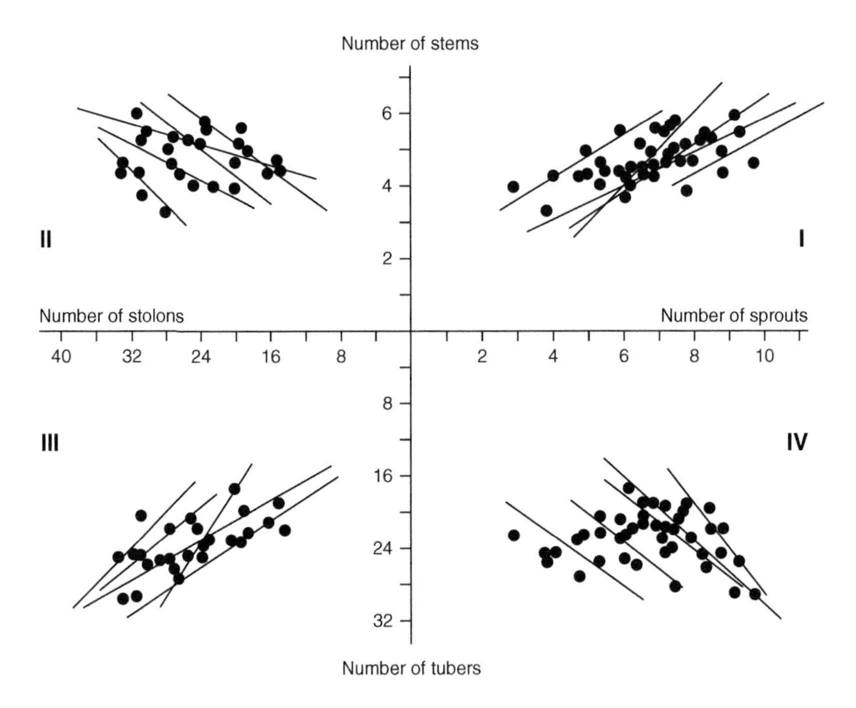

Fig. 11.13. An example of the interrelations between number of sprouts, number of stems, number of stolons and number of tubers. Within a specific year, the correlations are often high, but across years, the relations can differ substantially [*Source: Haverkort et al. (1990)*].

Table 11.2 Differences between three categories of genotypes of the same population varying in maturity type in the onset of plant development-related processes expressed as the proportion (%) of the total plant cycle.

Process	Classification of genotypes		
	Very early	Intermediate	Very late
Onset stolon formation	35	22	16
Onset stolon tip swelling	42	33	33
Onset stolon branching	53	36	27
Appearance first tubers >20 mm (onset of tuber bulking)	55	45	42
Onset flowering	44	31	22
Onset senescence	80	69	72

[*Based on*: Celis-Gamboa (2002)].

REFERENCES

Almekinders C.J.M, J.H. Neuteboom and P.C. Struik, 1995, *Sci. Hortic.* 61, 177.
Almekinders C.J.M. and P.C. Struik, 1994, *Neth. J. Agric. Sci.* 42, 311.
Almekinders C.J.M. and P.C. Struik, 1996, *Potato Res.* 39, 581.
Celis-Gamboa B.C., 2002, The life cycle of the potato (*Solanum tuberosum* L.): From crop physiology to genetics. Ph.D. thesis, Wageningen University, Wageningen, the Netherlands.
Cutter E.G., 1992, In: P.M. Harris (ed.), *The Potato Crop*, 2nd edition, p. 65. Chapman and Hill, London.
Ewing, E.E. and P.C. Struik, 1992, *Hortic. Rev.* 14, 89.
Firman D.M., P.J. O'Brien and E.J. Allen, 1991, *J. Agric. Sci. (Cambridge)* 117, 61.
Hannapel D.J., 1991, *Physiol. Plant.* 83, 568.
Haverkort A.J., M. van de Waart and K.B.A. Bodlaender, 1990, *Potato Res.* 33, 269.
Helder J., 1993, *Tuber Formation in the Wild Potato Species Solanum demissum Lindl.* Ph.D. thesis, Wageningen University, Wageningen, the Netherlands.
Helder J., A. van der Maarl, D. Vreugdenhil and P.C. Struik, 1993, *Potato Res.* 36, 317.
Park W.D., 1990, In: M.E. Vayda and W.D. Park (eds), *The Molecular and Cellular Biology of the Potato*, p. 43. CAB International, Wallingford, UK.
Schnieders B.J.M., L.H.L. Kerckhoffs and P.C. Struik, 1988, *Potato Res.* 312, 129.
Steckel J.R.A. and D. Gray, 1974, *J. Agric. Sci. (Cambridge)* 92, 375.
Struik P.C. and E.E. Ewing, 1995, In: A.J. Haverkort and D.K.L. MacKerron (eds), *Potato Ecology and Modelling of Crops Under Conditions Limiting Growth*, p. 19. Kluwer Academic Publishers, Dordrecht, the Netherlands.
Struik P.C., A.J. Haverkort, D. Vreugdenhil, C.B. Bus and R. Dankert, 1990, *Potato Res.* 33, 417.
Struik P.C., E. van Heusden and K. Burger-Meijer, 1988, *Neth. J. Agric. Sci.* 36, 11.
Struik P.C. and G. van Voorst, 1986, *Potato Res.* 29, 487.
Struik P.C., D. Vreugdenhil, A.J. Haverkort, C.B. Bus and R. Dankert, 1991, *Potato Res.* 34, 187.
Struik P.C. and S.G. Wiersema, 1999, *Seed Potato Technology*, 383 pp. Wageningen Pers, Wageningen, the Netherlands.

Vos J., 1995. In: P. Kabat, B.J. van den Broek, B. Marshall and J. Vos (eds), *Modelling and Parameterization of the Soil-Plant-Atmosphere System. A Comparison of Potato Growth Models*, p. 21. Wageningen Pers, Wageningen, the Netherlands.
Vos J. and H. Biemond, 1992, *Ann. Bot.* 70, 27.
Vreugdenhil D. and P.C. Struik, 1989, *Physiol. Plant.* 75, 525.
Vreugdenhil D., X. Xu, C.S. Jung, A.A.M. van Lammeren and E.E. Ewing, 1999, *Ann. Bot.* 84, 675.
Xu X., D. Vreugdenhil and A.M.M. van Lammeren, 1998, *J. Exp. Bot.* 49, 573.

Chapter 12

Signalling the Induction of Tuber Formation

David J. Hannapel

Department of Horticulture and Interdepartmental Plant Physiology Major, Iowa State University Ames, Ames, IA 50011-1100, USA

12.1 INTRODUCTION

The mechanism of tuberization has been the subject of considerable investigation by plant scientists in the past several decades. However, the precise controlling factors involved in this growth process are not entirely clear. A tuber is a shortened, thickened stem with leaves reduced to scales or scars subtending the axillary buds known as eyes. Potato tubers form from modified shoots called stolons. Tuberization is also used to describe the formation of the underground storage organs of dahlia, Jerusalem artichoke and begonia. Tuber formation in potatoes (*Solanum tuberosum* L.) is a complex developmental process that requires the interaction of environmental, biochemical and genetic factors. It involves many important biological processes, including carbon partitioning, signal transduction and meristem determination (reviewed by Ewing and Struik, 1992). Gregory (1965) has considered the development of potato tubers in three general stages: (1) tuber initiation, which is characterized by differentiation and the development of tuber primordium without evidence of any visible swelling; (2) tuber enlargement, characterized by rapid cell division, starch accumulation and visible swelling of the stolon tip and (3) tuber maturation, when the organ passes into the dormant period. This chapter will focus on those factors that affect the initial stages of tuber development and the signal that activates growth in the stolon tip.

In wild potato species such as *S. tuberosum* ssp. *andigena*, short-day (SD) photoperiods are required for tuber formation. This species tuberizes only under SD conditions (less than 12 h of light) and does not produce tubers when grown under long days (LDs) or SD supplemented with a night break. The main site of perception of the photoperiodic signal is in the leaves. Under inductive conditions, a transmissible signal is activated which initiates cell division and expansion and a change in the orientation of cell growth in the subapical region of the stolon tip (Xu et al., 1998a). In this signal transduction pathway, perception of the appropriate environmental cues is mediated by phytochrome and gibberellins (GAs) (Jackson and Prat, 1996; Jackson et al., 1996). High levels of GAs are correlated with the inhibition of tuberization, whereas low levels are associated with induction both at the site of perception (the leaf) and in the target organ, the stolon apex, site of the newly formed tuber (Jackson and Prat, 1996; Xu et al., 1998b). Although some excellent molecular work has been done characterizing the process of tuberization (Bachem et al., 1996; Macleod et al., 1999; Viola et al., 2001; Martinez-Garcia et al.,

Potato Biology and Biotechnology: Advances and Perspectives
D. Vreugdenhil (Editor)

2002b; Raices et al., 2003), the exact mechanism for inducing tuber formation is unknown. This chapter will focus on specific molecular switches that mediate the activation of the morphological events associated with the tuberizing stolon.

12.2 HISTORICAL BACKGROUND

One of the earliest studies of tuberization was made by Bernard (1902), who believed that tuberization in potatoes was caused by a symbiotic relationship between the plant and a mycorrhizal fungus. This was disproved by Magrou (1943) who produced tubers from seed-grown plants under sterile conditions and explained the process as an increase in osmotic pressure at the tip of stolons caused by a high carbohydrate concentration supplied by the culture medium or photosynthesis. Werner (1934) supported these conclusions by establishing that environmental and cultural factors, such as SDs, low temperatures and low supply of nitrogen, that raised the carbohydrate/nitrogen ratio also promoted tuberization. Driver and Hawkes (1943) suggested that increased respiration stimulated by LDs and high temperatures (both of which can be inhibitory for tuberization) decreases the quantity of carbohydrates available for translocation to underground parts and that this can prevent the onset of tuberization. Under these conditions, active vegetative growth is favoured and uses a greater proportion of photosynthates. Studies showed that by enhancing carbohydrate accumulation at stolon tips, initiation of tuberization is promoted by factors such as high radiation, low temperature and a low nutrient supply (Werner, 1934; Borah and Milthorpe 1962; Burt, 1964). Those that favoured the 'nutrient theory' postulated that the availability of carbohydrates and, consequently, the carbon/nitrogen ratio were more important than the action of a specific tuber-inducing growth substance.

Garner and Allard (1920, 1923) first established the significance of environmental factors in the regulation of tuberization. Their work demonstrated the effects of photoperiod on several tuberous species, and they suggested that tuberization is induced by a specific photoperiodic stimulus. The foliage was considered to be the receptor of this stimulus. Work by Gregory (1956) and Chapman (1958) suggested the existence of this specific tuber-forming stimulus. Gregory (1956) took cuttings from potato plants grown under induced (8-h day and 17 °C night temperature) and noninduced (16-h day and 20 °C night temperature) conditions and grafted them on to noninduced plants. Cuttings from induced plants produced tubers on scion plants, whereas those from noninduced plants produced no tubers. Using in vitro culture techniques, Gregory further showed that soluble sugars must be present in the medium for tuber formation from stem cuttings to occur. Using differential daylength treatment of plants with two stems, Chapman (1958) showed that unilateral formation of tubers occurred only on the SD half of the stolon system. Several other workers (Madec and Perennec 1959; Madec 1963; Kumar and Wareing 1972) have presented further evidence for a tuber-inducing stimulus. Scions from *S. tuberosum* ssp. *andigena* plants grown under SD and grafted on to potato stocks grown under LD can induce tuberization (Kumar and Wareing, 1973). In an eloquent grafting experiment (Fig. 12.1), a graft-transmissible substance was transported from the induced shoot tips of both LD- and SD-flowering tobacco plants to the potato stock grown under LDs to

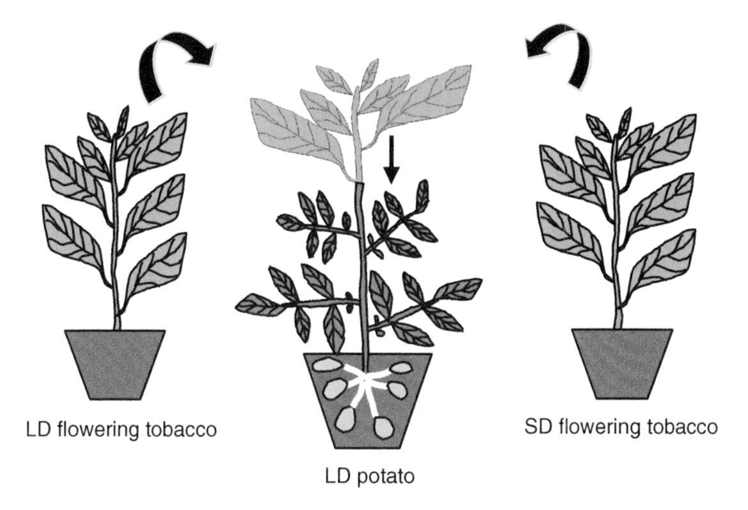

Fig. 12.1. Transmission of a flowering signal across a graft union to induce tuber formation. Scions from either long day (LD)- or short day (SD)-flowering tobacco plants grown under inductive conditions were grafted on to potato stock plants grown under noninductive conditions (LD) for tuber formation. After several days, tubers formed on the LD potato stocks (Chailakhyan et al., 1981). This graft-transmissible substance has not yet been identified.

induce tuber formation (Chailakhyan et al., 1981). These results imply that flowering and tuberization may be mediated by the same phloem-transported signal. The conclusions drawn from this work may be summarized as follows: (1) under conditions of low temperature and SDs, leaves undergo a change that results in the production of specific metabolites that stimulate tuber induction; (2) the tuber-inducing stimulus appears to be transmitted through a graft union and (3) soluble sugars appear to be essential for the initiation of tuberization but are not the only essential substance. What then is the exact nature and composition of this putative tuber-inducing stimulus?

Under the inductive conditions of an SD photoperiod and cool night temperatures, the leaf perceives the environmental cue and activates an inducing signal. This signal is then transmitted through the phloem to the target organ, an elongating stolon tip. The signal activates changes in cell growth activity in the subapical region of the stolon apex to create a strong sink organ. Informative studies of this latter process have been undertaken by using in vitro model systems (Xu et al., 1998a; Vreugdenhil et al., 1999). These changes include the induction of cell division, cell enlargement and a change in the orientation of cell division from transverse to longitudinal. Cells in the pith and cortex enlarged and divided longitudinally. Overall, these changes result in the retardation of stolon elongation and an increase in the radial expansion of the subapical region, leading to tuber formation (Fig. 12.2). Most of the cell growth that leads to an increase in tuber size occurs in the perimedullary region of the stolon (Fig. 12.3).

The creation of this strong sink, the newly formed tuber, results in the accumulation of a specific set of storage proteins and massive amounts of starch (reviewed by Fernie and Willmitzer, 2001).

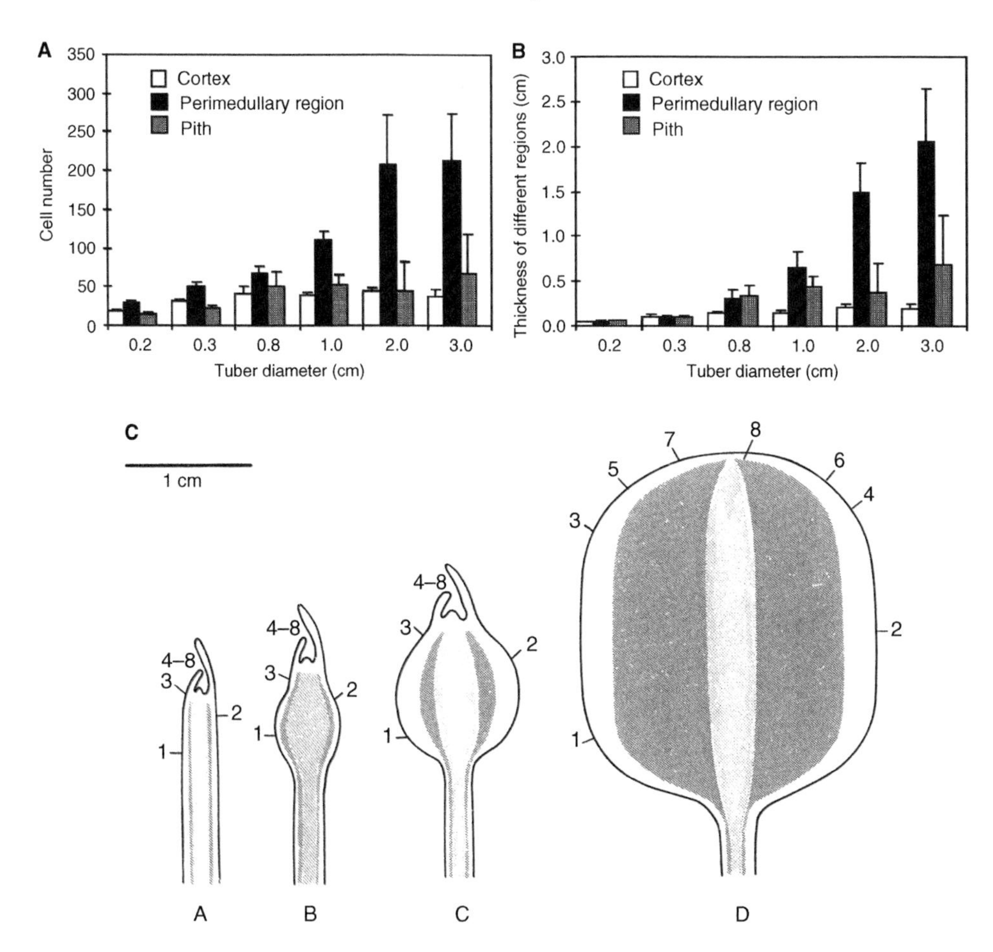

Fig. 12.2. Schematic of cell growth with cell numbers (A) and the thickness of different regions (B) along the transversal axis in a tuberizing stolon tip. Data are means with standard deviations of three independent measurements. (C) Diagram of longitudinal sections through in vivo grown potato tubers, showing the morphology of the stolon and tuber and the thickening of the perimedullary zone (dark-shaded area). From left to right are varying stages of the tuberizing stolon, from a 0.2-cm stolon to a 2.0-cm tuber. Throughout this development, the perimedullary region continues to expand. The positions of nodes are indicated numerically. Bar = 1.0 cm [*Source: Xu et al. (1998b)*].

12.2.1 Photoregulation

Phytochrome B (PHYB) is known to mediate the photoperiodic control of tuberization in a negative regulatory mechanism. Transgenic plants with suppressed levels of PHYB lose the capacity to respond to a LD photoperiod. These plants tuberize under LD conditions as efficiently as SD (Jackson et al., 1996). Grafting experiments with PHYB antisense plants

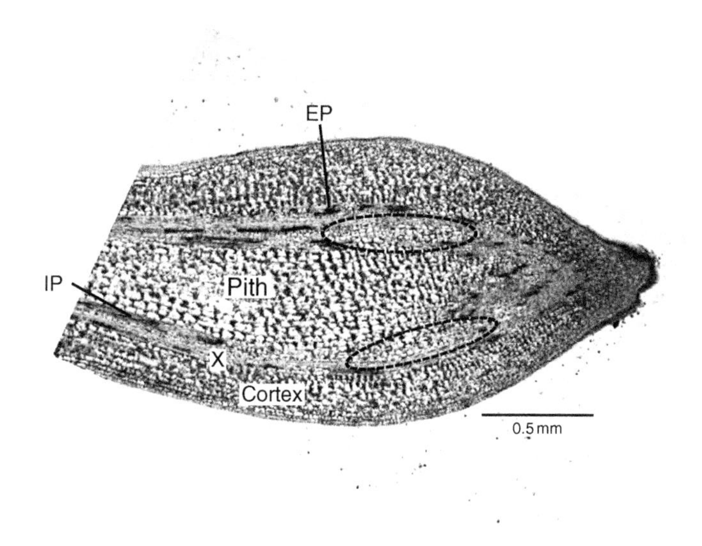

Fig. 12.3. A longitudinal section of the apex of a tuberizing stolon. The areas circled with dotted lines represent the expanding perimedullary region, representative of the most actively enlarging area of growth. IP, internal phloem; EP, external phloem and X, xylem.

and wild-type potato plants showed that PHYB regulates the transmission of an inhibitor of tuberization. Wild-type stocks grown under LD conditions tuberize when PHYB antisense scions were grafted on to them (Jackson et al., 1998). In Arabidopsis, CONSTANS is a putative transcription factor (TF) that accelerates flowering in response to LDs. Constitutive overexpression in potato of the Arabidopsis gene *constans* impairs tuberization under SD (inductive) conditions (Martinez-Garcia et al., 2002b). These transgenic lines exhibited a delayed tuberization phenotype, suggesting that CONSTANS has a function in the photoperiodic pathway controlling tuber formation. A *constans* orthologue exists in potato, which may mediate the photoperiod control of tuberization (González-Schain et al., 2003).

In an attempt to isolate genes with a role in photoperiodic control of tuber formation, a photoperiod responsive cDNA was isolated from leaves of subspecies *andigena* and designated *PHOR1* (Amador et al., 2001). *PHOR1* encodes a novel arm-repeat protein with homology to the *Drosophila* segment polarity gene, *armadillo*. Expression of *PHOR1* is enhanced in leaves of plants induced for tuberization. Transgenic antisense plants with reduced levels of the PHOR1 protein exhibited a phenotype characteristic of plants with an altered response to GAs. These lines exhibited reduced stem length, earlier tuberization under SDs and enhanced insensitivity to applied GAs. Subcellular localization studies showed that the PHOR1–green fluorescent protein (GFP) fusion had a cytoplasmic localization and a rapid migration of the fusion protein into the nucleus on treatment with GAs. PHOR1 is a novel component of the GA-signalling pathway with a function in the perception of the GA response.

12.3 THE ROLE OF GROWTH REGULATORS IN CONTROLLING TUBERIZATION

Similar to flower induction, numerous physiological studies have implicated both an inhibitor and a promoter working coordinately to control tuber induction. Early investigators presented evidence for phytohormones acting as the chemical regulators, with GAs acting as the inhibitor and cytokinins as the promoter (Hammes and Nel, 1975). This model envisions control by the regulatory effects of inhibitors on GA and subsequent activation of cytokinins leading to cell division and the mobilization of carbohydrates during tuber initiation.

Tuber initiation is accompanied by a reduction in dry weight of haulms, stolons and roots (Hammes and Nel, 1975) and an increase in stem height and the production of anthocyanin (Martinez-Garcia et al., 2002a). Such correlated growth reactions would seem to indicate that the induction stimulus is present throughout the plant and that interdependent control mechanisms stop stolon growth, induce tuber formation and increase the rate of photosynthesis, possibly by mobilizing carbohydrates from leaves to tubers. The possibility of a hormonal balance controlling tuberization has been proposed (Palmer and Barker, 1973; Tizio and Biain, 1973; Hammes and Nel, 1975). One explanation is the existence of a direct balance between endogenous GA and inhibitors. These inhibitors may readily control the level of active GA. As GA concentrations are lowered below a critical level, this stimulates the synthesis of cytokinins or another specific substance that activates tuberization. Let us examine what evidence exists to support this hypothesis.

12.3.1 Gibberellins

GAs have been implicated in inhibiting tuberization under LD photoperiods. GA levels in the leaf decreased under SD photoperiods and increased under LD conditions (Railton and Wareing, 1973). High levels of GA in the stolon tip favour elongation of stolon meristems, whereas decreasing levels of GA are required for the initiation of tuberization (Xu et al., 1998b). Several workers have shown that the noninduced state in potato plants is correlated with high endogenous GA levels (Okazawa, 1959, 1960; Pont-Lezica, 1970; Railton and Wareing, 1973) and that the exogenous application of GA inhibited tuberization (Kumar and Wareing, 1974; Hammes and Nel, 1975). Menzel (1980) demonstrated that the inhibitory effects of high temperatures (32 °C day and 28 °C night) and exogenous GA were similar: both promoted haulm growth and suppressed tuberization. He also showed that low temperature (22 °C day and 18 °C night), abscisic acid (ABA) and 2-chloroethyltrimethylammonium chloride (CCC) had the opposite effect. ABA and CCC are both inhibitors of GA (Moore, 1979). In this case, Menzel speculates that temperature exerts its influence by altering the balance between GA and inhibitors or other substances that act directly on the stolon tips. Photoperiod may play a similar role. Kumar and Wareing (1974) reported that GA levels were highest in LD plants and very low in SD plants. Applications of GA on SD cuttings inhibited tuberization, and CCC applications stimulated tuber formation on LD cuttings and lowered GA levels. GA is also reported (Palmer and Barker, 1972) to regulate enzyme activity. Invertase activity, associated with stolon elongation, is positively correlated with exogenous GA concentrations in vitro.

Enzymes involved in starch synthesis, characteristic of tuberization, were most active when GA levels were low (Obata-Sasamoto and Suzuki, 1979). GA can act as antagonist to the proposed mobilizing effect of natural cytokinins. Lovell and Booth (1967) showed that application of GA on potato plants at an early stage of tuber development resulted in greatly reduced transfer of assimilated ^{14}C to the stolons.

Although there are examples of both promotive (El-Antably et al., 1967; Zaib and Rafig, 1980) and inhibitory effects (Palmer and Smith, 1969) of ABA on tuberization, it is likely that promotive effects are due to inactivation of GA. Zaib and Rafig (1980) offered proof that ABA could reverse the inhibitory effects of GA on stem cuttings in vitro. In this study, ABA promoted tuberization, decreased total fresh weight and chlorophyll content and increased total protein. GAs have long been known to play an important role in regulating tuber formation. High levels of GA in leaves are associated with the noninduced state, whereas low levels of GA are associated with the induced state. High levels of GA are detected in the apices of noninduced, elongating stolons. These high levels of GA ($>500\,ng\,g^{-1}$) decrease precipitously on 4 days of culture on an inductive medium (8% sucrose). Concomitant with the formation of a new tuber, the reduction in GA in stolon apices can be as great as 50-fold (Xu et al., 1998b). Using transgenic *S. tuberosum* ssp. *andigena* lines, Carrera et al. (2000) showed that changing the expression level of the *GA20 oxidase1* gene, encoding for a key enzyme in the GA biosynthetic pathway, strongly affected tuber induction and yield. Suppressed levels of GA20 oxidase expression enhanced tuber production, whereas increased levels of GA20 oxidase expression retarded tuber development. The BEL/KNOX tandem of TFs discussed in Section 12.5 regulates tuber development by lowering bioactive GA levels through the repression of *GA20 oxidase1* gene activity (Rosin et al., 2003a; Chen et al., 2004)

In summary, support for a control role of GAs in tuberization comes from several lines of research: (1) the higher levels of GA in noninduced potato plants; (2) the inhibition of tuberization in whole plants, stem cuttings and stem segments treated with GA and (3) the promotion of tuberization by inhibitors and suppressors of GA activity. Although it is clear that GAs have an important function in controlling tuber formation, the site of synthesis for these GAs is not known. GAs contribute to the change in orientation of cell growth in the subapical region of the tuberizing stolon but also act in the leaf as a signalling mechanism.

12.3.2 Cytokinins

Early studies demonstrated that kinetin was incorporated into stolon tips prior to tuber formation (Smith and Palmer, 1969, 1970). Stolons cultured on medium with kinetin formed tubers, whereas those without kinetin formed none. Kinetin eliminated the need for photoperiod induction of tubers on stem sections of *S. tuberosum*, demonstrating the same propensity for tuberization as SD induction (Forsline and Langille, 1976). Using actinomycin D, chloramphenicol and 5-fluorodeoxyuridine, Palmer and Smith (1969) showed that kinetin-induced tuber formation was independent of protein and nucleic acid synthesis. They postulated the existence of a tuberization switch that induces cytokinin activity. The presence of zeatin riboside was positively correlated with the induced state in potato plants. In vitro studies produced 75% tuberization with zeatin riboside treatment compared with 0% for controls (Mauk and Langille, 1978).

Cytokinins indirectly affect tuberization by enhancing the activation of starch-synthesizing enzymes to support continuing starch deposition (Obata-Sasamoto and Suzuki, 1979). Mingo-Castel et al. (1976) demonstrated that applications of exogenous kinetin in vitro stimulated accumulation of starch, possibly by activation of ADP glucose phosphorylase and pyrophosphorylase during tuberization. In a study on the effects of kinetin and ethylene (Palmer and Barker, 1973), kinetin-treated stolons showed increased starch synthesis with decreased invertase activity, RNase activity and lower levels of soluble sugars at the stolon tip.

By activating cell division, cytokinins may be locally involved in the initial creation of the strong tuber sink. Li et al. (1992) showed that cytokinin activity can create a nutrient sink and that both ^{14}C-labelled sucrose and amino acids are mobilized to localized sites of high cytokinin accumulation. Amyloplast development and the increased transcription of starch biosynthesis enzymes are specifically induced by cytokinins in cultured tobacco cells (Miyazawa et al., 1999). Local synthesis of cytokinins in axillary buds of transgenic tobacco resulted in an increase in starch accumulation in the lateral shoots that formed (Guivarc'h et al., 2002). An increase in cytokinin levels mediated by suppression of the MADS box TF, *POTM1*, resulted in an increase in starch accumulation and active cell division in specific cells of meristems and leaves (Rosin et al., 2003b).

As a useful experimental approach, the *Agrobacterium ipt* gene has been introduced into plants to overproduce cytokinin (Barry et al., 1984). These cytokinin overproduction phenotypes include loss of apical dominance, decreased leaf size and internode length and poor root growth. Tuberization is also affected by the expression of the *Agrobacterium ipt* gene, with high levels of cytokinins inhibiting and moderate levels promoting tuber formation (Gális et al., 1995; Romanov et al., 2000). As illustrated by Sergeeva et al. (2000), the ratio of cytokinin to auxin is important for tuberization. A slight increase in the cytokinin to auxin ratio promotes tuberization. A large change in the ratio, however, inhibits tuberization, in favour of the formation of truncated shoots that accumulated high levels of starch. Local expression of the *ipt* gene in axillary buds of transgenic tobacco created a strong sink and resulted in the formation of truncated, tuberizing lateral branches (Guivarc'h et al., 2002). These swollen, lateral branches were similar in morphology to those produced from axillary buds of *POTM1* suppression lines (Rosin et al., 2003b).

The following points support the role of cytokinins as a regulatory factor for tuber induction: (1) There is a requirement for cell division during tuberization, a process known to be promoted by cytokinins (Skoog and Miller, 1957). Cytokinin accumulation is associated with the promotion of tuber formation (Gális et al., 1995; Romanov et al., 2000). (2) Tuber initiation requires the inhibition of cell elongation and the promotion of lateral growth or swelling (Vanderhoff and Key, 1968; Scott and Liverman, 1956). Local accumulation of cytokinins in axillary buds of transgenic tobacco produced truncated, tuberizing lateral branches (Guivarc'h et al., 2002). (3) Starch accumulation is an important component of tuberization, and cytokinins promote the mobilization of carbohydrates (Mothes, 1964). The transformation of stolon cells to tuber cells (Xu et al., 1998b) may be facilitated by the presence of cytokinins. TFs discussed in Section 12.5 enhance tuber formation and regulate cytokinin levels (Chen et al., 2003).

12.3.3 Lipoxygenase activity and the role of jasmonates

Plant lipoxygenases (LOXs) are a functionally diverse class of dioxygenases implicated in physiological processes such as growth, senescence and stress-related responses. LOXs incorporate oxygen into their fatty acid substrates and produce hydroperoxide fatty acids that are precursors of jasmonic acid and related compounds. Oxylipin products of this pathway, jasmonic and tuberonic acids, have been implicated as tuber-inducing compounds (Koda et al., 1991; Pelacho and Mingo-Castel, 1991). Specifically, jasmonic acid is involved in the induction of radial cell expansion in tuber cells (Takahashi et al., 1994) and tuber buds (Castro et al., 1999).

Studies by Kolomiets et al. (2001) showed that tuber-associated LOXs, designated the Lox1 class, are involved in the control of tuber growth. RNA hybridization analysis showed that accumulation of Lox1-class transcripts was restricted to developing tubers, stolons and roots and that mRNA accumulation and enzyme activity correlated positively with tuber initiation and growth. In situ hybridization showed that Lox1-class transcripts accumulated in the apical and subapical regions of the newly formed tuber and specifically in the vascular tissue of the perimedullary region, the site of the most active cell growth during tuber enlargement (Kolomiets et al., 2001). Suppression mutants produced by expressing antisense coding sequence of a specific tuber LOX, designated *POTLX-1*, exhibited a significant reduction in LOX activity in stolons and tubers. The suppression of LOX activity correlated with reduced tuber yield and a disruption of tuber formation. No morphological changes were observed in the vegetative shoots of any of the suppression lines. These results indicate that the pathway initiated by the expression of the Lox1-class genes of potato is involved in enhancing tuber enlargement.

Jasmonic and tuberonic acids are products of the 13-LOX pathway (Gardner, 1995). Because the tuber LOXs have predominately 9-LOX activity (Royo et al., 1996), it is possible that some of the 9-hydroperoxide derivatives may have a specific role, as yet unclear, in the regulation of tuber growth. Hamberg (2000) identified a novel cyclopentenone, 10-oxo-11-phytodienoic acid, produced by 9-LOX activity from young tubers, that is an isomer of 12-oxo-phytodienoic acid, the precursor to jasmonic acid. This new 9-LOX cyclopentenone could be the precursor of an undiscovered compound similar to tuberonic and jasmonic acids that regulates tuber growth. Low levels of expression of the 13-LOX of potato, designated H3, were detected in tubers (Royo et al., 1996). Transgenic antisense plants of LOX-H3, however, exhibited an increase in tuber yield in plants with suppressed H3 expression, indicating that LOX-H3 has no inductive role in the regulation of tuber development (Royo et al., 1999).

12.4 GENE ACTIVITY DURING EARLY TUBER FORMATION

To examine gene expression during the early stages of tuber formation, in conjunction with The Institute For Genomic Research (TIGR) Potato Genomics Project, microarray analysis was performed on RNAs extracted from stolon tips grown under SD photoperiodic conditions. Tissue culture stock plants of *S. tuberosum* ssp. *andigena* were grown in soil in growth chambers under either SD (inductive conditions) or SD plus a 60-min night break (noninductive conditions) with 22/18 °C day/night temperatures. Stolon tip samples, approximately 2.0 cm in length, from plants grown under both photoperiods were harvested at 2, 4, 8, 12 and 16 days after the beginning of the photoperiod regime. Visible tuber formation under these SD conditions was observed at 10–12 days. The experiment consisted of two biological replicates with two hybridizations per replicate. Lighting in the replicate growth chambers consisted of twelve 110-W fluorescent bulbs and four 25-W incandescent bulbs. RNA from plants exposed to the night break at each timepoint was used as the control hybridization probe and the SD samples as the test probe. Microarray data were scanned, normalized and catalogued by using the Gene Traffic software package (http://pga.swmed.edu/gene_traffic_2.htm). The potato microarray chip contained 12 000 cDNA clones and was made available by the Potato TIGR project. The gene list and microarray data are available at http://www.tigr.org/tdb/potato/.

There is very little difference in gene activity in stolon tips grown under inductive and noninductive conditions up to day 8 (Fig. 12.4). From day 8 to 16, over 600 genes are up- or down-regulated more than twofold. This coincides with the visible onset of swelling in the stolon tip. As expected, a number of genes involved in the synthesis and processing of starch are activated to very high levels (Fig. 12.5A). GA- and auxin-activated genes are repressed in their expression, whereas a cytokinin-induced gene exhibited enhanced expression (Fig. 12.5B and C). Numerous TFs are regulated, and an example of several

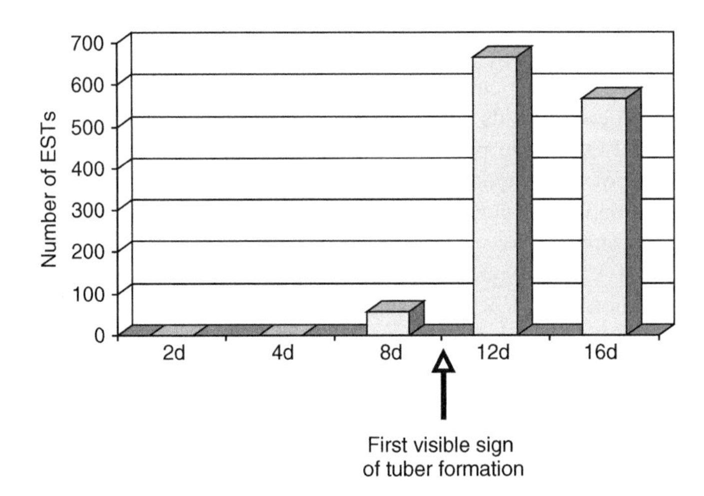

Fig. 12.4. Number of potato expressed sequence tags (ESTs) that exhibited a twofold or greater reduction or increase in RNA levels in stolon tips from short-day (SD) plants relative to control samples (SD + night break) over a 16-day period. Plants were generated from tissue culture stock plants of the photoperiod-responsive line *Solanum tuberosum* ssp. *andigena*. For the microarray hybridization, probes were generated from RNA extracted from stolon tips of plants grown under SD, inductive (8-h light and 16-h dark) conditions or SD plus a 15-min night break (noninductive conditions). Two centimeters of the stolon apex was harvested for RNA extraction.

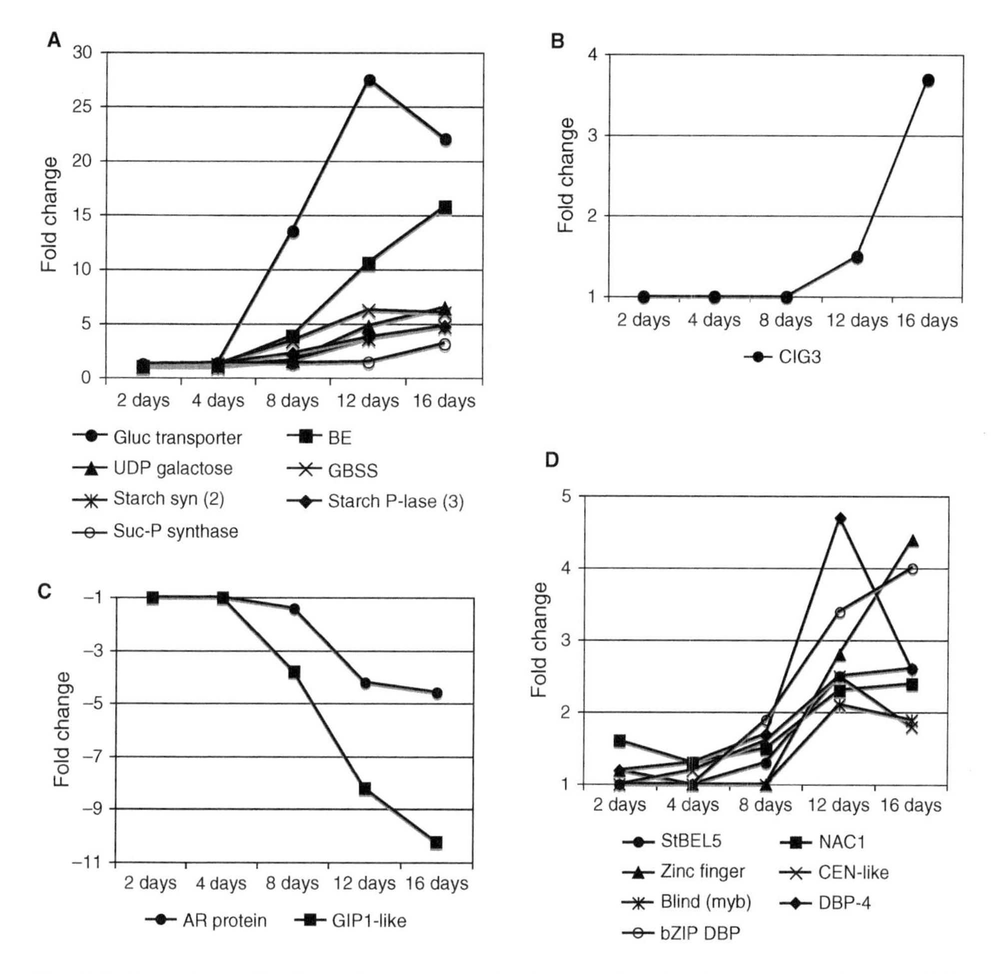

Fig. 12.5. Expression profiles from microarray data of select genes in stolons in response to short-day (SD) conditions in potato. All genes represented here exhibited a twofold or greater reduction or increase in RNA levels relative to control samples [long day (LD)] at day 12 or 16. The TC accession numbers are from the Potato The Institute For Genomic Research (TIGR) Genome website, http://www.tigr.org/tdb/potato/. (A) Gene-coding enzymes involved in the processing of starch: glucose transporter (TC93050), branching enzyme (BE, TC103098), uridine diphosphate (UDP)-galactose transporter (TC99045), granule-bound starch synthase (GBSS, TC102308), starch synthase (TC93629), starch phosphorylase (TC103326) and sucrose-phosphate synthase (TC107132). (B) A cytokinin-induced gene, *CIG3* (TC84728). (C) An auxin-responsive gene, *AR* (TC60904) and a gibberellin (GA)-induced protein, GIP1 (TC68380). (D) Transcription factors with increased RNA levels: StBEL5 (TC104412), NAC1 (no apical meristem, TC96473), a zinc finger protein (TC103019), a CENTRORADIALIS (CEN-like protein, TC98831), blind (a myb type, TC100120), DNA-binding protein-4 (DBP-4, TC62697) and a bZIP DBP (TC66586).

positively induced types is shown in Fig. 12.5D. Among those included are the BEL1 family, NAC1 types, a zinc finger type, a CENTRORADIALIS (CEN)-like protein, blind (a myb type), DNA-binding protein 4 and a bZIP DNA-binding protein. Genes involved in flowering time and floral development are both up- and down-regulated during early tuber

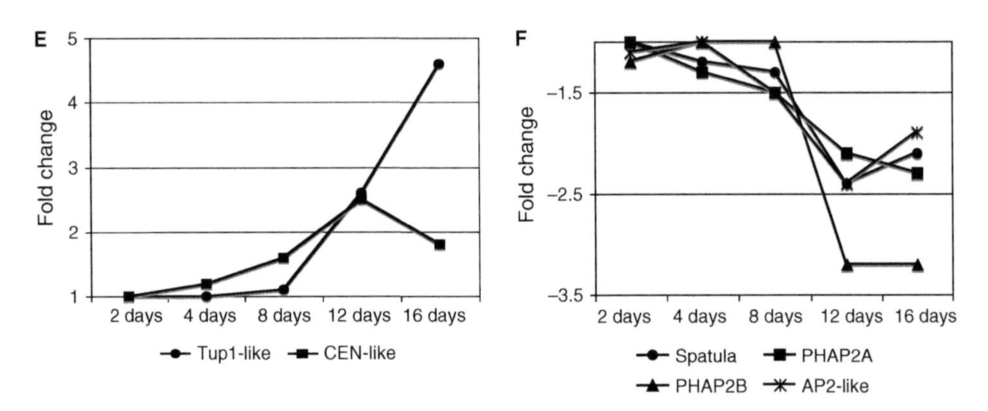

Fig. 12.5. *(Continued)* (E) Floral development genes with increased RNA levels, tup1-like protein (TC95867), and a CEN-like protein (TC98831). (F) Floral development genes with decreased RNA levels: SPATULA (TC67068), PHAP2A (AP2-like, TC60558), PHAP2B (AP2-like, TC59399) and an AP2-like protein (TC106729).

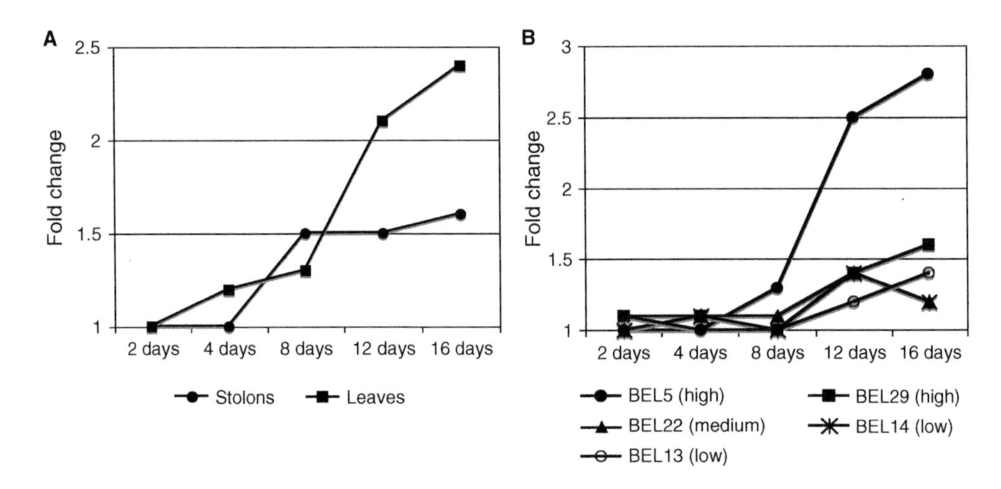

Fig. 12.6. Expression profiles from microarray data of the MADS box gene, *POTM1* (TC123022), in stolons and leaves (A) and five BEL1-like genes in stolons (B) in response to short-day conditions in potato. The five BEL1-like genes are *StBEL5* (TC104412), *StBEL13* (TC106745), *StBEL14* (TC96189), *StBEL22* (TC109648) and *StBEL29* (TC104712). Data were not available for the other two BEL1-like genes of potato, *StBEL11* and *StBEL30*.

formation (Fig. 12.5E and F). RNA of the potato BEL1-like protein, StBEL5, increases twofold through 16 days (Fig. 12.6B), and the potato MADS box gene, *POTM1*, is also developmentally regulated in both stolon tips and leaves grown under SD (Fig. 12.6A). These latter two TFs will be discussed in more detail in Section 12.5.

12.5 THE ROLE OF SPECIFIC TRANSCRIPTION FACTORS IN TUBER DEVELOPMENT

12.5.1 A MADS box protein that regulates axillary branching and affects tuber formation

MADS box genes are an example of a family of highly conserved TFs that have diverse roles in plant development. Although these TFs play a pivotal role in determining floral organ identity (Weigel and Meyerowitz, 1994; Theißen, 2001), they are also important regulators of vegetative development. *POTM1* (potato MADS box, GenBank accession number U23757) is a member of the *SQUA*-like family of plant MADS box genes isolated from an early-stage tuber cDNA library (Kang and Hannapel, 1995) and is expressed in apical and axillary meristems. Transgenic lines with reduced levels of *POTM1* mRNA exhibited decreased apical dominance accompanied by a compact growth habit and a reduction in leaf size. Suppression lines produced truncated, shoot clusters from stem buds (Fig. 12.7) and, in a model system, exhibited enhanced axillary bud growth instead of producing a tuber (Rosin et al., 2003b). Cytokinin levels in these axillary buds increased twofold to threefold, leading to the activation of cell division and starch accumulation (Rosin et al., 2003b). These results imply that *POTM1* mediates the control of axillary bud development by regulating cell growth in vegetative meristems, including stolon tips. Suppression of *POTM1* led to activation of branching and an approximately sevenfold reduction in tuber yield (Rosin et al., 2003b).

Fig. 12.7. Suppression of *POTM1* RNA accumulation alters shoot morphology. Suppression lines of POTM1 exhibited shoot clusters growing from the axillary buds of stems. Growth of these axillary buds was prolific, but limited, forming a dense cluster of truncated shoots and stolons as well as small tubers.

12.5.2 Transcription factors from the TALE superclass

Another important family of TFs involved in regulating the developmental events in apical meristems is the *knox* (knotted-like homeobox) gene family (Reiser et al., 2000). *Knox* genes belong to the group of TFs known as the three amino acid loop extension (TALE) superclass (Bürglin, 1997). These TFs are distinguished by a very high level of sequence conservation in the DNA-binding region, designated the homeodomain, and consist of three α-helices similar to the bacterial helix–loop–helix motif (Kerstetter et al., 1994). The third helix, the recognition helix, is involved in DNA-binding (Mann and Chan, 1996). TALE TFs contain a TALE, proline–tyrosine–proline, between helices I and II in the homeodomain that has been implicated in protein interactions (Passner et al., 1999). There are numerous TFs from plants and animals in the TALE superclass, and the two main groups in plants are the KNOX and BEL types (Bürglin, 1997). From early tuber cDNA libraries of potato, we have isolated two groups of TFs, KNOX and BEL types, that physically interact (Chen et al., 2003). BEL-type proteins are involved in floral architecture and development in several plant species (Smith and Hake, 2003; Bhatt et al., 2004; Smith et al., 2004).

12.5.3 Overexpression of POTH1 negatively regulates GA levels

The KNOX protein of potato, designated POTH1 (GenBank accession number U65648), regulates plant growth by controlling GA synthesis (Rosin et al., 2003a). Transgenic overexpression of POTH1 produced plants that were characterized by distorted, smaller leaves and reduced internode lengths (Rosin et al., 2003a). Both internode length and overall plant height were reduced approximately threefold in these mutant plants relative to controls. The mutant leaf traits are designated 'mouse-ear' or 'curled' phenotype as reported previously in other *knox* mutants (Parnis et al., 1997; Tamaoki et al., 1997). Application of GA_3 produced a partial reversal of the leaf phenotype and completely rescued the dwarf phenotype (Rosin et al., 2003a).

Because of the similarity of this POTH1 phenotype to those reported in tobacco (Tanaka-Ueguchi et al., 1998; Tamaoki et al., 1999), we examined the effect of GA20 oxidase mRNA accumulation in these POTH1 overexpressers. GA20 oxidase is a key biosynthetic enzyme in the GA pathway, catalyzing the conversion of GA_{53} to GA_{20} through GA_{44} and GA_{19} (Heddens and Kamiya, 1997). GA 20-oxidase1 mRNA was decreased in shoots of these overexpression lines. Biochemical analysis showed that the levels of GA_{53} and GA_{19} increased, whereas the levels of GA_{20} and GA_1 decreased in shoot tips of these plants (Rosin et al., 2003a).

12.5.4 POTH1 protein interacts with seven unique potato BEL transcription factors

Making use of the yeast two-hybrid system and immunoprecipitation, the interaction of the POTH1 protein with all seven members of the BEL1 family of TFs in potato was verified (Chen et al., 2003). Using deletion mutant analysis, we identified the first 80 amino acids of the BELL domain (Bellaoui et al., 2001) as the region involved in protein interaction with POTH1 (Chen et al., 2003). The existence of so many unique BEL partners that bind to POTH1 implies that they are involved in a complex system of developmental control in potato. One of the BEL1 partners, *StBEL5* (GenBank accession number AF406697), consistently exhibited enhanced RNA levels in stems, leaves and stolons (but not roots) in response to a SD photoperiod (Chen et al., 2003). In situ hybridization results have placed the RNA of both *POTH1* and *StBEL5* in the vascular tissue of tuberizing meristems (D.J. Hannapel and A.K. Banerjee, unpublished data).

12.5.5 Over-expression of POTH1 and StBEL5 produces an enhanced capacity to form tubers

Overexpression of *StBEL5* produced transgenic plants with an enhanced capacity to form tubers (Chen et al., 2003). These *StBEL5* sense lines were able to produce more tubers at a faster rate than controls under both inductive (SD) and noninductive (LD) conditions on soil- and in vitro-grown plants. *POTH1* also enhanced tuber formation when overexpressed in transgenic plants (Rosin et al., 2003a). *StBEL5* lines exhibited normal leaf and stem and actually grew faster than controls, with select lines exhibiting as much as a 50% increase in the rate of growth (Chen et al., 2003). These results indicate that POTH1 and StBEL5 may act through an identical signalling pathway to mediate tuber formation. The interaction between KNOX and BEL proteins appears to be ubiquitous in the plant kingdom (Bellaoui et al., 2001; Müller et al., 2001; Smith et al., 2002).

12.5.6 Mechanism for transcription factors in regulating tuberization

Our results indicate that these molecular switches control growth by modulating levels of phytohormones. For *POTM1*, suppression results in an increase in cytokinin levels. For *StBEL5* and *POTH1*, overexpression results in an increase in cytokinins (Table 12.1) and a reduction in GA (Fig. 12.8). Clearly the effect on cytokinin levels, for at least the *POTM1* suppression lines, resulted in disrupted source/sink relations in these lines. Starch accumulation occurred in leaves and axillary buds in preference to the normal sink of a forming tuber. This abnormal preferential redistribution of photosynthate resulted in reduced tuber yields. StBEL5 and POTH1 appear to affect tuber growth by lowering GA levels and activating cell growth with increased cytokinins in the stolon tip.

Table 12.1 Cytokinin content in POTH1 and StBEL5 overexpression lines

Plant	Zeatin Types (pm per g fr wt)	Isopentenyl Types (pm per g fr wt)
WT	10.5 ± 1.0	12.0 ± 1.5
POTH1-15	42.5 ± 15	35.5 ± 7.0
POTH1-29	34.0 ± 12	30.0 ± 6.0
STBEL5-11	55.5 ± 30	31.5 ± 11
STBEL5-20	30.0 ± 6.0	29.5 ± 6.5

Cytokinin was measured in the top 2 cm of the apical shoots from long-day plants. WT is *Solanum tuberosum* ssp. *andigena*. All samples were quantified in triplicate.

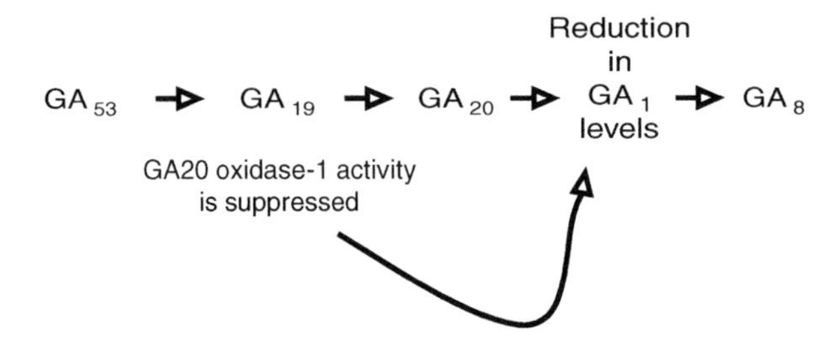

Fig. 12.8. Effect on the gibberellin (GA) biosynthetic pathway by overexpression of *StBEL5* and *POTH1* in transgenic potato lines.

Owing to the presence of conserved DNA-binding domains for the three families of TFs described here, it is very likely that these proteins target specific genes to regulate their expression. DNA-binding assays demonstrated that StBEL5 and POTH1 bind to the regulatory region of *ga20 oxidase1* from potato, a gene encoding a key enzyme in the GA biosynthetic pathway (Fig. 12.9). In tandem, StBEL5 and POTH1 had a

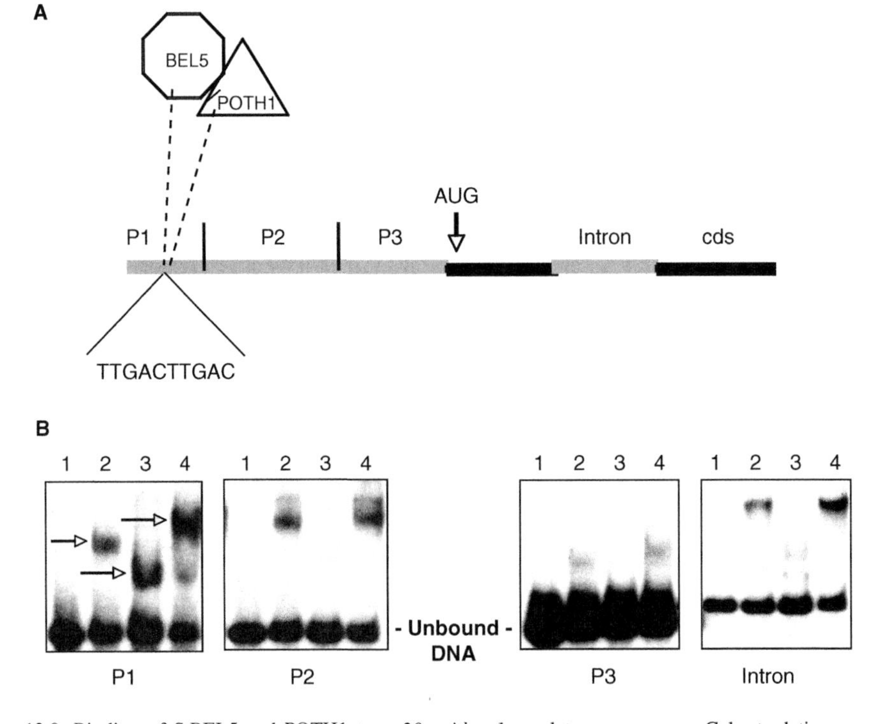

Fig. 12.9. Binding of StBEL5 and POTH1 to *ga20 oxidase1* regulatory sequences. Gel-retardation assay of (A) *ga20 oxidase1* promoter P1 and P2, P3 and first intron sequences with (B) StBEL5 (lanes 2), POTH1 (lanes 3) or both (lanes 4). Lanes 1 are labelled probe without protein to show the position of unbound probe. The DNA–protein complexes are indicated with arrowheads in the P1 panel. Both proteins bound only to sequence in the P1 region.

greater binding affinity for the *ga20 oxidase1* promoter than either protein alone (Chen et al., 2004). The StBEL5-POTH1 heterodimer bound specifically to a composite 10-bp sequence, containing two TTGAC cores in the P1 region of the promoter (Fig. 12.9). Transcription assays with the BEL and KNOX proteins indicate that, in tandem, they bind specific DNA sequences of *ga20 oxidase1* to repress its activity by more than 50% (Fig. 12.10). These results indicate that the tandem interaction of StBEL5 and POTH1 is essential for regulation of the expression of their target gene, *ga20 oxidase1*. In summary, POTH1/StBEL5 regulation on gene activity involves both protein/protein interaction and protein/DNA interaction (Fig. 12.11). The control of tuberization is mediated by a pathway that activates growth in the subapical region of the stolon tip. Cytokinins enhance cell division, and decreased GA levels favour radial growth over elongation of the stolon tip. POTM1 is putatively involved in regulating genes involved in cytokinin synthesis in specific meristematic cells, whereas the tandem StBEL5/POTH1 complex mediates an increase in cytokinin synthesis and a decrease in GA synthesis by targeting the promoters of specific genes. Based on the microarray results, however, it is clear that numerous other TFs and signalling molecules, such as micro-RNAs and flowering-time gene products, are likely involved in the pathways that regulate the initial stages of tuber formation.

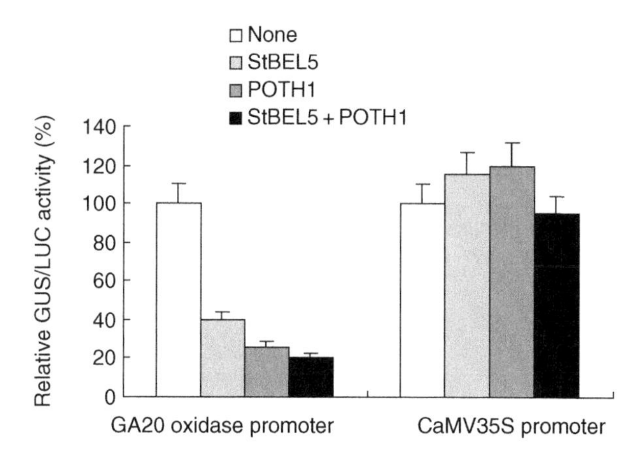

Fig. 12.10. The repression effect of StBEL5 and POTH1 on the *ga20 oxidase1* promoter in tobacco protoplasts. The *ga20 oxidase1* promoter (1028 nt) drives beta-glucuronidase (GUS) expression. The construct with the luciferase (LUC) gene under the control of cauliflower mosaic virus (CaMV) 35S promoter was used as control. Protein effector constructs were generated by using pBI221 vector as a backbone, with the GUS gene replaced by the full-length cDNAs of either *StBEL5* or *POTH1*. The protoplasts were subjected to electroporation with the various mixtures of plasmid DNA. Transfections were performed three times for each effector combination. Relative GUS-LUC activity was calculated by dividing the ratio of GUS activity to LUC activity from different effectors with the ratio from reporter plasmid alone. No suppression of transcription was observed with the CaMV 35S/GUS construct. Relative activities calculated from three transfection replications were presented as a mean ± SE[*Source: Chen et al. (2004)*].

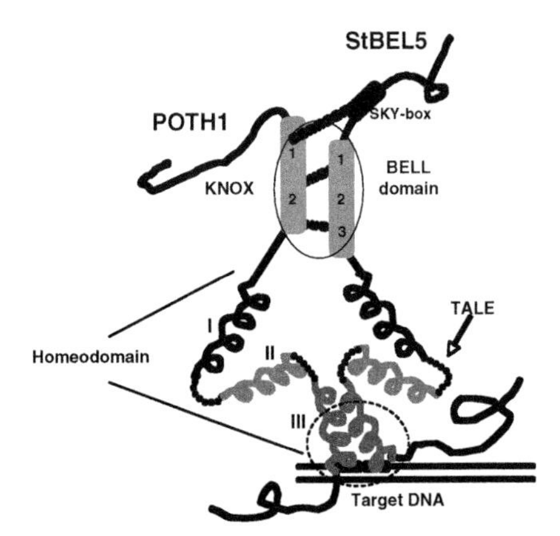

Fig. 12.11. POTH1/BEL5 regulation on gene activity involves both protein/protein interaction (solid oval) and protein/DNA interaction (dotted oval). The tandem complex of StBEL5 and POTH1 is essential to repress transcription of one of their target genes, *ga20 oxidase1*. This suppression leads to a reduction in enzyme activity and a subsequent decrease in bioactive gibberellin (GA) levels.

ACKNOWLEDGEMENTS

Many thanks to Professor Salomé Prat for providing the *ga20 oxidase1* promoter, to Dr. Hao Chen for assistance with the microarray data and his creative leadership in elucidating the mechanism of transcriptional control mediated by StBEL5/POTH1, to Drs Faye Rosin, Michael Kolomiets and Anjan Banerjee for their excellent contributions to the research described here and to Professor William D. Park for setting me on the right course. This work was supported by the National Science Foundation under Grant no. 0305647 and an award from the Plant Sciences Institute, Iowa State University.
Note: A recent report (Banerjee et al., 2006) has shown that a mobile RNA encoding StBEL5 plays an important role in signalling the onset of tuber formation.

REFERENCES

Amador V., E. Monte, J.L. Garcia-Martinez and S. Prat, 2001, *Cell* 106, 343.

Bachem C.W.B., R.S. van der Hoeven, S.M. de Bruijn, D. Vreugdenhil, M. Zabeau and R.G.F. Visser, 1996, *Plant J.* 9, 745.

Banerjee A.K., M. Chatterjee, Y.Yu, S.G. Suh, W.A. Miller and D.J. Hannapel, 2006, Plant Cell 18, 3443.

Barry G.F., S.G. Rogers, R.T. Fraley and L. Brand, 1984, *Proc. Natl. Acad. Sci. U.S.A.* 81, 4776.

Bellaoui M., M.S. Pidkowich, A. Samach, K. Kushalappa, S.E. Kohalmi, Z. Modrusan, W.L. Crosby and G.W. Haughn, 2001, *Plant Cell* 13, 2455.

Bernard N., 1902, *Rev. Gen. Bot.* 14, 5.

Bhatt A.M., J.P. Etchells, C. Canales, A. Lagodienko and H. Dickinson, 2004, *Gene* 328, 103.

Borah M.N. and F.L. Milthorpe, 1962, *Indian J. Plant Physiol.* 5, 53.

Bürglin T.R., 1997, *Nucleic Acids Res.* 25, 4173.

Burt R.L., 1964, *Eur. Potato J.* 7, 197.

Carrera E., J. Bou, J.L. Garcia-Martinez and S. Prat, 2000, *Plant J.* 22, 247.
Castro G., T. Kraus and G. Abdala, 1999, *J. Plant Physiol.* 155, 706.
Chailakhyan M.K., L.I. Yanina, A.G. Davedzhiyan and G.N. Lotova, 1981, *Dokl. Akad. Nauk SSSR* 257, 1276.
Chapman H.W., 1958, *Physiol. Plant.* 11, 215.
Chen H., A.K. Banerjee and D.J. Hannapel, 2004, *Plant J.* 38, 276.
Chen H., F. Rosin, S. Prat and D.J. Hannapel, 2003, *Plant Physiol.* 132, 1391.
Driver C.M. and J.G. Hawkes, 1943, *Imp. Bur. Plant Breed. Genet.*, Cambridge, England, (Tech. Comm.).
El-Antably H., P.F. Wareing and J. Hillman, 1967, *Planta* 73, 74.
Ewing E.E. and P.C. Struik, 1992, *Hortic. Rev.* 14, 89.
Fernie A.R. and L. Willmitzer, 2001, *Plant Physiol.* 127, 1459.
Forsline P.L. and A.R. Langille, 1976, *Can. J. Bot.* 54, 2513.
Gális I., J. Macas, J. Vlasák, M. Ondrej and H.A. Van Onckelen, 1995, *J. Plant Growth Reg.* 14, 143.
Gardner H.W., 1995, *Hortic. Sci.* 30, 197.
Garner W.W. and H.A. Allard, 1920, *J. Agric. Res.* 18, 553.
Garner W.W. and H.A. Allard, 1923, *J. Agric. Res.* 23, 871.
González-Schain N., S. Prat and P. Suárez-López, 2003, Abstract from *Proc. Int. Soc. Plant Mol.* Meeting, 24 June, Barcelona.
Gregory L.E., 1956, *Am. J. Bot.* 43, 281.
Gregory L.E., 1965, In: W. Ruhland (ed.), *Encyclopedia of Plant Physiol*ogy, p. 1328. Springer-Verlag, Berlin.
Guivarc'h A., J. Rembur, M. Goetz, T. Roitsch, M. Noin, T. Schmulling and D. Chriqui, 2002, *J. Exp. Bot.* 53, 621.
Hamberg M., 2000, *Lipids* 35, 353.
Hammes P.S. and P.C. Nel, 1975, *Potato Res.* 18, 262.
Hedden P. and Y. Kamiya, 1997, *Ann. Rev. Plant Physiol. Plant Mol. Biol.* 48, 431.
Jackson S.D., A. Heyer, J. Dietze and S. Prat, 1996, *Plant J.* 9, 159.
Jackson S.D., P. James, S. Prat and B. Thomas, 1998, *Plant Physiol.* 117, 29.
Jackson S.D. and S. Prat, 1996, *Physiol. Plant.* 98, 407.
Kang S.G. and D.J. Hannapel, 1995, *Gene* 166, 329.
Kerstetter R., E. Vollbrecht, B. Lowe, B. Veit, J. Yamaguchi and S. Hake, 1994, *Plant Cell* 6, 1877.
Koda Y., Y. Kikuta, H. Tazaki, Y. Tsujino, S. Sakamura and T. Yoshihara, 1991, *Phytochemistry* 30, 1435.
Kolomiets M.V., D.J. Hannapel, H. Chen, M. Tymeson and R.J. Gladon, 2001, *Plant Cell* 13, 613.
Kumar D. and P.F. Wareing, 1972, *New Phytol.* 71, 639.
Kumar D. and P.F. Wareing, 1973, *New Phytol.* 72, 283.
Kumar D. and P.F. Wareing, 1974, *New Phytol.* 73, 833.
Li Y., G. Hagen and T.J. Guilfoyle, 1992, *Dev. Biol.* 153, 386.
Lovell P.H. and A. Booth, 1967, *New Phytol.* 66, 525.
Macleod M.R., H.V. Davies, S.B. Jarvis and M.A. Taylor, 1999, *Potato Res.* 42, 31.
Madec P., 1963, In: J.D. Ivins and F.L. Milthorpe (eds), *The Growth of the Potato*, p. 121. Butterworths, London.
Madec P. and P. Perennec, 1959, *Eur. Pot. J.* 2, 22.
Magrou J., 1943, *Comp. Rend. Soc. Biol.* 138, 496.
Mann R.S. and S.K. Chan, 1996, *Trends Genet.* 12, 258.
Martinez-Garcia J.F., J.L. Garcia-Martinez, J. Bou and S. Prat, 2002a, *J. Plant Growth Reg.* 20, 377.
Martinez-Garcia J.F., A. Virgos-Soler and S. Prat, 2002b, *Proc. Natl. Acad. Sci. U.S.A.* 99, 15211.
Mauk C.S. and A.R. Langille, 1978, *Plant Physiol.* 62, 438.
Menzel C.M., 1980, *Ann. Bot.* 46, 259.
Mingo-Castel A.M., O.E. Smith and R.E. Young, 1976, *Plant Cell Physiol.* 17, 557.
Miyazawa Y., A. Sakai, S. Miyagishima, H. Takano, S. Kawano and T. Kuroiwa, 1999, *Plant Physiol.* 121, 461.
Moore T.C., 1979, *Biochemistry and Physiology of Plant Hormones*, Springer-Verlag, Brooklyn, NY.
Mothes K.L., 1964, *Reg. Nat. de la Croiss. Veg.* 131.
Müller J., Y. Wang, R. Franzen, L. Santi, F. Salamini and W. Rohde, 2001, *Plant J.* 27, 13.
Obata-Sasamoto H. and H. Suzuki, 1979, *Physiol. Plant.* 45, 320.
Okazawa Y., 1959, *J. Crop. Sci. Soc. Jpn.* 28, 129.
Okazawa Y., 1960, *J. Crop. Sci. Soc. Jpn.* 29, 121.

Palmer C.E. and W.G. Barker, 1972, *Plant Cell Physiol.* 13, 681.
Palmer C.E. and W.G. Barker, 1973, *Ann. Bot.* 37, 85.
Palmer C.E. and O.E. Smith, 1969, *Plant Cell Physiol.* 10, 657.
Parnis A., O. Cohen, T. Gutfinger, D. Hareven, D. Zamir and E. Lifschitz, 1997, *Plant Cell* 9, 2143.
Passner M., H.D. Ryoo, L. Shen, R.S. Mann and A.K. Aggarwal, 1999, *Nature* 397, 714.
Pelacho A.M. and A.M. Mingo-Castel, 1991, *Plant Physiol.* 97, 1253.
Pont-Lezica R.F., 1970, *Pot. Res.* 13, 323.
Raices M., P.R. Gargantini, D. Chinchilla, M. Crespi, M.T. Tellez-Inon and R.M. Ulloa, 2003, *Plant Mol. Biol.* 52, 1011.
Railton I.D. and P.F. Wareing, 1973, *Physiol. Plant.* 28, 88.
Reiser L., P. Sanchez-Baracaldo and S. Hake, 2000, *Plant Mol. Biol.* 42, 151.
Romanov G.A., N.P. Aksenova, T.N. Konstantinova, S.A. Golyanovskaya, J. Kossmann and L. Willmitzer, 2000, *Plant Growth Reg.* 32, 245.
Rosin F.M., J.K. Hart, H.T. Horner Jr, P.J. Davies and D.J. Hannapel, 2003a, *Plant Physiol.* 132, 106.
Rosin F.M., J.K. Hart, H. Van Onckelen and DJ. Hannapel, 2003b, *Plant Physiol.* 131, 1613.
Royo J., J. Leon, G. Vancanneyt, J.P. Albar, S. Rosahl, F. Ortego, P. Castañera and J.J. Sanchez-Serrano, 1999, *Proc. Natl. Acad. Sci. U.S.A.* 96, 1146.
Royo J., G. Vancanneyt, A.G. Perez, C. Sanz, K. Stormann, S. Rosahl and J.J. Sanchez-Serrano, 1996, *J. Biol. Chem.* 271, 21012.
Scott R.A. and J.L. Liverman, 1956, *Plant Physiol.* 31, 321.
Sergeeva L.I., S.M. de Bruijn, E.A.M. Koot-Gronsveld, O. Navratil and D. Vreudenhil, 2000, *Physiol. Plant.* 108, 435.
Skoog F. and C.O. Miller, 1957, *Symp. Soc. Exp. Biol.* 15, 118.
Smith H.M., I. Boschke and S. Hake, 2002, *Proc. Natl. Acad. Sci. U.S.A.* 99, 9579.
Smith H.M., B.C. Campbell and S. Hake, 2004, *Curr. Biol.* 14, 812.
Smith H.M. and S. Hake, 2003, *Plant Cell* 15, 1717.
Smith O.E. and C.E. Palmer, 1969, *Nature* 221, 279.
Smith O.E. and C.E. Palmer, 1970, *Physiol. Plant.* 23, 599.
Takahashi K., K. Fujino, Y. Kikuta and Y. Koda, 1994, *Plant Sci.* 100, 3.
Tamaoki M., S. Kusaba, Y. Kano-Murakami and M. Matsuoka, 1997, *Plant Cell Physiol.* 38, 917.
Tamaoki M., A. Nishimura, M. Aida, M. Tasaka and M. Matsuoka, 1999, *Plant Cell Physiol.* 40, 657.
Tanaka-Ueguchi M., H. Itoh, N. Oyama, M. Koshioka and M. Matsuoka, 1998, *Plant J.* 15, 391.
Theißen G., 2001, *Curr. Opin. Plant Biol.* 4, 75.
Tizio R. and M.M. Biain, 1973, *Phyton* 31, 3.
Vanderhoff L. and J.E. Key, 1968, *Plant Cell Physiol.* 9, 343.
Viola R., A.G. Roberts, S. Haupt, S. Gazzani, R.D. Hancock, N. Marmiroli, G.C. Machray and K.J. Oparka, 2001, *Plant Cell* 13, 385.
Vreugdenhil D., X. Xu, C.S. Jung, A.A.M. van Lammeren and E.E. Ewing, 1999, *Ann. Bot.* 84, 675.
Weigel D. and E.M. Meyerowitz, 1994, *Cell* 78, 203.
Werner H.O., 1934, *Neb. Agric. Exp. Station Res. Bull.* 75, 1.
Xu X., D. Vreugdenhil and A.A.M. van Lammeren, 1998a, *J. Exp. Bot.* 49, 573.
Xu X., A.A.M. van Lammeren, E. Vermeer and D. Vreugdenhil, 1998b, *Plant Physiol.* 117, 575.
Zaib N.A. and A. Rafig, 1980, *Plant Cell Physiol.* 21, 1343.

Chapter 13

Photosynthesis, carbohydrate metabolism and source–sink relations

Daniel Hofius[1] and Frederik A.J. Börnke[2]

[1]*Department of Molecular Biology, Copenhagen Biocenter, University of Copenhagen, Ole Maaloes Vej 5, 2200 Copenhagen N, Denmark;*
[2]*Lehrstuhl für Biochemie, Friedrich-Alexander-Universität Erlangen-Nürnberg, Staudtstr. 5, 91058 Erlangen, Germany*

13.1 INTRODUCTION

All biomass production depends on photosynthesis. Plants assimilate CO_2 from the atmosphere and reduce it to the level of triose phosphates, which can then be used to produce carbohydrates, mainly sucrose and starch. Photosynthetic carbon assimilation provides the driving force for biomass production but is only one of many factors influencing plant growth and development. There are other critical steps, such as sucrose synthesis and transport from the mesophyll tissue, phloem transport and carbon partitioning throughout the plant, that have considerable impact on these processes. Complex regulatory networks allow parallel responses to changes in environmental, metabolic and physiological limitations. The widespread adoption of transgenic approaches since the early 1990s has greatly advanced our understanding about the regulation of carbohydrate metabolism, and the potato (*Solanum tuberosum*) represents a particularly well-characterized example for this central part of plant metabolism.

Past improvements of yield potential have been derived largely from an increase in the proportion of accumulated dry weight which is invested into harvestable organs, i.e. the harvest index. In case of the potato, the harvest index has been increased from 0.09 in wild species to up to 0.81 in modern cultivars (Inoue and Tanaka, 1978). In plants, carbohydrates are produced in photosynthetically active tissues, primarily in the chloroplast containing cells of leaves. The conversion of photoassimilates into sucrose allows the transport via the phloem from these source tissues to support the growth of sink tissues such as roots or developing tubers which themselves are unable to produce assimilates. During development, sink-to-source ratios change, which implies that assimilate production must be adjusted to the changing needs of distant tissues. Past research has greatly advanced our understanding of the factors controlling the synthesis of carbohydrates and their partitioning within and between organs. In particular, the use of transgenic plants altered in the activity of single enzymes provided important information concerning metabolic control steps but also revealed the enormous flexibility of plant metabolism often bypassing single-site manipulations through the induction of

Potato Biology and Biotechnology: Advances and Perspectives
D. Vreugdenhil (Editor)

alternative pathways. Thus, a successful manipulation of biochemical pathways requires a thorough understanding of the regulatory properties of individual enzymes as well as of the regulatory networks linking entire pathways. In this chapter, we summarize the current understanding of the central pathways of carbohydrate metabolism, including current hypotheses concerning possible mechanisms underlying source-to-sink interactions involving sugars as signalling molecules.

13.2 PHOTOSYNTHETIC CARBON METABOLISM

13.2.1 CO_2 fixation

During photosynthesis, light energy is converted into chemical energy, namely ATP and NADPH, which is then used to reduce atmospheric CO_2 to carbohydrates via the reductive pentose phosphate (RPP) cycle (or Calvin cycle). The overall reaction can be described as the fixation of three molecules of CO_2 into a three-carbon sugar phosphate, triose-P, with the incorporation of one molecule of inorganic phosphate (Pi). To remobilize Pi, which is sequestered in the direct products of photosynthesis, assimilates are converted either to sucrose in the cytosol or to transitory starch, which is synthesized in the plastids and remobilized during periods when photosynthesis does not occur. The reactions of the Calvin cycle occur in the chloroplast stroma, and the cycle itself comprises 13 reactions catalysed by 11 enzymes (Fig. 13.1). Three phases can be distinguished:

(1) *Carboxylation*: Ribulose-1,5-bisphosphate carboxylase/oxygenase (Rubisco) catalyses the addition of CO_2 to ribulose-1,5-bisphosphate (RuBP) yielding two molecules of 3-phosphoglycerate (3-PGA).
(2) *Reduction* of the two molecules of 3-PGA to triose-P at the expense of two ATP and two NADPH molecules. This reaction is catalysed by the successive action of two enzymes, 3-PGA kinase (PGK) and NADP:glyceraldehyde-3-phosphate dehydrogenase (NADP:GAPDH).
(3) *Regeneration* of the primary acceptor, RuBP, from triose-P, by which five C_3 molecules are rearranged to three C_5 molecules. Each molecule of ribose-5-P must be converted to ribulose-5-P, which is, in turn, converted to RuBP at the expense of a molecule of ATP. Each molecule of CO_2 fixed in the Calvin cycle therefore requires three ATP and two NADPH to be provided by photosynthetic electron transport.

Although the activity of the Calvin cycle could in principle be merely governed by the availability of ATP and NADPH, provided by the light reaction, interaction between CO_2 fixation and photosynthetic electron transport occurs at multiple levels. Several enzymes of the Calvin cycle are subject to light/dark regulation. Light-dependent activation of these enzymes occurs rapidly upon illumination through a process that is coupled to photosynthetic electron transport via a ferredoxin/thioredoxin soluble electron transport system (Buchanan, 1980). Activation of target enzymes occurs when specific disulphide bonds (between sulphide groups contained in cysteine residues of the polypeptide) are reduced and hence cleaved. This is accomplished through reduced thioredoxin,

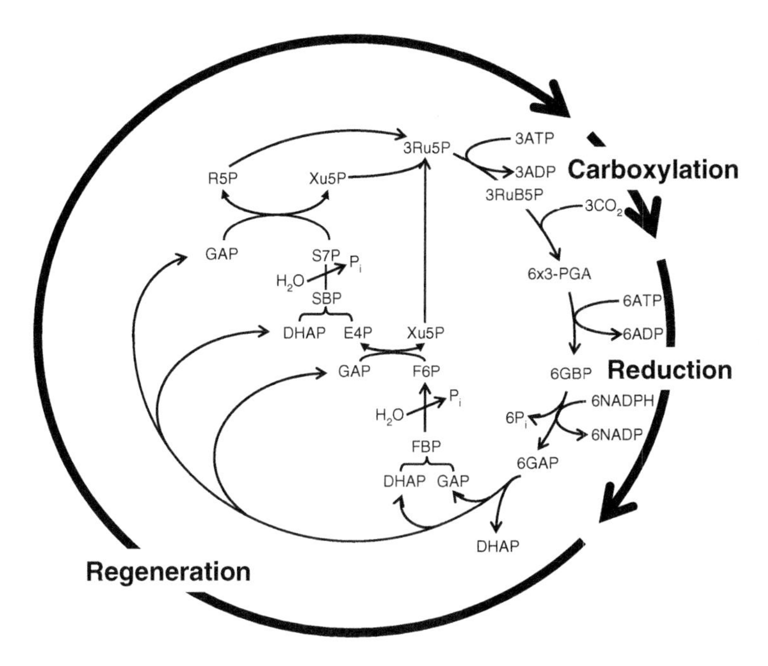

Fig. 13.1. The Calvin cycle. RuBP, ribulose-1,5-bisphosphate; 3-PGA, 3-phosphoglycerate; 1,3-PGA, 1,3-diphosphoglycerate; Ru5P, ribulose-5-phosphate; R5P, ribose-5-phosphate; Xu5P, xylulose-5-phosphate; GAP, glyceraldehyde-3-phosphate; DHAP, dihydroxyacetone-phosphate; E4P, erythrose-4-phosphate; F6P, fructose-6-phosphate; FBP, fructose-1,6-bisphosphate; SBP, sedoheptulose-1,7-bisphosphate; S7P, sedoheptulose-7-phosphate.

which in turn is reduced by reduced ferredoxin, and the reaction is catalysed by the enzyme ferredoxin–thioredoxin reductase. Ferredoxin itself is reduced by the electron transport chain, and hence, illumination brings about the rapid activation of several Calvin cycle enzymes including fructose-1,6-bisphosphatase (FBPase), sedoheptulose-1,7-bisphosphatase (SBPase), GAPDH and R5P kinase. Another level of regulation is represented by light-driven changes in stromal pH and Mg^{2+} concentration. For example, the pH of the darkened stroma is around 7 with an Mg^{2+} concentration of 1–3 mM, whereas the pH is around 8 and the Mg^{2+} concentration 3–6 mM in the illuminated stroma. These electron-transport-driven changes in the stromal environment are close to the optimum for the operation of the enzymes involved in the Calvin cycle. Photosynthetic carbon fixation is also regulated at the level of Rubisco activation. Enzyme activity requires that Mg^{2+} and CO_2 be bound to a lysine residue adjacent to the active site (carbamoylation). RuBP binds very tightly to the non-carbamoylated (inactive) form of Rubisco; this prevents carbamoylation of the enzyme and renders it effectively inactive until displaced (Portis, 1992). Furthermore, another naturally occurring inhibitor of Rubisco has been identified (2-carboxy-arabinitol-1-phosphate, CA1P) that is found in abundance during the night and is degraded during the photoperiod (Portis, 1992). This nocturnal inhibitor also binds tightly to the active site of Rubisco. Displacement of these inhibitory molecules is required to restore enzymatic activity. This is brought about by the activity of an enzyme called Rubisco activase (Portis, 1995). The process requires ATP and is inhibited by ADP; hence, the activation of Rubisco is in part dependent on the availability of ATP generated during photosynthetic electron transport (Portis, 1995).

13.2.2 Carbon partitioning in mesophyll cells

The principal product of photosynthetic carbon assimilation in the chloroplast is triose-P. This can be either exported to the cytosol to make soluble sugars or retained within the chloroplast to be fed into starch synthesis or to regenerate RuBP, respectively (Fig. 13.2). The rate of consumption of triose-P by either of these pathways has to be balanced with the momentary rate of photosynthesis to ensure that the correct proportion is cycled back into RuBP. Therefore, communication between the stromal compartment of the chloroplasts and the cytosol is essential for adjusting the rate of photosynthesis to the demands of various parts of the plants for photoassimilates. This communication is supposed to take place at the membrane system of the chloroplast. Two membranes enclose the plastidic stroma, an outer membrane that is permeable with an exclusion limit of 10 kDa and an

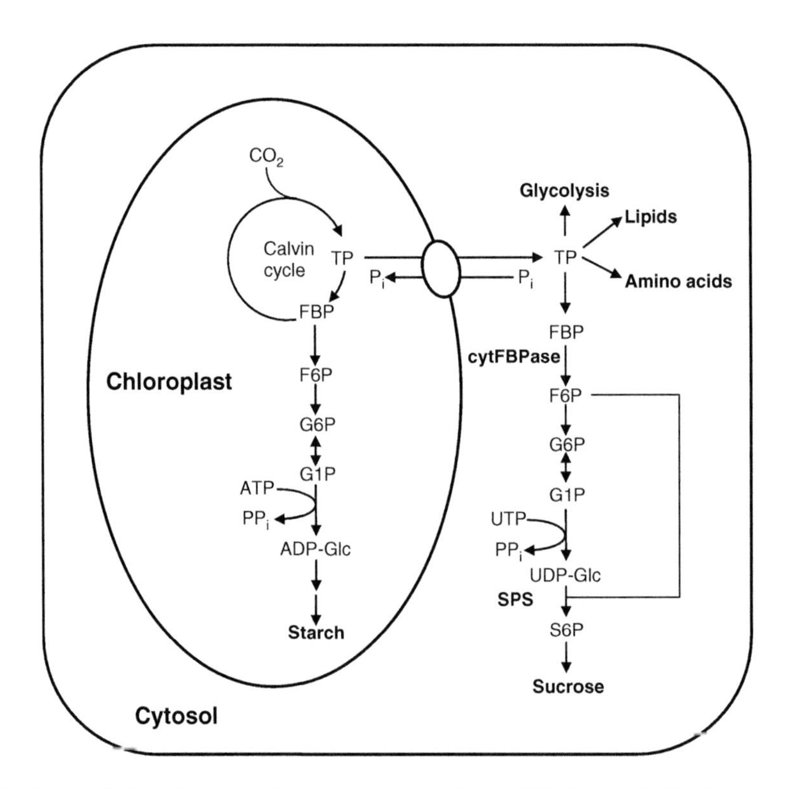

Fig. 13.2. Photosynthetic carbon metabolism in potato source leaves. FBP, fructose-1,6-bisphosphate; FBPase, fructose-1,6-bisphosphate phosphatase; F6P, fructose-6-phosphate; G6P, glucose-6-phosphate; G1P, glucose-1-phosphate; ADP-Glc, ADP-glucose; UDP-Glc, UDP-glucose; S6P, sucrose-6-phosphate; SPS, sucrose-6-phosphate synthase.

inner membrane that controls the substances entering and leaving the plastid. A specific transport system, the triose-P translocator (TPT), located in the inner membrane of the chloroplast envelope catalyses the export of triose-P into the cytosol in strict counter exchange for Pi (Flügge and Heldt, 1991). The rate of triose-P export catalysed by the TPT is thought to be primarily regulated by cytosolic processes liberating Pi from phosphorylated intermediates, particularly by sucrose synthesis (Stitt and Quick, 1989). When the export of triose-P from the chloroplasts is limited by the availability of cytosolic Pi, for example, due to decreased sucrose synthesis, fixed carbon can be deposited in the chloroplasts in the form of transitory starch. The TPT gene is nuclear encoded and the respective cDNA has been cloned from potato (Schulz et al., 1993), among other plant species. In accordance with the role of the protein in photosynthetic carbon metabolism, expression of the TPT gene was predominantly observed in green (chloroplast-containing) tissues and was shown to be light-responsive.

Transgenic potato plants transformed with the TPT cDNA in antisense orientation expressed under the control of the constitutive CaMV 35S promoter displayed a reduction in transport activity by approximately 30% (Riesmeier et al., 1993a). Under ambient light and CO_2 conditions, reduced TPT activity did neither impair photosynthesis nor affect growth and tuber yield. However, a detailed physiological analysis of the plants revealed profound changes in the allocation of carbohydrates and the distribution of metabolites between chloroplast and cytosol. The reduction of export of triose-P led to a three- to four-fold increase in starch accumulation during the day. This increase in starch synthesis and the resulting release of Pi within the chloroplast avoid Pi-limitation of photosynthesis and thus metabolically compensate for the reduced levels of TPT (Heineke et al., 1994). Moreover, the surplus starch, accumulated during the light period, was efficiently degraded during the night. Because starch degradation results partly in the release of glucose or maltose, sugars not requiring TPT activity to be exported from the chloroplast, the reduced availability of carbohydrates during the day is compensated for by an increased export rate during the night and thus prevents stronger phenotypic effects (Heineke et al., 1994). Taken together, the studies on transgenic potato plants with reduced TPT activity have established the central role of this translocator in the integration of assimilate partitioning during photosynthesis; however, they also demonstrate the enormous flexibility of plant metabolism, allowing the induction of metabolic bypasses in many cases.

13.2.3 Sucrose biosynthesis in source leaves

As outlined in the previous section, triose-P leaving the Calvin cycle can be exported from the plastid into the cytosol via the TPT and is then distributed between glycolysis and amino acid, lipid and sucrose biosynthesis (Fig. 13.2). Sucrose is the major transport form of carbohydrates in most higher plants and as such forms the interface between photosynthetically active source tissue and heterotrophic sink tissue, where it serves as an energy source for growth and development or feeds into the synthesis of storage compounds such as starch. Starting from triose-P, the sucrose biosynthetic pathway comprises seven enzymatic steps. The genes for all enzymes involved have been isolated from potato, and the *in planta* function of most of them has been assessed by reverse genetic studies. The entry molecule is fructose-1,6-bisphosphate (FBP), which is produced from the condensation of two molecules of triose-P catalysed by the enzyme aldolase. In a subsequent reaction, a phosphate group is cleaved from the C1 carbon by FBPase to yield fructose-6-phosphate (F6P). This reaction is essentially irreversible and is supposed to represent one of the key regulatory steps within the pathway of sucrose synthesis. Glucose-6-phosphate (G6P) and glucose-1-phosphate (G1P) are maintained in equilibrium with the F6P pool by the action of phosphoglucoseisomerase (PGI) and phosphoglucomutase (PGM), respectively. Uridine diphosphate glucose (UDP-Glc) and pyrophosphate are then formed from uridine triphosphate (UTP) and G1P catalysed by the enzyme UDP-glucose-pyrophosphorylase (UGPase). UDP-Glc is then combined with F6P to form sucrose-6-phosphate (Suc6P), catalysed by sucrose-6-phosphate synthase (SPS); subsequently, Suc6P is hydrolysed to form sucrose by a specific sucrose-6-phosphate phosphatase (SPP). The reaction catalysed by SPP is essentially irreversible and displaces the reversible SPS reaction from equilibrium into the direction of net sucrose synthesis (Stitt et al., 1987). SPS has been identified as the second major control point of the whole pathway, and the enzyme is subject to a complex regulation on multiple levels.

13.2.3.1 Regulatory enzymes of the pathway

In source leaves, sucrose synthesis has to be balanced with the momentary rate of photosynthesis. If sucrose synthesis is too fast, photosynthesis is inhibited because intermediates are withdrawn from the Calvin cycle too rapidly and RuBP regeneration is inhibited (Stitt et al., 1987). If sucrose synthesis is too slow, Pi is sequestered from the chloroplast into phosphorylated intermediates, and ATP synthesis, PGA production and photosynthesis are inhibited (Stitt et al., 1987).

Three enzymes of the pathway of sucrose synthesis are known to catalyse reactions removed from thermodynamic equilibrium in vivo. At these sites, the flux depends on the current activity of the enzyme, and such enzymes also often possess regulatory properties, including allosteric regulation and/or post-translational regulation, so they are viewed as potential control sites. The current view of how sucrose synthesis is regulated by coordination of these enzymatic activities is discussed below.

Cytosolic fructose-1,6-bisphosphatase. Cytosolic FBPase (cytFBPase) irreversibly converts F1,6BP into F6P and thus catalyses the first committed step in the pathway of sucrose synthesis. The enzyme is also active in non-photosynthetic tissues where it controls the rate of F6P production in the gluconeogenetic pathway. Its crucial role for photosynthetic carbon partitioning has been established in mutant *Flaveria linearis* plants with a reduction of cytFBPase activity of 75% (Sharkey et al., 1988) and in transgenic potato plants expressing an antisense FBPase gene under control of the CaMV 35S promoter (Zrenner et al., 1996). The analysis of the transgenic potato plants revealed that 45% reduction of the cytFBPase activity did not cause any measurable change in metabolite concentrations, growth behaviour or photosynthetic parameters of the transgenic plants (Zrenner et al., 1996). The inhibition of cytFBPase activity below 20% of the wild-type activity led to an accumulation of triose-P, 3-PGA and F1,6BP in source leaves. This resulted in a reduced light-saturated assimilation rate and a decreased photosynthetic rate under light- and CO_2-saturated conditions. The measurement of photosynthetic carbon fluxes under these conditions revealed a 53–65% reduction of sucrose synthesis, whereas starch synthesis decreased only around 20%. Despite these changes, steady-state sucrose levels were not affected in cytFBPase antisense plants. Starch accumulated by a factor of 3 as compared with the wild type and was degraded during the night. This situation very much resembles that observed in TPT antisense potato plants (see 13.2.2; Riesmeier et al., 1993a) in that the surplus of starch accumulated during the day was efficiently mobilized during the night, thereby circumventing the limitation of sucrose synthesis caused by a reduced cytFBPase activity (Zrenner et al., 1996).

The biochemical properties of the cytFBPase are highly complex and involve the action of several regulatory metabolites (Stitt, 1990; Daie, 1993). In the absence of effector molecules, the enzyme has a high affinity for its substrate F1,6BP (K_m = 4–6 μM). It is strictly Mg^{2+} dependent, weakly inhibited by F6P, Pi and adenosine monophosphate (AMP) and strongly inhibited by the regulatory metabolite fructose-2,6-bisphosphate (F2,6BP). F2,6BP is an important regulator of FBPase activity in mesophyll cells, and nano-to-micromolecular concentrations of the molecule are effective in substantially lowering the affinity of FBPase to its substrate F1,6BP. The concentration of F2,6BP is determined by the relative in vivo activities of fructose-6-phosphate,2-kinase (F6P,2-K) and fructose-2,6-bisphosphatase (F2,6BPase). Both of these activities reside on a bifunctional enzyme (F2KP), which in potato appears to be encoded by a single gene (Draborg et al., 1999; Nielsen et al., 2004). The ratio between the two enzymatic activities is allosterically regulated by metabolites representing intermediates of primary metabolism (Villadsen and Nielsen, 2001). F6P,2-K activity is enhanced by F6P and Pi and is inhibited by 3-PGA and triose-P, whereas F2,6BP activity is inhibited by F6P and Pi. Upon the onset of photosynthesis, rising 3-PGA and triose-P and decreasing Pi lead to a rapid decline in F2,6BP levels which, together with rising F1,6BP, stimulate cytFBPase. In addition to allosteric control by metabolites, F2KP from *Arabidopsis* has recently been shown to be subject to post-translational control by phosphorylation and binding of 14-3-3 proteins (Kulma et al., 2004). Phosphorylation and binding of 14-3-3 proteins regulate the activity of enzymes such as nitrate reductase and SPS. Although the mechanism of F2KP regulation by these post-translational events is currently unknown, it appears to link the regulation of key metabolic enzymes.

Studies on transgenic plants with modified F2,6BP levels have enabled a closer evaluation of the in vivo function of this signal metabolite. Potato and *Arabidopsis* plants harbouring F2KP antisense constructs have a decreased F2,6BP content and display changes in photosynthetic carbon partitioning in favour of sucrose, increasing the sucrose to starch ratio in $^{14}CO_2$-labelling experiments at least two-fold (Draborg et al., 2001; Rung et al., 2004). In *Arabidopsis*, this resulted in a 20–30% higher level of sucrose and a delay in diurnal starch accumulation (Draborg et al., 2001). However, in potato, the extra sucrose was mainly cleaved and hexoses accumulated in the leaves (Rung et al., 2004). All these studies prove that F2,6BP contributes significantly to the regulation of leaf carbon partitioning during photosynthesis.

SPS and SPP. The last two steps in the synthesis of sucrose are unique to the pathway; they are catalysed by SPS and SPP. Owing to the rapid removal of Suc6P by SPP, the reaction catalysed by SPS is considerably displaced from equilibrium in vivo (Stitt et al.,

1987), and thus, it is thought that SPS activity contributes to the control of flux into sucrose. SPS is regulated in a sophisticated manner by various mechanisms that operate at different levels and in different time frames (Stitt et al., 1987; Huber and Huber, 1996; Winter and Huber, 2000). Gene transcription is developmentally regulated and can be influenced by environmental factors such as light. Allosteric regulation, involving activation by G6P and inhibition by Pi, allows sucrose synthesis to be immediately increased in response to the increasing availability of precursors. SPS is also regulated by protein phosphorylation on multiple serine residues (McMichael et al., 1993; Toroser et al., 1998, 1999). Phosphorylation of serine-158 (Ser158) of the spinach enzyme leads to a reduced substrate (F6P) and effector (G6P) affinity and most likely is responsible for the light/dark regulation of the enzyme, which is thought to be one of the mechanisms to adjust the capacity for sucrose biosynthesis in relation to the rate of photosynthesis (McMichael et al., 1993). After the onset of photosynthesis, rising G6P and falling Pi activate SPS allosterically, and rising G6P inhibits SPS-kinase leading to dephosphorylation and post-translational activation of SPS (Huber and Huber, 1996). SPS dephosphorylation is further promoted by light-dependent stimulation of SPS-phosphatase (Weiner et al., 1992). Phosphorylation of Ser424 activates the enzyme and might be involved in the activation of SPS during osmotic stress (Toroser and Huber, 1997).

Besides the two phosphorylation sites discussed above, additional phosphorylation site(s) are proposed to be involved in the binding of 14-3-3 proteins to SPS. 14-3-3 proteins constitute a highly conserved eukaryotic family of regulatory proteins, binding as dimers to phosphorylated motifs in their diverse target proteins, generally functioning as adaptors, chaperones, activators or repressors (Aitken, 1996; Fu et al., 2000; Roberts, 2000). A phospho-peptide comprising the sequence surrounding Ser229 of SPS from spinach has been shown to bind 14-3-3s in vitro (Toroser et al., 1998), but it is not yet known whether this site is involved in the SPS:14-3-3 interaction in vivo, because substitution of Ser229 by alanine does not affect 14-3-3 binding to SPS in the yeast two-hybrid system (Börnke, 2005). The effect on SPS activity of binding of a 14-3-3 is also not clear. Both partial inhibition of the enzyme (Toroser et al., 1998) and activation (Moorhead et al., 1999) have been reported. Alternatively, 14-3-3 binding may modulate the rate of proteolytic degradation of SPS in response to changes in cellular carbohydrate status (Cotelle et al., 2000). Future experiments will have to clarify the function of 14-3-3 binding to SPS and whether binding is involved in signal transduction processes regulating sucrose synthesis in source leaves according to sink demand.

The role of SPP in the regulation of photosynthetic carbon partitioning has been relatively neglected. Earlier reports that SPP activity in vivo is substantially higher than SPS activity suggested that SPP is unlikely to play a regulatory role in sucrose synthesis (Hawker and Smith, 1984). However, recent evidence that SPS and SPP might form a complex in vivo (Echeverria et al., 1997) opened the possibility that SPP could have a role in metabolite channelling between the two enzymes and thus could contribute to control of flux through the pathway. It was not until recently that the enzyme has been purified to homogeneity from rice leaves and the gene cloned (Lunn et al., 2000). Genes encoding SPP-like sequences have as yet been described from several plant species, including *Arabidopsis*, maize, rice, tomato, potato, tobacco and wheat (Lunn, 2003; Chen et al., 2005), and it appears that they encode small gene families with, for example, four members in *Arabidopsis* and three in rice, respectively (Lunn, 2003). A recent study on transgenic tobacco plants expressing an SPP RNA-interference construct revealed that SPP is abundant in source leaves and does not exert significant control of sucrose synthesis under normal conditions (Chen et al., 2005). The SPP activity in those transformants could be reduced to less than 10% of wild-type SPP activity before any effect on sucrose biosynthesis and photosynthetic carbon partitioning was observed. Similar results have been obtained in potato plants with reduced SPP activity in their source leaves (Chen and Börnke, unpublished).

13.3 STARCH METABOLISM IN SOURCE LEAVES

13.3.1 Starch synthesis within the chloroplast

As much as half of the triose-P produced during photosynthesis may be retained in the chloroplast to synthesize starch. Leaf starch represents a transient store for assimilates, which is mobilized during the night to support leaf metabolism, and continued synthesis and export of sucrose (Geiger and Servaites, 1994). Regulation of sucrose synthesis in the cytosol is the critical factor in determining the supply of triose-P available for allocation to starch during photosynthesis (Stitt and Quick, 1989). A consensus has developed that leaf starch synthesis is regulated by changes in the levels of phosphorylated metabolites and Pi that are generated when the rate of photosynthesis increases or when rising levels of sugars lead to feedback regulation of sucrose synthesis.

Starch synthesis requires three consecutive enzymatic reactions catalysed by ADP-glucose pyrophosphorylase (AGPase), starch synthase (SS) and starch branching enzyme (SBE) (Martin and Smith, 1995). The first committed step in the pathway of starch synthesis is the formation of ADP-glucose (ADP-Glc) from G1P and ATP in a reaction that is catalysed by AGPase and liberates pyrophosphate. In chloroplasts, ATP is derived from photosynthesis, and G1P can be supplied by the Calvin cycle via the plastidial isoforms of PGI and PGM. The pyrophosphate that is produced during the reaction is removed by an inorganic alkaline pyrophosphatase (PPase) which effectively displaces the equilibrium of the AGPase reaction in favour of AGPGlc production.

Potato AGPase, as the enzyme from other higher plants, is a heterotetramer that contains two 'regulatory' (AGPS, 51 kDa) and two slightly smaller 'catalytic' (AGPB, 50 kDa) subunits (Okita et al., 1990). AGPase is exquisitely sensitive to allosteric regulation, with 3-PGA acting as an activator and Pi as an inhibitor (Preiss, 1988). This ensures that the rate of starch synthesis is always balanced with other pathways consuming triose-P, especially sucrose synthesis in the cytosol. For example, feedback regulation of sucrose synthesis will lead to the accumulation of phosphorylated intermediates, depletion of Pi and activation of AGPase by the rising 3-PGA:Pi ratio, resulting in a compensatory stimulation of starch synthesis (Herold, 1980; Stitt et al., 1987). A number of recent studies reported that AGPase is subject to post-translational regulation in non-photosynthetic and photosynthetic tissues in a range of plant species (Fu et al., 1998; Tiessen et al., 2002; Hendriks et al., 2003). This mechanism involves the redox-regulated formation of an intermolecular Cys bridge between the AGPB subunits of the AGPase heterotetramer.

Monomerization as a response to light or rising sugar levels leads to an activation of the enzyme. It is assumed that the main significance of the allosteric control of AGPase by the 3-PGA:Pi ratio is to rapidly increase Pi recycling in the stroma in case of a transient imbalance between photosynthesis and triose-P export (Hendriks et al., 2003). The redox regulation of AGPase seems more likely to participate in the direct light activation of starch synthesis and allows AGPase activity to be stimulated in response to rising sugar levels in leaves (Hendriks et al., 2003).

SSs catalyse the transfer of the glucosyl moiety of the soluble precursor AGPGlc to the reducing end of a pre-existing $\alpha(1 \rightarrow 4)$-linked glucan primer to synthesize the insoluble glucan polymers amylose and amylopectin. SBE then forms the $\alpha(1 \rightarrow 6)$ linkages found in amylopectin by cleaving internal $\alpha(1 \rightarrow 4)$ bonds and transferring the released reducing ends to other glucose residues (Martin and Smith, 1995; Tetlow et al., 2004).

13.3.2 Starch breakdown in leaves

The complex metabolic network regulating photoassimilate partitioning between the different pathways during the photoperiod and the regulatory properties of the enzymes involved is fairly well characterized. What is surprising is the lack of understanding how transitory starch is remobilized at night and how this process is regulated. Recently, considerable progress has been made in understanding starch degradation in leaves (for a recent review, see Lloyd et al., 2005; Smith et al., 2005), and the current understanding of the pathway is summarized below.

The process of starch degradation is initiated by the release of soluble glucans from the semi-crystalline starch granule surface. An essential prerequisite for starch mobilization is the phosphorylation of a small proportion of the glucosyl residues, catalysed by the enzyme glucan–water dikinase (GWD, formerly known as R1; Lorberth et al., 1998; Ritte et al., 2002). GWD transfers the β-phosphate of ATP to either the C3 or the C6 position of the glucosyl residue of amylopectin (Ritte et al., 2002). Reduction of GWD activity in transgenic potato and *Arabidopsis* plants, respectively, leads to decreased starch phosphate levels and causes a drastic starch-excess phenotype in leaves (Lorberth et al., 1998; Yu et al., 2001). However, it is not yet known whether it is the starch-bound phosphate or the GWD protein itself that is important in mediating starch breakdown.

Because of their role in cereal endosperm starch degradation, it has long been assumed that α-amylases are important in initiating granule degradation in leaves (Beck and Ziegler, 1989). α-Amylases are endoamylolytic enzymes that hydrolyse the α-1,4-glucosidic linkages of starch. However, recent evidence from *Arabidopsis* suggests that α-amylases are dispensable for this process. T-DNA knockout mutants of *AtAMY3*, the only α-amylase isoform localized to the plastid, have the same diurnal pattern of transitory starch metabolism as the wild type (Yu et al., 2005). Evidence from potato demonstrates a role for β-amylase in leaf starch degradation in this species (Scheidig et al., 2002). β-Amylases are exoamylases that yield maltose almost exclusively as a product. Down-regulation of a plastidial isoform of β-amylase in transgenic potato led to a reduction in leaf starch degradation and a starch-excess phenotype. In addition, the recombinant protein was shown to release maltose from starch granules (Scheidig et al., 2002).

α-Amylases and β-amylases cleave α-1,4 but not α-1,6 bonds which represent between 4 and 5% of the linkages in amylopectin. The involvement of the so-called debranching enzymes (DBEs) is therefore essential to ensure the complete breakdown of starch. DBEs can be divided into two classes, isoamylases and limit dextrinases, both being hydrolases whose activities lead to the release of linear malto-oligosaccharides (MOSs). These are probably further metabolized by β-amylases and a protein known as the disproportionating enzyme (D-enzyme). The D-enzyme catalyses the transfer of α-1,4-linked MOSs from the end of one glucan chain to the end of another. It preferentially acts on maltotriose transferring a maltosyl residue to an acceptor glucan while releasing glucose. According to the current model, the D-enzyme metabolizes glucans that are too short for the actions of other starch degrading enzymes, creating longer chains on which these enzymes can act (Smith et al., 2003).

Alternatively to being broken down hydrolytically via β-amylase, it has long been hypothesized that starch breakdown occurs phosphorolytically by the action of α-glucan phosphorylase releasing G1P (Beck and Ziegler, 1989). G1P can then be converted to triose-P and exported from the plastid via TPT (Häusler et al., 1998). Several lines of evidence suggest that this is not the case. For instance, antisense repression of TPT in potato and tobacco demonstrated that reduced TPT activity was correlated with increased transitory starch synthesis and increased capacity of the hydrolytic pathway (see 13.2.2; Riesmeier et al., 1993; Häusler et al., 1998). Moreover, transgenic potato plants with antisense expression of cytFBPase do not export carbon from the leaves during the day but store all assimilated carbon as starch. Starch is mobilized during the dark period using a pathway not involving cytFBPase (Zrenner et al., 1996). These results indicate that polyglucans are primarily broken down by amylolytic activity and that maltose might be an important product of leaf starch degradation.

There is now strong evidence that maltose is the major metabolite exported from chloroplasts during the night. *Arabidopsis* mutants lacking a plastidic maltose transporter accumulate high levels of maltose in their leaves, are severely impaired in growth and display a starch-excess phenotype (Niittylä et al., 2004). Genes encoding maltose exporter homologs appear to be present in a wide range of plant species (Niittylä et al., 2004) including potato (e.g. GenBank accession CX162346).

In addition to the maltose produced by β-amylase, glucose accumulates during starch breakdown, produced by either D-enzyme acting on MOSs or β-amylase acting on maltotriose. A specific transporter that could mediate glucose export from the plastid into the cytosol has been identified in potato and several other plant species (Weber et al., 2000), although studies concerning the *in planta* function of this protein are currently lacking.

In *Arabidopsis*, maltose imported into the cytosol is further metabolized by a transglucosidase named DPE2 because it shows the greatest similarity to the D-enzyme (Chia et al., 2004). DPE2 catalyses the transfer of one of the glucosyl moieties of maltose to branched polyglucans such as glycogen and releases the other as glucose. *Arabidospsis* mutants carrying a knockout for DPE2 accumulate high levels of maltose and show impaired starch breakdown, suggesting that metabolism via DPE2 is the major route for maltose in the cytosol (Chia et al., 2004). Similar results have been obtained from transgenic potato plants in which the DPE2 protein was repressed using an RNAi construct

(Lloyd et al., 2004). However, the DPE2 protein of potato was reported to reside within the chloroplast (Lloyd et al., 2004), whereas that of *Arabidopsis* was found to be in the cytosol (Chia et al., 2004). A chloroplastic location for this transglucosidase would imply a different, yet unknown, route for starch degradation to be present in potato.

13.4 CARBON EXPORT AND LONG-DISTANCE TRANSPORT

Sucrose is the major transported form of fixed carbon in plants. Its coordinated translocation to and partitioning between different heterotrophic tissues is essential for the supply of carbon energy and carbon skeletons required to support growth and storage. Within the leaf, sucrose is transported from the sites of synthesis across several cell layers towards the vascular tissue where it is loaded into the minor veins of the phloem and then distributed throughout the plant. Recent advance in our understanding of the underlying mechanisms associated with sucrose transport and allocation in solanaceous species has been mainly achieved by cloning and characterization of membrane transporters involved in phloem loading (for recent reviews, see Kühn, 2003; Lalonde et al., 2003, 2004). Additionally, new insights into the structure and function of plasmodesmata (PD) have expanded our knowledge as to how these intercellular cytoplasmic channels may contribute to the control of symplastic assimilate transport and the regulation of carbon distribution between source and sink (Lucas and Wolf, 1999; Schobert et al., 2000; Roberts and Oparka, 2003).

13.4.1 Pathway from the mesophyll to the phloem

Although direct evidence is still lacking, it is widely assumed that sucrose moves symplastically via PD from mesophyll cells towards the minor vein tissue. Electron microscopic observations of high plasmodesmal frequencies between cells along this route and microinjection studies with small fluorescent dyes suggested the potential for symplastic solute transport at least to the bundle sheath/vascular parenchyma (BS/VP) cells (Robards and Lucas, 1990). However, the presence of PD between individual cells does not necessarily indicate that these intercellular channels are constantly available for solute movement. It is now well established that PD undergo structural changes during sink–source transition and that the formation of secondary branched PD in the mesophyll and minor veins of exporting source tissue correlates with a decrease in plasmodesmal permeability and transport capacity compared with simple PD of importing sink tissue (Ding et al., 1992; Evert et al., 1996; Oparka et al., 1999; Oparka and Turgeon, 1999). Additionally, PD are known to alter their conductivity in response to various abiotic and biotic stimuli (summarized in Roberts and Oparka, 2003), which agrees with the recent finding that the leaf contains a heterogenous mixture of cells where PD apertures fluctuate according to changes in the environment (Crawford and Zambryski, 2001; Zambryski, 2004). Thus, it is likely that changes in plasmodesmal permeability act as an important control mechanism over the capacity of symplastic sucrose transport into the phloem.

Compelling evidence for this hypothesis has been provided by the analysis of viral movement proteins (MPs), which are well known to interact with PD components and to substantially increase the plasmodesmal size exclusion limits (SELs) (Lazarowitz and

Beachy, 1999; Roberts and Oparka, 2003; Oparka, 2004). Ectopic expression of MPs from various viruses in transgenic plants resulted in significant changes in carbohydrate metabolism and biomass partitioning (Olesinski et al., 1995; Lucas et al., 1996; Shalitin and Wolf, 2000; Hofius et al., 2001). For instance, sucrose transport and carbon allocation were specifically altered in potato by expression of the tobacco mosaic virus (TMV)-MP under the control of tissue-specific promoters. Phloem-restricted expression of TMV-MP in transgenic potato plants resulted in carbohydrate accumulation and decreased sucrose export from source leaves (Almon et al., 1997), whereas mesophyll-specific expression from the ST-LS1 (*S. tuberosum* Rubisco large subunit 1) promoter caused lower carbohydrate contents and higher sucrose export rates (Olesinski et al., 1996). Interestingly, these effects were in both cases under developmental control and observed only after tuber initiation, giving rise to a model, in that MPs interfere with the symplastic trafficking pathway for signal molecules involved in source–sink regulation (Lucas et al., 1996; Lucas and Wolf, 1999).

Additional support for an important role of the symplastic pathway for assimilate transport has previously been derived from the analysis of the *sucrose export defective1* maize mutant, which exhibits stunted growth and carbohydrate accumulation in source leaves paralleled by ultrastructural alterations and callose occlusion of PD between BS and VP cells (Russin et al., 1996; Botha et al., 2000). This suggested an inhibition of symplastic continuity at the BS/VP boundary leading to a block of sucrose transport into the phloem. Surprisingly, cloning of the maize *sxd1* locus and the respective orthologue in *Arabidopsis (vte1)* revealed that *sxd1* is not affected in a structural PD protein but disrupted in a gene encoding tocopherol cyclase, which resulted in a complete absence of tocopherols (Provencher et al., 2001; Porfirova et al., 2002; Sattler et al., 2003). Such unexpected link between tocopherol deficiency, source leaf-specific callose occlusion of PD and the subsequent blockage of carbohydrate export could also be demonstrated in transgenic potato plants expressing an RNAi construct targeted at the potato *sxd1* homolog (Hofius et al., 2004). On the basis of these results, it was proposed that tocopherol deficiency may indirectly affect callose synthesis and assimilate transport capacity by disturbance of cell-redox homeostasis and membrane lipid peroxidation (Hofius et al., 2004; Munné-Bosch, 2005). Although experimental proof for such mechanistic link is still lacking, the recent findings may give rise to a better understanding of the control of PD permeability and assimilate export, especially under conditions of environmental stress and leaf senescence (Munné-Bosch, 2005).

13.4.2 Phloem loading

In some plant families, a high degree of plasmodesmal connectivity extends even into the minor vein sieve element–companion cell complexes (SE-CCC), indicating the presence of an entirely symplastic route for sucrose transport into the phloem. In contrast to these so-called symplastic loaders, the abundance of PD has been described to drop remarkably between phloem parenchyma cells and the SE–CCC of potato leaves, suggesting the requirement of an apoplastic step in phloem loading (McCauley and Evert, 1989; van Bel et al., 1992) (Fig. 13.3). Since the beginning of the 1990s, extensive studies were performed to isolate and characterize membrane carriers providing transport activity for cellular efflux

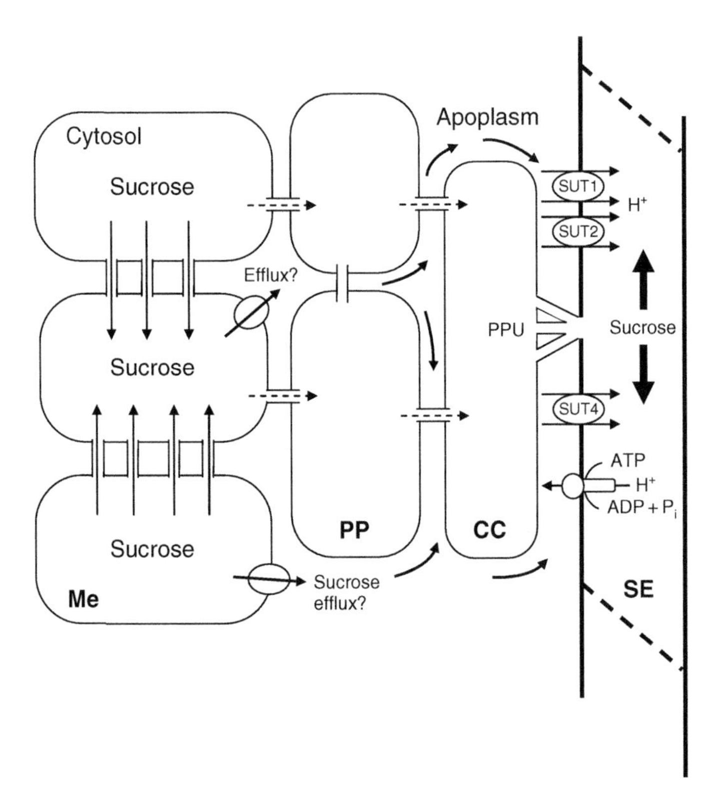

Fig. 13.3. Proposed model for phloem loading of sucrose in Solanaceae. CC, companion cell; Me, mesophyll; PP, phloem parenchyma; PPU, plasmodesmata pore units; SE, sieve element; SUT1, high-affinity sucrose transporter; SUT4, low-affinity transporter; SUT2, the putative sucrose sensor (modified after Kühn, 2003; for details, see text).

or vesicular transfer of sucrose into the apoplasm as well as uptake into the phloem cells. Although there is currently still insufficient knowledge regarding sucrose release into the extracellular space, the subsequent steps of phloem loading by sucrose transporters are well characterized (Williams et al., 2000; Lalonde et al., 2004). The first genes for sucrose transporters were identified by an artificial complementation system using a modified yeast strain, SuSy7, which was defective in all forms of invertase secretion but able to metabolize ingested sucrose due to expression of an introduced sucrose synthase (SuSy) (Frommer and Sonnewald, 1995). This strategy led to the identification of *su*crose *t*ransporters (*SUT1*) from spinach and potato (Riesmeier et al., 1992, 1993b), later followed by the isolation of *SUT1* paralogs from Solanaceae and homologs (also called SUCs) from many other plants including *Arabidopsis, Plantago* and *Ricinus* (summarized in Kühn, 2003). Transport studies in yeast and electrophysiology experiments in *Xenopus* oocytes provided persuasive evidence that SUT1 class transporters function as high-affinity H^+-cotransporters with a 1:1 proton/sucrose transport ratio (Zhou et al., 1997). Electron microscopic localization of solanaceous SUT1 protein and mRNA revealed

SE-specific localization of the protein, whereas SUT1 transcripts were detectable both in SEs and in CCs, especially at the orifices of the interconnecting PD (Kühn et al., 1997). This suggested trafficking of *SUT1* mRNA from CCs into SEs, which is supported by the detection of *SUT1* RNA in the phloem sap of a range of different plant species (Kühn et al., 1997; Ruiz-Medrano et al., 1999; Knop et al., 2001). However, it still needs to be verified whether this *SUT1* RNA is indeed translated on specialized ribosomes within the sieve tubes or rather subject to long-distance translocation and unloading in distant tissues. The provision of SUT1 protein to SE plasma membranes of phloem-loading sites could equally be obtained by symplastic trafficking of SUT1 protein as suggested for most of the proteins present in the sieve tube sap (Oparka and Santa Cruz, 2000; van Bel, 2003a).

Genetic proof for a critical role of SUT1 in phloem loading at the CC/SE boundary has been accumulated from studies on transgenic plants and mutants with reduced transporter activity. Antisense inhibition of SUT1 in tobacco and potato resulted in leaf carbohydrate accumulation and a curled bleached leaf phenotype due to a severe reduction of sucrose export capacity. These phenotypic alterations were coupled to a reduced assimilate supply of sink organs as evident by stunted growth and impaired root development as well as, in the case of potato, by a reduced tuber yield (Riesmeier et al., 1994; Kühn et al., 1996; Bürkle et al., 1998). Similar to the situation in solanaceous species, knockout mutation of the *Arabidopsis* SUT1 ortholog (*AtSUC2*) led to a starch-excess phenotype, impaired growth and sterility (Gottwald et al., 2000), which overall indicated that sucrose transport mediated by the SUT1/SUC2 type of transporters represents the major translocation pathway for photoassimilate supply of sink organs. However, plants are able to survive without this major loader, suggesting the existence of SUT1 paralogs, which may take over part of the function in mutant plants. Indeed, solanaceous plants contain at least two other distinct sucrose transporter proteins, SUT2 and SUT4 (Lalonde et al., 1999; Barker et al., 2000; Weise et al., 2000), that showed overlapping expression pattern and co-localized with SUT1 in the same enucleate SEs of tomato and potato (Barker et al., 2000; Reinders et al., 2002). SUT4 has been suggested to represent the low-affinity/high-capacity component of the sucrose transport and loading system in minor vein phloem (Weise et al., 2000), whereas SUT2 was postulated to function as putative sucrose sensor based on the specificity of extended cytoplasmic domains that structurally resemble the yeast sugar sensors SNF3 and RGT2 (Barker et al., 2000). Interestingly, all the three functionally different sucrose transporters have the capability to form homo- and hetero-oligomers as demonstrated by the use of the yeast split-ubiquitin system, suggesting that sucrose transporters may also function in higher order complexes (Reinders et al., 2002). However, the concept of oligomerization of sucrose transporter in plasma membranes, potentially with a regulatory role in phloem loading and assimilate transport, has not been confirmed *in planta*, which seems to be specially needed, because the situation in other species differs from the situation in Solanaceae. Schulze et al. (2003) showed in similar yeast experiments a potential interaction between AtSUC2, AtSUT2 (= AtSUC3) and AtSUT4 from *Arabidopsis*. However, it could recently be demonstrated that AtSUC3 is targeted to SEs (Wellmer et al., 2004) and, therefore, does not co-localize with the CC-confined AtSUC2 (Stadler and Sauer, 1996). Such spatial separation excludes the possibility of physical interaction in vivo and strongly questions the proposed regulation of AtSUC2-dependent phloem loading by AtSUC3 (Wellmer et al., 2004). In conclusion, the observed species-specific differences in localization and functionality of sucrose transporters suggest significant variations in the mechanisms and regulation of apoplastic phloem loading.

13.4.3 Long-distance transport in the phloem

Long-distance translocation of sucrose occurs in the sieves tubes, representing arrays of SE modules, each of them closely associated with one or more CCs. SEs are enucleate and devoid of almost all organelles, which implies that CCs provide the genetic and metabolic processes to ensure SE viability and transport functions (Oparka and Turgeon, 1999; van Bel and Knoblauch, 2000). The massive exchange of energy-carrying substances and macromolecules between CCs and SEs is largely mediated by a distinct class of PD, characterized by multiple branches on the CC side and a single branch on the SE, and commonly referred to as PD pore units (PPUs) (Fig. 13.3). It is now well established that most of the 150–200 proteins detected in the sieve tube sap are synthesized in the CCs and then are trafficked to SE via PPUs (Schobert et al., 1995, 1998; Ruiz-Medrano et al., 2001). Similarly, it is assumed that the large number of RNA species present in the phloem translocation stream, probably involved in long-distance signalling, are transported through PPUs with the help of specialized RNA-binding chaperones, resembling viral MPs (Xoconostle-Cázares et al., 1999; Lucas et al., 2001). The molecular SEL of PPUs was shown to be in the range of 20–30 kDa using fluorescence-tagged macromolecules (Kempers and van Bel, 1997) and transgenic tobacco and *Arabidopsis* plants expressing the freely diffusible 27-kDa green fluorescent protein (GFP) in the CCs (Imlau et al., 1999). It is likely that the unusually large plasmodesmal SEL of PPUs is mediated by the permanent gating activity of phloem-specific proteins, which were previously shown to enlarge the SEL of mesophyll PD up to 30 kDa after microinjection (Balachandran et al., 1997; Ishiwatari et al., 1998).

In addition to the supply with macromolecules, maintenance and functions of SEs fully rely on energy produced in CCs, which apparently is the reason for the particularly high number of mitochondria present in CCs (Oparka and Turgeon, 1999). Phloem loading requires energization of the plasma membrane by a proton gradient that is generated through the activity of phloem-specific H^+-ATPase to drive the proton-coupled sucrose uptake (Frommer and Sonnewald, 1995). Similar to the situation in sink organs, energy supply in the CC is based on the reversible turnover of sucrose into UDP-Glc and fructose, catalysed by a CC-specific SuSy (Nolte and Koch, 1993). Interestingly, blocking of the subsequent conversion of UDP-Glc to G1P by expression of a bacterial PPase in phloem cells resulted in impaired phloem loading and increased assimilate loss along the translocation pathway in transgenic tobacco plants (Lerchl et al., 1995; Geigenberger et al., 1996). This strongly suggested that a small proportion of the incoming sucrose is required as fuel to supply the ATP for the H^+-ATPase involved in sucrose uptake and retrieval. In agreement with this assumption, sugar metabolism could be restored in PPase-expressing plants by additional transformation with a cytosolic invertase, leading to similar phloem loading and sucrose loss as observed in wild-type controls (Lerchl et al., 1995).

The long-distance translocation in the phloem is generally thought to be driven by mass flow generated by a pressure gradient between sink tissues and source tissues. In the classical concept, the sieve tubes forming the translocation pathway between source and sink ends were considered to be osmotically isolated. However, this model has been adapted to a rather dynamic concept, suggesting that sieve tubes are essentially leaky and that the transported substances are released and retrieved along the phloem pathway (van Bel, 2003a). Sucrose, for example, is lost from the phloem at considerable rates but constantly retrieved from the apoplast by the activity of sucrose transporters decorating the phloem path (Kühn et al., 1997). Recent experimental data obtained by Ayre et al. (2003) support the view that small solutes in the CC enter the translocation stream indiscriminately but are then subject to mechanisms that control retention and/or reclamation along the transport pathway (for details, see van Bel, 2003b).

13.5 CARBON UNLOADING INTO SINK ORGANS

In contrast to the extensively investigated mechanisms of phloem loading, only a few studies have focussed on the processes involved in phloem unloading of the carbon translocated into sink organs. This might be due to the fact that the mode of unloading varies considerably not only between species but also in a tissue- and developmentally regulated manner. Therefore, several different unloading mechanisms were assumed to coexist in the plant (Frommer and Sonnewald, 1995). Since 1997, much knowledge was added to the field by a series of elegant studies that combined biochemical and/or molecular analyses with non-invasive imaging techniques using phloem-imported fluorescent dyes or transgenically expressed GFP (Roberts et al., 1997; Imlau et al., 1999; Oparka et al., 1999; Ruan et al., 2001; Viola et al., 2001). In the following section, we summarize the different models discussed for phloem unloading and then concentrate on the situation found in potato tubers.

13.5.1 Symplastic and apoplastic routes of unloading

In principle, four different pathways for SE unloading and post-phloem transport are proposed (Fig. 13.4), and the nature of the predominant route in sink tissues of many species is still under intense investigation and discussion (Herbers and Sonnewald, 1998; Lalonde et al., 2003). Evidence has accumulated that in many sink types such as root apices, expanding leaves or *Agrobacterium tumefaciens*-induced tumours, photoassimilates primarily follow a symplastic route formed by PD between the phloem and the sink parenchyma cells (Roberts et al., 1997; Imlau et al., 1999; Pradel et al., 1999; Haupt et al., 2001). It is assumed that symplastic unloading is controlled to a great extent by the frequency, structure and permeability of PD connecting the different cell types along the unloading pathway. Importantly, microscopic studies using transgenic tobacco or *Arabidopsis* plants that expressed GFP (27 kDa) or fusion proteins up to 50 kDa in CCs revealed free unloading of these proteins from sieve tubes and extensive movement along the post-phloem pathway in various sink organs including young leaves,

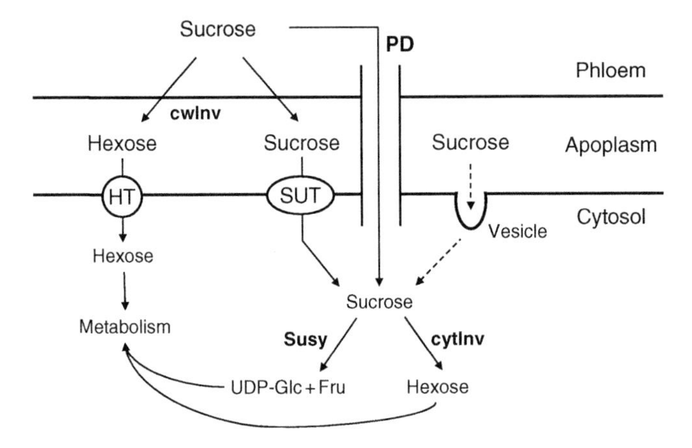

Fig. 13.4. Possible routes of phloem unloading and sucrose uptake into sink cells. CwInv, cell wall-bound invertase; HT, hexose transporter; SUT, sucrose transporter; cytInv, cytosolic invertase; SuSy, sucrose synthase; UDP-Glc, UDP-glucose; Fru, fructose; PD, plasmodesmata.

root tips and ovules (Imlau et al., 1999; Oparka et al., 1999). These observations indicated that PD involved in assimilate import possess an unusually high molecular SEL compared with other PD, most likely to accommodate high-assimilate fluxes but also non-specific import of macromolecules with potential signalling functions. However, comparable to the situation in source leaves, the genes responsible for the structural composition of sink PD and for the regulation of their conductance are still largely unknown. It has recently been suggested that phytohormones and calcium-regulated signalling may contribute to the regulation of sink PD aperture, especially by the modification of callose deposition and physicochemical states of cytoskeleton components located around and in the neck regions of PD (Baluška et al., 2001). In this respect, myosin VIII and endoplasmic reticulum (ER)-based calreticulin, both shown to specifically accumulate in sink PD, were proposed as candidate molecules to participate in the modulation of sink strength via plasmodesmal gating (Baluška et al., 1999, 2001; Reichelt et al., 1999).

In contrast to the symplastic route, all other pathways discussed for phloem unloading contain an apoplastic step either directly at the boundary between the SE–CC and parenchyma cells or at the boundary between different types of parenchyma cells during post-phloem transport (Patrick, 1997; Herbers and Sonnewald, 1998). The release of sucrose from the SE–CC into the apoplastic space might occur by leakage/diffusion down the concentration gradient but could likewise involve sucrose efflux proteins (Kühn et al., 2003). The subsequent uptake of sucrose into parenchyma cells could follow three different routes. Firstly, sucrose import is mediated by an active mechanism involving sucrose/proton symporters. Secondly, sucrose is cleaved by a cell wall invertase into hexoses, which are then transferred by sink-specific monosaccharide transporters into sink cells. Thirdly, sucrose could be taken up by endocytosis as indicated by earlier studies using fluorescent dyes in storage tissue of potato tuber (Oparka and Prior, 1988). Depending on the mode of phloem unloading, the incoming sugar is further metabolized by different enzymatic routes, such as sucrose breakdown by either SuSy or neutral invertase in the cytosol, or alternatively by a soluble acid invertase after sucrose import into the vacuole. Hexoses derived from apoplastic sucrose hydrolysis are either taken up into the vacuole or phosphorylated in the cytosol to enter glycolysis (for details, see Herbers and Sonnewald, 1998).

13.5.2 Phloem unloading in the tuber

Despite many studies on sink–source relations and carbon partitioning in potato plants, the mode of phloem unloading in the tuber remained unresolved for a long time. Early observations of numerous PD connections between SE–CCC and the surrounding storage parenchyma cells indicated a structural basis for the occurrence of a symplastic mode of phloem unloading (Oparka, 1986). This view was supported by subsequent work on growing tubers showing that plasmolysis had an inhibitory effect on the unloading of sucrose into the tuber (Oparka and Prior, 1987). Several experimental data, however, suggested the involvement and importance of an apoplastic step for sucrose delivery into the tuber. For instance, as already mentioned, studies with fluorescent dyes indicated that endocytosis could provide a potential route to translocate metabolites from the apoplast into the cytosol (Oparka and Prior, 1988). Also, isolated tuber discs were demonstrated to have the capacity for sugar uptake from the surrounding medium (Oparka and Wright, 1988; Fernie et al., 2001b), and finally, transgenic potato plants that expressed a yeast-derived invertase in the apoplast of growing tubers displayed significant alterations in tuber yield (Heineke et al., 1992; Sonnewald et al., 1997). Nevertheless, definite evidence for the mode(s) of phloem unloading during tuber development could only recently be provided by Viola and co-workers (2001). A combination of confocal imaging with the membrane impermeant marker carboxyfluorescein (CF) and autoradiography with ^{14}C-labelled assimilates was used to follow phloem unloading during different stages of tuber induction and growth. The authors convincingly showed that during very early stages of tuber development, i.e. during the elongation phase of stolon growth, the phloem-mobile tracer was confined to the phloem strands, whereas ^{14}C unloading was detectable on audioradiographs of the growing stolon, indicating that apoplastic phloem unloading predominated. However, during the first observable phases of tuberization, a transition occurred from apoplastic to symplastic transport as indicated by identical unloading pattern of both CF and ^{14}C sugar. Finally, during the remaining phases of tuber growth, the symplastic route of phloem unloading predominated. Apparently consistent with such a developmental switch in unloading mechanisms, the sucrose H^+-cotransporter *StSUT1* was recently found to be present in SEs of sink tubers (Kühn et al., 2003). Antisense inhibition of *StSUT1* using the B33 promoter suggested that SUT1 may play a crucial role in the unloading process at early stages of tuber development. Such a function might be related either to the retrieval of sucrose from the apoplast, thereby regulating the osmotic potential in the extracellular space, or to the transfer of sucrose from the SEs into the apoplast, thereby acting directly as a phloem exporter (Kühn et al., 2003).

13.6 SUCROSE TO STARCH CONVERSION IN THE TUBER

The major flux in potato tuber carbon metabolism is the conversion of sucrose through hexose phosphates to starch. The pathway by which this conversion occurs and the subcellular location of the enzymes involved are well documented (Fernie et al., 2002; Fernie and Willmitzer, 2004). An array of transgenic potato plants with reduced expression of most of the genes involved in this process has been generated. In this section, we intend to summarize the current understanding of this central pathway of carbohydrate metabolism while putting some stress on recent developments in understanding the regulation of starch synthesis.

13.6.1 Production of hexose phosphates in the cytosol

Sucrose delivered to the tuber can be cleaved in different ways. It can be either hydrolysed by apoplastic or cytosolic invertase, respectively, resulting in glucose and fructose, or converted into UDP-Glc and fructose by SuSy. The prevailing way of sucrose cleavage depends on the developmental stage of the tuber and is coordinated with the route of unloading. At the onset of tuberization, a switch from an invertase-dominated hydrolytic sucrose degradation route to a SuSy catalysed one occurs (Appeldoorn et al., 1997). The products of sucrose cleavage enter metabolism by the concerted action of UGPase or fructokinase and hexokinase (HK) in case of the SuSy and invertase pathways, respectively. As a result, the end products of sucrose degradation will enter the hexose-P pool, which consists of an equilibrium mixture of F6P, G6P and G1P, because of readily reversible reactions catalysed by the cytosolic isoforms of PGI and PGM (Fernie et al., 2001a).

The crucial role for SuSy in sucrose breakdown during the storage phase has been established from transgenic potato plants expressing a SuSy antisense construct driven by the CaMV 35S promoter (Zrenner et al., 1995). In transgenic tubers, a reduction in SuSy activity of up to 95% resulted in a reduction in starch and storage protein content of mature tubers, but surprisingly, sucrose levels remained unchanged. There was, however, a significant increase in the levels of hexoses, which was paralleled by a 40-fold increase in invertase activity. The fact that the induction of invertase activity could not compensate for the loss in SuSy activity argues for metabolic channelling of sucrose via the SuSy-dominated pathway into starch and thus indicates that SuSy plays a predominant role in determining potato tuber sink strength (Zrenner et al., 1995). Accordingly, expression of a yeast-derived invertase either in the apoplast or in the cytosol of transgenic tubers did not result in increased sink strength in these tubers despite their higher sucrolytic capacity (Sonnewald et al., 1997).

Although the net flux in the tuber is towards sucrose cleavage, the rate of sucrose (re)synthesis in this tissue is considerable (Geigenberger et al., 1997). This process can proceed either by the reversed SuSy reaction or by the reactions catalysed by SPS and SPP (Geigenberger and Stitt, 1993). It is thought that the combined operation of these pathways in concert with the degradative pathway allows the cell to respond sensitively to both variations in sucrose supply and the cellular demand of carbon for biosynthetic processes (Hatzfeld and Stitt, 1990). Several studies have shown that SPS is also present in potato tubers and that the tuber enzyme is regulated in an analogous manner to the leaf enzyme (Geigenberger and Stitt, 1993; Reimholz et al., 1994). The complex regulation of the enzyme (see discussion in 13.2.3.1) suggests that it plays an important metabolic role. The reduction of SPS activity in potato tubers by antisense transformation revealed only a minor influence of the enzyme on starch metabolism but a major role of sucrose synthesis via SPS in response to water stress (Geigenberger et al., 1999).

Beyond their role as intermediates in the conversion of sucrose to starch, hexose phosphates also serve as substrates for glycolysis and the oxidative pentose phosphate pathway in potato tubers (Neuhaus and Emes, 2000; for a recent review of the subject, see Stitt and Sonnewald, 1995).

13.6.2 Uptake of carbon into the amyloplast

Because starch synthesis in potato tubers is confined to amyloplasts, it relies entirely on translocation of metabolites from the cytosol through the amyloplast envelope. The form in which carbon enters the amyloplast has long been a matter of contention. Although potato tuber amyloplasts posses some TPT activity (Schünemann et al., 1996), it seems clear that triose-P is not the substrate taken up to support starch synthesis. In contrast to chloroplasts in photosynthetic tissues, they lack plastidial FBPase activity and thus cannot convert triose-P into F1,6BP (Entwistle and ap Rees, 1990). Consequently, other metabolites have been considered as cytosolic precursors for starch synthesis in the amyloplast. Studies on highly enriched amyloplasts from potato tubers provided evidence for uptake of carbon in the form of either G1P (Naeem et al., 1997) or G6P (Wischmann et al., 1999). However, the cloning of a hexose monophosphate transporter from potato and the finding that the cauliflower homologue is highly specific for G6P provide evidence for the latter being the major compound imported into the amyloplast (Kammerer et al., 1998). Furthermore, transgenic potato lines in which the activity of the plastidial isoform of PGM was reduced by antisense were characterized by a large reduction in starch content (Tauberger et al., 2000). This adds further weight to the view that interconversion of G6P to G1P for starch synthesis occurs inside the chloroplast, although G1P import might represent a minor route because the PGM antisense tubers were not completely starchless (Tauberger et al., 2000) (Fig. 13.5).

13.6.3 Starch synthesis in potato tubers

Following the uptake of carbon into the tuber amyloplast, starch synthesis proceeds via the concerted action of plastidial PGM, AGPase and the polymerizing reactions already described for the synthesis of transitory starch in chloroplasts (see Section 13.3.1). The important role of AGPase for starch synthesis in potato tubers has been proved by antisense studies in which a reduction of AGPase activity led to a dramatic reduction in the level of starch (Müller-Röber et al., 1992). Potato tuber AGPase resembles the leaf enzyme in being allosterically activated by 3-PGA and inhibited by Pi. In contrast to leaves, however, the levels of phosphorylated intermediates are remarkably constant on a diurnal basis and through development, respectively (Geigenberger and Stitt, 2000). During environmental perturbations such as wounding and water stress, this balance is disturbed, and consequently, changes in 3-PGA occur in the tubers that are well

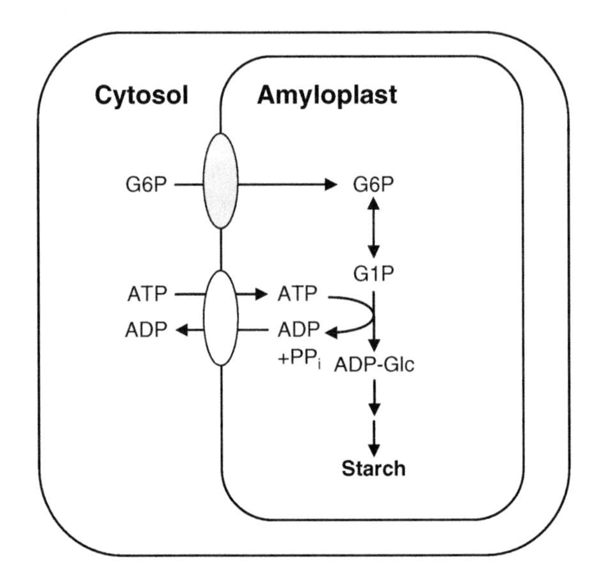

Fig. 13.5. Uptake of carbon into the amyloplast of potato tubers.

correlated with changes in the rate of starch synthesis and thus provide evidence for allosteric regulation of potato tuber AGPase in vivo (summarized in Geigenberger, 2003). Potato tuber AGPase is also subjected to post-translational redox regulation involving the mechanism described above for the leaf enzyme (Fu et al., 1998; Ballicora et al., 2000). This regulatory mechanism was found to be operating in vivo under a range of conditions that modify carbon supply and sucrose levels in growing potato tubers (Tiessen et al., 2002). Redox activation of AGPase was high at the beginning of tuber development and decreased as tubers matured. The decrease in AGPase redox state was paralleled by a decrease in sucrose levels and by a decline in the rate of starch synthesis in these tubers. At all events, a strong correlation was observed between the endogenous levels of sucrose and the redox activation state of AGPase (Tiessen et al., 2002), suggesting a link between sucrose levels and redox modulation of the enzyme.

Whereas in chloroplasts the ATP necessary for starch synthesis can be readily provided through photosynthesis, potato tuber amyloplasts have to import ATP from the cytosol via an ATP/ADP transport protein (AATP) located on the inner-envelope membrane (Neuhaus and Emes, 2000). Tjaden et al. (1998) recently showed that a relatively small decrease in ATP/ADP transporter activity leads to a reduced level of total starch content and a lower amylose-to-amylopectin ratio. By contrast, increased transporter activity correlated with higher starch contents and a higher amylose-to-amylopectin ratio. In total, these observations indicated that the rate of ATP import exerts considerable control on the rate of starch synthesis and affects the molecular composition of starch in potato tubers.

Although the later polymerizing reactions of starch synthesis catalysed by the various isoforms of SSs and SBEs do not appear to play a role in the control of starch accumulation, they are crucial in determining the structure of starch (Kossmann and Lloyd, 2000; Tetlow et al., 2004).

13.7 SOURCE–SINK REGULATION BY SUGARS

The plant life cycle is accompanied by source–sink transitions as well as changes with respect to sink strength of individual organs. Thus, complex mechanisms have to be assumed, which integrate the expression and activity of enzymes involved in carbohydrate synthesis in source tissues with those involved in carbon utilization in sink tissues on a whole plant level. The molecular mechanisms by which these complex interactions are achieved are far from understood, but many lines of evidence suggest the existence of sink signals that adjust source metabolism according to sink demand (Paul and Foyer, 2001). Mainly two concepts have been put forward to explain these sink signals: hormonal regulation and end-product inhibition of photosynthesis. According to the model of hormonal regulation, the variation in the activity of sink tissues will be translated into changes in the content of phytohormones which then will transmit appropriate messages to photosynthetic cells eventually adjusting photosynthesis to sink demand. This concept is supported by the observation that the application of cytokinin can partially overcome down-regulation of photosynthesis caused by the removal of sinks (Wareing et al., 1968).

According to the second model, assimilate accumulation in source tissues is supposed to be responsible for the inhibition of photosynthesis. In support of this hypothesis, a negative correlation between assimilate content and leaf photosynthetic rate has been observed in many plant species (Goldschmidt and Huber, 1992). Down-regulation of photosynthesis may be governed by fine regulation of biosynthetic enzymes or by the adjustment of gene expression. Fine regulation of photosynthetic sucrose biosynthesis may be initiated by the sucrose-mediated inhibition of SPS and the resulting changes in carbon partitioning between cytosol and chloroplast (see also Section 13.2.3.1). This fine regulation of carbon metabolism by metabolites is considered as short-term response to changing demands for assimilates. In case of continuous changes of assimilate demand, e.g. after the onset of tuberization, altered gene expression will accompany metabolic adjustments. The underlying molecular mechanisms are largely unknown, but several lines of evidence suggest that sugars play a role as signalling molecules in the down-regulation of photosynthetic genes. In agreement with this hypothesis, several genes have been shown to be regulated by external sugars (for a recent review, see Koch, 1996; Smeekens, 2000; Rolland et al., 2002). Depending on their response to sugars, plant genes have been grouped into three classes: down-regulated, up-regulated and non-responding. Photosynthetic genes belong to the class of down-regulated genes, which is in agreement with the hypothesis that sugars are involved in sink regulation of photosynthesis, whereas genes involved in carbon utilization, e.g. SuSy, are classified as up-regulated by sugars.

In general, mechanisms underlying sugar-sensing and sugar-mediated signal transduction in plants are poorly understood. It appears that separate sensing systems are present for hexoses and sucrose. Several lines of evidence suggest a prominent role of HKs in hexose-sensing. In *Arabidopsis*, two HK isoforms (AtHK1 and AtHK2) have been isolated (Jang et al., 1997). Transgenic *Arabidopsis* plants with reduced and elevated levels of AtHK1 or AtHK2 mRNAs have been constructed, and the sugar response of the respective seedlings was investigated. Overexpression of HK resulted in enhanced sensitivity to glucose, whereas reduced levels of HK led to reduced glucose sensitivity of seedlings, providing evidence for the involvement of HK in hexose-sensing (Jang et al., 1997).

In a recent study, using catalytically inactive AtHK1 mutants, Moore et al. (2003) were able to separate the metabolic and signalling function of AtHK1. For potato, however, the situation is far less clear. Transgenic potato plants transformed with *S. tuberosum* hexokinase-1 (StHK1) in sense or antisense orientation have been generated (Veramendi et al., 1999). Despite a wide range of HK activities, no evidence could be found that StHK1 is involved in sugar-sensing in potato. Interestingly, reduced StHK1 activity in leaves of potato plants led to a starch-excess phenotype, suggesting a role for this enzyme in the degradation of transitory starch rather than in hexose-sensing (Veramendi et al., 1999).

The SNF1 protein kinase has been shown to be an integral component of the sugar-sensing pathway in budding yeast (Carlson, 1999). SNF1-related protein kinases have also been identified in plants, and it has been proposed that they act as global regulators of carbon metabolism (Halford and Hardie, 1998). In support of this notion, antisense suppression of an SNF1-homologue in potato (SnRK1) prevents transcriptional activation of a sucrose-inducible *SuSy* gene and results in reduced SuSy activity in transgenic tubers (Purcell et al., 1998).

In addition to their importance in modulating gene expression, SNF-like proteins in plants also play a role in the post-translational regulation of metabolism by directly phosphorylating key enzymes such as 3-hydroxy-3-methyl glutaryl CoA reductase, nitrate reductase and SPS (Sugden et al., 1999). Recently, SnRK1 was also shown to be involved in the redox activation of AGPase in potato tubers (Tiessen et al., 2003). When tuber slices from either wild-type plants or SnRK1-antisense lines were supplied with exogenous sucrose, redox activation of AGPase was much stronger in the wild-type than in the antisense lines. Interestingly, SnRK1-antisense repression did not affect AGPase activation in response to external glucose feeding, which implies the existence of independent signalling pathways for glucose and sucrose possibly involving HK (Tiessen et al., 2003).

Yet an additional mechanism of sugar-sensing seems to occur at the plasma membrane and does not seem to require uptake or metabolism of the compound sensed. The feeding of the non-metabolizable and non-transportable sucrose analog palatinose has been shown to lead to a direct stimulation of starch synthesis in potato tuber slices without requiring an intervening increase in metabolite levels (Fernie et al., 2001b). Later, it was shown that this could at least partially be explained by a palatinose-mediated redox activation of AGPase, suggesting that metabolism of sucrose is not a necessary component of the sucrose-induced redox activation of AGPase (Tiessen et al., 2003). The mechanism by which sugar signals are perceived at the plasma membrane is currently completely unknown, but it has been suggested that sugar transport proteins might be involved in these processes (Lalonde et al., 1999; Xoconostle-Cázares et al., 1999).

Taken together, it seems evident that many independently operating sugar-sensing pathways exist in plants, and current models are shown in Fig. 13.6. Sucrose, synthesized in photosynthetically active tissues, is translocated to growing tubers via the phloem. In sink tissues, sucrose can either be hydrolysed by cell wall-bound invertases (cwInv) to glucose and fructose or be transported into sink cells by SUT proteins or through PD. Alternatively, hexoses derived from sucrose hydrolysis can be translocated into sink cells via hexose transporters (HT). In the cytosol, the hexoses trigger a signal via HK leading to transcriptional and post-translational regulation of several enzymes, including AGPase. A symplastic sucrose-specific signal is triggered by an SnRK1-dependent mechanism.

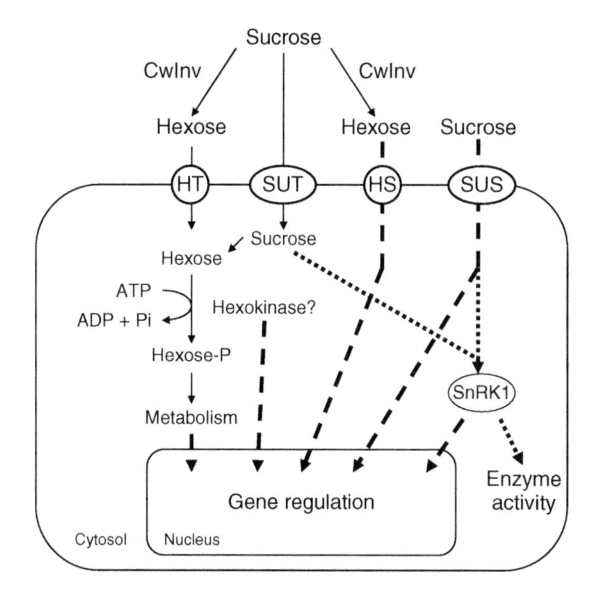

Fig. 13.6. Possible signal transduction pathways of sugar-sensing. CwInv, cell wall-bound invertase; HT, hexose transporter; SUT, sucrose transporter; HS, transmembrane hexose sensor; SUS, transmembrane sucrose sensor.

REFERENCES

Aitken A., 1996, *Trends Cell Biol.* 6, 341.

Almon E., M. Horowitz, H.L. Wang, W.J. Lucas, E. Zamski and S. Wolf, 1997, *Plant Physiol.* 115, 1599.

Appeldoorn N.J.G., S.M. de Bruijn, E.A.M. Koot-Gronsveld, R.G.F. Visser, D. Vreugdenhil and L.H.W. van der Plas, 1997, *Planta* 202, 220.

Ayre B.G., F. Keller and R. Turgeon, 2003, *Plant Physiol.* 131, 1518.

Balachandran S., V.M. Hurry, S.E. Kelley, C.B. Osmond, S.A. Robinson, J. Rohozinski, G.G.R. Seaton and D.A. Sims, 1997, *Physiol. Plantarum* 100, 203.

Ballicora M.A., J.B. Frueauf, Y.B. Fu, P. Schurmann and J. Preiss, 2000, *J. Biol. Chem.* 275, 1315.

Baluška F., F. Cvrčková, J. Kendrick-Jones and D. Volkmann, 2001, *Plant Physiol.* 126, 39.

Baluška F., J. Samaj, R. Napier and D. Volkmann, 1999, *Plant J.* 19, 481.

Barker L., C. Kuhn, A. Weise, A. Schulz, C. Gebhardt, B. Hirner, H. Hellmann, W. Schulze, J.M. Ward and W.B. Frommer, 2000, *Plant Cell* 12, 1153.

Beck E. and P. Ziegler, 1989, *Annu. Rev. Plant Phys.* 40, 95.

Börnke F., 2005, *J. Plant Physiol.* 162, 161.

Botha C.E.J., R.H.M. Cross, A.J.E. van Bel and C.I. Peter, 2000, *Protoplasma* 214, 65.

Buchanan B.B., 1980, *Annu. Rev. Plant Phys.* 31, 341.

Bürkle L., J.M. Hibberd, W.P. Quick, C. Kuhn, B. Hirner and W.B. Frommer, 1998, *Plant Physiol.* 118, 59.

Carlson M., 1999, *Curr. Opin. Microbiol.* 2, 202.

Chen S., M. Hajirezaei, M. Peisker, H. Tschiersch, U. Sonnewald and F. Börnke, 2005, *Planta* 221, 479.

Chia T., D. Thorneycroft, A. Chapple, G. Messerli, J. Chen, S.C. Zeeman, S.M. Smith and A.M. Smith, 2004, *Plant J.* 37, 853.

Cotelle V., S.E.M. Meek, F. Provan, F.C. Milne, N. Morrice and C. MacKintosh, 2000, *EMBO J.* 19, 2869.

Crawford K.M. and P.C. Zambryski, 2001, *Plant Physiol.* 125, 1802.

Daie J., 1993, *Photosynth. Res.* 38, 5.

Ding B., J.S. Haudenshield, R.J. Hull, S. Wolf, R.N. Beachy and W.J. Lucas, 1992, *Plant Cell* 4, 915.

Draborg H., D. Villadsen and T.H. Nielsen, 1999, *Plant Mol. Biol.* 39, 709.
Draborg H., D. Villadsen and T.H. Nielsen, 2001, *Plant Physiol.* 126, 750.
Echeverria E., M.E. Salvucci, P. Gonzalez, G. Paris and G. Salerno, 1997, *Plant Physiol.* 115, 223.
Entwistle G. and T. ap Rees, 1990, *Biochem. J.* 271, 467.
Evert R.F., W.A. Russin and A.M. Bosabalidis, 1996, *Int. J. Plant Sci.* 157, 247.
Fernie A.R., U. Roessner and P. Geigenberger, 2001b, *Plant Physiol.* 125, 1967.
Fernie A.R., U. Roessner, R.N. Trethewey and L. Willmitzer, 2001a, *Planta* 213, 418.
Fernie A.R. and L. Willmitzer, 2004, In: P. Christou and H. Klee (eds), *Handbook of Plant Biotechnology*. Wiley, Chichester, UK.
Fernie A.R., L. Willmitzer and R.N. Trethewey, 2002, *Trends Plant Sci.* 7, 35.
Flügge U.I. and H.W. Heldt, 1991, *Annu. Rev. Plant Phys.* 42, 129.
Frommer W.B. and U. Sonnewald, 1995, *J. Exp. Bot.* 46, 587.
Fu Y.B., M.A. Ballicora, J.F. Leykam and J. Preiss, 1998, *J. Biol. Chem.* 273, 25045.
Fu H.A., R.R. Subramanian and S.C. Masters, 2000, *Annu. Rev. Pharmacol.* 40, 617.
Geigenberger P., 2003, *J. Exp. Bot.* 54, 457.
Geigenberger P., J. Lerchl, M. Stitt and U. Sonnewald, 1996, *Plant Cell Environ.* 19, 43.
Geigenberger P., R. Reimholz, U. Deiting, U. Sonnewald and M. Stitt, 1999, *Plant J.* 19, 119.
Geigenberger P., R. Reimholz, M. Geiger, L. Merlo, V. Canale and M. Stitt, 1997, *Planta* 201, 502.
Geigenberger P. and M. Stitt, 1993, *Planta* 189, 329.
Geigenberger P. and M. Stitt, 2000, *Plant J.* 23, 795.
Geiger D.R. and J.C. Servaites, 1994, *Annu. Rev. Plant Phys.* 45, 235.
Goldschmidt E.E. and S.C. Huber, 1992, *Plant Physiol.* 99, 1443.
Gottwald J.R., P.J. Krysan, J.C. Young, R.F. Evert and M.R. Sussman, 2000, *Proc. Natl. Acad. Sci. U.S.A.* 97, 13979.
Halford N.G. and D.G. Hardie, 1998, *Plant Mol. Biol.* 37, 735.
Hatzfeld W.D. and M. Stitt, 1990, *Planta* 180, 198.
Haupt S., G.H. Duncan, S. Holzberg and K.J. Oparka, 2001, *Plant Physiol.* 125, 209.
Häusler R.E., N.H. Schlieben, B. Schulz and U.I. Flügge, 1998, *Planta* 204, 366.
Hawker J.S. and G.M. Smith, 1984, *Phytochemistry* 23, 245.
Heineke D., A. Kruse, U.I. Flügge, W.B. Frommer, J.W. Riesmeier, L. Willmitzer and H.W. Heldt, 1994, *Planta* 193, 174.
Heineke D., U. Sonnewald, D. Büssis, G. Günter, K. Leidreiter, I. Wilke, K. Raschke, L. Willmitzer and H.W. Heldt, 1992, *Plant Physiol.* 100, 301.
Hendriks J.H.M., A. Kolbe, Y. Gibon, M. Stitt and P. Geigenberger, 2003, *Plant Physiol.* 133, 838.
Herbers K. and U. Sonnewald, 1998, *Curr. Opin. Plant Biol.* 1, 207.
Herold A., 1980, *New Phytol.* 86, 131.
Hofius D., M.R. Hajirezaei, M. Geiger, H. Tschiersch, M. Melzer and U. Sonnewald, 2004, *Plant Physiol.* 135, 1256.
Hofius D., K. Herbers, M. Melzer, A. Omid, E. Tacke, S. Wolf and U. Sonnewald, 2001, *Plant J.* 28, 529.
Huber S.C. and J.L. Huber, 1996, *Annu. Rev. Plant Phys.* 47, 431.
Imlau A., E. Truernit and N. Sauer, 1999, *Plant Cell* 11, 309.
Inoue H. and A. Tanaka, 1978, *J. Sci. Soil Man. Jpn* 49, 321.
Ishiwatari Y., T. Fujiwara, K.C. McFarland, K. Nemoto, H. Hayashi, M. Chino and W.J. Lucas, 1998, *Planta* 205, 12.
Jang J.C., P. Leon, L. Zhou and J. Sheen, 1997, *Plant Cell* 9, 5.
Kammerer B., K. Fischer, B. Hilpert, S. Schubert, M. Gutensohn, A. Weber and U.I. Flügge, 1998, *Plant Cell* 10, 105.
Kempers R. and A.J.E. van Bel, 1997, *Planta* 201, 195.
Knop C., O. Voitsekhovskaja and G. Lohaus, 2001, *Planta* 213, 80.
Koch K.E., 1996, *Annu. Rev. Plant Phys.* 47, 509.
Kossmann J. and J. Lloyd, 2000, *Crit. Rev. Biochem. Mol. Biol.* 35, 141.
Kühn C., 2003, *Plant Biol.* 5, 215.
Kühn C., V.R. Franceschi, A. Schulz, R. Lemoine and W.B. Frommer, 1997, *Science* 275, 1298.

Kühn C., M.R. Hajirezaei, A.R. Fernie, U. Roessner-Tunali, T. Czechowski, B. Hirner and W.B. Frommer, 2003, *Plant Physiol.* 131, 102.
Kühn C., W.P. Quick, A. Schulz, J.W. Riesmeier, U. Sonnewald and W.B. Frommer, 1996, *Plant Cell Environ.* 19, 1115.
Kulma A., D. Villadsen, D.G. Campbell, S.E. Meek, J.E. Harthill, T.H. Nielsen and C. MacKintosh, 2004, *Plant J.* 37, 654.
Lalonde S., E. Boles, H. Hellmann, L. Barker, J.W. Patrick, W.B. Frommer and J.M. Ward, 1999, *Plant Cell* 11, 707.
Lalonde S., M. Tegeder, M. Throne-Holst, W.B. Frommer and J.W. Patrick, 2003, *Plant Cell Environ.* 26, 37.
Lalonde S., D. Wipf and W.B. Frommer, 2004, *Annu. Rev. Plant Biol.* 55, 341.
Lazarowitz S.G. and R.N. Beachy, 1999, *Plant Cell* 11, 535.
Lerchl J., P. Geigenberger, M. Stitt and U. Sonnewald, 1995, *Plant Cell* 7, 259.
Lloyd J.R., A. Blennow, K. Burhenne and J. Kossmann, 2004, *Plant Physiol.* 134, 1347.
Lloyd J.R., J. Kossmann and G. Ritte, 2005, *Trends Plant Sci.* 10, 130.
Lorberth R., G. Ritte, L. Willmitzer and J. Kossmann, 1998, *Nat. Biotechnol.* 16, 473.
Lucas W.J., S. Balachandran, J. Park and S. Wolf, 1996, *J. Exp. Bot.* 47, 1119.
Lucas W.J. and S. Wolf, 1999, *Curr. Opin. Plant Biol.* 2, 192.
Lucas W.J., B.C. Yoo and F. Kragler, 2001, *Nat. Rev. Mol. Cell Biol.* 2, 849.
Lunn J.E., 2003, *Gene* 303, 187.
Lunn J.E., A.R. Ashton, M.D. Hatch and H.W. Heldt, 2000, *Proc. Natl. Acad. Sci. U.S.A.* 97, 12914.
Martin C. and A.M. Smith, 1995, *Plant Cell* 7, 971.
McCauley M.M. and R.F. Evert, 1989, *Bot. Gaz.* 150, 351.
McMichael R.W., R.R. Klein, M.E. Salvucci and S.C. Huber, 1993, *Arch. Biochem. Biophys.* 307, 248.
Moore B., L. Zhou, F. Rolland, Q. Hall, W.H. Cheng, Y.X. Liu, I. Hwang, T. Jones and J. Sheen, 2003, *Science* 300, 332.
Moorhead G., P. Douglas, V. Cotelle, J. Harthill, N. Morrice, S. Meek, U. Deiting, M. Stitt, M. Scarabel, A. Aitken and C. MacKintosh, 1999, *Plant J.* 18, 1.
Müller-Röber B., U. Sonnewald and L. Willmitzer, 1992, *EMBO J.* 11, 1229.
Munné-Bosch S., 2005, *New Phytol.* 166, 363.
Naeem M., I.J. Tetlow and M.J. Emes, 1997, *Plant J.* 11, 1095.
Neuhaus H.E. and M.J. Emes, 2000, *Annu. Rev. Plant Phys.* 51, 111.
Nielsen T.H., J.H. Rung and D. Villadsen, 2004, *Trends Plant Sci.* 9, 556.
Niittylä T., G. Messerli, M. Trevisan, J. Chen, A.M. Smith and S.C. Zeeman, 2004, *Science* 303, 87.
Nolte K.D. and K.E. Koch, 1993, *Plant Physiol.* 101, 899.
Okita T.W., P.A. Nakata, J.M. Anderson, J. Sowokinos, M. Morell and J. Preiss, 1990, *Plant Physiol.* 93, 785.
Olesinski A.A., E. Almon, N. Navot, A. Perl, E. Galun, W.J. Lucas and S. Wolf, 1996, *Plant Physiol.* 111, 541.
Olesinski A.A., W.J. Lucas, E. Galun and S. Wolf, 1995, *Planta* 197, 118.
Oparka K.J., 1986, *Protoplasma* 131, 201.
Oparka K.J., 2004, *Trends Plant Sci.* 9, 33.
Oparka K.J. and D.A.M. Prior, 1987, *Plant Cell Environ.* 10, 667.
Oparka K.J. and D.A.M. Prior, 1988, *Planta* 176, 533.
Oparka K.J., A.G. Roberts, P. Boevink, S. Santa Cruz, L. Roberts, K.S. Pradel, A. Imlau, G. Kotlizky, N. Sauer and B. Epel, 1999, *Cell* 97, 743.
Oparka K.J. and S. Santa Cruz, 2000, *Annu. Rev. Plant Phys.* 51, 323.
Oparka K.J. and R. Turgeon, 1999, *Plant Cell* 11, 739.
Oparka K.J. and K.M. Wright, 1988, *Planta* 174, 123.
Patrick J.W., 1997, *Annu. Rev. Plant Phys.* 48, 191.
Paul M.J. and C.H. Foyer, 2001, *J. Exp. Bot.* 52, 1383.
Porfirova S., E. Bergmuller, S. Tropf, R. Lemke and P. Dormann, 2002, *Proc. Natl. Acad. Sci. U.S.A.* 99, 12495.
Portis A.R., 1992, *Annu. Rev. Plant Phys.* 43, 415.
Portis A.R., 1995, *J. Exp. Bot.* 46, 1285.
Pradel K.S., C.I. Ullrich, S. Santa Cruz and K.J. Oparka, 1999, *J. Exp. Bot.* 50, 183.
Preiss J., 1988, In: P.K. Stumpf and E.E. Conn (eds), *The Biochemistry of Plants*, Vol. 14, pp. 181–254. Academic Press, San Diego, CA.

Provencher L.M., L. Miao, N. Sinha and W.J. Lucas, 2001, *Plant Cell* 13, 1127.
Purcell P.C., A.M. Smith and N.G. Halford, 1998, *Plant J.* 14, 195.
Reichelt S., A.E. Knight, T.P. Hodge, F. Baluška, J. Samaj, D. Volkmann and J. Kendrick-Jones, 1999, *Plant J.* 19, 555.
Reimholz R., P. Geigenberger and M. Stitt, 1994, *Planta* 192, 480.
Reinders A., W. Schulze, C. Kuhn, L. Barker, A. Schulz, J.M. Ward and W.B. Frommer, 2002, *Plant Cell* 14, 1567.
Riesmeier J.W., U.I. Flügge, B. Schulz, D. Heineke, H.W. Heldt, L. Willmitzer and W.B. Frommer, 1993a, *Proc. Natl. Acad. Sci. U.S.A.* 90, 6160.
Riesmeier J.W., B. Hirner and W.B. Frommer, 1993b, *Plant Cell* 5, 1591.
Riesmeier J.W., L. Willmitzer and W.B. Frommer, 1992, *EMBO J.* 11, 4705.
Riesmeier J.W., L. Willmitzer and W.B. Frommer, 1994, *EMBO J.* 13, 1.
Ritte G., J.R. Lloyd, N. Eckermann, A. Rottmann, J. Kossmann and M. Steup, 2002, *Proc. Natl. Acad. Sci. U.S.A.* 99, 7166.
Robards A.W. and W.J. Lucas, 1990, *Annu. Rev. Plant Phys.* 41, 369.
Roberts M.R., 2000, *Curr. Opin. Plant Biol.* 3, 400.
Roberts A.G. and K.J. Oparka, 2003, *Plant Cell Environ.* 26, 103.
Roberts A.G., S. Santa Cruz, I.M. Roberts, D.A.M. Prior, R. Turgeon and K.J. Oparka, 1997, *Plant Cell* 9, 1381.
Rolland F., B. Moore and J. Sheen, 2002, *Plant Cell* 14, S185.
Ruan Y.L., D.J. Llewellyn and R.T. Furbank, 2001, *Plant Cell* 13, 47.
Ruiz-Medrano R., B. Xoconostle-Cázares and W.J. Lucas, 1999, *Development* 126, 4405.
Ruiz-Medrano R., B. Xoconostle-Cázares and W.J. Lucas, 2001, *Curr. Opin. Plant Biol.* 4, 202.
Rung J.H., H.H. Draborg, K. Jorgensen and T.H. Nielsen, 2004, *Physiol. Plant.* 121, 204.
Russin W.A., R.F. Evert, P.J. Vanderveer, T.D. Sharkey and S.P. Briggs, 1996, *Plant Cell* 8, 645.
Sattler S.E., E.B. Cahoon, S.J. Coughlan and D. DellaPenna, 2003, *Plant Physiol.* 132, 2184.
Scheidig A., A. Fröhlich, S. Schulze, J.R. Lloyd and J. Kossmann, 2002, *Plant J.* 30, 581.
Schobert C., L. Baker, J. Szederkenyi, P. Grossmann, E. Komor, H. Hayashi, M. Chino and W.J. Lucas, 1998, *Planta* 206, 245.
Schobert C., P. Grossmann, M. Gottschalk, E. Komor, A. Pecsvaradi and U. Zurnieden, 1995, *Planta* 196, 205.
Schobert C., W.J. Lucas, V.R. Franceschi and W.B. Frommer, 2000, In: R.C. Leegood, T.D. Sharkey and S. von Caemmerer (eds), *Photosynthesis: Physiology and Metabolism.* Kluwer Academic Press, The Netherlands.
Schulz B., W.B. Frommer, U.I. Flügge, S. Hummel, K. Fischer and L. Willmitzer, 1993, *Mol. Gen. Genet.* 238, 357.
Schulze W.X., A. Reinders, J. Ward, S. Lalonde and W.B. Frommer, 2003, *BMC Biochem.* 4, 3.
Schünemann D., K. Schott, S. Borchert and H.W. Heldt, 1996, *Plant Mol. Biol.* 31, 101.
Shalitin D. and S. Wolf, 2000, *Plant Physiol.* 123, 597.
Sharkey T.D., J. Kobza, J.R. Seemann and R.H. Brown, 1988, *Plant Physiol.* 86, 667.
Smeekens S., 2000, *Annu. Rev. Plant Phys.* 51, 49.
Smith A.M., S.C. Zeeman and S.M. Smith, 2005, *Annu. Rev. Plant Biol.* 56, 73.
Smith A.M., S.C. Zeeman, D. Thorneycroft and S.M. Smith, 2003, *J. Exp. Bot.* 54, 577.
Sonnewald U., M.R. Hajirezaei, J. Kossmann, A. Heyer, R.N. Trethewey and L. Willmitzer, 1997, *Nat. Biotechnol.* 15, 794.
Stadler R. and N. Sauer, 1996, *Bot. Acta* 109, 299.
Stitt M., 1990, *Annu. Rev. Plant Phys.* 41, 153.
Stitt M., S.C. Huber and P.S. Kerr, 1987, In: M.D. Hatch and N.K. Boardman (eds), *The Biochemistry of Plants.* Academic Press, New York.
Stitt M. and P. Quick, 1989, *Physiol. Plant.* 77, 633.
Stitt M. and U. Sonnewald, 1995, *Annu. Rev. Plant Phys.* 46, 341.
Sugden C., R.M. Crawford, N.G. Halford and D.G. Hardie, 1999, *Plant J.* 19, 433.
Tauberger E., A.R. Fernie, M. Emmermann, A. Renz, J. Kossmann, L. Willmitzer and R.N. Trethewey, 2000, *Plant J.* 23, 43.
Tetlow I.J., M.K. Morell and M.J. Emes, 2004, *J. Exp. Bot.* 55, 2131.
Tiessen A., J.H.M. Hendriks, M. Stitt, A. Branscheid, Y. Gibon, E.M. Farré and P. Geigenberger, 2002, *Plant Cell* 14, 2191.

Tiessen A., K. Prescha, A. Branscheid, N. Palacios, R. McKibbin, N.G. Halford and P. Geigenberger, 2003, *Plant J.* 35, 490.
Tjaden J., T. Möhlmann, K. Kampfenkel, G. Henrichs and H.E. Neuhaus, 1998, *Plant J.* 16, 531.
Toroser D., G.S. Athwal and S.C. Huber, 1998, *FEBS Lett.* 435, 110.
Toroser D. and S.C. Huber, 1997, *Plant Physiol.* 114, 947.
Toroser D., R. McMichael, K.P. Krause, J. Kurreck, U. Sonnewald, M. Stitt and S.C. Huber, 1999, *Plant J.* 17, 407.
van Bel A.J.E., 2003a, *Plant Cell Environ.* 26, 125.
van Bel A.J.E., 2003b, *Plant Physiol.* 131, 1509.
van Bel A.J.E., Y.V. Gamalei, A. Ammerlaan and L.P.M. Bik, 1992, *Planta* 186, 518.
van Bel A.J.E. and M. Knoblauch, 2000, *Aust. J. Plant Physiol.* 27, 477.
Veramendi J., U. Roessner, A. Renz, L. Willmitzer and R.N. Trethewey, 1999, *Plant Physiol.* 121, 123.
Villadsen D. and T.H. Nielsen, 2001, *Biochem. J.* 359, 591.
Viola R., A.G. Roberts, S. Haupt, S. Gazzani, R.D. Hancock, N. Marmiroli, G.C. Machray and K.J. Oparka, 2001, *Plant Cell* 13, 385.
Wareing P.F., M.M. Khalifa and K.J. Treharne, 1968, *Nature* 220, 453.
Weber A., J.C. Servaites, D.R. Geiger, H. Kofler, D. Hille, F. Groner, U. Hebbeker and U.I. Flügge, 2000, *Plant Cell* 12, 787.
Weiner H., R.W. McMichael Jr and S.C. Huber, 1992, *Plant Physiol.* 99, 1435.
Weise A., L. Barker, C. Kuhn, S. Lalonde, H. Buschmann, W.B. Frommer and J.M. Ward, 2000, *Plant Cell* 12, 1345.
Wellmer F., J.L. Riechmann, M. Alves-Ferreira and E.M. Meyerowitz, 2004, *Plant Cell* 16, 1314.
Williams L.E., R. Lemoine and N. Sauer, 2000, *Trends Plant Sci.* 5, 283.
Winter H. and S.C. Huber, 2000, *Crit. Rev. Plant Sci.* 19, 31.
Wischmann B., T.H. Nielsen and B.L. Möller, 1999, *Plant Physiol.* 119, 455.
Xoconostle-Cázares B., X. Yu, R. Ruiz-Medrano, H.L. Wang, J. Monzer, B.C. Yoo, K.C. McFarland, V.R. Franceschi and W.J. Lucas, 1999, *Science* 283, 94.
Yu T.S., H. Kofler, R.E. Häusler, D. Hille, U.I. Flügge, S.C. Zeeman, A.M. Smith, J. Kossmann, J. Lloyd, G. Ritte, M. Steup, W.L. Lue, J.C. Chen and A. Weber, 2001, *Plant Cell* 13, 1907.
Yu T.S., S.C. Zeeman, D. Thorneycroft, D.C. Fulton, H. Dunstan, W.L. Lue, B. Hegemann, S.Y. Tung, T. Umemoto, A. Chapple, D.L. Tsai, S.M. Wang, A.M. Smith, J. Chen and S.M. Smith, 2005, *J. Biol. Chem.* 280, 9773.
Zambryski P., 2004, *J. Cell Biol.* 164, 165.
Zhou J.J., F. Theodoulou, N. Sauer, D. Sanders and A.J. Miller, 1997, *J. Membrane Biol.* 159, 113.
Zrenner R., K.P. Krause, P. Apel and U. Sonnewald, 1996, *Plant J.* 9, 671.
Zrenner R., M. Salanoubat, L. Willmitzer and U. Sonnewald, 1995, *Plant J.* 7, 97.

Chapter 14

Dormancy and Sprouting

Jeffrey C. Suttle

USDA-ARS, Northern Crop Science Laboratory, 1307 18th Street N., P.O. Box 5677, Fargo, ND 58105, USA

14.1 INTRODUCTION

At some point during their life cycle, all sessile organisms must cope with adverse environmental conditions by adjusting their physiological/developmental status. The phenomenon of dormancy or developmental/metabolic arrest is a common strategy that enhances survival during times of environmental stress. Examples of dormancy can be found throughout the five kingdoms of life from single-cell bacteria to more complex multicellular organisms (Stearns, 1992). The existence of dormancy in primitive Archaean phyla suggests that dormancy was a very early adaptation to environmental extremes (Henis, 1987).

In plants, dormancy has been defined as 'the temporary suspension of visible growth of any plant structure containing a meristem' (Lang et al., 1987). Plant dormancy has been further subdivided into three distinct (but often overlapping) types: endodormancy, paradormancy, and ecodormancy (Lang et al., 1987). In endodormancy, a meristem is arrested by physiological factors arising from within the affected structure; in paradormancy, a meristem is arrested by external physiological factors; and in ecodormancy, a meristem is arrested by external environmental factors. During their life cycle, potato tubers can exhibit all three types of dormancy (Fig. 14.1). Immediately after formation and for an indeterminate period after harvest, tuber meristems (eyes) are endodormant and will not sprout. During extended storage, tubers exit endodormancy and begin to sprout, with (typically) one eye/sprout becoming dominant and inhibiting the growth of the other eyes that are paradormant. Tubers stored at temperatures of 3°C or lower will not sprout regardless of physiological dormancy status and are in a state of ecodormancy. In this review, unless specified otherwise, the term dormancy refers to the endodormant state.

As stated above, dormancy is thought to be a physiological adaptation to intermittent periods of environmental limitations and is therefore a survival mechanism. As such, many dormant organs or organisms exhibit enhanced resistance to otherwise lethal abiotic stress (Hand and Hardewig, 1996). For example, seeds of many frost-sensitive plants readily survive freezing temperatures that would certainly kill the entire plant (Osborne, 1981). Potato (*Solanum tuberosum* L.) plants are not frost tolerant (Li, 1977). No data have been published demonstrating enhanced low-temperature resistance in dormant tubers versus their non-dormant counterparts or parent plants. However, tuber dormancy prevents the emergence of sprouts in late summer/early fall when they would be exposed to the

Potato Biology and Biotechnology: Advances and Perspectives
D. Vreugdenhil (Editor)
2007 Published by Elsevier B.V.

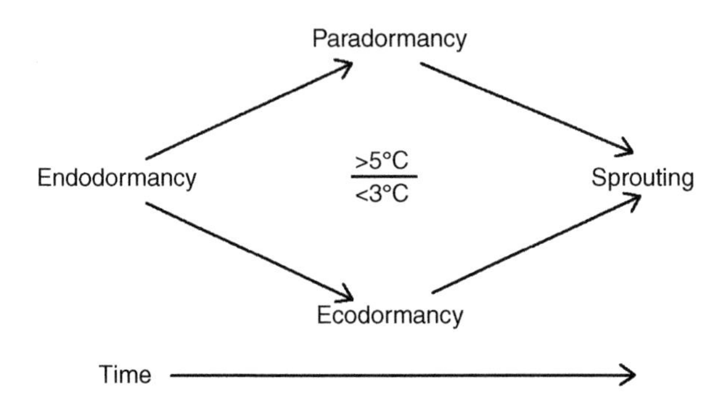

Fig. 14.1. Dormancy progression during tuber storage. At harvest, tubers will not sprout and are in a state of endodormancy. During storage above 5 °C (upper path), endodormancy weakens and tubers begin to sprout, with one eye becoming dominant and inhibiting the growth of the other eyes that are paradormant. Tubers stored at temperatures ≤3 °C (lower path) will not sprout regardless of physiological dormancy status and are in a state of ecodormancy.

subsequent killing temperatures of winter. This characteristic alone may confer a survival advantage in temperate climates. Interestingly, tuber dormancy has been associated with increased resistance to pathogen attack (Ladyzhenskaya and Protsenko, 2002).

In this review, the internal factors that accompany and possibly regulate dormancy progression and early sprout growth will be described.

14.2 TUBER DORMANCY CHARACTERISTICS

The tuber is a determinate organ with a finite functional lifespan. At what point during development does the tuber become dormant? In many studies, the length of tuber dormancy has been measured using harvest date as the starting point. Under field conditions, harvest date can vary widely and is dependent on many unrelated parameters including planting date, weather, field conditions, equipment availability, and intended use to name a few. Clearly, these factors have nothing to do with the physiological state of the tuber. Following a series of studies with several cultivars, Burton (1989) concluded that the inception of tuber dormancy occurs at or about the time of tuber initiation, and this view has been recently reemphasized (Claassens and Vreugdenhil, 2000). Thus, in indeterminate varieties such as Russet Burbank, the post-harvest duration of dormancy is longer for smaller (younger) tubers than it is for larger (older) tubers. The length of tuber dormancy in cultivars released for production is quite variable, ranging from essentially 0 to well over 9 months. Most cultivars grown for processing possess relatively short dormancy periods, necessitating the use of sprout-suppressing chemicals for successful long-term storage.

Tuber dormancy is affected by both genetic (i.e. genotype) and environmental influences. Protracted tuber dormancy is generally found in wild (non-domesticated) potato populations, whereas the reverse is often true in potato lines developed by modern breeding programs. One exception to this rule is the *Phureja* group, which, as a whole, is characterized by very short tuber and seed dormancy. Early studies by Simmons (1964) demonstrated that the inheritance of both (true) seed and tuber dormancy in potatoes was polygenic with at least three genes involved (Flewelling, 1987). Interestingly, a strong correlation between seed and tuber dormancy was observed in these studies (Simmons, 1964). Analyses for quantitative trait loci (QTL) have found between six (Freyre et al., 1994) and nine (Van den Berg et al., 1996) QTL that affect tuber dormancy, either alone or through epistatic interactions. Three of these dormancy-related QTL coincided with loci previously associated with photoperiodic sensitivity, suggesting that they may represent genes affecting tuber hormone content (Ewing et al., 2004). Further studies demonstrated that two of the dormancy-related QTL coincided with loci associated with tuber abscisic acid (ABA) content (Šimko et al., 1997).

Both pre- and post-harvest environmental factors can affect tuber dormancy duration. Unusually cold or hot weather during tuber development in the field often results in protracted or shortened (respectively) dormancy (Burton, 1989). Depending on the cultivar, extremely hot (>35°C) field temperatures can result in the immediate termination of tuber dormancy, a physiological disorder known as heat sprouting (Van den Berg et al., 1990). Tuber dormancy is restored as field temperatures moderate back to more seasonal norms. The overall effect of this loss and restoration of dormancy is malformed, knobby tubers with reduced starch content and lowered market value (Van den Berg et al., 1991). Varying the photoperiod over the range of 13–18 h during tuber development had no effect on the length of tuber dormancy (Emilsson, 1949). Under field conditions, the effects of photoperiod are difficult to determine, and any effects observed may relate more to the timing of tuber initiation or rate of maturation than to dormancy per se (Burton, 1989). Dormancy duration in microtubers was significantly reduced when microtubers were exposed to 8-h photoperiods instead of being held in complete darkness during tuber induction (Tovar et al., 1985).

Tuber dormancy is unlike that of other plant meristems in that there are no specific environmental cues required for dormancy exit. Nevertheless, post-harvest environmental conditions can have a dramatic quantitative effect on tuber dormancy. Between 3 and 25 °C, the length of tuber dormancy is inversely proportional to temperature (Burton, 1989). Prolonged exposure to temperatures $\leq$2 or $\geq$30°C abruptly terminates tuber dormancy, and sprouting commences upon return to moderate temperatures (Wurr and Allen, 1976). The presence or absence of light during post-harvest storage has little effect on dormancy duration but dramatically affects the morphology of emerging sprouts. Excluding biogenic volatiles that will be discussed later, the composition of the gas phase can affect tuber dormancy. As with temperature, storage gas phase composition has little effect on tuber dormancy, provided extremes are avoided. Both hypoxic and anoxic conditions that result in temporary anaerobiosis in the tuber can break tuber dormancy (Burton, 1989). Treatment of dormant tubers with non-physiologic concentrations of CO_2 (20%) and oxygen (40%) has been reported to effectively terminate tuber dormancy (Thornton, 1933; Reust and Gugerli, 1984; Coleman, 1998). By contrast, CO_2 concentrations $\leq$10% have no effect on dormancy duration per se but do stimulate subsequent sprout growth perhaps by antagonizing ethylene action (Burton, 1989).

14.3 CELL BIOLOGY OF DORMANCY

Metabolic depression is a general characteristic of dormant tissues (Hand and Hardewig, 1996; Guppy and Withers, 1999), and despite their fully hydrated condition, dormant potato tubers are no exception. Tuber respiration (a gross indicator of metabolic activity) is generally high immediately after harvest, declines rapidly thereafter, remains low during tuber dormancy, and rises dramatically as sprout growth is initiated (Burton, 1974). During storage, the contents of starch, amino acids, protein, nucleic acids, and lipids in tuber parenchyma either remain constant or decrease slightly (Burton, 1989). The onset of sprouting is accompanied by declines in most macromolecules and a corresponding increase in their monomeric constituents, which is especially evident in the starch/sugar ratio. These changes are consistent with the shift in tuber physiology from a sink/storage organ to a source tissue providing nutrients and biochemical precursors to the developing sprout.

Wholesale changes in macromolecular (DNA, RNA, and protein) syntheses also occur in tuber meristems during dormancy progression. When expressed on a weight basis, the total content of DNA, RNA, and protein increases in buds as dormancy ends and sprout growth commences (Tuan and Bonner, 1964; Désiré et al., 1995). Early studies with radiolabeled precursors indicated that eyes excised from dormant tubers exhibited very limited incorporation of labeled tracers into DNA and RNA (Tuan and Bonner, 1964; Rappaport and Wolf, 1969). Others have found that tuber dormancy is characterized by low (but measurable) rates of net RNA synthesis that increase substantially as dormancy ends (Korableva et al., 1976; MacDonald and Osborne, 1988). Sprouting is accompanied by increases in the synthesis of all classes of RNA, with the largest increases observed in rRNA (Korableva et al., 1976).

Although cell division is arrested in dormant tuber meristems, modest levels of thymidine incorporation into DNA have been observed that have been attributed to DNA repair processes (MacDonald and Osborne, 1988). By contrast, net protein synthesis is more active in dormant meristems and increases only modestly during dormancy exit (Korableva et al., 1976). As judged by electrophoretic mobility of the extracted proteins, the onset of sprouting is accompanied by wholesale qualitative changes in protein synthesis patterns (Désiré et al., 1995; Espen et al., 1999). Although the identity of many of these up- and down-regulated proteins has been determined, their role in tuber dormancy progression is largely unknown.

Chromatin preparations isolated from dormant eyes were essentially incapable of supporting net RNA synthesis either without supplement or when supplied with an excess of exogenous bacterial RNA polymerase (Tuan and Bonner, 1964). By contrast, chromatin prepared from tubers chemically forced to exit dormancy readily supported net RNA synthesis. This situation mirrored that found in dormant seeds (Deltour, 1985) and suggested that some component(s) in chromatin prepared from dormant tissues either inhibited polymerase activity or restricted template availability. Template availability of chromatin is known to be regulated by a number of reversible enzyme-catalyzed covalent modifications of both DNA and proteins collectively known as remodeling (Dobosy and Selker, 2001; Li et al., 2002). Two of the better characterized modifications are DNA–cytosine methylation and histone–lysine acetylation (Bender, 2004; Loidl, 2004).

Increased cytosine methylation and histone deacetylation inhibit transcription by altering nucleosome structure and inhibiting polymerase access to DNA, whereas increased template availability accompanies decreased DNA methylation and increased histone acetylation (Wu and Grunstein, 2000). Although genome-wide 5-methyl-cytosine (5mC) content in DNA isolated from tuber meristems was relatively constant during tuber dormancy progression, a significant transient decrease in 5mC levels was observed in certain sequence motifs as tubers exited dormancy (Law and Suttle, 2002). Similar declines in selected sequence-specific 5mC content were observed in both in vitro-generated microtubers and tubers chemically forced to prematurely terminate dormancy. In addition, tuber dormancy exit was accompanied by sustained increases in histone H3.1 and H3.2 multi-acetylation and transient increases in histone H4 multi-acetylation (Law and Suttle, 2004). The treatment of dormant tubers with bromoethane resulted in the rapid (<9 days) loss of dormancy and transient increases in histone H3.1, H3.2 and H4 multi-acetylation. Collectively, these results demonstrated that a defined sequence of epigenetic changes beginning with transient cytosine demethylation and followed by increased histone multi-acetylation accompanied the reactivation of tuber meristems during dormancy exit. Further studies are needed to identify changes in chromatin composition (nucleosome structure) in promoter regions of specific genes known to be up- or down-regulated during dormancy progression.

Tuber dormancy is defined by the absence of visible bud growth. Periodically, the cellular basis for the absence of macroscopic growth has been questioned (Coleman, 1987a; Burton, 1989). Some researchers (Leshem and Clowes, 1972) considered tuber dormancy to be a reflection of reduced (but not absent) cellular growth, whereas others concluded that visible tuber meristem dormancy was a result of total cell division arrest (Goodwin, 1967). A recent study using both destructive microscopic and non-destructive morphological observations concluded that tuber dormancy is indeed characterized by a cultivar-dependent period of meristematic inactivity (Van Ittersum et al., 1992). Cell division or mitosis is the final phase of a series of highly coordinated biochemical processes collectively referred to as the cell cycle (De Veylder et al., 2003; Dewitte and Murray, 2003). The absence of mitotic figures in dormant tuber meristems suggested that cell division was arrested in a pre-mitotic phase (Leshem and Clowes, 1972). Using histological methods, MacDonald and Osborne (1988) concluded that dormant tuber meristems were largely arrested prior to DNA synthesis in a 2C (Somatic cell DNA content) state. Cell cycle analysis by flow cytometry (Fig. 14.2) demonstrated that nuclei isolated from dormant tuber meristem cells were in the G1 phase of the cell cycle, and release from dormancy was accompanied by an increase in both S- and G2-phase nuclei (Campbell et al., 1996). Bud dormancy of aspen (*Populus tremula* L.) and tuber dormancy of Jerusalem artichoke (*Helianthus tuberosus* L.) are also characterized by cell cycle arrest in the G1 phase (Rohde et al., 1997; Freeman et al., 2003).

Cell cycle progression is controlled by both transcriptional and post-translational processes (De Veylder et al., 2003; Dewitte and Murray, 2003). Non-dividing cells can arrest at one of two checkpoints (G1/S or G2/M) of the cell cycle. Passage through these checkpoints is controlled in part by changes in activities of a group of heterodimeric, cyclin-dependent serine/threonine kinases (cdk). The activities of these enzymes are regulated by a combination of phosphorylation/dephosphorylation and association with other

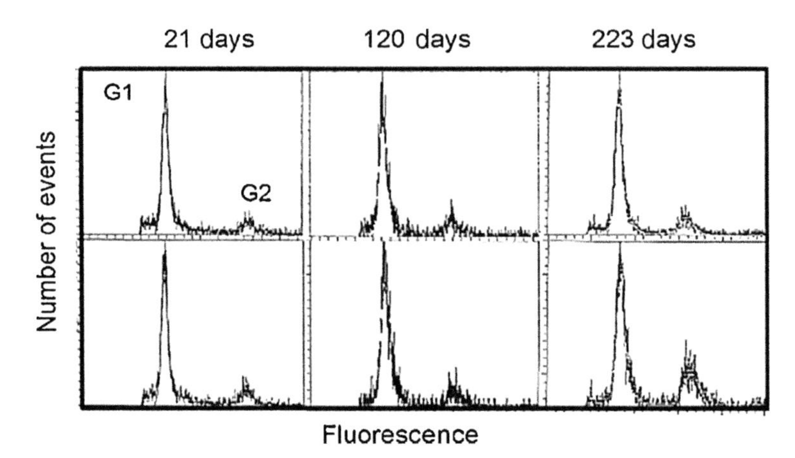

Fig. 14.2. Flow cytometry of nuclei isolated from tuber meristems after 21, 120, or 223 days of storage at continuous 3°C (upper panel) or after transfer to 20°C for 7 days (lower panel). Percentages of nuclei in G1/S/G2 phases in non-growing meristems isolated from tubers stored continuously at 3°C were 66/24/10, 73/16/11, and 67/19/14, whereas those of nuclei isolated from tubers transferred for 7 days to 20°C were 64/25/11, 69/19/12, and 55/17/29 after 21, 120, and 223 days of storage, respectively. In tubers transferred to 20°C, dormancy release was complete after 223 days of storage (Campbell et al., 1996).

regulatory proteins such as cyclins and inhibitors (Inzé, 2005). Studies from several plant systems have demonstrated that passage through the G1/S checkpoint is cytokinin dependent and requires an active cdc2a/cyclin D kinase complex that phosphorylates and inactivates a retinoblastoma-like protein which in turn de-represses the transcription of many S-phase genes (den Boer and Murray, 2000; Murray et al., 2001). A *cdc2a* gene has been cloned from potato tuber meristems (Campbell et al., 1996). Northern analysis indicated that no significant changes in *cdc2a* gene transcription occurred during tuber dormancy progression. Immunoblot (western) analysis demonstrated that an increase in $p34^{cdc2a}$ kinase protein occurred coincidently with the resumption of meristem growth in non-dormant tuber meristems. Preliminary studies found that an increase in tuber meristem histone H-1 kinase activity in PSTAIRE immunoprecipitates accompanied the onset of meristem growth following dormancy termination (Suttle, unpublished). The effects of dormancy status on cyclin D or cdk inhibitors in potato tuber meristems have not been reported. In the physiologically similar case of tuber dormancy in *H. tuberosus* L., an increase in cyclin D transcript levels was observed during re-initiation of cell cycle activity induced by hormone treatment (Freeman et al., 2003).

Although dormancy is defined by the absence of visible growth of meristematic tissues, it appears that non-growing tissues also display dormancy-related metabolic depression. The treatment of excised discs of potato tuber cortex with GA_3 increased the reducing sugar content only when discs were prepared from non-dormant tubers (Bailey et al., 1978). These results suggest that both the meristematic (bud) and storage tissues exhibit metabolic depression during dormancy.

14.4 GENE EXPRESSION DURING DORMANCY TRANSITION

The numerous reports describing qualitative and quantitative changes in tuber biochemistry and protein composition indicate that dormancy progression is accompanied by significant changes in gene expression. However, to date, there have been surprisingly few published reports describing such changes. In general, most of the published studies have been comparative survey analyses of global gene-expression patterns in cDNA libraries prepared from dormant and sprouting tissues. Because these studies are correlative in nature, generally encompass two end points (dormant and sprouting), and may use different tissues for each analysis, care must be exercised when interpreting these data.

Changes in gene expression from initiation through sprouting were assayed using whole microtubers generated in vitro from single-node explants and an amplified fragment-length polymorphism (AFLP) fingerprinting technique (Bachem et al., 2000). In total, an estimated 40 000 genes were expressed throughout the microtuber life cycle. Gene expression was highest during tuber initiation, was lowest during dormancy, and rose again as sprouting commenced. A total of 1300 transcript-derived cDNA fragments (TDFs) were followed during dormancy progression. Of these, only 13% were observed to change during dormancy, with 4% of the genes exhibiting decreased expression during early dormancy and 3% showing increased expression coincident with sprouting. The vast majority of genes either displayed no change during dormancy or displayed an irregular pattern of expression independent of physiological state. Following isolation and sequencing of 75 TDFs exhibiting dormancy-related changes in expression, it was found that 64% of these had no sequence similarity to any known genes in the databases but most were similar to anonymous cDNA sequences in expressed sequence tag (EST) databases. In a follow-up study (Trindade et al., 2004), putative promoters located upstream of the translation start site of 12 differentially expressed genes were isolated and analyzed. Response elements in the promoters identified confer light, sugar, and hormone responsiveness in other plant species and may do so in tuber tissues as well.

In another study involving EST sequencing from multiple libraries representing different potato tissues, Ronning et al. (2003) identified just under 20 000 unique expressed sequences, of which 44% were similar to known genes. Of these, approximately 2200 library-specific sequences from sprouting eyes and 851 from dormant tuber cortex (excluding buds and periderm) were identified. Just over 50 genes were more highly expressed in dormant tuber cortex versus actively growing microtubers, and over 40 genes were more highly expressed in sprouting eyes versus dormant tuber cortex. Because the libraries analyzed were prepared from very different tissues, it is not clear whether the expression patterns observed were the result of tissue or developmental specificity. Regardless, perhaps the most surprising aspect of the data presented is the paucity of differentially expressed genes. This was unexpected, given the wholesale changes in protein profiles noted in the earlier electrophoretic analyses (Désiré et al., 1995; Espen et al., 1999). Of the genes up-regulated in sprouting versus dormant tuber libraries, genes coding for ribosomal proteins constitute the largest group, and genes typically associated with cell cycle activity [cdk2, cyclins, proliferating cell nuclear antigen (PCNA), etc.] are notably absent.

More recently, a subtractive hybridization approach was used to generate a library enriched with genes up-regulated in tuber apical meristems during sprouting (Faivre-Rampant et al., 2004a). A total of 385 up-regulated ESTs were identified, of which 36% showed no significant similarity to known genes. Of the remainder, 12, 9, and 3% were similar to ribosomal protein genes, stress-related genes, and transcription factor genes, respectively. One gene coding for a member of the auxin-response factor (ARF6) family was identified and analyzed more closely. *StARF6* expression was greatest in tuber meristems during the initial stages of sprouting and *StARF6* expression seemed to parallel changes in meristem growth rate. These results are consistent with the role for endogenous auxins in tuber sprout growth (see 14.5.1). In a related study, a potato TCP-like transcription factor has also been identified and exhibited the inverse pattern of expression (Faivre-Rampant et al., 2004b). *StTCP* expression was highest in dormant meristems and could not be detected in libraries prepared from sprouting buds or in buds released from apical dominance (paradormancy). Interestingly, all members of the TCP family characterized seem to function as either positive or negative regulators of cell proliferation (Cubas et al., 1999).

Collectively, these studies are beginning to assemble an increasingly detailed description of differential gene expression during dormancy progression that will only become more refined as additional research is reported. In addition to these studies, sporadic reports describing changes in expression of various genes during tuber dormancy have appeared. However, the studies reported to date shed no light on the specific mechanisms of tuber dormancy control. It is possible that the 'master' control genes are very low abundance genes and therefore will escape detection using global survey techniques. The central question remains: will we know the control (master) genes when we find them?

14.5 HORMONAL REGULATION OF TUBER DORMANCY

Dormancy is a highly regulated phase of development that is crucial to plant survival. Successful progression from active growth to dormancy and, finally, to sprouting involves all tuber tissues and therefore requires physiological coordination. Integration of plant developmental cues is dependent on the synthesis and action of growth regulators or hormones. In Sections 14.5.1–14.5.7, the role(s) of all major classes of plant hormones in tuber dormancy control will be discussed. This topic has been the subject of several recent reviews (Coleman, 1987b; Wiltshire and Cobb, 1996; Suttle, 2004a). At first glance, much of the data reported in the literature appears fragmented and contradictory. This confusion is largely the result of disparate methods of hormone analysis (bioassay vs. physio-chemical analysis) and the choice of tissues for analysis. Where possible, these discrepancies will be explored and (hopefully) resolved.

14.5.1 Auxins

Using an *Avena* bioassay, Guthrie (1940) found that the content of auxin rose in tubers whose dormancy had been terminated by ethylene chlorohydrin treatment. However, this increase was observed only after the tubers began to visibly sprout, and he concluded that the increase in indole-3-acetic acid (IAA) content was a reflection of sprout growth rather than the cause. The content of bioactive acidic auxin during natural tuber dormancy progression was lowest during the initial period of dormancy, rose as storage was extended, and reached a maximum immediately prior to visible sprouting (Hemberg, 1942).

The introduction of more sophisticated methods of hormone analysis has not greatly clarified this situation. Using high-performance liquid chromatography (HPLC) coupled with fluorometric detection, Sukhova et al. (1993) found that the content of IAA remained constant during tuber dormancy rising only after sprouting commenced. In more recent studies using gas chromatography–mass spectrometry (GC–MS) together with internal standards, Sorce et al. (2000) determined that the content of free, ester- and amide-linked IAA in eye tissues isolated from tubers stored at 23°C increased during storage and reached a maximum coincident with 50% sprouting. In tubers stored at 3°C (where sprout growth is suppressed), loss of tuber dormancy (in the absence of sprout growth) was accompanied by only a small increase in free IAA, and maximum IAA content coincided with the commencement of sprout growth. These observations suggest that IAA content reflects sprout growth rather than bud dormancy.

Exogenous auxin has no apparent effect on dormant tubers (Hemberg, 1985). In non-dormant tubers, exogenous auxin elicits a bi-phasic response. The application of high concentrations of IAA to non-dormant tubers results in a dose-dependent temporary inhibition of sprout growth (Guthrie, 1940). Synthetic auxins (i.e. 1-naphthalene acetic acid and 2,4,-dichlorophenoxy acetic acid) elicit a more pronounced and protracted inhibition (Suttle, 2003). By contrast, the application of lower ($\leq 1\,\mathrm{mg\,L^{-1}}$) concentrations of IAA to single-eye tuber pieces resulted in a modest stimulation of sprout growth (Hemberg, 1949a). There have been no reports describing the effects of ectopic expression of bacterial auxin biosynthesis genes on tuber development or dormancy. However, ectopic expression of the *Agrobacterium rhizogenes rol* C gene under the control of a CaMV 35S RNA promoter in potato resulted in increased IAA content and altered tuber morphology, but no effects on tuber dormancy were noted (Fladung, 1990; Schmülling et al., 1993). In addition, expression of a bacterial IAA-lysine synthetase gene under the control of the CaMV 35S promoter in potatoes resulted in phenotypic changes and reduced IAA content, but again, no effects on tuber dormancy were noted (Spena et al., 1991).

Taken as a whole and as first proposed by Guthrie (1940), these results suggest that endogenous auxins are not cognate regulators of tuber dormancy but may play a role in the regulation of subsequent sprout growth. This is consistent with the proposed role of auxins as essential mediators of plant cell cycle progression at the G-2/M transition (Francis and Sorrell, 2001; Del Pozo et al., 2005), which is beyond the position of cell cycle arrest in dormant tubers (MacDonald and Osborne, 1988; Campbell et al., 1996).

14.5.2 Abscisic acid

Bioassay of plant extracts following paper chromatography typically results in zones of growth-promoting (auxin-like) and growth-inhibiting substances. Hemberg (1949a) was the first to recognize the potential significance of the growth-inhibiting substances in tuber dormancy. In a series of thorough studies, it was shown that the loss of tuber dormancy (through both natural progression and premature chemical termination) was accompanied by a dramatic decline in acidic (but not neutral) inhibitors (for review, see Hemberg, 1985). Similar observations were made in studies on winter bud dormancy in *Fraxinus* (Hemberg, 1949b). With the identification of ABA as a component of the inhibitor complex (Cornforth et al., 1966), attention was focused on the role of ABA in dormancy. The application of crude potato extracts containing ABA (and other inhibitors) temporarily inhibits sprout growth (Blumenthal-Goldschmidt and Rappaport, 1965; Franklin and Hemberg, 1980). Exogenous ABA transiently inhibits sprout growth in non-dormant tubers in a dose-dependent manner (El-Antably et al., 1967; Holst, 1971).

ABA content in potato tubers has been determined using both bioassay and, more recently, very sensitive and specific instrumental analyses. In one of the earliest investigations using instrumental analysis, Korableva et al. (1980) reported that tuber ABA content was highest in growing points (buds?) and periderm and lowest in storage tissues and that termination of dormancy was accompanied by substantial declines (>85%) in ABA content in all tissues examined. Similar results were reported by Coleman and King (1984) and Cvikrová et al. (1994). However, in the former study, significant declines in tuber ABA were observed in only two of three cultivars examined, whereas in the latter study, tuber ABA content was highest during mid-dormancy and declined dramatically thereafter. The ABA content of thin (1–2 mm) periderm discs containing the apical bud complex was highest at harvest and declined steadily during storage (Suttle, 1995). ABA content in stored tubers was inversely proportional to storage temperatures, whereas the rate of decline was lowest at cooler (3°C) storage temperatures. Consistent with earlier observations (Coleman and King, 1984), no definite minimum threshold level of ABA required for commencement of sprouting was identified. Similarly, in a study of six cultivars with varying lengths of dormancy, Biemelt et al. (2000) reported that ABA content declined in all varieties during storage but found no correlation between final ABA levels and sprouting behavior. Breakage of tuber dormancy with the synthetic cytokinin thidiazuron and heat stress was also accompanied by a decline in ABA content (Ji and Wang, 1988; Van den Berg et al., 1991). By contrast, Sorce et al. (1996) measured ABA content in several tissues of potato during storage at two temperatures and found that the concentration of ABA in 'eyes' was at a minimum at harvest and exhibited a modest rise coincident with sprouting. Recent studies from our laboratory using liquid chromatography (LC)–MS–selected ion monitoring (SIM) have demonstrated that a substantial decline in ABA content of tuber eyes occurred during dormancy progression, whereas the ABA content of the surrounding periderm and underlying cortex exhibited only modest reductions (Destefano-Beltrán et al., 2006a).

An ABA-deficient mutant of potato was identified and given the trivial name 'Droopy' (Quarrie, 1982). Biochemical analysis suggested that this mutant was impaired in the oxidation of ABA aldehyde to ABA (Duckham et al., 1989). Unfortunately, the effects of this mutation on tuber dormancy cannot be determined because this mutation occurs in *Solanum phureja*, a species known for its near-total lack of tuber dormancy (Hawkes, 1992).

Definitive proof of a role for endogenous ABA in potato tuber dormancy was obtained using a single-node, in vitro microtuber system and the phytoene desaturase inhibitor fluridone (Fig. 14.3). Under the culture conditions used, microtubers were harvested after 9 weeks of growth and remained dormant for approximately 15–20 weeks thereafter

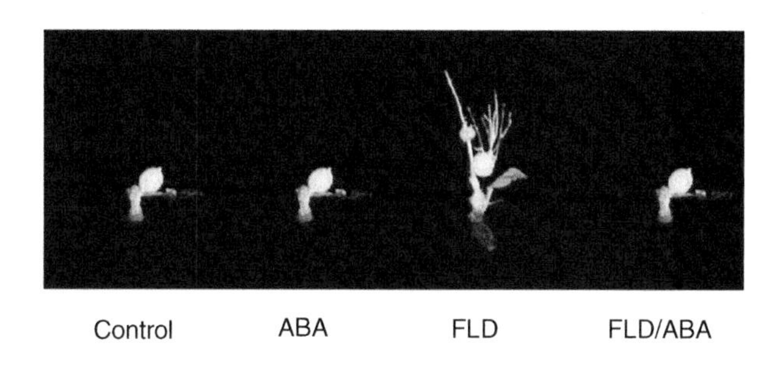

	Control	ABA	FLD	FLD/ABA
Sprouting (%)	0 ± 0	1 ± 1	95 ± 2	7 ± 3
ABA (pmol/g FW)	235 ± 20	420 ± 43	10 ± 5	90 ± 15

Fig. 14.3. Effects of the abscisic acid (ABA) biosynthesis inhibitor fluridone (FLD) on tuber morphology (upper photo), sprouting percentage, and endogenous ABA content in microtubers after 9 weeks of in vitro development (adapted from Suttle and Hultstrand, 1994).

(Suttle and Hultstrand, 1994). Continuous exposure to fluridone during development resulted in premature sprouting after only 3–6 weeks of in vitro culture. The inclusion of (±) ABA in the fluridone-containing medium restored endogenous ABA content to control levels and suppressed microtuber sprouting. Furthermore, the application of fluridone to fully developed and dormant microtubers after 9 weeks of culture also resulted in premature sprouting. These results are consistent with an essential role for endogenous ABA in the induction and maintenance of tuber dormancy.

ABA is synthesized from an oxygenated C_{40} carotenoid (xanthophyll) precursor through a series of cleavage and oxidation reactions (Seo and Koshiba, 2002; Nambara and Marion-Poll, 2005). In most plants, the rate-limiting step of ABA biosynthesis is catalyzed by either zeaxanthin epoxidase (ZEP) or 9-*cis*-epoxycarotenoid dioxygenase (NCED). Transgenic down-regulation of ZEP expression through either antisense or co-suppression resulted in numerous transformants with greatly elevated tuber contents of zeaxanthin (up to 134-fold), violaxanthin (approximately 4-fold), and neoxanthin (approximately 3-fold), all of which had normal ABA contents (Römer et al., 2002). Similarly, the transformation of potatoes with a bacterial phytoene synthase gene under the control of a tuber-specific promoter resulted in tubers with greatly elevated contents of zeaxanthin, violaxanthin, and neoxanthin and normal levels of ABA (Ducreux et al., 2005). These observations suggest that regulation of tuber ABA biosynthesis occurs downstream of ZEP activity.

Recent studies from this laboratory have begun to elucidate the molecular mechanisms regulating ABA content during tuber dormancy (Destefano-Beltrán et al., 2006a). Genes encoding proteins catalyzing the final post-zeaxanthin steps of ABA biosynthesis and the catabolic enzyme ABA-8′-hydroxylase have been cloned and characterized from tuber tissues (Fig. 14.4). Analyses using quantitative reverse transcription–polymerase chain reaction (qRT–PCR) have demonstrated a highly dynamic and tissue-specific pattern of expression during dormancy progression and have tentatively identified key regulatory steps. In particular, ABA biosynthesis in tuber meristems appears to be regulated at the level of *NCED 2* expression, whereas ABA catabolism is temporally correlated with expression of two of four putative ABA-8′-hydroxylase genes. Changes in meristem ABA content following chemically forced termination of tuber dormancy were also accompanied by altered expression of *NCED 2* and ABA-8′-hydroxylase genes (Destefano-Beltrán et al., 2006b).

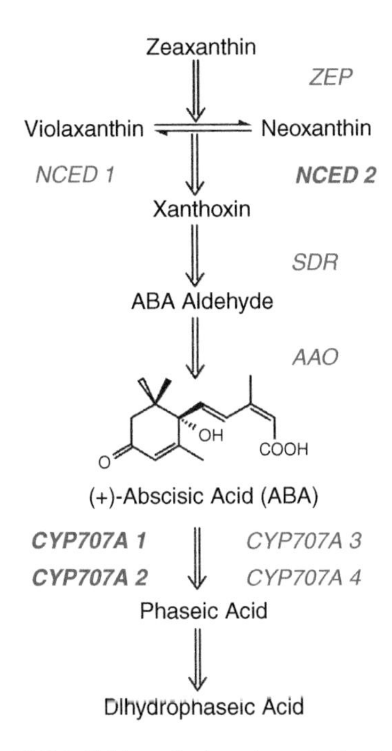

Fig. 14.4. Pathway of abscisic acid (ABA) biosynthesis and metabolism in potato tubers. Genes cloned and characterized from potato tuber tissues are indicated in red italics and putative regulatory genes are in red boldface. ZEP, zeaxanthin epoxidase; NCED, 9-*cis*-epoxycarotenoid dioxygenase; SDR, short-chain alcohol dehydrogenase/reductase; AAO, ABA aldehyde oxidase; CYP707A, ABA-8′-hydroxylase.

14.5.3 Ethylene

The involvement of ethylene in tuber dormancy is incompletely understood. The effects of exogenous ethylene on dormancy are dependent on both cultivar and duration of treatment. Long-term (approximately 4 weeks) treatment with ethylene hastened the sprouting of tubers once the gas was removed (Rosa, 1925). Subsequent studies failed to confirm this observation (Denny, 1926). Rylski et al. (1974) demonstrated that a 72-h treatment with ethylene shortened the length of dormancy, but continuous exposure to ethylene inhibited sprout growth. These observations were corroborated by Prange et al. (1998), who also showed that sprout number (per tuber) was increased by long-term ethylene treatment. In tubers stored at 20°C, reducing ethylene concentrations in storage below 0.01 ppm (v/v) increased the duration of dormancy and reduced total sprout number after 35 days (Wills et al., 2003). Post-harvest treatment with ethylene-releasing compounds has been reported to either hasten or delay tuber sprouting (Rama and Narasimham, 1982; Cvikrová et al., 1994).

Ethylene production is very low in intact dormant tubers but rises as sprout growth commences (Poapst et al., 1968; Creech et al., 1973; Okazawa, 1974). Wounding of tubers induced a substantial increase in ethylene production that persisted for 72 h (Lulai and Suttle, 2004). Because of this, ethylene production from excised (i.e. wounded) tuber tissues cannot be used to assay the situation in undamaged tubers during dormancy. Cvikrová et al. (1994) reported a transient surge in ethylene production in freshly harvested tubers followed by a very low rate of production that persisted for the duration of dormancy and then rose again as sprouting commenced. Although intact tubers were used in this study, it is possible that the initial transient rise in ethylene production was a response to the stresses and tissue damage that often accompany harvest. Exposure to various dormancy-breaking agents resulted in increased ethylene production coincident with the onset of sprout growth (Suttle, unpublished). Despite these observations, the significance (if any) of this sprouting-associated increase in ethylene production is unclear.

Potatoes (cultivar Russet Burbank) have been transformed with the Arabidopsis *ETR1* gene in both the sense and the antisense orientations (Haines et al., 2003). Regardless of transgene orientation, transformed plants exhibited many phenotypic alterations including modified leaf size and shape and malformed tubers with swollen eyes. When stored at 20 °C, no change in tuber dormancy was noted in any of the transformants, but when stored at 4 °C, two of the antisense lines failed to sprout after 2 years. It is not clear whether this indicates a change in tuber dormancy duration or reflects an exaggerated sensitivity to cold temperatures.

Ethylene has been shown to play a critical role in the induction of tuber dormancy. Following subculture of single-node explants under tuberizing conditions, the rate of ethylene production was highest during the initial 2 weeks of in vitro culture and declined precipitously thereafter (Suttle, 1998a). Continuous exposure of tuberizing explants at the time of transfer to the non-competitive ethylene antagonist silver nitrate or the competitive antagonist 2,5-norbornadiene (NBD) resulted in a dose-dependent increase in precocious sprouting (Fig. 14.5). Exposure of NBD-treated explants to ethylene inhibited precocious sprouting. Interestingly, the period of ethylene sensitivity was observed only

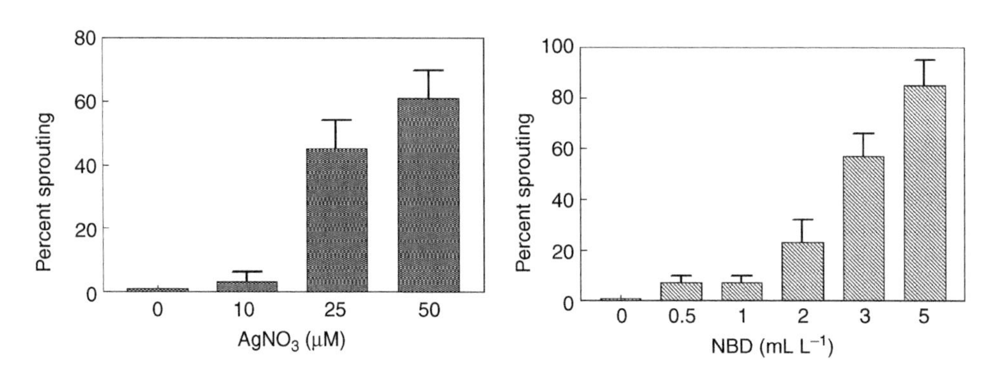

Fig. 14.5. Effects of the ethylene antagonists silver nitrate (left) and 2,5-norbornadiene (NBD; right) on percentage microtuber sprouting after 9 weeks of in vitro development (adapted from Suttle, 1998a).

during the initial period of in vitro culture. These results are consistent with a role for endogenous ethylene in the induction of tuber dormancy. Its role thereafter remains purely speculative.

14.5.4 Gibberellins

Together with ABA, gibberellins (GAs) have been posited to play a critical role in the processes of seed dormancy and germination (Raz et al., 2001). As first reported by Brian et al. (1955) and verified by others (Rappaport et al., 1958), application of GAs to tubers typically breaks dormancy and results in precocious sprouting. GA-like activity is low in dormant tubers and increases as sprouting commences (Smith and Rappaport, 1961; Boo, 1962; Bialek and Bielińska-Czarnecka, 1975). Interpretation of these bioassay results is complicated by the presence of large amounts of interfering phenolic acids in tuber peel extracts. Nevertheless, GAs have long been considered to play a role as cognate dormancy-terminating hormones in tubers (Hemberg, 1985).

Recent evidence does not support a role for endogenous GAs in tuber dormancy control. The endogenous GAs of potato are members of the early 13-hydroxylation pathway arranged in the following biosynthetic sequence: $GA_{19} \rightarrow GA_{20} \rightarrow GA_1 \rightarrow GA_8$ (Hedden and Phillips, 2000). GA_1 is considered to be the bioactive hormone and GA_8 an inactive metabolite (Jones et al., 1988; Van den Berg et al., 1995; Carrera et al., 2000). Consistent with this view, injection of GA_{19}, GA_{20}, or GA_1 but not GA_8 prematurely terminated tuber dormancy (Suttle, 2004b). The endogenous contents of GA_{19}, GA_{20}, and GA_1 were relatively high immediately after harvest, declined during storage (when dormancy was deepest), and rose to the highest levels during the period of robust sprout growth (Suttle, 2004b). Interestingly, at the time of initial sprouting, internal levels of these bioactive GAs were lower than those found in deeply dormant tubers. Tuber dormancy progression in a dwarf mutant of potato was similar to that of the normal phenotype despite the absence of detectable levels of tuber GA_1 at any stage of dormancy (Fig. 14.6), and the application of various GA biosynthesis inhibitors to developing microtubers was found to stimulate precocious sprouting (Suttle, 2004b). In addition, although artificial elevation of endogenous GA content by ectopic expression of GA 20-oxidase resulted in premature tuber sprouting, reduction of endogenous GAs in potato plants using an antisense approach had no effect on tuber dormancy but reduced subsequent sprout elongation (Carrera et al., 2000). Collectively, these results do not support a role for endogenous GAs in tuber dormancy per se but are consistent with a role in subsequent sprout growth.

14.5.5 Cytokinins

The role of cytokinins in seed and bud dormancy has not been clearly defined. Cytokinins are thought to play a permissive (rather than essential) role in seed dormancy especially under suboptimal conditions (Thomas, 1990). Although cytokinins have been implicated in the release of buds from apical dominance (Chatfield et al., 2000), their role in seasonal endodormancy is uncertain (Powell, 1987). The application of natural and synthetic cytokinins to dormant tubers rapidly breaks dormancy and results in sprouting of multiple eyes (Hemberg, 1970). Ectopic expression of the bacterial shoot-inducing Ti T_L-DNA in tubers resulted in massive wholesale increases in cytokinin content and increased premature sprouting (Ooms and Lenton, 1985). Cytokinin-like bioactivity is low in dormant tubers, and sprouting is accompanied by increased cytokinin-like activity (Engelbrecht and Bielińska-Czarnecka, 1972; Van Staden and Dimalla, 1978; Koda, 1982; Banas et al., 1984). As in the case of GAs, the interpretation of bioassay results using crude extracts of potato tubers is complicated by the presence of large amounts of interfering substances. Furthermore, endogenous cytokinins are chemically diverse and readily interconvertible and require significant purification prior to meaningful analysis.

Despite these concerns, more recent immunological techniques have largely confirmed the bioassay data and have provided a more thorough understanding of the complex metabolic dynamics of these hormones during dormancy progression. An increase in total immuno-reactive cytokinins in un-fractionated tuber extracts occurred coincident with sprouting (Turnbull and Hanke, 1985b; Sukhova et al., 1993). Additional studies with HPLC-fractionated extracts and side-chain-specific monoclonal antibodies found an increase in bioactive isopentenyl- and *trans*-zeatin-type cytokinins immediately prior to the onset of sprouting in tubers stored under both growth-permissive (20°C) and growth-inhibiting (3°C) temperatures (Fig. 14.7, left panel; Suttle, 1998b). Interestingly, the content of *cis*-zeatin also increased prior to sprout growth (Fig. 14.7, right panel; Suttle and Banowetz, 2000). Tuber dormancy was rapidly broken by treatment with either cytokinin isomer, suggesting a potential role for *cis*-cytokinins in dormancy regulation.

Cytokinins are rapidly metabolized in plant tissues (Auer, 2002). Immediately after harvest and during the initial period of storage, exogenous cytokinins have no effect on dormancy (Turnbull and Hanke, 1985a; Suttle, 2002). As storage was extended, tubers developed a time-dependent increase in cytokinin sensitivity. The acquisition of cytokinin sensitivity was not accompanied by significant changes in the qualitative or quantitative pattern of cytokinin metabolism, which suggested that other aspects of signal transduction were affected by tuber dormancy status (Suttle, 2002).

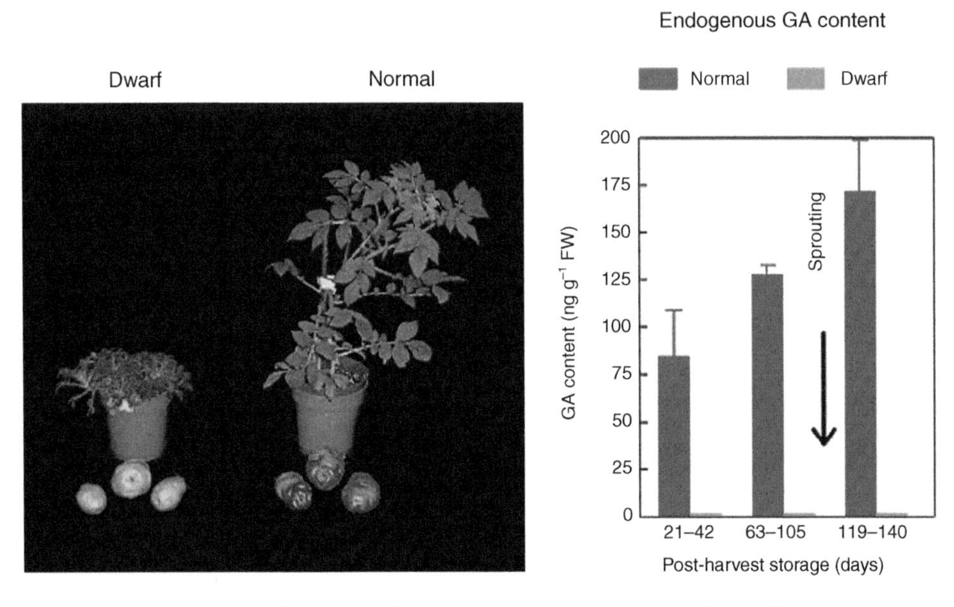

Fig. 14.6. Plant morphology and gibberellin (GA) content in dwarf and normal siblings of *Solanum tuberosum* L. ssp. andigena. Plant and tuber morphology in greenhouse-grown plants (left photo). The effect of post-harvest storage on GA_1 content in normal (red bars) and dwarf (green bars) tubers (right graph). Time of sprouting is denoted by downward arrow (adapted from Suttle, 2004b).

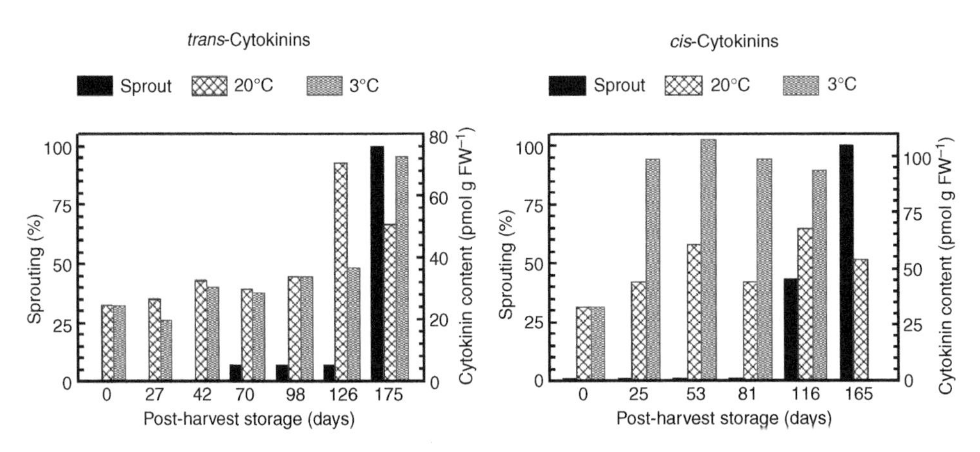

Fig. 14.7. Changes in endogenous cytokinin content during post-harvest storage. Black bars, percent sprouting; blue bars, tubers transferred to 20°C 7 days prior to extraction; red bars, tubers stored continuously at 3°C. Left panel: content of bioactive iso-pentenyl- and *trans*-zeatin type cytokinins. Right panel: combined content of *cis*-zeatin and *cis*-zeatin riboside. Data adapted from Suttle (1998b) and Suttle and Banowetz (2000).

Cytokinins are defined by their ability to stimulate cell division (Mok and Mok, 2001). In addition, cytokinins appear to be required for cell cycle progression through the G1/S transition (Francis and Sorrell, 2001; Del Pozo et al., 2005). These characteristics, together with the observed changes in cytokinin content during dormancy and the ability of exogenous cytokinins to terminate dormancy and stimulate sprouting, suggest that cytokinins are cognate regulators of tuber dormancy exit in potatoes.

14.5.6 Other endogenous growth substances

It has become clear that, in addition to the five original classes of hormones, plant development is regulated by a diverse array of endogenous signaling molecules (Davies, 2004). The role of these 'non-classical' hormones in tuber dormancy control has received only limited attention.

The original studies of Hemberg and associates found a correlation between tuber dormancy and the activity of a chromatographic zone known as inhibitor ß (Hemberg, 1985). Although ABA was shown to be a component of this fraction (Milborrow, 1967), inhibitor ß is a complex mixture of many compounds, most notably phenolic acids (Köves, 1957; Holst, 1971). Additional chromatographic fractionation demonstrated that the decline in growth-inhibiting activity of inhibitor ß isolated from tuber peels during storage could not solely be ascribed to a decline in ABA content (Holst, 1971; Franklin and Hemberg, 1980). Thus, other uncharacterized inhibitors in tuber extracts decline during storage and may play a role in dormancy control.

Jasmonates (JAs) and biosynthetically related compounds (oxylipins) have pronounced effects on potato development, and a hydroxylated derivative of JA (tuberonic acid) is one of the most potent inducers of in vitro tuberization and may play a role in tuberization *in planta* (Yoshihara et al., 1989). Although exogenous JA has been reported to inhibit sprout growth during storage and JA occurs in non-wounded tubers and sprouts (Abdala et al., 2000; Fauconnier et al., 2003), its role as a cognate regulator of tuber dormancy is uncertain. The endogenous content of JA in potato plants has been elevated by ectopic expression of a flax allene oxide synthetase gene and reduced by expression of an ω3 fatty acid desaturase in an antisense orientation (Harms et al., 1995; Martin et al., 1999). In neither case was there any mention of any effects of transformation on tuber JA content or dormancy progression. Similarly, brassinolides are a distinct class of bioregulators with numerous effects on plant development (Clouse and Sasse, 1998), and although treatment with brassinolide suppressed tuber sprouting (Korableva et al., 2002), the effects of tuber dormancy on endogenous levels of this class of bioregulators and their possible role(s) in tuber dormancy control are unknown.

Tubers (especially the periderm) contain a number of compounds that can affect sprout growth (Burton and Meigh, 1971). Several of these compounds are volatile and can accumulate in storage atmospheres at growth-inhibiting concentrations. In particular, tubers produce 1,4-dimethylnaphthalene and 1,6-dimethylnaphthalene, both of which are potent growth inhibitors (Meigh et al., 1973). The effects of dormancy status on the production of these inhibitors and the role of these compounds in tuber dormancy are unknown.

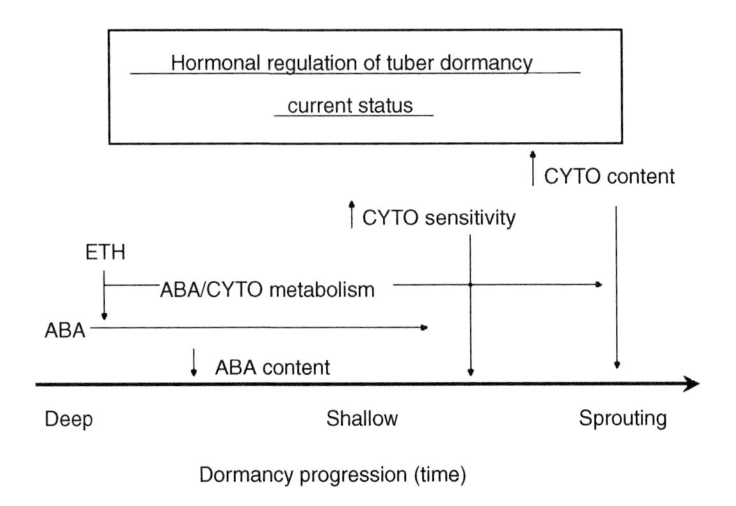

Fig. 14.8. Schematic overview of the hormonal regulation of potato tuber dormancy. Details are given in the text.

14.5.7 Hormonal regulation of tuber dormancy: an overview

A model incorporating current knowledge of the role of endogenous hormones in tuber dormancy is shown in Fig. 14.8. At formation and generally extending well-past harvest, tubers are highly dormant. During storage, dormancy gradually weakens and, ultimately, terminates resulting in sprouting. Both ABA and ethylene are required for the induction of dormancy, but only ABA appears necessary for the maintenance of dormancy. Endogenous contents of bioactive cytokinins are low in dormant tubers, and tubers are insensitive to exogenous cytokinins. Dormant tubers actively metabolize ABA and cytokinins throughout storage. As storage is extended and dormancy begins to weaken, ABA content in meristems declines and tubers acquire cytokinin sensitivity. An increase in cytokinin content slightly precedes or accompanies the onset of tuber sprouting. Sprout growth is accompanied by increases in GA and IAA contents. As currently envisioned, ABA and cytokinins are the principal regulators of tuber dormancy.

14.6 SPROUT GROWTH AND PHYSIOLOGICAL AGING

The rate and pattern of sprout growth (vigor) change with time (Krijthe, 1962). Immediately following the termination of dormancy, sprout growth rate is low, and this may reflect the lingering effects of dormancy-imposed growth inhibition. As storage is extended, the rate of sprout growth increases to a maximum, and typically, one sprout becomes dominant in an uncut tuber. At this stage, the inhibited buds are in a state of paradormancy or correlative inhibition.

From a biological standpoint, seed tuber longevity need not exceed 8–9 months, and during storage extending beyond this point, sprout number and vigor change dramatically.

With extended storage, initially correlative inhibition between eyes is lost and uncut tubers exhibit multiple sprouts. Later, apical dominance within an eye is weakened and individual eyes display multiple sprouts that branch profusely. In the final stage, sprout growth is supplanted by the formation of small tubers, a condition known as little tuber syndrome.

The entire process (from tuber formation to little tuber disorder) is known as physiological aging, and it has dramatic effects on subsequent plant growth and tuber yield (Van der Zaag and van Loon, 1987). Physiological aging of tubers is affected by both pre- and post-harvest conditions, most notably temperature (Jenkins et al., 1993). The application of GA dramatically increased the elongation of sprouts in tubers recently emerged from dormancy (Suttle, 2004b), and exogenous auxin restored apical dominance and reversed age-induced effects on subsequent plant field performance (Mikitzel and Knowles, 1990). The deteriorative changes seen with extended storage are accompanied by wholesale changes in tissue permeability, membrane lipid composition, tissue free radical content, and lipid peroxidation (Kumar and Knowles, 1993; Fauconnier et al., 2002; Zabrouskov and Knowles, 2002; Zabrouskov et al., 2002). Thus, the process of physiological aging can be divided into two phases: the initial changes in seed tuber performance and sprout growth, which appear to result from changes in hormone balance, and the latter changes leading to seed tuber deterioration and, ultimately, death, which resemble changes seen in other forms of plant senescence.

14.7 CONCLUSIONS

Despite tremendous progress in the past decade, the primary cognate processes controlling potato tuber dormancy remain an enigma. Much has been learned about the effects of tuber dormancy status on gene expression, protein levels, enzyme activities, and hormone content. However, the fundamental processes that control the transition between cell cycle arrest/metabolic depression and resumption of meristem growth and metabolic activity remain unknown. Unlike the situation in a number of other plant and non-plant systems whose life cycles include a period of metabolic depression and growth arrest, release from tuber dormancy is not dependent on specific environmental cue(s). Furthermore, the dormant tuber meristem is fully hydrated and metabolically competent. Thus, it is entirely possible that primary control of tuber meristem dormancy is metabolic in nature, with changes in gene expression secondary. In this scenario, dormancy exit would occur as a result of a progressive decline in one or more inhibitory factor(s) or the gradual accumulation of growth-promoting factor(s). Sprouting would occur once a specific threshold concentration was achieved. This hypothetical metabolic trigger could be a specific chemical entity (hormone, covalently modified protein/enzyme) or physiologic state such as redox potential (Claassens et al., 2005), energy charge (Bonhomme et al., 2000), intracellular pH (Robert et al., 1999), ion content, or membrane function (Aue et al., 2000; Rinne and van der Schoot, 2003).

The application of molecular genetics and biotechnology to dormancy research will undoubtedly reveal new insights into dormancy-control mechanisms. The selective over- and under-expression of candidate genes may greatly aid in this endeavor, but care must be taken in interpreting the results of these studies. The ectopic over-expression of genes has been shown to induce non-physiological responses in other tissues. Conversely, down-regulation of genes can lead to non-specific growth inhibition unrelated to dormancy-imposed arrest. Thus, pleiotropic effects of single-gene manipulations are common. For example, ectopic expression of a bacterial pyrophosphatase gene in potatoes results in dramatic increases in tuber pyrophosphatase activity, marked declines in pyrophosphate content, and either accelerated or completely inhibited sprouting behavior (cf. Hajirezaei and Sonnewald, 1999; Farré et al., 2001).

Regardless of the approaches taken, new information concerning the internal processes controlling potato tuber sprouting will result in improved storage technologies, enhanced market and nutritional value, and reduced post-harvest losses due to physiological and pathogen-induced deterioration.

REFERENCES

Abdala G., G. Castro, O. Miersch and D. Pearce, 2000, *Plant Growth Regul.* 36, 121.
Aue H.L., I. Lecomte and G. Pétel, 2000, *Biol. Plant.* 43, 25.
Auer C.A., 2002, *J. Plant Growth Regul.* 21, 24.
Bachem C., R. Van der Hoeven, J. Lucker, R. Oomen, E. Casarini, E. Jacobsen and R. Visser, 2000, *Potato Res.* 43, 297.
Bailey K.M., I.D.J. Phillips and D. Pitt, 1978, *Ann. Bot.* 42, 649.
Banas A., M. Bielińska-Czarnecka and J. Klocek, 1984, *Bull. Acad. Pol. Sci. Biol.* 23, 213.
Bender J., 2004, *Annu. Rev. Plant Biol.* 55, 31.
Bialek K. and M. Bielińska-Czarnecka, 1975, *Bull. Acad. Pol. Sci. Biol.* 23, 213.
Biemelt S., M. Hajirezaei, E. Hentschel and U. Sonnewald, 2000, *Potato Res.* 43, 371.
Blumenthal-Goldschmidt S. and L. Rappaport, 1965, *Plant Cell Physiol.* 6, 601.
Bonhomme M., R. Rageau and M. Gendraud, 2000, *Tree Physiol.* 20, 615.
Boo L., 1962, *Sven. Bot. Tidskr.* 56, 193.
Brian P.W., H.G. Hemming and M. Radley, 1955, *Physiol. Plant.* 8, 899.
Burton W.G., 1974, *Potato Res.* 17, 113.
Burton W.G., 1989, *The Potato*, 3rd edition, p. 470. Longman Scientific and Technical, Essex.
Burton W.B. and D.F. Meigh, 1971, *Potato Res.* 14, 96.
Campbell M.A., J.C. Suttle and T.W. Sell, 1996, *Physiol. Plant.* 98, 743.
Carrera E., J.L. Garcia-Martinez and S. Prat, 2000, *Plant J.* 22, 247.
Chatfield S.P., P. Stirnberg, B.G. Forde and O. Leyser, 2000, *Plant J.* 24, 159.
Claassens M.M.J., J. Verhees, L.H.W. van der Plas, A.R. van der Krol and D. Vreugdenhil, 2005, *J. Exp. Bot.* 56, 2515.
Claassens M.M.J. and D. Vreugdenhil, 2000, *Potato Res.* 43, 347.
Clouse S.D. and J.M. Sasse, 1998, *Annu. Rev. Plant Physiol. Plant Mol. Biol.* 49, 427.
Coleman W.K., 1987a, *Am. Potato J.*, 64, 57.
Coleman W.K., 1987b, *Potato Res.* 14, 96.
Coleman W.K., 1998, *Ann. Bot.* 82, 21.
Coleman W.K. and R.R. King, 1984, *Am. Potato J.* 61, 437.
Cornforth J.W., B.V. Milborrow and G. Ryback, 1966, *Nature (London)* 210, 627.
Creech D.L., M. Workman and M.D. Harrison, 1973, *Am. Potato J.* 64, 145.
Cubas P., N. Lauter, J. Doebley and E. Coen, 1999, *Plant J.* 18, 215.
Cvikrová M., L.S. Sukhova, J. Eder and N.P. Korableva, 1994, *Plant Physiol. Biochem.* 32, 685.
Davies P.J., 2004, In: P.J. Davies (ed.), *Plant Hormones. Biosynthesis, Signal Transduction, Action!*, p. 1. Kluwer Academic Publishers, Dordrecht.

De Veylder L., J. Joubes and D. Inzé, 2003, *Curr. Opin. Plant Biol.* 6, 536.
Del Pozo J.C., M.A. Lopez-Matas, E. Ramirez-Parra and C. Gutierrez, 2005, *Physiol. Plant.* 123, 173.
Deltour R., 1985, *J. Cell Sci.* 75, 43.
den Boer B.G. and J.A.H. Murray, 2000, *Trends Cell Biol.* 10, 245.
Désiré S., J.P. Couillerot, J.L. Hilbert and J. Vasseur, 1995, *Plant Physiol. Biochem.* 33, 479.
Destefano-Beltrán L., D. Knauber, L. Huckle and J.C. Suttle, 2006a, *Plant Mol. Biol.* 61, 687.
Destefano-Beltrán L., D. Knauber, L. Huckle and J.C. Suttle, 2006b, *J. Exp. Bot.*, 57, 2879.
Dewitte W. and J.A. Murray, 2003, *Annu. Rev. Plant Biol.* 54, 235.
Dobosy J.R. and E.U. Selker, 2001, *Cell. Mol. Life Sci.* 58, 721.
Duckham S.C., I.B. Taylor, R.S.T. Linforth, R.J. Al-Naieb, B.A. Marples and W.R. Bowman, 1989, *J. Exp. Bot.* 217, 901.
Ducreux L.J.M., W.L. Morris, P.E. Hedley, T. Shepherd, H.V. Davies, S. Millam and M.A. Taylor, 2005, *J. Exp. Bot.* 56, 81.
El-Antably H.M.M., P.F. Wareing and J. Hillman, 1967, *Planta* 73, 74.
Emilsson B., 1949, *Acta Agricul. Suec.* 3, 189–284.
Engelbrecht L. and M. Bielińska-Czarnecka, 1972, *Biochem. Physiol. Pflanzen.* 163, 499.
Espen L., S. Morgutti and S.M. Cocucci, 1999, *Potato Res.* 42, 203.
Ewing E.E., I. Simko, E.A. Omer and P.J. Davies, 2004, *Am. J. Potato Res.* 81, 281.
Faivre-Rampant O., G.J. Bryan, A.G. Roberts, D. Milbourne, R. Viola and M.A. Taylor, 2004b, *J. Exp. Bot.* 55, 951.
Faivre-Rampant O., L. Cardle, D. Marshall, R. Viola and M.A. Taylor, 2004a, *J. Exp. Bot.* 55, 613.
Farré E.M., A. Bachmann, L. Willmitzer and R.N. Trethewey, 2001, *Nat. Biotechnol.* 19, 268.
Fauconnier M.L., J. Rojas-Beltran, J. Delcarte, F. Dejaeghere, M. Marlier and P. du Jardin, 2002, *Plant Biol.* 4, 77.
Fauconnier M.L., R. Welti, E. Blée and M. Marlier, 2003, *Biochim. Biophys. Acta* 1633, 118.
Fladung M., 1990, *Plant Breed.* 104, 295.
Flewelling H.S., 1987, MS thesis, University of Wisconsin-Madison.
Francis D. and D.A. Sorrell, 2001, *Plant Growth Regul.* 33, 1.
Franklin J. and T. Hemberg, 1980, *Physiol. Plant.* 50, 227.
Freeman D., C. Riou-Khamlichi, E.A. Oakenfull and J.A.H. Murray, 2003, *J. Exp. Bot.* 54, 303.
Freyre R., S. Warnke, B. Sosinski and D.S. Douches, 1994, *Theor. Appl. Genet.* 89, 474.
Goodwin P.B., 1967, *J. Exp. Bot.* 18, 78.
Guppy M. and P. Withers, 1999, *Biol. Rev.* 74, 1.
Guthrie J.D., 1940, *Contrib. Boyce Thompson Inst.* 11, 29.
Haines M.M., P.J. Shiel, J.K. Fellman and P.H. Berger, 2003, *J. Agricul. Sci.* 141, 333.
Hajirezaei M. and U. Sonnewald, 1999, *Potato Res.* 42, 353.
Hand S.C. and I. Hardewig, 1996, *Annu. Rev. Physiol.* 58, 539.
Harms K., R. Atzorn, A. Brash, H. Kühn, C. Wasternack, L. Willmitzer and H. Peña-Cortés, 1995, *Plant Cell* 7, 1645.
Hawkes J.G., 1992, In: P. Harris (ed.), *The Potato Crop.* Chapman and Hall, London, p. 13.
Hedden P. and A.L. Phillips, 2000, *Trends Plant Sci.* 5, 523.
Hemberg T., 1942, *Sven. Bot. Tidskr.* 36, 467.
Hemberg T., 1949a, *Physiol. Plant.* 2, 24.
Hemberg T., 1949b, *Physiol. Plant.* 2, 37.
Hemberg T., 1970, *Physiol. Plant.* 23, 850.
Hemberg T. 1985, Potato rest. In: P.H. Li (ed.), *Potato Physiology.* Academic Press, New York, p. 353.
Henis Y., 1987, *Survival and Dormancy of Microorganisms.* John Wiley & Sons, New York, 335 pp.
Holst U.B., 1971, *Physiol. Plant.* 24, 392.
Inzé D., 2005, *EMBO J.* 24, 657.
Jenkins P.D., T.C. Gillison and A.D. Al-saidi, 1993, *Ann. Appl. Biol.* 122, 345.
Ji Z.L., Wang S.Y., 1988, *J. Plant Growth Regul.* 7, 37.
Jones M.G., R. Horgan and M.A. Hall, 1988, *Phytochemistry* 27, 7.
Koda Y., 1982, *Plant Cell Physiol.* 23, 851.
Korableva N.P., K.A. Karavaeva and L.V. Metlitskii, 1980, *Fiz. Rast.* 27, 585.
Korableva N.P., E.P. Ladyzhenskaya and L.V. Metlitskii, 1976, *Biokhimiya* 41, 971 (English translation).

Korableva N.P., T.A. Platonova, M.Z. Dogonadze and A.S. Evsunina, 2002, *Biol. Plant.* 45, 39.
Köves E., 1957, *Acta Univ. Szeged., Acta Biol.* 3, 179.
Krijthe N., 1962, *Eur. Potato J.* 5, 316.
Kumar G.N.M. and N.R. Knowles, 1993, *Plant Physiol.* 102, 115.
Ladyzhenskaya E.P. and M.A. Protsenko, 2002, *Biochemistry (Moscow)* 67, 151.
Lang G.A., G.C. Martin and R.L. Darnell, 1987, *HortScience* 22, 371.
Law R.D. and J.C. Suttle, 2002, *Plant Mol. Biol.* 51, 447.
Law R.D. and J.C. Suttle, 2004, *Physiol. Plant.* 120, 642.
Leshem B. and F.A.L. Clowes, 1972, *Ann. Bot.* 36, 687.
Li P.H., 1977, *Am. Potato J.* 54, 452.
Li G., T.C. Hall and R. Holmes-Davis, 2002, *Bioessays* 24, 234.
Loidl P., 2004, *Trends Plant Sci.* 6, 59.
Lulai E.C. and J.C. Suttle, 2004, *Postharv. Biol. Technol.* 34, 105.
MacDonald M.M. and D.J. Osborne, 1988, *Physiol. Plant.* 73, 392.
Martín M., J. León, C. Dammann, J. Albar, G. Griffiths and J.J. Sánchez-Serrano, 1999, *Eur. J. Biochem.* 262, 283.
Meigh D.F., A. Authur, E. Filmer and R. Self, 1973, *Phytochemistry* 12, 987.
Mikitzel L.J. and N.R. Knowles, 1990, *Plant Physiol.* 93, 967.
Milborrow B.V., 1967, *Planta* 76, 93.
Mok D.W.S. and M.C. Mok, 2001, *Annu. Rev. Plant Physiol. Mol. Biol.* 52, 89.
Murray J.A.H., J. Doonan, C. Riou-Khamlichi, M. Miejer and E.A. Oakenfull, 2001, In: D. Francis (ed.), *The Plant Cell Cycle and Its Interfaces.* CRC Press, Boca Raton, FL, p. 19.
Nambara E. and A. Marion-Poll, 2005, *Annu. Rev. Plant Biol.* 56, 165.
Okazawa Y. 1974, *J. Facul. Agricul. Hokkaido Univ.* 57, 443.
Ooms G. and J.R. Lenton, 1985, *Plant Mol. Biol.* 5, 205.
Osborne D.J., 1981, *Ann. Appl. Biol.* 98, 525.
Poapst P.A., A.B. Durkee, W.A. McGugan and F.B. Johnston, 1968, *J. Sci. Food Agric.* 19, 325.
Powell L.E., 1987, *HortScience* 22, 845.
Prange R.K., W. Kalt, B.J. Daniels-Lake, C.L. Liew and R.T. Page, 1998, *J. Am. Soc. Hort. Sci.* 123, 463.
Quarrie S.A., 1982, *Plant Cell Environ.* 5, 23.
Rama M.V. and P. Narasimham, 1982, *J. Food Sci. Technol.* 19, 144.
Rappaport L., H. Timm and L.F. Lippert, 1958, *Calif. Agric.* 12, 4.
Rappaport L. and N. Wolf, 1969, *Symp. Soc. Exp. Biol.* 23, 219.
Raz V., J.H.W. Bergervoet and M. Koornneef, 2001, *Development* 128, 243.
Reust W. and P. Gugerli, 1984, *Potato Res.* 27, 268.
Rinne P.L.H. and C. van der Schoot, 2003, *Can. J. Bot.* 81, 1182.
Robert F., M. Gendaraud and G. Pétel, 1999, *Plant Physiol. Biochem.* 37, 155.
Rohde A., M. van Montagu, D. Inzé and W. Boerjan, 1997, *Planta* 201, 43.
Römer S., J. Lübeck, F. Kauder, S. Steiger, C. Adomat and G. Sandman, 2002, *Met. Engineer.* 4, 263.
Ronning C.M., S.S. Stegalkina, R.A. Ascenzi, O. Bougri, A.L. Hart, T.R. Utterbach, S.E. Vanaken, S.B. Riedmuller, J.A. White, J. Cho, G.M. Pertea, Y. Lee, S. Karamycheva, R. Sultana, J. Tsai, J. Quakenbush, H.M. Giffiths, S. Retrepo, C.D. Smart, W.E. Fry, R. van der Hoeven, S. Tanksley, P. Zhang, H. Jin, M.L. Yamamoto, B.J. Baker and C.R. Buell, 2003, *Plant Physiol.* 131, 419.
Rosa J.T., 1925, *Potato Assoc. Am., Potato News Bull.* 2, 363.
Rylski I., L. Rappaport and H.K. Pratt, 1974, *Plant Physiol.* 53, 658.
Schmülling T., M. Fladung, K. Grossmann and J. Schell, 1993, *Plant J.* 3, 371.
Seo M. and T. Koshiba, 2002, *Trends Plant Sci.* 7, 41.
Šimko I., S. McMurry, H. Yang, A. Manschot, P.J. Davies and E.E. Ewing, 1997, *Plant Physiol.* 115, 1453
Simmons N.W., 1964, *Heredity (London)* 19, 489.
Smith O.E. and L. Rappaport, 1961, *Adv. Chem. Ser.* 28, 42.
Sorce C., R. Lorenzi, N. Ceccarelli and P. Ranalli, 2000, *Aust. J. Plant Physiol.* 27, 371.
Sorce C., A. Piaggesi, N. Ceccarelli and R. Lorenzi, 1996, *J. Plant Physiol.* 149, 548.
Spena A., E. Prinsen, M. Fladung, S.C. Schulze and H. Van Onkelen, 1991, *Mol. Gen. Genet.* 227, 205.
Stearns S.C., 1992, *The Evolution of Life Histories.* Oxford University Press, Oxford, 249 pp.

Sukhova L.S., I. Machackova, J. Eder, N.D. Bibik and N.P. Korableva, 1993, *Biol. Plant.* 35, 387.
Suttle J.C., 1995, *Physiol. Plant.* 95, 233.
Suttle J.C., 1998a, *Plant Physiol.* 118, 843.
Suttle J.C., 1998b, *Physiol. Plant.* 103, 59.
Suttle J.C., 2002, *Plant Growth Regul.* 35, 199.
Suttle J.C., 2003, *Am. J. Potato Res.* 80, 303.
Suttle J.C., 2004a, *Am. J. Potato Res.* 81, 253.
Suttle J.C., 2004b, *J. Plant Physiol.* 161, 157.
Suttle J.C. and G.M. Banowetz, 2000, *Physiol. Plant.* 109, 68.
Suttle J.C. and J.F. Hultstrand, 1994, *Plant Physiol.* 105, 891.
Thomas T.H., 1990, In: M. Kamínek, D.W.S. Mok and E. Zažimalová (eds), *Physiology and Biochemistry of Cytokinins in Plants*. SPB Academic Publishing, The Hague, p. 323.
Thornton N.C., 1933, *Contrib. Boyce Thompson Inst.* 5, 471.
Tovar P., R. Estrada, L. Schilde-Rentscher and J.H. Dodds, 1985, *C.I.P. Circul.* 13, 1.
Trindade L.M., B.M. Horvath, R. van Berloo and R.G.F. Visser, 2004, *Plant Sci.* 166, 423.
Tuan D.Y.H. and J. Bonner, 1964, *Plant Physiol.* 39, 768.
Turnbull C.G.N. and D.E. Hanke, 1985a, *Planta* 165, 359.
Turnbull C.G.N. and D.E. Hanke, 1985b, *Planta* 165, 365.
Van den Berg J.H., E.E. Ewing, R.L. Plaisted, S. McMurray and M.W. Bonierbale, 1996, *Theor. Appl. Genet.* 93, 317.
Van den Berg J.H., I. Simko, P.J. Davies, E.E. Ewing and A. Halinska, 1995, *J. Plant Physiol.* 146, 467.
Van den Berg J.H., P.C. Struik and E.E. Ewing, 1990, *Ann. Bot.* 66, 273.
Van den Berg J.H., D. Vreugdenhil, P.M. Ludford, L.L. Hillman and E.E. Ewing, 1991, *J. Plant Physiol.* 139, 86.
Van der Zaag D.E. and C.D. van Loon, 1987, *Potato Res.* 30, 451.
Van Ittersum M.K., F.C.B. Aben and C.J. Keijzer, 1992, *Potato Res.* 35, 249.
Van Staden J. and G.G. Dimalla, 1978, *J. Exp. Bot.* 29, 1077.
Wills R.B.H., M.A. Warton and K.K. Kim, 2003, *HortScience* 39, 136.
Wiltshire J.J.J. and A.H. Cobb, 1996, *Ann. Appl. Biol.* 129, 553.
Wu J. and M. Grunstein, 2000, *Trends Biochem. Sci.* 25, 619.
Wurr C.C.E. and E.J. Allen, 1976, *J. Agricul. Sci. (Cambridge)* 86, 221.
Yoshihara T., A. Omer, H. Koshino, S. Sakamura, Y. Kikuta and Y. Koda, 1989, *Agric. Biol. Chem.* 53, 2835.
Zabrouskov V. and N.R. Knowles, 2002, *Lipids* 37, 309.
Zabrouskov V., G.N.M. Kumar, J.P. Spychalla and N.R. Knowles, 2002, *Physiol. Plant.* 116, 172.

Chapter 15

Molecular Physiology of the Mineral Nutrition of the Potato

Marcel Bucher[1] and Jens Kossmann[2]

[1]*Federal Institute of Technology (ETH) Zuerich, Institute of Plant Sciences, Experimental Station Eschikon, Eschikon 33, 8315 Lindau, Switzerland*

[2]*Stellenbosch University, Institute of Plant Biotechnology, Department of Botany and Zoology, Private Bag X1, Matieland 7602, South Africa*

15.1 INTRODUCTION

Around 95% of the potato biomass is constituted of the elements carbon (C), hydrogen and oxygen that the plant absorbs through the assimilation of carbon dioxide and the uptake of water. The rest of the biomass is composed of a range of elements that have to be absorbed from the soil solution, where they are usually delivered through the application of fertilizers. The beneficial effect of adding mineral nutrients to soils to improve plant growth has been exploited in agriculture for more than 2000 years, but only in the nineteenth century was the scientific discipline of plant nutrition established by Justus von Liebig. Since then, large amounts of potash, superphosphate and inorganic nitrogen (N) have been applied to improve crop yield, and this practice contributes largely to the increase of crop productivity that was achieved in the past century. The excess use of fertilizers, however, also negatively impacts the environment, whereas developing countries usually cannot afford to provide sufficient nutrients to ensure food security. In addition, the production of N fertilizers through the Haber–Bosch procedure is extremely energy demanding, giving agricultural production systems usually a negative energy balance. For all these reasons, people are trying to breed or genetically engineer crops to optimize their nutrient acquisition and utilization efficiency. Especially some potato varieties are extremely nutrient inefficient due to their adaptation to high-input production systems (Sattelmacher et al., 1990). To achieve more nutrient-efficient plants, a general understanding of the molecular physiology of nutrient uptake, distribution and utilization is needed. In this chapter, these issues will be discussed with special emphasis on potato. An overview of plant nutrients, their occurrence and their functions is given in Table 15.1. They are grouped into macro- and micro-nutrients depending on the concentration that can be found in plant tissues. Focus will be given to our current knowledge on the uptake and utilization of the macro-nutrients N and phosphate, which are most often limiting potato yield.

Potato Biology and Biotechnology: Advances and Perspectives
D. Vreugdenhil (Editor)

Table 15.1 A summary of mineral nutrition in plants.

Element	Principal form in which element is absorbed	Usual concentration in healthy plants (% dry weight)	Important functions
Macronutrients			
Carbon	CO_2	~44%	Component of organic compounds
Oxygen	H_2O or O_2	~44%	Component of organic compounds
Hydrogen	H_2O	~6%	Component of organic compounds
Nitrogen	NO_3^- or NH_4^+	1–4%	Amino acids, proteins, nucleotides, nucleic acids, chlorophyll and coenzymes
Phosphorus	$H_2PO_4^-$ or HPO_4^{2-}	0.1–0.8%	Formation of 'high energy' phosphate compounds (ATP and ADP). Nucleic acids. Phosphorylation of sugars. Several essential enzymes. Phospholipids.
Sulfur	SO_4^{2-}	0.05–1%	Some amino acids and proteins. Coenzyme A Sulfolipids.
Potassium	K^+	0.5–6%	Enzymes, amino acids and protein synthesis. Activator of many enzymes. Opening and closing of stomata.
Calcium	Ca^{2+}	0.2–3.5%	Calcium of cell walls. Enzyme cofactor. Cell permeability.
Magnesium	Mg^{2+}	0.1–0.8%	Part of the chlorophyll molecule. Coenzyme A.
Micronutrients			
Iron	Fe^{2+} or Fe^{3+}	25–300 ppm	Chlorophyll synthesis, cytochromes and nitrogenase.
Chlorine	Cl^-	100–10 000 ppm	Osmosis and ionic balance; probably essential in photosynthetic reactions that produce oxygen.
Copper	Cu^{2-}	4–30 ppm	Activator of certain enzymes.
Manganese	Mn^{2-}	15–800 ppm	Activator of certain enzymes.
Zinc	Zn^{2+}	15–100 ppm	Activator of certain enzymes.
Molybdenum	MoO_4^{2-}	0.1–5.0 ppm	Nitrogen fixation. Nitrate reduction.
Boron	$BO3^-$ or B_4O7^{2-}	5–75 ppm	Influences Ca^{2+} utilization. Functions unknown cell wall stability.

15.2 NITROGEN

Nitrogen is the most abundant element after carbon, oxygen and hydrogen in biomass. The reason for this is that N is a constituent of amino acids, proteins, nucleotides and nucleic acids to name the most important molecules. Sub-optimal N supply immediately leads to growth retardation, whereas super-optimal supply leads to unwanted increases in the shoot/root ratio that might negatively impact the acquisition of other nutrients and water (Marschner, 1995). Fertilizer N is almost exclusively synthesized from N and hydrogen gas by the Haber–Bosch process that generates ammonia. Of the energy required to produce the three major fertilizer nutrients, 87% is consumed in ammonia synthesis. Fertilizer N represents the largest fossil-derived energy input for crop production in developed countries and represents one of the major barriers for energy crop production. However, if farmers stopped using fertilizer, crop yields would drop to a low level (Mengel, 1992).

The demand for fertilizer N could partially be covered by rotation or intercropping with N-fixing leguminous species (such as beans, clover, lupins or peas). Such sustainable cropping systems also are established for potato production, which could be optimized if it was possible to enhance the N acquisition and utilization efficiency of the potato varieties. Also potato plants are inefficient in their uptake and use of fertilizer N, with only about 50% of the applied N taken up.

The N not recovered by the crop may cause environmental problems such as pollution of ground water and run-off to surface waters.

15.2.1 Nitrogen uptake

Nitrogen fertilizers are supplied in three forms, ammonia, nitrate and urea. Urea usually is readily hydrolyzed by the ubiquitous enzyme urease to release ammonia. Uptake systems for all three forms of inorganic N have been described in higher plants (Williams and Miller, 2001; Liu et al., 2003). Most of the genes encoding such transporters have been biochemically characterized in the model organisms *Arabidopsis* and rice and, to a lesser extent, in tomato, but none of them in potato. However, because the uptake systems are in general quite similar between plant species, it can be assumed that similar systems are also operating in the potato, also given the close relationship of potato to tomato.

The nitrate uptake system of higher plants consists of a constitutive, low affinity transport system (LATS) and an inducible, high affinity transport system (HATS). The HATS are distinguished by whether they are substrate induced (iHATS) or constitutively active (cHATS) (Aslam et al., 1992). In *Arabidopsis* there are two genes encoding LATS, *AtNRT1.1* and *AtNRT1.2*, respectively. The *AtNRT1.1* gene (formerly called *CHL1*) was originally isolated by T-DNA tagging, but the mutant phenotype was described much earlier and had been selected by resistance to the toxic nitrate homologue chlorate (Doddema and Telkamp, 1979). Again in *Arabidopsis*, there are seven genes that encode proteins for HATS, designated *AtNRT2*.

Nitrite is also transported by two systems, of which the low-affinity system may play a greater role than in nitrate uptake. By uptake competition studies, the high-affinity system has been shown to be identical with that of nitrate transport, although not in all of the cases studied (Ullrich, 1992).

For ammonium again, HATS and LATS exist. Ammonium ions are taken up by a saturable, but apparently constitutive, carrier system with high substrate affinity, which may carry out ammonium uniport as long as H^+-ATPases establish a proton gradient. The low-affinity component of transport is stimulated by high external pH and probably reflects diffusion of uncharged NH_3 across the lipid phase of the plasmalemma. Three ammonium transporters have been identified in *Arabidopsis* roots: constitutive, diurnally regulated and starvation-induced (Gazzarrini et al., 1999).

Only recently a high-affinity uptake system for urea has been identified in *Arabidopsis* that was designated *AtDUR3* (Liu et al., 2003). This was identified based on homologies to a corresponding yeast gene *ScDUR3*, complementation of the corresponding yeast mutant and through expression in *Xenopus* oocytes, and was shown to be an urea/H^+ symporter.

15.2.2 Nitrogen assimilation

The nitrogenous compounds that are taken up from the soil solution need to be assimilated to enter metabolism. Ammonia is probably released from urea due to the action of urease, an enzyme that is present in sufficient amounts and not rate-limiting in the process of urea assimilation (Witte et al., 2002). Ammonia will then be incorporated into organic compounds by a range of enzymes (Section 15.2.2.2). Nitrate first needs to be reduced to form ammonia by the consecutive action of nitrate and nitrite reductases (NiRs).

15.2.2.1 Nitrate reduction

Nitrate reductase (NR) catalyzes the first reaction in nitrate assimilation, the reduction of nitrate to nitrite. Nitrate reductase requires molybdenum (Mo) as cofactor. Nitrate reductase in higher plants is proposed to be a homodimer, with two identical subunits joined and held together by the Mo cofactor (Kleinhofs et al., 1989).

Two cDNAs representing two alleles of the NR from potato have been described (Harris et al., 2000), which closely resemble the respective *Arabidopsis* encoding sequences and the ones from other plant species.

Ferredoxin-nitrite reductase (Fd-NiR) subsequently catalyzes the six-electron reduction of nitrite to ammonia, using reduced Fd as the electron donor. The Fd-dependent NiRs of plants, algae and cyanobacteria are monomeric proteins with molecular masses near 63 kDa which contain a single [4Fe-4S] cluster and a single siroheme (which serves as the binding site for nitrite) as prosthetic groups (Dose et al., 1997). Ferredoxin-NiR has been cloned from several higher plant species; however, no corresponding sequence so far has been isolated from potato.

In higher plants, nitrate reduction is highly regulated. A range of environmental factors influence the expression of the corresponding genes as well as the enzyme activity levels. NR activity expression and activity is controlled by light, temperature, pH, CO_2, O_2, water potential and N source. Drought for instance causes increased NR protein turnover and accelerated mRNA turnover (Ferrario-Mery et al., 1998; Foyer et al., 1998).

Spinach leaf NR undergoes a reversible phosphorylation in response to light/dark transitions, which leads to an inactivation of the enzyme (Huber et al., 1992). The low-activity, phosphorylated form of NR from darkened leaves of spinach is activated during purification, because it is separated from a approximately 110-kDa NR inhibitory protein (NIP). Re-addition of NIP inactivated the purified phosphorylated NR, but not the active dephosphorylated form of NR, indicating that the inactivation of NR requires both, phosphorylation and interaction with NIP. In addition, NR that had been inactivated in vitro through phosphorylation, and interaction with NIP could be reactivated either by dephosphorylation or by dissociation of NIP from NR (MacKintosh et al., 1995).

NIP has been shown to be a member of the family of 14-3-3 proteins. 14-3-3 proteins are chaperone proteins that modulate interactions between components of signal transduction cascades and enzymes, which finally results in the activation or inactivation of the interacting proteins (Aitken, 1996; Wu et al., 1997). 14-3-3-protein-binding proteins in plants amongst others also include NR and sucrose-phosphate synthase (SPS). Extensive work has been performed in potato to modulate the expression level of different isoforms of the 14-3-3 proteins (Zuk et al., 2003, 2005). Transgenic potato plants with lowered levels of five isoforms of the 14-3-3 proteins show increased NR as well as SPS activities, which ultimately leads to an increased productivity of the plants as they accumulated more starch in their tubers in greenhouse experiments. These experiments were done using a constitutive promoter. The authors therefore speculate that the increases in starch accumulation are a result of enhanced sucrose supply from the source leaves.

These results are in agreement with observations made with transgenic tobacco plants carrying an NR with an N-terminal deletion and hence an enzyme that is not subjected to phosphorylation, which results in the abolishment of post-transcriptional regulation of NR by light and higher NR activities (Nussaume et al., 1995). This could be due to a difference in dissociation of the NR–NIP complex (Lillo et al., 1997) or a different way of binding for 14-3-3 in the truncated NR (Provan et al., 2000). The truncated and deregulated form of NR also was over-expressed in transgenic potato plants. In agreement with the experiments where the 14-3-3 protein expression was reduced, under some environmental conditions the plants showed an increased productivity (Djennane et al., 2002, 2004). The only change that was seen, however, was a reduction of the nitrate contents of the transgenic potato tubers, which might support the speculation that rather the higher activation status of SPS is the reason for the enhanced productivity of plants with reduced 14-3-3 protein levels.

Our understanding of the regulation of the close coordination of N and carbon metabolism is only now emerging. Metabolite levels are often determined in order to provide information about responses to physiological or environmental changes. In the past, this has been applied to isolated aspects of metabolism, but today methodology is being developed that allows measuring increasing amounts of different metabolites simultaneously. In an example where a group of isolated metabolites were analyzed, e.g. in tobacco transformants with very low NR activity, it has been shown that nitrate acts as a signal to induce organic acid metabolism and repress starch metabolism (Scheible et al., 1997). These transformants show changes similar to those seen in nitrate-limited wildtype plants except that they accumulate large amounts of nitrate. The high levels of nitrate induce genes that encode NR, NiR, a cytosolic glutamate synthase (GS) (glutamine: 2-oxoglutarate amidotransferase or Fd-GOGAT), phosphoenolpyruvate carboxylase, cytosolic pyruvate kinase, citrate synthase and NADP-isocitrate dehydrogenase, whereas ADP-glucose pyrophosphorylase transcript abundance and activity decreases. In the plants with reduced NR activity, nitrate probably acts as a signal to initiate coordinated changes in carbon and N metabolism (Scheible et al., 1997). These kind of studies are now extended to a broader range of metabolites, transcripts and ideally also proteins (Stitt and Fernie, 2003; Usadel et al., 2005), mainly in the model *Arabidopsis*, but increasingly also in other species as the functional genomics toolboxes become available. In a comparison of potato tubers from several genotypes altered in the import or degradation of sucrose, many relationships between metabolites in carbon and N metabolism were uncovered (Roessner et al., 2001a,b). Some of the relations were expected, because they occurred between close neighbours in metabolic pathways. Other relations, however, were between less-related groups of metabolites or occurred only in certain genotypes indicating novel regulatory mechanisms. These techniques should now also be applied to transgenic potatoes with modifications in N metabolism, e.g. the plants with enhanced NR activity. It will be exciting to see how this will further our understanding of the interrelation of carbon and N metabolism.

Interestingly, it was recently shown that potato tubers are autotrophic for N under conditions where high nitrate levels are supplied (Mäck and Schjoerring, 2002; Lin et al., 2004). The tubers can take up nitrate through the skin and have the full capacity to reduce nitrate and assimilate ammonia for the biosynthesis of all amino acids and do not rely on the import of amino acids from the phloem sap.

15.2.2.2 Ammonia assimilation and recycling

The ammonia derived from direct uptake or through reduction of nitrate needs to be incorporated into organic compounds for further metabolism. The primary pathway of ammonia assimilation in higher plants occurs through the GS-GOGAT cycle that operates in leaves. Only minor portions of ammonia are assimilated through glutamate dehydrogenase (GDH). The reactions are the following:

GS-GOGAT

$$NH_3 + \text{glutamate} + ATP \longrightarrow \text{glutamine} + ADP + P_i$$

$$\text{glutamine} + \text{2-oxoglutarate} + NADPH + H^+ \ (\text{ferredoxin}) \longrightarrow \text{glutamate} + NADP^+$$

GDH

$$NH_3 + \text{2-oxoglutarate} + NADPH + H^+ \longleftrightarrow \text{glutamate} + NADP^+$$

In the GS-GOGAT pathway, glutamate acts as the acceptor for ammonia and glutamine is formed by GS. Glutamate synthase has a very high affinity to ammonia and is therefore able to incorporate ammonia into organic molecules also if it is present in very low concentrations. In a reductive transamination, one amino group from glutamine is then transferred to 2-oxoglutarate to form glutamate by the enzyme GOGAT. This process is more energy consuming than the reaction catalyzed by GDH, where ammonia is incorporated into 2-oxoglutarate, as an extra molecule of ATP is needed. However, the GS-GOGAT cycle is the major pathway in leaves, where energy balances do not play a major role in efficiently coordinating metabolic pathways. The high affinity of GS towards ammonia prevents the accumulation of ammonia to toxic concentrations, which could lead to an uncoupling of oxidative photophosphorylation and hence to the production of reactive oxygen species.

Higher plant GS is an octameric enzyme, as in yeast, but is apparently not regulated by dissociation/association (Stewart et al., 1980). The subunit size is approximately 40 kDa for all isoenzymes. Different isoenzymes can be resolved by isoelectric focusing and/or ion exchange chromatography. In leaves there are two forms of GS, a cytosolic form (GS1) and a chloroplastic form (GS2). Glutamate synthase 2 contains two additional cysteine residues per subunit – this may account for the higher susceptibility of this isoenzyme to sulfhydryl reagents (Stewart et al., 1980). Sequences of the chloroplastic GS genes indicate that they contain cysteines in the putative ATP and substrate binding sites, whereas in the cytosolic forms these positions are occupied by alanine residues (Baima et al., 1989).

Glutamate synthase from higher plant sources is subject to cumulative feedback inhibition by amino acids and nucleotides (e.g. ADP and AMP are inhibitory) (Stewart and Rhodes, 1977).

In potato, there is one gene encoding plastidic GS2 and two encoding cytosolic GS1 (*Stgs1a* and *Stgs1b*) that are differentially expressed at the cellular and organ level (Teixeira et al., 2005). Potato GS-encoding genes seem to be regulated at the transcriptional or mRNA stability level. Glutamate synthase 2 is expressed in leaves and decreases when the leaves are senescing, which is also true for *Stgs1a*. Glutamate synthase 1 transcripts and proteins are the only forms of GS present in non-photosynthetic tissues and senescing leaves (*Stgs1b*). Both of the GS1 proteins are also found in vascular tissues. *Stgs1a* is expressed only in the phloem companion cells, whereas *Stgs1b* is expressed in both phloem and xylem. These expression patterns indicate that GS2 is probably involved in ammonia assimilation in leaves, whereas GS1 fulfills this function in non-photosynthetic tissues and probably is mainly involved in the mobilization of N from leaves to other tissues.

Mutants of barley have been obtained which are deficient in GS2, the chloroplastic form of GS. In barley, GS1 is capable of maintaining normal growth under non-photorespiratory conditions. Mutants of barley lacking GS2 accumulate high levels of ammonia in the leaves when placed in air (Lea et al., 1989). Similar studies have not been undertaken with potato, but they could be simulated with compounds that are inhibitors of GS, e.g. methionine sulfoximine or phosphinothricin.

No genes encoding GOGAT have been described for potato. Two ESTs, which probably represent the identical allele, were found encoding NADH-GOGAT and one encoding Fd-GOGAT in tomato, but no further analysis has been performed so far. It seems likely also that *Solanaceous* species have similar genes for GOGAT as *Arabidopsis*, where two genes for Fd-GOGAT and one for NADH-GOGAT are present (Lam et al., 1995). Mutants of *Arabidopsis* defective in GOGAT exhibit markedly impaired ammonia assimilation, especially under photorespiratory conditions (Lea et al., 1992).

Again, no genes encoding GDH have been described for potato. One EST has been described from tomato. However, again it is probable that the situation for the genetic structure of the GDH-encoding genes in potato resemble the situation in *Arabidopsis*. Two distinct genes (designated *GDH1* and *GDH2*) encoding NADH-GDH have been identified in *Arabidopsis*. Both gene products contain putative mitochondrial transit polypeptides and NADH- and 2-oxoglutarate-binding domains. Subcellular fractionation confirms the mitochondrial location of the NADH-GDH isoenzymes. Glutamate dehydrogenase 1 encodes a 43.0-kDa polypeptide, designated alpha, and GDH2 encodes a 42.5-kDa polypeptide, designated beta. The two subunits combine in different ratios to form seven NADH-GDH isoenzymes (Turano et al., 1997). In wildtype *Arabidopsis*, GDH1 mRNA accumulates to high levels in dark-adapted or sucrose-starved plants; light or sucrose treatment each repress GDH1 mRNA accumulation. These results suggest that the GDH1 gene product functions in the direction of glutamate catabolism under carbon-limiting conditions. Low levels of GDH1 mRNA present in leaves of light-grown plants can be induced by exogenously supplied ammonia. Under such conditions of carbon and ammonia excess, GDH1 may function in the direction of glutamate biosynthesis. The recessive *Arabidopsis* GDH-deficient *gdh1-1* mutant displays only a conditional phenotype. Seedling growth is specifically retarded on media containing exogenously supplied inorganic N, suggesting that GDH1 plays a non-redundant role in ammonia assimilation under conditions of inorganic N excess (Melo-Oliveira et al., 1996). As in *Arabidopsis*, GDH in potato is localized to the mitochondria and seemingly fulfils similar functions in providing carbon skeletons under carbon-limiting conditions (Aubert et al., 2001).

15.2.3 Transport of organic N between source and sink

As N is usually limiting plant growth, it is a stringent necessity that, after assimilation, N is made available to the organs that are developing or that are deficient for N and therefore not able to fulfil their physiological function. Amino acids seem to represent the principal long-distance transport form for organic N in higher plants. Proteogenic amino acids accumulate to concentrations of 100–200 mM in the phloem sap and to approximately 10-fold less concentrations in the xylem, where amines and acidic amino acids predominate. Organic N transport in plants is highly complex, because the transport substrates are extremely diverse, encompassing the 20 proteogenic amino acids, their analogues GABA and oligopeptides, amines, as well as many uncharacterized N-containing compounds that are found in the phloem sap. Physiological studies using whole tissues, individual cells or plasma membrane vesicles indicated that amino acid transport is mediated by carriers with overlapping specificity, coupling electrochemical potential to active accumulation of amino acids (Lalonde et al., 2004).

As N metabolism is compartmentalized in cells, many transporter genes are needed. cDNAs encoding amino acid transporters for cellular import were originally isolated through functional expression in yeast uptake mutants (Frommer et al., 1993; Kwart et al., 1993). Excluding mitochondrial transporters, the *Arabidopsis* and rice genomes each encode more than 50 putative amino acid transporters as based on sequence similarity. For most of these genes, a precise function still needs to be established. By sequence similarities, they are grouped into at least four superfamilies: (a) the amino acid-polyamine-choline transporter superfamily (APC), (b) the amino acid transporter superfamily 1 (ATF1), (c) the amino acid transporters belonging to the major facilitator superfamily (MFS) and (d) members of the mitochondrial carrier (MC) family (Lalonde et al., 2004).

The only organic N transporter from potato that has been characterized on the molecular level, StAAP1, falls into the superfamily ATF1. Amino acid transporter superfamily 1 members contain 9–11 putative membrane-spanning domains with cytosolic N and extracellular C termini (Chang and Bush, 1997). The members of the best-studied amino acid permeases are preferentially expressed in vascular tissues and mediate Na^+-independent, proton-coupled uptake of a wide spectrum of amino acids (Fischer et al., 2002). The role for StAAP1 in mediating amino acid transport between organs is supported by the fact that potato tubers from plants with reduced expression of the transporter accumulate lowered levels of all free amino acids except aspartate (Koch et al., 2003). The reverse experiment, to overexpress this type of transporter to enhance amino acid levels in potato tubers, has so far not been undertaken. However, it seems more likely to achieve the enhancement of the amino acid composition of potato tubers by engineering specifically for the de novo synthesis of those essential amino acids that are not present in optimal concentrations.

15.3 PHOSPHORUS

Phosphorus (P) is an essential macro-nutrient for growth and development in all living organisms. It serves various basic biological functions as a structural element, in energy metabolism, in the activation of metabolic intermediates, in signal transduction and in the regulation of enzymes. The prevalent form of available P in the soil is the oxidized anion phosphate. Although the total amount of P in the soil may be high, the preferred form for assimilation, orthophosphate (P_i), is often not readily available, because of several reasons. P_i can be adsorbed to soil particles and clay minerals and surfaces of calcium (Ca) and magnesium (Mg) carbonates, or it is converted to organically bound forms or insoluble precipitates with common cations like iron (Fe), aluminium (Al) and Ca (Welp et al., 1983; Holford, 1997). More soluble minerals such as N move through the soil through bulk flow and diffusion, whereas P_i is moved mainly by diffusion in the soil solution. As the rate of diffusion of P_i in soils is slow, high plant P_i uptake rates create a P_i depletion zone around the root (Jungk, 1991; Marschner, 1995). Therefore, the low availability of P_i in the bulk soil affects its uptake into roots.

Furthermore, the way in which P_i is metabolized in storage organs is a key determinant for the accumulation of storage carbohydrates and the quality of the harvested organ. Therefore, P_i efficiency is of crucial importance for the growth and development of plants and a key component of the genetic yield potential of crops.

P_i efficiency can be divided into two different parameters, the P_i acquisition (PAE) and the P_i utilization efficiency (PUE). The PAE is characterized by the unit of P absorbed per unit of root length and relates to the different mechanisms by which plants are able to release P_i from insoluble fractions and to incorporate soluble forms available in the soil solution. The PAE is of major importance for plants growing on soils containing only low levels of P_i, which is the case for most of the soils under agricultural exploitation. But even in soils which are rich in P_i, e.g. in central Europe, P_i is not always available in sufficient amounts in the soil solution, due to the fact that most of it is present in an immobile form. The PUE, i.e. the unit of P per unit of dry matter, relates to the distribution of P_i to different organs, cells and subcellular compartments as well as the metabolism of P_i into different intermediate pools. The PUE must be optimal at the organismic, cellular and subcellular level for crop yield to be maximized. A plant that is supplied with sufficient P_i may still suffer from what is known as P_i limitation of photosynthesis, due to the fact that most of the P_i present in photosynthetically active cells is sequestered in the vacuole and not metabolically available during short-term limitations (Mimura, 1995, 1999). Besides allocation within the plant, another key process involved in PUE is the deficiency-induced remobilization of P_i from older leaves, critical for plants to overcome periods of malnutrition. A large variability of PAE and PUE has been observed between plant species and within varieties of the same species indicating that P efficiency is under genetic control (Sattelmacher et al., 1994; Marschner, 1995; Horst et al., 1996; Fageria and Baligar, 1997). In contrast to the crucial importance of P_i for plant growth and development, little is known about the underlying mechanisms determining P_i efficiency. However, improving the PUE of plants by breeding or genetic engineering strategies would require not only an appropriate knowledge on genes coding for P transport within the plant but also on genes controlling photosynthate allocation, growth, senescence, etc. This makes the PUE far more complicated and difficult to fully understand as opposed to the PAE of which key genes have meanwhile been identified. This chapter's focus is on a basic process involved in both PAE and PUE, i.e. the molecular and biochemical processes involved in P_i transport across membranes.

15.3.1 Phosphate uptake

Plants have developed numerous morphological, physiological, biochemical and molecular responses to cope with P_i limiting conditions, including changes in root morphology and architecture, accumulation of anthocyanins, secretion of phosphomonoesterases and organic acids into the rhizosphere, improved P_i uptake efficiency and changes in metabolism (Raghothama, 1999). The consequence of these adaptive changes is increased P_i availability in the rhizosphere, enhanced P_i uptake and maintenance of plant metabolism. To optimally exploit the soil P_i, plants respond to low P conditions with an increased root–shoot ratio and an enlargement of the absorptive surface area relative to the root volume (Marschner, 1995). The increase in root surface area can be attributed in part to the enhanced production of root hairs, their length being especially important for P_i uptake (Bates and Lynch, 1996; Jungk, 2001). Alternatively, arbuscular mycorrhizal (AM) associations between symbiotic soil fungi from the phylum Glomeromycota and plant roots have to be taken into consideration. In these associations, which are also established in potato plants (Bhattarai and Mishra, 1984; McArthur and Knowles, 1993), the absorbing surface area of the colonized roots is extended due to the large extraradical fungal mycelium exploiting soil volumes which are otherwise not accessible to the root (Smith and Read, 1997). Subsequently, available P_i is rapidly absorbed by an efficient uptake and transport system.

Among other mineral nutrients, plants are able to selectively and actively accumulate P_i in their cells. The transport of P_i across cellular and intracellular membranes is mediated by membrane-spanning transporter proteins with specific binding sites. P_i allocation on the cellular, tissue and organ level occurs through short- and long-distance transport including several membrane transport steps.

15.3.2 Molecular biological analysis of P_i transport systems

15.3.2.1 The Pht1 family of plant P_i transporters

Molecular biological studies in the past decade eventually made it possible to identify and investigate, similar to N-transport systems, membrane proteins exhibiting typical features of transport systems involved in P_i uptake from the soil solution.

The isolation of the first cDNA clones and genes encoding P_i transporters (P_iTs) from vascular plants was based on sequence information of an *Arabidopsis thaliana* expressed sequence tag clone (Muchhal et al., 1996; Leggewie et al., 1997; Mitsukawa et al., 1997; Smith et al., 1997; Daram et al., 1998; Liu et al., 1998b), which became available in 1995 from the Arabidopsis Biological Resource Center (Remy et al., 1994) and exhibited homology to the yeast high-affinity P_iT PHO84 (Bun-Ya et al., 1991). The Pht1 family presently consists of >80 proteins from both monocot and dicot species including *A. thaliana, Solanum tuberosum, Lycopersicon esculentum, Nicotiana tabacum, Medicago truncatula, Oryza sativa, Hordeum vulgare* (Karandashov and Bucher, 2005) and several others. Plant Pht1 transporters are integral membrane proteins predicted to contain 12 membrane-spanning domains separated into two groups of six by a large charged hydrophilic region (Muchhal et al., 1996). They belong to the phosphate: H^+ symporter (PHS) family within the MFS. The MFS consists of at least 29 distinct families present in bacteria, archaea and eukarya, each of which generally transports a single class of compounds (Saier et al., 1999). The three-dimensional structures of three MFS members, i.e. the oxalate (OxlT), lactose (LacY) and glycerol-3-phosphate/inorganic phosphate (GlpT) transporters from *Oxalobacter formigens* and *Escherichia coli*, respectively, have been elucidated; these demonstrate the presence of 12 transmembrane domains, substantiating the structural model of MFS proteins (Hirai et al., 2002; Abramson et al., 2003; Huang et al., 2003). Although it can be assumed that Pht1 proteins are structurally similar to these three well-characterized MFS proteins, experimental evidence clarifying the structure of Pht1 proteins is still lacking. Multiple alignments of Pht1 and homologous non-plant transporters have revealed the presence of several highly conserved sites for post-translational modification, as well as a highly conserved region, the Pht1 signature GGDYPLSATIxSE, in the fourth putative transmembrane domain (Karandashov and Bucher, 2005). Elucidating the function of these sequences is a future challenge. Functional complementation of different yeast mutants and overexpression of Pht1 proteins in plant cells have demonstrated great variability in affinity for P_i ranging from 3 to $\sim$700 μM. This also supports the presence of both high- and low-affinity P_iTs within the Pht1 family in plants (Leggewie et al., 1997; Harrison et al., 2002; Rausch and Bucher, 2002; Rae et al., 2003; Nagy et al., 2005).

Expression analysis and the proposed function of *Pht1* genes in different plant species have been published. Overall, the expression patterns of the respective genes in different organs suggest involvement of Pht1 proteins in P_i uptake at the root periphery from the soil solution and subsequent allocation to sink tissues in the plant (Schachtman et al., 1998; Raghothama, 1999, 2000; Smith et al., 2000; Bucher et al., 2001; Poirier and Bucher, 2002; Rausch and Bucher, 2002; Karandashov and Bucher, 2005). Clear evidence for a role of *Pht1* genes in P_i uptake was provided by detailed analysis of loss-of-function mutations in the model plant *A. thaliana* where mutations in Pht1;1 and Pht1;4 alone or in combination resulted in strongly reduced P_i uptake rates at both low and high P conditions (Misson et al., 2004; Shin et al., 2004). While the genomes of *A. thaliana* and rice contain 9 and 13 *Pht1* genes, respectively (Mudge et al., 2002; Paszkowski et al., 2002), five members of the Pht1 family have been identified so far in potato and tomato (Leggewie et al., 1997; Daram et al., 1998; Liu et al., 1998a; Nagy et al., 2005). The respective orthologous pairs of P_iTs LePT1 and StPT1, and LePT2 and StPT2, are predominantly expressed in the rhizodermis including root hairs, the former rather constitutively, the latter inducible by P_i deprivation. While mRNA of the respective orthologous pairs of P_iTs LePT1 and StPT1 was detectable in both root and shoot tissues (and transcript levels increased slightly in both tissues upon P_i starvation), LePT2 and StPT2 mRNA was exclusively detectable in roots, and its concentration was strongly enhanced during P_i deprivation (Daram et al., 1998; Liu et al., 1998a; Gordon-Weeks et al., 2003; Zimmermann et al., 2003). After resupply of P_i in the medium, transcript abundance decreased again to control levels (Liu et al., 1998a; Zimmermann et al., 2003). Using anti-LePT1 antibodies, Muchhal and Raghothama (1999) demonstrated that the fluctuations in *LePT1* transcripts under changing P_i supply to the plant were concurrent with the levels of the transporter protein. These data strongly suggest a predominantly transcriptional regulation of PAE, although post-translational modification of P_i transport activity cannot be excluded. In situ hybridization studies localized the *LePT1* transcript to rhizodermal cells including root hairs and root cap cells (Daram et al., 1998), and mRNA signals were also detected in leaf palisade parenchyma cells (Liu et al., 1998a). *LePT1* transcripts and protein accumulated along the entire root, and the respective protein was detected primarily in the plasma membrane (Daram et al., 1998; Muchhal and Raghothama, 1999). *LePT2* transcripts were primarily observed in root epidermal cells under P_i-limiting conditions (Liu et al., 1998a). Similarly, strong accumulation of StPT2 mRNA was shown to occur in root hairs of P_i-deprived potato roots (Zimmermann et al., 2003). Moreover, StPT2 protein localization in root transverse sections showed that StPT2 labelling is confined to the apical plasma membrane of epidermal cells indicating cellular polarity of P_i transport (Gordon-Weeks et al., 2003). Thus, similar to the situation in *Arabidopsis*, the expression pattern and localization of the two P_iTs from tomato and potato support a primary role of these Pht1 transporters in PAE at the root–soil interface. Enhanced P_iT gene expression under P_i-limiting conditions leads to an increased number of transport sites and thus most likely to an increase in V_{max} of P_i uptake per root area, which is in agreement with data obtained in physiological experiments.

15.3.2.2 P_i transport in mycorrhizas

A widespread form of adaptation to low P soil conditions is the formation of symbiotic associations with mycorrhizal fungi. Eighty per cent of vascular plants undergo interactions with AM fungi (Smith and Read, 1997). These are often mutualistic interactions, in which the plant acquires P_i and other nutrients from the fungus in exchange for photosynthetic carbohydrates. This symbiosis is especially effective for plant growth during P_i-limited conditions and is thought to have facilitated the colonization of terrestrial ecosystems by plants about 400 million years ago (Remy et al., 1994). In the years 1996–2006, the use of molecular physiological and genetic tools has greately promoted our understanding of the molecular and biochemical processes involved in PAE at the symbiotic root–fungus interface (Rausch et al., 2001; Harrison, 2005; Oldroyd et al., 2005). While root hairs represent important cellular structures involved in the 'direct P_i uptake pathway', the 'mycorrhizal P_i uptake pathway' expands from extraradical fungal hyphae to the root cortex cells of a mycorrhiza harbouring fungal structures, i.e. arbuscules and hyphal coils. It was demonstrated that the mycorrhizal P_i uptake pathway can dominate P_i supply to plants irrespective of whether colonized plants exhibited improved growth and/or total P uptake (Smith et al., 2003; Smith et al., 2004). Detailed expression studies performed with mycorrhized plants of different species allowed the identification of several AM fungus-inducible plant P_iT genes from the Pht1 family in potato (Rausch et al., 2001; Nagy et al., 2005), tomato (Nagy et al., 2005), *M. truncatula* (Harrison et al., 2002) and the cereals such as rice (Paszkowski et al., 2002), maize, wheat and barley (Glassop et al., 2005; Nagy et al., 2006). The cloning and characterization of the P_iT StPT3 from potato provided the first molecular and biochemical evidence for plant P_i uptake at the AM fungus–root interface in mycorrhizas (Rausch et al., 2001). Sequence analysis revealed the close relationship between the StPT3 protein and other members of the plant Pht1 family and high-affinity P_iTs of fungal origin. Biochemical evidence for the function of StPT3 was obtained by complementation of a yeast mutant defective in the two high-affinity P_iT genes *PHO84* and *PHO89*. P_i uptake mediated by StPT3 in the mutant yeast cells followed Michaelis–Menten kinetics, exhibiting an apparent K_m value of 64 μM (Rausch et al., 2001). In accordance with previous work (Daram et al., 1998), it was concluded that StPT3 functions as a high-affinity P_iT at the plasma membrane. *StPT3* expression was shown to be locally induced upon colonization by the AM fungus *Glomus intraradices*, and mRNA levels correlated with arbuscule formation in the roots. Additionally, in situ hybridization studies resulted in StPT3 mRNA labelling in arbuscule-containing cells, which is consistent with the presumed function of StPT3 as a mycorrhiza-inducible P_iT localized at the peri-arbuscular membrane. Subsequently, a second AM fungus-inducible P_iT, StPT4, has been described from potato (Karandashov et al., 2004). Orthologues of StPT4 were also found in other species, thus indicating the presence of two non-orthologous mycorrhiza-responsive P_i transport systems (Karandashov and Bucher, 2005). The cloned promoter regions from *StPT4* and its tomato orthologue *LePT4* exhibited a high degree of sequence identity and were shown to direct expression exclusively in colonized cells when fused to the GUS reporter gene, in accordance with the abundance of *LePT4* and *StPT4* transcripts in mycorrhized roots (Nagy et al., 2005). Furthermore, extensive sequencing of *StPT4*-like clones and subsequent expression analysis in both potato and tomato revealed the presence of a close paralogue to *StPT4* and *LePT4*, named *StPT5* and *LePT5*, representing a third P_i transport system in *Solanaceous* species which is upregulated upon AM fungal colonization of roots. The existence of the latter two genes can be explained by genome duplication. Knock-out of *LePT4* in the tomato variety MicroTom indicated considerable redundancy between LePT4 and presumably LePT5 and/or LePT3 (Nagy et al., 2005). Similar to StPT4/LePT4, the *M. truncatula* P_iT MtPT4 is expressed only in mycorrhizal roots, and the *MtPT4* gene promoter directs expression exclusively in cells containing arbuscules. MtPT4 protein is located in the membrane fraction of mycorrhizal roots, and immunolocalization revealed that MtPT4 colocalizes with the arbuscules, consistent with a location on the peri-arbuscular membrane. The transport properties and spatial expression patterns of MtPT4 are consistent with a role in the acquisition of P_i released by the fungus in the AM symbiosis (Harrison et al., 2002). Recently, experimental evidence was provided for a mycorrhizal P_i transport function of the *Lotus japonicus* P_iT LjPT3 (Maeda et al., 2006) that is orthologous to StPT3 from potato (Rausch et al., 2001). At a meeting on mycorrhizas in March 2006 on the Monte Verità (Ticino, Switzerland), it was reported that loss-of-function mutations in MtPT4 resulted in reduced shoot P_i content and premature degeneration of fungal structures in the mycorrhiza (Brachmann, 2006; Harrison et al., personal communication), a phenotype which was also described in the work on LjPT3 published by Maeda et al. (2006). This is strong evidence for an important role of mycorrhizal Pht1 P_iTs in the regulation of mycorrhiza development and functioning.

15.3.3 P_i translocation on the whole plant level: long-distance transport

After uptake of P_i in the root symplasm through rhizodermal cells, including root hairs (Drew et al., 1984; Sentenac and Grignon, 1985; Ullrich and Novacky, 1990), P_i is transported symplasmically to the xylem parenchyma cells and subsequently secreted into the xylem (apoplast) for long-distance translocation to the above-ground organs (Jeschke et al., 1997). Xylem transport is driven by the gradient in hydrostatic pressure and by the gradient in the water potential. The gradient in water potential between roots and shoots is particularly steep during the day when the stomata are open, and solute flow in the xylem is therefore unidirectional (Marschner, 1995). Several studies provide a picture of P_i distribution patterns on the whole plant level (Koontz and Biddulph, 1957; Biddulph et al., 1958; Jeschke et al., 1997). P_i is described as a phloem mobile element (Marschner, 1995) and it was suggested that P_i is the primary form in which P moves in the phloem (Bieleski, 1969; Jeschke et al., 1997). In P_i-sufficient plants, most of the P_i absorbed by the root is transported in the xylem to the young leaves. P_i mobilization, net export and retranslocation from older leaves to the root through the phloem were calculated to greatly exceed the requirements of the root under high P_i supply; thus, massive transfer of P_i from phloem to xylem for P_i transport back to the shoot was expected to occur in the root (Jeschke et al., 1997). In potato, sink–source relationships for P on the whole plant level have been demonstrated by feeding radioactive P_i to young and old leaves, respectively (Rausch et al., 2004). While the nutrient was retained in young sink leaves, it was rapidly distributed from old leaves to actively growing tissue indicating strong source capacity of old potato leaves. The question arises whether expression patterns of *Pht1* genes in potato suggest involvement of the encoded proteins in P-related source–sink interactions *in planta*. In fact, the data do not support this hypothesis. During high P conditions in fertilized soils, *StPT1* transcripts were more abundant in old leaves as compared with young leaves (Leggewie et al., 1997; Rausch et al., 2004). This can be explained by low P levels in old leaves triggering a local response to P deprivation.

Three *Arabidopsis* mutants impaired in P_i translocation within the shoot have been identified, i.e. *pho1, pho2* and *pht2;1-1*. The *pho1* mutant exhibited a defective mechanism of xylem loading with P_i. As a consequence, total P_i concentrations in the leaf tissue were strongly reduced, and root P_i uptake rates were similar to wildtype plants, whereas translocation rates to the shoot were reduced (Poirier et al., 1991). The *PHO1* gene was identified and characterized (Hamburger et al., 2002) and was shown to be a membrane protein involved either directly or indirectly in P_i transport across membranes.

Contrasting the phenotype of *pho1*, the *pho2* mutation affects a function normally involved in regulating the concentration of P_i in shoots of *Arabidopsis*. Compared to wildtype, *pho2* mutants had greater P_i concentrations in stems, siliques and seeds, but roots of *pho2* mutants had similar or lower P_i concentrations than either *pho1* mutants or wildtype seedlings (Delhaize and Randall, 1995). Later, it was shown that a mutation in the target gene of a micro-RNA (*miR399)* is responsible for the P_i overaccumulator phenotype in the *pho2* mutant (Aung et al., 2006; Bari et al., 2006). The *miR399* target gene encodes a putative ubiquitin-conjugating enzyme (UBC) which was also shown to be involved in the regulation of *Pht1;1* gene expression by P_i availability (Fujii et al., 2005). Ubiquitin-conjugating enzymes play an important role in the protein degradation pathway. Overexpression of *miR399* in *A. thaliana* resulted in downregulation of UBC and exagerated accumulation of P_i in the shoot (Chiou et al., 2006).

The *pht2;1-1* mutant is impaired in the expression of the Pht2;1 P_iT. Pht2;1 was the first member of the Pht2 family of plant P_iTs to be identified (Quigley et al., 1996; Daram et al., 1999). The protein belongs to the inorganic P_iT family of the MSF (Section 15.3.2.1). Functional analysis of the Pht2;1 protein in mutant yeast cells suggested H^+/P_i symport activity dependent on the electrochemical proton and potassium gradient across the yeast plasma membrane (Daram et al., 1999). Three orthologues from other plant species have been published, Pht2;1 from potato (Rausch et al., 2004), MtPHT2;1 from *M. truncatula* (Versaw and Harrison, 2002) and Phtc from spinach (*Spinacia oleracea*) (Ferro et al., 2002). Evidence was presented for developmental and light-dependent control of the *Pht2;1* gene and for localization of Pht2;1 protein to both photoautotrophic and heterotrophic plastids in *Arabidopsis* and potato and, moreover, to the inner membrane of spinach chloroplasts (Ferro et al., 2002; Versaw and Harrison, 2002; Rausch et al., 2004). Thus, molecular evidence for light-regulated unidirectional P_i transport in plastids mediated by Pht2;1 exists. Moreover, functional complementation studies in yeast indicated low-affinity P_i transport through Pht2;1 (Daram et al., 1999; Versaw and Harrison, 2002).

Steady-state transcript levels of *Pht2;1* slightly increased during growth at high P_i concentrations (Versaw and Harrison, 2002; Zhao et al., 2003; Rausch et al., 2004). In potato, *Pht2;1* expression levels were highest in young sink leaves correlating positively with sink strength in P_i allocation to growing tissues and inversely with expression levels of *StPT1* being highest in old P_i source leaves exhibiting reduced P content (Rausch et al., 2004). The T-DNA insertional *Pht2;1* loss-of-function mutant *pht2;1-1* exhibited an interesting phenotype with respect to interorgan P_i homeostasis (Versaw and Harrison,

2002). *Pht2;1-1* displayed reduced growth compared with the wild type, and the leaf P_i content was reduced by >20% at high external P conditions. At low P_i conditions, the root P_i content of *pht2;1-1* was strongly reduced as compared to control plants, whereas the shoot P_i level was markedly increased. Moreover, in contrast to the wild type, a redistribution of P_i from older to younger leaves under conditions of low external P_i was not observed. This finding indicated an important role of Pht2;1 in the allocation and partitioning of P_i at the whole plant level.

Overall, *Pht2;1* expression analysis suggested multifactorial regulation, and the physiological context in which Pht2;1 could play a role was difficult to imagine. With the *Arabidopsis* genome sequenced, high-throughput gene expression analysis has become a frequent and powerful research tool in plant biology. Using Genevestigator, a database and Web-browser data mining interface for Affymetrix GeneChip data (Zimmermann et al., 2004), the gene expression profiles of more than 22 000 *Arabidopsis* genes can be obtained. Meta-analysis of 228 microarrays from a large number of genome profiling studies covering 83 different experimental conditions in *Arabidopsis* allowed one to ask 'which genes are regulated in a similar manner to *Pht2;1* and could therefore be involved in the same physiological process?'. The analysis suggested a role of Pht2;1-mediated P_i transport in maintaining P_i homeostasis in plastids during periods of high demand for intermediates of carbohydrate metabolism (Rausch et al., 2004).

15.4 CONCLUSION AND OUTLOOK

The identification and molecular characterization of numerous genes encoding proteins for N and P_i transport and metabolism have largely furthered our understanding of resource utilization and allocation in plants and also in potato plants. A range of different transgenic potato plants that all are impaired in various critical steps in nutrient acquisition, allocation, intracellular transport or assimilation have been generated and analysed. Even more knowledge has been obtained from the work with functional genomics model species, which represents a rich resource for technologies that also could be applied to potatoes. More mutants will eventually be identified, and the study of interactions of individual genes will even contribute more to our understanding of mechanisms contributing to the nutrient efficiency of plants. The main challenge that remains is to elucidate the mechanisms that establish P_i homeostasis under variable P conditions, as no tonoplast P_iT protein has been identified until now. In the future, modifying the biochemical parameters of N and P_i transport systems and metabolizing enzymes may lead to enhanced N and PAE and utilization efficiency or changed P_i concentrations within cellular compartments, thus affecting for example plant development, photosynthetic capacity and starch synthesis or responses to stress (Marschner, 1995; Hurry et al., 2000). The ultimate goal will be to generate crops with high yield in low N and P soils to allow low input of N and P fertilizers. A more complete understanding of N and P_i uptake and utilization in plants will thus eventually have significant implications with respect to both the environment and world agriculture.

REFERENCES

Abramson J., I. Smirnova, V. Kasho, G. Verner, H.R. Kaback and S. Iwata, 2003, *Science* 301, 610.

Aitken A., 1996, *Trends Cell Biol.* 6, 341.

Aslam M., R.L. Travis and R.C. Huffaker, 1992, *Plant Physiol.* 99, 1124.

Aubert S., R. Bligny, R. Douce, E. Gout, R.G. Ratcliffe and J.K.M. Roberts, 2001, *J. Exp. Bot.* 52, 37.

Aung K., S.-I. Lin, C.-C. Wu, Y.-T. Huang, C.-L. Su and T.-J. Chiou, 2006, *Plant Physiol.* 141, 1000.

Baima S., A. Haegi, P. Stroman and G. Casadoro, 1989, *Carlsberg Res. Commun.* 54, 1.

Bari R., B. Datt Pant, M. Stitt and W.-R. Scheible, 2006, *Plant Physiol.* 141, 988.

Bates T. and J. Lynch, 1996, *Plant Cell Environ.* 19, 529.

Bhattarai I. and R. Mishra, 1984, *Plant Soil* 79, 299.

Biddulph O., S. Biddulph, R. Cory and H. Koontz, 1958, *Plant Physiol.* 33, 293.

Bieleski R.L., 1969, *Plant Physiol.* 44, 497.

Brachmann A., 2006, *New Phytol.* 171, 242.

Bucher M., C. Rausch and P. Daram, 2001, *J. Plant Nutr. Soil Sci.* 164, 209.

Bun-Ya M., M. Nishimura, S. Harashima and Y. Oshima, 1991, *Mol. Cell Biol.* 11, 3229.

Chang H.C. and D.R. Bush, 1997, *J. Biol. Chem.* 272, 30552.

Chiou T.J., K. Aung, S.I Lin, C.C Wu, S.F. Chiang and C.L. Su, 2006, Regulation of phosphate homeostasis by MicroRNA in Arabidopsis. Plant Cell 18, 412–421.

Daram P., S. Brunner, B.L. Persson, N. Amrhein and M. Bucher, 1998, *Planta* 206, 225.

Daram P., S. Brunner, C. Rausch, C. Steiner, N. Amrhein and M. Bucher, 1999, *Plant Cell* 11, 2153.

Delhaize E. and P.J. Randall, 1995, Characterization of a phosphate-accumulator mutant of *Arabidopsis thaliana. Plant Physiol.* 107, 207–213.

Djennane S., J.-E. Chauvin and C. Meyer, 2002, *J. Exp. Bot.* 53, 1037.

Djennane S., I. Quillere, M.T. Leydecker, C. Meyer and J.-E. Chauvin, 2004, *Planta* 219, 884.

Doddema H. and G.P. Telkamp, 1979, *Physiol. Plant.* 45, 332.

Dose M.M., M. Hirasawa, S. Kleis-SanFrancisco, E.L. Lew and D.B. Knaff, 1997, *Plant Physiol.* 114, 1047.

Drew M.C., L.R. Saker, S.A. Barber and W. Jenkins, 1984, *Planta* 160, 490.

Fageria N. and V. Baligar, 1997, *J. Plant Nutr.* 20, 1267.

Ferrario-Mery S., M.H. Valadier and C.H. Foyer, 1998, *Plant Physiol.* 117, 293.

Ferro M., D. Salvi, H. Rivière-Rolland, T. Vermat, D. Seigneurin-Berny, D. Grunwald, J. Garin, J. Joyard and N. Rolland, 2002, *Proc. Natl. Acad. Sci. U.S.A.* 99, 11487.

Fischer W.N., D.D.F. Loo, W. Koch, U. Ludewig, K.J. Boorer, M. Tegeder, D. Rentsch, E.M. Wright and W.B. Frommer, 2002, *Plant J.* 29, 717.

Foyer C.H., M.H. Valadier, A. Migge and T.W. Becker, 1998, *Plant Physiol.* 117, 283.

Frommer W.B., S. Hummel and J.W. Riesmeier, 1993, *Proc. Natl. Acad. Sci. U.S.A.* 90, 5944.

Fujii H., T.J. Chiou, S.I. Lin, K. Aung and J.K. Zhu, 2005, *Curr. Biol.* 15, 2038.

Gazzarrini S., L. Lejay, A. Gojon, O. Ninnemann, W.B. Frommer and N. von Wiren, 1999, *Plant Cell* 11, 937.

Glassop D., S.E. Smith and F.W. Smith, 2005, *Planta* 222, 688.

Gordon-Weeks R., Y. Tong, T.G.E. Davies and G. Leggewie, 2003, *J. Cell Sci.* 116, 3135.

Hamburger D., E. Rezzonico, J. MacDonald-Comber Petetot, C. Somerville and Y. Poirier, 2002, *Plant Cell* 14, 889.

Harris N., J.M. Foster, A. Kumar, H.V. Davies, C. Gebhardt and J.L. Wray, 2000, *J. Exp. Bot.* 51, 1017.

Harrison M.J., 2005, *Annu. Rev. Microbiol.* 59, 19.

Harrison M.J., G.R. Dewbre and J.Y. Liu, 2002, *Plant Cell* 14, 2413.

Hirai T., J.A. Heymann, D. Shi, R. Sarker, P.C. Maloney and S. Subramaniam, 2002, *Nat. Struct. Biol.* 9, 597.

Holford I.C.R., 1997, *Aust. J. Soil Res.* 35, 227.

Horst W., M. Abdou and F. Wiesler, 1996, *J. Plant Nutr. Soil Sci.* 159, 155.

Huang Y., M.J. Lemieux, J. Song, M. Auer and D.-N. Wang, 2003, Science 301, 616.

Huber J.L., S.C. Huber, W.H. Campbell and M.G. Redinbaugh, 1992, *Arch. Biochem. Biophys.* 296, 58.

Hurry V., A. Strand, R. Furbank and M. Stitt, 2000, Plant J. 24, 383.

Jeschke W., E. Kirkby, A. Peuke, J. Pate and W. Hartung, 1997, *J. Exp. Bot.* 48, 75.

Jungk A., 1991, In: J. Waisel, A. Eshel and U. Kafkafi (eds), *Plant Roots, the Hidden Half*, pp. 455–481. Marcel Dekker, New York, U.S.A.

Jungk A., 2001, *J. Plant Nutr. Soil Sci.* 164, 121.
Karandashov V. and M. Bucher, 2005, *Trends Plant Sci.* 10, 22.
Karandashov V., R. Nagy, S. Wegmüller, N. Amrhein and M. Bucher, 2004, *Proc. Natl. Acad. Sci. U.S.A.* 101, 6285.
Kleinhofs A., R.L. Warner and J.M. Melzer, 1989, In: J.E. Poulton, J.T. Romeo and E.E. Conn (eds), *Plant Nitrogen Metabolism, Recent Advances in Phytochemistry*, Vol 23, p. 117. Plenum Press, New York, U.S.A.
Koch W., M. Kwart, M. Laubner, D. Heineke, H. Stransky, W.B. Frommer and M. Tegeder, 2003, *Plant J.* 33, 211.
Koontz H. and O. Biddulph, 1957, *Plant Physiol.* 32, 463.
Kwart M., B. Hirner, S. Hummel and W.B. Frommer, 1993, *Plant J.* 4, 993.
Lalonde S., D. Wipf and W.B. Frommer, 2004, *Annu. Rev. Plant Biol.* 55, 341.
Lam H.M., K. Coschigano, C. Schultz, R. Melo-Oliveira, G. Tjaden, I. Oliveira, N. Ngai, M.H. Hsieh and G. Coruzzi, 1995, *Plant Cell* 7, 887.
Lea P.J., R.D. Blackwell, A.J.S. Murray and K.W. Joy, 1989, In: J.E. Poulton, J.T. Romeo and E.E. Conn (eds), *Plant Nitrogen Metabolism, Recent Advances in Phytochemistry*, Vol 23, p. 157. Plenum Press, New York, U.S.A.
Lea P.J., R.D. Blackwell and K.W. Joy, 1992, In: K. Mengel and D.J. $P_i$lbeam (eds), *Nitrogen Metabolism of Plants*, p. 153. Clarendon Press, Oxford, U.K.
Leggewie G., L. Willmitzer and J.W. Riesmeier, 1997, Plant Cell 9, 381.
Lillo C., S. Kazazaic, P. Ruoff and C. Meyer, 1997, Plant Physiol. 114, 1377.
Lin S., B. Sattelmacher, E. Kutzmutz, K.H. Muhling and K. Dittert, 2004, J. Plant Nutr. 27, 341.
Liu L.-H., U. Ludewig, W.B. Frommer and N. von Wiren, 2003, *Plant Cell* 15, 790.
Liu C., U.S. Muchhal, M. Uthappa, A.K. Kononowicz and K.G. Raghothama, 1998a, *Plant Physiol.* 116, 91.
Liu H., A.T. Trieu, L.A. Blaylock and M.J. Harrison, 1998b, *Mol. Plant Microbe Interact.* 11, 14.
Mäck G. and J.K. Schjoerring, 2002, *Plant Cell Environ.* 25, 999.
MacKintosh C., P. Douglas and C. Lillo, 1995, *Plant Physiol.* 107, 451.
Maeda D., K. Ashida, K. Iguchi, S.A. Chechetka, A. Hijikata, Y. Okusako, Y. Deguchi, K. Izui and S. Hata, 2006, *Plant Cell Physiol.* 47, 807.
Marschner H., 1995, *Mineral Nutrition of Higher Plants.* Academic Press London, U.K.
McArthur D.A.J. and N.R. Knowles, 1993, *Plant Physiol.* 102, 771.
Melo-Oliveira R., I.C. Oliveira and G.M. Coruzzi, 1996, *Proc. Natl. Acad. Sci. U.S.A.* 93, 4718.
Mengel K., 1992, In: K. Mengel and D.L. $P_i$lbeam (eds), *Nitrogen Metabolism of Plants*, p. 1. Clarendon Press, Oxford, U.K.
Mimura T., 1995, *Plant Cell Physiol.* 36, 1.
Mimura T., 1999, *Int. Rev. Cytol.* 191, 149.
Misson J., M.C. Thibaud, N. Bechtold, K. Raghothama and L. Nussaume, 2004, *Plant Mol. Biol.* 55, 727.
Mitsukawa N., S. Okumura, Y. Shirano, S. Sato, T. Kato, S. Harashima and D. Shibata, 1997, *Proc. Natl. Acad. Sci. U.S.A.* 94, 7098.
Muchhal U.S., J.M. Pardo and K.G. Raghothama, 1996, *Proc. Natl. Acad. Sci. U.S.A.* 93, 10519.
Muchhal U.S. and K.G. Raghothama, 1999, *Proc. Natl. Acad. Sci. U.S.A.* 96, 5868.
Mudge S.R., A.L. Rae, E. Diatloff and F.W. Smith, 2002, *Plant J.* 31, 341.
Nagy F., V. Karandashov, W. Chague, K. Kalinkevich, M. Tamasloukht, G.H. Xu, I. Jakobsen, A.A. Levy, N. Amrhein and M. Bucher, 2005, *Plant J.* 42, 236.
Nagy R., M.J. Vasconcelos, S. Zhao, J. McElver, W. Bruce, N. Amrhein, K.G. Raghothama and M. Bucher, 2006, *Plant Biol.* 8, 186.
Nussaume L., M. Vincentz, C. Meyer, J.P. Boutin and M. Caboche, 1995, *Plant Cell* 7, 611.
Oldroyd G.E.D., M.J. Harrison and M. Udvardi, 2005, *Plant Physiol.* 137, 1205.
Paszkowski U., S. Kroken, C. Roux and S.P. Briggs, 2002, *Proc. Natl. Acad. Sci. U.S.A.* 99, 13324.
Poirier Y. and M. Bucher, 2002, In: C. Somerville and E.M. Meyerowitz (eds), *The Arabidopsis Book.* American Society of Plant Biologists, Rockville, MD, doi/10.1199/tab.0024, http://www.aspb.org/publications/arabidopsis.
Poirier Y., S. Thoma, C. Somerville and J. Schiefelbein, 1991, *Plant Physiol.* 97, 1087.
Provan F., L.M. Aksland, C. Meyer and C. Lillo, 2000, *Plant Physiol.* 123, 757.
Quigley F., P. Dao, A. Cottet and R. Mache, 1996, *Nucleic Acids Res.* 24, 4313.

Rae A.L., D.H. Cybinski, J.M. Jarmey and F.W. Smith, 2003, *Plant Mol. Biol.* 53, 27.
Raghothama K.G., 1999, *Annu. Rev. Plant Physiol. Plant Mol. Biol.* 50, 665.
Raghothama K.G., 2000, *Curr. Opin. Plant Biol.* 3, 182.
Rausch C. and M. Bucher, 2002, *Planta* 216, 23.
Rausch C., P. Daram, S. Brunner, J. Jansa, M. Laloi, G. Leggewie, N. Amrhein and M. Bucher, 2001, *Nature* 414, 462.
Rausch C., P. Zimmermann, N. Amrhein and M. Bucher, 2004 *Plant J.* 39, 13.
Remy W., T.N. Taylor, H. Hass and H. Kerp, 1994, *Proc. Natl. Acad. Sci. U.S.A.* 91, 11841.
Roessner U., A. Luedemann, D. Brust, O. Fiehn, T. Linke, L. Willmitzer and A.R. Fernie, 2001b, *Plant Cell* 13, 11.
Roessner U., L. Willmitzer and A.R. Fernie, 2001a, *Plant Physiol.* 127, 749.
Saier M.H. Jr., J.T. Beatty, A. Goffeau, K.T. Harley, W.H. Heijne, S.C. Huang, D.L. Jack, P.S. Jahn, K. Lew, J. Liu, S.S. Pao, I.T. Paulsen, T.T. Tseng and P.S. Virk, 1999, *J. Mol. Microbiol. Biotechnol.* 1, 257.
Sattelmacher B., W. Horst and H. Becker, 1994, *J. Plant Nutr. Soil Sci.* 157, 215.
Sattelmacher B., F. Klotz and H. Marschner, 1990, *Plant Soil*, 123, 131.
Schachtman D.P., R.J. Reid and S.M. Ayling, 1998, *Plant Physiol.* 116, 447.
Scheible W.-R., A. Gonzalez-Fontes, M. Lauerer, B. Müller-Röber, M. Caboch and M. Stitt, 1997, *Plant Cell* 9, 783.
Sentenac H. and C. Grignon, 1985, *Plant Physiol.* 77, 136.
Shin H., H.-S. Shin, G.R. Dewbre and M.J. Harrison, 2004 *Plant J.* 39, 629.
Smith F.W., P.M. Ealing, B. Dong and E. Delhaize, 1997, *Plant J.* 11, 83.
Smith F.W., A.L. Rae and M.J. Hawkesford, 2000, *Biochim. Biophys. Acta* 1465, 236.
Smith S.E. and D.J. Read, 1997, *Mycorrhizal Symbiosis*. Academic Press, San Diego, CA, U.S.A.
Smith S.E., F.A. Smith and I. Jakobsen, 2003, *Plant Physiol.* 133, 16.
Smith S.E., F.A. Smith and I. Jakobsen, 2004, *New Phytol.* 162, 511.
Stewart G.R., A.F. Mann and P.A. Fentem, 1980, In: B.J. Miflin (ed.), *The Biochemistry of Plants*, Vol 5, p.6.9. Academic Press, New York, U.S.A.
Stewart G.R. and D. Rhodes, 1977, *New Phytol.* 79, 257.
Stitt M. and A.R. Fernie, 2003, *Curr. Opin. Biotechnol.* 14, 136.
Teixeira J., S. Pereira, F. Canovas, R. Salema, 2005, *J. Exp. Bot.* 56, 663.
Turano F.J., S.S. Thakkar, T. Fang and J.M. Weisemann, 1997, *Plant Physiol.* 113, 1329.
Ullrich W.R., 1992, In: K. Mengel and D.J. $P_i$lbeam (eds), *Nitrogen Metabolism of Plants*, p. 121. Clarendon Press, Oxford.
Ullrich C.I. and A.J. Novacky, 1990, *Plant Physiol.* 94, 1561.
Usadel B., A. Nagel, O. Thimm, H. Redestig, O.E. Blaesing, N. Palacios-Rojas, J. Selbig, J. Hannemann, M.C. $P_i$ques, D. Steinhauser, W.R. Scheible, Y. Gibon, R. Morcuende, D. Weicht, S. Meyer and M. Stitt, 2005, *Plant Physiol.* 138, 1195.
Versaw W.K. and M.J. Harrison, 2002, *Plant Cell* 14, 1751.
Welp G., U. Herms and G. Brümmer, 1983, *J. Plant Nutr. Soil Sci.* 146, 38.
Williams L.E. and A.J. Miller, 2001, *Annu. Rev. Plant Physiol. Plant Mol. Biol.* 52, 659.
Witte C.-P., S.A. Tiller, M.A. Taylor and H.V. Davies, 2002, *Plant Physiol.* 128, 1129.
Wu K., M.F. Rooney and R.J. Ferl, 1997, *Plant Physiol.* 114, 1421.
Zhao L.M., W.K. Versaw, J.Y. Liu and M.J. Harrison, 2003, *New Phytol.* 157, 291.
Zimmermann P., M. Hirsch-Hoffmann, L. Hennig and W. Gruissem, 2004, *Plant Physiol.* 136, 2621.
Zimmermann P., G. Zardi G., M. Lehmann, C. Zeder, N. Amrhein, E. Frossard and M. Bucher, 2003, *Plant Biotechnol. J.* 1, 353.
Zuk M., J. Skala, J. Biernat and J. Szopa, 2003, *Plant Sci.* 165, 731.
Zuk M., R. Weber and J. Szopa, 2005, *J. Agric. Food Chem.* 53, 3454.

Part IV

RESPONSE TO THE ENVIRONMENT

Chapter 16

Water Availability and Potato Crop Performance

J. Vos[1] and A. J. Haverkort[2]

[1]*Crop and Weed Ecology – WUR, PO Box 430, 6700 AK Wageningen, The Netherlands;*
[2]*Plant Research International – WUR, PO Box 16, 6700 AA Wageningen, The Netherlands*

16.1 INTRODUCTION

Globally, the demand for fresh water increases due to increased population pressure and altered food habits. Water is increasingly becoming a scarce resource. About 80 countries with 40% of the world population suffer from serious water shortage. Worldwide 85% of fresh water is used in agriculture. The efficiency of the use of available water, i.e. the proportion of available water directed towards plant transpiration, is often less than 50% (Hamdy et al., 2003), but there is a large variety of different options to reduce water losses and enhance water productivity, including micro-management such as drip irrigation (Chawla and Narda, 2001), improved varieties, growing alternative crop species and better timing of the crop cycle in the season.

Potato is a crop that uses water relatively efficiently. Its high harvest index of about 0.75, compared with approximately 0.5 for cereals, contributes to this property as does its agro-ecological character, characterized by relatively low evaporative water demand. Potato is best grown at places and in periods where average daily temperatures are above 5°C and below 21°C. Lower temperatures lead to risk of frost, and at higher temperatures, translocation of dry matter to the tubers is much reduced. Plant-available water primarily depends on amount and distribution of rainfall, water-holding capacity of the soil and the rooting depth. Limiting dependable rainfall is supplemented by irrigation in many potato-producing areas, but not all. Water shortage limits yield and also the associated use efficiencies (or productivities) of other limiting resources, including land, nutrients, fossil energy and labour; economic returns may also be reduced by poorer product quality. The importance of ensuring sufficient availability of water is illustrated in Fig. 16.1, comparing potential and water-limited potato yields in Africa.

Agricultural research can help mankind in several ways to cope judiciously with global problems of producing sufficient food from limiting supplies of water. Firstly, research generates insight into how the soil–plant–atmosphere system works. Such analyses generate ideas on the potential and the relative successes of options for improvement (e.g. soil improvement, irrigation and plant breeding). Secondly, such research provides reference standards for performance, including models of crop production in relation to environment, and helps to set realistic goals for improvement. While, lastly, research provides

Potato Biology and Biotechnology: Advances and Perspectives
D. Vreugdenhil (Editor)

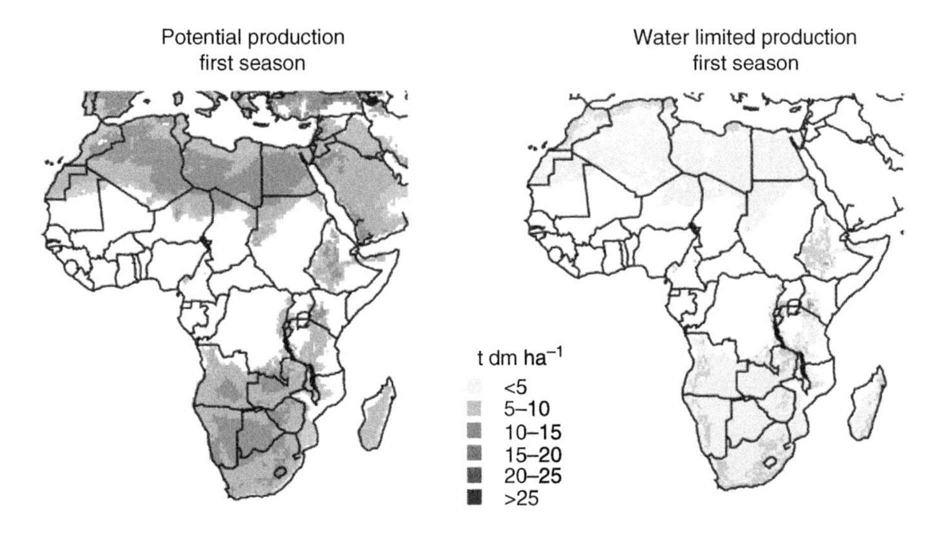

Fig. 16.1. Calculated potential (left) and water-limited tuber dry matter yields (right) for Africa (Haverkort et al., 2004).

the basis for the design of technical equipment for monitoring plant stress. This chapter (i) summarizes basic concepts on how the system works, (ii) addresses methods of monitoring the water status, (iii) presents the crop responses to drought and (iv) addresses environmental and genetic differences in water-use efficiency and product quality.

16.2 DETERMINANTS AND CONTROLS OF WATER MOVEMENT

16.2.1 The transport of water in the soil–plant–atmosphere continuum

Since the 1960s, the concepts have been well established describing the driving forces and controls governing the transport of water in the soil–plant–atmosphere continuum (SPAC). Plant transpiration can be described by analogy with Ohm's law on the flux of charge in an electrical circuit. According to this concept, the rate of flow of water through the entire SPAC is directly proportional to the total difference in pressure head between the soil–root interface and the leaf–atmosphere interface and is inversely proportional to the total water transport resistance of the system under steady-state conditions. Ignoring the very small proportion of water taken up by the plant that is stored, the general equation describing this relation can be repeated to describe the fluxes in the several parts of the SPAC [Eq. (16.1); Fig. 16.2]. The rules for summarizing an electrical system of resistors in parallel or in series can be applied to these transport resistances.

$$v = -\frac{\Delta h_{\text{total}}}{R_{\text{total}}} = -\frac{h_{\text{root}} - h_{\text{soil}}}{R_{\text{soil}}} = -\frac{h_{\text{leaf}} - h_{\text{root}}}{R_{\text{plant}}} = -\frac{h_{\text{leaf}} - h_{\text{soil}}}{R_{\text{soil}} + R_{\text{plant}}}, \quad (16.1)$$

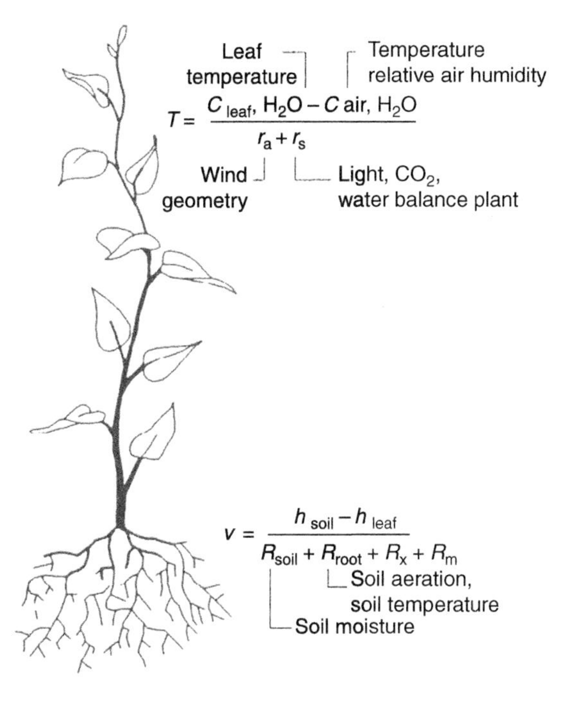

Fig. 16.2. Resistances in the soil plant air continuum and the effect of some soil and meteorological factors on transpiration (T) and water uptake (v). T = transpiration rate (M L^{-2} T^{-1}), v = water flux density (L^3 L^{-2} T^{-1}), C_{leaf,H_2O} = water concentration in the leaf (M L^{-3}), C_{air,H_2O} = water concentration in the air (M L^{-3}), R_{soil} and R_{root} = transport resistances in soil and in root cells (T), R_x and R_m = transport resistances in xylem and in leaf mesophyll (T) and r_s and r_a = resistance of stomata and laminar air layer (T L^{-1}). [*Source: Bierhuizen (1969).*]

where v = water flux density (L^3 L^{-2} T^{-1}); h_{soil}, h_{root} and h_{leaf} = pressure heads (L) in the soil, at the root surface and in the leaves; and R_{soil} and R_{plant} = water transport resistances (T) in the soil and the plant.

The common units of resistance to water flux are s cm^{-1} or s m^{-1}. In some studies, it is preferable to use conductance (the reciprocal of resistance, with units of mm s^{-1}) instead of resistance because the flux is directly proportional to the conductance. (The choice may be determined by the balance between resistances in series and in parallel in the system being studied.)

Hydraulic conductance is a common term to refer to the ‘ease’ of transport of water.

It is a limitation of Eq. (16.1) that steady-state conditions seldom exist in a plant. Throughout the day, the water content of the plant changes and the resistances to water flow are not constant either but may appear to depend on evaporative demand. It is difficult to determine the hydraulic conductance of different parts of roots of an extensive root system in a natural soil, although various techniques are available to measure hydraulic conductance, pressure heads and water uptake at detailed levels (Sanderson, 1983; Steudle, 1993). Again, it is difficult to relate these measurements to overall hydraulic conductance of the entire root system at a particular time or to translate laboratory findings to root systems grown under natural conditions (Steudle, 2000).

The soil water potential at the soil–root interface is not the same throughout the rooted soil volume. There are vertical gradients in average soil moisture potentials and gradients from the more remote points in the bulk soil to those nearest the root surface. The transport of water from the bulk soil to the root surface is also governed by the difference in pressure head and soil hydraulic conductance. The latter can be quantified as functions of soil water potential (Ψ) or soil water content, the form of these relations being strongly influenced by soil texture (Van Genuchten, 1980). Spatial heterogeneity of soil water and roots and the difficulty of measuring each of the resistances from root surface to stomatal cavity make it impractical to solve Eq. (16.1) to obtain the transpiration rate. Yet, the concept is sound and still very useful as a basis to guide thinking about the collection of water from the soil and water flow in the SPAC.

16.2.2 Plant water relations

Soil holds water against gravity. Plant tissue also binds water: if one cuts a potato shoot at the soil surface, water does not drip from the open ends of the xylem vessels. The water potential, Ψ, is a measure of the 'strength' with which the tissue holds the water. Basically, water potential is expressed in energy per unit water, but in practice, it is more convenient to use pressure units or equivalent lengths of columns of mercury or water. As the water potential of free water is 0 MPa, water in plants has negative water potential: one has to exert pressure to make the water exude from the cut end of a shoot. This is precisely what is done in a pressure chamber to measure the water potential (Scholander et al., 1965). A plant organ is fixed in the chamber with the cut end protruding through a hole in the lid into the open air. The chamber is pressurized, and the pressure required to make water appear at the cut surface, i.e. the pressure needed to compensate for the 'water binding force' of the tissue, is taken as a measure of the water potential. The water potential is the resultant of component potentials, the most of important of which are two opposing ones, namely the turgor, P (positive values), and the osmotic potential, π (negative values). In soil, the water retention curve describes the relation between soil water content and soil water potential. Plant water relations serve the same purpose for plant tissue. By definition, water relations are those between the relative water content (RWC), water potential (Ψ), osmotic potential (π) and turgor (P). Relative water content expresses the water content of a plant or an organ as a proportion of the water content that it could have if in equilibrium with free water at 0 MPa. At RWC = 1, the plant is maximally saturated with water, seen, for instance, pre-dawn in well-watered plants.

A representative, quantitative example of the relations between these variables is depicted in Fig. 16.3. Such associations have been explored from the working hypothesis that genetic differences in slopes and intercepts would be associated with drought tolerance, but such differences were not found.

Osmotic adjustment, i.e. active lowering of the osmotic potential because of the accumulation of solutes, seems to be absent in potato (Vos and Groenwold, 1988), whereas it is still expected that genetic variation in this property can be exploited to improve drought resistance in wheat (Skovmand et al., 2001). The osmotic potential at the point of full

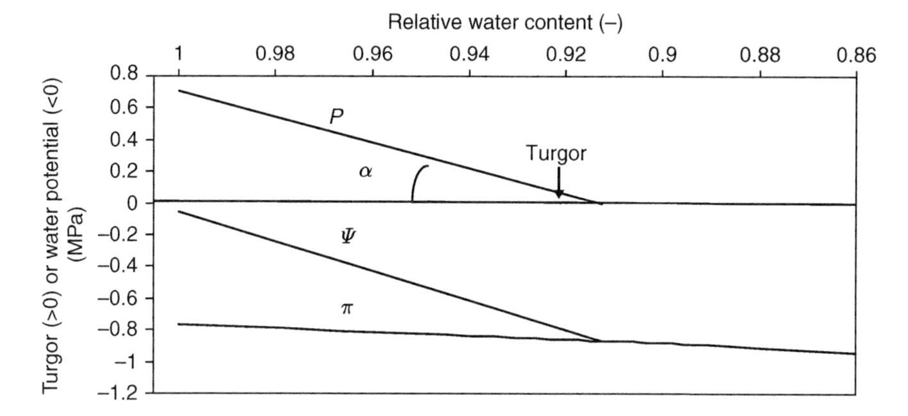

Fig. 16.3. Representative example for potato leaves of the relations between turgor and water potential as functions of relative water content (fraction). When $P > 0$ MPa, the osmotic potential, $\pi = \Psi - P$ (MPa), i.e. the negative value of the distance between the lines for P and Ψ. When RWC < 0.912 (i.e. less than point of turgor loss), $\pi \approx \Psi$. The slope α is a measure of the elasticity of the cell walls. [*Source: Vos and Oyarzún (1988).*]

turgor was found to decline (be more negative) with leaf insertion number on the stem (Vos and Groenwold, 1988), i.e. there is an osmotic gradient along the stem, spanning the range from approximately −0.6 to −1.0 MPa. Such a gradient in the plant explains why – even in full sunlight – the lower leaves on the plant and not the top leaves show wilting symptoms first when water supply falls short of evaporative demand. At the verge of turgor loss, which occurs at an RWC of approximately 0.91, the osmotic potential does not fall below −0.9 to −1.0 MPa at the top of the plant. There are three implications of these features.

(1) The permanent wilting point of potato crops is typically −0.6 MPa in young crops, falling to −1.0 MPa in older crops, both of which are higher than the commonly accepted value of −1.6 MPa. These figures imply that potato has a weaker capacity to dry out the soil than crops that wilt at −1.6 MPa or lower. However, the practical significance of this difference depends on the amount of water that is available to the plants in soils with water potentials between −0.6 and −1.6 MPa that, in turn, depends on soil texture and rooting depth.
(2) Relatively small imbalances between uptake of water and transpiration lead to loss of turgor. For each 1000 kg ha^{-1} of fresh leaves with 15% dry matter, wilting is reached when there is a net loss of approximately 77 kg water. Expressed in millimetres of water that loss is only 0.0077.
(3) The gradient in osmotic potential up the plant implies that drought resistance increases somewhat with ontogeny and that late varieties (producing higher levels of apical branches) acquire somewhat better drought resistance over time than early varieties that produce relatively fewer branches and leaves.

Since the mid-1980s, it has become clear that responses to drought, including stomatal reactions, are triggered by root signals rather than – or at least in addition to – leaf water potential per se. Gollan et al. (1986) pressurized root systems of wheat and sunflower and showed that drought reactions were not associated with leaf water status. Several lines of evidence support the idea that abscisic acid (ABA) plays a central role in stress signalling, inducing stomatal reactions (Wilkinson and Davies, 2002). Such reactions can be brought about even if only part of the root system is in dry soil (Gowing et al., 1990). Abscisic acid and root signalling has not been explored in potato. One could speculate on possible triggering of drought responses due to potato hills that dry out while the subsoil is still moist. However, in their extensive study on water uptake in potato in relation to the distribution of roots and water in the soil, Stalham and Allen (2004) did not find indications for such plant responses triggered in pockets of dry soil.

16.3 ASSESSING PLANT WATER STATUS

De Neve et al. (2000) reviewed several methods to assess water use and water status. Water potential, Ψ, can be measured directly by pressure chamber or indirectly by the use of psychrometers. Stem flow gauges are sensors that allow direct measurement of the water used by plants as it moves up the stem; however, the potato stem has a strongly angled cross-section that makes it difficult to effect and maintain a good thermal contact between stem and sensor. Stomatal aperture is the dominant factor in the stomatal conductance of leaf surfaces, which controls both the water loss from plant leaves and the uptake of CO_2 for photosynthesis. Measurements of stomatal conductance are therefore important indicators of plant water status and provide valuable insight into plant growth and plant adaptation to environmental variables. A porometer measures the stomatal conductance or resistance. There are two general designs, equilibrium and dynamic, that differ in their control philosophy. Both require a chamber to be clamped onto a leaflet. Water loss from the enclosed leaf then tends to increase the humidity in the chamber. In the equilibrium porometer, dry air is added at a measured rate to maintain a stable humidity close to the ambient conditions. In the dynamic porometer, the chamber is purged with dry air, and the transit time is measured while the leaf wets the air between two threshold values, dry and wet, either side of ambient conditions. Porometer estimates of stomatal conductance depend on the rate of transpiration, i.e. on radiation and humidity. Therefore, porometer readings can only be interpreted in relation to measurements from well-watered control plants.

The measurements just described are not useful for scheduling irrigation, but sensing of leaf temperature may be. Where stomata are open, water evaporates from the leaf and so cools it. The greater the evaporation rate, the greater the cooling of the leaf. Where the supply of soil water is limited, less water is evaporated and there is less cooling of the leaves, and the canopy temperature is higher than one would expect for a crop with an adequate water supply. That is, the difference in leaf–air temperature is less than it would be in an unstressed crop. Temperatures of air and canopy can be measured with infrared thermometers that are commercially available. Using the humidity of the air and the difference in leaf–air temperature, one can calculate a Crop Water Stress Index (CWSI) that can be used to schedule irrigation (Idso et al., 1981; Ben-Asher et al., 1989; McBurney, 2003). Critical values of the CWSI (i.e. values where yield reduction can be expected) are derived from field measurements on crops growing under a range of water stress. This technique is now reaching beyond development and is approaching the stage where it can be used in commercial applications (Shae et al., 1999).

16.4 POTATO PLANT RESPONSES TO DROUGHT AND BIOTIC STRESS

16.4.1 Leaf expansion

Water influx into a growing plant cell powers its expansion. Reduced expansion is one of the first symptoms of water limitation. As water moves down a gradient in potential, expanding cells attract water only if the local water potential is lower than elsewhere in the soil–plant system. As influx of water increases the water potential, movement of solutes into the cell expansion zones is a phenomenon that also powers cell expansion (Fricke and Flowers, 1998; Boyer and Silk, 2004).

Most early studies on cell and leaf expansion used water potential and its components as independent variables. The relative rate of cell expansion, R, has been described as a linear function of turgor (P) above a threshold value (P_t), below which there is no cell growth. This association is represented by

$$R = m(P - P_t), \quad (16.2)$$

where m = the extensibility factor of the cell wall. Limitation of water supply results in lower leaf water potentials (Vos and Groenwold, 1988), and Eq. (16.2) shows how reduced leaf growth is also an early symptom of water limitation. Early work by Gandar and Tanner (1976) indicated $P_t = 0.5\,\text{MPa}$ for potato. If P_t and m had conservative values, this would offer practical options for relating rates of extension processes to (soil) water potential. However, reality is more complex: according to Passioura and Fry (1992), these coefficients are not conservative at all, in several cases being functions of P, themselves. That makes doubtful the application of Eq. (16.2) to practical problems.

Leaves that expand during a period of limited water supply grow to smaller sizes and show lower specific leaf area (SLA) than leaves of well-watered controls. The magnitude of effects is illustrated in Fig. 16.4. Reduced SLA implies a larger effect of water limitation on area than on leaf dry weight.

Jefferies (1989) measured the rates of leaf length extension on drought stressed and control potato plants. The relative leaf extension rate (stressed/control) showed a quadratic association with soil moisture deficit (SMD) (Fig. 16.5A) and a linear relation with leaf water potential (Fig. 16.5B). Associations such as that in Fig. 16.5A are specific to soil type and root distribution and depend on weather conditions. Furthermore, one can apply the associations of Fig. 16.5 to particular potato crops only if the rate of leaf extension of well-watered plants is known or estimated in some way. Wider application of Fig. 16.5B seems justified (because leaf water potential reacts to both weather and soil conditions), provided leaf water potential is monitored exactly as described by Jefferies (1989): top leaves of plants measured between 1300 and 1400 h. This short review indicates that the

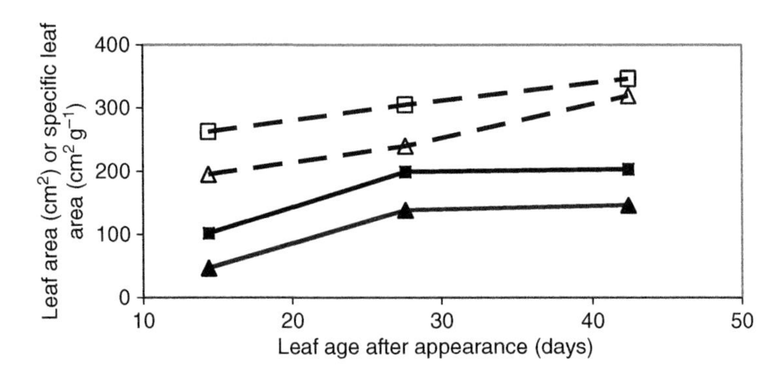

Fig. 16.4. Effect of water limitation on leaf area (full drawn lines) and specific leaf area (broken lines) for the first leaf on the first order of apical branches of plants of cultivar Saturna. Squares, well-watered controls; triangles, water supply rate 50% of control. [*Source: Vos and Groenwold (1989b).*]

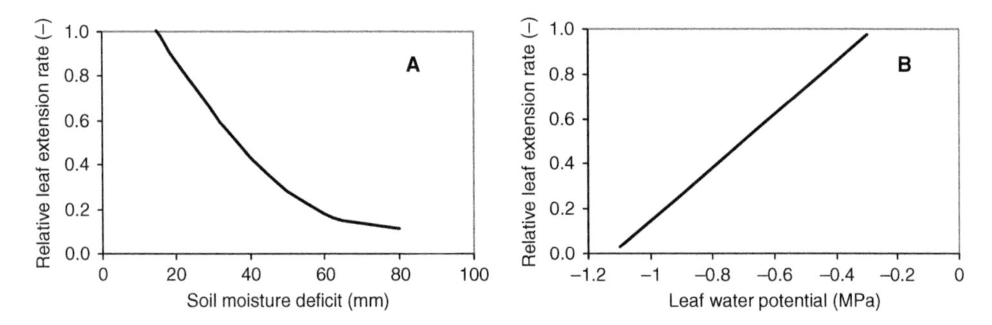

Fig. 16.5. Relative leaf extension of potato leaves (ratio of drought stressed/controls) as function of (A) the relation between soil moisture deficit [the amount of water (mm) required to refill the top 90 cm of the soil profile to field capacity] 0- to 90-cm layer and (B) leaf water potential [*Source: Jefferies (1989).*]

relation between water limitation and leaf expansion is not general and clear cut at all. As for stomatal reactions, evidence is accumulating that responses of extension growth to water limitation are not simply controlled by water potential.

16.4.2 Effect of drought on plant calcium and ^{13}C concentrations

Two indicators provide a cumulative reflection of the degree of drought stress a potato plant has been subjected to during its lifetime: its concentrations of calcium and ^{13}C.

The calcium concentration in plant parts is decreased when plants are subjected to drought. This is because the plant takes up less water per unit dry matter production (its water-use efficiency increases), whereas calcium is exclusively transported through the transpiration flow. Supply of calcium to plant roots can be accounted for by mass flow in most soils (Brewster and Tinker, 1970). Haverkort et al. (1996) tested dozens of genotypes varying in earliness in a dry and a wet year.

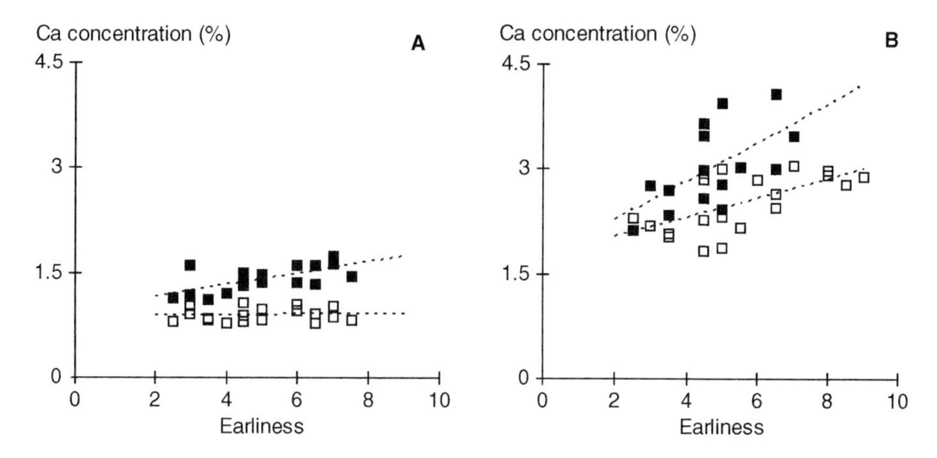

Fig. 16.6. Relations between calcium concentration in dry matter of the leaves and the earliness of genotypes according to the Netherlands List of Recommended Varieties. 2, very late; 10, very early. (A) Open symbol sampling date 21 June and closed symbol 29 July 1991. (B) Open symbol sampling date 17 July and closed symbol August 1992. [*Source: Haverkort et al. (1996).*]

The foliar calcium concentrations were between 0.5 and 2% in the dry year (1991) and more than double that level in the wet year (1992) (Fig. 16.6). Later sampling dates also showed higher calcium concentrations than at earlier sampling dates, indicating that more calcium-containing water was transpired, leaving calcium behind, as time progressed. Finally, the later the genotype, the longer it continues to form new leaves, and the younger the average age of such leaves is, hence the lower their calcium concentration.

Whereas potato plants subjected to drought have a lower foliar calcium concentration than unstressed plants, they have a higher concentration of ^{13}C. Like all higher plants with conventional C3 pathways, potato has a $^{13}C:{}^{12}C$ ratio about 20% less than carbon dioxide in the atmosphere (Farquhar et al., 1982). Potato plants discriminate against $^{13}CO_2$ during its diffusion from the air to the intercellular spaces. Also the primary carboxylation enzyme ribulose 1,5-bisphosphate carboxylase/oxygenase (Rubisco) discriminates against $^{13}CO_2$. If a plant is subjected to drought, discrimination is less because the ^{13}C accumulates in the intercellular cavities and more of it is carboxylated. Changes in ^{13}C content are expressed in isotope fractionation or isotope discrimination units, (^{13}€) (‰), commonly assuming negative values between −22 and −14‰ in potato, with less negative numbers indicating less discrimination (Vos and Groenwold, 1989a; Haverkort and Valkenburg, 1992).

Figure 16.7 shows the relation between ^{13}C discrimination and water-use efficiency: dry spells lead to increased water-use efficiencies and less ^{13}C discrimination. With time, the effect of an early drought wears off, especially in the leaves some of which were formed after the drought treatment was discontinued. The effect remained stronger in the tubers than in the green leaves. The authors reported that ^{13}C discrimination increased with plant age but that older plants were less efficient users of water because of increased respiration.

Changes in calcium and ^{13}C concentrations leave lasting marks in the plant's chemistry, revealing water limitation during (part of) the plant's life cycle.

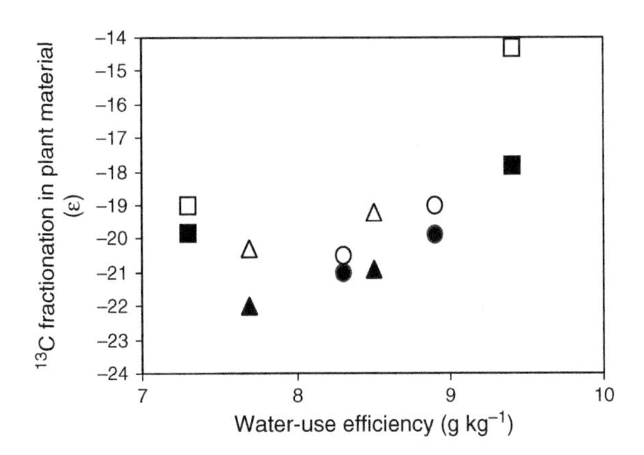

Fig. 16.7. Relationships between water-use efficiency (WUE) of the crop (g, dry matter per litre water used) and ^{13}C discrimination, expressed as isotope fractionation units (^{13}€) of the dry matter of green leaves (closed symbols) and tubers (open symbols) as measured on day 43 (■, □), day 70 (▲, Δ) and day 92 (○, ●) after planting. Droughted plants with the higher WUE of each data pair received 50% of the water supplied to controls. [*Source: Haverkort and Valkenburg (1992).*]

16.5 WATER USE, LEAF DYNAMICS AND POTATO PRODUCTIVITY

16.5.1 Water-use efficiency in different climates

Uptake and assimilation of carbon dioxide are intimately linked with transpirational loss of water, and a linear relation between potato production and water use is commonly found in very divergent agro-ecological conditions (Shalhevet et al., 1983; Feddes, 1987). Feddes (1987) found 43 kg ha^{-1}mm^{-1} (total dry matter production per millimetre water used) in sprinkler-irrigated potato in the Netherlands, whereas Shalhevet et al. (1983) observed a slope of 35 kg ha^{-1}mm^{-1} in sprinkler irrigation in the Negev, Israel.[1] The variation in the slopes of such relations that are observed among different studies can be attributed to systematic differences in the average water vapour pressure deficit (VPD) of the air. The effect of this climatic variable can be mediated by scaling the water use by the VPD, giving a fairly conservative estimate of the productivity per unit transpiration, 2740 kg ha^{-1}mm^{-1} kPa, as reported by Feddes (1987) (Fig. 16.8). That value is equal to the one quoted by Tanner (1981) and Tanner and Sinclair (1983) if one adjusts for the different units used. In those papers, Tanner and Sinclair set an upper limit for the transpiration efficiency in C3 crops, such as potato, equivalent to 6500 kg ha^{-1}mm^{-1} kPa. The equivalent value typical of a C4 crop would be 10 000 kg ha^{-1}mm^{-1} kPa The discrepancy between 2740 kg ha^{-1}mm^{-1} kPa (Fig. 16.8) and 6500kg ha^{-1}mm^{-1} kPa reported for potato could be due to a different calculation of the water VPD.

[1] The productivity on dry weight basis for the Israel case was obtained from the original data on fresh tuber production assuming that the dry matter content of tubers was 22% and the harvest index 0.75.

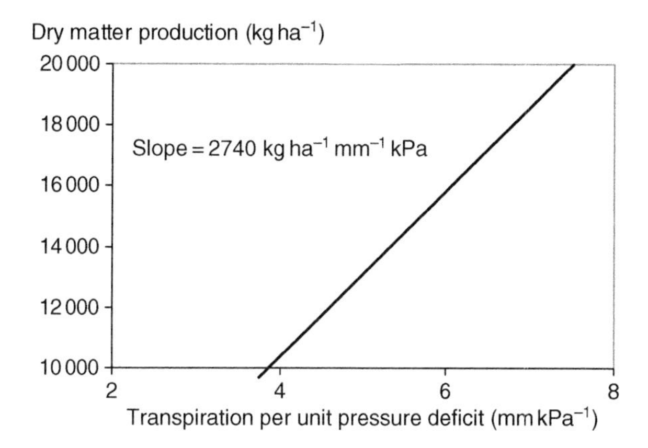

Fig. 16.8. Schematic representation of the relation between potato dry matter production and transpiration (millimetre) per unit water vapour pressure deficit of the air ($e^* - e$) (kPa), where e^* and e are the saturated water vapour pressure of air and the actual water vapour pressure, respectively. [*Source: Feddes (1987).*]

Plots of production versus water use commonly show a positive intercept with the x-axis, representing water use at zero production. The magnitude of that intercept is associated with non-transpirational, evaporative losses from the soil. For instance, Shalhevet et al. (1983) observed similar slopes but different intercepts between sprinkler- and drip-irrigation systems; the intercept with the x-axis, i.e. non-transpirational losses, was 112 mm for drip irrigation versus 197 mm for sprinkler irrigation. Especially when irrigation water is expensive, non-transpiration losses need to be minimized, e.g. by drip or trickle irrigation with subsurface tubes. Large evaporative losses should also be avoided on soils that are prone to salinity problems.

Under (mild) water limitation, transpiration is more reduced than photosynthesis resulting in 33% higher water-use efficiency under water limitation (50% regime) than in controls (Vos and Groenwold, 1989a). On a seasonal basis, the effect of water limitation on water-use efficiency depends on the fraction of the season over which plants were subjected to stress.

16.5.2 Relative transpiration and leaf dynamics

Although plant dry matter production per unit water transpired is up to 33% higher during water limitation than with optimal water supply, the following linear relation still holds as an approximation:

$$\frac{Y_a}{Y_p} = \frac{T_a}{T_p}, \tag{16.3}$$

where Y = dry matter production (kg ha^{-1}) and T = transpiration (mm) over a particular period of time or until maturity and the indices a = actual and p = potential (no water limitation).

Equation (16.3) implies that the loss of production is directly proportional to the deficit between actual (cumulative) transpiration and the transpiration that would be achieved by a crop provided with non-limiting supply of water. There are two categories of responses behind Eq. (16.3), namely the effects of water limitation on instantaneous gas exchange and the effects of water limitation on the dynamics of production and senescence of leaves. Equation (16.3) reflects that, at the stomatal level, entry of carbon dioxide is not possible without water loss. One simply cannot have the one without the other. Declining stomatal aperture is among the earliest signs of drought.

T_a depends on the amount of solar energy absorbed by the crop. Therefore, T_a depends on leaf dynamics over the season. There are three ways in which transient water limitation affects seasonal leaf dynamics. First, leaves that expand under water limitation grow to smaller sizes and show lower SLA than leaves of well-watered controls (Fig. 16.4). The immediate effect of smaller leaf sizes on radiation absorption depends on the amount of leaf already present and on whether the crop already intercepts all the solar radiation [leaf area index (LAI) >3–4]. Second, water limitation will slow down the rate of leaf appearance or will postpone the appearance of particular orders of apical branches (Fig. 16.9). Prolonged drought will not only hamper the production of new leaves but enhance leaf senescence in an acropetal direction. Unfortunately, as yet, accelerated leaf senescence in response to drought stress has not been quantified. Third, there is slowed maturation and compensatory leaf production upon recovery from transient water stress. Such effects are illustrated in Fig. 16.10. During water limitation [50–83 days after planting (DAP)], drought-treated plants produced much less leaf area on apical branches than did controls. However, at 131 DAP, i.e. some 50 days after returning to full water supply on 83 DAP, it turned out that treated plants had significantly more leaf area remaining on the main stem than the controls had, implying a reduced rate of senescence of existing leaves in plants subjected to transient stress. Between 83 and 131 DAP, leaf area on the first-order apical branches of treated plants caught up with controls and showed much more leaf area than the controls on the second-order apical branches, i.e. leaves that appeared after the treatment period. Slowed senescence of existing leaves

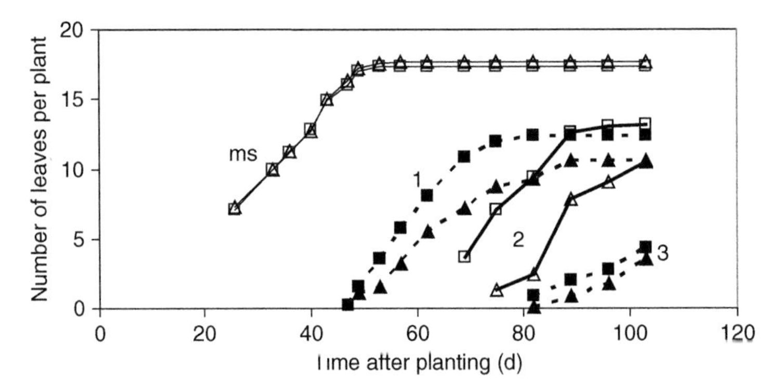

Fig. 16.9. Effect of transient water limitation on the number of leaves per stem. Squares, well-watered control; triangles, plants receiving 50% of the water of controls between day 50 and day 83 after planting; ms, main stem; 1–3, first- to third-order apical lateral branches, respectively. [*Source: Vos and Groenwold (1989b); cultivar Bintje.*]

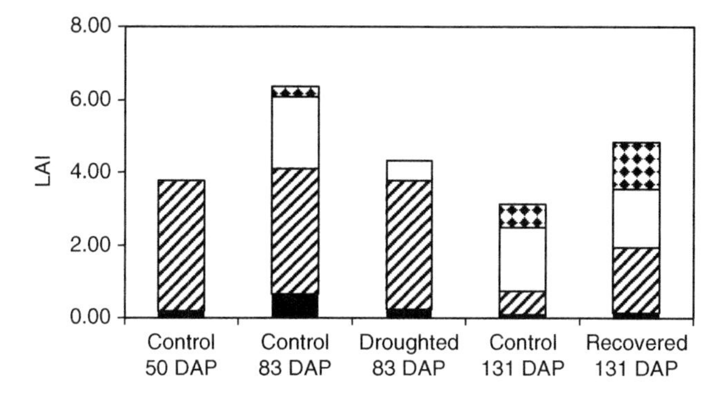

Fig. 16.10. Effect of transient water limitation on leaf area index (LAI) of cultivar Bintje comparing well-watered controls with plants receiving full water supply until 50 days after planting (DAP) and after 83 DAP, but only 50% of the water supplied to controls from 50 to 83 DAP. The figure compares the two sets of plants at the end of the drought period and again, 50 days later. Solid black part of stacked bar represents leaf area on basal lateral branches; dashed part, main stem leaves; white, first-order apical lateral branches; and diamonds, second-order apical lateral branches. [*Source: Vos and Groenwold (1989b).*]

and enhanced growth of new leaves after recovery from transient drought are features that are not well quantified, and further study is needed in this area, particularly in relation to maturity class.

The interaction between drought and temperature is another aspect that has had inadequate treatment in drought research. In rainout shelters, one can study effects of water regime on leaf dynamics for plant configurations as in a field canopy. Interruption of water supply will eventually induce stress in plants, even if the temperature is cool and the sky is overcast. However, it is questionable whether the effects of drought stress, observed in such systems, are the same as those occurring under the combination of drought stress, high temperature and exposure to full sunlight. Here again, there are no data available.

16.5.3 Interactions between drought and biotic stresses

Pests and diseases may reduce the growth and production of potato crops partly through their impact on the plant water economy. Haverkort et al. (1990a) and Bowden and Rouse (1991) studied the interaction between drought and effects of *Verticillium dahliae* (causing 'wilting disease'). Symptoms such as reduced stomatal conductance and reduced rates of transpiration and photosynthesis appeared 1 month after the inoculation of plantlets. Table 16.1 summarizes the results of a container experiment in which stomatal conductance and transpiration of individual leaves was measured with a handheld porometer. Wilting disease caused drought symptoms especially in the lower leaves, corroborated with high levels of *V. dahliae* infection in the corresponding nodes. Compared with the highest leaves along the stem, droughted control plants also showed reduced transpiration and stomatal conductance in the lower leaves as might be expected as they are subjected to less sunlight and they are older than the upper ones.

Table 16.1 Transpiration (TR, mg H_2O m^{-2} s^{-1}) and stomatal conductance (SC, mm s^{-1}) at different leaf positions in droughted and *Verticillium dahliae* infected plants (Haverkort et al., 1990a).

Leaf position	Drought treatment	*V. dahliae* treatment			
		Control		Treated	
		SC	TR	SC	TR
High	Control	15.6	174	19.2	205
Middle	Control	17.2	183	15.5	157
Low	Control	11.9	139	2.1	55
High	Droughted	11.1	173	12.5	149
Middle	Droughted	10.9	150	4.3	63
Low	Droughted	7.1	110	3.1	55

Potato cyst nematodes are also known to affect water collection as nematodes infest roots. Evans et al. (1975) reported a decrease of the leaf water potential and an increase of the stomatal resistance of leaves of potatoes infected with the potato cyst nematode *Globodera rostochiensis*. Haverkort et al. (1991) showed that both drought and potato cyst nematodes reduced plant water potential (Table 16.2) but that the effects were less than additive. A similar response was found in their trials for gas exchange and water-use efficiency. Both drought and nematodes led to increased dry matter concentrations in all plant components. Schans (1991) infected roots with juvenile 2 stages of *Globodera pallida* cyst nematodes and measured reduced gas exchange (CO_2 and H_2O) within 3 days, putatively through a signalling function of ABA. Schans and Arntzen (1991) reported that photosynthetic and transpiration rates were severely reduced 30 days after infection but that thereafter the difference between infected plants and control diminished, probably because a new equilibrium was reached between supply and consumption of water. Also,

Table 16.2 Pressure chamber readings (MPa) of potato plants grown in a container and subjected to drought and/or to potato cyst nematodes (Haverkort et al., 1991).

Treatment		Time after planting (days)		
Drought (d)	Nematodes (n)	34	37	42
Control	Control	0.537	0.602	0.694
Control	Infected	0.802	1.000	0.777
Droughted	Control	0.795	1.203	1.353
Droughted	Infected	0.793	1.045	1.155
P: d x n		<0.1	<0.001	<0.01
P: d		<0.1	<0.001	<0.001
P: n		<0.1	<0.001	NS

NS, non-significant; *P*, probabilities of main effects and interaction.

it should be noted, infected plants take up less water (Haverkort et al., 1991) depleting soil water at a lower rate than controls. Compared with controls, Haverkort et al. (1994) found root length density of infected plants to decline from about 2 to 1 cm cm^{-3} in the topsoil and from about 0.5 cm cm^{-3} to almost 0 in the subsoil.

Whereas drought and cyst nematodes or *V. dahliae* proved to exert less than additive negative impacts, free-living nematodes *(Pratylenchus penetrans)* and *Verticillium* showed synergism manifested by reduced gas exchange in potato (Saeed et al., 1997).

16.6 VARIETAL DIFFERENCES IN DROUGHT TOLERANCE

Varieties may vary in their tolerance of drought, expressed as the relative yield of crops subjected to drought compared with crops optimally supplied with water. Drought tolerance can also be defined as the relative ability of a variety to produce potato tubers from a limiting amount of water. Different aspects of tolerance may assume importance, depending on the timing of water limitation, e.g. early in the season before establishment of stem and tuber numbers, or in the middle of the season when shoots and tubers are growing, or during maturation when no more leaves are being produced and tuber bulking is the main growth process. Spitters and Schapendonk (1990) evaluated drought situations and crop behaviour in a simulation model. Characteristics that assumedly may vary when genotypes are subjected to water limitation include the light-use efficiency, SLA and leaf senescence rate (leaf longevity). The simulated instantaneous effects of drought on such characteristics were substantial, but the ultimate effect on tuber yield was much less and correlated best with the total amount of water transpired [Eq. (16.3)]. Short periods of early drought (up to mid-June) especially affected early genotypes that typically allocate more dry matter to the tubers early in the growth cycle and less to the foliage than late varieties. An early drought led to leaf shedding, and no recuperation was possible. Severe late droughts favoured the early varieties as the later varieties had not allocated similar proportions of dry matter to the tubers. Late genotypes generally tolerated short periods of transient drought relatively better than early ones. Haverkort et al. (1992) examined effects of transient drought on varieties differing in maturity class (Table 16.3)

Table 16.3 Relative values of yield components for unirrigated plants expressed as a percentage of values for fully irrigated controls.

Variety	Y	R	RUE	HI	DMC
Darwina	55	62	99	94	105
Désirée	77	88	99	94	105
Mentor	73	87	97	87	111
Elles	80	93	90	95	101

Fresh tuber yield (Y) equals the product of intercepted solar radiation (R), the radiation-use efficiency (RUE) and the harvest index (HI) divided by the tuber dry matter concentration (DMC). Varieties are ranked in increased lateness. [*Source: Haverkort et al. (1992).*]

and found that the factor affected most by transient drought switched from radiation interception to radiation-use efficiency across the spectrum of early to late varieties. They noted that varieties that shed their leaves (early varieties) before others (late ones) maintained a higher radiation-use efficiency, and vice versa. The results summarized in Table 16.3 were corroborated by Tourneux et al. (2003) who compared six *Solanum* and *Andigenum* genotypes that represented a great diversity in morphological characteristics such as number of leaves and stem length. These characteristics led to varying degrees of ground cover and hence intercepted radiation that was the main determinant of the yield differences because of drought.

That tolerance of drought depends mainly on the time of drought and on the lateness of the genotype was also concluded by DeBlonde et al. (1999). They found early genotypes to escape a late drought, suffering relatively less than late ones, whereas late varieties tolerated an early drought period better. Moreover, they found that the harvest index is reduced by drought indicating that it is one of the factors reducing yield as shown hypothetically by Spitters and Schapendonk (1990) in their modelling exercise.

Genetic differences in water-use efficiency are in the order of 15% between extremes and were found to correlate with ^{13}C discrimination (Vos and Groenwold, 1989a). Superior water-use efficiency would be a valuable trait, implying larger production from a limited amount of water, or conversely a lower requirement for irrigation water to achieve a particular level of production. In wheat, also, lower ^{13}C discrimination correlates with higher water-use efficiency. This led to research, exploring the benefit and feasibility of varieties with improved water-use efficiency. However, it appeared that in the absence of water limitation, water-efficient genotypes generally performed less than 'water spending' types (Condon et al., 2004). Yet, commercial wheat varieties were developed that combine good productivity and improved water-use efficiency. In potato, Tourneux et al. (2003) found the relative yield to correlate with ^{13}C discrimination, but the variation explained less than 42%. Jefferies and MacKerron (1997) arrived at the conclusion that assessing ^{13}C discrimination is not a useful breeding tool as differences among varieties did not significantly correspond with either water-use efficiency or final yield over a growing season.

Atmospheric CO_2 concentrations are expected to rise. Higher water-use efficiencies are found under raised CO_2 levels because higher CO_2 concentrations result in (transient) increase in the rate of leaf photosynthesis in potato and because stomatal conductance is lower (Lawson et al., 2001). The latter results in reduced transpiration. The varietal differences in water-use efficiency observed under current atmospheric CO_2 concentrations are associated with differences in stomatal conductance, too (Vos and Groenwold, 1989a,b), and it would be interesting to explore how these genetic differences are modulated by raised atmospheric CO_2 concentrations.

16.7 EFFECTS OF WATER AVAILABILITY ON QUALITY

The two factors most determining quality of potato are dry matter concentration and tuber size distribution. Dry matter concentration always increases with the duration of growth and ranges from about 10% at tuber initiation to anywhere between 15 and 25% at harvest. The latter range depends on variety of causes, such as length of the growing season, variety, average temperature during the growing season and the availability of water especially at the end of the growing season. The dry matter concentration of the tubers correlates well with the harvest index (DeBlonde et al., 1999). Too severe a drought reduces the harvest index, and the dry matter concentration then is lower than in crops that continue to grow to full maturity. A moderate drought at the end of the growing season usually leads to an increase of the dry matter concentration [Table 16.3 (Haverkort et al., 1992) and DeBlonde et al. (1999)].

The tuber size distribution of a potato crop depends on the yield and the number of tubers per square metre, which in turn depends on the number of stems per square metre and the number of tubers per stem. The stem number is independent of the water availability during the growing season as it is determined by factors prior to emergence such as variety, tuber size and the physiological age of the seed tubers. Haverkort et al. (1990b) subjected potato plants to early drought and counted the number of stems, stolons, tuber initials and tubers. A drought period before stolon appearance greatly reduced the number of stolons, but the subsequent number of tuber initials and tubers was unaffected (variety Radosa) or even increased (variety Bintje). A drought period after stolon and tuber initiation hardly affected tuber numbers in controlled conditions. The authors analysed 13-year experimental data of cultivar Bintje from a particular experimental station. Figure 16.11 shows the relation between the number of tubers per stem and precipitation during the first 60 days after planting. The relation can be represented by

$$\frac{\text{Tubers}}{\text{Stem}} = 0.0196\ P + 1.44 (R = 0.93), \qquad (16.4)$$

where P = precipitation in millimetre. Although the final yield did not correspond significantly with the amount of rainfall during the first 60 days, the final tuber size did: higher tuber numbers led to smaller average tuber size at final harvest. Especially, when high numbers are aimed for – in seed production – adequate moisture availability at stolon formation is essential.

The effects of water availability on quality are known but are less easily influenced by growers than is final yield, which correlates well with the amount of light intercepted

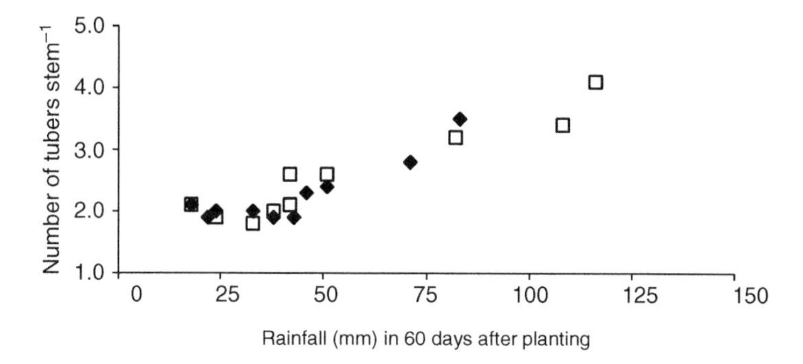

Fig. 16.11. Relationship between precipitation during the first 60 days after planting and the number of tubers per stem. Open symbols represent data from pre-sprouted seed and closed symbols from unsprouted seed at planting. [*Source: Haverkort et al. (1990b).*]

by the crop (if water is not limiting) and correlates well with the amount of water transpired by the crop. Dry soil conditions at the end of the growing season lead to higher dry matter concentrations of the tubers, which may be an advantage where tubers are produced for the processing industry that usually desires higher ranges of solids to increase the recovery rates in the production of chips and crisps. Growers then are given a bonus for high concentrations of solids. Jefferies et al. (1989) analysed data from several experiments with three varieties over several years to derive relations between tuber dry matter concentration and thermal time and water supply. Their model uses a quadratic relation in thermal time and a linear relation with SMD. The relation takes the form:

$$D = a + bT_{\Theta} + cT_{\Theta}^{2} + dSMD,$$

where $D =$ tuber dry matter concentration (%), $T_{\Theta} =$ thermal time (K days) from 50% emergence above a base temperature of 0°C and $a - d =$ all coefficients. The relation was validated against an independent data set where the relation accounted for 79.3% of the variance in the observed data. A sensitivity analysis showed that every change of 10 mm in SMD produces a change of 0.6 percentage points in dry matter concentration. Higher dry matter concentration is undesirable for the fresh potato market, and in such production conditions, an additional final irrigation may increase the tuber water content and the total fresh weight of the crops and thus their market value.

The number of tubers per plant is even more difficult to steer in practice than dry matter concentration. In general, larger seed tubers lead to more stems per plant, and moist soil conditions generally lead to increased tuber numbers per stem. How many stems are formed per seed tuber and how many tubers per stem cannot be predicted with certainty. Growers aiming at producing large tubers for processing using varieties that generally produce a lot of tubers (e.g. Bintje) cannot plan to withhold water around tuber initiation. Instead, they should plan their seed rates on the assumption that the soil will be moist between planting and tuber initiation. Then, if the rainfall is inadequate, irrigation can be used to return the soil to the planned condition. This approach is valid whether growing 'shy' varieties or ones that produce many tubers and whether growing for seed (small tubers required) or bakers (large tubers required) as it removes one variable from the procedure and allows the seed rate to be chosen to suit the variety and the market requirements.

REFERENCES

Ben-Asher J., D.W. Meek, R.B. Hutmacher and C.J. Phene, 1989, *Agron. J.* 81, 776.
Bierhuizen J.F., 1969, *Tuinbouwmeded* 32, 417.
Bowden R.L. and D.I. Rouse, 1991, *Phytopathology* 81, 293.
Boyer J.S. and W.K. Silk, 2004, *Funct. Plant Biol.* 31, 761.
Brewster J.L. and P.B. Tinker, 1970, *Soil Sci. Soc. Am. Proc.* 34, 421.
Chawla J.K. and N.K. Narda, 2001, *Irrig. Drain.* 50, 129.
Condon A.G., R.A. Richards, G.J. Rebetzke and D.G. Farquhar, 2004, *J. Exp. Bot.* 55, 2447.
De Neve S., D.K.L. MacKerron and J. Igras, 2000, In: A.J. Haverkort and D.K.L. MacKerron, (eds), *Management of Nitrogen and Water in Potato Production*, p. 188. Wageningen Pers, Wageningen, the Netherlands.

DeBlonde P.M.K., A.J. Haverkort and J.F. Ledent, 1999, *Eur. J. Agron.* 11, 91.
Evans K., K.J. Parkinson and D.L. Trudgill, 1975, *Nematologica* 21, 273.
Farquhar G.D., M.H. O'Leary and J.A. Berry, 1982, *Aust. J. Plant Physiol.* 9, 121.
Feddes R.A., 1987, *Acta Hortic.* 214, 45.
Fricke W. and T.J. Flowers, 1998, *Planta* 206, 53.
Gandar P.W. and C.B. Tanner, 1976, *Crop Sci.* 16, 534.
Gollan T., J.B. Passioura and R. Munns, 1986, *Aust. J. Plant Physiol.* 13, 459.
Gowing D.J., W.J. Davies and H.G. Jones, 1990, *J. Exp. Bot.* 41, 1535.
Hamdy A., R. Ragab and E. Scarlascia-Mugnozza, 2003, *Irrig. Drain.* 52, 3.
Haverkort A.J., M. Boerma, R. Velema and M. van de Waart, 1992, *Neth. J. Plant Pathol.* 98, 179.
Haverkort A.J., F.J. de Ruijter, M. Boerma and M. van de Waart, 1996, *Eur. J. Plant Pathol.* 102, 317.
Haverkort A.J., F. Fasan and M. van de Waart, 1991, *Neth. J. Plant Pathol.* 97, 162.
Haverkort A.J., J. Groenwold and M. van der Waart, 1994, *Eur. J. Plant Pathol.* 100, 381.
Haverkort A.J., D.I. Rouse and L.J. Turkensteen, 1990a, *Neth. J. Plant Pathol.* 96, 273.
Haverkort A.J. and G.W. Valkenburg, 1992, *Neth. J. Plant Pathol.* 98, 12.
Haverkort A.J., M. van der Waart and K.B.A. Bodlaender, 1990b, *Potato Res.* 33, 89.
Haverkort A.J., A. Verhagen, C. Grashoff and P.W.J. Uithol, 2004, In: D.K.L. MacKerron and A.J. Haverkort (eds), *Decision Support Systems in Potato Production: Bringing Models to Practice*, p 29. Wageningen Academic Publishers, Wageningen, the Netherlands.
Idso S.B., R.D. Jackson, P.J. Pinter Jr., R.J. Reginato and J.L. Hatfield, 1981, *Agric. Meteor.* 24, 45.
Jefferies R.A., 1989, *J. Exp. Bot.* 40, 1375.
Jefferies R.A., T.D. Heilbronn and D.K.L. MacKerron, 1989, *Potato Res.*, 32, 411.
Jefferies R.A. and D.K.L. MacKerron, 1997, *Plant Cell Environ.* 20, 124.
Lawson T., J. Graigon, A.-M. Tulloch, C.R. Black, J.J. Colls and G. Landon, 2001, *J. Plant Physiol.* 158, 309.
McBurney T., 2003, *Acta Hortic.* 619, 447.
Passioura J.B. and S.C. Fry, 1992, *Aust. J. Plant Physiol.* 19, 565.
Saeed I.A.M., A.E. MacGuidwin and D.I. Rouse, 1997, *Phytopathology* 87, 435.
Sanderson J., 1983, *J. Exp. Bot.* 34, 240.
Schans J., 1991, *Plant Cell Environ.* 14, 707.
Schans J. and F.K. Arntzen, 1991, *Neth. J. Plant Pathol.* 97, 297.
Scholander P.F., H.T. Hammel, E.D. Bradstreet and E.A. Hemmingsen, 1965, *Science* 148, 330.
Shae J.B., D.D. Steele and B.L. Gregor, 1999, *Trans. ASAE.* 42, 351.
Shalhevet J., D. Shimshi and T. Meir, 1983, *Agron. J.* 75, 13.
Skovmand B., M.P. Reynolds and I.H. DeLacy, 2001, *Euphytica* 119, 25.
Spitters C.J.T. and A.H.C.M. Schapendonk, 1990, *Plant Soil* 123, 193.
Stalham M.A. and E.J. Allen, 2004, *J. Agric. Sci.* 142, 373.
Steudle E., 1993, In: J.A.C. Smith and H. Griffith (eds), *Water Deficits: Plant Responses from Cell to Community*, p. 5. Bios Scientific Publications, Oxford.
Steudle E., 2000, *Plant Soil* 226, 45.
Tanner C.B., 1981, *Agron. J.* 73, 59.
Tanner C.B. and T.R. Sinclair, 1983, In: H.M. Taylor, W.R. Jordan and T.R. Sinclair (eds), *Limitations to Efficient Water Use in Crop Production*, p. 1. ASA/CSSA/SSSA, Madison, WI.
Tourneux C., A. Devaux, M.R. Camacho, P. Mamani and J.F. Ledent, 2003, *Aronomie* 23, 169.
Van Genuchten T.M., 1980, *Soil Sci. Soc. Am. J.* 44, 892.
Vos J. and J. Groenwold, 1988, *Ann. Bot.* 62, 363.
Vos J. and J. Groenwold, 1989a, *Potato Res.* 32, 113.
Vos J. and J. Groenwold, 1989b, *Field Crops Res.* 20, 237.
Vos J. and P.J. Oyarzún, 1988, *Ann. Bot.* 62, 449.
Wilkinson S. and W.J. Davies, 2002, *Plant Cell Environ.* 25, 195.

Chapter 17

Potato crop response to radiation and daylength

A.J. Haverkort

Wageningen University and Research Centre, P.O. Box 16, 6700 AA Wageningen, The Netherlands

As Monteith (1969) interpreted crop growth in terms of radiation interception, dry matter accumulation was described less and less as a function of actual dry matter and its relative increase over time. The fraction of light passing through the canopy without interception by the leaves depends on the leaf area index (LAI) which in turn determines the proportion of ground covered by green leaves. Intercepted radiation has been found to be linearly correlated with the quantity of dry matter produced which, in the case of potatoes, is distributed over the tubers and the other parts of the plant. Of the total incoming radiation, plants use only wavelengths between 400 and 700 nm for photosynthesis (McCree, 1972) which is about 0.5× global solar radiation (Monteith, 1969). Since Scott and Wilcockson (1978) first related measurement of light interception in potato crops to dry matter accumulation, there has been a steady flow of publications on how to measure intercepted radiation, on radiation use efficiency (RUE), and how it is influenced by biotic and abiotic environmental influences, for example, Khurana and McLaren (1982) to Shah et al. (2004). Moreover, it was recognized increasingly that development, driven by genotype, seed age, temperature and photoperiod, could be quantified partly in combination with radiation-driven growth. The concept also provided the basis for three international potato modelling conferences held in 1990, 1994 and 2003, of which the proceedings were published. In Kabat et al. (1995), the authors compared the performance of potato growth models, mainly simulating potential yields but also yields limited by water and nutrients. In Haverkort and MacKerron (1995), more environmental effects were simulated, and some applications (e.g. climatic change and agro-ecological zoning) were reported. In MacKerron and Haverkort (2004), the quantitative approaches and their use in decision support systems for policymakers and growers were shown.

17.1 RADIATION

17.1.1 Development of radiation interception

Radiation interception by the canopy can be described conveniently in four phases during the development of the crop over the growing season (Fig. 17.1). Throughout these phases, the crop growth rate can be calculated using the foliar expansion rate, whereas

Potato Biology and Biotechnology: Advances and Perspectives
D. Vreugdenhil (Editor)

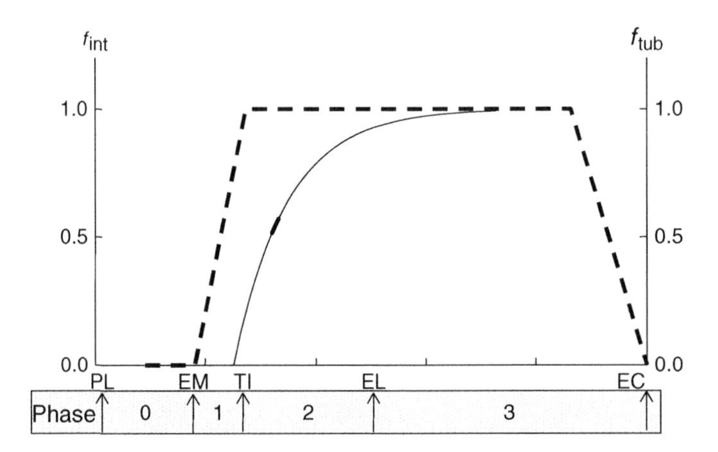

Fig. 17.1. Schematic representation of the fraction of solar radiation intercepted by the crop (dotted line, f_{int}) and the fraction of the daily assimilate that is partitioned to the tubers (drawn line, f_{tub}). PL, planting; EM, emergence; TI, tuber initiation; EL, end of leaf growth (Kooman and Haverkort, 1995).

the light-use efficiency determines the total amount of dry matter produced. During Phase 0 – between planting and emergence – the crop does not intercept radiation, and its development (sprout growth) only depends on soil temperature. There is no increase in dry matter. Phase 1 covers the period between emergence (usually defined as the moment when 50% of the plants have emerged) and tuber initiation. Its duration depends on the development rate until tuber initiation.

Phase 2 covers a period between tuber initiation and the stage when 90–100% of the daily acquired assimilates are partitioned to the tubers. This level of partitioning is only reached asymptotically and, as such, cannot be determined precisely. The length of Phase 2 depends on the relative tuber growth rate, which determines partitioning of dry matter between the tubers and the rest of the plant. The duration of Phase 3 is determined by the rate of leaf senescence and end of crop growth (EC) (Kooman and Haverkort, 1995).

As will be shown in Section 17.1.2, there is a close correlation between the proportion of intercepted solar radiation, the LAI and the proportion of the ground covered by green leaves. The LAI [area of leaves (m^2) per area of soil (m^2)] is determined by the amount of dry matter produced and partitioned to the leaves ($g\ m^{-2}$) and by the specific leaf area [area of leaf formed (m^2) per weight of leaf dry matter (g)]. When a potato crop emerges, its initial foliar growth is entirely due to temperature-dependent translocation of dry matter from the mother tuber to the emerging shoots. Usually potato crops have leaves with a specific leaf area of about $200\,cm^2 g^{-1}$. Prior to emergence, the leaves are much thicker, and they expand after emergence, leading to a greater specific leaf area and a strong increase of the early LAI. The rate of leaf expansion is also determined by temperature (Fig. 17.2).

Once the mother tuber is exhausted or when dry matter translocation stops for other reasons, all growth is autotrophic. To determine the transition from translocation-driven growth to autotrophic dry matter accumulation through intercepted radiation, Van Delden et al. (2001) compared the performance of a model simulating the increase in LAI (L)

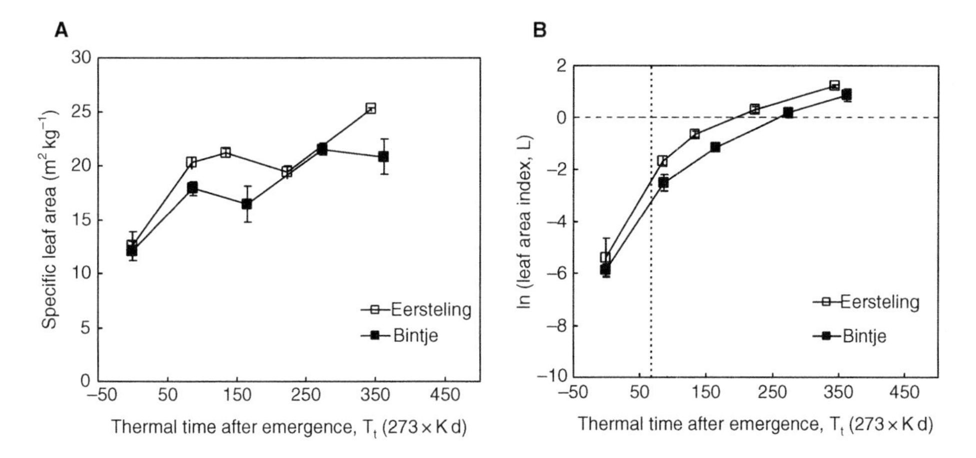

Fig. 17.2. (A) Specific leaf area (SLA) of two potato varieties as a function of thermal time (T_t, K d – Kelvin days) after 50% emergence and (B) the resulting leaf area index (LAI) (Van Delden et al., 2000).

against observations of several cultivars across a range of environments. From emergence ($L = 0$), L was assumed to increase in proportion to the increase in leaf dry weight which, in turn, depended on intercepted radiation, that is the model simulated radiation-limited leaf expansion. Figure 17.3 shows that the simulations that gave the least variation among environments (years and sites) and best reflected actual performance where L, the value of L at the switch to autotrophic growth, was equal to 1 ($R^2 = 0.91$). The sensitivity analysis in Fig. 17.3 also shows that the value of 1 for Ls gave close to the minimum rates of change in leaf area expansion and crop dry weight.

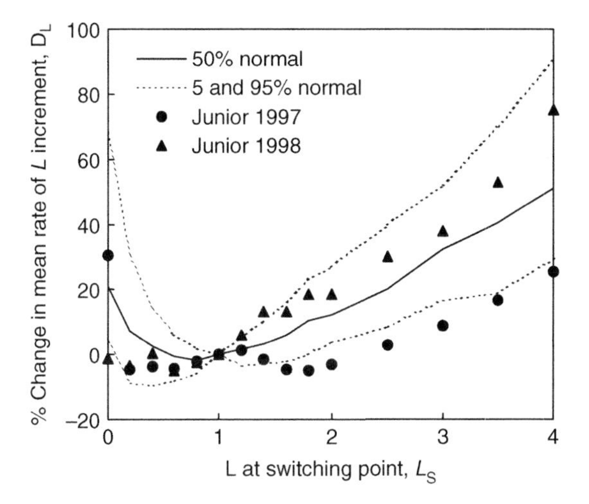

Fig. 17.3. Sensitivity of the mean rate of leaf increment of variety Junior in the Netherlands to a change in L_s (leaf area index at the point when its expansion is limited by radiation only) with an $L_S = 1$ as reference value. Lines represent 5, 50 and 95% of 30-year weather data.

17.1.2 Measurement of radiation interception

Solar radiation intercepted by green leaves is used for dry matter production. As explained in the introduction, photosynthetically active radiation (PAR) is only about half of the incident solar energy. Several methods exist to assess the proportion of solar radiation that is intercepted by the crop.

Figure 17.4 represents two courses in the development of ground cover and, hence, radiation interception. One where full ground cover is achieved and sustained and one where the crop does not achieve 100% ground cover. Proper measurement of radiation interception by the canopy is essential for interpretation of the influence of the environment on the crop, although measurement of ground cover by the canopy provides an approximation. Simulation of the effects of the environment on crop growth and development hinges on good estimates of both incident and intercepted solar radiation.

An appropriate estimate of the proportion of PAR that is intercepted by the foliage can be made using tube solarimeters above and below the canopy with a length equal to the distance between the rows of the crop so as to sample all parts of the row equally. Tube solarimeters measure total solar radiation, and so, an assumption is made that any changes in the spectral composition above and below the canopy are non-significant. Other instruments (e.g. Ceptometer) are available that will measure PAR rather than total solar radiation, but these are more expensive and tend to be used as portable devices to make spot measurements across a crop. Using the cheaper tube solarimeters in permanent locations allows measurements to be integrated across the whole day.

The proportion of the ground covered by green leaves can provide an acceptable estimate of intercepted radiation. Proportional ground cover (PGC) can be estimated using a grid divided in 100 equal sections viewed directly from above (Burstall and Harris, 1983). The dimensions of the frame should be a multiple of the planting pattern. When rows are 75 cm apart and plants are spaced at 30 cm within the row, a frame of 75×90 cm

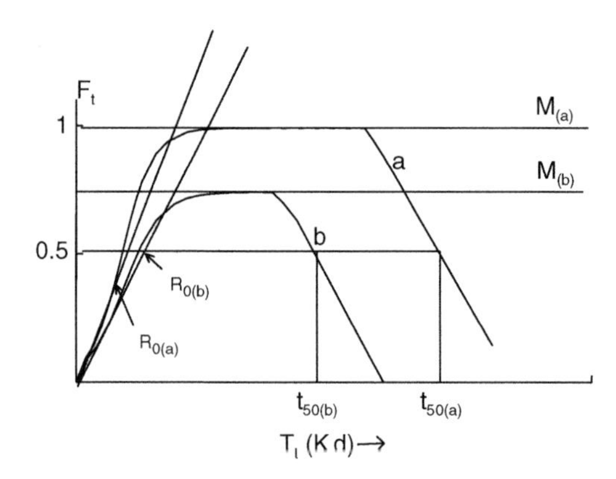

Fig. 17.4. The development of the fraction of the ground covered with green leaves (F_t) with thermal time during the growing season where the crop reaches full ground cover (a) and where it does not (b). R_0 indicates the relative rate of increase in light interception, M the maximum ground cover and t_{50} the thermal time when the ground cover is reduced to 50% of full ground cover (Kooman, 1995).

is appropriate. A measurement consists of counting the sections more than half filled with green leaves.

Measurements of the amount of incoming infrared radiation reflected by the canopy have been found to correlate well with the PGC (Birnie et al., 1987), offering a third non-destructive estimate. A radiometer may be multispectral, but for estimating canopy cover, the radiometer is fitted with filters allowing wavelengths between 836 and 846 nm to pass. The hemispherical irradiance and the crop reflectance are measured nearly simultaneously. The downward angle of view of the system is such that the estimate is based on about 1 m^2 of the canopy.

Light entering at the top of the canopy is extinguished with different extinction coefficients of the several radiation components (e.g. Spitters et al., 1986; Monteith, 2000). The light profile within the canopy can be characterized experimentally by destructive means in which light is measured above and below the canopy, and the leaf area is determined by detaching leaves and determining their area with a commercially available leaf area meter. The proportion of PAR intercepted is calculated as $P_{PAR} = 1 - e^{-kL}$ where L stands for LAI and k for the extinction coefficient that, typically, has a value of about 0.4 (Khurana and McLaren, 1982; Burstall and Harris, 1983; Haverkort et al., 1991).

Haverkort et al. (1991) compared PGC measured with the grid, LAI and infrared reflectance. They found highly significant correlations between the various methods (Fig. 17.5). The authors found that a disadvantage of the solarimeter is that it does not distinguish between green leaves and brown leaves and stems. The method tends to overestimate intercepted radiation in the second half of the growing season. The relationship between PGC measured with the grid and LAI was linear up to LAI = 3 after which full ground cover was reached.

17.1.3 Environmental effects on interception of solar radiation

The amount of photosynthetically active solar radiation intercepted by a potato crop depends on the amount of incident radiation and on the proportion intercepted by the

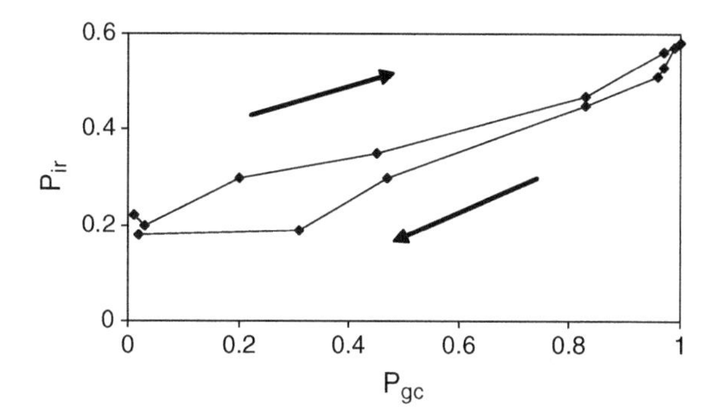

Fig. 17.5. Relationship between proportion ground cover with green leaves observed with the grid (P_{gc}) and the proportion of infrared reflected by the crop (P_{ir}). Arrows indicate the course of time from emergence until crop senescence (redrawn from Haverkort et al., 1991).

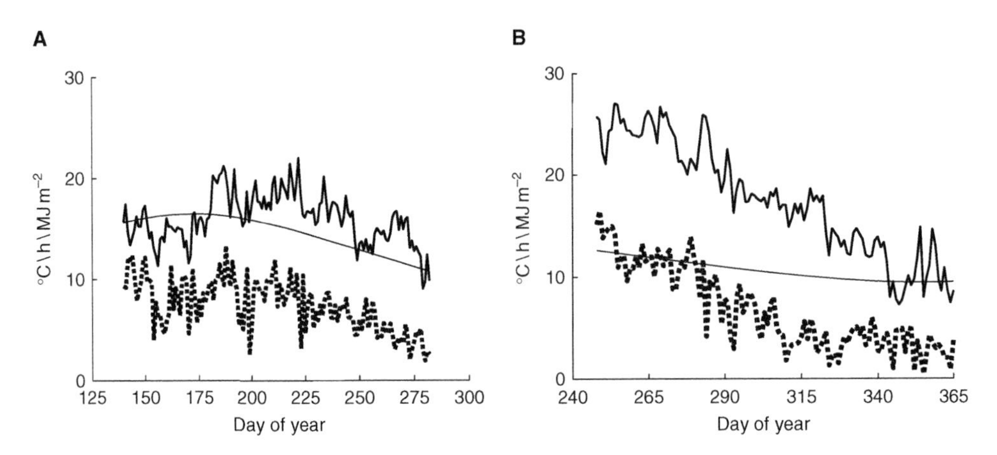

Fig. 17.6. The daily amount of photosynthetically active radiation in 1993 (MJ d^{-1}) (dotted line), daily average temperature in °C (drawn line) and the daylength in hours (thin drawn line) in summer in Wageningen (A) and in autumn in Tunisia (B) (Kooman et al., 1996a).

crop. Total incident PAR depends on the place on earth, time of the year and length of the growing season. Figure 17.6A shows that, in summer in a northern temperate climate, incident PAR varies widely around 9 MJ m^{-2} per day during the growing season, possibly between day 130 and 270, and then it declines. Crops in that part of the world will have been harvested by then. Figure 17.6B shows the development of PAR over time in a Mediterranean autumn growing season where its value declines strongly from about 15 to about 3 MJ m^{-2} per day between planting on day 250 and harvest on day 350.

The highest amounts of radiation intercepted by potato crops are achieved when the length of the crop cycle matches closely the available length of the growing season. The latter is usually determined by the suitable temperature window for potato production between about 5 and 30 °C (MacKerron and Haverkort, 2004). The most common instrument potato growers have to match cycle and season is the choice of the variety. The length of time to maturity (its earliness or lateness) should suit the length of the available season, and subsequently, growers take care that the period when solar radiation is intercepted is not shortened by lack of inputs such as water and nutrients nor by pests and diseases.

Many abiotic factors such as supply of nitrogen (Fig. 17.7) or of water or early or late frosts may reduce soil cover and so radiation interception to varying degrees. Similarly, biotic factors such as potato cyst nematodes (Shah et al., 2004) and late blight caused by *Phytophthora infestans* may cause drastic reduction of yield even as the sole cause of reduced radiation interception (Haverkort and Bicamumpaka, 1986; Van Oijen, 1991).

17.1.4 Radiation use efficiency

The term RUE (g MJ^{-1}) refers to the slope of the relation between total dry matter (g m^{-2}) and cumulative intercepted radiation (MJ m^{-2}) (e.g. Haverkort and Harris, 1987). Unlike cereal crops, where no leaf shedding takes place and no new leaves are formed

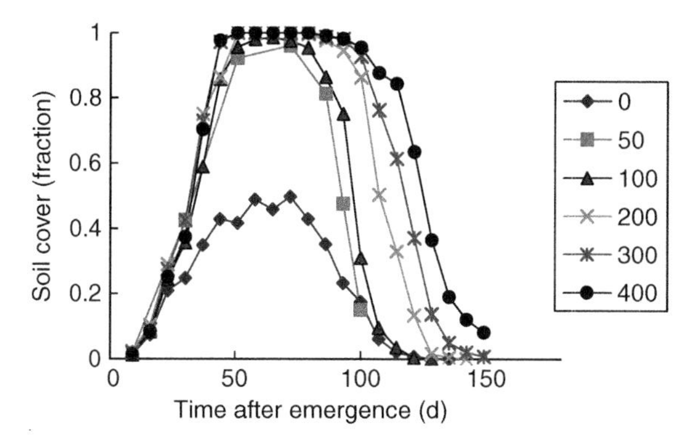

Fig. 17.7. Change in fractional soil cover with time for potatoes receiving 0, 50, 100, 200, 300 or 400 kg N ha^{-1} (Vos and MacKerron, 2000).

after the appearance of the flag leaf, potato crops may shed leaves when conditions are unfavourable (drought and exhaustion of nutrients) yet are able to form new leaves when conditions for growth improve. Therefore, the dynamics of the ability to intercept radiation (Fig. 17.7) is greater. Even so, a great spread in RUE values has been reported varying from 1.5 to 4 g MJ^{-1}. Part of this spread may be due to the methodologies used to measure PAR and its interception, but a large part is due to environmental, abiotic, and biotic factors (Sinclair and Muchow (1999)).

Manrique et al. (1991) referred to light use efficiency and showed these values to be negatively correlated with total solar radiation between 10 and 25 MJ m^{-2}d^{-1}, with increased mean daily temperature between 13 and 26 °C and with increased vapour pressure deficit between values of 0.4 and 1.5 kPa. In their model for potato growth under tropical highland conditions, based on experiments near the equator at elevations between 1300 and 2400 m above sea level, Haverkort and Harris (1987) corrected the conservative RUE for high temperatures and high levels of solar radiation with quantitative routines. Kooman and Haverkort (1995) elaborated this further to modify light use efficiency according to light intensity (roughly equivalent to PAR) as shown in Fig. 17.8.

RUE may or may not be affected by biotic factors such as pests and diseases. Late blight caused by *P. infestans* only reduces radiation interception but not the RUE as was shown conclusively by Haverkort and Bicamumpaka (1986) and by Van Oijen (1991). Inferences from that are that late blight did not influence photosynthesis or respiration. Some other pests and diseases influence the crop other than by removing foliage and may influence the RUE. For example, Haverkort et al. (1992) showed that potato cyst nematodes reduced the amount of intercepted radiation and the RUE. They observed disrupted water relations inferred from changes in water use efficiency and discrimination of ^{13}C.

Shah et al. (2004) reported that RUE increased from 1.8 to 2.8 g MJ^{-1} over a range of five nitrogen levels from 0 to 250 kg ha^{-1}($R^2 = 0.92$). They also reported that RUE declined from 3.07 to 1.9 g MJ^{-1} with level of potato cyst nematode infestation over a range of four nematode densities from 0 to 30 nematodes per millilitre of soil ($R^2 = 0.90$).

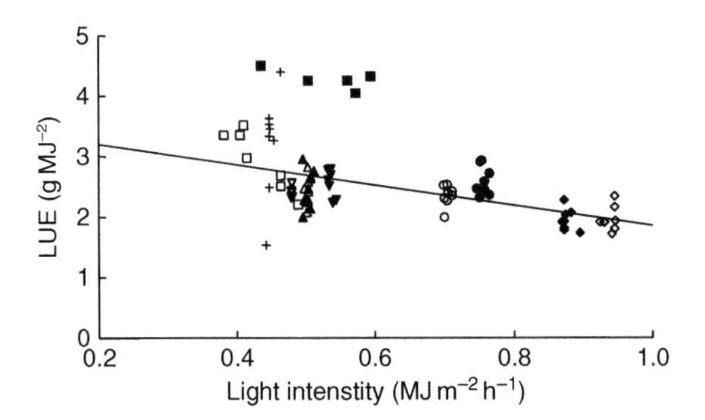

Fig. 17.8. The relation between average light intensity during the growing season and the light use efficiency (LUE) in various temperate and tropical conditions (Kooman et al., 1996b). Symbols refer to experiments in various seasons and sites between 0 and 53° latitude.

17.2 DAYLENGTH

17.2.1 Morphology

Menzel (1985a,b) cited several authors – mainly from the first half of the twentieth century – who reported that compared with short photoperiods, long photoperiods increase stem elongation and stem weight. The numbers of leaves and their weight are also increased while the size of the leaves and leaflets are less. Struik and Ewing (1995) reported a lot of research, both published and unpublished, on the influence of temperature and daylength on a number of crop characteristics. These are summarized in Table 17.1. Struik and Ewing (1995) demonstrated the effect of photoperiod (and temperature) on number of lateral stems per sympodium and per entire shoot (Table 17.2) for a short day variety, Atzimba, and a long day variety, Van Gogh. Long days stimulated the development of the sympodium, particularly in Atzimba and especially at the lower temperature. The higher temperature stimulated development at both photoperiods.

17.2.2 Tuber initiation

In Menzel's (1985a) review of previous literature, he defined a critical photoperiod (CPP) of a cultivar. The change from short to long days is associated with changed growth response in morphology and tuberization. As photoperiod increases towards the CPP, tuberization becomes irregular, then is retarded and finally is inhibited. Although all potatoes have a short-day reaction, there is a strong genotype × environment interaction. Bodlaender (1963) found the CPP to increase at high temperatures and at low irradiance. In general, *Solanum tuberosum* subsp. *andigena* varieties have a CPP of about 12 h and *S. tuberosum* subsp. *tuberosum* a CPP of about 16 h.

Table 17.1 Effects of increased photoperiod on selected crop characteristics (literature review by Struik and Ewing, 1995)

Effect on	Response
Sprout length	Reduced at pre-sprouting
Stem number	See Table 2
Stem branching	Increased
Number of leaves	Increased in first level of main stems only
Lifespan of shoot	Increased
Leaf characteristics	Specific leaf area decreased, leaf size decreased
Photosynthesis	Decreased
Stolon initiation	Delayed
Tuber initiation	Delayed
Tuber growth	Small and inconsistent
Tuber number	Decreased
Tuber yield	Increases when season is long enough as growth cycle increases

Table 17.2 Total number of stems (main stem plus lateral stems) with inflorescences per sympodium or per entire (single stem) shoot, as affected by photoperiod and temperature in cultivars Atzimba and van Gogh. From Struik and Ewing (1995) after Almekinders and Struik (1994)

	Photoperiod (h)			
	12		16	
	Temperature (°C)		(day/night)	
	20/10	30/20	20/10	30/20
Stems/sympodium				
Atzimba	3.2	4.6	3.8	4.8
Van Gogh	2.1	2.3	2.7	2.2
Stems/shoot				
Atzimba	14.0	27.1	27.0	33.1
Van Gogh	2.6	3.8	7.4	4.1

Following the early work on the timing of tuber initiation, emphasis shifted to the consequences of (early) tuberization and early tuber growth. Van Dam et al. (1996) subjected two varieties in controlled conditions with varying lengths of day and average daily temperatures (Table 17.3) and recorded the number of days until tubers appeared and the initial proportion of daily accumulated dry matter partitioned to the tubers.

Table 17.3 The influence of photoperiod and temperature on tuber initiation and dry matter partitioning (Van Dam et al., 1996)

Variety	Photoperiod (h)	Temperature (°C)	Days to tuberization	Relative partitioning rate (d^{-1})
Spunta	12	15	36	0.090
		19	38	0.090
	18	15	44	0.086
		19	47	0.060
Désirée	12	15	33	0.085
		19	36	0.060
	18	15	40	0.051
		19	45	0.035

Longer days retarded tuberization, especially at higher temperatures, and also decreased the relative partitioning rate of dry matter to the tubers. These findings are in line with other studies (Wolf et al., 1990) and Lorenzen and Ewing (1990) who subjected potato plants to changes in photoperiod and followed their dry matter partitioning through ^{14}C distribution and periodic harvests. Kooman et al. (1996a,b) carried out field trials on variety × temperature × daylength in the Tunisian spring (11 h daylength at emergence) and the northern European summer (17 h). They defined the development rate as the inverse of duration from emergence until tuberization. Figure 17.9 clearly shows

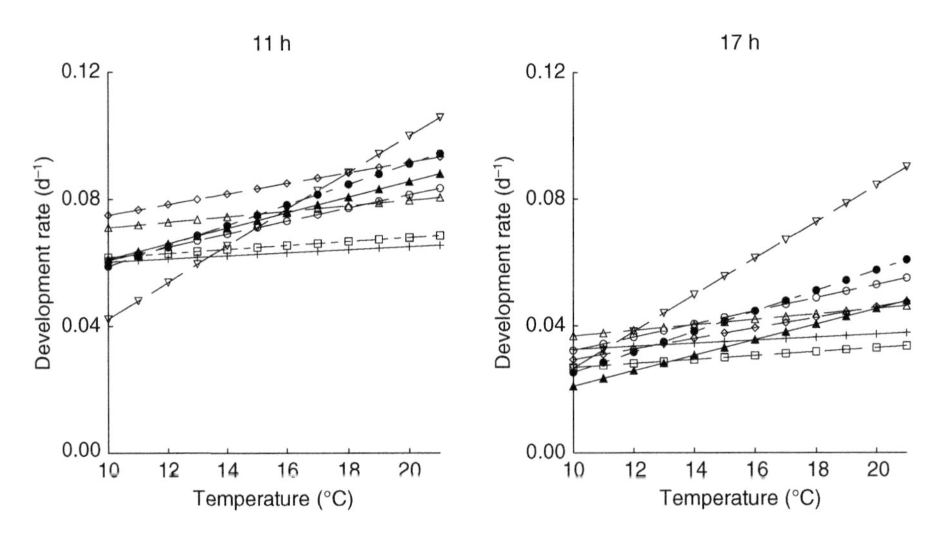

Fig. 17.9. The development rate (d^{-1}) during phase 1, from emergence to tuber initiation against temperature (°C) for Alpha (+), Bintje (Δ), Désirée (o), Diamant (▲), Escort (|), Première (∇), Producent (◊) and Spunta (□) at two daylengths (Kooman et al., 1996b).

higher rates of development (indicative of earlier tuberization) at the shortest daylength. The figures corroborate Van Dam's finding that higher temperatures enhance daylength effects.

17.2.3 Short day sensitivity

Tuberization in the wild-type potato is sensitive to daylength, requiring short days. The expression of the characteristic is less evident in the cultivated *S. tuberosum* but is not absent. Phytochrome is known to be involved in this because the effects of a night break of red light can be reversed by a far-red light treatment given immediately afterwards. Jackson et al. (1997) demonstrated which of the several phytochromes that are known to exist is responsible for the photoperiodic response in potato. They produced transgenic anti-sense short-day-requiring *S. tuberosum* subsp. *andigena* plants which had much lower levels of phytochrome B. These plants were able to tuberize under short and long days; the tuber inducing signal was present. In contrast, through grafting, the tuber-inducing signal in the wild-type was present only under short days.

Arabidopsis CONSTANS (AtCO) is a putative transcription factor known to accelerate flowering in long-day conditions. Martínez-García et al. (2002) have shown that constitutive overexpression of that gene in potato impairs tuberization under short-day inductive conditions. *AtCO* overexpressing lines of *S. tuberosum* subsp. *andigena* required longer exposure to short days before they could tuberize. Using grafts of these lines proved that *AtCO* exerts its action (prohibiting tuber formation) through the leaves. In potato, it delays tuberization. Development of improved understanding of the factors influencing tuberization promises to lead to the breeding of cultivars with the timing of tuberization matched to the growing season of diverse locations.

17.2.4 Earliness

Potato crops mature earlier if a greater proportion of daily dry matter is allocated to the tubers leaving less for the foliage. When all daily assimilates are allocated to the tuber, leaf growth ends and after a temperature-dependent period of time (Kooman et al., 1996b) the foliage dies. Hence, earlier tuber initiation and subsequent early dry matter partitioning to the tubers led to earlier maturing crops. So, tuberization induced by short photoperiods has an effect on the earliness of crop maturity. Kooman et al. (1996b) showed that the longer the duration between emergence and tuber initiation (Fig. 17.1, Phase 1) the longer the two subsequent phases last (from tuber initiation until end of leaf growth and end of leaf growth until crop death, Fig. 17.1). This is shown in Fig. 17.10: later tuber initiation is correlated with later maturing crops over a wide range of cultivars and environments. The authors found the length of Phase 1 (Fig. 17.9) to depend strongly on radiation and temperature, the length of Phase 2 also on these two factors but to a lesser extent, and the length of Phase 3 was shortened by higher temperatures and higher levels of solar radiation. Clearly, the photoperiod-induced timing of tuber initiation is the factor most determining the earliness or lateness of a potato crop.

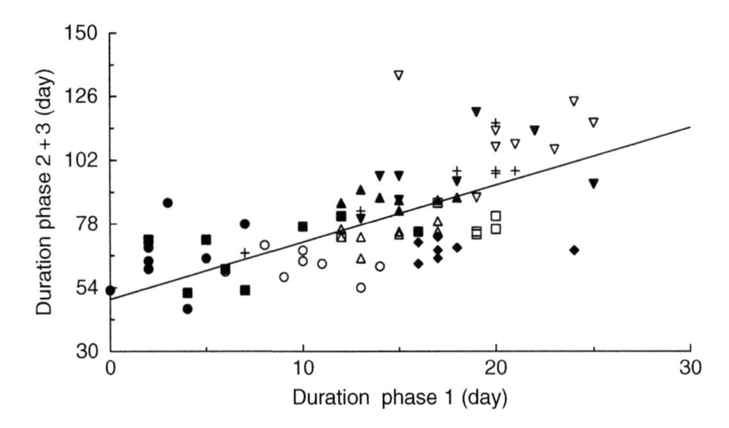

Fig. 17.10. The duration of phase 1 against the joined duration of phase 2 and 3 (Fig. 17.1) for cultivar (see Fig. 17.9 for cultivar names) × location combinations in field trials between 52 and 0° latitude (Kooman et al., 1996b).

REFERENCES

Almekinders C.J.M. and P.C. Struik, 1994, *Neth. J. Agric. Sci.* 42, 311.

Birnie R.V., P. Millard, M.J. Adams and G.G. Wright, 1987, *Res. Dev. Agric.* 4, 33.

Bodlaender K.B.A., 1963, Proceedings 10th Easter School Agricultural Science, University of Nottingham, 199.

Burstall L. and P.M. Harris, 1983, *J. Agr. Sci.* (Cambridge) 100, 241.

Haverkort A.J. and M. Bicamumpaka, 1986, *Neth. J. Plant Pathol.* 92, 239.

Haverkort A.J. and P.M. Harris, 1987, *Agr. Forest Meteorol.* 39, 271.

Haverkort A.J., D. Uenk, H. Veroude and M. Van de Waart, 1991, *Potato Res.* 34, 113.

Haverkort A.J., M. Boerma, R. Velema and M. van de Waart, 1992. *Neth. J. Plant Pathol.* 98, 179.

Haverkort A.J. and D.K.L. MacKerron (eds), 1995, *Potato ecology and modelling of crops grown under conditions limiting growth.* Kluwer Academic Publishers, Dordrecht, 380 pp.

Jackson S.D, S. Prat and B. Thomas, 1997, *Acta Hortic.* 435, 159.

Kabat P, B. Marshall, B.J. van den Broek, J. Vos and H. van Keulen (eds), 1995, *Modelling and parameterization of the soil-plant-atmosphere system: a comparison of potato growth models.* Wageningen Academic Publishers, the Netherlands, 513 pp.

Khurana S.C. and J.S. McLaren, 1982, *Potato Res.* 25, 329.

Kooman P.L., 1995, Yielding ability of potato crops as influenced by temperature and daylength. PhD thesis, Wageningen University, the Netherlands, 155 pp.

Kooman P.L. and A.J. Haverkort, 1995, In: A.J. Haverkort and D.K.L. MacKerron (eds), *Potato ecology and modelling of crops under conditions limiting growth,* p. 41. Kluwer Academic Publishers, Dordrecht.

Kooman P.L., M. Fahem, P. Tegera and A.J. Haverkort, 1996a, *Eur. J. Agron.* 5, 193.

Kooman P.L., M. Fahem, P. Tegera and A.J. Haverkort, 1996b, *Eur. J. Agron.* 5, 207.

Lorenzen J.H. and E.E. Ewing, 1990, *Ann. Bot.* 66, 457.

MacKerron D.K.L. and A.J. Haverkort (eds), 2004, *Decision support systems in potato production: bringing models to practice.* Wageningen Academic Publishers, Wageningen, 238 pp.

Manrique L.A., J.R. Kiniry, T. Hodges and D.S. Axness, 1991, *Crop Sci.* 31, 1044.

Martínez-García J.F., A. Virgós-Solerand and S. Prat, 2002, *Proc. Natl. Acad. Sci. U.S.A.* 99, 15211.

McCree K.J., 1972, *Agr. Meteorol.* 10, 443.

Menzel C.M., 1985a, *Ann. Bot.* 55, 35.

Menzel C.M., 1985b, *Field Crop Abst.* 38: 527.

Monteith J.L., 1969, In: J.D. Easton, F.A. Haskins, C.Y. Sullivan and C.H.M. van Bavel (eds), *Physiological aspects of crop yield*, p. 89. American Society of Agronomy, Madison, Wisconsin.
Monteith J.L., 2000, *Agr. Forest Meteorol.* 104, 5.
Scott R.K. and S.J. Wilcockson, 1978, In: P.M. Harris (ed.), *The potato crop; the scientific basis for improvement*, p. 678. Chapman and Hall, London.
Sinclair T.R. and R.C. Muchow, 1999, *Adv. Agron.* 61, 216.
Shah S.F., B.A. McKenzie, R.E. Gaunt, J.W. Marshall and C.M. Frampton, 2004, *N.Z. J. Crop Hort.* 32, 113.
Spitters C.J.T., H.A.J.M. Toussaint and J. Goudriaan, 1986, *Agr. Forest Meteorol.* 38, 217.
Struik P.C. and E.E. Ewing, 1995, In: A.J. Haverkort and D.K.L. MacKerron (eds), *Ecology and modelling of potato crops under conditions limiting growth*, p. 19. Kluwer, Dordrecht, the Netherlands.
Van Dam J., P.L. Kooman and P.C. Struik, 1996, *Potato Res.* 39, 51.
Van Delden A., A. Pecio and A.J. Haverkort, 2000, *Ann. Bot.* 86, 355.
Van Delden A., M.J. Kropff and A.J. Haverkort, 2001, *Field Crops Res.* 72, 119.
Van Oijen M., 1991, *Potato Res.* 34, 123.
Vos J. and D.K.L. MacKerron, 2000. In: A.J. Haverkort and D.K.L. MacKerron (eds), *Management of nitrogen and water in potato production*, p. 15. Wageningen Pers, Wageningen.
Wolf S., A. Marani and J. Rudich, 1990, *Ann. Bot.* 66, 513.

Chapter 18

Responses of the Potato Plant to Temperature

Paul C. Struik

Crop and Weed Ecology, Plant Sciences Group, Wageningen University, Haarweg 333, 6709 RZ Wageningen, The Netherlands

18.1 INTRODUCTION

The production and quality of a potato (*Solanum tuberosum* L.) plant is determined by the dynamics of the development of many organs (including stems, leaves, stolons, tubers and roots), the total dry matter production and the dry matter partitioning. Yield is determined by the number of plants per square metre, the number of tuber-bearing stems per plant, the number of stolons per tuber-bearing stem, the number of tubers per stolon and the average tuber weight. These parameters are all a result of a sequence of complex physiological events, many of them strongly influenced by temperature. The relative response of potato to temperature differs greatly between physiological processes.

18.1.1 Background and warnings

When discussing the literature on the effects of temperature, it is essential to discriminate between different parts of the plant exposed to certain temperatures, the period of exposure to specific temperature treatments and the range over which the temperature effects are investigated. For example, temperature effects on tuberization are very different for increases in temperature in the air, in the nutrient solution where the roots are grown or in the stolon environment where stolons and tubers are kept. Effects are also different for short or long exposure and for exposure during specific developmental stages. Finally, effects of increases in temperature can be opposite for different ranges.

The general literature on temperature on plant and crop physiology discriminates between a base temperature (i.e. the temperature below which a certain process does not occur), an optimal temperature (i.e. the temperature at which the process reaches its maximum rate) and a ceiling temperature (i.e. the temperature above which the process halts completely). In the range from the base temperature to the optimal temperature, an increase in temperature will result in an increase in the rate of the process. In the range from the optimal temperature to the ceiling temperature, the process will be slowed down by an increase in temperature. Below the base temperature and above the ceiling temperature, the process will not occur at a noticeable rate. Naturally, the changes in the rates of developmental processes with changes in temperature between the base temperature and the ceiling temperature are not always linear. Also, if the function is non-linear, then the development rates will differ between temperature regimes with the same

Potato Biology and Biotechnology: Advances and Perspectives
D. Vreugdenhil (Editor)

daily mean but with different diurnal fluctuations. The night temperature (usually being the lower one) is also considered to be more important for the rates of certain processes than the day temperature. This could simply be associated with the fact that the night temperature usually lies within the range of temperatures where the temperature response is still large and linear, whereas many of the day temperatures are in a range where the response to temperature is much smaller and curvilinear. This means that temperature sum (or thermal time, i.e. the integral of the product of time and temperature) is not always a useful concept. This is more so when the response to temperature also depends on plant age or stage of development.

18.1.2 Reader's guide

In this chapter, the focus will be on the temperature range between the base temperature and the ceiling temperature as experienced by a plant in the field or in facilities of climatic control. Temperatures at which the plant can no longer function or is even directly damaged (e.g. temperature at which the plant is killed by frost or scorched by heat) are not taken into account. The contribution starts with a description of the effects of temperature during the several phases of the potato plant (from seed tuber planted to seed tuber stored) and applying the morphological description of above-ground and below-ground development given elsewhere in this book (Chapter 11, Struik, this volume) as a guideline. For a proper understanding of the terminology, the readers are referred to that chapter. Subsequently, effects of temperature on the basic plant processes (photosynthesis, dry matter production and dry matter partitioning) are described, followed by sections in which partial exposure to specific temperatures, temporary exposure to certain temperature treatments and the effects of diurnal temperature fluctuations are described. Finally, we briefly describe the effects of temperature during storage on the physiology of seed tubers used to produce the next progeny.

18.2 SPROUT GROWTH, EMERGENCE AND CROP ESTABLISHMENT

The focus in this section is on soil temperatures.

Firman et al. (1991) showed that 20–40 leaf primordia may be produced on sprouts of ageing sprouted seed tubers kept in storage for a long period, before these sprouts start flower initiation. This suggests considerable variation in leaf number. However, in practice, the number of leaves on the main stem is rather constant, although the number of below-ground nodes may be variable. In normal agricultural practice, in most potato-growing areas, the seed tuber planted is no longer dormant and often has started sprout growth. Usually, the seed tubers have not been pre-sprouted for so long that there are already flower primordia present. This means that the final number of leaves of the main stem until the first inflorescence is influenced by conditions during storage, pre-conditioning of the seed, soil temperatures after planting and ambient conditions after emergence. However, the year-to-year variation in number of leaves of the main stem until the first inflorescence is much smaller than the variation in number of basal lateral branches, sympodial branches and number of leaves on those branches.

Soil temperature after planting drives sprout growth. Klemke and Moll (1990) stated on the basis of a literature review that the optimum growth rate of the sprout is about 20 °C or slightly higher. They produced a model for sprout growth which included the effect of physiological age of the seed tuber.

According to this model, emergence requires about 116 K d (Kelvin days, base temperature 2.9 °C) under optimal soil moisture. The model of MacKerron and Waister (1985) assumes that seed tubers require 125 K d (base temperature 5 °C) for sprouting. Once the seed tubers have sprouted, sprout extension proceeds at a rate of 1 mm per K d (base temperature 2 °C). Firman et al. (1992) claimed that, below 20 °C, lower soil temperatures delay and reduce sprout growth.

Squire (1995) used data of Sale (1979), Headford (1962) and Kirk et al. (1985) to analyse the relation between rate of emergence, rate of sprout extension and rate of leaf primordia formation. He showed clear maximum rates around 25 °C for 1/time (expressed in d^{-1}) to 50% emergence and sprout extension rate up to 20 mm in length. However, the optimum temperature for 1/time to 20 leaf primordia was above the highest temperature tested in the experiment he cited (i.e. above 30 °C). These optimal temperatures are well above normal soil temperatures in temperate regions, but, in semi-tropical and tropical zones, actual soil temperatures may be well above these optimum temperatures. Sale (1979) and Midmore (1984) showed that soil temperatures above 25 °C delay or even impede emergence, reduce plant survival and reduce the number of main stems per plant. Under such conditions, it might be advantageous to grow potato as an intercrop or relay-crop. The shading of the companion crop reduces soil temperature and damps temperature fluctuations and therefore advances and improves plant emergence (Midmore et al., 1988). Later during the growing season, the shading by the intercrop may also advance tuberization and enhance tuber bulking, although Midmore et al. (1988) observed a delay in tuberization despite the earlier emergence. Similar effects on soil temperature can be obtained by mulching (Midmore et al., 1986a,b).

Rapid emergence and rapid early shoot growth advance the time when maximum light interception starts. When the duration of the growing season is fixed by haulm killing or adverse conditions at the end of the growing season, this will result in higher yields. Rapid emergence and early growth may also increase yields by allowing a higher fraction of light intercepted during periods when light is abundantly available.

Table 18.1 summarizes data from Van Delden et al. (2000), indicating the thermal time required from planting until emergence under temperature conditions for different cultivars, mulch treatments, seed tuber treatments and planting depth.

18.3 THE SHOOT SYSTEM

In this section, the effects of temperature on the development of the shoot system are described. Effects include leaf appearance rate, the number of leaves per stem, the size of the leaves, the number of basal and sympodial lateral branches and other morphological and physiological aspects.

Table 18.1 Effects of cultivar, plastic mulch, pre-sprouting and planting depth on the soil thermal time (K d, Kelvin days) with a base temperature of 2 °C required from planting until emergence.

	Thermal time (K d)
Eersteling	249
Bintje	264
Plastic mulch	211
No plastic mulch	303
Pre-sprouted	211
Unsprouted	302
Planting depth 8 cm	227
Planting depth 18 cm	286

Source: Van Delden et al. (2000).

18.3.1 Leaf appearance

According to the older literature, the rate of leaf appearance peaks at a daytime temperature of 20 °C irrespective of night temperature (Borah and Milthorpe, 1962; Vos, 1995). Kirk and Marshall (1992) observed that the rate of leaf appearance increased linearly over the temperature range 9–25 °C. The temperature coefficient for the leaf appearance rate was 0.032 leaves per K d (base temperature 0 °C). Above 25 °C, they did not observe a further increase in rate. However, recent literature gives different results (Fleisher et al., 2006). Table 18.2 summarizes the results of a recent, thorough study on the effects of temperature on leaf appearance rate, in which also the leaves of the uppermost apical lateral stems, including all secondary and tertiary branches, were included. Note that Struik (Chapter 11, this volume) indicated that the rate of leaf appearance on apical, sympodial branches is lower than that on the main stem. Table 18.2 summarizes that leaf appearance shows optimal values at an average daily temperature of about 28 °C. Temperatures above that value show a longer phyllochron (measured in K d). The maximum rate of leaf appearance is 0.98 leaves per stem (the main stem and the uppermost apical lateral stems including all secondary and tertiary branches) per day.

18.3.2 Final leaf number

The final number of leaves of the main stem below the first inflorescence is affected by air temperature, but the effects are rather small within normal temperature ranges (Firman et al., 1991; Almekinders and Struik, 1994). High temperatures extend the period of leaf formation through their positive effects on sympodial growth (Marinus and Bodlaender, 1975). Although this temperature effect results in a lower harvest index under higher temperatures than under lower temperatures, it prolongs the life of the crop if weather conditions permit continued growth.

The final total number of leaves per plant may be increased greatly by an increase in temperature, because of the combined effects of an increase in number of basal and apical

Table 18.2 Leaf appearance rates in many temperature regimes, phyllochron (values for a base temperature of 4 °C) and relevant parameters from several regression models.

Assessment of linear leaf appearance rate		
Temperature regime (°C) day (16 h)/night (8 h)	**Leaf appearance rate (leaves per plant per day)**	**Phyllochron (K d per leaf)**
14/10	0.40	21.6
17/12	0.46	23.4
20/15	0.71	20.9
23/18	0.84	21.4
28/23	0.98	23.3
34/29	0.89	31.4
Common quadratic fit between leaf appearance rate and observed average daily temperature for all three atmospheric CO_2 treatments (no CO_2 effect or $CO_2 \times$ temperature interaction observed)		
Intercept (leaves per plant per day)		–1.41
Linear regression coefficient in leaves per plant per day per K		0.175
Quadratic regression coefficient in leaves per plant per day per $(K)^2$		–0.0033
Common phyllochron for two sets of temperature treatment data		
Phyllochron	28.2 K d per leaf[a]	
	23.5 K d per leaf[b]	
Modified beta response function parameters		
Tmax (°C)	39.9	
Topt (°C)	28.0	
Rmax (leaves per plant per day)	0.98	

Based on Fleisher et al. (2006); data are from the lowest atmospheric CO_2 treatment (370 μmol mol^{-1}), except when stated otherwise. Tmax is the ceiling temperature at which leaf appearance ceases, Topt is the temperature at which leaf appearance rate is maximal and Rmax is the maximum leaf appearance rate (i.e. leaf appearance rate at Topt).

[a] Value based on all treatments.

[b] Value based on all treatments except 34/29 °C.

Kelvin is the unit of temperature difference in both the Kelvin and the Celsius scales.

lateral branches (Struik et al., 1989a) and an increase in the number of leaves on those branches. The effect of temperature on the number of leaves per lateral branch at two photoperiods is summarized for the main stem and three types of branches in Table 18.3. Indeed, the relative effect of temperature on the number of leaves on the main stem is smaller than the relative effect on the number of leaves on either the low lateral branch or the total sympodium. The effects of temperature on the number of leaves of the entire shoot, however, are mainly brought about by the effect on the number of branches as the effect on the total shoot is much larger than the effect of one entire sympodium, at least in short days. Note the interaction between photoperiod and temperature for total leaf numbers: at higher temperature, there is a negative effect or only very small positive effect of long day on leaf number, whereas at low temperature, the long-day effect is large. This interaction is – again – mainly caused by the effects on number of branches (see also Section 18.3.7).

Increases in air temperature are particularly effective in increasing the number of leaves present by a given date, but, in some cultivars, higher root temperatures may

Table 18.3 The number of leaves produced by the main stem and completed secondary stems developing from nodes $n-1$, $n-2$ and $n-14$ on the main stem (n being the top leaf of the main stem) and the total number of leaves of the sympodium and of the entire shoot.

	Cv. Atzimba				**Cv. Van Gogh**			
	Temperature regime (°C)							
	20/10		**30/20**		**20/10**		**30/20**	
	Photoperiod (h)							
	12	**16**	**12**	**16**	**12**	**16**	**12**	**16**
Main stem	15.9	16.0	18.5	18.4	15.3	15.5	19.8	20.7
Secondary stem ($n-1$)	5.9	6.8	7.3	9.2	5.6	6.4	8.1	8.8
Secondary stem ($n-2$)	8.5	8.7	9.2	11.0	5.2	6.1	9.8	15.0
Secondary stem ($n-14$)	13.8	16.2	16.2	20.1	–	–	–	–
Total sympodium	31.0	41.8	48.0	48.3	25.5	32.3	54.0	56.5
Total entire shoot	169	348	392	415	30	104	90	85

For definitions of different stem parts, see Chapter 11. [*Source: Almekinders and Struik (1994).*]
–, no data available.

have an additional effect because they shift the dry matter partitioning to the haulm (Struik et al., 1989a). The final number of leaves per plant may be highest at about 30 °C (Mosille, 1985).

18.3.3 Leaf growth and leaf size

Kirk and Marshall (1992) showed a clear relationship between mature leaf size and leaf position on the main stem: the mature leaf size first increases with higher leaf insertion number and subsequently declines. The duration of leaf expansion is much more conservative than the rate of leaf expansion. Not only the time of appearance of each individual leaf but also the moment that an individual leaf reaches mature size can be calculated on the basis of thermal time (Jefferies, 1989; Kirk and Marshall, 1992). Leaf expansion is very sensitive to water stress (Jefferies, 1989; Jefferies and MacKerron, 1993).

Ewing and Struik (1992) claimed that when plants have been exposed to higher temperatures, leaves are smaller (shorter and narrower), more abundant (total dry weight increases), also the leaflets are smaller, and the angle of the leaf to the stem is more acute and the leaf: stem ratio decreases.

Kirk and Marshall (1992) illustrated the effect of temperature on individual leaf growth. According to Benoit et al. (1983), the rate and duration of individual leaf growth show an optimum at a temperature of 25 °C, that is slightly below the optimum temperature for the rate of leaf appearance reported by Fleisher et al. (2006). However, Vos (1995) reported a lower optimum temperature for rate and duration of leaf growth, in agreement with his lower optimum temperature for leaf appearance rate. Final leaf size (the product of rate and duration of individual growth) is optimal at relatively low temperatures, especially because the duration of leaf growth has a low optimum temperature. Individual leaf size tends to increase with a decrease in temperature well below the optimal temperatures for leaf appearance rate and leaf expansion rate, that is well below 25 °C. Despite the positive effect of high temperatures on leaf appearance and final leaf number, the negative effects of high temperature on individual leaf size result in a reduced total leaf area at high temperature.

18.3.4 Life span of leaves and specific leaf area

Vos (1995) inferred from literature data that leaves may have a life span of 60 to more than 80 days. Struik (Chapter 11, this volume) mentioned a life span between 30 and over 100 days. Vos (1995) also indicated that leaves of the main stems of cv. Bintje senesced (i.e. turned yellow) at a rate of 0.17 leaves per day when average temperature was 15 °C. Clear data on the effect of temperature on the life span of the individual leaf are scarce. However, it is generally assumed that the leaf ages physiologically faster under higher temperatures within a wide range of temperatures (Menzel, 1985). Optimum temperatures for life span of individual leaves are probably relatively low (Burton, 1972). However, the life span of the entire plant may be prolonged by an increase in air temperature (Struik et al., 1989a), certainly when expressed in terms of thermal time, because of the large positive effect on number of leaves. Stolon and root temperatures may also affect the life span of the canopy to some extent (Struik et al., 1989a). Under long days, the life span of the shoot is increased by high temperatures compared with low temperatures, but under short days, high temperatures may decrease the whole-shoot life span. The development of the shoot is mainly affected by the number of basal and apical lateral stems. Ewing and Struik (1992) indicated that senescence may be either accelerated or slowed down by increasing temperature, depending on photoperiod and other environmental conditions. Above 30 °C, however, senescence is usually accelerated by higher temperatures (Fahem and Haverkort, 1988; Midmore, 1990; Kooman et al., 1996).

Kooman et al. (1996) showed that the variation in the duration of the period of crop senescence is much larger than the variation in the duration of the period between emergence to tuber initiation or of the period between tuber initiation and the end of leaf growth. They confirmed that the effects of temperature during this period were contradictory, although the effect of higher temperature was mostly in the direction of faster senescence. They explained this contradiction by identifying two opposite effects of an increase in temperature: (1) higher temperatures may reduce the sink strength of the tubers, thus slowing down the rate of leaf senescence, and (2) higher temperatures hasten leaf senescence, thus advancing crop senescence.

Given the effect of temperature on the rate of haulm development, tuberization and sink strength, it is likely that temperature effects on whole-plant life span interact with maturity type and with the indeterminate or determinate behaviour of different cultivars. To some extent, the maturity class of a cultivar can be defined on the basis of the thermal time required to complete its life cycle naturally. However, the ranking of the cultivars based on the duration of their life cycle may not be consistent across different ecological zones. There are large differences between cultivars in their response to temperature

Table 18.4 Total number of stems (main stems plus lateral stems) with inflorescences per sympodium or per entire (single stemmed) shoot, as affected by photoperiod and temperature in cultivars Atzimba and Van Gogh.

	Photoperiod (h)			
	12		**16**	
	Temperature (°C)			
	20/10	**30/20**	**20/10**	**30/20**
Stems per sympodium				
Atzimba	3.2	4.6	3.8	4.8
Van Gogh	2.1	2.3	2.7	2.2
Stems per shoot				
Atzimba	14.0	27.1	27.0	33.1
Van Gogh	2.6	3.8	7.4	4.1

Source: Almekinders and Struik (1994) and Struik and Ewing (1995).

during different phases of development (Kooman et al., 1996), and there are strong three-way interactions between cultivar, photoperiod and temperature (Ewing and Struik, 1992; Almekinders and Struik, 1994, 1996) (Tables 18.3 and 18.4). This could result in the phenomenon that some cultivars that are indeterminate in the North are nearly determinate in the South.

Temperature affects the specific leaf area (SLA): SLA of plants grown under high temperatures is higher than that under cool temperatures (Midmore and Prange, 1991).

18.3.5 Number of stems

The number of stems is mainly determined by the seed size, the physiological age of the seed tubers and their pre-treatment (Struik and Wiersema, 1999) and by the soil conditions during emergence (see Section 18.2). High temperature during crop establishment can result in stem death before production takes place. Temperature, however, has a much larger effect on stem morphology and on the number of basal and apical lateral stems (see Sections 18.3.6 and 18.3.7).

18.3.6 Stem morphology

Stem morphology is affected by temperature: higher temperatures enhance branching, both lower branching and sympodial branching. Both root and shoot temperatures have an effect on these phenomena (Struik et al., 1989a).

Stem growth has a higher optimal temperature than leaf growth, resulting in a decline in the leaf: stem ratio with an increase in temperature above the optimum for leaf growth. Lafta and Lorenzen (1995) reported that supra-optimal temperatures resulted in taller plants, higher stem dry weights and smaller leaves compared with optimal temperatures for potato growth. Similar results were observed by Prange et al. (1990). Ewing and Struik (1992) stated that at higher temperatures (night temperatures above 20 °C and day temperatures above 30 °C), stems are taller, because the internodes are longer and because there are more sympodial branches. Furthermore, higher temperatures tend to enhance the auxiliary branching at lower nodes.

The optimum temperature for rate of stem elongation is above 25 °C (Borah and Milthorpe, 1962). Also stem dry matter production has a high optimum temperature, at least considerably higher than leaf production, resulting in a decline in leaf: stem ratio with an increase in temperature (Bodlaender, 1963; Wheeler et al., 1986; Ewing and Struik, 1992).

18.3.7 Stem branching

One of the most significant effects of temperature on the shoot system is the effect on the formation of basal lateral and apical branches. The shoot system is described in great detail by Struik (Chapter 11, this volume). Here, we describe the effects of temperature on the formation of branches. Table 18.4 illustrates the effect of temperature and photoperiod on the total number of stem segments (i.e. main stems plus lateral stems) with inflorescences (i.e. completed stem segments) for two cultivars. The effects of temperature on stems per symposium are relatively small, but the effects on the number of stems per shoot are large, especially for cv. Atzimba. Note the interactions between temperature and photoperiod and between temperature and cultivar at the longer photoperiod.

18.4 STOLONS

Usually the potato cultivars grown in temperate regions do not require a strong induction to produce stolons. The first stolons may already be produced shortly after or even before emergence, although this depends on the physiological status of the seed tuber. Stolon formation is usually enhanced by an increase in temperature over a wide range of temperatures. Higher temperatures not only stimulate the development of stolons (Bodlaender et al., 1964; Burt, 1964; Saha et al., 1974; Moorby and Milthorpe, 1975) but also favour stolon branching, thus increasing the number of potential tuber sites both per main stolon and per plant (Struik et al., 1989b,c). Very high temperatures, however, may partly or even almost completely impede the stolon formation (Struik et al., 1989b). Indeed, some reports indicate that high soil temperatures tend to reduce the number of stolons or the stolon yield (Randeni, 1980; Lemaga, 1986). Midmore (1984) showed that stolon development was delayed by higher temperature. However, the final stolon number was increased by higher temperature, probably because tuber initiation was more delayed than stolon initiation, thus allowing more time and providing more assimilates for further stolon formation.

Table 18.5 summarizes the effects of two levels of temperature in the environments of the shoot (air temperature), root (nutrient solution temperature) and stolons (temperature of a separate stolon chamber) on number of stolons, stolon weight and the potential number of tuber sites (tips of stolons or stolon branches). For stolon numbers, the effects

Table 18.5 Interaction between shoot, root and stolon temperature for cvs Bintje and Désirée with respect to stolon number, stolon weight of four stolons (numbers 4–7 when ranked in the order of date of initiation; mg dry weight per plant) and potential tuber sites per plant in a growth chamber experiment.

Temperature (°C) in environment of			Stolon number		Stolon weight		Potential tuber sites	
Shoot	Root	Stolon	Bintje	Désirée	Bintje	Désirée	Bintje	Désirée
18	18	Normal	25.9	22.6	131	96	153	186
18	18	Increased	26.7	25.0	418	199	250	161
18	28	Normal	26.4	21.8	345	145	272	149
18	28	Increased	19.0	19.0	346	134	202	121
28	18	Normal	26.0	15.8	416	32	309	65
28	18	Increased	21.3	12.7	92	8	140	31
28	28	Normal	12.4	11.2	0	0	99	35
28	28	Increased	6.5	8.9	0	0	24	15

Source: Struik et al. (1989b,c).

of increased temperatures in only one of the three compartments did not have a large effect on cv. Bintje, although increased air temperature alone decreased the stolon number of cv. Désirée. Combining high shoot temperature with a high root temperature had a detrimental effect on stolon number of cv. Bintje, an effect further aggravated by an increase in the stolon temperature. The weights of four selected stolons increased in response to an increase in temperature in any environment for cv. Bintje. Only when an increase in air temperature was combined with an increase in root temperature or stolon temperature (or both), a severe reduction of stolon weight was observed. In cv. Désirée, an increase in air temperature (alone but especially in combination with an increase in temperature in the root or stolon compartments or with an increase in both) strongly reduced stolon weight. A combination of high shoot and root temperatures (with its associated relatively high normal temperature in the stolon chamber) reduced the dry weights of the four selected stolons virtually to zero. In cv. Bintje, the number of potential tuber sites per plant was increased by an increase in the temperature of the shoot, root or stolon environments. A combination of high root and stolon temperatures however reduced the number of potential tuber sites, especially when the stolon temperature was increased also. In cv. Désirée, the control treatment (cool temperatures in all environments) gave the largest number of potential tuber sites, with the largest detrimental effects of high air temperatures.

Note that in these experiments, the increase in stolon temperature was not independent of the root temperature (Struik et al., 1989a).

Table 18.6 summarizes the stolon and tuber characteristics in a growth chamber experiment with cv. Désirée associated with two levels of root temperature. Although the number of tuber-bearing stolons was not affected over this range of root temperature, the number of tubers per tuber-bearing stolon and thus the total number of tubers per plant were higher for the higher root temperature. However, the total tuber yield was unaffected by the root temperature treatments. The combined result of no effect on total yield and

Table 18.6 Stolon and tuber characteristics in a growth chamber experiment (cv. Désirée) associated with two levels of root temperature (L.H.J. Kerckhoffs and P.C. Struik, unpublished data).

	Root temperature (°C)	
	15	20
Number of tuber-bearing stolons	13.9	14.0
Number of tubers per tuber-bearing stolon	2.79	3.72
Number of tubers per plant	39	52
Average fresh tuber weight (g)	42.8	31.3
Tuber fresh yield (kg per plant)	1.67	1.63

an increase in tuber number was that the tubers were smaller in size. In this experiment, a higher root temperature was probably associated with a higher stolon temperature.

18.5 TUBERS

The formation of tubers can be separated into different processes or steps, including tuber induction, tuber initiation, tuber set, tuber bulking and tuber maturation. During tuber induction, the plant translates the environmental and internal signals into a readiness to form tubers, without any visible signs of tuberization yet. Tuber initiation is the process underlying the first swelling of the stolon tips into tuber initials that may later grow out. In fact, they are the visible signs that tuber induction has taken place, and therefore, tuber induction and tuber initiation will be taken together in this contribution. During tuber set, those tuber initials that will grow out into marketable sizes are determined, although the actual number of tubers present may go down later during tuber bulking due to resorption and decay. Tuber maturation is characterized by suberization and an increase in dormancy. Temperature affects all these processes related to tuber formation. For many of them, there is a strong interaction between temperature and other environmental factors, especially photoperiod.

18.5.1 Tuber induction and tuber initiation

Photoperiod plays an important role in the induction and initiation of tuber formation in the potato plant. The traditional potato species grown worldwide is a quantitative short-day plant for this process, meaning that tuberization is faster when the photoperiod is shorter than a critical value. Temperature modifies the response of potato to photoperiod (Ewing and Struik, 1992; Struik and Ewing, 1995). Higher temperatures delay, impede or even inhibit tuberization in comparison with lower temperatures (Wheeler et al., 1986; Snyder and Ewing, 1989; Ewing and Struik, 1992; Jackson, 1999). In particular, night temperature has a strong influence (also see Section 18.11). In particular, high air temperatures impede induction and initiation. High soil temperatures do not prevent stolons from being formed (often seen as the first step towards tuberization) but prevent stolons from forming tubers (Ewing and Struik, 1992).

Kooman et al. (1996) reported that when temperatures increase to about 22 °C, the duration of the developmental phase from emergence to tuber initiation shortens. At temperatures above that value, the development will be slower (Ewing and Struik, 1992). As a consequence, high temperatures will prolong the period from emergence until tuber initiation and will result in a larger plant when tuberization starts.

Heat may completely impede tuber induction and the subsequent tuber formation. There is an enormous genetic variation in the ability to express tuber formation under heat (Ben Kheder and Ewing, 1985).

18.5.2 Tuber set

At the initial stages of tuberization, the balance of dry matter partitioning starts to shift from the shoot to the tubers. At the early stages of this shift, tuber initiation is most probably source regulated. Removal of tubers will enhance the tuberization of any remaining stolon tips. Presence of tubers lacking the ability to synthesize starch (e.g. in transgenic plants) will greatly increase the number of tuber initials. During tuber set, tuber formation will gradually become more determined by sink regulation and sink–sink competition. Struik et al. (1991) defined tuber set at the level of the individual tuber initial as the final phase of the early tuber development after which the small tuber definitely has the ability to produce a marketable tuber. Tuber set at the plant level may be defined as the onset of rapid bulking and is completed when the maximum number of tubers growing at a high rate is fixed. Beyond that moment, the number of tubers can still decline when (changes in) conditions do not allow the support of all growing tubers. Internal factors regulating tuber set and tuber size distribution have been extensively discussed by Struik et al. (1990, 1991).

Thermal time functions for the onset and rates of tuber bulking have been developed both for temperate regions (Jefferies and MacKerron, 1987; Spitters, 1990) and for tropical conditions (Manrique and Hodges, 1989). However, these relations are strongly cultivar-specific and are also affected by the photoperiod (Kooman and Haverkort, 1995).

Table 18.7 summarizes several parameters relating to the effect of temperature on tuber set based on curve fitting using data from frequent harvesting. Tuber set expressed in days after planting is delayed by higher temperatures, the effect being large above 23 °C, but certainly also present in the range 15–23 °C. The thermal time required to reach the onset of tuber growth (OTG) (base temperature 0 °C) doubles when the average temperature increases from 23 to 27 °C. Later during tuber growth, there is a shift from exponential to linear tuber growth. Linear tuber growth starts about 16–28 days after OTG, depending on temperature and cultivar. There is a caveat attaching to the data in Table 18.7: The data are based on growth cabinet experiments, and so, the light intensity may have been limiting at some temperatures.

Temperature may influence the competition among tuber initials and among tubers and thus may influence tuber set and sink-to-sink competition.

Table 18.7 Calculated values for time of the onset of tuber growth (OTG) either in days after planting (DAP) or in thermal time, Kelvin DAP (KDAP; base temperature 0 °C), relative tuber growth rate (RTGR), relative partitioning rate (RPR), time at which the tuber formation passes from exponential to linear growth (TLG), absolute tuber growth rate in the linear phase (LTGR), final dry matter tuber yield (yield) and final tuber number per plant (number).

Cultivar	Average temperature (°C)	OTG (DAP)	OTG (KDAP)	RTGR (d^{-1})	RPR (d^{-1})	TLG (DAP)	LTGR ($g\ pl^{-1}\ d^{-1}$)	Yield ($g\ pl^{-1}$)	Number (per plant)
Spunta	15	36	539	0.38	0.090	58	4.7	213	23
	19	38	722	0.37	0.090	54	3.9	191	23
	23	42	961	0.34	0.055	65	3.7	140	22
	27	68	1825	0.32	0.040	85	1.8	31	–
Désirée	15	35	525	0.38	0.085	55	3.6	205	22
	19	36	684	0.37	0.060	63	4.2	154	29
	23	40	913	0.34	0.045	68	3.4	42	42
	27	70	1895	0.32	0.035	–	–	0	–

Data are for cvs Spunta and Désirée grown at a photoperiod of 12 h. Temperature regimes had a day–night differential of 6 °C. [*Sources: Ingram and McCloud (1984) and Van Dam et al. (1996).*]
–, could not be assessed reliably.

18.5.3 Tuber bulking

Table 18.7 illustrates that the relative tuber growth rate (RTGR) is not very sensitive to temperature. These values are based on modelling studies by Ingram and McCloud (1984). Nevertheless, when average temperatures exceed 19 °C, the RTGR is reduced. Table 18.7 also lists the effects of temperature on the absolute linear tuber growth rates (LTGR). In cv. Spunta, there was a progressive decrease in the LTGR with an increase in temperature, whereas in cv. Désirée, there was an optimum at 19 °C. Effects became large at average daily temperatures higher than 23 °C.

18.5.4 Dry matter partitioning to tubers and harvest index

Table 18.7 summarizes the relative partitioning rate (RPR) to tubers. The RPR is highest at the cooler temperature regimes included in the experiment.

Other growth chamber studies carried out by Lafta and Lorenzen (1995) indicated that day/night temperatures of 31/29 °C reduced total plant dry matter by 44–72% after 4 weeks when compared with a temperature regime of 19/17 °C but that tuber dry weights were reduced even more, resulting in a lower harvest index under the high temperature regime. Wheeler et al. (1986) stated the harvest index increased with reducing temperatures in the range of 12–28 °C. Optimal temperatures for both total dry matter and tuber dry matter were 20 °C at 12 h and 16 °C at 24 h in their experiments.

Many other authors have reported similar results on the strongly negative effects of high temperature on harvest index (Marinus and Bodlaender, 1975; Wolf et al., 1990; Kooman et al., 1996).

Ben Kheder and Ewing (1985) investigated the dry matter partitioning of a set of cultivars under relatively cool temperatures and under heat in a greenhouse study. They showed that the effect of heat on the harvest index was large and that most cultivars responded in a very consistent way.

18.5.5 Tuber yield

The tuber yield of potato is very sensitive to temperature. Table 18.7 summarizes that the tuber yield decreased when average temperatures were increased in the range of 15–27 °C. Yandell et al. (1988) indicated that the optimum temperatures for tuber yield were 17.5 °C for Russet Burbank and 18.7 °C for Norland. Other authors report higher optimum temperatures or temperature ranges, for example 22 °C (Burton, 1981). Optimum temperatures reported, however, usually lie in the range of 14–22 °C (Yamaguchi et al., 1964; Marinus and Bodlaender, 1975; Sands et al., 1979; Timlin et al., 2006), a range that may be in agreement with the data reported in Table 18.7, although the lower end of the temperature range is lacking in that data set.

Optimal temperature, however, is strongly dependent on the photoperiod (Wheeler et al., 1986; Wheeler, 2006). This phenomenon is illustrated in Table 18.8. The optimum temperature was 20 °C under short day and 16 °C under long day, but both were within the range of 14–22 °C given earlier. However, the value for optimum temperature under a 12-h photoperiod is not consistent with the results in Table 18.7. There are interactions between not only temperature and photoperiod (Table 18.8) but also these and cultivar (and, possibly, irradiance) (Table 18.7 vs. Table 18.8). Menzel (1985) stated that tuber yield is also determined by a balance between temperature and irradiance.

18.5.6 Tuber number

Given the effects of temperature on the structure of the stolon system, and thus on the number of potential sites for tuberization, and given the effect of temperature on the size of the shoot system at first tuber initiation, tuber number may be enhanced by an

Table 18.8 Tuber yields (g fresh matter per plant) from 56-day-old plants of cultivar. Norland grown at two photoperiods (12 and 24 h) and at different temperatures.

Light temperature (°C)	12 h	24 h
12	321	446
16	433	755
20	460	559
24	289	15
28	1	0

Note that photoperiod extension was with full (photosynthetically active) light. [*Source: Wheeler et al. (1986).*]

increase in temperature even though tuber growth may be slowed down. Indeed, Borah and Milthorpe (1962) and Struik et al. (1989b,c) reported more tubers being formed at higher temperatures. Lafta and Lorenzen (1995), however, observed that high temperatures caused fewer tubers to be formed. Other reports indicate a low optimum temperature for number of tubers, especially under long days (Wheeler et al., 1986) and strong interactions between genotype and temperature (Van Dam et al., 1996). Some of the discrepancies may be associated with the definition of a tuber, that is the minimum size at which a swelling is still considered a tuber.

Table 18.7 provides some data on the effect of temperature on the number of tubers per plant. In cv. Spunta, at 23 °C and below, there was no effect of temperature on number of tubers. In cv. Désirée, there was an increase in number of tubers with an increase in temperatures within the range 15–23 °C, consistent with the prolonged stolon formation and enhanced stolon branching associated with the higher temperatures.

18.5.7 Tuber size distribution

Tuber size distribution is determined by the yield, the number of tubers and the relative variation in tuber size. High temperatures increase the within-plant variation in time of tuber initiation and in the rate and duration of tuber growth. In addition to the effects on tuber number, average tuber weight and tuber yield as discussed earlier, temperature may therefore have a large but complex effect on tuber size distribution.

18.5.8 Tuber quality

High temperatures cause an increase in abnormal tuber behaviour. Frequency and extent of knobbiness, chain-tuber formation and secondary growth and frequency of growth cracks, heat sprouts, heat necrosis and translucent ends are all increased by an increase in temperature (Lugt, 1960; Bodlaender et al., 1964; Lugt et al., 1964; Marinus and Bodlaender, 1975; Ben Kheder and Ewing, 1985; Hiller et al., 1985; Smith, 1987; Burton, 1989; Struik et al., 1989c; Van den Berg et al., 1989; Ewing and Struik, 1992; Ewing and Sandlan, 1995). The most important effects, however, are the reduction in starch content and the increase in sucrose content in tubers produced under high temperatures (Krauss and Marschner, 1984). Associated with this effect, the dry matter concentration of the tubers is also very low at high temperatures (Haynes et al., 1988; Van den Berg et al., 1989).

18.5.9 Tuber enzyme activity

The carbohydrate metabolism in potato tubers responds strongly to low temperature because of the cold-induced sweetening phenomenon (Kruger et al., 2003; Malone et al., 2006). This phenomenon is caused by a conversion of starch into sugars at low temperatures above freezing. The mechanism is not clear, especially given the commonly accepted theory that the cold lability of enzymes catalysing the conversion of fructose 6-phosphate to fructose 1,6-biphosphate is not a major factor in cold-induced sweetening in plants. Sugar accumulation in cold-stored tubers is not a direct consequence of a constraint in carbohydrate oxidation (Malone et al., 2006). Supra-optimal conditions for starch synthesis are also frequent in potato production, and such conditions may cause low dry matter concentration and thus poor tuber quality.

Wolf et al. (1991) observed that high temperatures reduced the incorporation of ^{14}C into starch but increased the amount of labelled sucrose. Obviously, temperature affects the carbon metabolism in different plant organs, and as a result, there is a change in the partitioning of assimilates (Wolf et al., 1990). Also within tubers, supra-optimal temperatures affect the carbohydrate metabolism. Mohabir and John (1988) observed a sharp temperature optimum for ^{14}C sucrose incorporation into starch. The optimum was 21.5 °C. This means that there is a relatively low temperature optimum for starch synthesis. It also means that the sink strength of tubers is already decreased at relatively low soil temperatures. This sharp optimum is also in agreement with the optimum temperature of tuber growth under adequate light of about 22 °C. Mohabir and John, however, assessed the optimum temperature by measuring carbohydrate metabolism in tuber slices. Their results are in contrast with the findings of Frydman and Cardini (1966) and Kennedy and Isherwood (1975) who observed temperature optima for adenosine diphosphate glucose (ADPG) pyrophosphorylase and starch synthase above 35 °C. Geigenberger et al. (1998) showed that the optimum temperature for starch synthesis was 25 °C, higher than the optimum temperature for tuber growth. At elevated temperature levels, hexose-phosphate levels were increased, and the levels of glycerate-3-phosphate and phosphoenolpyruvate were decreased. They suggested that elevated temperatures result in increased rates of respiration, and the resulting decline of glycerate-3-phosphate then inhibits the ADPG pyrophosphorylase and starch synthesis.

Lafta and Lorenzen (1995) reported enzyme activities in potato tubers after exposure to a heat treatment imposed 10 days after tuber initiation. The heat treatment (29/27 °C compared with the control 19/17 °C) showed much lower activities of sucrose synthase after 2 weeks of heat. A smaller but still significant reduction was observed for ADPG Pyrophosphorylase, but no effect was observed for uridine diphosphate pyrophosphorylase. Effects after 3 days of treatment were not statistically significant.

18.6 INFLORESCENCES AND FLOWERS

Potato (*Solanum tuberosum* spp. *tuberosum*) is a quantitative short-day plant for tuberization and also for flowering (Almekinders and Struik, 1996). However, the effect of temperature on flowering seems stronger for the lateral branches than for the main stem. Table 18.9 summarizes the thermal time (in K days) from planting until flowering of the primary inflorescence (i.e. the inflorescence of the main stem) and of three types of secondary inflorescence. The average effect of photoperiod on the timing of flowering is rather small, with a delay in flowering in long days compared with short days. Secondary stems need more thermal time to produce flowers than the main stem, and the effect of photoperiod is strongest for the apical lateral branches. Under higher temperatures, the several types of stem segments require more thermal time than under lower temperatures. The effect of lengthened photoperiod on time of flowering is very slightly greater at higher temperature than at lower temperature. The temperature effect is slightly larger at long days than at short days.

Table 18.9 Thermal time (Kelvin days; base temperature 0 °C) from planting until flowering of the primary inflorescence and the secondary inflorescences corresponding to the lateral branches developing from nodes $n-1$, $n-2$ and $n-14$ on the main stem.

	Temperature regime (°C)				Average
	20/10		30/20		
	Photoperiod (h)				
	12	16	12	16	
Main stem	687	717	848	875	782
Secondary stem ($n-1$)	854	920	1063	1145	996
Secondary stem ($n-2$)	923	995	1125	1190	1058
Secondary stem ($n-14$)	864	897	1120	1228	1027
Average	832	882	1039	1110	

Cv. Atzimba grown in a growth chamber. [*Source: Almekinders and Struik (1994).*]

The effects of temperature on the number of flowers per individual inflorescence are illustrated in Table 18.10. More flowers are produced at a longer photoperiod than at a shorter photoperiod and at a higher temperature than at a lower temperature on the primary inflorescence and the apical secondary inflorescences. For the lower secondary inflorescence, there was an interaction between temperature and photoperiod, and the higher temperature gave fewer flowers than the lower temperature.

The effects of temperature on number of flowers of the entire shoot are complex (Table 18.10). Temperature has a significant effect on the basal and apical branching and on the dry matter partitioning over shoot and tubers. Therefore, temperature affects

Table 18.10 The number of flowers per inflorescence of the main stem (primary inflorescence) and secondary stems developing from the nodes $n-1$, $n-2$ and $n-14$ on the main stem and the total number of flowers per plant.

	Temperature regime (°C)			
	20/10		30/20	
	Photoperiod (h)			
	12	16	12	16
Primary inflorescence	16.0	18.8	27.8	28.4
Secondary inflorescence ($n-1$)	11.4	15.6	16.7	18.0
Secondary inflorescence ($n-2$)	9.4	13.3	13.3	14.0
Secondary inflorescence ($n-14$)	8.5	11.0	6.0	5.3
Total flowers – entire shoot	65	184	215	352

Cv. Atzimba grown in a growth chamber. [*Source: Almekinders and Struik (1994).*]

the potential number of inflorescences. Furthermore, more flowers may be initiated per inflorescence, and flower bud abortion may be reduced by an increase in temperature within a certain temperature range. The overall effect of temperature on the number of flowers of the entire shoot is also summarized in Table 18.10. Both long days and high temperatures increase the number of flowers per shoot.

18.7 ROOT SYSTEM

If the temperature of the root zone is varied, but air temperature is kept constant at 20 °C, root growth has a distinct temperature optimum of 15–20 °C (Sattelmacher et al., 1990a). This agrees with findings of Saha et al. (1974): raising night temperatures from 0 to 20 °C increases root length. Higher temperatures, however, may strongly reduce or even inhibit root growth (Sattelmacher et al., 1990a) and root activity (Sattelmacher et al., 1990b). Root diameter and root hair formation, however, are enhanced by heat (Sattelmacher et al., 1990a). Root respiration is not severely affected by heat (Sattelmacher et al., 1990b). There is a negative relation between plant age and heat susceptibility of the roots, because of a decrease of the shoot: root ratio with age (Sattelmacher and Marschner 1990).

Supra-optimal temperatures in the root zone also limit the allocation of assimilates to the roots, especially in heat-sensitive cultivars (Sattelmacher et al., 1990b). This reduces root activity such as the export of nitrogen to the shoot.

18.8 PHOTOSYNTHESIS, DRY MATTER PRODUCTION AND DRY MATTER PARTITIONING

Leaf photosynthesis has an optimum temperature of about 24 °C (Ku et al., 1977). It rapidly decreases with an increase in temperature beyond that optimum (Leach et al., 1982; Wolf et al., 1990). Prange et al. (1990) stated that this temperature effect was mainly caused by a reduced efficiency of photosystem II.

Dwelle et al. (1981) found the optimum temperature of gross photosynthesis to be in the range of 24–30 °C. However, the optimum temperature for net photosynthesis is lower, not more than 25 °C (Winkler, 1971; Ku et al., 1977; Dwelle et al., 1981). Recent literature (Timlin et al., 2006) presents a more detailed view. Whole plant (canopy) photosynthesis showed no temperature effect during very early stages of plant growth (up to 12–14 days after emergence). Thereafter, 31 days after emergence, an optimum temperature was detectable at 24 °C, partly associated with a larger leaf area because of more (but smaller) leaves at higher temperatures. The optimum temperature for canopy photosynthesis appeared to be 24 °C for most of the growing period. The rate of photosynthesis at 24 °C declined during the growing period but remained more or less constant for lower temperatures. In mature leaves, a temperature increase reduces photosynthesis. Therefore, the optimum temperature shifted to 16–20 °C later during the growth period. Total end-of-season biomass had an optimum temperature of 20 °C. Tuber yields and harvest index declined when temperatures increased above 24 °C (Timlin et al., 2006).

A low photosynthetic rate at very high temperatures was the result of accelerated senescence, chlorophyll loss, reduced stomatal conductance, inhibition of dark reactions at high temperature (Reynolds et al., 1990) and also damage to photosystems and poorer sink–source relations. The photosynthesis system of potato seems to have a high adaptive capacity to high temperatures but not to very high temperatures. Photosynthesis is higher following tuber initiation than before (Moll and Henninger, 1978; Lorenzen and Ewing, 1990).

The rate of net photosynthesis has a relatively low optimum temperature, but leaf area duration is not reduced until much higher temperatures. At intermediate temperatures, the size of the aerial shoot may be much larger than that at cool temperatures. Even though there may be a positive effect of high temperature on the size of the shoot, this does not contribute to additional dry matter production as the leaf: stem ratio is reduced and the partitioning to the tubers is delayed and impeded at high temperatures compared with low temperatures (see Section 18.5.4). High temperatures will therefore reduce total dry matter production.

High temperatures also result in delayed and impeded tuberization and so in a later and lower production of sinks. This in itself may directly affect the photosynthetic rate later during the growth period (Prange et al., 1990). This inference is supported by the findings of Lafta and Lorenzen (1995) who observed that high temperatures caused fewer tubers to be formed, thus causing a reduction in sink size and in photosynthesis.

Even when the rate of dry matter production is sufficient and an adequate number of tubers are formed, tuber yield is much reduced by heat, because of the effect of air temperature on dry matter partitioning (Struik et al., 1989c). Even when only the tuber temperature is increased, tuber growth can be reduced through a reduction in the activity of the starch synthesizing enzymes (see Section 18.9).

18.9 PARTIAL EXPOSURE

Earlier in the text, frequent reference was made to the effects of partial exposure to certain temperature treatments. In this section, we therefore restrict the discussion to the effects of exposing different parts of the plant to certain temperature treatments on tuber growth and bulking.

Tuber initiation is sensitive to temperature in different plant organs. Growing plants at high temperatures gives poor tuberization in cuttings taken from these plants, independent of soil temperature during subsequent plant growth (Reynolds and Ewing, 1989). Exposing below-ground plant parts to high temperatures also inhibits tuber formation. Cuttings taken from similar plants when air temperatures are low may tuberize, even if soil temperatures are high, and there are no other tubers on the plant. Intact plants will not tuberize when stolons are exposed to high temperatures because the expression of the signal to tuberize is blocked (Reynolds and Ewing, 1989). Menzel (1983) however stated that exposing plants to high soil temperatures causes about the same reduction in tuber yield as increasing air temperature.

As described earlier in Section 18.4 and in Table 18.5, Struik et al. (1989a–c) varied the temperature in different compartments of plant growth, that is the shoot environment (air), the root environment (temperature of nutrient solution) and the stolon environment (temperature in special stolon chamber). The differences in stolon temperature were

Table 18.11 Fresh tuber yield (g per plant) in a growth chamber experiment illustrating the interaction between shoot, root and stolon temperatures for cvs Bintje and Désirée (Struik et al., 1989c).

Temperature (°C) in environment of			Bintje	Désirée	Average	
Shoot	Root	Stolon			Absolute	Relative
18	18	Normal	1863	1442	1653	100
18	18	Increased	1810	1499	1655	100
18	28	Normal	1160	661	911	55
18	28	Increased	603	262	433	26
28	18	Normal	999	243	621	38
28	18	Increased	560	78	319	19
28	28	Normal	5	8	7	0
28	28	Increased	0	0	0	0

smaller than the differences in temperature in the other compartments and not entirely independent of the root temperatures. The results indicated that high root and stolon temperatures may enhance the number of stolons, yet inhibit their diageotropic growth (data not shown), whereas high shoot and stolon temperatures may constrain stolon growth. Earliness of tuberization was particularly affected by shoot temperature, although the expression could be inhibited by a combination of high root and stolon temperatures (data not shown). Numbers of tubers, including very small ones, were much higher when stolon temperatures were high (data not shown), but tuber yields were reduced by a temperature increase in any compartment (Table 18.11). Second growth was enhanced by a high temperature in any compartment, but an increase in the stolon compartment was particularly effective (data not shown). Dry matter partitioning was mainly influenced by air temperature (data not shown).

Exposing a single tuber to high or low temperatures affects its growth but also the growth of the other tubers. Engels (1983) showed that cooling of a single tuber reduces its growth rate but increased the growth rate of other tubers. Krauss and Marschner (1984) showed that heating single tubers to 30 °C for 6 days reduced tuber growth and could finally result in a complete halt of the growth of that tuber. It also resulted in a faster growth of the other tubers or even in the initiation of bulking at other tuber sites.

18.10 EFFECTS OF SHORT PERIODS OF CHANGES IN TEMPERATURE

The complex nature of the yield formation in the potato crop and the fact that different plant processes show different sensitivities to temperature have consequences for the effects of the timing of a short period of high temperatures. Warm temperatures early during the growth followed by cool temperatures later during growth give better tuberization than the reverse treatment (Cao and Tibbitts, 1994). Tibbitts et al. (1984) showed that warm temperatures during the first weeks of growth, that is before tuber initiation, followed by cooler temperatures increased tuber yield compared with maintaining the cool temperatures during the entire growth period. This effect was probably caused by the larger plant size after warmer growth. Delaying the warm period to later phases of growth, however, reduced tuber yield, whereas delaying a short cool period from early growth until after the onset of tuber bulking greatly increased tuber yield.

Burt (1964) could advance tuber initiation by a 1-week period of low temperatures in the range of 3–9 °C.

Mild or short increases in temperature may enhance the number of tubers per stem, because of positive effects on the number of stolons, the rate and duration of longitudinal stolon growth – thus increasing the number of potential tuber sites –, the number of tubers per tuber-bearing stolon and on second growth. Under those temporary warm conditions, the period of tuber formation will be prolonged, resulting in larger variation in the initiation date of the individual tubers from one stem. Tuberization will also be delayed resulting in a larger canopy and a lower harvest index. The effects of a short period of heat on number of tubers, tuber yield and average tuber weight of cv. Désirée are summarized in Table 18.12. Short periods of temperature increases (from 20/15 to 32/27 °C for 1 week) usually increased tuber number, but especially so when these short periods were applied during or shortly after tuber set. These changes in tuber number were related to changes in shoot growth. When Désirée plants were exposed to heat during tuber set or during initial tuber growth, there were both lower dry matter production and

Table 18.12 Effect of the timing of a 1-week period of high temperature 32 °C (12 h)/27 °C (12 h) on the number and fresh yield of tubers of cv. Désirée.

		Tuber number (#/plants)	**Fresh tuber yield (g/plant)**	**Dry tuber yield (g/plant)**	**Average tuber weight (g)**
Control		27	923	176	34
Heat during, DAP	Phenological stage of control				
14–21	Onset stolonization	23	955	185	42
21–28	Onset tuberization	25	905	187	36
28–25	Tuber set	44	862	162	20
35–42	Initial tuber growth	45	856	152	19
42–49	Tuber bulking	37	942	166	25
49–56	Tuber bulking	36	932	173	26
14–56	Onset stolonization – bulking	33	456	84	14

Standard day (12 h)/night (12 h) temperature: 20 °C/15 °C; photoperiod 12 h. Greenhouse experiment, final harvest 98 days after planting. DAP = days after planting. Plant height was similar for all treatments except for heat during 14–56 DAP, which had slightly taller plants and many more leaves, because of more intensive branching and sympodial growth. The same treatment thus also had a lower harvest index. Tuber dry matter concentration ranged between 17.6 and 20.7% depending on treatment. [*Source: L.H.J. Kerckhoffs and P.C. Struik (unpublished data).*]

lower yield and a much smaller average tuber size (L.H.J. Kerckhoffs and P.C. Struik, unpublished data). However, in a different set of experiments, we observed only minor effects of the timing of the heat on the yield and the number of tubers in cv. Bintje.

18.11 DIURNAL TEMPERATURE FLUCTUATIONS

Providing a thermoperiod (i.e. diurnal fluctuations) may improve tuberization (Steward et al., 1981; Tibbitts et al., 1990; Bennett et al., 1991; Cao and Tibbitts, 1992). Night temperature is particularly important for tuberization (Slater, 1968). However, diurnal fluctuations are only beneficial for tuber initiation in certain cultivars (Steward et al., 1981; Bennett et al., 1991). Moreover, Bennett et al. (1991) also showed that diurnal fluctuations provide better growth and yield of tubers in some cultivars but not in others.

18.12 PHYSIOLOGICAL BEHAVIOUR OF SEED TUBERS

Temperature during seed tuber bulking has a small and cultivar-dependent effect on the physiological behaviour of the harvested seed tubers (Van Ittersum and Scholte, 1992a). Van Ittersum and Scholte (1992a) showed that even temperature fluctuation during the day was relevant: seed lots exposed to the same average temperature during their production, but with different diurnal patterns, gave different physiological behaviour of the seed tubers. The specific effects, however, depended on cultivar.

As long as tubers remain attached to the haulm, the effects of higher temperatures seem to be reduced by a dampening effect of the shoot. Once the tubers are detached (after haulm killing), there is a much larger effect of an increase in temperature; whether the tubers are still in the soil or already in store. Nevertheless, the number of sprouts produced shortly after the breaking of dormancy is larger when the tubers are grown under high temperatures. Such effects may disappear during the later stages of storage.

There is abundant information on the effects of temperature during storage and physiological ageing. For reviews, see Van der Zaag and van Loon (1987) and Struik and Wiersema (1999).

Dormancy can be shortened by warm storage (Struik et al., 2006) but also by a short treatment of heat (Van Ittersum, 1992) or in some cultivars also by a short treatment of cold storage (Van Ittersum and Scholte, 1992b). Optimum temperatures for dormancy break and sprout growth differ: the optimum temperature is higher for dormancy break than for sprout growth. Table 18.13 provides a classical example of the effects of storage temperatures on the duration of dormancy of seed tubers from several cultivars produced in 2 years. With an increase in storage temperature, the breaking of dormancy is faster.

Storage temperature also affects sprout growth on tubers that are no longer dormant. Prolonged storage at 4 °C results in more sprouts per seed tuber (Struik and

Table 18.13 Effect of temperature regime during storage on dormancy duration (in weeks).

Cultivar	Storage temperature (°C)							
	2		5		10		20	
	Year							
	1	2	1	2	1	2	1	2
Bintje	14	17	12	16	12	15	10	11.5
Eigenheimer	9.5	11	9.5	11	9.5	11	6.5	8
IJsselster	9.5	19	10	19	9.5	16	9	12
Libertas	12	23	12	19	14	19	10	15
Average	14.4		13.6		13.3		10.3	

Source: Struik and Wiersema (1999).

Wiersema, 1999). Also, the length of the sprouts increases with thermal time (O'Brien et al., 1983).

There is usually a close relation between thermal time from tuber initiation to use and the performance of the tuber. However, Struik et al. (2006) showed that the sensitivity to temperature may differ for the various stages of physiological development of the seed

Table 18.14 Effect of several storage regimes with constant or changing temperatures on accumulated thermal time (base temperature 0 °C) and its consequences for the growth vigour in a laboratory test (relative dry matter yield of sprouts per kilogram seed tuber) and in a field test (relative dry matter yield in a field test) for a group of cultivars with a low rate of physiological ageing (LRPA) and a group of cultivars with a high rate of physiological ageing (HRPA).

Storage interval (weeks)				Thermal time (Kelvin days)	Relative yield in laboratory test		Relative yield in field test	
0–8	8–16	16–24	24–32		LRPA	HRPA	LRPA	HRPA
Temperature (°C)								
4	4	4	4	896	100	100	100	100
12	12	12	12	2688	67	19	80	15
20	20	20	20	4480	42	14	47	8
4	4	20	20	2688	72	46	77	33
20	4	4	20	2688	114	106	107	75
20	20	4	4	2688	113	103	102	89
4	20	20	4	2688	74	31	89	24

Tubers were stored in darkness and de-sprouted before their growth vigour was tested. For more details on the performance in the field of some of the treatments, see Struik et al. (2006). [*Source: Struik and Wiersema (1999).*]

tuber. Similar thermal times differing in the timing of a short period of high temperatures result in different seed performance. A higher temperature after the end of dormancy advanced and accelerated the process of ageing of seed tubers, especially in certain cultivars sensitive to rapid ageing. Resulting differences in seed performance may be extreme, in terms of stem numbers emerged, tuber numbers produced and in yield and tuber size distribution (Struik et al., 2006). Table 18.14 summarizes some relevant results. The table summarizes that the relative yields in both the laboratory test and the field test were lower with an increase in thermal time for seed lots exposed to constant storage temperatures. When storage temperature was varied, the treatments with periods of storage at both 4 and 20 °C all performed better in the laboratory test than the seed tubers stored at 12 °C for the entire period. For the field test, a similar result was obtained. However, early exposure to 20 °C gave much better results than late exposure to 20 °C, despite the same value for thermal time during storage. When sprouts are already present on the seed tubers, warm storage can be very detrimental in certain cultivars.

Table 18.15 Estimates of optimal temperatures (in °C) for maximum rates of different processes in the potato plant.

Sprouting	16–20
Sprout growth	20–25
Emergence	20–25
Early shoot growth	24
Leaf primordial development	35
Leaf appearance	28
Individual leaf growth	25
Leaf area development	20–25
Stem elongation	> 25
Shoot growth	32
Progress to flowering	30?
Photosynthesis leaf	24
Whole plant photosynthesis	20–24[a]
Dry matter production	20
Stolon initiation	25
Stolon growth	25
Stolon branching	25
Tuber induction	15
Tuber initiation	22
Tuber set/onset of tuber growth	15
Dry matter partitioning to tubers	20
Tuber bulking	14–22
Tuber yield	20–24
Starch synthesis	21.5, 25 or > 35?
Breaking of dormancy	28

[a] Higher during early part of the growing season than during the late part (Timlin et al., 2006).

18.13 SUMMARY

Table 18.15 summarizes the optimal temperatures for various processes described in this chapter. The table summarizes that for the initial growth, optimum temperatures are rather low. For the creation of many leaves (leaf primordial development and leaf appearance rate), the optimum temperatures are rather high, but they are significantly lower for individual leaf growth and canopy development. Stem elongation, stem growth and flowering, however, have much higher optimal temperatures. Production is maximal at temperatures even below the best temperatures for leaf area development. The early processes in tuberization have rather high optimal temperatures, but the tuber formation processes themselves have low optima.

REFERENCES

Almekinders C.J.M. and P.C. Struik, 1994, *Neth. J. Agric. Sci.* 42, 311.
Almekinders C.J.M. and P.C. Struik, 1996, *Potato Res.* 39, 581.
Ben Kheder M. and E.E. Ewing, 1985, *Am. Potato J.* 62, 537.
Bennett S.M., T.W. Tibbitts and W. Cao, 1991, *Am. Potato J.* 68, 81.
Benoit G.R., C.D. Stanley, W.J. Grant and D.B. Torrey, 1983, *Am. Potato J.* 60, 489.
Bodlaender K.B.A., 1963, In: J.D. Ivins and F.L. Milthorpe (eds), *Growth of the Potato*, p. 199. Butterworths, London.
Bodlaender K.B.A., C. Lugt and J. Marinus, 1964, *Eur. Potato J.* 7, 57.
Borah M.N. and F.L. Milthorpe, 1962, *Indian J. Plant Physiol.* 5, 53.
Burt R.L., 1964, *Eur. Potato J.* 7, 197.
Burton W.G., 1972, In: A.R. Rees, K.E. Kockahull, D.W. Hand and G.R. Hurd (eds), *The Crop Processes in Controlled Environments*, p. 217. Academic Press, London.
Burton W.G., 1981, *Am. Potato J.* 58, 3.
Burton W.G., 1989, *The Potato.* Wiley, New York.
Cao W. and T.W. Tibbitts, 1992, *HortScience* 27, 344.
Cao W. and T.W. Tibbitts, 1994, *J. Am. Hort. Sci.* 119, 775.
Dwelle R.B., G.E. Kleinkopf and J.J. Pavek, 1981, *Potato Res.* 24, 49.
Engels Ch., 1983, Wachstumsrate der Knollen von *Solanum tuberosum* Var. Ostara in Abhängigkeit von exogenen und endogenen Faktoren – Konkurrenz zwischen Einzelknollen um Assimilate. Dissertation der Fakultät Pflanzenproduktion der Universität Hohenheim, Stuttgart-Hohenheim, Germany.
Ewing E.E. and K.P. Sandlan, 1995, In: P. Kabat, B. Marshall, B.J. van den Broek, J. Vos and H. van Keulen (eds), *Modelling and Parameterization of the Soil-Plant-Atmosphere System. A Comparison of Potato Growth Models*, p. 7. Wageningen Pers, Wageningen, The Netherlands.
Ewing E.E. and P.C. Struik, 1992, *Hort. Rev.* 14, 89.
Fahem M. and A.J. Haverkort, 1988, *Potato Res.* 31, 557.
Firman D.M., P.J. O'Brien and E.J. Allen, 1991, *J. Agric. Sci. Camb.* 117, 61.
Firman D.M., P.J. O'Brien and E.J. Allen, 1992, *J. Agric. Sci. Camb.* 118, 55.
Fleisher D.H., R.M. Shillito, D.J. Timlin, Soo-Hyung Kim and V.R. Reddy, 2006, *Agron. J.* 98, 522.
Frydman R.B. and C.E. Cardini, 1966, *Arch. Biochem. Biophys.* 116, 9.
Geigenberger P., M. Geiger and M. Stitt, 1998, *Plant Physiol.* 117, 1307.
Haynes K.G., F.L. Haynes and W.H. Swallow, 1988, *HortScience* 23, 562.
Headford D.W.R., 1962, *Eur. Potato J.* 5, 14.
Hiller L.K., D.C. Koller and R.E. Thornton, 1985, In: P.H. Li (ed.), *Potato Physiology*, p. 389. Academic Press, Orlando, FL.
Ingram K.T. and D.E. McCloud, 1984, *Crop Sci.* 24, 21.
Jackson S.D., 1999, *Plant Physiol.* 119, 1.

Jefferies R.A., 1989, *J. Exp. Bot.* 40, 1375.

Jefferies R.A. and D.K.L. MacKerron, 1987, *J. Agric. Sci. Camb.* 108, 249.

Jefferies R.A. and D.K.L. MacKerron, 1993, *Ann. Appl. Biol.* 122, 93.

Kennedy M.G.H. and F.A. Isherwood, 1975, *Phytochemistry* 14, 111.

Kirk W.W., H.V. Davies and B. Marshall, 1985, *J. Exp. Bot.* 36, 1634.

Kirk W.W. and B. Marshall, 1992, *Ann. Appl. Biol.* 120, 511.

Klemke T. and A. Moll, 1990, *Agric. Systems* 32, 295.

Kooman P.L., M. Fahem, P. Tegera and A.J. Haverkort, 1996, *Eur. J. Agron.* 5, 207.

Kooman P.L. and A.J. Haverkort, 1995, In: A.J. Haverkort and D.K.L. MacKerron (eds), *Potato Ecology and Modelling of Crops Under Conditions Limiting Growth*, p. 41. Kluwer Academic Publishers, Dordrecht, The Netherlands.

Krauss A. and H. Marschner, 1984, *Potato Res.* 27, 297.

Kruger N.J., R.G. Ratcliffe and A. Roscher, 2003, *Phytochem. Rev.* 2, 17.

Ku S.B., G.E. Edwards and C.B. Tanner, 1977, *Plant Physiol.* 24, 530.

Lafta A.M. and J.H. Lorenzen, 1995, *Plant Physiol.* 109, 637.

Leach J.E., K.J. Parkinson and T. Woodhead, 1982, *Ann. Appl. Biol.* 101, 377.

Lemaga B., 1986, The relationship between the number of main stems and tuber yield of potatoes (*Solanum tuberosum* L.) under the influence of different day length and soil temperature conditions. Thesis Institut für Nutzpflanzenforschung – Acker- und Pflanzenbau – der Technischen Universität Berlin, Berlin, Germany.

Lorenzen J.H. and E.E. Ewing, 1990, *Ann. Bot.* 69, 481.

Lugt C., 1960, *Eur. Potato J.* 3, 307.

Lugt C., K.B.A. Bodlaender and G. Goodijk, 1964, *Eur. Potato J.* 7, 219.

MacKerron D.K.L. and P.D. Waister, 1985, *Agric. Forest Meteorol.* 34, 241.

Malone J.G., V. Mittova, R.G. Ratcliffe and N. Kruger, 2006, *Plant Cell Physiol.* Doi: 10.1093/pcp/pcj101.

Manrique L.A. and T. Hodges, 1989, *Am. Potato J.* 66, 425.

Marinus J. and K.B.A. Bodlaender, 1975, *Potato Res.* 18, 189.

Menzel C.M., 1983, *Ann. Bot.* 52, 65.

Menzel C.M., 1985, *Ann. Bot.* 55, 35.

Midmore D.J., 1984, *Field Crops Res.* 8, 255.

Midmore D.J., 1990, *Potato Res.* 33, 293.

Midmore D.J., D. Berrios and J. Roca, 1986a, *Field Crops Res.* 15, 97.

Midmore D.J., D. Berrios and J. Rocca, 1988, *Field Crops Res.* 18, 159.

Midmore D.J. and R.K. Prange, 1991, *Euphytica* 55, 235.

Midmore D.J., J. Roca and D. Berrios, 1986b, *Field Crops Res.* 15, 109.

Mohabir G. and P. John, 1988, *Plant Physiol.* 88, 1222.

Moll A. and W. Henninger, 1978, *Photosynthetica* 12, 51.

Moorby J. and F.J. Milthorpe, 1975, In: L.T. Evans (ed.), *Crop Physiology: Some Case Histories*, p. 211. Cambridge University Press, London and New York.

Mosille E.A.W., 1985, Der Einfluss verschiedener Kombinationen von Wurzelraumtemperaturen und Tageslängen auf Wachstum und Ertrag der Kartoffelpflanze (*Solanum tuberosum* L.). Thesis Institut für Nutzpflanzenforschung – Acker- und Pflanzenbau – der Technischen Universität Berlin, Berlin, Germany.

O'Brien P.J., E.J. Allen, J.N. Bean, R.J. Griffith, S.A. Jones and J.L. Jones, 1983, *J. Agric. Sci. Camb.* 101, 613.

Prange R.K., K.B. McRae, D.J. Midmore and R. Deng, 1990, *Am. Potato J.* 67, 357.

Randeni G., 1980, Der Einfluss der Wurzeltemperatur auf die Entwicklung und Substanzbildung der Kartoffelpflanze (*Solanum tuberosum* L.). Thesis Institut für Nutzpflanzenforschung – Acker- und Pflanzenbau – der Technischen Universität Berlin, Berlin, Germany.

Reynolds M.P. and E.E. Ewing, 1989, *Ann. Bot.* 64, 241.

Reynolds M.P., E.E. Ewing and T.G. Owens, 1990, *Plant Physiol.* 83, 971.

Saha S.N., G.S.R. Murti, V.N. Banerjee, A.N. Purohit and M. Singh, 1974, *Indian J. Agric. Sci.* 44, 376.

Sale P.J.M., 1979, *Aust. J. Agric. Res.* 30, 667.

Sands P.J., C. Hackett and H.A. Nix, 1979, *Field Crops Res.* 2, 309.

Sattelmacher B. and H. Marschner, 1990, *J. Agron. Crop Sci.* 165, 190.

Sattelmacher B., H. Marschner and R. Küne, 1990a, *Ann. Bot.* 65, 27.

Sattelmacher B., H. Marschner and R. Küne, 1990b, *J. Agron. Crop Sci.* 165, 131.

Slater J.W., 1968, *Eur. Potato J.* 11, 14.
Smith O., 1987, In: W.F. Talburt and O. Smith (eds), *Potato Processing*, p. 305. Avi Publishing Co., Westport, CT.
Snyder R.G. and E.E. Ewing, 1989, *HortScience* 24, 336.
Spitters C.J.T., 1990, *Acta Horticult.* 267, 349.
Squire G.R., 1995, In: P. Kabat, B. Marshall, B.J. van den Broek, J. Vos and H. van Keulen (eds), *Modelling and Parameterization of the Soil-Plant-Atmosphere System. A Comparison of Potato Growth Models*, p. 57. Wageningen Pers, Wageningen, The Netherlands.
Steward F.C., V. Moreno and W.M. Roca, 1981, *Ann. Bot.* 48 (Suppl. 2), 1.
Struik P.C. and E.E. Ewing, 1995, In: A.J. Haverkort and D.K.L. MacKerron (eds), *Potato Ecology and Modelling of Crops Under Conditions Limiting Growth*, p. 19. Kluwer Academic Publishers, Dordrecht, The Netherlands.
Struik P.C., J. Geertsema and C.H.M.G. Custers, 1989a, *Potato Res.* 32, 133.
Struik P.C., J. Geertsema and C.H.M.G. Custers, 1989b, *Potato Res.* 32, 143.
Struik P.C., J. Geertsema and C.H.M.G. Custers, 1989c, *Potato Res.* 32, 151.
Struik P.C., A.J. Haverkort, D. Vreugdenhil, C.B. Bus and R. Dankert, 1990, *Potato Res.* 33, 417.
Struik P.C., P.E.L. van der Putten, D.O. Caldiz and K. Scholte, 2006, *Crop Sci.* 46, 1156.
Struik P.C., D. Vreugdenhil, A.J. Haverkort, C.B. Bus and R. Dankert, 1991, *Potato Res.* 34, 197.
Struik P.C. and S.G. Wiersema, 1999, *Seed Potato Technology*. Wageningen Pers, Wageningen, The Netherlands, 383 pp.
Tibbitts T.W., S.M. Bennett and W. Cao, 1990, *Plant Physiol.* 93, 409.
Tibbitts T.W., W. Cao and R.M. Wheeler, 1994, *Growth of Potatoes for CELLS*. NASA Coop Agreement Final Report NCC 2-301, Moffett Field, CA.
Timlin D., S.M. Lutfor Rahman, J. Baker, V.R. Reddy, D. Fleisher and B. Quebedeaux, 2006, *Agron. J.* 98, 1195.
Van Dam J., P.L. Kooman and P.C. Struik, 1996, *Potato Res.* 39, 51.
Van Delden A., A. Pecio and A.J. Haverkort, 2000, *Ann. Bot.* 86, 355.
Van den Berg J.H., P.C. Struik and E.E. Ewing, 1989, *Ann. Bot.* 66, 273.
Van der Zaag D.E. and C.D. van Loon, 1987, *Potato Res.* 30, 451.
Van Ittersum M.K., 1992, Dormancy and growth vigour of seed potatoes. PhD thesis, Wageningen University, Wageningen, The Netherlands, 187 pp.
Van Ittersum M.K. and K. Scholte, 1992a, *Potato Res.* 35, 365.
Van Ittersum M.K. and K. Scholte, 1992b, *Potato Res.* 35, 389.
Vos J., 1995, In: P. Kabat, B. Marshall, B.J. van den Broek, J. Vos and H. van Keulen (eds), *Modelling and Parameterization of the Soil-Plant-Atmosphere System. A Comparison of Potato Growth Models*, p. 21. Wageningen Pers, Wageningen, The Netherlands.
Wheeler R.M., 2006, *Potato Res.* 49, 67.
Wheeler R.M., K.L. Steffen, T.W. Tibbitts and J.P. Palta, 1986, *Am. Potato J.* 63, 639.
Winkler E., 1971, *Potato Res.* 14, 1.
Wolf S., A. Marani and J. Rudich, 1990, *Ann. Bot.* 66, 515.
Wolf S., A. Marani and J. Rudich, 1991, *J. Exp. Bot.* 42, 619.
Yamaguchi M., H. Timm and A.R. Spurr, 1964, *Proc. Am. Soc. Hortic. Sci.* 84, 412.
Yandell B.S., A. Najar, R.M. Wheeler and T.W. Tibbitts, 1988, *Crop Sci.* 28, 811.

Chapter 19

Response to the Environment: Carbon Dioxide

Ludwig De Temmerman[1], Karine Vandermeiren[1] and Marcel van Oijen[2]

[1]*Veterinary and Agrochemical Research Centre, Leuvensesteenweg 17, B-3080 Tervuren, Belgium;*
[2]*CEH-Edinburgh, Bush Estate, Penicuik EH26 0QB, United Kingdom*

19.1 INTRODUCTION

Potato (*Solanum tuberosum* L.) is of major importance for human nutrition in many parts of the world. At the end of the twentieth century, the area cropped with potato decreased slightly in Western Europe, but remained fairly stable in Asian countries and increased considerably in developing countries with exception of South America (Beukema and Van der Zaag, 1990). In the future, potato is likely to become increasingly important for human nutrition as the ratio of edible to non-edible components [harvest index (HI)] is much greater than in wheat, rice and maize and because of the high potential for increased production in Europe and especially in developing countries.

Potato production will be influenced by future climatic changes. The European climate of the twenty-first century is likely to change, resulting in warmer summers, wetter winters and more variable patterns of rainfall and temperature. The major cause of this climate change is the increasing concentration of atmospheric carbon dioxide. The ambient concentration has increased from 280 $\mu l\ l^{-1}$ in pre-industrial times to current levels of about 370 $\mu l\ l^{-1}$ and is expected to reach 540–970 $\mu l\ l^{-1}$ by the end of the twenty-first century if emissions continue at current rates (IPCC, 2001). These changes will affect the growth, development and yield of most important agronomic crops including potato. Positive effects of elevated CO_2 on the growth and yield of plants have been observed over many decades, and during this time, hundreds of experiments, under laboratory or controlled environmental conditions, have shown that crop yields are generally enhanced by elevated CO_2 (Kimball, 1983). However, studies so far have shown that the increases in yield achievable under elevated CO_2 are highly species-dependent (Kimball, 1983). This is not surprising as it is well documented that carbon flux to yield components is a whole-plant property, which depends on the co-ordination between the activity of source organs, transport and sink activity (Farrar, 1996). It has been shown that elevated CO_2 enhances photosynthetic rates in the leaves of almost all C3 species that are well supplied with nutrients (Long and Drake, 1992). Transport capacity and the presence of small or large sinks for assimilates may promote or impair the capacity of the plant to invest the extra carbon fixed under elevated CO_2 into more structural growth or long-term survival storage (Schnyder and Baum, 1992; Körner and Miglietta, 1994).

Potato Biology and Biotechnology: Advances and Perspectives
D. Vreugdenhil (Editor)

Experiments with potato in Italy and Germany (Miglietta et al., 1998; Craigon et al., 2002) have shown that such positive CO_2 effects occur when crops are grown in the field under realistic conditions in free air carbon dioxide enrichment (FACE) systems, in the absence of modifications caused by artificial enclosures. The positive effect of CO_2 on the yield of potato has recently been confirmed in a series of open-top field chamber (OTC) and FACE experiments in different countries during the European CHIP (Changing climate and potential impacts on potato yield and quality) project (Craigon et al., 2002; De Temmerman et al., 2002a). The results suggest that the expected increases in CO_2 concentration may enhance agricultural production in at least some areas of the world, although it must be emphasized that changes in other climatic factors have the potential to offset such a beneficial effect (Rosenzweig and Parry, 1994).

In this review, we report the effects of CO_2 on potato phenology, physiology and yield. We explain how the effects are caused, their interactions with other variables and how the acquired information can be integrated in predictive models.

19.2 EFFECTS OF INCREASED CO_2 ON CROP GROWTH AND DEVELOPMENT

Commonly reported responses of several plant species to a CO_2 doubling are earlier flowering, increased leaf area, specific leaf weight, canopy density and plant height, as well as accelerated crop maturation rate (Krupa and Kickert, 1993) and senescence that causes a shortened growth period (Manning and Tiedemann, 1995). Although it is difficult to distinguish any statistically significant changes in phenological development rate (Hacour et al., 2002), some changes have also been reported in potato. Crop phenological development was found to be advanced by elevated CO_2 in the two consecutive years of OTC experiments carried out in Belgium as part of the CHIP project (Hacour et al., 1998, 2000). Miglietta et al. (1998) observed that the date of flowering of potato plants (cultivar Primura) was progressively advanced by increasing CO_2 concentrations (460, 560 and 660 $\mu l\ l^{-1}$) compared to ambient air.

To clarify the impact of elevated CO_2 on the rate of crop growth and development, it is essential to make a distinction between the three main developmental stages: canopy expansion, during which tuber initiation (TI) occurs; full canopy, when maximum leaf area (MLA) is attained and the senescence phase when leaf area and chlorophyll concentration begin to decline. Although there was no indication of any changes in leaf area or leaf formation during initial growth and full canopy, the average number of green leaves was significantly reduced by elevated CO_2 during crop senescence. Owing to an earlier leaf shed, elevated CO_2 caused a more rapid decline of the leaf area index (LAI, leaf area per unit ground area) after MLA had been reached (Hacour et al., 2002). Interestingly, halfway through the growing season, at MLA, Lawson et al. (2001a) found an increase in green leaf number and LAI when plants had been growing at 550 $\mu l\ l^{-1}$ CO_2, but a decrease at 680 $\mu l\ l^{-1}$. This may be a consequence of the fact that the highest concentration induced a more serious temporal shift in crop development. Kimball et al. (2002) reported that in four of five FACE experiments with potato, elevated CO_2 caused a decrease in peak LAI (Bindi et al., 1998, 2000; Miglietta et al., 1998). Leaf area index peaked earlier and already started to decline, whereas at ambient CO_2, it continued to rise for another couple of weeks. Plant height and the number of leaves were not significantly affected in the FACE experiments of Miglietta et al. (1998). Earlier decline of LAI, as well as the decrease of final plant height (Donnelly et al., 2001; Hacour et al., 2002; Persson et al., 2003), indicated a premature ending of active vegetative, aboveground growth followed by an earlier senescence. Accelerated leaf senescence has also been reported for other crops and considered to be an effect of increased canopy temperature due to stomatal closure (Kimball et al., 1995; Miglietta et al., 1998; Vaccari et al., 2001).

Specific leaf area (SLA, leaf area per leaf mass) determines how much new leaf area to deploy for each unit of biomass produced. Under elevated CO_2, any storage of the extra carbohydrate in the leaves, or any reallocation of biomass to thicker leaves, would tend to increase leaf mass more than leaf area, thereby decreasing SLA (Kimball et al., 2002). Potato showed a fairly consistent reduction in SLA due to elevated CO_2 (Bindi et al., 1998, 2000; Miglietta et al., 1998; Schapendonk et al., 2000; Vaccari et al., 2001). Most probably, a decrease of SLA has also been the cause of a slight stimulation of aboveground dry biomass at MLA during the CHIP experiments, despite the height decrease and absence of any significant changes in leaf number and LAI (Donnelly et al., 2001; Hacour et al., 2002). Lawson et al. (2002) did indeed report an increase in leaf thickness and also in leaf area. The extra photosynthate, produced by an initial stimulation of CO_2 assimilation, is not used primarily to create a larger leaf area (thereby increasing light interception) but seems to be used in the development of thicker leaves, possibly by the accumulation of starch in the chloroplasts (Schapendonk et al., 2000).

A detailed assessment of the number of leaves, LAI and plant height, in combination with intermediate biomass measurements during the European CHIP experiments, made it clear that the influence of elevated CO_2 concentrations on plant growth was very dependent on the developmental stage of the crops. On the whole, elevated CO_2 induces an increase in growth rate that will lead to a biomass increase at full canopy, but at the same time, the active growing period is reduced, and there is an earlier onset of senescence that counteracts final biomass accumulation. The balance between these processes and changes in assimilate partitioning (influenced, e.g., by source–sink imbalance) is responsible for the final yield response.

19.3 EFFECTS OF INCREASED CO_2 ON POTATO PHYSIOLOGY

The effect of increased CO_2 on crop productivity is mediated through its interaction with photosynthetic CO_2 assimilation, assimilate partitioning and plant or leaf senescence. A plant's photosynthetic performance is determined by chlorophyll concentration, efficiency of the photosystems (light reactions), stomatal regulation of CO_2 uptake and the actual CO_2 incorporation through the Calvin cycle (dark reactions). (A detailed description of photosynthesis and related carbohydrate metabolism is given in Chapter 13, Hofius and Börnke, this volume). Ribulose 1,5-bisphosphate carboxylase/oxygenase (Rubisco) plays an essential regulatory role in the plant's CO_2 response because it is the entry point for inorganic C into the photosynthetic carbon reduction cycle. It is a bifunctional enzyme, and atmospheric O_2 is a competitive inhibitor with respect to CO_2 as a substrate for the mono-oxygenase activity of Rubisco to produce phosphoglycolate, which is partially metabolized to CO_2 in the photorespiratory carbon oxidation cycle (Bowes, 1991). Through O_2 inhibition and photorespiration, the present atmospheric CO_2/O_2 ratio causes a 35% reduction of the maximum photosynthetic capacity of C3 plants at 25 °C, and higher temperatures amplify this inhibition. Owing to competitive interaction, as CO_2 rises, it will diminish the inhibitory effects of O_2, a doubling of the present CO_2 concentration would more than halve photorespiration (Bowes, 1996). However, this enhancement of photosynthetic efficiency is mostly not maintained during long-term growth at elevated CO_2 (Allen, 1990; Long, 1994). A doubling of atmospheric CO_2 often induces an initial increase in net photosynthesis of approximately 50% in C3 species, followed by a decline to approximately 30% (Cure and Acock, 1986). This so-called photosynthetic acclimation is a common response to CO_2 enrichment (Sage et al., 1989; Barnes et al., 1995; Stirling et al., 1997) and has been defined as the difference in photosynthetic rates for ambient and elevated CO_2-grown plants when measured at the same CO_2 partial pressure (Long, 1991). A further increase of the ambient CO_2 concentration has also been reported to reduce stomatal conductance, enhance water use efficiency (WUE), increase C/N ratio and lower dark respiration. The impact on each of these processes is dependent on the plant and leaf developmental stage that may in turn interact with CO_2, increasing the complexity of predicting plant growth and performance in response to future global changes (Besford et al., 1990; Miller et al., 1997; Osborne et al., 1998; Backhausen and Sheibe, 1999; Adam et al., 2000; Kalina et al., 2001; Marek et al., 2001).

Despite the obvious agronomic importance of potato, only a limited number of studies deal with the effects of elevated CO_2 on the photosynthetic properties of this plant. Most of these studies do confirm an initial stimulation of CO_2 assimilation, followed by a steady decrease to the level of photosynthetic performance comparable to that at ambient CO_2 concentrations. So despite the indeterminate growth pattern of potato with its large sink capacity, acclimation still occurs, and a simulation model indicated that this phenomenon reduced the positive effect of elevated CO_2 on tuber yield by 50% (Schapendonk et al., 2000). Vandermeiren et al. (2002) reported that a season-long exposure of *S. tuberosum* cultivar Bintje to 680 $\mu l\ l^{-1}$ CO_2 resulted in an initial increase of light-saturated photosynthesis (A_{sat}) of the fully expanded upper canopy leaves by 40% at the time of TI. A 30-day exposure of high- and low-altitude potato species, *Solanum curtilobum* and *S. tuberosum*, respectively, to 720 $\mu l\ l^{-1}$ CO_2 from TI onwards induced an average increase in net photosynthesis of 56 and 53% (Olivo et al., 2002). Increases up to 80% have been established in cultivars Gloria and Elles (Schapendonk et al., 2000). Photosynthetic acclimation may set in from the early start of crop growth (TI), and A_{sat} is steadily reduced to that at the ambient CO_2 level as the crop reaches maturity. The acclimation mostly coincides with a parallel decline in stomatal conductance (Sicher and Bunce, 1999; Bunce, 2001; Donnelly et al., 2001; Finnan et al., 2002; Vandermeiren et al., 2002; Heagle et al., 2003), resulting primarily from reductions in stomatal aperture rather than stomatal number (Lawson et al., 2002). An increase in CO_2 assimilation combined with a relative decrease of stomatal transpiration leads to a higher WUE. Olivo et al. (2002) did indeed report an increase in instantaneous transpiration efficiency of 80% in *S. tuberosum* and 90% in *S. curtilobum*. Such an effect may have important beneficial implications for crop production under future climatic conditions, which are expected to be warmer and drier with increased atmospheric CO_2 concentration. On the other hand, the stomatal closure may increase leaf temperature which can cause a shift in leaf ontogeny and crop senescence. An increase of the non-photorespiratory mitochondrial respiration (R_d) after the crop reaches MLA may be considered as an increased demand to support maintenance processes resulting from this earlier senescence (Vandermeiren et al., 2002).

The decrease in net assimilation of the upper leaves was more rapid at 680 $\mu l\ l^{-1}$ CO_2 in comparison to 550 $\mu l\ l^{-1}$ CO_2 (Lawson et al., 2001b), but the acclimation response was not found for older shaded leaves farther down the canopy, demonstrating the existence of compensatory flexibility whereby the older leaves may further support tuber growth (Lawson et al., 2001b; Vandermeiren, 2003). Sage et al. (1989) and Ludewig et al. (1998) reported little or no photosynthetic acclimation of *S. tuberosum* under greenhouse or controlled environmental conditions, but the authors do not provide any details on the leaf position and/or plant age. In comparison to ambient conditions, the maximum carboxylation rate of Rubisco ($V_{c,max}$) and light-saturated rate of electron transport (J_{max}) are reduced at a similar rate during the early stages of the season. So the decrease in Rubisco activity is related to a comparable reduction in ribulose-1.5-biphosphate (RuBP) regeneration by a decreased rate of non-cyclic electron transport. These observations suggest that Rubisco activity and/or amount may be important in determining maximum photosynthetic capacity under elevated CO_2 and that adjustment in the quantity or activity of Rubisco is involved in acclimation (Donnelly et al., 2001; Lawson et al., 2001b; Vandermeiren et al., 2002). Sicher and Bunce (1999) did not find changes in Rubisco concentration during acclimation, but they proposed that the decrease in Rubisco activity was potentially due to the presence of inhibitory compounds bound to the active site of the enzyme. Vaccari et al. (2001) did not find any changes in $V_{c,max}$ and J_{max} of potato cultivar Primura during early growth, although elevated CO_2 did finally lead to a faster decrease in photosynthetic capacity of the leaves. The photosynthetic acclimation is mostly accompanied by a higher carbohydrate concentration in the leaves (Schapendonk et al., 1995, 2000; Miglietta et al., 1998; Sicher and Bunce, 2001; Vaccari et al., 2001), decrease in nitrogen concentration (Ludewig et al., 1998; Sicher and Bunce, 2001; Fangmeier et al., 2002) and lower levels of soluble proteins (Kauder et al., 2000), especially Rubisco (Sicher and Bunce, 2001).

A frequently proposed explanation for this photosynthetic acclimation is that CO_2 enrichment causes an imbalance between source and sink capacity, particularly when sink capacity is insufficient to cope with the additional carbohydrates produced (Farrar and Williams, 1991; Stitt, 1991). Schapendonk et al. (2000) reported that photosynthetic acclimation in potato was indeed accompanied by a reduced Rubisco concentration and was more correlated with the accumulation of sucrose than of starch. The surplus carbohydrates produced, which cannot be transported and metabolized in sink tissues, would lead to a feedback accumulation in source leaves and reduce the expression of photosynthetic genes by a mechanism involving sugar sensing. Several studies argue against the role of sugar accumulation itself for the reduction of Rubisco concentration and photosynthetic down-regulation because leaf starch synthesis functions as an efficient buffer for photoassimilates (Ludewig et al., 1998; Heineke et al., 1999). Potato leaves have a large capacity for starch synthesis to transiently store most of the carbon assimilated during the light period in the chloroplast (Heineke et al., 1994; Casanova, 2003), although this may not be sufficient under very high CO_2 concentrations (Kauder et al., 2000). In the dark period, starch is degraded and possibly exported by a hexose transporter (Schäfer et al., 1977). Large starch granules may however also cause chloroplast deformation (Cave et al., 1981). Potato also features very large sink organs for carbohydrates and uses an apoplastic mechanism for phloem loading based on a sucrose transporter (Riesmeier et al., 1994), all of which are important prerequisites for a large CO_2 response. Insufficient sink strength of the tuber can even be compensated for by stimulated growth of other sinks, i.e. the shoot (Zrenner et al., 1995).

Photosynthetic acclimation and associated metabolic changes have, on the other hand, also been related to an accelerated development and earlier senescence (Heineke et al., 1999; Kauder et al., 2000; Vaccari et al., 2001). In tobacco, Miller et al. (1997) observed that during acclimation, lower photosynthetic rates appear to be the result of a shift of the normal photosynthetic stages of leaf ontogeny to an earlier onset of natural decline associated with senescence. A similar conclusion was put forward with regard to tomato (Besford et al., 1990). This implies that the reduction in the Rubisco concentration and/or activity possibly reflects the developmental stage of the investigated leaves rather than the environmental conditions. A more rapid decrease of the leaf chlorophyll concentration confirms this hypothesis for potato (Lawson et al., 2001b; Bindi et al., 2002). In senescing leaves, a marked decrease of protein-containing components can be seen as the result of nutrient remobilization (Kolbe and Stephan-Beckmann, 1997). An earlier onset of such a reassignment caused by a prolonged exposure to elevated CO_2 would explain the more rapid decline in Rubisco content/concentration/activity after MLA. It has been argued, however, that the Rubisco concentration in general remains in excess of that required to support the measured photosynthetic rates, also at elevated CO_2, and that acclimation would be caused by other limiting factors such as the regeneration of orthophosphate in the chloroplasts (Sage et al., 1989). It has also been suggested that under elevated CO_2, earlier leaf senescence may be induced by nitrogen limitation (Nie et al., 1995; Stitt and Krapp, 1999).

On the basis of the present literature, it is clear that acclimation and senescence are difficult to separate. They are both regulated by ontogenetic and environmental factors, and it is often impossible to distinguish cause and consequence in these discussions.

19.4 EFFECTS OF INCREASED CO_2 ON YIELD AND QUALITY

Doubling of the ambient CO_2 level results in tuber yield (dry matter) increases ranging from 0 to 60% (Table 19.1). A negative value is ignored here but will be discussed in Section 19.4. Owing to its large sink capacity, it is clear that potato has a high potential to increase yield at more elevated CO_2 concentrations, but there appear to be many limitations to reaching those high yields. On average, the total, as well as the marketable (> 35 mm) tuber yield, increased by about 22% in OTC experiments and about 30% in FACE systems (Table 19.1). In the CHIP project, the yield increase was linked to a significantly higher total number of tubers as well as marketable ones (Craigon et al., 2002). The increase in tuber number is likely to be the result of the large increase in

Table 19.1 Percentage increases in potato tuber yield (marketable > 35 mm and total) relative to the response at ambient CO_2 and the climatic conditions during crop growth.

Country – year	Cultivar	Device	CO_2 $\mu L\,L^{-1}$	Yield increase (DM)		Crop duration (days)	Temperature average (°C)	VPD (kPa)	Radiation ($MJ\,m^{-2}d^{-1}$)	Reference
				> 35 mm (%)	Total (%)					
Belgium 1998	Bintje	OTC	680	2.2	2.9	103	17.4	0.55	13.7	1
Belgium 1999	Bintje	OTC	680	8.1	9.0	105	18.4	0.61	16.4	1
Finland 1999	Bintje	OTC	540	23.7	24.2	97	17.8	0.90	12.6	1
Germany 1998	Bintje	OTC	680	19.8	9.5	115	19.6	0.91	9.9	1
Germany 1998	Bintje	OTC	550	25.0	13.4	115	19.6	0.91	9.9	2
Germany 1999	Bintje	OTC	680	12.0	12.3	97	20.3	1.07	10.2	1
Germany 1999	Bintje	OTC	550	15.4	12.6	97	20.3	1.07	10.2	2
Ireland 1998	Bintje	OTC	680	45.1	39.8	141	16.6	0.47	10.4	1
Ireland 1999	Bintje	OTC	680	60.1	53.9	97	15.6	0.51	12.7	1
UK 1998	Bintje	OTC	680	6.6	10.7	105	14.8	0.33	12.1	1
UK 1998	Bintje	OTC	550	6.1	8.7	105	14.8	0.33	12.1	3
UK 1999	Bintje	OTC	680	41.9	41.4	105	15.6	0.35	11.8	1
UK 1999	Bintje	OTC	550	26.8	26.3	105	15.6	0.35	11.8	3
Sweden 1998	Bintje	OTC	680	−11.4	−7.3	96	13.5	0.39	14.5	1
Netherlands 1995	Elles	OTC	700	*	30.7	90	16.8	*	19.4	4
Netherlands 1995	Gloria	OTC	700	*	23.7	90	16.8	*	19.4	4
Netherlands 1996	Elles	OTC	700	*	48.9	91	15.1	*	16.8	4

* data not available

(*Continued*)

Table 19.1 (*Continued*)

Netherlands 1996	Gloria	OTC	700	*	0.0	91	15.1	*	16.8	4
USA 2001	Dark red Norland	OTC	540	13.6	13.9	86	22	0.82	14.3	5
USA 2001	Dark red Norland	OTC	715	33.3	34.8	86	22	0.82	14.3	5
USA 2001	Superior	OTC	540	34.1	33.1	86	22	0.82	14.3	5
USA 2001	Superior	OTC	715	29.5	29.0	86	22	0.82	14.3	5
Italy 1995	Primura	FACE	460	*	10.8	88	2.9–35.8[a]	*	21.5	6
Italy 1995	Primura	FACE	560	*	21.0	88	2.9–35.8[a]	*	21.5	6
Italy 1995	Primura	FACE	660	*	27.0	88	2.9–35.8[a]	*	21.5	6
Italy 1998	Bintje	FACE	520	*	45.4	82	19.9	0.94	21.3	1
Italy 1999	Bintje	FACE	520	54.9	53.7	83	19.5	0.86	21.4	1
Germany 1999	Bintje	FACE	490	8.7	8.8	86	17.2	0.65	19.7	1

1: Craigon et al., 2002; 2: Jäger et al., 2000; 3: Donnelly et al., 2000; 4: Schapendonk et al., 2000; 5: Heagle et al., 2003; 6: Miglietta et al., 1998.
OTC, open-top chambers; FACE, free air CO_2 enrichment; VPD, vapour pressure deficit.
[a] Min–max.

photosynthetic capacity of the young potato plant. However, there are large differences between individual sites. Up to 60% yield increase can only be reached under specific conditions that are largely beneficial to promote tuber yield differences between exposure to ambient and elevated CO_2. A deterministic simulation model of potato growth that considered the stimulatory effect of elevated CO_2 on photosynthesis predicted that yield might be increased by 20–30% by a doubling of atmospheric CO_2 depending on the earliness of the cultivars involved (Schapendonk et al., 1995). Experiments with potato cultivar Katahdin, grown in a glasshouse, did indeed confirm an increased tuber growth rate of 36% (Chen and Setter, 2003). These results are in agreement with the data presented in Table 19.1, but doubling the CO_2 concentration appears not to be necessary to reach the maximum yield increase. Craigon et al. (2002) did not find a significant difference in tuber yield between OTC exposures at 550 and 680 $\mu l\ l^{-1}$ CO_2. They concluded that the CO_2 effect had saturated or was close to its maximum at 550 $\mu l\ l^{-1}$. Mackowiak and Wheeler (1996) reported that tuber yield enhancement was at maximum at 1000 $\mu l\ l^{-1}$, but probably under more optimal conditions.

Predictions of potato yield stimulation by 20–30% (Schapendonk et al., 1995) have partly been based on the increase of light interception by CO_2-induced formation of more leaf area, which is obviously not always the case (Schapendonk et al., 2000). Consequently, the most important attribute to explain observed yield increases is the photosynthetic stimulation on a leaf area basis. Accordingly, the hypothesis can be made that increased sucrose availability under elevated CO_2 feeds forward to the production and the activity of the sink organs, thus resulting in a substantial enhancement of final tuber yield. Late and early cultivars respond differently to elevated CO_2. Stimulation of photosynthesis appeared to be maintained longer in late cultivars, which have greater sink activity than early cultivars because of their prolonged tubering capacity (Schapendonk et al., 2000).

The climate itself has a large influence on potato yield. The highest potential and actual tuber dry matter production are obtained in areas with temperate climates in north-western Europe and northwest of the USA (Van der Zaag, 1984). This was also clearly confirmed in the European CHIP project for field-grown potato exposed to ambient CO_2 concentrations. There was linear increase of yield from Italy to Finland as a function of decreasing daily total solar radiation during the growth period (De Temmerman et al., 2002b). This seems contradictory to the statement of Kooman et al. (1996) that potential yields of potato depend on the cumulative amount of radiation intercepted over the length of the growing season. However, as summarized in Table 19.1, the growing period in Italy is much shorter than in northern Europe. Early senescence of the potato crop is mainly temperature driven (Kooman et al., 1996; Vaccari et al., 2001), and because high radiation is linked to high temperatures, it coincides with the shorter growing period (Fig. 19.1). As such, light use efficiency (LUE) is lowest in warm environments and highest in cooler sites (Nelson and Midmore, 1986). Other essential prerequisites for obtaining high yield are clearly water and nitrogen (Haverkort et al., 2003). However, irrigation is not able to completely counteract drought stress. In spite of irrigation, the air humidity remains rather low in dry regions such as in Italy (Table 19.1), limiting stomatal opening and CO_2 uptake at present ambient concentrations.

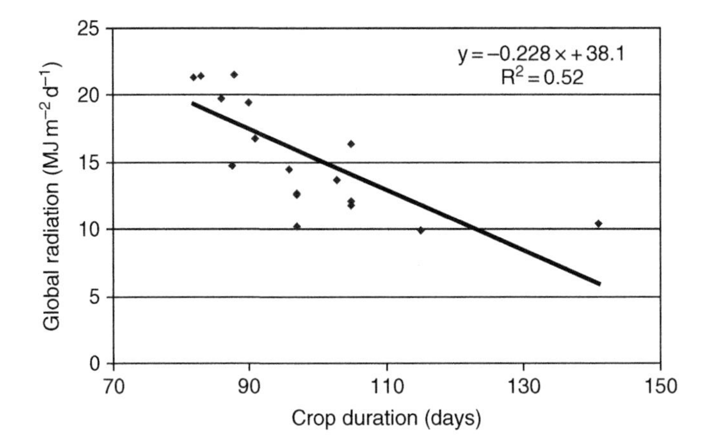

Fig. 19.1. Correlation between the radiation the potato crop experienced and the number of days between emergence and harvest (crop duration). Significance of regression: $P < 0.05$.

As the environmental conditions such as temperature (Lawlor and Mitchell, 1991), photoperiod and irradiance (Wheeler et al., 1991) vary between sites and years, the absolute as well as the relative CO_2 response will depend largely on those climatic differences. In addition, most of the CO_2 enrichment experiments have been carried out in OTCs, and typically, these facilities increase ambient temperature, reduce incident radiation and increase vapour pressure deficit (VPD). For wheat, van Oijen et al. (1999) found that cooling the air entering the OTC to obtain the ambient level delayed senescence, while the LUE was not affected. Cooling also increased the magnitude of the CO_2 effect. Moreover, increased CO_2 and VPD reduce stomatal opening, resulting in decreased losses of water and heat. In combination with increased OTC temperature, this additional stomatal closure causes increases in leaf temperature (Kimball et al., 1993) by a reduced transpirational cooling of the leaves. CO_2 itself causes partial stomatal closure, and consequently, increased canopy temperatures were measured in Italy on FACE plots with CO_2 enrichment (Magliulo et al., 2003). Increases in foliage temperature are probably beneficial if present temperatures are below the optimum for the crop but could be harmful if they exceed this optimum. Vaccari et al. (2001) observed that the increased canopy temperature in their FACE exposure of potato crops to CO_2 coincided with an accelerated chlorophyll breakdown and senescence (Bindi et al., 2002) and could limit yield. However, the beneficial effects of increased CO_2 supply on photosynthesis and WUE remain much larger and result in an important yield increase. Changes in radiation experienced by the plant growing in OTCs are probably even more important than temperature as a factor modifying the response to elevated CO_2 concentrations. Schapendonk et al. (2000) found a faster decline of LAI at elevated CO_2 under warm conditions and high irradiance, but other factors may also be responsible. Radiation also plays an important role in the allocation of assimilates, expressed as HI. The reduction of radiation in OTCs may increase the retention of assimilates in the aboveground biomass at the expense of the tuber biomass (De Temmerman et al., 2000). The effects on the aboveground biomass are also reflected in the canopy height. Reduced radiation increased height growth in the OTCs compared to the ambient air plots on average by more than 55% in the CHIP project (Hacour et al., 2002). Radiation measurements carried out in the Tervuren (Belgium) OTC system used for exposure of potato revealed a reduction of radiation by 30% below the air duct suspended along the chamber walls above the canopy, compared to the centre of the chamber where the reduction was almost zero for a large part of the day (unpublished results). It must be stressed, however, that OTCs are good tools for estimating CO_2 effects on crops, as homogeneous CO_2 enrichment is straightforward, but for potato, it is probably not the ideal method due to the important interactions of temperature and radiation on senescence and assimilate partitioning and probably also on yield.

Elevated CO_2 influences not only yield but also tuber quality: starch concentration was increased under elevated CO_2, but the reducing sugar concentration was not affected significantly (Vorne et al., 2002), whereas the nitrogen concentration was decreased (Fangmeier et al., 2002; Heagle et al., 2003). The increased starch concentration can be attributed to increased photosynthesis, which appeared to have more impact on starch than on reducing sugars. It is also a rather general CO_2 effect on many crops that the nitrogen concentration is decreased and consequently so is the protein concentration such as in wheat (Fangmeier et al., 1999).

19.5 INTERACTIONS BETWEEN YIELD AND STRESSES AT ELEVATED CO_2

In spite of a partial closure of the stomata, a rather general effect of increased CO_2 exposure, plants are able to maintain an increased photosynthetic rate during the main part of the growing season. Indeed, owing to the elevated ambient CO_2 concentration, the internal leaf concentration remains fairly high with limited effects on photosynthesis, whereas this partial closure can have important negative effects on photosynthesis at normal ambient levels (Van der Mescht et al., 1999). In addition, stomatal closure also has beneficial effects as the plants evaporate less water. As a consequence, it could be expected that dry soil conditions and/or high VPDs would be much more harmful for potato crops growing at ambient than at elevated CO_2. Apart from an overall decreased yield, the relative CO_2 effect will be much more important for water-stressed than for well-watered crops because of the better WUE. In the CHIP project, the option was taken to provide the plants with ample water to avoid drought stress. This may be the most important reason for the lack of CO_2 response in some experiments as reported by Craigon et al. (2002). As summarized in Table 19.1, yield increase due to elevated CO_2 exposure was high in Italy and rather low in Belgium, Sweden, Germany and Finland. Irrigation is important in dry conditions, but a high leaf to air water vapour pressure difference may continue to be a major limitation to CO_2 uptake even at twice the current concentration (Bunce, 2003). This is probably one of the reasons for a generally lower yield in Italy compared to the other sites, even at ambient levels of CO_2, but the relative yield increase at elevated CO_2 is higher. Indeed, in spite of ample water supply, the VPD remains high in Italy (Table 19.1). In temperate climate regions, the beneficial effect of ample water supply on stomatal opening is more important, and photosynthesis is less reduced compared to elevated CO_2. However, other stresses than drought might be involved, and moreover, if comparisons are made with other experiments than CHIP, the response to CO_2 increases appears to be highly cultivar dependent (Schapendonk et al., 2000) and largely variable from year to year.

Contrary to drought, stress caused by nutrient shortage will most likely reduce the effect of elevated CO_2. It appeared from several studies that elevated CO_2 decreases the nitrogen concentration in the plant, which is reflected in reduced protein concentrations (Fangmeier et al., 1999, 2002). It was suggested that earlier leaf senescence leading to a reduced yield may be induced by nitrogen limitation under elevated CO_2 conditions (Nie et al., 1995; Stitt and Krapp, 1999). This is probably the reason for the reduced yield at elevated CO_2 compared to ambient CO_2 in Sweden during the CHIP project (Table 19.1). To avoid any effect of drought stress on stomatal opening of the leaves, the well-drained OTC soils were abundantly irrigated, probably causing leaching of nitrates. The nitrogen shortage was reflected in the low yields at ambient CO_2 (De Temmerman et al., 2002b) and a negative yield effect of increased CO_2 supply. As CO_2 enrichment is increasing the nitrogen demand of the crop, the availability of the nutrient could become the limiting factor for the CO_2 effect on yield.

Elevated CO_2 is, to some extent, protecting plants from ozone effects (De Temmerman et al., 2002c), but elevated CO_2 did not directly protect against yield losses in the CHIP experiments, indicating that O_3-induced losses of potato yield will still occur in a future climate conditions even under elevated CO_2. The yield increases observed in response to elevated CO_2 far exceed ozone-induced losses (Craigon et al., 2002). Elevated CO_2 causes stomatal closure and thus protects plants from air pollution and water loss. Moreover, the increased carbohydrate status of the plants may benefit detoxification and repair mechanisms (van Oijen et al., 2004).

19.6 MODELLING FUTURE POTATO PRODUCTIVITY

The physiological and morphological responses of potato to elevated CO_2 have been discussed in the previous sections 2 to 5, as have some of the interactions with other environmental factors. In this section, we shall review how this knowledge has been incorporated in dynamic models of potato growth.

Crop modelling started in the 1960s (Bouman et al., 1996), and the earliest potato models were developed in the late 1970s (Ng and Loomis, 1984). Many models have been produced since that time; already in 1992, MacKerron (1992) was able to list 23 potato simulation models. However, none of these models was developed to account for elevated CO_2, their main purposes being to assist in potato breeding, assess the scope for growing potato in new areas or estimate the implications of limited water supply or high temperature (MacKerron, 1992; Haverkort and MacKerron, 1995). The earliest potato models attempted to represent the typical characteristics of the potato crop: indeterminate growth pattern, timing of TI dependent on temperature, day length and cultivar, a large carbohydrate sink in the tubers and only little allocation to seeds (Ng and Loomis, 1984). Later potato models tended to be simpler, re-parameterized instances of general crop growth models.

In the last decade, various reviews of crop modelling have appeared in the context of global change, including elevated CO_2 (Kickert et al., 1999). Here, we focus on modelling approaches that have been, or could be, applied to potato. The simplest vegetation models that represent the effect of CO_2 use the concept of the biotic growth factor β (Gifford, 1980) as follows:

$$\frac{G}{G_0} = 1 + \beta \ln\left(\frac{C_a}{C_{a,0}}\right), \tag{19.1}$$

where G and G_0 are growth rates at elevated and reference levels of ambient carbon dioxide (C_a, $C_{a,0}$). Goudriaan and de Ruiter (1983) varied C_a in greenhouses and found a value of 0.7 for β for different crops including potato.

The approach using the biotic growth factor is entirely empirical and, in contrast to more mechanistic approaches, cannot be used to study the variation in response to CO_2 among cultivars and growing conditions.

19.6.1 Source-driven potato growth models

In many potato models, growth rate is calculated as the product of light interception and LUE (Haverkort and MacKerron, 1995). Such models generally assume that LUE and allocation are independent of the carbohydrate status of the plants, growth thus being solely source-determined. This approach does not lend itself easily to incorporating complex effects of elevated CO_2. The effect of CO_2 on LUE could be represented using the biotic growth factor (Eq. 19.1), but no examples of this approach exist. Alternatively, the dependence of LUE on the photosynthetic capacity of the leaves (light-saturated photosynthetic rate, A_{sat}, and quantum yield, α) can be formulated as follows (van Oijen et al., 2004):

$$\text{LUE} = \frac{\gamma\alpha}{1 + kI_0/A_{sat}}, \tag{19.2}$$

where γ is the efficiency with which gross assimilated CO_2 is converted to biomass (typically about $0.4\,g\,g^{-1}$), k is the light extinction coefficient (about 1.0 for potato, Spitters and Schapendonk, 1990) and I_0 is photoperiod irradiation. Schapendonk et al. (2000) used Eq. (19.2) to quantify the reduction in LUE brought about by acclimation of A_{sat} to elevated CO_2. In their OTC experiments, CO_2 doubling increased the biomass yield of two potato cultivars by 20%, after initially increasing A_{sat} by about 80%. The increase in A_{sat} was only temporary – it gradually decreased to the level of the ambient CO_2 treatment. The observed time courses of A_{sat} were translated, using Eq. (19.2), into time courses of LUE, and these explained the observed yields well when used in an LUE-based potato model (Schapendonk et al., 2000). The model was then used to show that the yield increase would have been about 36% if acclimation had not taken place.

Equation (19.2) was also at the heart of the potato model used by Wolf and van Oijen (2002, 2003) to simulate the possible effects of elevated CO_2 and other environmental factors on potato yields in Europe. However, these authors had insufficient data on photosynthetic capacity to parameterize Eq. (19.2) directly, so they inferred changes in

A_{sat} and α from relationships between LUE and CO_2 and other factors, as observed in the CHIP project referred to above (De Temmerman et al., 2002a). Their modelling analysis suggests that environmental change expected for the year 2050 is likely to increase potato yields across the EU by 1000–4000 kg dry matter per hectare, mainly because of an expected 50% increase of atmospheric CO_2 concentration and with little interaction of CO_2 with other factors.

Instead of using an empirically determined dependence of A_{sat} on CO_2, we may employ a mechanistic equation that can be derived from the photosynthesis model of Farquhar et al. (1980):

$$A_{sat} = \lim_{I \to \infty} A = \frac{V_{c,max}\,(C_i - \Gamma^*)}{C_i + K_{mC}\left(\frac{O_2}{K_{mO}} + 1\right)}, \qquad (19.3)$$

where $V_{c,max}$ is maximum carboxylation velocity, C_i is the leaf internal CO_2 concentration, Γ^* is the CO_2 compensation point, O_2 is the oxygen concentration and K_{mC} and K_{mO} are the Michaelis–Menten constants of Rubisco for CO_2 and O_2. $V_{c,max}$ is proportional to leaf Rubisco concentration, so photosynthetic acclimation to elevated CO_2, but also damage to Rubisco by O_3, can be modelled using Eq. (19.3), by including Rubisco as a state variable in the model. Moreover, the parameters in Eq. (19.3) have shown temperature dependencies (Farquhar et al., 1980; Medlyn et al., 2002), so interactions of CO_2 and temperature can be assessed as well (van Oijen et al., 2004).

19.6.2 Source–sink-based potato growth models

The models described above do not represent sink strengths explicitly. The growth rates of leaves, stems, tubers and roots are assumed to be calculable by multiplying overall photosynthetic rate with organ-specific allocation factors that depend on thermal time and day length. In source-driven models, photosynthetic acclimation can only be represented as a direct reduction, by elevated CO_2, of growth, LUE, A_{sat} or Rubisco concentration. However, acclimation is the consequence of source strength exceeding sink strength, and mechanistic models of this phenomenon have been made. In fact, one of the earliest potato models (Ng and Loomis, 1984) explicitly simulated the accumulation of carbohydrates in case of sink limitation and a negative feedback of the carbohydrates on photosynthetic rate. This model and others that make growth a function of temperature-dependent, organ-specific sink strength (Ingram and McCloud, 1984) have not been very successful in practice because at the time of model development, data were lacking to support model parameterization and testing. However, the fact that photosynthetic acclimation to elevated CO_2 has recently been demonstrated in potato (Schapendonk et al., 2000; Vandermeiren et al., 2002), with sucrose accumulation the more likely signal as opposed to starch, renews the case for such source–sink-based modelling. Source–sink-based models have already proved successful in other crops such as grasses and maize (Höglind et al., 2001), although we still face difficulties in assessing and modelling the adaptability of sinks to prolonged excess source strength (Chen and Setter, 2003; van Oijen and Levy, 2005).

The leaves form a sink for carbohydrates whose strength is critical for the overall response to elevated CO_2 (Ewert, 2004). If leaf growth is enhanced by elevated CO_2, the capacity of the crop to intercept light will increase, thereby further enhancing growth. However, much recent evidence suggests that this positive feedback response to elevated CO_2 does not occur because of reduced allocation of carbohydrates to foliage (Ewert, 2004) or reduced SLA (Bindi et al., 1998, 2000; Miglietta et al., 1998; Schapendonk et al., 2000; Vaccari et al., 2001). For potato, recent modelling studies have found that, until LAI exceeds a value of 1.0, leaf growth rate is determined by temperature rather than source strength (Van Delden et al., 2001). This is likely to reduce the impact of elevated CO_2 on potato production.

19.6.3 Applications of potato models to CO_2-related issues: towards integrated assessment

As indicated above, potato models have mainly been developed to assist in plant breeding, scoping studies or estimating the effects of water stress or high temperature. The focus has been on modelling growth and yield, rather than the quality of the product. There have been efforts to model tuber size distribution (Nemecek et al., 1996), but tuber quality, and how it is affected by growing conditions such as CO_2, has not yet been modelled. Applications in breeding have focused on identifying potato cultivars with drought tolerance or disease resistance (Spitters and Schapendonk, 1990; van Oijen, 1992; Jefferies, 1993; Ellisèche and Hoogendoorn, 1995), but changes in atmospheric CO_2 concentration have not been considered. Potato models have been used to determine the scope for profitable potato production across the world (Van Keulen and Stol, 1995), but these studies have neglected global change including changes in atmospheric CO_2.

19.7 CONCLUSIONS

Potato has a high potential to increase yield at increasing ambient CO_2 concentrations. On average, 20–30% yield increase is reported by exposure of the crop to 1.5 or 2 times the actual ambient concentration. Under specific conditions such as interactions between environmental stress (drought) and increased CO_2, yield can increase by more than 60%. In temperate climate regions, where drought effects are less likely, fertilizer availability could become limiting to reach large yield differences between ambient and elevated CO_2 exposure. Accurate predictions of the effects of changing climatic variables on plant productivity will be dependent on the development of robust dynamic mechanistic models that are built on a sound understanding of the mechanisms by which environmental factors can influence photosynthetic processes (Baker, 1996). Although potato has some very specific features that would allow a large CO_2 response, the photosynthetic acclimation shows that this crop does not profit at full capacity from the increased CO_2 supply. Further investigations on the molecular background of this phenomenon may provide valuable information for further crop improvement. If photosynthetic acclimation can be decreased through breeding or management, potato could benefit more from an increase in the atmospheric CO_2 concentration than determinate crops such as cereals (Schapendonk et al., 2000).

The studies by Wolf and van Oijen (2002, 2003), referred to above, did include the effect of elevated CO_2, but focused only on one cultivar, 'Bintje'. In short, potato models have been used to determine what types of potato to grow, where to grow potatoes and how potatoes will respond to environmental change, but these questions have been studied in isolation. We suggest that we move towards integrated assessment. Potato breeding and cultivar selection may profoundly alter the geographical distribution of areas suitable for potato growth. Climate change may improve or worsen conditions, but elevated CO_2 is likely to increase yields and WUE (Wolf and van Oijen, 2002). It thus makes sense to address all these issues together, and the complexity of the combined problem will require models to identify solutions. The advances in mechanistic understanding of the physiological impact of elevated CO_2 on potato, outlined in this chapter, will be an essential ingredient in this integrated assessment of potato growth.

REFERENCES

Adam N.R., G.W. Wall, B.A. Kimball, P.J. Pinter, R.L. LaMorte, D.J. Hunsaker, F.J. Adamsen, T. Thompson, A.D. Matthias, S.W. Leavitt and A.N. Webber, 2000, *Photosynth. Res.* 66, 65.

Allen L.H. Jr, 1990, *J. Environ. Qual.* 19, 15.

Backhausen J.E. and R. Scheibe, 1999, *J. Exp. Bot.* 50, 665.

Baker N.R., 1996, In: N.R. Baker (ed.), *Environmental Constraints on Photosynthesis: An Overview of Some Future Prospects*, p. 469. Kluwer Academic Publishers, Dordrecht, The Netherlands.

Barnes J.D., J.H. Ollerenshaw and C.P. Whitfield, 1995, *Glob. Change Biol.* 1, 129.

Besford R.T., L.J. Ludwig and A.C. Withers, 1990, *J. Exp. Bot.* 41, 925.

Beukema H.P. and D.E. van der Zaag, 1990, *Introduction to Potato Production*. Pudoc, Wageningen.

Bindi M., L. Fibbi, A. Frabotta, M. Chiesi, G. Selvaggi and V. Magliulo, 2000, In: *Final Report CHIP*, contract ENV4-CT970489, C.E.U., Brussels, Belgium, 159.

Bindi M., L. Fibbi, A. Frabotta, G. Ottaviani and V. Magliulo, 1998, Free air CO_2 enrichment of potato (*Solanum tuberosum* L.). In: *Progress Report* contract ENV4-CT970489, Commission of the European Union, Brussels, Belgium, 132.

Bindi M., A. Hacour, K. Vandermeiren, J. Craigon, G. Selldén, P. Högy, J. Finnan and L. Fibbi, 2002, *Eur. J. Agron.* 17, 319.

Bouman B.A.M., H. Van Keulen, H.H. Van Laar and R. Rabbinge, 1996, *Agric. Syst.* 52, 171.

Bowes G., 1991, *Plant Cell Environ.* 14, 795.

Bowes G., 1996, In: N.R. Baker (ed.), *Photosynthesis and the Environment*, p. 387. Kluwer Academic Publishers, Dordrecht, The Netherlands.

Bunce J.A., 2001, *Glob. Change Biol.* 7, 323.

Bunce J.A., 2003, *Field Crops Res.* 82, 37.

Casanova M.A., 2003, Auswirkungen von erhöhter CO_2-Konzentration auf CO_2-Fixierung, Kohlenhydratmetabolismus und Kohlenhydratallokation in Kartoffel (*Solanum tuberosum* L.). PhD Thesis, Justus-Liebig-University, Giessen, Germany.

Cave G., L.C. Tolly and B.R. Strain, 1981, *Physiol. Plant.* 51, 171.

Chen C.T. and T.L. Setter, 2003, *Ann. Bot.* 91, 373.

Craigon J., A. Fangmeier, M. Jones, A. Donnelly, M. Bindi, L. De Temmerman, K. Persson and K. Ojanperä, 2002, *Eur. J. Agron.* 17, 273.

Cure J.D. and B. Acock, 1986, *Agric. Forest Meteorol.* 38, 127.

De Temmerman L., J. Craigon, A. Fangmeier, A. Hacour, H. Pleijel, K. Vandermeiren, V. Vorne and J. Wolf, 2000, Executive summary. In: *Final Report CHIP*, contract ENV4-CT970489, C.E.U., Brussels, Belgium.

De Temmerman L., A. Hacour and M. Guns, 2002a, *Eur. J. Agron.* 17, 233.

De Temmerman L., G. Pihl Karlsson, A. Donnelly, K. Ojanperä, H.-J. Jäger, J. Finnan, G. Ball, 2002c, *Eur. J. Agron.* 17, 291.

De Temmerman L., J. Wolf, J. Colls, M. Bindi, A. Fangmeier, J. Finnan, K. Ojanperä and H. Pleijel, 2002b, *Eur. J. Agron.* 17, 243.
Donnelly A., J. Craigon, C.R. Black, J.J. Colls and G. Landon, 2000, In: *Final Report CHIP*, contract ENV4-CT970489, C.E.U., Brussels, Belgium, 71.
Donnelly A., J. Craigon, C.R. Black, J.J. Colls and G. Landon, 2001, *Physiol. Plant.* 111, 501.
Ellisèche D. and J. Hoogendoorn, 1995, In: A.J. Haverkort and D.K.L. MacKerron (eds), *Potato Ecology and Modelling of Crops under Conditions Limiting Growth*, p. 341. Kluwer, Dordrecht.
Ewert F., 2004, *Ann. Bot. (Lond.)* 93, 619.
Fangmeier A., L. De Temmerman, C. Black, K. Persson and V. Vorne, 2002, *Eur. J. Agron.* 17, 353.
Fangmeier A., L. De Temmerman, L. Mortensen, K. Kemp, J. Burke, R. Mitchell, M. van Oijen and H.-J. Weigel, 1999, *Eur. J. Agron.* 10, 215.
Farquhar G.D., S. Von Caemmerer and J.A. Berry, 1980, *Planta* 149, 78.
Farrar J.F., 1996, *J. Exp. Bot.* 47, 1273.
Farrar J.F. and M.L. Williams, 1991, *Plant Cell Environ.* 14, 819.
Finnan J.M., A. Donnelly, J.L. Burke and M.B. Jones, 2002, *Agric. Ecosyst. Environ.* 88, 11.
Gifford R.M., 1980, In: G.I. Pearman (ed.), *Carbon Dioxide and Climate: Australian Research*, p. 167. Australian Academy of Science, Canberra.
Goudriaan J. and H.E. De Ruiter, 1983, *Neth. J. Agric. Sci.* 31, 157.
Hacour A., K. Vandermeiren and L. De Temmerman, 1998, In: *Progress Report CHIP*, contract ENV4-CT970489, Commission of the European Union, Brussels, Belgium, p. 1.
Hacour A., K. Vandermeiren and L. De Temmerman, 2000, In: *Final Report CHIP*, contract ENV4-CT970489, Commission of the European Union, Brussels, Belgium, p. 1.
Haverkort A.J. and D.K.L. MacKerron, 1995, *Potato Ecology and Modelling of Crops under Conditions Limiting Growth*. Kluwer, Dordrecht.
Haverkort A.J., J. Vos and R. Booij, 2003, *Acta Horticulturae* 619, 213.
Heagle A.S., J.E. Miller and W.A. Pursley, 2003, *J. Environ. Qual.* 32, 1603.
Heineke D., F. Kauder, W. Frommer, C. Kuhn, B. Gillissen, F. Ludewig and U. Sonnewald, 1999, *Plant Cell Environ.* 22, 623.
Heineke D., A. Kruse, U.-I. Flügge, W.B. Frommer, J.W. Reismeier, L. Willmitzer and H.-W. Heldt, 1994, *Planta* 193, 174.
Höglind M., A.H.C.M. Schapendonk and M. van Oijen, 2001, *New Phytol.* 151, 355.
Ingram K.T. and D.E. McCloud, 1984, *Crop Sci.* 24, 21.
IPCC, 2001, Climate change 2001. The Scientific Basis. Contribution of Working Group I – To the Third Assessment Report of the Intergovernmental Panel on Climate Change. IPCC, Cambridge University Press, Cambridge.
Jäger H.-J., A. Fangmeier and P. Högy, 2000, In: *Final Report CHIP*, contract ENV4-CT970489, C.E.U., Brussels, Belgium, 39.
Jefferies R.A., 1993, *Agric. Syst.* 41, 93.
Kalina J., O. Urban, M. Cajánek, I. Kurasová, V. Spunda and M.V. Marek, 2001, *Photosynthetica* 39, 369.
Kauder F., F. Ludewig and D. Heineke, 2000, *J. Exp. Bot.* 51, 429.
Kickert R.N., G. Tonella, A. Simonov and S.V. Krupa, 1999, *Environ. Pollut.* 100, 87.
Kimball B.A., 1983, *Agron. J.* 75, 779.
Kimball B.A., K. Kobayashi and M. Bindi, 2002, *Adv. Agron.* 77, 293.
Kimball B.A., J.R. Mauney, F.S. Nakayama and S.B. Idso, 1993, *Vegetatio* 104/105, 65.
Kimball B.A., P.J. Pinter Jr, R.L. Garcia, R.L. Lamorte, G.W. Wall, D.J. Hunsaker, G. Wechsung, F. Wechsung and T. Kartschall, 1995, *Glob. Change Biol.* 1, 429.
Kolbe H. and S. Stephan-Beckmann, 1997, *Potato Res.* 40, 111.
Kooman P.L., M. Fahem, P. Tegera and A.J. Haverkort, 1996, *Eur. J. Agron.* 5, 193.
Körner Ch. and F. Miglietta, 1994, *Oecologia* 99, 343.
Krupa S.V. and R.N. Kickert, 1993, *Vegetatio* 104/105, 223.
Lawlor D.W. and R.A.C. Mitchell, 1991, *Plant Cell Environ.* 14, 807.
Lawson T., J. Craigon, C.R. Black, J. Colls, G. Landon and J.D.B. Weyers, 2002, *J. Exp. Bot.* 53, 737.
Lawson T., J. Craigon, C.R. Black, J.J. Colls, A.-M. Tulloch and G. Landon, 2001a, *Environ. Pollut.* 111, 479.
Lawson T., J. Craigon, A.M. Tulloch, C.R. Black, J.J. Colls and G. Landon, 2001b, *J. Plant Physiol.* 158, 309.

Long S.P., 1991, *Plant Cell Environ.* 14, 729.

Long S.P., 1994, In: R.G. Alsher and A.R. Wellburn (eds), *Plant Responses to the Gaseous Environment*, p. 21. Chapman & Hall, London, UK.

Long S.P. and B.G. Drake, 1992, In: N.R. Baker and H. Thomas (eds), *Crop Photosynthesis: Spatial and Temporal Determinants*, p. 69. Elsevier, Amsterdam.

Ludewig F., U. Sonnewald, F. Kauder, D. Heineke, M. Geiger, M. Stitt, B.T. Müller-Röber, B. Gillissen, C. Kühn and W.B. Frommer, 1998, *Fed. Eur. Biochem. Lett.* 429, 147.

MacKerron D.K.L., 1992, Agrometeorological aspects of forecasting yields of potato within the E.C. Office for Official Publications of the European Communities, Luxembourg, 247 pp.

Mackowiak C.L. and R.M. Wheeler, 1996, *J. Plant Physiol.* 149, 205.

Magliulo V., M. Bindi and G. Rana, 2003, *Agric. Ecosyst. Environ.* 97, 65.

Manning W.J. and A. Tiedemann, 1995, *Environ. Pollut.* 88, 219.

Marek M.V., M. Sprtová, O. Urban and V. Spunda, 2001, *Photosynthetica* 39, 437.

Medlyn B.E., E. Dreyer, D. Ellsworth, M. Forstreuter, P.C. Harley, M.U.F. Kirschbaum, X. Le Roux, P. Montpied, J. Strassemeyer, A. Walcroft, K. Wang and D. Loustau, 2002, *Plant Cell Environ.* 25, 1167.

Miglietta F., V. Magliulo, M. Bindi, L. Cerio, F.P. Vaccari, V. Loduca and A. Peresotti, 1998, *Glob. Change Biol.* 4, 163.

Miller A., C.-H. Tsai, D. Hemphill, M. Endres, S. Rodermel and M. Spalding, 1997, *Plant Physiol.* 115, 1195.

Nelson D.G. and D.J. Midmore, 1986, *Potato Res.* 29, 258.

Nemecek T., J.O. Deroon, O. Roth and A. Fischlin, 1996, *Agric. Syst.* 52, 419.

Ng N. and R.S. Loomis, 1984, Simulation of Growth and Yield of the Potato Crop. Pudoc, Wageningen, 147 pp.

Nie G.Y., S.P. Long, R.L. Garcia, B.A. Kimball, R.L. Lamorte, P.J. Pinter, G.W. Wall and A.N. Webber, 1995, *Plant Cell Environ.* 18, 855.

Olivo N., C.A. Martinez and M.A. Oliva, 2002, *Photosynthetica* 40, 2, 309.

Osborne C.P., J. LaRoche, R.L. Garcia, B.A. Kimball, G.W. Wall, P.J. Pinter Jr, R.L. Lamorte, G.R. Hendry and S.P. Long, 1998, *PlantPhysiol.* 117, 1037.

Persson K., H. Danielsson, G. Sellden and H. Pleijel, 2003, *Sci. Total Environ.* 310, 1/3, 191.

Riesmeier J.W., L. Willmitzer, W.B. Frommer, 1994, *EMBO J.* 13, 1.

Rosenzweig C. and M.L. Parry, 1994, *Nature* 367, 133.

Sage R.F., T.D. Sharkey and J.R. Seemann, 1989, *Plant Physiol.* 89, 590.

Schäfer G., U. Heber, H.W. Heldt, 1977, *Plant Physiol.* 60, 286.

Schapendonk A.H.C.M., C.S. Pot and J. Goudriaan, 1995, In: A.J. Haverkort and D.K.L. MacKerron (eds), *Potato Ecology and Modelling of Crops under Conditions Limiting Growth*, p. 101. Kluwer Academic Publisher, Amsterdam.

Schapendonk A.H.C.M., M. van Oijen, P. Dijkstra, C.S. Pot, W.J.R.M. Jordi and G.M. Stoopen, 2000, *Aust. J. Plant Physiol.* 27, 1119.

Schnyder H. and U. Baum, 1992, *Eur. J. Agron.* 1, 51.

Sicher R.C. and J.A. Bunce, 1999, *Photosynth. Res.* 62, 155.

Sicher R.C. and J.A. Bunce, 2001, *Physiol. Plant.* 112, 55.

Spitters C.J.T. and A.H.C.M. Schapendonk, 1990, *Plant Soil* 123, 193.

Stirling C.M., P.A. Davey, T.G. Williams and S.P. Long, 1997, *Glob. Change Biol.* 3, 237.

Stitt M., 1991, *Plant Cell Environ.* 14, 741.

Stitt M. and A. Krapp, 1999, *Plant Cell Environ.* 22, 583.

Vaccari F.P., F. Miglietta, V. Magliulo, A. Giuntoli, L. Cerio and M. Bindi, 2001, *Ital. J. Agron.* 5, 3.

Van Delden A., M.J. Kropff and A.J. Haverkort, 2001, *Field Crops Res.* 72, 119.

Vandermeiren K., 2003, Global change and potatoes – impact of increased tropospheric CO_2 and O_3 on the physiological performance and tuber yield of *Solanum tuberosum* cv Bintje. PhD Thesis, Katholieke Universiteit Leuven, Belgium, pp. 234.

Vandermeiren K., C. Black, T. Lawson, M. Casanova and K. Ojanpera, 2002, *Eur. J. Agron.* 17, 337.

Van der Mescht A., J. De Ronde and F.T. Rossouw, 1999, *South Afr. J. Sci.* 95, 407.

Van der Zaag D.E., 1984, *Potato Res.* 27, 51.

Van Keulen H. and W. Stol, 1995, In: A.J. Haverkort and D.K.L. MacKerron (eds), *Potato Ecology and Modelling of Crops under Conditions Limiting Growth*, p. 357. Kluwer, Dordrecht.

van Oijen M., 1992, *Neth. J. Plant Pathol.* 98, 3.

van Oijen M., M.F. Dreccer, K.-H. Firsching and B.J. Schnieders, 2004, *Ecol. Model.* 179, 39.
van Oijen M. and P.E. Levy, 2005, In: S. Amâncio and I.D.H. Stulen (eds), *Nitrogen Acquisition and Assimilation in Higher Plants*, p. 133. Kluwer Academic Publishers, The Netherlands.
van Oijen M., A.H.C.M. Schapendonk, M.J.H. Jansen, C.S. Pot and R. Maciorowski, 1999, *Glob. Change Biol.* 5, 411.
Vorne V., K. Ojanperä, L. De Temmerman, M. Bindi, P. Högy, M.B. Jones, T. Lawson and K. Persson, 2002, *Eur. J. Agron.* 17, 369.
Wheeler R.M., T.W. Tibbitts and A.H. Fitzpatrick, 1991, *Crop Sci.* 31, 1209.
Wolf J. and M. van Oijen, 2002, *Agric. Forest Meteorol.* 112, 217.
Wolf J. and M. van Oijen, 2003, *Agric. Ecosyst. Environ.* 94, 141.
Zrenner R., M. Salanoubat, L. Willmitzer and U. Sonnewald, 1995, *Plant J.* 7, 97.

Chapter 20

Towards the Development of Salt-Tolerant Potato

D.J. Donnelly[1], S.O. Prasher[2] and R.M. Patel[2]

[1]*Department of Plant Science, Macdonald Campus of McGill University, 21,111 Lakeshore Rd., Ste Anne de Bellevue, QC, H9X 3V9 Canada;*
[2]*Department of Bioresource Engineering, Macdonald Campus of McGill University, 21,111 Lakeshore Rd., Ste Anne de Bellevue, QC, H9X 3V9 Canada*

20.1 INTRODUCTION

Agricultural sustainability, where salinity stress is a major problem, depends on a collaborative and integrative approach. This involves water management engineering and remedial cultural practices combined with crop improvement (Tester and Davenport, 2003). In 2006, there were no commercially important potato cultivars with outstanding salinity tolerance. However, significant variation for salinity tolerance exists among wild *Solanum* species and their accessions and among *Solanum tuberosum* cultivars. This salinity tolerance may be increased through conventional breeding efforts, tissue culture or molecular technologies. Improved or modified cultural and water management practices can enhance potato crop productivity. The application of these technologies to potato is reviewed, accomplishments to date are summarized, and possible strategies to accelerate the development of salinity tolerant potato are discussed.

Soil salinity, usually excess sodium chloride (NaCl) in soil, adversely affects the productivity of most crop plants. Although all crop plants tend to be affected by salt, they exhibit varying levels of yield depression in their response to saline soils or saline irrigation water. In this discussion, the relative terms 'sensitive' or 'tolerant' are used to suggest a continuum of decreasing injury caused by the exposure of crop species to Na^+. The uppermost spectrum of tolerance would approach the resistance demonstrated by halophytic species that have evolved in saline environments. For example, most crop plants cannot tolerate soil salinity levels that exceed $10\,dS\,m^{-1}$, whereas some halophytes are known to tolerate $30–40\,dS\,m^{-1}$ (Ahmad, 1997). This review is focussed on cultivated potato (*S. tuberosum* L.) and is not intended to encompass the multiplicity of salt effects on plants. The reader is referred to recent comprehensive review articles on sodium tolerance and salinity effects (Parida and Das, 2005), potential biochemical indicators of salinity tolerance (Ashraf and Harris, 2004), cellular basis of salinity tolerance (Mansour

Potato Biology and Biotechnology: Advances and Perspectives
D. Vreugdenhil (Editor)

and Salama, 2004), transport proteins and salt tolerance (Mansour et al., 2003), screening methods for salinity tolerance (Munns and James, 2003), sodium tolerance and transport (Tester and Davenport, 2003), comparative physiology of salt and water stress (Munns, 2002) and many others.

20.2 SALT-AFFECTED AGRICULTURAL LANDS – WHERE ARE THEY?

World population is increasing steadily, and as a result, demand for food is increasing. Meanwhile, world arable land is declining due to industrial development and settlements. These have put tremendous pressure on agricultural activities. Irrigated lands occupy about 15% of the total cultivated land area of the world but produce as much as 30% of the world's food (Flowers and Yeo, 1995; Munns, 2002). Production can be increased two-to-three-fold under irrigated land as compared with rain-fed lands. Therefore, it is necessary to develop irrigation potential by at least 2.25% per annum to cope with the food demand (FAO, 1992). Although land exploitation is becoming more imperative, the expansion of irrigated land is slow because most of the favourable land and water resources have already been exploited. Although more land areas are continually being put under irrigation, wide-scale abandonment of vast tracts of arable land in the order of 6–10 million hectares of irrigated land also occurs every year (Szabolcs, 1987, 1994; Umali 1993; FAO, 1995). This secondary salinization and waterlogging continues to expand due to inefficient execution of irrigation projects, improper or wasteful irrigation techniques and lack or mismanagement of drainage systems (FAO, 1995; Flowers and Yeo, 1995; Thomas and Middleton, 1993; Umali, 1993). Of the world's irrigated land, a frightening 30% overall (Munns, 2002) including 30% in the USA (Ward et al., 2003) and 50% in some countries (Thomas and Middleton, 1993; Ahmad, 1997) is affected by excess salinity. It is estimated that 6–10% of the total world land mass is now affected by excess salinity (Thomas and Middleton, 1993; Szabolcs, 1994; Flowers and Yeo, 1997; Munns, 2002). Salt-affected land is unevenly distributed around the globe. Most of these salt-affected lands are located in Australia and Asia (69%), and the rest are scattered around the world (Ahmad, 1997).

20.2.1 Is potato grown in salt-affected areas?

The major potato-growing countries, and their production areas, were listed by Hijmans (2001, 2003). Potato is grown all over the world, primarily in the Northern Hemisphere and in temperate regions; the top producers include China, Russia, Ukraine, Poland and India. It is estimated that 79% of countries in the world grow potato (Woolfe, 1986). Global distribution of potato is shown in the World Potato Atlas on the International Potato Center (CIP) website and is reproduced, with permission, as Fig. 20.1. For comparison purposes, the global distribution of stress-affected soils, including saline soils, is shown in Fig. 20.2. It is clear from these maps that most of the area under potato cultivation is in countries that are not overly affected by salinity. The exceptions are countries in southern and southeastern Asia, where coastal or inland salinity is more common (Ghosh et al., 2001). The weather has a direct impact on potato yield. Global climate

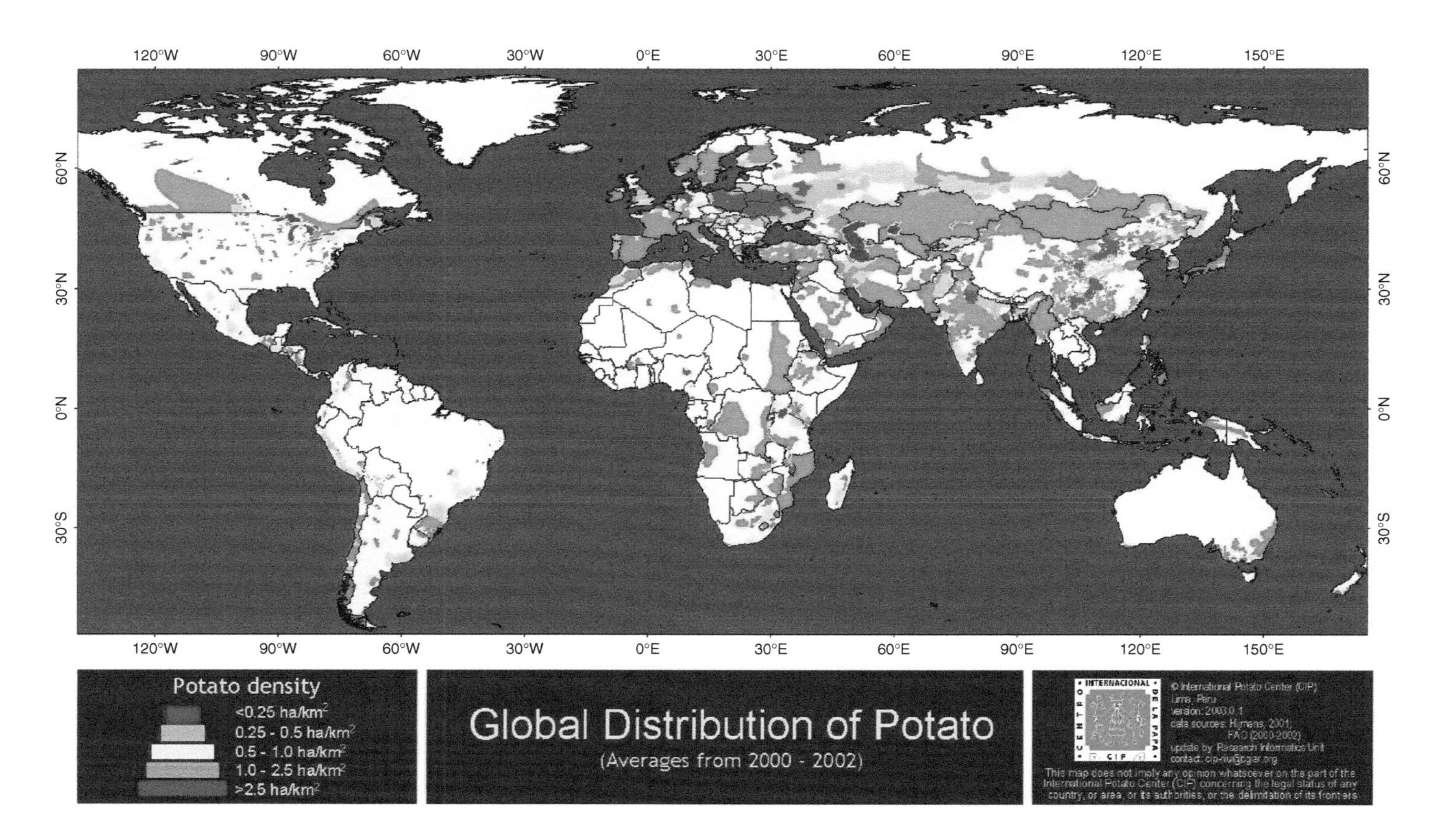

Fig. 20.1. The global distribution of potato map. Reprinted by permission from the CIP website (World Potato Atlas) (http://research.cip.cgiar.org/wpa/index.php).

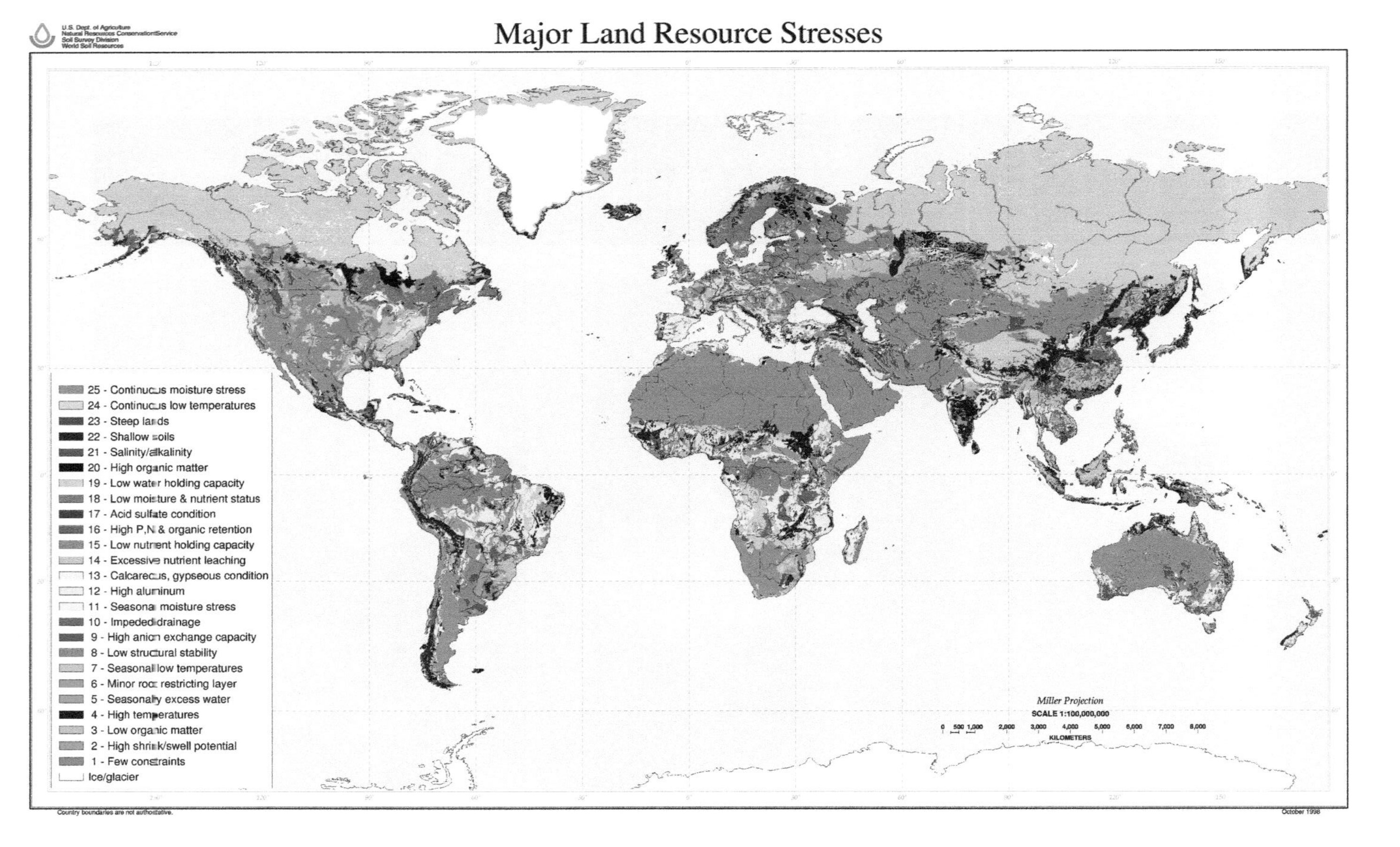

Fig. 20.2. The global distribution of stress-affected soils. Reprinted by permission from the U.S.D.A.'s Natural Resources Conservation Service (NRCS) Soils website (thematic maps) (http://soils.usda.gov/use/worldsoils/mapindex/stresses.html).

change predictions were recently applied to potato to estimate changes in growing area and yield by mid-century (Hijmans, 2003). Total potato yield is predicted to decrease by 18–32%. Most predicted yield reductions were attributed to temperature change alone, although changes in radiation (primarily due to shifts in planting time) could be responsible to some degree. However, higher temperatures will lead to greater water demand for agricultural production at a time when increased consumption for domestic and other use is expected due to industrial development and population growth (Elkhatib et al., 2004b). Scarcity of water would further decrease potato yield. Not only would there be scarcity of water, the quality of water would also deteriorate. To maintain potato production at the present level, an increase in cultivated area, the exploitation of brackish water sources and improved water management technology will be necessary. In this future scenario, potato with improved stress tolerance, including heat, drought and salinity tolerances, will be of increasing importance.

20.3 INTEGRATED APPROACH TO CROPPING SALINE SOILS

The case of potato is typical of other temperate crop species with moderate sensitivity to salinity. Cultural practices are important in determining the growth and productivity of such crops under salinity stress. Considerable engineering and cultural management expertise has emerged to enable the growth of crops such as potato under saline soil and brackish water conditions in developing countries [Ashraf, 1994; Flowers and Yeo, 1995; Ahmad, 1997; and many others, including field lysimeter studies (Patel et al., 1999, 2001; Katerji et al., 2000, 2003; Elkhatib et al., 2004a,b)].

The economic burden of soil reclamation, drainage and water desalination is increasing and often prohibitive. For this reason, it is important to improve farm management practices. This includes the use of drip irrigation to optimize water use where irrigation is practiced (Munns, 2002). Munns also suggests that rotation of annual crops with deep-rooted perennials could control rising water tables that bring salts up to the soil surface in the rain-fed zone. It is also essential to incorporate increased salinity tolerance into crops (Flowers and Yeo, 1995). However, both screening and breeding for salinity tolerance in crops have turned out to be intensely challenging.

To determine what real accomplishments had occurred in the development of salt-tolerant plants, Flowers and Yeo (1995) conducted a survey of the plant patent literature, the Science Citation Index and various scientific journals in which plant breeders register new varieties (such as Crop Science). Their survey showed that few salinity tolerant or resistant cultivars of any crop had ever been released. We repeated these searches in the summer of 2005. We were also disappointed; few crops were listed and no salinity tolerant potato releases had occurred. In 1987, Epstein and Rains (1987) observed, 'the number of breeders addressing the problem of salt tolerance is as yet minuscule'. Almost 20 years later, the observation of Epstein and Rains is still true for potato. Few potato breeders have considered salinity tolerance as a priority. In addition, the complexity of breeding for this multigenic trait may seem overwhelming to an individual breeder with limited resources; a team approach is essential (Flowers and Yeo, 1995).

20.4 MECHANISMS OF SALINITY TOLERANCE IN PLANTS

Salt stress imposes a complex metabolic disorder in glycophytic crop plants, such as potato, which is a combination of drought stress (lowering of water potential), specific ion toxicity, mineral (cation) deficiency, oxidative stress and other harmful effects which together reduce plant growth and productivity (Fidalgo et al., 2004; Netondo et al., 2004a,b). All major biochemical processes are depressed by salinity, including photosynthesis, protein synthesis, and energy and lipid metabolism (Parida and Das, 2005).

Salt-tolerant plants appear to use a wide spectrum of adaptive, interactive biochemical and molecular mechanisms that act additively and synergistically. These mechanisms involve ions (selective accumulation or exclusion, compartmentalization in cells or organs, control of root-uptake or transport into leaves), synthetic activities (compatible solutes, antioxidative enzymes and hormones) and altered photosynthetic pathways and membrane structure (Parida and Das, 2005). It has not been possible for plant physiologists to identify any one single trait for exploitation by plant breeders or geneticists (Ashraf and Harris, 2004; Mansour and Salama, 2004).

20.4.1 What is known of salinity tolerance mechanisms in potato?

Relatively little is known of salinity tolerance mechanisms specific to potato. The only physiological parameter that has been studied to any extent is in the category of compatible solutes. Correlations between relative salt tolerance and proline levels have been examined in potato plants grown in the field (Levy et al., 1988; Heuer and Nadler, 1995, 1998) and in vitro (van Swaaij et al., 1986; Potluri and Prasad, 1993; Martinez et al., 1996; Rahnama and Ebrahimzadeh, 2004) including in transformants tested in vitro (Evers et al., 1999). Findings have been inconsistent. No relationship was found between salt tolerance and proline levels in seven cultivars grown under salinity stress (Levy et al., 1988). However, in cv. Désirée, leaf water and osmotic potentials declined significantly with increased stress, and positive turgor was apparently maintained by osmotic adjustment to salinity stress through chloride and proline accumulation (Heuer and Nadler, 1995, 1998).

In vitro studies were also inconsistent with respect to proline accumulation in the presence of salinity stress. van Swaaij et al. (1986) used resistance to hydroxyproline to screen mutant cell lines of a diploid *S. tuberosum* clone for proline overproduction. Following growth for several months on non-selective medium, callus lines were tested for salinity tolerance and proline levels. Of four callus lines with elevated salinity tolerance, three showed greater concentrations of proline, whereas one line did not. No plants were regenerated from these calli. One variant line regenerated shoots that showed increased frost tolerance and elevated shoot proline levels, but its relative salinity tolerance was not reported. Of six potato varieties (from CIP) grown from axillary buds on MS basal medium (Murashige and Skoog, 1962) with various salt levels in vitro, some showed significantly increased proline concentration in stems and less in the leaves but not in the roots (Potluri and Prasad, 1993). Other varieties had increased protein but not proline content, and one showed increased shoot carbohydrate levels. Martinez et al. (1996) used four wild potato genotypes with varying frost tolerance (from CIP) to examine salinity tolerance in vitro. The genotypes included the highly frost-resistant *Solanum juzepczuckii* cv. Pinaza and *Solanum curtilobum* cv. Ugro shiri, the moderately frost-resistant *Solanum andigena* cv. Compis and the sensitive *S. tuberosum* cv. Baronesa. Based on the survival and shoot length, the two highly frost-tolerant genotypes performed better in vitro on elevated salinity levels than the other two genotypes. In addition, foliar proline concentrations increased in *S. juzepczuckii* and *S. curtilobum* but not the other two genotypes exposed to frost and NaCl stress in vitro (Martinez et al., 1996). In contrast, no relationship between proline levels and salinity tolerance occurred in shoots and callus from four cultivars exposed to salinity stress (Rahnama and Ebrahimzadeh, 2004).

In summary, salinity tolerance mechanisms in potato seem not to have been investigated in any detail, with the exception of proline, a compatible solute. Proline studies with potato showed inconsistent results in the field and in vitro. If proline overproduction is involved in some wild *Solanum* species, this would not be the only mechanism of salinity tolerance. Overproduction of osmo-protectants is insufficient to protect against salinity; regulation of uptake and compartmentalization is necessary to prevent accumulation and toxicity in photosynthesizing organs (Munns, 2002; Tester and Davenport, 2003).

20.5 CLASSIFICATION OF SALINITY TOLERANCE IN POTATO

The relative salinity tolerance of a long list of crop plants was compiled from early literature (Maas and Hoffman, 1977; Maas, 1985). Mediating factors in crop response to salinity included varietal differences, stage of growth, soil fertility, soil water and aeration, irrigation frequency and climate. Potato was described as moderately salt-sensitive, based on data from cv. Early Rose, which was cultivated under flood irrigation on loam soil (Bernstein et al., 1951). Reliant on substrate salinity and relative crop yield in field-based lysimeters, Katerji et al. (2000, 2003) also grouped potato (cv. Spunta) with the moderately salt-sensitive crops.

Katerji et al. (2000, 2003) proposed a more discriminating method of classification, the water stress day index (WSDI). This classification was based on measurement of pre-dawn leaf water potentials. The WSDI decreased as salinity levels increased. In this classification, potato was placed with the salt-tolerant crops; its sub-grouping was 'moderately sensitive'. What distinguished potato from species in the more sensitive group was its constant water-use efficiency as the salinity levels increased. Factors that affected transpiration did not determine salinity tolerance. No correlations occurred between leaf area, osmotic adjustments and yield reduction. For modelling crop response to salinity, it is preferable to use a reliable indicator of plant water status, such as pre-dawn leaf water potential, instead of stress coefficients that depend on the soil water status as the latter do not accurately express the water available to the plant (Katerji et al., 2000, 2003).

Potato could one day be classified on the basis of physiological mechanisms relating to salinity tolerance, including (1) concentration of organic solutes at the cellular level, such as glycinebetaine and proline (Ashraf and Harris, 2004), (2) transport proteins (Mansour et al., 2003) and (3) cytoplasmic viscosity (Mansour and Salama, 2004). However, these criteria are still under investigation for their effectiveness and efficiency. For this reason, classification of salt tolerance based on field performance and criteria such as WSDI appears to be more definitive.

20.6 EVALUATIONS OF SALINITY TOLERANCE IN POTATO

Determinations of relative salinity tolerance of crops (compiled by Maas and Hoffman, 1977; Maas 1985; reported by Katerji et al., 2000, 2003) were made using only one potato cultivar. It is important to note that, consistent with findings in other crops (Ashraf, 1994), significant variation in response to salinity has been found among *S. tuberosum* cultivars, wild *Solanum* species and among accessions from these wild species. Evaluation of salinity tolerance in cultivated and wild potato has been performed in the field, in the greenhouse and in tissue culture systems. In some cases, screening served to identify genotypes of potato that exhibited less impact of salt on growth and/or yield parameters and various physiological indices. As such, these studies are of interest to the present treatise and are summarized below.

20.6.1 Field and greenhouse evaluations of salinity tolerance in potato

Field trials (Bernstein et al., 1951; Barnes and Peele, 1958; Paliwal and Yadav, 1980; Levy, 1992; Heuer and Nadler, 1995; Nadler and Heuer, 1995), outdoor pot trials (Ahmad and Abdullah, 1979; Levy et al., 1988) and greenhouse studies (Bilski et al., 1988a, 1988b; Bruns and Caesar, 1990) explored the effects of salinity on potato plants or evaluated the salt tolerance of one or several cultivars. Few of these experiments used the same cultivars. Different studies used various salt mixtures, but NaCl was consistently used as the main salt. Salt tolerance was determined based on relative plant dry weight and/or relative reduction in tuber yield. Yield was assessed as mean tuber fresh weight per unit area (field trials) or mean fresh weight per plant (outdoor or greenhouse pot trials). These studies contributed much useful information on the response of potato to saline irrigation water. Salinity affected plant emergence, reduced haulm growth and accelerated maturity (Levy, 1992). However, no consistent relationship was found between maturation time and salinity tolerance (Levy, 1992). Salinity had a direct effect on tuber yield, as plants were most sensitive to salinity during tuber initiation and tuber bulking (Shannon and Grieve, 1999). This resulted in reduced tuber numbers overall or reduced number of over-large tubers. There were large differences in response to salinity stress among *S. tuberosum* cultivars. For example, relative yield increased for three of six cultivars as salt levels increased from 0.2 to 0.8%. However, at 1% mixed salts, relative yields decreased by 20–85% for all cultivars (Ahmad and Abdullah, 1979).

Wild *Solanum* species that evolved in dry South American habitats were suggested as potential sources of stress tolerance, with several listed as heat and drought tolerant (*Solanum acaule, Solanum bulbocastanum, Solanum chacoense, Solanum megistacrolobum, Solanum microdontum, Solanum papita, Solanum pinnatisectum* and *Solanum tarijense*) and a longer list as frost tolerant (Hawkes, 1990). In this regard, in vitro testing of some of the CIP selections was done by Elhag (1991). However, there are few published studies in which wild species were evaluated for salinity tolerance under field or greenhouse conditions. *Solanum kurzianum* shoot growth was superior to that of the two *S. tuberosum* cultivars Alpha and Russet Burbank in a salinized hydroponic trial (Sabbah and Tal, 1995). There were clear differences in response to salinity stress among *Solanum* species and between accessions of the same species. For example,

11 accessions of six species were evaluated for survival and growth (haulm dry weight) in greenhouse pot trials (Bilski et al., 1988b). *S. chacoense* (PI# 320285 was the only accession tested) was the most tolerant, *Solanum gourlayi, S. microdontum, Solanum sparsipilum* and *S. bulbocastanum* were moderately tolerant, and *S. papita* was the least tolerant. Although these studies provided information on salt tolerance from a vegetative growth perspective, inferences could not always be drawn on the effect on tuber yield.

20.6.2 In vitro evaluations of salinity tolerance in potato

Pioneering studies on in vitro screening for salinity tolerance evaluated a limited number of potato species and cultivars (Arslan et al., 1987; Morpurgo and Silva-Rodriguez, 1987; Morpurgo, 1991). In vitro assessments involved the addition of NaCl to MS basal medium. In most cases, stem cuttings with one or more axillary buds were taken from micropropagated plantlets and subjected to one or more levels of salt stress. The precise concentration of NaCl suitable for effective in vitro screening and useful growth parameters for evaluation were both uncertain. For example, one relatively high level of NaCl (103–154 mM) compared with MS control medium could not quantify differences in relative salinity tolerance between potato cultivars (Morpurgo and Silva-Rodriguez, 1987; Morpurgo, 1991).

The nodal-cutting bioassays were subsequently validated based on correlations between in vitro growth parameters and either haulm growth (Naik and Widholm, 1993) or yield parameters (Elhag, 1991; Morpurgo, 1991) in the field under salinized irrigation conditions. Elhag (1991) compared a large number of potato genotypes (86, including 11 cultivars and 36 wild species) at four NaCl levels (0, 40, 80 and 120 mM NaCl) in vitro and in the field. Evaluation of in vitro performance was based on the sum of the ranking of the relative growth parameters at 40, 80 and 120 mM NaCl. Two growth parameters (shoot length and fresh weight) decreased progressively with increase in salinity level. These two parameters were significantly correlated with each other in culture, in the field and with tuber yield in the field. However, Zhang and Donnelly (1997) did not find a consistent relationship between these two growth parameters at different salinity levels.

For thirteen genotypes tested in a similar manner, salinity exposure depressed growth parameters from 30 to 100% (Zhang et al., 1993). Single-node cuttings from six hybrids derived from *S. tuberosum* × three wild species (*S. chacoense, S. gourlayi* and *S. microdontum*; received from Dr. H. de Jong, AAFC, Fredericton, NB) were also tested. In preliminary trials, two *S. tuberosum* × *S. chacoense* hybrids were more tolerant than the other hybrids but less tolerant than the wild accessions and some of the cultivars.

Responses varied widely in different accessions of some wild species tested in vitro (Elhag, 1991; Khrais, 1996). The six top-ranked wild *Solanum* species in Elhag's study were *S. chacoense* (PI# 16979), *S. gourlayi, S. sparsipilum, Solanum spegazzinii, S. tarijense* and *Solanum vernei*. The response of different accessions from one species was sometimes dramatically different. For example, while one *S. chacoense* accession was the most tolerant, two others did not perform well. Thirteen *S. chacoense* accessions varied from sensitive to very tolerant (Khrais, 1996). Based on germination and early seedling growth under salinity stress, four of these *S. chacoense* accessions (PI# 275139, 320285, 320290 and 458309) were identified as outstanding. These studies highlighted the need to identify the accessions tested and to evaluate a representative number of accessions within each species. Many of the 36 wild species tested by Elhag (1991) were represented by only one to three accessions; some performed moderately well, suggesting that a greater number of accessions from these species should be evaluated.

In later studies, single-node cuttings were compared with other in vitro bioassays, including the growth of apical root segments (Naik and Widholm, 1993; Zhang and Donnelly, 1997), suspension cultures (Naik and Widholm, 1993) and microtubers (Kim et al., 1995; Zhang and Donnelly, 1997; Silva et al., 2001). Naik and Widholm (1993) used six levels of NaCl (0, 5, 100, 150, 200 and 250 mM) in the greenhouse (at 70% sprout emergence) or in vitro. Comparisons were based on vegetative growth, but not greenhouse yield parameters, and analyzed as a percentage of the control. Correlations were found between root length in the root segment bioassay and plant fresh weight *in vivo*, when averaged over the salt levels, excluding the control. Three types of in vitro bioassays (single-node, root tip segment and microtuberization) were verified using tuber yield criteria from field lysimeters under subsurface irrigation with salinized water (Zhang and Donnelly, 1997). Of these three bioassays, microtuberization is useful from a yield perspective. It is particularly important to extend understanding of salinity–nutrient relationships beyond plant shoot growth parameters to yield-related growth events. For this reason, effects of nutrients on microtuberization in salinized medium, including stolon formation and growth, induction and bulking steps, are of special interest. Microtuber fresh weights at any treatment level were not useful in predicting salinity tolerance, although the means across a range of salinity levels were predictive for the three genotypes tested (Zhang and Donnelly, 1997). The biggest drawback to the use of this bioassay was that not all wild species or hybrids could be tuberized in vitro (Zhang and Donnelly, 1997).

The single-node bioassay was simplest to perform and was applicable to *Solanum* species that were not readily microtuberized in vitro. However, of six growth parameters used in the single-node bioassay, none could be recommended individually for salinity tolerance ranking (Zhang and Donnelly, 1997). Accordingly, 130 European and North American potato cultivars were screened for relative salinity tolerance and vigour using the single-node bioassay for NaCl levels from 0 to 120 mM (Khrais et al., 1998). The sums of the relative values for six growth parameters (shoot and root length and fresh and dry weights), at each salinity level, were subjected to multivariate cluster analysis and used to partition the cultivars into eight groups. The cultivars with the greatest salinity tolerance included Amisk, BelRus, Bintje, Onaway, Sierra and Tobique. Tobique ranked as less vigorous than the other cultivars in this group. Chavez et al. (1997) evaluated ten advanced clones from CIP with potential for coastal Peru. These were examined using the single-node bioassay and under field salinity conditions. A moderate correlation between in vitro growth and yields in the field occurred. Cultivars with potential for growth in coastal soils of Peru were recommended, including Tacna (high tolerance), Basadre (medium-high tolerance), Costanera, Primavera and Desertica (medium tolerance).

In vitro evaluations of relative salinity tolerance have contributed useful knowledge about the various effects of salinity on growth parameters and the relative tolerances of different potato cultivars, species and accessions. The relative rankings of cultivars compared with these bioassays were confirmed in field or lysimeter studies (Elhag, 1991; Zhang and Donnelly, 1997), although correlations were not always strong (Chavez et al., 1997).

Reports by Elhag (1991) and Khrais et al. (1998) with the single-node cutting bioassay were particularly revealing. They showed that earlier studies, in which fewer genotypes were evaluated, had to be reinterpreted now and genotypes that were 'top ranked' among small populations needed to be benchmarked against these much larger populations. These studies also underlined the necessity of evaluation at a range of salinity levels in vitro; it was not possible to predict the growth response for individual genotypes as salinity levels varied. It has been suggested that differential gene expression may occur at different stress levels (Ashraf, 1994). Similar conclusions were drawn from work that involved short- and long-term screening of potato seedling populations (Jefferies, 1996). Jefferies found that potato seedling survival, even at 150 mM NaCl, was not predictive of long-term yield performance at 50 mM NaCl. Culture ventilation improved the accuracy of in vitro screening of tomato seedlings for salinity tolerance (Mills and Tal, 2004). Use of better ventilation may increase the effectiveness of in vitro screening for potato.

There are logistical problems in applying single-node cutting or microtuberization bioassays as screening tools on a large population of genotypes. This requires a significant amount of work in terms of tissue culture manipulations, measurement of growth parameters and analysis. As no single growth parameter accurately predicts salinity tolerance, labour-intensive multiple growth measurements are necessary. Despite the challenges, micropropagation systems have potential for screening genotypes capable of nutrient uptake under different environmental stress regimes, which apart from salinity could include water stress, high CO_2, high light or temperature extremes. Genotypes identified in vitro would need to be further tested in the greenhouse and/or field.

20.7 ENGINEERING AND CULTURAL MANAGEMENT PRACTICES FOR MODULATION OF SALINITY STRESS

There is little doubt that increasing the sustainability of irrigation and drainage systems will contribute most towards agricultural productivity in a future faced with increased pressure on water resources. This is best done in an integrated context where engineers and growers work together based on knowledge of the most vulnerable growth stages of each crop.

Crops are affected by salinity stress partly due to water stress and partly due to imbalance or deficiency of nutrients under osmotic stress conditions (Maas and Hoffman, 1977). A few studies have been done towards a better understanding of water and nutrient relationships of potato during NaCl stress. These may lead to improved productivity during the most vulnerable growth stages (Bruns and Caesar, 1990; Levy, 1992) and improved fertilisation to modulate salinity stress (Silberbush and Lips, 1991; Elkhatib et al., 2004a,b).

20.7.1 Water management for potato crops under salinity stress

It is apparent that at moderate or high salinity levels, certain potato growth stages are more vulnerable than others. Different water management strategies have been evaluated for potato growing under saline regimes in the field (Levy, 1992) and greenhouse

(Bruns and Caesar, 1990). Yield reductions were less when shoot emergence occurred before salt application (Levy, 1992). Applied 1 week after emergence, salinity delayed shoot development, especially at the higher salt concentrations (Bruns and Caesar, 1990). When salt was applied at the onset of tuberization, plants showed earlier senescence and reduced yields especially at the higher salt levels. Applied during tuber development, salinity treatments only marginally affected shoot development but downgraded tuber quality and accelerated plant senescence. Typical symptoms of salt stress on tubers include discolouration and fissures (Fig. 20.3).

Practices that may reduce salt stress to the crop could include green sprouting to promote early emergence and use of better-quality irrigation water during early-season growth, including during shoot emergence and tuber induction. If saline irrigation can be delayed to later in the season, impact on yield may be proportionally less.

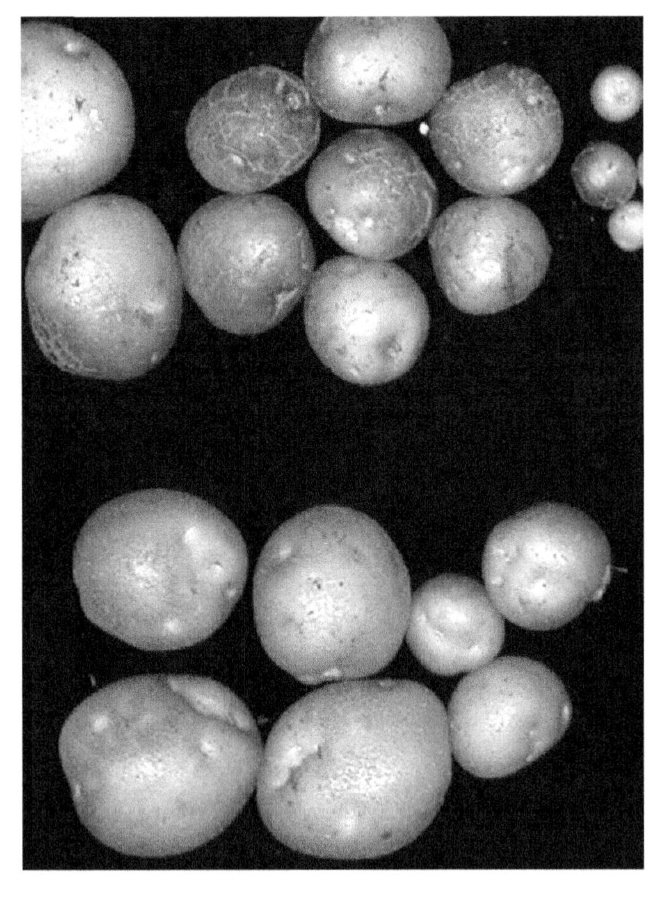

Fig. 20.3. Appearance of potato cv. Snowdon affected by soil salinity (top) compared with control tubers (bottom) (Dr. J. Abdulnour, Laval University, St. Foy, QC, Canada).

Irrigation methods play an important role in potato production. Although saline water can be applied using surface irrigation methods, it can cause leaf burn, depress yield and increase soil salinity due to greater saline water usage. Paliwal and Yadav (1980) reported that water with salinity as low as 20 meqL^{-1} can cause reduction in yield. They also found that salinity in the topsoil layer increased from 2.1 to 14.75 dS m^{-1} when irrigated with water having salinity of about 9 dS m^{-1}. Drip irrigation can be used to apply water with a salinity of 4.5–5.5 dS m^{-1} and a water saving of up to 30% compared with surface irrigation methods (Netafim, 2006). Frequent application of water with drip irrigation will also reduce water stress on the crop. Nadler and Heuer (1995) demonstrated that water with salinity levels of 1.5–6 dS m^{-1} could be used with drip irrigation without affecting potato yield.

In North America, sub-irrigation is quite often used to apply fresh water. In this method, water is applied from below the root zone through existing subsurface drainage systems. This method has not often been used to apply saline water for irrigation, although studies have shown that this is feasible. For example, Hanson and Carlton (1985) used saline water in open ditches for sub-irrigation purposes. Patel et al. (2004) used saline water in sub-irrigation for green peppers grown in lysimeters. Hanson and Carlton (1985) and Patel et al. (2004) reported that water present in the soil profile above sub-irrigation water was pushed upwards by sub-irrigation with brackish water. This is beneficial because water with lower salinity, present in the soil profile, can effectively be used during the early stages of crop growth. Patel et al. (1999, 2001) also used saline water to sub-irrigate lysimeter-grown potato. Yield of cultivars Russet Burbank and Atlantic was not affected by water table depth, but number of grade A tubers was increased for Russet Burbank (not Atlantic) when the water table was maintained at 40 cm compared with 80 cm below the soil surface (Patel et al., 1999). Yield of Norland, a short-season cultivar, was more affected compared with the two long-season cultivars, Kennebec and Russet Burbank (Patel et al., 2001). These studies showed that sub-irrigation with water having salinity as high as 9 dS m^{-1} could be effectively used for growing potato.

20.7.2 Fertiliser management for potato crops under salinity stress

Salinity–nutrient relationships are difficult to determine under field conditions but can be resolved more readily in vitro or through the use of lysimeters. Micropropagation and cell culture systems contribute a controlled environment in which nutrient relationships under salinity stress can be compared. Exploration of potassium, phosphorus and nitrate relationships of potato plantlets or cells in vitro was conducted using salinized medium. Wu et al. (1996) demonstrated that potassium (K) acquisition was critical to salt tolerance in glycophytes and suggested that increased K^+ fertilisation could potentially alleviate salt stress. This concept was tested using micropropagated potato (Alhagdow et al., 1999). Single-node cuttings were grown in MS basal medium containing 0, 40 or 80 mM NaCl and 6, 20 or 30 mM K. After 1 month of growth, tissue Na levels were greater when medium K^+ concentrations were 6 mM compared with 20 mM (MS control level). However, medium K^+ concentrations of 30 mM did not limit tissue Na levels or promote growth. This study concluded that while K^+ deficiency increases salt damage, its addition beyond the usual fertiliser recommendations was not likely to be beneficial in alleviating salinity stress. Similar results were found in field lysimeters that were salinized with NaCl (9.38 dS m^{-1}) (Elkhatib et al., 2004a,b). Elkhatib et al. characterized yield response to salinity stress using polynomial equations and used these to predict optimal K^+ fertilisation rates.

In vitro studies have also promoted understanding of sodium–phosphate relationships. High NaCl reduced inorganic phosphate uptake by several crops (summarized in Kalifa et al., 2000) including potato (Zhong, 1993). Using radiophosphorus (^{32}P), it was demonstrated that as MS basal medium was increasingly salinized, phosphorus translocation to micropropagated potato plantlet leaves was decreased (Kalifa et al., 2000). At high and constant NaCl levels (120 mM), translocation of phosphorus into the leaves improved as basal medium phosphorus levels were increased up to 2.5 mM.

Cell suspension cultures of Alpha and Désirée were depressed in biomass when NaCl was present in the medium at 75 mM, although the addition of 4 mM CaCl modulated this response (Hawkins and Lips, 1997). In salinized medium, salt-stressed cells took up only 45% of the amount of NO_3^- of control cells. This was not affected by addition of $CaCl_2$ to the medium. Reduced nitrate uptake during salinity stress was attributed to reduced plasma membrane ATPase activity in the presence of salinity (at least during the short term of these experiments), which reduced the proton gradient necessary for nitrate uptake.

20.7.3 Climatic conditions modulate salinity effects on potato

The response of plants to salinity is affected by climatic conditions. In a limited number of studies, potato response to salinity was better under cooler, more humid conditions (Bruns and Caesar, 1990) or under mild compared with high temperatures (Ferreyra et al., 1997; Bustan et al., 2004). When temperatures were higher, tuber yields decreased, and the differences were less between yields of controls and plants treated with salt (Bruns and Caesar, 1990). Potato and other crops performed well beyond predicted expectations on highly salinized (3–9 dS m^{-1}) drip-irrigated soils, even when boron levels were excessive, under mild daily temperatures and cool nights (Ferreyra et al., 1997). Similarly, Bustan et al. (2004) showed that current classifications of potato crop response to salinity underestimated potato crop yields under mild climatic conditions. In drip-irrigated (pulsed) fields in Israel, when the climate was not otherwise stressful, relatively high potato yields were possible even when irrigation water reached salinity levels as high as 6.2 dS m^{-1}. However, heat wave events, and especially their timing with respect to tuber development, had an increasingly critical impact on yield as salinity stress increased. Following a heat wave, plants irrigated with saline water were slower to recover; leaves were more likely to wilt permanently, and new leaf regeneration was slower or absent compared with plants irrigated with fresh water. While sodium levels tended to be greater in stems and older leaves than in young leaves, this ion compartmentalization appeared to break down, and sodium levels increased dramatically in young leaves compared with older tissues following a heat wave. Injury to young expanding leaves was less critical before tuber initiation, at a time when these leaves could be regenerated, or during declining tuber growth at the end of the season. Yields collapsed when heat waves caused leaf injury during early tuber formation (40–60 days after planting in cv. Désirée).

20.8 PRODUCING SALINITY TOLERANT POTATO

Salinity tolerance is complex and multigenic. Breeding for salinity tolerant crop plants can be difficult and slow for many reasons (Munns, 1993). Breeding for salinity tolerance in potato seems to be in its infancy. By comparison, understanding of salinity tolerance in some model species such as *Arabidopsis* is relatively advanced. Great progress has also been made in certain crop species related to potato, such as tomato. These works may contribute strategic insights towards breeding for potato with salinity tolerance.

20.8.1 Salinity-tolerant wild and/or primitive potato species

Conventional breeding efforts for development of salinity tolerance in potato will require parental material with documented salinity tolerance or resistance.

20.8.1.1 Where can a plant breeder find wild and/or primitive potato species with potential salinity tolerance?

A list of genebanks currently holding collections of wild and/or primitive potato species is posted on the website of the Centre for Genetic Resources, The Netherlands (CGN) under 'Major Genebanks' (http://www.cgn.wageningen-ur.nl/pgr/collections/crops/potato/genbanks.htm). On this website are listed databases held in Europe and The Americas as well as Inter-genebank databases. The Association of Potato Inter-genebank Collaborators (APIC) links the major potato genebanks from around the world (Europe, Russia, USA and South America) and maintains the Inter-genebank Potato Database (http://www.potgenebank.org). In 1996, there were no descriptors relating to salinity tolerance in potato. Some databases, such as GRIN, show evaluation data that include 'frost' and 'heat' in their list of stress descriptors (http://www.ars-grin.gov/ars/MidWest/NR6/evaluation.html), but salinity is not listed. Only the CGN has 'Evaluation data' for salt tolerance (http://www.genebank.nl/collections/crops/potato/). This website has tabular data on the best clones identified through in vitro screening and greenhouse pot trials by Elhag (1991). Salinity screening at 80 and 120 mM NaCl identified four wild species (one accession each) as tolerant from among 58 accessions of 36 wild species and 11 cultivars of *S. tuberosum*. The most tolerant genotypes were *S. chacoense* PI#16979 (CGN#17898), *S. sparsipilum* PI#24687 (CGN#18099), *S. gourlayi* PI#16837 (CGN#17853) and *S. vernei* PI#15451 (CGN#17836). When retested at 120 and 160 mM NaCl, the *S. chacoense* and *S. sparsipilum* accessions were tolerant, but *S. tarijense* was only moderately tolerant.

20.8.2 Domestication of wild salt-tolerant potato

Flowers and Yeo (1995, 1997) suggested that domestication of a wild salt-tolerant species might be less expensive than breeding existing cultivars for salinity tolerance.

20.8.2.1 Are there salt-resistant wild relatives of potato?

There are no apparent halophytic relatives of potato, although many accessions have been collected (Huamán et al., 2000) from countries where saline soils are prevalent such as Argentina, Bolivia, Chile, Mexico, Peru, to name only a few. A few field and greenhouse studies (Section 20.6.1) and several in vitro trials (Section 20.6.2) have compared the relative salinity tolerance of wild species or accessions. Only the in vitro screening trials of Elhag (1991), Khrais (1996) and Khrais et al. (1998) have compared large numbers of species, many accessions of an individual species, or many cultivars, respectively. Despite the limited number of comprehensive investigations of this subject, there is no doubt that considerable variation exists among the wild species and their accessions and among the cultivars of *S. tuberosum*.

20.8.2.2 Do drought or temperature tolerance contribute salt tolerance?

A two-phase growth response characterizes the relative yield response (percentage of the yield under control conditions) of any crop to salinity levels (Maas and Hoffman, 1977; Munns and Termaat, 1986; Munns, 1993, 2002). The first phase (with a time frame of minutes or days) results from low water potential of the medium and results in osmotic effects on roots, root hormonal signals and reduced plant growth. During this phase, compartmentalization of ions (Na^+ and Cl^-) in cell vacuoles occurs, and direct toxic effects are avoided. In the second phase, salt toxicity is involved; vacuoles are saturated, salts accumulate in the cytoplasm and cell walls and may result in tissue death. Long-term rather than short-term testing was advocated as a means of detecting the real salt tolerant not just osmotically tolerant selections as it was felt that all glycophytes reacted similarly to short-term water stress. Long-term effects (with a time-frame of days or weeks) include features of leaf injury and plant growth or biomass production that correlate with yield and are more likely to distinguish genotypes with salinity tolerance (Munns, 1993, 2002). Neumann (1997) argued that it is desirable to consider the osmotic effects of external salinity separately and to select for this in the short term, in addition to screening for increased tolerance to salt accumulation. The rational presented is that diversity has often been found in the first phase of crop response. This diversity may have particular value in areas of intermittent irrigation when irrigation water is moderately saline. For this reason, response to both salinity and polyethylene glycol (PEG) or another osmotic agent should be evaluated.

No relationship was found between salt tolerance and heat or drought tolerance in a greenhouse pot trial of several wild potato species, although few accessions were tested (Bilski, 1988b). However, frost tolerance and salt tolerance appeared related in two wild species tested in vitro (Martinez et al., 1996). As few studies of this sort have occurred with potato, independent screening for osmotic and temperature stresses will be necessary.

20.8.3 Breeding for increased vigour and yield

As maximal plant productivity is affected by quantitative traits, there were some concerns that yield potential would decrease if a trait was introduced to improve salinity tolerance

(Bohnert and Jensen, 1996; Winicov, 1998). This has not been discussed in any detail for potato. Breeding for increased vigour, rather than salinity tolerance per se, is one strategy that has been advocated for lands where salinity levels vary (Munns, 1993).

It is possible to screen in vitro for both salinity tolerance and vigour at the same time. Screening activities indicated clear differences in relative vigour among *S. tuberosum* cultivars in the most salinity tolerant group, where one of the six top-ranked cultivars was less vigorous (Khrais et al., 1998).

20.8.4 Obtaining salinity tolerance through cell and tissue culture techniques

Screening at the cell level cannot identify whole-plant response mechanisms (Rains et al., 1986). Increased salinity tolerance of cultured plant cells does not necessarily lead to expression of salt tolerance in regenerated plants, primarily due to failure to understand the tolerance mechanisms operating at the cellular and whole-plant levels (Dracup, 1991). This was true for potato, where salinity tolerance relationships did not correspond between callus cultures and greenhouse-tested plants (Sabbah and Tal, 1995). However, in some cases, the salt-sensitive potato cultivar and the less-sensitive cultivar may have been too similar in response to salinity at the whole-plant level to expect major differences at the cellular level (Hawkins and Lips, 1997).

Another attempt to derive salinity tolerant plants from potato callus was described by Hannachi et al. (2004). Adventitious shoots of cv. Claustar were collected from control callus from MS basal medium or callus adapted to growth medium containing $3\,g\,L^{-1}$ NaCl. Plants derived from control or salt-adapted callus were field-evaluated under conditions of clean or saline ($4\,g\,L^{-1}$ NaCl) irrigation water. Yield parameters (number and mean weight of tubers per plant) were similar in both groups. Reduced number of eyes and reduced tuber diameter in the salt-adapted plants occurred. Deep fissuring of tubers resulted from exposure to saline irrigation water and was worse in the salt-adapted plants than in the adventitiously produced control plants. It is not always clear from such studies whether adventitious shoots (caliclones) produced on callus exposed to elevated salt levels are 'escapes' from the high salt environment or truly produced under conditions of elevated salt. It is also unclear to what extent these caliclones might vary from the original cultivar and how stable this variation might be.

In a unique report of salt tolerance at the cellular level translating to tolerance at the whole-plant level, a salt-tolerant caliclone (clone 150) from potato, cv. Kennebec was developed using a callus culture system (Ochatt et al., 1999). Callus cells were subjected to recurrent selection on medium salinized with 150 mM NaCl followed by adventitious regeneration in the continued presence of NaCl. A sensitive caliclone was produced through recurrent selection and regeneration on control medium without NaCl. Selected tolerant and sensitive caliclones were multiplied on medium without salt for 8 years prior to transfer to the greenhouse. Clone 150 that performed well when exposed to 150 mM NaCl in culture did not grow well when exposed to iso-osmotic levels of PEG-600 in culture but performed well when watered with 90 mM NaCl in the greenhouse. Clone 150 demonstrated stable salinity tolerance maintained over many years. It was different both phenotypically (more lobed leaves, oval tubers with reddish periderm and flesh compared with yellow periderm and flesh of Kennebec or the sensitive clone) and genotypically (polymorphic for 1 of 70 primers used for RAPDs) from the sensitive clone. Comparative growth analysis using salinized (100 mM NaCl) hydroponic culture distinguished the two clones after 28 but not 14 days. Clone 150 had greater fresh and dry weights than the sensitive clone and greater relative accumulation rate of water at all salt levels tested. The sensitive clone accumulated very high levels of Na^+ and high levels of K^+, whereas clone 150 accumulated less Na^+ and K^+ into both roots and shoots (Marconi et al., 2001).

Later work determined that chlorophyll concentration was reduced in the salt-sensitive clone but not clone 150 (Benavides et al., 2000). Glutathione concentration and all enzymatic activities measured (SOD, APOX, GR, DHAR and CAT) were constitutively elevated in clone 150 compared with the sensitive clone grown at 0 mM NaCl. Under salt challenge (100 or 150 mM NaCl) chlorophyll concentration and antioxidant parameters were unaffected in clone 150. However, in the sensitive clone, chlorophyll levels were reduced, antioxidant defence enzyme levels were reduced or unaltered, and salinity damage was apparent. Acquired salt tolerance appears to have resulted from improved resistance to oxidative stress through increased soluble and enzymatic antioxidant systems. The relationship between antioxidant defence systems and salinity tolerance could possibly be exploited as a screening tool to further increase salinity tolerance. In 1999, clone 150 was under evaluation for economic relevance. To the best of our knowledge this clone has not been released.

Spontaneously occurring somaclonal variation has led to many reports of salt-tolerant or salt-resistant plant cells, usually with no increase in the salinity tolerance of plants derived from these cells (Tal, 1990). There are a few reports of acquisition of salinity tolerance in potato cells through somaclonal techniques. Potato cells selected for resistance to proline analogues were relatively more resistant to stress, including cold and salt stress (van Swaaij et al., 1986). Potato callus and suspension cultures were selected with improved resistance to NaCl and mannitol (Sabbah and Tal, 1990). However, these studies did not lead to recovery of salt-tolerant plants (Tal, 1996). More recently, somatic embryos were reliably produced from a range of potato tissues, genotypes and ploidy levels (Seabrook and Douglass, 2001; Seabrook et al., 2001). These techniques should have application in the development of commercially interesting salinity tolerant intraclones of established cultivars or wild species.

Shepard et al. (1980) were first to demonstrate the possibilities of wide hybridization using protoplast fusion in potato breeding. Only recently, this technology has been applied in the development of potato plants with stress tolerance. Plantlets of salt-tolerant *S. vernei* (origin unstated), the drought-tolerant *S. tuberosum* cv. Sahel, drought- and salt sensitive *S. tuberosum* cv. Belle de Fontenay and disease-resistant *S. stenotomum* (PI#1234013), as well as two somatic hybrids made from fusing cells of cv. Belle de Fontenay and *S. stenotomum*, were challenged in vitro with salt (200 mM) or PEG (10% PEG 8000) for 2, 24 or 48 h (Akossiwoa Quashie et al., 2004). Various enzymes and aquaporins were examined that might distinguish biochemically or physiologically between osmotic and salinity stress in micropropagated plantlets. Preliminary results suggested that aquaporin expression may be useful in distinguishing stress responses. Aquaporin expression is initially reduced under salinity stress and is thought to explain resistance to osmotic stress – as a mechanism of reducing water loss and excluding sodium. In halophytes, aquaporin expressin is less affected. In the presence of salt stress, the salinity tolerant *S. vernei* initially showed little disturbance of plasma membrane aquaporins of either 26 or 19 kDa but after some time, overexpression occurred in the 19-kDa group. Hybrids with intermediate tolerance over-expressed aquaporins in the 17–19-kDa range, but the 26-kDa aquaporins were not affected. In response to osmotic stress, it was the 26-kDa aquaporin that was affected; less in the drought-tolerant cv. Sahel than in the other genotypes.

Nabors (1990) provides a comprehensive discussion of the constraints slowing the use of tissue culture selection. He pointed out the need for relative growth measurements compared with control cells for putatively tolerant cells grown on salinized medium and the necessity for regeneration of putatively tolerant cells to plantlets for demonstration of stable and inheritable tolerance. The latter does not apply to many cultivated potato. He also suggested that a working link be established between laboratories that identify stress-tolerant germplasm and plant breeders capable of field-testing this material. Without this link, potentially useful germplasm and much scientific effort will be wasted. Fifteen years later, there is still no obvious mechanism in place for this type of communication or help available for testing putatively salinity tolerant potato. It would certainly seem appropriate that scientists who feel they have produced a salinity tolerant plant communicate with a germplasm repository to assist them in preserving this material. In addition, plant breeders working on salinity tolerance in potato should identify themselves to one of the potato germplasm repositories. Coordinates of these plant breeders could be clearly posted on the website to facilitate transfer of the material for suitable field-testing and incorporation into a breeding program.

20.8.5 Obtaining salinity tolerant potato through genetic engineering

Some scientists believe that salinity tolerance of crop plants must be accompanied by plant transformation (Bohnert and Jensen, 1996). The basis for this argument is that lack of success through conventional breeding may have resulted from outdated concepts. For potato, we can safely conclude that the basis for failure in achieving salinity tolerance has involved the complexity of this multigenic characteristic, the apparent lack of outstanding salinity tolerance in wild *Solanum* species, the lack of coordinated effort and the modest effort invested to date.

Molecular engineering has resulted in major improvements in salinity tolerance in model plants such as *Arabidopsis* (reviewed in Apse and Blumwald, 2002; Ward et al., 2003). Some of these traits, such as overexpression of a vacuolar Na^+/H^+ antiporter protein has now been transferred to tomato, which has become a model species for pyramiding quantitative trait loci (QTL) through marker-assisted selection (reviewed by Foolad, 2004). The development of commercial cultivars of tomato with salinity tolerance is almost at hand; breeding efforts are far ahead of studies with potato and may serve as a beacon of hope in the development of potato with salinity tolerance.

20.9 SUMMARY

Salinity is a serious threat to world agriculture that may intensify under proposed global warming scenarios. Potato is widely grown around the world and some of these growing areas are threatened by increase in salinity levels and increase in the saline land area. An integrated approach involving engineering and improved cultural management as well as increased salinity tolerance of potato genotypes is necessary. This chapter summarizes the little that is known of water and fertiliser management for potato growing under salinity stress and how climatic conditions modulate salinity effects on potato. In 2006, there were no known commercially important *S. tuberosum* cultivars with outstanding salinity tolerance. For this reason, emphasis must be placed on improved water management techniques for potato production. This will become increasingly important in the context of predicted climate change.

Efforts to classify and evaluate the salinity tolerance of cultivars and wild species in the field and in vitro are reviewed as is the scanty literature on salinity tolerance mechanisms specific to potato. The one physiological parameter that has been studied to any extent in potato is proline accumulation. However, correlations between relative salinity tolerance and proline levels were inconsistent both in field and in in vitro studies.

Breeding for salinity tolerance in potato may draw on a wealth of wild *Solanum* species and accessions held in germplasm repositories around the world. The web coordinates for the major repositories of potato germplasm are listed; only the CGN (Netherlands) provides evaluation criteria for salt tolerance and lists salt-tolerant genotypes. Some wild species with relative salinity tolerance have been suggested as potentially useful to breeding efforts, but only a few accessions of these were tested. However, it should be noted that no wild *Solanum* species have been found with salinity tolerance levels beyond that of some cultivars. In vitro testing has been well correlated with results from field or lysimeter studies but is labour intensive and time consuming. Numerous growth parameters must be measured at various growth stages and salt levels; tolerance at one growth stage or salt level was not necessarily predictive of tolerance at other growth stages or salt concentrations.

Few studies have examined the relationship between salt tolerance and other abiotic stress tolerances in potato. Independent screening for various abiotic stresses will be necessary. Screening at the cell level has not usually led to expression of salt tolerance in regenerated plants. A notable exception involved a somatic variant with excellent salinity tolerance regenerated from salt-tolerant potato callus (Ochatt et al., 1999). Some mechanisms of tolerance in this clone involved improved resistance to oxidative stress (Benavides et al., 2000) and less accumulation of Na^+ and K^+ in shoots and roots (Marconi et al., 2001). Recent improvements in somatic cell technology (Seabrook and Douglass, 2001; Seabrook et al., 2001) may increase the likelihood of identifying a somaclonal variant with improved salinity tolerance that also has suitable commercial characteristics. The same may be said for protoplast fusion technologies that have recently led to the development of potato plants with increased stress tolerance (Akossiwoa Quashie et al., 2004). In the near future, we are likely to see these two in vitro technologies more fully exploited for the development of salinity tolerance in potato. Ideally, we would see them coupled to in vitro screening systems that are less cumbersome than at present. Molecular engineering has resulted in major improvements in salinity tolerance of *Arabidopsis* (Apse and Blumwald, 2002; Ward et al., 2003) and tomato; commercial cultivars of tomato with salinity tolerance will soon be available (Foolad, 2004). These major advances will also lend impetus to the development of potato with salinity tolerance.

REFERENCES

Akossiwoa Quashie M.-L., A. Nato and K. Akpagana, 2004, *Acta Bot. Gallica* 151, 127.

Ahmad R., 1997, In: P.K. Jaiwal, R.P. Singh and A. Gulati (eds), *Strategies for Improving Salt Tolerance in Higher Plants*, p. 403. Science Publishers Inc., U.S.A.

Ahmad R. and Z. Abdullah, 1979, *Pak. J. Bot.* 11, 103.

Alhagdow M.M., N.N. Barthakur and D.J. Donnelly, 1999, *Potato Res.* 42, 73.

Apse M.P. and E. Blumwald, 2002, *Curr. Opin. Biotech.* 13, 146.

Arslan N., G. Mix and N. El Bassam, 1987, *Landbauforschung Völkenrode* 37, 128.

Ashraf M., 1994, *CRC Crit. Rev. Plant Sci.* 13, 17.

Ashraf M. and P.J.C. Harris, 2004, *Plant Sci.* 166, 3.

Barnes W.C. and T.C. Peele, 1958, *Proc. Am. Soc. Hort. Sci.* 72, 339.

Benavides M.P., P.L. Marconi, S.M. Gallego, M.E. Comba and M.L. Tomaro, 2000, *Aust. J. Plant Physiol.* 27, 273.

Bernstein L., A.D. Ayers and C.H. Wadleigh, 1951, *Proc. Amer. Soc. Hort. Sci.* 57, 231.

Bilski J.J., D.C. Nelson and R.L. Conlon, 1988a, *Am. Potato J.* 65, 85.

Bilski J.J., D.C. Nelson and R.L. Conlon, 1988b, *Am. Potato J.* 65, 605.

Bohnert H.J. and R.G. Jensen, 1996, *Aust. J. Plant Physiol.* 23, 661.

Bruns S. and K. Caesar, 1990, *Potato Res.* 33, 23.

Bustan A., M. Sagi, Y. De Malach and D. Pasternak, 2004, *Field Crops Res.* 90, 275.

Chavez R., A. Wijntje, R. Berríos, M. Upadhya, P. Zúniga, S. Colque, R. Cabello, J. Espinoza, G. Cueva, H. Mendoza, W. Amoros, G. Bollo, P. Siles, K. Monasterio and M. Huacollo, 1997, *Ciencia y Desarrollo* 5, 60.

Dracup M., 1991, *Aust. J. Plant Physiol.* 18, 1.

Elhag A.Z., 1991, *Eignung von in vitro verfahreh zur Charakterisierung der Salztoleranz bei Solanum-arten.* Vom Fachbereich Gartenbau der Universitat, Hannover. 148 pp.

Elkhatib H.A., E.A. Elkhatib, A.M. Khalaf-Allah and A.M. El-Sharkawy, 2004a, *J. Plant Nutr.* 27, 111.

Elkhatib H.A., E.A. Elkhatib, A.M. Khalaf-Allah and A.M. El-Sharkawy, 2004b, *J. Plant Nutr.* 27, 1575.

Epstein E. and D.W. Rains, 1987, *Plant Soil* 99, 17.

Evers D., S. Overney, P. Hausman, H. Simon and J.F. Greppin, 1999, *Biol. Plantarum* 42, 105.

FAO, 1992, *Irrigation and Drainage Papers, 48.* FAO, Rome, pp. 184.

FAO, 1995, *Irrigation and Drainage Papers, 53.* FAO, Rome, pp. 75.

Ferreyra R.E., A.U. Aljaro, R.S. Ruiz, L.P. Rojas and J.D. Oster, 1997, *Agr. Water Manage.* 34, 111.

Fidalgo F., A. Santos, I. Santos and R. Salema, 2004, *Ann. Appl. Biol.* 145, 185.

Flowers T. J. and A.R. Yeo, 1995, *Aust. J. Plant Physiol.* 22, 875.

Flowers T.J. and A.R. Yeo, 1997, In: P.K. Jaiwal, R.P. Singh and A. Gulati (eds), *Strategies for Improving Salt Tolerance in Higher Plants*, p. 247. Science Publishers Inc., U.S.A.

Foolad M.R., 2004, *Plant Cell Tiss. Org. Cult.* 76, 101.

Ghosh S.C., K. Asanuma, A. Kusutani and M. Toyota, 2001, *Soil Sci. Plant Nutr.* 47, 467.

Hannachi C., P. Debergh, E. Zid, A. Messaï and T. Mehouachi, 2004, *Biotechnol. Agron. Soc. Environ.* 8, 9.

Hanson B.R. and A.B. Carlton, 1985, *Trans. ASAE* 28(3), 815.

Hawkes J.G., 1990, *The Potato. Evolution, Biodiversity and Genetic Resources.* Belhaven Press, Pinter Publications, London, pp. 209.

Hawkins H.J. and S.H. Lips, 1997, *J. Plant Physiol.* 150, 103.

Heuer B. and A. Nadler, 1995, *Aust. J. Agr. Res.* 46, 1477.

Heuer B. and A. Nadler, 1998, *Plant Sci.* 137, 43.

Hijmans R.J., 2001, *Am. J. Potato Res.* 78, 403.
Hijmans R.J., 2003, *Am. J. Potato Res.* 80, 271.
Huamán Z., R. Hoekstra and J.B. Bamberg, 2000, *Am. J. Potato Res.* 77, 353.
Jefferies R.A. 1996, *Euphytica* 88, 207.
Kalifa A., N.N. Barthakur and D.J. Donnelly, 2000, *Am. J. Potato Res.* 77, 179.
Katerji N., J.W. van Hoorn, A. Hamdy and M. Mastrorilli, 2000, *Agr. Water Manage.* 43, 99.
Katerji N., J.W. van Hoorn, A. Hamdy and M. Mastrorilli, 2003, *Agr. Water Manage.* 62, 37.
Khrais T., 1996, Evaluation of salt tolerance in potato (*Solanum* spp.). M.Sc. Thesis, Faculty of Graduate Studies, McGill University, 127 pp.
Khrais T., Y. Leclerc and D.J. Donnelly, 1998, *Am. J. Pot. Res.* 75, 1.
Kim H.S., J.H. Jeon, Y.H. Jeung and H. Joung, 1995, *J. Kor. Soc. Hort. Sci.* 36, 172.
Levy D., E. Fogelman and Y. Itzhak, 1988, *Potato Res.* 31, 601.
Levy D., 1992, *Ann. Appl. Biol.* 120, 547.
Maas E.V. and G.J. Hoffman, 1977, *J. Irrig. Drain. Div., ASCE* 103, 115.
Maas E.V., 1985, *Plant Soil* 89, 273.
Mansour M.M.F., K.H.A. Salama and M.M. Al-Mutawa, 2003, *Plant Sci.* 164, 891.
Mansour M.M.F. and K.H.A. Salama, 2004, *Env. Exp. Bot.* 52, 113.
Marconi P.L., M.P. Benavides and O.H. Caso, 2001, *N.Z. J. Crop Hort. Sci.* 29, 45.
Martinez C.A., M. Maestri and E.G. Lani, 1996, *Plant Sci.* 116, 177.
Mills D. and M. Tal, 2004, *Plant Cell Tiss. Org. Cult.* 78, 209.
Morpurgo R., 1991, *Plant Breed.* 107, 80.
Morpurgo R. and D. Silva-Rodriguez, 1987, *Riv. Agric. Subtrop. Trop.* 81, 73.
Munns R. and A. Termaat, 1986, *Aust. J. Plant Physiol.* 13, 143.
Munns R., 1993, *Plant Cell Environ.* 16, 15.
Munns R., 2002, *Plant Cell Environ.* 25, 239.
Munns R. and R.A. James, 2003, *Plant Soil* 253, 201.
Murashige T. and F. Skoog, 1962, *Physiol. Plant.* 15, 473.
Nabors M.W., 1990, In: P.J. Dix (ed.), *Plant Cell Line Selection: Procedures and Applications*, p. 167. Weinheim, New York, U.S.A.
Nadler A. and B. Heuer, 1995, *Potato Res.* 38, 119.
Naik P.S. and J.M. Widholm, 1993, *Plant Cell Tiss. Org. Cult.* 33, 273.
Netafim, 2006, http://www.netafim.com/Business_Divisions/Agriculture/New_Trends/Growing Potatoes_ Under_Drip_Irrigation.
Netondo G.W., J.C. Onyango and E. Beck, 2004a, *Crop Sci.* 44, 797.
Netondo G.W., J.C. Onyango and E. Beck, 2004b, *Crop Sci.* 44, 806.
Neumann P., 1997, *Plant Cell Environ.* 20, 1193.
Ochatt S.J., P.L. Marconi, S. Radice, P.A. Arnozis and O.H. Caso, 1999, *Plant Cell Tiss. Org. Cult.* 55, 1.
Paliwal K.V. and B.R. Yadav, 1980, *Indian J. Agric. Sci.* 50, 31.
Parida A.K. and A.B. Das, 2005, *Ecotox. Env. Safety* 60, 324.
Patel R.M., S.O. Prasher, C.A. Madramootoo, P.K. Goel, R.S. Broughton, K. Stewart, and R.B. Bonnell, 2004, *J. Veg. Crop Prod.* 10, 57.
Patel R.M., S.O. Prasher, D. Donnelly, R.B. Bonnell and R.S. Broughton, 1999, *Bioresource Technol.* 70, 33.
Patel R.M., S.O. Prasher, D. Donnelly and R.B., Bonnell, 2001, *Agric. Water Manage.* 46, 231.
Potluri S.D.P. and P.V.D. Prasad, 1993, *Plant Cell Tiss. Org. Cult.* 32, 185.
Rahnama H. and H. Ebrahimzadeh, 2004, *Acta Physiol. Plant.* 26, 263.
Rains D.W., S.S. Croughan and T.P. Croughan, 1986, In: I.K. Vasil (ed.), *Cell Culture and Somatic Cell Genetics of Plants. vol. 3. Plant Regeneration and Genetic Variability*, p. 537 Academic Press Inc., Orlando, FL.
Sabbah S. and M. Tal, 1990, *Plant Cell Tiss. Org. Cult.* 21, 119.
Sabbah S. and M. Tal, 1995, *Potato Res.* 38, 319.
Seabrook J.E.A. and L.K. Douglass, 2001, *Plant Cell Rep.* 20, 175.
Seabrook J.E.A., L.K. Douglass and G.C.C. Tai, 2001, *Plant Cell Tiss. Org. Cult.* 65, 69.
Shannon M.C. and C.M. Grieve, 1999, *Scientia Hortic.* 78, 5.
Shepard J.F., D. Bidney and E. Shahin, 1980, *Science* 208, 17.
Silberbush M. and S.H. Lips, 1991, *J. Plant Nutr.* 14, 765.

Silva J.A.B., W.C. Otoni, C.A. Martinez, L.M. Dias and M.A.P. Silva, 2001, *Sci. Hortic.* 89, 91.
Szabolcs I., 1987, *Acta Agron. Hung.* 36, 159.
Szabolcs I., 1994, In: M. Pessarakali (ed.), *Handbook of Plant and Crop Stress*, p. 3. M. Dekker, New York.
Tal M., 1990, In: Y.P.S. Bajaj, (ed.), *Biotechnology in Agriculture and Forestry, vol. 2, Somaclonal Variation in Crop Improvement II*, p. 236. Springer-Verlag, Germany.
Tal M., 1996, In: Y.P.S. Bajaj, (ed.), *Biotechnology in Agriculture and Forestry, vol. 36, Somaclonal Variation in Crop Improvement II*, p. 132. Springer-Verlag, Germany.
Tester M. and R. Davenport, 2003, *Ann. Bot.* 91, 503.
Thomas D.S.G. and N.J. Middleton, 1993, *J. Arid Environ.* 24, 95.
Umali D.L., 1993, *World Bank Technical Paper No. 215*. The World Bank, Washington, D.C., pp. 78.
van Swaaij A.C., E. Jacobsen, J.A.K.W. Kiel and W.J. Feenstra, 1986, *Physiol. Plant.* 68, 359.
Ward J.M., K.D. Hirschi and H. Sze, 2003, *Trends Plant Sci.* 8, 200.
Winicov I., 1998, *Ann. Bot.* 82, 703.
Woolfe J.A., 1986, *The Potato in the Human Diet*. Cambridge University Press. Cambridge, U.K., pp. 57.
Wu S.J., L. Ding and J.K. Zhu, 1996, *Plant Cell* 8, 617.
Zhang Y., M. Brault, V. Chalavi and D. Donnelly, 1993, Proceedings of 13th International Congress of Biometeorology, September 12–18, 1993. Calgary, AB, Canada, Part 2, p. 491.
Zhang Y. and D.J. Donnelly, 1997, *Potato Res.* 40, 285.
Zhong H, 1993, *J. Plant Nutr.* 16, 1733.

Part V

TUBER QUALITY

Chapter 21

The Harvested Crop

Michael Storey

British Potato Council, 4300 Nash Court, Oxford Business Park South, Oxford OX4 2RT, United Kingdom

21.1 INTRODUCTION

The importance of the potato as one of the world's major staple crops is increasingly being recognized because it produces more dry matter (DM) and protein per hectare than the major cereal crops (Burton, 1989).

Globally, production (2005) is an estimated 321 million tonne per annum (FAOSTAT, 2006) from an area of 18.6 million hectares. Average production in the 5-year period 2001–05 was 16.76 tonnes per hectare compared with 12.34 tonnes per hectare for the period 1961–65. The largest producer in the world is China with 73 million tonnes (2005); Europe and the Russian Federation account for about 40% of the world hectares harvested and about 40% production. Within the EU, Germany and Poland are the largest producers, each with about 11 million tonnes, but there is a large amount used for animal stock-feed. Ireland, Portugal and Great Britain have per capita consumption about 100 kg per annum. Overall, in Europe, consumption of fresh potatoes is greater than that of processed potato products. By contrast, North America has a higher proportion of processed consumption, and although it has less than 2.5% of global area, the USA accounts for 6% of production. Despite a global production increase over the past 45 years of some 50 million tonnes, the area has declined from an average of approximately 21.9 million hectare in 1960–65 to a current area of 18.6 million hectare. This increased productivity per unit area, however, has been offset by a population increase from 3.1 to 6.4 billion people, and hence globally, per capita availability of potatoes has decreased by almost 40%.

With a burgeoning global population, the International Potato Centre (CIP) aims to increase potato production further, particularly in subtropical and highland tropical regions throughout the world and also in the developing countries to meet UN Millenium Targets (CIP, 2003). The nutritional value of the crop is the key driver for this growth, linked to the economic benefits it can bring to developing economies, and provides the international impetus for breeding programmes and improving production technologies.

In contrast to many developing countries where the nutritional status of the potato is the key driver, there is increasing sophistication in developed economies in the market for fresh potatoes and increasing consumption of processed potato products including chips (crisps in the UK), French fries and an increasingly wide range of frozen and chilled products. This is being driven by the changing lifestyles of consumers who are looking for greater convenience. The opportunities presented by new culinary experiences

Potato Biology and Biotechnology: Advances and Perspectives
D. Vreugdenhil (Editor)

and changed eating patterns, more snacking, are resulting in a higher proportion of out-of-home consumption. The food service sector is of increasing importance, and in the UK, it represents approximately 50%, with the balance from retail sales. Whilst the consumer drive for convenience is happening, there is also greater awareness of issues such as obesity and the increased risk of type 2 diabetes, and the value of potatoes in a healthy balanced diet is being promoted. Several national programmes have been developed around the 'five-a-day' type initiative, some of them including potatoes in the count, others categorizing potatoes as a starchy carbohydrate within the 'balanced plate' (Produce for Better Health Foundation, USA, http://www.5aday.org; UK Department of Health, http://www.5aday.nhs.uk).

Consumers' requirements for fresh market potatoes are often associated with the visual characteristics such as the shape and appearance of the tuber with good freedom from defects and disorders. Blemish diseases such as silver scurf (*Helminthosporium solani*) and black dot (*Colletotrichum coccodes*) may be superficial and detract from the appearance but not necessarily the culinary acceptability of the potato. The culinary qualities are increasingly important in differentiating potato products in a commodity market. For the processed potato crop, the size and uniformity of the harvested crop, in terms of for example tuber shape, dry matter (DM) content or reducing sugar level, is crucial to product-manufacturing efficiency. Defects resulting from harvest and handling – damage and bruising – detract from crop quality for all markets.

End-use acceptability for markets is also influenced by quality characteristics associated with preparation and cooking attributes. These include enzymic browning resulting from the oxidation of tyrosine (Storey and Davies, 1992) and the non-enzymic, after-cooking blackening, also known as stem-end blackening. This results from chlorogenic acid combining with iron during cooking to form a complex that is oxidized on cooling to give the coloured ferri-dichlorogenic acid responsible for the blue-grey discolouration at the normal pH (6.0–6.5) of cooked potatoes (Hughes and Swain, 1962a,b). These attributes, together with taste and flavour, which are increasingly important to the consumer, are covered in detail elsewhere (Chapter 23, Sowokinos, this volume).

The nutritional value of the crop is a major impetus for breeding and development programmes in developing countries and also from a consumer's perspective in more sophisticated markets. Improved protein content and nutritional value can be affected by growth and development and by subsequent storage (Chapter 24, Taylor et al., this volume). This chapter therefore considers the nutritionally important characteristics of the harvested crop and some of the morphological and biochemical aspects associated with quality and defects that affect suitability for the market place.

21.2 NUTRITIONAL VALUE

The chemical composition of the potato is fairly conservative, and variation depends largely on the genetic features of the variety, although tubers of the same variety, and even tubers from the same plant, can vary in levels of particular contents. The chemical composition is also affected by environmental conditions during production and subsequent storage of the crop.

Analyses of the nutritional value have been carried out for many years and have been reported in several reviews (e.g. Woolfe, 1987; Burton, 1989; Lisińska and Leszczyński, 1989; Burton et al., 1992).

Carbohydrates, which constitute about 75% of the total DM, are the main energy source, and the potato provides significant amounts of protein, vitamins and dietary fibre. Overall, the harvested potato is a very good source of vitamin C (ascorbic acid), with the average 175 g serving providing 44% of Recommended Daily Allowance (RDA), a good source of vitamins B6 (29%) and B1 (16%), a source of folate (16%), and potato also provides potassium, iron and magnesium for the diet (BPC, 2004).

The energy value of boiled potatoes, 72–75 kcal, is about only half of that of rice (138 kcal) and pasta (159 kcal) and about one-third of that of bread (207–219 kcal). The almost total absence of fat in potatoes, which has twice the calorific value of carbohydrate, means that a considerable amount of boiled potato would need to be eaten to meet total daily energy requirements. However, the potato's total contribution to diet depends on the cooking method. Roast and chipped potatoes act as a carrier for absorbed and surface fat, and with the dehydration that results from the cooking process, these preparations have a higher calorific value. A summary of nutritional information (FSA, 2002) used for potatoes in dietary assessments in the UK is given in Table 21.1.

The potato is one of the richest sources of antioxidants in the human diet (Lachman and Hamouz, 2005). The main antioxidants are polyphenols (123–441 mg/100 g), ascorbic acid (8–54 mg/100 g), carotenoids (up to 0.4 mg/100 g) and tocopherols (up to 0.3 mg/100 g). L-Tyrosine, caffeic acid, chlorogenic acid and ferulic acid are amongst the main polyphenols, which have about twice the level in the skin compared with the flesh of the potato. Flavonoids are especially effective antioxidants (Bors and Saran, 1987). In the USA, potato consumption ensures a daily intake of about 64 mg polyphenols, which places potatoes second to tomatoes (Al-Saikhan et al., 1995).

Potato is also an important source of dietary fibre, and a 175 g serving contributes about 14% of UK's daily requirement. This, together with the energy that potato provides, the important vitamin contribution, the quality of the tuber protein and the nutritional benefits of antioxidants, means that potato, as a recognized starchy carbohydrate, is an important dietary component in many parts of the world.

Table 21.1 Proximate composition of raw and cooked potatoes (g/100 g) (FSA, 2002).

	Dry matter	Protein	Starch	Sugars	Fat	Dietary fibre
Uncooked[a] (flesh only)	21.0	2.1	16.6	0.6	0.2	1.3
Boiled (flesh only)	19.7	1.8	16.3	0.7	0.1	1.2
Baked (without skin)	21.1	2.2	17.3	0.7	0.1	1.4
Roast (in corn oil)	35.3	2.9	25.3	0.6	4.5	1.8
Chipped (in blended oil)	46.5	3.9	29.5	0.6	6.7	2.2

[a] Four varieties sampled over 2 years.

21.3 DRY MATTER

The DM content of most varieties selected for commercial usage ranges from about 18 to 26% (Burton, 1989); it is an important aspect of tuber quality and is affected by a wide range of factors, including most importantly, environmental factors during growth of the crop and development of the tuber, such as intercepted solar radiation, soil temperatures, available soil moisture and cultural treatments. See Chapters 17 (Haverkort, this volume) and 16 (Vos and Haverkort, this volume) and the study by Allen and Scott (2001) for a description of the agronomic principles involved in DM production and accumulation and the practical management of DM content in potato crops. In general, cool dull years and short growing seasons reduce DM production, whilst the reverse occurs in warm, sunny locations with long growing seasons and an adequate water supply. DM accumulates during growth of the crop, and tuber DM content often declines from a peak after defoliation of the haulm if the root systems remain active in moist soils (Wilcoxon et al., 1985). There is variation in the level of DM accumulated between tubers within a crop and within individual tubers (Gaze et al., 1998; Kumar and Ezekiel, 2004).

The variation can have a considerable impact on the end-user acceptability for both the fresh and the processed markets. The DM content and distribution has implications for bruise susceptibility during harvest and effects on cooking type, e.g. a waxy or mealy texture when boiled, organoleptic characteristics and, in processed potatoes, the final product texture (Chapter 24, Taylor et al., this volume).

The level of DM is generally highest in the storage parenchyma between the cortex and the vascular ring, is lower for the periderm and cortex and decreases towards the pith, and longitudinally, the stolon end is usually higher than the rose end (Fig. 21.1).

The specific gravity (SG) of tubers, determined from the underwater weight of a sample or by hydrometer readings, is extensively used as a measure of DM, and this often provides an indication of the suitability of the crop for a particular, e.g. processing, usage. With the methods used, anything that affects the density of the water or solution (e.g. temperature or suspended solids) must be accounted for. The relationship between SG and percentage of DM is usually described as linear, but several authors have obtained different regression equations for this relationship (Lisińska and Leszczyński, 1989). Differences are reported for cultivars (Vanesse et al., 1951) and on crops grown under different environmental conditions (Schippers, 1976). Varietal differences may, in part, be the result of differences in the intercellular spaces or the composition of the DM, and internal disorders such as hollow heart can affect measurements. The SG can also provide an estimate of the starch content of the tubers, which is the major component of the tuber DM and usually constitutes about 65–75% of total solids.

The DM content of early maturing cultivars is usually lower than that of mature full season crops. The total DM content of a tuber is made up of different components.

21.3.1 Carbohydrates

21.3.1.1 Starch

Starch is the main component of the tuber DM content. The starch is a polymer of glucose molecules and occurs in two main forms, amylose, which is a linear chain of

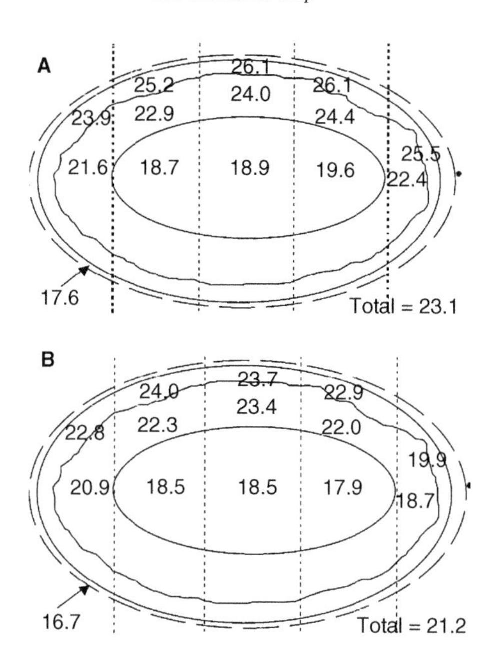

Fig. 21.1. Dry matter percentage distribution within 40–50 mm tubers of (A) Saturna and (B) Russet Burbank. Stolon attachment point (•) is on right hand side. Total refers to total tuber dry matter (Gaze et al., 1998).

glucose monomers linked by 1,4-glycosidic bonds, and amylopectin, in which the chains are branched by the addition of 1,6-glycosidic bonds. Potato starch is about 20% amylose, the balance being amylopectin. There are several forms of starch synthase, and these are grouped into two classes of soluble and granule-bound starch synthase. Shewmaker and Stalker (1992) reported that the granule-bound starch synthase is responsible for the synthesis of amylose. The pathways for sucrose degradation and starch synthesis are well known, the enzymes mediating it are well characterized and most genes encoding them have been cloned (Geigenberger, 2003). Teissen et al. (2002) provided evidence for a novel mechanism involving post-translational regulation acting through redox-modification of ADP-glucose pyrophosphorylase that regulates the rate of starch synthesis in response to sugars.

During crop development, starch accumulates in amyloplasts within tuber cells forming single or complex granules with size ranging from approximately 15 to 75 μm. The starch granules contain alternating zones of semi-crystalline and amorphous material known as growth rings. Both diurnal and circadian rhythms have previously been suggested as contributing to the development of the rings (Buttrose, 1962) by changing the relative activities of the starch-synthesizing enzymes, but Pilling and Smith (2003) recently suggested that there was a more complex interplay with physical mechanisms and speculated that this was due to the structured packing arrangements of the branched amylopectin molecules. It has been suggested that differences in starch rheological properties may partly explain differences in organoleptic qualities (Dobson et al., 2004), and the starch structure will have a bearing on these properties. The starch distribution in the tuber is uneven, generally following that of the total DM content with the lowest levels in the medullary tissue of the pith. The influence of starch levels on the quality parameters may also be mediated through cell size as higher starch content is associated with larger cell sizes (Chen and Liao, 1993), and these follow trends in DM concentration with tuber size and the increase from apical to stolon end.

The structural composition of the starch and the grain number, size and distribution within the tuber cells have implications for several quality factors, including bruise susceptibility of tubers and crop acceptability in both the fresh and the processed sectors (Chapter 6, van Eck, this volume) where authors make associations with the mealiness of cooked potato, the mash quality and the texture of finished product, e.g. bite and mouth feel of chips (crisps). Modification of the amylose: amylopectin ratio, of chain lengths, or increasing the level of starch biosynthesis offers the potential to modify these characteristics for both consumer and industrial attributes.

With recent attention (BPC, 2004) on the nutritional quality of the diet, and in particular the contribution of carbohydrates, the focus is on the potato because of its relatively high starch content and its Glycaemic Index (GI) contribution.

21.3.1.2 Glycaemic Index

The GI (Jenkins et al., 1981) is a classification of the blood glucose-raising potential of carbohydrate foods. It is defined as the incremental area under the blood glucose curve of a 50-g carbohydrate portion of a test food expressed as a percentage of the response to 50 g of carbohydrate of a standard (reference) food taken by the same subject, on a different day (FAO/WHO, 1998).

Carbohydrate foods produce different glycaemic responses depending on many factors, such as particle size, cooking and food processing, other food components (e.g. fat, protein and dietary fibre) and starch structure (Bjorck et al., 1994). The principle is that the slower the rate of carbohydrate absorption, the lower the rise of blood glucose level and the lower the GI value (Augustin et al., 2002). Indeed, high-GI foods are characterized by fast-release carbohydrate and higher blood glucose levels. A GI value ≤ 55 is low, a GI value 56–69 inclusive is medium and a GI value ≥ 70 is considered high (where glucose = 100).

The GI of foods may have important implications for the prevention and treatment of the major causes of morbidity and mortality in Western countries, and recent data support the preventive potential of a low-GI diet against the development of type 2 diabetes and cardiovascular disease (Frost et al., 1999; Meyer et al., 2000).

Potatoes generally have one of the highest GI values of any food, although some varieties appear to be lower than others (Table 21.2). Published GI values for boiled potatoes are highly variable. A GI value range of 56–101 for boiled potatoes, where the type of potato has been specified, and a GI value range of 23–76 for unspecified potato varieties have been reported (Foster-Powell et al., 2002). GI values related to preparation also need to be interpreted with care as variety is usually not identified, with reports of GI ranging across new potatoes 57, white skinned mash 70, French fries 75, baked 85, instant mash 86 and red-skinned boiled 88 (K. Brown, 2006, personal communication).

As varieties have different starch levels, it is necessary to clarify the GI of different varieties of potatoes as they make a considerable contribution to total starch intake in the

Table 21.2 Glycaemic Index for commercially available potato varieties in UK.

Potato variety	GI value	Classification
Maris Peer	94 ± 16	High
Maris Piper	85 ± 4	High
Desiree	77 ± 17	High
Estima	66 ± 5	Medium
Charlotte	66 ± 15	Medium
Marfona	56 ± 13	Medium
King Edward	75 ± 10	High

diet. The GI values of seven commercially available potato varieties in the UK ranged from 56 to 94 (medium to high), and all were similar in their satiating capacities (Henry et al., 2005). Other workers found no difference in GI values between three varieties (Sebago, Pontiac and Desiree) of potato (Soh and Brand-Miller, 1999). However, it has been demonstrated that young or 'new' potatoes have a lower GI than more mature potatoes, which may be attributed to differences in starch structure. If a low-GI potato variety can be identified, it could be used to lower the overall glycaemic load of the western diet and thus decrease the risk of type 2 diabetes, coronary heart disease and obesity.

The high-GI rating for potato has been seen as a disadvantage in comparison with other carbohydrate foods, but as potatoes are rarely eaten alone, the GI rating of a meal is more appropriate, and the addition of toppings to baked potatoes, e.g. cheese, can reduce GI classification from high (93 ± 8) to low (39 ± 5) (Henry et al., 2006).

21.3.1.3 Sugars

The sugar content of potatoes is variable depending on variety and physiological status of the tuber, and the relative contributions and levels of the different sugars change during crop development and subsequent storage. The main sugars present are the monosaccharides, glucose (0.15–1.5%) and fructose (0.15–1.5%), which are reducing sugars, and sucrose (0.4–6.6%), a non-reducing disaccharide. The dynamics of the change in sugar levels during tuber development vary between varieties, but when tubers reach maturity, the ratio of sucrose to reducing sugars reaches a minimum value. The management of sugars for harvest and during storage is a particular concern for the processing sector, where high levels of reducing sugars can make crops unacceptable for processing. This is due to a typical Maillard reaction between the reducing sugars and the α-amino groups of nitrogenous compounds (Schallenberger et al., 1959). The level of reducing sugars that are generally acceptable for processing for chips is 0.2–0.3% and for French fries is 0.3–0.5%, although fry colour charts, e.g. USDA, IBVL or British Potato Council (BPC), are used to determine colour acceptability. The sugar content of the harvested crop is also important for the fresh market, and sucrose levels above 1% fresh weight (FW) are reported to give an unacceptably sweet taste to the boiled potatoes. This sweetening is more likely to occur after a period of storage, particularly at holding temperatures (below 4 °C) that promote low-temperature sweetening. Conversely, it has been reported for cultivars Carlingford and Maris Peer that the increased sucrose levels after low temperature (3.5 °C) and modified atmosphere 5% O_2/5% CO_2 storage (up to 1.8% FW for cultivar Carlingford) are not reflected in tasting assessments (Mawson, 1998). However, with all taste perceptions, a range of genetic, physiological and metabolic variables can influence consumer acceptability (Drewnowski, 1997), and it is important to relate any preferences to the characteristics of the test group. The low-temperature accumulation of sugars may be reduced by reconditioning the crop for a period of time (2–6 weeks) at an elevated holding temperature (Storey and Davies, 1992). Senescent sweetening is an irreversible accumulation of reducing sugars resulting from prolonged storage and cannot be ameliorated by reconditioning. Variety selection and management of the sucrose levels in the crop at harvest, with monitoring of conditions in store, minimize the risk of adverse effects (Chapter 23, Sowokinos, this volume).

21.3.1.4 Crude fibre and dietary fibre

The cell wall and intercellular components of the tuber, which are mainly cellulose, hemicellulose, pentosans and pectic substances, are often termed crude fibre and contribute about 2.3% of the tuber weight.

Dietary fibre consists of the insoluble and soluble polysaccharides associated with the cell wall, but also lignin and resistant starch. The dietary fibre is not digested by enzymes of the small intestine, and so is 'non-available'; however, some components may be fermented by micro-organisms in the large intestine. The distribution of dietary fibre is not uniform, and cooking the potatoes and the resulting pectin breakdown may change some of the fibre properties and the amount of resistant starch. During cooking, the apparent level of dietary fibre increases. The skin of the tuber contains more dietary fibre than the flesh, for example 22.6% dry basis for micro-waved skin compared with 7.5% dry basis for micro-waved flesh (Mullin and Smith, 1991).

21.3.2 Protein

The protein content of potatoes is in the range 1.7–2.1 g per 100 g of FW for boiled new (in skin) and maincrop potatoes. Because there is a low total protein content on a FW basis, potatoes are not usually considered as a protein source compared with other foods; however, in countries with a high per capita consumption of total potato, they can make a significant nutritional contribution to the diet because of its high nutritional value. This high quality of the dietary nitrogen means that potatoes can supply a high proportion of international RDA for children and adults and 100 g of boiled potato supplies 8–13% of FAO/WHO RDA of nitrogen for children and 6–7% for adults, respectively. In the UK, potatoes have been estimated to contribute 3.4% of protein intake compared with 4.6, 4.8 and 5.8% from eggs, fish and cheese.

The major potato tuber protein, patatin (Racusen and Foote, 1980) is a family of 40–42 kDa glycoproteins that constitute 40% of the soluble protein in tubers but is generally undetectable in other tissues. It has been characterized as an esterase enzyme complex (Park, 1983). There are also three abundant proteins with approximate molecular

Table 21.3 Amino acid composition of potato tuber protein.

Amino acid	Range (%)	Amino acid	Range (%)
Alanine	4.62–5.32	Lysine	6.70–10.1
Arginine	4.74–5.70	Methionine	1.20–2.15
Aspartic acid	11.9–13.9	Phenylalanine	4.80–6.53
Cysteine	0.20–1.25	Proline	4.70–4.83
Glutamic acid	10.2–11.8	Serine	4.90–5.92
Glycine	4.30–6.05	Threonine	4.60–6.50
Histidine	2.10–2.50	Tryptophan	0.30–1.85
Isoleucine	3.73–5.80	Tyrosine	4.50–5.68
Leucine	9.70–10.3	Valine	4.88–7.40

Source: Lisińska and Leszczyński, 1989.

masses of 22, 23 and 24 kDa (Suh et al., 1990). Potato tuber protein contains high levels of the essential amino acids – lysine, leucine, threonine, phenylalanine and valine (Table 21.3). However, there are lower concentrations of sulphur amino acids – methionine and cysteine (Woolfe, 1987). This means that foods low in lysine, e.g. pasta and rice, can be supplemented in mixed diets with potatoes to provide an improved amino acid balance in the diet.

21.3.2.1 Amino acids

The quality of the protein depends on the amount and quantity of different amino acids and can be measured by its biological value. This is a measure of how much body nitrogen can be formed if renewed by 100 g of absorbed food nitrogen. If the protein provides all that is required for protein synthesis at a rate equal to protein turnover, then the biological value is 100. The standard is whole chicken egg protein. A comparison of potato protein with other sources is summarized in Table 21.4, and this also shows that because of the

Table 21.4 Biological value of different sources of protein.

Protein source	Biological value
Whole egg and potato (35%:65%)	137
Whole egg and milk (71%:29%)	122
Whole chicken egg	100
Potato	90–100
Cow milk	84–88
Beef	83–92
Edam cheese	85
Rice	83
Fish	83
Maize	72–26
Beans	73

Source: Kasper, 2004.

high quality of potato protein, combination with other protein sources can improve the overall biological value of the food.

Manipulation of amino acid levels for the harvested crop can potentially enhance the nutritional quality of tubers. Increased levels of free methionine (2.66 times that of Russet Burbank control) have been produced in tubers of selected protoclones grown from regenerated callus, which was derived from leaf protoplasts cultured in the amino acid analogue ethionine (Langille et al., 1998). Labelling approaches, such as those used by Roessner-Tunali et al. (2004) using incubation in ^{13}C isotopes with GC-MS, have provided significant information on the pathways and kinetics of sugars, organic acids and amino acids that will help understanding the effects of manipulating particular systems either through conventional breeding or transgenic approaches. Comparisons using transgenic lines exhibiting enhanced sucrose mobilization or where sucrose supply to tubers was restricted by environmental perturbation has provided evidence of partitioning towards several amino acids, with de novo biosynthesis being regulated by sucrose levels (Roessner-Tunali et al., 2003). Analysis of proteome diversity (Lehesranta et al., 2005), investigated with the application of 'omic' technologies, has demonstrated clear and qualitative differences in protein patterns of a range of varieties and landraces with over 95% of 1111 protein spots analysed showing significant differences. The potential of such methods to establish intended or detect unintentional effects of modifications is apparent, as in the same analysis much less variation was found between genetically modified (GM) lines and their non-GM controls, where only nine proteins of 730 showed significant differences.

The organic nitrogen fraction in tubers contains the free amino acids and amides, glutamine and asparagine. The free amino acid pool accounts for about 40–60% of the total nitrogen, and the factors affecting protein content, such as variety and crop maturity and fertiliser usage, also affect the levels of free amino acids and amides. Attention has been given to these effects on asparagine levels and more particularly on reducing sugar levels in the harvested crop and during subsequent storage because of their key role in the Maillard reaction as precursors for acrylamide formation (Mottram et al., 2002; Stadler et al., 2002).

21.3.2.2 Acrylamide

Acrylamide was reported as occurring in baked and fried foods by the Swedish National Food Administration (Tareke et al., 2002), and as a probable human carcinogen (IARC, 1994), there has been concern about the implication of acrylamide in the diet. However, population-based case–control studies have found no positive association between acrylamide food exposure and cancer risk, nor have any increases in cancer been identified in cohort studies of humans occupationally exposed to acrylamide, with only one possible exception of an increase in pancreatic cancer (Marsh et al., 1999). Acrylamide forms in carbohydrate-rich foods cooked (baked and fried) at temperatures of above 120 °C, but it has not been detected in boiled foods (Table 21.5).

Although the mechanism is not yet fully understood, the Maillard reaction, in particular a reaction between the amino acid asparagine and reducing sugars, accounts for most acrylamide production (Mottram et al., 2002; Stadler et al., 2002). However, in potato, more acrylamide is formed for a given concentration of sugars and asparagine than with

Table 21.5 Summary of reported levels of acrylamide in potato food groups.

Product type	Acrylamide levels (μg/kg)	
	Minimum	Maximum
Potatoes – raw	<10	<50
Potato chips (crisps – UK)	117	3770
French fries	59	5200
Potato fritters/rosti (fried)	42	2779

Source: Codex and FAO/WHO (2006), discussion paper on acrylamide.

wheat flour or cornstarch, which led Mottram et al. (2002) to suggest that other factors, such as availability of ammonia, may be involved. In the harvested potato crop, as the concentration of reducing sugars is the most important factor influencing acrylamide formation, limiting the levels is an option for reducing cooking-mediated acrylamide formation. Selection of cultivars with low sugar levels and subsequent storage management to limit sugar accumulation provides a major opportunity to reduce acrylamide formation (BPC, 2001). Other mitigation measures can include reducing concentrations of the reactants during early stages of food processing, minimizing heating conditions that result in excessive heat/low moisture conditions and modifying the thermal profile of the cooking process, all of which are being actively adopted by the potato industry (CIAA, 2005).

21.3.3 Vitamins

Vitamin C is the main vitamin in potatoes. Global dietary contribution of vitamin C from potatoes is important with an estimated 40% of daily-recommended intake (OECD, 2002). It is present in both the reduced state as ascorbic acid and in the oxidized state as dehydroascorbic acid (12–15%). There is a wide range of vitamin C in tubers, with the usual range for freshly harvested tubers reported as 10–25 mg/100 g FW. Brown (2005) reported an average level of 20 mg/100 g FW, which may account for up to 13% of the total antioxidant capacity of the tuber. The highest levels of vitamin C are found in the central pith region (25–29 mg/100 g FW) rather than in the outer cortex (22–26 mg/100 g FW) (Munshi and Mondy, 1989).

There are significant differences between cultivars and germplasm, and Tarn (2005), reviewing the heritability estimates that range from 0.45 for narrow-sense heritability (Sinden et al., 1978) to 0.71 for broad-sense heritability (Love et al., 2004), suggested that as the interaction between genotype and environment is low, there is potential to exploit this and increase ascorbic acid levels and improve the nutritional status of tubers.

Storage is known to reduce vitamin C content, but not all varieties show the same rate of decline during holding (Dale et al., 2003). Furthermore, it has been demonstrated by Davies et al. (2002) that a *Solanum phureja* × *Solanum tuberosum* hybrid retained a two-fold higher level of ascorbic acid after storage, and this identifies the opportunity that could be exploited using novel germplasm in breeding programmes.

Table 21.6 Vitamin content of potato (for maincrop potatoes, per 100 g edible portion) (FSA, 2002).

Vitamin E (mg)	Thiamin B_1 (mg)	Riboflavin B_2 (mg)	Niacin B_2 (mg)	Trypt/60 (mg)	Vitamin B_6 (mg)	Folate (μg)	Panto-thenate (mg)	Biotin (μg)	Vitamin C (mg)
0.06	0.21	0.02	0.6	0.5	0.44	35	0.37	0.3	11[a]

[a] Freshly dug potatoes contain average 21 mg/100 g FW. This falls to 9 mg/100 g FW, after 3 months storage and to 7 mg after 9 months storage.

The GM modification of the starch component (a higher amylopectin: amylose ratio) of cultivar Prevalent has also resulted in increased levels of vitamin C [67 mg/100 g in the altered line (EH92-527-1) compared with 49 mg/100 g in the parental line (EFSA, 2006)]. Although this difference is significant, a change is not unexpected given the alteration in carbohydrate metabolism and it is within the background ranges reported in the literature.

The other vitamins that occur in the potato, mostly in the vitamin B group, are at much lower concentrations than vitamin C (Table 21.6), but they can make a significant contribution to daily nutritional requirements.

21.3.4 Allergens and anti-nutritionals

21.3.4.1 Allergens

Potatoes are not usually considered a source of allergens. Cooked potato is usually well tolerated, but adverse reactions have been documented for both raw and cooked potatoes (Quirce et al., 1989; Majamaa et al., 2001). This is because both heat-labile and heat-stable allergens can be present, and although thermal processes largely inactivate the protease inhibitors, anti-nutritional reactions can occur if raw or inadequately cooked potatoes are consumed or fed. Allergic reactions, including atopic dermatitis, induced by patatin have also been reported in sensitive children (Seppälä et al., 1999).

21.3.4.2 Protease inhibitors

The protease inhibitors found in potatoes inhibit the activity of trypsin, chymotrypsin and other proteases, decreasing the digestibility and biological value of ingested protein. They represent approximately 50% of the total soluble proteins in potato juice of cultivar Elkana (Pouvreau et al., 2003) and have been classified into seven different families: potato inhibitor (PI)-1, PI-2, potato cysteine protease inhibitor (PCPI), potato aspartate protease inhibitor (PAPI), potato Kunitz-type protease inhibitor (PKPI), potato carboxypeptidase inhibitor (PCI) and 'other serine protease inhibitors'. The most abundant families were the PI-2 and PCPI families, representing 22 and 12% of all proteins in potato juice, respectively. Exposure to light, sufficient to induce greening over a 20-day period, did not alter the levels of inhibitors in cultivar White Rose (Dao and Friedman, 1994).

21.3.4.3 Lectins

Potato lectin, sourced from cultivar King Edward, is a glycoprotein that contains about 47% (by weight) L-arabinose, 3% D-galactose and 11% hydroxyproline (Allen et al., 1978). The lectins have a common property of binding to specific carbohydrate structures on cell surfaces, e.g. on intestinal or blood cells, and some, e.g. from beans, are known to cause serious health effects when ingested by humans or animals. However, heating inactivates them and only consumption of raw or inadequately cooked potato may have an adverse effect. Based on the sequencing of a complete cDNA encoding a potato lectin, the molecular structure has been determined and has provided insight into the physiological role of Solanaceae lectins (Van Damme et al., 2004).

21.3.5 Glycoalkaloids

Glycoalkaloids occur in nearly all potato tissues. Structurally, one or more sugar molecules (usually three) are linked to the steroidal alkaloid solanidine. The principal glycoalkaloids are α-chaconine (solanidine-glucose-rhamnose-rhamnose) and α-solanine (solanidine-galactose-glucose-rhamnose) and generally contribute about 90–95% total glycoalkaloids (TGAs) (Maga, 1980; Jadhev et al., 1981). Other glycoalkaloids that occur in smaller quantities include β-chaconine, γ-chaconine, β1-solanine, β2-solanine and γ-solanine. TGA levels in most commercial crops range from 2 to 10 mg/100 g FW. Because of the association of TGA with a bitter astringent taste and their relative toxicity, a generally accepted safe level is 20 mg/100 g FW (Sinden and Webb, 1972), although some individuals can detect a bitter taste in tubers with levels as low as 10 mg/100 g FW. The glycoalkaloids are not evenly distributed throughout the tuber. Higher concentrations are at the periphery (Smith et al., 1996) and in the vicinity of the eyes. The skin, comprising about 2–3% of the tuber, contains 30–80% of TGAs. Preparation by peeling can considerably reduce levels (Haase, 2002). However, the TGA are not broken down during cooking and frying.

Early potato varieties have reportedly higher levels of TGA, which may reflect growing conditions, the relative immaturity and smaller size tubers in these crops (Verbist and Monnet, 1979). Variation occurs in TGA levels between varieties, and this can be affected by conditions during production (Sinden and Webb, 1972) and subsequently in store (Love et al., 1994; Griffiths et al., 1997). However, in a comparison for two varieties, Saba and Asterix, grown either conventionally or organically, no significant effects of the type of production system were found in the TGA level of the crops (Strömberg et al., 2004.), although Hajšlová et al. (2005) reported that elevated TGA levels occurred in some potato varieties throughout a 4-year series of experiments grown under an organic farming system. The α-solanine level was about 40% of TGA content in the freshly harvested organic crops, and this is similar to 40:60 ratio for α-solanine: α-chaconine reported for a range of controlled growing conditions for cultivars Norland, Russet Burbank and Denali (Nitithamyong et al., 1999).

High levels of TGA are found in the tuber sprouts, and there are enhanced levels in greened tubers where they have been exposed to light. After harvest, the TGA levels may be increased, and this is dependent on duration of exposure to sunlight

(Percival et al., 1996) or artificial light (Dale et al., 1993), with the magnitude of increase being cultivar dependent (Griffiths et al., 1994). However, the TGA formation is independent of greening (Gull and Isenberg, 1960), so greened tubers do not always have high levels of TGA and vice versa. Patchett et al. (1977) report that immature tubers form larger amounts of glycoalkaloids when exposed to light compared with mature tubers. At harvest, bruising of tubers also increases TGA levels, and Dale et al. (1998) demonstrated that the rates of TGA synthesis for the varieties after bruising were in good agreement with their response to other stresses, light and low temperature. However, for one variety, cultivar Torridon, which exhibited severe internal damage in response to the impacts, the extensive cell death curtailed TGA and chlorogenic acid synthesis.

21.3.6 Other tuber metabolites

A review of ethno-botanical databases suggests that about 700 entries are chemicals documented as being present in potatoes. The levels of individual metabolites are variable and are likely to be linked to cultivar and production management. Although there is no evidence of high-value chemicals being extracted from potato, it has the potential to be a good source of high-value products for both food and non-food uses. Metabolite profiling is identifying new compounds, and non-targeted metabolomics have recently identified the occurrence of kukoamine and its allies in tubers (Parr et al., 2005). Kukoamines, found previously only in *Lycium chinense*, have been associated with reduced blood pressure, and they selectively affect a chemotherapeutic target for trypanosome diseases such as sleeping sickness (Ponasik et al., 1995).

The compositional characteristics of the tuber affect the nutritional quality, but to fully meet end-user requirements, morphological characteristics are also important for consumer and processor acceptability.

21.3.7 Minerals

Minerals contribute about 1.1% of total tuber weight. Potassium is the major cation, and the potato is a major source for the diet (Table 21.7).

Higher concentrations of potassium are found in the skin and the cell layers immediately below the skin and levels increase during the growing season. Potassium levels in tubers are known to have a consistent effect on levels of bruising, with crops grown with low potassium more susceptible to bruising (Van der Zaag and Meijers, 1970).

Phosphorus has the second highest level in tubers and is present in various forms with large amounts in inorganic compounds (12–39%) and contained within starch (20–48%). Although there are lower levels of iron and zinc in the tuber, the potato provides an important contribution of these elements to the diet. The iron content is also involved in the development of after-cooking blackening, as it forms a complex during cooking with chlorogenic acid, which then oxidizes on cooling to give the tuber tissue a blue-grey discolouration (Hughes and Swain, 1962a,b). Selenium is present at very low levels in tubers but does contribute to the antioxidant potential of potato (Lachman and Hamouz, 2005). Supplementing fertilisers with sodium selenate is used in Finland to increase selenium levels in vegetables (Eurola et al., 1989), and the use of foliar selenium sprays

Table 21.7 Mineral content of the potato (for maincrop potatoes, per 100 g edible portion) (FSA, 2002).

	New potatoes[a]	Old potatoes[b]	Units
Calcium	13.0	5.0	mg
Phosphorus	54.0	31.0	mg
Magnesium	18.0	14.0	mg
Sodium	10.0	7.0	mg
Potassium	430.0	280.0	mg
Chloride	43.0	45.0	mg
Iron	1.6	0.4	mg
Zinc	0.3	0.3	mg
Copper	0.06	0.07	mg
Selenium	1.0	1.0	μg

[a] New potatoes, in skins, boiled in unsalted water.
[b] Old potatoes, boiled in unsalted water.

(up to 150 g Se/ha) with humic acid was reported (Poggi et al., 2000) to increase selenium content of tubers approximately five-fold for cultivar Primura.

21.4 FLESH AND SKIN COLOUR

In many markets, consumers are becoming less discriminating about the flesh colour of the raw potato with greater emphasis being placed on the appropriateness for its intended end use (BPC, 2004), although there are invariably local or regional preferences. In most of Europe, the preference for fresh market potatoes is for varieties with yellow/light-yellow-coloured flesh. There are also very definite preferences for the processing industry, depending on the type of product being produced. For the chip (crisp) sector, varieties with light-yellow and yellow flesh predominate, e.g. in the UK this includes cultivars Hermes, Saturna and Lady Rosetta, whereas for French fries, cultivars Maris Piper, Russet Burbank and Shepody with white or cream flesh colour are preferred.

Skin colour of potatoes is also a varietal characteristic, and preferences are evident in many countries. The skin finish is also important, and russet varieties with a netted reddish brown skin are popular for most uses in the USA and whites and small reds are used for boiling. In most of Europe, a cream to light-yellow skin colour is preferred, whereas in many Arabic countries, varieties with a red skin colour are popular.

21.4.1 Carotenoids

Carotenoids are present in the flesh of all potatoes, and carotene contents in the literature (reported by Brown, 2005) range from 50–100 μg per 100 g FW for white-fleshed varieties to 2000 μg per 100 g FW in deep yellow-orange-fleshed varieties. Lu et al. (2001), examining 11 diploid clones of a hybrid breeding population and two tetraploid cultivars, Yukon

Gold (yellow flesh) and Superior (white flesh), showed that both total and individual carotenoid contents are positively correlated with tuber yellow intensity. Yellow-fleshed varieties obtain their colour from xanthophyll carotenoids. Many have been identified, including lutein, violaxanthin, neoxanthin, antheraxanthin and β-cryptoxanthin, with deep yellow pigmentation due to the presence of zeaxanthin (Nesterenko and Sink, 2003; Morris et al., 2004; Brown, 2005). There are usually just trace levels of α- or β-carotene meaning that potato is not a source of pro-vitamin A carotenes.

Antioxidant activity, measured by the oxygen radical absorbance capacity (ORAC), in a range of material with differing degrees of yellowness, with total carotenoid content ranging from 35 to 795 μg per 100 g FW, has ORAC values from 4.6 to 15.3 nmoles α-tocopherol equivalents per 100 g FW with the total carotenoid content correlated with lipophilic ORAC values ($r = 0.77$) (Brown et al., 2005). The antioxidant values of potato can be enhanced by exploiting conventional breeding (Bradshaw and Ramsay, 2005) or using transgenic approaches. Ducreux et al. (2005) produced transformed *S. tuberosum* L. cultivar Desiree × *phureja* group cultivar Mayan Gold crosses that expressed an *Erwinia uredovora crtB* gene encoding phytoene synthase. The carotenoid content increased from approximately 5.6 μg carotenoid/g dry weight (DW) in the untransformed Desiree to 35 μg carotenoid/g DW in the transgenic Desiree lines; with Mayan Gold, the increase was typically from 20 μg carotenoid/g DW to 78 μg carotenoid/g DW in the most affected transgenic line. There were also significant changes in the balance of carotenoids, with β-carotene in the transformed Desiree reaching approximately 11 μg/g DW and lutein 19-fold higher than in the untransformed control. The identification of genes that are consistently up- or down-regulated opens opportunities for improved selection of material with higher carotenoid contents and antioxidant levels.

During tuber development and DM accumulation, carotenoids are reported to decrease (Haynes et al., 1994), but this does not occur in all varieties, as reported by Morris et al. (2004) for the *S. tuberosum* group *phureja* variety, Inca Dawn. During storage, levels are stable, but there is a change in relative components.

21.4.2 Anthocyanins

Like ascorbic acid and carotenoids, anthocyanins are antioxidants and if their level can be increased, this can add to the nutritional quality of the potato. Diets rich in antioxidant flavonoids and carotenoids have been associated with lower incidence of atherosclerotic heart disease, certain cancers, macular degeneration and severity of cataracts (Hertog et al., 1993; Knekt et al., 1996; Stintzing and Carle, 2004).

The anthocyanins are responsible for the red, blue and purple colours of the skin and flesh. In the Andes, highly coloured flesh and skin is common in potato germplasm but deep coloured flesh has been unusual in most modern cultivated potatoes. However, because of increasing interest in antioxidants (Espin et al., 2000), this is changing and information has been comprehensively reviewed by Lachman and Hamouz (2005). Antioxidant levels are about two to three times higher in red- and purple-coloured flesh varieties than in white-fleshed potato. Indeed, highly coloured flesh is exploited in selection programmes, for example in China, where it is traditionally considered to have medicinal benefits and also within developed market economies, where flesh colours are used to differentiate some fresh and processed potato products, particularly chips (crisps). Colour is retained during baking and microwaving but leaches out during boiling, as anthocyanins are water-soluble.

Anthocyanin contents have been found to range from 2 to 40 mg/100 g FW in 33 red-fleshed breeding lines and cultivars (Rodriguez-Saona et al., 1998), and Brown et al. (2003) reported 6.9 mg/100 g FW in red-fleshed and 5.5–17.1 mg/100 g FW in purple-fleshed tubers, with predominantly acylated glycosides of pelargonidin in the red-fleshed and in addition acylated glycosides of petunidin and peonidin but also malvidin and delphinidin in the purple-fleshed tubers. The colour of their pigment depends on the presence and number of hydroxyl and methyl groups on the molecule, and the number of free hydroxyl groups determines the antioxidant activity, so petunidin has greater antioxidant effects in comparison with malvidin, peonidin or pelargonidin, respectively (Lachman and Hamouz, 2005). The hydrophilic antioxidant activity of solidly pigmented red or purple potatoes is comparable with Brussels sprouts or spinach (Brown, 2005). In red and purple potatoes with solidly pigmented flesh with levels of total anthocyanin ranging from 9 to 38 mg per 100 g FW, ORAC ranged from 7.6 to 14.2 μmole/g FW of Trolox equivalents.

Changes occur in total anthocyanins during tuber development, with levels during growth reported to decrease in the cultivar Norland (Hung et al., 1997), whereas Lewis et al. (1999), for an intensely coloured variety, Urenika, reported an almost constant level of anthocyanin as biosynthesis matched increase in tuber weight, whereas with a less intensely coloured variety Desiree, anthocyanin concentrations increased gradually. In store, an increase in anthocyanin level occurred during holding of cultivars Desiree and Arran Victory at 4 °C but a decrease at 10 °C.

Phenolic acids, such as chlorogenic acid, which can constitute about 80% of total phenolic acids, caffeic acid, protocatechuic acid and *p*-coumaric acid amongst several others have been identified in purple- and red-fleshed potatoes (Lewis et al., 1998; Brown, 2005) and contribute to the antioxidant capacity of potatoes. There is a similarity in phenolic composition among potato varieties irrespective of periderm colour. High positive correlations have been found between antioxidant capacity ($R^2 = 0.879$) and total anthocyanins ($R^2 = 0.910$) with total phenolic content (Reyes et al., 2005), and this is important for researchers and breeders to increase antioxidant capacity and functional value of coloured varieties for food and nutraceutical industries.

Anthocyanins and total phenolics have higher concentrations in the peel than in the flesh, but the overall contribution of the peel is estimated at approximately 20% of the total, as the peel only accounts for about 10% of tuber weight. There is also a longitudinal distribution with higher anthocyanins at the stem end of the tuber during development, but on reaching maturity, the concentration becomes approximately the same (Lewis et al., 1999). By contrast, after cold storage, tubers had a higher concentration of anthocyanins at the bud end.

Changes in the colour of tubers also occur in store and may be related to deterioration in skin bloom, where an increase in periderm phenolics and decrease in anthocyanins per given surface area may lead to darkening of skin during storage (Anderson et al., 2002).

21.5 GREENING

Crops may contain greened tubers at harvest, leading to grading rejections of the harvested crop by processors, pre-packers and merchants. Amongst the main reasons for this are insufficient soil cover at planting and exposure of the tubers due to heavy rainfall, which lead to the synthesis of chlorophyll at the exposed tuber surface. Modification of potato-ridge profiles needs to be considered for higher yielding varieties, those with larger tuber cluster size or longer tubers. Increasing the internal ridge height and top width can minimize the risk of greening to these crops (Kouwenhoven et al., 2003). Whilst rejections from the harvested crop can be a problem, exposure to light during subsequent storage and retail distribution can also result in greening, and this may also be associated with increased levels of glycoalkaloids in the tubers (Section 21.3.5). Pigment synthesis is influenced by the wavelength of the light, with blue light being more effective than red in starting the greening process in cultivars King Edward and Bintje (Virgin and Sundqvist, 1992). The pigment formation is greatest in the cells just below the periderm, and maximum chlorophyll level accumulated over a 20-day period in cultivar White Rose was 0.5 mg/100 g FW (Dao and Friedman, 1994). However, differences occur between varieties in the rate of greening (Dale et al., 1993) with, for example, significantly lower rates of chlorophyll accumulation in cultivars Ailsa and Eden than in Brodick and Torridon, the latter varieties accumulating approximately three-fold more chlorophyll over a 7-day period when all varieties were stored in bright light (140 μmol/m^2/s) at 20 °C. Grunenfelder et al. (2006) also reported differences in chlorophyll development for several US varieties, White Rose, Yukon Gold, Dark Red Norland, Russet Norkotah and Reba over a 5- to 7-day interval and have quantified the colour differences objectively using optical measurements. The changes in chlorophyll content affected darkness (L value) and colour (hue angle) of the periderm producing a characteristic off-colour for each variety.

Although Dale et al. (1993) reported some apparent relationship between chlorophyll and glycoalkaloid synthesis in light and darkness, this may be attributed to damage and heat stress. However, there is no direct metabolic link between chlorophyll biosynthesis and TGAs, and Edwards et al. (1998) have demonstrated that the chlorophyll biosynthesis inhibitor 4-amino-5-fluoropentanoic acid (AFPA) reduced total chlorophyll synthesis by 50–70% in cultivars Pentland Dell and Record and that there was no effect of the inhibitor on light-enhanced TGA accumulation. Temperature also influences rate of chlorophyll accumulation, and during holding cultivar King Edward for 8 days at temperatures between 5 and 25 °C, chlorophyll accumulation was greatest at 20 °C (Edwards and Cobb, 1997), and conversely, lower rates of synthesis occurred in cultivars Kerrs Pink, Peik and Troll when holding at 6 °C compared with either 18 or 24 °C (Kaaber, 1993). Although optical measurements can differentiate the development of unique colours between varieties experimentally, the commercial detection and elimination of greening defects during grading and potato preparation is essential to maintain the quality of fresh or prepared potato products for consumers. Visual grading and enhanced optical systems provide that facility for fresh potatoes, and Noordam et al. (2005) demonstrate the benefits of multi-spectral imaging in detecting latent greening in prepared French fries.

21.6 MECHANICAL DAMAGE AND BRUISING

Mechanical damage, including bruising, is a major concern for the potato industry worldwide. For example, Brook (1996) estimated that a 1% reduction in impact damage was worth approximately $7.5 million annually for the US industry, and in the UK, a survey by the BPC in 2004 estimated that the cost of bruising to growers was £26 million in an average year, equivalent to £200 per hectare. Packers and processors incur additional costs when loads are rejected or downgraded, and several businesses reported the annual cost of bruising to them at £1–2 million.

Peters (1996) highlighted the extent of damage caused during harvesting and grading operations. Improvements in field preparation and potato crop agronomy can contribute to reductions in overall damage levels of the harvested crop. Mechanization methods have also improved significantly, with increased emphasis on better-designed and operated harvesters, where cumulative damage can be minimized with, for example, better share and web design and managed horizontal web agitation (Witney and McRae, 1992; Baheri and de Baerdemaeker, 1997; Kang and Wen, 2005). Although improved management at store loading and grading has also contributed to overall damage reduction, by minimizing drop heights and use of cushioning materials, there are still opportunities for improvement, as Molema et al. (2000), monitoring instrumented sphere (IS) impacts during different phases of representative Dutch ware potato-handling chains, found that of the 340 impacts above 30 g, 79% of the impacts occurred during packaging.

The distinction between external damage and internal damage is important because the internal defects cannot be readily seen. The visibly damaged tubers, which includes scuffing of the skin, cuts or gouges and splits or cracking, can be physically removed at harvest, but handling during store loading, unloading and grading can add to the internal damage incurred at harvest.

Various classifications of damage types have been published (Hughes, 1980; Hiller et al., 1985; Bouman, 1996; Baritelle et al., 2000), the latter presenting a modification (Fig. 21.2) of the comprehensive classification proposed by Hughes (1980) that included a description of the tuber characteristics that influenced the type of damage. Consistent and accurate descriptions of damage and bruise types are important for comparative assessments and for understanding the relationships between the type of impact, the energy absorbed and the interaction of the physical condition of the tuber with the biochemical propensity to discolouration of the tuber tissue.

In the Washington State University (WSU) classification system (Baritelle et al., 2000), six different bruise types were identified: (1) blackspot, typically showing no visible cell separation, although the cells are often damaged, and a typical blue-black discolouration; (2) internal crush damage, where there is obvious cell wall or cell separation in addition to blue-black discolouration; (3) white spot, which is similar to a crush, except that the damaged tissues are not discoloured; (4) internal shatter, which is normally in the perimedullary tissue, has distinct failure planes where tissue has sheared and where damage to cells results in discolouration; (5) external shatter, where failure planes extend through the cortex to the skin and the discolouration is brown rather than blue-black because air enters the shatter and dries tissue before complete oxidation occurs and (6) external cracking, which is mostly the result of cell separation, with little or no tissue discolouration.

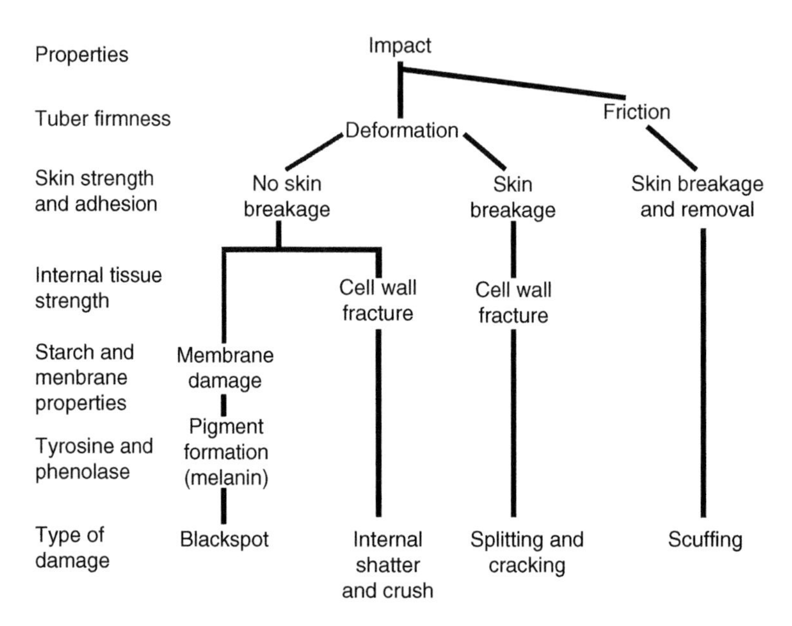

Fig. 21.2. Influence of tuber properties on types of impact damage (Baritelle et al., 2000 – adapted from Hughes, 1980).

The shatter bruise described by Hughes (1980) was further differentiated into star shatter and ring shatter (Noble, 1985). The WSU description from type (1) blackspot through to type (6) external cracking largely reflects an increasing degree of hydration of the tuber tissue. The understanding of skin set and factors affecting the scuffing damage of potatoes are dealt with in detail in Chapter 22 (Lulai, this volume).

21.6.1 Enzymic browning

The discolouration of potato flesh arising from mechanical impacts through enzymic browning is well documented (Walker, 1977; Matheis, 1987). The pigment melanin formed in bruised tissue results from initial oxidation of the substrate tyrosine by the enzyme polyphenol oxidase (PPO E.C. 1.10.3.1) to 3,4-dihydroxyphenylalanine (DOPA). Intracellular membrane disruption is needed to allow the substrate in the vacuole to come into contact with the enzyme that Cobb (1998) suggests is located in the amyloplast. DOPA then oxidizes to DOPA quinone. The polymerization to the blue-black melanin pigment occurs over about 24 hours, although some of the intermediary dopachromes, with reddish-orange pigments, are evident within a few hours of impact. PPO is a doublet of molecular weights of 60 and 69 kDa and has a reported michaelis constant (Km) of 4.3 mM using L-DOPA as a substrate (Partington and Bolwell, 1996). This implies the need for considerable substrate for the melanin formation to proceed and is an important consideration in relation to crop maturity and bruise development for the harvested crop. Laerke et al. (2002b) reported increasing abundance of the red- and yellow-coloured intermediaries during tuber growth and a high correlation coefficient (r) of the final black-coloured compounds with free tyrosine levels for cultivars Dali (0.69) and Oleva (0.67). Tyrosine is synthesized in the tuber through the shikimic acid pathway, and Sabba and Dean (1994, 1996) demonstrated that the principal determinant of raised tyrosine concentrations in the tuber results from increased proteinase activity and that this activity was more apparent following prolonged storage at 3 °C. The detrimental effect of low temperature is not explained by changes in PPO activity, catalase activity or ascorbic acid content (Brierley et al., 1998). Chlorogenic acid and caffeic acid are also potential substrates for PPO and in vivo they are rapidly oxidized, but their concentrations are too low to influence the bruising process (Cobb, 1998).

The chemical control of enzymic browning, which also occurs in cut or prepared potatoes, involves inhibition of PPO activity or prevention of tyrosine oxidation, most commonly by adding bisulphite or hypochlorite compounds (Brecht et al., 1993). Alternatives such as citric acid or ascorbic acid have been assessed for minimizing tuber discolouration (Molnar-Perl and Friedman, 1990). The recent classification of sulphite as an allergen has added impetus to this interest, particularly as a number of countries require the level of sulphur dioxide and sulphite in foods to be declared, for example in the UK, if the content exceeds 10 mg/kg of SO_2 in the finished product. Muneta (1981) has shown that sodium sulphite and diethyldithiocarbamic acid (DIECA) inhibit browning by restricting oxidation of tyrosine and that DIECA also causes enzyme inactivation. Ascorbate and dihydroxyfumarate inhibit discolouration by reducing DOPA quinone to DOPA. Reducing synthesis of phenols such as chlorogenic acid, which may be oxidized to produce lighter brown *o*-quinones, may also limit discolouration, and there is evidence that phenoxyacetic acids such as 3,5-dichlorophenoxyacetic acid inhibit browning (Burrell, 1984).

21.6.2 Structural and cellular changes

The ultra-structural sequence of events in tubers following a bruising impact have been observed in cultivar Pentland Dell using a falling bolt with known energy (0.7 J) (Edgell et al., 1998). Typically, there was collapse of intracellular compartmentation, increased abundance of ribosomes and mitochondria in the cytoplasm, an increase in granular density of the cytoplasm adjacent to cell walls and amyloplasts and, as a result of mixing of enzymes and substrates, development of melanin as a dark amorphous layer adjacent to cell walls and amyloplasts. This intercellular membrane disruption may occur as an immediate effect of impact, or the membranes may undergo physiological deterioration over time (Partington et al., 1999; Laerke et al., 2000). Changes occur in cell membrane permeability and lipid content following harvest. Marked changes in glycolipid composition take place during storage at 5 °C (Spychalla, 1994), and oxygen free radical build up in stored tubers contributes to a progressive deterioration of the cell membranes (Kumar and Knowles, 1993). The effects of any impact on the tissue mechanical properties and internal structural changes are influenced by the tuber-cell size and packing arrangements, cell wall strength and cell–cell cohesion. At the stolon end of the tuber, a dense zone of lignified xylem and tracheids are positioned close to the tuber surface, near to the point of stolon attachment, and discontinuities in starch packing in parenchymatous cells may concentrate pressure shocks in this zone (Croy et al., 1998). By contrast, the sides of the tuber have less sub-surface lignified tissue and possible greater dissipation of impact.

Externally, the strength of the periderm will also have an effect on the type of resulting damage depending on the size and site of the impact on the tuber. The speed and propagation of a shock wave created by a pendulum impact is influenced by the tuber turgor (Bajema et al., 1998c), and that affects the type and degree of damage at both tissue and cellular levels, but the effects on blackspot development differ depending on the stage of crop growth (Laerke et al., 2002a) and the biochemical potential of the tuber. The size, number and angularity of starch granules in the cells may also contribute to the disruption of membranes and cell walls (Gray and Hughes, 1978), but the magnitude of these effects on membrane and cell walls could be modified depending on tissue turgor, as Scanlon et al. (1998) suggested that for flaccid tubers that had been osmotically adjusted, starch granule movement in viscous protoplasm dissipated shear strain energy.

At the cellular level, an early response to mechanical stress on tuber tissue is a significant and rapid synthesis of superoxide radicals. This burst of radical production distinctively displays a reproducible biphasic pattern over time (Johnson et al., 2003) with differences in superoxide production between impacted bruise-susceptible and bruise-resistant cultivars (Fig. 21.3).

A consequence of the free radical generation is elevated levels of oxidatively modified tuber proteins. Both radical generation and protein modification vary between cultivars, but both are directly proportional to the amount of melanin pigments produced. This proportionate generation of superoxide radicals has provided the basis for the development of a diagnostic kit that has the potential to predict, prior to harvest, the susceptibility of tubers to bruising. A standard 0.7-J impact is imposed on tubers and, using a colorimeter, the test quantitatively measures the level of oxygen free radical production in the tissue that can be related to crop bruise susceptibility (Fig. 21.4).

21.6.3 Field factors and tuber water status

Practical application of such technology requires an appreciation of the status of tubers at the time of the assessment and the ability to identify and implement crop management options for remediation.

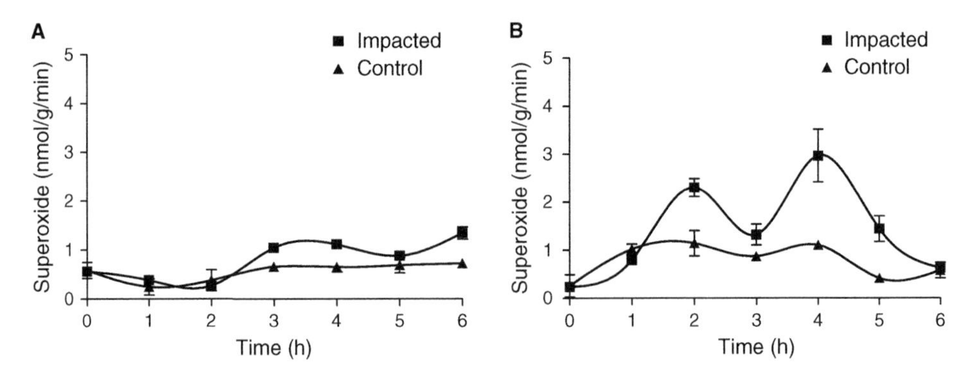

Fig. 21.3. Superoxide production after impact in (A) bruise-resistant and (B) bruise-susceptible variety (BPC Bruise forum 2006 – courtesy of R.R.M. Croy and S.M. Johnson).

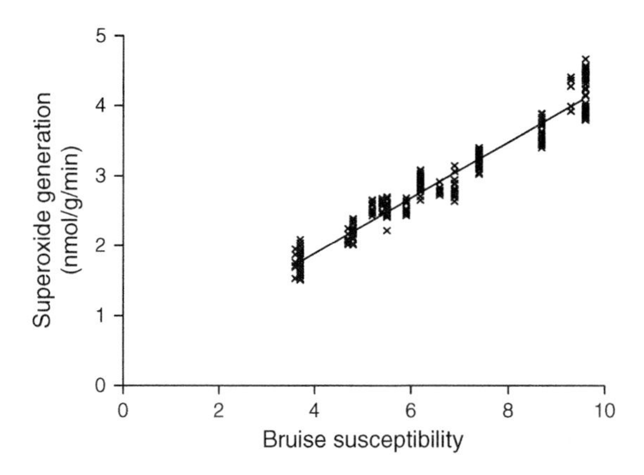

Fig. 21.4. Relationship between levels of superoxide generation and bruise susceptibility of potato tissue (BPC Bruise forum 2006 – courtesy of R.R.D. Croy and S.M. Johnson).

Commercially, the susceptibility of a crop to bruising is often not established until an assessment is made of damage and bruise levels at harvest, and that assessment reflects the soil conditions at the time and efficiency of the harvesting operation as well as the physical and biochemical condition of the tuber. Hot-box treatments, where a harvested tuber sample is held at an elevated temperature for 8–12 hours to enhance bruise development prior to assessment, are used frequently to inform harvesting operations. However, in attempts to characterize the bruise susceptibility of varieties and determine experimentally the effects of different field factors that may predispose crops to damage, a wide range of other tests have been adopted, including impacting tubers with a known energy using pendulums and falling bolts, using a shaking table or alternatively dropping tubers from a given height. The use of different bruise tests in the literature makes comparisons and interpretation between evaluations difficult (Aeppli et al., 1981; Pavek et al., 1985; Dean et al., 1993; Laerke et al., 2002a). Genotypic variation does exist in the extent of enzymic browning and has been widely exploited through the adoption of new varieties, but the incidence of bruising is affected by climatic and edaphic factors. Information on soil-related factors affecting bruise susceptibility has been widely reported, but most of the effects are inconsistent. However, there is fairly compelling evidence of a reduction in susceptibility with increasing application rate of potassium fertiliser, but the effect is not large and is primarily observed for the potassium-deficient range of concentrations (Chapman et al., 1992; McGarry et al., 1996). The effect of potassium on bruise susceptibility is not mediated through changes in tuber DM content, as many studies demonstrate that there is no relationship between tuber DM and bruise susceptibility (Rogers-Lewis, 1980; McGarry et al., 1996; Craighead and Martin, 2002). However, it has been suggested the effect may be mediated through the hydration status of the tuber (Van der Zaag and Meijers, 1970) or lower phenol levels and PPO activity (McMabnay et al., 1999).

The effect of soil temperatures at harvest on bruise susceptibility has been widely reported, with lower temperatures increasing the bruise susceptibility of the tuber and recommendations that soil temperature should be approximately above 10 °C and preferably 12 °C before harvesting begins (Allen and Scott, 2001).

The importance of soil condition at the time of lifting has also been recognized, but the emphasis has usually placed on having sufficient soil on the harvester webs to provide a degree of cushioning to the tubers being lifted. Surveys reviewing field crop agronomy have been related to information on the incidence of bruising assessed post-harvest, but the challenge has been identifying specific field factors pre-harvest that affect susceptibility. One approach has involved discriminative analysis of a total of 180 commercial crops of cultivars Cara, Marfona and Maris Piper over a 3-year period (2001–03). This work identified that dry soil conditions at crop defoliation were consistently associated with greater bruising susceptibility at harvest (Fellows, 2004). The soil moisture level in the period prior to harvest will affect the hydration status of the tuber and potentially affect damage susceptibility.

There has been an appreciation since the 1960s that tuber water status has a significant influence on tuber damage and the occurrence of a shatter or blackspot bruise. Smittle et al. (1974) hypothesized that highly turgid tubers were more susceptible to shatter damage than the flaccid tubers that were susceptible to bruising (Fig. 21.5). The relationship suggested that there was an optimum level of tuber hydration at which total damage was minimal.

Despite the knowledge that tuber hydration was implicated in bruise and damage susceptibility, the practical application for active crop management has been limited because of the complexity of the inter-relationships between the different parameters that affect tissue susceptibility and blackspot colour development.

Bruising is known to be affected by cultivar and the crop maturity, tuber mass and tuber shape, temperature, tuber turgor but not DM content per se, although this is frequently incorrectly associated with bruise susceptibility by many researchers, advisors and potato growers. Integrating this information so that hypotheses can be tested has been a crucial step, and studies at WSU have examined the different components: tuber impact thresholds (Mathew and Hyde, 1997), temperature and tissue strain rate (Bajema et al., 1998a), turgor and temperature (Bajema et al., 1998b), turgor and failure stress and strain

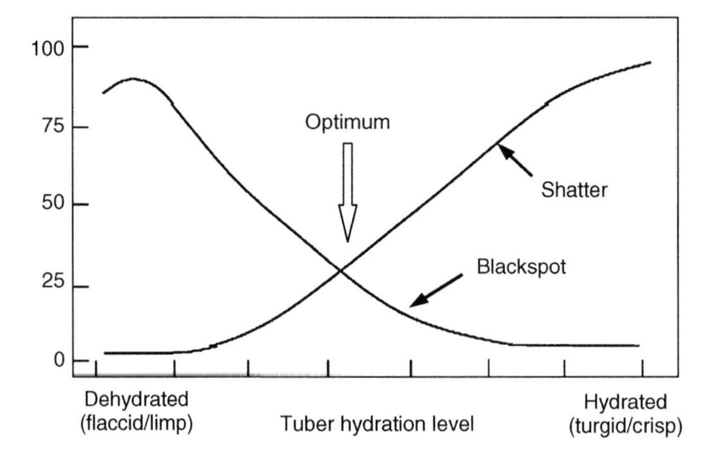

Fig. 21.5. Proposed effect of tuber water status on blackspot and shatter bruise at 7–10 °C (redrawn from Smittle et al., 1974). [Source: Codex and FAO/WHO (2006).]

(Bajema et al., 1998c), tuber size (Baritelle and Hyde, 1999), SG and cultivar (Baritelle and Hyde, 2003) within a biological and mechanical context (Baritelle and Hyde, 2001).

Experimentally, work on cultivars Russet Burbank and Atlantic (Bajema et al., 1998b) had demonstrated that progressive dehydration of tubers to reduce turgor affected the tissue stress–strain relationship and that there was an optimum level for tissue hydration that was linked to the physical characteristics determining bruise susceptibility.

At a threshold when bruising is about to occur, the induced stress (σ_i) on the tuber will be equal to the tissue failure stress (σ_f); when the induced stress is then increased, the threshold is exceeded, tissue failure happens and bruising occurs. The amount of stress that the tissue can withstand is related to the elasticity of the tissue, and the following elasticity-based equation has been developed for potato to explain the relationship between the bruise threshold (h) and tissue failure stress (σ_f) and strain (ε_f), the radius at the site of impact (R) and the force of the impact (m and g):

$$h \approx \frac{7600\sigma_f \sigma_f^4 R^3}{mg}$$

where h = potato tuber bruise threshold (the drop height at which bruising just begins to occur);
σ_f = tissue failure stress (strength);
ε_f = tissue failure strain (elasticity – ability to deform before failure);
R = radius of curvature at point of impact (mm);
m = tuber mass;
g = acceleration due to gravity.

The equation has provided an indication of the relative importance of the different components as they relate to the condition of a tissue and its bruise threshold. Baritelle and Hyde (2001) report that, in general, higher temperature and/or lower relative turgor (within limits) in potato tubers resulted in improved bruise threshold. Preliminary validation work using the model predicted apple and potato bruise thresholds within 23% of experimental values or better for assumed values of Poisson's ratio (Baritelle and Hyde, 2001). The importance of small changes in tissue elasticity is identified by the equation and the understanding of where, and to what extent, changes in tissue turgor impact on the relationship with the bruise threshold have been central to making progress on crop management.

At the time of harvest, the DM concentration in the tubers will have been influenced by length and conditions during the growing season. By managing soil moisture levels, a range of different tuber turgor levels has been generated experimentally for varieties of differing bruise susceptibility. Stalham and Allen (2006) manipulated the tuber hydration status during the development of field grown crops and have shown that internal impact damage could be predicted from measurements of tuber water potential with reasonable accuracy in cultivar Lady Rosetta but not in cultivars Maris Piper or Smith's Comet. Actively altering the tuber turgor level prior to harvest to improve the level of bruising will be dependent on the hydration status of the tuber at the given point as it will have been affected by soil water availability over the growth period to that time; the degree of change in turgor that can occur in the tuber will depend on whether the crop has senesced, whether it has been defoliated and whether it has an active rooting system that can take up water. Whether the active intervention in tuber hydration management to minimize blackspot will be effective will also depend on the maturity of the crop, as immature tubers are less susceptible to enzymic browning (Pavek et al., 1985; Corsini et al., 1999). However, there are also reports that some young, small tubers, e.g. of cultivar Maris Peer, can be very susceptible to blackspot bruising (E.J. Allen, personal communication). If the crop does not have the biochemical propensity for discolouration, there may be no observed effect on blackspot levels, but other forms of internal damage such as whitespot may still occur (Baritelle et al., 2000). Active intervention may be more readily achieved in other situations, e.g. with cultivar Smith's Comet; as the tuber became more turgid it was increasingly susceptible to cracking, and Stalham and Allen (2006) have suggested this could be reduced by maintaining higher Soil Moisture Deficits (SMDs) late in the season.

21.7 CONCLUDING COMMENTS

Storey and Davies (1992) considered that although little could be done that affected tuber tissue structure, a range of cultural practices adopted during growth and subsequent storage could have a significant effect on damage levels. The maintenance of optimum tuber turgor, avoidance of unfavourable harvesting temperatures and reducing impacts were the three key areas identified that would produce marked benefits. Practical progress has been achieved since then by greater attention to temperatures at harvesting and reducing numbers and size of damaging impacts. More could still be done in this respect in relation to improving soil condition for crop growth and harvest (Stalham et al., 2005), and in many production scenarios by improving seed selection and appropriate fertiliser use to manage the harvest window. The understanding of the relationship between bruising and tuber turgor is beginning to lead to a better understanding of the practical potential that could be available to growers to minimize bruising through crop water management. If this could be integrated, in the future, with diagnostic systems that reliably predict the biochemical potential of crop to bruise susceptibility, then there are real opportunities to improve the quality of potatoes delivered to end users. This improvement in defect management will need to be applied within both existing and new cropping situations to enable potato to meet the nutritional demands that will be required of it as a crucial food resource in developing economies and for consumers within more developed markets who are demanding a wider range of quality potatoes and potato products.

REFERENCES

Aeppli A., E.R. Keller and F. Schwendlmann, 1981, *Zeit. Acker. Pflanzen.* 150, 372.
Allen K.A., N.N. Desai and A. Neuberger, 1978, *Biochem. J.* 171(3), 665.
Allen E.J. and R.K. Scott, 2001, *BPC Research Review – Potato Agronomy.* British Potato Council, Oxford, 137 pp.
Al-Saikhan M.S., L.R. Howard and J.C. Miller Jr, 1995, *J. Food Sci.* 60, 341.
Anderson A.A., C.B.S. Tong and D.E. Kreuger, 2002, *Am. J. Potato Res.* 79, 249.

Augustin L.S., S. Franceschi, D.J.A. Jenkins, C.W.C. Kendall and C. La Vecchia, 2002, *Eur. J. Clin. Nutr.* 56, 1049.

Baheri M. and J. de Baerdemaeker, 1997, ASEA meeting 1997, paper 971101, 14 pp.

Bajema R.W., G.M. Hyde and A.L. Baritelle, 1998a, *Trans. ASAE* 41, 733.

Bajema R.W., G.M. Hyde and A.L. Baritelle, 1998b, *Trans. ASAE* 41, 741.

Bajema R.W., G.M. Hyde and A.L. Baritelle, 1998c, *Postharvest Biol. Technol.* 14, 199.

Baritelle A. and G. Hyde, 1999, *Trans. ASAE* 42, 159.

Baritelle A. and G. Hyde, 2001, *Postharvest Biol. Technol.* 21, 331.

Baritelle A. and G. Hyde, 2003, *Postharvest Biol. Technol.* 29, 279.

Baritelle A., G. Hyde, R. Thornton and R. Bajema, 2000, *Am. J. Potato Res.* 77, 143.

Bjorck I., Y. Granfeldt, H. Liljeberg, J. Tovar and N.G. Asp, 1994, *Am. J. Clin. Nutr.* 59, S699.

Bors W. and M. Saran, 1987, *Free Radic. Res. Commun.* 2, 289.

Bouman A., 1996, Abstracts of the 13th Triennial Conference of the European Association for Potato Research, p. 215.

BPC (British Potato Council), 2001, In: A.C. Cunnington (ed.), *Store Managers' Guide*, British Potato Council, Oxford, 67 pp.

BPC (British Potato Council), 2004, *Potatoes, A Healthy Market*, British Potato Council, Oxford, 52 pp.

Bradshaw J.E. and G. Ramsay, 2005, *Euphytica* 146, 9.

Brecht J.K., A.U.O. Sabaa-Srur, S.A. Sargent and R.J. Bender, 1993, *Acta Hortic.* 343, 341.

Brierley E.R., T. Edgell, J.J.J. Wiltshire and A.H. Cobb, 1998, *Aspects Appl. Biol.* 52, 309.

Brook R.C., 1996, *Potato Bruising: How and why, emphasising black spot bruise.* Running Water Publishing, Haslett, Michigan, 117 pp.

Brown C.R., 2005, *Am. J. Potato Res.* 82, 163.

Brown C.R., D. Culley, C.P. Yang, R. Durst and R. Wrolstad, 2005, *J. Am. Soc. Hortic. Sci.* 130(2), 174.

Brown C.R., R. Wrolstad, R. Durst, C. P. Yang and B. Clevidence, 2003, *Am. J. Potato Res.* 80, 241.

Burrell M.M., 1984, *Plant Pathol.* 33, 325.

Burton W.G., 1989. *The Potato*, 3rd edition. Longman Scientific and Technical, London.

Burton W.G., A. van Es and K.J. Hartmans, 1992, In: P.M. Harris (ed.), *The Potato Crop: Scientific basis for improvement*, p. 608. 2nd edition. Chapman & Hall, London.

Buttrose M.S., 1962, *J. Cell Biol.* 14, 159.

Chapman K.S.R., L.A. Sparrow, P.R. Hardman, D.N. Wright and J.R.A. Thorp, 1992, *Aust. J. Exp. Agric.* 32, 521.

Chen J.J. and Y.J. Liao, 1993, *J. Am. Soc. Hortic. Sci.* 118, 831.

CIAA (Confederation of the Food and Drink Industries in the EU), 2005, The CIAA toolbox (23 September 2005). http://europa.eu.intl

CIP (International Potato Centre), 2003, *Preserving the Core, Stimulating Progress: CIP's Vision Statement.* CIP, Lima. 64 pp.

Cobb A.H., 1998, *Aspects Appl. Biol.* 52, 199.

Codex and FAO/WHO (Codex Alimentarius and Food and Agriculture Organisation/World Health Organisation), 2006, Discussion Paper on Acrylamide. FAO, Rome.

Corsini D., J. Stark and M. Thornton, 1999, *Am. J. Potato Res.* 76, 221.

Craighead M.D. and R.J. Martin, 2002, *Agron. NZ* 32, 15.

Croy R.R.D., P. Baxter, W. Deakin, R. Edwards, J.A. Gatehouse, P. Gates, N. Harris, C. Hole, S.M. Johnson and R. Raemaekers, 1998, *Aspects Appl. Biol.* 52, 207.

Dale M.B.F., D.W. Griffiths and H. Bain, 1998, *J. Sci. Food Agric.* 77, 499.

Dale M.B.F., D.W. Griffiths, H. Bain and D. Todd, 1993, *Ann. Appl. Biol.* 123, 411.

Dale M.B.F., D.W. Griffiths and D.T. Todd, 2003, *J. Agric. Food Chem.* 51, 244.

Dao L. and M. Friedman, 1994, *J. Agric. Food Chem.* 42, 633.

Davies C.S., M.J. Ottman and S.J. Peloquin, 2002, *Am. J. Potato Res.* 79, 295.

Dean B.B., N. Jackowiak, M. Nagle, J. Pavek and D. Corsini, 1993, *Am. Potato J.* 70, 201.

Dobson G., D.W. Griffiths, H.V. Davies and J.W. McNicol, 2004, *J. Agric. Food Chem.* 52, 6306.

Drewnowski A., 1997, *Annu. Rev. Nutr.* 17, 237.

Ducreux L.J.M., W.L. Morris, P.E. Hedley, T. Shepherd, H.V. Davies, S. Millam and M.A. Taylor, 2005, *J. Exp. Bot.* 56, 81.

Edgell T., E.R. Brierley and A.H. Cobb, 1998, *Ann. Appl. Biol.* 132, 143.
Edwards E.J and A.H. Cobb, 1997, *J. Agric. Food Chem.* 45, 1032.
Edwards E.J., R.E. Saint and A.H. Cobb, 1998, *J. Sci. Food Agric.* 76, 327.
EFSA (European Food Safety Authority), 2006, *EFSA J.* 324, 1.
Espin J.C., C. Soler-Rivas, H.J. Wichers and C. Garcia-Viguera, 2000, *J. Agric. Food Chem.* 44, 1588.
Eurola M., P. Ekholm, M. Ylinen, P. Koivistoinen and P. Varo, 1989, *Acta Agric. Scand.* 39, 345.
FAOSTAT, 2006, Agricultural database. http://faostat.fao.org
FAO/WHO, 1998, Carbohydrates in Human Nutrition. Report of a Joint FAO/WHO Expert Consultation. Food and Agriculture Organisation of the United Nations, Rome.
Fellows J., 2004, BPC project report 807/227 (final). British Potato Council, Oxford.
Foster-Powell K., S.H.A. Holt and J.C. Brand-Miller, 2002, *Am. J. Clin. Nutr.* 76, 5.
Frost G., A.A. Leeds, C.J. Dore, S. Madeiros, S. Brading and A. Dornhorst, 1999, *Lancet* 353, 1045.
FSA (Food Standards Agency), 2002. *McCance and Widdowson's The Composition of Foods*, 6th summary edition, p. 237. Royal Society of Chemistry, Cambridge.
Gaze S.R., M.A. Stalham, R.M. Newbery and E.J. Allen, 1998, BPC project report 807/182 (1997). British Potato Council, Oxford.
Geigenberger P., 2003, *J. Exp. Bot.* 54(382), 457.
Gray D. and J.C. Hughes, 1978, In: P.M. Harris (ed.), *The Potato Crop*, p. 504. Chapman & Hall, London.
Griffiths D.W., H. Bain and M.F.B. Dale, 1997, *J. Sci. Food Agric.* 74, 301.
Griffiths D.W., M.F.B. Dale and H. Bain, 1994, *Plant Sci.* 98, 103.
Grunenfelder L., L.K. Hiller and N.R. Knowles, 2006, *Postharvest Biol. Technol.* 40, 73.
Gull and Isenberg, 1960, *Proc. Am. Soc. Hortic. Sci.* 75, 545–556.
Haase N.U., 2002, *Kartoffelbau* 7, 284.
Hajšlová J., V. Schulzová, P. Slanina, K. Janné, K.E. Hellenäs and C. Andersson, 2005, *Food Addit. Contam.* 22, 514.
Haynes K.G., W.E. Potts, J.L. Chittams and D.L. Fleck, 1994, *J. Am. Soc. Hortic. Sci.* 119, 1057.
Henry C.J.K., H.J. Lightowler, F.L. Kendall and M. Storey, 2006, *Eur. J. Clin. Nutr.* 60, 763.
Henry C.J.K., H.J. Lightowler, C.M. Struik and M. Storey, 2005, *Br. J. Nutr.* 94, 917.
Hertog M.G.L., E. Feskens, P. Hollman, M. Katan and D. Kromhout, 1993, *Lancet* 324, 1007.
Hiller L.K., D.C. Koller and R.E. Thornton, 1985, In: P.H. Li (ed.), *Potato Physiology*, p. 389. Academic Press, Orlando, Florida.
Hughes J.C., 1980, *Span* 23, 65.
Hughes J.C. and T. Swain, 1962a, *J. Sci. Food Agric.* 13, 229.
Hughes J.C. and T. Swain, 1962b, *J. Sci. Food Agric.* 13, 358.
Hung C., J.R. Murray, S.M. Ohmann and C.B.S. Tong, 1997, *J. Am. Soc. Hort. Sci.* 122, 20.
IARC (International Agency for Research on Cancer), 1994, IARC Monographs on the Evaluation of the Carcinogenic Risk of Chemicals to Humans: Acrylamide. 60, 389. IARC, Lyon, France.
Jadhev S.J., R.P. Sharma and D.K. Salunkhe, 1981, *CRC Crit. Rev. Toxicol.* 9, 21.
Jenkins D.J.A., T.M.S. Wolever, R.H. Taylor, H. Barker, H. Fielden, J.M. Baldwin, A.C. Bowling, H.C. Newman, A.L. Jenkins and D.V. Goff, 1981, *Am. J. Clin. Nutr.* 34, 362.
Johnson S.M., S.J. Doherty and R.R.D. Croy, 2003, *Plant Physiol.* 131, 1440.
Kaaber L., 1993, *Nor. J. Agric. Sci.* 7, 221.
Kang W.S. and X.Z. Wen, 2005, *Appl. Eng. Agric.* 25, 807.
Kasper H., 2004, *Ernährungsmedizin und diätetik*, 10th edition. Elsevier, Munich. 634 pp.
Knekt P., R. Jarvinen, A. Reunanen and J. Matatela, 1996, *Br. Med. J.* 312, 478.
Kouwenhoven K.J., U.D. Perdok, E.C. Jonkheer, P.K. Sikkema and A. Wieringa, 2003, *Soil Till. Res.* 74(2), 125.
Kumar D. and R. Ezekiel, 2004, *Potato J.* 31(3/4), 129.
Kumar G.N.M. and N.R. Knowles, 1993, *Plant Physiol.* 102, 115.
Lachman J. and K. Hamouz, 2005, *Plant Soil Environ.* 51(11), 477.
Laerke P.E., E.R. Brierley and A.H. Cobb, 2000 *J. Sci. Food Agric.* 80, 1332.
Laerke P.E., J. Christiansen, M.N. Andersen and B. Veierskov, 2002a, *Potato Res.* 45, 187.
Laerke P.E., J. Christiansen and B. Veierskov, 2002b, *Postharvest Biol. Technol.* 26, 99.
Langille A.R., Y. Lan and D.L. Gustine, 1998, *Am. J. Potato Res.* 75, 201.

Lehesranta S.J., H.V. Davies, L.V.T. Shepherd, N. Nunan, J.W. McNicol, S. Auriola, K.M. Koistinen, S. Suomalainen, H.I. Kokko and S.O. Kärenlampi, 2005, *Plant Physiol.* 138, 1690.

Lewis C.E., J.R.L. Walker and J.E. Lancaster, 1999, *J. Sci. Food Agric.* 79, 311.

Lewis C.E., J.R.L. Walker, J.E. Lancaster and K.H. Sutton, 1998, *J. Sci. Food Agric.* 77, 45.

Lisińska G. and W. Leszczyński, 1989, *Potato Science and Technology*. Elsevier Applied Science, London.

Love S.L., T.J. Herrman, A. Thompson-Johns and T.P. Baker, 1994, *Potato Res.* 37, 77.

Love S.L., T. Salaiz, A.R. Mosley and R.E. Thornton, 2004, *HortScience* 39, 156.

Lu W.-H., K. Haynes, E. Wiley and B. Clevidence, 2001, *J. Am. Soc. Hortic. Sci.* 126(6), 722.

Maga J.A., 1980, *CRC Crit. Rev. Food Sci. Nutr.* 12, 371.

Majamaa H., U. Seppälä, T. Palosuo, K. Turjanmaa, N. Kalkkinen and T. Reunala, 2001, *Pediatr. Allergy Immunol.* 12(5), 283.

Marsh G.M., L.J. Lucas, A.O. Youk and L.C. Schall, 1999, *Occup. Environ. Med.* 56, 181.

Matheis G., 1987, *Chem. Mikrobiol. Technol. Lebensm* 11, 5.

Mathew R. and G.M. Hyde, 1997, *Trans. ASAE* 40, 705.

Mawson K., 1998, *Aspects Appl. Biol.* 52, 321.

McGarry A., C.C. Hole, R.L.K. Drew and N. Parsons, 1996, *Postharvest Biol. Technol.* 8, 239.

McMabnay M., B.B. Dean, R.W. Bajema and G.M. Hyde, 1999, *Am. J. Potato Res.* 76, 53.

Meyer K.A., L.H. Kushi, D.R. Jacobs, J. Slavin, T.A. Sellers and A.R. Folsom, 2000, *Am. J. Clin. Nutr.* 71, 921.

Molema G.J., P.C. Struik, B.R. Verwijs, A. Bouman and J.J. Klooster, 2000, *Potato Res.* 43, 225.

Molnar-Perl I. and M. Friedman, 1990, *J. Agric. Food Chem.* 38, 1652.

Morris W.L., L. Ducreux, D.W. Griffiths, D. Stewart, H.V. Davies and M.A. Taylor, 2004, *J. Exp. Bot.* 55, 975.

Mottram D.S., B.L. Wedzicha and A.T. Dodson, 2002, *Nature* 419, 448.

Mullin W.J. and J.M. Smith, 1991, *J. Food Compost. Anal.* 4(2), 100.

Muneta P., 1981, *Am. Potato J.* 58, 85.

Munshi C.B. and N.I. Mondy, 1989, *J. Food Sci.* 54, 220.

Nesterenko S. and K.C. Sink, 2003, *HortScience* 38(6), 1173.

Nitithamyong A., J.H. Vonelbe, R.M. Wheeler and T.W. Tibbitts, 1999, *Am. J. Potato Res.* 76, 337.

Noble R., 1985, *J. Agric. Eng. Res.* 32, 237.

Noordam J.C., W.H.A.M. van den Broek and L.M.C. Buydens, 2005, *J. Sci. Food Agric.* 85, 2249.

OECD (Organisation for Economic Co-operation and Development), 2002, Series on the safety of novel foods and feeds, No.4. Consensus document on compositional considerations for new varieties of potatoes: key food and feed nutrients, anti-nutrients and toxicants. 20 pp.

Park W.D., 1983, *Plant Mol. Biol. Rep.* 1, 61.

Parr A., F. Mellon, I. Colquhoun and H. Davies, 2005, *J. Agric. Food Chem.* 53(13), 5461.

Partington J.C. and G.P. Bolwell, 1996, *Phytochemistry* 42, 1499.

Partington J.C., C. Smith and G.P. Bolwell, 1999, *Planta* 207, 449.

Patchett B., P.S. Cunningham and R.E. Lill, 1977, *NZ J. Exp. Agric.* 5, 55.

Pavek J., D. Corsini and F. Nissley, 1985, *Am. Potato J.* 62, 511.

Percival G., G.R. Dixon and A. Sword, 1996, *J. Sci. Food Agric.* 71, 59.

Peters R., 1996, *Potato Res.* 39, 479.

Pilling E. and A.M. Smith, 2003, *Plant Physiol.* 132, 365.

Poggi V., A. Arcioni, P. Filippini and P.G. Pifferi, 2000, *J. Agric. Food Chem.* 48, 4749.

Ponasik J.A., C. Strickland, C. Faerman, S. Savvides, P.A. Karplus and B. Ganem, 1995, *Biochem. J.* 311, 371.

Pouvreau L, H. Gruppen, G.A. Van Koningsveld, L.A. Van Den Broek and A.G. Voragen, 2003, *J. Agric. Food Chem.* 51, 5001.

Quirce S., M.L. Diez Gomez, M. Hinjosa, M. Cuevas, V. Urena, M.F. Rivas, J. Puyana, J. Cuesta and E. Losada, 1989, *Allergy* 44(8), 532.

Racusen D. and M. Foote, 1980, *J. Food Biochem.* 4, 43.

Reyes L.F., J.C. Miller Jr and L. Cisneros-Zevallos, 2005, *Am. J. Potato Res.* 82, 271.

Rodriguez-Saona L.E., M.M. Giustiand and R.E. Wrolstad, 1998, *J. Food Sci.* 63, 458.

Roessner-Tunali U., J. Liu, A. Leisse, I. Balbo, A. Perez-Melis, L. Willmitzer and A.R. Fernie, 2004, *Plant J.* 39, 668.

Roessner-Tunali U., E. Urbanczyk-Wochniak, T. Czechowski, A. Kolbe, L. Willmitzer and A.R. Fernie, 2003, *Plant Physiol.* 133, 683.

Rogers-Lewis D.S., 1980, *Ann. Appl. Biol.* 96, 345.
Sabba R.P. and B.B. Dean, 1994, *J. Am. Soc. Hortic. Sci.* 119, 770.
Sabba R.P. and B.B. Dean, 1996, *Am. Potato J.* 73, 113.
Scanlon M.G., A.J. Day and M.J.W. Povey, 1998, *Int. J. Food Sci. Tech.* 33, 461.
Schallenberger R.S., O. Smith and R.H. Treadaway, 1959, *J. Agric. Food Chem.* 7, 274.
Schippers P.A., 1976, *Am. Potato J.* 53, 111.
Seppälä U., H. Alenius, K. Turjanmaa, T. Reunala, T. Palosuo and N. Kalkkinen, 1999, *J. Allergy Clin. Immunol.* 101, 165.
Shewmaker C.K. and D.M. Stalker, 1992, *Plant Physiol.* 100, 1083.
Sinden S.L and R.E. Webb, 1972, *Am. Potato J.* 49, 334.
Sinden S.L., R.E. Webb and L.L. Sandford, 1978, *Am. Potato J.* 55, 394.
Smith D.B., J.G. Roddick and J.L. Jones, 1996, *Trends Food Sci. Tech.* 7, 126.
Smittle D.A., R.E. Thornton, C.L. Peterson and B.B. Dean, 1974, *Am. Potato J.* 51, 152.
Soh N.L. and J. Brand-Miller, 1999, *Eur. J. Clin. Nutr.* 53, 249.
Spychalla J.P., 1994. In: W.R. Belknap, M.E. Vayda and W.D. Parks (eds), *The Molecular and Cellular Biology of the Potato*, p. 107. CABI, Wallingford.
Stadler R.H., I. Blank, N. Varga, F. Robert, J. Hau, P.A. Guy, M.-C. Robert and S. Riediker, 2002, *Nature* 419, 449.
Stalham M.A. and E.J. Allen, 2006, BPC Project Report R263 (interim). British Potato Council, Oxford.
Stalham M.A., E.J. Allen and F.X. Henry, 2005, *BPC Research Review – Effects of Soil Compaction on Potato Growth and its Removal by Cultivation.* British Potato Council, Oxford. 61 pp.
Stintzing F.C. and R. Carle, 2004, *Trends Food Sci. Tech.* 15, 19.
Storey R.M.J. and H.V. Davies, 1992, In: P.M. Harris (ed.), *The Potato Crop: Scientific basis for improvement*, 2nd edition, p. 507. Chapman & Hall, London.
Strömberg A., C. Branzell, C. Andersson and K.E. Hellenäs, 2004, *Vår Föda* 56(2), 24.
Suh S.-G., J.E. Peterson, W.K. Stiekema and D.J. Hannapel, 1990, *Plant Physiol.* 94, 40.
Tareke E., P. Rydberg, P. Karisson, P. Eriksson and M. Tornquvist, 2002, *J. Agric. Food Chem.* 51(17), 4998.
Tarn T.R., 2005, In: A.J. Haverkort and P.C. Struik (eds), *Potato in Progress. Science Meets Practice*, p. 66. Wageningen Academic Publishers, Wageningen.
Teissen A., J.H.M. Hendriks, M. Stitt, A. Branscheid, Y. Gibon, E.M. Farre and P. Geigenberger, 2002, *Plant Cell* 14, 2191.
Van Damme E.J.M., A. Barre, P. Rouge and W.J. Peumans, 2004, *Plant J.* 37, 34.
Van der Zaag D. E. and C.P. Meijers, 1970, *Proc. Trienn. Conf. Eur. Assoc. Potato Res.* 1969 4, 93.
Vanesse A., I.D. Jones and H.L. Lucas, 1951, *Am. Potato J.* 28, 781.
Verbist J.F. and R. Monnet, 1979, *Potato Res.* 22, 239.
Virgin H.I. and C. Sundqvist, 1992, *Physiol. Plant.* 86, 587.
Walker J.R.L., 1977, *Food Technol. NZ* 12, 19.
Wilcoxon S.J., E.J. Allen, R.K. Scott and D.C.E. Wurr, 1985, *J. Agric. Sci.* 105, 413.
Witney B.D. and D.C. McRae, 1992, In: P.M. Harris (ed.), *The Potato Crop: Scientific basis for improvement*, 2nd edition, p. 570. Chapman & Hall, London.
Woolfe J.A., 1987, *The Potato in the Human Diet.* Cambridge University Press, Cambridge.

Chapter 22

Skin-Set, Wound Healing, and Related Defects

Edward C. Lulai

USDA-ARS, Northern Crop Science Laboratory, 1307 18 Street N, Fargo, ND, USA

22.1 INTRODUCTION

The physiology and biochemistry of the development of resistance to tuber skinning injury (idiom = skin-set), wound healing, and wound-related defects are of global importance because of the magnitude of food and financial losses impacted by these processes. The amount of these losses is difficult to determine because of the large range of infections, bruise defects, water vapor loss, and various quality issues that are affected by inadequate skin-set and slow wound healing. Collectively, minor to serious wounding and bruising can average 40%, resulting in serious food quality and loss problems and the creation of grower–processor contracts with stringent incentives to reduce these losses (Hampson et al., 1980; Brook, 1996). This chapter will discuss important physiological and biochemical research that impacts these costly wound-related problems.

This chapter will discuss the formation and maturation of native tuber periderm and its relationship to tuber skin-set development in section 22.2. Skinning wounds are difficult to control during harvest unless the tuber periderm has matured so that the skin is set and resistant to skinning injury. The structure of tuber periderm and the maturational changes that result in resistance to tuber skinning injury are of fundamental importance in developing physiological approaches to enhance skin-set and reduce associated losses. The current status of this relatively young research area is described.

The following section, 22.3, discusses tuber wound-healing in detail. The process of wound-induced suberization to heal skinned, cut, and so-called bruised areas covers a vast research plane. Sections 22.3 includes information on the induction and regulation of suberization, and the composition, biosynthetic pathways, assembly, and molecular structure of suberin. These areas of suberin physiology and biochemistry are not fully understood, but they are of great importance in mitigating infection, defect development, and other losses. Suberin is somewhat of an enigma that is often misunderstood and poorly described in conjunction with wound healing and wound periderm development. Consequently, the section covers what is currently considered appropriate terminology and description of suberin.

The final section, 22.4, discusses wound-related defects including shatter bruise, blackspot bruise, growth cracks, and skinning. These defects directly affect food losses and market quality. The physiology and biochemistry of these defects are discussed in relation to wounding and suberization.

Potato Biology and Biotechnology: Advances and Perspectives
D. Vreugdenhil (Editor)
2007 Published by Elsevier B.V.

22.2 NATIVE PERIDERM AND SKIN-SET

22.2.1 Native periderm formation

A well-developed intact periderm with its suberin biopolymer provides the primary barrier against disease, insects, dehydration, and physical intrusions for the potato tuber (Lulai, 2001a). These critical roles of protecting and preserving the tuber are indicative of the importance of the native periderm and wound periderm development.

Prior to development of the native periderm, an epidermis exists for a short time on the youngest tubers of approximately 1 cm or less in diameter. Epidermal tissues are created as the underground stem of the potato plant, i.e. the stolon, swells to form the nascent potato tuber (Artschwager, 1918; Peterson and Barker, 1979). Stomata are scattered in the epidermis and permit gas exchange.

The native periderm forms from epidermal tissues of the nascent potato tuber. Tuber periderm consists of three distinct types of cells; each cell type is grouped into separate layers: (1) phellem, (2) phellogen, and (3) phelloderm (Fig. 22.1). Periderm formation in the emergent tuber is initiated by periclinal division of the epidermal and subepidermal cells, but the meristematic layer of periderm cells, referred to as the phellogen or cork cambium, is formed from the hypodermis (Peterson and Barker, 1979). The phellogen then divides outwardly forming the phellem of the periderm. The phellem is a corky material consisting of several layers (approximately 4–10, depending on genotype, environment, and stage of growth) of well-organized, rectangular suberized cells located at the very surface of the tuber. The innermost layer of periderm cells, the phelloderm, neighbors the phellogen which utilizes substrate sources and starch collectively found in the phelloderm and neighboring cortical cells as is evidenced by the lack of starch granules in these cells. The phellem and phelloderm cells are derivatives of the phellogen and as such are organized into the same rectangular file as that of the original phellogen cell from which the file originates. During periderm development, the meristematic action of the

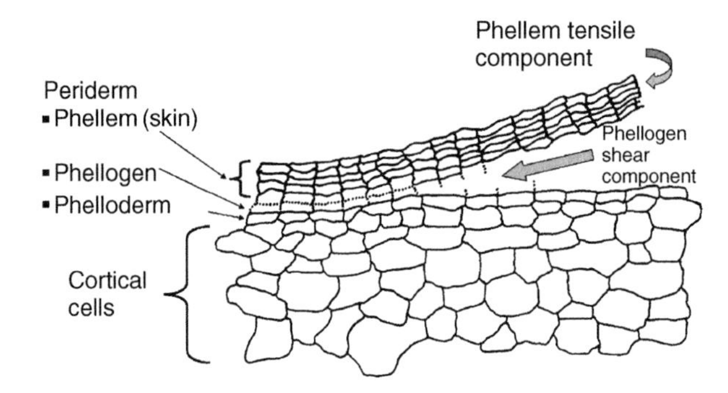

Fig. 22.1. Outline of the cell walls comprising tuber native periderm (immature) and neighboring cortical cells illustrating the two types of fractures that occur on tuber skinning/excoriation injury. Skinning injury occurs by (1) exceeding the tensile strength of the phellem resulting in a fracturing of the phellem cell walls and tearing of the fabric-like skin (phellem tensile component) and (2) a fracture of the phellogen cell walls (phellogen shear component). (Courtesy: E.C. Lulai, *Am J. Potato Res.* 2002).

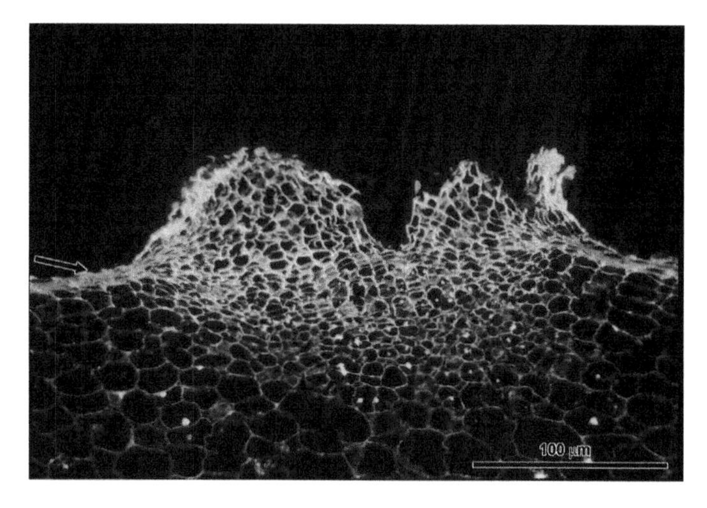

Fig. 22.2. A fully suberized tuber lenticel visualized by treatment with Sudan III/IV. The arrow on the left side points to the red coloration created by the partitioning of Sudan III/IV into the suberin poly(aliphatics) domain. Other cell wall material is visible because of autofluorescence. (Courtesy: E.C. Lulai, S. Goecke and T.J. Wirta).

phellogen also forms lenticels below stomata that existed in the previous epidermal tissues (Artschwager, 1918; Peterson and Barker, 1979) (Fig. 22.2).

As the tuber grows, the radial and longitudinal dimensions increase, thereby requiring commensurate growth of the periderm tissues to maintain the protective suberized covering on the tuber surface. The meristematic action of the phellogen layer continues to form cells necessary to replace sloughed cells and maintain the native periderm on the expanding surface of the growing tuber. During tuber growth, the surface of the periderm may remain smooth, as it does in red- and white-skinned genotypes, or it may develop a rough surface. These rough surfaces of the potato tuber are referred to as russeted or netted skins.

The biology of native periderm and wound-induced suberization to heal wounds and bruises that breach or damage the native periderm are of great importance in minimizing disease, tuber rot during storage, and the development of various defects. Among the most common and yet problematic types of tuber wounds are those created by tuber skinning injury.

22.2.2 Skin-set: a part of native periderm maturation

During growth and for a period during potato plant senescence, the periderm covering the tuber is immature and as such is fragile and susceptible to abrasion. These abrasion-type wounds are frequently referred to as scuffing, skinning injuries, or skinning wounds. This type of periderm damage can be extensive during harvest and handling operations, and results in costly disease, dehydration, and defect development that adversely affects all sectors of the potato industry including processing, fresh market, and seed (Hampson et al., 1980; Lulai and Orr, 1993, 1994). The ensuing losses are consistent with the

250- to 1000-fold increase in the initial rate of water vapor loss from skinned areas compared with non-skinned areas of freshly harvested tubers (Lulai and Orr, 1995). Water vapor loss from freshly harvested tubers that are damage-free is up to 28 times greater than that from mature tubers (Lulai and Orr, 1994). The process of periderm maturation is initiated after growth ceases, and the potato vines die or begin to die in the field (Murphy, 1968). The decades-old approach employing vine/haulm killing or desiccation remains the standard means of promoting periderm maturation and related development of resistance to skinning injury. However, results are varied, and reasonably good skin-set development generally requires about 3 weeks of periderm maturation following vine-killing treatment (Lulai and Orr, 1993). Cultural practices and conditions, such as excessive amounts of fertilizer applied or remaining in the soil because of dry growing conditions followed by rain near the time of harvest, may drastically increase the time required for skin-set (Stark and Love, 2003). In addition to cultural conditions, potato genotype is a major factor in skin-set development; this can be broadly illustrated where russeted genotypes generally mature more quickly whereas red-skinned genotypes often develop skin-set slowly (Lulai and Orr, 1993). Some potato genotypes are deemed unacceptable for commercial production because inherent skin-set development is exceedingly slow, leading to excessive skinning wounds, disease, and development of related defects.

The term 'skinning' has often been loosely used to imply that the entire periderm becomes physically detached from the underlying tuber cells; this is not the case as will be discussed in this chapter. The term excoriation (Latin: ex – out of; corium – skin) has been used to describe this same type of skinning wound but with emphasis that the entire periderm does not become detached (Lulai and Freeman, 2001). The terms excoriation and skinning will be used interchangeably to streamline this discussion. The development of resistance to excoriation, or skin-set, is an important part of periderm maturation that occurs during and after potato vine senescence as well as postharvest (Lulai and Orr, 1993, 1995; Bowen et al., 1996; Pavlista, 2002). Although slow or hindered development of resistance to excoriation during periderm maturation is a serious and costly problem, little is known about the physiology of susceptibility and resistance of the potato tuber to skinning injury.

22.2.3 Skin-set and native periderm physiology

Tuber periderm is composed of (1) phellem (suberized cells), (2) phellogen (cork cambium), and (3) phelloderm (parenchyma-like cells derived from the phellogen) tissues (Reeve et al., 1969). Analysis of mature tuber periderm, however, may not produce easily identifiable phellogen or phelloderm (Lyshede, 1977). A study of all three periderm cell types in immature and mature periderm was needed to determine maturational changes. Only recently have these periderm cell structures been clearly illustrated for easier identification and associated morphological description (Lulai and Freeman, 2001; Lulai, 2002). The suberization processes involved in phellem development are only partially characterized (Kolattukudy, 1980, 2001; Lulai and Morgan, 1992; Thomson et al., 1995; Bernards and Lewis, 1998; Lulai and Corsini, 1998; Lulai, 2001a; Bernards, 2002).

Considering the long history of potato cropping and the breadth and depth of global potato research, it is surprising that earlier identification was not made of the type of periderm cells and cellular changes involved in susceptibility and resistance to tuber excoriation. The lack of fundamental information, particularly at the cellular level, describing the simplest aspects of susceptibility and resistance to excoriation, has hampered the development of effective, rational approaches to describe periderm maturation and associated skin-set development. In turn, there has been a lack of technological advancements necessary to move toward solving the costly problem of tuber skinning injuries that occur during harvest and that hinder successful long-term storage of tubers. The lack of research led to non-scientific explanations for skinning and skin-set, which resulted in postulates incorrectly ascribing skin thickness, periderm thickness, and suberization as determinants of susceptibility and resistance to tuber skinning in immature and mature tubers. These characterizations of skin-set often incorrectly refer to the skin, i.e. phellem, as the periderm of the potato tuber even though the skin constitutes but one of the three types of cells that make up the periderm (Reeve et al., 1969). Because these postulates and idioms arose without scientific investigation or verification, they have become entrenched as descriptive vernaculars and they have been appropriately found in various reviews (Hiller et al., 1985; Peterson et al., 1985; de Haan, 1987; Hiller and Thornton, 1993). Research advancements have moved toward new information and hypotheses describing periderm maturation and excoriation. The ability to objectively measure the status of skin-set development is an important requisite for this research.

A few techniques have been developed to objectively measure the total resistance to skinning during periderm maturation (Ostby et al., 1990; Halderson and Henning, 1993; Lulai and Orr, 1993; Muir and Bowen, 1994; Bowen et al., 1996). All of these techniques rely on measurement of the physical resistance to skinning injury, i.e. the tangential or torsional force required to mechanically shear the phellem from the tuber. Results obtained using the basic principle for these techniques were quantitatively related to observed tuber skinning injury (Pavlista, 2002). The ability to objectively measure the development of resistance to skinning injury is essential for assessing the effectiveness of cultural practices intended to address skin-set development and for uncovering physiological factors associated with susceptibility and resistance to excoriation. However, a uniformly acceptable means of objectively measuring skin-set has not been adopted.

Postharvest controlled environment studies, in conjunction with objective measurement of skin-set, have shown that for some genotypes low relative humidity may hasten periderm maturation and the development of resistance to excoriation in freshly harvested tubers (Lulai and Orr, 1993). Periderm maturation was more rapid in tubers from cultivars with characteristically higher water vapor loss, particularly russeted genotypes (Lulai and Orr, 1994). Periderm maturation and skin-set development did not relate to phellem/skin thickness, phellem/skin weight, or phellem histology. Also, the method of haulm destruction did not influence skin morphology (Lulai and Orr, 1993, 1994; Bowen et al., 1996). These results suggested that the first layer of fully hydrated cells within the periderm, i.e. the phellogen, should play an important role in tuber periderm maturation and skin-set development. Until recently, there was no published information available on the changes that occur within the cork cambium/phellogen of potato tuber periderm as growth ceases and as the periderm matures (Lulai and Freeman, 2001). Extensive studies had been conducted on the structure, ultrastructure, cytology, and biochemistry of the vascular cambium of perennial woody plants and taproots as the plants cycle through growth and dormancy. These are periods when the vascular cambium correspondingly cycles from being meristematically active to inactive (Catesson, 1994; Catesson et al., 1994; Chaffey et al., 1998; Lachaud et al., 1999). The changes in cell wall architecture of the vascular cambium from perennial plants may be a poor model for the changes in cork cambium/phellogen from periderm tissues of annual plants such as potato tubers. In potato tuber, the cells of the lateral meristem irreversibly change from meristematically active to inactive. However, as noted in section 22.2.4, the overall changes in cell wall morphology occurring in the vascular cambium as it enters dormancy are very similar to those found in tuber phellogen as it becomes meristematically inactive upon periderm maturation.

22.2.4 Periderm architecture and skinning injury

Lulai and Freeman (2001) investigated the cellular architecture of immature and mature tuber periderm at the light and electron microscope levels and developed a new paradigm for susceptibility and resistance to tuber skinning injury. They confirmed that the periderm consisted of phellem, phellogen, and phelloderm cells and defined the physiology of skinning by showing that the phellem portion of the periderm comprises what had been loosely referred to as the skin. They further showed that skin thickening and suberization are not the source of development of resistance to skinning injury upon periderm maturation. Instead, the tissue separation responsible for skinning injury occurred solely within the phellogen layer (cork cambium) of immature periderm. As discussed earlier, in section 22.2.1, the phellogen is a lateral meristem from which the periderm is formed (Artschwager, 1918, 1924; Peterson and Barker, 1979). The cell walls of the phellogen mechanically connect and hold the phellem, i.e. skin, to the underlying phelloderm cells, which are tough and rigidly connected to the neighboring cortical tissues. The phellogen cells of immature periderm are meristematic and have thin walls that are easily fractured. This fracturing was found to be synonymous with skinning injury. The walls of these cells were shown to strengthen and thicken considerably and were no longer susceptible to fracture upon development of full resistance to skinning injury and periderm maturation. Although the changes that occur in tuber phellogen cells during periderm maturation and the susceptibility of immature phellogen cell walls to fracture prior to periderm maturation are only beginning to be studied, the biomechanics and fracture of plant cell walls in general have been described (Niklas, 1992a,b).

During tuber skinning, two physical fractures or breakages occur within the periderm. The modulus associated with each of these fractures consists of the stress and strain related to (1) tearing of the fabric-like phellem/skin when the tensile strength is exceeded and (2) shearing type of fracture incurred by the radial cell walls in the phellogen. Lulai and Freeman (2001) showed that the phellogen is the specific area of contiguous fracture upon skinning and that phellogen cell wall strengthening is a determining factor for skin-set development. Lulai (2002) quantitatively determined the role of phellem/skin tensile-related fractures and shear-related fractures in susceptibility and resistance to skinning injury and in skin-set development (Fig. 22.1). The relative strength of the phellem tensile component was nearly constant for the time points tested during periderm maturation for each cultivar and did not measurably increase as the periderm approached maturation. These results indicated that the phellem tensile component did not significantly contribute to skin-set development. However, the force required for fracture of the phellogen shear component did increase upon periderm maturation. Results indicated phellogen shear was the major determinant for development of resistance to skinning injury and confirmed phellogen cell wall strength as the determinant for susceptibility and resistance to tuber excoriation. Collectively, this information showed that the nature of tuber phellogen cell wall strengthening is of major importance to tuber excoriation and skin-set.

22.2.5 Cellular changes associated with skin-set

Research has been conducted to determine maturational changes that occurred in phellem, phellogen, and phelloderm cell walls in native periderm in comparison with changes observed in wound periderm (Sabba and Lulai, 2002, 2004, 2005). As the wounded tuber tissue began to heal, it formed a closing layer (Fig. 22.3) where the walls of existing parenchyma cells became suberized. In conjunction with the formation of the closing layer, a wound phellogen formed under the closing layer. The meristematic action of the wound phellogen formed a wound phellem and a wound phelloderm. This newly formed wound periderm was subject to similar susceptibility and later resistance to excoriation as that of the native periderm. Thickening of wound phellogen cell walls, after meristematic activity ceased, was evident at the light microscope level. Both the phellogen and the phelloderm are often difficult to discern. Sabba and Lulai (2004, 2005) resolved this difficulty by developing a technique employing toluidine blue O, which differentially stained periderm cell walls. The technique proved to be useful in immunolocalization

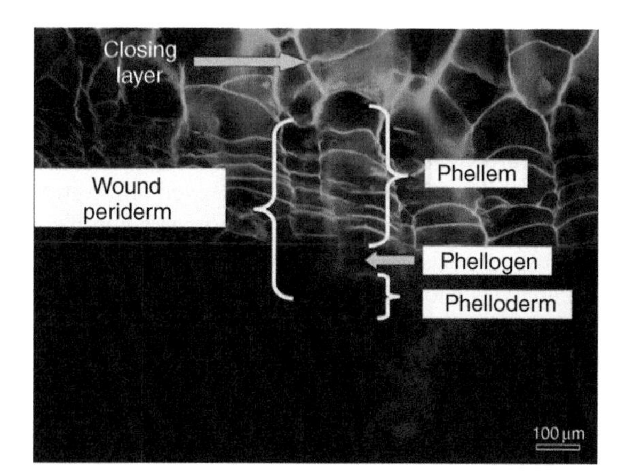

Fig. 22.3. Suberized closing layer and wound periderm viewed using epifluorescence microscopy 31 days after wounding. Autofluorescence (AF) reveals only those walls that have accumulated poly(phenolic) material and will not show poly(aliphatic) material without further cytochemical treatment. Note suberized cells of the closing layer, i.e. existing parenchyma cells that were induced to suberize prior to development of a wound periderm. The wound periderm consists of (1) phellem cells, suberized cells derived from phellogen meristematic action, visualized by poly(phenolic) AF; (2) phellogen, cork cambium or meristematic layer, not suberized and poorly visible under these AF conditions; and (3) phelloderm, derived from phellogen, does not suberize and is weakly AF, and therefore not visible. Under the wound periderm are parenchyma cells; in this case, cortical parenchyma cells that weakly AF and are therefore not visible. (Courtesy R.P. Sabba and E.C. Lulai).

studies where, in separately stained parallel sections, the orthochromatic staining (blue) of phellem walls could be used to identify these cells (Sabba and Lulai, 2004, 2005). Adjoining phellogen and neighboring phelloderm layers, which stained metachromatically (violet) in native and wound periderm, could then be separately identified based on their location within the file of periderm cells and their morphology. Identification of these cells in native and wound periderm is essential in the interpretation of immunolocalization data and other biological investigations that involve wound healing and periderm maturation.

Results from histological and immunolabeling studies of immature and mature native periderm indicated that pectin and extensin depositions are associated with the processes involved in the thickening and related strengthening of phellogen walls upon meristematic inactivation of the phellogen and development of resistance to skinning injury (Sabba and Lulai, 2004). Histological and immunolocalization data indicated that native periderm maturation and associated development of resistance to skinning injury are accompanied by an increase in unesterified pectin in the walls of phellogen cells (Sabba and Lulai, 2002, 2004, 2005). Unesterified pectin can impart rigidity to the cell walls through calcium bridges. Similar processes have been shown to occur in the cambium of aspen in association with cessation of meristematic activity and the onset of dormancy and thickening of cambial walls (Micheli et al., 2000). Pectins are methyl-esterified before they are transported into the wall, and as such, immature cells tend to be characterized by highly esterified pectins that prevent calcium pectate cross-linking and associated strengthening of the cell wall. Older, fully differentiated cells have more rigid walls that are characterized by less esterification and more calcium pectate cross-linking (Goldberg et al., 1989). Interestingly, pre- and postharvest skin-set measurements of maturing tubers in some instances actually show a decrease and later an increase in the resistance to skinning injury (Lulai, 2002; Rolf Peters, personal communication, EAPR, 2005). These mechanical measurements indicated that phellogen cell wall strength may at times decrease prior to cell wall thickening and further skin-set development. The biochemical nature of these anomalies is not known, but they are likely related to reversible compositional changes in the thin cell walls of the phellogen as the tuber undergoes some form of stress imparted by pre- and/or postharvest practices or conditions.

Immunocytological analysis showed that some changes occurred in the localization of pectin and extensin epitopes in periderm cell walls during native periderm maturation (Sabba and Lulai, 2005). The antibodies JIM5 and JIM7 recognize a range of homogalacturonan (HG) epitopes (50% and less esterified, and up to 90% esterified, respectively); LM5 and LM6 recognize rhamnogalacturonan (RG)-I epitopes (β-galactan tetrasaccharide and α-L-arabinan pentasaccharide, respectively); and LM1 recognizes an extensin epitope. The walls of the three types of potato periderm cells labeled differentially for these pectin and extensin epitopes. While the phelloderm labeled equally well for all the epitopes tested, most of the phellem only labeled abundantly for the HG epitope recognized by JIM7 and was lacking in the HG epitope recognized by JIM5, as well as the RG-I and extensin epitopes. Most significantly, labeling of the phellogen layer varied between immature and mature periderm. Cell walls of meristematically active phellogen were lacking in (1,4)-β-galactan and extensin epitopes, as well as those for HG. Upon maturation of the periderm and development of resistance to excoriation, labeling for all these epitopes increased dramatically in phellogen cell walls. These results imply that an increase in the presence of HG, RG-I, and extensin polymers in these walls coincide with meristematic inactivation and periderm maturation. HG would be expected to form calcium pectate and extensin would be expected to become cross-linked in phellogen walls to provide increased strength and reinforcement once meristematic activity is terminated (Sabba and Lulai, 2005). The (1,4)-β-galactan epitope recognized by LM5 in phellogen walls at the end of periderm maturation may also be involved in the strengthening of these walls. By contrast, the (1,5)-α-L-arabinan epitope recognized by LM6 is present in phellogen walls before and after periderm maturation, and is apparently not specifically associated with periderm maturation or a discernable role in the thickening of phellogen walls upon meristematic inactivation. The changes in the epitopes associated with these cell wall polymers provide special insight into the maturational processes that occur during periderm maturation and skin-set development. Collectively, these immunolocalization responses for the targeted cell wall polymers indicate pectin deposition and de-esterification, and extensin depositions are associated with phellogen cell wall strengthening upon inactivation of the phellogen layer as a lateral meristem and maturation of the periderm in potato tuber.

Interestingly, both histological and immunolabeling analyses, distinctly different chemistries, indicated differences between phellogen cell walls of wound versus native periderm. Results from both immunolocalization and histochemical techniques indicated that in wound periderm there is no increase in phellogen cell wall pectin upon wound periderm maturation (Sabba and Lulai, 2004). The reason for these apparent differences remains to be determined.

Although various studies have been conducted to characterize tuber parenchyma cell walls (Li and Showalter, 1996; Bush and McCann, 1999; Oomen et al., 2002; Oomen, 2003; Obro et al., 2004), information on the physiology of periderm maturation remains sparse. Determination of the biochemical nature and regulatory mechanisms for strengthening of radial phellogen cell walls is important in the development of technologies to hasten skin-set and reduce associated market quality problems. The mechanisms regulating development of resistance to excoriation and biological markers for progress have not been determined. In conjunction with these knowledge gaps, the physiological and biological basis for slow/hindered skin-set development is not known, nor are the mechanisms known for the unpredictable reversal of skin-set development. Much more needs to be learned about the biology of potato periderm maturation so that effective technologies may be created to enhance skin-set development.

22.3 WOUND HEALING

22.3.1 The process of tuber wound healing

Tubers that are skinned, nicked, or bruised during harvest or cut for seed lack the robust protection provided by the suberized layer of the native periderm, i.e. native phellem. Rapid wound healing is essential to avoid infection, desiccation/shrinkage, and defect development after injury. Tuber wound responses and wound healing involve many biological processes, perhaps the most important of which is wound-induced suberization (suberin biosynthesis). Regardless of the type of cells and associated polymeric structures covering the plant surface, wounding induces suberization to protect the damaged area (Dean and Kolattukudy, 1976). Wound-induced suberization of tuber tissue involves two stages during which two types of cells are suberized. The first stage involves formation of a 'closing layer' whereby the walls of existing cells at the wound site suberize, i.e. accumulate suberin biopolymers; 'closing layer formation' is also often referred to as 'primary suberization'. Following the formation of the closing layer, the second stage involves development of a 'wound periderm' whereby files of new cells are formed and suberized below the closing layer; 'wound periderm development' is also often referred to as 'secondary suberization' (Section 22.3.6). A fully suberized tuber wound is resistant to infection by both bacteria and fungi (Lulai and Corsini, 1998). Suberin is a complex of suberin poly(phenolics) (SPP) and suberin poly(aliphatics) (SPA) that are cross-linked by glycerol, embedded with soluble waxes, and laminated to the interior side of the plant cell wall (Bernards, 2002). The terms wound healing and suberization are often loosely used interchangeably. Other forms of resistance are required to provide some degree of protection against infection and rot until suberization of the wound has been completed. These resistance processes and associated mechanisms are not fully understood, are generally temporal, and include various hypersensitive responses, the oxidative bursts, phytoalexin production, pathogenesis-related proteins, and other mechanisms (Lyon, 1989; Beckman, 2000; Pérombelon, 2002). Some resistance responses at the wound site appear to be related to the appearance of phenolic compounds (Lyon, 1989; Nolte et al., 1993; Beckman, 2000), some of which may be precursors of SPP (Bostock and Stermer, 1989). This observation is consistent with the correlative relationship found between strong temporal resistance and the rapidity of suberization across diverse genotypes (Lulai and Corsini, 1998, unpublished results). Importantly, a fully suberized surface provides the final, most durable, and broad ranging barrier for protectin against pests, desiccation, and infection (Lulai and Corsini, 1998). Rapid suberization is essential for protecting tubers that are injured during growth, during harvest and handling, and upon seed cutting. Suberization is also responsible for closure of lenticels as orifices for gas exchange and as a portal of entry for pests (Wigginton, 1973; Banks and Kays, 1988; Scott et al., 1996; Tyner et al., 1997) (Fig. 22.2). The soluble waxes embedded in the suberin matrix control water loss and prevent desiccation (Soliday et al., 1979; Lulai and Orr, 1994; Schreiber et al., 2005a). For the sake of conciseness, the discussion in section 22.3.2 will be primarily directed to potato tuber suberin and suberization.

22.3.2 Induction of suberization

Upon cellular damage incurred during wounding or stress, a wide range of signals is perceived, which induces various protective or wound-related responses including suberization (de Bruxelles and Roberts, 2001; Kolattukudy, 2001; Leon et al., 2001; Schilmiller and Howe, 2005; Wasternack et al., 2006). A four-fold increase in translational activity is induced within 1 h of wounding, and a range of transcripts appears including those involved in primary suberization (Vayda and Morelli, 1994). Despite their importance, the signal(s) that induce suberization are enigmas that require further study. Several mitosis-related signals were hypothesized to be involved (Rosenstock and Kahl, 1978). Although induction of cell division is an essential part of wound periderm formation, it does not address the critical induction of suberin accumulation on the existing cell walls in the closing layer, i.e. during primary suberization resulting from a wound or from other stresses. Suberization of existing tuber parenchyma cells, i.e. similar to closing layer formation/primary suberization, has been demonstrated without mechanical wounding (Lulai, 2005; Lulai et al., 2006). The hollow heart disorder is a different type of example; it may be described as an internal growth-related wound with no cells exposed to the external environment; yet, cells neighboring the hollow heart area are suberized and are compositionally and ultrastructurally similar to that of wound periderm (Dean and Kolattukudy, 1977; see Chapter 23, Sowokinos, this volume, for more information on hollow heart). The signals involved in wound-, pathogen-, and other stress-induced suberization have not been determined. Rosenstock and Kahl (1978) asked the fundamental question 'what is the primary event after mechanical disturbance of a tissue?' and whether a primary event triggers the sequence of reactions in a wound cell. They hypothesized that one of the first primary changes upon wounding is that of a change in osmotic potential of the cell at the wound site. Many wound-related or wound-induced signals have been studied (de Bruxelles and Roberts, 2001; Leon et al., 2001; Schilmiller and Howe, 2005). However, the signal(s) and related mechanism(s) that induce suberization in potato tuber have not been identified.

22.3.3 Regulation of suberization

In association with induction, the regulation of suberization is of equal importance, yet poorly understood. The plant hormone ethylene has been shown to be involved with various kinds of plant stress including wound response (Abeles et al., 1992; Bleecker and Kende, 2000; Ciardi and Klee, 2001; de Bruxelles and Roberts, 2001). However, the involvement of ethylene in wound-induced suberization had not been determined until recently. Using various inhibitors of ethylene biosynthesis and action, Lulai and Suttle (2004) determined the involvement of ethylene in wound-induced suberization. Ethylene biosynthesis was found to be stimulated by tuber wounding. Separate analysis for accumulation of SPP and SPA on wound-induced cell walls showed that ethylene is not required for wound-induced suberization of the closing layer or subsequent suberization associated with wound periderm development.

Soliday et al. (1978) employed a gravimetric technique, diffusive resistance, to measure water vapor loss in wound-healing tissue. Results from this technique reflected wax deposition as an indirect assessment of wound-healing responses and the effect of various hormone treatments. They found that the development of resistance to water vapor loss during wound healing was inhibited by indole-3-acetic acid (IAA) and cytokinin treatments, but stimulated by abscisic acid (ABA) treatments. They further found that ABA and other 'suberization induction factors' could be washed from the tuber tissue, thereby inhibiting suberization. ABA treatment of cultured potato cells also resulted in increased accumulation of suberin components, waxes, and enzymes involved in suberin biosynthesis including peroxidase, phenylalanine ammonia lyase

(PAL), and ω-hydroxy-fatty acid dehydrogenase (Cottle and Kolattukudy, 1982b). Lulai and Orr (1995), using a sensitive porometric technique for direct electronic measurement of water vapor loss, found that ABA treatment hastened development of resistance to water vapor loss during the first day of wound healing but made little difference after that time point when SPP and later SPA began to accumulate. Schreiber et al. (2005a) conducted detailed analysis of wax deposition during periderm maturation and obtained results consistent with that of Lulai and Orr (1995). Schreiber et al. (2005a) also found that even though periderm permeability quickly approached a nearly constant value during periderm development and maturation, significant wax formation continued. These results suggest that regulation of wax accumulation was still active even though water vapor loss had been minimized. Collectively, the above results indicate that ABA is involved in the regulation of wax accumulation and that water vapor loss is controlled by wax accumulation during wound healing, but water vapor loss may not be quantitatively used for assessing suberization. Lulai and Suttle (2005) employed other technologies to determine the involvement of ABA in tuber wound-induced suberization. Liquid chromotagraphy-mass spectrometry (LC-MS) was used to determine changes in tuber ABA content during wound healing. A specific xenobiotic inhibitor of carotenoid (ABA precursor) biosynthesis was used to determine the effect of metabolically blocking ABA biosynthesis during wound healing. Specific cytological techniques were used to directly assess accumulation of SPP and SPA on suberizing cell walls of inhibitor-treated tuber tissue during wound healing. The ability to block ABA biosynthesis provided a non-correlative approach to help determine the role of ABA in the regulation of wound responses. Tuber wounding resulted in an increase in ABA concentration. Inhibition of ABA biosynthesis resulted in a diminution of tuber wound-healing responses including reduced PAL activity, a mild delay in SPP accumulation, a more noticeable delay in SPA accumulation on suberizing cell walls, and a significant hampering in the reduction of water vapor loss. These results clearly indicate that ABA has a role in wound-induced suberization but that other signaling factors could be involved.

Jasmonic acid (JA) is included in the wide range of signaling/regulatory compounds induced upon wounding and is involved in a range of wound responses such as insect and disease resistance (Choi et al., 1994; Koda and Kikuta, 1994; Negrel et al., 1995; de Bruxelles and Roberts, 2001; Schilmiller and Howe, 2005; Wasternack et al., 2006). Yet, information on the involvement of JA in wound-induced suberization is sparse. Two hydroxycinnamoyl transferases involved in suberin biosynthesis were shown to be influenced by treatment with ABA; however, JA treatment did not significantly modify the time course or intensity of the induction of these enzymes during wound healing (Negrel et al., 1995). Wound-induced PAL activity and the associated accumulation of phenolics in purple-flesh tubers with varying anthocyanin concentrations did not show a response to methyl jasmonate treatment unless anthocyanin concentrations were low (Reyes and Cisneros-Zevallos, 2003). Although under current investigation, there is little other information available concerning the involvement or regulatory roles of JA in tuber wound healing and suberization. Many signaling compounds, in addition to those discussed in this section, are induced upon wounding, yet their involvement in the regulation of suberization has not been determined.

22.3.4 Environmental effects on suberization

After induction, the rate of suberization is influenced by genotype (Lulai and Corsini, 1998), type or severity of wound or bruise (Lulai and Orr, 1995; Thomson et al., 1995), physiological age (Thomson et al., 1995; Kumar and Knowles, 2003), a range of possible chemical treatments (Nolte et al., 1987; Gronwald, 1991; Oosterhaven et al., 1995), and environmental conditions (e.g. relative humidity, temperature, and aeration including oxygen/carbon dioxide concentrations) (Artschwager, 1927; Wigginton, 1974; Dean, 1989; Morris et al., 1989; Schaper et al., 1993). Relative humidity above 80%, preferably 90–95%, is needed to ensure that the cells at the wound site do not desiccate and die. Low relative humidity after wounding may lead to tissue dehydration and result in a layer of dead non-suberized cells over the wound. A layer of dead desiccated cells will not resist penetration by pathogens or prevent water vapor loss and should not be confused with suberized cells. Excessive humidity may result in a film of water over the wounds, thereby restricting oxygen supply and causing cell proliferation both of which inhibit suberization. If relative humidity and oxygen supply are favorable, the most important environmental factor affecting suberization is temperature. Suberization increases from 2.5 to 25 °C. The rate of suberization increases approximately three-fold between 5 and 10 °C, increases another three-fold between 10 and 20 °C, and is greatest near 25 °C (Artschwager, 1927; Wigginton, 1974; Dean, 1989; Morris et al., 1989). Suberization is prevented at temperatures of 35 °C and above. Excessively warm storage temperatures directly after harvest can hamper suberization and result in infection and rapid decay.

Ample oxygen (air is approximately 21% oxygen) is required for suberization. Low oxygen concentrations and carbon dioxide concentrations above ambient (air is approximately 0.03% carbon dioxide) inhibit suberization and wound periderm formation. Tuber wounding from harvest and seed-cutting operations can quickly result in a three- to five-fold increase in respiration, which within 24 h can increase another three- to five-fold (Laties, 1978). Wound respiration may be 25 times that of the intact tuber by 24 h after wounding. Fresh wound respiration is derived primarily from lipid. Within 24 h after wounding, carbohydrate becomes the respiratory substrate. Carbon dioxide concentrations increase while oxygen concentrations decrease within the potato storage; both conditions inhibit suberization. Proper ventilation of freshly harvested tubers or cut seed is essential in maintaining oxygen and carbon dioxide concentrations that promote suberization. Unless properly ventilated, the carbon dioxide concentration in a storage bin of freshly harvested skinned and bruised tubers can increase to well over 100 times that of ambient air (Schaper et al., 1993).

22.3.5 Characteristics of the biopolymers that form suberin

The identification and description of suberin has been confusing in part because of its complex and poorly understood composition and ultrastructure, and its similarities to the plant polymers lignin and cutin (Kolattukudy, 1980; Cottle and Kolattukudy, 1982a; Kolattukudy, 2001; Bernards, 2002). Lignin is composed of a dense polymeric matrix of aromatic/phenolic monomers, and cutin is composed of hydroxy and epoxy aliphatic monomers (Kolattukudy, 1980; Lewis et al., 1999). The biopolymer suberin is composed of both phenolic/aromatic and aliphatic/hydrophobic polymeric domains (Kolattukudy, 1980, 2001), which are spatially separate (Lulai and Morgan, 1992) and cross-linked by glycerol (Moire et al., 1999; Graca and Pereira, 2000; Bernards, 2002). Because of the polymeric and compositional nature of the phenolic/aromatic and aliphatic/hydrophobic domains, they are now referred to as the suberin poly(phenolic) domain (SPPD) and suberin poly (aliphatic) domain (SPAD) (Bernards, 2002). The SPPD had been referred to as lignin-like (Kolattukudy, 1980; Cottle and Kolattukudy, 1982a). The complex biosynthesis and macromolecular assembly of the SPP biopolymer on the cell wall may have similarities to that of lignin and involve dirigent proteins and sites (Lewis et al., 1999; Davin and Lewis, 2000). However, the SPPD lacks the dense matrix of guaiacyl and syringyl phenylpropane units characteristic of lignin (Lapierre et al., 1996). Detailed labeling and spectroscopic studies clearly showed that the phenolic domain of suberin has a very low monolignol content, about one-tenth that of wood or straw, and instead is largely composed of hydroxycinnamic acids and associated derivatives (Bernards et al., 1995; Lapierre et al., 1996; Bernards, 2002). Also, confusion arises because lignin and the SPPD react similarly to histological analyses, such as phloroglucinol and other treatments, and both autofluoresce under commonly used epifluorescent microscopy. Results of Lulai (2005) may be interpreted to indicate that, in some cases, pathogen-induced accumulation of SPP on the cell wall may have been mistaken for lignification in other published reports. SPAs, like cutin, are composed of fatty acid monomers. However, unlike cutin, suberin aliphatics are made up of ω-hydroxy and α, ω-dicarboxylic acids, which can be of equal or longer chain length than cuticular aliphatics and lack the characteristic cuticular epoxy and internal chain hydroxy groups (Kolattukudy and Dean, 1974; Dean and Kolattukudy, 1977; Kolattukudy, 1980). Bernards (2002) indicated a possible role for epoxy fatty acids in hydroxylation reactions involved in SPA biosynthesis. Although SPP and SPA accumulations are tightly coupled, if conditions are not favorable, SPP accumulation on the cell wall may terminate before completion and SPA accumulation will not occur as has been demonstrated in tuber pink-eye tissues (Lulai et al., 2006). There have been no reports of wound-induced SPA accumulation on the primary cell wall without first accumulating SPP about the entire cell wall nor have there been demonstrations of SPA accumulation on a polyphenolic matrix that had been fully characterized as lignin. These features help define the two biopolymers, SPP and SPA, that form the phenolic/hydrophilic and aliphatic/hydrophobic domains of suberin. Molecular genetic similarities and differences between cutin and suberin continue to be uncovered (Yephremov and Schreiber, 2005).

22.3.6 Suberization: closing layer and wound periderm formation

The hierarchy for wound-induced accumulation of SPP and SPA on the suberizing cell walls also differs. The process of closing layer formation, i.e. primary suberization, is cell specific, occurring one cell layer at a time beginning with the first layer of viable cells neighboring the wound site. SPP accumulates on the cell walls in a segmented fashion, first on the outer tangential walls of the cells in the first layer followed by accumulation on the radial cell walls and then the inner tangential cell walls (Lulai and Corsini, 1998). After SPP accumulation on the cell wall is complete, SPA begins to accumulate and does so relatively uniformly about the inner perimeter of the cell. SPA does not accumulate in a pronounced segmented fashion like that of SPP. As SPA accumulates, presumably in conjunction with glycerol, over the SPP in the cells of the first layer, SPP accumulates in a segmented fashion on the walls of the cells in the second cell layer. SPA accumulation follows completion of SPP deposition on the second layer and so on. Generally, under favorable conditions, primary suberization will continue until there are about two or three layers of suberized parenchyma cells forming the closing layer. At this time, the wound phellogen will have sufficiently developed and will form additional suberized cells, i.e. wound phellem, directly beneath the closing layer. In addition, a layer of wound phelloderm cells is formed beneath the wound phellogen (Lulai and Corsini, 1998; Lulai, 2001a). These three cell types, i.e. phellem, phellogen, and phelloderm, comprise the wound periderm (Reeve et al., 1969), but it is the suberized cells of the closing layer and wound phellem that provide the durable protective barrier (Lulai and Corsini, 1998).

It is important to note that primary suberization, i.e. formation of the closing layer, involves suberization of existing cell walls. Therefore, these suberized cells do not derive from progenitor cells in meristems, i.e. not from a distinct cambium/meristematic layer. Arguably, these suberized cells may play a more important role than those of the wound phellem because they provide the first suberized form of durable and broad-ranging resistance to infection over the wound (Lulai and Corsini, 1998). Interestingly, the biological differences involved in suberization of closing layer cells and that of wound phellem cells derived from the wound periderm, i.e. secondary suberization, are not known. Development of the closing layer provides a unique perspective compared with that of wound phellem development because SPP accumulation must occur in conjunction with the existing primary cell wall. Whether there are fine biosynthetic and structural differences between suberized cell walls of the closing layer and wound phellem is not certain.

22.3.7 Suberin biosynthesis and structure

Suberization requires biosynthesis of phenolic, aliphatic, and glycerol monomers and assembly of these monomers into the two separate biopolymeric domains, the SPPD and the SPAD. Together, the SPPD and the SPAD comprise the suberin barrier. The composition, biosynthesis, and associated macromolecular assembly of these phenolic and aliphatic biopolymers into their respective domains on the cell wall continue to be described through hypothetical models and schemes because the pathways and mechanisms of assembly are only partially elucidated. Suberin biosynthesis is an important area of research that is complicated by the distinctly different biochemical pathways required for SPP and SPA syntheses, assembly, glycerol cross-linking, and placement on/in the cell wall. Detailed issues associated with SPP and SPA biosyntheses have been reviewed and will not be repeated in this chapter (Bernards and Lewis, 1998; Bernards, 2002). The SPPD and SPAD have been shown to be spatially separate and their time courses for wound-induced accumulation on the cell wall initiated at distinctly different times (Lulai and Morgan, 1992; Lulai and Corsini, 1998). SPP biosynthesis occurs prior to that of SPA (Kolattukudy, 1980). Furthermore, nuclear magnetic resonance (NMR) data (Stark and Garbow, 1992) and ultrastructural analysis also show that the SPAD is distinct

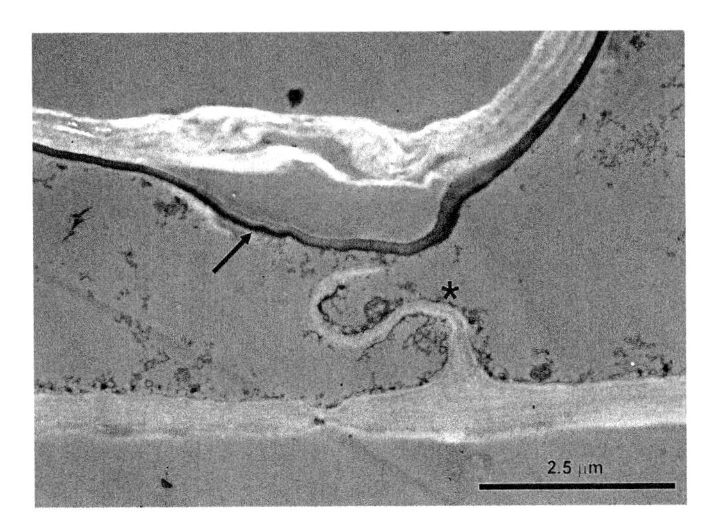

Fig. 22.4. Transmission electron micrograph of a suberized cell wall neighboring an active phellogen layer. The upper primary cell wall is suberized and includes suberin poly(phenolic) material and darkly stained suberin poly(aliphatic) material. The suberin poly(aliphatics) domain (SPAD) is spatially separate from the primary cell wall. The delamination of the SPAD from the primary cell wall (arrow) illustrates the spatial separation of the two suberin polymeric domains. The thin and fragile radial cell wall (*) of the active phellogen has fractured. (Courtesy: E.C. Lulai and T.P. Freeman).

and spatially separate from the SPPD and primary cell wall (Fig. 22.4). Although spatial features and architecture of the SPPD accumulated on/into the primary cell wall are not clear, spectroscopic data indicate that the aromatic carbons of the SPPD and glycosidic carbons of cell wall polysaccharides are in proximity, perhaps 0.05 nm (Yan and Stark, 2000). Wound signals induce enzymes of the phenylpropanoid pathway to provide the phenolic precursors for the biosynthesis of SPP (Kolattukudy, 1980) (Fig. 22.5A). The first enzyme of this pathway, PAL, which produces cinnamic acid for further biosynthesis, is essential and its inhibition prevents formation of the SPP barrier (Cottle and Kolattukudy, 1982a; Hammerschmidt, 1984; Bostock and Stermer, 1989; Bernards and Lewis, 1998). Hydroxycinnamic acid derivatives, primarily ferulic acid and *N*-feruloyltyramine in wound-induced suberin, are channeled to comprise a major part of the SPPD (Bernards and Lewis, 1992; Bernards et al., 1995, 2000; Negrel et al., 1996). Ferulate esters of long-chain fatty alcohols, hypothesized to act as bioplasticizers, are found in the wound periderm (Bernards and Lewis, 1992; Bernards, 2002); however, little has been indicated about the presence of these alkyl ferulates in native periderm. Information on the composition of the SPPD has largely been obtained by analysis of end products from harsh chemical degradation treatments, which inherently alter the structure of the final monomers to be identified. NMR analysis of wound periderm confirmed that sinapyl and guaiacyl structures are part of the SPPD (Yan and Stark, 2000), and LC-MS techniques have been developed, which quantitatively show that there are differences in the biosynthetic flux of phenylpropanoids, i.e. formation and conversion of specific monomers into SPP during tuber wound healing (Matsuda et al., 2003).

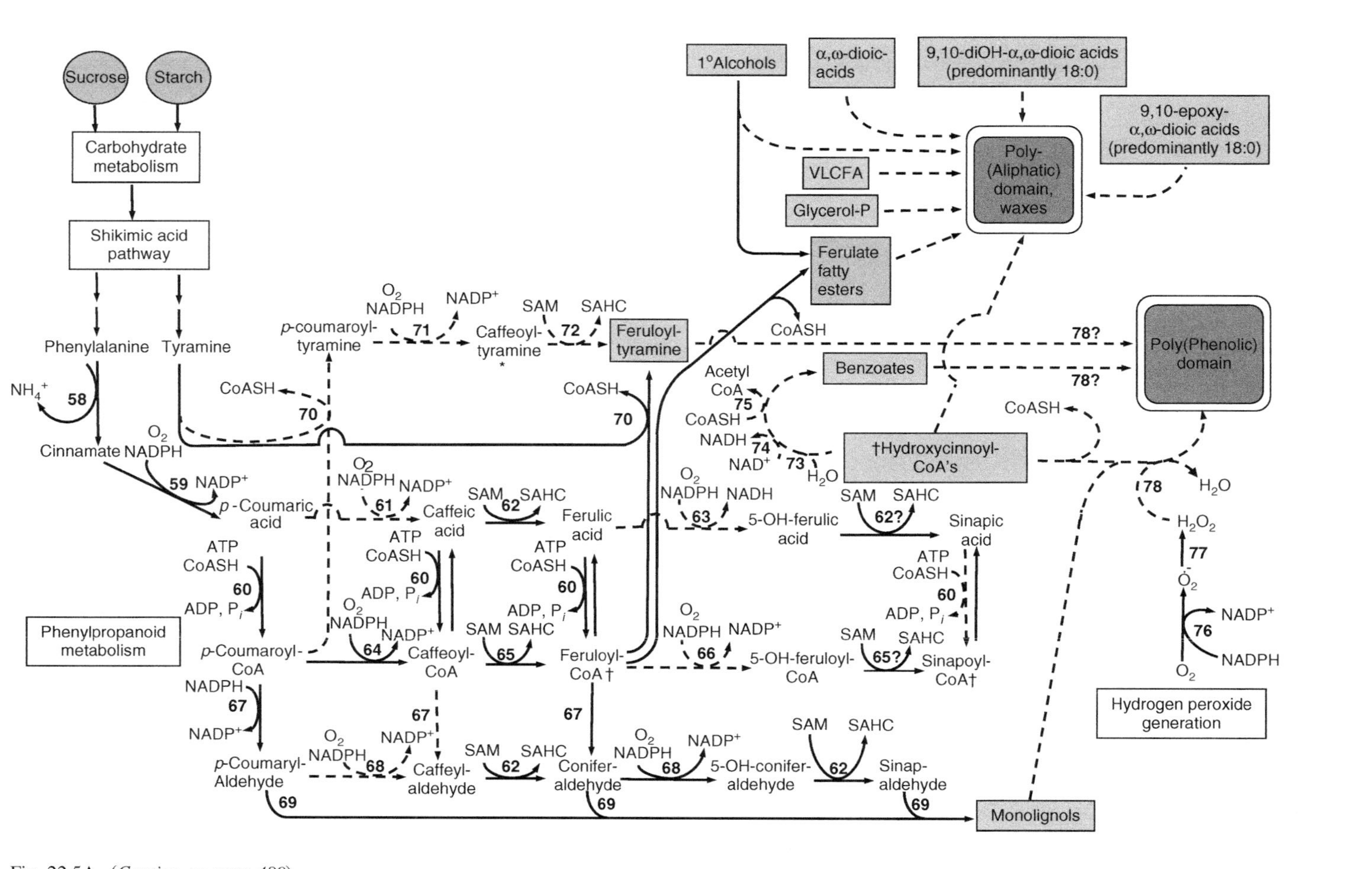

Fig. 22.5A. (*Caption on page 489*)

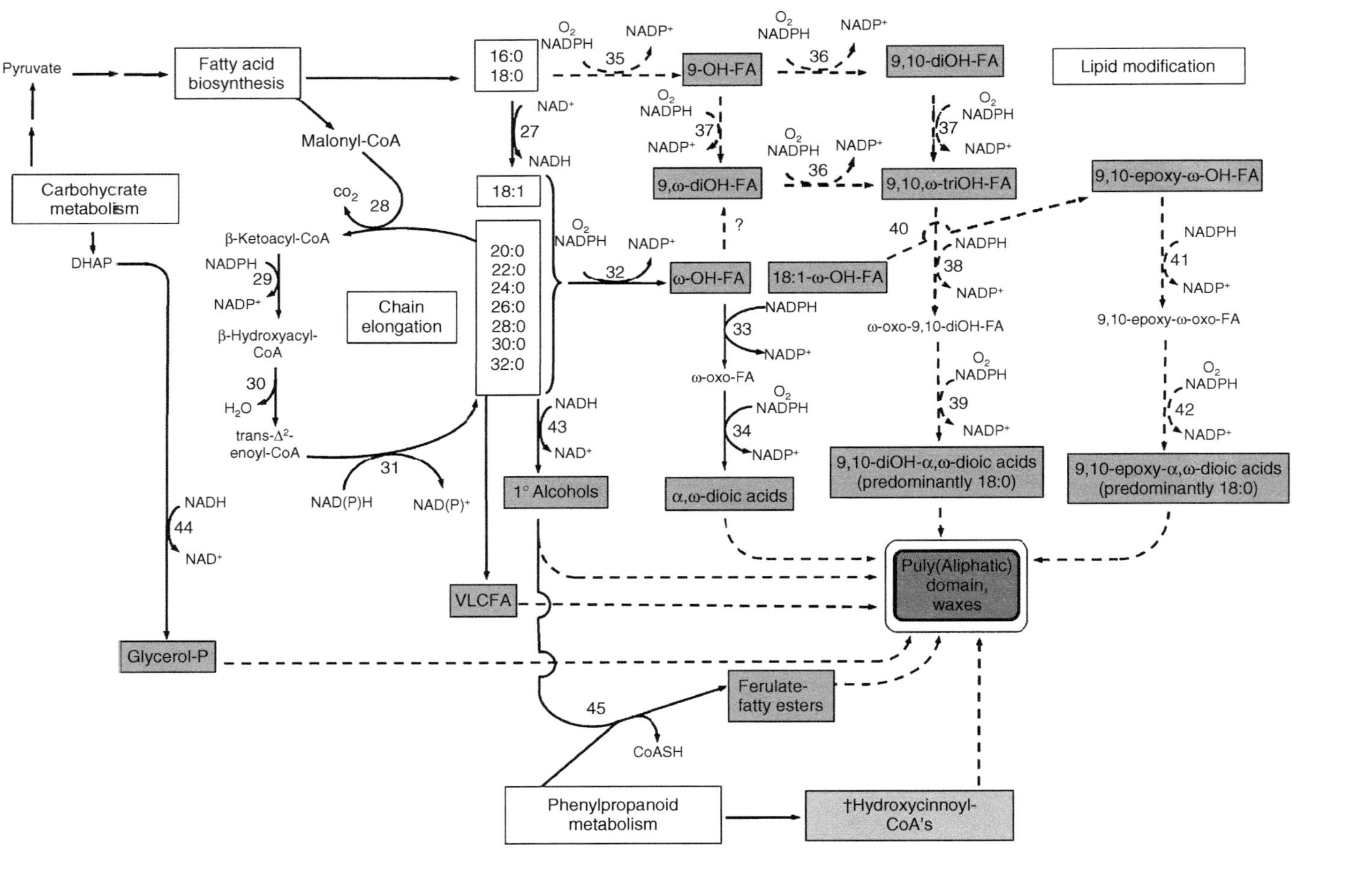

Fig. 22.5B. (*Caption on page 489*)

Fig. 22.5. Scheme for the biosynthesis of the suberin poly(phenolic) (SPPD) and poly(aliphatic) (SPAD) domains, respectively. Figure 5A and B is part of a total scheme developed by Bernards (2002) and represents a composite of knowledge derived from a number of different plant species, but principally derived from potato and maize. Subcellular localization of the processes is not depicted. The enzymes involved are identified by number as in the original scheme; a few enzymes within the scheme are hypothetical (*): **1–26**, **46–57**, and **84–85** involve carbohydrate metabolism, fatty acid biosynthesis, shikimic acid pathway, and ammonia recovery and may be found in Bernards (2002); **27**, stearoyl-ACP Δ^9-desaturase; **28**, β-ketoacyl-CoA synthase III; **29**, β-ketoacyl-CoA reductase; **30**, β-hydroxyacyl-CoA dehydratase; **31**, enoyl-CoA reductase; **32**, fatty acyl-ω-hydroxylase; **33**, ω-hydroxyacid dehydrogenase; **34**, ω-oxoacid dehydrogenase; **35**, fatty acyl-9-hydroxylase*; **36**, 9(ω)-hydroxy fatty acyl-10-hydroxylase*; **37**, 9(10)-hydroxy fatty acyl-ω-hydroxylase*; **38**, 9,10,ω-trihydroxyacid dehydrogenase*; **39**, 9,10-dihydroxy-ω-oxoacid dehydrogenase*; **40**, ω-hydroxyacid-9,10-epoxide synthase*; **41**, 9,10-epoxy-ω-hydroxyacid dehydrogenase*; **42**, 9,10-epoxy-ω-oxoacid dehydrogenase*; **43**, reductases*; **44**, glycerol-3-phosphate dehydrogenase; **45**, hydroxycinnamoyl-CoA; 1-alkanol hydroxycinnamoyl transferase; **58**, phenylalanine ammonia-lyase; **59**, cinnamate-4-hydroxylase; **60**, 4-coumaroyl-CoA ligase; **61**, *p*-coumaric acid 3-hydroxylase; **62**, caffeic acid 3-Î-methyltransferase; **63**, ferulic acid 5-hydroxylase; **64**, *p*-coumaroyl-CoA-3-hydroxylase; **65**, caffeoyl-CoA-3-O-methyltransferase; **66**, hydroxycinnamoyl-CoA-5-hydroxylase*; **67**, cinnamoyl-CoA oxidoreductase; **68**, hydroxycinnamaldehyde hydroxylase*; **69**, coniferyl alcohol dehydrogenase; **70**, hydroxycinnamoyl-CoA : tyramine hydroxycinnamoyltransferase; **71**, *p*-coumaroyltyramine-3-hydroxylase*; **72**, caffeoyltyramine-O-methyltransferase*; **73**, hydroxycinnamoyl-CoA-7-hydroxylase*; **74**, (7-hydroxy)-hydroxycinnamoyl-CoA reductase*; thiolase*; **76**, NAD(P)H-dependent oxidase; **77**, superoxide dismutase or spontaneous; **78**, peroxidase. Reactions denoted by solid lines are known while those denoted by broken lines are hypothetical or assumed. Shaded boxes denote known precursors incorporated into the suberin poly(aliphatic) domain (SPAD) and suberin poly(phenolic) domain (SPPD). (Courtesy: M.A. Bernards and the *Can. J. Bot.*, 2002, revised).

Peroxidases and H_2O_2 formation are induced upon wounding and are required for suberization, possibly in a peroxidase-mediated coupling process cross-linking phenolics within the SPPD and attachment to the cell wall (Espelie et al., 1986; Bernards et al., 1999, 2004; Bernards and Razem, 2001; Razem and Bernards, 2003). A model has been constructed whereby phenolics are synthesized in the cytoplasm, transported to the cell wall (by an unknown mechanism), and polymerized by a wall-associated peroxidase into the SPPD (Bernards and Razem, 2001). The polymeric attachments to the cell wall are at discrete sites (Stark and Garbow, 1992). Specific peroxidases have been determined to preferentially cross-link suberin phenolics but not monolignols (Bernards et al., 1999). In vitro studies have shown that there are several specific combinations of cross-coupling products formed by peroxidase-catalyzed polymerization of hydroxycinnamic acids (Arrieta-Baez and Stark, 2006). The identity of these coupling products provides important insight into the molecular architecture of suberin and the structural requisites that impart durable protective properties to the barrier. The construction of the SPPD is particularly perplexing when the ambiguities of SPP biosynthesis, phenolic transport, and polymeric assembly are combined with polymeric attachment at discrete sites on the cell wall in conjunction with the segmented hierarchy for SPP accumulation on the cell wall.

Biosynthesis of the SPAD involves fatty acid synthesis and the creation of aliphatic monomers from palmitic (16:0), stearic (18:0), and oleic (18:1) acids (Fig. 22.5B). The suberin aliphatics are composed primarily of ω-hydroxy acids and the corresponding dicarboxylic acids and smaller amounts of long ($>C_{20}$) chain acids and corresponding alcohols (Kolattukudy, 1978). Fatty acid monomers are substrate to a chain elongation system (Kolattukudy, 1978), which now appears to be microsomal and

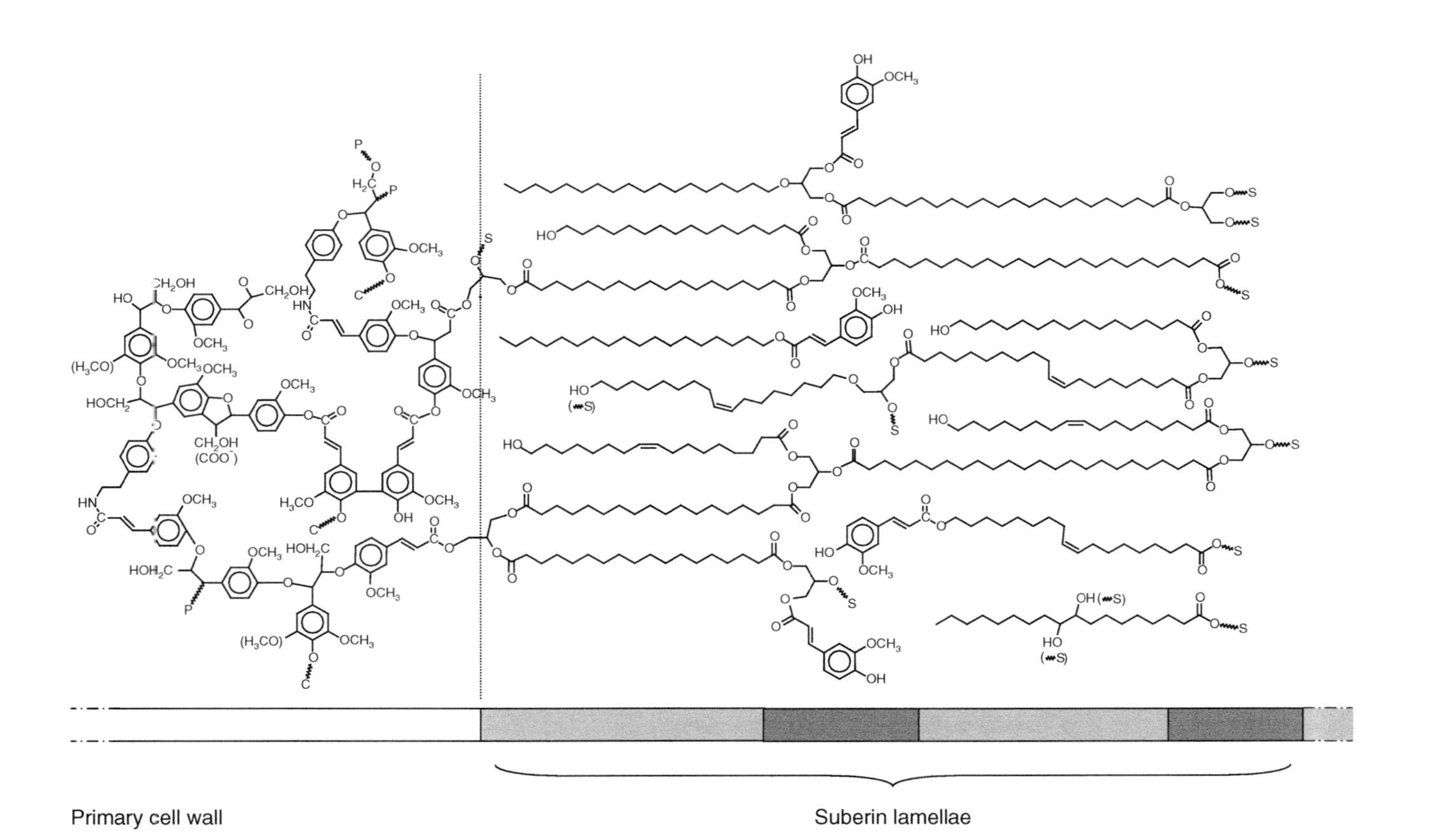

Fig. 22.6. (*Caption on page 491*)

Fig. 22.6. Tentative model for the structure of potato suberin. The suberin poly(phenolic) domain (SPPD) is covalently attached to carbohydrate units of the cell wall. Glycerol units are used to cross-link suberin poly(aliphatic) monomers and account for the lamellae found within the suberin poly(aliphatic) domain SPAD. The SPAD is illustrated using two lamellae and is covalently attached to the SPPD through easily hydrolyzed ester linkages. Very long-chain fatty acids and waxes are not illustrated within this diagram but are thought to be embedded in the aliphatic domain. Ferulates esterified to long-chain fatty alcohols are integrated into the aliphatic domain and are thought to act as plasticizers. The constituents are connected to C, carbohydrate; P, phenolic; S, suberin (phenolic or aliphatic). (Courtesy: M.A. Bernards and the *Can. J. Bot.*, 2002, revised).

elongate through malonyl-CoA as substrate (Schreiber et al., 2000, 2005b). The fatty acids are presumed to undergo ω-hydroxylation, similar to cutin, and then further conversion to dicarboxylic acids in reactions that may have some uniqueness to SPA biosynthesis (Kolattukudy, 1980). The ω-hydroxy fatty acids are converted to ω-oxo fatty acids through an NADP-dependent oxidoreductase (i.e. ω-hydroxy fatty acid dehydrogenase) that is wound inducible. A constitutive ω-oxoacid dehydrogenase catalyzes the conversion of the ω-oxoacid to the dicarboxylic acid. These enzymes have been separated and the dehydrogenase enzyme purified and characterized. The catalytic properties of the ω-hydroxy fatty acid dehydrogenase appear to be distinct from alcohol dehydrogenase (Agrawal and Kolattukudy, 1977, 1978a,b; Kolattukudy, 2001). Kinetic studies involving chain-length substrate specificity suggest that the pocket of the active site of the ω-fatty acid dehydrogenase enzyme limits binding of longer chain ($>C_{18}$) substrate, which may explain the presence of dicarboxylic acids in the range of C_{18} and ω-hydroxy fatty acids of longer chain length (Kolattukudy, 2001). Tracking the flux of aliphatic monomers of the SPAD during wound healing further demonstrated that the newly formed acids had two possible metabolic fates: (1) desaturation and oxidation to form suberin C_{18} ω-hydroxy and dioic acids, and (2) elongation to form very long-chain fatty acids (C_{20}–C_{28}) associated with reduction to 1-alkanols, decarboxylation to n-alkanes, and minor amounts of hydroxylation (Yang and Bernards, 2006). In some plant tissues, fatty 9,10-epoxides may be hydroxylated during suberization, thereby creating a hypothetical role for these compounds in SPA biosynthesis (Bernards, 2002). Glycerol is brought into the SPA biosynthetic scheme and incorporated into the polymer at about the same rate as that of the major/diagnostic suberin monomers, α,ω-alkanedoic acids (dicarboxylic acids), thus suggesting cross-linking of aliphatic and aromatic domains (Moire et al., 1999). The cross-linking hypothesis is supported by ultrastructural data, which show that the alternating electron-transparent regions found within the electron-dense areas of the suberin aliphatic domain have dimensions similar to that created by a C-22 hydrocarbon located between two glycerol molecules (Schmutz et al., 1996; Moire et al., 1999). The lamellae dimensions match with the tentative model for the structure of potato suberin (Fig. 22.6). The source for glycerol in the suberin model is hypothesized to be dihydroxyacetone phosphate (Fig. 22.5B) (Bernards, 2002). However, the complete pathway for these processes is not fully known including how these specialized fatty acids and glycerol are assembled to form the SPAD. Nor is it clear how these monomers or oligomers are transported on to the cell wall after some apparent resident time elsewhere while construction of the SPPD is being completed. As with the SPPD, the construction of the SPAD becomes more complicated and perplexing when attempting to integrate the biosynthesis and coordinated macromolecular assembly of the SPA biopolymer as it accumulates in a non-segmented fashion over the SPPD on the cell wall.

22.3.8 Suberization and resistance to infection

Suberization and loosely defined wound periderm development have been suggested to play various roles in resistance to tuber infection (Nnodu et al., 1982; O'Brien and Leach, 1983; Nolte et al., 1987; Bostock and Stermer, 1989; Lyon, 1989; Vaughn and Lulai, 1991; Pérombelon, 2002). However, resistance to bacterial and fungal infections was often poorly related to the various definitions for suberin and wound periderm; consequently, the roles of suberin in resisting bacterial and fungal infections had not been clearly resolved. Lulai and Corsini (1998) determined that the hierarchy for SPP and SPA accumulations during wound healing is responsible for the differential development of resistance to bacterial and then fungal penetration. Total resistance to bacterial infection occurred after the SPP accumulation was complete along the contiguous outer tangential walls of the first layer of cells (2–3 days) in the closing layer. However, the SPP accumulation offered no protection against fungal infection even after SPP accumulation advanced to the adjoining radial and inner tangential cell walls. Fungi could breach the SPP barrier and thereby provide an entry for bacterial infections such as *Erwinia carotovora* ssp. *carotovora*. Lulai and Corsini (1998) indicated that this process is responsible for the confusing development of resistance to bacteria, followed by fungal infection and subsequent susceptibility to bacterial infection. Resistance to infection by the fungal agent that causes dry rot, *Fusarium sambucinum*, began to develop after accumulation of SPA was initiated. The final form of full resistance to fungal infection occurred upon completion of SPA accumulation on the first layer of cells. The specificity of the SPA domain as a barrier to fungi has been further demonstrated at the cellular level by blockage of advancement of fungal hyphae within pink-eye-afflicted tubers (Lulai et al., 2006).

The knowledge gaps and ambiguities associated with wound healing and wound-induced suberization are of great importance because of the critical roles that rapid suberization plays in minimizing food-quality deterioration, shrinkage, defects, and perhaps the most easily and widely recognized problem of infection by a wide range of bacterial and fungal pathogens.

22.4 RELATED DEFECTS

22.4.1 Wound-related tuber defects

Various forms of wound-related defects, i.e. those caused by different types of bruising and growth anomalies, are responsible for a range of tuber quality problems and losses. Reviews indicate that bruising and associated defects may range from 9 to 40% and that they are influenced by genetics, cultural conditions/practices, and handling procedures (Storey and Davies, 1992; Brook, 1996; McGarry et al., 1996). The types of defects and associated disorders that may fit this general category are variable. Many of these defects and disorders are poorly described because of the dearth of published information and the frequent use of untested hypotheses as fact. The discussion herein will primarily deal with the wound-related issues of these defects and will not delve deeply into other details.

22.4.2 Shatter bruising and tuber cracking

Shatter bruising and cracking results from mechanical impact stresses incurred during potato harvest and handling operations (Thornton, 2001). The internal shattering or cracking of tuber tissue to the surface is a serious problem because of the impaired suberization inherent with this type of wound. This type of bruise damage results in a combination of problems including disease, severe defects created by discoloration, and the development of scars from the wound. The impairment of suberization in these cracks is inherent and is a significant cause for the exacerbation of the defect. The impaired suberization of these bruise wounds exemplifies the differences in wound healing created by irregular versus smooth wound surfaces, and altered oxygen and carbon dioxide concentrations at the wound site (Section 22.3.4). Wound healing in the microenvironment of a shatter bruise or tuber crack can be especially hindered because of the irregular wound surface and the confinement of the narrow wound crevasse compared with that of a smooth cut. The amount of cell damage and respiration increases proportionally with the larger wound surface area created by the irregular fracture of the shatter bruise. The carbon dioxide concentrations would be expected to increase and the oxygen concentrations decrease in the confined area formed by the wound crevice of a shatter bruise. Ventilation into the confined area of a tuber crack is difficult. Roughened or irregular wound surface areas severely hamper development of contiguous SPP accumulation in the formation of a closing layer and consequently also impair formation of a wound periderm. The irregular surface area and likelihood of a hypoxic environment predispose shatter bruises and cracks to hindered suberization and increased susceptibility to infection (Vayda et al., 1992; Lulai and Corsini, 1998; Lulai, 2001a,b).

Higher tuber turgor appears to increase susceptibility to dynamic failure, i.e. shattering and cracking, whereas higher temperature decreases failure rates (Smittle et al., 1974; Bajema et al., 1998). Tuber size also influences failure properties; tubers weighing more than 340 g are significantly more likely to undergo dynamic failure in the form of shatter bruise/cracking whereas tubers of 170–340 g showed little difference in failure rates throughout their range (Baritelle and Hyde, 1999). Managing tuber turgor, especially on rain-fed land, and managing harvest temperature can create unworkable approaches to minimize shattering. Further improvement of mechanical harvesting procedures and biologically improving the rate of suberization and cell–cell adhesion are feasible approaches to develop technologies to reduce shatter bruise/cracking defects. For example, polygalacutronic acid of low ester content has been shown to play a role in cell–cell adhesion at 'edge-of-face regions' of tuber parenchyma cells (Parker et al., 2001).

22.4.3 Blackspot and pressure/crush bruising

Blackspot is an internal discoloration of the tuber cortical parenchyma found beneath an impact or pressure bruise. A significant amount of research has been conducted on impact-induced blackspot bruise development, while there has been less research on pressure bruising and pressure bruise-induced blackspot development. In general, tubers of higher specific gravity have a lower bruise threshold, a lower bruise resistance, or both (Baritelle and Hyde, 2003). Depending on the genotype, less severe impact bruising and impact bruising sustained by tubers at warmer temperatures or by tubers with lower turgor tend not to result in tuber shattering or cracking (Smittle et al., 1974; Brook, 1996; McGarry et al., 1996; Bajema et al., 1998; Baritelle and Hyde, 1999, 2001). Instead, these impact bruises result in cellular damage or physiological disruption and the development of blue-black or gray-black discoloration in tissues beneath the tuber surface. These dark bruise discolorations are referred to as blackspot. Although other secondary mechanisms have been implicated, further evidence indicates that the discolorations are primarily a result of oxidation of tyrosine by polyphenol oxidase and subsequent reactions to form melanin, which in combination with cysteine influences coloration of the pigment (Dean et al., 1992, 1993; Stevens and Davelaar, 1996, 1997; Edgell et al., 1998; Stevens et al., 1998; Partington et al., 1999; Laerke et al., 2002). Antioxidants normally found within the tuber, such as ascorbic acid, may be involved in limiting the enzymatic formation of melanin (Delgado et al., 2001b; Pawelzik et al., 2005). Intracellular compartmentation is an important factor in determining susceptibility to blackspot development (Laerke et al., 2002). McGarry et al. (1996) pointed out that these wound-related discolorations involve compromising the integrity of intracellular membranes and bringing the polyphenol oxidase enzyme, located in the plastid membrane, i.e. amyloplast, into contact with phenolic substrates that are putatively in the vacuole. There is evidence that certain field applications of soluble calcium during plant growth will enhance tuber calcium concentrations and reduce blackspot bruising presumably through improved cell membrane stability and stronger cell walls (Karlsson et al., 2006). Although blackspot development is indicative of cellular disruption within the tuber, indications of wounding on the surface of the tuber are often difficult to detect. Blackspot detection is accomplished by peeling to expose the darkened areas within the cortical parenchyma. Surface detection of wounds, including less obvious bruised areas that involve damaged periderm, which may develop into blackspot or a surface blemish, may be detected in freshly harvested tubers by testing with catechol or other peroxidizable substrate that produces a detectable color difference. Freshly bruised tubers are submerged in a solution of catechol and then visually inspected for oxidative color change of the catechol on the tuber surface (Gould, 1989). Methods have been developed to assess the biochemical potential of potato tubers to discolor and darken in vitro. However, it appears that the conditions for in vitro analysis, possibly including oxygen concentration, are not always fully consistent with the biochemical potential of tubers to discolor and develop blackspot (Delgado et al., 2001a; Laerke et al., 2002). Although the mechanisms associated with the development of blackspot discoloration have been partially characterized, little is known about other wound responses that may or may not be induced by bruising including suberization of cells in the proximity of this type of internal wound.

Pressure bruising or crushing occurs at the tuber–tuber contact points or tuber–storage structure contact points during storage and is caused by excessive pressures created by potato pile heights of 3.5–4.0 m or greater (Schaper and Yaeger, 1982). The constant pressure created by the weight of the pile on tubers located at lower depths in a potato storage bin may cause the tuber surfaces at the points of contact to become flattened or indented creating a surface blemish. Cellular damage in the cortical parenchyma under the pressure bruise area may result in the development of blackspot discolorations (Lulai et al., 1996). Although this periderm damage may not be visible to the eye, the pressure often creates a layer of crushed cells and debris, presumably from the phellogen and possibly phelloderm. This layer is found under the tough phellem cells that are also often damaged. High water vapor conductance resulting from wounding normally rapidly declines during the first 24–36 h after the tuber is cut (Lulai and Orr, 1995). However, water vapor conductance through the pressure bruised areas was approximately three times higher than that of control areas and remained high, without decline, for up to 4 days after removal from the pressures created by the storage environment. These results indicate that there was no induction of wax biosynthesis/accumulation as these cells were damaged by storage pressures. Nor was there induction of wax biosynthesis/accumulation at these bruise points after the tubers were removed from the pressures created by the storage environment and after atmospheric oxygen was readily available. Most, but not all, of the pressure bruise areas showed no induction of suberization. Wound healing responses, including suberization and reduced water vapor conductance, as an indication of wax deposition, were inhibited in tuber cells damaged by pressure bruising and in neighboring cells (Lulai et al., 1996). The continued pressure and tight tuber contact at the pressure bruise site would restrict the supply of oxygen from the ventilation stream to the damaged cells. However, oxygen would be available to the pressure bruise areas after removal from the bin. The changes in oxygen availability to cells compromised by pressure bruise could influence blackspot development during and after removal from storage. Proper humidification to reduce loss of turgor is important in minimizing cellular damage and defect development associated with pressure bruising.

22.4.4 Growth cracks

Tuber growth cracks are fully healed wounds that develop during tuber growth. These cracks occur after rain or irrigation prompts rapid water uptake resulting in increased tuber turgor pressure and growth (Storey and Davies, 1992; Hiller and Thornton, 1993; Thornton, 2001). Internal pressure exceeds the tensile strength of the surface tissues during tuber enlargement; the resulting tissue failure results in a physical cracking of the tuber surface. The cracks frequently occur on the bud end of the tuber, usually extend lengthwise in the tubers, and vary in length and depth. There is little published information describing the physiology of growth cracks and associated wound healing. The apparent lack of disease associated with growth cracks suggests that these wounds heal readily as the tubers grow in the soil. Growth cracks may also result from various infections and chemical treatments.

22.4.5 Skinning

Tuber skinning injury is a superficial wound generally inflicted by mechanical forces during harvest and handling, which results in the fracture of fragile radial phellogen cell walls of the native periderm and loss of the protective layer of phellem cells (Section 22.2). Radial phellogen cell walls connect and physically hold the tuber phellem (skin) in place.

These cell walls are thin and fragile while the phellogen of the native periderm is meristematically active during growth; tubers are susceptible to skinning/excoriation during this time. As the potato plant senesces and growth terminates, phellogen meristematic activity ceases, the phellogen cell walls strengthen/thicken, and the tuber becomes resistant to skinning, i.e. skin-set develops. Skinning results in discoloration of the wounded area, release of water sequestered within tuber cells, loss of turgor and increased susceptibility to associated pressure bruising and blackspot development, and creation of areas that are open to infection. Red-skinned genotypes are generally most susceptible to skinning injury and require more time for periderm maturation and proper skin-set development. The reason for red-skinned genotypes being predisposed to hindered skin-set development is not known. The red-skin pigment is not synthesized and replaced during wound healing, thus leaving a light off-color blemish amid the red skin. Depending on the maturity of the potato plant at the end of the growing season, no less than 10 days and generally 21 days of periderm maturation are routinely recommended for proper skin-set before harvest (Lulai and Orr, 1993; Struik and Wiersema, 1999; Olsen et al., 2003; Stark and Love, 2003). The wounding caused by skinning injury is frequently categorized as a bruise and can contribute to significant increases in carbon dioxide concentrations in storage bins directly after harvest (Mazza and Siemens, 1990; Schaper et al., 1993). The increased carbon dioxide concentration adversely affects suberization and can increase the reducing sugar concentrations and darkening of processed food products (Mazza and Siemens, 1990). Skinning injuring is among the most common yet troublesome and costly problems of the potato industry.

22.5 SUMMARY

A healthy intact periderm is essential for protecting the potato tuber from desiccation, infection, pests, and other intrusions that cause food quality and supply problems and financial losses. A comprehensive understanding of the physiological and biochemical processes associated with skin-set, wound healing, and wound-related responses to bruising is essential for the development of technologies to minimize these problems. Advances have been made; however, the understanding of these processes is far from complete.

Research has shown that skin-set development is not caused by skin thickening, suberization, or increased tensile strength of the skin. Instead, phellogen cell wall strengthening and cell wall thickening within the periderm have been shown to be responsible for the development of resistance to tuber skinning injury. Immunolocalization data indicate that accumulation and modification of specific cell wall polymers are among those processes involved in skin-set development. Further information is needed on the biological processes that are involved in tuber phellogen cell wall strengthening and skin-set development. The biological signals that initiate and regulate skin-set development have not been elucidated. This lack of biological information hinders development of technologies to predict, initiate, and/or enhance the development of resistance to skinning injury.

Wound-induced suberization is perhaps the most important facet of wound healing. Tubers cut for seed, wounded by various means including harvest/handling processes, and tubers challenged by various biotic and abiotic stresses utilize suberization as the most durable form of protection. A competent suberized barrier protects the tuber from infection, dehydration, and various intrusions. The SPPD has been shown to provide resistance to bacterial but not fungal infection. The SPAD is architecturally distinct and spatially separate from the phenolic domain and is required to provide the final barrier to fungal infection. Associated waxes are part of the protective mechanism that controls water vapor loss. The induction and regulation of suberin biosynthesis, the biosynthetic pathways as well as the assembly and molecular architecture of the suberin biopolymer are of fundamental importance in enhancing wound healing and solving wound-related problems. A better understanding of these processes and the structure of suberin is on the horizon. However, it is obvious that further research is required in all these areas to provide the basis for the development of new technologies that will enhance suberization and mitigate associated food quality and loss problems. Lastly, wound-related responses are crucial elements of the various bruise-related problems and losses described in section 22.4.

ACKNOWLEDGEMENTS

The author attempted to include references pertinent to the topics and apologies to those authors who published papers that were cited indirectly through review articles or for other reasons were not directly included in this chapter.

REFERENCES

Abeles F.B., P.W. Morgan and M.E. Saltveit (eds), 1992, In: *Ethylene in Plant Biology*, p. 26. Academic Press, N.Y.
Agrawal V.P. and P.E. Kolattukudy, 1977, *Plant Physiol.* 59, 667.
Agrawal V.P. and P.E. Kolattukudy, 1978a, *Arch. Biochem. Biophys.* 191, 452.
Agrawal V.P. and P.E. Kolattukudy, 1978b, *Arch. Biochem. Biophys.* 191, 466.
Arrieta-Baez D. and R.E. Stark, 2006, *Phytochemistry*, 67, 743.
Artschwager E.F., 1918, *J. Agric. Res.* 14, 221.
Artschwager E., 1924, *J. Agric. Res.* 27, 809.
Artschwager E., 1927, *J. Agric. Res.* 27, 995.
Bajema R.W., G.M. Hyde and A.L. Baritelle, 1998, *Trans. ASAE* 41, 741.
Banks N.H. and S.J. Kays, 1988, *J. Am. Soc. Hortic. Sci.* 113, 577.
Baritelle A.L. and G.M. Hyde, 1999, *Trans. ASAE* 42, 159.
Baritelle A.L. and G.M. Hyde, 2001, *Postharvest Biol. Technol.* 21, 331.
Baritelle A.L. and G.M. Hyde, 2003, *Postharvest Biol. Technol.* 29, 279.
Beckman C.H., 2000, *Physiol. Mol. Plant Pathol.* 57, 101.
Bernards M.A., 2002, *Can. J. Bot.* 80, 227.
Bernards M.A. and N.G. Lewis, 1992, *Phytochemistry* 31, 3409.
Bernards M.A. and N.G. Lewis, 1998, *Phytochemistry* 47, 915.
Bernards M.A. and F.A. Razem, 2001, *Phytochemistry* 57, 1115.
Bernards M.A., W.D. Fleming, D.B. Llewellyn, R. Priefer, X. Yang, A. Sabatino and G.L. Pluourde, 1999, *Plant Physiol.* 121, 135.
Bernards M.A., M.L. Lopez, J. Zajicek and N.G. Lewis, 1995, *J. Biol. Chem.* 270, 7382.
Bernards M.A., D.K. Summerhurst and F.A. Razem, 2004, *Phytochem. Rev.* 3, 113.
Bernards M.A., L.M. Susag, D.L. Bedgar, A.M. Anterola and N.G. Lewis, 2000, *J. Plant Physiol.* 157, 601.
Bleecker R.C. and H. Kende, 2000, *Annu. Rev. Cell Dev. Biol.* 16, 1.

Bostock R.M. and B.A. Stermer, 1989, *Annu. Rev. Phytopathol.* 27, 343.
Bowen S.A., A.Y. Muir and C.T. Dewar, 1996, *Potato Res.* 39, 313.
Brook R.C. (ed.), 1996, In: *Potato Bruising, How and Why, Emphasizing Blackspot Bruise*, p. 1. Running Water Publishing, Haslett, MI.
Bush M.S. and M.C. McCann, 1999, *Physiol. Plant.* 107, p. 201.
Catesson A.M., 1994, *Int. J. Plant Sci.* 155, 251.
Catesson A.M., R. Funada, D. Robert-Baby, M. Quinet-Szel, J. Cu-Ba and R. Goldberg, 1994, *IAWA J.* 15, 91.
Chaffey N.J., P.W. Barlow and J.R. Barnett, 1998, *New Phytol.* 139, 623.
Choi D., R.M. Bostock, S. Avdiushko and D.F. Hildebrand, 1994, *Proc. Natl. Acad. Sci. U.S.A.* 91, 2329.
Ciardi J. and H. Klee, 2001. *Ann. Bot.* 88, 813.
Cottle W. and P.E. Kolattukudy, 1982a, *Plant Physiol.* 69, 393.
Cottle W. and P.E. Kolattukudy, 1982b, *Plant Physiol.* 70, 775.
Davin L.B. and N.G. Lewis, 2000, *Plant Physiol.* 123, 453.
de Bruxelles G.L. and M.R. Roberts, 2001, *Crit. Rev. Plant Sci.* 20, 487.
de Haan P.H., 1987, In: Rastovski, A., A. van Es. et al. (eds), *Storage of Potatoes: Post-harvest Behavior, Store Design, Storage Practice, Handling*, p. 371. Pudoc, Wageningen, The Netherlands.
Dean B.B., 1989, *Plant Physiol.* 89, 1021.
Dean B.B., N. Jackowiack and S. Munck, 1992, *Potato Res.* 35, 49.
Dean B.B., N. Jackowiak, M. Nagle, J. Pavek and D. Corsini, 1993, *Am. Potato J.* 70, 201.
Dean B.B. and P.E. Kolattukudy, 1976, *Plant Physiol.* 58, 411.
Dean B.B. and P.E. Kolattukudy, 1977, *Plant Physiol.* 59, 48.
Dean B.B., P.E. Kolattukudy and R.W. Davis, 1977, *Plant Physiol.* 59, 1008.
Delgado E., J. Poberezny, E. Pawelzik and I. Rogozinska, 2001a, *Am. J. Potato Res.* 78, 389.
Delgado E., M.I. Sulaiman and E. Pawelzik, 2001b, *Potato Res.* 44, 207.
Edgell T., E.R. Brierley and A.H. Cobb, 1998, *Ann. Appl. Biol.* 132, 143.
Espelie K.E., V. Franceschi and P.E. Kolattukudy, 1986, *Plant Physiol.* 81, 487.
Graca J. and H. Pereira, 2000, *J. Agric. Food Chem.* 48, 5476.
Goldberg R., P. Devillers, R. Prat, C. Morvan, V. Michon and C. Hervé du Penhoat, 1989, In: N.G. Lewis and M.G. Paice (eds), *Plant Cell Wall Polymers – Biogenesis and Biodegradation*, p. 312. American Chemical Society, Washington D.C.
Gould W.A. 1989, In: W.A. Gould, J.R. Sowokinos, E. Banttari, P.H. Orr and D.A. Preston (eds), *Chipping Potato Handbook*, p. 15. The Snack Food Association, Alexandria VA.
Gronwald J.W., 1991, *Weed Sci.* 39, 435.
Hammerschmidt R., 1984, *Physiol. Plant Pathol.* 24, 33.
Hampson C.P., T.J. Dent and M.W. Ginger, 1980, Proceedings Conference Potato Tuber Damage, *Ann. Appl. Biol.* 96, 366.
Halderson J.L. and R.C. Henning, 1993, *Am. Potato J.* 70, 131.
Hiller L.K., D.C. Koller and R.E. Thornton, 1985, In: P.H. Li (ed.), *Potato Physiology*, p. 389. Academic Press Inc., New York.
Hiller L.K. and R.E. Thornton, 1993, In: R.C. Rowe (ed.), *Potato Health Management*, p. 87. APS Press, The American Phytopathological Society, St. Paul, MN, U.S.A.
Karlsson B.H., J.P. Palta and P.M. Crump, 2006, *HortScience*, 41, 1213.
Koda Y. and Y. Kikuta, 1994, *Plant Cell Physiol.* 35, 751.
Kolattukudy P.E., 1978, In: G. Kahl (ed.), *Biochemistry of Wounded Plant Tissues*, pp. 43. Walter de Gruyter, Berlin, New York.
Kolattukudy P.E., 1980, *Science* 208, 990.
Kolattukudy P.E., 2001, Biopolyesters. In: T. Scheper (ed.), *Advances in Biochemical Engineering Biotechnology*, Vol. 71, p. 1. Springer-Verlag, Berlin.
Kolattukudy P.E. and B.B. Dean, 1974, *Plant Physiol.* 54, 116.
Kumar G.N.M. and N.R. Knowles, 2003, *Physiol. Plant.* 117, 108.
Lachaud S., A.M. Catesson and J.L. Bonnemain, 1999, *C.R. Acad. Sci. III-Vie* 322, 633.
Laerke P.E., J. Christiansen and B. Veierskov, 2002, *Postharvest Biol. Technol.* 26, 99.
Lapierre C., B. Pollet and J. Negrel, 1996, *Phytochemistry* 42, 949.
Laties G.G., 1978, In: G. Kahl (ed.), *Biochemistry of Wounded Plant Tissues*, p. 421. W. de Gruyter, Berlin.

Leon J., E. Rojo and J.J. Sanchez-Serrano, 2001, *J. Exp. Bot.*, 52, 1.

Lewis N.G., L.B. Davin and S. Sarkanen, 1999, Carbohydrates and their derivatives including tannins, cellulose and related lignins. In: Sir D.H.R. Barton and K. Nakanishi (eds-in-chief), *Comprehensive Natural Products Chemistry*, Vol. 3, p. 617. Elsevier Science Ltd, Oxford, U.K.

Li S.-X. and A.M. Showalter, 1996, *Physiol. Plant.* 97, 718.

Lulai E.C., 2001a. In: W.R. Stevenson, R. Loria, G.D. Franc and D.P. Weingartner (eds), *Compendium of Potato Diseases*, p. 3. APS Press, St. Paul, MN, U.S.A.

Lulai E.C., 2001b, In: W.R. Stevenson, R. Loria, G.D. Franc and D.P. Weingartner (eds), *Compendium of Potato Diseases*, p. 6. APS Press, St. Paul, MN, U.S.A.

Lulai E.C., 2002, *Am. J. Potato Res.* 79, 241.

Lulai E.C., 2005, *Am. J. Potato Res.* 82, 433.

Lulai E.C. and D.L. Corsini, 1998, *Physiol. Mol. Plant Pathol.* 53, 209.

Lulai E.C and T.P. Freeman, 2001, *Ann. Bot.* 88, 555.

Lulai E.C. and W.C. Morgan, 1992, *Biotech. Histochem.* 67, 185.

Lulai E.C. and P.H. Orr, 1993, *Am. Potato J.* 70, 599.

Lulai E.C. and P.H. Orr, 1994, *Am. Potato J.* 71, 489.

Lulai E.C. and P.H. Orr, 1995, *Am. Potato J.* 72, 225.

Lulai E.C. and J.C. Suttle, 2004, *Postharvest Biol. Technol.* 34, 105.

Lulai E.C. and J.C. Suttle, 2005, 16th Triennial Conference of the EAPR, Abstracts of papers and posters, II poster presentations, p. 775.

Lulai E.C., M.T. Glynn and P.H. Orr, 1996, *Am. Potato J.* 73, 197.

Lulai E.C., J.J. Weiland, J.C. Suttle, R.P. Sabba and A.J. Bussan, 2006, *Am. J. Potato Res.*, 83, 409.

Lyon G.D., 1989, *Plant Pathol.* 38, 313.

Lyshede O.B., 1977, In: *Yearbook of the Royal Veterinary and Agricultural University*, p. 68. Copenhagen, Denmark.

Matsuda F., K. Morino, M. Miyashita and H. Miyagawa, 2003, *Plant Cell Physiol.* 44, 510.

Mazza G. and A.J. Siemens, 1990, *Am. Potato J.* 76, 121.

McGarry A., C.C. Hole, R.L.K. Drew and N. Parsons, 1996, *Postharvest Biol. Technol.* 8, 239.

Micheli F., B. Sundberg, R. Goldberg and L. Richard, 2000, *Plant Physiol.* 124, 191.

Moire L., A. Schmutz, A. Buchala, B. Yan, R.E. Stark and U. Ryser, 1999, *Plant Physiol.* 119, 1137.

Morris S.C., M.R. Forbes-Smith and F.M. Scriven, 1989, *Physiol. Mol. Plant Pathol.* 35, 177.

Muir A.Y. and S.A. Bowen, 1994, *Int. Agrophysics* 8, 531.

Murphy H.J., 1968, *Am. Potato J.* 45, 472.

Negrel J., S. Lotfy and F. Javelle, 1995, *J. Plant Physiol.* 146, 318.

Negrel J., B. Pollet and C. Lapierre, 1996, *Phytochemistry* 43, 1195.

Niklas K.J., 1992a, In: K.J. Niklas (ed.), *Plant Biomechanics. An Engineering Approach to Plant Form and Function*, p. 48. The University of Chicago Press, Chicago.

Niklas K.J., 1992b, In: K.J. Niklas (ed.), *Plant Biomechanics. An Engineering Approach to Plant Form and Function*, p. 234. The University of Chicago Press, Chicago.

Nnodu E.C., M.D. Harrison and R.V. Park, 1982, *Am. Potato J.* 59, 297.

Nolte P., G.A. Secor and N.C. Gudmestad, 1987, *Am. Potato J.* 64, 1.

Nolte P., G.A. Secor, N.C. Gudmestad and P.J. Henningson, 1993, *Am. Potato J.* 70, 649.

O'Brien V.J. and S.S. Leach, 1983, *Am. Potato J.* 60. 227.

Obro J., J. Harholt, H.V. Scheller and C. Orfila, 2004, *Phytochemistry* 65, 1429.

Olsen N., G.E. Kleinkopf and J.C. Stark, 2003, In: J.C. Stark and S.L. Love (eds), *Potato Production Systems*, p. 309. The University of Idaho Extension/The University of Idaho Agricultural Communications, Moscow, ID, U.S.A.

Oosterhaven K., K.J. Hartmans, J.J.C. Scheffer and L.H.W. van der Plas, 1995, *Physiol. Plant.* 93, 225.

Ostby P.B., A.Y. Muir and F.N. Zender, 1990, Abstracts 11th Triennial Conference, European Association for Potato Research, Edinburgh, Scotland, p. 176.

Oomen R.J.F.J. 2003, *In planta* modification of the potato tuber cell wall. Thesis, Wageningen University, The Netherlands.

Oomen R., C.H.L. Doeswijk-Voragen, M.S. Bush, J.-P. Vincken, B. Borkhardt, L.A.M. van den, Broek, J. Cosar, P. Ulvskov, A.G.J. Voragen, M.C. McCann and R.G.F. Visser, 2002, *Plant J.* 30, 403.

Parker C.C., M.L Parker, A.C. Smith and K.W. Waldron, 2001, *J. Agric. Food Chem.* 49, 4364.
Partington J.C., C. Smith and G.P. Bolwell, 1999, *Planta* 207, 449.
Pavlista A.D., 2002, *Am. J. Potato Res.* 79, 301.
Pawelzik E., N.U. Haase, A. Schulze and R. Peters, 2005, Abstracts 16th Triennial Conference, European Association for Potato Research, Bilbao, Spain, p. 375.
Pérombelon M.C.M., 2002, *Plant Pathol.* 51, 1.
Peterson R.L. and W.G. Barker, 1979, *Bot. Gaz.* 140, 398.
Peterson R.L., W.G. Barker and M.J. Howarth, 1985, In: P.H. Li (ed.), *Potato Physiology*, p. 123. Academic Press Inc., New York.
Razem F.A. and M.A. Bernards, 2003, *J. Exp. Bot.* 54, 935.
Reeve R.M., E. Hautala and M.L. Weaver, 1969, *Am. Potato J.* 46, 361.
Reyes L.F. and L. Cisneros-Zevallos, 2003, *J. Agric. Food Chem.* 51, 5296.
Rosenstock G. and G. Kahl, 1978, In: G. Kahl (ed.), *Biochemistry of Wounded Plant Tissues*, p. 623. Walter de Gruyter & Co., Berlin, New York.
Sabba R.P. and E.C. Lulai, 2002, *Ann. Bot.* 90, 1.
Sabba R.P. and E.C. Lulai, 2004, *Am. J. Potato Res.* 81, 119.
Sabba R.P. and E.C. Lulai, 2005, *J. Am. Soc. Hort. Sci.* 130, 936.
Schaper L.A., M.T. Glynn, and J.L. Varns, 1993, *Appl. Eng. Agric.* 10, 89.
Schaper L.A. and E.C. Yaeger, 1982, *Trans. ASAE* 25, 719.
Schilmiller A.L. and G.A. Howe, 2005, *Curr. Opin. Plant Biol.* 8, 369.
Schmutz A., A.J. Buchala and U. Ryser, 1996, *Plant Physiol.* 110, 403.
Schreiber L., M. Skrabs, K. Hartmann, D. Becker, C. Corsagne and R. Lessire, 2000, *Biochem. Soc. Trans.* 28, 647.
Schreiber L., R. Franke and K. Hartmann, 2005a, *Planta* 220, 520.
Schreiber L., R. Franke and R. Lessire, 2005b, *Phytochemistry* 66, 131.
Scott R.I., J.M. Chard, M.J. Hocar, J.H. Lennard and D.C. Graham, 1996. *Potato Res.* 39, 333.
Smittle D.A., R.E. Thornton, C.L. Peterson and B.B. Dean, 1974, *Am. Potato J.* 51, 152.
Soliday C.L., B.B. Dean and P.E. Kolattukudy, 1978, *Plant Physiol.* 61, 170.
Soliday C.L., P.E. Kolattukudy and R.W. Davis, 1979, *Planta* 146, 607.
Stark R.E. and J.R. Garbow, 1992, *Macromolecules* 25, 149.
Stark J.C. and S.L Love, 2003, In: J.C. Stark and S.L. Love (eds), *Potato Production Systems*, p. 329. The University of Idaho Extension/The University of Idaho Agricultural Communications, Moscow, ID, U.S.A.
Stevens L.H. and E. Davelaar, 1996, *Phytochemistry* 42, 941.
Stevens L.H. and E. Davelaar, 1997, *J Agric. Food Chem.* 45, 4221.
Stevens L.H., E. Davelaar, R.M. Kolb, E.J.M. Pennings and N.P.M. Smit, 1998, *Phytochemistry*, 49, 703.
Storey R.M.J. and H.V. Davies, 1992, In: P.M Harris (ed.), *The Potato Crop, The scientific basis for improvement*, p. 507. Chapman and Hall, London.
Struik P.C. and S.G. Wiersema (eds), 1999, In: *Seed Potato Technology*, p. 219. Wageningen Pers, Wageningen, The Netherlands.
Thomson N., R.F. Evert and A. Kelman, 1995, *Can. J. Bot.* 73, 1436.
Thornton R.E., 2001, In: W.R. Stevenson, R. Loria, G.D. Franc and D.P. Weingartner (eds), *Compendium of Potato Diseases*, p. 84. APS Press, St. Paul, MN, U.S.A.
Tyner D.N., M.J. Hocart, J.H. Lennard and D.C. Graham, 1997, *Potato Res.* 40, 181.
Vaughn S.F. and E.C. Lulai, 1991, *Physiol. Mol. Plant Pathol.* 38, 455.
Vayda M.E., L.S. Antonov, Z. Yang, W.O. Butler and G.H. Lacy, 1992, *Am. Potato J.* 69, 239.
Vayda M.E. and J.K. Morelli, 1994, In: W.R. Belknap, M.E. Vayda and W.D. Park (eds), *The Molecular and Cellular Biology of the Potato*, p. 188. CAB International, Wallingford, U.K.
Wasternack C., I. Stenzel, B. Hause, G. Hause, C. Kutter, H. Maucher, J. Neumerkel, I. Feussner and O. Miersch, 2006, *J. Plant Physiol.*, 163, 297.
Wigginton M.J., 1973, *Potato Res.* 16, 85.
Wigginton M.J., 1974, *Potato Res.* 17, 200.
Yan B. and R.E. Stark, 2000, *J. Agric. Food Chem.* 48, 3298.
Yang W.L. and M.A. Bernards, 2006, *Plant Signal. Behav.* 1, el.
Yephremov A. and L. Schreiber, 2005, *Plant Biosyst.* 139, 74.

Chapter 23

Internal Physiological Disorders and Nutritional and Compositional Factors that Affect Market Quality

Joseph R. Sowokinos

Department of Horticultural Science, University of Minnesota, 305 Alderman Hall, 1970 Folwell Avenue, St. Paul, Minnesota 55108, USA

23.1 INTRODUCTION

'Potato quality' is a diverse term covering features such as texture, colour of raw and processed product, external and internal tuber morphology, general appearance and nutritive value. Potato tubers are made up of hydrated, highly respiring, mainly parenchymatous cells that are metabolically responsive to their environment (Peterson et al., 1985). Internal physiological tuber disorders that reduce the market quality of potatoes originate from abiotic stresses that are non-infectious in nature. Although the qualities of tubers are preset by their genetic limits, their final compositional and nutritional value is influenced by variables such as fertility, soil type and agronomic practices, as well as by environmental conditions experienced during growth and development in the field. Just as biotic stress (e.g. bacteria, fungi, viruses, phytoplasms and nematodes) can severely affect crop yield and tuber quality, internal physiological disorders can impact the salability of fresh and processed potato products. Economic loss from biotic stress can range from minor to complete loss dependent on the severity of the disease. Losses resulting from internal physiological disorders are more indeterminate in nature as they are influenced not only by severity and percentage of tubers affected but also by supply and demand. Economic loss due to internal physiological disorders may be greatest during years when potatoes are in abundant supply. Although tubers with external defects can be graded-out going into storage, internal damage is not visible until the tuber is peeled or sliced. Generally, internal physiological disorders are relatively slow in expression. Considerable time may elapse between the initiation or induction stage and the manifestation of symptoms, making it difficult to identify the causal factors and the time of initiation (Hiller et al., 1985).

Although there is extreme genotypic variation in a cultivar's ability to resist internal physiological disorders (Collier et al., 1980; Rex and Mazza, 1989; Davies and Monk-Talbot, 1990; Wannamaker and Collins, 1992; Lambert et al., 1996; McGarry et al., 1996), such variation has not been previously exploited fully to dissect the genetics and biochemistry of the key processes affected (Davies, 1998). Since 1985, some advances have been made into the biological features of brown centre (BC), hollow heart (HH), internal brown/rust spot, internal heat necrosis (IHN), stem-end discolouration (SED),

Potato Biology and Biotechnology: Advances and Perspectives
D. Vreugdenhil (Editor)

translucency and mottling disorders. Black heart is a condition where the supply of oxygen to the tuber is deficient or impaired leading to dark, necrotic cavities that develop within its centre. As this disorder is easily diagnosed and chiefly leads to cellular disruption and death (Olsen et al., 2003), it will not be discussed further in this chapter.

Primarily, this chapter will attempt to identify key biological and molecular changes that accompany the establishment of certain internal physiological disorders in potatoes used for crisp (chip) and French fry production. An effort will be made to look for possible biological traits peculiar to resistant cultivars to aid in the selection of plants with superior genotypes and/or with which existing genotypes can be enhanced for improved tolerance. Secondarily, gross tuber compositional and nutritional elements that have a major effect on post-harvest market quality of potatoes will be discussed along with a description of a commercially used method to manage the 'chemical maturity' (sugar levels) of potatoes from harvest through storage. Owing to the vast amount of literature on some of the above topics, appropriate reviews are cited for the sake of brevity.

23.2 GENERAL NATURE, INCIDENCE AND SEVERITY OF INTERNAL PHYSIOLOGICAL DISORDERS

Many of the internal physiological disorders in potatoes are related in that they produce coloured (rust or brown) or translucent (TransL) (devoid of starch) areas in distinct parts of the tubers, which in time become necrotic resulting in cellular death. It is obvious from the literature that some, if not all, of these internal physiological disorders are induced, in part, or exacerbated by high temperature. A fluctuation in moisture availability is also a recurring factor that is suggested to initiate or increase the magnitude of many internal physiological disorders. Inducing conditions are complex and result from uneven tuber growth affected by irregular temperature/moisture relationships (Sowokinos et al., 1985; Lulai et al., 1986; Olsen et al., 2003), substandard fertilization regimes and/or soil types (Nelson, 1970; Simmons and Kelling, 1987; Lambert et al., 1996), excess nitrogen (Kallio, 1960; Rex and Mazza, 1989) and an inadequate supply of calcium (Ca^{2+}) (Collier et al., 1980; Kratzke and Palta, 1985, 1986; Monk and Davies, 1989; Davies and Monk-Talbot, 1990; Olsen et al., 1996; Kleinhenz et al., 1999; Ozgen et al., 2006). The type of soil, rainfall, planting date and seedling rate have more influence on the magnitude of the disorder than on its incidence (Iritani et al., 1984; Lambert et al., 1996). The impact of these factors on the magnitude of the disorders also varies widely across cultivars. In this chapter, key internal physiological disorders of potatoes are discussed in an attempt to clarify major causal factors as well as to identify metabolic changes that may be distinctive to their development.

23.2.1 Calcium nutrition and tuber quality

Several recent studies have demonstrated that the severity and incidence of some internal physiological disorders can be mitigated by in-season calcium application (Tawfik and Palta, 1992; Olsen et al., 1996; Kleinhenz et al., 1999; Karlsson et al., 2001; Ozgen et al., 2006). Lack of calcium has been documented to cause severe deficiency symptoms in the tissues and organs of numerous plants (Bangerth, 1979). Calcium is involved not only in maintaining the structural integrity of the cell but also in regulating metabolic processes (Burstrom, 1968). Over 30 physiological plant disorders have been associated with calcium deficiency as originally reviewed by Shear (1975) and Maynard (1979). The important role of calcium as a second messenger in the signal transduction process in plants has been reviewed by Poovaiah and Reddy (1993). Calcium contributes to the maintenance of cell membrane permeability and cell wall structure by forming stable but reversible linkages between the polar head groups in the membranes and pectic acid fraction in the cell wall (Clarkson and Hanson, 1980; Hanson, 1984; Palta, 1996).

Recently, it has been suggested that calcium moves predominately (if not exclusively) with water through the zylem (Palta, 1996). The discovery of functional roots present on the tuber at the tuber–stolon junction and on the stolon associated with the tuber has helped to modify the view of calcium nutrition of potato tubers (Fig. 23.1) (Kratzke and Palta, 1985; Struckmeyer and Palta, 1986). By feeding a water-soluble dye, these authors demonstrated that these tiny roots were able to supply water to the tuber. In a follow-up study using a divided pot system, where supplemental calcium could be applied either in the tuber area or to the main root system, Kratzke and Palta (1986) found that adding calcium to the main root system did not increase calcium concentration of the tuber tissue. However, applying calcium to the tuber and stolon area resulted in a three-fold increase in calcium concentration in the tuber peel and medullary tissue. More recently, by studying simultaneous transport of a water-soluble dye and $^{45}Ca^{2+}$, Busse and Palta (in press) have demonstrated that tiny roots on the stolons can supply water and calcium to the tuber. In support of these studies, Simmons and Kelling (1987), Simmons et al. (1988) and Kleinhenz et al. (1999) found that maximum increase in tuber calcium occurred when

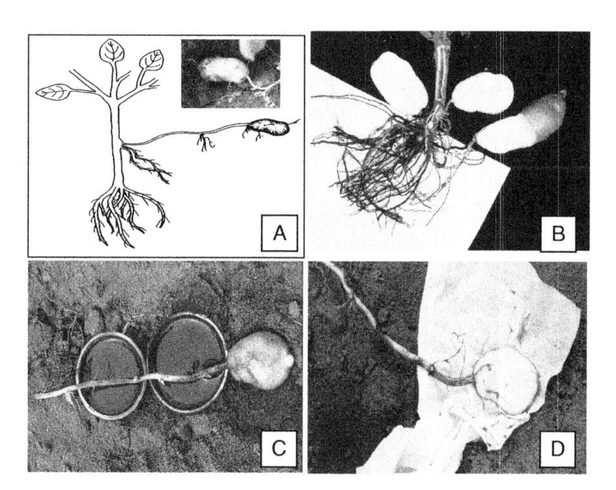

Fig. 23.1. Transport of a water-soluble dye (Safranin O) from various roots present on the potato plant. (A) Description of various types of roots present on a potato plant. Inset showing tuber roots. (B) Dye is transported from the main root to the shoot and not to the tuber. (C) Uptake of dye from the stolon roots closely associated with a tuber. (D) Tuber from C following 30 minutes of dye exposure. The dye is transported from the stolon roots into the tuber (Source: Kratzke and Palta, 1985).

it was applied in the hill where the tubers develop. Similarly, using an artificial system, Krauss and Marschner (1971) reported higher calcium content in potato tubers when it was applied to the tuber and stolon area as opposed to the basal root system. Using this concept, a number of studies have also demonstrated that in-season application of calcium around the tuber and stolon area not only increased tuber calcium but also reduced the incidence of internal physiological disorders (Tawfik and Palta, 1992; Kleinhenz et al., 1999; Karlsson et al., 2001).

Internal physiological disorders due to calcium deficiency are usually associated with tissues that experience a low rate of transpiration, such as fruits, underground storage organs and enclosed leafy structures (Bangerth, 1979). Potato tubers, being storage organs and surrounded by moist soil, also have very low transpiration. Thus, a very small amount of water moves to the tubers resulting in relatively low calcium concentration in the tuber tissue as compared with leaves and above-ground stems (Palta, 1996). Calcium deficiency in tuber tissue is even greater for potatoes grown in sandy soil because of the low level of exchangeable calcium in these soils. Internal physiological disorders are induced, in part, and/or are exacerbated by temperature stress. There is considerable evidence that calcium can mitigate plant responses to temperature stress (Palta, 1996). This is especially true for heat stress in potatoes (Tawfik et al., 1996; Kleinhenz and Palta, 2002). Furthermore, Vega et al. (1996) found an improvement in freezing stress tolerance of wild potato germplasm by supplemental calcium fertilization.

Selection of potato genotypes with superior stolon and tuber root systems may also help to increase post-harvest potato quality. Bamberg and coworkers (1993) screened tuber-bearing *Solanum* germplasm for their ability to accumulate tuber calcium when a control level is available and for ability to respond when higher levels are available. *Solanum* species *Solanum gourlayi* and *Solanum microdontum* had the ability to have higher tuber calcium content than *Solanum tuberosum* in one or both of these measurements. This material may provide a valuable resource for breeding cultivars that are more efficient calcium accumulators. Variation in resultant progeny may provide the germplasm necessary to clarify the biological, genetic and molecular traits directly responsible for tuber resistance to internal physiological potato disorders. In addition to bringing desired genes from the wild germplasm, it appears that variation in cultivated germplasm could also be exploited to improve calcium concentration of desired genotypes (Vega et al., 1996).

23.2.2 Brown centre and internal brown spot

Necrotic lesions of BC are localized in the central area of the potato tuber, whereas internal brown spot (IBS) is primarily found in the phloem-rich perimedullary region between the pith and vascular ring (Hiller et al., 1985; Davies, 1998). The BC disorder can also be a precursor to HH development as cells collapse and, in certain types, are layered with suberin (Dean et al., 1977). Non-wound-induced suberization of parenchyma cells can also develop in response to pathogen infection, e.g. that caused by *Verticillium dahliae* (Lulai, 2005). van Denburgh et al. (1980) demonstrated that there was an optimal soil temperature (i.e. 10–15°C) for the expression of BC in cultivar Russet Burbank potatoes. Tubers grown at temperatures below or above this range had decreased incidence of the

BC disorder. IBS has been referred to as internal brown fleck, internal browning, internal rust spot, internal necrosis and chocolate spot (Iritani et al., 1984; Hiller et al., 1985; Olsen et al., 1996).

Calcium availability during the early tuber growth of cultivar Russet Burbank potatoes was shown to be an influential factor in reducing the development of IBS (Olsen et al., 1996). In-season application of calcium during tuber development and bulking has been shown to reduce BC (Kleinhenz et al., 1999) and IBS (Ozgen et al., 2006). In a recent study, an inverse relationship between calcium and IBS was found (Fig. 23.2, Ozgen et al., 2006), suggesting a linkage between tuber calcium concentration and IBS. If calcium was withheld during tuber initiation, symptoms of BC were observed (Davies, 1998), whereas symptoms of IBS were evident if calcium was reduced from 20 mM to 1 mM after tubers reached 5 cm in diameter (Davies and Monk-Talbot, 1990). Calcium deficiency in potato tubers can lead to rapid metabolic changes leading to a loss of membrane integrity and a series of uncontrolled oxidative reactions (Monk et al., 1989). These investigators used electron spin resonance (ESR) to explore the formation of paramagnetic species during the development of IBS in potato tubers. Disruption of cellular organization causes leakiness of deleterious reactive species from mitochondria that may expose membrane lipids to peroxidation. The resulting changes in membrane permeability and subcellular compartmentalization could expose membrane lipids to endogenous lipase and lipoxygenase (LOX) action, resulting in the generation of free radicals (Davies, 1998). Using optical and electron microscopy, Baruzzini et al. (1989) observed groups of IBS cells to be spongy, suberized and absent of starch. Separation and degeneration of the plasmalemma and the death of the cells were observed in tissues in the final stages of damage. Osmiophilic granules attached to the tonoplast were suggested to be catabolic products of a disturbed metabolism (Table 23.1).

Differences in antioxidant status of tubers could not unequivocally explain their differences in susceptibility to IBS (Davies, 1998). Superoxide dismutases (SOD: EC 1.15.1.1) are a group of metallo-enzymes that catalyse the disproportionation of superoxide free

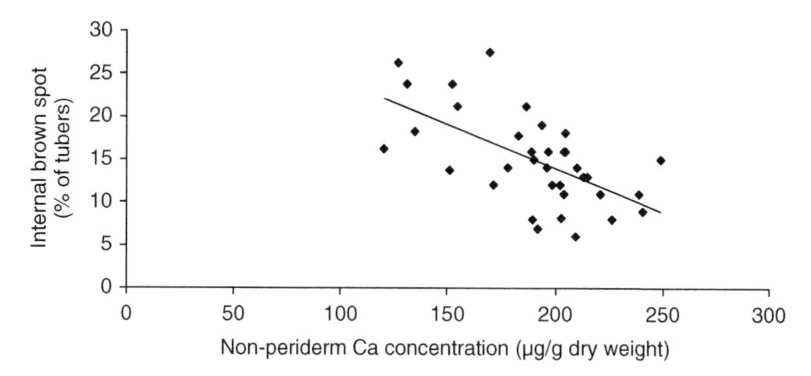

Fig. 23.2. Scatter plot of the relationship between non-periderm tissue calcium (Ca) and internal brown spot (IBS). Each point represents the mean Ca level in 15 tubers within a given replication and the corresponding IBS in that replication in 1997. Model for the graph is $y = -0.1022x + 34.368$, $R^2 = 0.3677$. Significant at $P < 0.05$ (Source: Ozgen et al., 2006).

Table 23.1 Changes in composition, enzyme activities and subcellular organization in potato tubers experiencing abiotic stress that are characteristic of a particular internal psychological disorder.

Parameters	Internal physiological disorder (fold change from healthy tissue)			
	TransL[a,b]	**SED[c]**	**Mottling[d]**	**IBS[e,f]**
Composition				
Starch	–2.3	–1.7	No change	absent
Sucrose	+7.0	–1.8		
Glucose	+4.2	+14.1	+1.6	
Sucrose/glucose ratio	2.1	0.2		
Phosphate	+1.5	+1.1	No change	
Total Solids	–1.5	–1.4	No change	
Glutathione				+2.0[g]
Enzyme activity				
ADP-Glc pyrophosphorylase	ND	–2.1		
Starch phosphorylase	+5.4	+1.7		
UDP-Glc pyrophosphorylase	+1.6	–2.8		
Sucrose synthase	No change	–1.7		
Sucrose 6-phosphate synthase	+3.8	No change		
Acid invertase (basal)	ND	+7.0		
Acid invertase (total)	+1.5	+4.2		
α-Amylase	+2.0	ND		
Acid phosphatase	+2.0	ND		
Glutathione reductase				+1.5[g]
Superoxide dismutase				Higher[h]
Dehydroascorbate reductase				+2.0[g]
Lipolytic acyl hydrolase	–3.7	ND		
Lipoxygenase	–1.4	ND		
Membrane and microstructural changes	Yes	Yes	No change	Yes
Starch granule	Degrading[i]			
Mitochondrial	Swollen			
Cell walls			Suberized	
Plasmalemma				Disrupted
Tonoplast				Osmiophilic granules[j]

*IBS, internal brown spot; ND, not determined; SED, Stem-end discoloration; TransL, translucency.
[a] Sowokinos, et al., 1985.
[b] Lulai, et al., 1986.
[c] Sowokinos, et al., 2000.
[d] Jankowski, et al., 1997.
[e] Davies, H.V., 1998.
[f] Baruzzini, et al., 1989.
[g] In *IBS* resistant cultivars.
[h] Most *IBS* resistant cultivars possessed the maximum number of isoforms.
[i] Starch granules few in number; those remaining contained a large electron-dense stroma area surrounding the senescing starch granule located inside a physically intact amyloplast membrane.
[j] Attached to the tonoplast.

radicals to oxygen and hydrogen peroxide. A negative correlation ($R > -0.73$) between the percentage of tubers affected by necrotic lesions and the maximum catalytic activity of SOD was observed (Monk-Talbot et al., 1991). Isozyme analysis revealed extensive polymorphism for SOD among cultivars. However, there was no indication that restricted supplies of calcium induced or repressed the synthesis of specific isoforms. They concluded that although SOD may contribute to resistance for IBS, it is not the primary mechanism involved.

Resistance to IBS may be related to a plant's ability to distribute and maintain calcium in soluble form for physiological processes in fast tuber growing areas that have a minimum of vascularized connections (Peterson et al., 1985). Certain environmental conditions may cause a disruption in the calcium supply needed for a cell's metabolic requirements resulting in tissue necrosis. Collier and coworkers (1980) showed in tubers of 10 potato cultivars that the incidence of IBS increased when the concentration of calcium chloride supplied to the plants fell from 9 mM to 1 mM. They suggested that in the field, factors that reduced the availability and uptake of calcium would enhance the development of IBS. In any case, a cultivar's susceptibility to IBS did not depend simply on tuber calcium concentration as other cellular factors are apparently implicated.

Climatic/cultural factors that regulate calcium homeostasis within the growing tuber cell could play an important role in BC and IBS (Davies, 1998). It has been reported that over 90% of the calcium in potato tuber cells is in a physiologically active form and very little is in the form of insoluble components, such as calcium oxalate (Davies and Millard, 1985). Autoradiographs of potato tubers (cultivar Maris Piper) grown in the presence of $^{45}Ca^{2+}$ revealed high activity in the periderm, vascular ring (predominantly xylem), phloem bundles and pith. They reported that 40% of tuber calcium entered directly through the periderm and that the localization of calcium in the region of phloem bundles in unsprouted tubers raised the possibility of import through the phloem. Their data also indicated that $^{45}Ca^{2+}$ is associated with both xylem and phloem when tubers and only a small section of a stolon are directed into moist compost containing this isotope. Calcium uptake through the roots present on the stolon could account for this result. Other investigators have suggested that calcium moves predominately (if not exclusively) with water through the xylem (Wiersum, 1966; Tibbits and Palzkill, 1979; Clarkson, 1984; Palta, 1996). The discovery of functional roots present on the tuber at the tuber–stolon junction and on the stolon associated with the tuber has helped to modify the view of calcium nutrition of potato tubers (for more details, see Section 2.1). Therefore, very low calcium content in the tuber may be due to a relatively small amount of water that is transported to the tuber as compared with leaves (Palta, 1996). This would be particularly critical during drought conditions.

23.2.3 Hollow heart

Hollow heart is characterized by the formation of an irregular cavity in the flesh (pith) of the tuber, typically surrounded with brown, discoloured tissue that causes serious losses in crop quality and economic return to the grower and processor following harvest. Numerous factors have been implicated in the initiation and manifestation of HH as reviewed by Hiller et al. (1985). Investigations into the causal factors of HH, however, are often conflicting or inconclusive. HH has frequently been reported to be most prevalent in large tubers (Kallio, 1960; Nelson and Thoreson, 1986; Jansky and Thompson, 1990). It was suggested that HH was due to tissue tension associated with tuber enlargement (Nelson and Thoreson, 1986). Crumbly and coworkers (1973) observed that tuber growth rate was related to HH occurring within a variety but could not account for differences among varieties. They also reported that small tubers can experience the disorder under certain growing conditions. These conditions were (1) moisture stress resulting in conversion of starch to sugar followed by a rapid influx of water leading to a rapid enlargement and separation of pith cells and (2) injury to some cells in the tuber due to reabsorption of minerals and carbohydrates or depletion of food reserves followed by tuber enlargement. Early initiation of HH appears shortly after tuber set and can follow from the same factors that cause BC (Olsen et al., 2003). HH in larger tubers may primarily be associated with excessive rapid tuber enlargement, without the involvement of reabsorption or depletion of food reserves. Conditions favouring rapid vine growth (excess nitrogen) during rapid tuber bulking lead to a physiological imbalance in the plant increasing the incidence of HH (Kallio, 1960; Rex and Mazza, 1989). It was suggested that important nutrients may be temporally insufficient resulting from cellular instability of pith cells that eventually collapse and/or separate as growth conditions and moisture availability improve. Excess moisture during field growth in August and September, which reduces specific gravity, tends to increase the incidence of HH, and in some years IBS (Silva et al., 1991). Growers in the Pacific Northwest of the USA limit HH development in russeted potatoes by subjecting plants to slight water stress during the last part of the growing season. Researchers in Idaho concluded that late season irrigation practices have a significant impact on HH in varieties susceptible to late-season development. An experiment was conducted comparing irrigation regimes in which full evapo-transpiration (ET) was either met or irrigation was reduced to 75% of ET replacement on July 20 and further reduced to 50% replacement on August 10. For the cultivars Russet Norkotah and Gemstar Russet, limiting late-season water supply in a step-down fashion reduced HH from 16 to 5% and from 18 to 2%, respectively, while having minimal effect on yield (S.L. Love, personal communication).

Although the incidence of HH is influenced strongly by the environment, there is also an effect by genotype. High estimates of general combining ability in both male and female parents for expression of HH have been reported (Veilleux and Lauer, 1981). Jansky and Thompson (1990) conducted an experiment to study HH in segregating populations from parents in a breeding programme. The highest levels of HH were found in families with susceptible parents, but susceptible clones were obtained in families with two resistant parents. Regardless of the type of cross, selection of clones with large tubers increased the probability of selecting for susceptibility to HH. Ehlenfeldt (1992) evaluated HH frequency and leaf transpiration rates in nine potato cultivars from an irrigated and dry-land location. He reported that 88% of the total variation in HH was explained as a quadratic function of transpiration. The positive correlation suggested that varieties that have high levels of transpiration may be more likely to undergo severe cycles of dehydration and hydration, which can result in rapid and severe expansion stress in tubers. It was suggested that measurement of leaf transpiration rates might be used as a tool to select for advanced potato selections that have the genetic capability of resisting HH development.

Early studies by Levitt (1942) showed reduced concentration of many nutrients, including calcium in tubers with HH. Later, electron microprobe and neutron activation analysis were used to evaluate the distribution of elements within potato tubers having the physiological disorder HH (Arteca et al., 1980). Both methods of analysis correlated well for all elements tested. In HH tubers, there were significantly higher calcium levels in the stem end than in the bud end; however, a calcium gradient in the control was not evident. In support of the role of calcium in HH, Kleinhenz et al. (1999) found a significant reduction in this incidence with in-season application of soluble calcium fertiliser during bulking. This reduction in HH was associated with an increase in non-periderm tissue calcium concentration. A Cl^- gradient from stem to bud end was present in both the HH and control tubers. The low levels of calcium in HH tubers may be involved in the weakening or destruction of cell walls and membranes, subsequently leading to the HH disorder. These observations are consistent with the concept that tubers are generally low in calcium and that the incidence of internal defects can be reduced by increasing its concentration in the tuber (Palta, 1996; Kleinhenz et al., 1999).

In general, conflicting reports are widespread in the literature regarding the effects that nutrients have on the occurrence of HH. These differences are likely because of variation in cultivar selection, soil types, cultural practices and environmental conditions (Nelson, 1970; Rex and Mazza, 1989; Lambert et al., 1996). Excess nitrogen has been reported to increase HH, whereas higher applications of potassium reduce its occurrence (Kallio, 1960; Nelson, 1970). Varying phosphorus applications showed no effect on HH occurrence (Kallio, 1960). As there are many factors that act synergistically to cause variation in the incidence of HH outside the scope of this chapter, the reader is referred to a comprehensive review by Rex and Mazza (1989), who discuss the formation, methods of control and means of detecting HH in potatoes.

23.2.4 Internal heat necrosis

There are several synonyms for IHN, including internal necrosis, IBS and others as listed by Sterrett et al. (2003). Here, however, IHN will be described as a distinct internal physiological disorder as it can vary in form, severity and time of injury from IBS. Symptoms of IHN in the tuber include round to irregular, light tan to reddish-brown spots or blotches that appear first towards the apical end of the tuber (Olsen et al., 2003; Sterrett et al., 2003). This necrotic tissue occurs primarily in the parenchymal tissue associated closely with the vascular ring. Hiller et al. (1985) indicated that location of the necrotic areas across the tuber is used by some authors to distinguish IHN from IBS. IBS lesions are generally more localized throughout the parenchymal tissue internal to the vascular ring.

It is generally accepted that IHN tissue occurs at/or near the time of harvest in the largest tubers (Sterrett and Henninger, 1997). Both the colour intensity and the tuber area affected increase over time (Sterrett et al., 1991a). On the contrary, IBS has been reported to occur at various times throughout the period of potato growth (Hiller et al., 1985). The influence of environment and cultural management practices on the development of

IHN has been discussed in detail by Sterrett and Henninger (1997) for the IHN-sensitive cultivar Atlantic. The genetic and molecular factors involved in the development of IHN, however, have been difficult to determine. Although cultivar selection is important, the soil type and soil composition influence the effectiveness of nitrogen and calcium treatments. Minimum and maximum temperatures exceeding the optimum for potato growth, rainfall events and tuber size distribution have been used in modelling to predict occurrence and development/severity of IHN in cultivar Atlantic in the mid-Atlantic states (Sterrett et al., 1991b). The variables appearing in the regression models for incidence, rating and distribution suggested that the development of IHN is not necessarily a simple response to high temperature (Sterrett et al., 1991a). Sterrett et al. (1991b) developed a predictive model for onset and development of IHN. This model was designed to allow farmers to predict when their potato crop is estimated to go off-grade because of IHN and thus suggest an earlier harvest to minimize the economic impact.

Studies of cultural and management practices to alleviate IHN have included delayed planting, use of straw mulch to reduce soil temperature, decreased in-row spacing, application of additional calcium and application of various substances to prolong vine growth (Wannamaker and Collins, 1992; Sieczka and Thornton, 1993). Because IHN is associated with heat stress and calcium, nutrition has been shown to mitigate heat stress in potatoes (Tawfik et al., 1996; Kleinhenz and Palta, 2002), it may be interesting to investigate whether the incidence of IHN can be reduced by targeted calcium nutrition. Regardless, cultural and management practices have been limited in their success in reducing the occurrence of IHN. This is, in large part, due to the general lack of understanding of the biology responsible for IHN resistance. Wannamaker and Collins (1992) suggested that the most economical practice for a potato grower would be to plant a cultivar resistant to IHN. They cite a number of researchers who have demonstrated genetic variability to IHN in *S. tuberosum*. In a 3-year study of 19 genotypes at two locations, Henninger et al. (1994) found that broad-sense heritability (additive + dominance components) was 86% for incidence and 88% for severity of IHN with most of the remaining variation in this population due to genotype × environment interaction. Henninger et al. (2000) reported that the genetic potential exists for developing new cultivars that are high in specific gravity and yield and free of IHN. In retrospect, it appears that the best avenue for attaining resistant genotypes to IHN in the future may be through the development of more efficient breeding strategies (Wannamaker and Collins, 1992).

23.2.5 Stem-end discolouration

A serious problem for French fry manufactures is an internal physiological disorder known as sugar ends, also referred to as dark ends, jelly ends, TransL ends and/or glassy ends (Olsen et al., 2003). The losses due to this disorder have never been quantified but can be serious for growers because of crop rejection and trim loss and product downgrades by processors. The most common occurring type of SED is located at the stem or basal end of 'russeted' tubers and is characterized by reduced starch content along with a high concentration of reducing sugar (RS) (Sowokinos et al., 2000) (Table 23.1). A more rare form of the SED condition leads to the development of dumbbell-shaped tubers induced by a mid-season moisture stress (Iritani and Weller, 1980).

As with most internal potato physiological disorders, heat and drought stress at specific growth stages are implicated in causing SED. Although temperature exacerbates the physiological and biochemical changes characteristic of the SED condition, research has suggested that a transient moisture deficiency during tuberization may be the primary initiator (Shock et al., 1993; Eldredge et al., 1996). Timing of moisture stress appeared to be critical to the incidence of SED development. Plants stressed before tuber initiation had fewer tubers with dark stem-end fry colours and demonstrated an increased percentage of US No. 1 potatoes (e.g. smooth, uniform-shaped tubers and independent of size) (Shock et al., 1992). A single transitory soil water potential stress (i.e. –69 kPa to –80 kPa) episode during early tuber bulking has been associated with an increased incidence of SED at harvest (Eldredge et al., 1996). The incidence and severity of dark ends increased with the duration and intensity of transitory soil water potential stress. The concentration of RSs in the stressed tubers continued to increase after storage (9°C) for 6 weeks as did the darkness of the fries. Owing to the latency of the tuber in developing a noticeable decrease in solids and an increased concentration of RSs following a transient moisture stress period, the successful use of these parameters to screen commercial potato fields for SED potential is unavailable at the present time.

Research has shown that heat and moisture-deficit stress lead to a plethora of changes affecting whole plant metabolism and phenotypic characteristics including (1) changes in photosynthetic and respiration rates (Owings et al., 1978; Dwelle, 1985; Lulai et al., 1986; Wolf et al., 1990, 1991), (2) aberrations in source–sink relationships that affect tuberization, tuber number, yield and tuber quality (Rufty and Huber, 1983; Sowokinos et al., 1985; Levy, 1986) and (3) development of irregular-shaped tubers (see Shock et al., 1993, and references therein). A study was conducted to determine how moisture deficit, during early tuberization, affected tuber composition and activity of enzymes involved with the synthesis and degradation of starch and sucrose (Sowokinos et al., 2000). It was apparent that compositional and enzymatic changes associated with SED development had, in time, shifted the tuber from a starch-storing tissue to one actively involved in starch mobilization. A decrease in the activity of adenosine diphosphate ADP-Glc pyrophosphorylase (−2.1-fold) compared with an increase in starch phosphorylase (+1.7-fold) seemed to favour an environment for starch mobilization rather than starch synthesis. This shift in carbohydrate metabolism was supported by the observation that total solids and starch decreased in SED tissue (Table 23.1). The vacuolar enzyme, acid invertase, demonstrated the largest change in activity (i.e. increase) in SED tissue. Although its activity increased gradually, prior to harvest, it continued to increase (seven-fold) in storage (Sowokinos et al., 2000). Kinetic analysis suggested that following SED induction, apparent invertase activity increased and/or the level or effectiveness of the endogenous invertase inhibitor decreased. There was no significant difference in the specific activity of sucrose 6-P synthase (SPS) whereas that of UDP-Glc pyrophosphorylase (UGPase) and sucrose synthase (Susy) both decreased from that observed in normal, healthy tissue. Other changes in tissue composition and enzyme activities in the SED tissue compared with data available with other internal physiological disorders are summarized in Table 23.1.

A genetic component to the expression of sugar-ends under stressful environments has been demonstrated. Following severe moisture stress, cultivar Lemhi Russet tubers showed little to no sugar gradient from basal to stem end of tubers, whereas Russet

Burbank developed a high incidence of SED (Love et al., 1991). The numbered selection A082260-8 was relatively SED free compared with Russet Burbank following a severe moisture-deficit condition (Shock et al., 1993). Although several wild species have demonstrated the ability to tolerate heat and drought stress (Reynolds and Ewing, 1989), the biological parameters responsible for this increased resistance were not investigated. SED resistance is likely inherited in a polygenic, additive manner, and through the selection of appropriate parents, breeding progress for SED resistance should be fairly rapid (Thompson-Johns, 1998).

23.2.6 Translucency

In principally non-irrigated growing areas, a 'translucent disorder' has periodically been observed that seems to be primarily induced by extreme growing temperatures. Although SED translucency occurs in oblong, russeted potatoes, this translucency defect occurs in more spherically shaped potatoes of a non-russeted skin type. These TransL areas occur at random, internal to the vascular ring. When a thin tuber slice is held to a light source, a definite TransL area devoid of starch can be seen. Upon frying, this area becomes dark brown to black in colour indicating a high concentration of RSs (Sowokinos et al., 1985). In this respect, this TransL disorder is similar to SED tissue described above, but differs from BC, IBS or IHN, where discolouration is evident prior to frying.

Starch phosphorylase in TransL tissue experienced a marked 5.4-fold increase in specific activity, and this was reflected by a significant drop in total solids and starch. Although some compositional and enzymatic changes in TransL are common to SED tissues, obvious differences are apparent (Table 23.1). In the TransL tissue, total acid invertase activity did not experience the significant change as observed in SED tissue, and UGPase showed a positive 1.6-fold increase in activity. This may explain, in part, the major difference observed in the sucrose/glucose ratio between TransL and SED tissue of 2.1 and 0.2, respectively (Sowokinos et al., 1985, 2000) (Table 23.1). The respective increase in both UGPase (+1.6-fold) and SPS (+3.8-fold) (coupled with only a slight increase in invertase activity) may account for the large positive change in sucrose content (+seven-fold) observed in the TransL tissue (Table 23.1).

Lipid content of potato is only 0.1% on a fresh weight basis and is primarily membrane in nature (Galliard, 1973). Potato tubers contain sufficient lipolytic acyl hydrolase (LAH) and LOX to rapidly hydrolyse all membrane lipids (phospholipids and galactolipid) and to rapidly peroxidize the endogenous liberated linoleic and linolenic acids (Galliard, 1970). Normally, LAH and LOX are necessarily sequestered, probably in the tuber vacuole (Galliard, 1978) or LAH may exist in an inactive form in vivo (Walcotot et al., 1982). Changes in lipid-metabolizing enzymes along with changes in mitochondria and starch granule morphology were examined in relation to the loss of tuber quality associated with the TransL disorder (Lulai et al., 1986). It was shown that a reduction in LAH (−3.7-fold) and LOX (−1.4-fold) activities occurred in TransL tissue suggesting a possible alteration in lipid metabolism compared with that in healthy tubers. The effects of these reduced activities are uncertain, but they may have influenced composition, turnover and/or permeability of cellular membranes. Electron micrographs of the TransL tissue revealed a discernible decrease in the number of starch granules (amyloplasts) compared

Fig. 23.3. Electron micrograph of translucent potato tuber tissue showing an atypical amyloplast with an enlarged stroma surrounding the remainder of a degraded starch granule (Lulai et al., 1986).

with healthy tissue. The few remaining amyloplasts possessed a large electron-dense stroma area that surrounded the remainder of the starch granule (Lulai et al., 1986) (Fig. 23.3). Mitochondria from TransL tissue were aggregated in large numbers, were irregular is size and shape and had fewer cristae than mitochondria found in healthy tuber tissue (Fig. 23.4). The defective mitochondria resembled those present in aged tissue slices of beetroot and rutabaga as described by van Steveninck and Jackman (1967). It appeared that TransL tissue was experiencing an exaggerated or accelerated form of senescence/ageing.

23.2.7 Mottling

Since 1990, a less severe colour defect has been recognized in the French fried potato industry that has been termed 'mottling'. This disorder is expressed as a non-uniform surface browning (i.e. streaking) of French fries. Mottling appears following storage (i.e. 2–4 months). The incidence and severity of mottling is sporadic from year to year but leads to losses of several hundred thousand dollars annually for major processing plants. Increased temperature with time intensifies the disorder, and the streaking is not evident until the product is finished-fried at the foodservice establishment. The only study addressing the nature of this internal physiological disorder was conducted by Jankowski et al. (1997). Mottling appeared to lead to the heterogeneous distribution of RSs along the potato tissue strips, giving rise to enhanced Maillard browning in localized areas during finished-frying (Fig. 23.5). Fresh and processed tissue areas that are prone to mottling have significantly ($P < 0.01$ to $P < 0.05$) greater levels of glucose and fructose than do areas lighter in colour. Scanning electron microscopy (SEM) revealed no apparent

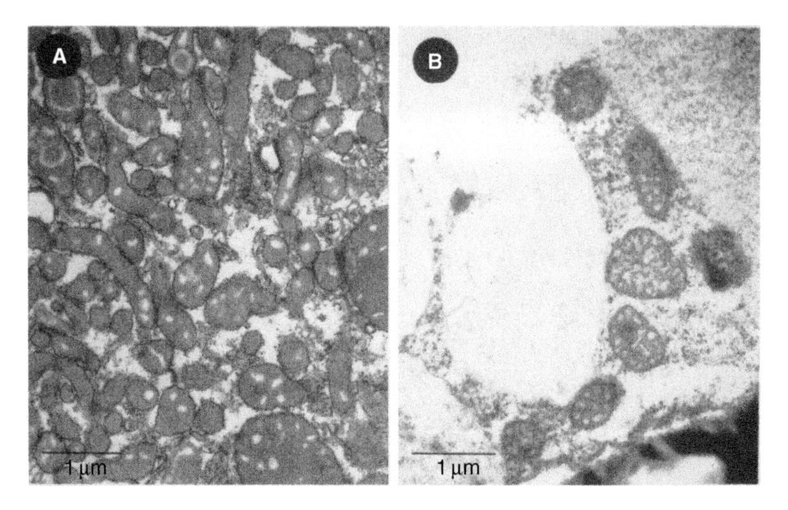

Fig. 23.4. Electron micrograph showing mitochondria from translucent tissue of defective tubers (A) and from normal/healthy tuber tissue (B) (Lulai et al., 1986).

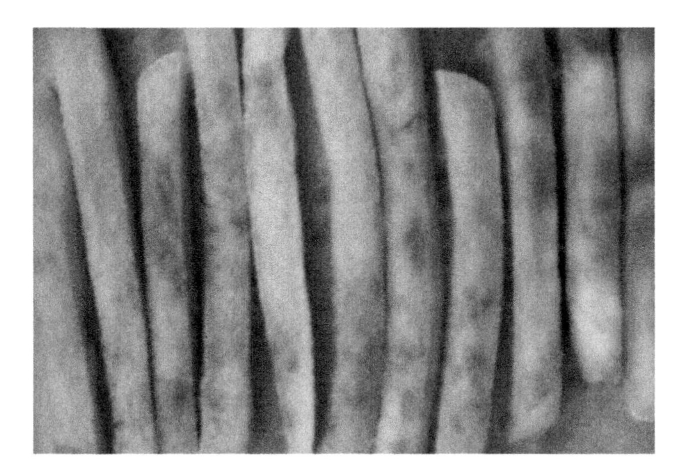

Fig. 23.5. The nature of mottling in commercial French fries (Jankowski et al., 1997).

differences in microstructure of potato tuber tissue sections prone to mottling as compared with that described above for TransL (Sowokinos et al., 1985; Lulai et al., 1986) and SED tissues (Sowokinos et al., 2000) (Table 23.1). This disorder was similar to TransL, in that they both appeared to involve localized differences in RS concentration that increased in storage. In addition to a lack of microstructure changes, mottling did not lead to any significant decrease and increase in total solids and inorganic phosphate, respectively (Table 23.1). At this point in time, the cause of mottling cannot be speculated on. It appears to be a unique disorder when both tissue appearance and constituent changes are compared. Jankowski et al. (1997) suggested that staining of raw potato strips with a solution of glucose oxidase/peroxidase enzymes might be used as a predictor of the incidence of mottling.

23.3 SUMMARY OF INTERNAL PHYSIOLOGICAL DISORDERS

Collectively, internal physiological disorders may be initiated by abiotic stresses that affect singly or in combination: (1) hormonal levels, (2) membrane composition, structure and function, (3) gene induction or repression, (4) endogenous enzyme activity and (5) the compartmentalization and/or concentration of key metabolites, ions/minerals, enzymes or other effectors (i.e. antioxidants) (Lulai et al., 1986; Sowokinos, 1990; Davies, 1998; Sowokinos et al., 2000). Although cultural and environmental factors giving rise to such disorders have been studied extensively, little information is available on the underpinning physiological, molecular and biochemical mechanisms involved (Davies, 1998). Of those internal physiological disorders where a few biological changes have been examined (i.e. TransL, SED, Mottling and IBS), distinct differences are noted in their expression (Table 23.1). Less biological information is known about BC, HH and IHN related to their 'causal factors' or in relation to their 'putative' metabolic response profiles. Considering internal physiological disorders in their entirety, it is expected that several 'biological alterations' that lead to necrosis and eventual death of the tissue are 'shared' among these disorders. Future research is needed to better understand internal physiological disorders on a molecular basis. Until then, it is apparent that developing resistant lines of potatoes may be the best initial step to minimize the impact that internal physiological disorders have on post-harvest potato quality. Such resistant germplasm could be an important source for the identification of the key genes involved. Meanwhile, the potential for developing most internal physiological disorders may be decreased by (1) managing the choice of cultivar, (2) establishing uniform plant spacing and planting depth to encourage uniform emergence, (3) managing the rate and timing of fertiliser application, (4) controlling soil moisture and temperature and (5) conducting other cultural practices in an effort to promote steady, uniform growth rates. For details of the symptoms, cultural variation, distribution and control of internal physiological disorders in potatoes beyond the scope of this chapter, see reviews presented by Hiller et al. (1985), Rex and Mazza (1989), McGarry et al. (1996) and Olsen et al. (2003).

23.4 COMPOSITIONAL AND NUTRITIONAL CHANGES AFFECTING END-USE QUALITY

The potato is one of most nutritious crops that man currently grows. In addition to supplying abundant starch reserves as an energy source, the potato is an important contributor of fibre, nitrogen and high-quality protein, vitamins and minerals, which support human growth and health (Storey and Davies, 1992). A medium potato (150 g/5.3 oz) with the skin contains a mere 100 calories and provides 45% of the daily value (DV) for vitamin C, has more potassium than bananas (720 mg or 21% of DV) and contains trace amounts of calcium, thiamine, riboflavin, niacin, pyridoxine, folate, magnesium, phosphorus and iron

(0.12–24 mg/150 g potato) (Beals, 2004). Lesser amounts of copper, cadmium, lead and selenium (0.02–0.07 mg/150 g potato) are also present (Dugo et al., 2004). Moreover, the fibre content of a potato with skin (3 g) equals or exceeds that of many whole grain breads, pastas and cereals (Beals, 2004).

The potato is an important component of healthy diet plans created by many heart and dietetic associations. The cooking method used, however, can significantly impact the nutrient availability of the potato. Significant losses of water-soluble vitamins (particularly Vitamin C), amino acids, potassium and calcium can occur. A substantial loss of vitamin C occurs when potatoes are fried in oil (55–79%), which is second only to losses experienced when potatoes are boiled in water (77–88%) (Han et al., 2004). Potatoes cooked in a microwave retain more of the vitamin C content compared with baking or boiling (Beals, 2004). In general, baking a potato with the skin is the best overall cooking method to retain the nutrition of the potato (Beals, 2004).

The remainder of this chapter will be concerned with factors that influence changes in post-harvest carbohydrate balance (i.e. starch–sugar) as they affect the final colour quality of crisps and French fries. A commercial method to manage the sugar levels in potatoes before harvest through delivery to the processor is also described.

23.4.1 Carbohydrates – starch

Starch comprises approximately 80% of the total solids found in potato tubers. Specific gravity is an indirect measure of the potato solids content (Stark and Love, 2003). High specific gravity (HSG) potatoes are important to the processing industry as they yield more product per unit of raw material and the product obtained is more rigid, crisp and less oily (Lulai and Orr, 1979; Storey and Davies 1992). It has been reported that HSG potatoes lose less amino acids during processing than low SG potatoes (Jaswal, 1973). One disadvantage of HSG potatoes is that they are more susceptible to bruising during harvest handling and transport (Gray and Hughes, 1978; Killick and Macarthur 1980). Although starch is the major component of a potato and the source of free RSs that interfere with the final colour of fried potato products, variation in its post-harvest concentration is not an efficient predictor of the final colour quality (O'Donoghue et al., 1996).

23.4.2 Carbohydrates – sugars

The 12-carbon, non-RS 'sucrose' is the major form of carbon translocated to the tuber from the leaves and therefore becomes the predominant sugar found in potatoes during growth and development (Burton, 1965; Sowokinos, 1978, 1990; Mares et al., 1985; ap Rees and Morrell, 1990). The presence of a specific inhibitor of the enzyme 'acid invertase' prevents sucrose from being hydrolysed into its two 6-carbon RS components, glucose and fructose, during tuber growth (Pressey, 1969). As a result, the high energy of the glucosidic bond of sucrose (6.6 kcal/mole) is conserved as 'chemical energy' to drive all biosynthetic reactions that occur in the developing potato rather than being dissipated as useless heat energy (Mares et al., 1985). As the foliage dies (either induced or naturally), or as tubers are harvested, acid invertase commences to convert any excess sucrose to the RS. Frying of potatoes with excess RS produces a brown to black colour pigmented product due to the Maillard reaction with free amino acids in the potato cell (Shallenberger et al., 1959). These dark-coloured products are unacceptable to the consumer because of their appearance and bitter taste (Roe et al., 1990). Substantial economic loss occurs for both potato growers and processors when RS accumulates in a potato resulting in the manufacture of unacceptably dark-coloured products. Although several factors contribute to a slight variation in crisp or French fry colour (Habib and Brown, 1956, 1957; Ashoor and Zent, 1984; Leszkowiat et al., 1990; Khanbari and Thompson, 1993; Rodriguez-Saona and Wrolstad, 1997; Rodriguez-Saona et al., 1997), variation in the content of RS is the most critical determinant of the colour quality of fried potato products (Fuller and Hughes, 1984; Roe et al., 1990; Mann et al., 1991; Roe and Faulks 1991; Pritchard and Adam, 1994; Putz and Lindhauer, 1994). Blenkinsop and coworkers (2002) studied changes in sucrose, RS, nitrogen, protein, vitamin C and dry matter content in relation to processing colour from low-temperature storage. Using Pearson correlation analysis of the data, they found that crisp colour was most closely correlated with RS concentration. Although fructose and glucose are equally formed as the result of sucrose hydrolysis, glucose is the main sugar measured as an indicator of crisp colour (Brown et al., 1990). Light crisp colour has been found to be directly related to tuber glucose content regardless of the detection method, cultivar, growing site or storage temperature (Coleman et al., 1993).

23.4.3 Factors affecting RS concentration in stored potatoes

It is commonly known by crisp processors that the more chemically and physically mature potatoes are when harvested, the easier they are to handle, store and condition to a light-coloured product. Immature tubers not only lose more weight and shrivel in storage, but their chemical composition is different (Yamaguchi et al., 1966). The concentration of RS is genetically determined (Lauer and Shaw, 1970; Coffin et al., 1987; Ehlenfeldt et al., 1990; Loiselle et al., 1990; Jakuczun et al., 1995), and selection for low RS germplasm is possible through breeding evaluation. The final content of sugars in a potato, however, is influenced by planting date, growing location, biotic and abiotic stresses experienced during development, fertility, vine-kill and harvesting conditions, handling and storage practices, e.g. see reviews by Smith (1967, 1977), Burton (1989) and Sowokinos (2001b).

It is important for potatoes to be 'physically mature' before harvest (i.e. adequate yield, high solids and firm skin-set), but attainment of 'chemical maturity' is a necessary feature of a successful processing potato. Chemical maturity is reached when a potato crop reaches its minimum free-sugar (sucrose) content (Burton, 1965). Following vine death or harvest, excess sucrose is efficiently converted to RS through the activity of the enzyme, acid invertase (Pressey, 1969). The more chemically immature a potato is at harvest (i.e. higher the sucrose concentration), the greater is the potential for RS accumulation in storage for crisps (Sowokinos, 1978) and French fry potatoes (Pritchard and Adam, 1992, 1994). Santerre et al. (1986) indicated that changes in sucrose levels of more recently developed cultivars, as tubers matured, were helpful in evaluating their chemical maturity compared with those already established. A reciprocal relationship has been shown between a cultivar's harvest sucrose content and its ability to produce acceptably coloured crisps from 8.9°C storage (Sowokinos, 1978). This relationship was determined, however, with cultivars previously screened for a similar range of invertase activity, e.g. low to moderate. Variability in sucrose content of 'immature tubers' explained greater than 70% of the variation in crisp colour observed among cultivars following 9°C for 4–6 months (Nelson and Sowokinos, 1983). Sucrose values, obtained following harvest, have not been a reliable indicator of fry colour with potatoes grown in the UK (Briddon and Storey, 1996) or Ireland (Burke et al., 2005). Variation in invertase activity, however, is known to be an over-riding factor in determining the hexose: sucrose ratio in stored potatoes (Zrenner et al., 1996; McKenzie et al., 2005). Sucrose concentration was shown to be a better predictor of crisping quality following moderate to long-term storage when measured during early tuber development, when the differences in sucrose concentration among cultivars was the greatest (Nelson and Sowokinos, 1983).

It is well known that the RS content of the tuber is affected by storage temperature and by time in storage (Isherwood, 1973, 1976; Dwelle and Stallknecht, 1978). Although there are several advantages for storing potatoes at colder temperatures (e.g. 3–5°C), a process known as cold-induced sweetening (CIS) occurs (ap Rees et al., 1981; Storey and Davies, 1992; Sowokinos, 2001b), which renders processed potato products unacceptably dark. The normal compromise is to store processing potatoes for the short term at 10°C, or warmer, where CIS is avoided. Holding potatoes at warmer temperature, however, decreases the time to rapid sprout growth and irreversible senescent sweetening (Burton and Wilson, 1978). Although biochemical mechanisms have been suggested to explain why different potato genotypes accumulate RS at differential rates in storage, the molecular events leading to the initiation and subsequent regulation of CIS process in potato tubers needs further investigation (Sowokinos, 2001b). A potato's ability to sweeten in storage varies markedly among genotypes. The advanced selection ND3828-15 differed 100-fold in its sweetening potential following 7 months storage at 5°C compared with cultivar Picasso [i.e. 0.4 μmol/g fresh weight (FW) and 47.1 μmol/g FW, respectively] (Sowokinos, 2001b). It is therefore critical to evaluate a new potato cultivar for CIS potential before being selected for storage and crisp processing.

23.4.4 Chemical maturity monitoring

A commercial system has been developed, which has permitted the monitoring of the free-sugar content (i.e. sucrose and glucose) of processing potatoes. The theory, technique and application of chemical maturity monitoring (CMM) were initially described by Sowokinos and Preston (1988). This system has proved valuable in assisting the potato industry in predicting and maintaining processing quality of potatoes from harvest and throughout the storage period. CMM involves the measurement of the sucrose rating (SR = mg sucrose/g tuber FW) and glucose (mg/g tuber FW) content of potatoes using the Yellow Springs Instrument (YSI), Model 2700 Select, Industrial Sugar Analyzer (Sowokinos, 2001a). CMM has also been utilized to maintain the long-term storage quality of potatoes used for the production of French fries (Pritchard and Adam, 1994). The proper chemical maturity necessary to be reached before harvest has an upper limit for sucrose of 1.0 mg/g (0.1% FW) or less for both crisps and French fries (Santerre et al., 1986; Sowokinos and Preston, 1988). The upper limits for glucose are 0.35 mg/g (0.035% FW) and 1.0 mg/g (0.1% FW), or less, for crisp and French fry production, respectively

(Sowokinos and Preston, 1988; Coleman et al., 1993). Higher glucose values are generally tolerated with French fry potatoes due, in part, to a blanching process that takes place before cooking. Experience has shown that superior processing cultivars have the genetic ability to reach and maintain glucose concentrations in storage markedly lower than the 'upper limits' described above, e.g. in the range of 0.01–0.1 mg/g (0.001–0.1% FW). For a reliable prediction of processing performance from storage, the concentration of both sucrose (i.e. source for new RS following harvest or vine death) and the RS currently available must be considered.

If 'chemically mature' potatoes are placed into storage and stressed (principally influenced by temperature and ventilation parameters), RS can accumulate from the vast starch reserves resulting in the production of unacceptable, dark-coloured products. Additionally, if 'chemically immature/high sugar' potatoes are placed into storage and suberized at temperatures of 10°C, or below, they will maintain their sugar level and process with a dark colour long into the storage season. To minimize this colour problem, potatoes are usually subjected to 'pre-conditioning temperatures' early in storage ranging from 13 to 16°C. At these temperatures, RS levels usually drop to acceptable levels within a 2- to 4-week period following which bin temperatures can be slowly decreased (0.5°/day) for long-term holding (Sowokinos and Preston, 1988). Prolonged exposure to high temperature will decrease the storage life of the potato by increasing potential for disease loss and by decreasing the time to 'irreversible' senescent sweetening. Sprouting can be delayed by applying a natural or synthetic sprout inhibitor, but an unnecessarily prolonged period at a high temperature can diminish their effectiveness. Outside air must be intermittently circulated through the pile (to replenish oxygen supply) each day to allow for this sugar-removal process to occur successfully. At pre-conditioning temperatures, RS levels drop as carbons are recycled back into starch, whereas others are directed into respiratory pathways (Isherwood, 1976; Smith, 1977). The temperature used for pre-conditioning must be compromised (i.e. decreased slightly) if soft-rot or other infectious organisms are detected. Intermediate storage temperatures can be maintained (10–12°C) if the potatoes are to be stored short term before processing (1–3 months). Long-term holding of potatoes (4–10 months) requires temperatures ranging between 5.5 and 8.9°C. The actual temperature tolerated depends on the CIS potential of the cultivar used (Sowokinos and Glynn, 2001).

During a period of extended storage, the concentration of sucrose in potatoes can serve as an 'indicator' of their holding potential (Sowokinos and Preston, 1988). In a stress-free storage environment, the sucrose content should remain well below the upper limit of 1.0 mg/g FW. Potatoes maintaining a continuous low level of sucrose display the potential for long-term holding. This is due to the fact that starch must first be converted into sucrose before RS can accumulate. Experience has shown that sucrose levels need to increase to a concentration of near 1.5 mg/g FW in storage before RS accumulates, leading to a significant degree of visual browning of crisps or French fries. If sucrose is monitored often enough (bimonthly or weekly), this may allow time for the storer/processor to move potatoes in a more timely manner. During the storage period, outside air must be intermittently circulated through the pile to allow respiratory processes to continue. If sugars increase because of a decrease in oxygen content or an increase in carbon dioxide, relieving these stresses will aid the reversal of this condition and will allow colour quality to return. Success for regaining colour quality is limited, in part, by the duration and severity of the stress, the physiological age of the potato and the cultivar used. If tuber sucrose begins to approach and surpass 1.0 mg/g FW while the bin environment is relatively stress free, this could be an indication that the potatoes are approaching senescence and should be processed as quickly as possible.

By minimizing the harvest of chemically immature or stressed potatoes and with a careful assessment of their sugar concentration in storage, the quality of fried products can be controlled with a minimum loss of raw material (Sowokinos and Preston, 1988). CMM has not eliminated all the colour problems that the industry faces in maintaining the post-harvest quality of processing potatoes, but it has helped the industry fine-tune its efforts to minimize the detrimental effects that RS levels have on final colour quality.

REFERENCES

ap Rees T., D.L. Dixon, C.J. Pollock and F. Franks, 1981, In: J. Friend and M.J.C. Rhodes (eds), *Recent Advances in the Biochemistry of Fruits and Vegetables*, p. 41. Academic Press, New York.

ap Rees T. and S. Morrell, 1990, *Am. Potato J.* 67, 835.

Arteca R.N., B.W. Poovaiah and L.K. Hiller, 1980, *Am. Potato J.* 57, 241.

Ashoor S.H. and J.B. Zent, 1984, *J. Food Sci.* 49, 1206

Bamberg J.B., J.P. Palta, L.A. Peterson, M. Martin and A.R. Krueger, 1993, *Am. Potato J.* 70, 219c

Bangerth F., 1979, *Annu. Rev. Phytopathol.* 17, 97.

Baruzzini L., L.A. Ghirardelli and E. Honsell, 1989, *Potato Res.* 32, 405.

Beals K., 2004, *Potato Literature Review*, United States Potato Promotion Board, Internal Document, pp. 1–91.

Blenkinsop R.W., L.J. Copp, R.Y. Yada and A.G. Marangoni, 2002, *J. Agric. Food Chem.* 50, 4545.

Briddon A. and R.M.J. Storey, 1996, In: *Abstracts of Conference Papers of the 13th Triennial Conference of the European Association of Potato Research*, pp. 565.

Brown J., G.R. Mackay, H. Bain, D.W. Griffith and M.J. Allison, 1990, *Potato Res.* 33, 219.

Burke J.J., T. O'Donavan and P. Barry, 2005, *Potato Res.* 48, 69.

Burstrom H., 1968, *Biol. Rev.* 43, 278.

Burton W.G., 1965, *Eur. Potato J.*, 8, 80.

Burton W.G., 1989, In: W.G. Burton (ed.), *The Potato*, pp. 423–522. John Wiley & Sons, New York.

Burton W.G. and A.R. Wilson, 1978, *Potato Res.* 21, 145.

Busse J.S. and J.P. Palta, 2006, *Physiol. Plantarum* 128, 313.

Clarkson D.T., 1984, *Plant Cell Environ.* 7, 449.

Clarkson D.T. and J.B. Hanson, 1980, *Annu. Rev. Plant Physiol.* 31, 239.

Coffin R.H., R.Y. Yada, K.L. Parkin, B. Grodzinski and D.W. Stanley, 1987, *J. Food Sci.* 52, 639.

Coleman W.K., G.C.C. Tai, S. Clayton, M. Howie and A. Pereira, 1993, *Am. Potato J.* 70, 909.

Collier G.F., D.C.E. Wurr and C. Huntington, 1980, *J. Agric. Sci. (Camb.)* 94, 407.

Crumbly I.J., D.C. Nelson and M.E. Duysen, 1973, *Am. Potato J.* 50, 266.

Davies H.V., 1998, *Am. J. Potato Res.* 75, 37.

Davies H.V. and P. Millard, 1985, *Ann. Bot.* 56, 745.

Davies H.V. and L.S. Monk-Talbot, 1990, *Phytochemistry* 29, 2833.

Dean B.B., P.E. Kolattukudy and R.W. Davis, 1977, *Plant Physiol.* 59, 1008.

Dugo G., L. La Pera, V. Lo Turco, D. Giuffrida and S. Restuccia, 2004, *Food Addit. Contam.* 21, 649.

Dwelle R.P., 1985, In: P.H. Li (ed.), *Potato Physiology*, pp. 35–58. Academic Press, New York.

Dwelle R.B. and B.F. Stallknecht, 1978, *Am. Potato J.* 55, 561.

Ehlenfeldt M.K., 1992, *Am. Potato J.* 69, 537.

Ehlenfeldt M.K., D.F. Lopez-Portilla, A.A. Boe and R.H. Johansen, 1990, *Am. Potato J.* 67, 83.

Eldredge E.P., Z.A. Holmes, A.R. Mosley, C.C. Shock and T.D. Stieber, 1996, *Am. J. Potato Res.* 73, 517.

Fuller T.J. and J.C. Hughes, 1984, *J. Food Technol.* 19, 455.

Galliard T., 1970, *Phytochemistry* 9, 1725.

Galliard T., 1973, *J. Sci. Food Agric.* 24, 617.
Galliard T., 1978, In: Kahl G (ed.), *Biochemistry of Wounded Plant Tissue*, p. 155. Walter de Gruyter, Berlin.
Gray D. and J.C. Hughes, 1978, In: P.M. Harris (ed.), *The Potato Crop*, p. 504. Chapman and Hall, London.
Habib A.T. and H.D. Brown, 1956, *Food Technol.* 10, 332.
Habib A.T. and H.D. Brown, 1957, *Food Technol.* 11, 85.
Han J.S., N. Kozukue, K.S. Young, K.R. Lee and M. Friedman, 2004, *J. Agric. Food Chem.* 52, 6516.
Hanson J.B., 1984, In: P.B. Tinker and A. Lauchli (eds), *Advances in Plant Nutrition*, Vol. 1, p. 149. Prager Press, New York.
Henninger M.R., K.G. Haynes and S.B. Sterrett, 1994, *Am. Potato J.* 71, 677.
Henninger M.R., S.B. Sterrett and K.G. Haynes, 2000, *Crop Sci.* 40, 977.
Hiller L.K., D.C. Koller and R.E.Thornton, 1985, In: P.H. Li (ed.), *Potato Physiology*, p. 389. Academic Press, New York.
Iritani W.M. and L.D. Weller, 1980, *Washington State University Cooperative Extension Bulletin* 0717, 3.
Iritani W.M., L.D. Weller and N.R. Knowles, 1984, *Am. Potato J.* 61, 335.
Isherwood F.A., 1973, *Phytochemistry* 12, 2579.
Isherwood F.A., 1976, *Phytochemistry* 15, 33.
Jakuczun H., K. Zgorska and E. Zimnoch-Guzowska, 1995, *Potato Res.* 38, 331.
Jankowski K.M., K.L. Parkin and J.H. von Elbe, 1997, *J. Food Process. Pres.* 21, 33.
Jansky S.H. and D.M. Thompson, 1990, *Am. Potato J.* 67, 695.
Jaswal A.S., 1973, *Am. Potato J.* 50, 86.
Kallio A., 1960, *Am. Potato J.* 37, 338.
Karlsson B.H., J.P. Palta and S. Ozgen, 2001, *Am. J. Potato Res.* 78,462.
Khanbari O.S. and A.K. Thompson, 1993, *Potato Res.* 36, 359.
Killick R.J. and A.W. Macarthur, 1980, *Potato Res.* 23, 457.
Kleinhenz M.D. and J.P. Palta, 2002, *Physiol. Plantarum* 115,111.
Kleinhenz M.D., J.P. Palta, C.G. Gunter and K.A. Kelling, 1999, *J. Am. Soc. Hortic. Sci.* 124, 498.
Kratzke M.G. and J.P. Palta, 1985, *Am. Potato J.* 62, 27.
Kratzke M.G. and J.P. Palta, 1986, *HortScience* 21,1022.
Krauss A. and H. Marschner, 1971, *Pflanzenernahr Dang BodenKunde* 129, 1.
Lambert R., R. Michaud, R. Romain and S. Yelle, 1996, *Acta Hortic.* 429, 481.
Lauer F. and R. Shaw, 1970, *Am. Potato J.* 47, 275.
Leszkowiat M.J., V. Barichello, R.Y. Yada, R.H. Coffin, E.C. Lougheed and D.W. Stanley, 1990, *J. Food Sci.* 55, 281.
Levitt J., 1942, *Am. Potato J.* 19, 134.
Levy D., 1986, *Potato Res.* 26, 315.
Loiselle F., G.C.C. Tai and B.R. Christie, 1990, *Am. Potato J.* 67, 633.
Love S.L., M.K. Thornton, G. Beaver and A.L. Thompson, 1991, In: A.L. Thompson and S.L. Love (eds), *Tri-State Potato Trials*, p. 73, University of Idaho Misc. Ser. No. 149. Moscow, Idaho.
Lulai E.C., 2005, *Am. J. Potato Res.* 82, 9.
Lulai E.C and P.H. Orr, 1979, *Am. Potato J.* 56, 379.
Lulai E.C., J.R. Sowokinos and J.A. Knoper, 1986, *Plant Physiol.* 80, 424.
Mann J.D., J.P. Lammerink and G.D. Coles, 1991, *N.Z. J. Crop Hort.* 19, 199.
Mares D.J., J.R. Sowokinos and J.S. Hawker, 1985, In: P.H. Li (ed.), *Potato Physiology*, p. 279. Academic Press, New York.
Maynard D.N., 1979, *J. Plant Nutr.* 1, 1.
McGarry A., C.C. Hole, R.L.K. Drew and N. Parsons, 1996, *Postharvest Biol. Technol.* 8, 239.
McKenzie M.J., J.R. Sowokinos, I.M. Shea, S.K. Gupta, R.R. Lindlauf and J.A.D. Anderson, 2005, *Am. J. Potato Res.* 82, 231.
Monk L.S. and H.V. Davies, 1989, *Physiol. Plantarum* 75, 411.
Monk L.S., D.B. McPhail, B.A. Goodman and H.V. Davies, 1989, *Free Radical Res. Com.* 5, 345.
Monk-Talbot L.S., H.V. Davies, M. Macaulay and B.P. Forster, 1991, *J. Plant Physiol.* 137, 499.
Nelson D.C., 1970, *Am. Potato J.* 47, 130.
Nelson D.C. and J.R. Sowokinos, 1983, *Am. Potato J.* 60, 949.
Nelson D.C. and M.C. Thoreson, 1986, *Am. Potato J.* 63, 155.

O'Donoghue E.P., A.G. Marangoni and R.Y. Yada, 1996, *Am. Potato J.* 73, 545.
Olsen N.L., L.K. Hiller and E.J. Mikitzel, 1996, *Potato Res.* 39, 165.
Olsen N.L., G.E. Kleinkopf and J.C. Stark, 2003, In: J.C. Stark and S.L. Love (eds), *Potato Production Systems*, p. 309. University of Idaho Agricultural Communications, Moscow, Idaho.
Owings T.R., W.M. Iritani and C.W. Nagel, 1978, *Am. Potato J.* 55, 211.
Ozgen S., B.H. Karlsson and J.P. Palta, 2006, *Am. J. Potato Res.* 83,73.
Palta J.P., 1996, *HortScience* 31, 51.
Peterson R.L., W.G. Barker and M.J. Howarth, 1985, In: P.H. Li (ed.), *Potato Physiology*, p. 123, Academic Press, New York.
Poovaiah B.W. and A.S.N. Reddy, 1993, *Crit. Rev. Plant Sci.* 12, 185.
Pressey R., 1969, *Am. Potato J.* 46, 291.
Pritchard M.K. and L.R. Adam, 1992, *Am. Potato J.* 69, 805.
Pritchard M.K. and L.R. Adam, 1994, *Am. Potato J.* 71, 59.
Putz B. and M.G. Lindhauer, 1994, *Agric. Biol. Res.* 47, 335.
Rex B.L. and G. Mazza, 1989, *Am. Potato J.* 66, 165.
Reynolds M.P. and E.E. Ewing, 1989, *Am. Potato J.* 66, 63
Rodriguez-Saona L.E. and R.E. Wrolstad, 1997, *Am. Potato J.* 74, 87.
Rodriguez-Saona L.E., R.E. Wrolstad and C. Pereira, 1997, *J. Food Sci.* 62, 1001.
Roe M.A. and R.M. Faulks, 1991, *J. Food Sci.* 56, 1711.
Roe M.A., R.M. Faulks and J.L. Belsten, 1990, *J. Sci. Food Agric.* 52, 207.
Rufty T.W. and S.C. Huber, 1983, *Plant Physiol.* 72, 474.
Santerre C.R., J.N. Cash and R.W. Chase, 1986, *Am. Potato J.* 63, 99.
Shallenberger R.S., O. Smith and R.H. Treadway, 1959, *J. Agric. Food Chem.* 7, 274.
Shear C.B., 1975, *HortScience* 10, 361.
Shock C.C., Z.A. Holmes, T.D. Stieber, E.P. Eldredge and P. Zhang, 1993, *Am. J. Potato Res.* 70, 227.
Shock C.C., J.C. Zalewski, T.D. Stieber and D.S. Burnett, 1992, *Am. J. Potato Res.* 69, 793.
Sieczka J.B. and R.E. Thornton, 1993, *USDA Handbook 267*, University of Maine, Orono, Maine.
Silva G.H., R.W. Chase, R. Hammerschmidt, M.L. Vitosh and R.B. Kitchen, 1991, *Am. Potato J.* 68, 751.
Simmons K.E. and K.A. Kelling, 1987, *Am. Potato J.* 64, 119.
Simmons K.E., K.A. Kelling, R.P. Wolkowski and A. Kelman, 1988, *Agron. J.* 80, 13.
Smith, O., 1967, In. W.F. Tallburt and O. Smith, (eds), *Potato Processing*, p. 262. AVI Publishing, Westport.
Smith, O., 1977, In: O. Smith (ed.), *Potatoes: Production, Storing, Processing*, p. 436. AVI Publishing, Westport.
Sowokinos J.R., 1978, *Am. Potato J.* 55, 333.
Sowokinos J.R., 1990, In: M.E. Vayda and W.D. Park (eds), *The Molecular and Cellular Biology of the Potato*, p. 137. C.A.B. International, Wallingford, Oxon.
Sowokinos J.R., 2001a, *Am. J. Potato Res.* 78, 57.
Sowokinos J.R., 2001b, *Am. J. Potato Res.* 78, 221.
Sowokinos J.R. and M. Glynn, 2001, *Potato Grower*, 66, 14.
Sowokinos J.R., E.C. Lulai and J.A. Knoper, 1985, *Plant Physiol.* 78, 489.
Sowokinos J.R. and D.A. Preston, 1988, *Minnesota Agric. Exp. Station Bulletin 586-1988*, Item No. AD-SB-3441, p. 1.
Sowokinos J.R., C.C. Shock, T.D. Stieber and E.P. Eldredge, 2000, *Am. J. Potato Res.* 77, 47.
Stark J.S. and S.L. Love, 2003, In: J.C. Stark and S.L. Love (eds), *Potato Production Systems*, p. 329. University of Idaho, Moscow, Idaho.
Sterrett S.B. and M.R. Henninger, 1997, *Am. Potato J.* 74, 233.
Sterrett S.B., M.R. Henninger and G.S. Lee, 1991a, *J. Am. Soc. Hortic. Sci.* 116, 697.
Sterrett S.B., M.R. Henninger, G.C Yencho, W. Lu, B.T. Vinyard and K.G. Haynes, 2003, *Crop Sci.* 43, 790.
Sterrett S.B., G.S. Lee, M.R. Henninger and M. Lentner, 1991b, *J. Am. Soc. Hortic. Sci.* 116, 701.
Storey R.M.J. and H.V. Davies, 1992, In: P. Harris (ed.), *The Potato Crop*, 2nd edition, pp. 507. Chapman and Hall, London.
Struckmeyer B.E. and J. P. Palta, 1986, *Am. Potato J.* 63, 57.
Tawfik A.A., M.D. Kleinhenz and J.P. Palta, 1996, *Am. Potato J.* 73, 261.
Tawfik A.A. and J.P. Palta, 1992, *HortScience* 27, 665.

Thompson-Johns A.L., 1998. *Inheritance of the sugar-end disorder in potato (Solanum tuberosum, L.).* Ph.D. Dissertation, University of Idaho, Moscow.

Tibbits T.W. and D.A. Palzkill, 1979, *Commun. Soil Sci. Plant* 10, 251.

van Denburgh R.W., L.K. Hiller and D.C. Koller, 1980, *Am. Potato J.* 57, 371.

van Steveninck R.F.M. and M.E. Jackman, 1967, *Aust. J. Biol. Sci.* 20, 749.

Vega S.E., J.P. Palta and J.B. Bamberg, 1996, *Am. Potato J.* 73, 397.

Veilleux R.E. and F.I. Lauer, 1981, *Euphytica* 30, 547.

Walcotot P.J., J.R. Kenrick and D.G. Bishop, 1982, In: J.F.G.M. Wintermans and P.J.C. Kuiper (eds), *Biochemistry and Metabolism of Plant Lipids*, p. 297. Elsevier, New York.

Wannamaker M.J. and W. Collins, 1992, *Am. Potato J.* 69, 221.

Wiersum L.K., 1966, *Acta Bot. Neerl.* 15, 406.

Wolf S., A. Marani and J. Rudich, 1990, *Ann. Bot.* 66, 513.

Wolf S., A. Marani and J. Rudich, 1991, *J. Expt. Bot.* 42, 619.

Yamaguchi M., H. Timm, M.D. Clegg and F.D. Howard, 1966, *Proc. Am. Soc. Hortic. Sci.* 89, 456.

Zrenner R., K. Schuler and U. Sonnewald, 1996, *Planta* 198, 246.

Chapter 24

Potato Flavour and Texture

Mark A. Taylor, Gordon J. McDougall and Derek Stewart

Quality, Health and Nutrition, Scottish Crop Research Institute, Invergowrie, Dundee DD2 5DA, UK

24.1 INTRODUCTION

Increasingly tuber quality traits are assuming a greater importance in breeding programmes, as consumers demand greater variety and retailers wish to market cultivars that have distinctive commercial advantages. However, as with many food crops, potato flavour and texture are difficult to assess in breeding programmes. Assessments are highly subjective and require trained sensory panels. These have a low sample throughput and are consequently expensive. As a result, flavour and texture are generally only assessed in the later stages of a breeding programme after selection for more easily quantifiable traits. In fact, most of the potential flavour and texture improvements are likely to be discarded, and to a large extent, the market place determines whether a new cultivar is acceptable to consumers (Wang and Kays, 2003).

Means of analytically determining flavour and texture that can be efficiently applied to germplasm are obvious requirements for improvement of the potato crop. In this chapter, we shall review progress that has been made in understanding these complex traits and opportunities that are arising, particularly from the Omic technologies that are now being applied to potato.

24.2 POTATO FLAVOUR

24.2.1 Non-volatile components

Flavour is a complex trait and depends on the interaction of volatile compounds and soluble cellular constituents. The soluble compounds define the basic taste parameters, sweet, sour, salty or bitter. In the early twentieth century, a glutamate-like taste was proposed as a fifth basic taste quality (Ikeda, 1912). There is a growing recognition of this taste as umami (from the Japanese word meaning delicious). Compounds including monosodium glutamate, several process-derived glutamate glycoconjugates, adenosine-5′-monophosphate, inosine-5′-monophosphate and guanosine-5′-monophosphate (GMP) are well known to show umami-like sensory characteristics (Bellisle, 1998). Umami compounds generally enhance flavour and mouthfeel, giving the impression of creaminess and viscosity to savoury dishes (Halpern, 2000).

Potato Biology and Biotechnology: Advances and Perspectives
D. Vreugdenhil (Editor)

As early as 1971, it was suggested that the flavour of boiled potato was largely due to the natural mixture of glutamic acid and other amino acids in combination with GMP and other 5′-ribonucleotides produced on cooking (Solms, 1971). Thus chemicals that impart an umami flavour are likely to be an important component of potato flavour. Indeed, some authors claim that there is only a small contribution from volatile (olfactory) components and that chemicals representing the so-called sweet, sour, salty and bitter tastes do not make a major contribution to potato flavour (Solms, 1971; Solms and Wyler, 1979). Thus, the presence of salt, sugars or alkaloids does not enhance potato flavour although their presence at high levels may decrease palatability. With the discovery of umami-tasting glutamate glycoconjugates, produced by a Maillard reaction (Beksan et al., 2003), the potential indirect effects of sugars on potato flavour should not be overlooked.

Data on the levels of several umami compounds in potato tubers have been published (Solms and Wyler, 1979). Although raw potatoes contain only very small amounts of 5′-nucleotides and no 5′ GMP, cooked potatoes contain appreciable levels, higher than most other plant foods examined. Thus, these nucleotides accumulate because of the action of nucleases during cooking processes, particularly due to RNA degradation. Ribonucleases are active under the pH and temperature conditions that occur during heating, being particularly active at around 50°C. As the temperature of potato tissues increases slowly from 40 to 60°C during some cooking processes (e.g. boiling), nuclease activity may be significant (Solms and Wyler, 1979).

Several studies have addressed the levels of free amino acids (including glutamic acid) in potato tubers and also examined changes in amino acid levels during storage (Brierley et al., 1996, 1997). It is clear that the soluble protein and amino acid contents of potato tubers are modified substantially during storage, with larger effects observed after 3 months of storage at 10°C than at 4°C (Brierley et al., 1997). At harvest, the major free amino acids are asparagine, glutamine, glutamic acid, arginine and aspartic acid, with some cultivar-dependent variations in the contribution of these amino acids (Brierley et al., 1997; Davids et al., 2004). Additionally, the soil nitrogen fertilization regime impacts on amino acid content, a high level of nitrogen fertilization being associated with increased glutamine content (Burton, 1989). After storage, the profile of free amino acids that can be extracted in potato juices changes significantly depending on the storage temperature. For example, in one study, the combined levels of glutamine and asparagine contributed 34% of the amino acid pool at harvest but up to 90% after 25 weeks of storage at 10°C (Brierley et al., 1997). In fact, it has been suggested recently that flavourings may be extracted from potato juice and that the amino acid content and hence the flavour characteristics of the juice may be manipulated by controlling the storage temperature of the potato tubers prior to extraction (Davids et al., 2004). Treatment of the extracted juice with glutaminase may increase the glutamic acid content and, consequently, the umami character of the extract.

Only limited taste panel data are available to support the importance of umami compounds in defining potato flavour. However, the effects of supplementing boiled potato with glutamic acid and glutamic acid and nucleotides both resulted in a 'stronger' potato taste as assessed by a trained panel of 18 individuals. Additionally, an aqueous mixture of amino acids and nucleotides that reproduces the levels found in boiled potatoes 'had practically no odor, but an agreeable basic potato-like taste' (Solms, 1971). Furthermore, a mixture with components replicating the amino acid and nucleotide concentrations of a preferred boiled potato was judged to taste better than that replicating a less preferred boiled potato (Buri and Solms, 1971).

Although potato starch probably has no taste of its own, the interaction of volatile and non-volatile components with starch may influence flavour. In a recent study, it has been suggested that starch-thickened products have a preferable taste to those prepared with other thickeners because of an 'in-mouth' viscosity decrease caused by amylase action (Ferry et al., 2004). The decrease in viscosity per se may enhance flavour perception or, alternatively, volatile components complexed with starch may be released during this process providing additional stimulation of the olfactory receptors. Interestingly, starch can form inclusion complexes with decanal, a volatile derived from fatty acids that is produced during cooking (Solms and Wyler, 1979).

24.2.2 Glycoalkaloids and flavour

Potato tubers are well known to contain glycoalkaloids, the major forms being α-chaconine and α-solanine, which represent approximately 95% of the total (Slanina, 1990). The level of tuber glycoalkaloids varies according to cultivar (Ramsay et al., 2004) and is also influenced by growth and storage conditions (Sengul et al., 2004). Tuber glycoalkaloids are toxic to humans if present at levels in excess of 20 mg/100 g (Osman, 1983). Additionally, high levels of glycoalkaloids are responsible for off-flavours in potato tubers (Zitnak, 1961; Sinden et al., 1976; Ross et al., 1978). The emerging consensus is that potato glycoalkaloids at elevated levels are responsible for flavours described as bitter, burning, scratchy or acrid and are thus generally undesirable components of flavour. In one study, taste panels correlated glycoalkaloid content in excess of 10 mg/100 g fresh weight with a burning taste. However, at levels between 0.76 and 4.98 mg/100 g, no correlation was found (Ross et al., 1978). Other taste panel studies described how tubers with glycoalkaloid contents greater than 14 mg/100 g had a bitter taste (Sinden et al., 1976).

24.2.3 Volatile compounds

Recent studies on potato flavour have focused almost exclusively on the volatile components released from raw and cooked potato tubers, despite earlier claims that potato flavour was largely due to soluble matrix-associated components and the lack of convincing sensory panel data in support of the volatile components being of primary importance to flavour. The volatiles produced by raw and cooked potatoes have been studied extensively (reviewed by Maga, 1994), and over 250 compounds have been identified in potato volatile fractions. Since then, attempts have been made to discriminate which of these components are important for potato flavour, which are specific to the method of cooking, cultivar differences, the effects of agronomic conditions and effects of storage (Duckham et al., 2001, 2002; Oruna-Concha et al., 2001, 2002a,b; Dobson et al., 2004).

Although potato tubers are only consumed after cooking, there is an advantage to determining analytically the volatiles arising from the raw product. The determination of a favourable analytical profile in raw potatoes would enable a rapid evaluation of a given product free from artefacts arising from differences in cooking profiles. Raw potato volatiles are derived from biosynthetic processes in the tuber or associated microorganisms rather than thermally catalyzed reactions and include compounds such as methoxypyrazines (Buttery and Ling, 1973). The presence of relatively high amounts of methoxypyrazines in peeled tubers may argue against this compound being synthesized by soil microflora associated with the tuber and indicate that a biosynthetic route for their production is present in the potato tuber. The most prevalent methoxypyrazine present in raw potato volatiles was 2-methoxy-3-isopropylpyrazine (Murray and Whitfield, 1975). Sesquiterpenes are also present in raw potato volatiles, and in general, their levels are much higher following tuber damage or microbial attack (Desjardins et al., 1995).

Lipid oxidation products contribute to the spectrum of volatiles from raw potato – on slicing or shredding potato tubers, there is a large increase in classical lipid oxidation products such as hexanal, octenal and isomeric forms of 2,4-decadienal (Josephson and Lindsay, 1987).

The volatile profile obtained from cooked potatoes contains many process-derived compounds that undoubtedly contribute to the overall potato flavour (Maarse, 1991). Important thermally driven reactions include the Maillard reaction between reducing sugars and amino acids, the Strecker degradation of methionine to produce methional and the thermal and enzymatic degradation of fatty acids (Fig. 24.1). The Maillard reaction takes place when compounds possessing a carbonyl group, typically reducing sugars, react with components with a free amino group such as amino acids. The Strecker degradation of methionine involves the interaction of α-dicarbonyl compounds, intermediates in the Maillard reaction, with methionine, resulting in the formation of methional (Lindsay, 1996). Autooxidation (Frankel, 1998) and enzymic (hydroperoxide lyase) action on hydroperoxides derived from linoleic and α-linolenic acids (Gardner, 1995) produce a range of flavour-active volatile aldehydes, ketones, alcohols and alkyl furans. The other major classes of cooked potato volatiles include methoxypyrazines and terpenes. In

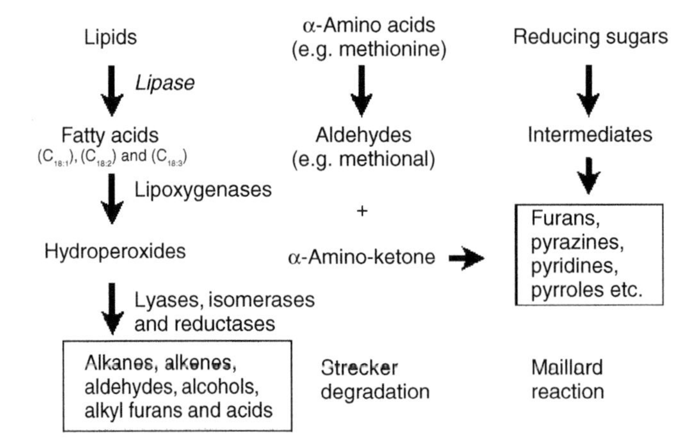

Fig. 24.1. Origin of flavour volatiles released from cooked potato. Precursor metabolites are shown at the top, and the main products formed on boiling are shown in solid boxes.

addition, several specific compounds have been directly related to desirable flavour and aroma characteristics with methional (Lindsay, 1996), methoxypyrazines (Murray and Whitfield, 1975) and the lipid degradation product, *cis*-4-heptenal (Josephson and Lindsay, 1987), all reported to exhibit 'a cooked potato odour'. This provides some evidence that these compounds may be significant contributors to potato flavour. To generalize the findings of several studies, in boiled potatoes, the most important volatile constituents are likely to be those derived from lipid degradation, the Maillard reaction and the Strecker reaction and include methional, aliphatic alcohols and aldehydes, thiols and sulphides and methoxypyrazines.

As the volatiles derived from potato tubers are greatly altered on cooking, it is important to assess the changes in profiles associated with different cooking methods. Distinctive quantitative and qualitative differences in volatile profile were observed in a comparison of tubers (cultivars Estima and Maris Piper) cooked by boiling, baking or microwaving (Oruna-Concha et al., 2002b). Although for all three methods of cooking the main sources of flavour compounds were from lipid degradation and the Maillard reaction and/or sugar degradation, there was a characteristically different ratio of the contribution of each class to the total volatile fraction. For boiled potatoes, volatiles formed by lipid degradation were judged to be the most important, whereas for baked potatoes, those arising from amino acids contributed more significantly (Oruna-Concha et al., 2002b). A much lower amount of volatile aroma compounds was produced from microwave-cooked potatoes than for the other cooking methods, possibly indicating that microwave-cooked potato products are blander in flavour than potato products produced by boiling or conventional baking. The cooking-dependent differences in volatile profiles have been attributed to variations in heat and mass transfer processes associated with the different cooking methods (Wilson et al., 2002a,b). For example, during conventional baking, initially the surface temperature of the tuber increases to 100°C with significant water evaporation (Wilson et al., 2002a). As baking progresses, the surface temperature increases above 100°C, a crust develops as evaporation continues and the 100°C isotherm moves towards the centre of the tuber. Water loss during baking is typically 20% but with most of the moisture loss (approximately 57%) from the outer 3 mm of the tuber. Boiling tubers resulted in negligible water loss and a more rapid migration of the 100°C isotherm to the centre of the tuber than conventional baking (Oruna-Concha et al., 2002b). Microwave cooking proceeds by a different heating mechanism (Fryer, 1997); all parts of the tuber reach 100°C within a few minutes except for the outer surface that remains cooler because of evaporative cooling. Starch gelatinization is complete throughout the tuber after 2–2.5 min of cooking. Although water loss is similar during conventional and microwave baking, the loss was more uniform with microwave cooking and an outer crust was not formed. During microwave cooking, water boils explosively in the intercellular spaces rupturing the surrounding tissue. It has been hypothesized that the greater proportion of lipid degradation products measured in boiled potato volatiles may be because of more significant lipoxygenase activity during this mode of cooking (Oruna-Concha et al., 2002b). As tuber slices were boiled, the lipoxygenases may have been able to come into contact with their substrates more readily. Additionally, as the tubers were placed in cold water and heated to boiling, lipoxygenase activity may have persisted longer on boiling than on baking. For baked potatoes, lipid degradation compounds arise probably as a result of thermal degradation and so account for a smaller proportion of the total volatiles produced.

In view of the large differences in volatile profiles arising from different cooking methods, it is significant that trained sensory panels can detect differences between microwave and conventionally baked potatoes (Maga and Twomey, 1977; Brittin and Trevino, 1980).

Significant effects of storage time on volatile flavour components of baked potatoes have been investigated (Duckham et al., 2002). As described above for amino acids, other flavour precursors, including fatty acids and sugars, change in levels during storage. Although potato lipids account for only 0.8–1.3 mg/g dry weight (Galliard, 1973), 70–75% of lipids are the relatively reactive polyunsaturated linoleic and linolenic acids, precursors of a wide range of volatile compounds (Galliard, 1973). Some studies report an increase in total fatty acid levels on storage (Cotrufo and Lunsetter, 1964; Cherif and Ben Abdelkader, 1970), whereas other studies report genotypic variations in fatty acid levels depending on the duration of storage (Dobson et al., 2004). Changes in tuber-reducing sugar during storage are well characterized (Finglas and Faulks, 1984; Brown et al., 1990) and generally increase during storage although the extent of the increase is cultivar-dependent (Blenkinsop et al., 2002).

24.2.4 Molecular and genetic approaches to the study of potato flavour

From this brief summary of our knowledge of potato flavour, it is clear that there are still large gaps in our knowledge. Although substantial progress has been made in describing the chemical composition of raw and cooked potato volatiles and of raw potato composition, studies that correlate extremes in these profiles with taste-trial data are sparse. Consequently, it is still uncertain which volatile and non-volatile components should be targeted in an attempt to manipulate potato flavour for particular consumer groups. It would be useful to correlate profiles with flavour preferences in populations that segregate for diverse potato flavour and produce quantitative trait locus (QTL) maps from these data. Potato germplasm, particularly diploid accessions of *Solanum phureja*, has been identified with a distinctive flavour. Compared with *Solanum tuberosum* cultivars, sensory analysis revealed that *S. phureja* tubers have a distinctive mouthfeel (high in smooth and low in grainy and floury traits) and a higher intensity of flavour attributes (De Maine et al., 2000). In an attempt to explore reasons for the difference in flavour, the fatty acid and polar lipid contents in tubers from *S. phureja* and *S. tuberosum* have been compared (Dobson et al., 2004). The absolute levels of linoleic and α-linolenic acids were significantly higher in four *S. phureja* accessions than in four *S. tuberosum* accessions, perhaps providing a partial explanation for the taste-trial data although differences in amino acid and sugar content may also contribute to the different taste. Genetic crosses using *S. phureja* accessions, in combination with analytical and sensory analysis, may be particularly valuable in determining the relative importance of aspects of the flavour trait. Identifying QTL in potato is becoming more routine (Bradshaw et al., 2004; Gebhardt et al., 2004) and may lead to the discovery of genes underlying the QTL.

24.2.5 Molecular approaches to dissecting key constituents of tuber flavour

Although the precise contribution of individual volatile and non-volatile components to overall potato flavour remains to be clarified, circumstantial evidence implicates several metabolites. In some cases, the biosynthetic routes for the production of the flavour compound or its precursor is known. For example, the key enzymatic steps that regulate methionine biosynthesis are now known (Hesse et al., 2004, Fig. 24.2), and using a transgenic approach, the level of soluble tuber methionine can be manipulated. In one example, antisense inhibition of threonine synthase leads to a 30-fold increase in tuber methionine content (Zeh et al., 2001), and in another, over-expression of an *Arabidopsis* cystathionine γ-synthase gene in potato tubers leads to a six-fold increase in tuber methionine content (Di et al., 2003, Fig. 24.2). Furthermore, in the latter study, the increased methionine content resulted in an increase in methional content (up to 4.4-fold) in baked tubers.

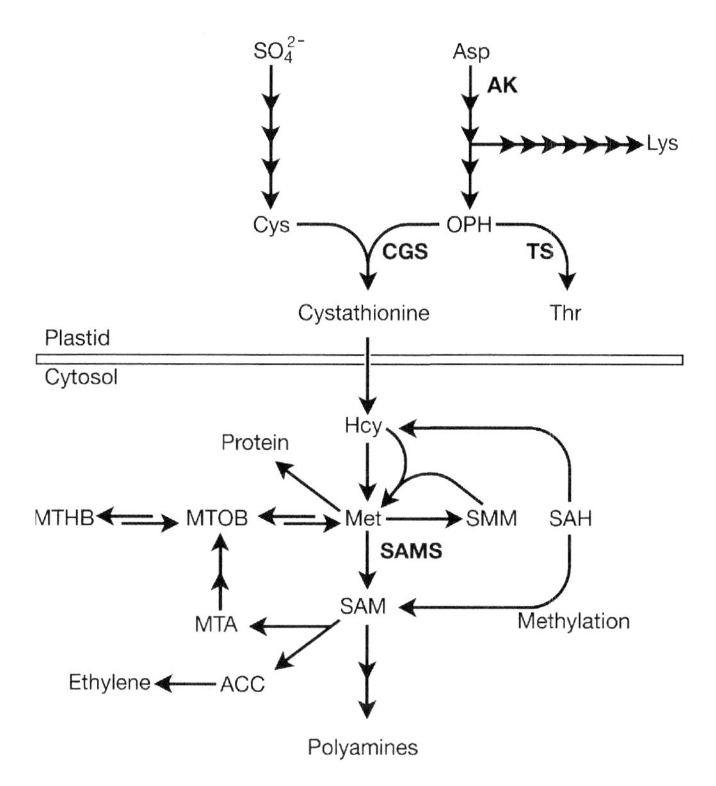

Fig. 24.2. Biosynthesis and metabolism of methionine in plants [*Source: Di et al. (2003)*]. Dashed arrows indicate that steps are omitted for clarity. ACC, 1-aminocyclopropane-1-carboxylic acid; AK, aspartate kinase; Asp, aspartate; CGS, cystathionine γ-synthase; Cys, cysteine; Hcy, homocysteine; Lys, lysine; Met, methionine; MTA, 5′-methylthioadenosine; MTHB, 4-methylthio-2-hydroxybutyrate; MTOB, 4-methylthio-2-oxobutanate OPH, *O*-phosphohomoserine; SAM, *S*-adenosylmethionine; SMM, *S*-methylmethionine; SAMS, SAM synthetase; Thr, threonine; TS, threonine synthase.

Other transgenic experiments have demonstrated that tuber fatty acid levels can be increased by over-expressing an *Arabidopsis* acetyl-CoA carboxylase gene (Klaus et al., 2004). In this study, over-expression led to approximately 40% increase in total fatty acid content and changes in the ratio of linoleic and α-linolenic acid. A third class of volatile metabolites, the sesquiterpenes, have also been manipulated transgenically in *Arabidopsis*, again demonstrating the potential of metabolic engineering for re-designing the profile of flavour compounds (reviewed in Dudareva and Negre, 2005). The transgenic approach clearly offers an opportunity to dissect the various elements contributing to overall flavour; however, this approach has not been carried out in conjunction with sensory analysis, making calibration of analytically derived profiles impossible.

Although transgenic routes may provide a means to manipulating potato tuber flavour, acceptability with consumers remains a problem. This may make molecular breeding strategies a more commercially attractive route to flavour improvement. Such an approach could exploit germplasm with diverse flavour characteristics and aim to identify alleles of naturally occurring genes that may account for different flavour profiles. These alleles can then be introgressed into germplasm with good agronomic traits. With our increased knowledge of the biochemical pathways leading to flavour compounds and the genes encoding the key enzymes in these pathways, it may be possible to identify such alleles in the near future. The combination of metabolically profiling raw and cooked tuber material with microarray analysis of germplasm, exploiting germplasm with wide phenotypic variation, has great potential for making further progress in understanding this complex trait. Indeed some headway has been made with regard to achieving this aim. For example, Dobson et al. (2005) have shown some correlations between the metabolic profiles of the raw tuber and that of the volatiles from the cooked material of potatoes already reported to exhibit distinct flavours (*S. tuberosum* and *S. phureja*, Fig. 24.3). Significant species differences were evident with regard to the relatively elevated levels of leucine, isoleucine and valine in *S. phureja*, and in turn, these were shown to be related to the subsequently increased levels of branched chain aldehydes, esters and alcohols. Such compounds are associated with desirable potato flavour and aromas (Oruna-Concha et al., 2002a,b). This was supported by further studies that showed that as the metabolites changed during storage, so did the corresponding putatively related volatiles.

24.3 POTATO TUBER TEXTURE

Potato tuber texture is a key quality determinant of cooked potato and a major trait that influences consumer preference. As with flavour, texture is a complex trait to analyse as it depends on the interaction of many factors, and defining texture that is attractive to consumers currently depends on sensory panel analysis rather than analytical measurements. Lugt (1961) proposed a scheme for describing cooked potato texture (Table 24.1), including aspects related to the degree of disintegration, the consistency, mealiness and structure of the cooked potato. This scheme provides a basis for comparison between subjective assessments. Other terms are also commonly used to describe

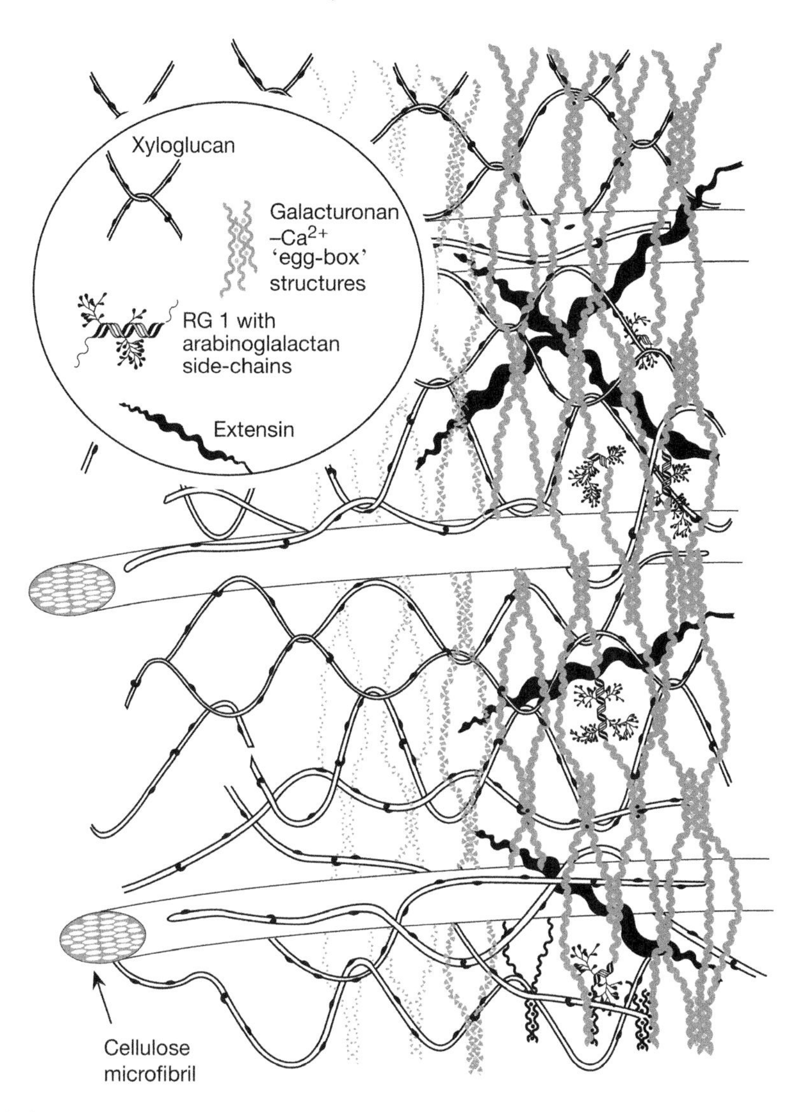

Fig. 24.3. Schematic representation of primary cell wall structure. A framework composed of insoluble extensin and cellulose microfibrils is coated with hydrogen bonded xyloglucan and filled with a matrix of pectic homogalacturonans and rhamnogalacturonans. The matrix has been progressively filled from left to right to show interactions. Microfibrils and extensin rods can be extensively overlapped but this has been omitted for clarity. Adapted from Carpita and Gibeaut (1993) with permission.

potato texture: flouriness, synonymous with mealiness, describing a dry, soft texture and waxiness describing a moist firm texture (Jarvis and Duncan, 1992).

Different consumer groups prefer different textures. For example, a dry boiled potato texture is preferred in Scotland, whereas a waxy texture is preferred in the Netherlands.

Table 24.1 Recommended vocabulary for describing potato structure (Lugt, 1961)

Texture	**Expression of characteristics**			
Disintegration	None	Slight	Moderate	Complete
Consistency	Firm	Fairly firm	Soft	Soft and uneven
Mealiness	Not mealy	Slightly mealy	Slightly dry	Dry
Dryness	Humid	Slightly humid	Slightly dry	Dry
Structure	Fine	Fairly fine	Slightly coarse	Coarse

Furthermore, boiled potatoes should not disintegrate even if slightly overcooked, a property that is difficult to combine with a dry texture (Jarvis and Duncan, 1992).

Factors that have an impact on cooked potato texture include starch content and distribution within the tuber (Barrios et al., 1963; Burton, 1989; Storey and Davies, 1992; Matsuura-Endo et al., 2002a), starch swelling pressure (Hoff, 1973; Jarvis et al., 1992; Shomer, 1995a,b), cell size (Barrios et al., 1963; Linehan et al., 1968; Hoff, 1972), cell-wall structure and composition (Hoff, 1973) and the breakdown of the cell wall middle lamella (Hughes et al., 1975a,b,c; Burton, 1989; Van Marle et al., 1994; Matsuura-Endo et al., 2002b) during cooking.

The specific gravity and dry weight of tubers, parameters that are easy to measure, are often used as a predictor of the degree of cooking disintegration. Within a variety, there is often a good correlation between specific gravity and cell separation on cooking, but this relationship does not extend between varieties (Storey and Davies, 1992; Matsuura-Endo et al., 2002b), indicating other factors are involved. Specific gravity and dry weight measurements in tubers are indicators of starch content. The gelatinization of starch on cooking may be a factor in cell separation; gelatinized starch swells resulting in an increase in the internal pressure within tuber cells that may lead to cell separation (reviewed by Reeve, 1977). Although some studies concluded that starch could not generate pressure when it swelled (Linehan et al., 1968; Hoff, 1972), Jarvis et al. (1992) quantified starch swelling pressure directly and demonstrated that this pressure could contribute to cell separation following weakening of the middle lamella by pectin degradation and subsequent cell wall deformation.

The uncertain nature of the relationship between starch content and degree of disintegration on cooking was illustrated by the study of Matsuura-Endo et al. (2002a,b). Although the degree of tuber disintegration generally increased with starch content within each of three cultivars tested, there was a wide difference in the degree of disintegration between the cultivars for a given starch content. A cultivar that contained high starch content in the area outside the vascular ring was more prone to disintegration than a cultivar with high starch associated with the vascular ring despite similar overall starch contents. Larger cell size and intracellular spaces were also evident in tubers from the cultivar that disintegrated most readily suggesting that cellular architecture plays an overarching role in cell disintegration (Matsuura-Endo et al., 2002a). Ormerod et al. (2002) measured thermal weakening of potato parenchyma tissue from tubers containing a range of high amylose contents and reduced starch-swelling properties. No amylose content-dependent differences in thermal weakening could be detected leading these authors to conclude that thermal degradation of the middle lamella exerts the major influence on disintegration during cooking. To summarize, the relationship between starch content and composition and cooked texture is not simple. It may be that most commercially grown tubers contain enough starch to generate sufficient swelling pressure to drive or initiate cell separation, and other factors such as starch distribution within the tuber may be more important, but further study is required.

Cell size has also been implicated as a determinant of cooked potato texture as larger cells appear to separate more readily during cooking (Barrios et al., 1963; Linehan et al., 1968; Gray, 1972). Tuber cell size varies according to variety (Linehan et al., 1968) and increases during tuber development (Gray, 1972; Reeve et al., 1973). As with starch and amylose content, the correlation between cell size and cooking texture is not always strong but is another factor that should be considered as contributing to overall texture (Gray, 1972).

The primary cell walls of the dicotyledonae have been modelled as a composite of two phases. An insoluble framework of cellulose micro-fibrils hydrogen-bonded by xyloglucans, which coat and partly span the spaces between the microfibrils (Selvendran et al., 1987; Carpita and Gibeaut, 1993), provides rigidity and strength. The cellulose–xyloglucan framework is embedded in a matrix of pectic polysaccharides (Fig. 24.4). The pectic polysaccharides are thought to perform many functions, such as the definition of cell wall porosity and the provision of charged surfaces capable of modulating the cell wall pH and ion balance (Carpita and Gibeaut, 1993). The middle lamella is the interstitial layer of the cell walls of two adjacent cells and is characterized by its relatively high level of pectin (Jarvis et al., 1981). Three classes of pectin have been described (Carpita and Gibeaut, 1993), homogalacturonan, rhamnogalacturonan I and rhamnogalacturonan II. The distribution of pectic epitopes in the cell walls of potato tubers has been studied in detail using monoclonal antibodies specific for different pectic epitopes (Bush and McCann, 1999; Bush et al., 2001; Parker et al., 2001). The distribution of the pectic epitopes changes both during tuber development and during storage (Bush et al., 2001). Homogalacturonans with low levels of methyl esters are capable of binding divalent cations (such as Ca^{2+} and Mg^{2+}) and are particularly abundant in the middle lamella of potato parenchyma cells (Bush and McCann, 1999; Parker et al., 2001). These cation-binding homogalacturonan polysaccharides can form electrostatic bridges with multiple divalent cations producing *egg-box* structures that can lead to the formation of pectin gels (Carpita and Gibeaut, 1993). The dissociation of these electrostatic bridges by treatment with calcium chelators, a method which is used to selectively extract pectins (Selvendran, 1985; Brett and Waldron, 1996), may explain the separation of cells along the middle lamella (Jarvis et al., 1981; Burton, 1989). It is also notable that Ca^{2+} ions are specifically located in the middle lamella of potato parenchyma cells (Bush et al., 2001).

During cooking, it is generally considered that pectin in the middle lamella is solubilized leading to softening of the potato (Storey and Davies, 1992). Release of pectic polysaccharides into the cooking medium is closely correlated with a decrease in cell firmness (Van Marle et al., 1994). The release of pectic material during cooking was compared in tubers with different textural properties (mealy cultivar Irene compared with non-mealy cultivar Nicola). With both cultivars having equivalent cell size, the release of pectic material and cell sloughing were higher for cultivar Irene. Similar results were obtained

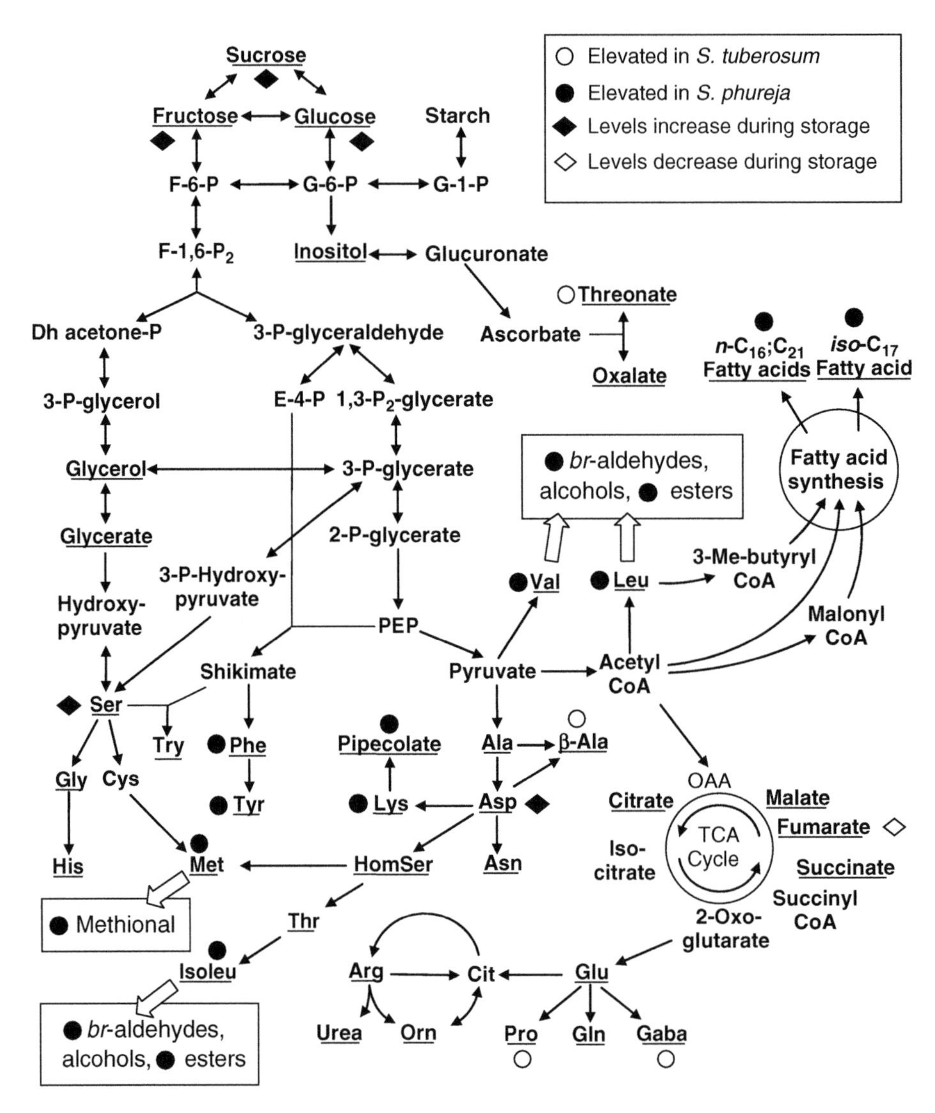

Fig. 24.4. Simplified schematic representation of interrelationship between polar and non-polar metabolites isolated from tubers of *Solanum tuberosum* and *Solanum phureja*. Individual metabolites showing differences in abundance between *S. tuberosum* and *S. phureja* are indicated by open and closed circles. Metabolites which show changes in abundance during storage at low temperature are indicated by closed and open diamonds. Several of the metabolites shown are the source of a number of volatile flavour-related compounds generated when tubers are boiled. These volatiles, shown in the outlined boxes, and their precursor metabolites were relatively more abundant when sampled from tubers of *S. phureja*.

by Hughes et al. (1975a,b), who demonstrated that reduced cell firmness was associated with the release of pectic polysaccharides into the cooking medium when potato discs of a uniform starch content and cell size had been cooked for varying times. The degree of cell separation on cooking is also dependent on the nature of the cooking medium as this will affect the degree and rate of leaching of depolymerized pectic substances. Conditions that disrupt interactions of pectic polysaccharides with calcium and magnesium bridges by ion exchange, hydrogen bond breakage or chelation will reduce cell adhesion. Thus, inclusion of chelating agents such as sodium hexametaphosphate (Hughes et al., 1975c) or EDTA (Matsuura-Endo et al., 2002b) induces cell separation. Conversely, soaking tubers in a solution of calcium chloride prior to cooking reduces cell separation probably as a result of stabilizing or enhancing linkages between homogalacturonan acid chains in the middle lamella (Storey and Davies, 1992).

Attempts have been made to dissect the various factors that contribute to cell separation during cooking (Matsuura-Endo et al., 2002b). This study compared three cultivars with tubers that disintegrated to different extents on cooking. The swelling power of isolated starch was determined and found to be greatest in the cultivar that exhibited the lowest degree of cooking-dependent cell separation, indicating that this parameter was not of primary importance in the process. Tuber cell separation was induced by treatment of tuber slices for 32 h in the presence of EDTA at 50°C. At this temperature, starch is not gelatinized and cannot contribute to cell separation. The cultivar that exhibited the greatest degree of cell separation under these conditions also exhibited the highest degree of boiling-induced cell separation. Additionally, the cultivar that exhibited the highest degree of cooking-dependent cell separation contained significantly lower levels of calcium and galacturonic acids, possibly reflecting differences in middle lamella structure. However, studies by Van Dijk et al. (2002) have suggested that there was minimal relationship between cell wall and texture. They studied two potato cultivars with very distinct cooked textures that had been segregated into small, medium and large tubers with low and high dry matter and which were stored for set periods. They found that there were no significant differences in pectin composition and amount between the cultivars and that despite large changes in pectin during storage there were minimal changes in sensory-perceived texture of the cooked potatoes following storage. Indeed, they found that dry matter (starch content) was the dominant factor with regard to texture.

Cell cleavage is also an important factor in determining texture after cooking (Jarvis and Duncan, 1992). A soft texture arises from both cell separation and cell cleavage, whereas a dry texture results from a high degree of cell separation with lesser degree of cell cleavage.

With the advent of metabolic profiling technologies, potato microarrays, transgenic technologies and improved knowledge of potato genetics, we are poised to build on the studies outlined above in order to produce potato cultivars with improved texture. As with flavour, potato germplasm that exhibits major differences in tuber texture exists (De Maine et al., 2000). It is now possible to establish genetic populations that segregate for aspects of tuber texture and to carry out QTL analysis of these populations. As many of the genes involved in starch and cell wall biosynthesis and turnover are known, it may be possible to use microarray technologies to identify candidate genes by correlating their expression patterns in tubers that show differences in texture. Candidate genes can then be function-tested in transgenic models by modulating their expression levels and examining the effects on tuber texture. Already, attempts have been made to engineer potato cell-wall structure by the down-regulation of endogenous genes (Pilling et al., 2004) or the over-expression of bacterial genes (Oomen et al., 2002). Ultimately, it may be possible to identify alleles of genes associated with particular aspects of texture and to design molecular markers for such alleles to assist their introgression into commercial germplasm.

ACKNOWLEDGEMENTS

The authors are grateful to Dr. Tom Shepherd for providing Fig. 24.1 and to many colleagues at SCRI for discussions on potato flavour and texture. This work was funded by the Scottish Executive Environment and Rural Affairs Department.

REFERENCES

Barrios E.P., D.W. Newsom and J.C. Miller, 1963, *Am. Potato J.* 40, 200.
Beksan E., P. Schieberle, F. Robert, I. Blank, L.B. Fay, H. Schlichtherle-Cerny and T. Hofmann, 2003, *J. Agric. Food Chem.* 51, 5428.
Bellisle F., 1998, *Food Rev. Int.* 14, 309.
Blenkinsop R.W., L.J. Copp, R.Y. Yada and A.G. Marangoni, 2002, *J. Agric. Food Chem.* 50, 4545.
Bradshaw J.E., B. Pande, G.J. Bryan, C.A. Hackett, K. McLean, H.E. Stewart and R. Waugh, 2004, *Genetics* 168, 983.
Brett C.T. and K.W. Waldron, 1996, *The Physiology and Biochemistry of Plant Cell Walls*, 2nd ed. Chapman and Hall: London.
Brierley E.R., P.L.R. Bonner and A.H. Cobb, 1996, *J. Sci. Food Agric.* 70, 515.
Brierley E.R., P.L.R. Bonner and A.H. Cobb, 1997, *Plant Sci.* 127, 17.
Brittin H.C. and J.E. Trevino, 1980, *J. Food Sci.* 45, 1425.
Brown J., G.R. Mackay, H. Bain, D.W. Griffith and M.J. Allison, 1990, *Potato Res.* 33, 219.
Buri R. and J. Solms, 1971, *Naturwissenschaften* 58, 56.
Burton W.G., 1989, *The Potato*, 3rd ed. Longman Scientific and Technical. Harlow, UK.
Bush M.S., M. Marry, I.M. Huxham, M.C. Jarvis and M.C. McCann, 2001, *Planta* 213, 869.
Bush M.S. and M.C. McCann, 1999, *Physiol. Plant.* 107, 201.
Buttery R.G. and L.C. Ling, 1973, *J. Agric. Food Chem.* 21, 745.
Carpita N.C. and D.M. Gibeaut, 1993, *Plant J.* 3, 1.
Cherif A. and A. Ben Abdelkader, 1970, *Potato Res.* 13, 284.
Cotrufo C. and P. Lunsetter, 1964, *Am. Potato J.* 41, 18.
Davids S.J., V.A. Yaylayan, G. Turcotte, 2004, *Lebensm. Wiss. U. Technol.* 37, 619.
De Maine M.J., A.K. Lees, D.D. Muir, J.E. Bradshaw and G.R. Mackay, 2000, In: S.M.P. Khurana, G.S. Shekhawat, B.P. Singh and S.K. Pandey (eds), *Potato, Global Research and Development*, Volume 1. Indian Potato Association, Shimla, India.
Desjardins A.E., S.P. McCormick and D.L. Corsini, 1995, *J. Agric. Food Chem.* 43, 2267.
Di R., J. Kim, M.N. Martin, T. Leustek, J. Jhoo, C.T. Ho and N.E. Tumer, 2003, *J. Agric. Food Chem.* 51, 5695.
Dobson G., D.W. Griffiths, H.V. Davies. and J.W. McNicol, 2004, *J. Agric. Food Chem.* 52, 6306.
Dobson G., T. Shepherd, R. Marshall, S.R. Verrall, S. Conner, D.W. Griffiths, J.W. McNicol, D. Stewart and H.V. Davies, 2007, In: B.J. Nicolau and E.S. Wurtele (eds), *Concepts in Plant Metabolomics*. Springer, The Netherlands.
Duckham S.C., A.T. Dodson, J. Bakker and J.M. Ames, 2001, *Nahrung* 45, 317.
Duckham S.C., A.T. Dodson, J. Bakker and J.M. Ames, 2002, *J. Agric. Food Chem.* 50, 5640.
Dudareva N. and F. Negre, 2005, *Curr. Opin. Plant Biol.* 8, 113.
Ferry A.L., J. Hort, J.R. Mitchell, S. Lagarrigue, and B. Pamies, 2004, *J. Textural Stud.* 35, 511.
Finglas P.M. and R.M. Faulks, 1984, *J. Agric. Food Chem.* 35, 1347.
Frankel E.N., 1998, *Lipid Oxidation*, The Oily Press. Bridgwater, UK.

Fryer P.J., 1997, In: P.J. Fyrer, D.L. Pyle and C.D. Rielly (eds), *Thermal Treatment of Foods, in Chemical Engineering for the Food Industry*. Blackie Academic and Professional: London.
Galliard T., 1973, *J. Agric. Food Chem.* 24, 617.
Gardner H.W., 1995, *HortScience* 30, 197.
Gebhardt C., A. Ballvora, B. Walkemeier, P. Oberhagemann and K. Schuler, 2004, *Mol. Breeding* 13, 93.
Gray D., 1972, *Potato Res.* 15, 317.
Halpern B.P., 2000, *J. Nutr.* 130, 910.
Hesse H., O. Kreft, S. Maimann, M. Zeh and R. Hoefgen, 2004, *J. Exp. Bot.* 55, 1799.
Hoff J.E., 1972, *J. Agric. Food Chem.* 20, 1283.
Hoff J.E., 1973, *HortScience* 8, 108.
Hughes J.C., R.M. Faulks and A. Grant, 1975a, *J. Agric. Food Chem.* 26, 731.
Hughes J.C., A. Grant and R.M. Faulks, 1975b, *J. Sci. Food Agric.* 26, 739.
Hughes J.C., R.M. Faulks and A. Grant, 1975c, *Potato Res.* 18, 495.
Ikeda K., 1912, *On the Taste of the Salt of Glutamic Acid.* Proceedings of the 8th International Congress of Applied Chemistry Volume 38, 147.
Jarvis M.C., E. Mackenzie and H.J. Duncan, 1992, *Potato Res.* 35, 93.
Jarvis M.C. and H.J. Duncan, 1992, *Potato Res.* 35, 83.
Jarvis M.C., M.A. Hall, D.R. Threlfall and J. Friend, 1981, *Planta* 152, 93.
Josephson D.B. and R.C. Lindsay, 1987, *J. Food Sci.* 52, 328.
Klaus D., J.B. Ohlrogge, H.E. Neuhaus and P. Dormann, 2004, *Planta* 219, 389.
Lindsay R.C., 1996, In: O.R. Fennema (ed.), Food Chemistry, 3rd ed. Dekker: New York.
Linehan D.J., C.E. Stooke and J.C. Hughes, 1968, *Eur. Potato Res.* 11, 221.
Lugt C., 1961, *Results of the Assessment of the Cooking Quality of Internationally Exchanged Potato Samples.* Proceedings of the 1st Triennial Conference of the European Association of Potato Research, Braunschweig-Volkenrode, FDR, p. 321.
Maarse H. (ed.),1991, *Volatile Compounds in Foods and Beverages.* Marcel Dekker: New York.
Maga J.A., 1994, *Food Rev. Int.* 10, 1.
Maga J.A. and J.A. Twomey, 1977, *J. Food Sci.* 42, 541.
Matsuura-Endo C., A. Ohara-Takada, H. Yamauchi, M. Mori and S. Fujikawa, 2002a, *Food Sci. Technol. Res.* 8, 252.
Matsuura-Endo C., A. Ohara-Takada, H. Yamauchi, Y. Mukasa, M. Mori. and K. Ishibashi, 2002b, *Food Sci. Technol. Res.* 8, 323.
Murray K.E. and F.B. Whitfield, 1975, *J. Sci. Food Agric.* 26, 973.
Oomen R.J.F.J., C.H.L. Doeswijk-Voragen, M.S. Bush, J.P. Vincken, B. Borkhardt, L.A.M. van den Broek, J. Corsar, P. Ulvskov, A.G.J. Voragen, M.C. McCann and R.G.F. Visser, 2002, *Plant J.* 30, 403.
Ormerod A., J. Ralfs, S. Jobling and M. Gidley, 2002, *J. Mater. Sci.* 37, 1667.
Oruna-Concha M.J., J. Bakker and J. M. Ames. 2002a, *Lebensm. Wiss. Technol.* 35, 80.
Oruna-Concha M.J., J. Bakker and J.M. Ames, 2002b, *J. Sci. Food Agric.* 82, 1080.
Oruna-Concha M.J., S.C. Duckham and J.M. Ames, 2001, *J. Agric. Food Chem.* 49, 2414.
Osman S.F., 1983, *Food Chem.* 11, 235.
Parker C.C., M.L. Parker, A.C. Smith and K.W. Waldron, 2001, *J. Agric. Food Chem.* 49, 4364.
Pilling J., L. Willmitzer, H. Bucking and J. Fisahn, 2004, *Planta* 219, 32.
Ramsay G., D.W. Griffiths and N. Deighton, 2004, *Genet. Resour. Crop. Ev.* 51, 805.
Reeve R.M., 1977, *J. Texture Stud.* 8, 1.
Reeve R.M., H. Timm and M.L. Weaver, 1973, *Am. Potato J.* 50, 49.
Ross H., P. Pasemann and W. Nitzsche, 1978, *Z. Pflanzenzuchtung* 80, 64.
Selvendran R.R., 1985, *J. Cell Sci. Suppl.* 2, 51.
Selvendran R.R., B.J.H. Stevens and M.S. Dupont, 1987, *Adv. Food Res.* 31, 117.
Sengul M., F. Keles and M.S. Keles, 2004, *Food Control* 15, 281.
Shomer I., 1995a, *Carbohydr. Polym.* 26, 47.
Shomer I., 1995b, *Carbohydr. Polym.* 26, 47.
Sinden S.L., K. Deahl and B.B. Aulenbach, 1976, *J. Food Sci.* 41, 520.
Slanina P., 1990, *Food Chem. Toxicol.* 28, 759.
Solms J., 1971, In: G. Olhloff and A.F. Thomas (eds), *Gustation and Olfaction.* Academic Press, London, UK.

Solms J. and R. Wyler, 1979, In: J. Boudreau (ed.), *Taste Components of Potatoes in Food Taste Chemistry*. ACS Symposium Series, Washington, D.C., USA.
Storey R.M.J. and H.V. Davies, 1992, In: P. Harris (ed.), *The Potato Crop*, 2nd ed. Chapman and Hall: London.
Van Dijk C, J.G. Beekhuizen, T. Gibcens, C. Boeriu, M. Fischer and T. Stolle-Smits, 2002, *J. Agric. Food Chem.* 50, 5089.
Van Marle J.T., C. Van Dijk, A.G.J. Voragen and E.S.A. Biekman, 1994, *Potato Res.* 37, 183.
Wang Y. and S.J. Kays, 2003, *J. Am. Soc. Hortic. Sci.* 128, 711.
Wilson W.D., I.M.R. Mackinnon and M.C. Jarvis, 2002a, *J. Sci. Food Agric.* 82, 1074.
Wilson W.D., I.M.R. Mackinnon and M.C. Jarvis, 2002b, *J. Sci. Food Agric.* 82, 1070.
Zeh M., A.P. Casazza, O. Kreft, U. Roessner, K. Bieberich, L. Willmitzer, R. Hoefgen and H. Hesse, 2001, *Plant Physiol.* 127, 792.
Zitnak A., 1961, *Can. J. Biochem. Physiol.* 39, 1257.

Part VI

PESTS AND DISEASES

Chapter 25

Insect Pests in Potato

Edward B. Radcliffe[1] and Abdelaziz Lagnaoui[2]

[1]*Department of Entomology, University of Minnesota, 219 Hodson Hall, 1980 Folwell Ave., St. Paul, MN 55108, USA;*

[2]*The World Bank, 1818 H. Street, NW, Washington, DC 20433, USA*

There are various insect pests of potato capable of causing tuber yield or quality reductions of 30–70% if not routinely controlled (Raman and Radcliffe, 1992). Losses of this magnitude can occur in potatoes unprotected from Colorado potato beetle, *Leptinotarsa decemlineata* (Say) (Chrysomelidae); aphid (Aphididae)-transmitted viruses, e.g. potato leafroll virus (PLRV) (Luteoviridae, Genus *Polerovirus*) and potato virus Y (PVY) (Potyviridae, Genus *Potyvirus*); potato tuber moths (potato tuberworm), *Phthorimaea operculella* (Zeller) (Gelechiidae); *Symmetrischema tangolias (Plaesiosema)* (Gyen) (Gelechiidae); Guatemalan moth, *Tecia (Scrobipalpopsis) Solanivora* (Polyvoný) (Gelechiidae) and tomato leafminer, *Tuta (Scrobipalpula) absoluta* (Meyrick); pea leafminer, *Liriomyza huidobrensis* (Blanchard) (Agromyzidae); Andean potato weevils, *Premnotrypes* spp. (Curculionidae) and potato leafhopper, *Empoasca fabae* (Harris) (Cicadellidae).

25.1 YIELD AND QUALITY EFFECTS

25.1.1 Defoliators

Of the multitude of leaf-feeding insects attacking potato, the best known and most destructive is Colorado potato beetle. Numerous researchers have reported on the effects of defoliation imposed at various stages of potato growth (Elkinton et al., 1985). These studies were consistent in that plant damage inflicted before tuber bulking delayed haulm and tuber growth rates but plants could recover, whereas damage during tuber bulking permanently reduced growth rates indicating decreased assimilation efficiency (Dripps and Smilowitz, 1989). Ratio of leaf area to plant dry matter weight appeared to explain this temporal shift in sensitivity to damage, with vegetative plants having greater leaf area to dry weight ratios than tuber-bulking plants. Early season damage to upper younger portions of the plant, typical of most defoliating insects, can actually increase assimilation efficiency and tuber yields, presumably because opening the canopy promotes development of axillary buds which permits greater light interception by lower leaves. In general, potato growers tend to have a lower tolerance threshold for insect-inflicted defoliation than does potato, often resulting in unnecessary application of insecticide.

Potato Biology and Biotechnology: Advances and Perspectives
D. Vreugdenhil (Editor)

25.1.2 Sap feeders

Insects with piercing-sucking mouthparts damage potatoes by direct physical injury, extracting phloem, injecting toxic saliva and transmitting pathogens or facilitating their establishment on the host. Although potato has considerable ability to compensate for early season loss of foliage by defoliators, there is little compensation for adverse effects on plant health, e.g. disruption of nutrient transport, reduced photosynthetic efficiency or infection with plant pathogens. Direct physical injury by sap-feeding insects often results in premature leaf senescence.

25.1.3 Pathogen transmission

Many insect pests are potential vectors of potato pathogens. This association can be casual with the insect serving as a mechanical vector or providing entry for invasion by pathogens present in the environment. Of greatest importance are instances in which the insect is the principal or sole means of pathogen transmission. Aphid-transmitted viruses probably cause greater economic loss in potato production than all other insect-related damage. Some 40 virus species are known to infect potato, and of these, 13 are aphid-transmitted (Salazar, 1996, Chapter 28, Valkonen, this volume).

The aphid-transmitted potato viruses of most common occurrence and greatest economic importance are PLRV and PVY. Planting seed tubers with a high incidence of secondary (tuber borne) PLRV or PVY infection can result in yield losses exceeding 50–80% (Beemster and de Bokx, 1987). Primary infection (current season) with either PLRV or most PVY strains tends to have little effect on tuber yield. However, primary infection with PLRV can affect tuber quality in certain cultivars, most notably Russet Burbank, by causing a phloem defect termed 'net necrosis' that develops in storage.

Other aphid-transmitted potato viruses include potato virus A (PVA, genus *Potyvirus*), potato virus M [PVM (paracrinkle), genus *Carlavirus*], potato virus S (PVS, genus *Carlavirus*), potato latent virus (PLV, genus *Carlavirus*), Alfalfa mosaic virus [AMV (calico), genus *Alfamovirus*] and cucumber mosaic virus (CMV, genus *Cucumovirus*).

Most plant viruses occur in nature as variants, which if they differ sufficiently from type are designated as strains. A number of PVY strains have been described that vary in symptomology in potato. PVY^{N} (the 'tobacco Veinal necrotic strain' of PVY) appears increasingly common worldwide and is predominant in Europe (Weidemann, 1988). PVY^{N} causes severe necrosis in tobacco but in potato is usually mild compared with PVY^{O} ('ordinary' PVY) and is essentially asymptomatic in many cultivars. PVY^{NTN} (tuber necrosis subgroup) infection can induce a rapid and severe systemic veinal necrosis and severely damaged tubers that cannot be marketed or stored. This disease, potato tuber necrotic ringspot disease (PTNRD), is now distributed worldwide (Nie and Singh, 2002). Co-infection of PVY with other viruses, e.g. potato virus X (PVX, genus *Potexvirus*) or PVS, may influence the severity of symptom expression. In North America, recombinants of PVY^{N} and PVY^{O} (designated as $PVY^{N:O}$) are now more common than PVY^{O} (Piche et al., 2004). All PVY strains are transmitted by aphids, but PVY^{C}, stipple streak strain, is transmitted only in the presence of a helper component the vector acquires feeding on PVY- or PVA-infected plants. Within vector species, populations differ in ability to transmit (Gray and Gildow, 2003; Basky and Almási, 2005). Virus infections not readily detected by visual symptoms because of virus strain or asymptomatic cultivars make it difficult for seed certification programmes to purge virus inoculum (Mollov and Thill, 2004).

25.1.4 Root and tuber feeding

Insect damage to potato roots and tubers is usually caused by feeding of larval stages with adults generally causing little or no subterranean damage. Soil insects tend to have minor effects on potato yields, but the damage caused can greatly reduce tuber quality. Damaged tubers may necessitate post-harvest sorting, reduce saleability or market price and create storage problems. Some insects damage potato seed pieces or roots of developing seedlings resulting in weak plants or stand loss. Others, e.g. seedcorn maggot, *Delia platura* (Meig.) (Anthomyiidae), can transmit bacterial cells of *Erwinia* spp. (Enterobacteriaceae), causal agents of such potato pathogens as blackleg, aerial stem rot and tuber soft rot (see Chapter 27, Van der Wolf and De Boer, this volume).

Each region and potato production system has its own complex of tuber-infesting pests. In the high altitudes of South America, the Andean weevil complex, primarily *Premnotrypes* spp. (Curculionidae), can be extraordinarily destructive. Other regions contend with white grubs (Scarabaeidae), wireworms (Elateridae), weevils, subterranean cutworms (Noctuidae) or others of a host of potential tuber pests. In worldwide economic importance, no tuber pest equals potato tuber moth. This insect damages both foliage and tubers, but potential for loss is greatest in storage. Larvae entering storage in infested tubers can continue their development filling the tubers with frass and enabling entry of decay organisms.

Larvae of most tuber-infesting potato pests are generalist herbivores. Thus, knowledge of previous crop rotation is helpful in assessing risk and need for insecticidal control. Planting into warm soil to assure rapid plant growth is recommended. To protect developing tubers from damage by soil insects, insecticides must be applied at-planting, either in-furrow or as seed treatments, but control is often incomplete. There is little that can be done to protect the potato crop after planting other than to assure that seed pieces and developing tubers are well covered with soil.

25.2 INSECT PESTS OF WORLDWIDE IMPORTANCE

25.2.1 Aphids

25.2.1.1 Occurrence and damage

High aphid densities can cause significant yield losses and even direct plant injury, but because of their role as virus vectors, aphids are of paramount concern to seed potato producers. Severe losses of tuber yield and quality in ware production can result when potato plants are propagated from virus-infected seed pieces (Marshall et al., 1988; Hamm and Hane, 1999; Hane and Hamm, 1999).

In plants grown from infected tubers, PLRV is found almost exclusively in phloem tissues and most abundantly in companion cells (van den Heuvel et al., 1995). Thus, vectors must feed on phloem to acquire the virus. Potato leafroll virus is only transmitted by aphids that colonize potato and not by all colonizing aphids. Green peach aphid (peach–potato aphid), *Myzus persicae* (Sulzer), is the most efficient and important vector of PLRV. Most populations of potato aphid (green and pink potato aphid), *Macrosiphum euphorbiae* (Thomas), transmit PLRV poorly if at all, but this species occasionally has been implicated in early season spread (Woodford et al., 1995). Other potential PLRV vectors include black bean aphid, *Aphis fabae* Scopoli; buckthorn aphid (buckthorn–potato aphid), *Aphis nasturtii* (Kaltenbach); the melon (cotton) aphid complex, *Aphis gossypii* Glover [including *Aphis frangulae* (Kaltenbach)]; shallot aphid, *Myzus ascalonicus* Doncaster; violet aphid, *Myzus ornatus* Laing; mottled arum aphid, *Aulacorthum* (*Neomyzus*) *circumflexum* (Buckton); foxglove aphid (glasshouse–potato aphid), *Aulacorthum solani* (Kaltenbach); damson–hop aphid, *Phorodon humuli* (Schrank); the stolon-infesting bulb and potato aphid, *Rhopalosiphoninus latysiphon* (Davidson) and mangold aphid, *Rhopalosiphoninus staphyleae* (Koch). It should be noted that *M. persicae* is a species complex with two other described species, *Myzus nicotianae* (Blackman) and *Myzus antirrhinii* (Macchiati). Genetic marker research has cast doubt on validity of *M. nicotianae* as a distinct species but confirmed validity of *M. antirrhinii* (Blackman and Spence, 1992).

Over 50 aphid species are known to transmit PVY (Ragsdale et al., 2001). Green peach aphid, the most important vector of PLRV, is also the most efficient vector of PVY (Bourdin et al., 1998), but the greater abundance of some less-efficient vector species, or their propensity to develop alatae, can make them more important in PVY epidemiology (Harrington and Gibson, 1989; DiFonzo et al., 1997; Heimbach et al., 1998; Davis et al., 2005). Most potential PVY vectors have only transient association with potato, alighting and probing, but not colonizing. Aphid species implicated as important PVY vectors include potato and pea aphids, *Acyrthosiphon pisum* (Harris); black bean aphid; melon aphid; buckthorn aphid; soybean aphid, *Acyrthosiphon glycines* (Matsumura); leaf-curling plum aphid, *Brachycaudus helichrysi* (Kaltenbach) and several species of cereal aphids, the most important being bird cherry–oat aphid, *Rhopalosiphum padi* (L.).

25.2.1.2 Life cycles

Aphid life cycles often involve clonal polymorphism (morphs). Polymorphism allows aphids to allocate resources efficiently to accomplish specific functions but imposes constraints, e.g. resources allocated to flight muscles are not available for reproduction. In 97% of aphid species, reproduction is both asexual (parthenogenetic) and sexual, with multiple generations of asexual, viviparous reproduction between each occurrence of sexual reproduction (Blackman, 1980). Most aphid species have four larval instars (nymphs) before becoming adults. In temperate regions, most aphid species show seasonal alternation of hosts (are heteroecious) and overwinter as fertilized eggs (are holocyclic). The overwintering (primary) host is often a woody plant, whereas summer (secondary) hosts are usually short-lived herbaceous species. The primary host range typically is highly specific, whereas the secondary host range may be broad. Alternation of hosts is an evolutionary adaptation that has permitted aphids to exploit agricultural crops more successfully perhaps than any other insect group (Moran, 1992).

Green peach aphid eggs are deposited in fall on the primary host, peach, *Prunus persica* (L.) Batsch (Rosaceae), and closely related *Prunus* species. In spring, eggs hatch producing nonwinged (apterae), parthenogenetic, viviparous stem mothers (fundatrices). Winged progeny (alatae) appear by second generation, and the number and proportion of winged progeny peak in the third generation on the primary host (Boiteau and Parry, 1985). Winged green peach aphid invading potato can be true spring migrants from the primary host or alatae produced after several generations of parthenogenetic, viviparous reproduction (virginoparae) on secondary hosts, which include many common crops and weeds. Progeny of spring migrants are invariably apterous, but some proportion of each succeeding generation is winged. In late season, fall migrants, male and female, are produced. Female migrants that reproduce asexually on the primary host (gynoparae) give birth to apterous, sexual female progeny (oviparae). Oviparae mate with male migrants (androparae) and produce fertilized eggs. In harsh climates, wintering outdoors is not always possible. In milder climates and glasshouses, continuous asexual reproduction can occur. In some parts of the world, particularly in the tropics, clones that have lost the ability to produce sexually can persist. The parthenogenetic life style of aphids coupled with a rapid completion of generations (typically 7–10 days) allows for spectacular rates of population increase. Under ideal field conditions, green peach aphid populations can double in as little as 1.7 days.

25.2.1.3 Transmission of viruses

Numerous authors have reviewed the role of aphids in transmitting potato viruses (Ragsdale et al., 2001; Robert and Bourdin, 2001; Radcliffe and Ragsdale, 2002). A common theme of these reviews is that aphids possess biological attributes that make them especially effective in acquiring and transmitting plant viruses. Among the most important of these attributes are complex life cycles with specialized morphs, alternation of host plants and exceptionally short life cycles facilitated by parthenogenetic reproduction. Movement of virus into a potato field from an outside inoculum source is almost exclusively by winged aphids (Boiteau, 1997). Within field, spread of PLRV is often by apterae walking from plant to plant (Hanafi et al., 1989), but movement of PVY is thought to be almost exclusively by alatae (Ragsdale et al., 1994). Green peach aphid apterae tend to be more efficient vectors of PLRV than are alatae, and nymphs are more efficient than adults (Robert, 1971).

Aphid-transmitted viruses differ in how they are acquired, whether they circulate within the body of the vector, how they are transmitted to healthy plants and how long a vector remains infective following virus acquisition (Nault, 1997). Potato leafroll virus is circulative in the vector and transmitted persistently. A latent period (typically 24 h or more) exists between acquisition of PLRV by the vector and onset of ability to transmit. This lag occurs because PLRV must pass through the gut into the hemocoel and enter accessory salivary glands before transmission can occur. Potato leafroll virus does not replicate in aphids. Both acquisition and transmission take minimum feeding times of minutes to hours with maximum acquisition requiring about 12 h. Once a vector acquires PLRV, it usually remains infective for life and capable of transmission even following moults. Green peach aphid acquires more PLRV and tends to transmit sooner when it feeds at higher temperatures (Syller, 1994). Potato leafroll virus is acquired more readily from the top leaves of young, secondarily infected plants than from older plants, but acquisition is not correlated with virus titre or feeding activity (van den Heuvel et al., 1993). Bacterial endosymbionts, *Buchnera* spp. (Lepraliellidae), play a role in PLRV transmission (van den Heuvel et al., 1994). Potato leafroll virus has a strong affinity for symbionin, the major protein synthesized and released into the hemolymph by *Buchnera*. Potato leafroll virus cannot be transmitted mechanically.

All known aphid-transmitted potato viruses other than PLRV are transmitted in a nonpersistent manner. Transmission of nonpersistent viruses is unique to aphids and occurs during short duration surface probes required for host recognition (Pollard, 1973). With nonpersistent viruses, e.g. PVY, no latent period occurs, and the aphid remains viruliferous only for a few feeding probes following acquisition and never following a moult. Nonpersistent viruses are described as stylet borne, but clearly, more dynamic processes are involved than mere stylet contamination as not all aphid species can transmit nonpersistent viruses. Electrical penetration graph (EPG) recording has been used to associate specific aphid-feeding activities with *Potyvirus* transmission. Results show that *Potyvirus* inoculation occurs during the first subphase of stylet intracellular puncture (Martín et al., 1997), whereas acquisition occurs primarily in the last subphase (Powell, 2005).

Association of aphid flight activity and spread of potato viruses is well documented. Aphid trapping networks have been established in many potato seed-producing areas to monitor aphid flight activity and issue risk advisories. A European consortium operates 12-m suction traps at 73 sites in 19 countries 'to provide an integrated, standardised, long term, consolidated, Europe-wide database on aphid incidence' (Harrington and EXAMINE Consortium, 2004). Trap captures were found to be representative of green peach aphid abundance patterns over large areas of Northwestern Europe and to be associated with climatic gradients and land use patterns (Cocu et al., 2005b). High temperatures and low rainfall had a weak-positive association with abundance (Cocu et al., 2005a). Mild winters and early springs favour green peach aphid, e.g. an increase of 1 °C in mean January–February temperatures advances spring migration by 2 weeks in the UK (Harrington et al., 1995). However, local crop landscapes were the factors most strongly associated with local abundance, e.g. area of oilseed rape (canola), *Brassica napus* L. (Cruciferae), was positively correlated with green peach aphid abundance, whereas first capture was associated with rainfall (Cocu et al., 2005c).

25.2.1.4 Control

Insecticides are the only practical means of suppressing colonizing aphids on crops but are of inconsistent benefit in controlling virus spread. Among reported successes in controlling virus spread with insecticides (all crops and insect vectors), 94 of 119 cases involved persistent and semi-persistent viruses (Perring et al., 1999), whereas most failures, 32 of 48 cases, involved nonpersistent viruses. Viruliferous alatae are seldom killed quickly enough to prevent virus transmission, even of PLRV. In contrast, spread of PLRV from within-field sources can be interrupted because of the extended post-acquisition latent period before an aphid can transmit. Systemic insecticides applied at time of planting or plant emergence can reduce within field spread of PLRV significantly (Woodford et al., 1988; Boiteau and Singh, 1999).

Insecticide resistance is a worldwide problem in green peach aphid (Devonshire et al., 1998; Dewar et al., 1998), and resistance has developed to all major insecticide classes except neonicotinoids and pymetrozine (Nauen and Elbert, 2003). Resistance to organophosphates was first documented in green peach aphid from Washington State in 1952. Resistant populations spread rapidly, and within 10 years, aphids with the identical resistance genes, *E4* and *FE4*, were found in Europe and elsewhere leading to the conclusion that migration of winged aphids and transport of aphid-infested plants had spread resistance throughout the world. Impairment of aphid control has not been reported to neonicotinoids, but susceptibility of green peach aphid can vary 20-fold among populations (Nauen and Denholm, 2005). Most organophosphate, carbamate and pyrethroid insecticides registered for use on potato tend to flare outbreaks of green peach aphid (ffrench-Constant et al., 1988; Harrington et al., 1989). Secondary outbreaks are usually attributed to disruption of natural enemies and insecticide resistance in the secondary pest, but in some instances, there is a direct stimulation of reproduction (Gordon and McEwen, 1984).

Mineral oils applied to plants can substantially reduce PVY transmission. Limitations to crop oils include weathering of oil deposits, new plant growth between applications and incomplete coverage. The mechanism by which oils inhibit aphid transmission of viruses is still not known. Bradley et al. (1966) suggested that oil interferes with adherence of the virus to receptors on the mouthparts. Crop oils can be mildly repellent to aphids, and some cause direct mortality, but the latter effect is unlikely to occur quickly enough to prevent transmission of a nonpersistently transmitted virus (Martín et al., 2004).

Current potato cultivars differ too little in aphid susceptibility for host plant resistance to be useful in preventing virus spread. Through *Agrobacterium*-mediated transformation, various potato lines were generated expressing genes conferring resistance to viruses (Berger and German, 2001). Transgenic lines have been developed that are highly resistant, but not immune to infection by PLRV, PVY and PVX. Although aphids can still acquire virus from low-titre plants, efficiency of transmission is greatly reduced (Thomas et al., 1997). This technology was far more effective than any previously used tactic, but these cultivars have been withdrawn because of concerns over a public backlash against genetically modified food (Thornton, 2003).

Ideally, seed potato increase should occur in localities well isolated from potential sources of virus inoculum. However, in many countries, there is little isolation of late-generation seed and commercial potato production. The question then is what is the minimum separation from sources of virus inoculum or crops producing large vector populations required to reduce risk of virus spread to acceptable levels? Halbert et al. (1990) suggested that 400 m–5 km could provide effective isolation from known PVY sources but that 32 km might be required for isolation from PLRV sources, because the vector remains infective for life. In England, minimum separation of 800 m from potential sources of PVY is recommended (Harrington et al., 1986), although just 40 m reduced spread of PVY in Denmark (Hiddema, 1972). Isolation can also be achieved by early planting and haulm destruction ('vine-kill').

Physical removal of symptomatic plants from seed potato fields has long been an important virus management tactic and is most practical when virus incidence is low and the field is small enough that every plant can be inspected several times during the growing season. Roguing is easiest to accomplish before the canopy closes and must be accomplished before secondary infection occurs. If the seed field is heavily infected (>1% virus), roguing is often ineffective because some infected plants are almost certain to be missed and will remain as sources of inoculum. When vector pressure is high, even initially low levels of seed-borne infection can result in explosive spread (Sigvald, 1989).

Polymer webs can provide a high degree of protection against aphid-transmitted viruses (Harrewijn et al., 1991). However, the cost and inconvenience of row covers limit their application to seed potato fields of very high value and small size (e.g., first-generation field increase after tissue culture). Barrier crops are more widely adaptable than mulches or floating row covers; they are easier to install and keep in place and do not lose effectiveness because of weathering or when the crop canopy closes. Barrier crops should have a fallow outside border with no gap between barrier crop and the potato field, because winged aphids tend to alight at the interface of fallow ground and crop canopy. If immigrating alatae carrying PVY feed first on the border crop, they will probably lose their virus inoculum before moving into the potatoes (DiFonzo et al., 1996). Barrier crops need be only a few metres wide to be effective. In Lower Saxony, oat borders just 1 m wide reduced the number of winged aphids, e.g. *R. padi*, caught in potato fields and were more effective in reducing PVY spread than intensive use of insecticides (Thieme et al., 1998). However, crop borders will not be effective if there is virus inoculum within the field being protected.

In temperate climates, most known weed hosts of PVY or PLRV are annuals, and as transmission through true seed does not occur, these species are not virus sources. In the Souss Valley of Morocco, jimsonweed, *Datura stramonium* L. (Solanaceae), supports green peach aphid before the winter potato crop emerges. This weed and volunteer potatoes were implicated as being principal sources of both viruliferous aphids and PLRV inoculum (Hanafi et al., 1995). Volunteer potatoes that emerge as weeds in rotation crops in the Columbia Basin are important PLRV reservoirs (Thomas, 1983), as are potatoes sprouting in cull piles in Canada (Frazer, 1987). Annual weed hairy nightshade, *Solanum sarrachoides* (Sendtner), is both a reservoir of PLRV and a preferred host of green peach aphid in the US Pacific Northwest (Alvarez and Srinivasan, 2005). However, because ware producers often plant tubers with 1–5% virus, potato generally overshadows weeds as a source of potato viruses. In regions where winter frost extends below the root zone of potatoes, volunteer potatoes are rare and usually inconsequential sources of potato viruses.

Static action thresholds in the range of 20–100 green peach aphid apterae per 100 leaves have been proposed to control PLRV spread in potatoes grown for fresh market or processing. The threshold concept may be less applicable to seed potato production because of the biological complexities involved and stringent phytosanitary standards that must be achieved, but thresholds in the range of 1–10 apterae per 100 leaves have been proposed for cultivars susceptible to net necrosis and in seed production (Flanders et al., 1991).

25.2.2 Colorado potato beetle

25.2.2.1 *Origin, occurrence and damage*

The genus *Leptinotarsa* apparently originated in central Mexico, where the primary host of Colorado potato beetle is still buffalobur, *Solanum rostratum* Dunal, and *Solanum angustifolium* Mill. Westward expansion of agriculture into the US central plains introduced potato production within the range of Colorado potato beetle distribution by 1820, but the beetle was unknown as a potato pest until 1859. Over the next 25 years, Colorado potato beetle extended its range north into Canada east to the Atlantic and became permanently established in Europe prior to 1922 (de Wilde and Hsiao, 1981). Its spread eastward was rapid, reaching Germany in 1936, Poland in the mid-1950s, the western borders of the USSR by 1959 and the Caucasus and Turkey by 1976. Colorado potato beetle has since spread to Iran, the Central Asian Republics, western China, Siberia and Russian far east.

In temperate climates, Colorado potato beetle passes winter as adults in the soil. The developmental threshold for this species is about 10 °C, but a considerable range of values have been reported for different populations. Emergence in spring usually occurs before potatoes have broken ground. The insect is remarkably fecund, with females typically laying 300–500 eggs in masses of 20 or more. At constant temperatures, development is optimal at 28 °C, with the beetle going from egg to adults in as little as 20.7 days (Ferro et al., 1985). At optimal temperatures, larval development can be completed in 7.1 days. There are four larval instars, with 75% of total foliage consumption by larvae occurring in the final instar. The insects are voracious feeders, and complete defoliation and even haulm destruction can result.

Throughout most of its range, Colorado potato beetle has one or two generations per year, but in southern latitudes and mild climates, three or even four are possible. Even where two generations are the rule, a significant portion of newly emerged first-generation adults burrow directly into the soil and enter diapause (Tauber et al., 1988; Senanayake et al., 2000). Photoperiod at time of emergence from pupation is critical in determining the onset of diapause in summer adults. In northern latitudes and at high elevations, Colorado potato beetle is a typical long-day insect. Photoperiodic-induced adult diapause appears to be a recent adaptation to the insect's expanded distribution range (Hsiao, 1981). In its southern range, Colorado potato beetle populations are not photoresponsive.

25.2.2.2 Control

Colorado potato beetle has been the object of numerous innovations in crop protection. The first success of the agricultural insecticide era was the use of Paris green, cuprous acetoarsenite, against Colorado potato beetle in midwestern United States in the 1860s. Many advances in insecticide application equipment were in response to this pest, including first hand-operated compression sprayers, first wheel-drawn sprayers, first traction-operated dusters, first engine-operated sprayer and first air-blast sprayer (Gauthier et al., 1981). Potatoes were also one of the first crops to be treated by airplane.

Colorado potato beetle is legendary for its capacity to develop insecticide resistance (Forgash, 1981). Colorado potato beetle has developed resistance to all major classes of synthetic insecticidal classes, including the neonicotinoid imidacloprid (Zhao et al., 2000). Imidacloprid resistance is still localized in the United States, but the trend is of concern as worldwide neonicotinoids are the insecticide class most used on potato.

Recommended thresholds for application of insecticidal sprays to control Colorado potato beetle are often based either on beetle densities or on defoliation. For example, treatment has been recommended if beetle densities reach 0.5 adults, 4 small larvae (1st and 2nd instar) or 1.5 large larvae per stem (Ferro, 1985), exceed 20 larvae per plant (Martel et al., 1986) or if defoliation exceeds 10% (Holliday and Parry, 1987). However, treatments at or shortly after egg hatching tend to give best control. Sprays applied later may not control large larvae that are much more tolerant of exposure. Also, if spray applications are delayed, some larvae may have already quit feeding or dropped to the ground to pupate.

With the introduction of the neonicotinoids, a common approach to Colorado potato beetle control has been to apply the insecticide in-furrow or as seed piece treatments. Treating only border rows or some proportion of rows within fields could make for more efficient use of at-planting treatments and conserve susceptibility in the population (Dively et al., 1998).

Crop rotation can be effective in delaying establishment of Colorado potato beetle in new plantings (Wright, 1984; Sexson and Wyman, 2005). Overwintered Colorado potato beetles need to regenerate their flight muscles and do not oviposit until they have fed on a nutritionally adequate host plant. Rotating a planting by as little as 200–400 m from last year's location can delay colonization of the new planting by 1–2 weeks, generally resulting in lower population densities than in nonrotated fields and can save 1–2 sprays for first-generation larvae. Reducing the need for repeated insecticide applications has the indirect benefit of reducing selection for insecticide resistance. In some situations, field rotation will sufficiently delay emergence of most summer adults until day lengths are short enough to induce diapause.

25.2.3 Potato tuber moths

25.2.3.1 Origin, occurrence and damage

Tuber moths are the most damaging lepidopteran pests of potato throughout the tropics and subtropics. The tuber moth complex consists of four species that are pests of potato: potato tuber moth; Andean potato tuber moth, *Symmetrischema tangolias (plaesiosema)* (Gyen); Guatemalan moth, *Tecia (Scrobipalpopsis) solanivora* (Polvoný) and tomato leafminer, *Tuta (Scrobipalpula) absoluta* (Meyrick). Tuber moths are believed to have originated in the humid subtropics of South America where their principal host plants, potato and tobacco, *Nicotiana tabacum* L. (Solanaceae), are indigenous and where potato tuber moth is associated with a rich parasite complex and is of minor economic importance.

Potato tuber moth is the most widely distributed species of the complex, and despite quarantine efforts, it has spread worldwide. In warm climates, potato tuber moth can complete 8–14 generations per year (Trivedi and Rajagopal, 1992). Survival is optimal from 20 to 30 °C, and optimal population increase occurs at 28–30 °C (Sporleder et al., 2004). Potato appears to be the preferred host, but many plant species, cultivated and wild, are attacked. Potato tuber moth damages both foliage and tubers, but the major damage occurs when larva feed on stored tubers.

Potato tuber moth eggs are deposited on upper leaf surfaces and emerging larvae mine leaflets. As leaflets are consumed, larvae may leave their tunnels and web together adjacent leaflets. Larvae sometimes mine stems causing entire terminals to die. Partially grown larvae sometimes invade tubers in the field, particularly when the soil is dry and deeply cracked.

Potato tuber moth is particularly damaging in North Africa. In Tunisia, 60–70% of the main season potato production is kept for 3–4 months in rustic, nonrefrigerated stores. In the absence of control measures, tuber infestations can reach 100% within 2–3 months of storage (von Arx et al., 1990). In Peru, tuber infestation of over 90% was recorded following 4 months of rustic storage (Raman et al., 1987).

Guatemala moth is a devastating and increasingly important potato pest in Central America, Colombia, Ecuador and Venezuela. The risk is great that Guatemala moth will move southward to Peru and Bolivia. Larvae of this species primarily attack tubers. Ecuador had severe infestations in 2001 with tens of thousands of hectares of potato unfit to harvest and 0.5 million bags of harvested potatoes that had to be discarded (IRD, 2002). The insect is an aggressive invader. In 2000, Guatemala moth was found in the Canary Islands and is now on the red list of the European and Mediterranean Plant Protection Organization. It is considered to be a major threat to potato crops throughout southern Europe.

Andean tuber moth causes significant economic losses in potato stores in Peru and Bolivia. It has been reported in Tasmania, Australia and more recently in Indonesia and New Zealand. In the field, Andean tuber moth larvae mine foliage and bore into stems making chemical control extremely difficult. In storage, tubers are destroyed by larval mining, have a foul odour and are unfit for consumption or seed. Tuber damage in 51 farm stores surveyed in Peru ranged from 2 to 78% (Raman, 1988b).

Tomato leafminer damages only above ground portions of the plant. It is a pest of both potato and tomato, *Solanum lycopersicon* L., in temperate coastal locations of Peru, Chile, Argentina, Colombia and Brazil (Moore, 1983). Susceptible potato genotypes can be completely destroyed by this pest. Tomato leafminer has developed resistance to a wide range of insecticides, including methamidophos, deltamethrin and even abamectin (Lietti et al., 2005).

25.2.3.2 Control

Sex pheromones have provided a useful tool for managing populations of potato tuber moth. In severe field infestations, mating disruption can be effective (Bosa et al., 2005). Water pan and funnel traps were effective in mass trapping to reduce potato tuber moth damage in storage (Raman, 1988a). Thresholds are cultivar dependent, but a general recommendation is for insecticide application when moth captures exceed 40 moths in any one night or a cumulative mean of 15–20. In Israel, trap captures of potato tuber moth proved useful in predicting first appearance in the field and correlated with infestations in foliage, but the latter only on dry sandy soils (Coll et al., 2000).

Potato tuber moth pheromone consists of two components, *E*-(trans-)4, *Z*-(cis-)7-tridecadien-1-ol acetate (PTM-1) (Roelofs et al., 1975) and trans-4, cis-7, cis-10-tridecatrien-1-ol acetate (PTM-2) (Persoons et al., 1976). Blends of PTM-1 and PTM-2 were more effective than either alone, but the ratio was not critical (Voerman and Rothschild, 1978). For Andean tuber moth, the pheromone consists of a 2:1 mixture of (*E*, *Z*)-3,7 tetradecadienyl acetate and (*E*)-3-tetradecenyl acetate (Griepink et al., 1995), for Guatemala moth, a blend of (*E*)-3-dodecenyl acetate and (*Z*)-3-dodecenyl acetate (Nesbitt et al., 1985) and for tomato leafminer, a blend of (*E*, *Z*)-3,8,14 acetate and (*E*)-3, (*Z*)-8,11,14 acetate (Michereff Filho et al., 2000).

Extensive efforts have been undertaken to identify potential sources of genetic resistance to the tuber moth complex. Lack of knowledge about resistance mechanisms and their genetic expression has obstructed efficient selection of wild potato genotypes with promising traits. Also, selection of hybrid progenies for the expression of desirable agronomic traits often leads to loss of resistance inherited from wild progenitors. The glandular trichomes of *Solanum berthaultii* (Hawkes) confer resistance against potato tuber moth egg-laying and larval mobility (Malakar and Tingey, 2000). This trait appears to have a great potential, but to date, resistant selections have not been widely adopted.

Genetic transformation with, for example, *Bacillus thuringiensis (Bt)* genes has opened new opportunities for pest control (Cañedo et al., 1999). Expression of *Bt* genes encoding different proteins (Cry1A(b) and Cry5) conferred resistance to both potato tuber moth and Andean tuber moth (Jansens et al., 1995; Lagnaoui et al., 2000). *Bt*-Cry5 Spunta potato lines were highly resistant to potato tuber moth under field and storage conditions, with over 90% damage-free tubers (Douches et al., 2004). The same study demonstrated that potato clone NYL235-4.13 combining glandular trichomes with a *Bt*-Cry5/gus fusion construct had field resistance to potato tuber moth infestation. Toxic crystal proteins from *Bt* have proved highly effective against potato tuber moth larvae. The challenge is to manage this transgenic material to avoid selection for resistant moth genotypes and to preserve *Bt* insecticides as a tool for managing potato tuber moth in storage where control is most needed.

Cultural practices can greatly reduce potato tuber moth infestations. Sprinkler irrigation, deep planting, reshaping planting beds (hilling), preharvest defoliation and prompt harvest all reduce infestations in tubers entering storage (Shelton and Wyman, 1979; von Arx et al., 1987). Timely sanitation is effective in control of Andean tuber moth as the pest survives in crevices and crop residues (Palacios et al., 1999). Modern storage technology can essentially eliminate losses in storage to potato tuber moth. In Tunisia, producers were found to benefit most from investments in technologies that reduced marketing inputs (e.g. labour, depreciation and interest costs), whereas consumers benefited most from reduction of losses in storage (Fuglie, 1995).

Exposing male potato tuber moths to substerilizing doses of gamma irradiation has shown promise for management. In follow-up studies, Saour and Makee (2004) showed that gamma irradiation induced a high level of sterility and a reduction in female mating ability and fecundity.

25.2.4 Leafminers

25.2.4.1 Occurrence and damage

Leafminers (Agromyzidae) are important insect pests worldwide. Of greatest economic importance are the highly polyphagous pea leafminer; vegetable leafminer, *Liriomyza trifolii* (Burgess) and serpentine leafminer, *Liriomyza sativae* (Blanchard) (Parrella, 1987). These species tend to have overlapping host ranges, and several species may occur within the same field. *Liriomyza* species attack a wide variety of vegetable and horticultural crops, including potatoes and many other field and greenhouse crops. Pea leafminer is the most destructive leafminer on potato, but American serpentine leafminer is an important potato pest in Africa, from Senegal to Kenya and Mauritius (Bourdouxhe, 1982). *Liriomyza* species are morphologically similar and extremely difficult to tell apart. DNA sequence data of putative pea leafminer populations from the United States and elsewhere suggest that *L. huidobrensis* consists of at least two cryptic species (Scheffer, 2000).

Pea leafminer is indigenous to South America but was not considered to be an important potato pest there before the 1970s as it was generally held in check by natural enemies. More than 40 species of endo- and ecto-parasitoids have been reported from *Liriomyza* larvae and pupae, but most are not highly host specific (Izhevskiy, 1985). Pea leafminer populations developed insecticide resistance when exposed to intensive spraying for tomato leafminer control in coastal locations of Peru and Chile. Spray costs for pea leafminer became the most expensive component of potato production in the Cañete Valley, Peru (Cisneros and Mujica, 1999). Insecticide-resistant populations of pea leafminer since have spread to Europe and Israel (Weintraub and Horowitz, 1995). Pea leafminer is now present in China, Indonesia and has a foothold in Africa. Distribution of pea leafminer in China ranges from southern subtropical to northern temperate zones, but apparently, the species cannot persist through winter in nonsheltered environments beyond ~35 °N (Chen and Kang, 2004).

25.2.4.2 Control

Liriomyza pest species have been the subject of management efforts using a variety of tactics. Pea leafminer control has traditionally depended on chemical insecticides; however, in many places, leafminer flies show significant levels of resistance to all major classes of insecticides commonly used against them. In early season while plants are still actively growing, natural enemies may provide sufficient control. As plant growth slows, greater levels of protection are required, and selective use of pesticides is often necessary.

In the tropics, potatoes are commonly intercropped with maize, beans or other crops. By the choice of intercropped species and appropriate agronomic practices, it is possible to reduce populations of various potato pests. Potato intercropped with wheat, *Triticum* spp. (Poacea), is less damaged by pea leafminer because wheat attracts leafminer parasitoids (Raymundo and Alcázar, 1983). Yellow sticky traps have been used to achieve direct control of adult pea leafminer in both Asia and South America. In other parts of the world, sticky traps are used mainly for monitoring purposes. Trapping will reduce fly populations, but more importantly, when fly populations are low, trapping can convince farmers that insecticides are not necessary.

Pea leafminer attacks potato shortly after plant emergence. Adult females puncture the leaves to feed or lay eggs. Eggs are inserted individually on lower leaf surfaces and hatched within 3 or 4 days. Larvae tunnel into the leaflets making characteristic serpentine mines. Pea leafminer eggs laid in growing leaves often produce a hypersensitivity reaction causing many neonate larvae to be unsuccessful in burrowing into the leaf tissue. Tissue surrounding the encrusted egg starts an abnormal multiplication of cells that push the egg to the surface exposing them to greater risk of dehydration and predation. Breeding at

Centro Internacional de la Papa (CIP) resulted in selection of potato genotypes with good levels of tolerance to leafminer fly infestations, high yields and good processing qualities. Clone CIP-282 gave high yields in both low elevations and highlands of Peru and when officially released as 'Maria Tambena' was quickly adopted by growers (Mujica and Cisneros, 1997).

25.3 REGIONAL PESTS

25.3.1 Leafhoppers

Leafhoppers [Cicadellidae (Jassidae)] are abundant in the tropics and subtropics. Reproduction is usually sexual, and most species are multivoltine. Eggs are typically inserted into the main vein or petiole on the underside of leaves. In warm weather, a generation typically takes about 30 days. The most destructive leafhoppers are those whose feeding causes 'burning' ('hopperburn') of foliage (Backus et al., 2005). Among the leafhoppers, most destructive to potato are *Amrasca biguttala biguttala* Ishida and *Amrasca devastans* Ghauri in India, Pakistan and China, *Austroasca virigrisea* (Paoli) in Australia, *Empoasca decipiens* (Paoli) in Europe, Central Asia and North Africa and potato leafhopper in North America. Most other leafhopper species cause only mild stippling in potato.

Leafhoppers employ a 'lacerate-and-flush-feeding' strategy. Feeding leafhoppers move their stylets continuously or intermittently, secreting watery saliva, rupturing plant cells and then ingesting the resulting slurry. Burners tend to make repeated, contiguous, short duration probes in the vascular bundle, whereas stipplers tend to make fewer probes of longer duration and mostly in mesophyll-parenchyma cells (Backus et al., 2005).

Potato leafhopper is a consistently important pest in central and eastern North America. Potato leafhopper overwinters along the Gulf Coast. In spring, flying adults are caught in updrafts and transported north on upper level airstreams. Northward influxes of leafhoppers are associated with southerly winds of 36 h or more duration and precipitation in the fallout area. These long-distance migrants often arrive before potatoes have emerged. Significant populations seldom appear on potato until a generation of new adults has been produced locally on alternative hosts (Flanders and Radcliffe, 1989). For potato leafhopper, treatment has been recommended if densities exceed 10–30 nymphs per 100 leaves (Johnson and Radcliffe, 1991).

Yield loss in potato can occur before visual symptoms of potato leafhopper damage are obvious. Damaged plants show large increases in respiration that deplete photosynthates required for haulm and tuber development. These initial effects are reversible if leafhoppers are controlled before leaf tissue is destroyed, but any impairment of photosynthetic efficiency is irreversible. Hopperburn affects plant biomass accumulation by reducing green leaf area and efficiency of solar radiation capture. Potato is most susceptible to damage by leafhopper in early tuber bulking.

25.3.1.1 Phytoplasma transmission by leafhoppers

Certain leafhopper species transmit phytoplasma (Class Mollicutes). Various phytoplasma occur on potato, and these are difficult to separate or group. In North America, aster leafhopper, *Macrosteles quadrilineatus (fascifrons)* Forbes, transmits a phytoplasma belonging to the aster yellows complex that causes potato purple-top wilt (purple top) (Banttari et al., 1993). Potato stolbur phytoplasma (aster yellows group), vectored primarily by the planthopper *Hyalesthis obsoletus* Signoret (Cixiidae), occurs on potato in southern and central Europe (Cousin and Moreau, 1977). In India, potato marginal flavescence phytoplasma and potato toproll phytoplasma are important leafhopper-vectored diseases (Khurana et al., 1988). Phytoplasma diseases on potato appear to increase in importance worldwide. An emerging problem in the Pacific Northwest USA is a 'purple-top' disease caused by a clover proliferation group phytoplasma vectored primarily, but not exclusively, by beet leafhopper, *Circulifer tenellus* (Baker) (Crosslin et al., 2005). Vector relationships are not well known for the phytoplasma on potato, but all appear to be leafhopper transmitted.

25.3.2 Potato psyllid

Potato psyllid, *Bactericerca (Paratrioza) cockerelli* (Sulc.) (Psyllidae), occurs throughout the western United States, especially the Mountain states. Losses caused by potato psyllid sometimes are catastrophic. The saliva of psyllids is toxicogenic and induces a condition called psyllid yellows. The primary symptoms diagnostic of this disorder in potato include stunting, increasing chlorosis, basal cupping and erectness of leaflets, and reddish colouration of new foliage and aerial tubers (Cranshaw, 1994). Affected plants tend to set excessive number of tubers, most close to the stem, often misshapen tubers attributed to internal necrosis and premature sprouting. This pathology has been attributed to disrupted hormonal regulation of plant growth. Progress of psyllid yellows can be interrupted and even reversed if psyllid infestations are controlled early. Only the nymphs cause psyllid yellows. Symptoms are detected with as few as 3–5 nymphs per plant.

Psyllids overwinter as adults in Texas and New Mexico and migrate northward in early spring. Psyllid adults are attracted to yellow traps. When adults are detected, monitoring should begin for the nymphs. Nymphs are found mostly on middle to lower leaves. Control is recommended when nymphs exceed one per 100 sweeps. Psyllids are readily controlled by most insecticides.

25.3.3 Thrips

Palm (melon) thrips, *Thrips palmi* Karny (Thripidae), is highly polyphagous, and populations can build quickly on potato. Both larvae and adults feed gregariously along midribs, veins and stems, particularly at bud terminals, leaving numerous scars and deformities and finally killing the plant. Palm thrips was first found in Indonesia in 1925 and has since spread widely throughout the Pacific. In Indonesia and the Philippines, severe infestations of palm thrips and *Megalurothrips usitatus* (Bagnall) have been reported in lowland potato production. It is likely that palm thrips is present in most countries in south and southeastern Asia. Palm thrips has been present in the Caribbean region since 1985, and there have been several limited outbreaks in the Netherlands. Neonicotinoid insecticides are effective but can disrupt natural enemies (Nemoto, 1995).

25.3.4 White grubs

White grubs are larvae of scarab beetles (Scarabaeidae). White grubs are creamy white, medium to large, with a distinct brown head capsule, C-shaped body with prominent legs and a slightly enlarged abdomen. Scarabs have long life cycles, typically 1–4 years. Species that are potato pests cut roots and stems and make large, clean, shallow circular surface wounds in tubers. Infestations are usually worse when potatoes are planted in weedy fields or fields previously in sod. White grubs have become increasingly troublesome in Asia and Central America since the phasing out of chlorinated hydrocarbon insecticides. Major pest genera in India include *Anomala* spp. and *Melolontha* spp. In the Americas, *Phyllophaga* is the most economically important genus.

25.3.5 Wireworms

Wireworms are the larval stage of click beetles (Elateridae). Wireworm larvae are slender, yellowish to brown orange, with short legs. The tip of the abdomen is flattened and has a pair of short hooks. Mature larvae range from 12 to 25 mm in length, depending on species. Wireworms typically have life cycles of 1–3 years, so it is important to know the cropping history of a particular field. Wireworms are usually present following pasture, small grain cereals, maize, beans or weedy fields. Wireworm infestations tend to be spotty and often reoccur in the same locations within fields for several years. Wireworms feed on seed pieces in the spring and occasionally damage emerging shoots. Damaged seed pieces can become secondarily infected with bacteria or fungi and either fail to emerge or produce weak plants. Later in the summer, wireworms bore into the new tubers, making a clean, round entry hole, about 2–3 mm in diameter. The tunnel often enters straight into the tuber and is lined with periderm. Wireworm damage does not reduce yields but lowers tuber value and can make the crop unmarketable.

There is great local variation in the extent to which wireworm species are important. Usually, no one wireworm species can be regarded as a key pest, but collectively wireworms can represent a serious problem (Jansson and Seal, 1994). In many potato-producing regions, relative importance of the various wireworm species has not been documented.

It is possible to sample for wireworms either by putting soil samples through screens or by baiting traps with carrot. Unfortunately, action thresholds usually are not based on empirical data nor are they species specific. Wireworms tend to be worse on light soils. In dry years, wireworms tend to be more destructive because the larvae will burrow into tubers seeking moisture. Insecticides applied in-furrow give fair to good control but seldom are as effective as the broadcast-chlorinated hydrocarbon insecticides they replaced. Treating cereal crops grown in rotation preceding potato can reduce wireworm populations in potato.

25.3.6 Ladybird beetles

In India, Epilachna beetles (Coccinellidae), e.g. twenty-eight-spotted potato lady beetle *Henosepilachna (Epilachna) vigintioctopunctata* Motschulsky and *Henosepilachna ocellata* Jadwiszczak and Wegrzynowicz, are important defoliators that often limit successful potato cultivation. Both adults and grubs skeletonize leaves.

25.3.7 Flea beetles

Flea beetles (Chrysomelidae) are economic pests in the Americas and Asia. In Peru, there are at least five species important on potato, *Epitrix parvula* (Fab.), *Epitrix subcrinita* (Le Conte), *Epitrix ubaquensis* Haarold, *Epitrix harilana rubia* Bech. and Bech. and *Epitrix yanazara* Bech. In North America, the most important species are potato flea beetle, *Epitrix cucumeris* (Harris), common throughout the eastern half of the continent, and tuber flea beetle, *Epitrix tuberis* Gentner, which occurs primarily in the Northwest.

Flea beetles overwinter as adults in or near potato fields. Potato flea beetle has one generation per year, whereas tuber flea beetle generally has two. Adult flea beetles are 1–2 mm long and feed on foliage, producing numerous minute circular holes. Damage is most serious on young plants. Heavy flea beetle infestations may cause substantial yield losses, but flea beetles are usually controlled incidentally by insecticides targeted against other potato insect pests. Flea beetle-damaged leaves often become desiccated. The feeding wounds can serve as entry point for potato pathogens, including bacterial diseases and early blight, *Alternaria solani* (Ell. and Mart.) Jones and Grout (Pleosporaceae). Larvae of all flea beetle species feed on roots, stolons and tubers, but most species cause only superficial injury. Tuber flea beetle makes deep tunnels; one or two larvae can do enough damage for a tuber to be unmarketable. New adults emerging from the soil may carry Verticillium wilt, *Verticillium dahliae* Kleb. (Clavicipitaceae) and *Fusarium* spp. (Hypocreaceae). Common scab, *Streptomyces scabies* (Thaxter) Waksman and Henrici (Streptomycetaceae), infection is also associated with flea beetle injury.

25.3.8 Andean potato weevils

The Andean potato weevil complex is a group of closely related curculionids attacking potatoes grown above 2500 m. The most important species are *Premnotrypes latithorax* (Pierce) limited to Chile, Bolivia and southern Peru, *Premnotrypes suturicallus* Kuschel in central Peru and *Premnotrypes vorax* (Hustache) distributed over the northern Andes (Venezuela, Colombia, Ecuador and northern Peru) (Alcázar and Cisneros, 1999). Less abundant species include *Premnotrypes fractirostris* Marshall, *Premnotrypes piercei* Alcalá, *Premnotrypes pusillus* Kuschel, *Premnotrypes sanfordi* (Pierce), *Premnotrypes solani* Pierce, *Premnotrypes solanivorax* (Heller), *Premnotrypes clivosus* Kuschel, *Premnotrypes zischkai* Kuschel, *Premnotrypes solaniperda* Kuschel and *Amitrus jelskyi* (Kirsh). Other potato weevils of economic importance, but primarily in Bolivia, are *Rhigopsidius tucumanus* Heller occurring above 2500 m and *Phyrdenus muriceus* Germar in mesothermic valleys.

Adult Andean weevils feed on young potato plants, and severe infestations may completely denude fields. Life cycles of the three common species differ only in detail. There is usually one generation per year, except for *P. vorax* that has two generations per year in parts of Ecuador and Colombia where potato is produced year round. Eggs are laid in plant debris. Neonate larvae bore into potato and remain there feeding until ready to pupate. Larvae leave the tubers to make their pupal cells in the soil where they pass the winter. *P. muriceus* also pupates in the soil, but *R. tucumanus* pupates inside potato tubers. When rains start, adults emerge and feed mainly on leaves and stems of potato plants. Eggs are laid on potato, weeds or in the soil. In fields not protected by insecticides, loss of marketable yield can be total (Cálvache, 1986).

No significant sources of host plant resistance have been found in primitive cultivars or wild potato species. Similarly, there are no known parasitoids, and predators are few and seemingly ineffective. Management practices are mainly based on cultural control measures, including selection of clean seed, deep planting, hilling, careful soil cultivation, border crops (lupine or small grains) and chemical barriers, removal of weed hosts, early harvest and ditches to protect fields from invasion by migrating adults.

Adult weevils are nocturnal and tend to take refuge under plant debris during the day. This behaviour is exploited for monitoring and control purposes. Weevils can be monitored using pit fall traps and insecticides applied upon first capture of adult weevils. However, the most widely used trapping technique is 'shelter traps' baited with insecticide-drenched potato foliage (Alcázar and Cisneros, 1997). Shelter traps are particularly effective just prior to the emergence of potato plants in new fields.

As Andean potato weevil larvae leave tubers to pupate, infested tubers can be stored on plastic sheets to collect emerging larvae that can then be destroyed or fed to chickens. An entomopathogenic fungus, *Beauveria brongniartii* (Sacc.) (Clavicipitaceae), is mass-produced as a biopesticide in cottage-type facilities in several Andean communities of Peru, Bolivia and Colombia (Cisneros and Vera, 2001). Use of *B. brongniartii* in potato stores can achieve up to 90% weevil mortality under moist conditions.

25.3.9 Cutworms

Numerous cutworm (Noctuidae) species attack potato. Cutworm larvae are nocturnal, living just below the ground surface. Larvae can cut the stalks of young growing plants, with severely infested fields appearing grazed. On more mature plants, cutworms can damage tender stems. After tuberization, cutworm damage is largely confined to the tubers. Among the most cosmopolitan and destructive species attacking potato are *Agrotis* spp., e.g *Agrotis ipsilon* (Hufnagel) and *Agrotis segetum* (Denis and Schiffermüller).

25.4 INSECT CONTROL TACTICS

Potatoes are grown in most countries of the world and under an extraordinary range of environmental conditions. Production systems and pest issues vary greatly among and within geographic regions. This complex of pest problems is dynamic, changing within and across years. The relative importance of particular pest problems can vary with cultivar, cultural practices, weather, intended market, storage conditions and other variables. Given such management complexities, it is understandable that pesticides continue to play a dominant role in potato pest management. There seems little question that the most pressing need in potato pest management worldwide is to find ways to reduce the present dependency on pesticides (Yanggen et al., 2004).

25.4.1 Insecticides

Worldwide, economic return on investment in potato insecticides (1988–1990) was estimated to be about 6:1, but overall effectiveness of crop protection was less than 44% (Oerke et al., 1994). Effectiveness of crop protection tended to be greatest in production systems that use insecticides most intensively, e.g. 60% in North America. Estimated potato yield losses to insects in 1988–1990 were more than 10% higher worldwide than corresponding estimates 25 years earlier (Cramer, 1967), despite much greater insecticide inputs in terms of applications per hectare and product cost.

Development of insecticide resistance can severely limit insecticide options. Two key potato pests, Colorado potato beetle and green peach aphid, rank among the most notorious of all insects for their capacity to develop insecticide resistance, with resistance reported to 41 and 68 insecticide chemistries, respectively (Whalon et al., 2004). Another important vector of potato viruses, melon aphid, is reported to have developed resistance to 39 different insecticides, including methamidophos, the insecticide of choice for green peach aphid control in many seed potato production systems. The alarming rate at which insecticide resistance has evolved, the consolidation of agrochemical companies worldwide and a slowing rate of development of new insecticides are persuasive arguments for active development and promotion of sound insecticide resistance management (IRM) programmes.

To slow the development of insecticide resistance, it is desirable to avoid sequential use of products with the same mode of action that exposes more than one generation of the target pest to selection. To assist crop professionals practicing IRM, the Insecticide Resistance Action Committee (IRAC), an interagrochemical company committee whose mission is to 'promote the development of resistance management strategies in crop protection' (IRAC, 2005), has developed a classification system and tentatively assigned all insecticides to one of 28 unique groups based on their mode of action. This numbering system when included on the label for insecticides (fungicide and herbicide labels already indicate mode of action codes) will make it easier to implement IRM programmes as practitioners will not have to be conversant in insecticide toxicology to make appropriate decisions on product rotation.

25.4.2 Host plant resistance

Potato is unique among crop plants in the diversity of its wild relatives. More than 150 species of potato have been described, and many present potentially useful sources of insect resistance. Insect resistance appears to be a primitive trait in the wild potatoes but is also characteristic of the most advanced species, whereas susceptibility is characteristic of cultivated species and their nearest relatives (Flanders et al., 1992). Wild potatoes possess a wide variety of defence mechanisms that can protect against insects. These include glandular trichomes (Gregory et al., 1986), foliar and tuber glycoalkaloids (Tingey, 1984) and a hypersensitivity response (Balbyshev and Lorenzen, 1997).

European workers in the 1930s discovered that resistance to Colorado potato beetle was associated with presence of foliar glycoalkaloids, as in *Solanum chacoense* Bitter, but breeding efforts to exploit this resistance met little success. Two major problems were encountered: levels of tuber glycoalkaloids tended to be highly correlated with levels in the foliage and resistance always declined in backcrosses. Colorado potato beetle resistance is now known to be associated with specific glycoalkaloids, especially leptines, commersonine and dehydrocommersonine (Sinden et al., 1986; Lorenzen et al., 2001), suggesting that it may be possible to make greater progress.

Breeding for trichome-conferred resistance began at Cornell University in 1977 when *S. berthaultii* accessions were crossed with cultivar Hudson. Hybrids have since been selected with high levels of resistance to Colorado potato beetle and potato leafhopper. When the Cornell University potato-breeding programme released an advanced breeding line, NYL235-4, it was said to be the first near cultivar quality insect-resistant potato clone developed by traditional plant breeding (Plaisted et al., 1992). In 1992, CIP was awarded the King Baudouin International Agricultural Research Award for 'innovative research in integrated pest management' and the creation of a hybrid potato population (the 'hairy potato') with resistance to a range of insect pests.

25.4.3 Biological control

Despite the extensive research that has been done on insect pests of potato, comparatively scant attention has been given to exploiting natural enemies and microbial agents. The biological control information available has in only a few instances been integrated into successful pest management strategies. Intuitively, it might seem unlikely that biological control agents could be effective given the intensive use of pesticides in potato production. However, the tremendous outbreaks of pests, such as aphids, leafminers, spider mites and thrips occasionally induced by inappropriate use of pesticides provide indirect evidence of the importance of natural enemies. Future success with biological control for potato insect control is likely to depend on finding ways that insecticides can be used selectively.

25.5 CONCLUSIONS

With each passing year, potato pest management becomes more sophisticated. Considerable progress has been made in understanding pest biology and ecology and applying this knowledge to management of the crop. Examples include establishment of economic thresholds for key pests, development of strategies for managing insecticide resistance, increased understanding and quantification of multiple pest interactions and elucidation of vector–virus relationships.

Integrated pest management (IPM) has been defined variously, but the common theme is that reliance on pesticides should be reduced by use of alternative control measures substituting a coordinated, systems approach. For potato, IPM measures may include use of resistant or tolerant varieties, introduction or enhancement of natural enemies, cultural manipulations of the crop and pest environment and careful selection and timing of chemical applications (Cisneros, 1984). The argument is often advanced that in developing countries there usually is not enough information available on the various components to formulate IPM programmes. A counterpoint to that argument is that IPM implementation efforts often reveal knowledge gaps that in turn serve to focus research resources where they are most needed. Integrated pest management in the tropics will become reality only when the component technologies are taken by farmers and adapted to their conditions (Morse and Buhler, 1997). Sponsored pest surveys and on-farm research can play important roles in facilitating this process. National and international programmes should encourage such activities (Ezeta, 1988). Farmers' perceptions and knowledge of pests are an important source of information and can be very useful in formulating IPM programmes. There will always be need for additional research. However, in most situations, considerable knowledge of pest management components already exists, and practical IPM programmes at some level of complexity already are in place or could be formulated. Modifications to Integrated Post Management programmes to incorporate more or better information and to integrate more complex interactions can proceed as the knowledge base expands.

REFERENCES

Alcázar J. and F. Cisneros, 1997, In: *The International Potato Center Program Report 1995–1996*, p. 169. International Potato Center (CIP), Lima, Peru.

Alcázar J. and F. Cisneros, 1999, In: *Impact on a Changing World, Program Report 1997–1998*, p. 141. International Potato Center (CIP), Lima, Peru.

Alvarez J.M. and R. Srinivasan, 2005, *J. Econ. Entomol.* 98, 1101.

Andow D.A., G.P. Fitt, E.E. Grafius, R. Jackson, E.B. Radcliffe, D.W. Ragsdale and L. Rossiter, In: R. Hollingworth, D. Mota-Sanchez and M. Whalen (eds), *Global Pesticide Resistance in Arthropods*. CAB International, Wallingford (in press).

Backus E.A., M.S. Serrano and C.M. Ranger, 2005, *Annu. Rev. Entomol.* 50, 125.

Balbyshev N.F. and J.H. Lorenzen, 1997, *J. Econ. Entomol.* 90, 652.

Banttari E.E., P.J. Ellis and S.M.P. Khurana, 1993, In: R.C. Rowe (ed.), *Potato Health Management*, p. 127. APS Press, St. Paul, MN.

Basky Z. and A. Almási, 2005, *J. Pest Sci.* 78, 67.

Beemster A.B.R. and J.A. de Bokx, 1987, In: J.A. de Bokx and J.P.H. van der Want (eds), *Viruses of Potatoes and Seed-Potato Production*, p. 84. Pudoc, Wageningen, the Netherlands.

Berger P and T. German, 2001, In: G. Lobenstein, P.H. Berger, A.A. Brunt and R.H. Lawson (eds), *Virus and Virus-Like Diseases of Potatoes and Production of Seed-Potatoes*, p. 341. Kluwer Academic Publishers, Dordrecht, the Netherlands.

Blackman R.L., 1980, In: R.L Blackman, G.M. Hewitt and M. Ashburner (eds), *Insect Cytogenetics*, p. 133, 10th Symp. R. Entomol. Soc. London. Blackwell, Oxford.

Blackman R.L. and J.M. Spence, 1992, *Bull. Entomol. Res.* 11, 267.

Boiteau G., 1997, *Can. J. Zool.* 75, 1396.

Boiteau G. and R.H. Parry, 1985, *Am. Potato J.* 62, 489.

Boiteau G. and R.P. Singh, 1999, *Am. J. Potato Res.* 76, 31.

Bosa C.F., A.M. Cotes Prado, T. Fukumoto, M. Bengtsson and P.Witzgall, 2005, *Entomol. Exp. Appl.* 114, 137.

Bourdin D., J. Rouzé, S. Tanguy and Y. Robert, 1998, *Plant Pathol.* 47, 794.

Bourdouxhe L., 1982, *Bull. Phytosanitaire FAO* 30, 81.

Bradley R.H.E., C.A. Moore and D.D. Pond, 1966, *Nature* 209, 1370.

Cálvache H.G, 1986, In: L. Valencia (ed.), *Memorias del Curso Sabre Control Integrado de Plagas de Papa*, p. 18. International Potato Center (CIP) and Instituto Colombiano Agropecuario (ICA), Bogotá, Colombia.

Cañedo V., J. Benavides, A. Golmirzaie, F. Cisneros, M. Ghislain and A. Lagnaoui, 1999, In: *Impact on a Changing World, Program Report 1997–98*, p. 161. International Potato Center (CIP), Lima, Peru.

Chen B. and L. Kang, 2004, *Environ. Entomol.* 33, 155.

Cisneros F.M, 1984, *Report of the XXVII Planning Conference on Integrated Pest Management*, June 4–8, p. 19–30. International Potato Center (CIP), Lima, Peru.

Cisneros F.M. and A. Vera, 2001, *Scientist and Farmer: Partners in Research for the 21st Century, Program Report 1999–2000*, p. 155. International Potato Center (CIP), Lima, Peru.
Cisneros F.M. and N. Mujica, 1999, *Impact on a Changing World: Program Report 1997–98*, p. 129. International Potato Center (CIP), Lima, Peru.
Cocu N., K. Conrad, R. Harrington and M.D.A. Rounsevell, 2005a, *Bull. Entomol. Res.* 95, 47.
Cocu N., R. Harrington, M. Hullé and M.D.A. Rounsevell, 2005b, *Agric. Forest Entomol.* 7, 31.
Cocu N., R. Harrington, M.D.A, Rounsevell, S.P. Worner, M. Hullé and the EXAMINE project participants, 2005c, *J. Biogeogr.* 32, 615.
Coll M., S. Gavish and I. Dori, 2000, *Bull. Entomol. Res.* 90, 309.
Cousin M.T. and J.P. Moreau, 1977, *Phytoma*, 291, 15.
Cramer H.H., 1967, *Plant Protection and World Crop Production*. Pflanzenschutz-Nachr, Leverkusen.
Cranshaw W.S, 1994, In: G.W. Zehnder, M.L. Powelson, R.K. Jansson and K.V. Raman (eds), *Advances in Potato Pest Biology and Management*, p. 83. APS Press, St. Paul, MN.
Crosslin J., J.E. Munyaneza, A. Jensen and P.B. Hamm, 2005, *J. Econ. Entomol.* 98, 279.
Davis J.A., E.B. Radcliffe and D.W. Ragsdale, 2005, *Am. J. Potato Res.* 82, 197.
de Wilde J. and T. Hsiao, 1981, In: J.H. Lashomb and R. Casagrande (eds), *Advances in Potato Pest Management*, p. 47. Hutchinson Ross Publ. Co., Stroudsburg, PA.
Devonshire A.L., L.M. Field, S.P. Foster, G.D. Moores, M.S. Williamson and R.L. Blackman, 1998, *Phil. Trans. Roy. Soc. Lond. Ser. B* 353, 1677.
Dewar A., L. Haylock, S. Foster, A. Devonshire and R. Harrington, 1998, *Br. Sugar Beet Rev.* 66, 14.
DiFonzo C.D., D.W. Ragsdale, E.B. Radcliffe, N.C. Gudmestad and G.A. Secor, 1996, *Ann. Appl. Biol.* 129, 289.
DiFonzo C.D., D.W. Ragsdale, E.B. Radcliffe, N.C. Gudmestad and G.A. Secor, 1997, *J. Econ. Entomol.* 90, 824.
Dively G.P., P.A. Follett, J.J. Linduska and G.K. Roderick, 1998, *J. Econ. Entomol.* 91, 376.
Douches D.S., W. Pett, F. Santos, J. Coombs, E. Grafius, W. Li, E.A. Metry, T. Nasr El-Din, M. Madkour, 2004, *J. Econ. Entomol.* 97, 1425.
Dripps J.E. and Z. Smilowitz, 1989, *Environ. Entomol.* 18, 854.
Elkinton J.S., D.N. Ferro and E. Ng, 1985, In: D.N. Ferro and R.H. Voss (eds), *Mass. Agric. Exp. Sta. Bull.* 704, p. 9, Proc. Symp. on the Colorado Potato Beetle, XVIIth Int. Congr. Entomol., Hamburg, Germany. Amherst, MA.
Ezeta F.N., 1988, *Collaborative Country Research Networks*. International Potato Center (CIP), Lima, Peru.
Ferro D.N., 1985, In: D.N. Ferro and R.H. Voss (eds), *Mass. Agric. Exp. Sta. Bull.* 704, p. 1, Proc. Symp. on the Colorado Potato Beetle, XVIIth Int. Congr. Entomol., Hamburg, Germany. Amherst, MA.
Ferro D.N., J.A. Logan, R.H. Voss and J.S. Elkinton, 1985, *Environ. Entomol.* 14, 343.
ffrench-Constant R.H., R. Harrington and A.L. Devonshire, 1988, *Crop Prot.* 7, 55.
Flanders K.L. and E.B. Radcliffe, 1989, *Environ. Entomol.* 18, 1015.
Flanders K.L., E.B. Radcliffe and D.W. Ragsdale, 1991, *J. Econ. Entomol.* 84, 1028.
Flanders K.L., J.G. Hawkes, E.B. Radcliffe and F.I. Lauer, 1992, *Euphytica* 61, 83.
Forgash A.J., 1981, In: J.H. Lashomb and R. Casagrande (eds), *Advances in Potato Pest Management*, p. 34. Hutchinson Ross Publ. Co., Stroudsburg, PA.
Frazer B.D., 1987, In: G. Boiteau, R.P. Singh and R.H. Parry (eds), *Potato Pest Management in Canada*, p. 23, Proceedings Symposium on Improving Potato Pest Prot., January 27–29. Canada Agri-Food, Fredericton, NB, Canada.
Fuglie K.O., 1995, *Am. J. Agric. Econ.* 77, 162.
Gauthier N.L., R.N. Hofmaster and M. Semel, 1981, In: J.H. Lashomb and R. Casagrande (eds), *Advances in Potato Pest Management*, p. 13. Hutchinson Ross Publ. Co., Stroudsburg, PA.
Gordon P.L. and F.L. McEwen, 1984, *Can. Entomol.* 116, 783.
Gray S. and F.E. Gildow, 2003, *Annu. Rev. Phytopathol.* 41, 539.
Gregory P., W.M. Tingey, D.A. Avé and P.Y. Bouthyette, 1986, In: M.B. Green and P.A. Hedin (eds), *Natural Resistance of Plants to Pests – Roles of Allelochemicals*, p. 160. American Chemical Society, Washington, DC.
Griepink F.C., T.A. van Beek, J.H. Visser, S. Voerman and A. de Groot, 1995, *J. Chem. Ecol.* 21, 2003.
Halbert S., J. Connelly and L. Sandvol, 1990, *Acta Phytopathol. Entomol. Hun.* 25, 411.

Hamm P.B. and D.C. Hane, 1999, *Plant Dis.* 83, 1122.
Hanafi A., E.B. Radcliffe and D.W. Ragsdale, 1989, *J. Econ. Entomol.* 82, 1201.
Hanafi A., E.B. Radcliffe and D.W. Ragsdale, 1995, *Crop Prot.* 14, 145.
Hane D.C. and P.B. Hamm, 1999, *Plant Dis.* 83, 43.
Harrewijn P., H. den Ouden and P.G.M. Piron, 1991, *Entomol. Exp. Appl.* 58, 101.
Harrington R., E. Bartlet, D.K. Riley, R.H. ffrench-Constant and S.J. Clark, 1989, *Crop Prot.* 8, 340.
Harrington R. and EXAMINE Consortium, 2004, *EXploitation of Aphid Monitoring Systems IN Europe*, Environment Project EVK2-1999-00151, http://www.rothamsted.ac.uk/examine/.
Harrington R., J.S. Bale and G.M. Tatchell, 1995, In: R. Harrington and N.E. Stork (eds), *Insects in a Changing Environment*, 17th Symp. Roy. Entomol. Soc, September 7–10, p. 125. Academic Press, London.
Harrington R., N. Katis and R.W. Gibson, 1986, *Potato Res.* 29, 67.
Harrington R. and R.W. Gibson, 1989, *Potato Res.* 32, 167.
Heimbach U., T. Thieme, H.-L. Weidemann and R. Thieme, 1998, In: J.M. Nieto Nafría and A.F.G. Dixon (eds), *Aphids in Natural and Managed Ecosystems*, p. 555. Universidad de León, León, Spain.
Hiddema J, 1972, In: J.A. de Bokx (ed.), *Viruses of Potatoes and Seed-Potato Production*, p. 206. Pudoc, Wageningen, the Netherlands.
Holliday N.J. and R.H. Parry, 1987, In: G. Boiteau, R.P. Singh and R.H. Parry (eds), *Potato Pest Management in Canada*, p. 77, Proc. Symp. on Improving Potato Pest Prot., January 27–29. Canada Agri-Food, Fredericton, NB, Canada.
Hsiao T, 1981, In: J.H. Lashomb and R. Casagrande (eds), *Advances in Potato Pest Management*, p. 69. Hutchinson Ross Publ. Co., Stroudsburg, PA.
Insecticide Resistance Action Committee (IRAC), 2005, *Resistance Management from IRAC*, http://www.irac-online.org.
Institut de Recherche pour le Développement (IRD), 2002, *Insect Pest of Potatoes Tecia solanivora is Devastating Crops in Latin America and has Reached the Canary Islands*, http://www.innovations-report.de/html/berichte/agrar_forstwissenschaften/bericht-9922.html.
Izhevskiy S.S., 1985, *Entomol. Rev.* 64, 148.
Jansens S., M. Cornelissen, R. De Clerco, A. Reynaerts and M. Peferoen, 1995, *J. Econ. Entomol.* 88, 1469.
Jansson R.K. and D.R. Seal, 1994, In: G.W. Zehnder, M.L. Powelson, R.K. Jansson and K.V. Raman (eds), *Advances in Potato Pest Biology and Management*, p 31. APS Press, St. Paul, MN.
Johnson K.B. and E.B. Radcliffe, 1991, *Crop Prot.* 10, 416.
Khurana S.M.P., R.A. Singh and D.M. Kalay, 1988, In: K. Maramorosch and J.P Raychaudhuri (eds), *Mycoplasma Diseases of Crops, Basic and Applied Aspects*, p. 285. Springer-Verlag, New York, NY.
Lagnaoui A., V. Cañedo and D. Douches, 2000, *Am. J. Potato Res.* 77, 406.
Lietti M.M.M., E. Botto, E. and R.A. Alzogaray, 2005, *Neotrop. Entomol.* 34, 113.
Lorenzen J.H., N.F. Balbyshev, A.M. Lafta, H. Casper, X. Tian, B. Sagredo, 2001, *J. Econ. Entomol.* 94, 1260.
Malakar R. and M.W. Tingey, 2000, *Entomol. Exp. Appl.* 94, 249.
Marshall B., H. Barker and S.R. Verrall, 1988, *Ann. Appl. Biol.* 113, 297.
Martel P., J. Belcourt, D. Choquette and G. Boivin, 1986, *J. Econ. Entomol.* 79, 414.
Martín B., I. Varela and C. Cabaleiro, 2004, *J. Hortic. Sci. Biotechnol.* 79, 855.
Martín B., J.L. Collar, W.F. Tjallingii and A. Fereres, 1997, *J. Gen. Virol.* 78, 2701.
Michereff Filho M., E.F. Vilela, A.B. Attygalle, J. Meinwald, A. Svatoš and G.N. Jham, 2000, *J. Chem. Ecol.* 26, 875.
Mollov D.S. and C.A. Thill, 2004, *Am. J. Potato Res.* 81, 317.
Moore J.E, 1983, *Trop. Pest Manage.* 29, 231.
Moran N.A., 1992, *Annu. Rev. Entomol.* 37, 321.
Morse S. and W. Buhler, 1997, *Integrated Pest Management, Ideals and Realities in Developing Countries.* Lynne Rienner Publishers, London.
Mujica N. and F. Cisneros, 1997, In: *The International Potato Center Program Report 1995–1996*, p. 177. International Potato Center (CIP), Lima, Peru.
Nauen R. and A. Elbert, 2003, *Bull. Entomol. Res.* 93, 47.
Nauen R. and I. Denholm, 2005, *Arch. Insect Biochem. Physiol.* 58, 200.
Nault L.R., 1997, *Ann. Entomol. Soc. Am.* 90, 521.
Nemoto H., 1995, *Jpn. Agric. Res. Q.* 29, 25.

Nesbitt B.F., P.S. Beevor, A. Cork, D.R. Hall, R.M. Murillo and H.R. Leal, 1985, *Entomol. Exp. Appl.* 38, 81.
Nie X. and R.P. Singh, 2002, *J. Virol. Methods* 103, 145.
Oerke E.-C., A. Weber, D.W. Dehne and F. Schönbeck, 1994, In: E.-C. Oerke, H.-W. Dehne, F. Schönbeck and A. Weber (eds), *Crop Production and Crop Protection – Estimated Losses in Major Food and Cash Crops*, p. 742. Elsevier, Amsterdam, the Netherlands.
Palacios M., J. Tenorio, M. Vera, F.Y. Zevallos and A. Lagnaoui, 1999, In: *Impact on a Changing World, Program Report 1997–1998*, p. 153. International Potato Center (CIP), Lima, Peru.
Parrella M.P., 1987, *Annu. Rev. Entomol.* 32, 201.
Perring T.M., N.M. Gruenhagen and C.A. Farrar, 1999, *Annu. Rev. Entomol.* 44, 457.
Persoons C.J., S. Voerman, P.E.J., Verwiel, F.J. Ritter, W.J. Nooyen and A.K. Minks, 1976, *Entomol. Exp. Appl.* 20, 289.
Piche L.M., R.P. Singh, X. Nie and N.C. Gudmestad, 2004, *Phytopathology* 94, 1368.
Plaisted R.L. W.M. Tingey and J.C. Steffens, 1992, *Am. Potato J.* 69, 843.
Pollard D.G., 1973, *Bull. Ent. Res.* 62, 631.
Powell G., 2005, *J. Gen. Virol.* 86, 469.
Radcliffe E.B. and D.W. Ragsdale, 2002, *Am. J. Potato Res.* 79, 353.
Ragsdale D.W., E.B. Radcliffe and C.D. DiFonzo, 2001, In: G. Lobenstein, P.H. Berger, A.A. Brunt and R.H. Lawson (eds), *Virus and Virus-Like Diseases of Potatoes and Production of Seed-Potatoes*, p. 237. Kluwer Academic Publishers, Dordrecht, the Netherlands.
Ragsdale D.W., E.B. Radcliffe, C.D. DiFonzo and M.S. Connelly, 1994, In: G.W. Zehnder, M.L. Powelson, R.K. Jansson and K.V. Raman (eds), *Advances in Potato Pest Biology and Management*, p. 99. APS Press, St. Paul, MN.
Raman K.V, 1988a, *Agric. Ecosys. Environ.* 21, 85.
Raman K.V, 1988b, *Integrated Insect Pest Management for Potatoes in Developing Countries*. CIP Circular 16, p. 1. International Potato Center (CIP), Lima, Peru.
Raman K.V. and E.B. Radcliffe, 1992, In: P.M. Harris (ed.), *The Potato Crop, The Scientific Basis for Improvement*, 2nd ed., p. 476. Chapman and Hall, London.
Raman K.V., R.H. Booth and M. Palacios, 1987, *Trop. Sci.* 27, 175.
Raymundo S.A. and J. Alcázar, 1983, In: W.J. Hooker (ed.), *Research for the Potato in the Year 2000*, p. 159, Proc. Int. Congr. in Celebration of the Tenth Anniversary of the Int. Potato Cent., February 22–27. International Potato Center (CIP), Lima, Peru.
Robert Y., 1971, *Potato Res.* 14, 130.
Robert Y. and D. Bourdin, 2001, In: G. Lobenstein, P.H. Berger, A.A. Brunt and R.H. Lawson (eds), *Virus and Virus-like Diseases of Potatoes and Production of Seed-potatoes*, p. 195. Kluwer Academic Publishers, Dordrecht, the Netherlands.
Roelofs W.L., J.P. Kochansky, R.T. Cardé, G.G. Kennedy, C.A. Henrick, J.N. Labovitz and V.L. Corbin, 1975, *Life Sci.* 17, 699.
Salazar L.F, 1996, *Potato Viruses and Their Control*. International Potato Center (CIP), Lima, Peru.
Saour G. and H. Makee, 2004, *J. Econ. Entomol.* 97, 711.
Scheffer S.J., 2000, *J. Econ. Entomol.* 93, 1146.
Senanayake D.G., E.B. Radcliffe and N.J. Holliday, 2000, *Environ. Entomol.* 29, 1123.
Sexson D.L. and J.A. Wyman, 2005, *J. Econ. Entomol.* 98, 716.
Shelton A.M. and J.A. Wyman, 1979, *J. Econ. Entomol.* 72, 261.
Sigvald R., 1989, *J. Appl. Entomol.* 108, 35.
Sinden S.L., L.L. Sanford, W.W. Cantelo and K.L. Deahl, 1986, *Environ. Entomol.* 15, 1057.
Sporleder M., J. Kroschel, M. R. Gutierrez Quispe and A. Lagnaoui, 2004, *Environ. Entomol.* 33, 477.
Syller J, 1994, *Ann. Appl. Biol.* 125, 141.
Tauber M.J., C.A. Tauber, J.J. Obrycki, B. Gollands and R.J. Wright, 1988, *Ann. Entomol. Soc. Am.* 81, 748.
Thieme T., U. Heimbach, R. Thieme and H.-L. Weidemann, 1998, *Aspects Appl. Biol.* 52, 25.
Thomas P.E, 1983, *Plant Dis.* 67, 744.
Thomas P.E., W.K. Kaniewski and E.C. Lawson, 1997, *Plant Dis.* 81, 1447.
Thornton M., 2003, In: S. Ristow, E.A. Rosa and M.J. Burke (eds), *Biotechnology: Science and Society at a Crossroad*, p. 235, Nat. Agric. Biotechnol. Council. Boyce Thompson Inst., Ithaca, NY.
Tingey W.M., 1984, *Am. Potato J.* 61, 157.

Trivedi T.P. and D. Rajagopal, 1992, *Trop. Pest Manage.* 38, 279.
van den Heuvel J.F.J.M., C.M. de Blank, D. Peters and J.W.M. van Lent, 1995, *Eur. J. Plant Pathol.* 101, 567.
van den Heuvel J.F.J.M., J.A.A.M. Dirven, G.J. van Os and D. Peters, 1993, *Potato Res.* 36, 89.
van den Heuvel J.F.J.M., M. Verbeek and F. van der Wilk, 1994, *J. Gen. Virol.* 75, 2559.
Voerman S. and G.H.L. Rothschild, 1978, *J. Chem. Ecol.* 4, 531.
von Arx R., J. Goueder, M. Cheikh and A. Ben Temine, 1987, *Insect Sci. Appl.* 8, 989.
von Arx R., O. Roux and J. Baumgärtner, 1990, *Agric. Ecosyst. Environ.* 31, 277.
Weidemann H.-L, 1988, *Potato Res.* 31, 85.
Weintraub P.G. and A.R. Horowitz, 1995, *Phytoparasitica* 23, 177.
Whalon M.E., D. Sanchez and L. Duynslager, 2004, The MSU Arthropod Pesticide Resistance Database, http://www.pesticideresistance.org/.
Woodford J.A.T., C.A. Jolly and C.S. Aveyard, 1995, *Potato Res.* 38, 133.
Woodford J.A.T., S.C. Gordon and G.N. Foster, 1988, *Crop Prot.* 7, 96.
Wright R.J, 1984, *J. Econ. Entomol.* 77, 1254.
Yanggen D., D.C. Cole, C. Crissman and S. Sherwood, 2004, *EcoHealth* 1, SU72.
Zhao J.-Z., B.A. Bishop and E.J. Grafius, 2000, *J. Econ. Entomol.* 93, 1508.

Chapter 26

The Nematode Parasites of Potato

Didier Mugniéry[1] and Mark S. Phillips[2]

[1]*INRA, UMR BiO3P, Domaine de la Motte-au-Vicomte, BP 35327, 35653 Le Rheu, France*
[2]*Scottish Crop Research Institute, Invergowrie, Dundee, DD2 5DA, Scotland, UK*

Numerous nematodes are able to feed and to reproduce on potato. The majority withdraw nutrients from cells in the epidermis of the roots, without inducing any measurable damage. However, some of these nematodes, for example *Trichodorus* and *Paratrichodorus* species, are extremely noxious, firstly because their saliva is toxic to the roots and secondly because they transmit viruses. Some species penetrate and migrate into the roots, destroying the cells by removing the cytoplasm. Some nematodes may cause little direct damage but may facilitate secondary infections with fungi, which can be extremely destructive. Some genera are highly specialized and induce cellular changes resulting in feeding sites that are called giant cells for the genera *Nacobbus* and *Meloidogyne* and syncytia for the genus *Globodera*. The first two genera are extremely polyphagous, whereas *Globodera* species are highly specific for Solanaceous species. The last type of nematode causing damage in potato is the polyphagous species, *Ditylenchus destructor*, which directly damages the tubers by causing a specific dry decay. All these principal genera have specific parasitic methods (Fig. 26.1) and belong to different classes in the phylogenetic tree of the Nematoda (Baldwin et al., 2004).

26.1 POTATO CYST NEMATODES (*GLOBODERA ROSTOCHIENSIS* AND *GLOBODERA PALLIDA*)

Among the four species of *Globodera* that develop on Solanaceous species, *G. rostochiensis, G. pallida, G. tabacum* and *G. mexicana*, only the first two develop on potato. Present in Central and North America, *G. tabacum* spreads to Europe (France and Italy) where it may be sympatric with *G. rostochiensis* and *G. pallida*. *G. mexicana* is found only in Mexico. Listed as quarantine pests in almost all countries, *G. rostochiensis* and *G. pallida* are specific parasites of the Solanaceae. It is assumed that these species originate from South America where *G. pallida* is found more to the North of Lake Titicaca, whereas *G. rostochiensis* is found more to the South. They were probably introduced into Europe during the nineteenth century where they multiplied and have subsequently been spread all around the world with infested tubers. They are known to be present in North, Central and South America, North and South Africa, Asia, Australia and New Zealand. The only exceptions are the warm tropical areas. The European populations of *G. pallida* are likely to have originated from the south of Peru (Picard, 2005).

Potato Biology and Biotechnology: Advances and Perspectives
D. Vreugdenhil (Editor)

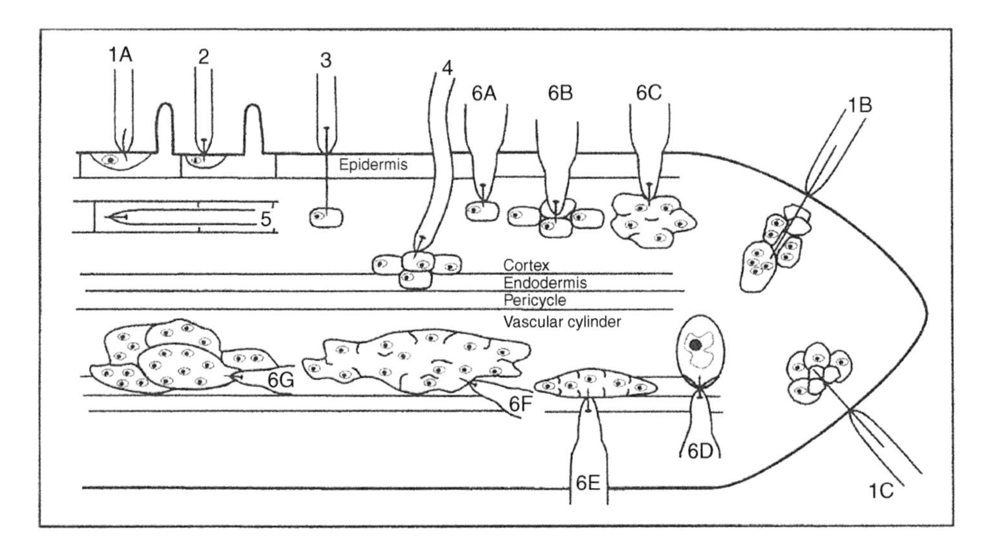

Fig. 26.1. Schematic representation of food sites for several plant parasitic nematodes according to Wyss (1997). 1, migrating ectoparasites Dorylaimidae; 1A, ***Trichodorus*** sp.; 1B, *Xiphinema index* and 1C, *Longidorus elongatus*. 2–6, Tylenchidae: 2, migrating ectoparasites, *Tylenchorhynchus dubius*; 3, sedentary ectoparasites, *Criconemella xenoplax*; 4, migrating ectoectoparasites, *Helicotylenchus* sp.; 5, migrating endoparasites, ***Pratylenchus*** sp.; 6, sedentary endoparasites; 6A, *Trophotylenchus obscurus*; 6B, *Tylenchulus semipenetrans*; 6C, *Verutus volvigentis*; 6D, *Cryphodera utahensis*; 6E, *Rotylenchulus reniformis* and ***Nacobbus aberrans***; 6F, *Heterodera* sp. and ***Globodera*** sp. and 6G, ***Meloidogyne*** sp. Potato nematodes are bold faced.

Characterization of *G. rostochiensis* and *G. pallida* based on morphological and morphometric traits of the juveniles and cysts is difficult and only feasible for specialists. Monoclonal antibodies recognizing species-specific thermostable proteins are used as a routine method in the Netherlands to distinguish the two species (Schots et al., 1989, 1992). Burrows (1990) and Fullaondo et al. (1999) developed diagnostic protocols using the polymerase chain reaction (PCR) to amplify species-specific DNA sequences. In order to differentiate these two species from each other and from the related species *G. tabacum* and *G. mexicana*, Thiéry and Mugniéry (1996) developed a PCR-RFLP (Restriction Fagment Length Polymorphism) methodology. Marshall and Crawford (1992) and Mulholland et al. (1996) differentiated *G. pallida* from *G. rostochiensis* by a multiplex PCR diagnostics. Two-dimensional electrophoresis (2DE) may be used to distinguish different populations of *G. rostochiensis* (Bakker and Gommers, 1982).

All these species reproduce by obligate amphimixis. However, whatever the criteria used, their physiological, biochemical or molecular variability is extremely high.

26.1.1 Host range

G. rostochiensis and *G. pallida* have the same host range. They are able to develop on all tuberous species of *Solanum*. Among the cultivated plants, they freely develop on *Lycopersicon esculentum* (tomato) and are able to reproduce on *Solanum melongena*

Table 26.1 Differential host range of the Solanaceous *Globodera*

	***Solanum tuberosum* (potato)**	***Lycopersicon esculentum* (tomato)**	***Nicotiana tabacum* (tobacco)**	***Solanum nigrum* (nightshade)**	***Solanum dulcamara* (bittersweet)**
Globodera rostochiensis	+	+	–	–	+
Globodera pallida	+	+	–	–	+
Globodera tabacum	–	+	+	+	+
Globodera mexicana	–	+	–	+	+

(egg-plant). Among the wild Solanaceae, bittersweet (*Solanum dulcamara*) is a fairly good host, but the nightshade *Solanum nigrum* allows the juveniles to hatch and to penetrate into the roots but not to develop. Stelter (1971) gives a list of putative hosts. In Table 26.1, common and differential hosts of the four species of *Globodera* are given. It is shown that the two potato cyst-nematode species have the same host range. This specificity is due to two main factors. The first is that only the root exudates of Solanaceous plants are able to induce the juveniles to hatch from the cysts. But there also exists a strong interaction between type of exudates and species of *Globodera*, which partly explains this specificity (Table 26.2). So, egg-plant is never attacked in natural conditions by the two potato cyst nematodes, although artificially hatched juveniles penetrate and develop freely on this plant. The second factor is a result of the specific recognition of the roots by the hatched juveniles because of specific carbohydrate-binding lectins. The ability of both *G. rostochiensis* and *G. pallida* to penetrate the roots of potato, tomato and egg-plant is extremely strong, whereas it is extremely low in roots of pepper. In all cases, the penetration of juveniles occurs in the root zone of elongation and in the area where there is formation of auxiliary rootlets.

Table 26.2 Efficiency of the root exudates on the hatching of *Globodera* species.

	Globodera rostochiensis	***Globodera pallida***	***Globodera tabacum***	***Globodera mexicana***
Potato	+	+	ε	+
Tomato	+	+	+	+
Tobacco	–	–	+	–
Nightshade	±	±	+	+
Egg-plant	ε	ε	ε	ε

+, strong efficiency; ±, efficiency more or less important depending of the populations; ε, weak efficiency and –, no effect.

26.1.2 Diseases

The growth of attacked plants is retarded. In the field, it is common to observe foci of smaller plants, which correspond with areas having high nematode densities. The symptoms are not specific to nematode damage and may be confused with many other causes, such as soil structure, effects of herbicides, lack of fertilizers and other biotic or abiotic stress factors. The consequence of nematode attack is that tuber yield losses are sustained as a result of a reduction in the size of tubers and therefore total weight per hectare. Different mathematical models describing the relationship between the reduction of the yield and the density of the soil population of nematodes before planting have been described (Seinhorst, 1965; Oostenbrink, 1966; Elston et al., 1991). Some cultivars are more or less tolerant, that is to say that at similar levels of nematode population density some cultivars are damaged more than others irrespective of their resistance. Some tolerant cultivars such as cultivar Cara develop a very large leaf canopy that enables the crop to achieve 100% ground cover and thus maximum light interception despite damage (Trudgill et al., 1990). Consequently, the amount of actual yield loss is not as high as that if light interception is reduced. Commercial yield is also less affected in cultivars that produce few but very large tubers. Qualitative damage to the tuber skin can also occur and is seen as numerous small spots on the tuber surface due to penetration of juveniles into the epidermis of potato cyst nematodes (Hide and Read, 1991), mainly due to *G. pallida* (Fig. 26.2).

As for practically all phytoparasitic nematodes, the amount of disease is proportional to the log of the initial population density expressed as the number of second stage juveniles (J2s) per gram of soil. Total plant yield reduction cannot be detected below a certain minimum density called the tolerance threshold. Above this threshold, the yield reduction may be considered as proportional to the log (J2/g) until a yield minimum lower than the asymptote is reached. Many models have been developed: sigmoid (Seinhorst, 1965), linear (Oostenbrink, 1966) and inverse linear (Elston et al., 1991). The general slope of the curves may be modified by many factors. In general, all agronomic factors favourable to the crop (water supply, fertilizers, etc.) reduce the nematode damage. Some cultivars as Multa possess a tolerance threshold higher than others, whereas other cultivars are characterized by having a high relative minimum yield. Finally, some potato cyst populations are more or less pathogenic.

Association with fungi such as *Verticillium dalhiae* (Harrison, 1970) or *Rhizoctonia solani* (Grainger and Clark, 1963) may increase the amount of damage. The wilt due to *V. dalhiae* appears sooner than it otherwise would in the presence of *G. rostochiensis*.

Quarantine regulations are of essential economic importance for farmers. In most countries, potato seed production is strictly forbidden on infested land. This prohibition is generally maintained for 5–7 years. All potato seed has to be certified to have been produced from non-infested fields, and a phytosanitary passport must accompany all shipping. For consumption and ware potatoes, a potato crop may be authorized if soil disinfection, resistant cultivars or early harvest are practised.

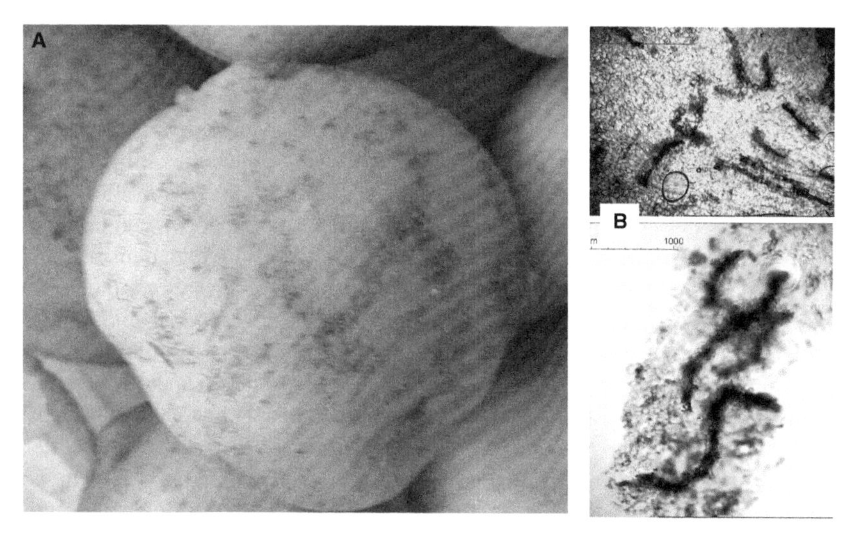

Fig. 26.2. Skin diseases caused by *Globodera pallida* [A, whole tuber and B, second stage juveniles (J2s) in the cuticle].

26.1.3 Biology

The J2 hatch and mechanically penetrate the elongation zone of the roots. They migrate into the roots to the vascular vessels. With their stylet, they pierce a cell and induce a syncytium that is characterized by thickened cell walls, a dense cytoplasm, a reduced vacuole and multilobate nuclei (Fig. 26.3A). This syncytium, which may be formed from up to 15 cells, acts as a nurse cell providing a metabolic sink from which the nematode feeds and develops.

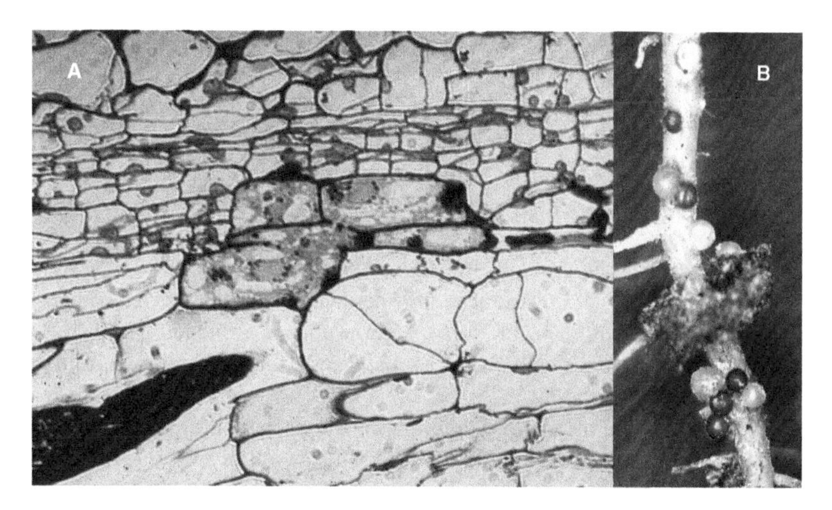

Fig. 26.3. (A) Syncytium induced by *Globodera* and (B) females and cysts of *Globodera rostochiensis* on potato roots.

Induction and maintenance of such syncytia is presumed to be regulated by proteins produced by the nematode. Some enzymes have been detected, but the real mechanism of induction is as yet unknown. Once the syncytium has been induced, the J2 looses its mobility and grows through three successive moults [third stage juveniles (J3s), fourth stage juveniles (J4s) and adult]. Sexual determination is under the influence of environmental factors and is apparent at the J4 stage. Factors unfavourable to syncytium development and thus nematode development result in a change in sex ratio with an increase in the production of males. The factors that affect this are high densities of nematodes in the roots resulting in high rates of competition and damage and the production of syncytia in small lateral roots where development is limited by space and in resistant plants where syncytial development is restricted by the resistance response.

On reaching J4, the immature female swells and bursts through the root epidermis. Finally, the females are visible on the root surface as small spheres (Ø 500–750 μ), white for *G. pallida* and yellow to golden for *G. rostochiensis* (Fig. 26.3B). Adult males emerge from the roots and are attracted to females by the female cuticle pheromones. After copulation, the females lay numerous eggs inside their uterus and subsequently die. The female's cuticle darkens and becomes a cyst containing up to 1000 eggs. Embryogenesis occurs inside the female/cyst: a first moult occurs inside the chorion, producing a new J2 that enters into diapause. At 20 °C, the life cycle needs 15 days from penetration to copulation and 16 days for embryogenesis. In the natural conditions of Western Europe, a generation requires about 3 months from potato planting to the formation of cysts. Generally, one generation occurs per potato crop, but a partial second one is observed on late potatoes.

After harvest, the cysts remain in the soil. The diapause, generally stronger for *G. rostochiensis* than for *G. pallida*, requires cold conditions for it to be broken but shows large variability among populations. Some populations are able to develop on volunteers that may start to grow from September and achieve their cycle during winter. If a second potato crop is grown in the same season, i.e. in July after the harvest of a crop in June, the new J2s of some populations are able to hatch immediately while others do not.

26.1.4 Dormant stage

The cyst protects the eggs and J2s for many years. In the absence of potatoes, the J2s remain alive in the cysts. Annually, a small proportion hatch freely and die in the soil. The climatic conditions may act to decrease the survival of the J2. Adverse conditions are mainly high temperatures and soil dryness. Generally, the annual decline has been found to be extremely variable. In Ireland, the Plant Protection Service considers an annual decline not more than 10%. In Brittany, this percentage reaches 18%, whereas in the South of France, it is about 50%. In North Algeria, it was estimated as 80%. In Agadir (South Morocco), the annual decline of 95% balances exactly the annual multiplication on the potato crop (Schlüter, 1976).

The cysts are an unusual means of dissemination. Very light, they can be spread locally by water, wind, machinery (wheels) and man (shoes) as well as in soil attached to potato tubers. On a larger scale, they accompany potato seeds from country to country and from continent to continent. These characteristics explain the very wide distribution of these species despite the very severe quarantine regulations.

26.2 ROOT-KNOT NEMATODES (*MELOIDOGYNE* SPP.)

In contrast to the potato cyst nematodes, root-knot nematodes are extremely polyphagous. They are able to develop on thousands of species, cultivated or wild, belonging to all botanical families. Approximately 80 species of *Meloidogyne* have been described, of which only a few are pathogenic on potato. In warm and tropical areas, as well as in glasshouses, the economically important species are *Meloidogyne incognita, M. arenaria, M. javanica* and *M. mayaguensis*, whereas in temperate areas, the most important species are *M. hapla, M. chitwoodi* and *M. fallax*.

Specific identification of *Meloidogyne* species is extremely difficult. Various criteria have been used including the host range (Hartman and Sasser, 1985), the morphology of females, males and juveniles (Jepson, 1983; Eisenback, 1985;), isoenzyme patterns of soluble proteins from females (Esbenshade and Triantaphyllou, 1985) and specific PCR markers (Zijlstra, 2000; Zijlstra et al., 2000; Wishart et al., 2002). No system exists other than biological host range tests to differentiate pathotypes (Table 26.3) and biological races (Table 26.4).

Genetically, *M. incognita, M. arenaria* and *M. javanica* are a very closely related group and more distantly related to *M. mayaguensis, M. hapla* and the closely related group of *M. chitwoodi* and *M. fallax*.

Reproduction of the tropical species and of the race B of *M. hapla* is by mitotic parthenogenesis, with between 41 and 48 chromosomes. The other species reproduce by facultative meiotic parthenogenesis and have between 14 and 18 chromosomes. This means that males are potentially able, but probably only in a small proportion of cases, to contribute to the genetic flow (Triantaphyllou, 1985). As a consequence, the variability, worldwide, of the tropical species is extremely small in contrast to the temperate species.

Table 26.3 Host range and pathotypes of the four main species of *Meloidogyne* (according to Hartman and Sasser, 1985).

	Meloidogyne incognita				***Meloidogyne javanica***	***Meloidogyne arenaria***		***Meloidogyne hapla***
	Pathotype					**Pathotype**		
	1	**2**	**3**	**4**		**1**	**2**	
Tobacco[a]	–	+	–	+	+	+	+	+
Cotton[b]	–	–	+	+	–	–	–	–
Pepper[c]	+	+	+	+	–	+	–	+
Watermelon[d]	+	+	+	+	+	+	+	–
Peanut[e]	–	–	–	–	–	+	–	+
Tomato[f]	+	+	+	+	+	+	+	+

[a]cultivar NC95; [b]cultivar Deltapine; [c]cultivar Early California Wonder; [d]cultivar Charleston Gray; [e]cultivar Florunner; [f]cultivar Rutgers.

Table 26.4 Races of *Meloidogyne chitwoodi* and difference with *Meloidogyne hapla* (according to Mojtahedi et al., 1988)

	Meloidogyne hapla	*Meloidogyne chitwoodi*	
		Race	
		1	2
Wheat[a]	–	+	+
Pepper[b]	+	–	–
Alfalfa[c]	+	–	+
Carrot[d]	+	+	–
Tomato[e]	+	+	+

[a]cultivar Nugaines; [b]cultivar California wonder; [c]cultivar Thor; [d]cultivar Red Cored Chantenay; [e]cultivar Columbian.

26.2.1 Disease

All *Meloidogyne* species induce galls on the root system. On potato, they are able to induce galls on tubers (Fig. 26.4A). The presence of *Meloidogyne* in the roots limits the water supply of the plants, disrupts their physiology and, as a direct consequence, decreases yield. Galled tubers are not marketable. When tubers are used for processing, e.g. for French fry production, the presence of females, mainly of *M. chitwoodi*, on the tubers leads to holes reducing quality and can result in the rejection of a whole crop (Fig. 26.4B).

Yield reduction is correlated with the initial density of nematodes before planting. Disease on the tubers tends to be more correlated to the soil temperature, which influences the number of generations and then the number of juveniles present at tuber induction. Disease symptoms on tubers can be very variable (Griffin, 1985; Pinkerton et al., 1991).

Disease may increase when soil fungi are present. This is the case with *Streptomyces scabies* (Mehiar et al., 1984) and with *R. solani* (Scholte and s'Jacob, 1990). The interaction with the bacterium *Ralstonia solanacearum* has been demonstrated for tomato but is not so clear with potato.

26.2.2 Biology

In contrast to *Globodera*, no real resting or dormant stage exists. All eggs are deposited in an external gelatinous matrix, in which the embryos develop rapidly to hatch at the J2 stage whereupon they invade the roots. Mechanical and biochemical penetration occurs in the elongation zone of a growing root. Intercellular migration, facilitated by the breakdown of the intercellular membrane, takes place through the cortical cells to the root apex at which point the J2s turn round and migrate back up the root deep inside the vascular cylinder (Wyss et al., 1992). The J2s stop and pierce up to 6–7 adjacent cells that are then induced to form the giant cells from which the nematode feeds.

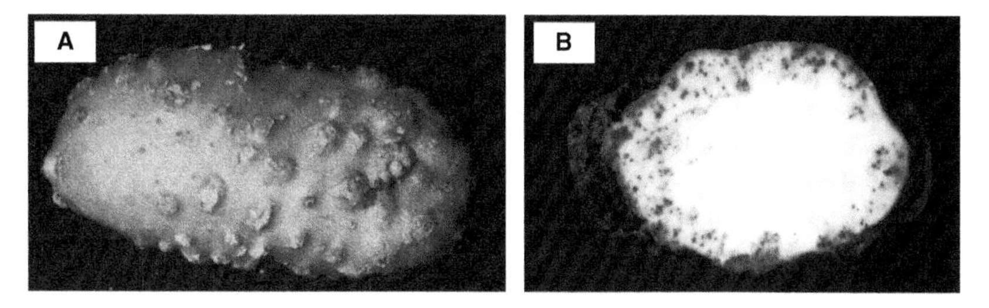

Fig. 26.4. Galls on tubers due to *Meloidogyne javanica* (A) and internal spots due to *Meloidogyne chitwoodi* (B).

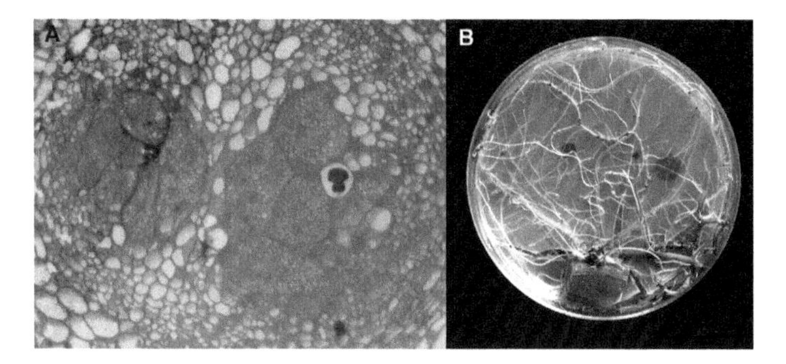

Fig. 26.5. Giant cells (A) and root galls in Petri dish (B) induced by *Meloidogyne incognita* in a susceptible potato genotype.

In each one, the cell walls thicken, the vacuole disappears, the cytoplasm becomes extremely dense and nuclear division occurs without cell wall formation (Fig. 26.5A). Around these nurse cells, the adjacent cells divide rapidly, which results in the formation of a root gall (Fig. 26.5B). Three moults occur and the mature female starts to deposit eggs outside the root if the gall is small, inside the roots if the gall is large and in the tubers.

Sex determination is strictly determined by environmental factors. Males are produced under unfavourable conditions. Again in contrast to *Globodera*, reversal of sex is possible throughout the developmental cycle, producing normal males with one testis, males with two testes or with two testes and a vestigial vulva when unfavourable factors act during the third or the fourth stage. Males of meiotic species may mate and participate in producing the new generation.

The number of annual generations depends on the temperature. At 20 °C, the time taken from penetration to the deposit of the first eggs is about 40 days for *M. chitwoodi* and 30 days for *M. incognita*. In glasshouses and in tropical countries, three to four annual generations may be observed. In temperate areas, *M. chitwoodi* may produce three generations on late potatoes.

Some tropical populations, mainly those of the Sahelian area, have an embryonic diapause (de Guiran, 1979). Eggs enter diapause at an early stage and remain viable during the dry season.

In the stored tubers, all tropical species develop and produce multiple generations. This has not been observed with *M. chitwoodi* and *M. fallax*.

26.2.3 Spread

Meloidogyne species are spread over short distances by water and wind. Long-distance spread is facilitated by exchange of contaminated soil, root-stocks and tubers. In Europe, *M. chitwoodi* and *M. fallax* are listed as quarantine pests on potato. That means that tuber seeds must be certified free of these species. It is essential to keep in mind that in case of a low infestation, without any symptoms on tubers, these nematodes are quite undetectable. This means also that, in Europe, root-stocks, ornamental species, etc. in infested soil could enable undetected spread of these species.

26.3 THE FALSE ROOT-KNOT NEMATODE *NACOBBUS ABERRANS*

This species (or this species complex) originates from America. Some populations originating from North America develop on sugar beet. The populations originating from South America develop on potatoes. Named 'El Rosario' by the Andean farmers, it damages potato crops in the Peruvian, Bolivian and Argentinean mountains from 2000 m above sea level. The populations found at lower altitudes do not develop on potatoes and are mainly found on tomatoes. It was introduced into Russia through the Netherlands and Finland but disappeared later. This species is listed on the European quarantine list. A PCR diagnosis has been described, which is able to recognize all the South American populations (Anthoine and Mugniéry, 2005).

26.3.1 Host range

Nacobbus aberrans is extremely polyphagous. Among the cultivated plants, it may be found on potato, tomato, sugar beet, carrot, lettuce, spinach, pepper and cucumber. Numerous wild species are hosts, among them *Chenopodium album* and *Amaranthus hybridus*. Other hosts, such as *Brassica campestris* and *Erodium cicutarium*, are hosts in the Central Peruvian Mountains but not in the Altiplano of Puno (Gomez Tovar, 1973).

Physiological races have been described (Manzanilla-López et al., 2002) according to their ability to complete their cycle on differential hosts (Table 26.5). These races, all amphimictic, are able to interbreed and to give viable and fertile progenies. Reid et al. (2003) suggested the existence of a species complex and advocated separating *N. aberrans* into three species.

Table 26.5 Differential hosts able to discriminate the different races of *Nacobbus aberrans*

	Potato	Tomato	Sugar beet	Chilli pepper	Bean
Race 1 (sugar beet)	–	+	+	±	
Race 2 (potato)	+	+	+	±	
Race 3 (bean)	–	+	–	+	+

26.3.2 Disease

N. aberrans may be considered as the most damaging nematode of the potato in the high Andes. The symptoms are typically the development on the root system of numerous small and spherical galls, very often accompanied by small rootlets (Fig. 26.6). Reduction of yield may reach up to 60–90% (Gomez Tovar, 1973).

26.3.3 Biology

The biology of this species is extremely complex and partly unknown. The J2s penetrate everywhere along the root. They migrate into the cortex where they create a large cavity in which they develop to the pre-adult stage. The cells around the cavity are extremely enlarged, with a dense cytoplasm, and contain numerous starch particles. The internal cells are close to the pericycle and the vascular vessels (Fig. 26.7). The form and distribution of the starch particles depend on the host and nematode race.

On potato, the pre-adult stages enter a strong diapause. The occurrence of this is rather uncommon and is characterized by nematodes having a black colouration, showing no movement and with their internal organs difficult to distinguish. This stage may be encountered in soil and dry roots. Cold conditions are necessary to break the diapause. After winter, the pre-adult stage becomes male or pre-female. The pre-females penetrate into the roots and induce a giant cell. This giant cell is characterized by a dense cytoplasm, hypertrophied nuclei and nucleoli with adjacent cells storing numerous starch particles (Manzanilla-Lopéz et al., 2002). The females enlarge while at the same time the cortical and vascular systems show hypertrophy resulting in gall formation. With the shape of an elongate lemon, the female remains embedded in the root, except for its posterior part. After mating, eggs are deposited in an external gelatinous matrix where embryogenesis occurs. Up to 500 J2s are produced, which then hatch, though not all of them immediately. However, a proportion of the eggs is able to stop their development at the morula stage. The hatched J2s penetrate the roots, but in the absence of host plants, they are able to survive for a long time. At 25 °C, embryogenesis takes 9–10 days and 51 days at 15 °C. The whole cycle requires 37–48 days at 22–24 °C.

Whatever the race, the J2s are able to develop to the pre-adult stage. The race specificity is measurable only at the adult stage. The tomato race pre-female does not penetrate potato roots.

26.3.4 Spread

This nematode is mainly spread in infested tubers. All the juvenile stages are able to penetrate the developing tubers and develop just under the epidermis. During storage, the majority of the nematodes found in tubers are J2 and J3, seldom females. In Argentina, infested seed tubers, produced in the mountains, are mainly transported to the Pampas, i.e. where the tomato race occurs. Despite this fact, the potato race has been unable to establish at low altitudes.

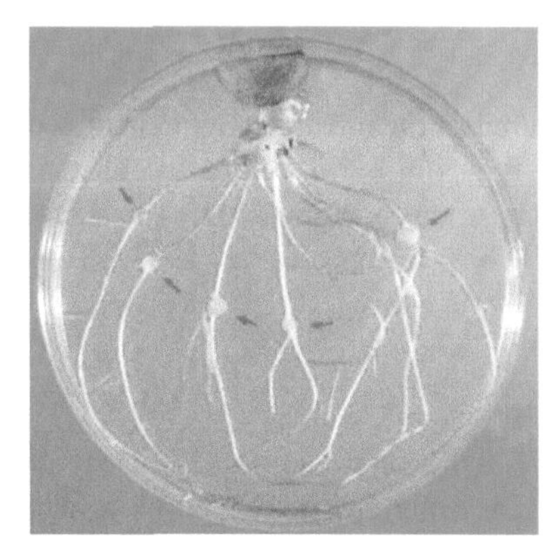

Fig. 26.6. Galls developed in Petri dish by, each one, a pre-female stage of *Nacobbus aberrans.*

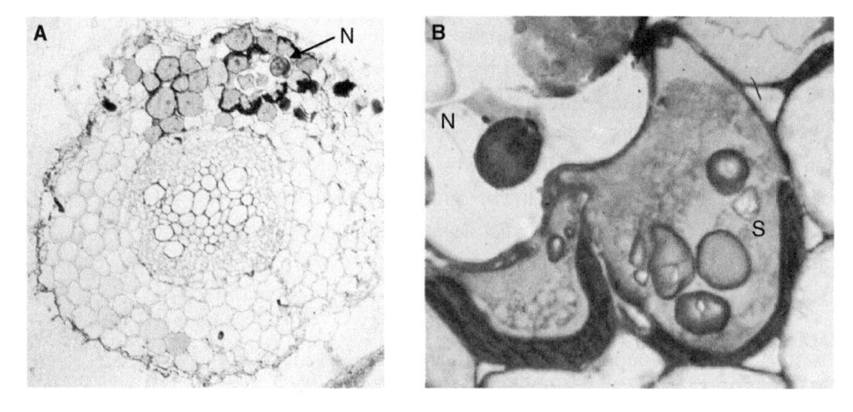

Fig. 26.7. Cavity surrounded by transformed cortical cells (A) and (B) particles of starch (S) in the transformed cortical cells around the pre-juvenile nematode (N).

26.4 VIRUS VECTOR NEMATODES (*TRICHODORUS* SPP.)

Numerous species of the genera *Trichodorus* and *Paratrichodorus*, belonging to the Dorylaimidae, feed on potato roots. When population densities are high, they can cause damage to the root system by inhibiting growth.

However, their main damaging effect on potato is due to their ability to transmit tobacco rattle virus (TRV), with the typical symptoms of a cork ring in the tuber flesh.

According to Decraemer (1991), the main species are *Trichodorus primitivus*, mainly found in Europe but also found in the USA and New Zealand; *T. similis*, a European species, found in Michigan; *T. viruliferus*, found in Europe and in Florida; *Paratrichodorus minor*, worldwide species; *P. pachydermus*, a European species, found in Canada and in the Northwest of the USA and *P. teres*, a European species, found in South Africa and in Oregon. All these species are parasites in temperate climates, except *P. minor*.

They live preferentially in sandy soils though, where they can colonize all the arable soil. Each species lives in the soil at its own preferred depth with some living as far as 70 cm below the soil surface.

26.4.1 Disease

The nematodes ingest viral particles when feeding on the roots. The attachment of the particles on the inner part of the oesophageal lumen and viral transmission are specific for the nematode species and the isolates of TRV (Brown et al., 1989a).

If the population level is low, damage is only visible on tuber slices. In case of heavy infestation, the potato skin may appear cracked, and the tuber symptoms may be confused with those of mop top viruses (Fig. 26.8). When virus-infected potato plants are growing at low temperatures, necrotic striations are visible on leaves, veins and stems (Joubert and Dalmasso, 1974). Some cultivars are relatively tolerant and show few symptoms (Dale et al., 2004).

There is no direct relationship between the density of *Trichodorus* and *Paratrichodorus* and the percentage of infected tubers (Fritzsche et al., 1986). If the nematodes are not viruliferous and if the seed tubers and other plants in the same soil are virus free, then the only disease will be the direct nematode damage to the roots, which can reduce yield.

26.4.2 Biology

Except *T. primitivus*, all species prefer very sandy soils. They are extremely susceptible to desiccation and migrate only to near the soil surface during rainfall or other favourable moisture conditions. The majority of species are amphimictic: the eggs are deposited in soil and hatching occurs without diapause. The newly hatched juveniles are infective and may survive for up to a year.

All these species are highly polyphagous and can feed on cultivated plants, such as potato, sugar beet, cauliflower, clover, alfalfa, pea, barley, oats, maize, tobacco and carrot. They are also able to feed and reproduce on numerous weeds, and the population densities are often extremely high in grassland.

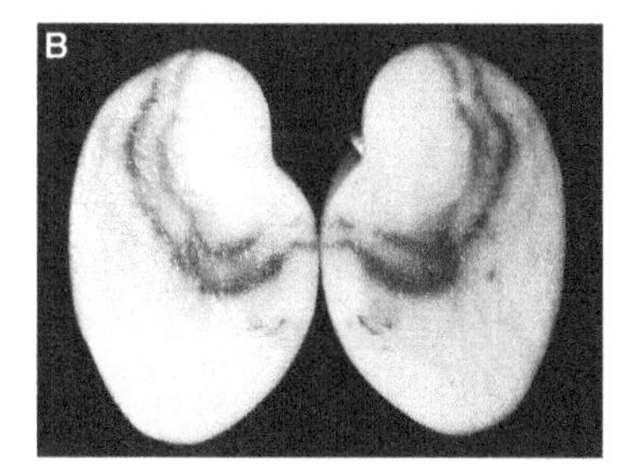

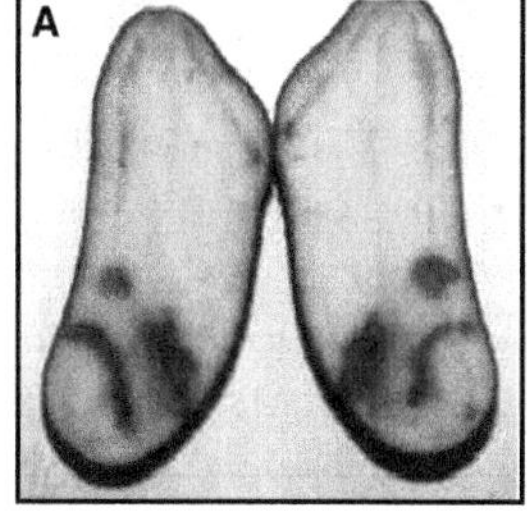

Fig. 26.8. Symptoms of Rattle (A) compared with those from Mop-Top (B).

26.5 THE ROOT LESION NEMATODES (*PRATYLENCHUS* SPP.)

Numerous species of the genus *Pratylenchus*, mainly *P. coffeae, P. scribneri, P. brachyurus, P. penetrans* and *P. neglectus*, feed and reproduce on potato roots and may be responsible for damage, especially in the USA. Their identification, mainly based on morphological characters, is difficult. All these species, among 13 others, may be identified by PCR-RFLP of the ITS (Internal Transcribed Spacer) region (Waeyenberge et al., 2000).

26.5.1 Disease

In case of a heavy soil infestation, potato growth and consequently total tuber yield may be strongly adversely affected. Depending on the environmental and soil conditions, yield reduction may reach 19% (Olthof, 1990). Furthermore, the presence of multiple pimples on the tuber surface, mainly attributed to *P. scribneri*, depreciates the product that becomes unmarketable.

The most damaging effects seem to be due to fungal interactions. The early dying of potato crops in the USA is attributed to the interactions between *Verticillium dahliae* and *P. penetrans* (Kotcon et al., 1985) or with *P. neglectus* (Scholte and s'Jacob, 1990).

26.5.2 Biology

All the above species of *Pratylenchus* are polyphagous. Juveniles and adults penetrate into the roots without any preferential locus. They feed on the cortical cells and migrate in the roots, destroying cell after cell. Necrotic lesions appear along the roots and are visible in case of high infestation levels that may reach as high as 30,000 nematodes per gram of root.

Egg laying occurs either inside the root or in the soil. As the root system becomes invaded by soil microorganisms, it becomes unfavourable for *Pratylenchus*. They leave the roots and migrate to the rhizosphere. After crop harvest, they survive in the soil, in the remaining root debris or feed on weeds. During winter, population decline varies between 15 and 86% (MacGuidwin and Forage, 1991). This winter decline is generally insufficient to decrease population levels to a point at which in the following season potato or other host plants can be grown without risks.

26.6 *DITYLENCHUS DESTRUCTOR* AND *DITYLENCHUS DIPSACI*

The genus *Ditylenchus* consists of about 80 species with a range of food sources including fungi and plants, with some species being both plant parasitic and mycophagous. Among them, two species are extremely polyphagous, namely *Ditylenchus dipsaci* and *D. destructor*, and able to damage potatoes. *D. dipsaci* develops mainly in stems and leaves, whereas *D. destructor* develops only in underground organs. Correct identification, mainly based on morphological traits, is important but difficult (Viscardi and Brzeski, 1993). Molecular tools are not yet available to distinguish all the known species (Wendt et al., 1994).

D. dipsaci is an occasional parasite of aerial parts of the potato. In tubers, it may be found in the presence of *Phoma exigua* var. *foveata*. The growth of potato is reduced, the leaves remain small and the stems become thickened. Brown pustules are visible on attacked tubers. This type of infection is exceptional and has been observed in the Netherlands, Russia and Romania.

D. destructor is much more serious and a frequent pest on potato. It has been observed in Europe, (mainly in Russia), Asia (Iran, Pakistan, China and Japan), America (Canada, USA and Peru), New Zealand and South Africa. In Europe, it is listed as a quarantine pest on potato and bulbs.

26.6.1 Host range

D. destructor is able to cause diseases on numerous cultivated plants, including potato, sugar beet, carrot, hop and rhubarb. It is also able to multiply on an extremely large number of cultivated plants (sweet potato, lupin, clover, tomato, wheat, barley, etc.) and weeds but without disease symptoms. In the laboratory, it can be multiplied on fungi, but it is not clear whether it can survive on fungi in natural conditions.

26.6.2 Disease

D. destructor penetrates directly into the tubers through the stolons, wounds and eyes. The first symptoms are initially only visible on the tubers where on the skin small translucent spots are seen. Later, these spots darken, enlarge, coalesce and the tuber skin cracks. Finally the typical symptom is a dry rot (Fig. 26.9).

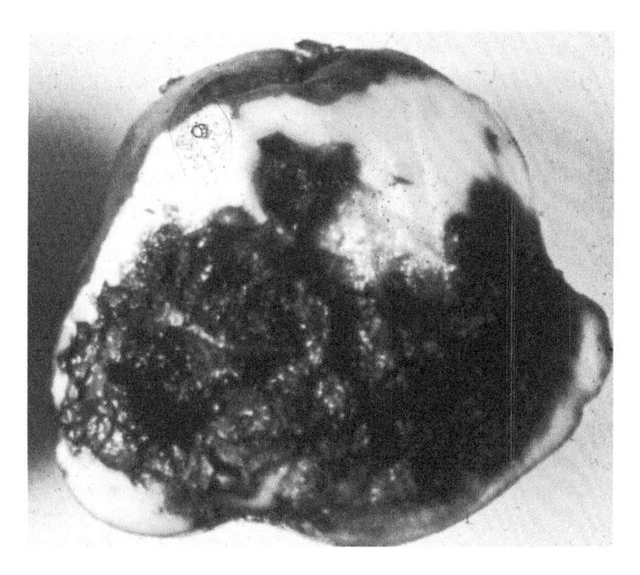

Fig. 26.9. Dry decay caused by *Ditylenchus destructor*.

Very often, microorganisms such as fungi – *Fusarium* sp. (Rojankovskii and Ciuera, 1986) – and bacteria – *Clavibacter sepedonicus, Erwinia atroseptica, Erwinia carotovora* (Polozhenets, 1984) – colonize the infected tubers, increasing the disease. Cold conditions during storage limit the aggravation of diseases.

26.6.3 Biology

During winter, *D. destructor* stays in the soil and remaining roots as juveniles and eggs (Andersson, 1971). It does not survive temperatures below 4.5 °C. In a host plant, cultivated or not, its generation time is less than a month at 20 °C. Its thermal optimum is between 20 and 27 °C with 5 and 34 °C as minimum and maximum. In natural conditions, the annual number of generations may be six to nine.

During tuber storage, it continues its cycle with a rate of multiplication of about 6. Curiously, these nematodes are unable to multiply on carrot, beet and onion during storage (Ivanova, 1975).

26.7 CONTROL

26.7.1 Prophylaxis

The key point is to avoid infestation of healthy areas and where they already exist to avoid the introduction of new populations. This last point is extremely important to consider for potato cyst nematodes, characterized by a large gene pool: the introduction of new genotypes in an infested area may increase the heterogeneity with regard to virulence, which may cause serious problems to control of nematode populations by means of resistant cultivars. The first step in avoiding spread and infestation is to use only certified seed tubers, accompanied by a phytosanitary passport. This passport certifies that the shipment is free of quarantine nematodes and/or that the tubers originate from soil in which, after official analyses, nematodes could not be detected. Despite the phytosanitary passport, it is extremely important that the grower themselves ensure that the seed tubers meet high phytosanitary standards. Tubers may be disinfected using sodium hypochlorite, for example, in order to destroy the *Globodera* cysts (Sarakoski, 1977) or by ethoprophos against *N. aberrans* (Costilla and Basco, 1985). In the areas where potato nematodes are known to be present, the publication of infested farms should be available to avoid spread, e.g. by exchange of infested soil from different fields or on machinery that is lent or hired. Within a farm, because of the risk of carrying infested soil on cultivation machinery, it is highly recommended to plough the uninfected fields first followed by the fields that may be or are known to be infested. Following cultivation of the latter, all machinery should be cleaned.

Rotation and control of volunteers are probably the most efficient preventative methods that can be used against the monophagous potato cyst nematodes. In case of accidental introduction of cysts of *Globodera*, it prevents the population from increasing if no host is present. When light infestations are present, growing seasons without a suitable host will limit the build-up of the population both to the detection level and to the point at which the tolerance threshold is reached. As mentioned above, the length of an effective rotation will depend on the climatic conditions. It is generally recognized that a minimum period of 7 years without a potato crop is required, but this duration may be shortened under warm conditions or if resistant varieties and soil disinfestation are included as part of the control strategy. The efficiency of rotation *sensu stricto* is much more questionable against all the other polyphagous species because of their large host range. In these cases, the farmers will incorporate in the rotation cultivated plants that are a non-host or, better, resistant.

To control the Southern root-knot nematodes (*M. incognita, M. arenaria* and *M. javanica*), non-hosts, such as cereals, are grown. In Europe, where only race 1 of *M. chitwoodi* is found, 1 year of alfalfa greatly lowers the *Meloidogyne* population. Against these two types of *Meloidogyne*, resistant cultivars may be used, like, respectively, the Mi tomato and some fodder radishes. Growers need to be aware that some cultivated crops, while suffering no noticeable damage, are in fact hosts. Examples of these are pea for *D. dipsaci* and wheat for *M. chitwoodi* and *M. fallax*. In contrast, some crop rotations are extremely efficient even when the intermediate plants are considered as good hosts. For instance, *D. destructor* seems not to produce symptoms on potato when grown following a carrot, lupin or buckwheat crop. Three or four years rotations with rye, alfalfa and rye-grass result in 70–100% decrease of the diseases (Andersson, 1971).

Weed control is another crucial factor to consider. Firstly, because they allow the multiplication of polyphagous species, such as *Meloidogyne, N. aberrans, Ditylenchus* and *Pratylenchus*, and consequently reduce the positive effect of the rotation. Secondly, many of them are hosts of TRV on which the different species of *Trichodorus* are able to feed and recover the viral particles.

26.7.2 Cultural methods

Three main types of cultural methods for control may be distinguished. The first is to obtain the maximum vigour of the plants in order that they are able to obviate the damaging effects of nematode attack. Increasing the levels of soil fertilization may mask the effect of the nematodes, and heavy dressing of nitrogen is generally used to balance the decrease of the yield. Although this strategy may result in little or no yield loss, it has the consequence that unless the cultivar used is resistant, there will be a higher rate of nematode reproduction as larger amounts of roots will be available for colonization. Fertilizers may in certain cases act directly or indirectly on the nematodes themselves. It has been reported that there may be a possible lethal effect of calcium ammonium nitrate on the juveniles of *D. destructor* (Sepselev and Glez, 1973). Soil pH can affect populations of *P. penetrans* that are higher in acidic soils than in basic ones. Increasing the pH by liming may be a possibility (1) to allow the potatoes to grow in more favourable pH conditions, (2) to reduce nematode populations and (3) to attenuate the damage, when they are present alone or in association with *Verticillium* (de Pelsmaeker and Coomans, 1987).

The second type is aimed at decreasing the soil population before growing potatoes. Deep ploughing, mainly during summer, can greatly reduce the soil populations of *Trichodorus* sp. nematodes, which are extremely susceptible to desiccation. Intercropping with nematicidal plants has been advocated as a means of decreasing the soil populations.

Against the polyphagous species of *Pratylenchus*, *Meloidogyne* and *Trichodorus*, numerous plants have been evaluated for their nematicide activities and reviewed by Gommers (1973). Among them, the best known plant is the marigold with three species *Tagetes minuta*, *T. patula* and *T. erecta*. Heterocyclic sulphur-containing molecules occur in their roots and are thought to be toxic to the nematodes. Compounds such as alpha-terthienyl accumulate in nematodes that have penetrated the roots. After 2–3 months of cultivation followed by burying the aerial parts, decreases in populations of *Pratylenchus* and *Meloidogyne* may reach 70–95% (Lung et al., 1997). It is important to note that the efficiency increases with the length of cultivation.

The third type Manipulating the potato plants directly which can have a significant effect on the nematode populations. To control *Globodera* spp., harvesting the crop before female maturity may result in a huge decrease of the population. Using the potato as trap-crop may be valuable (1) in seed production areas in order to avoid any new generation of the nematodes to develop by destruction of the plants at mating time (Mugniéry and Balandras, 1984) and (2) in early potato crops when harvest is done during the period of nematode embryogenesis, which prevents a new generation being produced and for the grower to obtain a reasonable yield (Grainger, 1964). Predicting optimal time of harvest is possible, using accumulated day degrees (Mugniéry, 1982). However, there is a potential risk that where such methods are used intensively it may be possible to select populations. It has been reported, for example, that intensive growing of early potatoes resulted in populations able to hatch at lower temperatures, i.e. sooner than normal, and develop faster thus completing their life cycle by harvest time (Hominick, 1979, 1982).

26.7.3 Physical methods

The main physical methods for nematode control are solarization and flooding. In warmer locations, solarization has been tested against *Meloidogyne* with efficiency found to be good if the treatment duration is at least 4–8 weeks. When flooding is used, the nematodes die of asphyxiation during long periods of soil flooding, which induces anaerobic conditions. In certain type of tropical soils, flooding favours the activity of sulphur-reducing anaerobic bacteria, which product sulphurous compounds toxic to the nematodes (Jacq and Fortuner, 1978).

26.7.4 Chemical treatments

Three groups of chemicals may be used to disinfest soils. The first are the fumigants (or the precursors of fumigants). These products belong to the organo-halogens, and the most commonly used on potato is 1,3-dichloropropene.

These chemicals kill nematodes by contact, but as well as being nematicidal, they are also bactericidal, fungicidal and herbicidal; with regard to the latter, they must be injected into the soil a long time before sowing or planting. As they are strongly adsorbed by organic matter, they are generally used to disinfect low organic matter soils. A consequence of their biocidal effect on the microflora and fauna is that yields may be noticeably increased even in absence of nematodes because of the availability of extra nutrients.

The second group belongs to the organophosphorus family of chemicals with the most commonly used being Ethoprophos. The third group belongs to the carbamate family with three well-known representatives: Oxamyl, Carbofuran and Aldicarb. These chemicals, which are also insecticidal, act directly on the acetylcholinesterase receptors and lead to the paralysis of the nematodes. They are partly or completely systemic, but their systemic activities are only effective against insects. They have to be applied at planting or sowing time either in the totality of the soil or in the rows. They limit the penetration of nematodes into the roots.

All these chemicals can be extremely efficient. When they are normally used, they lead to a direct (fumigants) or indirect (organophophorus and carbamate) mortality of between 80 and 90%. However, this rate of mortality is seldom able to decrease the multiplication rate of the nematodes, except *Trichodorus* spp., because at low-population densities, reproduction rates can be very high resulting in a level of infestation, after harvest, very often higher than before sowing. Their use for population control, therefore, needs to be in conjunction with other methods, such as rotation and resistant varieties.

Furthermore, in Europe, the regulations now limit or suppress their utilization. In numerous European countries, the carbamates are now prohibited on many food plants (and on potatoes). The programmed prohibition of methyl bromide (probably in 2007) will most likely lead to the banning of 1,3-dichloropropene, and we speculate that in a very near future, the only nematicide authorized on potato will be Ethoprophos.

26.7.5 Biological methods

Meloidogyne are, in practical terms, the only nematodes that may be controlled by microorganisms. Among them, two fungi, one predator, *Arthrobotrys irregularis* (Cayrol, 1983), and the other parasitic, *Paecilomyces lilacinus* (Jatala et al., 1981), may be used. *A. irregularis* traps the juveniles before penetration into the roots. Therefore, it must be applied a long time before sowing to give it time to colonize the soil before the hatching of the juveniles. *P. lilacinus* parasitizes the females and the egg masses, destroying the eggs and limiting their hatching. This fungus is also used against *Globodera* spp. in Philippines. These two fungi are manufactured in some countries.

The most promising biological agent is probably the bacteria (or the actinomycete) *Bacillus* sp. The spores of these bacteria bind to the cuticle of the nematode, and some species are known to parasitize *Meloidogyne, Globodera* and *Pratylenchus*. The most famous is *B. penetrans*, a specific parasite of *Meloidogyne* species. The specificity of this species is variable, from very strong to weak. Once the juvenile is inside the roots, the spores germinate and invade the juvenile. The spores multiply inside the developed female of *Meloidogyne*, which eventually bursts producing millions of spores. The main practical problem of this antagonist is that its culture in artificial medium is not yet possible on an industrial scale. Consequently, practical application is limited to small areas such as glasshouses.

The fungi *P. lilacinus* and *Verticillium chlamydosporium* have been investigated for the control of potato cyst nematodes, but for many reasons, difficulty of application, price, variable results and commercial sensitivity, little information is available.

26.7.6 Resistant varieties

Host resistance may be considered as the most valuable form of control. A plant is considered as resistant when the number of nematodes able to develop to female is reduced compared with a control. Then all degrees of resistance may exist from absolute, i.e. zero female development to a few females reaching maturity. Resistance has been found in a number of wild potato species against the sedentary nematodes, i.e. those that have a very close relationship with their host.

26.7.6.1 Globodera *spp.*

Numerous genes and quantitative trait loci (QTL) of resistance are known, and some have been mapped (see Chapter 7, Simko et al., this volume). Against *G. rostochiensis*, the first discovered was the *H1* gene from *Solanum tuberosum* ssp. *andigena*. Found by Ellenby in 1954 in the genotype CPC 1673, this dominant gene was introgressed into *S. tuberosum* ssp. *tuberosum* and is now present in many commercial varieties. It acts when the J2s are inducing the syncytium and around which a strong necrotic reaction occurs. As a consequence, the J2s cannot develop and die or become male. Thus, the resistant varieties act as a trap-crop, and the annual decrease of the nematode population is directly related to the hatching rate that is usually between 70 and 90%.

According to the Flor's theory, a dominant resistance gene corresponds with a recessive virulence gene in the nematode. So virulent pathotypes – Ro2/3/5 versus Ro1/4 (Janssen et al., 1991) – were observed, mainly in the Netherlands and in Germany. Curiously, this was not observed in the UK despite the selection pressure exercised by the very popular resistant cultivar Maris Piper. As the *H1* gene does not act against *G. pallida*, the main consequence of such selection pressure has been largely to eliminate *G. rostochiensis* but to allow the multiplication of *G. pallida*, which is now the major problem.

Following the discovery of the *H1* gene, a second major gene *(H2)*, which confers resistance to a limited number of *G. pallida* populations in Northern Ireland and Scotland, was found. As a consequence, some pathotypes were described by their ability to develop on these major genes. In Europe, *G. rostochiensis* and *G. pallida* consist of two pathotypes (Table 26.6) different from those described in Peru (Canto and Scurrah, 1977).

Table 26.6 European pathotypes of *Globodera* spp. and corresponding resistance genes

Species	**Gene**	**Pathotype**			
		Globodera rostochiensis		***Globodera pallida***	
		Ro1/4	**Ro2/3/5**	**Pa1**	**Pa2/3**
Solanum tuberosum		+	+	+	+
Solanum andigena	*H1*	–	–	+	+
Solanum multidissectum	*H2*	+	+	–	+

Other sources of resistance which have mainly been derived from *Solanum vernei* and *Solanum spegazzinii* tend to exhibit quantitative resistance that is measured by the number of females that develop and that is never absolute but dependent on a number of environmental factors, including the initial population density. It acts, firstly, by restricting the development of the J2s and, secondly, by limiting the development of the juveniles, whereby they are only able to mature as males. The second effect is much more important than the first. The resistance is partial and some females always succeed in developing to maturity. Faced with this quantitative resistance, we observe the existence of diverse degrees of aggressiveness as measured by the number of females that can develop. Thus, it is possible to rank populations, and the ranking obtained is extremely consistent (Phillips et al, 1989). However, different sources of resistance, e.g. *S. vernei, S. spegazzinii or S. tuberosum* ssp. *andigena*, will rank populations in different ways.

Breeding against potato cyst nematodes has mainly been conducted using polygenic resistance from *S. vernei* and has led to release of partially resistant varieties, which are mainly starch potatoes.

More recently, resistance has been found in numerous wild *Solanum* species, such as *S. spegazzinii* and *Solanum sparsipilum.* Genes and QTL are currently being mapped, and the use of PCR-specific markers will greatly facilitate breeding for resistance. The durability of such resistance is yet unknown. Against major genes, the speed of selection mainly depends on the relative proportion of virulent nematodes in the populations (Franco and Gonzales, 1990). Against polygenic resistance, circumventing resistance should be slower (Forrest and Phillips, 1984). However, results from Turner et al. (1983) seem to invalidate this theory. Recent and not yet published results show that this phenomenon may be extremely slow if very resistant varieties are used.

26.7.6.2 Meloidogyne *spp.*

Numerous resistance genes are known and used against root-knot nematodes, mainly the tropical species. Commercial varieties of cotton, peanut, watermelon and beans (Hartmann, 1970; Ginoux et al., 1979; Castagnone-Sereno et al., 1996) are available. Breeding is in progress for sweet potato (Jones and Dukes, 1980). Some varieties of fodder radish appear resistant to both *M. chitwoodi* and *M. fallax*, although the gene(s) characterization is still unknown. The first rape, resistant to *M. chitwoodi*, has just been released in France in 2004.

In the Solanaceous family, numerous resistance genes have been identified. Seven genes are found in pepper (Hendy et al., 1985) and two are mapped. The gene *Rk* of tobacco acts against the tropical species and is available in numerous varieties, although it loses its efficiency in mixed infestations with *M. hapla* (Eisenback, 1983). The most famous gene of resistance is the *Mi* gene from *Lycopersicon peruvianum*, discovered in the 1940s and subsequently introgressed in many cultivars of commercial tomatoes. As with many resistance genes, *Mi* acts during penetration and migration. The J2s remain embedded in necrotic cells and are unable to reach the vascular cylinder and induce nurse cells. This hypersensitive reaction is lost at high temperatures above 25–28 °C. The wild aubergines bear resistance against the tropical species, and *Solanum torvum* is used as root-stock for commercial egg-plant in Africa.

Fig. 26.10. Resistance from *Solanum sparsipilum* (gene *Mh*) to Southern root-knot nematodes.

In potato, the resistance gene *Mh* (Berthou et al., 1993) was found in *S. sparsipilum* and introgressed into tetraploid progenitors (Fig. 26.10). This gene acts in the same way as the *Mi* gene. Against *M. chitwoodi*, total resistance was found by Brown et al. (1989b, 1998) both in *S. bulbocastanum* and in *S. hougasii*, by Janssen et al. (1998) in *S. fendleri* and by Berthou (unpublished) in *S. schenckii*. Against *M. fallax*, the main source of resistance has been found in *S. sparsipilum*, and a QTL associated with this resistance has now been mapped (Kouassi et al., 2005). The mechanism of resistance appears very different against these temperate species. The main effect of the *S. sparsipilum* resistance, and to some extent that found in *S. schenkii*, is that there is a much later reaction, which appears not during penetration and migration but after the induction of the nurse cells around which a necrotic reaction appears and isolates them from the vascular cylinder (Fig. 26.11).

The durability of such resistance is unknown but has to be discussed species by species. With the mitotic species, there is evidence that few natural populations exist, which are naturally fully virulent and are called race B. They were found in Europe and elsewhere in Africa, but there is no evidence that they were always selected by natural selection pressure. Notwithstanding, it has been possible to artificially select some populations for

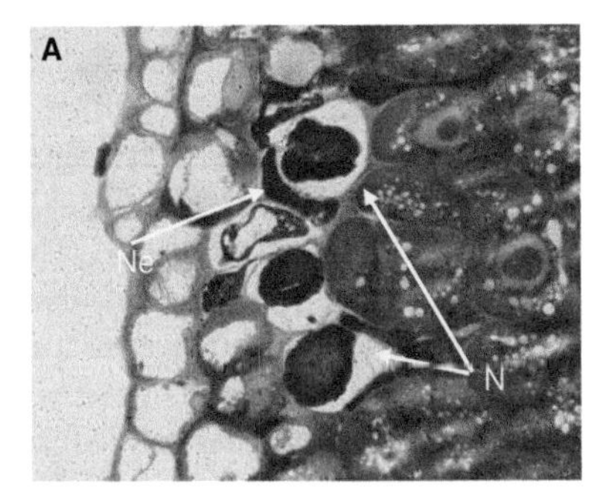

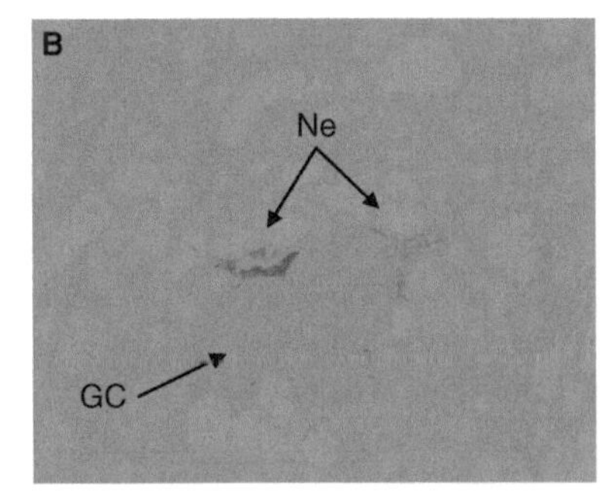

Fig. 26.11. Effect of resistance from *Solanum sparsipilum* to Southern root-knot nematodes during migration (A) and to *Meloidogyne fallax* after induction of giant cells (B). GC, giant cell; N, nematode; and Ne, necrosis.

virulence by continuous rearing on hosts with the *Mi* gene. Experimental results show that this is not possible from some populations but that in other populations it was possible but needs 50–60 generations before virulence is selected.

At the other end of the spectrum, with the highly variable temperate species such as *M. chitwoodi*, it has been shown that few females are able to develop on the resistance gene from *S. bulbocastanum* and subsequently produce fully virulent progenies. Thus there is a paradox in that the resistance is much more durable against the mitotic species than against the meiotic species.

A consequence of the use of resistant cultivars, mainly in Africa, with the *Mi* gene has been the decrease of *M. incognita, M. arenaria* or *M. javanica* in favour of the virulent *M. mayaguensis*, where this species has now become predominant.

26.7.6.3 Nacobbus aberrans

Some native Andean cultivars with different diverse degrees of ploidy seem to have fairly good resistance. According to Alarcón and Jatala (1977), the native variety Huaca Laira (*Solanum andigena*) is resistant to the Peruvian populations of *N. aberrans*. More recently, Finetti-Sialer (1990) bred a hybrid in which the nematode is unable to induce syncytia. Necrosis appears soon around the female's cuticle. Up until now, no modern resistant tetraploid cultivar is yet available.

26.8 CONCLUSIONS

As the potato is cultivated at all latitudes and in many different climatic regions, numerous and very different nematodes are known to cause much damage in the crop.

The potato cyst nematode, in temperate areas, is the main problem as a consequence of intensive and modern agriculture. Long rotations of 7 years in northern Europe to 3 years in Mediterranean Europe are generally long enough to solve or contain the problem, but this often seems incompatible with modern intensive agriculture. The use of chemical soil treatment has to be strictly limited and will probably be prohibited in the near future. The success of the *G. rostochiensis*-resistant varieties grown in nearly all infested areas has led to its replacement by *G. pallida* against which the first resistant varieties have only recently become available and for which the durability of resistance is still uncertain. As numerous new resistance genes have recently been found and genetically mapped, it is hoped that, with the help of marker-assisted breeding, combinations of genes can be constructed and limit or delay the nematodes' ability to circumvent these resistances. If and when accepted, genetically modified varieties would be a supplementary hope (Urwin et al., 2003).

The problems caused by *Pratylenchus* sp. and *Verticillium* sp. seem to be restricted to the USA. In Europe, where these two pests are often found, disease does not occur. Probably, soil improvement would be enough to reduce the problem.

M. hapla is common but, although this species is known to attack numerous cultivated plants, disease on potato is rare.

The vector virus nematodes are generally present in sandy soils, in which the efficiency of fumigants is optimal. Weed control, summer ploughing and use of certified seed and varieties unsusceptible to TRV are enough to reduce the problem. Against *D. destructor*, prophylactic methods have reduced the problem to a minimum.

Now in 2007, the most potentially dangerous nematodes are *M. chitwoodi* and *M. fallax*. Fortunately, they are not widely spread, but were they to become so, their importance is great. Due to their high variability, it is unlikely that resistant potato varieties will be durable. The most promising control method would probably be the use of horizontal resistance as found in fodder radish.

In tropical and Mediterranean areas, the two main problems are due to the southern root-knot nematodes and *N. aberrans*. Faced with the highly polyphagous *Meloidogyne* species, control must favour biological antagonists such as *B. penetrans* integrated with the use of resistant cultivated plants. The recent introgression of the *Mh* gene in modern tetraploid potatoes would be a supplementary tool not only to protect the potato but also to protect the other susceptible crops of the rotations. Although we have no estimate of the durability of this new resistance, the fact that it behaves as the very durable *Mi* gene is extremely encouraging.

In the mountains of tropical South America, the problem caused by *N. aberrans* is not yet resolved. This nematode is serious, because it occurs in very poor areas and few studies had been made. It is an urgent requirement that resistant varieties be developed and released in order to help the Andean farmers.

REFERENCES

Alarcón C. and Jatala P., 1977, *Nematrópica* 7, 2.

Andersson S., 1971. Potatisrötnematoden, *Ditylenchus destructor*, Thorne, som skadegörare i potatis. Dissertation, Agricultural College, Uppsala, Sweden.

Anthoine G. and D. Mugniéry, 2005, *Nematology* 7, 503.

Bakker J. and F.J. Gommers, 1982, *Proc. K. Ned. Akad. Wet.*, C85, 309.

Baldwin J.G., S.A. Nadler and B.J. Adams, 2004, *Annu. Rev. Phytopathol.* 42, 83.

Berthou F., P. Rousselle, H.A. Mendoza and D. Mugniéry, 1993, Caractérisation des mécanismes de résistance à *Meloidogyne* spp., chez diverses Solanacées, *Lycopersicon* spp. et *Solanum tuberosum*: intérêt pour l'analyse du déterminisme génétique des caractères de la résistance de la pomme de terre à ce nématode. C.R., 12th EAPR Trien. Conf., Paris, 141–142.

Brown D.J.F., A.T. Ploeg and D.J. Robinson, 1989a, *Rev. Nématol.* 12, 235.

Brown C.R., H. Mojtahedi and G.S. Santo, 1989b, *Plant Dis.* 73, 957.

Brown C.R., H. Mojtahedi and G.S. Santo, 1991, *Am. Potato J.* 68, 445.

Burrows P.R., 1990, *Rev. Nématol.* 13, 185.

Canto M. and M. Scurrah, 1977, *Nematologica* 23, 340.

Castagnone-Sereno P., M. Bongiovanni, A. Palloix and A. Dalmasso, 1996, *Eur. J. Plant Pathol.* 102, 585.

Cayrol J.C., 1983, *Rev. Nématol.* 6, 265.

Costilla M.A. and H. Basco, 1985. In: J. Franco and H. Rincon (eds), *Investigaciones Nematologicas en Programas Latinoamericanos de Papa*, Centro Internacional de la papa, Lima, Peru.

Dale M.F.B., D.J. Robinson and D. Todd, 2004, *Plant Pathol.* 53, 788.

de Guiran G., 1979, *Rev. Nématol.* 2, 223.

de Pelsmaeker M. and A. Coomans 1987, *Meded. Faculteit Landbouwwetensch. Rijksuniversiteit Gent.* 52, 561.

Decraemer W., 1991, In: W.R. Nickle (ed.), *Manual of Agricultural Nematology*, p. 587. Marcel Dekker Inc., New York.

Eisenback J.D., 1983, *J. Nematol.*, 15, 478.

Eisenback J.D., 1985, In: J.N. Sasser and C.C. Carter (eds), *An Advanced Treatise on Meloidogyne, vol. I. Biology and control*, p. 95. North Carolina State University Graphics, Raleigh, NC.
Ellenby C., 1954, *Euphytica* 3, 195.
Elston D.A., M.S. Phillips and D.L. Trudgill, 1991, *Rev. Nématol.* 14, 213.
Esbenshade P.R. and A.C. Triantaphyllou, 1985, *J. Nematol.*, 17, 6.
Finetti-Sialer M.S., 1990, *Rev. Nématol.* 13, 155.
Forrest J.M.S. and M.S. Phillips, 1984, *Plant Pathol.* 33, 53.
Franco J. and A. Gonzalez, 1990, *Rev. Nématol.* 13, 181.
Fritzsche R., S. Thiele and G. Barchend, 1986, *Arch. Phytopathol. Pfl.* 22, 329.
Fullaondo A., E. Barrena, M. Viribay, I. Barrena, A. Salazar and E. Ritter, 1999, *Nematology* 1, 157.
Ginoux J-P., G. Jaqua, A. Fougeroux and C.M. Messiaen, 1979, *Ann. Amélior. Plant.* 29, 717.
Gomez Tovar J., 1973, *Nematrópica* 3, 4.
Gommers F.J., 1973, *Nematicidal Principles in Compositae, Meded. Landbouwhogesch. Wageningen*, p. 71.
Grainger J., 1964, *Nematologica* 10, 5.
Grainger J. and M.R. Clark, 1963, *Eur. Potato J.* 6, 131.
Griffin G.D., 1985, *J. Nematol.* 17, 396.
Harrison J.A.C., 1970, *Ann. Appl. Biol.* 67, 185.
Hartman K.M. and J.N. Sasser, 1985, In: K.R. Barker, C.C. Carter and J.N. Sasser (eds), *An Advanced Treatise on Meloidogyne, vol. II. Methodology*, p. 69, North Carolina State University Graphics, Raleigh, NC.
Hartmann R.W., 1970, *Circ. Hawaii Exp. Stn.* 67, 5.
Hendy H., E. Pochard and A. Dalmasso, 1985, *Agronomie* 5, 93.
Hide G.A. and P.J. Read, 1991, *Ann. Appl. Biol.* 119, 77.
Hominick W.M., 1979, *Nematologica* 25, 322.
Hominick W.M., 1982, *Ann. Appl. Biol.* 100, 345.
Ivanova I.V., 1975, *Byull. Vses. Inst. Gel'mintol. Im. K.I. Skryabina* 15, 67.
Jacq V.A. and R. Fortuner, 1978, *C.R. Acad. Agric.* 64, 1248.
Janssen G.J.W., O.E. Scholten, A. van Norel and C.J. Hoogendoorn, 1998, *Eur. J. Plant Pathol.* 104, 645.
Janssen R., J. Bakker and F.J. Gommers, 1991, *Rev. Nématol.* 14, 213.
Jatala P., R. Salas, R. Kaltenbach and M. Bocangel, 1981, *J. Nematol.* 13, 445.
Jepson S.B., 1983, *Rev. Nématol.* 6, 291.
Jones A. and P.D. Dukes, 1980, *J. Am. Soc. Hortic. Sci.* 105, 154.
Joubert J. and A. Dalmasso, 1974, In: I.T.P.T (ed.), *Maladies et parasites animaux de la pomme de terre*, p. 26. ITPT, Paris.
Kotcon J.B., D.I. Rouse and J.E. Mitchelle, 1985, *Phytopathology* 75, 68.
Kouassi A., M.C. Kerlan, B. Caromel, J-P. Dantec, D. Fouville, D. Ellissèche and D. Mugniéry, 2005, *Theor. Appl. Genet.* 112, 699.
Lung G., A. Fried and U. Schmidt, 1997, *Gesunde Pflanzen.* 49, 111.
MacGuidwin A.E. and T.A Forage, 1991, *J. Nematol.* 23, 198.
Manzanilla-López R.H., M.A. Costilla, M.E. Doucet, J. Franco, R.N. Inserra, P.S. Lehman, I.V. Cid Del Prado, R.M. Souza and K. Evans, 2002, *Nematrópica* 32, 149.
Marshall J.W. and A.M. Crawford, 1992, *N.Z. J. Zool.* 19, 133.
Mehiar F.F., R.A. Omar, S.F. Mashaall, M.A. Abdel-Hadi and M.A. Gabf, 1984, *Ann. Agric. Sci.* 29, 493.
Mojtahedi H., G.S. Santo and J.H. Wilson, 1988, *J. Nematol.* 20, 468.
Mugniéry D., 1982, *Agronomie* 2, 629.
Mugniéry D. and C. Balandras, 1984, *Agronomie* 4, 773.
Mulholland V., L. Carde, K.J. O'Donnell, C.C. Fleming and T.O. Powers, 1996, *Diagnostics in Crop Production*, BCPC Symp. Proc. 65, 247.
Olthof T.H.A., 1990, *J. Nematol.* 22, 303.
Oostenbrink M., 1966, *Meded. Landbouwhogesch. Wageningen* 66–4, 46.
Phillips M.S., H.J. Rumpenhorst and D.L. Trudgill, 1989, *Nematologica* 35, 207.
Picard D., 2005, *Génétique des populations et phylogéographie du nématode à kyste de la pomme de terre (Globodera pallida) au Pérou.* Thèse ENSAR, Rennes, France.
Pinkerton J.N., G.S. Santo and H. Mojtahedi, 1991, *J. Nematol.* 23, 283.
Polozhenets V.M., 1984, *Kartofel. Boleznei Vreditelei (Nauch. Trud.)*, Moskva, 146.

Reid A., R.H. Manzanilla-López and D.J. Hunt, 2003, *Nematology* 5, 441.
Rojankovskii E. and A. Ciuera, 1986, *Arch. Phytopathol. Pfl.* 22, 101.
Sarakoski M.L., 1977, *Ann. Agric. Fenn.* 16, 137.
Schlüter K., 1976, *Z. Pflanzenkr. Pflanzenschutz.* 83, 401.
Scholte K. and J.J. s'Jacob, 1990, *Potato Res.* 33, 191.
Schots A., F.J. Gommers and E. Egberts, 1992, *Fund. Appl. Nematol.* 15, 55.
Schots A., T. Hermsen, S. Schouten, F.J. Gommers and E. Egberts, 1989, *Hybridoma* 8, 401.
Seinhorst J.W., 1965, *Nematologica* 11, 137.
Sepselev Z.G. and V.M. Glez, 1973, *Sel. Semenovod. Kartofel.*, Moskva, 157.
Stelter H., 1971, *Der Kartoffelnematode* (Heterodera rostochiensis *Wollenweber*) *Deutsche Demokratische Republik.* Deutsche Akademie, Der Landwirtschaftswissenschaften zu Berlin. Akademie-Verlag, Berlin.
Thiéry M. and D. Mugniéry, 1996, *Fund. Appl. Nematol.* 19, 471.
Triantaphyllou A.C., 1985, In: J.N. Sasser and C.C. Carter (eds), *An Advanced Treatise on Meloidogyne, vol. I. Biology and Control*, p 113. North Carolina State University Graphics, Raleigh, NC.
Trudgill D.L., B. Marshall and M.S. Phillips, 1990, *Ann. Appl. Biol.* 117, 107.
Turner S.J., A.R. Stone and J.N. Perry, 1983, *Euphytica* 32, 911.
Urwin P. E., J. Green and H.J. Atkinson, 2003, *Mol. Breeding* 12, 263.
Viscardi T. and M. Brzeski, 1993, *Fund. Appl. Nematol.* 16, 389.
Waeyenberge L., A. Ryss, M. Moens, J. Pinochet and T. Vrain, 2000, *Nematology* 2, 135.
Wendt K., T. Vrain and J. Webster, 1994, *J. Nematol.* 25, 555.
Wishart J., M.S. Phillips and V.C. Blok, 2002, *Phytopathology* 92, 884.
Wyss U., 1997, In: C. Fenoll, F.M.W Grundler and S.A. Ohl (eds), *Cellular and Molecular Aspects of Plant-Nematode Interactions*, p. 5. Kluwer Academic Publishers, Dordrecht, the Netherlands.
Wyss U., F.M.W. Grundler and A. Munch, 1992, *Nematologica* 38, 98.
Zijlstra C., 2000, *Eur. J. Plant Pathol.* 106, 283.
Zijlstra C., T.H.M. Donkers-Venne and M. Fargette, 2000, *Nematology* 2, 847.

Chapter 27

Bacterial Pathogens of Potato

Jan M. van der Wolf[1] and Solke H. De Boer[2]

[1]*Plant Research International, PO Box 16, 6700 AA Wageningen, The Netherlands;*
[2]*Canadian Food Inspection Agency, Charlottetown Laboratory, 93 Mount Edward Road, Charlottetown, Prince Edward Island, Canada C1A 5T1*

27.1 INTRODUCTION

Diseases caused by bacterial pathogens are major threats to potato production. Pathogenic bacteria, once introduced, may persist and spread in agricultural environments. Moreover, their inadvertent and unnoticed spread as contamination or in latent (asymptomatic) infections of seed potatoes is particularly important. Under conditions favourable for disease development, bacterial phytopathogens multiply rapidly and cause significant yield loss and economic damage. In general, bactericides are ineffective as crop protection agents, and their usefulness for disinfecting seed tubers is limited. Immunity to bacterial diseases is generally not available, although some cultivars show tolerance and various levels of resistance. Control of some bacterial diseases such as those caused by the quarantine pathogens, *Ralstonia solanacearum* and *Clavibacter michiganensis* ssp. *sepedonicus*, can only be achieved by costly statutory measures.

In this chapter, the biology and pathology of the major bacterial diseases including blackleg, stem rot, bacterial soft rot, bacterial wilt, bacterial ring rot and common scab will be discussed. Thereafter, measures to control bacterial diseases by production of pathogen-tested seed, reduction of inoculum and helpful agronomic practices are addressed.

27.2 PATHOGEN BIOLOGY

27.2.1 *Ralstonia solanacearum*

R. solanacearum (Smith) Yabuuchi et al. (formerly *Pseudomonas* and *Burkholderia solanacearum*) causes bacterial wilt, also known as potato brown rot (Hayward, 1991). This disease is a major constraint for potato production in tropical and subtropical regions, but it has also become a threat in temperate European countries (Janse, 1996). The bacterium is Gram-negative non-spore-forming, aerobic, motile and rod-shaped. It grows relatively fast, with colonies becoming visible within 2 days when cultured on agar medium. Upon isolation, colonies are mucoid, but when subcultured they readily convert

Potato Biology and Biotechnology: Advances and Perspectives
D. Vreugdenhil (Editor)

into non-mucoid forms that have reduced virulence (Wallis and Truter, 1978; Hayward, 1991). The bacterium has a diverse and broad host range. Traditionally, it has been informally classified into five races on the basis of differences in host range and into six biovars on the basis of biochemical and physiological characteristics (He and Hua, 1985; Fegan and Prior, 2005). Biovars 1–4 and race 1 and 3 strains cause bacterial wilt in potato. Biovar 2 is nearly synonymous with race 3 and includes strains adapted to temperate and cooler regions at high altitudes in (sub)tropical areas. Biovar 2 strains are mainly seed-borne on potato and are genetically and phenotypically homogeneous. Potato strains of biovars 1, 3 and 4 are grouped into race 1 and are mainly soil-borne in tropical regions. They show wide variations in host range and have high genetic and phenotypic diversity. Biovar 1 is primarily found in North and South America and Africa, whereas biovar 3 is primarily found in Asia. Recently, *R. solanacearum* has been classified into four phytotypes (subspecies level) on the basis of the nucleotide sequence of the internal spacer region of the ribosomal gene operon. Twenty-three sequevars (infrasubspecific groups) have also been identified on the basis of genetic fingerprinting methods and the sequence of an endoglucanase gene (Fegan and Prior, 2005). This classification offers a more stable and meaningful taxonomy than the division into races.

The typical wilting of plants caused by *R. solanacearum* seems to be largely due to clogging of the vascular system by large amounts of extracellular polysaccharides (EPSs) produced by the bacterium (Schell, 1996; Genin and Boucher, 2002). EPS-deficient mutants are much less virulent, although they are not entirely non-pathogenic. It has been suggested that EPS, apart from interfering with water transport, also contributes to protecting bacterial surface structures from the plant defence system. Extracellular cell-wall degrading enzymes may also be involved in the induction of wilt symptoms, as they cause disintegration of vascular tissue, limiting water transport. The role of cell-wall degrading enzymes, such as endopolygalacturonase, pectin methylesterase (PME) and glucanase, in pathogenesis of *R. solanacearum*, however, is minor compared with the role these enzymes play in pathogenesis of pectolytic erwinias. Extracellular enzyme-deficient mutants of *R. solanacearum* are pathogenic although symptom expression is delayed when they are inoculated into host plants.

Pathogenicity factors are under control of two cell-density–dependent quorum sensing systems controlled by different signal molecules, 3-hydroxypalmitic acid methyl ester and acylhomoserine lactone, that interact in complex ways (Von Bodman et al., 2003). The quorum sensing mechanism allows bacteria to respond to population density and thereby delay production of the extracellular enzymes until a critical cell density has been achieved. It has been suggested that the genetic regulatory apparatus postpones enzyme production until bacterial cell numbers are high enough to overwhelm the plant's defence response. The quorum sensing mechanism allows the bacterium to modulate expression of factors important in the early phase of pathogenesis such as motility and production of siderophores when bacterial density is low and, at high densities, to produce factors such as EPS, PME and glucanases that are required in a later phase of disease development. It is anticipated that further analysis of the complete *R. solanacearum* genome will further unveil the secrets of its pathogenesis (Salanoubat et al., 2002).

27.2.2 *Clavibacter michiganensis* ssp. *sepedonicus*

C. michiganensis ssp. *sepedonicus* (Spiekermann and Kotthoff) Davis et al. is the causative agent of bacterial ring rot. The disease was reported for the first time by Appel (1906) after an outbreak in Germany. The pathogen has been described under the following names: *Corynebacterium sepedonicum, Bacterium sepedonicum, Aplanobacter sepedonicum, Phytomonas sepedonica, Mycobacterium sepedonicum* and *Pseudobacter sepedonicum*. *C. michiganensis* ssp. *sepedonicus* is a Gram-positive, non-motile, slow-growing, aerobic, coryneform bacterium (Slack, 1987). On laboratory media, colonies become visible only after 3–7 days and remain small. *C. michiganensis* ssp. *sepedonicus* does not form spores and cannot resist high temperatures but survives well under cool and dry conditions. It forms a relatively homogeneous group. In BOX-PCR analyses, a genetic fingerprinting technique based on primers that anneal to repetitive sequences dispersed over the entire bacterial genome, *C. michiganensis* ssp. *sepedonicus* strains form a tight group with a similarity of over 80% (Smith et al., 2001). Also in analyses of fatty acid methyl esters, *C. michiganensis* ssp. *sepedonicus* forms a tight group with profiles that overlap with profiles of related subspecies (Henningson and Gudmestad, 1991). There is little variation in the serological specificity among *C. michiganensis* ssp. *sepedonicus* strains. Polyclonal antibodies generated against whole cells and a monoclonal antibody selected against a cell wall component reacted in immunofluorescence antibody-staining procedures with all strains tested (De Boer and Wieczorek, 1984). A monoclonal antibody produced against the soluble EPS, suitable for use in ELISA, reacted with all strains except those that were non-mucoid. The poor reaction of non-mucoid strains is presumably due to a low level of the polysaccharide antigen (De Boer et al., 1988).

The pathogenicity factors of *C. michiganensis* ssp. *sepedonicus* and their genetic control are largely unknown, in part, because the genetics of coryneform, Gram-positive bacteria remain unexplored (Metzler et al., 1997). Thus far, only the plant cell-wall degrading enzyme, cellulase, and evidence for a hypersensitivity gene operon have been identified as pathogenicity factors. A cellulose-negative wild-type strain was nearly non-virulent, but virulence could be restored by transformation with a cloned cellulase gene (Laine et al., 2000). Virulent *C. michiganensis* ssp. *sepedonicus* strains produce a hypersensitive response in tobacco, a non-host plant for *C. michiganensis* ssp. *sepedonicus*, and used widely to bioassay for the presence of *hrp* genes (Nissinen et al., 2001). With the recent availability of the genome sequence for *C. michiganensis* ssp. *sepedonicus*, it may now be possible to explore similarities and differences of pathogenicity factors of Gram-positive and Gram-negative bacteria on a molecular basis (http://www.sanger.ac.uk/Projects/C_michiganensis).

27.2.3 Pectolytic erwinias

Pectolytic erwinias are responsible for the bacterial foliage diseases known as blackleg, aerial stem rot and stem wet rot, as well as bacterial soft rot in tubers. Until recently, the causative agents were distinguished as *Erwinia carotovora* ssp. *atroseptica* (van Hall) Dye, *Erwinia carotovora* ssp. *carotovora* (Jones) Bergey et al. and *Erwinia chrysanthemi*

Burkholder et al. (Pérombelon and Kelman, 1980). In 2004, atypical strains of *E. carotovora* causing blackleg in Brazil were described as *Erwinia carotovora* ssp. *brasiliensis* (Duarte et al., 2004). *E. carotovora* ssp. *brasiliensis* strains grow at higher temperatures than *E. carotovora* ssp. *atroseptica* and are considerably more virulent. They are distinguishable by characteristic serological, genetic and biochemical features. On the basis of DNA–DNA hybridization studies, serology and numerical taxonomy, *E. carotovora* ssp. *atroseptica* and *E. carotovora* ssp. *carotovora* were recently reclassified, respectively, as *Pectobacterium carotovorum* ssp. *atrosepticum* and *Pectobacterium carotovorum* ssp. *carotovorum* (Hauben et al., 1998). In a further study, *E. carotovora* ssp. *atroseptica* was given species status as *Pectobacterium atrosepticum* (Gardan et al., 2003). In this chapter, we have retained the familiar *Erwinia* taxons used in most recent scientific publications.

The host range of *E. carotovora* ssp. *atroseptica* is mostly limited to potato in which it incites the blackleg disease. The subspecies is genetically and phenotypically homogeneous. In contrast, *E. carotovora* ssp. *carotovora* is a far more heterogeneous subspecies. It causes aerial stem rot and is often responsible for bacterial soft rot of tubers during storage (De Boer and Kelman, 1978; Bartz and Kelman, 1984). *E. carotovora* ssp. *carotovora* also has a broad host range, reflecting its polyphagous nature. Analysis of restriction fragment length polymorphisms (RFLPs) of a *recA* gene fragment revealed 18 groups among 56 *E. carotovora* ssp. *carotovora* strains compared with 2 RFLP groups among 13 *E. carotovora* ssp. *atroseptica* strains (Waléron et al., 2002). Similarly, RFLP analysis of the *pel* gene revealed 16 groups among 40 *E. carotovora* ssp. *carotovora* strains and 2 groups among 32 *E. carotovora* ssp. *atroseptica* strains. In the field, *E. carotovora* ssp. *carotovora* strains are often secondary invaders, colonizing and macerating plants wounded by other pathogens.

E. chrysanthemi, the causal agent of stem wet rot, like *E. carotovora* ssp. *carotovora*, has a wide host range and is phenotypically diverse. On the basis of biochemical and serological features, *E. chrysanthemi* has been divided into nine biovars (Samson et al., 1987; Ngwira and Samson 1990). Biovars 1, 5 and 7 have been found associated with disease in potato (Samson et al., 1987; Janse and Ruissen, 1988). Some relationship was found between biovar and host specificity, but most biovars induce symptoms of decay in potato under conditions favourable for disease expression. *E. chrysanthemi* has recently been redistributed among six genomic species in a new genus, *Dickeya* (Samson et al., 2005). Potato strains of *E. chrysanthemi* are found within the new species *Dickeya zeae, Dickeya chrysanthemi* and *Dickeya dianthicola* (Samson et al., 2005; unpublished results). *D. dianthicola* (biovar 7) grows relatively well at low temperatures and is mainly found in temperate zones. It is unknown to what extent the species further differ in their biology and pathogenesis.

Production of cell-wall degrading enzymes, including pectinases, cellulases, proteases and xylanases, is a dominant factor in pathogenesis of erwinias. The pectinases, that degrade pectin in the middle lamella between plant cells, are considered to be the most important for pathogenicity (Collmer and Keen, 1986; Pérombelon, 2002; De Boer, 2003). Of the various pectolytic enzymes that are produced, many are present in multiple forms (isoenzymes), each with different optimal pH values and active on substrates with different degrees of methylation. The multiplicity of isoenzymes allows plasticity in metabolism and thereby enhances the erwinias' ability to survive in and colonize different environments. The production of pectolytic enzymes is regulated in a cell-density–dependent manner (Pirhonen et al., 1993; Von Bodman et al., 2003). Only when populations reach a particular threshold level does enzyme production commence. It has been suggested that in this way the bacteria can thrive unnoticed at lower densities without triggering plant defence mechanisms. Subsequent bacterial multiplication and initiation of decay commences when host resistance is impaired in some way by physiological or environmental changes. Several other determinants have been implicated as important in pathogenesis, although they may not be clearly distinguished from functions of general housekeeping processes. For instance, flagellar motility may be important during tuber lenticel infection when free soil water is available. Lipopolysaccharides (LPSs), a major component in the bacterial outer cell wall, inhibit the hypersensitivity response of plants; LPS mutants of *E. chrysanthemi* have considerably reduced virulence. The plant-inducible peptide methionine sulphoxide reductase protects *E. chrysanthemi* against active oxygen species. The exact role of some pathogenicity-related genes such as *hrp* and *avr* is only partially understood. Recently, the genome sequence of *E. carotovora* ssp. *atroseptica* has become available and has already revealed the presence of novel pathogenicity determinants, including a putative synthetase system similar to that associated with phytotoxin production in other plant pathogenic bacteria (Bell et al., 2004).

27.2.4 *Streptomyces scabies*

Streptomyces scabies is also to be reckoned among the significant prokaryote pathogens of potato although it differs from the other phytopathogenic bacteria in its hyphal-like filamentous growth pattern. *S. scabies* is the principal causal agent of the common scab disease of potato, but other species of streptomycetes including *Streptomyces acidiscabies* and *Streptomyces turgidiscabies* also cause the disease under varying conditions (Lambert and Loria, 1989a,b; Miyajima et al., 1998). Streptomycetes are atypical, high G + C, Gram-positive, filamentous bacteria that produce vegetative spores by fragmentation of aerial filaments. The grey, smooth spores of *S. scabies* are borne in spiral chains. This phenotypic characteristic plus utilization of specific sugars characterize the species (Lambert and Loria, 1989a). *S. scabies* itself is a heterogeneous species, and other scab-inducing streptomycetes are only distantly related on the basis of genotypic and phenotypic criteria (Bouchek-Mechiche et al., 2000). The main virulence factor common to all the pathogenic streptomycetes is the phytotoxin thaxtomin, which is a nitrated dipeptide toxin (Lawrence et al., 1990). Most strains carry a second virulence factor, the *nec1* gene, which encodes a necrogenic protein (Bukhalid et al., 1998). Both the *nec1* and the thaxtomin synthesis genes, *txtABC*, along with other virulence-associated genes, are carried on a 325-kb chromosomal element known as a pathogenicity island (PAI). PAIs are known to carry clusters of pathogenicity-related genes in other plant and animal pathogenic bacteria (Kers et al., 2005). Pathogenicity can in some cases be passed on to related strains by horizontal gene transfer involving the entire PAI. Mobilization of the PAI gene cluster into *Streptomyces diastatochromogenes* transferred the scab-inducing phenotype to this non-pathogenic species, whereas transconjugants of *Streptomyces coelicolor* did not produce thaxtomin or necrotize potato tuber tissue (Kers et al., 2005). Horizontal transfer of the PAI from *S. scabies* into specific receptive *Streptomyces* spp. perhaps accounts for the variability of apparent species associated with induction of the scab disease.

27.3 PATHOLOGY

27.3.1 Symptoms and factors favouring symptom expression

Field symptoms caused by *R. solanacearum* are wilting, stunting, yellowing of foliage and collapse of stems (Fig. 27.1A). Virulent strains of *R. solanacearum* colonize the vascular tissue in which they produce large amounts of EPS (Fig. 27.1B). The occlusion of vascular tissue interferes with the movement of water and results in wilting. By placing sections of infected stems in water, fine white strands of bacterial slime can be observed to ooze from cut vascular bundles, a diagnostic characteristic of the disease. Vascular bundles of infected cross-sectioned tubers also exude drops of greyish slime without the application of pressure (Fig. 27.1C). In advanced infections, parenchymatous tuber tissue disintegrates, resulting in a progressing rot (Fig. 27.1D). Symptoms of bacterial wilt develop with high bacterial densities at high humidity and temperatures exceeding 25°C. Resistance of some cultivars to bacterial wilt is temperature dependent and varies with bacterial strain. Some potato cultivars that show resistance at low temperatures become susceptible at higher temperatures (French, 1994).

Early symptoms of bacterial ring rot caused by *C. michiganensis* ssp. *sepedonicus* are characterized by wilting of leaves and slight rolling of leaf margins (Fig. 27.2A). Interveinal spaces of the leaves become light green to pale yellow. As the disease progresses, leaves become necrotic, starting at the margins or in interveinal tissue and continues to spread until the entire stem collapses. Development of necrotic lesions has been described at high inoculum levels that expand to cover entire leaves in susceptible cultivars, whereas they are arrested in resistant ones, which may suggest a hypersensitive response (Romanenko et al., 2002). Tuber symptoms are characterized by decay in the vascular ring (Fig. 27.2D). Prior to the onset of decay, infected vascular tissue may have a glassy appearance and take on a dark colour at the heel end (De Boer and Slack, 1984). In advanced infections, the vascular tissue becomes creamy and cheesy in texture, and milky exudate can sometimes be squeezed from it (Fig. 27.2C). Eventually, the entire tuber usually decays. As rot progresses, surface cracks and dark blotches immediately beneath the periderm sometimes become visible (Fig. 27.2B). Secondary infections by saprophytic bacteria can mask typical ring rot symptoms. Recently in Europe, it was observed that the central cortex decays at the same time as the vascular ring in infected tubers, perhaps a characteristic of some European cultivars. Symptom development follows the acquisition of high bacterial densities in potato tissue and varies with cultivar susceptibility and temperature (van der Wolf et al., 2005a). Symptoms in susceptible cultivars are generally observed 90–100 days after planting (Hukkanen et al., 2005) while they rarely develop in tolerant cultivars. To avoid aphid transmission of virus diseases, haulms of seed potato crops are often destroyed prior to the appearance of ring rot symptoms. This practice leads to the persistence of latent ring rot infections and the inevitable consequences for

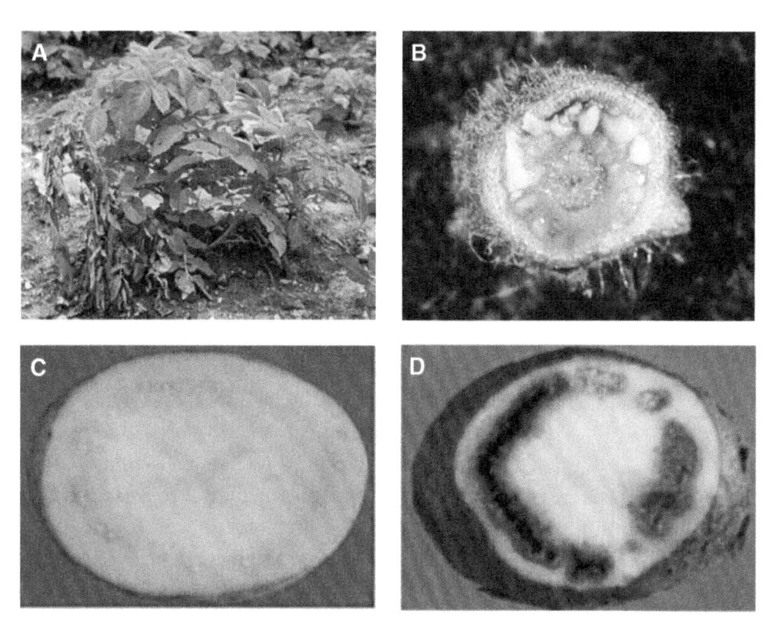

Fig. 27.1. Symptoms caused by *Ralstonia solancearum*. (A) Wilted foliage. (B) Bacterial slime oozing from a cross-section of an infected stem. (C) Bacterial slime droplets spontaneously emerging from the vascular ring of a cross-sectioned tuber. (D) Brown rot symptoms in the vascular ring of a cross-sectioned tuber. (Pictures generously provided by J.G. Elphinstone, CSL, York, UK.)

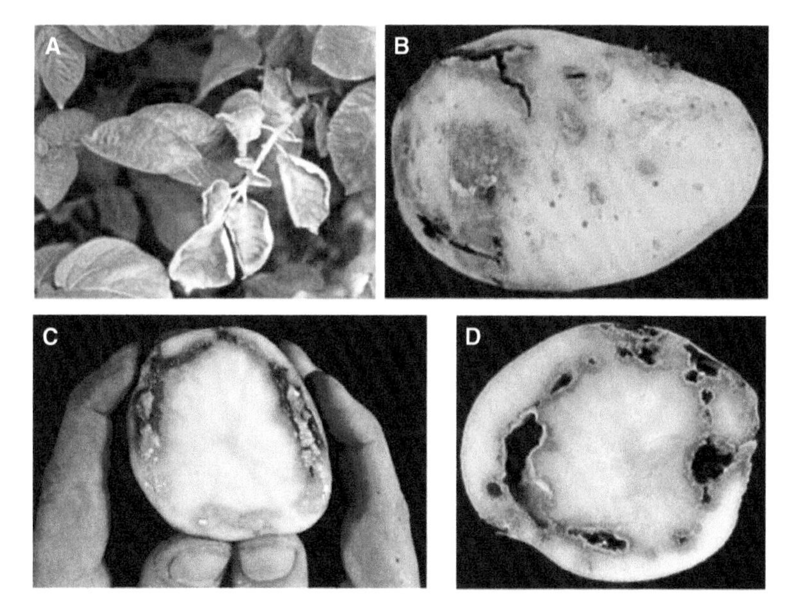

Fig. 27.2. Symptoms caused by *Clavibacter michiganensis* ssp. *sepedonicus*. (A) Upward rolling of leave margins and interveinal chlorosis. (B) Surface cracks in a heavily infected tuber. (C) Soft cheesy-textured exudate expressed from the vascular ring of an infected tuber upon squeezing. (D) Late stage tuber decay showing discolouration and hollow areas in the vascular ring.

disease dissemination. Ring rot foliar symptoms can also be masked by other diseases and natural senescence (van der Wolf et al., 2005a).

When *Erwinia*-infected seed tubers are used for planting, they sometimes decay before sprouts emerge or a plant is established creating a condition known as non-emergence or seed piece decay (Pérombelon and Kelman, 1980). Non-emergence is exacerbated by heavy rainfall that causes soil compaction and saturation leading to anaerobic conditions that favour bacterial soft rot. Post-emergence symptoms of *E. carotovora* ssp. *atroseptica* and *E. chrysanthemi* infections often start with dark colouring of upper leaves, followed by chlorosis and wilting (Fig. 27.3A and B). As the disease progresses, stems wilt or show blackleg symptoms and ultimately the entire stem dies (Fig. 27.3C). Contamination of foliage and subsequent crop damage due to heavy rainfall, hail, insects or human activities can result in aerial stem rot. The identity of the *Erwinia* species or subspecies cannot usually be reliably established on the basis of symptomology alone. At high field temperatures, *E. carotovora* ssp. *carotovora* can be the primary source of infection and result in wilting of plants. In temperate zones, some *E. carotovora* ssp. *carotovora* strains can cause blackleg-like symptoms (D. Stead, personal communication).

Bacterial soft rot of potato tubers is favoured by anaerobic conditions and a high relative humidity (Fig. 27.3D and E). The temperature determines largely whether erwinias are able to infect tubers and which (sub)species will dominate (Pérombelon and Kelman, 1980). At temperatures below 15°C, *E. carotovora* ssp. *atroseptica* is primarily involved in decay, whereas at higher temperatures, it is primarily *E. chrysanthemi* and *E. carotovora* ssp. *carotovora*, possibly in association with other pectolytic bacteria.

Although the erwinias are the primary causal agents of bacterial soft rot of tubers, other pectolytic bacteria from the soil environment contribute to the development and progression of post-harvest decay. In particular, anaerobic *Clostridium* spp. are effective soft rotters of oxygen-deprived potato tissues and are responsible for the slimy, foul-smelling characteristic of tuber decay (Campos et al., 1982). Although there are many pectolytic clostridial species, only one species, *Clostridium puniceum*, has been well characterized as a component of bacterial soft rot (Lund et al., 1981). *Clostridium* spp. produce long-living endospores that are commonly present in soil and on potato tuber surfaces. Water on tubers in storage resulting from inappropriate storage conditions or leakage from decaying tubers creates hypoxic conditions favourable to the growth of pectolytic clostridia.

Pectolytic fluorescent pseudomonads also form a component of the soil microbial flora and are associated with soft rot of fleshy vegetables including potato (Cuppels and Kelman, 1980). These bacteria, a biochemically diverse group identified as *Pseudomonas fluorescens*, have also been reported to be the causal agent of the pink eye disease of potato (Folsom and Friedman, 1959). The pink eye disease is characterized by discolouration of cortex cells that appear as pink blotches on the skin. Pink eyes may break down into watery lesions that turn brown and become sunken and cracked when dry. Underlying tissue may develop soft rot with concomitant infection by other organisms (Hagar and McIntyre, 1972). Symptoms of pink eye occur only when soil moisture content is high and the disease is particularly enhanced by the presence of *Verticillium* wilt (Goth et al., 1993). Although pink eye symptoms were induced by artificial inoculation with *P.*

Fig. 27.3. Symptoms caused by pectinolytic *Erwinia* spp. (A) Initial symptoms showing dark coloring and upward curling of leaves. (B) Wilting of foliage. (C) Blackleg symptom in the lower stem. (D) Decay of vascular tuber tissue. (E) Tuber decay proceeding from the stolon end.

Fig. 27.4. Common scab symptoms caused by *Streptomyces scabies*. (A) Superficial rough, circular or irregular lesions on the skin of young tubers. (B) Scab symptoms on mature tubers. (C) Tubers with deep-pitted scab lesions. (Pictures generously provided by George Lazarovits, AAFC, London, Canada.)

fluorescens (Folsom and Friedman, 1959), a subsequent attempt to confirm the association of the bacterium with the disease by isolation and completion of Koch's postulates was unsuccessful (Nolte et al., 1993). Because mature pink eye lesions are similar in appearance to corky patch syndrome that has multiple aetiologies, it has been suggested that pink eye is just one manifestation of corky patch characterized by the accumulation of fluorescent phenolic metabolic products of plant defence and wound healing processes (Nolte et al., 1993).

Common scab is characterized by superficial rough, circular or irregular lesions on the tuber skin (Fig. 27.4A and B). Symptoms may vary with the particular strain of *Streptomyces* involved. When lesions are particularly superficial, the disease is sometimes called russet scab, and when particularly deep called pitted scab (Fig. 27.4C). *S. turgidiscabies* is specifically associated with erumpent scab in which the lesion perimeters are slightly raised above surrounding peridermal tissue. The pathogen invades lenticels or wounds on the surface of developing tubers. Initially, infections appear as tan or brown-coloured circular lesions of 5–8 mm in diameter. As the lesions mature, they may increase in size to 10 mm and coalesce into irregular patches that, in severe infections, cover a significant portion of the tuber surface. Although tuber infections are the most common manifestation of the disease, root infection can occur but is not known to decrease yield significantly.

27.3.2 Economic importance

R. solanacearum and *C. michiganensis* ssp. *sepedonicus* are considered of quarantine significance throughout the old world, e.g. by EPPO, APPPC and IAPSC, and also in the new world by NAPPO and COSAVE. A zero tolerance has been established for these pathogens. They cause economic damage in three different ways: (i) by direct crop loss during growth and storage, (ii) by rejection of infected seed lots and the cost of control measures and (iii) by loss of export markets and difficulty in developing new markets. The finding of only a single infected tuber can result in enormous economic consequences, not only for an individual grower but for an entire seed potato-producing area.

From a global perspective, *R. solanacearum* is, after *Phytophthora infestans* (late blight), probably the most economically important pathogen of potato. In temperate zones of Europe, the economic consequence is mainly a result of statutory measures that are imposed upon the detection of the bacterium in asymptomatic tubers. Symptoms of bacterial wilt or brown rot rarely occur in commercial potato production in temperate regions of Europe. In contrast, in tropical and subtropical regions, the incidence of bacterial wilt disease can be very high and result in high yield losses. There is a linear relationship between yield loss and wilt incidence in these areas (Elphinstone, 1989). In Nepal, the average incidence of bacterial wilt was 33% of potato crops raised from farmer-saved seed. The incidence of latent infections in seed was found to be as high as 15–19% (Pradhanang et al., 2000). Also in African countries, wilt disease incidences of up to 100% have been reported in heavily infested soils planted with farmer-saved seed (Lemaga et al., 2001). Before certified seed was used, extensive crop losses were also reported in subtropical areas such as Israel and Greece (Zachos, 1957; Volcani and Palti, 1960).

It is claimed that statutory control measures for control of *C. michiganensis* ssp. *sepedonicus* cost more than the actual yield losses due to the ring rot disease (De Boer, 1987; Stead and Wilson, 1996). However, the potential for direct losses caused by *C. michiganensis* ssp. *sepedonicus* should not be underestimated. Without appropriate precautions, *C. michiganensis* ssp. *sepedonicus* can rapidly disseminate through seed tubers and cause considerable yield loss. Even without visible symptom development, tuber yield (number and weight) of *C. michiganensis* ssp. *sepedonicus*-infected plants is considerably lower than of healthy plants (R. Karjalainen, unpublished results). In the 1940s, the pathogen was present in all major potato-producing districts of Canada and the USA. In particular, the use of picker planters and the practice of cutting seed were largely responsible for the high incidence of bacterial ring rot. The lack of adequate detection methods for asymptomatic infections hampered control of the disease in those years. Incidentally, during that time, up to 80% of seed lot rejections in parts of the USA were due to bacterial ring rot (Eddins, 1939; Kreutzer and McLean, 1943). In Maine (USA), in 1939 and in 1940, respectively, 11.5 and 7.5% of all certified seed lots were rejected due to bacterial ring rot (Baribeau, 1948). In Canadian surveys conducted during 1943–47, 9–16% of the farms in Ontario and Quebec were found to be infected with ring rot (Richardson and Goodin, 1949). As recently as 1978, 60% of the total seed rejected for certification in the USA was rejected due to infections with *C. michiganensis* ssp. *sepedonicus* (Slack et al., 1979). Only by implementation of an intensive eradication programme could the incidence of ring rot be reduced. The incidence of bacterial ring rot in eastern Canada dropped from 5.1% in 1983 to only 0.04% in 1992 (De Boer, 1999). Also in Western Europe, where *C. michiganensis* ssp. *sepedonicus* was frequently found in Scandinavian countries and in Germany, functional eradication is being achieved by targeted campaigns. Only an occasional reoccurrence happens, which is now followed by immediate action to mitigate its impact.

Blackleg and stem wet rot cause large annual economic losses in some geographic areas due to rejection and declassification to a lower grade of seed. In the Netherlands, for example, declassification to a lower grade costs on average €2/100 kg of seed. From 2002 to 2004, on average 9% of the seed potato hectarage was declassified and 3% was rejected due to blackleg or stem wet rot. In 2003, when weather conditions were very favourable for disease expression in the Netherlands, the economic loss due to erwinia infections was estimated at €17 million. Thirty years ago, blackleg was estimated to cause losses of 5–10% in some potato-growing regions of the USA with losses as high as 40% in individual fields (Stanghellini and Meneley, 1975). Today, however, the incidence of blackleg, in particular, has been significantly decreased by the use of erwinia-tested minitubers for the initiation of seed stocks.

Loss of agricultural produce due to bacterial soft rots is difficult to estimate but is known to be large. An estimate of 10% loss caused by post-harvest decay of root crops is realistic, and much of it is due to decay initiated by the pectolytic erwinias. In 1980, potato crop loss due to bacterial soft rot on a worldwide basis was estimated to be in the range of $US50–100 million (Pérombelon and Kelman, 1980). Improvements in harvesting equipment and better controlled storage have undoubtedly curbed losses in storage over the last 25 years, which likely still exceed $US100 million at today's currency levels. Seed piece decay causing failure in crop stands have been recorded as high as 50% (Pérombelon et al., 1988).

Although common scab does not generally affect yield, it is an economically important disease on account of its impact on marketability of affected tubers. Scab-infected tubers have decreased value on the fresh market because infections are unsightly and decrease value for processing due to decreased quality. Grade-out of scabby tubers can be significant and seriously diminish profit margins for both seed and ware potatoes. Actual estimates of economic losses due to common scab are not available, however.

27.3.3 Geographic distribution

R. solanacearum is widespread in tropical and subtropical regions throughout the world. Biovars 1, 3 and 4 are present in the warmer areas, whereas biovar 2 is present in temperate zones and cooler areas at high altitudes in tropical regions (Anonymous, 2005a). Recently, biovar 2 has become widespread in temperate regions of Europe. *C. michiganensis* ssp. *sepedonicus* occurs chiefly in temperate zones and is not known to occur in tropical or subtropical areas. Bacterial ring rot has been reported from at least 31 countries distributed over five continents (Stead and Wilson, 1996; Anonymous, 2005b). Some reports are unclear, however, on whether the disease reported is due to *R. solanacearum* or *C. michiganensis* ssp. *sepedonicus* because of similarities in symptomologies and inadequacy of microbiological resources in developing countries. The disease has never been found in Australia. It was first described in Germany in 1906 and first found in North America in 1931 in Quebec (Canada). Subsequently, it has been reported in most potato-growing areas of Canada and the United States. *C. michiganensis* ssp. *sepedonicus* occurs in the Scandinavian countries and Eastern Europe. It is occasionally reported in Western European countries. The disease has fortunately never been found in Scotland, a major producer of seed. A few occurrences have recently been reported in the Netherlands, the other major seed potato-producing area of Europe. *E. chrysanthemi* and *E. carotovora* ssp. *carotovora*, having a wide host range, are worldwide in distribution. *E. carotovora* ssp. *atroseptica* is found in all potato-producing areas in the Northern Hemisphere, but the finding of *E. carotovora* ssp. *brasiliensis* in Brazil as the causal agent of blackleg raises questions about the identity of blackleg-causing strains elsewhere. Scab-inducing streptomycetes are widespread and occur wherever potatoes are grown. Specific species or metabolic types may vary from region to region.

27.4 ECOLOGY

27.4.1 Plant colonization

Bacterial pathogens lack the enzymatic repertoire to penetrate intact plant tissue directly. They depend on wounds or natural entrances, such as stomata, lenticels and hydathodes, or regions of secondary root emergence to enter plant tissues. Once inside the plant, however, bacterial pathogens can remain latent in asymptomatic infections and be spread in vegetative plant propagules. In this way, seed potato tubers serve as the major primary inoculum source for most of the bacterial potato diseases. After planting, bacteria from the seed tuber are transported through the xylem vessels of the vascular tissue and colonize sprouts and the growing plants. The vascular pathogens such as *C. michiganensis* ssp. *sepedonicus* and *R. solanacearum*, in particular, infect successive generations of potato plants from infected tubers that are often symptomless and used inadvertently for seed. This may also be true to a lesser extent for *E. carotovora* ssp. *atroseptica* (De Boer, 2002). In tropical areas, *R. solanacearum* is known to be soil-borne, and disease can be initiated directly from root infection.

In active infections caused by *C. michiganensis* ssp. *sepedonicus, R. solanacearum* or *E. carotovora* ssp. *atroseptica*, bacterial populations achieve very high levels and result in decay of the seed tuber or early death of the plant. Alternatively, bacteria from the seed tuber may invade and colonize the stem and proceed to infect progeny tubers, only causing symptoms of disease when cultivar characteristics and growing conditions are appropriate for their manifestation. Bacterial slime, consisting of EPSs, occludes the plant's water transport system, resulting in wilting symptoms typical of vascular diseases.

Scab-inducing streptomycetes are not systemic invaders of potato plants. Rather they infect tubers, and to a lesser extent roots, through lenticels and remain localized to superficial tissues. Some strains do, however, cause necrosis deep into tuber tissue to cause a condition know as deep-pitted scab.

27.4.2 Survival

The bacterial pathogens of potato are weak competitors among the microbial flora in soil and other natural habitats. Although they survive while confined in infected potato debris, in general they do not survive for long periods of time in unprotected environments with high microbial activity. *C. michiganensis* ssp. *sepedonicus* and some of the erwinias such as *E. carotovora* ssp. *atroseptica* survive for less than a year in soil at temperatures higher than 10°C. *E. carotovora* ssp. *carotovora* survives better in soil than does *E. carotovora* ssp. *atroseptica* and can often be detected in potato field soil with appropriate enrichment techniques. Survival data for European potato strains of *E. chrysanthemi* are not available although the bacterium survived in association with detritus in alpine streams in Australia (Cother and Gilbert, 1990). However, it was not detected in soil and bog samples near *E. chrysanthemi*-positive streams. *C. michiganensis* ssp. *sepedonicus* only survives in soil for prolonged periods in a dried state or at temperatures below −10°C. *R. solanacearum* is a soil resident in tropical areas, where in many soils relatively high densities are found. In temperate regions, *R. solanacearum* survival periods never exceeded 2 years, even in the presence of high numbers of overwintering tubers (Elphinstone, 1996). It survived longest when protected from microbial antagonism and desiccation in deep soil layers (75 cm) and in association with plant host tissue. Soils that are flooded annually (e.g. sugar cane fields) and river flood plains generally do not harbour populations of *R. solanacearum* (Hartman and Elphinstone, 1994).

Alternate hosts, including weeds and field crops, play a role in the survival of *R. solanacearum* and erwinias between two successive potato crops. However, they seem to be of little importance in the epidemiology of *C. michiganensis* ssp. *sepedonicus*, which has a restricted host range (van der Wolf et al., 2005b). In particular, biovars 1, 3 and 4 of *R. solanacearum* have a broad host range and infect many crops grown in rotation with potato and in a number of solanaceous weeds, such as *Solanum carolinense* in the USA and *Portulaca oleracea* in Australia (Moffett and Hayward, 1980). Biovar 2, which is responsible for brown rot in Europe, survives during the winter in roots of bittersweet (*Solanum dulcamara*), a perennial riparian weed (Janse, 1996). During the spring when temperatures rise and exceed 15°C, *R. solanacearum* cells in infected *S. dulcamara* roots multiply and are released into adjacent streams at high densities. In Australia, *Solanum cinereum*, another perennial weed, may be involved in overwintering of biovar 2 (Graham and Lloyd, 1978). In Scotland and the USA, both *E. carotovora* ssp. *atroseptica* and *E. carotovora* ssp. *carotovora* were found in the rhizosphere of different crops and weeds (Mendoca and Stanghellini, 1979; McCarter-Zorner et al., 1985). In Scotland, infected weeds were even found in fields where no potatoes were grown for 2 years or more, although at low frequency.

C. michiganensis ssp. *sepedonicus*, but not the pectolytic erwinias and *R. solanacearum*, survives for long periods on the surfaces of farm equipment, tools and storages (Nelson, 1980). Survival of *C. michiganensis* ssp. *sepedonicus* is particularly enhanced by low humidity and low temperature. Survival on various surface materials for more than 10 months under conditions of widely fluctuating temperatures and relatively low humidity has been recorded. Consequently, contaminated storage bins and farm equipment that are not adequately cleaned and disinfected serve as sources of inoculum for disease spread. The erwinias and *R. solanacearum* do not persist for long periods on surfaces of farm equipment but do survive for sufficient periods of time to spread infections during potato handling operations such as harvesting and grading.

Streptomyces spp. are common soil inhabitants and also the plant pathogenic strains survive in soil and plant debris for more than 10 years. However, potato tubers are also an important source of inoculum. Propagules of pathogenic *Streptomyces* spp. have been estimated to be as high as 10^9 colony forming units/g of scab lesion and at 10^5 colony forming units/g of healthy potato tuber periderm (Wang and Lazarovits, 2004). Tuber-borne inoculum is an important means by which the pathogen is introduced to new areas and also contributes directly to disease severity (Wilson et al., 1999).

27.4.3 Dissemination

27.4.3.1 Through potato tubers

The chief means by which the bacterial potato pathogens are dispersed over long distances are in or on infected and contaminated potato tubers. Dissemination of pathogens occurs mostly on seed potatoes, but there are also examples of ware potatoes being used as tablestock for fresh consumption or production for processing into starch or other products that served to introduce the bacteria into formerly pathogen-free regions. There is good evidence that during the mid-1970s surface water in Sweden became infested with *R. solanacearum* through waste water from a potato processing factory (Olsson, 1976). Similar scenarios may have happened more recently in other European countries, contributing to the widespread occurrence of *R. solanacearum* in temperate European countries today.

27.4.3.2 Through soil

The erwinias and *R. solanacearum* are motile bacteria and thus readily move through water films in soil and contaminate progeny tubers from decaying seed tubers and other infected plant tissues. The spread of the pectolytic erwinias, released by decaying seed tubers, to progeny tubers resulting in lenticel infections, has been well documented as an important infection route. Erwinias have been detected in soil at distances of 1–3 m from the source of a decaying seed tuber. There is no evidence that dissemination of *C. michiganensis* ssp. *sepedonicus*, being non-motile, occurs through soil to play a role in the epidemiology of the ring rot disease. Infections largely, if not entirely, are a result of internal colonization from infected seed tubers or from contaminated equipment used for planting, harvesting, grading, transporting or storing the crop.

27.4.3.3 Through surface water

There is good evidence that overhead irrigation with contaminated surface water has resulted in initial infections of seed lots with *R. solanacearum* in several countries in Europe. Erwinias have also been found in both salt and fresh surface water, although they are found at relatively low densities of 1–1000 cells/ml. Of the erwinias, *E. carotovora* ssp. *carotovora* is most commonly found, but in one study *E. carotovora* ssp. *atroseptica*- and *E. chrysanthemi* strains together comprised up to 10% of the erwinias in surface water. It is speculated that some aquatic plants might serve as reservoirs for the erwinias and perhaps their survival is enhanced in natural biofilms that form on rocks and other surfaces within bodies of surface water. As with *R. solanacearum* populations detectable in water, the population of erwinias was higher at water temperatures above 10–15 °C. Contamination of surface water with erwinias may result from entry of drainage water from fields harbouring decayed tubers. In contrast to *R. solanacearum*, no data are available on whether there is a relationship between irrigation with contaminated surface water and incidence of blackleg, aerial soft rot or stem wet rot. *C. michiganensis* ssp. *sepedonicus* survives for less than 2 weeks in surface water (van der Wolf and van Beckhoven, 2004). Contaminated water will only create a risk when there is a source, such as a susceptible host, that constantly releases high densities of the bacterium into the water. The existence of such a source is unlikely as *C. michiganensis* ssp. *sepedonicus* has never been found in surface water.

27.4.3.4 Through aerosols and rain

Heavy rainfall on infected crops and mechanical destruction of haulms before harvest cause aerosolization of the erwinia bacteria. Even at low densities of diseased plants, the density of erwinias in aerosols can be as high as 10^8 cells/ha. Erwinas have been detected in aerosols at great distances from infected hosts, indicating that if relative humidity is high, aerosols persist and can transport the bacterium over long distances. Erwinias were present in more than 90% of rain samples collected at different times during the year in Oregon (USA) (Franc et al., 1985). There is no evidence to suggest that the other bacterial pathogens of potato are also dispersed by aerosols or rain.

Erwinas deposited on leaf surfaces from aerosols may be further distributed to the below-ground portion of the plant during summer rains. Even symptomless plants may carry relatively high concentrations of erwinias of 4×10^6 cells/g of plant material (N.J.M. Roozen, personal communication). The bacteria may multiply to even higher populations during subsequent decay of contaminated leaves after senescence or after intentional haulm destruction. As rain water percolates through the soil, the bacteria are washed on to progeny tubers and serve as an inoculum source for subsequent generations of potatoes.

27.4.3.5 Through insects

Insects transmit various bacterial pathogens to initiate new crop infections, although their role in disease epidemiology is limited. Insects acquire contamination with bacterial phytopathogens simply by mechanical contact, but the bacteria may survive on insects for several days. There is no indication that persistent infections of insects with plant pathogenic bacteria occurs in the same way as persistent plant viral infections of insect vectors. Insects can travel large distances in a few days and are specifically attracted to plant wounds that serve as ideal infection loci. Plant wounds are created by insect feeding and if the insect mouth parts are contaminated with pathogenic bacteria, it is an ideal means of spreading infections. In Scotland and Colorado (USA), 15% of flies in potato-growing areas carried pectolytic erwinias, and those captured near potato cull dump piles contained up to 10^6 cells/insect.

27.5 CONTROL

The vascular bacterial diseases of potato can only be controlled by integrating measures based on exclusion and reduction of seed- or soil-borne inoculum. These measures include the use of clean seed and implementation of appropriate agronomic practices during planting, growing, harvesting and storage of potato crops. No chemical compounds or treatments are available that effectively cure plants from bacterial diseases, but some physical treatments or bactericides may be useful for reducing inoculum load. Potato cultivars differ in susceptibility to the various bacterial diseases, but none are immune. The use of resistant cultivars, therefore, has not been an important component of strategies to control these diseases. Avoidance of particularly susceptible cultivars is recommended. Disease-tolerant cultivars present a special hazard in that they may serve as a reservoir of inoculum unless specific measures are taken to prevent infection.

27.5.1 Use of clean seed

The use of pathogen-free certified seed is the most important element in strategies for controlling bacterial diseases. Potatoes are not usually propagated from true botanical seed but rather are propagated from vegetative tubers. It is these vegetative propagules that are much more likely to carry pathogens from the parent plant than true seed. Not only are the vascular pathogens present in parent plants likely to be transferred to tubers, but also are the surface pathogens such as *S. scabies* migrate from parent plant to progeny tubers.

Because all the bacterial pathogens of potato may be borne on seed tubers, seed production should preferably start with pathogen-tested stem cuttings or minitubers, for which risks of contamination with pathogens are considerably reduced. The number of subsequent field generations should also be limited. In an 8-year survey in the USA, seed stocks from stem cuttings resulted in 54% of crops being free of symptoms of bacterial disease compared with an average of 8% for lots not produced from stem cuttings (Knutson, 1985). When the two groups were compared, 6% of lots grown from stem cuttings contained at least 0.5% infected tubers, compared with 41% of lots produced from tubers. With the advent of minituber production technologies, stem cuttings are rarely used today, but the starting of seed stocks from pathogen-free material remains an important principle. Because bacterial infections may be latent, it is important that all plant material used to initiate seed stocks be tested. Many protocols based on serological or molecular tests are available for each of the bacterial pathogens of potato.

Seed stocks become contaminated with bacterial pathogens during the growing season in agricultural fields to varying degrees depending on the presence of the pathogen in the geographic area and inoculum load if present. The level of contamination and infection increases with each field generation. Hence, field inspections combined with laboratory assays are required to identify lots that become contaminated or infected beyond tolerance levels. Elimination of positive lots minimizes the risks for dissemination of bacterial pathogens through seed.

R. solanacearum and *C. michiganensis* ssp. *sepedonicus* are considered quarantine pathogens in many countries and are under strict statutory control. There is a zero tolerance for these pathogens in most seed potato certification programmes. Export of potatoes into controlled areas usually requires laboratory testing of tubers to ensure freedom from these pathogens. For detection and identification of *R. solanacearum* in samples of potato tubers within the European Union, both dilution plating on a semi-selective agar (SMSA) medium and immunofluorescence (IF) cell staining can be used for initial screening. The use of ELISA and TaqMan-PCR tests are optional (EU Council Directive 98/57/EC). For final verification, *R. solanacearum* should be isolated and characterized by physiological, biochemical and pathogenicity tests.

In the European Union, IF is used for initial screening of tubers for *C. michiganensis* ssp. *sepedonicus*, whereas the ELISA test is used in North America. In North America, IF or PCR is used to confirm the presence of *C. michiganensis* ssp. *sepedonicus* in ELISA positive lots. However, in the European Union, IF positive samples need to be confirmed by bioassay in eggplant. Subsequent isolation of *C. michiganensis* ssp. *sepedonicus* from eggplant and characterization of isolates by serology, PCR using TaqMan chemistry and a pathogenicity test are required to verify the presence of *C. michiganensis* ssp. *sepedonicus*.

In Europe, the standard sample size for detecting *C. michiganensis* ssp. *sepedonicus* or *R. solanacearum* in a lot of seed potatoes is 200 tubers/25 tonnes of potatoes. In North America, tests for *C. michiganensis* ssp. *sepedonicus* (*R. solanacearum* does not occur in the North American potato industry) is usually based on 400 tubers/seed lot. The probability of detecting *C. michiganensis* ssp. *sepedonicus* or *R. solanacearum*, assuming random distribution of diseased tubers, is a function of sensitivity of the laboratory test, sample size and incidence of infection. If the sensitivity of the laboratory test is considered to be 100%, even then there is only a 0.95 probability of detecting these bacteria in a potato lot with a 200 tuber sample size if more than 1.5% of the tubers in the lot are infected. Freedom from bacterial infections, therefore, cannot be guaranteed based on laboratory testing alone (Janse and Wenneker, 2002).

With few exceptions, the pectolytic erwinias are not considered to be quarantine pathogens. Most seed certification inspection programmes include only an obligatory field inspection for the foliage manifestations of infections. In some instances, however, laboratory tests are used to evaluate the health of seed potatoes. Available tests for specific *Erwinia* (sub)species include plating on a SMSA medium (e.g. crystal violet pectate medium), ELISA, enrichment ELISA, immunofluorescence cell or colony staining and (enrichment) PCR techniques. Serological testing for the erwinias is limited by their serological heterogeneity; only subpopulations are detected with any one antiserum or antibody preparation.

27.5.2 Inoculum reduction

Seed tuber treatment for reducing seed-borne inoculum has been considered. Tuber-borne inoculum can be reduced but not eliminated by heat treatment using hot water, steam or air or exposure to ultraviolet light. Such treatments are often not practical, however, and only address surface contamination. Bacteria located internally in tuber tissue are protected from disinfection. Seed treatment to reduce *R. solanacearum* and *C. michiganensis* ssp. *sepedonicus* contamination levels is not recommended, but such treatments have been explored for reducing erwinia inoculum. Treatment with hot water for 5 min at 55°C followed by forced air-drying resulted in a 99.9% reduction of *E. carotovora* ssp. *atroseptica* in artificially contaminated seed tubers (Pérombelon et al., 1989). Similarly, steam treatment reduced blackleg incidence from 46 to 3% (Afek and Orenstein, 2002). Longer exposures to high temperature resulted in sprout damage. It is unknown whether heat treatment can be used for *E. chrysanthemi*, which is more heat resistant. Treatment with steam and hot air are preferred to hot water treatment because there is less risk for cross-contamination and avoids the need for drying after treatment. Efficacy of heat treatment may be due to the direct eradication of surface located pathogens, but it may also stimulate the production of antibacterial compounds in potato tubers and thereby increase tuber resistance. Ultraviolet light, even at low dosages, effectively reduced storage rot in erwinia-inoculated tubers (Ranganna et al., 1997).

Treatment of seed tubers with fungicides provides some control of seed piece decay and common scab. The fungicides may directly reduce the inoculum level of streptomyces as well as pathogenic fungi that provide infection sites for the pectolytic erwinia.

In tropical countries, reduction of soil-borne inoculum is particularly important as a control strategy for *R. solanacearum* (French, 1994). Soil populations of *R. solanacearum* can be reduced by a number of measures including crop rotation, fumigation, solarization, application of soil amendments, flooding and the roguing of volunteers and weeds that serve as hosts to the bacterium. To achieve effective control, a combination of different measures may be integrated.

Crop rotation is used to promote natural decline of soil inoculum. The effect of crop rotation on bacterial wilt is dependent on the level of inoculum in the soil and the crops used in the rotation scheme. In Uganda, a 1-year crop rotation with wheat or maize reduced the bacterial wilt incidence in mildly infested fields from 62 to 13%, but for heavily infested fields, a two-season rotation was required (Lemaga et al., 2001). In this study, the greatest wilt reduction was obtained with a potato–beans–maize–potato rotation. Similarly, in India, a 5-year rotation with wheat, lupin and maize reduced the wilt incidence considerably (Verma and Shekhawat, 1991), whereas, in Indonesia, a two-season rotation with pulses, cereals and root crops was effective (Gunadi et al., 1998). Crop rotation also reduces inoculum levels of scab-inducing streptomycetes. Rotations of 3–4 years reduce the incidence of scab to a relatively constant level but seldom completely eliminate soil populations of the pathogen.

Some soils are naturally suppressive, due to the presence of a specific, possibly diversified, microbial community, that results in low incidences of bacterial wilt despite the presence of the bacterium (Shiomi et al., 1999). The use of soil organic amendments such as pig manure (Gorissen et al., 2004), sewage sludge (Prior and Beramis, 1990) and mixtures of agricultural and industrial wastes (Sun and Huang, 1985; Anith et al., 2004), which contain active compounds such as urea and mineral ash, helps to accelerate the decline of *R. solanacearum* populations. Organic amendments have particular potential for controlling common scab (Lazarovits et al., 2001). Incorporation of high nitrogen-containing amendments such as soy and bone meal reduced the severity of scab by 90%, but the disease rebounded in subsequent years. Evidently, soil amendments of this type not only kills pathogenic bacteria but also enhances the overall soil microbial population that selectively displaces some organisms and enriches for others.

Soil fumigation with synthetic chemical crop protection agents can reduce soil-borne *R. solanacearum*, although results may be variable (Chellemi et al., 1994). Jones et al. (1995) reported that soils treated with methyl bromide/chloropicrin, chloropicrin, metam-sodium, dazomet or dichloropropene reduced bacterial wilt incidence in tomato from 92 to 20–48%. Various attempts have been made to reduce soil populations of *R. solanacearum* with natural compounds. Thyme, palmorosa and lemongrass oil, volatile essential oils with antimicrobial properties, are reported to have eliminated *R. solanacearum* in soil treated with a concentration of 700 μg/l of soil (Pradhanang et al., 2003).

Solarization of soil prior to planting has also been used successfully in reducing *R. solanacearum* populations, but it does not achieve eradication of the pathogen (Chellemi et al., 1994; Pradeep and Sood, 2001). Soil temperature increases of 5–12°C under polyethylene film used for the solarization treatment have been reported. Reduction in *R. solanacearum* populations may be caused, in part, by direct heat killing of the bacterium, but increased activity of microbial antagonists may also be a factor in reducing pathogen populations (Pradeep and Sood, 2001). An impact of solarization was found up to a depth of 15 cm.

27.5.3 Agronomic practices

As there is a zero tolerance for bacterial ring rot and bacterial wilt (brown rot), seed stocks in which these diseases occur are normally not used for planting. However, good agronomic practices are recommended to avoid losses because of pectolytic erwinias and scab-inducing streptomycetes. Potato production should start with small undamaged tubers, carefully checked by an experienced inspector. Small tubers disintegrate more readily than large tubers, thus reducing risks for contaminating progeny tubers during harvest. When cut seed is used, tuber size is less important as seed pieces usually disintegrate during the growing season if they contain pectolytic bacteria.

Potatoes should be planted in soils that are well drained and aerated as anaerobic conditions in the soil negatively affect tuber resistance to decay (Pérombelon and Lowe, 1975; Bain and Pérombelon, 1988). Maintenance of high soil moisture, however, helps to suppress the incidence of common scab.

Fertilization should be optimized to promote healthy plant growth. Low nitrogen content will make the crop susceptible to infection, whereas high nitrogen levels will result in a high nitrate content that promotes multiplication of erwinias (Smid et al., 1993). High nitrogen levels also stimulates growth of foliage that in turn increases leaf wetness in the canopy providing conditions that are favourable for development of aerial stem rot. Soil amendments that increase soil pH should be avoided to control scab.

Damage to potato haulms must be avoided during hilling and spraying operations as all injuries or wounds to potato haulms provide places of entry for pathogenic bacteria. As it is usual to apply fungicides to control late blight up to 4–12 times/season, considerable crop damage can be done if extreme care is not exercised. Pest control is important, however, as fungal infections, nematode damage and insect feeding sites allow entry of bacterial pathogens.

Spread of bacterial contamination within seed stocks often occurs during harvesting and grading of tubers. The presence of decayed mother or progeny tubers can hardly be avoided and results in the contamination of equipment with bacteria-laden tuber debris. A single decayed tuber can contaminate up to 100 kg of potatoes during mechanical grading (Elphinstone and Pérombelon, 1986). During harvesting and grading, a high percentage of tubers are damaged or bruised and become particularly susceptible to infection from contaminated equipment (Pérombelon and Kelman, 1980). Wet harvest conditions exacerbate tuber damage and spread of inoculum.

For the first 10–14 days after harvest, potato storage temperature should be modulated to promote wound healing. Subsequently, the temperature should be decreased to below 10°C to prevent the growth of fungal and bacterial pathogens. Storages should be well ventilated to prevent formation of condensation on tubers and provide adequate aeration. Oxygen depletion can be favourable for growth of soft rot bacteria and be inhibitory to the resistance response of tubers.

27.6 PERSPECTIVES

Globalization has resulted in an increase in international trade of seed potatoes with concomitant, albeit inadvertent, dissemination of bacterial pathogens in latently infected tubers. The chances for control and eradication of bacterial diseases after introduction into new areas largely depend on the available infrastructure enabling their detection and trace back to source. Availability of adequate instruments to detect and identify the pathogens along with the means to contain and eradicate infection sources is an important consideration for addressing new pathogen incursions.

To minimize the impact of bacterial diseases in potato, producers of seed need to specialize by separating seed production from ware production, optimize agronomic practices and use strict hygienic measures in both the field and storage components of their operation. The use of pathogen-tested minitubers should be favoured, and the number of field generations reduced to limit the build-up of bacterial populations. Monitoring of inoculum carried in or on seed by laboratory testing is recommended if it can be done in a cost-effective manner. For seed testing, sampling strategies and robust sensitive detection methods need to be designed and optimized within quality control systems.

It is anticipated that advances in knowledge of the genomics, proteomics and metabolomics of potato and bacterial pathogens will further enhance the possibilities for control. Genome sequencing has already resulted in the identification of several hitherto unknown pathogenicity factors (Salanoubat et al., 2002; Bell et al., 2004). Microarray systems for analysing gene expression and high-throughput systems for protein analysis will increase understanding of pathogenesis, including factors involved in bacterial survival strategies, host specificity and the transition from latent to active phase. Deciphering of genome sequences and extending knowledge of plant pathogen interactions will facilitate marker-assisted resistance breeding. Additionally, increased knowledge may suggest new strategies for designing more effective and perhaps systemic bactericides to combat the bacterial diseases that are so harmful to optimizing potato production.

REFERENCES

Afek U. and J. Orenstein, 2002, *Can. J. Plant Pathol.* 24, 36.

Anith K.N., M.T. Momol, J.W. Kloepper, J.J. Marois, S.M. Olson and J.B. Jones, 2004, *Plant Dis.* 88, 669.

Anonymous, 2005a, *Ralstonia solanacearum*. Data Sheets and Quarantine Pests, Prepared by CABI and EPPO for the EU Under Contract 90/399003.

Anonymous, 2005b, *Clavibacter michiganensis* subsp. *sepedonicus*. Data Sheets and Quarantine Pests, Prepared by CABI and EPPO for the EU Under Contract 90/399003.

Appel O., 1906, *Botanik* 3, 122.

Bain R.A. and M.C.M. Pérombelon, 1988, *Plant Pathol.* 37, 431.

Baribeau B., 1948, *Am. Potato J.* 25, 71.

Bartz J.A. and A. Kelman, 1984, *Am. Potato J.* 61, 485.

Bell K.S., M. Sebaihia, L. Pritchard, M.T.G. Holden, L.J. Hyman, M.C. Holeva, N.R. Thomson, S.D. Bentley, L.J.C. Churcher, K. Mungall, R. Atkin, N. Bason, K. Brooks, T. Chillingworth, K. Clark, J. Doggett, A. Fraser, Z. Hance, H. Hauser, K. Jagels, S. Moule, H. Norbertczak, D. Ormond, C. Price, M.A. Quail, M. Sanders, D. Walker, S. Whitehead, G.P.C. Salmond, P.R.J. Birch, J. Parkhill and I.K. Toth, 2004, *Proc. Natl. Acad. Sci. U.S.A.* 101, 11105.

Bouchek-Mechiche K., L. Gardan, P. Normand and B. Jouan, 2000, *Int. J. Syst. Evol. Microbiol.* 50, 91.

Bukhalid R.A., S.Y. Chung and R. Loria, 1998, *Mol. Plant Microbe Interact.* 11, 960.

Campos E., E.A. Maher and A. Kelman, 1982, *Plant Dis.* 66, 543.

Chellemi D.O., S.M. Olson and D.J. Mitchell, 1994, *Plant Dis.* 78, 1167.

Collmer A. and N.T. Keen, 1986, *Annu. Rev. Phytopathol.* 24, 383.

Cother E.J. and R.L Gilbert, 1990, *J. Appl. Bacteriol.* 69, 729.

Cuppels D.A. and A. Kelman, 1980, *Phytopathology* 70, 1110.

De Boer S.H., 1987, *Am. Potato J.* 64, 683.

De Boer S.H., 1999, In: *Abstracts of the 14th Triennial Conference of the European Association for Potato Research (EAPR '99)*, 2–7 May, Sorrento, Italy, p. 537.

De Boer S.H., 2002, *Plant Dis.* 86, 960.

De Boer S.H., 2003, *Eur. J. Plant Pathol.* 108, 893.

De Boer S.H. and A. Kelman, 1978, *Potato Res.* 21, 65.

De Boer S.H. and S.A. Slack, 1984, *Plant Dis.* 68, 841.

De Boer S. H. and A. Wieczorek, 1984, *Phytopathology* 74, 1431.

De Boer S.H., A. Wieczorek and A. Kummer, 1988, *Plant Dis.* 72, 874.

Duarte V., S.H. De Boer, L.J. Ward and A.M.R. De Oliveira, 2004, *J. Appl. Microbiol.* 96, 535.

Eddins A.H., 1939, *Am. Potato J.* 16, 309.

Elphinstone J.G, 1989, *Tropic. Agric. Res. Ser.* 22, 120.

Elphinstone J.G, 1996, *Potato Res.* 39, 403.

Elphinstone J.G. and M.C.M. Pérombelon, 1986, *Plant Pathol.* 35, 25.

Fegan M. and P. Prior, 2005, In: C. Allen, P. Prior and A.C. Hayward (eds), *Bacterial Wilt Disease and the Ralstonia solanacearum Species Complex*, p. 449. APS Press, St Paul, MN, USA.

Folsom D. and B.A. Friedman, 1959, *Am. Potato J.* 36, 90.

Franc G.D., M.D. Harrison and M.L. Powelson, 1985, In: D.C. Graham and M.D. Harrison (eds), *Report of the International Conference on Potato Blackleg Disease*, 26–29 June 1984, Edinburgh, UK, p. 48.
French E.R., 1994, In: A.C. Hayward and G.L. Hartman (eds), *Bacterial Wilt: The Disease and Its Causative Agent*, p. 199. CAB International, Wallingford, UK.
Gardan L., C. Gouy, R. Christen and R. Samson, 2003, *Int. J. Syst. Evol. Microbiol.* 53, 381.
Genin S. and C. Boucher, 2002, *Mol. Plant Pathol.* 3, 111.
Gorissen A., L.S. van Overbeek and J.D. van Elsas, 2004, *Can. J. Microbiol.* 50, 587.
Goth R.W., K.G. Hynes and D.R. Wilson, 1993, *Plant Dis.* 77, 402.
Graham J. and A.B. Lloyd, 1978, *J. Aust. I. Agr. Sci.* 44, 124.
Gunadi N., E. Chujoy, M. Kusmana, I. Surviani, O.S. Gunawan and Sinung-Basuki, 1998, In: *Potato Research in Indonesia: Results in Working Paper Series*, p. 56. CIP/RIV, Lima, Peru.
Hagar S.S. and G.A. McIntyre, 1972, *Can. J. Bot.* 50, 2479.
Hartman G.L. and J.G. Elphinstone, 1994, In: A.C. Hayward and G.L. Hartman (eds), *Bacterial Wilt: The Disease and Its Causative Agent*, p. 157. CAB International, Wallingford, UK.
Hauben L., E.R.B. Moore, L. Vauterin, M. Steenackers, J. Mergaert, C. Verdonck and J. Swings, 1998, *Syst. Appl. Microbiol.* 21, 384.
Hayward A.C., 1991, *Annu. Rev. Phytopathol.* 29, 65.
He L.Y. and J.Y. Hua, 1985, *Plant Prot.* 11, 10.
Henningson P.J. and N.C. Gudmestad, 1991, *J. Gen. Microbiol.* 137, 427.
Hukkanen A., R. Karjalainen, S. Nielsen and J.M. van der Wolf, 2005, *J. Plant Dis. Protect.* 112, 88.
Janse J.D, 1996, *EPPO Bull.* 26, 679.
Janse J.D. and M.A. Ruissen, 1988, *Phytopathology* 78, 800.
Janse J.D. and M. Wenneker, 2002, *Plant Pathol.* 51, 523.
Jones J.P., J.P. Gilreath, A.J. Overman and J.W. Noling, 1995, *Proc. Fla. State Hort. Soc.* 108, 201.
Kers J.A., K.D. Cameron, M.V. Joshi, R.A. Bukhalid, J.E. Merello, M.M. Wach, D.M. Bibson and R. Loria, 2005, *Mol. Microbiol.* 55, 1025.
Knutson K.W., 1985, In: D.C. Graham and M.D. Harrison (eds), *Report of the International Conference on Potato Blackleg Disease*, 26–29 June 1984, Edinburgh, UK, p. 67.
Kreutzer W.A. and J.G. McLean, 1943, *Colorado Agricultural Experiment Station – Technical Bulletin 30.*
Laine M.J., M. Haapalainen, T. Wahlroos, K. Kankare, R. Nissinen, S. Kassuwi and M.C. Metzler, 2000, *Physiol. Mol. Plant Pathol.* 57, 221.
Lambert D.H. and R. Loria, 1989a, *Int. J. Syst. Bacteriol.* 39, 387.
Lambert D.H. and R. Loria, 1989b, *Int. J. Syst. Bacteriol.* 39, 393.
Lawrence C.H., M.C. Clark and R.R. King, 1990, *Phytopathology* 80, 606.
Lazarovits G., K. Conn, M. Tenuta and N. Soltaini, 2001, In: S.H. De Boer (ed.), *Plant Pathogenic Bacteria*, p. 291. Kluwer Academic Publishers, Dordrecht, The Netherlands.
Lemaga B., R. Kanzikwera, R. Kakuhenzire, J.J. Hakiza and G. Maniz, 2001, *Afric. Crop Sci. J.* 9, 257.
Lund B.M., T.F. Brocklehurst and G.M. Wyatt, 1981, *J. Gen. Microbiol.* 122, 17.
McCarter-Zorner N.J., M.D. Harrison, C.D. Frank, C.E. Quinn, I.A. Sells and D.C. Graham, 1985, *J. Appl. Bacteriol.* 59, 357.
Mendoca M. and M.E. Stanghellini, 1979, *Phytopathology* 69, 1096.
Metzler M.C., M.J. Laine and S.H. De Boer, 1997, *FEMS Microbiol. Lett.* 150, 1.
Miyajima K., F. Tanaka, T. Takeuchi and S. Kuninaga, 1998, *Int. J. Syst. Bacteriol.* 48, 495.
Moffett M.L. and A.C. Hayward, 1980, *Aust. Plant Pathol.* 9, 6.
Nelson G.A., 1980, *Am. Potato J.* 57, 595.
Ngwira N. and R. Samson, 1990, *Agronomie* 10, 341.
Nissinen R., S. Kassuwi, R. Peltola and M.C. Metzler, 2001, *Eur. J. Plant Pathol.* 107, 175.
Nolte P., G.A. Secor, N.C. Gudmestad and P.J. Henningson, 1993, *Am. Potato J.* 70, 649.
Olsson K., 1976, *EPPO Bull.* 6, 199.
Pérombelon M.C.M., 2002, *Plant Pathol.* 51, 1.
Pérombelon M.C.M., E.M. Burnett, J.S. Melvin and S. Black, 1989, In: E.C. Tjamos and C. Beckman (eds), *Vascular Wilt Diseases of Plants*, Vol. H28, p. 557. NATO ASI Series.
Pérombelon M.C.M. and A. Kelman, 1980, *Annu. Rev. Phytopathol.* 18, 361.
Pérombelon M.C.M., M.M. Lopez, J. Carbonell and L.J. Hyman, 1988, *Potato Res.* 31, 591.

Pérombelon M.C.M. and R. Lowe, 1975, *Potato Res.* 18, 64.
Pirhonen M., D. Flego, R. Heikinheimo and E.T. Palma, 1993, *EMBO J.* 12, 2467.
Pradeep K. and A.K. Sood, 2001, *Indian Phytopathol.* 54, 12.
Pradhanang P.M., J.G. Elphinstone and R.T.V. Fox, 2000, *Plant Pathol.* 49, 403.
Pradhanang P.M., M.T. Momol, S.M. Olson and J.B. Jones, 2003, *Plant Dis.* 87, 423.
Prior P. and M. Beramis, 1990, *Agronomie* 10, 391.
Ranganna B., A.C. Kushalappa and G.S.V. Raghavan, 1997, *Can. J. Plant Pathol.* 19, 30.
Richardson L.R. and R.E. Goodin, 1949, *Am. Potato J.* 26, 85.
Romanenko A.S., L.A. Lomovatskaya, I.A. Graskova and R.K. Salyaev, 2002, *Russ. J. Plant Physiol.* 49, 690.
Salanoubat M., S. Genin, F. Artiguenave, J. Gouzy, S. Mangenot, M. Arlat, A. Billault, P. Brottier, J.C. Camus, L. Cattolico, M. Chandler, N. Choisne, C. Claudel Renard, S. Cunnac, N. Demange, C. Gaspin, M. Lavie, A. Moisan, C. Robert, W. Saurin, T. Schiex, P. Siguier, P. Thebault, M. Whalen, P. Wincker and M. Levy, 2002, *Nature* 415, 497.
Samson R., J.B. Legendre, R. Christen, M. Fischer-Le Saux, W. Achouak and L. Gardan, 2005, *Int. J. Syst. Evol. Microbiol.* 55, 1415.
Samson R., F. Poutier, M. Sailly and B. Jouan, 1987, *EPPO Bull.* 17, 11.
Schell M.A., 1996, *Eur. J. Plant Pathol.* 102, 459.
Shiomi Y., M. Nishiyama, S. Suzuki and T. Marimoto, 1999, *Appl. Environ. Microbiol.* 65, 3996.
Slack S.A., 1987, *Am. Potato J.* 64, 665.
Slack S.A., A. Kelman and J.B. Perry, 1979, *Am. Potato J.* 56, 441.
Smid E.J., A.H.J. Jansen and C.J. Tuijn, 1993, *Appl. Environ. Microbiol.* 59, 3648.
Smith N.C., J. Hennessy and D.E. Stead, 2001, *Eur. J. Plant Pathol.* 107, 739.
Stanghellini M.E. and J.C. Meneley, 1975, *Phytopathology* 65, 86.
Stead D.E. and J. Wilson, 1996, *A Review of the Risks and Yield Losses Caused by the Potato Ring Rot Pathogen Clavibacter michiganensis ssp. sepedonicus.* Internal Report Central Science Laboratory, Harpenden, UK.
Sun S.K. and J.W. Huang, 1985, *Plant Dis.* 69, 917.
van der Wolf J.M and J.R.C.M. van Beckhoven, 2004, *J. Phytopathol.* 152, 161.
van der Wolf J.M., J.G. Elphinstone, D.E. Stead, M. Metzler, P. Müller, A. Hukkanen and R. Karjalainen, 2005a, *Plant Research International*, Internal Report 95, Wageningen, The Netherlands.
van der Wolf J.M., J.R.C.M. van Beckhoven, A. Hukkanen, R. Karjalainen and P. Müller, 2005b, *J. Phytopathol.* 153, 358.
Verma R.K. and G.S. Shekhawat, 1991, *Indian Phytopathol.* 44, 5.
Volcani Z. and J. Palti, 1960, *Plant Dis. Rep.* 44, 448.
Von Bodman S.B., W.D. Bauer and D.L. Coplin, 2003, *Annu. Rev. Phytopathol.* 41, 455.
Waléron M., K. Waléron and E. Lojkowska, 2002, *Plant Protec. Sci.* 38, 288.
Wallis F.M. and S.J. Truter, 1978, *Physiol. Plant Pathol.* 13, 307.
Wang A. and G. Lazarovits, 2004, *Can. J. Plant Pathol.* 26, 563.
Wilson C.R., L.M. Ransom and B.M. Pemberton, 1999, *Phytopathology* 147, 13.
Zachos D.G., 1957, *Annales de l'Institut Phytopathologique Benaki, New Series* 1, 115.

Chapter 28

Viruses: Economical Losses and Biotechnological Potential

Jari P.T. Valkonen

Department of Applied Biology, University of Helsinki, PO Box 27, FIN-00014 Helsinki, Finland

28.1 INTRODUCTION

'Viruses, because of their tuber-perpetuation and the heavy yield losses they cause, are the most dangerous parasites of the potato' (Ross, 1986). The threat of economic losses caused by viruses is still there, although programmes for the production of certified seed have drastically reduced the prevalence of viruses in potato crops in many areas. Detection and identification of viruses have improved with the application of molecular methods. Efforts to breed for virus-resistant potato cultivars utilizing genes derived from wild potato germplasm were strongly enforced by Ross, Cockerham and others since the 1950s. New cultivars with resistance to the most important viruses have helped to reduce yield losses. However, there are new emerging problems and threats that need attention. The current negative impact of potato viruses can be illustrated using *Potato virus Y* (PVY) as an example. A positive aspect, however, is that potato viruses can be converted to gene vectors and utilized as tools possessing economic potential. Thus, the new methods of biotechnology provide these pathogens with positive features, of which exploitation has just begun.

28.2 VIRUSES INFECTING POTATO

There are some 40 viruses infecting cultivated potatoes in the field. Most of them are listed in Table 28.1. The polerovirus *Potato leafroll virus* (PLRV), the potyviruses PVY and PVA (Fig. 28.1), the carlaviruses PVM and PVS and the potexvirus PVX are widely distributed in potato-growing areas in different parts of the world. Many other viruses listed in Table 28.1 also occur in many parts of the world, but in potato only occasionally or locally. *Wild potato mosaic virus* (genus *Potyvirus*, family *Potyviridae*) is not included in Table 28.1, because it is not known to infect cultivated potatoes (Jones and Fribourg, 1979; Spetz et al., 2003). The vectors and means of transmission of all viruses have not been studied in potato. Table 28.1 provides information referring to studies carried out on potato or presumes that the vectors known for a virus also transmit it on potato. Symptoms caused by viruses in potato are illustrated elsewhere (Jeffries, 1998).

Potato Biology and Biotechnology: Advances and Perspectives
D. Vreugdenhil (Editor)

Table 28.1 Viruses infecting cultivated potatoes[a]

Species (acronym), genus, family[b]	Occurrence in potato	Transmission by
Alfalfa mosaic virus (AMV), *Alfamovirus, Bromoviridae*	Worldwide (uncommon)	Aphids
Andean potato latent virus (APLV), *Tymovirus, Tymoviridae*	S-America	Beetles, TPS[c]
Andean potato mottle virus (APMV), *Comovirus, Comoviridae*	S-America	Contact
Arracacha virus B (AVB), tentative *Cheravirus, Sequiviridae*	Peru, Bolivia	TPS
Beet curly top virus (BCTV), *Curtovirus, Geminiviridae*	Arid areas worldwide	Leafhoppers
Cucumber mosaic virus (CMV), *Cucumovirus, Bromoviridae*	Worldwide (uncommon)	Aphids
Eggplant mottled dwarf virus (EMDV), *Nucleorhabdovirus, Rhabdoviridae*	Iran	Aphids
Potato aucuba mosaic virus (PAMV), *Potexvirus, Flexiviridae*	Worldwide (uncommon)	Contact, aphids
Potato black ringspot virus (PBRSV), *Nepovirus, Comoviridae*	Peru	(nematodes?)
Potato deforming mosaic virus = *Tomato yellow vein streak virus* (ToYVSV), *Begomovirus, Geminiviridae*[d]	Brazil	Whiteflies
Potato latent virus (PotLV), *Carlavirus, Flexiviridae*	N-America	Aphids
Potato leafroll virus (PLRV), *Polerovirus, Luteoviridae*	Worldwide	Aphids
Potato mop-top virus (PMTV), *Pomovirus,* –	N- & C-Europe, Peru	*Spongospora subterranea*
Potato rough dwarf virus = Potato virus P, tentative *Carlavirus*[e] –	Argentina, Uruguay	Aphids
Pototo virus A (PVA), *Potyvirus, Potyviridae*	Worldwide	Aphids
Pototo virus M (PVM), *Carlavirus, Flexiviridae*	Worldwide	Aphids
Pototo virus S (PVS), *Carlavirus, Flexiviridae*	Worldwide	Aphids
Pototo virus T (PVT), *Trichovirus, Flexiviridae*	S-America	Contact, TPS, pollen
Pototo virus U (PVU), *Nepovirus, Comoviridae*	Peru	(nematodes?)
Pototo virus V (PVV), *Potyvirus, Potyviridae*	N-Europe, S-America	Aphids
Pototo virus X (PVX), *Potexvirus, Flexiviridae*	Worldwide	Contact
Pototo virus Y (PVY), *Potyvirus, Potyviridae*	Worldwide	Aphids

Table 28.1 (*Continued*)

Species (acronym), genus, family[b]	Occurrence in potato	Transmission by
Potato yellow dwarf virus (PYDV) *Nucleorhabdovirus, Rhabdoviridae*	N-America	Leafhoppers
Potato yellow mosaic virus (PYMV), *Begomovirus, Geminiviridae*	Caribbean region	Whiteflies
Potato yellow vein virus (PYVV), tentative *Crinivirus, Closteroviridae*	S-America	Whiteflies
Potato yellowing virus (PYV), tentative *Alfamovirus*[f]	S-America	aphids, TPS
Solanum apical leaf curl virus (SALCV), tentative *Begomovirus*	Peru	?
Sowbane mosaic virus (SoMV), *Sobemovirus*, –	Worldwide (uncommon)	?
Tobacco mosaic virus (TMV), *Tobamovirus*, –	Worldwide (uncommon)	Contact
Tobacco necrosis virus (TNV), *Necrovirus, Tombusviridae*	Europe, N-America, Tunisia	*Olpidium brassicae*
Tobacco rattle virus (TRV), *Tobravirus*, –	Worldwide	Nematodes
Tobacco ringspot virus (TRSV), *Nepovirus, Comoviridae*	S-America	TPS (nematodes?)[g]
Tobacco streak virus (TSV), *Ilarvirus, Bromoviridae*	S-America	?
Tomato black ring virus (TBRV), *Nepovirus, Comoviridae*	Europe	Nematodes, TPS
Tomato mosaic virus (ToMV), *Tobamovirus*, –	Hungary	Contact
Tomato mottle Taino virus (ToMoTV), *Begomovirus, Geminiviridae*	Cuba	Whiteflies
Tomato spotted wilt virus (TSWV), *Tospovirus, Bunyaviridae*	Hot climates worldwide	Thrips

Please note that the number of viruses is likely to increase, especially by the inclusion of additional tospo- and begomoviruses.

[a] For the illustration of symptoms, see Jeffries (1998) and Loebenstein et al. (2001). For references to original detection and description in potato, see Valkonen (1994) and Loebenstein et al. (2001).

[b] Taxonomic treatment according to Fauquet et al. (2005). Note that only species names approved by the International Committee on Taxonomy of Viruses (ICTVs) are italixed.

[c] TPS, true potato seed.

[d] Ribeiro et al. (2005).

[e] Nisbet et al. (2005).

[f] Missing from Fauquet et al. (2005). See Valkonen et al.(1992).

[g] Jones (1982).

The taxonomy and genome organization of viruses are explained and illustrated in Fauquet et al. (2005) and at the official home page of the International Committee on Taxonomy of Viruses (ICTV). Most potato-infecting viruses (Table 28.1) have a positive-sense, single-stranded RNA (ssRNA) genome that is replicated in cytoplasm by the viral RNA-dependent RNA polymerase. Virions are assembled from hundreds to thousands of viral coat protein (CP) molecules that encapsidate the viral RNA. However, tospoviruses exemplified by *Tomato spotted wilt virus* (TSWV) (genus *Tospovirus*, family

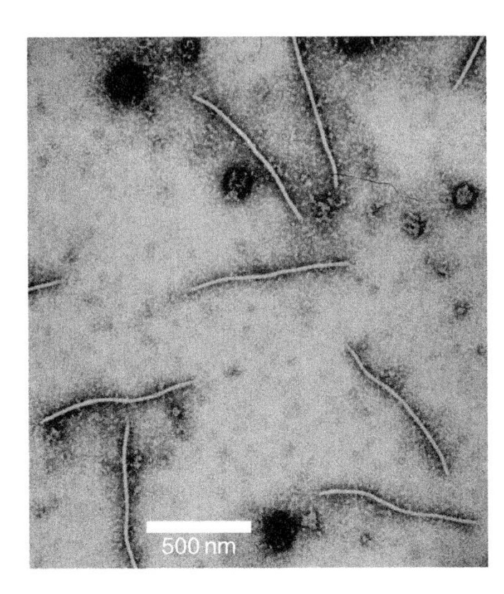

Fig. 28.1. Particles of *Potato virus A* (PVA). (Photo: J. Valkonen.)

Bunyaviridae) contain a negative-sense ssRNA genome of an ambisense nature: open-reading frames (genes) exist in both reading directions. A third type of RNA genome is possessed by nucleorhabdoviruses (family *Rhabdoviridae*) such as *Potato yellow dwarf virus* that has a negative-sense ssRNA genome replicated in the nucleus. Tospo- and nucleorhabdoviruses differ from most RNA viruses also in that their virions are enveloped by a lipid membrane. A few viruses such as curtoviruses and begomoviruses (family Geminiviridae) have a single-stranded ambisense DNA genome replicated in the nucleus by the host DNA polymerase. Their virions are not enveloped.

Potato spindle tuber viroid (PSTVd, *Pospiviroidae*) is one of the 27 known viroids (Flores et al., 2005). It is a severe pathogen of potato distributed worldwide. Although no vector is known for PSTVd, it is readily transmitted by contact and in true potato seed (Jeffries, 1998). Viroids are the smallest known pathogens and only found infecting plants. Molecular details of PSTVd will be discussed at the end of this chapter.

28.3 NEW AND EMERGING VIRUSES AND THEIR DETECTION

28.3.1 Molecular detection and identification

Because most potato viruses originate in the Andes, the evolutionary centre of potato species (Hawkes, 1990; Ochoa, 1999), movement of virus-infected cultivated and wild potato germplasm constitutes a risk for the dissemination of yet unknown viruses to other parts of the world. Systematic detection and characterization of viruses infecting potato species in their natural habitats in Central and South America would be important to continue, to develop antibodies, primers and probes needed for virus detection during international movement of germplasm. The International Potato Center (CIP) in Peru has worked for the characterization of many potato viruses since its establishment 30 years ago (Salazar, 1996, 2005). Biology and aetiology of the new viruses are best studied in the native geographic area, not least for reducing the risk of inadvertant dispersal of the viruses. The past and continuing efforts of CIP and other institutes in South America are therefore important.

Contributions of gene banks and virology laboratories in different parts of the world are needed for the characterization of the less-studied potato viruses. The latest technologies available for virus detection and characterization have made it increasingly feasible. Sequence analysis of large numbers of virus isolates is feasible using robotics-assisted automated sequencing facilities (Ala-Poikela et al., 2005; Allander et al., 2005). Microarray-based approaches offer the possibility to detect many viruses and virus isolates simultaneously and to distinguish them (Boonham et al., 2003; Bystricka et al., 2003; Zhang et al., 2005).

Nucleotide and deduced amino acid sequences of viral genes and genomes have become the most important single criteria used in viral taxonomy (Fauquet et al., 2003, 2005; Adams et al., 2005). Sequence data alone are sufficient to group virus isolates and often to determine whether they should be considered the same virus species. Accordingly, Nisbet et al. (2005) have recently concluded that potato rough dwarf virus and PVP, independently described in Argentina and Brazil, are isolates of the same carlavirus. However, sequence data cannot yet be used to predict more than a few biological properties in a limited number of viruses. Basic research needs to be continued even with the best characterized potato viruses before the sequence–function relationships are adequately understood. Therefore, biological characterization of viruses and virus isolates in conjunction with molecular studies remains an important topic.

28.3.2 New viruses

Only a few reports have been published on new viruses in potatoes during the past 10 years. Sequences homologous to *Apple latent spherical virus*, a nematode-transmitted virus with a wide host range, were found while analysing a potato cDNA library (Thompson et al., 2004). The host range or other properties of the virus isolate have not yet been reported. A new isometric virus designated as SB26/29 was found in potatoes in the Andes in 1990s. It is transmitted by a psyllid vector (*Russeliana solanicola*) (Salazar, 2005).

One may claim that the reason for lack of reports on new viruses from potato crops grown outside the Andean region is that most viruses causing economically significant losses in potato crops have already been detected and described. Also, the economically insignificant viruses that cause conspicuous symptoms, such as the calico mosaic symptoms in potato leaves infected with *Alfalfa mosaic virus* (Fig. 28.2), have probably been noticed.

The few reports on new potato viruses may, however, also be a sign of little effort put on the detection of new viruses in potato crops. There may be viruses that cause mild or no symptoms and are hard to notice when occasionally infecting potatoes. However, such viruses could become a considerable problem if agricultural practices, cultivars or climatic conditions or other agroecological factors change. One possible threat could be *Columbian Datura virus* (CDV) (Verhoeven et al., 1996) that was recently found in PVY-resistant tobacco cultivars in Poland, Germany and Hungary. It has also been found

Fig. 28.2. Yellow mosaic caused by *Alfalfa mosaic virus* in potato. An alfalfa plant growing as a volunteer from the previous crop has served as the virus source for aphids. (Photo: J. Valkonen.)

in *Datura* plants in gardens in Poland. Of more than 50 potato cultivars tested by Prof. M. Chrzanowska at IHAR, Poland, 30 were systemically infected, displaying necrosis, yellowing and conspicuous collapse or breaking of stems, depending on the cultivar. The virus is tuber-transmitted (M. Chrzanowska, personal communication). As tobacco and potato cultivation coincide temporally and spatially in Poland and many other areas, virus transmission between the crops is possible. It would be worthwhile to pay more attention to viruses that infect potatoes occasionally.

28.3.3 Emerging viruses

Viruses transmitted by whiteflies or thrips have become more and more prevalent in crops grown in tropical and subtropical parts of the world (Anderson et al., 2004). Reasons are not fully understood, but the increased abundance of viruses is associated with increasing prevalence of their whitefly vectors. Another example of emerging diseases, albeit not caused by a virus, is the relatively fast expansion of insect-transmitted phytoplasma in potato crops grown in the USA, for which reasons are unknown (Secor et al., 2005).

28.3.3.1 DNA viruses

Members of the genus *Begomovirus* (family *Geminiviridae*) cause vast yield losses in solanaceous and leguminous crops, such as tomato and beans, in Central and South America. They are transmitted by the whitefly, *Bemisia tabaci* (Gennadius). The number of approved and tentative virus species in genus *Begomovirus* has increased to nearly 200 and approaches the number of potyviruses, the largest group of plant-infecting viruses

(Fauquet et al., 2005). Taxonomy of begomoviruses is complicated by recombination between virus strains and species in co-infected plants. Recombination results in new strains and species that are sometimes hard to classify taxonomically because they show the closest sequence homology with different viruses, depending on which genomic region is used for comparison (Rojas et al., 2005). New virus recombinants may acquire an ability to infect new hosts (Monci et al., 2002).

In Cuba, the begomovirus *Tomato mottle Taino virus* infects potatoes in the field and is tuber-transmitted (Cordero et al., 2003). *Potato yellow mosaic virus* (PYMV), another begomovirus, causes yellowing and leaf malformation in potatoes in Venezuela. It is probably synonymous to tomato yellow mosaic virus described in the Caribbean region (Morales et al., 2001; Fauquet et al., 2003; Urbino et al., 2004). Potato deforming mosaic virus is a tentative begomovirus that was described in potato in Brazil. It is probably synonymous to tomato yellow vein streak virus (Ribeiro et al., 2005). Solanum apical leaf curl virus is a tentative begomovirus infecting potatoes in Peru (Table 28.1).

Until now, there are relatively few reports on begomoviruses infecting potato crops. For example, no begomovirus was detected in potatoes in a recent survey in Nicaragua where tomato production has been devastated by begomoviruses (Ala-Poikela et al., 2005). The native wild vegetation acts as a reservoir of the begomoviruses, but specific hosts are unknown. Some of the begomoviruses in tomato infect also potatoes and are tuber-transmitted when experimentally inoculated (Rojas et al., 2005).

Taken together, recent data suggest that begomoviruses may threaten potato crops in warm climates where potato growing is currently expanding. It would be important to understand why begomoviruses have not yet caused epidemics in potato, despite an abundance of the same whiteflies in potato crops and the adjacent begomovirus-infected tomato cultivations in potato growing areas, for example in Nicaragua. Understanding the reasons might help to plan precautionary actions to prevent begomoviruses from becoming a problem in potato.

28.3.3.2 RNA viruses

Potato yellow vein is a disease known for a long time in cultivated potatoes in the Andean region, but the disease has only recently become widespread and devastating (Salazar et al., 2000). *Potato yellow vein virus* (PYVV) is the causal agent and was identified as a member of family Closteroviridae, most closely resembling members of genus *Crinivirus* (Salazar et al., 2000; Livieratos et al., 2004). PYVV is transmitted by the whitefly *Trialeurodes vaporariorum* Westwood. It is also tuber-transmitted, albeit not very efficiently. Many infected tubers produce symptomless plants (Salazar et al., 2000).

Tomato spotted wilt virus has the largest host range of known plant viruses and occurs in all continents (EPPO, 1997). It is transmitted by western flower thrips, *Frankliniella occidentalis*, and infects many solanaceous hosts, including potato. Introduction of this vector species to new areas has led to problems also with TSWV. The virus is one of the most important potato pathogens in Australia (R.A.C. Jones, personal communication) and infects potatoes also in many other parts of the world (e.g. Al Shahwan et al., 1997; Abad et al., 2005). Other tospoviruses have also recently begun to occur in potato crops in warm climates (C. Jeffries, personal communication).

28.4 ECONOMIC IMPACT OF PVY

PVY is the type member of genus *Potyvirus*, the largest group of plant viruses containing 128 approved and 89 tentative species (Rajamäki et al., 2004; Fauquet et al., 2005). Many potato viruses, including PVY, are transmitted non-persistently by many aphid species. In non-persistent transmission, only a few seconds to minutes are needed for virus acquisition and inoculation. Therefore, the most important vectors for these viruses in potato crops are often those that do not colonize potato. The aphids only land to probe, reject the leaf as a site for feeding and move further and hence disseminate the viruses efficiently. Aphicides do not efficiently prevent transmission but may increase the restless behaviour of aphids and thereby virus transmission. Therefore, the spread of PVY is difficult to control in the field unless resistant cultivars (Fig. 28.3) are grown (De Bokx and van der Want, 1987). Repeated treatment of foliage with mineral oil reduces virus transmission but may also reduce plant growth (Kurppa and Hassi, 1989). Helpful guidelines for control of aphid-transmitted viruses in potatoes have been provided by Evans (2000).

PVY is considered to be economically the most harmful virus in cultivated potatoes, a position that used to be held by PLRV. Relative importance of PLRV has decreased owing to the statutory production schemes for healthy seed and breeding for PLRV-resistant varieties. In the northernmost areas of Europe, PVY is the only aphid-transmitted virus still causing yield losses in potato crops (Fig. 28.4) (Tapio et al., 1997). Also in South America, the origin of cultivated potato, relative importance of PLRV has decreased as compared with PVY (L.F. Salazar, personal communication). At the same time, new strains of PVY have evolved or spread to new geographical areas (Beczner et al., 1984; McDonald and Kristijansson, 1993; Le Romancer et al., 1994; Piche et al., 2004). Australia is exceptional in that PVY has never caused significant problems to potato production there (Bald and Norris, 1945; R.A.C. Jones, personal communication).

The evidence for viruses being an important cause of cultivar degeneration (Salaman, 1921) gave a shove to systematic breeding for new, virus-resistant potato varieties (Davidson, 1980; Ross, 1986; Bradshaw and Mackay, 1994). Crosses made by J. Wilson in St. Andrews, Scotland, in 1912–1919 resulted in cv. Crusader that marked the beginning of a new generation of PLRV- and PVY-resistant potato varieties bred using genes from wild potato species (Davidson, 1980). PLRV and PVY (and also PVX) have been ever since the main targets because of their high ability to reduce yields (Andersen, 1956; Arenz and Hunnius, 1959; Kolbe, 1981), but combining resistance to them with the desired agronomic traits has been challenging (Davidson, 1980; Ross, 1986).

Only a few recent studies provide quantitative data on the yield losses caused by viruses in potato and even fewer on losses specifically caused by PVY in the field. Perhaps, once the negative impact of viruses was generally accepted, it became gradually less important to do trials for demonstrating the yield losses. However, there are also well-known difficulties in quantifying yield losses caused by viruses in the field (Waterworth and Hadidi, 1998). To fill in the gap of recent data on the economic impact of PVY, 30 potato virologists and seed potato experts were contacted in different parts of the

Fig. 28.3. Differing hypersensitive responses to PVY^O visualized by development of necrotic lesions on inoculated leaves of four potato cultivars. (Photo: J. Valkonen.)

Fig. 28.4. Chlorosis, reduced growth and leaf-drop symptoms following seed-borne (secondary) infection with PVY^O in potato cv. Sieglinde. (Photo: J. Valkonen.)

world. The information obtained is quoted in the following sections, with permission. In a few countries, no detailed data were available, but as such, even this was significant to know because it revealed that more information is needed about the local impact of PVY in some areas.

28.4.1 Mixed infections

Mixed virus infections are common in the field, which complicates the estimation of the proportion of losses caused by PVY (Lihnell et al., 1975; Kolbe, 1981), because PVY and other potyviruses may synergize unrelated viruses in co-infected plants. Consequently, symptoms and yield losses caused by the other viruses may be enhanced (Anandalakshmi et al., 1998; Moissiard and Voinnet, 2004). Interactions between PVY and PVX in tobacco are the classical examples of viral synergism (Rochow and Ross, 1955; Vance, 1991) and have been reported also in potato plants (Stone, 1936; Bald, 1945).

28.4.2 Impact of primary and secondary infection

Growth reduction, yield losses and the proportion of progeny tubers that becomes infected following the current season (primary) infection with PVY depend on the time of infection in relation to the growth stage. The impacts of cultivar and virus strain are discussed in Section 28.4.4.

Multiplication and subsequent translocation from the initially infected cell are the steps of viral infection cycle needed for significant symptom development (Fig. 28.3). Virus transport occurs according to the source–sink partitioning of photoassimilates and results in systemic infection of the sink tissues (Haywood et al., 2002). Consequently, symptoms such as mosaic, rugosity, epinasty and malformation are induced in developing leaves and leaf tissues. The symptoms are a manifestation of the disturbed physiology and development, which potyviruses cause by interfering with host gene expression (Wang and Maule, 1995; Teycheney and Tepfer, 2001; Kasschau et al., 2003). Young plants are most susceptible and severely affected in their growth and development. Indeed, infection at an early growth stage results in the biggest yield losses. During maturation, plants become gradually less susceptible to infection, less suitable for virus propagation and less efficient in transporting the virus from leaves to tubers (Gibson, 1991). The phenomenon has been coined as mature plant resistance (Beemster, 1987). Thus, delayed infection in relation to the developmental stage reduces yield losses.

The biggest losses are experienced when the crop is grown from PVY-infected seed (secondary infection; Fig. 28.4) (Kurppa and Hassi, 1989; Anonymous, 1999; Locke et al., 1999; Hane and Hamm, 1999). Combined data from 31 cultivars indicated that the yield obtained using PVY-infected seed (incidence of infected tubers not reported) was 50–85% less than the yield of non–PVY-infected plants in a pot trial (Arenz and Hunnius, 1959). If all seed tubers of cvs Bintje, Matilda and Sabina were infected with PVY^{O} or PVY^{N}, yield reduction was 29–59% depending on the cultivar–virus strain combination (Kurppa and Hassi, 1989). Cultivars Shepody and Russet Norkotah show only mild symptoms when infected with PVY; however, plants grown from PVY-infected tubers suffered approximately 30–50 and 45–50% yield reduction, respectively, depending on the trial and year (PVY strain was not specified). Marketable yield was reduced by approximately 55–80 and 65%, respectively (Hane and Hamm, 1999).

In the field, a low PVY incidence does not always result in yield reduction because the poor growth of infected plants may be compensated for by the neighbouring healthy plants that gain from the additional space (Kurppa and Hassi, 1989; Anonymous, 1999). Hence, in Sweden, it is considered that yield reduction (%) caused by PVY $= 0.5 \times (\beta - 10)$, where β is the percentage of PVY-infected plants in the crop (H. Bång, personal communication). This agrees well with the estimated 10–15% yield reduction that would be expected if the incidence of PVY-infected seed tubers was 30% in Spain (J. Legorburu, personal communication). It also agrees with the negligible yield loss with 10–20% PVY-infected seed tubers in Finland (Kurppa and Hassi, 1989). However, no yield compensation was observed in a study carried out in North America where PVY was found to spread quickly to the neighbouring virus-free plants. The neighbouring plants that were infected also suffered from yield reduction, albeit to a much lesser extent than the plants that grew from PVY-infected seed (Hane and Hamm, 1999). With a high incidence of secondary infection (poor-quality seed), no compensation is expected.

Taken all aforementioned variables to consideration, PVY is estimated to account for 45% of the losses because of diseases in potato production in the UK (Anonymous, 1999). Many authors have independently arrived at an estimated annual loss of £30–40 million there (Anonymous, 1999; I. Barker, personal communication).

28.4.3 Costs to seed production

Owing to the devastating effect of secondary infections on yield, there is a threshold for incidence of PVY (or 'severe viruses' including PVY) allowed in certified seed. The threshold value is set according to the anticipated yield loss caused by PVY under the local conditions. Infection pressure and possibilities to control the spread of PVY also need to be considered. For example, the naturally low virus infection pressure in the seed production zone close to the polar circle in Finland allows setting the limit to 2% PVY-infected tubers in the highest class (class A) of certified seed. Despite the stringent threshold value, less than 1% of the seed crops are usually rejected due to virus infections (statistics of the Finnish Seed Testing Institute; H. Kortemaa, personal communication). On the contrary, in the Czech Republic, 5–20% of the seed potato crops are rejected because of a too high incidence of viruses, PVY being the predominant one (P. Dědič, personal communication). If the crop cannot be certified for seed, it may be used for ware potato or other commercial purposes, depending on the market situation. However, the price depends on the purpose of use and is the highest for seed potatoes. Therefore, to avoid rejection, seed crops need to be inspected and symptomatic plants rogued during the growing season. Symptomless infections in some cultivars, especially by PVY^N, hamper this practice. Haulm killing may also be needed to decrease virus translocation from foliage to tubers (Beemster, 1987). The costs related to virus-control measures (roguing, mineral oil sprays and haulm killing) may be €300 per hectare in France (Y. Le Hingrat, personal communication). Following harvest, tuber lots must be virus-tested and certified in accordance with official regulations, causing additional costs to the seed grower.

Taken together, certification of seed crops for sufficient freedom from PVY (and the other important potyvirus, PVA, and other severe viruses) causes significant costs in the potato production chain before the actual main crop is grown for other purposes. Although these costly effectors are necessary for maintaining high productivity of potato crops, they are a burden to the economic profitability and accumulate in product prices.

28.4.4 Yield loss depending on cultivar resistance and PVY strain

Potato varieties expressing extreme resistance (ER) that inhibits multiplication of all PVY strains (Barker and Harrison, 1984; Ross, 1986), or strain-specific hypersensitive resistance (HR) (Fig. 28.3) that restricts infection to the inoculated leaves (Jones, 1990), are protected against growth reduction and transmission of PVY to progeny tubers (Ross, 1986; Valkonen, 1994; Anonymous, 1999). Cultivars that develop mature plant resistance quickly are less prone to suffer from a high incidence of virus-infected progeny tubers (Beemster, 1987; Gibson, 1991). Therefore, production of qualified, certified seed is more feasible from resistant than susceptible varieties.

However, in many varieties and under elevated temperatures, HR may fail to restrict the movement of PVY. Consequently, systemic infection with strains PVY^{O}, PVY^{C} or PVY^{Z}, recognized by different resistance genes (Jones, 1990), will cause extensive necrosis, leaf-drop symptoms (Fig. 28.4) and even plant death (Dykstra, 1939; Bald and Norris, 1945; Valkonen et al., 1998). Potato cultivars do not express HR to the tobacco veinal necrosis strain of PVY (PVY^{N}), which, in general, causes mild or no foliar symptoms in potato plants. However, growth reduction and yield losses caused by PVY^{N} may be substantial, similar to those observed with other PVY strains that induce more severe foliar symptoms (Kurppa and Hassi, 1989; Anonymous, 1999).

Variants of PVY^{N} designated as PVY^{NTN} can induce superficial necrotic symptoms on tubers, which causes quality losses (Beczner et al., 1984). The symptoms are different from those of two soil-borne viruses: the corky ringspot symptoms caused by the nematode-transmitted *Tobacco rattle virus* (TRV) and the spraing symptoms caused by *Potato mop-top virus* (PMTV) that is transmitted by the powdery scab pathogen, *Spongospora subterranea* (Jeffries, 1998). Symptom development following PVY^{NTN} infection depends on many factors, all of which are not yet fully elucidated (Le Romancer et al., 1994; Browning et al., 2004). The biological and genetic variability of PVY has been recently reviewed (Glais et al., 2002a), but until now, it is unclear which genomic region of PVY is responsible for the tuber necrosis symptoms. Thus, specific detection of PVY^{NTN} using molecular methods is not yet possible.

Certain cultivars typically respond with tuber necrosis to infection with PVY^{NTN} (Browning et al., 2004). Tolerance to PVY^{NTN} has become an important criterion in the selection of new cultivars. Sensitive varieties are avoided (Y. Hingrat, personal communication). For example, cv. Yukon Gold was abandoned from use in France because of its high sensitivity to PVY^{NTN} (C. Kerlan, personal communication). The physiological basis of tuber necrosis waits to be elucidated. A host mechanism that resembles interferon-mediated virus defence in mammals recognizes dsRNA (replicative form of virus) and triggers death of virus-infected cells in plants. However, potyviruses can suppress this mechanism with their helicase (CI) protein, and, consequently, infected plants develop only mild symptoms (Bilgin et al., 2003). PVY^N typically causes only mild symptoms in potato plants. Therefore, one possibility is that PVY^{NTN} contains mutations in CI and cannot inhibit host defence and cell death, unlike PVY^N. The second possibility is that potatoes recognize dsRNA and defend against viruses also by a mechanism called RNA silencing (Voinnet, 2005). RNA silencing targets viral RNA to degradation without disturbing cellular physiology. However, viruses suppress RNA silencing with specific proteins, such as the helper component proteinase (HCpro) of potyviruses (Brigneti et al., 1998). The viral silencing suppressors sometimes induce necrosis when over-expressed or expressed from a heterologous viral genome in plants (Yelina et al., 2002; Kreuze et al., 2005). Indeed, tobacco veinal necrosis symptoms caused by PVY^N are attributable to HCpro-mediated physiological effects (Tribodet et al., 2005). It is therefore not excluded that HCpro is the virulence factor of PVY^{NTN} causing tuber necrosis symptoms. The third possibility is that the necrotic symptoms are associated with an HR-like response controlled by as yet unknown resistance-like genes recognizing PVY^{NTN}. This would not be unexpected because many potato cultivars currently grown are offspring from inter-specific hybrids (Davidson, 1980; Ross, 1986; Bradshaw and Mackay, 1994) including wild potato species that respond with HR to PVY^N (Singh et al., 1994; Valkonen, 1997).

The unexpected occurrence of PVY^{NTN} in imported potato seed in northern Italy in 1997 illustrated the economic losses caused by this virus strain (Tomassoli et al., 1998; L. Tomassoli and A. Sonnino, personal communication). The crop was grown for chips. Yield was reduced by 28% (6000 tonnes) because of the secondary virus infection (percentage infected seed tubers unknown). Increased waste during processing resulted in an additional 5% loss of the yield. Owing to poor quality, 17% of the yield was processed to flour or flakes instead of chips. In total, economic losses were estimated to be €1.4 million.

28.4.5 Aetiology and evolutionary perspectives

PVY is hard to eradicate because potatoes are also widely grown in home gardens. Subsistence crops may act as reservoirs of PVY in many production areas. However, the most important initial PVY sources for vectors are the PVY-infected potato plants within the crop. These plants may originate from infected seed tubers or the volunteers remaining in the field from the previous crop. Other solanaceous crops and weeds infectible with PVY are potential reservoirs (Glais et al., 2002a). Transmission of PVY from tobacco crops to potato and *vice versa* may be a significant problem in some areas (Glais et al., 2002a; M. Chrzanowska, personal communication). However, usually, the PVY isolates infecting, for example tomato and pepper, show reduced virulence in potato (Gebre Selassie et al., 1985; Marte et al., 1991).

Exchange of PVY isolates between host species and genotypes is likely to increase genetic and biological variability of PVY, because host-driven selection (host adaptation) is one of the major forces shaping virus populations (García-Arenal et al., 2001). Mutations occur at a high rate during RNA virus replication, which results in a population of nearly identical genotypes, the so-called quasispecies (Smith et al., 1997). The viral genotypes most fit in the host will become predominant over time (Drake and Holland, 1999;

García-Arenal et al., 2001). Recombination between nearly identical viral mutants and genotypes of a viral quasispecies may result in new virus strains that cause distinctly different symptoms, as shown with PVA (Paalme et al., 2004). This is also evidenced with the PVY variant 'Wilga' found in the field (Glais et al., 2002b) and the artificial chimeras of PVY^N and PVY^O (Tribodet et al., 2005).

Clonal propagation of potato by tubers and tuber transmissibility of viruses allow viruses to remain in the same potato plant for years, and viral quasispecies may be genetically stabilized. However, the use of virus-certified seed enforces the more typical potyvirus infection cycle involving aphid-transmission from plant to plant. Exchange of PVY isolates between potato cultivars (genotypes) exposes the virus to host selection, which is accelerated by the introduction of new cultivars.

Taken together, properly executed potato production reduces the prevalence of PVY, which is crucial for economical profitability, but it also includes elements that may increase the variability of PVY. Evolution of PVY in cultivated potato could be largely prevented by breeding and using potato cultivars that carry genes for ER and inhibit PVY multiplication in the initially infected cells (Barker and Harrison, 1984; Ross, 1986; Jones, 1990; Valkonen, 2002).

28.5 INFECTIOUS cDNA CLONES OF POTATO VIRUSES AND THEIR USE AS BIOTECHNOLOGICAL TOOLS

Potato viruses are an economical burden and a risk factor in potato production. However, they also possess some potential as biotechnological tools. They can be used to express heterologous proteins in plants, for example for pharmaceutical purposes. They can also be used to reveal functions of plant genes based on a reverse-genetics approach, which is achieved by virus-induced gene silencing (VIGS) (see chapter 30, Millam, this volume). Thus, potato viruses can be considered as a valuable natural resource for certain purposes.

28.5.1 Use of potato viruses as gene vectors

PVX was the first potato virus tagged with marker genes. A bacterial β-glucuronidase (β-GUS) gene and an *Aquorea victoria* (jellyfish) gene for green fluorescent protein (GFP) (Chapman et al., 1992; Baulcombe et al., 1995) were inserted in the PVX genome to study the viral infection process. Similarly, PLRV (Nurkiyanova et al., 2000), PVA (Ivanov et al., 2003; Rajamäki et al., 2005) and PMTV (Zamyatnin et al., 2004) have been tagged with GFP to monitor the viral infection cycle non-invasively. Hence, cells and tissues infected with the GFP-tagged virus can be detected and monitored under hand-held UV light or using fluorescence microscopy during the course of infection (Roberts et al., 1997). Sub-cellular localization of the virus or specific viral proteins can be detected with confocal microscopy (Zamyatnin et al., 2004).

Expression of non-viral proteins from viral genomes demonstrates the possibility of using potato viruses as biotechnological tools. For example, production of biopharmaceutical proteins in plants reduces production costs and allows utilization of a eukaryotic cellular environment free from human pathogens (Gleba et al., 2004). Biopharmaceuticals can be produced in transgenic plants, but utilization of a viral vector allows much faster production (a few days to weeks) and is versatile because production can be carried out in different species and cultivars, depending on the needs.

PVX has been used for many applications, for example production of human epidermal growth factor (Wirth et al., 2004) and a partial lactoferrin protein (Li et al., 2004). The foreign protein can be expressed on the surface of the PVX particle, which allows harvesting the recombinant protein in high yields using virus purification protocols (Santa Cruz et al., 1996). PVA has been utilized to express human catechol-*O*-methyltransferase (S-COMT) in plants, yielding 0.9% of S-COMT in the total leaf proteins (Kelloniemi et al., 2006).

Vector viruses can be used in transgenic plants to remove selectable marker genes that are unnecessary after plant regeneration and sometimes risky for environmental reasons. This is achieved using the Cre–*lox* recombination system. Expression of the Cre recombinase from a PVX vector in transgenic plants catalysed recombination between the repeated *lox* sequences, which removed the *bar* marker gene inserted between the *lox* sites (Kopertekh et al., 2004).

PVX-based vectors are used in studies on RNA silencing (e.g. Brigneti et al., 1998; Thomas et al., 2001; Kreuze et al., 2005) and functional genomics (Lu et al., 2003). VIGS is used for the inhibition of gene expression, which allows elucidating functions of the genes. PVX suits for this purpose in potato (Faivre-Rampant et al., 2004). Infectious cDNA of TRV is used for similar purposes (Ratcliff et al., 2001). Using the TRV vector, silencing of endogenous potato genes, including several disease resistance genes, was achieved (Brigneti et al., 2004).

One of the future applications of potato viruses and PSTVd sequences may be their utilization in the regulation of plant gene expression, owing to the ability of certain viral proteins to suppress or interfere with specific stages of RNA silencing (Voinnet, 2005) and the ability of specific PSTVd sequences to direct RNA silencing against host genes (Wang et al., 2004).

28.5.2 Studies on viral infection cycle using infectious cDNAs of potato viruses

The primary reason to prepare infectious cDNA clones of viruses is to construct chimeric, point-mutated and/or marker gene-tagged variants that can be used to critically show which functional roles are played by the different genomic regions and viral proteins during the infection cycle. The knowledge on these molecular details will enable combating the infection cycle at different stages using natural or engineered virus resistance and may also turn out to be useful for control of virus transmission.

Preparation of an infectious cDNA of new viruses is no trivial task, despite the experience gained on many unrelated viruses over the past 20 years. A significant invention was made by Johansen (1996) who inserted a plant intron to the viral cDNA, to interrupt an open-reading frame and to alleviate problems during cloning in bacterial cells. This strategy is now used for cloning large viral cDNAs.

28.5.2.1 Potato virus Y

The important questions on avirulence factors, symptom determinants and virus–host interactions have recently become possible to be resolved in PVY from which an infectious viral cDNA is now available (Moury et al., 2004). Many attempts with limited success were made by many researchers in the past. The construct that finally marked a break through was relatively cumbersome in use because it was difficult to clone in the bacterial vectors (Jakab et al., 1997). However, it was used to elucidate the viral determinants associated with veinal necrosis caused by PVY^N in tobacco plants. This was accomplished using chimeric viruses made of PVY^N and PVY^O, which revealed that amino acids 400 and 419 of HCpro were critical for the induction of vein necrosis in tobacco (Tribodet et al., 2005).

This intron insertion strategy worked to solve the problems with cloning of PVY cDNA. Three introns were required for overcoming cloning problems in *Escherichia coli* (Moury et al., 2004).

Two infectious PVY cDNAs were made by Moury et al. (2004), one from a PVY isolate obtained from tomato and another one from pepper. Subsequently, mutagenesis of the cDNAs identified the viral genome-linked protein (VPg) as an avirulence determinant corresponding to the recessive PVY resistance genes *pot-1* and *pvr2* in *Lycopersicon hirsutum* and pepper (*Capsicum annuum*), respectively (Moury et al., 2004). No studies on potato plants have yet been published using the infectious PVY cDNA.

28.5.2.2 Potato virus X

Potato virus X was the first potato virus, of which an infectious cDNA was made (Chapman et al., 1992). The infectious cDNA was instrumental in identifying the PVX avirulence genes corresponding to the potato resistance genes *Nx, Nb* and *Rx*. The CP of PVX elicits *Nx* and *Rx* (Goulden and Baulcombe, 1993; Santa Cruz and Baulcombe, 1995), whereas the 25K movement protein triggers *Nb* (Malcuit et al., 1999). The data provided a molecular explanation from the viral side to the gene-for-gene resistance conferred by these resistance genes (Cockerham, 1970) and also to the PVX strain group concept (Cockerham, 1955).

The non–protein-encoding regions of PVX control viral replication, which was revealed using mutated infectious PVX cDNAs (Kim and Hemenway, 1996; Pillai-Nair et al., 2003). Cell-to-cell movement of PVX requires co-ordinated activities of three viral proteins encoded by the so-called triple gene block (TGB), as shown with infectious chimera of PVX (Morozov et al., 1999; Yang et al., 2000; Tamai and Meshi, 2001; Schepetilnikov et al., 2005).

PVX is contact-transmitted between plants and has no vector. However, only a few amino acid substitutions in PVX CP would be needed for aphid transmissibility from plants also infected with PVY. The CP DAG motif is known for its essential role in transmission of potyvirus particles by aphids (Atreya et al., 1991). However, for particles to be transmitted, the potyviral HCpro has to be acquired before the virions (Govier and Kassanis, 1974), which materializes in PVY-infected cells. The N-proximal end of PVX CP was substituted for that of *Potato aucuba mosaic virus* (PAMV), a potexvirus which contains a DAG amino acid motif. Consequently, the chimeric PVX (Baulcombe et al.,

1993), like non-engineered PAMV (Kassanis and Govier, 1971), was aphid transmissible if aphids were first fed on a PVY-infected plant.

28.5.2.3 Potato virus A

Mutagenesis of the infectious cDNA of PVA (Puurand et al., 1996) confirmed that the DAG motif in CP is necessary for aphid transmissibility (Andrejeva et al., 1999). The VPg of PVA regulates accumulation and movement of PVA in plants, but the amino acids that are significant differ depending on the host species (Rajamäki and Valkonen, 1999, 2002). Several PVA proteins have co-ordinated activities during viral propagation and movement (Andrejeva et al., 1999; Hämäläinen et al., 2000; Paalme et al., 2004). On the contrary, the small break 6-kDa protein (6K2) of PVA controls viral long distance movement and symptom induction functions independently (Spetz and Valkonen, 2004). All protein-encoding and non-coding regions of PVA contain essential and non-essential sites for virus propagation, which was revealed by the analysis of 1125 mutants of the infectious PVA cDNA (Kekarainen et al., 2002).

28.5.2.4 Potato leafroll virus

Infectious cDNAs of PLRV are available (Nurkiyanova et al., 2000; Kawchuk et al., 2002). However, relatively little has been reported about transmission or virus–host interactions of PLRV using them. Recently, Lee et al. (2005) showed that a loop protruding from the surface of PLRV CP controls many viral functions, such as virus assembly, long-distance movement in cultivated potato (cv. Russet Burbank) and indicator plants and movement through the aphid vector.

Spread of PLRV and other members of genus *Polerovirus* in the field is dependent on aphids that transmit PLRV in a persistent manner (i.e. for their lifetime) (Taliansky et al., 2003; Stevens et al., 2005). Vector aphids are specifically attracted to PLRV-infected potato plants, probably because of volatiles that are induced following PLRV infection. In contrast, aphids are not attracted to plants infected with PVY or PVX (Eigenbrode et al., 2002). It would be worthwhile to resolve, using the infectious cDNA of PLRV, which PLRV proteins are responsible for induction of the aphid attractant. It would help identification of the host factors involved and their removal or suppression using methods of molecular breeding.

28.5.2.5 Potato mop-top virus

PMTV is a soil-borne virus that induces necrotic symptoms (spraing) in potato tubers, causes heavy yield losses in northern Europe, especially in Scandinavia, and is very difficult to control (Sandgren et al., 2002). An infectious cDNA made of the tripartite genome of PMTV has been used to show that PMTV can spread cell-to-cell and systemically in plants without CP, virions or even the CP-encoding RNA2 (Savenkov et al., 2003). The TGB proteins 2 and 3 of PMTV show co-ordinated functions, as they mediate the subcellular localization and cell-to-cell movement of TGB protein 1. These results were validated using the infectious PMTV cDNA from which a GFP-tagged TGB protein 1 was expressed in the infection foci (Zamyatnin et al., 2004). Hence, experimental artefacts sometimes experienced with over-expressed viral proteins could be excluded. The necrosis-inducing capacity of the PMTV 8K protein in *Nicotiana* spp. (Lukhovitskaya et al., 2005) suggests that this protein may be a virulence factor involved in the induction of necrotic symptoms also in potato tubers.

28.5.2.6 Potato spindle tuber viroid

The viroid PSTVd was the first potato pathogen for which the whole genomic sequence was cloned and determined (Gross et al., 1978; Dickson et al., 1979). The genome of PSTVd consists of a rod-shaped, circular ssRNA molecule of 356–360 nucleotides that contains characteristic secondary structures with functional importance (Flores et al., 1999). Viroids are the smallest known plant pathogens, but still many of their functions remain unknown. One of their most puzzling features is that they do not encode proteins but mediate their functions through direct viroid RNA–cellular factor interactions. Mutational analysis of PSTVd (cDNA) has revealed several genomic regions important for replication and symptom induction (Qi and Ding, 2003; Wang et al., 2004; and references therein).

The use of a PSTVd cDNA allows the initiation of infection with a single viroid genotype. Thus, the factors affecting viroid variability during replication can be addressed critically. Studies on influence of the host and temperature on variability of PSTVd showed that the viroid quasispecies is dynamic: higher temperatures increased the number of mutants, and a new population of viroid quasispecies developed following transmission of the old population to a new host species (Matoušek et al., 2004).

Potato spindle tuber viroid is transmitted in true potato seed by contact, besides the infected seed tubers. However, there is evidence indicating that PSTVd RNA can get encapsidated in the particles of velvet tobacco mottle virus (genus *Sobemovirus*) (Francki et al., 1986) and PLRV (Querci et al., 1997) in plants co-infected with PSTVd and the virus. Heteroencapsidation of PSTVd in virus particles would allow the transmission of PSTVd with the virus vectors, which in turn has important epidemiological implications. It would seem worthwhile to re-address these questions as infectious cDNAs of PLRV are currently available and determinants for virion assembly are being elucidated, as discussed above.

Taken together, the application of molecular methods on research of potato viruses using infectious viral cDNAs continues to reveal many important details and mechanisms, which can be used to enhance virus control and exploitation of the viruses in biotechnological applications.

ACKNOWLEDGEMENTS

I am grateful to colleagues for sharing their unpublished information. In addition to those mentioned in the text, I thank R. Sigvald, T. Munthe, J.P. Palohuhta, G. Clover, J. Schubert, H. v.d. Heuvel, F. Ponz, N. Katis, S. Horváth, A. Tsahkna, R.P. Singh, S. Slack, G. Thompson, B. Anoshenko, G. Yakovleva, Z. Polgar, F. Tanaka and M. Sato for their efforts in collecting and forwarding information. The studies on potato and viruses in the author's laboratory are financially supported by the Academy of Finland, Ministry of Agriculture and Forestry, National Technology Agency of Finland (TEKES), Finnish Private Enterprises and the Swedish Research Council for Environment, Agricultural Sciences and Spatial Planning (Formas).

REFERENCES

Abad J.A., J.W. Moyer, G.G. Kennedy, G.A. Holmes and M.A. Cubeta, 2005, *Am. J. Potato Res.* 82, 255.

Adams M.J., J.F. Antoniew and C.M. Fauquet, 2005, *Arch. Virol.* 150, 459.

Al Shahwan I.M., O.A. Abdalla and M.A. Al Saleh, 1997, *Plant Pathol.* 46, 91.

Ala-Poikela M., E. Svensson, A. Rojas, T. Horko, L. Paulin, J.P.T. Valkonen and A. Kvarnheden, 2005, *Plant Pathol.* 54, 448.

Allander T., M.T. Tammi, M. Eriksson, A Bjerkner, A. Tiveljung-Lindell and B. Andersson, 2005, *Proc. Natl. Acad. Sci. U.S.A.* 102, 12891.

Anandalakshmi R., G.J. Pruss, X. Ge, R. Marathe, A.C. Mallory, T.H. Smith and V.B. Vance, 1998, *Proc. Natl. Acad. Sci. U.S.A.* 95, 1379.

Andersen T., 1956, *Tidskr. Planteavl* 59, 571.

Anderson P.K., A.A. Cunningham, N.G. Patel, F.J. Morales, P.R. Epstein and P. Daszak, 2004, *Trends Ecol. Evol.* 19, 535.

Andrejeva J., Ü. Puurand, A. Merits, F. Rabenstein, L. Järvekülg and J.P.T. Valkonen, 1999, *J. Gen. Virol.* 80, 1133.

Anonymous, 1999, *The Effect of Primary and Secondary Infection of PVY Strains on Potato Yield.* Final Project Report, CSG 15, MAFF Project Code HP0128, London, U.K.

Arenz, B. and W. Hunnius, 1959, *Bayerische Landw. Jahrbuch* 36, 163.

Atreya P.L., C.D. Atreya and T.P. Pirone, 1991, *Proc. Natl. Acad. Sci. U.S.A.* 88, 7887.

Bald J.G., 1945, *Phytopathology* 35, 585.

Bald J.G. and D.O. Norris, 1945, *Phytopathology* 35, 591.

Barker H. and B.D. Harrison, 1984, *Ann. Appl. Biol.* 105, 539.

Baulcombe D.C., S. Chapman and S. Santa Cruz, 1995, *Plant J.* 7, 1045.

Baulcombe D.C., J. Lloyd, I.N. Manossopoulos, I.M. Roberts and B.D. Harrison, 1993, *J. Gen. Virol.* 74, 1245.

Beczner L., H. Horvath, I. Romhanyi and H. Foster, 1984, *Potato Res.* 27, 339.

Beemster A.B.R., 1987, In: J.A. De Bokx and J.P.H. van der Want (eds), *Viruses of Potatoes and Seed-Potato Production*, p. 116. Pudoc, Wageningen, The Netherlands.

Bilgin D.D., Y. Liu, M. Schiff and S.P. Dinesh-Kumar, 2003, *Dev. Cell* 4, 651.

Boonham N., K. Walsh, P. Smith, K. Madagan, I. Graham and I. Barker, 2003, *J. Virol. Methods* 108, 181.

Bradshaw J.E. and G.R. Mackay, 1994, In: J.E. Bradshaw and G.R. Mackay (eds), *Potato Genetics*, p. 467. CAB International, Wallingford, U.K.

Brigneti G., O. Voinnet, W.X. Li, L.H. Ji, S.W. Ding and D.C. Baulcombe, 1998, *EMBO J.* 17, 6739.

Brigneti G., A.M. Martin-Hernandez, H.L. Jin, J. Chen, D.C. Baulcombe, B. Baker and J.D.G Jones, 2004, *Plant J.* 39, 264.

Browning I., K. Charlet, M. Chrzanowska, P. Dědič, C. Kerlan, A. Kryszczuk, J. Schubert, C. Varveri, A. Werkman and I. Wolf, 2004, *What is PVY^{NTN}? The Reaction of Potato Cultivars to Inoculation with a Range of PVY Isolates*, p. 151. 12th EAPR Virology Section Meeting, Rennes, France.

Bystricka D., O. Lenz, I. Mraz, P. Dědič and M. Sip, 2003, *Acta Virol.* 47, 41.

Chapman S., T. Kavanagh and D. Baulcombe, 1992, *Plant J.* 2, 549.

Cockerham G., 1955, *Strains of Potato Virus X*, p. 89. Proc. 2nd Conference on Potato Virus Diseases. Lisse-Wageningen, The Netherlands.

Cockerham G., 1970, *Heredity* 25, 309.

Cordero M., P.L. Ramos, L. Hernandez, A.I. Fernandez, A.L. Echemendia, R. Peral, G. Gonzalez, D. Gracia, S. Valdes, A. Estevez and K. Hernandez, 2003, *Phytoparasitica* 31, 478.

Davidson T.M.W., 1980, *Breeding for Resistance to Virus Disease of the Potato (Solanum tuberosum) at the Scottish Plant Breeding Station*, p. 100. 59th Ann. Rep. Scottish Plant Breed. Inst. Dundee, UK.
De Bokx J.A. and J.P.H. van der Want (eds), 1987, *Viruses of Potatoes and Seed-Potato Production*. Pudoc, Wageningen, The Netherlands.
Dickson E., H.D. Robertson, C.L. Niblett, R.K. Horst and M. Zaitlin, 1979, *Nature* 277, 60.
Drake J.W. and J.J. Holland, 1999, *Proc. Natl. Acad. Sci. U.S.A.* 96, 13910.
Dykstra T.P., 1939, *Phytopathology* 29, 40.
Eigenbrode S.D., H.J. Ding, P. Shiel and P.H. Berger, 2002, *Proc. R. Soc. Lond. B Biol. Sci.* 269, 455.
EPPO, 1997, In: *Quarantine Pests for Europe*, 2nd ed. European and Mediterranean Plant Protection Organisation (EPPO). Oxford, UK.
Evans A., 2000, *Aphids and Aphid-Borne Virus Disease in Potatoes*. Technical Note T492, The Scottish Agricultural College, Edinburgh, UK.
Faivre-Rampant O., E.M. Gilroy, K. Hrubikova, I. Hein, S. Millam, G.J. Loake, P. Birch, M. Taylor and C. Lacomme, 2004, *Plant Physiol.* 134, 1308.
Fauquet C.M., D.M. Bisaro, R.W. Briddon, J.K. Brown, B.D. Harrison, E.P. Rybicki, D.C. Stenger and J. Stanley, 2003, *Arch. Virol.* 148, 405.
Fauquet C.M., M.A. Mayo, J. Maniloff, U. Desselberger and L.A. Ball, 2005, *Virus Taxonomy. Eighth Report of the International Committee on Taxonomy of Viruses*. Elsevier, San Diego, CA, U.S.A.
Flores R., J.-A. Navarro, M. De La Pena, B. Navarro, S. Ambros and A. Vera, 1999, *Biol. Chem.* 380, 849.
Flores R., C. Hernandez, A.E.M. de Alba, J.A. Daros and F. Di Serio, 2005, *Annu. Rev. Phytopathol.* 43, 117.
Francki R.I.B., M. Zaitlin and P. Palukaitis, 1986, *Virology* 155, 469.
García-Arenal F., A. Fraile and J.M. Malpica, 2001, *Annu. Rev. Phytopathol.* 39, 157.
Gebre Selassie K., G. Marchoux, B. Delecolle and E. Pochard, 1985, *Agronomie* 5, 621.
Gibson R.W., 1991, *Potato Res.* 34, 205.
Glais L., C. Kerlan and C. Robaglia, 2002a, In: J.A. Khan and J. Dijkstra (eds), *Plant Viruses as Molecular Pathogens*, p. 225. Haworth Press, Binghampton, NY, U.S.A.
Glais L., M. Tribodet and C. Kerlan, 2002b, *Arch. Virol.* 147, 363.
Gleba Y., S. Marillonnet and V. Klimyuk, 2004, *Curr. Opin. Plant Biol.* 7, 182.
Goulden M.G. and D.C. Baulcombe, 1993, *Plant Cell* 5, 921.
Govier D.A. and B. Kassanis, 1974, *Virology* 61, 420.
Gross H.J., H. Domdey, C. Lossow, P. Jank, M. Raba, H. Alberty and H.L. Sänger, 1978, *Nature* 273, 203.
Hämäläinen J.H., T. Kekarainen, C. Gebhardt, K.N. Watanabe and J.P.T. Valkonen, 2000, *Mol. Plant Microbe Interact.* 13, 402.
Hane D.C. and P.B. Hamm, 1999, *Plant Dis.* 83, 43.
Hawkes J.G., 1990, *The Potato: Evolution, Biodiversity and Genetic Resources*. Belhaven Press, London, U.K.
Haywood V., F. Kragler and W.J. Lucas, 2002, *Plant Cell* 14, Suppl., 303.
Ivanov K.I., P. Puustinen, R. Gabrenaite, H. Vihinen, L. Rönnstrand, L. Valmu, N. Kalkkinen and K. Mäkinen, 2003, *Plant Cell* 15, 2124.
Jakab G., E. Droz, G. Brigneti, D. Baulcombe and P. Malnoë, 1997, *J. Gen. Virol.* 78, 3141.
Jeffries C.J., 1998, *Potato*. FAO/IPGRI Technical Guidelines for the Safe Movement of Germplasm no. 19. Plant Production and Protection Division of the Food and Agriculture Organization of the United Nations (FAO) and the International Plant Genetic Resources Institute (IPGRI), Rome, Italy.
Johansen E.I., 1996, *Proc. Natl. Acad. Sci. U.S.A.* 93, 12400.
Jones R.A.C., 1982, *Ann. Appl. Biol.* 100, 315.
Jones R.A.C., 1990, *Ann. Appl. Biol.* 117, 93.
Jones R.A.C. and C.E. Fribourg, 1979, *Phytopathology* 69, 446.
Kassanis B. and D.A. Govier, 1971, *J. Gen. Virol.* 13, 221.
Kasschau K.D., Z.X. Xie, E. Allen, C. Llave, E.J. Chapman, K.A. Krizan and J.C. Carrington, 2003, *Dev. Cell* 4, 205.
Kawchuk L., H.M. Jaag, K. Toohey, R. Martin, W. Rohde and D. Prüfer, 2002, *Can. J. Plant Pathol.* 24, 239.
Kekarainen T., H. Savilahti and J.P.T. Valkonen, 2002, *Genome Res.* 12, 584.
Kelloniemi J., K. Mäkinen and J.P.T. Valkonen, 2006, *Biochimie* 88, 505.
Kim K.-H. and C. Hemenway, 1996, *J. Virol.* 70, 5533.
Kolbe W., 1981, *Der Kartoffelbau* 32, 352.

Kopertekh L., G. Jüttner and J. Schiemann, 2004, *Plant Mol. Biol.* 55, 491.
Kreuze J.F., E.I. Savenkov, X. Li, W. Cuellar and J.P.T. Valkonen, 2005, *J. Virol.* 79, 7227.
Kurppa A. and A. Hassi, 1989, *Ann. Agric. Fenn.* 28, 297.
Le Romancer M., C. Kerlan and M. Nedellec, 1994, *Plant Pathol.* 43, 138.
Lee L., I.B. Kaplan, D.R. Ripoll, D. Liang, P. Palukaitis and S.M. Gray, 2005, *J. Virol.* 79, 1207.
Li Y., Y.F. Geng, H.H. Song, G.Y. Zheng, L.D. Huan and B.S. Qiu, 2004, *Biotechnol. Lett.* 26, 953.
Lihnell D., H. Bång, K. Kvist and C. Nilsson, 1975, *Nat. Swed. Inst. Plant Prot. Contr.* 16, 149.
Livieratos I.C., E. Eliasco, G. Müller, R.C.L. Olsthoorn, L.F. Salazar, C.W.A. Pleij and R.H.A. Coutts, 2004, *J. Gen. Virol.* 85, 2065.
Locke T., K.P.A. Wheeler, I. Barker, J. Morris, J.P. North, S.A. Weller and A.N. Phillips, 1999, *Effect of PVY on Yield of UK Potato Cultivars*, p 553. Proc. 14th Triennial Conf. Eur. Assoc. Potato Res., Sorrento, Italy.
Loebenstein G., P.H. Berger, A.A. Brunt and R.H. Lawson (eds), 2001, *Virus and Virus-Like Diseases of Potatoes and Production of Seed Potatoes*. Kluwer Academic Publishers, Dordrecht, The Netherlands.
Lu R., I. Malcuit, P. Moffett, M.T. Ruiz, J. Peart, A.J. Wu, J.P. Rathjen, A. Bendahmane, L. Day and D.C. Baulcombe, 2003, *EMBO J.* 22, 5690.
Lukhovitskaya N.I., N.E. Yelina, A.A. Zamyatnin Jr., M.V. Schepetilnikov, A.G. Solovyev, M. Sandgren, S.Y. Morozov, J.P.T. Valkonen and E.I. Savenkov, 2005, *J. Gen. Virol.* 86, 2879.
Malcuit I., M.R. Marano, T.A. Kavanagh, W. De Jong, A. Forsyth and D.C. Baulcombe, 1999, *Mol. Plant Microbe Interact.* 12, 536.
Marte M., G. Bellezza and A. Polverari, 1991, *Ann. Appl. Biol.* 118, 309.
Matoušek J., L. Otrová, G. Steger, J. Škopek, M. Moors, P. Dědič and D. Riesner, 2004, *Virology* 323, 9.
McDonald J.G. and G.T. Kristijansson, 1993, *Plant Dis.* 77, 87.
Moissiard G. and O. Voinnet, 2004, *Mol. Plant Pathol.* 5, 71.
Monci F., S. Sanchez-Campos, J. Navas-Castillo and E. Moriones, 2002, *Virology* 303, 317.
Morales F.J., R. Lastra, R.C. de Uzcátequi and L. Calvert, 2001, *Arch. Virol.* 146, 2249.
Morozov S.Y., A.G. Solovyev, N.O. Kalinina, O.N. Fedorkin, O.V. Samuilova, J. Schiemann and J.G. Atabekov, 1999, *Virology* 260, 55.
Moury B., C. Morel, E. Johansen, L. Guilbaud, S. Souche, V. Ayme, C. Caranta, A. Palloix and M. Jacquemond, 2004, *Mol. Plant Microbe Interact.* 17, 322.
Nisbet C., I. Butzonitch, M. Colavita, J. Daniels, J. Martin, R. Burns, E. George, V. Mulholland and C.J. Jeffries, 2005, *Characterisation of Potato Rough Dwarf Virus and Potato Virus P*, p. 301. 16th Int. Conf. Eur. Assoc. Potato Res. (EAPR), Bilbao, Spain.
Nurkiyanova K.M., E.V. Ryabov, U. Commandeur, G.H. Duncan, T. Canto, S.M. Gray, M.A. Mayo and M.E. Taliansky, 2000, *J. Gen. Virol.* 81, 617.
Ochoa C.M., 1999, *Las Papas de Sudamérica: Perú*. International Potato Center, Lima, Peru.
Paalme V., E. Gammelgård, L. Järvekülg and J.P.T. Valkonen, 2004, *J. Gen. Virol.* 85, 739.
Piche L.M., R.P. Singh, X. Nie and N.C. Gudmestad, 2004, *Phytopathology* 94, 1368.
Pillai-Nair N., K.H. Kim and C. Hemenway, 2003, *J. Mol. Biol.* 326, 701.
Puurand Ü., J.P.T. Valkonen, K. Mäkinen, F. Rabenstein and M. Saarma, 1996, *Virus Res.* 40, 135.
Qi Y. and B. Ding, 2003, *Plant Cell* 15, 1360.
Querci M., R.A. Owens, I. Bartolini, V. Lazarte and L.F. Salazar, 1997, *J. Gen. Virol.* 78, 1207.
Rajamäki M.-L., J. Kelloniemi, A. Alminaite, T. Kekarainen, F. Rabenstein and J.P.T. Valkonen, 2005, *Virology* 342, 88.
Rajamäki M.L., T. Mäki-Valkama, K. Mäkinen and J.P.T. Valkonen, 2004, In: N.J. Talbot (ed.), *Plant-Pathogen Interactions*, p. 68. Blackwell Publishing, Sheffield, U.K.
Rajamäki M.L. and J.P.T. Valkonen, 1999, *Mol. Plant Microbe Interact.* 12, 1074.
Rajamäki M.L. and J.P.T. Valkonen, 2002, *Mol. Plant Microbe Interact.* 15, 138.
Ratcliff F., A.M. Martin-Hernandez and D.C. Baulcombe, 2001, *Plant J.* 25, 237.
Ribeiro S.G., A.K. Inoue-Nagata, J. Daniels and A.C. Avila, 2005, *Potato Deforming Mosaic Disease in Caused by an Isolate of Tomato Yellow Vein Streak Virus*. New Disease Reports Vol. 12, The British Society for Plant Pathology. Reading, UK.
Roberts A.G., S. Santa Cruz, I.M. Roberts, D.A.M. Prior, R. Turgeon and K.J. Oparka, 1997, *Plant Cell* 9, 1381.
Rochow W.F. and A.F. Ross, 1955, *Virology* 1, 10.
Rojas A., A. Kvarnheden, D. Marcenaro and J.P.T. Valkonen, 2005, *Arch. Virol.* 150, 1281.

Ross H., 1986, *J. Plant Breed.* Suppl. 13, 1.
Salaman R.N., 1921, *Degeneration of Potatoes.* In: Report on the International Potato Conference, pp. 79–91. Royal Horticultural Society, London.
Salazar L.F., 1996, *Potato Viruses and Their Control.* International Potato Center, Lima, Peru.
Salazar L.F., 2005, *Emerging and Re-Emerging Potato Virus Diseases in the Andes*, p. 44. 16th Intern. Conf. Eur. Assoc. Potato Res. (EAPR), Bilbao, Spain.
Salazar L.F., G. Müller, M. Querci, J.L. Zapata and R.A. Owens, 2000, *Ann. Appl. Biol.* 137, 7.
Sandgren M., R.L. Plaisted, K.N. Watanabe, S. Olsson and J.P.T. Valkonen, 2002, *Am. J. Potato Res.* 79, 205.
Santa Cruz S. and D.C. Baulcombe, 1995, *J. Gen. Virol.* 76, 2057.
Santa Cruz S., S. Chapman, A.G. Roberts, I.M. Roberts, D.A.M. Prior and K.J. Oparka, 1996, *Proc. Natl. Acad. Sci. U.S.A.* 93, 6286.
Savenkov E.I., A. Germundsson, A.A. Zamyatnin Jr., M. Sandgren and J.P.T. Valkonen, 2003, *J. Gen. Virol.* 84, 1001.
Schepetilnikov M.V., U. Manske, A.G. Solovyev, A.A. Zamyatnin, J. Schiemann and S.Y. Morozov, 2005, *J. Gen. Virol.* 86, 2379.
Secor G.A., I.M. Lee, K.D. Bottner, V.V. Rivera and N.C. Gudmestad, 2005, *A New Potato Disease Causing Extensive Discoloration of Processed Crisps Associated with a Phytoplasma Infection*, p. 117. 16th Intern. Conf. Eur. Assoc. Potato Res. (EAPR), Bilbao, Spain.
Singh M., R.P. Singh and T.H. Somerville, 1994, *Am. Potato J.* 71, 567.
Smith D.B., J. McAllister, C. Casino and P. Simmonds, 1997, *J. Gen. Virol.* 78, 1511.
Spetz C., A.M. Taboada, S. Darwich, J. Ramsell, L.F. Salazar and J.P.T. Valkonen, 2003, *J. Gen. Virol.* 84, 2565.
Spetz C. and J.P.T. Valkonen, 2004, *Mol. Plant Microbe Interact.* 17, 502.
Stevens M., B. Freeman, H.Y. Liu, E. Herrbach and O. Lemaire, 2005, *Mol. Plant Pathol.* 6, 1.
Stone W.E., 1936, *J. Agric. Res.* 52, 295.
Taliansky M., M.A. Mayo and H. Barker, 2003, *Mol. Plant Pathol.* 4, 81.
Tamai A. and T. Meshi, 2001, *Mol. Plant Microbe Interact.* 14, 1158.
Tapio E., K. Bremer and J.P.T. Valkonen, 1997, *Agr. Food Sci. Finl.* 6, 323.
Teycheney P.Y. and M. Tepfer, 2001, *J. Gen. Virol.* 82, 1239.
Thomas C.L., L. Jones, D.C. Baulcombe and A.J. Maule, 2001, *Plant J.* 25, 417.
Thompson J.R., K.L. Perry and W. De Jong, 2004, *Arch. Virol.* 149, 2141.
Tomassoli L., V. Lumia, C. Cerato and R. Ghedini, 1998, *Plant Dis.* 82, 350.
Tribodet M., L. Glais, C. Kerlan and E. Jacquot, 2005, *J. Gen. Virol.* 86, 2101.
Urbino C., J.E. Polston, C.P. Patte and M.L. Caruana, 2004, *Arch. Virol.* 149, 417.
Valkonen J.P.T., 1994, *Plant Breed.* 112, 1.
Valkonen J.P.T., 1997, *Ann. Appl. Biol.* 130, 91.
Valkonen J.P.T., 2002, In: J.A. Khan and J. Dijkstra (eds), *Plant Viruses as Molecular Pathogens*, p. 367. Haworth Press, Binghampton, NY, USA.
Valkonen J.P.T., E. Pehu and K.N. Watanabe, 1992, *Potato Res.* 35, 401.
Valkonen J.P.T., V.M. Rokka and K.N. Watanabe, 1998, *Phytopathology* 88, 1073.
Vance V.B., 1991, *Virology* 182, 486.
Verhoeven J.T.J., D.E. Lesemann and J.W. Roenhorst, 1996, *Eur. J. Plant Pathol.* 102, 895.
Voinnet O., 2005, *Nat. Rev. Genet.* 6, 206.
Wang D.W. and A.J. Maule, 1995, *Science* 267, 229.
Wang M.B., X.Y. Bian, L.M. Wu, L.X. Liu, N.A. Smith, D. Isenegger, R.M. Wu, C. Masuta, V.B. Vance, J.M. Watson, A. Rezaian, E.S. Dennis and P.M. Waterhouse, 2004, *Proc. Natl. Acad. Sci. U.S.A.* 101, 3275.
Waterworth H.E. and A. Hadidi, 1998, In: *Plant Virus Disease Control*, p. 1. APS Press, St. Paul, MN, USA.
Wirth S., G. Calamante, A. Mentaberry, L. Bussmann, M. Lattanzi, L. Baranao and F. Bravo-Almonacid, 2004, *Mol. Breed.* 13, 23.
Yang Y., B. Ding, D.C. Baulcombe and J. Verchot, 2000, *Mol. Plant Microbe Interact.* 13, 599.

Yelina N.E., E.I. Savenkov, A.G. Solovyev, S.Y. Morozov and J.P.T Valkonen, 2002, *J. Virol.* 76, 12981.
Zamyatnin A.A., A.G. Solovyev, E.I. Savenkov, A. Germundsson, M. Sandgren, J.P.T. Valkonen and S.Y. Morozov, 2004, *Mol. Plant Microbe Interact.* 17, 921.
Zhang D.-Y., P. Willingmann, C. Heinze, G. Adam, M. Pfunder, B. Frey and J.E. Frey, 2005, *J. Virol. Methods* 123, 101.

Chapter 29

Fungal and Fungus-Like Pathogens of Potato

Aad J. Termorshuizen

Biological Farming Systems Group, Wageningen University, Marijkeweg 22, 6709 PG Wageningen, The Netherlands

There are several fine and recent overviews of fungal and fungus-like pathogens of potato (Stevenson et al., 2001; Mulder and Turkensteen, 2005). In addition, several monographic assays of the major pathogens are available (referred to in this chapter). Here, I intend to provide a concise overview of these pathogens including the correct names, names used as synonyms, major references and a short characterization of each pathogen.

Among the 33 species listed, *Puccinia piettieriana, Phoma andigena, Thecaphora solani* and *Septoria lycopersici* var. *malagutii* are considered A1 and *Synchytrium endobioticum* A2 quarantine organisms according to EPPO (2004). All A1 pathogens are limited to the region of origin of the potato. Among these quarantine organisms, *P. andigena* and *S. endobioticum* are soil-borne pathogens and the other species are air-borne. It is remarkable that only little research has been performed on the A1 quarantine species. Two other species occurring only in South America might receive quarantine status, namely *Aecidium cantense* and *Phoma crystalliniformis*, although current regulations for the already quarantined species probably will prevent spread of these pathogens too.

A minority of eight (24%) of the described species is air-borne. Of these, five species inflict great actual (*Alternaria solani* and *Phytophthora infestans*) or potential (*P. piettieriana, S. lycopersici* var. *malagutii* and *T. solani*) damage. Among the 25 species listed as soil-borne, the majority has a worldwide distribution (only two *Phoma* species, *P. andigena* and *P. crystalliniformis*, are restricted to South America). Of the 25 soil-borne species with a worldwide distribution, nine are restricted to solanaceous hosts (*Fusarium coeruleum, Fusarium sambucinum, Phoma foveata, Polyscytalum pustulans, Rhizoctonia solani* anastomosis group-3 (AG-3), *Spongospora subterranea* and *S. endobioticum*), and these species have probably been spread by infested soil or infected seed tubers. No correlation seems to exist between the magnitude of actual damage incited by a pathogen and its host specificity. So, pathogens that probably have not co-evolved with Solanum spp. may inflict great damage on potato (e.g. *Phytophthora erythroseptica* and *Verticillium dahliae*) or not (e.g. *Helicobasidium brebissonii* and *Sclerotinia sclerotiorum*), whereas other pathogens that are specifically linked to solanaceous hosts may inflict great damage (e.g. *P. infestans*) or not (e.g. *Passalora concors*).

For the correct names of pathogens, I mainly followed the publications of Boerema et al. (1993; published in different issues in *Netherlands Journal of Plant Pathology* as

Potato Biology and Biotechnology: Advances and Perspectives
D. Vreugdenhil (Editor)

indicated in the text and collected in one monograph). Under the heading 'synonyms', those names are mentioned that may still be found in present-day literature, irrespective of whether they are used as real synonyms or whether their status is taxonomically illegitimate.

In the overview given below, references to taxonomic literature are made. Frequently used literature is listed in Table 29.1.

Several pathogens that are able to infect nearly any weakened tissue, for example due to extraordinary weather circumstances, severe nutrient deficiencies or other diseases, are excluded. These include *Alternaria alternata* (Fr. : Fr.) Keissl. (infecting shoots and tubers; distribution worldwide), *Botrytis cinerea* Pers. : Fr. [Teleomorph *Botryotinia fuckeliana* (de Bary) Whetzel] (grey mould; infecting all tissues; distribution worldwide), *Choanephora cucurbitarum* (Berk. and Rav.) Thaxter (shoot rot; distribution worldwide), *Phoma huancayensis* Turkenst. (probably a saprophyte on potato leaves; South America and New Zealand; Boerema et al., 2004: 88), *Phoma subherbarum* Gruyter et al. (probably a saprophyte occurring on potato leaves and other plant species; North and South America; Boerema et al., 2004: 113), *Pleospora herbarum* (Fr. : Fr.) Rabenh. (anam. *Stemphylium herbarum* E.G. Simmons) (leaf lesions; distribution worldwide), *Rhizopus stolonifer* (Fr.) Lind (soft rot of potato tubers; distribution worldwide) and *Ulocladium atrum* Preuss [≡ *Stemphylium atrum* (Preuss) Sacc.] (leaf lesions; South America). Also excluded are some anomalous diseases, occurring only under extraordinary circumstances and even then inciting only limited damage, such as *Armillaria* spp. The species dealt with in this chapter are listed in Table 29.2 and described below in alphabetical order.

1 *Aecidium cantense* Arthur (Common disease names: Deforming potato rust; Peruvian rust)

ORIGINAL DESCRIPTION in *Estac. Exp. Agric. Soc. Nac. Agrar. Lima, Peru, Bol.* 2 (1929) (as *A. 'cantensis'*, an error to be corrected according to Art. 32.5 of the International Code of Botanical Nomenclature). CLASSIFICATION: Uredinales, Basidiomycota, Fungi (further

Table 29.1 Frequently cited literature.

Abbreviation	Title	Reference
Agrios	*Plant Pathology*, 5th ed.	Agrios, 2004
Comp. APS	*Compendium of Potato Diseases*, 2nd ed.	Stevenson et al., 2001
CMI Descr.	*CMI Description of Fungi*	–
EHPD	*European Handbook of Plant Diseases*	Smith et al., 1988
M&T	*Potato Diseases*	Mulder and Turkensteen, 2005

Table 29.2 Fungal and fungus-like pathogens of potato dealt with in this chapter.

Scientific name	Entry	Common disease name	Remarks
Aecidium cantense	1	Deforming potato rust and Peruvian rust	
Alternaria solani	2	Early blight	
Athelia rolfsii	26	Southern blight and stem rot	Teleomorph of *Sclerotium rolfsii*
Colletotrichum coccodes	3	Black dot	
Erysiphe cichoracearum	4	Powdery mildew	
Fusarium coeruleum	5	Fusarium dry rot	
Fusarium eumartii	6	Potato wilt	
Fusarium javanicum	7	Jelly end rot and potato wilt	
Fusarium sambucinum	8	Dry rot	
Galactomyces candidus	9	Rubbery rot	Teleomorph of *Geotrichum candidum*
Geotrichum candidum	9	Rubbery rot	
Helicobasidium brebissonii	10	Violet root rot	
Helminthosporium solani	11	Silver scurf	
Macrophomina phaseolina	12	Charcoal rot	
Mycovellosiella concors	13	Yellow leaf blotch and Cercospora leaf blotch	See *Passalora concors*
Passalora concors	13	Yellow leaf blotch and Cercospora leaf blotch	
Phoma andigena	14	Potato black blight and leaf spot and Phoma leaf spot	
Phoma crystalliniformis	15	Potato and tomato black blight and leaf spot	
Phoma eupyrena	16	Dry rot and Damping-off	
Phoma exigua var. *exigua*	17	Thumbmark rot	
Phoma foveata	18	Gangrene	
Phytophthora erythroseptica	19	Pink rot	
Phytophthora infestans	20	Late blight	
Polyscytalum pustulans	21	Tuber skin spot and skin spot	
Puccinia pittieriana	22	Common potato rust	

(*Continued*)

Table 29.2 (*Continued*)

Scientific name	Entry	Common disease name	Remarks
Pythium spp.	23	Leak and watery wound rot	
Rhizoctonia crocorum	10	Violet root rot	Anamorph of *H. brebissonii*
Rhizoctonia solani	24	Black scurf and Rhizoctonia canker	
Rosellinia necatrix	25	Rosellinia wilt and Rosellinia black rot	
Sclerotinia sclerotiorum	26	White mould; Sclerotinia rot and Sclerotinia stalk rot	
Sclerotium rolfsii var. *rolfsii*	27	Southern blight and stem rot	
Septoria lycopersici var. *malagutii*	28	Septoria leaf spot and annular leaf spot	
Spongospora subterranea f. sp. *subterranea*	29	Powdery scab	
Synchytrium endobioticum	30	Wart disease and Wart	
Thanatephorus cucumeris	24	Black scurf and Rhizoctonia canker	Anamorph of *R. solani*
Thecaphora solani	31	Potato smut and Thecaphora smut	
Verticillium alboatrum	32	Potato early dying and Verticillium wilt	
Verticillium dahliae	33	Potato early dying and Verticillium wilt	

taxonomic position not clear); forms aecia containing aeciospores. For distinguishing from the only other rust species occurring on potatoes, see *Puccinia pittieriana*. MAJOR LITERATURE: Comp. APS: 21. SELECTED ILLUSTRATIONS: Comp. APS: pl. 28.

Hosts and distribution. Potatoes, although research may reveal other solanaceous hosts. Host specialization unknown. Reported only from Peru and more recently from northern Argentina at high altitudes (2300–3200 m).

Epidemiology. Established in areas where potatoes are grown the year round. It is unknown to what extent the fungus can persist in the absence of a host.

Management. Unknown.

2 *Alternaria solani* Sorauer (Common disease name: Early blight)

ORIGINAL DESCRIPTION in *Z. PflKrankh.* 6 (1896): 6. SYNONYMS [Boerema and Verhoeven in *Neth. J. Plant Pathol.* 82 (1976): 197]: *A. solani* Sorauer, *Macrosporium solani* Ellis & G. Martin, *Alternaria porri* (Ell.) Cif. f. sp. *solani* (Ellis & G. Martin) Neergaard 1945. For a discussion on the correct citation, see Simmons (2000: 48). CLASSIFICATION: Anamorphic *Lewia*, Pleosporaceae, Pleosporales, Ascomycota, Fungi. Sexual stage unknown; forms hyphae and conidia. MAJOR LITERATURE: CMI Descr. 475 (1975), Comp. APS: 22, EHPD: 368, M&T: 19, Chelkowski and Visconti (1992), Rotem (1994; monograph on *Alternaria*), Simmons (2000: 36). SELECTED ILLUSTRATIONS: Agrios: 454, Comp. APS: pl. 31–34, M&T 19, 20.

Hosts and distribution. Solanaceous hosts; other hosts such as *Brassica* may become infected but are probably of limited importance. No host specialization. Considerable variation in genetic composition and virulence exists but could not be related to geographic origin of isolates (van der Waals et al., 2004). Distribution worldwide but important mainly in warm and dry regions.

Epidemiology. Surviving in infected plant material as mycelium and conidia (infected seed, crop residues left on the field or solanaceous weeds) and in soil for approximately 1 year as chlamydospores (Basu, 1971). Dispersal mainly by wind or rain splash. Tuber infections occur post-harvest.

Management. Use of tolerant cultivars; avoid tuber wounding; application of fungicides.

3 *Colletotrichum coccodes* (Wallr.) S. Hughes (Common disease name: Black dot)

ORIGINAL DESCRIPTION in *Fl. Crypt. Germ.* 2 (1833): 265 as *Chaetomium coccodes* by Wallroth. SYNONYMS [Boerema and Verhoeven in *Neth. J. Plant Pathol.* 82 (1976): 75]: *Colletotrichum atramentarium* (Berk. & Broome) Taubenh. 1916, *Vermicularia atramentaria* Berk. & Broome 1850. CLASSIFICATION: Anamorphic fungi, anamorphs of *Glomerella*, Ascomycota, Fungi. Forms hyphae, sclerotia and conidia in acervuli. MAJOR LITERATURE: CMI Descr. 131 (1967), Comp. APS: 16, Bailey and Jeger (1992), EHPD: 330, Jeger et al. (1996; review). SELECTED ILLUSTRATIONS: Comp. APS: pl. 24, M&T 13.

Hosts and distribution. Mainly on Solanaceae, Cucurbitaceae and Leguminosae. Distribution worldwide. Large variation in pathogenicity between isolates, but no 'formal' infraspecific classification has been proposed.

Epidemiology. The pathogen is an opportunist that incites problems only under suboptimal growing conditions or in the presence of other pathogens. The sclerotia can survive for a few years in soil. Potato seed can be infected.

Management. The pathogen can be avoided by creating optimal growing conditions and using disease-free planting material.

4 *Erysiphe cichoracearum* DC. (Common disease name: Powdery mildew)

ORIGINAL DESCRIPTION in *Fl. fr.*, 3rd ed., 2 (1805): 274 as *Erysiphe cichoracearum* by de Candolle. SYNONYMS: see Braun (1995: 130). CLASSIFICATION: Erysiphaceae,

Erysiphales, Ascomycota, Fungi; forms hyphae, conidia and rarely ascomata. MAJOR LITERATURE: Comp. APS: 34, EHPD: 256, Braun (1995: 130). SELECTED ILLUSTRATIONS: Comp. APS: 35, pl. 57, 58.

Hosts and distribution. Wide host range (approximately 230 hosts in nine families). Some host specialization occurs: for example strains from Compositae do not infect other hosts, but those on cucurbits do not specialize. Strains of potato tend to specialize as well (Hopkins, 1956).

Epidemiology. The disease develops strongly under dry conditions, at moderate to high temperatures and high levels of available nitrogen. If ascomata are formed, the fungus may survive a limited period of time in soil.

Management. The disease of minor importance only. Application of fungicides.

5 *Fusarium coeruleum* Lib. ex Sacc. (Common disease name: Fusarium dry rot)

ORIGINAL DESCRIPTION in *Syll. Fung.* 4 (1886): 705 (short description). SYNONYMS: *Fusarium caeruleum* (Lib.) ex Sacc., *F. coeruleum* (Lib.) ex Sacc., *Fusarium solani* (Mart.) Sacc. var. *coeruleum* (Lib. ex Sacc) C. Booth. CLASSIFICATION: Anamorphic Nectriaceae, Hypocreales, Ascomycota, Fungi; forms chlamydospores. MAJOR LITERATURE: Comp. APS: 23, Comp. Soil Fungi: 499, EHPD: 282. SELECTED ILLUSTRATIONS: Comp. APS: 24 and pl. 35, Gerlach and Nirenberg (1982: 371; also monograph).

Hosts and distribution. Reported to occur specifically on potato in central and northern Europe and North America.

Epidemiology. Dry rot is often followed by infection by secondary pathogens such as *Erwinia* spp., but *F. coeruleum* can also invade lesions caused by other pathogens. Mechanical damage at harvesting increases damage. Most proliferous development at 15–20 °C and high humidity.

Management. Harvest tubers from dead stolons, optimize storage conditions, reduce mechanical damage in combination with usage of cultivars that are tolerant to mechanical damage, reduce tuber handling at low temperatures. Application of fungicides.

Note: *Fusarium solani* is an aggregate of very similar but genetically and pathologically distinct species. *Fusarium coeruleum* is often confused with purple colonies of *Fusarium oxysporum*.

6 *Fusarium eumartii* Carpenter (Common disease name: Potato wilt)

ORIGINAL DESCRIPTION in *J. Agric. Res.* 5 (1915): 204. SYNONYMS: *Fusarium solani* var. *eumartii* (Carp.) Wollenw. Teleomorph *Haematonectria haematococca* (Berk. & Broome) Samuels & Nirenberg in *Stud. Mycol.* 42 (1999): 135 [originally described as *Nectria haematococca* Berk. & Broome in *J. Linn. Soc., Bot.* 14 (1873): 116; synonyms

Hypomyces haematococcus (Berk. & Broome) Wollenw., *Nectria citri* Henn., *Nectria asperata* Rehm]. CLASSIFICATION: Anamorphic Nectriaceae, Hypocreales, Ascomycota, Fungi; forms chlamydospores. MAJOR LITERATURE: Comp. Soil Fungi: 499, EHPD: 282, Gerlach and Nirenberg (1982: 375; also monograph).

Hosts and distribution. Distribution worldwide but limited to areas with a hot climate, including greenhouses. Mainly or only of importance on potato but infecting also tropical crops.

Epidemiology. Insufficiently known.

Management. Insufficiently known.

Note: The taxonomy requires more study.

7 *Fusarium javanicum* Koord. (Common disease names: Jelly end rot; Potato wilt)

ORIGINAL DESCRIPTION in *Verh. K. Akad. Wet.*, 2nd sect., 13 (1907): 247 by Koorders. CLASSIFICATION: Anamorphic Nectriaceae, Hypocreales, Ascomycota, Fungi; forms chlamydospores. The purported teleomorph connection with *Haematonectria ipomoeae* was not confirmed (Rossman et al., 1999). MAJOR LITERATURE: Comp. Soil Fungi: 499, EHPD: 282, Gerlach and Nirenberg (1982: 361; also monograph).

Hosts and distribution. Distribution worldwide but limited to areas with a hot climate, including greenhouses. Very wide host range.

Epidemiology. Insufficiently known.

Management. Insufficiently known.

Note: The taxonomy requires more study.

8 *Fusarium sambucinum* Fuckel (Common disease name: Dry rot)

ORIGINAL DESCRIPTION in *Symb. Mycol.* 167: 1869. SYNONYMS: *F. sambucinum* Fuckel f. 3 Wollenw., *F. sambucinum* f. 3 Raillo, *Fusarium trichothecioides* Wollenw. apud Jam. and Wollenw., *F. sambucinum* var. *trichothecioides* (Wollenw.) Bilai, *Fusarium sulphureum* Schlecht., *Fusarium tuberivorum* Wilcox and Link. Teleomorph *Gibberella pulicaris* (Fr. : Fr.) Sacc. [originally described as *Sphaeria cyanogena* Desm. in *Ann. Sci. Nat. (Bot.) III*, 10 (1848): 352]; synonym *Fusarium cyanogena* (Desm.) Sacc. CLASSIFICATION: Anamorphic Nectriaceae, Hypocreales, Ascomycota, Fungi; forms chlamydospores. MAJOR LITERATURE: Comp. APS: 23, Hering and Nirenberg (1995), Nirenberg (1995). SELECTED ILLUSTRATIONS: Comp. APS: pl. 36–38, Gerlach and Nirenberg (1982: 211; also monograph).

Hosts and distribution. Mainly occurring in Europe and North America. See further *F. coeruleum*.

Epidemiology. See *F. coeruleum*.

Management. See *F. coeruleum*.

9 Geotrichum candidum Link : Fr. (Common disease name: Rubbery rot)

ORIGINAL DESCRIPTION in *Mag. Naturf. Freunde, Berlin* 9 (1809): 17. SYNONYMS: see de Hoog & Smith (2004). Teleomorph *Galactomyces candidus* de Hoog & Smith in *Stud. Mycol.* 50 (2004): 503–505. CLASSIFICATION: Anamorphic *Galactomyces*, Dipodascaceae, Saccharomycetales, Ascomycota, Fungi; forms ascospores in asci, hyphae and arthroconidia. MAJOR LITERATURE: Comp. Soil Fungi: 348, EHPD: 250, M&T: 54, de Hoog and Smith (2004; taxonomy). SELECTED ILLUSTRATIONS: M&T: 54.

Hosts and distribution. Worldwide distributed soil-borne fungus. Usually a saprophyte, occurring opportunistically on for example diverse fruits and potato tubers. Virulent and avirulent strains are known.

Epidemiology. Develops well under wet and warm (30 °C) conditions.

Management. Usually not needed.

10 *Helicobasidium brebissonii* (Desm.) Donk (Common disease name: Violet root rot)

ORIGINAL DESCRIPTION in *Pl. Cryptog. N. France*, 1st ed., 14 (651) (1834) as *Protonema brebissonii* by Desmazières. SYNONYMS [Boerema et al. in *Neth. J. Plant Pathol.* 93 (1987): 195]: *Helicobasidium purpureum* Pat. 1885. Anamorph: *Thanatophytum crocorum* (Pers. : Fr.) Nees [see Taxon 47 (1998): 725]; originally described as *Rhizoctonia crococum* D.C. in de Candolle and de Lamarck, *Fl. fr.*, 3rd ed., 5 (1815): 111; synonyms [Boerema et al. in *Neth. J. Plant Pathol.* 93 (1987): 195]: *Rhizoctonia crocorum* (Pers. : Fr.) D.C., *Rhizoctonia violacea* Tul. and C. Tul. 1851, *Sclerotium crocorum* Pers. : Fr. 1801. CLASSIFICATION: Platygloeaceae, Platygloeales, Basidiomycota, Fungi; forms hyphae and sclerotia. MAJOR LITERATURE: EHPD: 504, M&T: 63. SELECTED ILLUSTRATIONS: M&T: 63.

Hosts and distribution. Worldwide distributed soil-borne pathogen with a wide host range. The disease is mostly noted on tuber crops. Severe outbreaks are rare and unpredictable.

Epidemiology. The pathogen may well develop on many kinds of weeds.

Management. Unknown.

Note: This is the earliest pathogen known, which was studied (by Duhamel) experimentally [see Zadoks in *Meded. Landbouwhogesch. Wageningen* 81 (1981): 1].

11 *Helminthosporium solani* Dur. & Mont. (Common disease name: Silver scurf)

ORIGINAL DESCRIPTION in *Fl. d'Algérie, Crypt.* 1 (Sect. 4): 356 as *H. solani* by Durieu de Maisonneuve and Montagne. SYNONYMS [Boerema and Verhoeven in *Neth. J. Plant Pathol.* 82 (1976): 201]: *Dematium atrovirens* Harz 1871, *Helminthosporium atrovirens* (Harz) E.W. Mason & S. Hughes apud Hughes 1953, *Sphondylocladium atrovirens* (Harz) Harz ex Sacc. 1886. CLASSIFICATION: Anamorphic Massarinaceae, Pleosporales, Ascomycota, Fungi; sexual stage unknown; forms hyphae and multicellular conidia. MAJOR LITERATURE: CMI Descr. 166 (1968), Comp. APS: 40, EHPD: 407, M&T: 55, Errampalli et al. in *Plant Pathol.* 50 (2001): 141 (review), Jeger et al. (1996; review). SELECTED ILLUSTRATIONS: Comp. APS: 69–71, M&T: 55, 56.

Hosts and distribution. Very common in Europe and North America, and probably also elsewhere, occurring only on potatoes. Tuber quality rather than yield is affected.

Epidemiology. Inoculum originates from infected seed tubers; only the skin is infected, mainly post-harvest. The pathogen spreads especially fast during warm (15–20 °C) and humid storage.

Management. Use of clean seed tubers. Slightly infected seed tubers usually result in relatively heavy infections, as these nearly healthy seed tubers can produce much inoculum. Cool and dry storage of tubers.

12 *Macrophomina phaseolina* (Tassi) Goid. (Common disease name: Charcoal rot)

ORIGINAL DESCRIPTION in *Boll. R. Orto Bot. Siena* 4 (1901): 9 as *Macrophoma phaseolina* by Tassi. SYNONYMS [Boerema and Hamers in *Neth. J. Plant Pathol.* 95 (1989) Suppl. 3: 13]: *Macrophoma phaseoli* Maubl. 1927, *Macrophomina phaseoli* (Maubl.) Ashby 1927, *Rhizoctonia bataticola* (Taubenh.) Butler 1925, *Sclerotium bataticola* Taubenh. 1913, *Tiarosporella phaseoli* (Maubl.) van der Aa 1977, *Tiarosporella phaseolina* (Tassi) van der Aa 1981. CLASSIFICATION: Anamorphic Ascomycetes, Ascomycota, Fungi; sexual stage unknown; forms dark brown, one-celled conidia in pycnidia and microsclerotia. MAJOR LITERATURE: CMI Descr. 275 (1970), Comp. APS: 19, EHPD 404, M&T: 36, Dhingra and Sinclair 1978. SELECTED ILLUSTRATIONS: M&T: 36.

Hosts and distribution. Extremely wide host range. Limited to countries with a hot climate, mainly tropical countries. Infraspecific groups not known.

Epidemiology. Microsclerotia infect the plant roots under the influence of root exudates, from where it spreads up into the shoot, where it forms numerous new microsclerotia in decaying tissue. The pathogen is strongly favoured by predisposing factors such as a hot, dry periods after periods of good growing conditions. Microsclerotia survive in soil for several years.

Management. Elimination of predisposing factors, especially maintenance of irrigation during periods of hot weather may be an effective management option. Rotation with unsusceptible crops. N-rich soil amendments may suppress the pathogen. Use of disease-free planting material.

13 *Passalora concors* (Casp.) U. Braun & Crous (Common disease names: Yellow leaf blotch; Cercospora leaf blotch)

ORIGINAL DESCRIPTION in *Ber. Verh. K. Preuss. Akad. Wiss. Berl.* (1855): 314 as *Fusisporium concors* by Caspary. SYNONYMS [Boerema et al. in *Neth. J. Plant Pathol.* 93 (1987) Suppl. 1: 9]: *Cercospora concors* (Casp.) Sacc. 1886, *Carex heterosperma* Bres. 1903, *Mycovellosiella concors* (Casp.) Deighton 1974 (invalidly published). CLASSIFICATION: Anamorphic Mycosphaerellaceae, Mycosphaerellales, Ascomycota, Fungi; forms septate hyphae and multi-celled conidia. MAJOR LITERATURE: CMI Descr. 724 (1982), Comp. APS: 19, M&T: 15, Crous and Braun (2003: 134; taxonomy). SELECTED ILLUSTRATIONS: M&T: 15.

Hosts and distribution. Reported from the northern hemisphere from potatoes, in notably Europe and Russia, as well as some parts of India and Africa.

Epidemiology. Prefers cool and humid climate.

Management. Fungicide application; it is likely that cultivars differ in tolerance.

Note: A leaf spot pathogen with lesions constricted by the veins and forming conidia and mycelium underneath the leaves. Of minor importance only but becoming more prevalent according to Mulder and Turkensteen (2005).

14 *Phoma andigena* Turkensteen (Common disease names: Potato black blight and Leaf spot; Phoma leaf spot)

ORIGINAL DESCRIPTION in *Fitopatología* 13 (1978): 67 as *Phoma andina.* This name was an illegitimate homonym because this binomial had already been introduced in 1904 by Saccardo and Sydow in *Ann. Mycol.* 2: 170 for another fungus (Boerema et al., 2004). CLASSIFICATION: Anamorphic Pleosporaceae, Pleosporales, Ascomycota, Fungi; sexual stage unknown; forms pycnidia containing unicellular conidia of two different sizes and chlamydospores. MAJOR LITERATURE: Comp. APS: 32, Boerema et al. (2004): 130, EPPO (2005a). SELECTED ILLUSTRATIONS: Comp. APS: pl. 54 (as *P. andina*), Boerema et al. (2004): 121 (pathogen), EPPO (2005b), M&T: 70.

Hosts and distribution. Restricted to wild and cultivated *Solanum* spp. in Bolivia and Peru (EPPO, 2005a) at altitudes more than 2000 m. It is unclear whether tomato is a host. Considered an A1 quarantine organism by EPPO (2004).

Epidemiology. Survival in the form of pycnidia on infected plant debris, but persistence time unknown. The importance of chlamydospores is unknown. Infection takes place under moist conditions below 15 °C. Only the larger conidia have been shown to be infective. The fungus is splash-dispersed over short distances. Tubers are not infected.

Management. Not well investigated. The application of fungicides and the use of resistant cultivars are recommended (Comp. APS: 47).

15 *Phoma crystalliniformis* (Loer. et al.) Noordel. & Gruyter (Common disease name: Potato and tomato black blight and leaf spot)

ORIGINAL DESCRIPTION in *Fitopatología* 21 (1986): 100 as *P. andina* var. *crystalliniformis* by Loerakker, Navarro, Lobo and Turkensteen. CLASSIFICATION: Anamorphic Pleosporaceae, Pleosporales, Ascomycota, Fungi; sexual stage unknown; forms pycnidia containing unicellular conidia of two different sizes. Formally described as a variety of *P. andigena* but differing in host range (primarily occurring on tomato), morphology (the large-type conidia are larger) and culture characteristics (growing faster) (Boerema et al., 2004). MAJOR LITERATURE: Boerema et al. (2004: 136).

Hosts and distribution. Reported from tomato and *Solanum tuberosum* ssp. *andigenum* in Colombia and Venezuela at altitudes more than 1500 m.

Epidemiology. No details known.

Management. No details known.

16 *Phoma eupyrena* Sacc. (Common disease names: Dry rot; Damping-off)

ORIGINAL DESCRIPTION in *Michelia* 1 (1879): 525 by Saccardo. CLASSIFICATION: Anamorphic Pleosporaceae, Pleosporales, Ascomycota, Fungi; sexual stage unknown; forms pycnidia containing unicellular conidia and chlamydospores. MAJOR LITERATURE: Comp. Soil Fungi: 632, M&T: 37 (as *Phoma eupyrina*), Boerema et al. (2004: 75). SELECTED ILLUSTRATIONS: Comp. Soil Fungi: 633, M&T: 37, Boerema et al. (2004: 66).

Hosts and distribution. Wide-spread (probably cosmopolitan) soil-borne fungus. Formerly considered to be restricted to potato, but currently at least 30 hosts are known (including various monocotyledonous hosts).

Epidemiology. The disease is favoured under wet conditions. The fungus colonizes quickly disinfested soils.

Management. Usually not needed, but the fungus is sensitive to fungicides.

17 *Phoma exigua* Desm. var. exigua (Common disease name: Thumbmark rot)

Note 1: For *P. exigua* var. *foveata*, see *P. foveata*.
Note 2: The common disease name 'gangrene' is reserved for the symptoms inflicted by *P. foveata*.

ORIGINAL DESCRIPTION in *Ann. Sci. Nat. (Bot.) Sér. III* 11 (1849): 282–283 as *P. exigua* Desm. The first notion of a strain of *P. exigua* pathogenic on potato was made by Prillieux & Delacroix in *Bull. Soc. Mycol. Fr.* 6 (1890): 179 (as *Phoma solanicola*). SYNONYMS: The synonymy of this species is very complicated and extensive (>300 synonyms), see Boerema et al. (2004: 240). CLASSIFICATION: Anamorphic Pleosporaceae, Pleosporales, Ascomycota, Fungi; forms hyphae, pycnidia and unicellular and 1–2-septate conidia. MAJOR LITERATURE: Comp. APS: 25, Comp. Soil Fungi: 634, EHPD: 399, M&T: 23, Boerema et al. (2004: 240). SELECTED ILLUSTRATIONS: M&T: 24.

Hosts and distribution. Cosmopolitan, opportunistic plant pathogen isolated from more than 200 different host genera.

Epidemiology. The disease is promoted under wet and cool conditions but may occur especially when wounding occurs during or after harvesting.

Management. Promote wound healing by maintaining the tubers after harvest for 1 week at elevated temperature (~15–18 °C). Avoid tuber wounding. This pathogen is much less important than *P. foveata*.

18 *Phoma foveata* Foister (Common disease name: Gangrene)

ORIGINAL DESCRIPTION in *Trans. Proc. Bot. Soc. Edinb.* 33 (1940): 66–68 (volume dated 1943). SYNONYMS [Boerema et al. (2004: 266)]: *P. exigua* f. sp. *foveata* (Foister) Malc. 1958, *P. exigua* var. *foveata* (Foister) Boerema 1967, *P. solanicola* f. *foveata* (Foister) Malc. 1958. CLASSIFICATION: Anamorphic Pleosporaceae, Pleosporales, Ascomycota, Fungi; sexual stage unknown; forming hyphae, pycnidia, uni-, sometimes bicellular conidia. Formation of chlamydospores and pseudosclerotia is induced by the bacterium *Serratia plymuthica* (Camyon and Gerhardson, 1997). MAJOR LITERATURE: Comp. APS: 57, EHPD: 399, M&T: 23, Boerema et al. (2004: 266). SELECTED ILLUSTRATIONS: M&T: 24.

Hosts and distribution. Occurs wherever potatoes are grown, mostly in areas with a cool and wet climate. Originates from South America, where it causes brown stalk rot of *Chenopodium quinoa*.

Epidemiology. Stems, leaves and tubers may become infected. Harvested tubers have latent infections. Gangrene of tubers develops during tuber storage and is promoted when tubers are wounded and at low temperatures (<4 °C). Generally persists in soil no longer than 2 years.

Management. Pathogen-free soil is more important than disease-free seed tubers. Promote wound healing by maintaining the tubers after harvest for 1 week at elevated temperature

(~15–18 °C). Avoid tuber wounding. Fungicide treatment of harvested tubers to control contamination is possible. Avoid highly susceptible cultivars.

19 *Phytophthora erythroseptica* Pethybr. (Common disease name: Pink rot)

ORIGINAL DESCRIPTION in *Scient. Proc. R. Dubl. Soc. II [New Series]* 13 (35): 547–548 by Pethybridge. CLASSIFICATION: Pythiaceae, Pythiales, Oomycetes, Oomycota, Chromista; forms hyphae (without septa), oospores and zoospores in sporangia. MAJOR LITERATURE: CMI Descr. 593 (1978), Comp. APS: 33, EHPD: 204, M&T: 39, Erwin and Ribeiro (1996; monograph on *Phytophthora*). SELECTED ILLUSTRATIONS: Comp. APS: pl. 55, M&T: 39, 40.

Hosts and distribution. Various hosts including non-solanaceous plants but economically significant only on potatoes. Distribution worldwide.

Epidemiology. May survive several years in the form of oospores. Can be transmitted in apparently healthy planting material. Disease progress is favoured by circumstances that promote soil wetness.

Management. Cultivars vary in tolerance (Peters et al., 2004) but no complete resistance. Improve soil drainage and avoid potato growing on infested soils.

Note: Also other soil-borne *Phytophthora* spp. are able to cause similar disease symptoms (*Phytophthora cryptogea*) Pethybr. and Laff. (very wide host range; distribution worldwide; of minor importance for potatoes), *Phytophthora drechsleri* Tucker (like *P. cryptogea* but preferring higher temperatures), *Phytophthora megasperma* Drechsler (very wide host range; limited to warm countries) and *Phytophthora nicotianae* Breda de Haan (= *Phytophthora parasitica* Dastur) (very wide host range, limited to areas with a hot climate; of minor importance for potatoes).

20 *Phytophthora infestans* (Mont.) de Bary (Common disease name: Late blight)

ORIGINAL DESCRIPTION in *Bull. Soc. Philomath. Paris* (1845): 98 as *Botrytis infestans* by Montagne. SYNONYMS [Boerema et al. in *Neth. J. Plant Pathol.* 93 (1987) Suppl. 1: 197]: *Peronospora infestans* (Mont.) Casp. 1948. CLASSIFICATION: Pythiaceae, Pythiales, Oomycetes, Oomycota, Chromista; forms hyphae (without septa), oospores and zoospores in sporangia. MAJOR LITERATURE: CMI Descr. 838 (1985), Comp. APS: 28, EHPD: 202, M&T: 28, Erwin and Ribeiro (1996; monograph on *Phytophthora*). SELECTED ILLUSTRATIONS: Comp. APS: 28, 29, pl. 43–47, M&T: 28, 29.

Hosts and distribution. Various members of the Solanaceae, most notably potato and tomato. Heterothallic; the A1 mating type has a wide distribution since the 1840s. The A2 mating type has become common in many areas since the 1980s (Hohl and Iselin, 1984). Since then, the sexual spores (oospores) are found, and as a result, the diversity of genotypes has drastically increased and epidemics have become more severe (Smart and Fry, 2001). Distribution worldwide.

Epidemiology. Onset of an epidemic is by spread of sporangia from infected volunteers or uncovered potato heaps. Infection may also occur from soil-borne oospores, but its importance for the onset of disease is unknown. Sporangia may germinate either directly (18–24 °C) or first form zoospores (<18 °C). In either case, free water is necessary for infection. Appearance of new sporangiophores halts under dry weather conditions. Under optimal weather conditions, 5 days are needed from infection till sporulation.

Management. Partial resistance can reduce disease spread. Early planting and cropping of early cultivars may limit disease. Conventional growers apply metalaxyl and dithiocarbamates on the basis of disease forecasting systems. Before harvest, an infected potato crop can be burned to prevent *Phytophthora* infection of tubers.

21 *Polyscytalum pustulans* (Owen & Wakef.) M.B. Ellis (Common disease names: Tuber skin spot; Skin spot)

ORIGINAL DESCRIPTION in *Kew Bull.* (1919): 297 as *Oospora pustulans* by Owen and Wakefield. CLASSIFICATION: Anamorphic Ascomycetes, Ascomycota, Fungi; forms septate hyphae, mainly unseptate conidia and microsclerotia. MAJOR LITERATURE: Comp. APS: 41, EHPD: 408, M&T: 7. SELECTED ILLUSTRATIONS: Comp. APS: 41, M&T: 57.

Hosts and distribution. Only of importance on potato, but infecting other solanaceous hosts, including tomato, as well. Distribution worldwide but important mainly in Europe and Russia. Limited to areas with a cool temperate climate.

Epidemiology. The disease is tuber-borne, but microsclerotia, formed on decaying plant material, may survive in soil up to 7 years.

Management. Use of disease-free planting material; fungicide treatment of planting material; avoidance of humid conditions during tuber storage.

22 *Puccinia pittieriana* Hennings (Common disease name: Common potato rust)

ORIGINAL DESCRIPTION in *Hedwigia* 43 (1904): 147 by Hennings. CLASSIFICATION: Pucciniaceae, Uredinales, Urediniomycetes, Basidiomycota, Fungi; forms teliospores and basidiospores. MAJOR LITERATURE: CMI Descr. 286 (1971), Comp. APS: 20, EHPD: 497, *Annu. Rev. Phytopathol.* 11 (Thurston, 1973): 40, EPPO (2005c). SELECTED ILLUSTRATIONS: Comp. APS: pl. 29, M&T: 69.

Hosts and distribution. Solanaceae, including potato. Also reported from tomato but without heavy damage. Reported only from South America (EPPO, 2005c) at high elevations (2700–4300 m). Considered an A1 quarantine organism by EPPO (2004).

Epidemiology. Teliospores are formed in infected leaf tissue. They germinate at temperatures below 15 °C to give rise to basidiospores at temperatures below 10 °C, which become easily air-borne. As basidiospores are generally more sensitive to desiccation than uredospores (the spore type mostly associated with dispersal in the rusts), transcontinental transport is not expected by this means. The persistence of teliospores is unknown, but they may very well persist for 1 year or so on infected debris, as many other rust fungi do. The fungus does not infect belowground tissue.

Management. Application of carbamate fungicides (Diaz and Echeverria, 1963). Destruction of infected shoot tissue and control of solanaceous weeds.

Note: The only other rust species occurring on potatoes is *A. cantense* (see entry 1), which forms aecia (5–10 mm diameter) on the leaves, whereas *P. pittieriana* forms smaller (3–4 mm diameter) telia.

23 *Pythium* spp. (Common disease names: Leak or Watery wound rot)

Various species have been associated with diseased potato tubers, including *Pythium splendens* Hans Braun (warm areas) and *Pythium ultimum* Trow (temperate areas), but given the complex taxonomy and problems associated with the identification of *Pythium* spp., other species may be involved as well. Classification: Pythiaceae, Pythiales, Oomycetes, Oomycota, Chromista. Forms hyphae (without septa), oospores and sporangia. Many, but not all, species form zoospores. Major literature: Comp. APS: 30, EHPD: 200, M&T: 34, van der Plaats-Niterink (1981; monograph including illustrations). Selected illustrations: Comp. APS: pl. 48–50, M&T: 34, 35.

Hosts and distribution. Distribution worldwide but mostly a rare disease.

Epidemiology. Opportunistic pathogens, infecting only weakened, usually wounded, tubers.

Management. Avoid mechanical injury during tuber handling and avoid humid storage conditions.

24 *Rhizoctonia solani* Kühn, nom. cons. (Common disease names: Black scurf; Rhizoctonia canker)

Original description in *Krankh. Kulturgew.* (1858): 224 as *R. solani* by Kühn. Synonyms [Boerema and Hamers in *Neth. J. Plant Pathol.* 94 (1988): 228]: *Moniliopsis aderholdii* Ruhland 1908, *M. solani* (Kühn) R.T. Moore 1987. Teleomorph *Thanatephorus cucumeris* (Frank) Donk [originally described as *Hypochnus cucumeris* Frank in *Ber. Dt. Bot. Ges.* 1 (1883): 62]; synonyms [Boerema and Hamers in *Neth. J. Plant Pathol.* 94 (1988) Suppl. 1: 228]: *Corticium solani* (Prill. & Delacr.) Bourdot and Galzin 1911, *Corticium vagum* var. *solani* Burt apud Rolfs 1903, *H. solani* Prill. & Delacr. 1891. Classification (*T. cucumeris*): Ceratobasidiaceae, Ceratobasidiales, Basidiomycota, Fungi. Forms hyphae, basidiospores and sclerotia. Major literature: CMI Descr. 406 (1974), Comp. APS: 36, EHPD: 505, M&T: 47, Jeger et al. (1996; review), Sneh et al. (1996; proceedings). Selected illustrations: Comp. APS: pl. 65–67, M&T: 47–49.

Hosts and distribution. *Rhizoctonia solani* has an extremely wide host range. Currently 12 AGs are known, each exhibiting its specific ecology including host range. At least AG-3, -5 and -8 may infect potato, of which AG-3 is the major cause of black scurf and the only AG inciting stem canker. Contrary to most other AGs, AG-3 is fairly limited to potatoes. Vegetative compatibility groups (VCGs) within AG-3 have been reported (Jeger et al., 1996). Cosmopolitic.

Epidemiology. *Rhizoctonia solani* is able to translocate its cytoplasm through its hyphal system, enabling the fungus to grow very fast even under nutrient-deprived circumstances. In addition, the fungus may colonize fresh organic matter. The function of basidiospores that are formed at the stem base is not understood. Arthropods grazing on hyphae of *R. solani* are significant antagonists (Lartey et al., 1994).

Management. Use disease-free planting material. Fresh organic matter should not be applied shortly before planting potato tubers. There exist various selective fungicides against *R. solani* including pencycuron. Separation of the haulms from the tubers before harvesting limits production of sclerotia on the tubers. This method, called 'Green Crop Lifting' (Mulder et al., 1992), may be combined with the application of the hyperparasite *Verticillium biguttatum*, which is effective against *R. solani* (van den Boogert and Luttikholt, 2004).

25 *Rosellinia necatrix* Prill. (Common disease names: Rosellinia wilt; Rosellinia black rot)

ORIGINAL DESCRIPTION in *Bull. Soc. Mycol. Fr.* 20 (1904): 34 by Prillieux. Asexual stage: *Dematophora necatrix* Hartig. CLASSIFICATION: Xylariaceae, Xylariales, Ascomycota, Fungi; forms hyphae, rhizomorphs and sclerotia; in culture, it also forms conidia on synnemata and ascospores in perithecia. MAJOR LITERATURE: Comp. APS: 37, EHPD: 333, M&T: 52. SELECTED ILLUSTRATIONS: Comp. APS: 38, 39, pl. 68, M&T: 52, 53.

Hosts and distribution. No host specialization. Occurs worldwide but mainly in areas with higher temperature and humid conditions.

Epidemiology. May occur severely on cleared forest or pasture.

Management. Use of uninfected planting material; proper drainage.

26 *Sclerotinia sclerotiorum* (Lib.) de Bary (Common disease names: White mould; Sclerotinia rot; Sclerotinia stalk rot)

ORIGINAL DESCRIPTION in *Pl. Cryptog. Ard.* 4 (326) (1837) as *Peziza sclerotiorum* by Libert. SYNONYMS [Boerema et al. in *Neth. J. Plant Pathol.* 93 (1987): 202]: *Whetzelinia sclerotiorum* (Lib.) Korf & Dumont 1972. Asexual stage: *Sclerotium varium* Pers. : Fr. 1801. CLASSIFICATION: Sclerotiniaceae, Helotiales, Ascomycota, Fungi; forms sclerotia and ascospores. MAJOR LITERATURE: CMI Descr. 513 (1976), Comp. APS: 47, EHPD: 443, M&T: 66. SELECTED ILLUSTRATIONS: Comp. APS: pl. 86–90, M&T: 66–68.

Hosts and distribution. Very wide host range, limited to dicotyledonous hosts. Special forms are not known. Infections in potatoes are not common, and the resulting damage is limited. Distribution worldwide.

Epidemiology. Ecologically obligate, unspecialized and soil-borne pathogen. Survives in the soil in the form of black sclerotia that germinate and give rise to an aboveground-formed apothecium. The resulting ascospores become air-borne and may infect host tissue, notably the stems. The disease progresses most under cool and moist conditions.

Management. Heavy attacks rarely occur in potatoes. Soil inoculum build-up can be avoided by proper rotations. Certain crops such as lettuce, beans and chicory can contribute to inoculum build-up significantly. Soil infestations can also be reduced by metam sodium and biological soil disinfestation (Blok et al., 2000). The hyperparasite *Coniothyrium minitans* may provide biological control, and a commercialized product 'Contans' is based on this (de Vrije, 2001).

27. *Sclerotium rolfsii* Sacc. var. *rolfsii* (Common disease names: Southern blight; Stem rot)

ORIGINAL DESCRIPTION in *Ann. Mycol.* 9 (1911): 257 as *Sclerotium rolfsii* by Saccardo. Teleomorph: *Athelia rolfsii* (Curzi) C. Tu & Kimbrough 1978 [originally described as *Corticium rolfsii* Curzi in *Boll. Staz. Patol. Veg. Roma II* 11 (1931): 306]. CLASSIFICATION (*A. rolfsii*): Atheliaceae, Polyporales, Basidiomycota, Fungi; forms hyphae and brown sclerotia; in culture, rarely in nature, also basidiospores. MAJOR LITERATURE: CMI Descr. 410 (1974) (as *C. rolfsii*), Comp. APS: 42, EHPD: 509, M&T: 58. SELECTED ILLUSTRATIONS: Comp. APS: 42, 43, pl. 72, 73, M&T: 58.

Hosts and distribution. Very wide host range. Worldwide occurrence but limited to areas with a hot climate.

Epidemiology. Ecologically obligate and unspecialized plant pathogen. Sclerotia can survive many years in soil.

Management. Management of fields infested with *S. rolfsii* is difficult. Removal of plant residues and weed control; use of disease-free planting material. Also deep-ploughing and soil fumigation have been advised.

28 *Septoria lycopersici* Speg. var. *malagutii* (Common disease names: Septoria leaf spot; Annular leaf spot)

ORIGINAL DESCRIPTION in *An. Soc. Cient. Argent.* 12 (1881): 115 by Spegazzini. The variety *malagutii* was described in *Phytopathologia Mediterranea* 17 (1978): 87 by Ciccarone and Boerema. CLASSIFICATION: Anamorphic *Mycosphaerella*, Mycosphaerellales, Ascomycota, Fungi; sexual stage unknown; forms pycnidia containing unicellular conidia.

MAJOR LITERATURE: CMI Descr. 89 (1966), Comp. APS: 39, EPPO (2005f). SELECTED ILLUSTRATIONS: Comp. APS: 39, EPPO (2005g).

Hosts and distribution. Occurs on cultivated and wild potatoes. Limited to Central and South America. The variety *malagutii* infects mainly, or only, *Solanum* spp. var. *lycopersici*, causes leaf spot on tomato and is more widely spread. Var. *malagutii* is considered an A1 quarantine organism by EPPO (2004).

Epidemiology. The pathogen thrives well at cool temperatures and moist conditions.

Management. Use of resistant cultivars and application of fungicides [Carrera and Orellana in *Fitopatología* 36 (2001): 37].

29 *Spongospora subterranea* (Wallr.) Lagerh. f. sp. *subterranea* (Common disease name: Powdery scab)

ORIGINAL DESCRIPTION in Linnaea, *Halle* 16 (1842): 332 as *Eryisiphe subterranea* by Wallroth (but there erroneously written as *Erysibe subterranea*). CLASSIFICATION: Plasmodiophoraceae, Plasmodiophorales, Plasmodiophoromycota, Protozoa; forms zoospores, cystosori (= spore balls) consisting of cysts (= resting spores) and zoosporangia. MAJOR LITERATURE: CMI Descr. 477 (1975), Comp. APS: 35, EHPD: 245, M&T: 42, Harrison et al. (1997; review), Jeger et al. (1996; review), Karling (1968: 71), Merz et al. (2005; detection). SELECTED ILLUSTRATIONS: Agrios: 407, Comp. APS: 35, pl. 59–65, Karling (1968: 73; microscopic drawings).

Hosts and distribution. There are two formae speciales known of *S. subterranea*: f. sp. *nasturtii* specific to watercress and f. sp. *subterranea* specific to potato. The f. sp. *subterranea* infects a wide range of hosts, but resting spores are formed only in various solanaceous hosts. Economic damage occurs only in potatoes. Probably occurring wherever potatoes are grown in cool and wet regions.

Epidemiology. Resting spores may survive 6–10 years in the absence of a host. Infection is favoured strongly by cool temperatures and wet soil conditions. In addition to causing powdery scab, *S. subterranea* vectors potato mop top virus.

Management. Avoidance by using uninfected potatoes, no transport of infested soil. There is variation between cultivars in susceptibility, but new cultivars are not necessarily tolerant. Complete resistance seems non-existent.

30 *Synchytrium endobioticum* (Schilb.) Perc. (Common disease names: Wart disease; Wart)

ORIGINAL DESCRIPTION in *Ber. Dt. Bot. Ges.* 14 (1896): 36 as *Chrysophlyctis endobiotica* by Schilbersky. SYNONYMS [Boerema and Verhoeven in *Neth. J. Plant Pathol.* 82 (1976): 86]: *Synchytrium solani* Massee 1910. CLASSIFICATION: Synchytriaceae, Chytridiales, Chytridiomycota, Fungi; forms resting spores (= secondary sporangia = winter sporangia), zoospores and summer sporangia. MAJOR LITERATURE: Baayen et al. (2004; in Dutch, with reference list), CMI Descr. 755 (1983), Comp. APS: 46, EHPD: 242, M&T: 64, Karling (1964; monograph on *Synchytrium*). SELECTED ILLUSTRATIONS: Agrios (2004: 434), Comp. APS: pl. 83–85, EPPO (2005e), M&T: 64, 65.

Hosts and distribution. Infects only solanaceous hosts, in particular tuber-forming species, but also tomato. No host specialization. About 40 pathotypes have been distinguished that differentially affect potato cultivars of which pathotypes 1, 2, 6, 8 and 18 are internationally of great significance. Cosmopolitan but significant damage is mainly limited to regions with moderate temperatures and moderate to high rainfall. Considered an A1 quarantine organism by EPPO (2004).

Epidemiology. The persistence of the resting spores in the absence of hosts is extremely long: 10–20 years. In the presence of hosts, zoospores arising from the resting spores infect the plant and form summer sporangia that subsequently form zoospores. This may occur repeatedly in one growing season. For their transport, zoospores depend on free water.

Management. Against pathotype 1, fully resistant cultivars are widely available, and for all other pathotypes, this is not or hardly the case. For these other pathotypes, refraining from cropping potato for several years is necessary. In addition, cropping of agricultural products that bring about transport of soil should also be forbidden.

31 *Thecaphora solani* (Thirum & O'Brien) Mordue (Common disease names: Potato smut; Thecaphora smut)

ORIGINAL DESCRIPTION in *Phytopathology* 34 (1944): 714 by Barras, but without Latin description. Thirum and O'Brien published the taxon under the name *Angiosorus solani* in Sydowia 26 (1972): 201 (published in 1974). SYNONYMS: *Angiosorus solani* (Thirum and O'Brien). CLASSIFICATION: Glomosporiaceae, Ustilaginales, Ustomycetes, Basidiomycota, Fungi; forms smut spores (= ustilospores = ustospores). MAJOR LITERATURE: Andrade et al. in *Phytopathology* 94 (2004): 875, Comp. APS: 43, EPPO (2005d). SELECTED ILLUSTRATIONS: Comp. APS: pl. 74–76, M&T: 69.

Hosts and distribution. Potatoes and other solanaceous hosts in northern regions of South America. Considered an A1 quarantine organism by EPPO (2004).

Epidemiology. A smut converting tubers into masses of smut spores. The smut spores are likely to persist for several years in soil, although research is needed on this. Dispersal only by infected tubers or infested soil.

Management. Use of resistant cultivars, planting uninfected seed, long rotation and elimination of solanaceous weeds (notably *Datura stramonium* has been mentioned).

32 *Verticillium alboatrum* Reinke & Berth. (Common disease names: Potato Early Dying; Verticillium wilt)

Note: *Verticillium dahliae* (entry 33) and *V. alboatrum*, although genetically distinct species, have not been differentiated for some time when *V. dahliae* was designated as *V. alboatrum* 'microsclerotial form'.

ORIGINAL DESCRIPTION in *Unters. Bot. Lab. Univ. Göttingen* 1 (1879): 75 as *V. alboatrum* by Reinke and Berthold. CLASSIFICATION: Anamorphic Ascomycetes, Ascomycota, Fungi, related to *Glomerella*; sexual stage unknown; forms hyaline hyphae, dark hyphae (in contrast to *V. dahliae*), conidia and yeast cells. MAJOR LITERATURE: CMI Descr. 255 (1970), Comp. APS: 62, Comp. Soil Fungi: 830; EHPD: 299, Pegg and Brady (2002; extensive monograph). SELECTED ILLUSTRATIONS: M&T: 62.

Hosts and distribution. Host range not precisely known but much narrower than *V. dahliae*, causing damage mainly in lucerne, hop and tomato. Damage in potatoes is generally less than that caused by *V. dahliae*. Distribution worldwide in temperate zones. A strain highly pathogenic to hop has been reported (Sewell and Wilson, 1984).

Epidemiology. Prefers cooler temperatures than *V. dahliae*. An interaction with root-infecting nematodes has been reported (see *V. dahliae*, entry 33). For a discussion on significance of air-borne inoculum, see Pegg and Brady (2002: 68).

Management. Cultivars of many crops evidently differ in tolerance to the disease but less so in susceptibility. The black mycelium is able to survive a short host-free period. A proper rotation with non-hosts adequately controls the pathogen.

33 *Verticillium dahliae* Kleb. (Common disease names: Potato Early Dying (PED; see note 2); Verticillium wilt)

Note 1: See also under *V. alboatrum* (entry 32).
Note 2: The common disease name Potato Early Dying (PED) is often used when both root infecting nematodes and *V. dahliae* or *V. alboatrum* are involved.
Note 3: *Verticillium longisporum* (Stark) Karapapa et al. (= *V. dahliae* var. *longisporum* Stark) causes wilt exclusively on crucifers and is most likely a hybrid (amphihaploid) of *V. dahliae* × *V. alboatrum* (Barbara and Clewes, 2003; Collins et al., 2003).

ORIGINAL DESCRIPTION in *Mycol. Centbl.* 3 (1913): 66 as *V. dahliae* by Klebahn. SYNONYMS: *Verticillium alboatrum* Reinke & Berth. 1879 'microsclerotial form'. CLASSIFICATION: Anamorphic Ascomycetes, Ascomycota, Fungi, related to *Glomerella*; sexual stage unknown; forms hyphae, microsclerotia, conidia and yeast cells. MAJOR LITERATURE: CMI Descr. 256 (1970), Comp. APS: 62, Comp. Soil Fungi: 836, EHPD: 299, M&T: 60, Jeger et al. (1996; review), Pegg and Brady (2002; extensive monograph), Rowe et al. [*Plant Dis.* 71 (1987): 482; review]. SELECTED ILLUSTRATIONS: Comp. APS: 45, pl. 77–82, M&T: 61.

Hosts and distribution. Very wide host range but monocotyledonous plants (except barley) are immune. Limited host specialization. VCGs have been described from many parts in the world, and there are differences in pathogenicity to potato (Rowe et al., 1997). Some VCGs do not infect potatoes, but VCGs pathogenic to potato are very common. In Europe, two VCGs have been recognized, and both are pathogenic to potato (Hiemstra and Rataj-Guranowska, 2003). Distribution worldwide but absent in regions of high temperatures, because its development is suppressed at temperatures above 28–30 °C.

Epidemiology. Damage because of Verticillium wilt is strongly increased by nematode infection (*Globodera rostochiensis, Meloidogyne* spp. and *Pratylenchus* spp.). Presence of nematodes doubles the number of infections and the degree of colonization because of biochemical rather than mechanical damage (Bowers et al., 1996; and more references in Pegg and Brady, 2002: 85–86). For a discussion on significance of air-borne inoculum, see Pegg and Brady (2002: 68).

Management. There are evident differences in tolerance of cultivars of many crops to the disease but less so in susceptibility. Wide rotation with non-hosts or hosts that do not build up inoculum, such as sugar beets, is necessary. Soil disinfestation with nematicides or metam sodium is effective. Irrigation and fertilization affects disease severity.

ACKNOWLEDGEMENT

I thank W. Gams (Centraalbureau voor Schimmelcultures, Baarn) for comments on an earlier version of this manuscript.

REFERENCES

Agrios G.N., 2004, *Plant Pathology*, 5th ed. Elsevier Academic Press, Amsterdam, The Netherlands, pp. 922.

Baayen R.P., G. Cochius, H. Hendriks, G.C.M. van Leeuwen, J.P. Meffert and F.J.A. Janssen, 2004, *Gewasbescherming* 35, 160.

Bailey J.A. and M.J. Jeger (eds), 1992, *Colletotrichum: Biology, Pathology and Control*. CAB International, Wallingford, U.K., pp. 388.

Barbara D.J. and E. Clewes, 2003, *Molec. Plant Pathol.* 4, 297.

Basu E., 1971, *Phytopathology* 61, 1347.

Blok W.J., J.G. Lamers, A.J. Termorshuizen and G.J. Bollen, 2000, *Phytopathology* 90, 253.

Boerema G.H. et al., 1993, *Check-Lists for Scientific Names of Common Parasitic Fungi. Libri Botanici Vol. 10*. IHW-Verlag, Eching, Germany, pp. 370.

Boerema G.H., J. de Gruyter, M.E. Noordeloos and M.E.C. Hamers, 2004, *Phoma Identification Manual*. CABI Publishing, Wallingford, U.K., pp. 470.

Bowers J.H., S.T. Nameth, R.M. Riedel and R.C. Rowe, 1996, *Phytopathology* 86, 614.

Braun U., 1995, *The Powdery Mildews (Erysiphales) of Europe*. Jena, Germany, pp. 337.

Camyon S. and B. Gerhardson, 1997, *Eur. J. Plant Pathol.* 103, 467.

Chelkowski J. and A. Visconti (eds), 1992, *Alternaria: Biology, Plant Diseases and Metabolites*. Elsevier Academic Press, Amsterdam, The Netherlands, pp. 573.

Collins A., C.A.N. Okoli, A. Morton, D. Parry, S.G. Edwards and D.J. Barbara, 2003, *Phytopathology* 93, 364.

Crous P.W. and U. Braun, 2003, *Mycosphaerella and Its Anamorphs: 1. Names Published in Cercospora and Passalora*. Centraalbureau voor Schimmelcultures, Utrecht, The Netherlands, pp. 571.

de Hoog G.S. and M.Th. Smith, 2004, *Stud. Mycol.* 50, 489.
de Vrije T., N. Antoine, R.M. Buitelaar, S. Bruckner, M. Dissevelt, A. Durand, M. Gerlagh, E.E. Jones, P. Luth, J. Oostra, W.J. Ravensberg, R. Renaud, A. Rinzema, F.J. Weber and J.M. Whipps, 2001, *Appl. Microbiol. Biotechnol.* 56, 58.
Diaz M.J. and J. Echeverria, 1963, *Turrialba* 13, 152.
EPPO, 2004, *EPPO A1 and A2 Lists of Pests Recommended for Regulation as Quarantine Pests.* http://www.eppo.org/QUARANTINE/pm1-02(13).pdf. European and Mediterranean Plant Protection Organization.
EPPO, 2005a, *Data Sheets on Quarantine Pests. Phoma andina*, pp. 2. http://www.eppo.org/QUARANTINE/fungi/Phoma_andigena/PHOMAN_ds.pdf.
EPPO, 2005b, *Phoma Andina. Colour Picture of Symptoms.* http://www.eppo.org/QUARANTINE/fungi/Phoma_andigena/PHOMAN_images.htm.
EPPO, 2005c, *Data Sheets on Quarantine Pests. Puccinia pittieriana.* http://www.eppo.org/QUARANTINE/fungi/Puccinia_pittieriana/PUCCPT_ds.pdf.
EPPO, 2005d, *Data Sheets on Quarantine Pests. Thecaphora solani.* http://www.eppo.org/QUARANTINE/fungi/Thecaphora_solani/THPHSO_ds.pdf.
EPPO, 2005e, *Synchytrium endobioticum. Colour Pictures of Symptoms and Fungus.* http://www.eppo.org/QUARANTINE/fungi/Synchytrium_endobioticum/SYNCEN_images.htm.
EPPO, 2005f, *Data Sheets on Quarantine Pests. Septoria lycopersici.* http://www.eppo.org/QUARANTINE/fungi/Septoria_lycop_malagutii/SEPTLM_ds.pdf.
EPPO, 2005g, *Colour Picture of Symptoms.* http://www.eppo.org/QUARANTINE/fungi/Septoria_lycop_malagutii/SEPTLM_images.htm.
Erwin D.C. and O.K. Ribeiro, 1996, *Phytophthora Disease Worldwide.* APS, St. Paul, MN, U.S.A., pp. 562.
Gerlach W. and H. Nirenberg, 1982, *The Genus Fusarium, a Pictorial Atlas.* Parey, Berlin, Germany, pp. 406.
Harrison J.G., R.J. Searle and N.A. Williams, 1997, *Plant Pathol.* 46, 1.
Hering O. and H.I. Nirenberg, 1995, *Mycopathologia* 129, 159.
Hiemstra J.A. and M. Rataj-Guranowska, 2003, *Eur. J. Plant Pathol.* 109, 827.
Hohl H.R. and K. Iselin, 1984, *Trans. Br. Mycol. Soc.* 83, 529.
Hopkins J.F.C. (ed.), 1956, *Tobacco Diseases.* CMI, Kew, U.K., pp. 93.
Jeger M.J., G.A. Hide, P.H.J.F. van den Boogert, A.J. Termorshuizen and P. van Baarlen, 1996, *Potato Res.* 39, 437.
Karling J.S., 1964, *Synchytrium.* Academic Press, New York, U.S.A., pp. 470.
Karling J.S., 1968, *The Plasmodiophorales*, 2nd ed. Hafner Publishing Company, New York, U.S.A., pp. 256.
Lartey R.T., E.A. Curl and C.M. Peterson, 1994, *Soil Biol. Biochem.* 26, 81.
Merz U., J.A. Walsh, K. Bouchek-Mechiche, Th. Oberhänsli and W. Bitterlin, 2005, *Eur. J. Plant Pathol.* 111, 371.
Mulder A. and L.J. Turkensteen, 2005, *Potato Diseases.* Aardappelwereld BV & NIVAP, Wageningen, The Netherlands, pp. 280.
Mulder A., L.J. Turkensteen and A. Bouman, 1992, *Neth. J. Plant Pathol.* 98 (Suppl. 2), 103.
Nirenberg H.I., 1995, *Mycopathologia* 129, 131.
Pegg G.F. and B.L. Brady, 2002, *Verticillium Wilts.* CABI Publishing, Wallingford, U.K., pp. 416.
Peters R.D., A.V. Sturz and W. Arsenault, 2004, *Can. J. Plant Pathol.* 26, 63.
Rotem J., 1994, *The Genus Alternaria: Biology, Epidemiology, and Pathogenicity.* APS, St. Paul, MN, U.S.A., pp. 221.
Rowe R.C., D.A. Johnson, W.R. Beery and M.A. Omer, 1997, Proc. 7th Int. Verticillium Symp. Cape Sounion, Athens, Greece, pp. 23.
Sewell G.W.F. and J.F. Wilson, 1984, *Plant Pathol.* 33, 39.
Simmons E.G., 2000, *Mycotaxon* 75, 1.
Smart C.D. and W.E. Fry, 2001, *Biol. Invasions* 3, 234.
Smith I.M., J. Dunez, R.A. Lelliott, D.H. Phillips and S.A. Archer (eds), 1988, *European Handbook of Plant Diseases.* Blackwell Scientific Publications, Oxford, U.K., pp. 583.
Sneh B., S. Jabaji-Hare, S. Neate and G. Dijst (eds), 1996, *Rhizoctonia Species: Taxonomy, Molecular Biology, Ecology, Pathology and Disease Control.* Kluwer Academic Publishers, Dordrecht, The Netherlands, pp. 578.

Stevenson W.R., R. Loria, G.D. Franc and D.P. Weingartner (eds), 2001, *Compendium of Potato Diseases*, 2nd ed. American Phytopathological Society, St. Paul, MN, U.S.A., pp. 106.
Thurston H.D., 1973, *Annu. Rev. Phytopathol.* 11, 27.
van den Boogert P.H.J.F and A.J.G. Luttikholt, 2004, *Eur. J. Plant Pathol.* 110, 111.
van der Plaats-Niterink A.J., 1981, *Pythium. Stud. Mycol.* 21, 242.
van der Waals J.E., L. Korsten and B. Slippers, 2004, *Plant Dis.* 88, 959.

Part VII

BIOTECHNOLOGY

Chapter 30

Developments in Transgenic Biology and the Genetic Engineering of Useful Traits

Steve Millam

Institute of Molecular Plant Sciences, University of Edinburgh, Kings Buildings, Mayfield Road, Edinburgh EH9 3JH, United Kingdom

30.1 INTRODUCTION

Plant biotechnology has, in the last 20 years, made a significant impact on world agriculture, with over 81 million hectares of genetically modified (GM) crops being grown worldwide in 2005 (http://www.isaaa.org). The transformed traits of the GM crops currently cultivated can be generalized as 'first-generation' targets, largely based on single-gene modifications. This is due to the substantial timescale from laboratory to field-scale commercial cultivation. Recent rapid developments in technology may allow much more sophisticated and complex transgenic crops to be grown in the future. The major component of the current total is herbicide-resistant soya; indeed, over 50% of all soya cultivated is GM. In addition to the commercial applications of GM technology, which are increasing in both total area and the number of countries growing GM crops, gene transfer methods can be used for a wide range of fundamental studies, contributing to a better understanding of the mechanisms of plant : pathogen interactions and biosynthetic pathways in plants.

Examples of in vitro manipulation technology in potato (*Solanum tuberosum* L.) vary from the low-input and widely adopted (and non-GM) technology of micropropagation to the complex manipulation of multi-gene biosynthetic pathways through transgenic intervention. Potato is considered, due to its high in vitro regeneration capacity, a model species for methods such as somatic hybridization and *Agrobacterium*-mediated transformation. In addition, the related technology of micropropagation is an important component of many breeding schemes, and anther culture methodology has been adapted recently to enable the production of dihaploid lines from an increasing number of lines (Millam et al., 2005). Furthermore, recent advances in somatic embryogenesis in this species have increased the scope for the rapid propagation of novel material (Sharma and Millam, 2004).

Clearly, the lengthy breeding programmes and the problems of tetrasomic inheritance involved in the creation of potato lines with improved characters establish potato as a leading target for enhancement through *Agrobacterium*-mediated transformation. Indeed, historically, potato was one of the first crop plants to be successfully transformed (Ooms et al., 1986). Considerable scientific and commercial progress has been made since these early reports, and a large number of single-gene and multiple-gene traits have been

Potato Biology and Biotechnology: Advances and Perspectives
D. Vreugdenhil (Editor)

engineered into a wide range of potato and related germplasm (Millam, 2005). The development and large-scale adoption of GM technology has not, however, been trouble-free due to both scientific and ethical issues. Problems in scientific areas include alleged problems with gene flow (Snow, 2002) and the toxicity of introduced traits (Malarkey, 2003). There has also been considerable public concern over GM technology in general, and to allay many of the perceived problems, substantial scientific advances have been made in such areas as developing 'clean-gene' technology (see section 30.3.3.2), studies on resistance breakdown (Bates et al., 2005), large-scale studies of gene flow in the environment (Conner et al., 2003) and the actual composition and safety of introduced traits (Rischer and Oksman-Caldentey, 2006).

The changing emphasis in target traits has been a key feature in the development of transgenic technology in potato, which has been at the forefront of progress in crop plants due to the relative ease of transformation of this species. Early targets were mainly *input traits*, e.g. herbicide resistance (Eberlein et al., 1988), pest resistance (Cheng et al., 1992) and virus resistance (Kawchuk et al., 1991). Fungus/bacteria resistance was introduced at a later stage (Lyapkova et al., 2001), largely due to the complexities of developing durable systems for such traits. The trend today is towards the generation of crops with *output traits*, e.g. modified starch (Vardy et al., 2002), carotenoids (Ducreux et al., 2005a) and the production of pharmaceuticals in tubers (Park and Cheong, 2002). These more complex traits have been facilitated by the development of increasingly efficient and sophisticated approaches (Section 30.3.2), including the vastly increased knowledge of genomes and gene function. Furthermore, such output traits are considered to be more acceptable to consumers and have thus been the focus of commercially funded as well as academic projects.

30.2 GENETIC TRANSFORMATION OF POTATO

Since the development of the first transgenic plant (tobacco) reported in 1983, by four groups simultaneously (although historically the actual first publication was in May of that year by Herrera-Estrella et al., 1983), over 120 plant species have been successfully transformed with a wide and increasingly sophisticated range of traits. The historical development of transgenic technology in potato follows a familiar route for most of the first plant species investigated (Table 30.1).

Table 30.1 Historical development of transformation technology in crop plants

1. *Agrobacterium rhizogenes* infection of model cultivars
2. The extension of the system to other cultivars
3. The development of more efficient *Agrobacterium tumefaciens* systems
4. Further extension to a wider range of cultivars
5. Protocol refinements using model systems (e.g. optimized selectable markers)
6. The application of developed technology to commercial targets
7. The application of novel technologies

One of the key requirements for an efficient transformation system is the ability to rapidly regenerate plants from an isolated explant, and in potato, a number of strategies have been successfully employed for a wide range of germplasm (Wheeler et al., 1985). Early work on potato transformation focused on infection of potato tissue by strains of *Agrobacterium rhizogenes*, and the first direct evidence of the uptake of foreign DNA by transformation was obtained by Ooms et al. (1986), where clones of the potato cultivar Desirée were regenerated from *A. rhizogenes*-infected tissue. This work was extended to include several other cultivars by Ooms et al. (1987) and further to monohaploids and diploids of *S. tuberosum* (Devries and Gilissen, 1987). The use of *A. rhizogenes* has a number of technical drawbacks, however, not least the requirement to be able to regenerate intact plants from isolated root tissues. Thus, the use of *Agrobacterium tumefaciens* became more widespread in the mid-1980s, as this approach facilitates regeneration from more amenable explants such as leaf, tuber and stem. Cultivars Desiree and Bintje were transformed using *A. tumefaciens* by Stiekema et al. (1988) (interestingly, these remain the 'model' cultivars for potato transformation), and DeBlock (1988) reported a largely genotype-independent method for transformation, using leaf discs as the target tissue. This work was extended by Visser et al. (1989) who published a two-stage regeneration/transformation method using stem and leaf explants and binary *A. tumefaciens* vectors, which is the basis for many protocols in use today.

These early reports were devoted to the establishment and development of protocols, but as technology became established more widely, reports appeared describing significant scientific and commercial applications of transformation technology in potato, e.g. resistance to potato leafroll virus (PLRV) in the major variety Russet Burbank (Kawchuk et al., 1991). The ultimate objective of a transformation project would be to transfer the transgene into an existing cultivar in order to produce an enhanced version. However, the directed integration of an introduced gene into a unique and characterized position on the genome, despite many advances (section 30.3.2), has proved to be virtually impossible. Thus, many transformation projects have created large populations of independent transformants which have been screened (a time-consuming and expensive process) for the lines of interest. The first stage of screening would be the use of a selectable marker during the regeneration process. Following this, the resultant population of independent lines could be grown on and analysed further for the gene/phenotype of interest.

Historically, a survey of the literature reveals rapid application of the novel transformation technologies of the time, e.g. the use of antisense approaches (Kawchuk et al., 1991). In reports of that time, antisense could be defined as the insertion, by genetic engineering, of a gene in a reverse or backwards orientation. The antisense gene interferes with the operation of the same gene naturally found in the organism. This approach was used to switch off, delay or slow down the operation of a natural process such as softening in tomatoes and other fruits (Smith et al., 2002). In terms of tuber characters, Visser et al. (1991) reported inhibition of expression of the gene for granule-bound starch synthase (GBSS) in potato by antisense constructs, with implications for the starch : sugar balance in the tuber of transgenic lines and the production of amylose-free starch similar to that found in waxy mutants. For another key target, that of virus resistance, Mackenzie et al. (1991) reported genetically engineered resistance to potato virus-S in the US processing cultivar Russet Burbank. Virus resistance was among the first traits investigated for the genetic enhancement of potato, and the first report of antisense RNA-mediated resistance (to PLRV) was made by Kawchuk et al. (1991) again using the variety Russet Burbank.

It became apparent, however, that *Agrobacterium*-mediated transformation was not a simple 'one plus one' approach, and one of the first reports of the effect of transgene insertion was made by Brown et al. (1991), citing findings on insert copy number where it was shown that the range of T-DNA insert number could be from one to seven among a small population of transgenics. The same authors also showed that around one-third of the diploid regenerants had doubled their chromosome number (see also Ducreux et al., 2005b). Pollen stainability revealed that the tetraploid regenerants were male-sterile, thus affecting the crossability of the *Agrobacterium*-transformed diploid potato. This was important information, and because of the random nature of insertion, many projects aim to produce large numbers of independent transformants, undertake preliminary screenings and select a small subset of the desired lines for further analysis. The downside to this strategy is that many of the large number of transgenic lines created are discarded, some of which may be of great biological interest. Further refinements in improving the efficiency and applicability of potato transformation methodology, mainly by adjustments to in vitro conditions and modifications to the plant tissue culture media formulations, were made by Hulme et al. (1992) who reported an efficient genotype-independent method for the regeneration and transformation of potato plants from leaf tissue explants. This was extended to the report of a total of over 35 potato varieties that were assessed for their morphogenic and transformation efficiency in a review by Dale and Hampson (1995). A range of tissue explants have been successfully used for regeneration and transformation in potato. These include leaf discs (DeBlock, 1988), the advantages of which are the ease and rapidity of production and reproducibility and the disadvantage being the high incidence of damage during handling, internodal stem sections (Visser et al., 1989) that are also easy to produce and more robust than leaf sections for handling purposes and microtubers (Snyder and Belknap, 1993) that are technically more complicated and time-consuming to produce. The target of much of this work was to increase the efficiency of transformation across a wide range of cultivars. A widely applicable method has recently been published (Millam, 2005) based on the use of internodal sections as an explant source and successfully demonstrated for model cultivars such as Desiree, more recalcitrant cultivars such as Saturna and Stirling, breeding lines and diploid material. With regard to the plant tissue culture media formulations, nearly all methods utilize the widely used Murashige and Skoog (1962) plant tissue culture medium or minor modifications. There has been a range of growth regulator components commonly used in regeneration and transformation systems, but most protocols utilize a two-stage regeneration system. The first stage (approximately 10–14 days duration) stimulates callus formation, and the second stage is designed to induce de novo shoot outgrowth. In the first media, cytokinin (often in the form of zeatin riboside or less commonly benzylaminopurine) ratio to auxin (often naphthalene acetic acid) is between 20 and 200:1. In many protocols, gibberellic acid is used in the second media to enhance shoot outgrowth. Many variations exist in precise media formulations, and if introducing novel germplasm into a transformation system, some degree of optimization may be required.

Variation in transgene expression among the progeny is a commonly reported problem in potato transformation and has been attributed to the random integration of the transgene into different sites of the plant genome. A detailed analysis of significant populations of transgenic lines of potato has additionally revealed a number of phenotypic changes and substantially reduced tuber yields in field trials. Such alterations have been attributed to epigenetic and genetic events occurring during the regeneration phase of transformation. Depending on the potato cultivar, the frequency of these off-types has been recorded as 15–80%, and these often do not become apparent until plants are field grown (Kuipers et al., 1991; Jongedijk et al., 1992). It is apparent that data derived from glasshouse trials or first-generation tubers derived from microplants need to be treated with caution.

Thus, for the unequivocal analysis of introduced traits, it is important to generate sufficient experimental material and trial it under field conditions. The first report of a field evaluation of transformed potato was by Kuipers et al. (1991), who described an evaluation of the antisense RNA-mediated inhibition of GBSS gene expression in potato. Conner et al. (1994) reported findings from a field trial of transgenic potatoes undertaken in New Zealand. On a commercial basis, Monsanto's 'New Leaf' potato, which was initially the variety Russet Burbank (but later included the varieties Atlantic and Superior), was transformed to contain a *Bacillus thuringiensis (Bt) CryIIIA* gene (Perlak et al., 1993) conferring resistance to the Colorado potato beetle (CPB) and was first approved by US regulatory agencies in early 1995. This first line was followed, in late 1998, by a second type of transgenic potato, a Russet Burbank marketed as 'New Leaf Plus', which combined the *Bt* resistance trait with resistance to the PLRV (Thomas et al., 1997), and a third version, 'New Leaf Y', combining *Bt* with resistance to the potato virus Y (PVY). Despite being grown commercially between 1995 and 2001, plantings of these lines never amounted to more than 2–3% of the total US potato market and were discontinued, one reason cited being the premium price of the seed tubers (http://www.geo-pie.cornell.edu/crops/potato.html). In our own work, we successfully field-trialled over 10 000 transgenic potato plants of the model cultivar Desiree as well as the commercial crisping cultivar Saturna on two sites in Scotland between 1996 and 1998 (http://www.biomatnet.org/secure/Fair/S1140.htm). This study also included 'double-transformants' where selected lines transformed for one gene were re-transformed with another (using a second selectable marker), and populations comprising a range of promoters, sense and antisense constructs and substantial somaclonal controls were grown in replicated field plots. Material was assessed in the field for phenotypic differences, and in one line, it was found that the tetraploid had doubled to octoploid, which is in line with previous findings (Brown et al., 1991). Generally, the vast majority of the transgenic material was phenotypically indistinguishable from the control plants and was stable over the several tuber generations tested. Although the outcome of the work was the generation of a number of improved crisping quality lines, due to patent and ownership issues and a reluctance of the industry to commit to the use of GM potatoes, this work was discontinued.

The first report of transgenic potato was in 1986, and by 1995, extensive field trials in several countries and commercialization of GM potato crops had occurred. This rapid development was largely attributable to the amenability of potato to first-generation transformation methodology. Some of the important traits that have been engineered will be discussed in Section 30.5. However, there have been a number of advances (see Sections 30.3.2.2 and 30.3.2.3) and breakthroughs in gene transfer technology, many of which have been applied to potato. Additionally, the extensive information that has become available from genomics has enabled more precise and advanced transformation targets to be addressed. These will be discussed in the rest of this chapter.

30.3 DEVELOPMENTS IN TRANSGENIC BIOLOGY

Many transformation protocols, in a number of species including potato, are still largely based on the methods detailed in the original papers (DeBlock, 1988; Visser et al., 1989). There have been some developments to enable a wider range of related germplasm to be transformed, for example to the cultivated diploid species *Solanum phureja* (Ducreux et al., 2005b), and to increase overall transformation efficiencies. These improvements have been principally based on protocol refinements (Section 30.3.1). Gene transfer strategies have widened since the early work on *Agrobacterium*, and alternative systems have been developed for introducing novel genetic material into plants and for the transfer of multi-gene traits (Section 30.3.2).

30.3.1 Protocol refinements

30.3.1.1 Improved vectors

Early vectors used for transformation were closely based on wild-type strains of *Agrobacterium* and were deletion derivatives of Ti plasmids (co-integrative vectors tended to be large and were problematic in enabling efficient plant regeneration and for assembling the transgene constructs). It was found that the only features of the Ti plasmid actually necessary for integration into the host genome were the short border sequences and also that the removal of the natural oncogenic features enabled plants to be more easily regenerated using plant growth regulators. Several key features are necessary for the efficient design of a vector. The plasmid needs to be able to replicate in *Escherichia coli* (thus enabling easy DNA 'cut and paste' manipulations), additional selectable markers need to be incorporated for plant selection, border sequences need to be incorporated into the design of plasmid vectors to ensure efficient integration into the plant's genome and the 'foreign' genes (even though they may be from potato itself) need to have appropriate terminators and promoters to ensure correct expression. The development of binary vectors (Bevan, 1984) enabled the transfer system and the T-DNA to be located on separate plasmids. This enabled the vectors to become much smaller and easier to manipulate. A wide range of *Agrobacterium* vectors has been used for transformation of potato (for an extensive guide to currently available *Agrobacterium* vectors, see Hellens et al., 2000). There have been very recent reports that bacteria other than *Agrobacterium* may be used as vectors for gene transfer (Broothaerts et al., 2005), although this has not, at the time of writing, been applied to potato yet.

30.3.1.2 In vitro conditions

This is an aspect that has been strangely neglected over the years. Most protocols utilize plant tissue culture media formulations developed in the 1950s and 1960s, and there have been no apparent technical advances on these. The development of new growth regulators has also been slow. Possibly, the only developments in this section are the use of refined or alternative gelling agents such as Phytagel™ (Sigma Aldrich) in place of agar and in using vented culture vessels such as Magenta® (Sigma Aldrich) containers. This is an area where it is quite conceivable that improvements to conditions, e.g. light spectra and modified day–night conditions, could make a significant impact on the efficiency of transformation.

30.3.1.3 Improvements in selection

The first-generation transformation vectors utilized a number of antibiotic resistance genes for selection, and such methods have a number of advantages and disadvantages. In the majority of transgenic potato plants developed to date, kanamycin resistance, which is cheap to use and, in potato, proves to be highly discriminatory (Millam, 2005), has been used for selection. Other selectable markers that have been successfully used for potato transformation include methotrexate resistance, hygromycin resistance and, interestingly, 'benign' markers such as galactose mediated by xylose isomerase (Haldrup et al., 2001) or a UDP-glucose: galactose-1-phosphate uridyltransferase gene (Joersbo et al., 2003). Herbicide-resistant genes have been used but have been found to be less effective (DeBlock, 1988). However, owing to public opinion, the use of antibiotic-resistant genes is being phased out (Section 30.3.2.2). Practically, after the initial transformation and regeneration, selectable markers are no longer required for transgene expression.

30.3.2 Enhanced or alternative transformation strategies

30.3.2.1 Strategies for transgene pyramiding

Although single-gene traits have been successfully introduced into potato by transformation, the range of characters (e.g. quality, yield and drought resistance) and their consequent uptake would be significantly enhanced if multiple genes could be expressed or manipulated. Considerable progress has been made in recent years in elucidating biosynthetic pathways in plants by the modern tools of metabolomics and gene sequencing. Thus, this information could be used in transformation programmes if methodologies were available for the stable transfer of multiple-gene traits into plants. The simplest strategy for combining two or more transgenes would be *crossing* one transgenic with another, but this would be a time-consuming approach in potato. Another strategy could be *re-transformation* of an existing transgenic with another gene construct. This is again a time-consuming approach and necessitates the use of two different selectable markers. Such an approach has other limitations, in terms of potential co-suppression of transgenes, but has been used successfully in a number of commercial potato projects (http://www.nf-2000.org/secure/Fair/S1140.htm). Many methods for combining transgenes have disadvantages. One of the main problems is that transgenes introduced by the above methods will not be linked and will be sited at different loci in the plant's genome. In a commercial crop-breeding programme, each unlinked transgenic locus to be introduced would double the size of the breeding population needed, and the introduction of a third gene would potentially require eight times the population (Hitz, 1999). The second key constraint would be the need to use a number of selectable markers. Although methods exist for their removal, they are time-consuming and inefficient (Hare and Chua, 2002). The approach of *co-transformation with multiple transgenes* has a number of variants: mixing two (or more) different *Agrobacteria* each carrying a different binary vector, combining two T-DNAs on a single plasmid within an *Agrobacterium* strain or using one *Agrobacterium* carrying two compatible plasmids each with a different T-DNA. These approaches were investigated in our group using potato, with some success in generating transgenic lines at expected rates containing both genes of interest. These approaches have been discussed by Komari et al. (1996) and Daley et al. (1998).

30.3.2.2 Clean gene technologies

There has been considerable interest in the development of systems for the production of marker-free transgenic plants, which are considered more acceptable to consumers. Indeed, EU legislation will necessitate the removal of such 'extra' DNA sequences in transgenic plants in the near future. The range of options used is wide, ranging from attempts to transform without using selectable markers and screening the regenerants by phenotypic or molecular markers to highly sophisticated transposon-based gene removal systems. The first approach relies on a very high transformation efficiency and low cost of screening and has yet to be widely adopted. Transformation rates, in potato, using such an approach have been cited at 0.2% (De Vetten et al., 2003) compared with 40% using selection (Millam, 2005). Marker removal systems have been developed, but problems include the complexity of methods such as inducible recombination systems (Zuo et al., 2001). Other approaches tend to be scientifically complex and not yet suitable for widespread adoption (Halpin, 2005).

30.3.2.3 'Vectorless' transformation systems – vectors assembled only from potato-derived sequences

There have been several recent reports of a potentially useful variant of transformation technology, whereby native genes and regulatory elements can be reintroduced into plants, in this instance, potato, without the need to use selectable markers (Rommens, 2004; Rommens et al., 2004). These findings offer some new approaches. By combining the rapid increase in the use of native plant genes for transformation targets (rather than those derived from bacterial or viral sources), the use of native regulatory elements (promoters and terminators) and the avoidance of foreign markers, the system offers a number of advantages over existing transformation methodologies. A problem in conventional transformation methodologies is the unwanted transfer of 'backbone' DNA from the *Agrobacterium* vector, containing bacterial genetic elements such as origins of replication and antibiotic-resistant genes used for bacterial selection. Rommens (2004) describes the discovery of plant transfer DNAs (P-DNAs) that resemble *Agrobacterium* T-DNA borders and were demonstrated to support DNA transfer from *Agrobacterium* to plant cells creating 'intragenic' plants. The second report by the same authors describes the first example of genetically engineered plants that contain only native DNA, including potatoes engineered to have reduced the expression of a tuber-specific polyphenol oxidase (PPO) gene. This work clearly has great potential for future studies.

30.3.2.4 *Plastid transformation*

Most transformation methodologies are targeted towards nuclear transformation although some elegant work has been done using chloroplast transformation, largely limited to tobacco, although one earlier report exists in potato using a selected breeding line (Sidorov et al., 1999). Plastid transformation offers some advantages in that there would be no position effects as the genes of interest are introduced into the plastome through homologous recombination. This approach is an alternative method for the generation of marker-free lines, but is complex. The approach transforms plants with constructs that contain reconstitution elements and a gene of interest cloned within flanks used for homologous recombination and marker gene outwith these flanks. With the removal of selection pressure, a process known as 'loop out' recombination leads to the loss of co-integrates (Klaus et al., 2004). The generation of homoplastic transformants of a commercial cultivar (Desiree) of potato was reported by Nguyen et al. (2005). In this work, integration and expression of foreign genes into the plastid genome of potato was achieved.

30.3.2.5 *Particle bombardment*

The technique of particle bombardment (syn. biolistics) was developed in the late 1980s and is widely used in the transformation of cereals and legumes. The principal rationale for using biolistics for plant transformation is in its use for species that are problematic to regenerate in vitro, and potato does not fall into such a category. However, there have been reports of its use to transform potato (Romano et al., 2001, 2003), but such a system has not been widely adopted due to the more complex integration patterns that occur. These, of course, could be simplified if a passage of meiosis occurs, but due to the vegetative propagation system in potato, this is not possible. Furthermore, efficiency of transformation by this method is 0.1–10%, whereas by *Agrobacterium*, it can be up to 100%. However, a recent report (Romano et al., 2005) describes how the synthesis of microbial polyesters has been attempted in transgenic potato using this approach, offering possibilities for a novel application of transgenic potatoes.

30.3.2.6 *Gene silencing*

Potato viruses cause significant economic losses to crops, and the avoidance of virus infection is a major factor in the design and application of seed potato production programmes. However, potato viruses also have the potential to be used as biotechnological tools for the introduction of novel traits into the crop (Chapter 28, Valkonen, this volume). Virus-induced gene silencing (VIGS) is a tool that exploits an RNA-mediated anti-viral defence mechanism, triggered by incoming viruses, to target individual genes for silencing. In plants infected with unmodified viruses, the process is specifically targeted against the viral genome. However, with modified virus vectors carrying inserts derived from the host genes, the process can be additionally targeted against the corresponding mRNAs. The discoveries, that plants co-suppress their own genes if they are transformed with homologous transgenes and that plants recognize and degrade invading viral RNA, have resulted in new models for plant intercellular communication and defence mechanisms (Robertson, 2004). Historically, transgene expression in plants was not found to be predictable, and in some cases, transgene expression was lost during development or in subsequent generations (Allen et al., 2000). It was further found that not only the transgene but also the endogenous gene expression was subject to silencing, and despite active transcription, mRNA levels for both genes were found to have decreased (Napoli et al., 1990). This process was termed post-transcriptional gene silencing (PTGS). Both antisense and sense-mediated suppression of gene expression were commonly used as a tool for the down-regulation of target genes in plants, including potato (Romer at al., 2002).

Two recent developments associated with such technology are RNAi (interfering RNA from dsRNA) and siRNA-mediated (the use of 22-nt oligonucleotide dimers) silencing in plants. Virus-induced gene silencing has been used to silence a wide variety of genes in plants (Robertson, 2004). It has been used to generate transient loss-of-function assays and also has potential as a powerful reverse-genetics tool in functional genomic programmes, as a more rapid alternative to stable transformation. There have been recent reports of the use of VIGS for dissecting signalling factors in disease resistance, such as the in N-mediated resistance against tobacco mosaic virus (Peart et al., 2002). Until recently, most applications of VIGS have been in *Nicotiana benthamiana*; however, a potato virus X (PVX) VIGS vector was shown to be effective in triggering a VIGS response in both diploid and cultivated tetraploid *Solanum* species. It was shown (Faivre-Rampant et al., 2004) that a systemic silencing of a phytoene desaturase gene was observed and maintained throughout the foliar tissues of potato plants and, significantly, was also observed in tubers. It was further reported that VIGS can be triggered and sustained on in vitro micropropagated tetraploid potato for several cycles and on in vitro-generated microtubers. Another paper (Brigneti et al., 2004) produced similar findings but used a tobacco rattle virus (TRV) vector. In terms of risk assessment, the use of recombinant viruses requires approval from the relevant plant pathogen licensing authorities. Notably, the use of entire RNA viruses requires a high level of containment to avoid accidental transmission. However, in the future, VIGS approaches will facilitate large-scale functional analysis of potato expressed sequence tags and provide a non-invasive reverse-genetic approach to study mechanisms involved in developmental processes in potato.

30.4 THE GENETIC ENGINEERING OF USEFUL TRAITS

The genetic modification of potato cultivars by *Agrobacterium* (or any of the other methods described above in Sections 30.3.2.3, 30.3.2.4 and 30.3.2.5) offers the possibility of introducing genes into the potato genome that are not present in cultivated potatoes, and their wild relatives, and hence of the introduction of novel biochemical and desirable traits into this important crop plant. Since the first reports of the genetic manipulation of potato in 1986 and the development of more efficient model systems, the range of introduced genes has significantly increased, as has the complexity and durability of the traits altered.

The genes used to date in transformation studies have mainly coded for proteins that are toxic to pests and pathogens, sequences whose expression interferes with virus multiplication in host cells and genes that code for key enzymes in biochemical pathways in other organisms, often, but not always, derived from other plant species. With the increasing amount of metabolomic and genomic information becoming available, the precise mechanisms of biochemical and host plant: pathogen interactions are becoming more clearly understood, and thus, transgenic intervention strategies can be designed more precisely.

30.4.1 Resistance to major pests and diseases

The major driver for research and development into this aspect is evident, due to the substantial crop losses attributable to the effect of pathogens, not only during the growing season but, in the case of potato tubers, also during storage. However, a secondary factor in the substantial research effort made in this area is that many of the resistance mechanisms could be introduced by a single gene, which was the only available strategy until the development of methods for introducing multiple genes was demonstrated on an efficient scale in the late 1990s. A further consideration is that much of the (expensive) research has been undertaken in laboratories associated directly, or indirectly, with the major agrochemical manufacturers.

30.4.1.1 Colorado potato beetle

Colorado potato beetle (*Leptinotarsa decemlineata* Say) is a major pest in North America and elsewhere. The CPB is notorious for its ability to rapidly develop resistance to insecticides that are used repeatedly for control. Insecticides having the same class of chemical structure will often have the same mode of action. Consequently, resistance develops more rapidly to an insecticide when used repeatedly as the only control measure. Additionally, the repeated use of one class kills the susceptible beetles, leaving those that are resistant, i.e. selecting for the resistant population. Consequently, to delay or prevent resistance, a complicated and expensive crop protection strategy involving insecticide rotation is necessary. Transgenic resistance was developed with the introduction of a gene that encodes the Cry3A protein derived from the bacterium *B. thuringiensis* var. *tenebrionis* and expressed in potato using the constitutive 35S cauliflower mosaic virus (CaMV 35S) promoter. The strategy was so successful that such plants were the first GM potato varieties to be commercialized by Monsanto using the varieties Russet Burbank, Atlantic, Snowden and Superior in North America from 1995 to 2001 (Duncan et al., 2002). Extensive testing of the *Bt*-protected crops had established their safety for humans, animals and the environment, but (Section 30.2) the product was withdrawn in 2001 mainly for commercial rather than agronomic reasons.

30.4.1.2 Potato tuber moth

Potato tuber moth (*Phthorimaea operculella*) is a troublesome pest of potatoes and is found in warm tropical and subtropical climates. It is the most damaging pest of potatoes in fields and stores in warm, dry areas of the world, such as North Africa and the Middle East, Mexico, Central America and the inter-Andean valleys of South America. Transgenic resistance has been provided by the Bt protein encoded by the *cry5* gene (Mohammed et al., 2000) and by the *cry*1Ac9 gene (Davidson et al., 2004), again using the constitutive 35S CaMV promoter. Davidson et al. (2004) demonstrated that their transgenic potato lines exhibited stable resistance to larvae across field seasons, between affected plant organs and between plant organs of different ages.

30.4.1.3 Potato cyst nematodes

Globodera rostochiensis and *Globodera pallida* [potato cyst nematodes (PCN)] are the major pests of potatoes in Great Britain and elsewhere. A recent survey of soil samples taken by UK growers revealed that up to 60% of the land cropped with potatoes in the UK is infested with PCN. The annual loss to the UK potato industry as a direct consequence of PCN has been estimated at approximately 9% of crop value, worth around £43 million in 2004. It has been found that in highly infested soil, there may be several thousand nematodes per gram of potato root. Furthermore, even in low to medium infestations, some degree of crop loss will result and the numbers of PCN will multiply, thus increasing the problem for subsequent potato crops. In the worst case scenario, in fields with high PCN infestations, complete crop loss is possible (http://www.syngenta-crop.co.uk). A number of transgenic strategies have been proposed to control nematodes, but due to the complexity of the life cycle and infestation methods, control has been difficult. Urwin et al. (2003) were able to demonstrate that constructs based on a cysteine proteinase inhibitor (cystatin) from sunflower and a protein-engineered variant of a rice cystatin conferred similar levels of resistance to PCN as chicken egg white cystatin (CEWC) under the control of CaMV35S, and in a field trial, these levels of resistance were similar to that provided by the natural partial resistance of cultivar Sante. Transformation of Sante and the South American cultivar Maria Huanca with CaMV35S/CEWC raised the status of both cultivars from partial to full resistance.

30.4.1.4 Viruses PLRV and PVY

Potato leafroll (PLRV) and potato Y (PVY) are the two most serious virus diseases of potatoes worldwide. PLRV causes both qualitative and quantitative damage and is transmitted in a persistent manner by several aphid species. Also, the virus infects other Solanaceous crops and weeds. Potato virus Y is readily spread by aphids in a non-persistent manner as well as mechanically by human activity and may result in severely depressed yields. Sense and antisense RNA-mediated resistance to PLRV was engineered into Russet Burbank potato plants in 1991 (Kawchuk et al., 1991), and this work was expanded to generate Russet Burbank potatoes in which CPB resistance was combined with resistance to PLRV provided by a construct designed to prevent virus replication using the constitutive Figwort mosaic virus (FMV) promoter. Russet Burbank and Shepody potatoes have also been produced with combined CPB and PVY resistance, the latter provided by the PVY coat protein gene, again using the FMV promoter. In both these examples, the process started with about 3000 original transgenic potato clones in 1991 from which six were finally selected for commercialization in 1998 (Davies, 2002). Trait stability has been demonstrated in field trials over a number of years, as has the greatly reduced use of pesticides (Duncan et al., 2002).

30.4.1.5 Bacteria and fungi

Owing to the complexity of the host: pathogen response, progress in engineering bacterial and fungal resistance into potato (and other crop plants) has been less rapid than with other traits. Genes coding for lytic enzymes from bacteriophage to humans are being evaluated in a number of laboratories worldwide as a method to achieve transgenic resistance to a number of bacteria and fungi. For example, the gene *chly* encoding the enzyme lysozyme from chicken has been introduced into cultivar Desiree through *Agrobacterium*-mediated transformation and shown to enhance resistance to blackleg and soft rot caused by infection with *Erwinia carotovora* subsp. *atroseptica* (Serrano et al., 2000).

30.4.2 Tuber quality traits

30.4.2.1 Anti-bruise potatoes

Bruise resistance is important in potatoes, as mechanical damage initiates enzymic browning which results in the production of black, brown and red pigments and either crop rejection by processors or waste during processing. A transgenic solution to the problem can be provided by the down-regulation of PPO gene expression so that the reaction leading to pigment production is no longer catalysed by the enzyme PPO (Bachem et al., 1994). The results of a 4-year field trial of transgenic Desiree showed that the average rotting caused by Ec-bacteria was diminished in tubers of pectic lyase (PL)-transgenic lines by 34.1%, and the resistance of tubers to Ec soft rot was significantly correlated with the PPO activity in tuber tissue (Wegener, 2001).

30.4.2.2 Reduced glycoalkaloid content

Steroidal glycoalkaloids are a class of potentially toxic compounds with a bitter taste, which are found throughout the family Solanaceae. Cultivars vary with regard to their inherent tuber glycoalkaloid content. Levels above 20 mg per 100 g fresh weight are considered unsuitable for human consumption as they can cause various symptoms typically associated with food poisoning. Although breeders check potential cultivars for unacceptably high levels, particularly where pedigrees involve wild species, a transgenic option for further reduction would be useful. Initial reports of down-regulating a gene encoding a sterol alkaloid glycosyltransferase (*Sgt1*) and an almost complete inhibition of α-solanine accumulation which was compensated by elevated levels of α-chaconine (McCue et al., 2005) have been made; however, further transformation will be required to inhibit chaconine accumulation. Of more general interest, transgenic potato plants overexpressing a soybean [(type 1 sterol methyltransferase (GmSMT1)] cDNA were generated and used to study sterol biosynthesis in relation to the production of toxic glycoalkaloids (Arnqvist et al., 2003). The results show that glycoalkaloid biosynthesis can be down-regulated in transgenic potato plants by reducing the content of free non-alkylated sterols and support the view of cholesterol as a precursor in glycoalkaloid biosynthesis.

30.4.3 Nutritional value

30.4.3.1 Protein and amino acid content

Chakraborty et al. (2000) reported improvements in the nutritive value of potato through transformation with a non-allergenic seed albumin gene (*AmA1*) from *Amaranthus hypochondriacus*. The seed protein has a well-balanced amino acid composition with no known allergenic properties. Five- to ten-fold increases in transcript levels in tubers were achieved using the tuber-specific GBSS promoter compared with the 35S CaMV promoter. Significant two- to four-fold increases were achieved in the lysine, methionine, cysteine and tyrosine content of the protein amino acids, and a 35–45% increase was achieved in total protein content.

30.4.3.2 Inulin

Inulin is a mixture of linear fructose polymers with different chain length and a glucose molecule at each C2 end. Inulin belongs to the fructan group of polysaccharides and serves as a carbohydrate storage in many plant species. Compounds such as inulin reduce the energy density of food and are used to enrich food with dietary fibre or to replace sugar and fat. Hellwege et al. (2000) have developed transgenic potato tubers which synthesize the full range of inulin molecules naturally occurring in globe artichoke (*Cynara scolymus*). High-molecular-weight inulins have been produced by expressing the sucrose : sucrose 1-fructosyl transferase and the fructan : fructan 1-fructosylhydrolase genes from globe artichoke. Inulin made up 5% of the dry weight (DW) of the transgenic tuber.

30.4.3.3 Carotenoids

Yellow and orange flesh colour comes from a class of pigments known as carotenoids (e.g. zeaxanthin) which are also antioxidants with health-promoting attributes. Hence, their enhancement by breeding provides an opportunity to improve the nutritive value of potatoes and processed foods made from potatoes (Brown, 2005). Transgenic approaches are also possible as Romer et al. (2002) discovered that tuber-specific down-regulation of the zeaxanthin epoxidase gene in *S. tuberosum* increased not only the amount of zeaxanthin that accumulated but also the total carotenoid level by up to 5.7-fold of the controls. Recent work on the underlying mechanisms of carotenogenesis during tuber development and storage in potato has been reported (Morris et al., 2004), and in further work by the same group, it was found that in developing tubers of transgenic Desiree (using an *Erwinia uredovora crtB* gene encoding phytoene synthase) lines, carotenoid levels reached 35 μg carotenoid g DW, and the balance of carotenoids changed radically compared with controls (Ducreux et al., 2005a). The *crtB* gene was also transformed into *S. phureja* (cultivar Mayan Gold), again resulting in an increase in total carotenoid content to 78 μg carotenoid g DW in the most affected transgenic line (Ducreux et al., 2005a,b).

30.4.3.4 Starch

Starch is the primary storage compound in tubers. It is also widely used for a range of industrial processes. The physical properties of starch vary with plant source, but there are considerable opportunities to generate novel starches for use in food and non-food market sectors (Davis et al., 2003). Genetic engineering has already generated novel potato starches of which the two extremes are high amylopectin starch and high amylose starch. High amylopectin starch was produced by the down-regulation of the *GBSS* gene that controls amylose synthesis (Visser et al., 1991). In contrast, to produce high amylose starch, it was necessary to concurrently down-regulate two starch branching enzymes, A and B (Schwall et al., 2000). Field trialling confirmed the stability of the modification over years and demonstrated an increased tuber yield, reduced starch content, smaller granule size and an increase in reducing sugars (Hofvander et al., 2004). Potatoes containing starch with a very low degree of branching, such as 0.3%, were not suitable for commercial cultivation due to severe starch yield reduction and other effects. However, with slightly more branching, the effects were much reduced, and the modified starch was considered suitable for biodegradable plastics, expanded products and film-forming operations.

30.4.3.5 Reducing sugars

Ideally, the potato industry would like to store tubers at low temperatures (about 4°C) to minimize sprout growth and eliminate the need to use chemicals to suppress the sprouting process. However, low temperatures induce glucose and fructose accumulation, and these reducing sugars are primarily responsible for non-enzymic browning through a typical Maillard reaction that occurs at temperatures required to generate potato chips (crisps) and French fries. Whilst breeders have been able to select for lower levels of reducing sugars out of cold storage, transgenic approaches are also possible based on an understanding of primary carbohydrate metabolism. Stark et al. (1992) increased tuber starch content and lowered the levels of reducing sugars by expressing an *E. coli glgC16* mutant gene that encodes for the enzyme ADPglucose pyrophosphorylase and increases the production of ADPglucose, which in turn becomes incorporated into the growing starch granule. Greiner et al. (1999) were able to minimize the conversion of sucrose to glucose and fructose by expressing a putative vacuolar invertase inhibitor protein from tobacco, called Nt-inhh, in potato plants under the control of the CaMV35S promoter. See also http://www.nf-2000.org/secure/crops/potato.html for a report on the field trialling of a range of transgenic potato lines engineered for modified reducing sugar content.

30.5 SUMMARY AND FUTURE DEVELOPMENTS

Since the first report of transformation in potato in 1986, the technology has developed considerably, both in terms of the range of germplasm and the number and complexity of traits successfully introduced into the potato. The potato has been a model crop for transformation, due to the relative ease of *A. tumefaciens*-mediated transformation, but developments in areas such as VIGS may lead to hitherto unconsidered possibilities for testing genes and host: pathogen interactions. Potato is an important staple food crop, the fourth most important in the world, and one with great potential for development in developing countries. Transformation projects have been designed to enhance the nutritional quality of potato tubers with some success in the lab and glasshouse (Chakraborty et al., 2000; Ducreux et al., 2005a), but many field trials of such material have been precluded due to the constrictive legislation regarding transgenic crops in Europe. Nevertheless, such approaches towards the nutritional enhancement of such an important crop will continue in future as knowledge of the potato genome and the processes controlling pathways increase.

Other areas for future transgenic progress will be in using tubers as biorefineries for valuable products, either by conventional transformation, for example the expression and production of recombinant human interleukin-2 in potato plants (Park and Cheong, 2002) and the expression of antibodies and Fab fragments in transgenic potato plants (De Wilde et al., 2002), or by biolistics (Romano et al., 2005) or plastid transformation (Nguyen et al., 2005). Biological processes and biosynthetic pathway elucidation will also be enhanced, and further insights into the mechanisms of disease resistance such as the role of mitogen-activated protein kinase cascades in blight resistance (Yamamizo et al., 2006), tuber life cycle and dormancy will be mediated by the transformation of this important species.

Potato will continue to be at the forefront of transformation technology, and developments in this area offer exciting challenges for future crop improvement, sustainability and scientific advancement.

REFERENCES

Allen G.C., S. Spiker and W.F. Thompson, 2000, *Plant Mol. Biol.* 43, 361.

Arnqvist L., P.C. Dutta, L. Jonsson and F. Sitbon, 2003, *Plant Physiol.* 131, 1792.

Bachem C.W.B., G.J. Speckmann, P.C.G. Van Der Linde, F.T.M. Verheggen, M.D. Hunt, J.C. Steffens and M. Zabeau, 1994, *Nat. Biotechnol.* 12, 1101.

Bates S.L., J.-Z. Zhao, R.T. Roush and A.M. Shelton, 2005, *Nat. Biotechnol.* 23, 57.

Bevan M., 1984, *Nucleic Acids Res.* 12, 8711.

Brigneti G., A.M. Martin-Hernandez, H.L. Jin, J. Chen, D.C. Baulcombe, B. Baker and J.D.G. Jones, 2004, *Plant J.* 39, 264.

Broothaerts W., H.J. Mitchell, B. Weir, S. Kaines, L.M.A. Smith, W. Yang, J.E. Mayer, C. Roa-Rodriguez and R.A. Jefferson, 2005, *Nature* 133, 629.

Brown C.R., 2005, *Am. J. Potato Res.* 82, 163.

Brown C.R., C.P. Yang, S. Kwiatkowski and K.D. Adiwiliga, 1991, *Am. Pot. J.* 68, 317.

Chakraborty S., N. Chakraborty and A. Datta, 2000, *Proc. Natl. Acad. Sci. U.S.A.* 97, 3724.

Cheng J., M.G. Bolyard, R.C. Saxena and M.B. Sticklen, 1992, *Plant Sci.* 81, 83.

Conner A.J., T.R. Glare and J.P. Nap, 2003, *Plant J.* 33, 19.

Conner A.J., M.K. Williams, D.J. Abernethy, P.J. Fletcher and R.A. Genet, 1994, *N.Z. J. Crop Hort.* 22, 361.

Dale P.J. and K.K. Hampson, 1995, *Euphytica* 85, 101.

Daley M., V.C. Knauf, K.R. Summerfelt and J.C. Turner, 1998, *Plant Cell. Rep.* 17, 459.

Davidson M.M., R.C. Butler, S.D. Wratten and A.J. Conner, 2004, *Ann. Appl. Biol.* 145, 271.

Davies H.V., 2002, In: V. Valpuesta (ed.), *Fruit and Vegetable Biotechnology*, pp. 222–249. Woodhead Publishing Limited, Cambridge.

Davis J.P., N. Supatcharee, R.L. Khandelwal and R.N. Chibbar, 2003, *Starch–Starke* 55, 107.

DeBlock M., 1988, *Theor. Appl. Genet.* 76, 767.

Ducreux L.J.M., W.L. Morris, P.E. Hedley, T. Shepherd, H.V. Davies, S. Millam and M.A. Taylor, 2005a, *J. Exp. Bot.* 56, 81.

De Vetten N., A.M. Wolters, K. Raemakers, I. Van Der Meer, R. Ter Stege, E. Heeres, P. Heeres and R. Visser, 2003, *Nat. Biotechnol.* 21, 439.

Devries E. and L.J.W. Gilissen, 1987, *Acta Bot. Neerl.* 36, 182.

Ducreux L., W.L. Morris, M.A. Taylor and S. Millam, 2005b, *Plant Cell Rep.* 24, 10.

Duncan D.R., D. Hammond, J. Zalewski, J. Cudnohufsky, W. Kaniewski, M. Thornton, J.T. Bookout, P. Lavrik, G.J. Rogan and J. Feldman-Riebe, 2002, *HortScience* 37, 275.

De Wilde C., K. Peeters, A. Jacobs, I. Peck and A. Depicker, 2002, *Mol. Breed.* 9, 271.

Eberlein C.V, M.J. Guttieri and J. Steffen-Campbell, 1988, *Weed Sci.* 46, 150.

Faivre-Rampant O., E.M. Gilroy, K. Hrubikova, I. Hein, S. Millam, G.J. Loake, P. Birch, M. Taylor and C. Lacomme, 2004, *Plant Physiol.* 134, 1308.
Greiner S., T. Rausch, U. Sonnewald and K. Herbers, 1999, *Nat. Biotechnol.* 17, 708.
Haldrup A., M. Noerremark and F.T. Okkels, 2001, *In Vitro Cell. Dev. Biol.-Plant* 37, 114.
Halpin C., 2005, *Plant Biotechnol. J.* 3, 141.
Hare P.D. and N.H. Chua, 2002, *Nat. Biotechnol.* 20, 575.
Hellens R., P. Mullineaux and H. Klee, 2000, *Trends Plant Sci.* 5, 446.
Hellwege E.M., S. Czapla, A. Jahnke, L. Willmitzer and A.G. Heyer, 2000, *Proc. Natl. Acad. Sci. U.S.A.* 97, 8699.
Herrera-Estrella L., A. Depicker, M. Van Montagu and J. Schell, 1983, *Nature* 303, 209.
Hitz B., 1999, *Curr. Opin. Plant Biol.* 2, 135.
Hofvander P., M. Andersson, C.-T. Larsson and H. Larsson, 2004, *Plant Biotechnol. J.* 2, 311.
Hulme J.S., E.S. Higgins and R. Shields, 1992, *Plant Cell Tiss. Org. Cult.* 31, 161.
Joersbo M., K. Jorgensen and J. Brunstedt, 2003, *Mol. Breed.* 11, 315.
Jongedijk E., A.J.M. Deschutter, T. Stolte, P.J.M. Vandenelzen and B.J.C. Cornelissen, 1992, *Nat. Biotechnol.* 10, 422.
Kawchuk L.M., R.R. Martin and J. McPherson, 1991, *Mol. Plant Microbe Inter.* 4, 247.
Klaus S.M.J., F.C. Huang, T.J. Golds and H.U. Koop, 2004, *Nat. Biotechnol.* 22, 225.
Komari T., Y. Hiei, Y. Saito, N. Murai and T. Kumashiro, 1996, *Plant J.* 10, 165.
Kuipers G.J., J.T.M. Vreem, H. Meyer, E. Jacobsen, W.J. Feenstra and R.G.F. Visser, 1991, *Euphytica* 59, 83.
Lyapkova N.S., N.A. Loskutova, A.N. Maisuryan, V.V. Mazin, N.P. Korableva, T.A. Platonova, E.P. Ladyzhenskaya and A.S. Evsyunina, 2001, *Appl. Biochem. Microbiol.* 37, 301.
Mackenzie D.J., J.H. Tremaine and J. Mcpherson, 1991, *Mol. Plant Microbe Inter.* 4, 95.
Malarkey T., 2003, *Mutat. Res.* 544, 217.
McCue K.F., L.V.T. Shepherd, P.V. Allen, M.M. Maccree, D.R. Rockhold, D.L. Corsini, H.V. Davies and W.R. Belknap, 2005, *Plant Sci.* 168, 267.
Millam S., 2005, In: I. Curtis (ed.), *Transgenic Crops of the World – Essential Protocols*, pp. 257–270. Kluwer Academic, Dordrecht.
Millam S., S.K. Sharma, G. Bryan, V. Matti-Rokka and J. Middlefell-Williams, 2005, Annual Report Scottish Crop Research Institute 2004, pp. 124–125.
Mohammed A., D.S. Douches, W. Pett, E. Grafius, J. Coombs, W. Liswidowati, W. Li and M.A. Madkour, 2000, *J. Econ. Entomol.* 93, 472.
Morris W.L., L. Ducreux, D.W. Griffiths, D. Stewart, H.V. Davies and M.A. Taylor, 2004, *J. Exp. Bot.* 55, 975.
Murashige T. and F. Skoog, 1962, *Physiol. Plant.* 15, 473.
Napoli C., C. Lemieux and R. Jorgensen, 1990, *Plant Cell* 2, 279.
Nguyen T.T., G. Nugent, T. Cardi and P.J. Dix, 2005, *Plant Sci.* 168, 1495.
Ooms G., M.E. Bossen, M.M. Burrell and A. Karp, 1986, *Potato Res.* 29, 367.
Ooms G., M.M. Burrell, A. Karp, M. Bevan and J. Hille, 1987, *Theor. Appl. Genet.* 73, 744.
Park Y. and H. Cheong, 2002, *Protein Express Purif.* 25, 160.
Peart J.R., G. Cook, B.J. Feys, J.E. Parker and D.C. Baulcombe, 2002, *Plant J.* 27, 569.
Perlak F.J., T.B. Stone, Y.M. Muskopf, L.J. Petersen, G.B. Parker, S.A. McPherson, J. Wyman, S. Love, G. Reed, D. Biever and D.A. Fischhoff, 1993, *Plant Mol. Biol.* 22, 313.
Rischer H. and K.-M. Oksman-Caldentey, 2006, *Trends Biotechnol.* 24, 102.
Robertson D., 2004, *Annu. Rev. Plant Biol.* 55, 495.
Romano A., K. Raemakers, J. Bernardi, R. Visser and H. Mooibroek, 2003, *Transgenic Res.* 12, 461.
Romano A., K. Raemakers, R. Visser and H. Mooibroek, 2001, *Plant Cell Rep.* 20, 198.
Romano A., L.H.W. van der Plas, B. Witholt, G. Eggink and H. Mooibroek, 2005, *Planta* 220, 455.
Romer S., J. Lubeck, F. Kauder, S. Steiger, C. Adomat and G. Sandmann, 2002, *Metab. Eng.* 4, 263.
Rommens C.M., 2004, *Tr. Plant Sci.* 9, 457.
Rommens C.M., J.M. Humara, J. Ye, H. Yan, C. Richael, L. Zhang, R. Perry and K. Swords, 2004, *Plant Physiol.* 135, 421.
Schwall G.P., R. Safford, R.J. Westcott, R. Jeffcoat, A. Tayal, Y.C. Shi, M.J. Gidley and S.A. Jobling, 2000, *Nat. Biotechnol.* 18, 551.

Serrano C., P. Arce-Johnson, H. Torres, M. Gebauer, M. Gutierrez, M. Moreno, X. Jordana, A. Venegas, J. Kalazich and L. Holuigue, 2000, *Am. J. Potato Res.* 77, 191.
Sharma S.K. and S. Millam, 2004, *Plant Cell Rep.* 23, 115.
Sidorov V.A., D. Kasten, S.Z. Pang, P.T.J. Hajdukiewiecz, J.M. Staub and N.S. Hehra, 1999, *Plant J.* 19, 209.
Smith C.J.S., C.F. Watson, J. Ray, C.R. Bird, P.C. Morris, W. Schuch and D. Grierson, 2002, *Nature* 334, 724.
Snow A.A., 2002, *Nat. Biotechnol.* 20, 542.
Snyder G.W. and W.R. Belknap, 1993, *Plant Cell Rep.* 12, 324.
Stark D.M., K.P. Timmerman, G.F. Barry, J. Preiss and G.M. Kishore, 1992, *Science* 258, 287.
Stiekema W.J., F. Heidekamp, J.D. Louwerse, H.A. Verhoeven and P. Dijkhuis, 1988, *Plant Cell Rep.* 7, 47.
Thomas P.E., W. Kaniewski and C. Lawson, 1997, *Plant Disease* 81, 1447.
Urwin P.E., J. Green and H.J. Atkinson, 2003, *Mol. Breed.* 12, 263.
Vardy K.A., M.J. Emes and M.M. Burrell, 2002, *Funct. Plant Biol.* 29, 975.
Visser R.G.F., E. Jacobsen, A. Hesselingmeinders, M.J. Schans, B. Witholt and W.J. Feenstra, 1989, *Plant Mol. Biol.* 12, 329.
Visser R.G.F., I. Somhorst, G.I. Kuipers, N.J. Ruys, W.J. Feenstra and E. Jacobsen, 1991, *Mol. Gen. Genet.* 225, 289.
Wegener C.B., 2001, *Potato Res.* 44, 401.
Wheeler V.A., N.E. Evans, D. Foulger, K.J. Webb, A. Karp, J. Franklin and S.W.J. Bright, 1985, *Ann. Bot.* 55, 309.
Yamamizo C., K. Kuchimura, A. Kobayashi, S. Katou, K. Kawakita, J.D.G. Jones, N. Doke and H. Yoshioka, 2006, *Plant Physiol.* 140, 681.
Zuo J.R., Q.W. Niu, S.G. Moller and N.H. Chua, 2001, *Nat. Biotechnol.* 19, 157.

Chapter 31

Field-Testing of Transgenic Potatoes

A.J. Conner

New Zealand Institute for Crop and Food Research, Private Bag 4704, Christchurch, New Zealand and National Centre for Advanced Bio-Protection Technologies, Lincoln University, P.O. Box 84, Canterbury, New Zealand

31.1 INTRODUCTION

For many years, the cultivated potato (*Solanum tuberosum* L.) has been at the forefront of genetic engineering developments in crop plants. Many research institutes and private companies have targeted potato improvement through the transformation of existing cultivars with specific genes. This is a consequence of the importance of the potato crop throughout the world, the relative ease with which the crop can be transformed and genetic limitations associated with traditional potato breeding (Conner et al., 1997). Improvements in characters such as resistance to pests and diseases and specific quality attributes are widely anticipated to allow potato breeders to respond much more quickly to the market need for new and improved cultivars. The anticipated result is higher quality, blemish-free tubers with reduced chemical residues as demanded by the processors and consumers (Conner et al., 1997).

Despite the considerable potential for highly targeted improvement of potato cultivars, the release and agricultural deployment of transgenic crops have raised public concerns on environmental issues and food safety (Nap et al., 2003). This is well reflected in potatoes, with only a very limited number of transgenic events receiving full approval for commercial release (Nap et al., 2003). Two of these events involve potatoes with resistance to Colorado potato beetle (*cry*3A gene), with other events involving this trait plus resistance to potato leafroll virus (putative viral helicase and replicase genes) or resistance to potato virus Y (PVY) (viral coat protein gene) (Nap et al., 2003). Despite the considerable promise of potatoes possessing these traits and the anticipated environmental benefits from reduced pesticide applications, these events have essentially failed in the market place due to consumer perception issues.

The limited commercial success of transgenic potatoes to date has not deterred the ongoing development of transgenic potato cultivars with improved agronomic and quality traits. Such research continues throughout the world, primarily in research institutes. This is well evident by the ongoing activity associated with the field-testing of transgenic potatoes. Records of approvals/permits issued for experimental field tests of transgenic crops around the world can be found in various databases, although there are many discrepancies between different sources of information. The Biotrack database of the Organization for Economic Cooperation and Development (OECD)

Potato Biology and Biotechnology: Advances and Perspectives
D. Vreugdenhil (Editor)

Table 31.1 Field tests on transgenic potatoes in OECD countries until 1999.

Country	Number of field tests
Australia	10
Austria	1
Belgium	7
Canada	93
Denmark	6
Finland	2
France	13
Germany	74
Italy	6
Japan	23
Netherlands	54
New Zealand	18
Portugal	2
Russian Federation	3
Spain	6
Sweden	14
Switzerland	2
United Kingdom	21
United States of America	830

Data from http://www.webdominol.oecd.org/ehs/biotrack.nsf/.OECD, Organization for Economic Cooperation and Development.

(http://www.webdominol.oecd.org/ehs/biotrack.nsf) contains approvals/permits issued for experimental field tests of genetically modified (GM) organisms in its 30-member countries until 1999. It currently records over 1185 permits issued for field-testing of transgenic potatoes between 1986 and 1999 (Table 31.1). The European Summary Notification Information Format (SNIF) database on Biotechnology and genetically modified organisms (GMOs) (http://biotech.jrc.it/deliberate/gmo.asp) lists 256 field tests on potato within the EU, whereas the database of the Information Systems for Biotechnology (http://www.isb.vt.edu/cfdocs/fieldtests1.cfm) lists 764 field tests in the USA till May 2006. This latter figure includes 162 field tests on potato from 2000 and is therefore notably inconsistent with the 830 field tests listed in the OECD database for the USA before 2000.

The numbers of approvals/permits in databases do not necessarily reflect the actual number of field releases performed due to the varying approval procedures in different countries (Nap et al., 2003). Some countries require separate applications for every specific modification in a specific plant for each location and year. In contrast, other countries will issue a single approval/permit for applications involving groups of GM crops with a range of different transgenes over multiple sites and/or years. In many non-OECD countries of Asia, Africa, South America and eastern Europe, the number of field tests on transgenic crops, including potatoes, has rapidly increased in recent years (Nap et al., 2003). The total number of field tests on transgenic potatoes is likely to exceed two thousand between 1986 and 2006. The vast majority of field tests on transgenic potatoes has occurred in the USA, with considerable activity also occurring in Canada, Germany and the Netherlands (Table 31.1).

The motivation for the ongoing development of transgenic potato cultivars lies in the immense opportunities provided by genetic engineering for the efficient genetic improvement of potatoes to mitigate many global problems associated with the potato industry. The field-testing of transgenic potatoes is a key component of all research programmes investigating genetic engineering for potato improvement. This chapter summarizes important considerations for field-testing transgenic potatoes. The intention is to illustrate these issues with key examples, rather than to provide an exhaustive survey of the scientific literature.

31.2 TRANSGENIC POTATOES IN THE CONTEXT OF POTATO BREEDING

Potato breeding has been highly successful over the past century. Traditional breeding methods involve the hybridization of parental clones and the subsequent selection among large seedling populations for superior individuals with the desired combination of traits (Plaisted et al., 1994). Single-plant selections are then propagated vegetatively and evaluated as clones for relevant agronomic and quality attributes. This general approach to potato breeding has resulted in the development of many elite clones, which have become highly successful potato cultivars. More information can be found in Chapter 8 (Bradshaw, this volume).

The autotetraploid status ($2n = 4x = 48$) of virtually all commercial potato cultivars presents one of the major difficulties associated with traditional potato breeding. The resulting tetrasomic inheritance, in conjunction with the associated high heterozygosity, adds considerable complexity to potato breeding (Conner et al., 1997). These difficulties require exceptionally large populations of potato seedlings to be screened to recover individuals for evaluation as potential cultivars. Initial selection for many desirable characters can often be inefficient and/or time consuming. Potato breeders often have to screen up to a million seedlings to find one clonal line that survives through to the release of a successful cultivar (Plaisted et al., 1994). The main objectives of most potato breeding programmes involve the improvement of specific processing attributes and resistance to pests and diseases, while maintaining or improving traits such as tuber colour, shape and yield. Since the 1950s, achieving many of these traits has been greatly assisted by the transfer of genes from related *Solanum* species (Ross, 1986).

Potatoes were one of the first crop plants in which transgenic plants were successfully regenerated (An et al., 1986; Shanin and Simpson, 1986). Potato transformation has since become routine in many laboratories and offers new opportunities to transfer genes for cultivar improvement. The introgression of new genes through the development of transgenic potatoes has several advantages over other breeding methods for potato improvement:

1. the widening of the germplasm base from which new characters can be transferred to any source of DNA;
2. the effective transfer of new genes directly into existing elite cultivars without many generations of additional crosses;
3. the ability to limit gene transfer to discrete genes without many unknown closely linked genes often associated with negative traits (linkage drag); and
4. the opportunity to develop new gene formulations involving promoter-swapping to target gene expression in a specific manner and/or using gene knock-down strategies such as RNA interference.

Using genetic engineering approaches has allowed the successful transfer of numerous transgenes into elite potato cultivars. These transgenes confer a wide range of traits, including pest and disease resistance, quality attributes for improved processing and nutrition and appearance and novel protein production for biopharming (Vayda and Belknap, 1992; Conner et al., 1997). Many further examples of transgene expression conferring specific traits, too numerous to mention here, have been published since 1997 (Chapter 30, Millam, this volume).

Plant transformation is highly unpredictable with respect to the nature of transgene integration, the magnitude, specificity and stability of transgene expression and the frequency of off-types observed within populations of independently derived transformed plants (Conner and Christey, 1994). Potatoes are no exception (Conner et al., 1997). In this context, the use of transformation for crop improvement and the intended release of superior cultivars involve 'playing the numbers game'. It is important to select a large population of independently transformed individuals in the laboratory and then screen the population to identify the lines with the desired expression of the transgene, while maintaining the phenotype and yield performance of the parental cultivar. In this respect, plant improvement through transformation is very similar to traditional breeding. A greater chance of success results from a larger programme with thorough evaluation of the lines.

If the 'numbers game' is played correctly, potato transformation offers a highly effective means of adding single genes to existing elite clones with no, or very minimal, disturbances to their genetic background. This is virtually impossible through traditional breeding because the genetic integrity of potato clones is instantly lost due to allele segregation upon sexual reproduction. This is a direct consequence of the high heterozygosity in the tetraploid potato genome. As a consequence, in potato and other clonal crops, transformation represents the only effective way to produce isogenic lines of specific genotypes/cultivars (Conner and Christey, 1994). For these reasons, transgenic potato cultivars will most likely be maintained as clonal lines from the primary transformants initially regenerated from tissue culture through to commercial release. Such cultivars will therefore be hemizygous for transgene insertion events. For tetraploid potato cultivars, this means that the transgenes will be represented in the simplex state (Aaaa) (Conner et al., 1997).

Although transformation provides a mechanism for gene transfer to existing potato cultivars without compromising existing attributes, in reality, it is not quite that simple. The insertion and expression of transgenes, as well as the tissue culture processes associated with gene transfer, can all impose important constraints that limit the performance of the resulting transgenic line.

31.3 THE IMPORTANCE OF FIELD-TESTING TRANSGENIC POTATOES

31.3.1 Field confirmation of transgenic phenotype

Many of the first reports on field tests of transgenic potatoes were largely focused on confirming that the phenotypic performance conferred by the transgene was maintained under field conditions. A high correlation between greenhouse and field performance of transgenic traits was frequently reported. For example, for transgenic potato lines expressing a *cry*1Ac9 gene conferring tuber moth resistance, a correlation coefficient of $r = 0.98$ was obtained for larval growth indices from bioassays on foliage from greenhouse and field-grown plants (Davidson et al., 2002a). High correlations between the greenhouse and field are especially common when a large population of independently derived transgenic lines are compared. However, such high correlations represent a summary of the response at a population level and provide no guarantee that the greenhouse performance will translate to field conditions for any specific transgenic line. For example, when evaluating a large number of transgenic potato lines expressing a *cry*1Ac9 gene conferring tuber moth resistance, exceptions were noted where insect resistance on greenhouse foliage was not matched with field resistance (Davidson et al., 2002b).

The failure of specific transgenic lines to exhibit the expected transgenic phenotype under field conditions has been frequently reported across a wide range of transgenic phenotypes. The frequency of plants developing viral infection was greater under field conditions, compared with prior experiments in growth chamber, in three of four transgenic lines expressing the viral coat protein genes of potato virus X (PVX) and PVY (Kaniewski et al., 1990). Similarly, one of three transgenic lines with high herbicide (chlorsulfuron) resistance in laboratory and greenhouse studies exhibited slight toxicity symptoms following herbicide applications in the field (Moses et al., 1993). Stable expression in field-grown tubers of an antisense granule-bound starch synthase gene conferring amylose-free starch (red staining with iodine) was only apparent in one of four transgenic lines selected from in vitro and greenhouse assays (Kuipers et al., 1992; Heeres et al., 2002).

These results illustrate the critical importance of verifying that transgenes confer the desired phenotype under field conditions. This is especially important when identifying specific transgenic lines for further research. The failure of some transgenic lines to perform under field conditions may be related to position effects associated with specific integration events (Conner and Christey, 1994). Because transgene expression can vary with environmental conditions (Broer, 1996), the choice of promoter for transcriptional regulation of the transgene may be an important consideration. It is also possible that the greater fluctuations of environmental conditions associated with growth in the field result in subtle dynamic changes in the flux of biochemical pathways towards the biosynthesis of many interacting metabolites. As a consequence, a higher threshold of transgene expression at the molecular level may be necessary to maintain a specific phenotype in the field.

In some cases, it is not possible to perform meaningful assessment of transgenic phenotypes under laboratory or greenhouse conditions. For example, when expressing transgenes encoding antimicrobial peptides in potatoes for soft rot resistance, it was necessary to use field-grown tuber for bioassays against *Erwinia* (Barrell, 2002). New Zealand regulatory guidelines require transgenic plants to be container-grown in containment greenhouses. Potato plants grown in pots or bags are expected to produce tubers that vary markedly in physiological properties within and between tubers as a consequence of the physical constraint imposed on tuber growth and expansion, especially for tubers developing near the sides of the container. In the case of tuber bioassays against *Erwinia*, this resulted in a high variance within and between plants of the same clone, which overshadowed any statistical differences between transgenic lines and non-transgenic parental controls. It is likely that analysis of other transgenic traits in tubers is also obscured by such limitations.

31.3.2 Occurrence of off-types

Most of the early studies reporting yield performance of transgenic potato in field tests involved only a small number of transgenic lines, for example, four lines expressing viral coat protein genes of PVX and PVY (Kaniewski et al., 1990); four lines expressing a herbicide (glufosinate/bialaphos) resistance gene (de Greef et al., 1989); four lines expressing an antisense granule-bound starch synthase gene (Kuipers et al., 1992); three lines with herbicide (chlorsulfuron) resistance (Moses et al., 1993) and one line expressing PVY coat protein gene (Malnoë et al., 1994). These studies generally report tuber yield of transgenic lines similar to the non-transgenic parental cultivar, although in some cases, a few transgenic lines had slightly reduced yield (Kuipers et al., 1992; Moses et al., 1993). It is not clear in these studies the extent to which the transgenic lines tested in the field had been pre-selected for phenotypic appearance similar to the non-transgenic control under either greenhouse conditions or previous field trials. Although these studies were important to establish that transgenic potatoes could be developed without changing the intrinsic properties of elite commercial potato cultivars, the results presented do not represent the performance of most transgenic lines developed in the laboratory.

The first field evaluations of transgenic potatoes involving large numbers of independently derived transgenic lines commonly reported marked phenotypic changes in plant appearance and/or significantly reduced tuber yield in many transgenic potato lines (Dale and McPartlan, 1992; Jongedijk et al., 1992; Belknap et al., 1994; Conner et al., 1994). In some instances, dramatic changes in plant appearance involving stunted growth, altered shoot morphology, markedly reduced yield and/or plant death were reported (e.g. Dale and McPartlan, 1992; Belknap et al., 1994; Conner et al., 1994). Typical foliage off-types include short internodes, pronounced rolling of apical leaves, small cupped leaves, wrinkled leaf surface, smaller rugose leaflets and twisted petioles, whereas typical tuber off-types include a large number of small tubers, small and elongated tubers, deep eyes, knobbly tubers from secondary growths, loss of skin colour and russetting (Belknap et al., 1994; Conner et al., 1994; Davidson et al., 2002b). In general terms, the appearance of aberrant foliage is usually accompanied by reduced tuber yield (Fig. 31.1).

At the time, it was important to understand the basis for these unexpected phenotypes in transgenic potatoes. Although many of the phenotypic changes in foliage were similar to symptoms of viral infection, virus testing quickly established that they occurred independent of the presence of viruses (Conner et al., 1994). When field-tested for the first time, transgenic potatoes are usually established from transplants micropropagated in tissue culture. The observed off-types were consistent when transgenic lines were replanted in the second year from tubers harvested in the first year (Belknap et al., 1994; Conner et al., 1994), thereby eliminating a transient response to tissue culture. Other possible explanations for the unexpected phenotypic changes included pleiotropic effects of transgene expression, insertional mutagenesis resulting from random integration into the plant genome and/or somaclonal variation arising during tissue culture. It is often difficult to discern between these options when phenotypic changes are observed in transgenic plants such as potato, which are highly heterozygous and already segregate for a multitude of other genes affecting phenotypic appearance (Conner and Christey, 1994).

If the unexpected phenotypes result from pleiotropic effects of transgene expression, then all the transgenic lines expressing the same transgene are expected to show a similar phenotypic change. The recovery of some transgenic lines with high transgene expression in the field as well as a phenotypic appearance identical to the 'parental' cultivar suggests that phenotypic changes in other lines are unlikely to result from pleiotropic effects of transgene expression (Dale and McPartlan, 1992; Conner et al., 1994). Transgenic potato cultivars are usually maintained as clonal lines from the primary transformants through to commercial release (Section 31.2). Deviant phenotypes resulting from gene disruption generally require the transgene to be in a homozygous state, which does not occur in the initial transformants regenerated from tissue culture. Therefore, insertional mutagenic events are not expected to be observed among the plants originating as the primary transformants regenerated from tissue culture, except in very rare instances of gene activation (Conner and Christey, 1994).

The appearance of off-type transgenic potato lines in field tests has been commonly attributed to 'somaclonal variation' (Dale and McPartlan, 1992; Belknap et al., 1994; Conner et al., 1994). The occurrence of somaclonal variation among plants regenerated from cell culture has been well documented for many years, including field-grown potatoes (Shepard et al., 1980; Secor and Shepard, 1981; Potter and Jones, 1991). These events arise from random epigenetic or genetic events during the cell culture and plant regeneration phase of transformation and are often apparent in the primary transgenic plants initially regenerated following transformation (Phillips et al., 1994; Veilleux and Johnson, 1998). Therefore, the occurrence of some transgenic potato lines that differ in the phenotypic appearance of the 'parental' cultivar is to be expected. The spectrum of off-type appearances is expected to vary markedly between different independently selected transgenic lines (Conner and Christey, 1994).

A comprehensive study on the frequency of off-types among field-grown transgenic potato lines was undertaken by Jongedijk et al. (1992). They compared the field performance of tuber-grown progeny from independently derived transgenic lines of cultivars

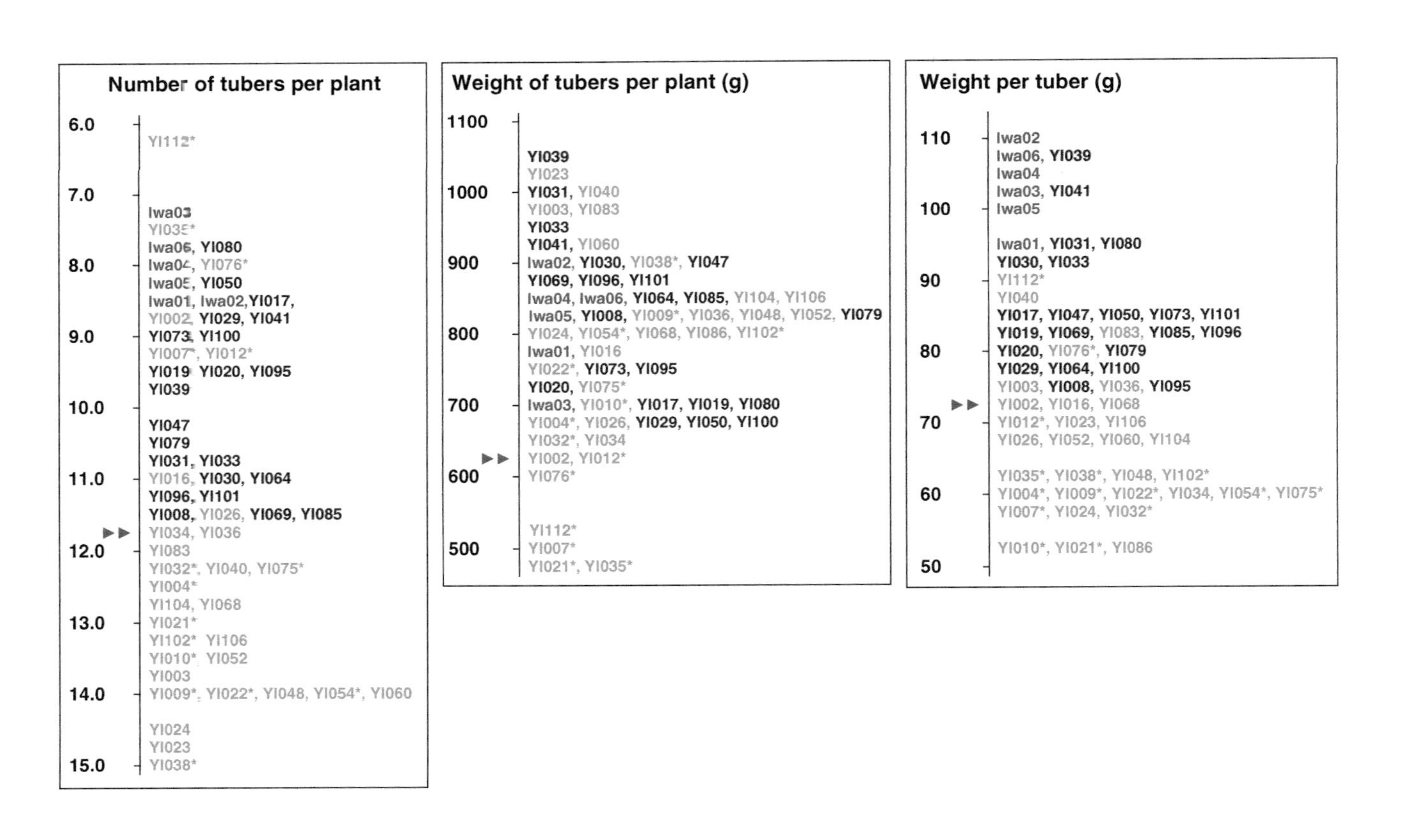

Number of tubers per plant
6.0
YI112*
7.0
Iwa03
YI035*
Iwa06, YI080
8.0
Iwa04, YI076*
Iwa05, YI050
Iwa01, Iwa02, YI017,
YI002, YI029, YI041
9.0
YI073, YI100
YI007*, YI012*
YI019, YI020, YI095
YI039
10.0
YI047
YI079
YI031, YI033
11.0
YI016, YI030, YI064
YI096, YI101
YI008, YI026, YI069, YI085
YI034, YI036
12.0
YI083
YI032*, YI040, YI075*
YI004*
YI104, YI068
13.0
YI021*
YI102*, YI106
YI010*, YI052
YI003
14.0
YI009*, YI022*, YI048, YI054*, YI060
YI024
YI023
15.0
YI038*
Weight of tubers per plant (g)
1100
YI039
YI023
1000
YI031, YI040
YI003, YI083
YI033
YI041, YI060
900
Iwa02, YI030, YI038*, YI047
YI069, YI096, YI101
Iwa04, Iwa06, YI064, YI085, YI104, YI106
Iwa05, YI008, YI009*, YI036, YI048, YI052, YI079
800
YI024, YI054*, YI068, YI086, YI102*
Iwa01, YI016
YI022*, YI073, YI095
YI020, YI075*
700
Iwa03, YI010*, YI017, YI019, YI080
YI004*, YI026, YI029, YI050, YI100
YI032*, YI034
YI002, YI012*
600
YI076*
YI112*
500
YI007*
YI021*, YI035*
Weight per tuber (g)
110
Iwa02
Iwa06, YI039
Iwa04
Iwa03, YI041
100
Iwa05
Iwa01, YI031, YI080
YI030, YI033
90
YI112*
YI040
YI017, YI047, YI050, YI073, YI101
YI019, YI069, YI083, YI085, YI096
80
YI020, YI076*, YI079
YI029, YI064, YI100
YI003, YI008, YI036, YI095
YI002, YI016, YI068
70
YI012*, YI023, YI106
YI026, YI052, YI060, YI104
YI035*, YI038*, YI048, YI102*
60
YI004*, YI009*, YI022*, YI034, YI054*, YI075*
YI007*, YI024, YI032*
YI010*, YI021*, YI086
50

Fig. 31.1. Summary of yield performance for transgenic potato lines resulting from *Agrobacterium*-mediated transformation using a binary vector containing a component of the coat protein gene from potato virus Y (PVY) and a kanamycin-resistant selectable marker gene. The 54 transgenic lines were field-tested in single plots of 10 plants, randomly interspersed with replicated plots of the non-transgenic control (cultivar Iwa). All data points represent the mean of the 10 plants in each plot. The red arrowheads mark the point on the scale for each yield parameter at which all lines below the arrowheads are significantly different from the control at the 5% probability level as determined by Duncan's multiple range test. Data points displayed in red represent the six non-transgenic control plots (cultivar Iwa). Data points in green represent 32 transgenic Iwa lines, with at least one of the three yield parameters significantly deviating from the control. Data points in blue represent 22 transgenic lines determined to be equivalent to the control for all three yield parameters. Lines indicated with an asterisk exhibited an off-type appearance in foliage traits.

Bintje (39 lines) and Escort (22 lines), all constitutively expressing the coat protein of PVX, with non-transgenic control cultivars. Each was evaluated for total yield and grading, plus 50 different UPOV (Union pour la Protection des Obtentions Végétales) morphological traits used in official cultivar registration procedures. The frequency of 'true-to-type' transgenic lines was 18% for Bintje and 82% for Escort (Jongedijk et al., 1992). Other studies have also reported the frequency of off-types among transgenic lines varying between potato cultivars (Belknap et al., 1994; Davidson et al., 2002b; Heeres et al., 2002).

The elimination of transgenic lines with an off-type appearance is therefore one of the main objectives of initial small-scale field trials with a new series of transgenic potato lines. Some studies on transgenic potato lines have reported a good correlation of yield attributes and phenotypic appearance between greenhouse and field conditions (e.g. Bones et al., 1997). However, this association does not always hold. In our experience, transgenic lines that exhibit off-type appearances in the greenhouse generally exhibit a more severe off-type in the field. Furthermore, it has been commonly reported that transgenic lines that appear 'true-to-type' under greenhouse conditions can show phenotypic deviations when grown in the field (Conner et al., 1994; Davidson et al., 2002a,b; Heeres et al., 2002). Because off-types cannot be reliably identified under laboratory or greenhouse conditions, the critical importance of field-testing transgenic potato lines is further illustrated.

31.4 THE DESIGN OF A FIELD-TESTING PROGRAMME

When developing transgenic potatoes for cultivar improvement, the aim is to recover a line with the desired expression of the transgene, while retaining the elite genetic attributes of the parental clone. Clearly, these attributes must be maintained under field conditions. Consequently, field-testing of transgenic potato lines is a highly important and critical part of the research programme. Furthermore, such field-testing must be conducted at an early stage in the programme to eliminate all the lines performing at a sub-optimal level. Because research space and resources are usually finite, this allows efficient use of resources.

In any new breeding cycle of a traditional potato breeding programme, field evaluation begins with screening one to a few plants from many hundreds or thousands of progeny lines (Plaisted et al., 1994). Many lines are discarded after one season and retained tubers from a reduced number of selected lines used to establish replicated trials the next year. As selection further reduces the number of lines over successive years, the plot size and replication are increased, until a few highly promising lines are incorporated into regional trials (Plaisted et al., 1994).

The strategy for field-testing transgenic potato lines is very similar. In the first year, a well-designed field-testing programme will plant many transgenic lines in small plots, followed by increasingly larger, replicated plots in subsequent years. For any combination of potato cultivar with a specific vector or transgene, a large number of independently derived transformed lines generated in the laboratory are established as single-field plots in the first year. In our field-testing programme, we routinely establish single 3-m plots of 10 plants for each transgenic line (e.g. Davidson et al., 2002b). For convenience, plots are separated by 1-m gaps within rows, with rows a standard 0.75 m apart. Lines derived from the same cultivar are always planted in a block of neighbouring plots, with replicated non-transgenic plots of the parental cultivars positioned randomly among the single plots of numerous independently derived transgenic clones. We generally establish one non-transgenic control plot for every 5–10 plots of transgenic lines. In this manner, any statistically significant variance components between plots of the same cultivar can be determined. Our first-generation field tests are usually established from in vitro micropropagated plants initially hardened-off under greenhouse, then screen-house, conditions.

The field expression of the transgene and the phenotypic value of the conferred transgenic trait as well as the appearance of deviant phenotypes can be readily determined from such small-plot field tests. Consistent results among transgenic lines have been reported over successive field seasons using this approach (Davidson et al., 2004; Douches et al., 2004). Such initial field assessments could be determined with only 3–5 plant plots; indeed, others have successfully conducted such trials with only three plants per plot (Heeres et al., 2002). However, the key advantage of 10 plants per plot is that when promising lines are identified, more tubers are harvested and are therefore available for establishing large-scale replicated field trials in the subsequent season. This provides a seed source in which tubers are treated in the same manner as commercial crops and represents a more realistic cycle of storage and dormancy breaking for more advanced field tests.

From these first-season field tests, selections are made of high-performing lines for incorporation into larger-scale field trials with more rigorous experimental plot design and statistical analysis. For the efficient use of resources, it is important to define how many lines should be produced for initial field evaluation and how many plants per plot are required to allow meaningful selection between the transgenic lines. The number of required lines is controlled by two independent components: the frequency of lines with the required transgene expression and the frequency of off-types. Another important consideration is to determine whether the potato cultivar influences either of these estimates.

In a thorough study by Heeres et al. (2002), field-testing was conducted on 995 transgenic lines from 14 potato cultivars that expressed an antisense granule-bound starch synthase gene conferring amylose-free starch. After small-plot field tests assessing both expression of the transgenic trait and true-to-type appearance, it was estimated that, depending on the cultivar, between 70 and 500 transgenic lines are required for initial field tests to identify 10 transgenic lines in each cultivar worthy of subsequent scale-up field tests (Heeres et al., 2002). Likewise, we have field-tested transgenic potato lines expressing a *cry*1Ac9 gene determined to exhibit resistance to tuber moth and have a normal phenotypic appearance under greenhouse conditions. Of 56 transgenic lines from seven cultivars, only 13 lines from five of the cultivars were identified with sufficient insect resistance as well as true-to-type appearance and yield to merit planting in the following year (Davidson et al., 2002b).

Promising lines identified from the first-season field tests are established in replicated plots the following season. We usually aim to evaluate up to 10 high-performing transgenic lines and the non-transgenic control for each combination of cultivar and transgene or vector. At this stage, we increase plot size to five rows of 10 plants, with plots replicated three to four times (e.g. Davidson et al., 2006). Often, tubers harvested from the 10 plants of the previous season need to be divided to produce sufficient 'seed tubers', especially when some tubers are sacrificed for analysis of the transgenic trait the previous season. If necessary, it is also possible to establish field tests of this size using tissue culture transplants. This experimental design permits more rigorous statistical analysis of performance. For each replicated plot, we record performance measurements of the transgenic phenotype and yield from only the three inner rows of each replicated plot. In this manner, plants are evaluated when surrounded by other plants of the same line. However, the tubers from all five rows are harvested to provide multiplied 'seed tubers' for further scale-up field trials of lines worthy of further evaluation. Transgenic lines recorded as being equivalent to the parental cultivar in the single-plot field tests of the previous season often exhibit slightly reduced yield in larger-scale field trials when evaluated using more rigorous experimental plot design and statistical analysis (Moses et al., 1993; Davidson et al., 2006).

31.5 STRATEGIES TO REDUCE THE FREQUENCY OF OFF-TYPES

It is clear from Section 31.3.2 that the appearance of off-types among populations of independently derived transgenic lines is an inherent feature of potato genetic engineering. In highly heterozygous clonal crops such as potato, the genetic integrity of cultivars is instantly lost on self- or cross-pollination and sexual crossing. Therefore, in most instances, transgenic potato cultivars will be clonally propagated from the primary transformant arising in tissue culture through to commercial release (Section 31.2). Consequently, the possibility of segregating the transgenic insertion event away from any independently arising somaclonal event following sexual crosses is not appropriate. From the perspective of potato improvement, any strategy to reduce the frequency of off-types among primary transformants is important.

The extent and frequency of phenotypically variant plants from tissue culture varies markedly under different culture conditions and is generally understood to increase when the time to plant regeneration is longer (Cullis, 1983; Karp, 1991; Veilleux and Johnson, 1998). Efficient plant regeneration systems are therefore a key component of plant transformation. During our first field tests in 1988, all the transgenic potato lines exhibited deviant phenotypes for at least one morphological trait (Conner et al., 1994). In subsequent field tests, the frequency of true-to-type plants substantially increased following improvements in our transformation protocol (Fig. 31.1). This emphasizes the importance of initiating field-testing early in a research programme to ensure that the laboratory protocols being employed are capable of producing true-to-type plants in the field.

When recovering any transgenic line in tissue culture, the first shoot to regenerate from each transformation event is usually selected. This is based on a widely held premise that the frequency of somaclonal variation increases with longer periods in cell culture and the resulting assumption that the longer time a shoot takes to regenerate, the greater chance that an 'off-type' will appear. However, an important experiment by Barrell (2002) established that this assumption is not necessarily correct. Using our standard potato transformation protocol, Barrell (2002) regenerated multiple shoots from a series of transformation events and used Southern analysis to confirm that each of the multiple shoots arose from a single transformed somatic potato cell. The analysis of phenotypic variation in field tests demonstrated marked differences between these multiple regeneration events, the origin of which must have occurred after T-DNA insertion and consequently during the tissue culture phase. This result unequivocally demonstrates that somaclonal variation occurs during tissue culture and happens independent of transgene insertion. Most importantly, the first shoots recovered did not necessarily exhibit less deviant phenotypes (Fig. 31.2). Plants exhibiting a true-to-type appearance were occasionally derived from later regeneration events. Therefore, one strategy to assist recovery of true-to-type plants when developing transgenic potato lines is to regenerate and evaluate multiple shoots from each transformation event.

Tissue culture is generally recognized as a stress environment (Madlung and Comai, 2004), and the extent and frequency of off-type plants arising from tissue culture increases when stress is applied in cell culture (Cullis, 1983; Karp, 1991). Selectable marker genes are routinely used during the recovery of transformed plants in tissue culture to facilitate

Fig. 31.2. An extreme example illustrating the value of evaluating multiple shoots regenerated from each potato transformation event. Southern analysis established that both clones were derived from the same transformation event. The poorly growing clone on the right was derived from the first regenerated shoot, whereas the clone on the left was derived from a later regeneration event.

the identification of rare transformation events among a large population of wild-type cells. For potato transformation, kanamycin resistance is almost universally used (Conner et al., 1997), although other possibilities are available (Barrell et al., 2002). Under such selection systems, transformed cells are selected as the rare survivors among many dying or dead cells. This can result in greater stress in the culture environment as a consequence of the release of toxic metabolites and volatiles from the large population of dying wild-type cells (Conner, 1986). It is therefore not surprising that the frequency of off-types in field tests is greater among populations of transgenic potato plants than that among non-transgenic plants regenerated without selection (Dale and McPartlan, 1992; Belknap et al., 1994). A recent alternative approach for the recovery of transgenic potato lines, without the use of selectable marker genes, involves the PCR screening of plants for the gene-of-interest following regeneration from potato tissue co-cultivated with *Agrobacterium* (de Vetten et al., 2003). It is to be expected that the recovery of transgenic lines in this manner may substantially reduce the frequency of off-types during field tests of transgenic potatoes.

31.6 ASSESSMENT OF BIOSAFETY ISSUES

The deployment of transgenic crops in global agriculture has raised a diverse range of perceived environmental and food safety issues (Conner et al., 2003). Many of these issues are relevant for the commercial release of transgenic potatoes (Conner, 1994) and include

1. the incidence of transgenic volunteer or ground-keeper potatoes and their potential invasiveness or weediness;
2. the incidence of pollen-mediated gene flow to other potatoes and related species;
3. the impact on non-target invertebrates and microflora, especially on species considered beneficial;
4. the potential increased selection pressure on pest and disease populations to become more virulent;
5. the evaluation of transgene expression products on food safety of the edible tubers; and
6. the impact on nutritional composition and concentration of natural toxins (e.g. glycoalkaloids).

Plants growing under field conditions are the most appropriate environment in which to assess such biosafety issues. Consequently, initiating biosafety assessments in conjunction with field tests of transgenic potatoes must be considered important. This represents an opportunity to gather valuable data to help address issues of public concern over the deployment of transgenic potatoes. For example, during some of the early field tests on transgenic potatoes, data were recorded on the frequencies and distance of pollen-mediated transgene dispersal (Tynan et al., 1990; McPartlan and Dale, 1994), as well as nutritional composition and animal feeding studies (Monro et al., 1993). Examples of various biosafety evaluations conducted on field-grown transgenic potatoes can be for transgenic lines with herbicide (chlorsulfuron) resistance (Conner, 1995), antisense granule-bound starch synthase gene conferring amylose-free starch (Heeres et al., 1995) and cystatin expression for nematode resistance (Celis et al., 2004).

The large number of field tests on transgenic potatoes conducted around the globe over the past 20 years (Section 31.1) provided an opportunity to repeatedly verify the validity of the biosafety data over a wide range of environments and with a wide range of potato cultivars. In some respects, this has been a wasted opportunity. In many cases, a relevant set of data has not been recorded, or if it has, it has not been published or made available to the wider scientific community. This would have greatly increased the confidence in the conclusions made to date and facilitate future applications for field-testing transgenic potatoes. A good example is pollen-mediated gene flow from field tests of transgenic potato. Early studies established that the frequency of transgenic progeny from non-transgenic potatoes growing in rows adjacent to transgenic potatoes ranges from 1 to 24% and rapidly falls to negligible within 3–5 m (Tynan et al., 1990; McPartlan and Dale, 1994). Based on this and similar data, it was recommended that a distance of 20 m is generally adequate for containing novel gene constructs during initial field tests of transgenic potato (Conner and Dale, 1996). If this data had been repeatedly verified in many environments and for additional potato cultivars, some regulatory bodies may no longer require the removal of flower buds during field tests of transgenic potato or set more realistic isolation distances.

There has been a much greater emphasis on gaining knowledge on the impacts of transgenic crops since the late 1990s. However, too many of these studies are still performed under laboratory or greenhouse conditions, rather than under a more appropriate field environment. It is unfortunate that field tests on transgenic crops are often limited by the containment controls imposed by regulatory authorities to manage the perceived risks. The resulting field designs are not always conducive to critical experimental evaluation of biosafety issues. This is ironic because the information that could be obtained would be especially valuable to assist regulatory bodies to make better informed decisions.

When conducting experiments to assess the ecological risk or food safety evaluation of transgenic potatoes, it is critical to use appropriate controls (Conner et al., 2003). The transgenic lines should be compared to the isogenic non-transgenic 'parental' cultivar. Ideally, such transgenic lines should be selected from previous field trials and established to be 'true-to-type' with the desired expression of the transgene. This allows unambiguous assessment of the impact of the transgene during biosafety assessment, because the comparison is between lines that differ only by the presence/absence of the transgene. In highly heterozygous crops such as potato, transformation represents the only effective way to produce such isogenic lines (Section 31.2).

It is also important that biosafety assessments include other non-transgenic potato cultivars as additional controls to determine the relevance of any slight changes in the performance of the transgenic line. For example, in conjunction with field tests on transgenic potatoes secreting an antimicrobial peptide (magainin II) for improved resistance to *Erwinia carotovora* isolates that incite soft rot diseases, we have investigated the impacts of the transgenic plants on microbial communities on the root, tuber and leaf surfaces and during decomposition of the plant material (O'Callaghan et al., 2004; Conner and O'Callaghan, 2005). In all experiments, comparisons were made between transgenic line, the non-transgenic parental cultivar and another unrelated non-transgenic cultivar. In some instances, minor differences were apparent between the transgenic lines and the non-transgenic parent cultivar, but these were usually negligible relative to the differences between the two non-transgenic cultivars. Similarly, when comparing the compositional similarity of tubers from transgenic potato and the isogenic non-transgenic parental cultivar, metabolomic profiling revealed subtle differences (Catchpole et al., 2005). However, these were very minor relative to the large variation in metabolite profile between conventional non-transgenic cultivars (Catchpole et al., 2005).

31.7 CONCLUSIONS

Transgenic potatoes have been widely field-tested around the world since 1986. Although most of the activity initially occurred in North America and western Europe, since the turn of the century this has become a global activity extending into many Asian, African, South American and eastern European countries. In general terms, field-testing of transgenic potatoes has three key components:

1. Verifying that the transgene confers the desired phenotype under field conditions;
2. Identifying transgenic lines that retain all the phenotypic traits, including yield and quality attributes characteristic of the parental cultivar and
3. Initiating biosafety assessments on transgenic potatoes under realistic experimental conditions.

When developing transgenic potatoes for cultivar improvement, the aim is to recover a high-performing line with the desired expression of the transgene, while retaining all the elite genetic attributes of the parental clone. From the perspective of potato improvement, an important limitation is the frequency of off-types due to somaclonal variation arising during the tissue culture phase of plant transformation. Consequently, field-testing must be conducted at an early stage in the research programme to eliminate transgenic lines performing at a sub-optimal level. The majority of these off-types can be easily eliminated using single-field plots with up to 10 plants. A smaller number of high-performing lines are subsequently established in larger-scale field trials with more rigorous experimental plot design and statistical analysis. These latter field tests are the most appropriate stage to initiate the assessment of biosafety issues.

ACKNOWLEDGEMENTS

Many thanks to the numerous colleagues for their enthusiastic assistance to plant, maintain, harvest and record data during our field-testing programme on transgenic potatoes since 1988. The assistance of Jeanne Jacobs in preparing this manuscript is also much appreciated.

REFERENCES

An G., B.D. Watson and C.C. Chiang, 1986, *Plant Physiol.* 81, 301.

Barrell P.J., 2002, Expression of synthetic magainin genes in potato. PhD thesis, Lincoln University, New Zealand, 153 pp.

Barrell P.J., Y.J. Shang, P.A. Cooper and A.J. Conner, 2002, *Plant Cell Tiss. Org.Cult.* 70, 61.

Belknap W.R., D. Corsini, J.J. Pavek, G.W. Snyder, D.R. Rockhold and M.E. Vayda, 1994, *Am. Potato J.* 71, 285.

Bones A.M., B.I. Honne, K.M. Nielson, S. Visvalingam, S. Ponnampalam, P. Winge and O.P. Thangstad, 1997, *Acta Agric. Scand. Sect. B Soil Plant Sci.* 47, 156.

Broer I., 1996, *Field Crops Res.* 45, 19.

Catchpole G.S., M. Beckman, D.P. Enot, M. Mondhe, B. Zywicki, J. Taylor, N. Hardy, A. Smith, R.D. King, D.B. Kell, O. Fiehn and J. Draper, 2005, *Proc. Natl. Acad. Sci. U.S.A.* 102, 14458.

Celis C., M. Scurrah, S. Cowgill, S. Chumbiauca, J. Green, J. Franco, G. Main, D. Kiezebrink, R.G.F. Visser and H. Atkinson, 2004, *Nature* 432, 222.

Conner A.J., 1986. *N.Z. J. Technol.* 2, 83.

Conner A.J., 1994, In: W.R. Belknap, M.E. Vayda and W.D. Park (eds), *The Cellular and Molecular Biology of Potatoes*, 2nd edition, p. 245. CAB International, Wallingford.

Conner A.J., 1995, In: D.D. Jones (ed.), Proceedings of the 3rd International Symposium on the Biosafety Results of Field Tests of Genetically Modified Plants and Microorganisms, p. 245. University of California, Oakland.

Conner A.J. and M.C. Christey, 1994, *Biocontrol Sci. Technol.* 4, 463.

Conner A.J. and P.J. Dale, 1996, *Theor. Appl. Genet.* 92, 505.

Conner A.J., T.R. Glare and J.P. Nap, 2003, *Plant J.* 33, 19.

Conner A.J., J.M.E. Jacobs and R.A. Genet, 1997, In: G.D. McLean, P.M. Waterhouse, G. Evans and M.J. Gibbs (eds), *Commercialisation of Transgenic Crops: Risk, Benefit and Trade Considerations*, p. 23. Cooperative Research Centre for Plant Science and Bureau of Resource Sciences, Canberra.

Conner T. and M. O'Callaghan, 2005, *Grower* 60 (10), 52.

Conner A.J., M.K. Williams, D.J. Abernethy, P.J. Fletcher, P.J. and R.A. Genet, 1994, *N.Z. J. Crop Hort. Sci.* 22, 361.

Cullis C.A., 1983, *CRC Crit. Rev. Plant Sci.* 1, 117.

Dale P.J. and H.C. McPartlan, 1992, *Theor. Appl. Genet.* 84, 585.

Davidson M.M., R.C. Butler, S.D. Wratten and A.J. Conner, 2004, *Ann. Appl. Biol.* 145, 271.

Davidson M.M., R.C. Butler, S.D. Wratten and A.J. Conner, 2006, *Crop Prot.* 25, 216.

Davidson M.M., J.M.E. Jacobs, J.K. Reader, R.C. Butler, C.M. Frater, N.P. Markwick, S.D. Wratten and A.J. Conner, 2002a, *J. Am. Soc. Hort. Sci.* 127, 590.

Davidson M.M., M.F.G. Takla, J.K. Reader, R.C. Butler, S.D. Wratten and A.J. Conner, 2002b, *N.Z. Plant Protection* 55, 405.

de Greef W., R. Delon, M. De Block, J. Leemans and J. Botterman, 1989, *Bio/Technology* 7, 61.

de Vetten N., A. Wolters, K. Raemakers, I. van der Meer, R. ter Stege, E. Heeres, P. Heeres and R. Visser, 2003, *Nat. Biotechnol.* 21, 439.

Douches D.S., W. Pett, F. Santos, J. Coombes, E. Grafius, W. Li, E.A. Metry, T.N. El-din and M. Madkour, 2004, *J. Econ. Entomol.* 97, 1425.

Heeres P., M. Schippers-Rozenboom, E. Jacobsen and R.G.F. Visser, 2002, *Euphytica* 124, 13.

Heeres P., A.C. van Swaaij, P.M. Bruinenberg, A.G.J. Kuipers, R.G.F. Visser and E. Jacobsen, 1995, In: D.D. Jones (ed.), Proceedings of the 3rd International Symposium on the Biosafety Results of Field Tests of Genetically Modified Plants and Microorganisms, p. 271. University of California, Oakland.

Jongedijk E., A.A.J.M. de Schutter, T. Stolte, P.J.M. van den Elzen and B.J.C. Cornelissen, 1992, *Bio/Technology* 10, 422.

Kaniewski W., C. Lawson, B. Sammons, L. Haley, J. Hart, X. Delannay and N.E. Turner, 1990, *Bio/Technology* 8, 750.

Karp A., 1991, *Oxf. Surv. Plant Mol. Cell Biol.* 7, 1.

Kuipers G.J., J.T.M. Vreem, H. Meyer, E. Jacobsen, W.J. Feenstra and R.G.F. Visser, 1992, *Euphytica* 59, 83.

Madlung A. and L. Comai, 2004, *Ann. Bot.* 94, 481.
Malnoë P., L. Farinelli, G.F. Collet and W. Reust, 1994, *Plant Mol. Biol.* 25, 963.
McPartan H.C. and P.J. Dale, 1994, *Transgenic Res.* 3, 216.
Monro J.A., K.A.C. James and A.J. Conner, 1993, *FoodInfo Rep.* 6, 1.
Moses T.J., R.J. Field and A.J. Conner, 1993, In: Proceedings I: 10th Australian Weeds Conference and 14th Asian Pacific Weed Science Society Conference, p. 319. Weed Society of Queensland, Brisbane.
Nap J.P., P.L.J. Metz, M. Escaler and A.J. Conner, 2003, *Plant J.* 33, 1.
O'Callaghan M., E.M. Gerard, N.W. Waipara, S.D. Young, T.R. Glare, P.J. Barrell and A.J. Conner, 2004, *Plant Soil* 266, 47.
Phillips R.L., S.M. Kaeppler and P. Olhoft, 1994, *Proc. Natl. Acad. Sci. U.S.A.* 91, 5222.
Plaisted R.L., M. Bonierbale, G.C. Yencho, O. Pineda, W.M. Tingey, J. van den Berg, E.E. Ewing and B.B. Brodie, 1994, In: W.R. Belknap, M.E. Vayda and W.D. Park (eds), *The Cellular and Molecular Biology of Potatoes*, 2nd edition, p. 1. CAB International, Wallingford.
Potter R. and M.G.K. Jones, 1991, *Plant Sci.* 76, 239.
Ross H., 1986, *J. Plant Breed. Suppl.* 13, 1.
Secor G.A. and J.F. Shepard, 1981, *Crop Sci.* 21, 102.
Shanin E.A. and R.B. Simpson, 1986, *HortScience* 21, 1199.
Shepard J.F., D. Bidney and E. Shahin, 1980, *Science* 208, 17.
Tynan J.L., M.K. Williams and A.J. Conner, 1990, *J. Genet. Breed.* 44, 303.
Vayda M.E. and W.R. Belknap, 1992, *Transgenic Res.* 1, 149.
Veilleux R.E. and A.A.T. Johnson, 1998, *Plant Breed. Rev.* 16, 229.

Chapter 32

Soil-Free Techniques

Steve Millam[1] and Sanjeev K. Sharma[2]

[1]*Institute of Molecular Plant Sciences, University of Edinburgh, Kings Buildings, Mayfield Road, Edinburgh EH9 3JH, UK;*
[2]*Gene Expression, Scottish Crop Research Institute, Invergowrie, Dundee DD2 5DA, UK*

32.1 INTRODUCTION

The use of soil-free techniques provides unique opportunities for producing seed potatoes at enhanced rates in a controlled environment with no, or a highly reduced, incidence of pests and diseases.

The most important applications of soil-free methodologies can be considered to be in the commercial production of disease-free potato seed tubers. Potato is infected by more than 30 viruses and virus-like agents. Being systemic pathogens, potato viruses are able to disseminate through the tubers thus leading to substantial levels of crop degeneracy with every increase in the number of crop cycles. Thus, the yield losses attributed to virus infection are not only confined to a particular year but also accumulated with each successive propagation cycle for as long as the infected tubers are used as seed. To check this spread of viruses and other pathogens, most seed potato certification schemes in use today are largely based on a uni-directional principle, whereby a certified nuclear stock is used to initiate a seed multiplication scheme, and after many in vitro, semi-in vivo and field multiplications, a much larger quantity of certified seed is subsequently generated for further use. Enhanced propagation rates using fewer field generations reduce the seed's exposure to field-borne diseases, minimize the accumulation of tuber-borne diseases and, additionally, facilitate the rapid introduction of new varieties enabling early initiative market gains for the producer. Soil-free techniques play an increasingly important role in these activities, and the critical in vitro and semi-in vivo phases almost invariably rely on the use of one or more of the range of soil-free techniques that have been developed.

Historically, many of the techniques related to the production of high-quality planting material have evolved concurrently with the increasing sophistication of the potato industry and of scientific advances in related areas. Potato seed certification was initiated in Europe in the early 1900s (Appel, 1934). In the USA, the National Plant Quarantine Act was adopted in 1912, and the importance of disease resistance among potato varieties was also recognized as a crucial factor towards improving crop productivity. The history of seed potato certification has been extensively reviewed (Shepard and Claflin, 1975). It was demonstrated that virus diseases could be transmitted from plant to plant, and the crucial role of aphids in transmitting plant viruses was further established (Schultz and Folsom, 1923). In 1926, the transmission of spindle tuber viroid was demonstrated (Goss,

Potato Biology and Biotechnology: Advances and Perspectives
D. Vreugdenhil (Editor)

1926). In the 1920s, the first greenhouse indexing services were set up, to assist growers in isolating disease-infected seed sources, and this was a significant step in improving seed quality because of the subsequent reduction in mosaic virus disease and potato bacterial ring rot. The first truly 'virus-free' potato programmes were developed in Scotland and the Netherlands in the 1950s.

Meristem tip culture (first reported by Morel and Martin, 1955) is a fundamental tool in the production of virus-free potatoes. An apical meristem, together with one to three adjacent leaf primordia measuring 0.1–0.5 mm, can be referred to as a meristem tip. The typical meristem dome is generally free of viruses or, in some cases, may carry a very low concentration of the viruses. Although not proven unequivocally, this concept arises from several different hypotheses, including that the mitotic chromosome replication and high-auxin content found in the meristem interfere with viral nucleic acid metabolism and, thus, are inhibitory to virus multiplication. Possibly, as compared with other regions in the plant, the apical region possess more active virus-inactivating systems, and a further hypothesis is to suggest that although virus particles proliferate through the plant vascular system, the meristem, being a zone of actively dividing cells, is devoid of vascular system. In the production of virus-free material, the meristem tip from the infected plants is excised aseptically and cultured on a defined media to obtain virus-free mericlones. Although the process is principally used for virus elimination, it often also has the additional benefits of the simultaneous elimination of other pathogens such as mycoplasmas, bacteria and fungi (http://cropandsoil.oregonstate.edu/fpsp/psf5.htm). Viruses can also be inactivated by thermotherapy, a process of growing the plants at higher temperatures (35–37°C) for a period of up to 4–6 weeks. The process of thermotherapy disrupts the production and activity of virus-encoded movement proteins and coat proteins, thus restricting their movement through plasmodesmata and the plant vascular system, and their reconstitution from replicated viral nucleic acids, respectively. In potato, the apical meristems are often free of viruses, but some viruses actually invade the meristematic region of the growing tips. Moreover, potato virus S (PVS) and potato virus X are difficult to eliminate by thermotherapy or meristem-tip culture alone. In these cases, it has been possible to obtain virus-free plants by combining meristem-tip culture with thermotherapy. Additionally, chemotherapy, i.e. the treatment of plants with anti-viral chemicals such as ribavirin or acyclovir, has also been used to eliminate viruses, either alone or in combination with meristem culture and/or heat treatment (Sanchez et al., 1991).

The virus-indexed and pathogen-free mericlones are subjected to a rapid tissue culture plant propagation method to generate abundant clean (pathogen-free) cultures. Several facets of the generic science of plant tissue culture have both current and potential future applications. These include micropropagation, i.e. clonal propagation through axillary shoot proliferation using explants (isolated plant tissues, e.g. leaf sections or stem internodes used in in vitro culture) containing pre-existing meristems, de novo shoot production following the induction of adventitious meristems by the application of plant growth regulators and somatic embryogenesis, which is the process of embryo initiation and development from vegetative or non-gametic cells. All these processes hold significant applications in potato biotechnology. The widespread introduction of plant tissue culture to seed potato production during the 1970s enabled the rapid multiplication of disease-free seed material and resulted in more productive seed stock. Micropropagation now underpins many seed potato production systems and specifically provides the nuclear stock material, in the form of microplants (plants derived through in vitro axillary bud proliferation) or micro-tubers (in vitro produced tubers) for their subsequent use in a chain of potato seed production programmes. The microplants and/or micro-tubers are specifically used to increase the number of first-year clones. These can be either planted directly in the field or more preferably used as the planting stock for the production of pre-basic seed comprising small seed potato tubers (mini-tubers). For producing mini-tubers, the in vitro propagated propagules are planted at a high density, and the production can be carried out all year round in a glasshouse environment. Although micro-tubers were first described in the mid-1950s (Barker, 1953; Mes and Menge, 1954), they have generally been under-utilized as a propagule, mainly due to physiological constraints, but offer a fascinating and highly informative system for a number of basic studies on tuber formation. Micro-tubers are resting structures, and their packaging requires much less space when compared with microplants. They can be conveniently used for long-distance shipments without any need for specialist light conditions, growth medium and immediate/intermittent subculturings. These features are useful in meeting quarantine restrictions from different countries and thus make micro-tubers a favourable option for international germplasm exchange.

The term hydroponics is used to describe many different types of system for growing plants without soil but is far from being a new concept. It has been suggested that the hanging gardens of Babylon and the floating gardens of the Aztecs of Mexico were early examples of 'Hydroponic' culture (http://en.wikipedia.org/wiki/Hydroponic). The scientific principles of hydroponics were established in the 1930s, when researchers of plant metabolism discovered that plants absorbed nutrients as simple ions in water. The definition of hydroponics has been extended to include the use of an inert medium, such as gravel, sand, peat, vermiculite, perlite, rock wool or sawdust, to which is added a nutrient solution, containing all the essential elements needed by the plant for its normal growth and development. The productivity of these technologies can be greatly advanced by increasingly sophisticated methods that have been developed for the accurate control of both the gaseous atmosphere and the light intensity in enclosed hydroponic systems to ensure optimal production. Due to the very high density of planting, increased yields per unit area can be obtained, and pest and disease measures tightly controlled. Additionally, the previous limitations of seasonality no longer become a problem, as production can be maintained all year round. The use of a continuous flowing nutrient film technique was demonstrated for actual tuber production with acceptable yields in studies at NASA (Wheeler et al., 1990) as part of the Controlled Ecological Life Support Systems (CELSS) studies.

Aeroponics is a variation (generally attributed to work by NASA in the 1960s and 1970s, where research was undertaken on food production in low-gravity situations) on the fundamental hydroponic technique that has been enhanced by the use of nebulizers, foggers or other devices to create a fine mist of solution to enable nutrient delivery to plant roots. In many systems, the plant roots are suspended above a reservoir of nutrient solution or inside a channel connected to a reservoir. Nutrient solution is delivered to the roots through sprayer nozzles, which then drips or drains back into the reservoir. It is generally considered that such systems have advantages due to the increased aeration of nutrient solution and its consequent effect on increased oxygen availability to plant roots, stimulating growth and preventing algae formation.

It will be apparent that many of the above techniques are closely inter-related, and in many cases complementary. Technical developments are constantly being evolved and introduced into commercial seed production programmes worldwide. With the increasing emphasis on energy conservation, the need to reduce production costs per unit area and to maintain output quality, the techniques of soil-free culture of potato are likely to increase in importance.

32.2 MINI-TUBER PRODUCTION

Although the definition given above that mini-tubers are small seed potato tubers that can be produced year round in glasshouses from in vitro propagated plantlets planted at a high density is fundamentally correct, a more modern definition would include that the propagule itself would be grown in a soil-free medium, within a facility for which there are appropriate procedures and physical barriers to prevent the entry of plant pathogens and insects. This technology package is a fundamental component of many seed potato production companies.

The basis of mini-tuber production is to rapidly facilitate the stage between the delivery of virus-free material derived from meristem culture nuclear stock and the production of tubers destined for field planting. Although schemes can vary between countries and markets, the format remains constant. The basic starting point is the nuclear stock that comprises in vitro pathogen-tested microplants to ensure that the starting material is pathogen-free, according to a programme of official testing for indigenous and non-native pathogens. Microplants or micro-tubers derived from the nuclear stock can then be grown in a pathogen-free medium to produce mini-tubers or, less commonly, can be grown in the field to produce tubers under protected environments. Production rates vary between systems and the method of harvesting. Some commercial companies quote rates of up to 1000 mini-tubers per square metre with tuber weight ranging from 1 to 5 g following a non-destructive harvesting every 40–50 days from a crop derived from a single microplant under optimum glasshouse conditions (http://www.quantumtubers.com/techinfo.htm).

The effectiveness of using mini-tubers for the selection of agronomic characters was reported by Gopal et al. (2002), suggesting that the selection for tuber yield can be practised at the mini-tuber crop level in potato-breeding programmes. The effectiveness of an early stage selection for quality traits in the potato-breeding population was also demonstrated by the selection for low-glucose content in mini-tubers by Xiong et al. (2002).

In a theme that will recur throughout this chapter, the production of mini-tubers under low-gravity situations has also been the subject of considerable interest. Kordyum et al. (1997) first reported the cultivation of microplants grown for 8 days on board the 'Mir' orbital space station. Spherical mini-tubers were formed with no statistically significant differences in either the frequency of tuber formation or the tuber size during this period under space flight and stationary conditions. The authors reported, however, that the grain size of starch was decreased and lamellae within the amyloplasts were locally enlarged.

In summary, although many mini-tuber production systems exist, they are all based on the rapid propagation of nuclear stock microplants in vitro, described below in Section 32.3.1, and the subsequent planting out, often in enhanced environmental conditions (including hydroponic or derived technologies) for the rapid production of mini-tubers, which are generally used as the starting point for a field multiplication system.

32.3 IN VITRO MULTIPLICATION TECHNIQUES

32.3.1 Axillary-bud proliferation

Micropropagation is the rapid vegetative propagation of plants under in vitro conditions of high light intensity, controlled temperature and a defined nutrient medium. The technique has been applied to a substantial number of commercial vegetatively propagated plant species. The cost and efficiency of production are the key issues in the commercial micropropagation of potato, and different modes adopted for potato micropropagation have different implications over these issues. In vitro propagation of potato by the serial culture of single-node cuttings (containing axillary buds) has been widely used in the rapid multiplication of disease-free material in elite seed potato programmes (Goodwin et al., 1980). The in vitro produced microplants are uniform, true-to-parent type and reliable propagules of choice. They are extensively used by the industry for the production of mini-tubers, almost exclusively in greenhouses, although direct planting into the field can also be employed. In many situations, a relatively low-technology facility can be assembled and utilized – costs can be cut by using generic systems such as supermarket shelving for lighting racks, the use of household sugar rather than laboratory-grade sucrose, replacing distilled water by de-mineralized water and the use of inexpensive food containers as culture vessels. These, although not technically sterile, are produced as food quality items and can reduce operational costs considerably. However, an area that requires careful attention is the matrix used for culture, which, when expensive laboratory-grade agar is used, accounts for more than 75% of the total media cost. Regarding the use of less-expensive agar alternatives, one report cited the use of sago, a processed (gelatinized) edible starch, as a replacement for agar in potato in vitro systems (Naik and Sarkar, 2001), but the economics of this approach need to be determined. Another critical factor is the need for some form of aeration for the growing cultures. The use of cheaper food-grade vessels does not allow proper aeration, so a form of 'venting' using micropore-tape may be needed. Several commercial micropropagation companies have devised ingenious systems for carbon dioxide enrichment of the in vitro culture vessels. Additionally, there are low-cost lighting systems on the market. A recent technical report on manipulation of the culture environment on in vitro air movement and its impact on the process of photosynthesis carried out by potato plantlets (Kitaya et al., 2005) described how enhancement of the air movement in the culture vessel promoted photosynthesis of the in vitro plantlets. Photoautotrophic micropropagation using sugar-free medium is an area of potential utility to scale-up potato micropropagation for commercialization, and recent advancements in the development and utilization of large culture vessels and case studies concerning photoautotrophic micropropagation were described by Zobayed et al.

(2004). Manual handling is a major requirement for all conventional micropropagation systems, and it has been estimated that labour costs may represent 65–85% of the total costs; consequently, it has been proposed that the introduction of methods for the automation of specific stages in the whole micropropagation process may assist in reducing total operating costs (Kondo and Ting, 1998). One specific area that may in the future lead to a reduction in the total number of handling steps is the use of aeroponic or bioreactor systems. The use of temporary immersion techniques to enhance production in vitro has been the subject of recent interest and was first reported in potato by Teisson and Alvard (1999) and further advanced by Piao et al. (2003). This methodology has seen widespread uptake in many other species, and this may be an area for further study and refinement in potato.

Microplants, however, tend to be non-functional in terms of normal physiological processes. Certainly, microplants have malfunctioning photosynthetic apparatus, reduced epidermal wax, are fragile and photoheterotrophic in nature and lack completely functional roots. To overcome these limitations, the weaning of plantlets under an intermediary ex vitro hardening phase involving conditions of high humidity and reduced light intensity, until the emergence of true and photosynthetically active leaves, is critical to the success of this stage. This hardening phase of microplants requires careful execution to minimize or eliminate losses at this crucial stage of micropropagation. The maintenance of the high health status of the material is paramount, and the microplants are vulnerable at this stage. The use of sterilized media, or non–soil-based substrates, greatly facilitates the avoidance of subsequent pathogen problems. The careful design of the facilities used for these procedures is important, and attention is needed to assure aseptic conditions as far as possible. Many production facilities use protective clothing, filtered air and detailed hygiene procedures, as clearly any contamination at this stage would prove to be costly for the company in terms of loss, or the disruption of production.

32.3.2 Micro-tuber production

Micro-tubers are miniature tubers and differ from mini-tubers in that they are produced under aseptic conditions designed to simulate the normal conditions of tuber-induction in vitro. The subject was previously extensively reviewed by Donnelly et al. (2003). Although there is some uncertainty regarding the uptake and utility of micro-tubers for any reliable evaluation of agronomic characters, the application of micro-tubers in the area of germplasm conservation is widely accepted, and, as a research tool, this biological system has many further applications (Coleman et al., 2001). A recent example was the use of an in vitro micro-tuber system as an important component of a virus-induced gene-silencing method in potato (Faivre-Rampant et al., 2004). Micro-tubers may be utilized commercially for mini-tuber production in greenhouses or screenhouses and, less commonly, can be directly field planted. One problem with the wider commercial uptake of micro-tubers is the requirement for a minimum fresh weight of 0.50 g to enable the tuber to perform adequately. Although most protocols fail to generate micro-tubers that meet this criteria, several reports do state that micro-tubers of 0.5–1.0 g can be consistently generated (Akita and Takayama, 1994; Leclerc et al., 1994). Large micro-tubers are generally easier to handle, are less subject to excessive shrinkage and degeneration in cold storage, also have a shorter dormancy period and exhibit greater survival rates when planted out directly in the field (Leclerc et al., 1994). Most commonly used micro-tuber production systems are based on the use of single-node cultures, from which generally one tuber per node is obtained. However, larger micro-tubers can be obtained by culturing whole plantlets or cuttings composed of several nodes on a semi-solid medium, but this takes more resources in terms of cultured plant tissues, media and glassware. It has been shown that larger micro-tubers can be produced by growing potato plantlets in liquid culture (Leclerc et al., 1994). However, as compared with their aerial counterparts, the micro-tubers remaining submerged in liquid cultures are considered to store poorly and tend to be very soft, with open lenticels that become the site of entry for pathogenic microorganisms.

The biological basis for the induction of micro-tuberization in potato is strongly related to the artificial simulation of the natural tuber-inducing conditions found in the field. Factors employed in an in vitro system in many cases include a high sucrose and cytokinin supply in the medium, the provision of short days (or complete darkness), lower temperatures or the application of anti-gibberellins such as chlorocholine chloride (CCC). The specific carbon source and its concentration within the tissue culture medium have great significance, with sucrose being the favoured substrate, largely because of its translocation to the developing micro-tubers (Khuri and Moorby, 1995). The optimum sucrose concentration for micro-tuber production ranges from 60 to 80 g/l, which is up to three to four times the normal amount used for micropropagation. It has been suggested that sucrose plays a dual role in micro-tuber development by providing a favourable osmolarity for tuber development in addition to its role as a carbon source (Khuri and Moorby, 1995). Jasmonate is a growth regulator normally produced by plants that have been exposed to stress and has previously been found to be highly effective in the induction of micro-tubers (Koda et al., 1991; Van den Berg and Ewing, 1991). However, although the use of jasmonate has been widely advocated, the response was shown to vary with cultivar (Pruski et al., 2003).

The process of micro-tuberization can be adapted to a culture production system within bioreactors (Yu et al., 2000), and methods for the automation of the technology have been further described (http://www.osmotek.com/PotatoProtocol.html). Also, there have been proposals to adapt micro-tuber technology to some form of large-scale mechanized planting process (Ranalli et al., 1994; Struik and Lommen, 1999). The growth and the yield of potato plants grown from micro-tubers in field trials has been investigated (Kawakami et al., 2003), and it was found that the micro-tuber-derived plants had a lower initial increase in root and leaf area index than conventional seed tuber plants. The first tuber formation in micro-tuber plants was about 7 days later than in conventional seed tuber plants, and tuber bulking occurred about 14 days later in micro-tuber plants. At harvest, the tuber fresh weight of micro-tuber plants was 82% that of conventional tuber plants, suggesting the potential for using micro-tubers for direct field planting, although with certain limitations.

Micro-tubers are also convenient for handling, storage and transport of germ plasm (Estrada et al., 1986) and in comparison with in vitro plantlets do not need a hardening period in the greenhouse or field (Ranalli, 1997; Coleman et al., 2001). However, a major constraint to the wider uptake of micro-tubers is that micro-tubers, after in vitro tuberization, are generally very dormant (Tabori et al., 1999) and will not sprout unless stored for 4 months or more at low temperatures.

As previously stated, micro-tubers have great utility as an experimental tool for research studies, reviewed by Coleman et al. (2001). These applications have been extended to several biochemical studies on the temperature effects on starches (Debon et al., 1998; Tester et al., 2004). There are obvious constraints to the widespread uptake of micro-tubers as a commercial tool, but significant research is ongoing on the optimization of systems and future applications.

32.3.3 Somatic embryogenesis

A potentially novel method of producing high-quality potato (*Solanum tuberosum* L.) nuclear seeds is through the process of somatic embryogenesis. Somatic embryo formation has been successfully reported in many plant species, but in potato, reliable systems were, until recently, only reported at the experimental stage (De Garcia and Martinez, 1995). The process of somatic embryogenesis can be defined as the development of a bipolar structure, with both root and shoot poles, from any sporophytic part of the plant occurring, through the same key stages of embryo development as zygotic embryogenesis (i.e. globular, heart and torpedo stages). This is yet another example of the potential of totipotency that can be exhibited by plant cells, whereby the cells first de-differentiate and then re-determine towards the embryogenic pathway. Recent reports (Seabrook and Douglass, 2001; Sharma and Millam, 2004) offer the prospect of developing and utilizing an efficient somatic embryogenesis system in potato. In the somatic embryogenesis system demonstrated by these authors, the in vitro potato internodal segments were subjected to a three-stage culturing regime of shoot multiplication, induction of somatic embryogenesis and regeneration/expression of somatic embryos using modified culture media. Somatic embryos were visible within 3 weeks of explant transfer to auxin-free medium, after an initial incubation for 2 weeks on medium containing auxin (alone or in combination with a cytokinin). The characteristic developmental stages observed during potato somatic embryogenesis were unequivocally confirmed using histology, and the development of somatic embryos was suggested to be through a unicellular mode of origin (Sharma and Millam, 2004). The transfer of potato somatic embryos to a plant growth regulator-free medium resulted in the observation of growth patterns similar to those expected from potato seedlings, in contrast with those observed for in vitro propagated shoots. The successful transplantation of emblings (plantlets derived from somatic embryos, a term similar to seedlings for zygotic embryos) to glasshouse conditions resulted in good quality potato plants and tubers with normal morphology (Fig. 32.1).

The somatic embryogenesis system in potato offers a potentially novel method for producing nuclear seed material, the propagation of transgenic plants through a unicellular mode and also as a novel biological system for many studies on gene expression and regulation. As a comparison of the potential of this mode of propagation versus other methods previously discussed (Sections 32.3.1 and 32.3.2), micro-tubers, theoretically, can be produced throughout the year in the controlled and non-seasonal environment of a laboratory, but the dormancy problems of micro-tubers and their physiological age limit their continual production and utilization. Practically, micro-tubers are produced in such a way that their dormancy period coincides with, or ends just before, the normal crop season. Microplants are free of such dormancy and of physiological age-related constraints,

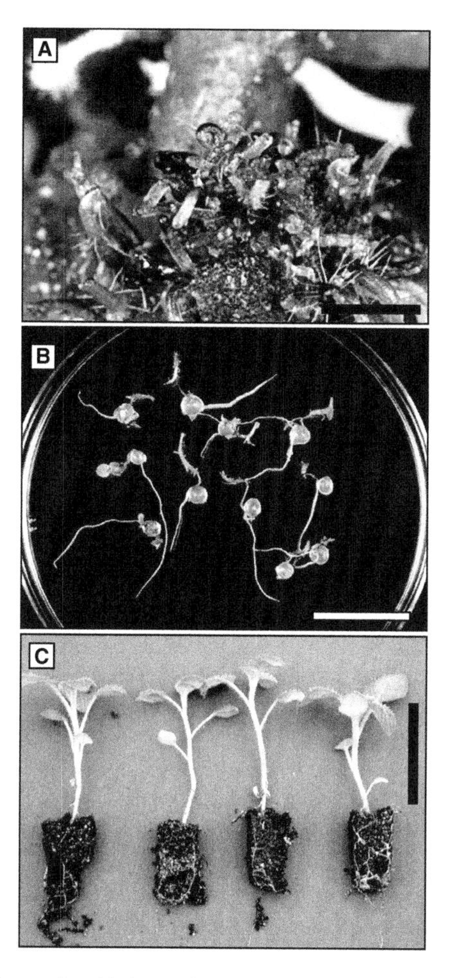

Fig. 32.1. (a) Emerging embryos after 21 days culture of internodal sections of potato, scale bar 1 mm. (b) Encapsulated embryos (synthetic seeds) with emerging shoots and roots, scale bar 20 mm. (c) Embryo-derived plantlets 2 weeks post-glasshouse planting, scale bar 25 mm.

but are fragile, and existing production methods are labour-intensive and need substantial in vitro and ex vitro storage space for bulking-up. Additionally, upon prolonged storage, the plants become tall, thin and etiolated which adversely affects transplantation success rates. However, somatic embryos have the significant advantages that they are also free from dormancy and physiological constraints and require substantially less manual labour input and specialized storage space; dormancy can be induced and released at will (under optimized systems) aiding the bulking and harvesting of embryos in a programmed manner. Additionally, somatic embryos could be encapsulated as synthetic seeds thus greatly, and advantageously, facilitating their handling, storage and transportation. Thus, somatic embryos combine many of the distinct advantages of microplants, micro-tubers and seed propagation systems. However, they do require a hardening phase before transplantation, and, for the initial post-hardening period, growth is uneven and less vigorous when compared with micro-tuber- and microplant-derived plants. At the time of maturity, these differences tend to disappear, and all the three described systems of micropropagation are generally comparable in terms of overall and individual mini-tuber yield.

32.4 HYDROPONICS AND AEROPONICS

These techniques offer many interesting opportunities for developing enhanced production systems, mainly for mini-tubers. Although requiring a degree of technical sophistication to design, establish and run, the benefits offered are sufficient for such systems to have been widely adopted by seed production companies worldwide.

One aspect of both types of systems is the potential to continually harvest tubers of the required size. However, both types of system operate in high-humidity conditions, and problems associated with open lenticels such as increased penetration of pathogens may need special attention and additional costs. Specific details of production rates have been cited in a number of papers, from a range of modifications of the basic protocols. Two contrasting culture systems for propagating first-generation potatoes were compared by Muro et al. (1997), a conventional system using a peat/sand mixture with mineral fertilizer and a hydroponic culture method using perlite as the matrix and nutrient solution. The authors stated that the total production and number of tubers produced through the hydroponic system were significantly higher than using the conventional system. It was further reported that the tuber yields from in vitro plants and mini-tubers were variable depending upon the time of year. Interestingly, it was shown that during the autumn/winter cycle, yield from mini-tubers was approximately double that from in vitro plants, whereas the reverse was reported during the spring/summer cycle. The work was extended to assess the effects of four hydroponic test cultures to study the influence of seed density, and although the number of tubers obtained increased significantly with seed density, there was no concurrent decrease in the number of large-diameter tubers. A soil-less production technique using clay balls as a substrate has been previously reported (Rolot and Seutin, 1999), and the system further employed a recycled and disinfected nutritive solution as the source of nutrients but was set up using micro-tubers as an in vitro starting material. Results varied according to cultivar but were reported to be in the magnitude of 1:8–1:13, giving 470–760 mini-tubers per square metre instead of a multiplication rate of 4–8 as observed in the conventionally grown controls (230–470 mini-tubers per square metre).

It has been claimed by many companies that an aeroponic-based multiplication system (see Section 32.2) has the potential to eliminate all but one generation of seed potato multiplication in the field, thus, lowering costs and raising the plant health quality of the first field-generation clones, although these claims have not been published in detail. Clearly, however, there are several obvious advantages of applying the aeroponic culture system, mainly for the avoidance of soil pathogens or contaminants, but importantly by increasing the availability of water and minerals in the root zone, and optimizing aeration around roots. Returning to the theme of generating plant material in low-gravity situations, it was shown that plant-derived nutrients could be successfully recycled in a presumably closed-cycle CELSS set up using biological methods (Mackowiak et al., 1997). The majority of the essential nutrients were recovered by microbiologically treating the plant biomass in an aerobic bioreactor. Effluent containing the nutrients was then returned to the biomass production component through a recirculating hydroponic system. All nutrient solutions were continually recirculated during the entire 418-day study. In general, tuber yields with reclaimed minerals were within 10% of control solutions.

A direct comparison of the hydroponic and aeroponic systems for the production of mini-tubers in greenhouse beds was made by Ritter et al. (2001). It was demonstrated that plants raised in the aeroponic system showed increased vegetative growth, delayed tuber formation and an extended vegetative cycle of about 7 months after transplanting. Their study further reported the accomplishment of two production cycles using the hydroponic system as compared with only one with the aeroponic system in a definite time span. Significantly, however, compared with the total production in hydroponics, the per plant tuber yield and number in the aeroponic system were almost 70% and 2.5-fold higher, respectively. Thus, the aeroponic system provides a higher overall tuber yield and increased efficiency per cropping cycle. This positively compensates for the loss of one crop cycle, which in itself means reduced resource inputs while at the same time maintaining a higher output. However, the mean tuber weight from tubers produced in the aeroponic system was reported to be reduced by 33%. Further advantages and possible problems with the aeroponic system for mini-tuber production were also discussed in this report, and there has been substantial, but largely unpublished, commercial research activity in this area.

32.5 FUTURE PROSPECTS

The development of methods for the soil-free production of potato has a long and interesting history. The technological developments described have enabled an increase in mini-tuber production, with substantial effects on reducing the cycle of seed tuber production and increasing seed quality while reducing the incidence of disease. Clearly, in terms of the need for a greater efficiency of seed potato production and for a reduced energy input, research into soil-free techniques will continue to be the subject of focus in both established and developing potato-producing areas, in the near and distant future. Advances in engineering technology will also assist in the development of more automated and controlled seed propagation systems. However, there are also options for simplifying the seed potato production systems for adaptation to low-technology situations, which has greater scope and relevance towards the increasing trends of potato production in developing countries.

REFERENCES

Akita M. and S. Takayama, 1994, *Plant Cell Tissue Org.* 36, 177.
Appel O., 1934, *Phytopathology* 24, 482.
Barker W.G., 1953, *Science* 118, 384.
Coleman W.K., D.J. Donnelly and S.E. Coleman, 2001, *Am. J. Potato Res.* 78, 47.

De Garcia E. and S. Martinez, 1995, *J. Plant Phys.* 145, 526.
Debon S.J.J., R.F. Tester, H.V. Davies and S. Millam, 1998, *J. Food Sci. Agric.* 76, 599.
Donnelly D.J., W.K. Coleman and S.E. Coleman, 2003, *Am. J. Potato Res.* 80, 103.
Estrada R., P. Tovar and J.H. Dodds, 1986, *Plant Cell Tissue Org.* 7, 3.
Faivre-Rampant O., E.M. Gilroy, K. Hrubikova, I. Hein, S. Millam, G.J. Loake, P. Birch, M. Taylor and C. Lacomme, 2004, *Plant Physiol.* 134, 1308.
Goodwin P.B., Y.C. Kim and T. Adisarwanto, 1980, *Potato Res.* 23, 9.
Gopal J., R. Kumar and G.S. Kang, 2002, *Potato Res.* 45, 145.
Goss R.W., 1926, *Phytopathology* 16, 299.
Kawakami J., K. Iwame, T. Hasegawa and Y. Jitsuyama, 2003, *Am. J. Potato Res.* 80, 371.
Khuri S. and J. Moorby, 1995, *Ann. Bot.* 75, 295.
Kitaya Y., Y. Ohmura, C. Kubota and T. Kozai, 2005, *Plant Cell Tissue Org.* 83, 251.
Koda Y., Y. Kikuta, H. Tazaki, Y. Tsuhino, S. Sakamura and T. Yoshihara, 1991, *Phytochemistry* 30, 1435.
Kondo N. and K.C. Ting, 1998, *Artif. Intell. Rev.* 12, 227.
Kordyum E., V. Baranenko, E. Nedukha and V. Samoilov, 1997, *Plant Cell Physiol.* 38, 1111.
Leclerc Y., D.J. Donnelly and J.E.A. Seabrook, 1994, *Plant Cell Tissue Org.* 37, 113.
Mackowiak C.L., G.W. Stutte, J.L. Garland, B.W. Finger and L.M. Ruffe, 1997, *Adv. Space Res.* 20, 2017.
Mes M.G. and I. Menge, 1954, *Physiol. Plant.* 7, 637.
Morel G. and C. Martin, 1955, *CR Acad. Agric. Fr.* 41, 54.
Muro J., V. Diaz, J.L. Goni and C. Lamsfus, 1997, *Potato Res.* 40, 431.
Naik P.S. and D. Sarkar, 2001, *Biol. Plant.* 44, 293.
Piao X.C., D. Chakrabarty, E.J. Hahn and K.Y. Paek, 2003, *Curr. Sci.* 84, 1129.
Pruski K., T. Astatkie, P. Duplessis, T. Lewis, J. Nowak and P.C. Struik, 2003, *Am. J. Potato Res.* 80, 183.
Ranalli P., 1997, *Potato Res.* 40, 439.
Ranalli P., F. Bassi, G. Ruaro, P. del Re, M. di Candilo and G. Mandolino, 1994, *Potato Res.* 42, 559.
Ritter E., B. Angulo, P. Riga, C. Herran, J. Relloso and M. San Jose, 2001, *Potato Res.* 44, 127.
Rolot J.-L. and H. Seutin, 1999, *Potato Res.* 42, 457.
Sanchez G.E., S.A. Slack and J.H. Dodds, 1991, *Am. J. Potato Res.* 68, 299.
Schultz E.S. and D. Folsom, 1923, *J. Agric. Res.* 25, 117.
Seabrook J.E.A. and L.K. Douglass, 2001, *Plant Cell Rep.* 20, 175.
Sharma S.K. and S. Millam, 2004, *Plant Cell Rep.* 23, 115.
Shepard J.F. and L.E. Claflin, 1975, *Annu. Rev. Phytopathol.* 13, 271.
Struik P.C. and W.J.M. Lommen, 1999, *Potato Res.* 37, 383.
Tabori K.M., J. Dobranszki and A. Ferenczy, 1999, *Potato Res.* 42, 611.
Teisson C. and D. Alvard, 1999, *Potato Res.* 42, 499.
Tester R., M. Yusuph, S. Millam and H.V. Davies, 2004, *J. Food Sci. Agric.* 84, 1397.
Van den Berg J.H. and E.E. Ewing, 1991, *Am. Potato J.* 68, 781.
Wheeler R.M., C.L. Mackowiak, J.C. Sager, W.M. Knott and C.R. Hinkle, 1990, *Am. Potato J.* 67, 177.
Xiong X., G.C.C. Tai and J.E.A. Seabrook, 2002, *Plant Breed.* 121, 441.
Yu W.-C., P.J. Joyce, D.C. Cameron and B.H. McCown, 2000, *Plant Cell Rep.* 19, 407.
Zobayed S.M.A., F. Afreen, Y. Xiao and T. Kozai, 2004, *In Vitro Cell Dev. Plant.* 40, 450.

Part VIII

CROP MANAGEMENT

Chapter 33

Agronomic Practices

D.M. Firman and E.J. Allen

Cambridge University Farm, Huntingdon Road, Cambridge, CB3 0LH, United Kingdom

33.1 INTRODUCTION

Production of potatoes for the demanding markets of developed countries in the twenty-first century increasingly requires growers to revise agronomic practices frequently to provide potatoes of suitable specification at low cost. The trend towards larger growers and producer groups in many countries (e.g. British Potato Council, 2006) has aided rapid adoption of new techniques, but traditional husbandry remains important and policies that provide short-term gains can place sustainable future production at risk. The value of potato crops has become more dependent on meeting quality requirements than on the total yield, and many practices are adopted or revised in an attempt to achieve these requirements. The diverse markets for which crops are grown often have contrasting requirements, so that all aspects of crop production must be tailored to the particular market, and this may require individual growers to specialize in production for a single market to remain profitable. Potato production has often been highly profitable, albeit subject to variable yields and crop prices year to year, and high input costs have been accepted, but currently, economic and environmental pressures require the use of all inputs to be constantly reviewed. The information that growers need to make rational decisions in relation to many of their production practices is not always known or available, and although crop growth models and decision support systems (DSS) offer opportunities to aid in this, the trend to turn the research focus away from experimental data and towards simulation is not entirely wise. It presents growers with a challenging future for they increasingly require accurate knowledge of the major individual disciplines that contribute to the agronomic management of the potato crop, and this demands improved understanding of growth and development.

33.2 PLANNING AND PREPARATION

Many of the opportunities to influence a potato crop occur prior to planting as there is limited opportunity during growth to influence key crop parameters; so, effective planning is vital to ensure that these are not compromised. Inputs and interventions to the growing crop then enable the potential of the planned crop to be achieved. In practice, planning individual crops is usually not made in isolation but as part of a target production

Potato Biology and Biotechnology: Advances and Perspectives
D. Vreugdenhil (Editor)

comprising many crops, and this increases the complexity of the task. The scope of crop planning is considerable as it includes everything from choice of variety and procurement of seed through site selection, nutrient requirements and establishing a sequence of operations from soil preparation until the end of crop storage.

33.2.1 Market

The choice of market for a potato crop determines the varieties that are suitable, quality traits that affect crop value and other related aspects of planning. The main markets (Chapter 1, McGregor, this volume) for processed products, fresh potatoes, seed and starch have very different requirements, and within these sectors, distinct crops with particular specifications exist. Although some varieties can be used for both fresh and processed products, the more sophisticated markets often dictate a narrow choice of suitable varieties, so that traits of disease and drought resistance are generally secondary. Varieties are chosen also on the basis of their suitability for processing or marketing immediately following harvest or after various periods of storage, this being a particularly important consideration for chip (French fry) and crisp production. Size specifications of particular products for fresh market production are an important influence on variety choice with premium markets for both very large and very small tubers. Newly bred varieties generally account for a small proportion of the planted area, and several widely grown varieties were bred more than 40 years ago (Russet Burbank was introduced before 1890) (Table 33.1).

Table 33.1 The main potato varieties and the percentage of the potato area planted in Great Britain and the USA.

Variety	Area (%)
Great Britain	
Maris Piper	20
Estima	13
Lady Rosetta	6
Maris Peer	5
Saturna	4
Nadine	4
Total	52
The USA	
Russet Burbank	44
Russet Norkotah	14
Ranger Russet	10
Shepody	5
Fritolay[a]	5
Norland	4
Total	82

Source: Great Britain 2005 data, British Potato Council, 2006. United States 2004 data for fall potatoes from eight major states, United States Department of Agriculture, 2005.
[a] A group of varieties grown for processing.

Market choice may dictate site selection, the requirement to control particular diseases and the calendar of crop production. Sites where irrigation is available may be essential for certain markets, and soils differ in suitability for particular crops as, for example skin appearance is adversely affected at harvest by abrasive soils. Soil type also affects the period over which crops may be grown as sands and light soils remain workable in wetter conditions than heavier soils; so, clay soils may be unsuitable for early-spring production or for crops destined for late harvest dates in wet climates. Rigorous control of seed-borne diseases is often required for production of seed crops to meet certification standards and also for fresh market crops where infection detracts from the appearance and thus from crop value. Control of diseases and disorders causing internal discolouration is particularly important for some processed products. The high value of fresh crops for some premium markets (e.g. salad and baking potatoes) can justify the cost of practices and inputs of agrochemicals and irrigation that are not economically viable for other crops.

The tonnage of crop to be produced affects other important planning considerations, and estimating this may require a combination of historical data and predictions based on a range of factors influencing yield. Significant over-production may be unprofitable if product cannot be sold to the target market, whereas under-production can lead to contractual and other problems.

33.2.2 Calendar

In some locations, potato crops can be planted and harvested at almost any time of year, but for most parts of the world, there are climatic limitations mainly associated with frost damage or heat stress, but excessive or inadequate seasonal rainfall may also be important in affecting crop growth directly and in limiting opportunities to prepare seedbeds or to harvest. Although the interval from emergence to tuber initiation in some varieties may be slightly less in short days than in long days (e.g. 12 versus 16 h, Demagante and Van der Zaag, 1988a) at high temperatures, the effect of photoperiod on tuber initiation in field-grown potatoes is generally limited (O'Brien et al., 1998), and daylength per se is not an important constraint to potato production in most of the areas where the crop is produced. Successful potato production can be carried out in latitudes above 60° N in Europe and North America (e.g. De Temmerman et al., 2002; Walworth and Carling, 2002) and 45° S but also <15° N and S in Africa, South America and Asia (e.g. Hay and Allen, 1978; Midmore, 1988; Demagante and Van der Zaag 1988b; Ifenkwe and Odurukwe, 1990). As these environments have major differences in almost all aspects of climate, pathology and infrastructure, a great range of agronomic practices have been adopted to accommodate these. Potential and actual yields differ considerably, but commercial varieties of *Solanum tuberosum* show remarkable adaptability.

In practice, limitation to the window for growing the crop may result from the requirement to plant and then to harvest a large area of crop with the same machinery or labour. Full season crops of some varieties have the potential to grow far longer than many climates allow, so that even crops planted in early spring can retain near complete crop cover up to normal harvest dates in the autumn (>180 days after planting) when grown in fertile conditions (Fig. 33.1). Although such long-season crops maximize the opportunity to grow the largest crop biomass, in practice, many crops are grown for shorter periods,

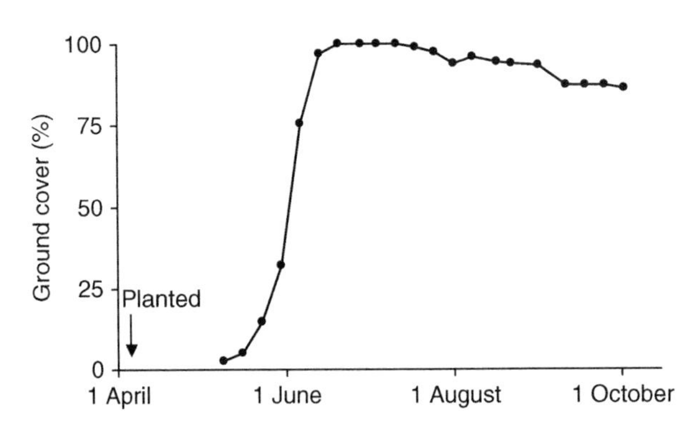

Fig. 33.1. Ground cover over the season for Cara grown in Cambridge, UK (Firman, unpublished).

and very profitable crops of small potatoes can be harvested within 60 days of planting. Cropping plans for a single market may consist of a sequence of plantings suitable for sequential harvesting or combinations of plantings and varieties differing in growth pattern to provide a staggered harvest. Particularly early or late season availability often attracts premium prices, and such crops often require special husbandry. Crops destined for use without storage require production cycles appropriate to achieve market criteria by the planned harvest date, whereas crops destined for storage are best grown, so that the period of growth, skin set and date of harvest do not compromise crop quality at the intended date of use. Sophisticated production systems can be integrated in such a way that seed and ware production cycles are matched, so that seed of the most appropriate physiological state is used for ware production.

33.2.3 Seed

Seed is one of the most expensive and important inputs, and effective planning should make best use of seed by accurately establishing the quantity required and ensuring it is suitable for the intended crop. Although seed certification standards in many countries may simply restrict marketing of seed affected by diseases only to the extent that substantial yield loss does not occur, procurement of seed with lower disease thresholds may be justified for premium markets. Some blemishing diseases (e.g. silver scurf) are ignored for seed certification purposes, and for other diseases (e.g. black scurf), tolerances may allow up to a quarter of the surface area of tubers to be affected (Great Britain, 2006), although the incidence of less severe infection on seed may provide sufficient inoculum for blemishing diseases to affect the value of the ware. An increasing range of diagnostic tests (Barker et al., 2005) can provide data on the incidence and severity of infection by bacteria, fungi and viruses to guide the choice of seed.

33.2.3.1 Seed rate

The quantity of seed required varies considerably according to different markets, expected yields and seed size (mean weight) with a range from <1 to >6 t/ha being used. Although

Table 33.2 Example of recommended seed populations (sets/ha) for Estima for a range of seed sizes and cost relative to ware (Ministry of Agriculture, Fisheries and Food, 1982).

Seed size (sets/50 kg)	Ratio of cost of seed to value of ware		
	1:1	2:1	3:1
400	48 000	39 000	29 000
600	56 000	49 000	39 000
800	62 000	56 000	46 000
1000	66 500	62 000	51 500
1200	69 000	66 500	56 000
1400	70 500	70 000	60 000

Note these Ministry of Agriculture, Fisheries and Food recommendations do not account for differences in ware yield or the market-specified size of the ware and presume a single price for all of the ware.

seed requirements are generally expressed as 'seed rate' (tons of seed per hectares planted), the effective unit of density for the potato crop is the stem (Holliday, 1960; Bleasdale, 1965). Each seed tuber may produce from one to eight or more stems depending on the variety, the weight and physiological state of the tuber. The number of stems produced tends to increase with increase in seed weight but smaller seed produces more stems per unit weight than large seed, so that a lower seed rate of small seed is required to achieve a given stem population. In some countries, most seed is used whole, but in others, seed tubers may be cut into two or more pieces prior to planting, and this generally results in a greater stem population than from planting whole seed at the same rate. A very wide range of seed tuber size, from <20 to >120 g, is used, but grading by size or weight into several divisions is recommended to reduce the variation in stem production within lots that would otherwise result. The mean tuber weight or tuber count (i.e. number of tubers per 50 kg) of seed lots may vary considerably, especially if they are divided into few size classes, and because seed weight is an important determinant of the stem population likely to be produced, a wide range of values compromises accurate determination of seed requirements.

Very low stem populations result in low yield as ground cover is slow to develop (Chapter 17, Haverkort, this volume) but over a wide range in stem density, there is little or no difference in total yield (Wurr, 1974) as shown in Fig. 33.2A. In practice, an important effect of increasing stem density is an increase in the number of tubers formed per square metre (Fig. 33.2B), resulting in a reduction in the mean tuber size (Fig. 33.2C). Tuber populations generally continue to increase up to greater stem densities than yield, but the rate of increase declines with increase in stem density (Fig. 33.2). For crops with no requirement to limit tuber size, the optimum stem density is the minimum required to achieve maximum yield, although the most economic yield may be somewhat less than the maximum, particularly where seed is costly. For markets where very large tubers are of low value or are unmarketable, and particularly for crops where a narrow range of quite small tuber sizes are required, optimum stem densities are greater than the

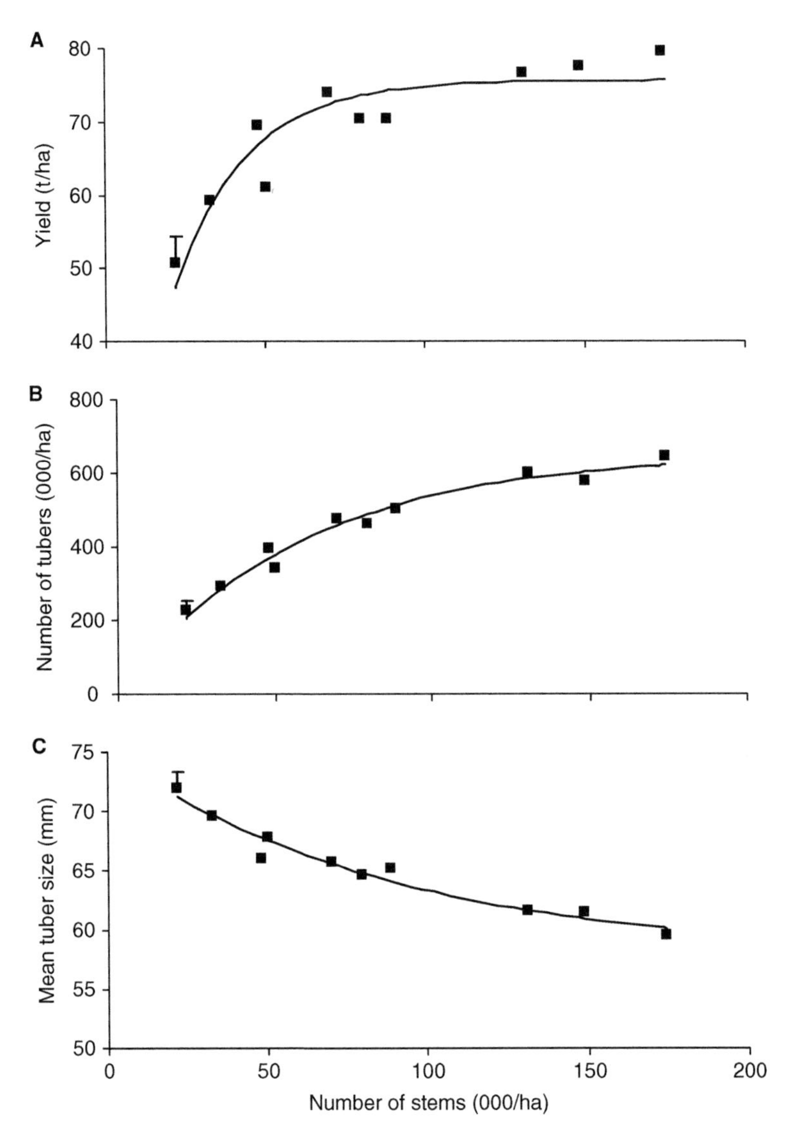

Fig. 33.2. Relation between (A) yield, (B) number of tubers (>10 mm) and (C) mean tuber size and stem density in Estima (Firman et al., 2004)

minimum required to produce maximum yield (Fig. 33.3). For these crops, the optimum stem density increases with increase in yield as more tubers are required to limit mean tuber size. The optimum seed rate is thus determined by the number of stems expected per seed tuber and the optimum stem density required for the target crop and expected yield.

In practice, seed rates are planned for many crops without direct reference to the principles in Section 33.2.3.1 but based on the average economic return from prior

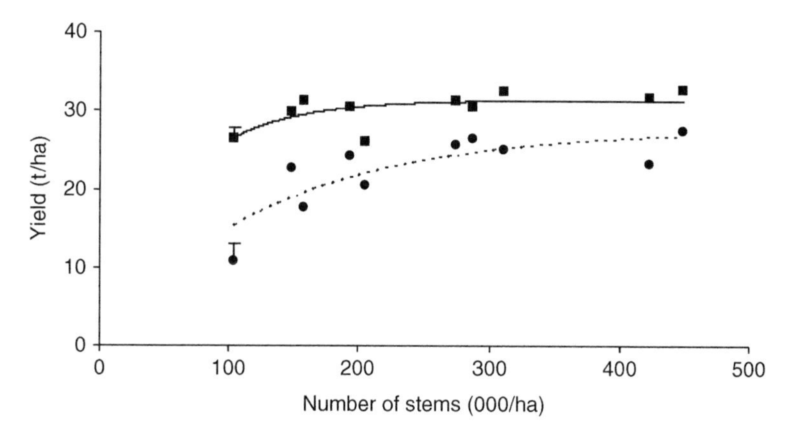

Fig. 33.3. Relation between total yield (■ and solid line), yield of 20–40 mm tubers (• and broken line) and stem density in Maris Peer (Firman and Shearman, unpublished).

experience or according to empirical recommendations from experiments where a range of seed rates have been examined (e.g. Ministry of Agriculture, Fisheries and Food, 1982). In the future, optimum seed rates are increasingly likely to be determined by DSS where mechanistic models based on the key principles can provide more accurate estimates for specific crops, but to date, systems that have been developed (e.g. MacKerron et al., 2004) have not been widely adopted by growers. Quantification of the changes in numbers of stems with seed age has enabled better prediction of stem populations (Firman et al., 2004), so that recommendations for planting densities can be improved.

33.2.3.2 Seed storage, handling and grading

Seed production in many countries is carried out in regions distinct and geographically separate from those mainly used for ware (e.g. British Potato Council, 2006), with seed often stored in the area of production until shortly before planting. For ware producers with limited facilities for seed storage, co-ordinating delivery of seed with planting requires effective logistics, and the ability to store seed locally is an advantage in unpredictable climates where planting may be interrupted. Seed is often stored for long periods after the natural dormant period has ended, and sprout growth must be prevented by keeping the seed cold (typically 2–4°C) unless sprouting is desired and suitable systems implemented (see Section 33.2.3.4). Unintended sprouting of seed in bulk often causes difficulty in handling and planting, and damage to the fragile sprouts causes variable emergence and is associated with loss of yield and increased susceptibility to some diseases (e.g. *Erwinia* spp.).

Some preparation of seed for ware production by size grading may be carried out prior to storage, or seed may be stored initially ungraded and then be prepared during the storage period or immediately prior to dispatch for planting. Grading may be used to divide seed into two or more fractions for better control of seed rates, and the process also provides an opportunity to remove diseased or damaged tubers and to apply fungicidal seed treatments. Inspection of seed may be required for certification prior to sale and labels or plant passports issued to identify seed lots when prepared for transport. Seed may be transported in bulk, large wooden boxes or polypropylene bags (often 1 t) or smaller hessian bags (often 25–50 kg).

33.2.3.3 Seed fungicides

Seed fungicides are usually applied in a liquid spray onto tubers carried on a roller table or as powders applied to tubers during loading into hoppers at planting. Soil adhering to tubers adversely affects the efficacy of these treatments, and an even coverage is best achieved with powered rollers, so that tubers are rotated to expose all surfaces to the spray. Early application of fungicide (i.e. prior to storage) is beneficial for controlling diseases that develop in store (e.g. *Fusarium* spp., *Polyscytalum pustulans*, Carnegie et al., 1990), but later application can provide effective control of diseases that develop after planting (e.g. *Rhizoctonia solani*, Adam and Malcolm, 1988). Many seed fungicides can inhibit emergence particularly when applied to seed tubers after sprout growth has just begun; so, indiscriminate use of such treatments should be avoided.

33.2.3.4 Seed sprouting

Sprouting seed prior to planting results in more rapid emergence (Headford, 1962) and, in some varieties, production of fewer stems (Toosey, 1964), so that economic yields can be obtained earlier than from unsprouted seed (Allen et al., 1979). Substantial sprout development limits the maximum yield, however, as early development is at the expense of canopy longevity. The increased cost and constraints that sprouting imposes on mechanization now limit its widespread adoption, but it is practised where the effects lead to advantages for production. Typically, seed is loaded into wooden or plastic boxes each holding <20 kg and placed in a warm well-illuminated environment (e.g. cool glasshouse or shed with vertical hanging fluorescent strip lighting). Semi-mechanized systems for sprouting have been developed allowing sprouting of tubers held in large suspended nets. Insufficient light results in weak etiolated sprouts particularly at high temperatures (>15°C), and such sprouts are readily damaged during handling and planting, reducing the advantages of sprouting and increasing variation between plants (Jarvis and Palmer, 1973). Recording the accumulation of heat units (Kelvin days) after sprout growth begins provides a quantitative measure of sprout development proportional to the effects on crop growth (O'Brien et al., 1983).

33.2.3.5 Seed cutting

Cutting seed is widely practised in some countries (e.g. the USA) but is not favoured in others (e.g. the UK) as, although seed costs can be reduced by cutting, there is a cost associated with the process, and crop establishment is often more variable particularly if cutting allows rotting diseases to develop. Seed may be cut by hand or by machinery, some of which is fitted with hot knives to reduce the risk of disease spread. Normally, only large seed is cut, and the largest seed may have up to three cuts to produce as many as six pieces, although cutting such large seed results in a wide size range of seed pieces which increases variation in numbers of stems as some may have no eyes. Seed is usually warmed prior to cutting, and coating the cut face of seed with lime or fungicides can inhibit disease development and aid handling. Cutting a few weeks prior to planting spreads the workload, but seed must be cured to allow cut surfaces to suberize, and, instead, seed is often planted immediately following cutting (the same day).

33.2.4 Site selection

Potatoes can be grown on a wide range of soils, but heavy clay soils are generally unsuitable for seedbed preparation and difficult for harvesting, whereas soils with a high stone content result in excessive damage at harvest. Shallow soils may be unsuitable unless ample irrigation is available, and for premium fresh market products where skin appearance is important, abrasive soils are unsuitable with clay loams and silt soils often giving best results. Short intervals (<4 years) between successive potato crops risk effects of soil-borne pathogens on the succeeding crop (e.g. Hide and Read, 1991), and longer rotations are usually practised in the UK (Minnis et al., 2002), although monoculture of potatoes is not unknown. Inadequate intervals between crops may lead to such high populations of some organisms, for example potato cyst nematodes (PCN), that potato production is unsustainable. For seed and high-value crops, land with no history of potato cropping may be sought to minimize the risk of contamination with soil-borne pathogens. Other site considerations include availability of irrigation, access for farm operations and the distance through which produce must be transported for storage and marketing. The high cost of haulage is an increasingly important consideration for processing factories and packhouses, and changes in the location of potato production may accompany the building or expansion of facilities.

33.2.5 Soil analysis

Soil samples are often taken to provide an estimate of nutrient availability for crop growth, and soil may also be analysed for pH, soil texture and presence of pathogens. The amount of P, K and Mg required as fertiliser is dependent on the availability of these elements in the soil, and analysis may also indicate a requirement for other minor elements. In contrast, the content of organic matter affects the mineralization of N, but soil mineral N content varies with date of sampling, and a practical, predictive relation with N requirement has not been demonstrated. For nutrient analysis, soil in the top 30 cm is usually sampled from several points in each field and sub-sampled to provide a single composite sample for analysis. Texture analysis of the soil to a depth of a metre or more may be carried out to determine the water holding capacity of the soil (Hall et al., 1977) for irrigation scheduling in addition to determining characteristics of the topsoil which have wider importance. Sampling for PCN is routinely practised where this pathogen is endemic and the detection and quantification of other soil-borne pathogens and diseases are of increasing interest as diagnostic techniques improve (Barker et al., 2005). Detecting a low incidence of infestation requires rigorous and costly sampling particularly where distribution is patchy and the relation between measures of soil inoculum and their consequences for crops are currently unknown for many diseases.

33.2.6 Fertiliser

Fertiliser requirements are affected by soil type and nutrient content, previous cropping, the expected length of the growing season and variety grown (e.g. Ministry of Agriculture, Fisheries and Food, 2000). Where the cost of excess fertiliser is marginal in relation to any loss of yield from inadequate application, fertilisers may be applied at a rate equal to or above that required for maximum yield. However, excess fertiliser N can decrease yield, make haulm destruction more difficult and reduce quality as well as increasing N residues susceptible to leaching. Fertilisers can affect many aspects of the crop other than yield (e.g. number of tubers, dry matter concentration, susceptibility to bruising and disorders), but there is little evidence that such effects are of great importance where crops are grown with levels of nutrients at or near that required for optimum yield. Fertiliser application may also be made to maintain or improve soil reserves, so that removal of nutrients (particularly P and K) over time does not deplete the soil, although the slow rate of change of soil reserves is only important over long time frames (decades). On many soils in Great Britain, there is little or no effect of P or K fertiliser on potato yield (Allison et al., 2001a,b), and although rates of application have declined over the 1990s, these remain high (Fig. 33.4) and greater than required for maximizing potato yield or maintaining soil indices in the medium term.

Although requirements for P, K, Mg and minor elements can be determined quite accurately from soil analysis, requirement for N generally cannot be determined, even though it often has the greatest effect on crop growth. Nitrogen requirements are least following crops where there is much residual N (e.g. lucerne) or where large amounts of farmyard manure or slurry were applied to the preceding crop. Indeterminate varieties require less N than determinate varieties to produce and maintain a persistent leaf canopy (Firman et al., 1995), and short-season crops require relatively little applied N or none. Nitrogen requirements are greatest for crops following cereals, poor quality permanent pasture or where N has been depleted by the previous crop and high winter rainfall.

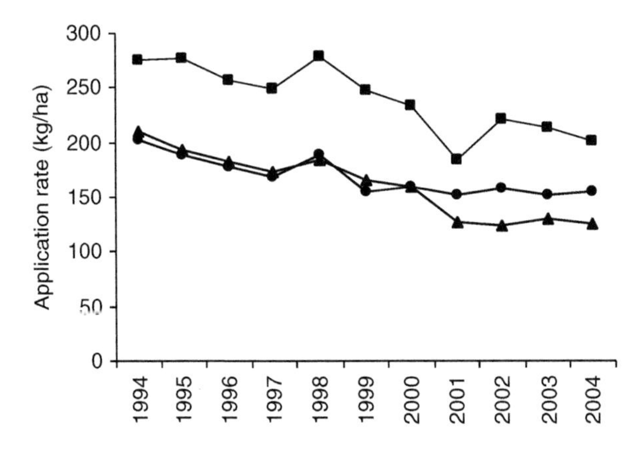

Fig. 33.4. Overall application rates of nitrogen (•), phosphate (▲) and potash (■) on potato crops in Great Britain (Department for Environment, Food and Rural Affairs, 2004).

More N is required for crops where a long period of growth is anticipated and for determinate varieties that produce relatively few leaves (Allen and Scott, 2001).

Fertiliser is often applied by broadcasting just prior to planting and then incorporating it as part of the soil preparation. Despite the conclusion of Cooke et al. (1957) that there was no justification for applying top-dressings of nitrogen to potatoes that received adequate amounts at planting, there remains much interest in sequential applications of N during crop growth especially in North America. Even on light soils where leaching might be expected, there are few data to support the use of split N applications. Fertiliser can be applied around the seed tubers, but such placement is unlikely to be beneficial unless root growth is restricted and high local concentrations of fertiliser may be deleterious. Liquid application of major and minor nutrients to growing crops is practised, sometimes using irrigation water (fertigation), and is often combined with systems for establishing nutrient requirements from analysis of plant nutrient status. The uptake of nutrients through foliage is often limited particularly as crops mature, and the relation between various measurements of plant nutrient status (e.g. petiole N, see MacKerron et al., 1995) and optimal status at any time is complex and may not be amenable to augmentation, so that such systems may be of limited benefit.

33.3 SOIL MANAGEMENT

33.3.1 Cultivation

Soil cultivation is a very important aspect of potato production as effects of the tilth of the seedbed and resistance of soil to root penetration may be very significant for many aspects of production including growth of the crop, tuber quality and incidence of disease (Van Loon and Bouma, 1978; Rosenfeld, 1997). Planting may be preceded by four cultivations, ploughing, bed-forming, bed-tilling and destoning or declodding, so that a large volume of soil may be moved although fewer cultivations are preferable where possible, saving energy and reducing risk of damage to soil structure. Between 1978 and 2007, there has been almost universal adoption of this sequence of cultivations in the UK with a great increase in the depth of working. In many other countries, levelling and cultivating to c. 30 cm suffice prior to planting (Stalham et al., in press).

Ploughing is often carried out in winter, but on light soils, ploughing immediately prior to planting is possible unless there is appreciable trash to bury and decompose. Excessively deep ploughing or cultivation of wet soil results in compaction and smearing. Compaction may impede drainage leading to poor root growth, increase diseases favoured by wet conditions (e.g. blackleg and many tuber-rotting diseases) and can lead to lenticel eruption and difficult harvesting conditions. Compacted soil also inhibits root penetration directly, adversely affecting the rate of plant development and the availability of nutrients and water.

Bed-forming produces a mound of soil between furrows where the tractor wheels run, and depending on the soil type, this bed may require further cultivation to produce a suitable seedbed for planting, but rotary bed-tillers can produce an over-fine tilth and ridges with unstable, structureless soil. Destoning or declodding separates stones and large clods from the top soil through a web or series of fingered 'stars' and deposits them in windrows (furrows), whereas smaller clods are broken up through a series of oscillating blades, stars or fine webs. This can reduce severe tuber damage during harvest by up to 50% and allow up to a 40% increase in the spot rate of harvesting (Whitney and McRae, 1992), although mechanisms for separation have improved on modern harvesters. Separation and breakdown of clods also enables a fine seedbed to be produced, so that close contact between soil and tubers occurs. This is beneficial for early growth of seed tubers, and as dry pockets of soil are less likely to occur, control of common scab may be improved. A fine seedbed may also aid efficacy of herbicides and reduce greening of tubers from penetration of light into the soil, but on some soils, slumping may occur if excessive cultivation destroys soil structure.

33.3.2 Control of soil-borne pests and diseases

The survival of several important pests and diseases between potato crops allows these to increase to economically damaging populations with the relatively short rotations often practised. For soil pests, fumigant chemicals (e.g. metham sodium) are occasionally used to treat soil some weeks prior to planting using specialist machinery. More commonly, non-fumigant chemicals (e.g. aldicarb and fozthiazate) are incorporated immediately prior to planting as part of seedbed preparation using bed-tilling machinery to provide partial control of free-living nematodes and PCN as well as wireworm larvae (*Agriotes* spp.). For control of free-living nematodes, in-furrow application is sometimes recommended as the aim is to inhibit transmission of tobacco rattle virus to daughter tubers by the nematode vectors rather than overall control of the nematode population as is the case with PCN. Use of soil-applied fungicides is generally less common than seed treatments, but these are sometimes used for soil-borne *R. solani* (e.g. pencycuron) and *Colletotrichum coccodes* (e.g. azoxystrobin) and in-furrow application with specialist equipment is practised to provide the most efficient control.

33.3.3 Weed control

Control of weeds by mechanical means is rarely practised except by producers of 'organic' crops as in most situations very effective control can be achieved by use of herbicides. Mechanical control by harrowing down of ridges and re-ridging two or three times can be effective during early growth, and weeds are subsequently suppressed by the crop, provided adequate foliage is produced, but damage to plants and soil compaction can reduce yield. Pre-planting cultivations usually obviate the requirement for pre-planting herbicides, but these may be used (generally systemic products such as glyphosate) in autumn or spring where perennial species are present and for control of potato weeds from previous crops. Soil-acting pre-emergence treatments applied after planting, but before potatoes and weeds have emerged, are generally the most important herbicides. These remain active for several weeks after application in undisturbed soil and are absorbed by shallow roots of weed species. Soil-acting pre-emergence herbicides do not usually affect potatoes, but shallow-planted minitubers (used for seed multiplication) may be damaged by some products particularly if they are washed down to rooting depth. Contact pre-emergence treatment may be required along with soil-acting products where weed seedlings have grown sufficiently to be insensitive to the residual herbicides. Application of pre-emergence herbicides is often delayed, so that products remain effective until weeds are suppressed by the crop. Applications may be made after some plants have emerged resulting in good weed control but at the expense of some check to early crop establishment. Post-emergence treatments may be active through foliage and soil and depend on physiological selectivity between potatoes and the weeds but can generally only be used on relatively small potato plants without causing significant damage.

33.3.4 Irrigation

Irrigation is widely practised both to increase yield and to improve the quality of potatoes. Irrigation affects the severity of several diseases (e.g. Adams et al., 1987) with some increased (e.g. black dot caused by *C. coccodes*) and others reduced (e.g. silver scurf caused by *Helminthosporium solani*), so that irrigation is generally applied mainly to increase yield. For premium markets, irrigation is applied for control of common scab (*Streptomyces scabiei*) as very little disease develops if a low soil moisture deficit is maintained in the soil surrounding developing tubers for approximately 6 weeks following tuber initiation (Lapwood et al., 1973). Irrigation to achieve this is often in excess of that required to maintain soil moisture deficits less than those limiting crop growth. Irrigation may also be applied prior to harvest to reduce the tuber damage that can result from harvesting in dry conditions when insufficient soil is present on the harvester web to cushion tubers. Irrigation can also be used to protect crops against frost damage when short periods of frost occur (e.g. in Mediterranean countries).

For large fields, irrigation is usually carried out using either centre pivot sprinkler systems or hose reels coupled to rain guns or booms, but several other systems are used particularly for smaller fields, including grids of sprinklers, drip tape and flood irrigation. Centre pivot systems require a high initial investment, and as these are fixed, the other crops in the rotation should require irrigation so as to justify the expense. Sprinklers mounted on booms that are not fixed are also used but, for mobile irrigation systems, readily transported rain-guns are more common, although their distribution of water is less uniform (Department for Environment, Food and Rural Affairs, 2005), and they are unsuitable for the application of small quantities of water (approximately <15 mm).

Efficient use of irrigation depends on supplementing soil water and rainfall with the minimum amount of water to ensure that crop growth rates are not limited by soil water availability. In arid environments (e.g. parts of South Africa and the USA) where potatoes are grown on light soils, the frequency of irrigation is high, and stored soil water makes only a limited contribution to crop growth. In wetter environments, a soil water deficit is allowed to accumulate, so that soils contribute more to the overall water need and can accept some rainfall without exceeding field capacity. Scheduling of irrigation aims to balance crop demand (evapotranspiration) with the supply of water from the soil, and the basic calculations are well documented (Penman, 1948; Monteith, 1965; Thom and Oliver, 1977). More recently, irrigation requirements have been determined from measurement of soil moisture, but adequate accurate instrumentation (e.g. using neutron probes) is difficult to achieve (Gaze et al., 2002), and so calculating soil moisture

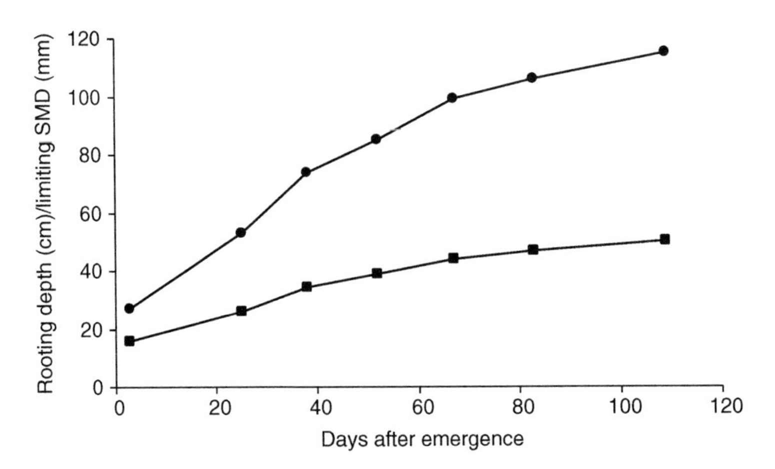

Fig. 33.5. Change in the limiting soil moisture deficit (SMD) (■) with increase in rooting depth (•) in the variety Cara (Stalham, 1993, unpublished).

from inputs and evapotranspiration using meteorological and crop data is generally more successful. The trigger for irrigation applications may be based on the requirement for control of common scab (i.e. soil moisture in the ridge surrounding daughter tubers) or to maintain the soil moisture deficit above a critical value, so that plant growth is not adversely affected. The limiting soil moisture deficit for plant growth increases with rooting depth (Fig. 33.5) and scheduling systems making allowance for this enable the most effective use of water (Stalham et al., 1999). Where irrigation is based on average crop requirements (e.g. 25 mm per week) rather than scheduling, crops may become dry or be over-irrigated during periods according to prevailing meteorological conditions and crop development. In practice, the ability to meet irrigation demand may be limited by the available equipment or by water supply. Improved efficiency in use of irrigation is increasingly likely to be required in Europe and elsewhere to conform with regulations such as the European Water Framework Directive (European Parliament and Council, 2000) that recognizes the importance of controlling abstraction of surface and groundwater. As a significant amount of the irrigation of potatoes is not scheduled quantitatively, there is much improvement to be made.

33.4 CROP ESTABLISHMENT AND MANAGEMENT

33.4.1 Planting

Although early spring planting has the potential to maximize the growing season in some climates, the start of planting may be delayed to avoid soil damage from working wet soil or the risk of frost damage after emergence. Seed planted early in cold soil emerges slowly, and small canopies may result, but the requirement to plant large areas may necessitate planting to begin earlier than is optimal for crop growth.

Potatoes are usually planted into relatively widely spaced ridges (typically approximately 0.9 m apart) with two to four rows planted at a time by placing tubers into a furrow created by an opening body after which a closing body forms the ridge of soil over the tuber. Planting into beds (typically 1.8 m wide) comprising three or more rows is also carried out, reducing the rectangularity of plant spacing and allowing for staggered spacing to reduce early competition between plants. Use of beds is often adopted where large numbers of tubers are required (seed and salad crops) or to improve conservation of soil moisture. Planters require systems for singulating seed tubers and adjusting the spacing to plant single tubers or seed pieces at the required spacing. Singulation is achieved by using cups, by forming a stream of touching tubers in belt planters or by pick machines. Shallow planting can delay emergence if soils dry near the surface (Firman et al., 1992) and can increase the proportion of green tubers at harvest, whereas overly deep planting also delays emergence and may result in harvesting problems, so that tubers are generally planted approximately 15 cm deep with allowance made in some soils for slumping or erosion of the ridge.

33.4.2 Crop protection

Subsequent to application of seed fungicides (Section 33.2.3.3) and products to control soil-borne pests and diseases (Section 33.3.2), fungicides, insecticides and molluscicides may be applied throughout crop growth. The number of sprays and active substances applied to potato crops has tended to increase although rates of application (kg/ha) have generally decreased (e.g. Garthwaite et al., 2005). Control of late blight (*Phytophthora infestans*) is often the most important crop protection consideration for potato crops worldwide, and with 15 or more foliar applications on some crops (e.g. Bradshaw et al., 2000), it represents a significant input of active substances. Products used for protection against late blight also control other diseases that would otherwise be important (e.g. early blight, *Alternaria solani*). Application of fungicides for late blight may begin shortly after emergence and be repeated at intervals of 7 days or less where conditions are highly conducive for blight. In many countries, a proportion of *Phytophthora* isolates are resistant to widely used phenylamide fungicides (Gisi and Cohen, 1996), but many other very effective protectant and systemic products are available (Whitehead, 2005), and application usually prevents significant development of foliar blight, whereas without fungicides total crop loss may occur. Even with only limited development of foliar blight, there is a risk of tuber blight developing in store particularly where tubers are harvested, while active spores remain in the soil, and a delay in harvest following crop desiccation may be justified to reduce this.

Applications of foliar sprays are often conducted with tractor-mounted sprayers but aerial spraying is practised in some regions. Sprays cannot be applied effectively during rainfall or high winds, and when soils are very wet, spraying by tractor may be difficult. Prolonged periods of unsuitable spraying conditions may allow infection to occur, and some appreciation of practical constraints to application of pesticides is necessary for effective application of DSS for crop protection. DSS are used for scheduling blight sprays (Chapter 36, Marshall, this volume) in the Netherlands, UK and elsewhere, and adoption may enable a reduction in the number of applications and input of active ingredients (Bouma, 2003).

Insecticides are used to control pests that defoliate the crop (notably Colorado beetle) in some regions but also may be applied to control pests causing tuber damage or vectors of virus diseases such as aphids. Control of virus spread is particularly important in seed crops, but use of aphicides to limit virus spread may be of limited efficacy where non-colonizing vectors are important (e.g. with potato virus Y). Molluscicides may be applied as pellets broadcast by spinning disc applicators with some growers making repeat applications at the same time as application of blight sprays.

33.4.3 Covers, mulches, soil amendments and intercropping

Crop covers are used in some climates to avoid frost damage and advance crop growth by raising soil and air temperatures. Non-woven agricultural fleece generally allows for greater penetration of water than perforated polythene that may also be subject to condensation. Covers are generally applied at planting in strips covering one or more beds each and held in place by burying the edges with soil. Covers of fleece or fine net may also be used as a barrier to aphid vectors of viruses for seed multiplication.

Mulching and incorporation of soil amendments are not widely practised but can be used to retain moisture and reduce temperature (Midmore et al., 1986) for soil amelioration and to provide some control of pests and pathogens. Intercropping with maize (*Zea mays*) or other crops is practised in some countries, and although contrasting growth habits of potato and other plants can enable more efficient utilization of resources, intercropping is not necessarily more profitable than a sole potato crop (Ifenkwe and Odurukwe, 1990; Liu and Midmore, 1990).

33.4.4 Defoliation

Crops left to senesce naturally may be harvested without defoliation but, where yield and tuber size are sufficient, defoliation is frequently carried out to end growth. Defoliation may also aid harvesting and advance the date of skin set, enabling earlier harvesting of undamaged tubers. Large quantities of haulm on harvesters may interfere with tuber separation and, where tubers remained attached to stolons, damage can be increased. Green top harvesting is carried out where set skins are not required, but it may still require haulm to be mechanically pulverized immediately prior to harvest. Following defoliation, the tuber periderm becomes progressively more resistant to removal (Bowen et al., 1996), but depending on variety and conditions, an interval of 2–6 weeks may be required before significant damage can be avoided by harvesters.

Defoliation may be achieved mechanically, chemically, by burning or indirectly by undercutting the roots. Desiccation with sulphuric acid, often as two sequential applications, is usually very effective but requires specialist contractors to apply it, whereas herbicides such as diquat and glufosinate ammonium can be applied by farm sprayers. Risk of stem-end necrosis and vascular discolouration of tubers resulting from desiccation can be minimized by sequential applications of relatively low doses or by adopting combinations of mechanical and chemical treatment.

33.4.5 Harvesting

Mechanized harvesters separate tubers from haulm, soil, clods and stones on a series of rollers and webs and collect the tubers for transport from the field. The design and settings of the harvester affect the efficiency with which tubers of the required size are harvested with minimal damage. Failure to optimize the harvesting process can result in damage in numerous ways; tubers may be crushed by over-sized tractor wheels, cut by inadequate depth of the share, pinched by haulm removal rollers and scuffed and bruised by bouncing and rolling on the webs. It is important that the forward speed and main web speed are sufficient to carry some soil to the top of the main web, so that tubers are cushioned (e.g. Bentini et al., 2006). Separation is improved by agitation of the webs, but excessive agitation results in bouncing or rolling of tubers. Significant bruising may result where tubers fall from the elevator into a box or trailer, and the height of this drop should be minimized. Regular monitoring of potatoes for scuffing and bruising during the harvesting period allows problems to be identified and appropriate action to be taken. Samples placed in a 'hot box' (high humidity and approximately 35°C) give early warning of bruising, so that changes can be made to avoid future storage problems, and electronic potatoes with impact sensors can be used to identify the position on the harvester where tubers are most likely to be damaged. Some harvesting problems can be related to the beginning of the season with poor seedbed preparation resulting in cloddy soils (Campbell, 1982) and excessive planting depth necessitating digging shares to be set deeply. Two-stage harvesting, where the crop is dug and two or more rows deposited in a windrow prior to lifting, may offer advantages over single-stage harvesting under some conditions including faster lifting rates of low yielding crops, and is practised in parts of the USA but use elsewhere is limited.

33.5 POST-HARVEST HANDLING AND STORAGE

A significant quantity of potatoes is processed or consumed soon after harvest (see Section 33.2.1), but most crops are stored prior to use. For almost all markets, some grading and sorting of the crop is carried out, and for premium markets, sophisticated facilities are used to prepare and package 'pre-pack' potatoes. Potatoes are washed to remove soil and graded by size (diameter) or weight according to the particular market requirements. Damaged, green and diseased tubers can be removed by manual or computerized detection systems to meet high quality-thresholds provided the proportion of tubers rejected is relatively low. Some preliminary grading and application of fungicide may be carried out prior to storage, but generally this is not done.

Scuffed surfaces of potatoes harvested without set skins cure within a few days of harvest, so that for the fresh market, efficient supply chains are used to ensure products reach retail outlets without losing quality. For some products, hydro-cooling is used to reduce the temperature of the potatoes within hours of harvest before transport to packers and retailers. Washing of potatoes prior to transport is advantageous for processors, and waste remains near the site of production.

Potatoes are stored in bulk and box stores, and although bulk stores are generally cheaper, they do not readily allow segregation of different lots. The risk of pressure bruising can limit the depth of piles in bulk stores to 4 m although in the USA piles 7 m deep are stored. Fans are used to provide controlled ventilation through the crop (positive ventilation), and air can be vented through apertures for cooling and be re-circulated to prevent condensation. Automated controls optimize the use of re-circulated and fresh air for storage in relation to external conditions (e.g. use of cool night air for cooling), but refrigeration and heating may be required depending on the climate.

Stores are ideally loaded within 7–10 days to limit differences in initial storage conditions for early and later harvested tubers. Within the first few hours of potatoes entering the store, ventilation is important to remove surface moisture. Increased moisture loss and risk of disease resulting from unset skins and wounds can be reduced by holding tubers above 10°C for 10–14 days (Wigginton, 1974) before storage at lower temperatures, but this curing period may be reduced or eliminated where loss of quality through development of blemishing disease is more important. Once wound healing of the last crop into store is complete, store temperatures are reduced to holding temperatures at approximately 0.5°C per day.

For stored potatoes, maintaining quality requires control of sprouting, dehydration and disease. Over long periods of storage, changes in tuber composition, notably in reducing sugars, also occur affecting taste and processing characteristics, and these changes may limit the duration of storage even in the absence of deterioration from other causes. Storage temperature, relative humidity, ventilation and composition of the storage atmosphere may all be controlled to optimize storage for periods of 10 months or more. Once the dormant period has passed, sprout growth will occur unless storage is below approximately 2°C or control measures are used to suppress sprouting. For table potatoes, storage for more than 6 months is possible without the use of control measures where temperatures of approximately 3°C are used, and this has been widely adopted to avoid use of chemical sprout suppressants such as chlorpropham. Modifying storage atmospheres with ethylene and other volatile compounds can also be used to limit sprouting in table potatoes, although their use is not currently widespread. For processing crops, potatoes are generally stored between 6 and 10°C with lower temperatures used for the longer periods of storage (8 months or more), and use of chlorpropham to suppress sprouting is normal (e.g. Anderson et al., 2004). Usually, products are first applied within a few weeks of filling stores and several applications may be made, although long-term storage without sprouting can be achieved from a single application as is widely used in the USA. Foliar application of growth regulators such as maleic hydrazide may be made several weeks prior to harvest to suppress sprouting during storage although these are frequently used in conjunction with products applied in store. In well-sealed stores, a bulk of potatoes maintains a high humidity (approximately 96%), but humidification is used to limit moisture loss particularly in dry climates.

33.5.1 Crop monitoring

During the life of a potato crop, considerable information is recorded, particularly in order to meet the requirements of various national and commercial crop assurance schemes (e.g. Assured Produce, 2006). Most of these records are documentations of the practices and agro-chemicals used and contribute nothing to the management of the crop. A body of crop-specific meteorological data is collected by some growers, and this may be used in DSS to support disease forecasting. There is, as yet, generally only limited recording of the growth of the crop and of the key parameters determining success – number of stems and tubers, time course of leaf canopy and quality attributes. As a result, many growers do not know the results of their planning and the reasons for their success or failure, and no managerial decisions are taken during the crops' life based on measurement of the crops' true state. This is likely to change as understanding of crop growth and development has improved to allow prediction of future events – yield and tuber size distribution (e.g. MacKerron et al., 2004) – to be made with sufficient accuracy to have commercial value. Electronic transmission of images and data allows remote access to crop production intelligence for an increasing number of growers to facilitate this.

REFERENCES

Adam N.M. and A.J. Malcolm, 1988, Pests and diseases, p. 959, Proceedings of the Brighton Crop Protection Conference, Brighton, U.K.

Adams M.J., P.J. Read, D.H. Lapwood, G.R. Cayley and G.A. Hide, 1987, *Ann. Appl. Biol.* 110, 287.

Allen E.J. and R.K. Scott, 2001, *Potato Agronomy*, Research Review. British Potato Council, Oxford.

Allen E.J., J.N. Bean, R.L. Griffith and P.J. O'Brien, 1979, *J. Agric. Sci. (Camb.)* 92, 151.

Allison M.F., J.H. Fowler and E.J Allen, 2001a, *J. Agric. Sci. (Camb.)* 136, 407.

Allison M.F., J.H. Fowler and E.J. Allen, 2001b, *J. Agric. Sci. (Camb.)* 137, 397.

Anderson H.M., D.G. Garthwaite and M.R. Thomas, 2004, *Pesticide Usage Survey Report 189*, Potato Stores in Great Britain 2002. DEFRA/SEERAD, London.

Assured Produce, 2006, *Crop Specific Protocol – Potatoes*. Assured Produce, Cobham.

Barker I., M. Hims, N. Boonham, T. Fisher, J. Elmore, H. Swan and R. Mumford, 2005, *Aspects Appl. Biol.* 76, 147.

Bentini M., C. Caprara and R. Martelli, 2006, *Biosyst. Eng.* 94, 75.

Bleasdale J.K.A., 1965, *J. Agric. Sci. (Camb.)* 64, 361.

Bouma E., 2003, *EPPO Bull.* 33, 461.

Bowen S.A., A.Y. Muir and C. Dewar, 1996, *Potato Res.* 39, 313.

Bradshaw N.J., S.J. Elcock, J.A. Turner and N.V. Hardwick, 2000, *Proceedings British Crop Protection Council Conference* 3, 847.

British Potato Council, 2006, *Yearbook of Potato Statistics in Great Britain*. British Potato Council, Oxford.

Campbell D.J., 1982, *J. Agric. Eng. Res.* 27, 373.

Carnegie S.F., A.D. Ruthven, D.A. Lindsay and T.D. Hall, 1990, *Ann. Appl. Biol.* 116, 61.

Cooke G.W., F.V. Widdowson and J.C. Wilcox, 1957, *J. Agric. Sci. (Camb.)* 49, 81.

De Temmerman L., J. Wolf, J. Colls, M. Bindi, A. Fangmeier, J. Finnan, K. Ojanperä and H. Pleijel, 2002, *Eur. J. Agron.* 17, 243.

Demagante A.L. and P. Van der Zaag, 1988a, *Potato Res.* 31, 73.

Demagante A.L. and P. Van der Zaag, 1988b, *Field Crops Res.* 19, 153.

Department for Environment, Food and Rural Affairs, 2004, *British Survey of Fertiliser Practice*. The Stationery Office, London.

Department for Environment, Food and Rural Affairs, 2005, *Irrigation Best Practice – Water Management for Potatoes. A Guide for Growers*. The Stationery Office, London.

European Parliament and Council, 2000, EU Water Framework Directive 2000/60/EC.

Firman D.M., E.J. Allen and V.J. Shearman, 2004, Production practices, storage and sprouting conditions affecting number of stems per seed tuber and the grading of potato crops. British Potato Council Project Report 2004/14. British Potato Council, Oxford.

Firman D.M., P.J. O'Brien and E.J. Allen, 1992, *J. Agric. Sci. (Camb.)* 118, 55.
Firman D.M., P.J. O'Brien and E.J Allen, 1995, *J. Agric. Sci. (Camb.)* 125, 379.
Garthwaite D.G., M.R. Thomas, H. Anderson and H. Stoddart, 2005, Pesticide usage survey report 202, Arable crops in Great Britain 2004. DEFRA, London.
Gaze S.R., M.A. Stalham and E.J. Allen, 2002, *J. Agric. Sci. (Camb.)* 138, 135.
Gisi U. and Y. Cohen, 1996, *Annu. Rev. Phytopathol.* 34, 549.
Great Britain, 2006, *The Seed Potatoes (England) Regulations 2006*, Statutory Instrument 2006 No. 1161. The Stationery Office.
Hall D.G.M., M.J. Reeve, A.J. Thomason and V.F. Wright, 1977, Water retention, porosity and density of field soils, soil survey technical monograph No. 9, 75 pp.
Hay R.K.M. and E.J. Allen, 1978, *Trop. Agric.* 55, 289.
Headford D.W.R., 1962, *Eur. Potato J.* 5, 14.
Hide G.A. and P.J. Read, 1991, *Ann. Appl. Biol.* 119, 77.
Holliday R., 1960, *Field Crops Abst.* 13, 247.
Ifenkwe O.P. and S.O. Odurukwe, 1990, *Field Crops Res.* 25, 73.
Jarvis R.H. and G.M. Palmer, 1973, *Exp. Husbandry* 24, 29.
Lapwood D.H., L.W. Wellings and J.H. Hawkins, 1973, *Plant Pathol.* 22, 34.
Liu J. and D.J. Midmore, 1990, *Field Crops Res.* 25, 41.
MacKerron D.K.L., B. Marshall and J.W. McNicol, 2004, In: D.K.L MacKerron and A.J. Haverkort (eds), *Decision Support Systems in Potato Production*, p. 119. Wageningen Academic Publishers, Wageningen, the Netherlands.
MacKerron D.K.L., M.W. Young and H.V. Davies, 1995, *Plant Soil* 172, 247.
Midmore D.J. 1988, *Field Crops Res.* 19, 183.
Midmore D.J., J. Roca and D. Berrios, 1986, *Field Crops Res.* 15, 109.
Ministry of Agriculture, Fisheries and Food, 1982, *Seed Rate for Potatoes Grown as Maincrop*, Leaflet 653, 19 pp.
Ministry of Agriculture, Fisheries and Food, 2000, *Fertiliser Recommendations for Agricultural and Horticultural Crops* (Reference Book 209), 7th edition. HMSO, London.
Minnis S.T., P.P.J. Haydock, S.K. Ibrahim, I.G. Grove, K. Evans and M.D. Russell, 2002, *Ann. Appl. Biol.* 140, 187.
Monteith J.L., 1965, *The State and Movement of Water in Living Organisms*, p. 205, XIXth Symposium, Society for Experimental Biology, Swansea. Cambridge University Press, Cambridge.
O'Brien P.J., E.J. Allen and D.M. Firman, 1998, *J. Agric. Sci. (Camb.)* 130, 251.
O'Brien P.J., E.J. Allen, J.N. Bean, R.L. Griffith, S.A. Jones and J.L. Jones, 1983, *J. Agric. Sci. (Camb.)* 101, 613.
Penman H.L., 1948, *P. Roy. Soc. Lond. Ser. A* 193, 120.
Rosenfeld A.B., 1997, Effects of nitrogen and soil conditions on growth, development and yield in potatoes, PhD thesis, University of Cambridge, Cambridge.
Stalham M.A., E.J. Allen and S.R. Gaze, 1999, *Irrigation Scheduling and Efficient Use of Water in Potato Crops*, Research Review 1999/2, British Potato Council, Oxford.
Stalham M.A., E.J. Allen, A.B. Rosenfeld and F.X. Herry, 2007, *J. Agric. Sci. (Camb.)* 145, 1.
Thom A.S. and H.R. Oliver, 1977, *J. Roy. Meteorol. Soc.* 103, 345.
Toosey R.D., 1964, *Field Crops Abst.* 17, 239.
United States Department of Agriculture, 2005, *National Agricultural Statistics Service*, Potatoes 2004 Summary.
Van Loon C. D. and J. Bouma, 1978, *Neth. J. Agric. Sci.* 26, 421.
Walworth J.L. and D.E. Carling, 2002, *Am. J. Potato Res.* 79, 387.
Whitehead R., 2005, *The UK Pesticide Guide 2005*. CABI Publishing, CAB International, Wallingford, Oxon, U.K.
Whitney B.D. and D.C. McRae, 1992, In: P.M. Harris (ed.), *The Potato Crop – The Scientific Basis for Improvement*, 2nd edition, p. 570. Chapman and Hall, London.
Wigginton M.J., 1974, *Potato Res.* 17, 200.
Wurr D.C.E., 1974, *J. Agric. Sci. (Camb.)* 82, 37.

Chapter 34

Minerals, Soils and Roots

Philip J. White[1], Ron E. Wheatley[1], John P. Hammond[2] and Kefeng Zhang[2]

[1]*Scottish Crop Research Institute, Invergowrie, Dundee DD2 5DA, Scotland, United Kingdom;*
[2]*Warwick HRI, University of Warwick, Warwick CV35 9EF, United Kingdom*

34.1 INTRODUCTION

Potato plants require over 14 essential mineral elements (Table 34.1), which include both macronutrients (N, P, K, Ca, Mg and S) and trace elements (Cl, Fe, Mn, B, Zn, Cu, Mo and Ni). These are generally acquired from the soil solution by the root system. For adequate nutrition and, to avoid mineral toxicities, the concentrations of these elements in plant tissues must be maintained within certain limits. Insufficient tissue mineral concentrations limit potential growth and can affect tuber quality, whereas excess mineral concentrations may inhibit growth through toxicity.

Often, the soil solution lacks sufficient mineral elements for optimal plant growth, and to achieve maximal tuber yields, fertilisers are added to soils. Typical responses in tuber yield to N, P and K fertilisation are shown in Fig. 34.1. The amounts of fertiliser required for maximal yields will depend on environmental factors such as soil type, soil mineral

Table 34.1 The minimal concentrations of mineral elements in diagnostic leaves (fourth leaf from the top) of potato plants during tuber bulking identified with optimal growth.

Mineral element	Minimal tissue concentration
N	2.50–3.50 %
K	2.25–3.50 %
P	0.15–0.25 %
Ca	0.30–0.60 %
Mg	0.15–0.25 %
S	0.12–0.20 %
Fe	11–30 ppm
Zn	15–20 ppm
Mn	10–20 ppm
B	10–20 ppm
Cu	2–5 ppm
Mo	ND

Potato Biology and Biotechnology: Advances and Perspectives
D. Vreugdenhil (Editor)

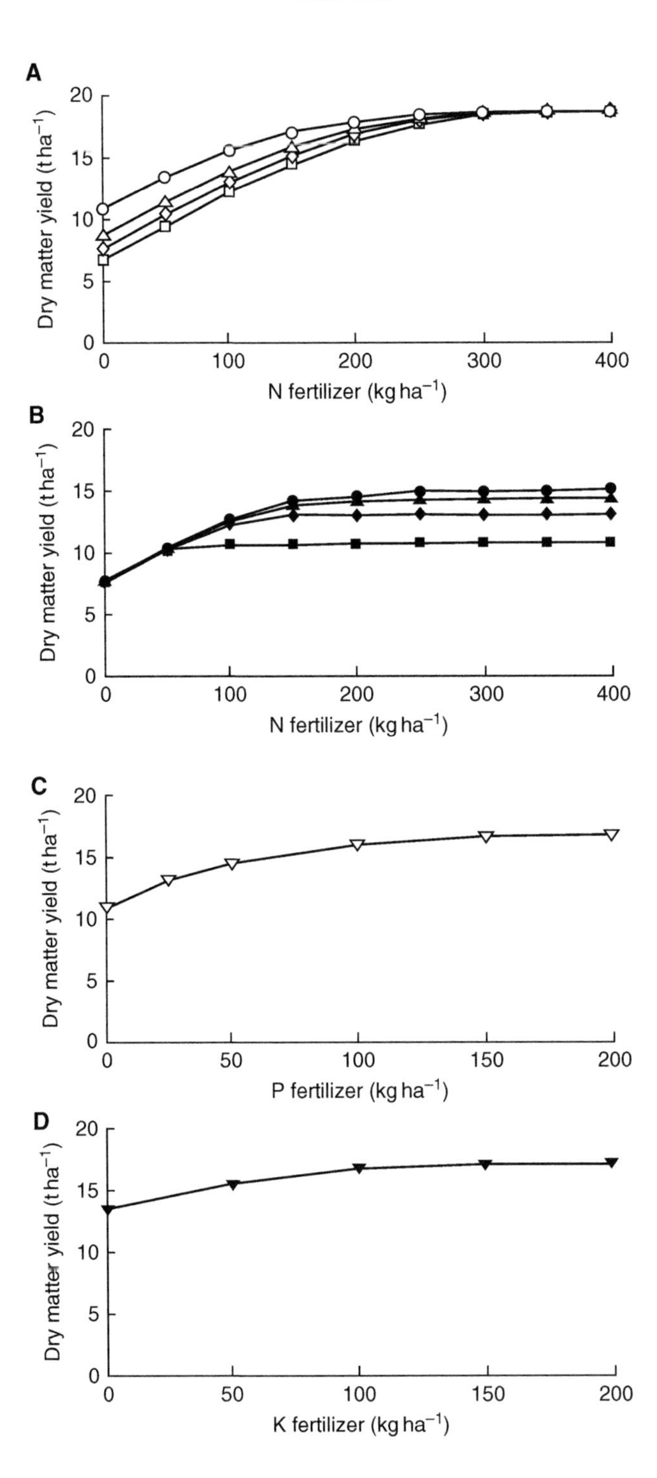
A
20
15
10
5
0
Dry matter yield (t ha^{-1})
0
100
200
300
400
N fertilizer (kg ha^{-1})
B
20
15
10
5
0
Dry matter yield (t ha^{-1})
0
100
200
300
400
N fertilizer (kg ha^{-1})
C
20
15
10
5
0
Dry matter yield (t ha^{-1})
0
50
100
150
200
P fertilizer (kg ha^{-1})
D
20
15
10
5
0
Dry matter yield (t ha^{-1})
0
50
100
150
200
K fertilizer (kg ha^{-1})

Fig. 34.1. Predicted responses of potato yields to N, P and K fertilisation. Data were obtained using a modification of the simulation models of Greenwood and colleagues (Greenwood et al. 1996, 2001; Greenwood and Karpinets, 1997) available at http://www.qpais.co.uk/. The predictions assumed that potatoes were planted at a density of 500 kg dry weight (DW) per hectare on 15 April 2006 at Wellesbourne, UK. The soil type was assumed to be a sandy loam with 20% clay, with an initial exchangeable P concentration of 21 kg ha^{-1}, an exchangeable K concentration of 100 kg ha^{-1} and mineral N concentrations of 30 kg ha^{-1} to a depth of 30 cm, 20 kg ha^{-1} at depths between 30 and 60 cm and 10 kg ha^{-1} at depths between 60 and 90 cm on 1 April 2006, unless otherwise specified. The soil water capacity was assumed to be 0.26, with a soil moisture deficit of 30 mm on 15 June 2006 and irrigation being applied at 25 mm weekly. It was assumed that fertilisers were applied on 8 April 2006 and that the crop was harvested on 1 October 2006. The yield potential of the crop was assumed to be 20 t DW ha^{-1}. Panel A shows predicted yield responses to N fertiliser applications when the initial mineral N concentrations were 0 (squares), 20 (diamonds), 50 (triangles) and 100 (circles) kg N ha^{-1} and both 600 kg P ha^{-1} and 600 kg K ha^{-1} were also applied. Panel B shows predicted yield responses to N fertiliser applications when 50 kg K ha^{-1} and 0 (squares), 25 (diamonds), 50 (triangles) and 100 (circles) kg P ha^{-1} were also added. Panel C shows predicted yield responses to P fertiliser applications when 500 kg N ha^{-1} and 100 kg K ha^{-1} were also applied. Panel D shows predicted yield responses to K fertiliser applications when 500 kg N ha^{-1} and 200 kg P ha^{-1} were also applied.

concentrations and weather conditions, agronomic factors such as timing, location and chemical form of the fertiliser applied and genetic factors such as longevity, growth rate and tissue mineral requirements of a potato variety (Defra, 2000; Dampney et al., 2002; White et al., 2005b).

Currently, potato production in Great Britain occupies 141 000 hectares, representing about 3% of its arable land (Defra, 2005), and about 18 600 000 hectares worldwide (FAO, 2006), representing approximately 1.2% of all cultivated land. However, although N fertilisation rates are similar to other tillage crops, a disproportionate amount of P fertiliser and K fertiliser is applied to the potato crop, which, for example, consumes about 9.4% of the inorganic P fertiliser and 11.8% of the inorganic K fertiliser applied to tillage crops in Great Britain (Defra, 2005). Although appropriate mineral fertilisation is necessary to maintain tuber yields, excessive fertiliser applications are not only an unnecessary financial burden but can also lead to reduced yields, due to mineral toxicities, and environmental pollution. The contribution to eutrophication of watercourses through the use of N and P fertilisers in agriculture is of particular concern for the implementation of the Water Framework Directive (2000/60/EC) in the EU, which aims to deliver good ecological status in European waters by 2015, and similar legislation in other countries, such as the USA where the Clean Water Act (PL-95-217; 1977) is enforced. Unfortunately, inefficient N (Zebarth et al., 2004) and P fertilisation (Dampney et al., 2002) is a particular problem for the potato crop, which generally recovers $<70\%$ of broadcast N fertiliser and $<10\%$ of broadcast P fertiliser in the year it is applied (Tran and Giroux, 1991; Dampney et al., 2002). To address this problem, agronomists aim to devise protocols to reduce fertiliser losses to the environment, and plant breeders aim to develop varieties that acquire and utilize mineral elements more efficiently (Davenport et al., 2005; White et al., 2005a). In addition, microbiologists aim to identify and foster soil microbial communities that improve mineral acquisition by crops (McArthur and Knowles, 1993; Morgan et al., 2005).

34.2 OPTIMIZING THE APPLICATION OF FERTILISERS

Plant roots take up most mineral elements in their ionic forms (White, 2003, Chapter 15, this volume). Because the relative concentrations of different ions in the soil solution vary greatly, their speed of movement in the soil and availability to plant root systems differ considerably. Some ions, such as nitrate and potassium, are sufficiently available and mobile in fertilised soils to supply plant demand, but the root system must forage for less readily available ions, such as phosphate (Barber, 1995). Conversely, leaching through the soil profile is of concern for nitrate and potassium, but not for phosphate. Most P is lost from fields in infrequent, discrete runoff events (Sharpley et al., 2002). Thus, agronomic strategies to optimize the use of N, P and K fertilisers differ.

34.2.1 Nitrogen

The initial N requirement of the potato crop can usually be met from the mineral-N present in the soil at planting and further N made available from transformation processes in the soil. However, although the surface layers of most cultivated soils generally contain between 0.1 and 0.5% N, which means that the 3500 tonnes of soil in the top 25-cm layer of a 1-ha field contains between 3.5 and 17.5 tonnes of N, mineralization rates of this N into forms that can be acquired by the plant (NH_4^+ and NO_3^-) are too low to sustain growth throughout the season and certainly not enough to provide maximal yield. At planting, this N is present in soils as soil organic matter (SOM), at various degrees of humification, and in the incorporated residues of the previous crop. The amount of N in these residues is dependent on the identity of the previous crop. For example, potato residues may contain the equivalent of 35 kg N ha^{-1}, sugarbeet 120 kg N ha^{-1} and wheat residues 25 kg N ha^{-1} (Velthof et al., 1999). Although the N in these residues will be gradually mineralized to plant available forms, not all of this N will become available to the following crop. Significant losses are associated with the transformation of soil organic N to mineral N, both as NO_x gasses during nitrification and through leaching of the NO_3^- formed. Immobilization of both NH_4^+ and NO_3^- and uptake by soil microorganisms will also reduce the amount of N available to plants. The rates of N mineralization and immobilization respond to cultivation and soil disturbance: maximal rates of both N mineralization and N immobilization follow ploughing or formation of ridges. Hence, predicting the amounts of N fertiliser to apply to a particular crop requires an accurate estimate of the N that will be available from previous residues and the SOM (Fig. 34.1). Soil type and winter rainfall are also important parameters in this calculation.

To optimize tuber production, N must be added in the form of organic or inorganic fertilisers. Organic manures used as a source of N will also contain P, K and other mineral elements. Some of the N present in such manures will be as NH_4^+, but most requires microbial transformation from an organic form to NH_4^+ and possibly also conversion to NO_3^- before it becomes available to the crop. Interestingly, the crop itself appears to promote the availability of mineral N (Fig. 34.2). On most soils, fertilisation with inorganic NH_4^+ and/or NO_3^- salts at a rate of 150–250 kg N ha^{-1} is recommended for maximal tuber yields. This can be broadcast over a field before furrow formation, placed

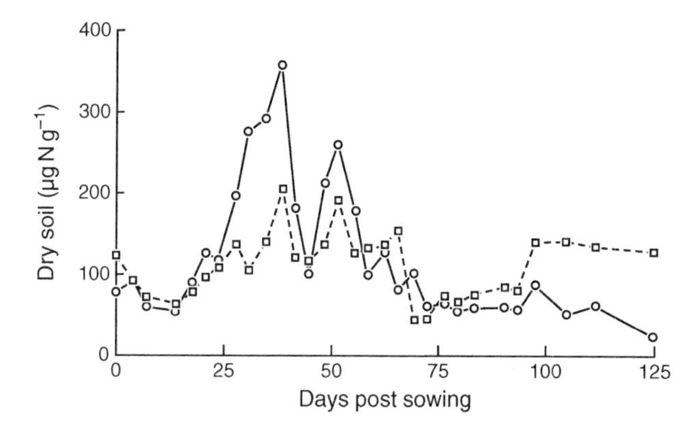

Fig. 34.2. Average mineral N concentration in soil cultivated to grow potatoes that were either planted (circles) or left unplanted (squares) and were fertilised with 187 kg N ha^{-1}. Day 0 was 17 April 1990. Data from Wheatley and Ritz (1995).

at the bottom of the drills or applied in a split application. The timing of fertiliser application is critical to achieve optimal plant uptake and subsequent yield, with minimal negative environmental impact. For example, more N is lost to the environment, and yield may be reduced, if N fertilisers are applied unnecessarily early (Fig. 34.3).

34.2.2 Potassium

Total K in agricultural soils may range from 1 to 2%, but only a small proportion of this is exchangeable, water soluble and available for uptake by plant roots. Typical K concentrations in the soil solution lie between 0.01 and 0.2 mg K ml^{-1} (0.25–5.0 mM). However, because K is comparatively mobile in the soil solution, its movement through the soil rarely limits plant growth. Hence, K fertilisation is an effective method to supply the demand of a potato crop for K, and soil analysis is an effective guide for K-fertiliser application rates (Vos, 1996). It has been observed that the K concentration in the soil solution required for optimal growth is much greater for potato than for many other crops. For example, sugar beet and wheat achieved 90% of their maximum growth at a K concentration of about 5 μM under conditions when potato required at least 50 μM for maximum growth (Steingrobe and Claassen, 2000). Thus, K fertilisation of potato is generally greater than that of other crops (Allison et al., 2001b; Defra, 2005). The large applications of K fertiliser to the potato crop, coupled with the mobility of K through the soil profile, may present an environmental challenge. Under continuous potato cultivation, K inputs are significantly greater and the K-input/K-offtake quotient is much higher than in a conventional rotation (Vos, 1996). However, absolute K offtake is greater under continuous potato cultivation than in a conventional rotation, because potato tubers have higher K concentrations than tissues of other crops, such as wheat (Vos, 1996). The optimum K-fertiliser application rate in most soils is 170–210 kg K ha^{-1}. Over-application of K fertiliser may cause a reduction in tuber dry matter concentrations. In a conventional rotation, the excess K inputs to the potato crop will probably be used by the following

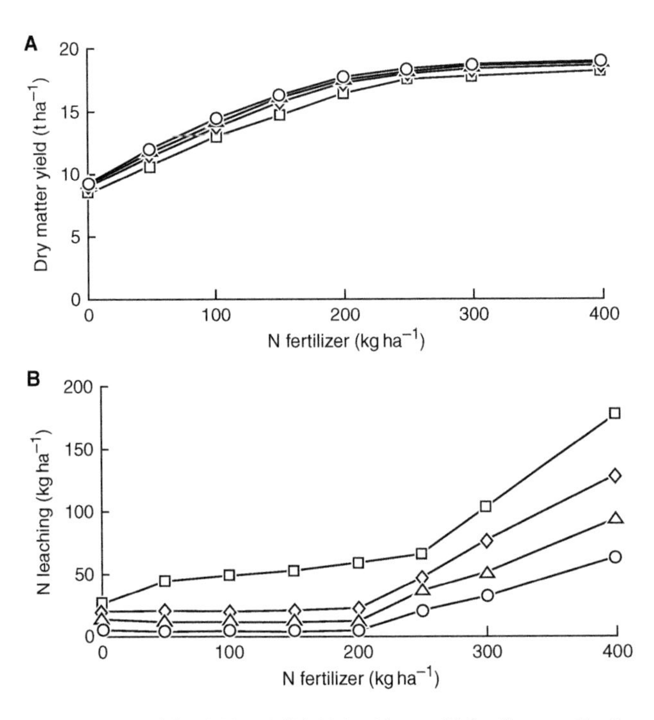

Fig. 34.3. Predicted responses in (A) yield and (B) N leaching to N fertiliser applications 30 (squares), 20 (diamonds), 10 (triangles) and 0 (circles) days before planting a potato crop. Data were obtained using a modification of the simulation models of Greenwood and colleagues (Greenwood et al., 1996, 2001; Greenwood and Karpinets, 1997), with the parameter values listed in the legend to Fig. 34.1. It was assumed that 600 kg P ha^{-1} and 600 kg K ha^{-1} were also applied

crops, but it would be more desirable to balance the inputs better during potato cultivation. This might be achieved by better positioning of K fertilisers to improve their availability and uptake by roots of young plants. The timing of the application of K fertilisers may also be significant, but it will be much more difficult to modify cultivation techniques to accommodate this.

34.2.3 Phosphorus

The effects of P fertilisation on tuber yield are thought to be a direct consequence of increased leaf area index, ground cover and radiation absorption (Harris, 1992; Jenkins and Ali, 1999, 2000; Allison et al., 2001a; Dampney et al., 2002; White et al., 2005b). The application of P fertiliser results in increased ground cover at tuber initiation and, often, also throughout the season. Because yield is dependent on photoassimilate and radiation absorption during the period of tuber initiation is one of the factors influencing the number of tubers found at harvest, P fertilisation produces not only greater yields but also more tubers.

Plant roots take up P from the soil solution as phosphate. Because phosphate is present at low concentrations in the soil solution and is relatively immobile in the soil, agronomic strategies to improve P-fertiliser use efficiency (PUE) have focused on placing concentrated P-fertilisers close to seed tubers at planting. A typical potato plant will have access only to the P fertiliser placed close to its roots during its initial growth period, when most P is accumulated and P is in greatest demand for tuber production (Simpson, 1962; Asfary et al., 1983; Van Noordwijk et al., 1990; Kolbe and Stephan-Beckmann, 1997; Dampney et al., 2002). Thus, placing P fertilisers close to the seed tuber often results in more efficient capture of fertiliser P, accelerated early growth and a small increase in final tuber yield (Prummel, 1957; Verma and Grewal, 1979; Harris, 1992; Lewis and Kettlewell, 1992; Sparrow et al., 1992), although care must be taken to avoid high salt concentrations in the rhizosphere (Harris, 1992; Hegney and McPharlin, 1999). In a recent field-plot experiment at Warwick HRI (Wellesbourne, UK), equivalent yields of tubers to those achieved by broadcasting the amount of P fertiliser recommended in the UK (Defra, 2000) were obtained by placing half this amount at a distance 3 inches below and 3 inches to the side of the seed tuber (P.J. White, H.C. Bowen, R. Hayden, M.C. Meacham, W.P. Spracklen and J.P. Hammond, Warwick HRI, UK, unpublished data). It is thought that P fertiliser placed close to the seed tuber will be acquired most efficiently by a vigorous plant with a rapidly developing, extensive root system and that this phenotype will subsequently exploit the entire soil volume most effectively (White et al., 2005b). Indeed, higher root length density (RLD = root length per unit volume soil) and specific root length (SRL = root length per unit root mass) are correlated with improved soil exploration, increased P acquisition, faster canopy development and, ultimately, a greater tuber yield (Opena and Porter, 1999). Eventually, the ability of the potato root system to exploit the soil volume will be limited by competition with its neighbours and by physical constraints. In particular, the depth to which the potato root system can explore the soil is restricted by its inability to penetrate beneath the plough pan (Lesczynski and Tanner, 1976; Gregory and Simmonds, 1992). Thus, agronomic practices such as deep tillage of compacted soils have been advocated that might increase P acquisition and, thereby, increase tuber yields (Gregory and Simmonds, 1992).

34.3 OPTIMIZING PLANT PHYSIOLOGY FOR MINERAL ACQUISITION AND UTILIZATION

One strategy to reduce fertiliser inputs and the loss of minerals to the environment is to develop potato varieties that acquire or utilize mineral elements more efficiently. There are many definitions for mineral use efficiency in the literature, which can be divided into three conceptual categories (Gourley et al., 1994; Baligar et al., 2001; Greenwood et al., 2005; White et al., 2005a).

The first category is concerned with productivity. Definitions include (1) yield on low-mineral soils, (2) yield obtained with insufficient fertilisation divided by the yield obtained with sufficient fertilisation, (3) increase in yield per unit of added fertiliser, which is often referred to as agronomic efficiency, and (4) amount, or concentration, of a mineral element in the rooting medium required for a given percentage of maximum yield, which may be expressed as either the 'Km' value required for half-maximal yield or the 'critical' value required for 90% yield. Crops and varieties with low values for Km and critical mineral supply will achieve their maximal yield with minimal fertilisation.

The second category is concerned with mineral acquisition from the soil. Definitions include (1) amount of a mineral element in a plant at a given concentration of the mineral element in the rooting medium, (2) amount of a mineral element in a plant divided by root biomass, both of which are referred to as mineral acquisition efficiency in the literature, (3) amount of a mineral element in a plant divided by the amount of the element applied in fertiliser and (4) percentage of the element added in fertiliser that is present in the crop at harvest, which is referred to as the fertiliser recovery efficiency. Crops with high mineral acquisition efficiencies generally grow well on soils with low mineral availability, and crops with high fertiliser recovery efficiencies reduce pollution.

The third category is concerned with the internal mineral economy of plants. Definitions include (1) yield divided by the amount of a mineral in the plant, which is referred to as the mineral efficiency ratio, (2) yield divided by tissue mineral concentration, which is referred to as tissue, or physiological, mineral use efficiency, and (3) tissue mineral concentration required for a given percentage of maximum yield, which is referred to as the 'critical' tissue mineral concentration if this is 90% of maximum yield. Crops and varieties with lower critical tissue mineral concentrations are likely to tolerate growth on soils with low mineral availability.

In the context of this section, reference will be made to varietal differences in (1) *agronomic efficiency*, as yield with low fertiliser inputs or per unit fertiliser supply, (2) *acquisition efficiency*, as the amount of an element in a plant divided by the root biomass, and (3) *tissue utilization efficiency*, expressed as the reciprocal of tissue mineral concentration.

34.3.1 Nitrogen

Potato varieties differ in their agronomic N-fertiliser use efficiency, their N acquisition efficiency and their tissue nitrogen utilization efficiency (Kleinkopf et al., 1981; Sattelmacher et al., 1990a; Errebhi et al., 1999; Govindakrishnan et al., 1999; Saluzzo et al., 1999; Zebarth et al., 2004; Love et al., 2005).

In general, early maturing (determinate) varieties have the lowest agronomic N-fertiliser use efficiency (Zebarth et al., 2004), which has been attributed to their shorter growth period. This is reflected in the recommendation by Defra that more N fertiliser be applied to determinate varieties than indeterminate varieties (Defra, 2000). However, potato varieties capable of good yields with low N inputs have been identified within groups of varieties with similar longevity. For example, the early maturing variety Kufri Ashoka required only 65 kg N ha^{-1} to yield 33.4 t ha^{-1} compared with the standard variety Kufri Chandramukh, which needed 194 kg N ha^{-1}, and the medium maturing variety MS 78-46 required only 133 kg N ha^{-1} to yield 41.7 t ha^{-1} compared with the standard variety Kufri Bahar, which required 211 kg N ha^{-1} (Govindakrishnan et al., 1999). In the experiments of Saluzzo et al. (1999), the agronomic efficiency (here defined as the yield without N fertilisation/maximal yield) of Jaerli potatoes was lower than that of Spunta, Huinkel and Mailén potatoes. Similarly, when four related cultivars, Bannock Russet,

Gem Russet, Summit Russet and Russet Burbank, were grown with four N rates, 0, 100, 200 and 300 kg N ha^{-1} using two different application timings, each of the four showed a unique response (Love et al., 2005). Maximum yields and net returns for Bannock Russet were achieved with relatively small N applications, but Gem Russet required a larger application of N fertiliser for maximum returns and had a similar N requirement to Russet Burbank. Differences between varieties in their tissue N concentrations can be illustrated by data on petiole N concentrations in Russet Burbank and Shepody (Fig. 34.4) (Porter and Sissons, 1993). Although petiole N concentrations declined through the growth period in both varieties and responded to both the amount and the timing of N fertilisation, absolute petiole N concentrations at a particular time point and N fertiliser treatment were generally lower in Shepody (Porter and Sisson, 1993). Similarly, tissue N concentrations differ among potato varieties grown under identical conditions.

34.3.2 Potassium

In general, potato is less efficient in acquiring and utilizing K fertiliser than many other crops (Trehan and Claassen, 1998; Steingrobe and Claassen, 2000) and, as a consequence,

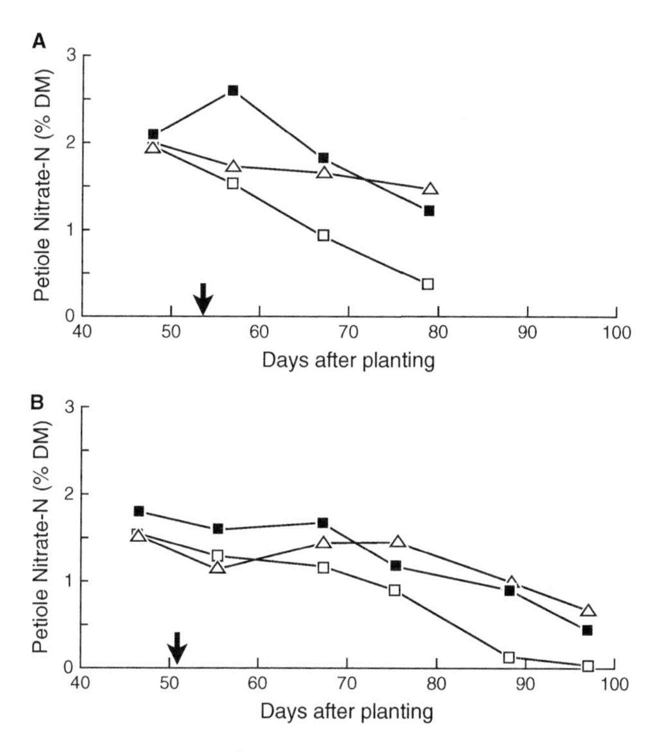

Fig. 34.4. The influence of the rate (kg N ha^{-1}) and timing of N application on the petiole nitrate-N concentration of (A) Russet Burbank and (B) Shepody potatoes. Nitrogen was applied at planting (90 N, open squares, and 180 N, closed squares) or split half upon planting and half after tuber initiation (90 + 90 N, triangles). The arrows indicate the timing of the second N application. Adapted from Porter and Sisson (1993).

is fertilised heavily with K. Nevertheless, potato varieties do differ in their dry matter accumulation when grown on low K soils, and there is, therefore, scope to reduce the K-fertiliser requirement of commercial varieties (Trehan and Sharma, 2002; Elkhatib et al., 2004; Moinuddin et al., 2004; Trehan et al., 2005). It is observed that varieties yielding most on low K soils maintain higher shoot K concentrations and acquire K more effectively by releasing more non-exchangeable K from the soil (Trehan and Sharma, 2002; Trehan et al., 2005). It is noteworthy that an increased ability to acquire K is not attributed to an increase in root length, which is consistent with the high mobility of K in the soil solution.

34.3.3 Phosphorus

Potato is less efficient than many crops in acquiring P and does not yield well on low P soils (Sattelmacher et al., 1990b; White et al., 2005b). The yield of a particular variety on soils with low P availability depends greatly on its maximum yield potential (Sattelmacher et al., 1990b; Trehan and Sharma, 2005). Although potato varieties differ in their tissue PUE (Dampney et al., 2002; Trehan and Sharma, 2003), which appears to correlate with haulm longevity (White et al., 2005b), differences in their yield responses to P supply are associated largely with differences in their P acquisition efficiency and correlate with root growth parameters rather than with the phosphate uptake capacity of root cells (Sattelmacher et al., 1990b; Trehan and Sharma, 2005). In addition, potato varieties with greater P acquisition and yield on low P soils are better able to increase P availability in soils (Trehan and Sharma, 2005).

Because phosphate is relatively immobile in the soil and is taken up effectively by all parts of the potato root system, the P nutrition of the crop will be governed by the volume of soil explored by the root system and the distribution of phosphate within the soil profile. Thus, the development of a rapidly growing, extensive and deep root system should improve P acquisition, plant growth and tuber yield. Consistent with this hypothesis, simulation models indicate that P acquisition by a potato plant is dominated by the morphological characteristics of its root system and the maximal rate of P uptake by root cells (Dechassa et al., 2003), and field trials of 36 tuber-bearing *Solanum* species have shown that tuber yield is significantly correlated with root dry weight (Sattelmacher et al., 1990b). Incidentally, the root system of the potato displays a surface-foraging phenotype (White et al., 2005b) that is considered to be most effective for the acquisition of mineral elements located in the topsoil (White et al., 2005a).

Breeding strategies to improve the response in potato yield to P fertilisation might focus on the development of varieties with the greatest vigour and root growth. Greater root growth rates per se are likely to benefit early maturing (determinate) varieties, and deeper rooting will benefit later maturing (indeterminate) varieties (White et al., 2005b). Although commercial potato varieties show little variation in their maximal root growth rates, the ultimate depth of rooting differs between varieties because the duration of root growth varies and is particularly extended in indeterminate varieties (Allen and Scott, 1992; Stalham and Allen, 2001). For example, Cara, a variety with exceedingly long haulm longevity, produces a deeper root system than the indeterminate varieties Maris Piper or Hermes, which, in turn, have deeper root systems than the partially determinate varieties

Estima or Wilja (Harris, 1992; Stalham and Allen, 2001). Morphological characteristics also differ between commercial potato varieties. For example, King Edward produces large numbers of fibrous roots, and Pentland Crown produces fewer thicker and longer roots (Steckel and Gray, 1979). Thus, there is genetic potential for the selection or breeding of genotypes with appropriate rooting phenotypes to exploit the soil volume and acquire P more effectively.

34.4 SOIL CONDITIONS AND POTATO GROWTH

Crop yield is affected by many factors such as crop type, soil type, degree of compaction and moisture status, drought and waterlogging. Rooting volume is the major factor, because, as this increases, so will the ability of a plant to take up the water and both major and minor mineral elements necessary for plant growth and tuber formation. Different cultivars have different potential rooting volumes, but achievement of these is dependent on the soil type and its physical condition. Soils can frequently become compacted because of the use of heavy machinery and powered cultivation implements, particularly when the soil is too wet, in an attempt to plant earlier and extend the growing season. During such practices, the soil is sheared by compressive rather than brittle failure, causing a smeared profile at the cultivation depth. Compaction reduces the number and size of macropores and increases their tortuosity, consequently reducing the movements of air and water through soil. An increase in soil bulk density restricts the rate of downward and lateral extension of roots, so effectively reducing the volume of soil from which the potato plant can acquire water and nutrients, resulting in a reduction in yield. A lighter textured soil may not compact so readily, and so, the potato roots can extend more easily. However, as these soils may hold less water, this increase in rooting volume may not result in increased crop yields. Attempts to relieve the depression of plant yield caused by compaction by sub-soiling below the plough depth are not always successful. Frequently, the compaction occurs at a relatively shallow depth during seed-bed preparation, an operation that has to occur after sub-soiling. Shallow compaction is frequently caused by bed-tilling and de-stoning. As de-stoning is usually the rate-limiting step during seed-bed preparation, it frequently occurs when conditions for cultivation are marginal. The subsequent compaction, at about 30 cm in wet soil, will slow early root growth and development. Soil compaction is a serious issue in potato cultivation, potentially causing severe reductions in crop yield. Such losses are a result of the reduced efficiency in the use of resources such as water and nutrients.

Soil compaction can also affect senescence in the potato crop. More rapid senescence was reported in a deeply compacted soil (Van Loon and Bouma, 1978) than in a loose soil as root growth had been restricted later in the season. However, Van Oijen et al. (1995) reported that compaction did not significantly advance senescence but that senescence was delayed because of the smaller number of leaves in the canopy and, consequently, less self-shading. Such compaction effects on leaf development can cause significant reductions in both tuber number and yield, together with other unwanted effects, such as secondary growth cracking (Table 34.2) (Van Loon and Bouma, 1978; Stalham et al., 1997).

Table 34.2 Effect of soil compaction depth and irrigation regime on the number of tubers (thousands per hectare) and total tuber yield (t ha^{-1}) of Maris Piper potatoes.

	Uncompacted	Compacted (10 cm)	Compacted (40 cm)	Compacted (10 + 40 cm)
Tuber number				
Unirrigated	813	685	715	591
Irrigated	676	591	699	611
Tuber yield				
Unirrigated	73.8	46.4	65.3	47.3
Irrigated	87.9	59.4	79.0	56.5

Source: Stalham et al. (1997).

34.5 SUMMARY

Potatoes need at least 14 essential mineral elements. Six of these (K, N, P, Ca, Mg and S) are required in large amounts and are often supplied as inorganic fertilisers. However, although fertilisation with these elements may be necessary for adequate tuber yields, there is concern that the use of fertilisers, especially those containing N, P and K, can lead to environmental pollution. The N fertiliser requirement of the potato is similar to that of many other field crops. However, the potato crop is fertilised heavily with K and P, because potato plants acquire both P and K extremely inefficiently. For this reason, agronomists are working to optimize the application of N, P and K fertilisers for uptake by the crop, and plant breeders are searching for varieties that acquire and utilize these elements more efficiently.

Agronomic strategies to optimize N and K fertilisation include the use of computer models to inform the amount and timing of organic and inorganic fertilisation, whereas recent strategies to optimize P fertilisation have focused on techniques for placing P fertiliser close to the roots of the seed tuber. In addition, practices that increase rooting depth and, thereby, the availability and acquisition of both major and minor mineral elements are being promoted. Plant breeders are seeking to quantify the genetic diversity and heritability of traits in potato impacting on mineral acquisition and utilization and their effects on plant growth and tuber yield (Davenport et al., 2005; White et al., 2005b). In tandem, knowledge of chromosomal loci that impact on the acquisition and utilization of mineral elements could be used to develop molecular markers for use in breeding programmes and/or to screen existing germplasm for mineral-efficient varieties. Such chromosomal loci could be identified using genetic mapping populations of *Solanum tuberosum* (Bradshaw et al., 2004) or of more distant interspecific crosses (Brouwer et al., 2004). Alternatively, trait-to-genotype association mapping using germplasm collections could be used (Simko, 2004). Indeed, it is already possible to genotype *Solanum* accessions using markers for genes known to improve the acquisition and utilization of mineral elements in other plant species (Vreugdenhil et al., 2005; Amtmann et al., 2006; Bucher and Kossmann, 2007, Chapter 15, this volume) and to test for allelic associations with the acquisition and utilization of mineral elements, plant growth and tuber yield. In the future, it is expected that the fertiliser requirements of the potato crop will be reduced through both improved agronomy and the development of mineral-efficient genotypes.

ACKNOWLEDGEMENTS

Work at SCRI is supported by the Scottish Executive Environment and Rural Affairs Department (SEERAD). The work of P.J.W., J.P.H. and K.Z. on potatoes, fertiliser placement and fertiliser response models is supported by grants HH3504SPO, HH3507SFV and HH3509SFV from the UK Department for Environment, Food and Rural Affairs. We thank Helen Bowen, Duncan Greenwood and Martin Broadley for their continued support, time, insight and enthusiasm for this work.

REFERENCES

Allen E.J. and R.K. Scott, 1992, In: P.M. Harris (ed.), *The Potato Crop: The Scientific Basis for Improvement*, p. 816. Chapman and Hall, London, U.K.

Allison M.F., J.H. Fowler and E.J. Allen, 2001a, *J. Agric. Sci.* 137, 379.

Allison M.F., J.H. Fowler and E.J. Allen, 2001b, *J. Agric. Sci.* 136, 407.

Amtmann A., J.P. Hammond, P. Armengaud and P.J. White, 2006, *Adv. Bot. Res.* 43, 209.

Asfary A.F., A. Wild and P.M. Harris, 1983, *J. Agric. Sci.* 100, 87.

Baligar V.C., N.K. Fageria and Z.L. He, 2001, *Comm. Soil Sci. Plant Anal.* 32, 921.

Barber S.A., 1995, *Soil Nutrient Bioavailability: A Mechanistic Approach.* John Wiley and Sons Inc., New York, U.S.A.

Bradshaw J.E., G.J. Bryan, C.A. Hackett, K. McLean, B. Pande, H.E. Stewart and R. Waugh, 2004, *Euphytica* 137, 13.

Brouwer D.J., E.S. Jones and D.A. St Clair, 2004, *Genome* 47, 475.

Dampney P., P. Johnson, G. Goodlass, C. Dyer, A. Sinclair and T. Edwards, 2002, Review of the Response of Potatoes to Phosphate. Final Report on Defra Project PE0108, Department for Environment, Food and Rural Affairs, London, U.K.

Davenport J.R., P.H. Milburn, C.J. Rosen and R.E. Thornton, 2005, *Am. J. Potato Res.* 82, 321.

Dechassa N., M.K. Shenk, N. Claassen and B. Steingrobe, 2003, *Plant Soil* 250, 215.

Defra, 2000, *Fertiliser Recommendations for Agricultural and Horticultural Crops (RB209)*, 7th edition. H.M.S.O., Norwich, U.K.

Defra, 2005, *The British Survey of Fertiliser Practice. Fertiliser Use on Farm Crops for Crop Year 2004.* Department for Environment, Food and Rural Affairs, London, and Scottish Executive, Environment and Rural Affairs Department, Edinburgh, U.K.

Elkhatib H.A., E.A. Elkhatib, A.M. Khalaf Allah and A.M. El-Sharkawy, 2004, *J. Plant Nutr.* 27, 111.

Errebhi M., C.J. Rosen, F.L. Lauer, M.W. Martin and J.B. Bamberg, 1999, *Am. J. Potato Res.* 76, 143.

FAO, 2006, FAOSTAT database: http://faostat.fao.org/faostat/, last accessed July 2006.

Gourley C.J.P., D.L. Allan and M.P. Russelle, 1994, *Plant Soil* 158, 29.

Govindakrishnan P.M., R.C. Sharma, I.A. Khan and H.C. Sharma, 1999, *Indian J. Agric. Sci.* 69, 350.

Greenwood D.J. and T.V. Karpinets, 1997, *Soil Use Manage.* 13, 178.

Greenwood D.J., T.V. Karpinets and D.A. Stone, 2001, *Ann. Bot.* 88, 279.

Greenwood D.J., C.R. Rahn, A. Draycott, L.V. Vaidyanathan and C.D. Paterson, 1996, *Soil Use Manage.* 12, 13.

Greenwood D.J., A.M. Stellacci, M.C. Meacham, M.R. Broadley and P.J. White, 2005, *Crop Sci.* 45, 1728.

Gregory P.J. and L.P. Simmonds, 1992, In: P.M. Harris (ed.), *The Potato Crop: The Scientific Basis for Improvement*, p. 214. Chapman and Hall, London, U.K.

Harris P.M., 1992, In: P.M. Harris (ed.), *The Potato Crop: The Scientific Basis for Improvement*, p. 162. Chapman and Hall, London, U.K.
Hegney M.A. and I.R. McPharlin, 1999, *Aust. J. Exp. Agric.* 39, 495.
Jenkins P.D. and H. Ali, 1999, *Ann. Appl. Biol.* 135, 431.
Jenkins P.D. and H. Ali, 2000, *Ann. Appl. Biol.* 136, 41.
Kleinkopf G.E., D.T. Westermann and R.B. Dwelle, 1981, *Agron. J.* 73, 799.
Kolbe H. and S. Stephan-Beckmann, 1997, *Potato Res.* 40, 135.
Lesczynski D.B. and C.B. Tanner, 1976, *Am. Potato J.* 53, 69.
Lewis D.J. and P.S. Kettlewell, 1992, *Aspects Appl. Biol.* 33, 29.
Love S.L., J.C. Stark and T. Salaiz, 2005, *Am. J. Potato Res.* 82, 21.
McArthur D.A.J. and N.R. Knowles, 1993, *Plant Physiol.* 102, 771.
Moinuddin K.S., S.K. Bansal and N.S. Pasricha, 2004, *J. Plant Nutr.* 27, 239.
Morgan J.A.W., G.D. Bending and P.J. White, 2005, *J. Exp. Bot.* 56, 1729.
Opena G.B. and G.A. Porter, 1999, *Agron. J.* 91, 426.
Porter G.A. and J.A. Sisson, 1993, *Am. Potato J.* 70, 101.
Prummel J., 1957, *Plant Soil* 8, 231.
Saluzzo J.A., H.E. Echeverria, F.H. Andrade and M. Huarte, 1999, *J. Agron. Crop Sci.* 183, 157.
Sattelmacher B., F. Klotz and H. Marschner, 1990a, *Plant Soil* 123, 131.
Sattelmacher B., R. Kuene, P. Malagamba and U. Moreno, 1990b, *Plant Soil* 129, 227.
Sharpley A.N., P.J.A. Kleinman and R.B. Bryant, 2002, *J. Soil Water Conserv.* 57, 425.
Simko I., 2004, *Trends Plant Sci.* 9, 441.
Simpson K., 1962, *J. Sci. Food Agric.* 13, 236.
Sparrow L.A., K.S.R. Chapman, D. Parsley, P.R. Hardman and B. Cullen, 1992, *Aust. J. Exp. Agric.* 32, 113.
Stalham M.A. and E.J. Allen, 2001, *J. Agric. Sci.* 137, 251.
Stalham M.A., A.B. Rosenfield and E.J. Allen, 1997, Abstracts of 81st Meeting of the Potato Association of America, Charlestown, PEI, Canada, 3–7 August 1997.
Steckel J.R.A. and D. Gray, 1979, *J. Agric. Sci.* 92, 375.
Steingrobe B. and N. Claassen, 2000, *J. Plant Nutr. Soil Sci.* 163, 101.
Tran T.S. and M. Giroux, 1991, *Can. J. Soil Sci.* 71, 519.
Trehan S.P. and N. Claassen, 1998, *Potato Res.* 41, 229.
Trehan S.P., H. El Dessougi and N. Claassen, 2005, *Comm. Soil Sci. Plant Anal.* 36, 1809.
Trehan S.P. and R.C. Sharma, 2002, *Comm. Soil Sci. Plant Anal.* 33, 1813.
Trehan S.P. and R.C. Sharma, 2003, *Indian J. Agric. Sci.* 73, 54.
Trehan S.P. and R.C. Sharma, 2005, *Adv. Horticult. Sci.* 19, 13.
Van Loon C.D. and J. Bouma, 1978, *Neth. J. Agric. Sci.* 26, 421.
Van Noordwijk M., P. De Willigen, P.A.I. Ehlert and W.J. Chardon, 1990, *Neth. J. Agric. Sci.* 38, 317.
Van Oijen M., F.J. De Ruijter and R.J.F. Van Haren, 1995, *Ann. Appl. Biol.* 127, 499.
Velthof G.L., P.J. van Erp, J.C.A. Steevens, 1999, Meststoffen 1999.
Verma R.S. and J.S. Grewal, 1979, *J. Indian Potato Assoc.* 5, 76.
Vos J., 1996, *Eur. J. Agron.* 5, 105.
Vreugdenhil D., M.G.M. Aarts and M. Koornneef, 2005, In: M.R. Broadley and P.J. White (eds), *Plant Nutritional Genomics*, p 201. Blackwell, Oxford, U.K.
Wheatley R.E. and K. Ritz, 1995, *Biol. Fert. Soils* 19, 36.
White P.J., 2003, In: B. Thomas, D.J. Murphy and B.G. Murray (eds), *Encyclopaedia of Applied Plant Sciences*, p. 625. Academic Press, London, U.K.
White P.J., M.R. Broadley, D.J. Greenwood and J.P. Hammond, 2005a, Proceedings of the International Fertiliser Society 568. Genetic modifications to improve phosphorus acquisition by roots. IFS, York, U.K. ISBN 0853102058.
White P.J., M.R. Broadley, J.P. Hammond and A.J. Thompson, 2005b, *Aspects Appl. Biol.* 73, 111.
Zebarth B.J., G. Tai, R. Tarn, H. de Jong and P.H. Milburn, 2004, *Can. J. Plant Sci.* 84, 589.

Chapter 35

Mathematical Models of Plant Growth and Development

D.K.L. MacKerron

Scottish Crop Research Institute, Invergowrie, Dundee DD2 5DA, United Kingdom

'Wisdom is the principal thing; therefore get wisdom: and with all thy getting get understanding. Exalt her... (and) she shall bring thee to honour, ...'
The Book of Proverbs. Chap. 4: verses 7, 8

35.1 INTRODUCTION

Models are not reality. Properly, they do not even represent reality. Rather, models represent abstractions taken from reality. Their primary functions are to formalize our knowledge of a subject in a manner that allows us to understand and quantify some aspects of the system that we are studying, to test that understanding and hence make inferences beyond our first studies and to show us where there are gaps in our knowledge. A worked example is seen in MacKerron and Waister (1985) where gaps found in the existing knowledge stimulated new experimental work to allow the generation of functions necessary for the model and in MacKerron (1985) where among other things the model was used to quantify the sensitivity of the potential yield of the crop to changes in the timing of management operations. These points are among those considered by Sinclair and Seligman (2000) in their debate on criteria for publishing papers on crop modelling. As a bonus, models allow us also to communicate our understandings with others.

For the purposes of this chapter, I propose to dismiss or ignore all 'models' that are based on a single statistical relation. An example of those could be one to estimate yield as a single defined function of air temperature in June, solar radiation in June and solar radiation in July. Such a 'model' has no explanatory powers and, quite simply, only describes the data set from which it was derived. Further data from the same area in further years would almost certainly not agree with the same statistical relation. I shall not cite examples. Sinclair and Seligman (2000), in their article, give a more detailed critique of types of growth models, whereas Karvonen et al. (2000) consider the role of simulation and other modelling approaches in decision making.

The models that we will consider here treat the potato crop, its growth and development and its interactions with its environment and with other organisms. Largely, they are mechanistic. That is, they describe the operation of the system at one level of organization by providing an estimate of processes operating at a lower level (Haverkort and

Potato Biology and Biotechnology: Advances and Perspectives
D. Vreugdenhil (Editor)

MacKerron, 1995; Kabat et al., 1995b). No models attempt to describe the whole system of the crop and all the influences on it, although some decision support systems (DSSs) (cf. Chapter 36, Marshall, this volume) have been designed to draw on two or more models within the one system to suggest responses of the crop to a wider range of input variables than are normally treated by single models. Instead, mechanistic models are usually designed to quantify a few specified output variables in terms of another specified set of input variables that have appeared to be relevant to the modellers and, if done properly, that have been shown experimentally to modify relevant processes towards the output variables (Kabat et al., 1995b).

The boundaries between mathematical models and DSSs may be blurred and hard to define, but for the purposes of this chapter, we will restrict ourselves to simulation models that take a maximum of two classes of inputs to describe the behaviour of the crop: e.g. environmental factors and management decisions or management decisions and economic variables. A mathematical system that handles three or more such sets of variables will be considered a DSS and will be considered in Chapter 36 (Marshall, this volume).

35.2 AIMS AND APPROACHES

The aims of simulation models of the type being considered here are to simulate and explain crop development and behaviour, yield and quality as functions of environmental and management conditions or disease pressure or (possibly) of genetic variation. Then, if the values of the variables are known, output variables of the model – e.g. yield (e.g. Fishman et al., 1985; Jefferies and Heilbronn, 1991; Kooman, 1995), quality (e.g. Jefferies and Heilbronn, 1991; Hertog et al., 1997a,b) or risk of disease (e.g. Spitters and Ward, 1988; Phillips et al., 1998; Elliott et al., 2004) – may be estimated rather than measured. The important point in that declaration is 'to explain'. By 'explain', I mean to provide a description of the operation at one level of organization through the quantification of operations at another lower level. If the design and parameterization of the model are successful, then it may be used to analyse certain questions and provide answers, on a specific crop in the field, for example. What constitutes a successful design will be considered later.

Several different approaches have been taken since the 1960s on how best to tackle the simulation of aspects of the crop. Most crop modelling within that time has been what I will call 'definitive' – that is, one set of values for the input variables produces one answer for each of the output variables. These models do not offer estimates of natural variation or of sampling error, and most of those to be considered in this chapter are of that nature. More recently developed kinds of models are able to deal specifically with uncertain values in both the input and the output variables. Examples include rule-based models with flexible querying and causal models based on Bayesian probability theory (e.g. Gu et al., 1994, 1996; Marshall et al., 1995). Problems inherent in this latter style of model include how to assemble a chain of uncertain processes in a way that avoids unlikely outcomes and how to present answers in a style that is meaningful to others than scientists engaged in the same exercise. To inform a grower or an agronomist that a particular well-grown crop has a 50% chance of exceeding 60 t/ha, a 75% chance of exceeding 50 t/ha and a 95% chance of exceeding 40 t/ha is to invite the response,

'I could have told you that myself!'. For these reasons, it is likely that developments in those kinds of models will be targeted towards economists and policy makers rather than growers (e.g. Schans, 1991; Abbaspour et al., 1992; Abbaspour, 1994). These models will not be discussed further.

Another basic division in approach to modelling is between what can be called 'top-down' and 'bottom-up' approaches. Within these subdivisions, approaches can be classified as 'source-driven' or 'sink-based'.

35.2.1 Top-down/bottom-up

Some of the earliest mechanistic models of crop behaviour were of the bottom-up kind: for example, those of De Wit (1965) in which understandings of photosynthesis and of canopy structure, each at the organizational level of the leaf, were tested by aggregating and organizing that knowledge to simulate the behaviour of a canopy of leaves and hence the level of production in their canopy. Given the underlying generality of photosynthesis, it can be difficult to introduce meaningful factors that constrain the next level of the system to reflect differences between species, let alone cultivars. However, an early example of modelling the production of potatoes was provided by Rijtema and Endrödi (1970) based on the work of De Wit (1965), coupled with some necessary but notably arbitrary assumptions such as those on diffusive resistances and that the respiration rate is 25% of the production rate. Some time later, the development of a universal mechanistic model applicable across many species of crop was still the aim of some workers, e.g. Van Keulen and Wolf (1986) and Spitters et al. (1989).

A more recent development of a bottom-up model was developed by Fishman et al. (1984, 1985, 1995), in which the basic process is the photosynthesis of individual leaves as a function of the intensity of photosynthetically active radiation. However, the levels of organization that need to be climbed impose unrealistic demands on the user of this model. For example, this model does not use even daily totals of radiation but requires daily integration of hourly calculations of the rate of photosynthesis per unit ground area for each of 10 layers. Furthermore, to derive daily dry matter production from photosynthesis, the model has embedded in it several functions that are not readily assessed independently to estimate the effects of temperature, respiration, water stress and nitrogen supply on reducing the contribution from photosynthesis (Fishman et al., 1995). The performance of potato crops growing in three fields in the spring season was used to 'calibrate' the parameters in the model. The model was then 'verified' fairly successfully against an autumn-sown crop in the same area.

For models intended to produce values for output variables at the upper level of organization, plant population or field, it has generally been more productive to take the top-down approach. Then, the modeller, knowing the target level for his/her answers, has the task of identifying the important variables within the system that is to be studied and what factors modify these from a lower level. For example, Spitters and colleagues developed a model to analyse variation in yield between cultivars in terms of differences in light absorption and use and dry matter partitioning (Spitters, 1987; Spitters and Schapendonk, 1990), whereas MacKerron and colleagues did the same for potential yields as functions of management operations – planting dates and depths and times of harvesting – and standard measurements of weather variables (MacKerron, 1985; MacKerron and Waister, 1985) with the important design criterion that all the input variables should be readily available. If such models are then successful, then they may be developed for further applications by addition of further relations. So, Kooman (1995) and Kooman and Haverkort (1995) developed the model of Spitters (1987) by the addition of relations between tuberization and both day length and temperature. Jefferies and Heilbronn (1991) developed MacKerron and Waister's model (1985) by redefining canopy expansion to permit the inclusion of new experimentally derived functions describing the effect of constraints in water supply on leaf expansion. Other functions newly derived from experimental data were also included to improve the generality of the model (Jefferies et al., 1991).

35.2.2 Source-driven/sink-based

The subdivision between source-driven and sink-based models is not always clear-cut. (Source-driven implies that the more of a resource is captured the faster the growth rate, whereas sink-based suggests that growth will be determined by the rate at which the plant can use the resources that are potentially available.) Some models may be source-driven, essentially, and yet include one or more sink-based functions or, more confusingly, some may include functions that become sink-based as the simulated crop develops. In these circumstances, it is better to examine critically the balance between source and sink and the simulations of them. It is not part of the purpose of this chapter to offer a firm conclusion on what is the correct balance. Rather, it is the intention to draw attention to the question and to encourage readers to be aware of the inherent problems and to come to an informed opinion on what is appropriate in each circumstance.

For example, some models simulate leaf expansion as a function of assimilate supply and specific leaf area (SLA, the amount of leaf area per unit leaf biomass) (De Wit et al., 1970; Van Keulen and Seligman, 1987). Other models use functions of temperature to determine leaf expansion in early season then limit leaf area expansion by assimilate supply (i.e. radiation interception) and a function of SLA, later in the growing season, e.g. Spitters et al. (1989). So, Van Delden et al. (2001) examined the sensitivity of the timing or developmental stage of switching between temperature and radiation as the limitation to leaf expansion. For two data sets on potato, each based on one site, 2 years and two cultivars in each year, they found the leaf area index (LAI) at which the switch was best introduced (Ls) to be 0.26 and 1.41. The disparity between these numbers suggests that the variation in Ls would need to be explored further if that approach were to be generally adopted.

In the model of Spitters and Schapendonk (1990), incident solar radiation is conditioned by fractional interception and efficiency, which then drives the increase in dry matter. Some of that dry matter is partitioned into leaves, and so, through an estimate of SLA, conditions the estimated leaf area, which in turn influences fractional interception of radiation. This does indeed reflect one's understanding of crop growth, but the process, if not circular, is certainly helical so that if, say, some intermediate functions give a wrong answer at an earlier time in the season, then the answers will be wrong later even if these functions are correct later. That is, the rate of assimilation now is heavily dependent on the rate of assimilation then.

Models that drive canopy expansion using assimilate can be contrasted with models that treat leaf expansion as being independent of assimilate supply – that is, they assume that assimilate supply is ample for the purposes of leaf expansion, at least until canopy closure – and drive canopy expansion by functions of temperature (MacKerron, 1985) or temperature and water supply (Jefferies and Heilbronn, 1991; MacKerron and Lewis, 1995; Rijneveld, 1997). A characteristic of this class of model is that it is entirely independent of assumptions about SLA, which is recognized to be a most variable ratio, differing within the plant at any one time between younger and older leaves, being influenced by the radiation environment and being highly sensitive to water status.

In Kooman's model (Kooman, 1995; Kooman and Haverkort, 1995), there is a clearly defined progression after tuber initiation in which tuber growth shifts from being sink-limited, taking a small proportion of daily assimilate production, to being source-limited, taking all the daily assimilate production. The progression caused by this device is sensitive to the differing degrees of sink strength attributed to biomass in leaf and tuber. The shift away from the leaves in partitioning assimilate provides a rational means to simulate the death of the leaf canopy in the later stages of crop growth.

35.3 APPLICATIONS

The most commonly modelled processes are the growth and the development of the crop. This is not only because the final outputs are of considerable economic importance but also because such a model may often offer a basis on which to develop the simulation of other effects or constraints such as the effects of pests and diseases or of a wider range of environmental factors such as CO_2 concentration, $[CO_2]$, or nutrient supply.

In almost all models of growth and development, growth is driven by intercepted solar radiation. However, in some models derived in lower latitudes – e.g. Sands et al. (1979) in Australia and Ingram and McCloud (1984) in Florida – solar radiation plays a less important role than temperature. Perhaps, this is because at the higher levels of insolation experienced at these latitudes, the canopy of a potato crop becomes light saturated (Sale, 1974) while high temperatures may be a serious constraint.

Among top-down models, the relation between biomass and intercepted radiation is taken generally as the principal function driving growth. There have been occasional questions posed as to the validity of using these relations, e.g. Demetriades-Shah et al. (1992), but most criticisms of the principle were answered by Monteith (1994). That still leaves the condition that the values used in the relation between intercepted solar radiation and biomass should be properly determined with due regard to errors of measurement.

35.3.1 Potential yields

A significant proportion of the models of growth and yield have begun as models of potential yield. That is, they began by assuming that the crops are free of pests and diseases and that mineral nutrition is adequate for growth (Ng and Loomis, 1979; Van der Zaag, 1984). Some even assume that water supply is non-limiting (MacKerron and Waister, 1985). These are seen as areas of potential development of the model. However, models of potential yield should not be seen simply as drafts for later more detailed models. They have their own uses.

An early and significant model of the growth of the potato crop was produced by Ng and Loomis (1979, 1984) in which they set out clear objectives for their study. Their model included representations of a wide range of influences and responses from the basic effects of radiation and temperature on photosynthesis to the 'feed-forward' effects of the mother tuber and the 'feedback' effects of age on capacity for growth and of senescence, initiation of branches and the growth of fibrous roots. Ng and Loomis clearly declared most of the physiological relations that they used as 'physiological opinions' and recognized that their model comprises 'a large set of hypotheses'. Many were derived from the literature and others from a field experiment. They were clear also that they were unable to give a true validation of their model, as the data used in the model and those from the field experiment were not completely independent. Their model was successful, however, in that it advanced the understanding of integrative crop physiology and offered a tool to study the action of environmental variables on the crop.

In complete contrast, Van der Zaag (1984) presented an extremely simple model for the estimation of potential yields based on solar radiation and length of growing season to estimate potential tuber yield at various locations throughout the world. That, too, had a useful role in focusing the attention of others on differences in potential yields and in actual yields and between these two at various locations across the world.

A weakness of the model of Ng and Loomis (1979) is that, in trying to be 'explanatory' and to have a 'high degree of physiological and morphological detail', it incorporated so many variables and tried to simulate too much. Many subsequent models have been set similar targets but with tighter boundaries on the domain within which they apply. The key to the top-down approach is not to ask 'What else do we know about?' but rather 'Is there anything else we *need* to include?'

A successful model of potential yield does not provide an 'airy-fairy' estimate of yield but one which the best of crops in a locality will approximate to (MacKerron, 1985; Kunkel and Campbell, 1987). MacKerron (1985) took the opportunity presented by a local competition organized by growers in Angus, Scotland, and test-dug by staff of the Scottish Agricultural College to assess the comparison between achieved yields in the best crops and the potential yield calculated for each of them. The four best crops, in order, achieved 102, 101, 94 and 85% of the calculated potential yields, respectively. The four achieved yields ranged from 80 to 64 t/ha at a time when the average UK yield of potatoes was in the order of 40 t/ha.

Kunkel and Campbell (1987) examined achieved yields from 134 field plots from experiments performed in the Columbia Basin, Washington, over 15 years from 1959 through 1973 and, recognizing that most plots had not given potential yields, used Webb's (1972) 'Boundary Line' method to indicate the progress of potential yields with time from planting in the relevant years. They considered a number of models for the estimation of potential yields and were best satisfied with the relatively simple model of MacKerron and Waister (1985). They also reported that Van der Zaag's (1984) estimate of 140 t/ha potential yield had been confirmed by the models and that actual yields close to that figure had been achieved. This emphasizes that models do not need to be complex. What is needed in a model will depend on the task in hand.

35.3.2 Actual yields

The principal difference between a model of potential yield and a model of actual yields is that the modeller has included certain constraints on the growth and development of the crop. The purpose of such developments has to be that the modeller is looking to provide answers that better reflect actual experience more often. The most widely treated constraint is water supply. There are, of course, a wide range of models for which the sole purpose is scheduling irrigation. These may require values for crop-specific parameters and for crop cover, but these will not be considered in this chapter. Rather, we will consider those models where it is the crop that is of primary importance.

35.3.2.1 Effects of water supply/water use/water stress/irrigation

The interest in the influence of water supply on growth and development in potatoes can be subdivided broadly into two categories. First, there are attempts to answer the question of how the crop will grow in a field soil under rain-fed conditions and then there is the question of which thresholds need to be avoided, using irrigation if necessary, if growth and development are not to fall substantially below their potential rates.

Campbell et al. (1976) developed a model of water flow, from soil to plant, to determine an optimal soil water potential to maximize potato growth. The factors that they included as variables were transpiration, leaf osmotic potential, root density and distribution and the hydraulic conductivity of the soil. Their model was developed using growth chamber experiments and then tested against measurements in field-grown plants. A characteristic to be noted from the model is that the plant water potential at which transpiration is first reduced is not constant but is dependent on soil hydraulic properties, transpiration rate and several plant properties including rooting density. Although the model produced a fairly adequate description of plant water potential during the day and described the maximum transpiration rates that could be sustained as a function of soil water potential, the nature of other plant variables that needed to be defined possibly limited its value. An additional problem recognized by the authors is that the growth of the crop may not be limited by stomatal closure but by turgor (Hsiao, 1973). Jefferies (1989) showed that even under relatively low evaporative conditions, differences in leaf extension rates were detectable with soil moisture deficits (SMD) as low as 16 mm, and the reductions increased rapidly with SMD to a minimum when SMD equalled 77 mm. These changes corresponded to reductions in turgor. In contrast, the conversion coefficient for intercepted radiation into dry matter (arguably a function of stomatal aperture) was unaffected by SMD less than 47 mm (Jefferies and MacKerron, 1989).

When simulating the growth of potatoes under a constraint such as water supply, an important advantage of the top-down models becomes apparent – that there are fewer relations to be modified. However, an adequate treatment of the problem may necessitate invoking another lower level of organization. So, MacKerron and Waister (1985) simulated the development of LAI, but in addressing the problem of water as a constraint on growth, Jefferies and Heilbronn (1991) had to recognize that the stress could be imposed and relieved at various times and that increases in leaf area did not occur proportionately over all the existing leaf area. Rather, individual leaves appear, grow and stop growing; other leaves grow in their turn. Therefore, the simulation of the dependence of canopy expansion on water supply had to treat the growth of individual leaves in turn. A leaf that has matured before water stress is imposed is unaffected. A leaf that has matured during a period of water stress is not available for further growth. Jefferies and Heilbronn (1991) therefore related the extension of individual leaves to thermal time constrained by the level of available soil water. Among the criticisms of that model are that root growth is not simulated, maximum root depth is defined, and potentially, all the plant-available water in that depth of soil is available from the start of growth. Another is that by including a constant water-use efficiency term, normalized for vapour pressure deficit, dry matter production and water use were firmly coupled together (Jefferies et al., 1991). There are many sources of information supporting the relation between these two variables, but the relation is not constant and varies with site and season. Separate estimation of evaporation by a 'Penman'-type approach would have avoided that feature but only at the price of added constants, crop coefficients and meteorological data.

In later developments of that model (MacKerron and Lewis, 1995; Rijneveld, 1997), potential evaporation was included as an input variable to the model to avoid the problem yet minimize the number of inputs.

Others have found it necessary to take equivalent approaches. So, Kooman and Spitters (1995) for example noted that in their model with differing levels of complexity, the simpler version was the best for estimating potential yield, but where water supply was a factor, the more complex version was needed.

All the models presented in the book by Kabat et al. (1995b) treated potato growth as potential constrained by water supply (De Koning et al., 1995; Fishman et al., 1995; Jefferies and Heilbronn, 1995; Karvonen and Kleemola, 1995; Kooman and Spitters, 1995; Ritchie et al., 1995; Roth et al., 1995; Van den Broek and Kabat, 1995). Fortunately, the comparisons and evaluations were made most thoroughly by Kabat et al. (1995a). In conclusion, they noted that crop simulation models are critically dependent on the procedures defining partitioning of carbon and its redistribution during senescence. They noted the particular sensitivity of models where the relative sink sizes of different plant organs are used to quantify coefficients. They doubted the utility of most models beyond the environments for which they had been calibrated. Kabat et al. (1995a) not only considered that the understanding of potato crop development had reached the stage where yields limited by temperature and radiation alone can be estimated with reasonable confidence but also considered that proper simulation of the effects of changing water supply needed better handling.

Unusually, Hamer et al. (1994) modelled the development of the potato crop as a means to simulate water use by the crop. Tuber growth and yield were not a consideration. In particular, root growth was calculated in a simple manner and the soil moisture volume fraction at four levels of matric potential (5, 40, 200 and 1500 kPa) were used as inputs to the model. Tested against data from four widely separated sites in the UK, a regression between observed and measured values of soil moisture data accounted for 94% of the variance in the data.

Other efforts to take account of the growth of roots will be in Section 35.3.3.

35.3.2.2 The effects of nitrogen and the combined effects of water and nitrogen

The nitrogen requirement of a potato crop depends on many factors. One group of these affects the rate of nitrogen uptake and the total nitrogen uptake by a potato crop. A second group affects the supply of nitrogen to the crop from the soil. Among the latter group, the effects of previous cropping and manuring can be very variable, and so, fertilizer recommendations ought to be field specific. Attempts to develop simulation models (and DSSs) have been handicapped by that variability, and, at the practical level, there have been strong arguments for measurements to be made both before and during crop growth to guide the use of nitrogen fertilizer. Yet, these very difficulties have continued to drive a perceived need for simulation modelling or DSSs that will be capable of giving realistic and timely advice.

Greenwood et al. (1985b) produced a model said to be 'dynamic' and for 'practical purposes rather than understanding' of the response of potatoes to nitrogen fertilizer. Although the description of the nitrogen uptake may have been dynamic and although the model formalized the data collected from 11 experiments (Greenwood et al. 1985a) and served to explain the response of the crops to nitrogen, the simulation of the underlying growth of the potato crop was not even semi-mechanistic but was driven by a standard daily increment. That feature implies that without 'calibration' (see Section 35.4.2), the model cannot be expected to be transportable much beyond the originating data. Neeteson et al. (1987) did extend the work by testing the model against 61 experiments on nitrogen fertilizer with potatoes, but there were 99 such experiments in total and those with yields of less than 55 t/ha were excluded from the study.

In the comparative exercise led by Kabat et al. (1995b), only two models, CROPWATN (Karvonen and Kleemola, 1995) and SUBSTOR (Ritchie et al., 1995), were applied to the question of the effects of differing levels of nitrogen supply.

In the model of Karvonen and Kleemola (1995), special attention was paid to combining physically based sub-models of soil water balance and soil nitrogen balance with the sub-models of crop growth and crop nitrogen. Being clearly described, the model offers a good example of how complexity develops as functions are added to a model. Transpiration is computed following the Penman–Monteith equation, and potential evapotranspiration and evaporation from soil are calculated following Feddes et al. (1978) and Jensen (1983). In the soil nitrogen balance sub-model, the uptake of nitrogen by plant roots not only follows the method of Van Keulen and Seligman (1987) but also includes procedures for transport of mineral nitrogen and the fates of mineral fertilizers. The sub-model on potential crop production requires the computation of daily leaf net assimilation after Goudriaan (1986), which is then integrated over the canopy, adjusted for conversion from CO_2 to glucose, modified for maintenance respiration, partitioned between sinks and further modified for leaf senescence (Spitters, 1986; Spitters et al., 1986). The sub-model on nitrogen balance in the crop uses the method of Van Keulen and Seligman (1987), but that is a model of growth of a crop of spring wheat. The greatest difficulties were met in predicting the development of LAI as SLA differs under both water and nitrogen stresses. There were also similar problems of partitioning between root and shoot.

35.3.2.3 *Effects of day length*

In the model of Sands et al. (1979), developed under Australian conditions, solar radiation and day length were inputs for only the first 3 weeks after emergence. Thereafter, tuber bulking was driven by temperature. However, that model did not transfer to the higher, European, latitudes because the physiological time to the start of tuber growth was 'an unbounded function of day length'. Regel and Sands (1983) then modified the model to accommodate what they called early and late tuberization so that the start of tuber growth was independent of day length while it was less than 14.1 hours for cultivar Russet Burbank and 15.1 hours for cultivar Kennebec. However, in neither version of the model was day length an explanatory variable. Rather, it was adjusted to set the bounds for numerical summation.

Possibly, the first versions of a model to treat day length as an input variable were those of Kooman (1995) and Kooman and Haverkort (1995) in which partitioning of assimilate to the foliage is a function of day length and temperature until tuber initiation. Tuber growth rate is independent of day length in this model, but it was used as input to a function describing development rate until tuber initiation. Once the simulation of tuber growth has started, the partitioning of dry matter is determined by the demand from the tubers – sink determined – and only the balance of assimilate is available for the growth of the foliage. Once the tuber growth rate matches the daily assimilation rate, foliage growth is stopped.

That approach (Kooman, 1995) where day length influences development rather than growth directly and so influences a progressive transition between phases of crop growth is almost certainly a better one than that of the earlier model. Certainly, its developers have gone on to apply it and its developments to several applications for which timing of tuber initiation and length of the period for tuber growth are critical (Haverkort and Kooman, 1997; Haverkort and Grashoff, 2004).

35.3.2.4 *Effects of CO_2 concentrations*

In a comparatively early study of the effects of elevated [CO_2] Yandell et al. (1988) conducted to establish what might be achievable in space systems, response surfaces for tuber weight were fitted as functions of temperature and [CO_2] and simulation studies were promised but do not seem to have been realized.

Van de Geijn and Dijkstra (1995) discussed the expected changes of increased photosynthetic rates, partially offset by lower stomatal conductance, and the many feedback systems in the soil–plant–atmosphere system. Schapendonk et al. (1995) assembled a mechanistic model from several sources including a biochemical model of photosynthesis described by Farquhar et al. (1980) and presented simulations of some effects of [CO_2]. However, their experimental data, which both fed and tested the model, were all derived from crops grown under the ambient [CO_2] of 1988. It follows that none of the developmental processes, e.g. partitioning to tubers and the progress of dry matter concentration following the conversion of sugars to starch, could have been tested.

The simpler of the top-down models in which dry matter accumulation is driven by experimentally derived relations between biomass and intercepted radiation (e.g. Jefferies and Heilbronn, 1991; Kooman, 1995) are unable to be applied to the question of the effects of elevated [CO_2] on the growth of the potato. Or, at least, it is not valid to do so as the necessary underlying experimental information for their primary driver is lacking. However, Schapendonk et al. (2000) devised a modification to the light conversion coefficient (LCC) that is central to this type of model. The validity of the device remains to be seen. On the contrary, bottom-up models driven by photosynthetic rates, e.g. Schapendonk et al. (1995), do not have that limitation and there is quite extensive experimental evidence available of the effect of [CO_2] on photosynthetic rates. A serious finding of recent studies of photosynthesis in crops introduced to sustained elevated [CO_2] is that although the saturated rate of photosynthesis is initially increased, it later declines or acclimatizes until it is near to the rate achieved under current ambient conditions. That change is associated with reductions in stomatal conductance. These observations suggest that neither kind of simulation model is yet suited to addressing the questions raised by an increasing ambient [CO_2]. What are needed are quantities of good, reproducible experimental data (cf. Chapter 19, De Temmerman et al., this volume).

Even if the problem of quantifying the primary driver of growth at consistently elevated [CO_2] were to be solved, the problem of assembling an adequate mechanistic model of growth would not have been solved, as such evidence as is available at present suggests that there may be changes in a whole range of developmental functions and stages. For example, leaves may be more dense, SLA may be lower, and partitioning may be modified. Any model intended to analyse the questions posed by rising [CO_2] will need to incorporate modified developmental functions of the kind that are currently being established (cf. Chapter 19, De Temmerman et al., this volume).

35.3.3 Outstanding difficulties and inadequacies – root growth

The growth of roots presents growth modellers with an apparently intractable problem. Sampling roots is tedious, and when done, generally, it has not been coupled with other extensive measurements that might facilitate establishment of relations between their growth and environmental factors. For example, there have been only a few attempts to couple observations of root growth with progressive changes in soil water profiles observed using neutron probes or other instruments making equivalent measurements (Durant et al., 1973; Belmans et al., 1982).

Many modellers, from Ng and Loomis (1979) onwards, have sought to simulate root mass (Fishman et al., 1995; Kooman and Spitters, 1995). But one should ask why. How can the mass of roots be derived and what is to be done with that figure once it has been estimated? Some years later, Kooman (1995) agreed with this point, stating (p. 102) that the model Lintul-Potato did not take into account root biomass, as precise measurements of roots in field experiments are rare. In models that work up from photosynthesis, it may be acceptable to assign a share of the net assimilates to root growth (e.g. Dewar, 1993) but how big a share? Generally, there is little or no good field data on partitioning that include mass of root. In those models that estimate the net dry matter production by a simple relation between biomass and intercepted radiation, there is an even more basic question to be asked. One ought to ask how the conversion coefficient from intercepted radiation to biomass was determined. Almost certainly, the observations of biomass did not include good quality estimates of root biomass. If one returns to the original experimental investigations that gave rise to these relations, then one will find that if root growth was measured in any of them, they are the exception (e.g. the data used by MacKerron and Waister, 1985, to derive their LCC did not include root mass, nor did the data set of Allen and Scott, 1980). Some may have included estimates of the roots lifted from within the soil ridge, but few if any have seriously extracted roots throughout the rooting profile. Then, the coefficient used does not include the production that went into the root biomass, and the estimate of biomass in these cases is 'biomass without roots'. To 'adjust' the value of the light use efficiency or conversion coefficient to give a slightly higher biomass production that can then be diverted to root biomass is at best self-deceiving. In such models, the practice of assigning a proportion of assimilate to root production is quite invalid.

Presuming that an appropriate value of LCC could be found, the problems of speculative partitioning between shoot, root and tubers (MacKerron and Peng, 1989, showed that dry mass of roots could vary between 3 and 17% of the total), converting total root dry matter into length (specific root weight and length are not constants; Allmaras et al., 1975; Jefferies, 1993), and of distributing it in the soil remain.

Roots do not function in proportion to their mass (contrary to the assumption in Fishman et al., 1995; and in Kooman and Spitters, 1995) but in proportion to their length (Tinker, 1976). When trying to quantify the uptake function of roots, it is not wise even to try to estimate the length of roots from the (estimate of) root biomass, as the greater length of roots is in fine roots that represent an almost insignificant fraction of the whole root biomass. The important feature of a root system is its absorptive surface area. Often, the total length of root is used as a surrogate for surface area (Hillel and Talpaz, 1976; Taylor and Klepper, 1978), and this appears to be satisfactory for those plants, such as potato, that do not have tapering roots (Hamza and Aylmore, 1992). So, the question that should be addressed is 'What length of roots is being produced and where?' Van den Broek and Kabat (1995) estimated depth of rooting as a function of time, following Van Wijk and Feddes (1986), whereas Karvonen and Kleemola (1995) avoided the need to estimate the quantity of roots. An approach that avoids the problems of attributing biomass to the roots is to simulate root length directly from estimates of numbers of roots initiated and of root extension rates (Hackett and Rose, 1972) and death rates (Hillel and Talpaz, 1976). This is the approach taken by MacKerron and Lewis (1995). Monteith (1986) used root length density (cm of root per cm^3 of soil) to estimate the rate at which a growing root system could extract water from soil. Following Gregory et al. (1978a,b) and Gregory and Squire (1979), Monteith assumed that root growth was limited to an advancing front and that there was little change in the root length at a given depth once that front had moved deeper. (Strictly, these observations were made on cereal crops.)

In the potato crop, a few authors have reported the development of the root system at only a few stages through the season, determined by sequential sampling (Lesczynski and Tanner, 1976; Vos and Groenwold, 1986). Parker et al. (1989, 1991) used minirhizotrons to observe rooting depth on as many as 11 occasions in a growing season. In their experiments, potato roots extended continuously until the end of July, more than 100 days after planting.

The questions of what lengths of roots were in the soil and when do they get there were among the questions addressed by MacKerron and Lewis (1995) who divorced root growth from assimilate production and, instead, calculated root length densities from number of roots per stem internode below ground (each stem is rooted independently) and from root extension calculated as a function of soil temperature. They considered the root and soil moisture profiles to be distributed over 10 horizons in the soil, the upper two being in the ridge.

Two limitations were imposed on root extension (MacKerron and Lewis, 1995; Rijneveld, 1997). One was designed to produce a profile of root length density that declines linearly with depth from a maximum in the ridge (Vos and Groenwold, 1986; Parker et al., 1989) to a limiting potential root depth. (Feddes and Rijtema (1972) used an exponential function to describe the change in root length density with depth, although they found evidence of a change to a linear pattern with prolonged drought. Van Bavel and Ahmed (1976) used a linear distribution.) That was achieved (Rijneveld, 1997) by arbitrarily starting the death of individual, randomly selected roots from 55 days after emergence (Vos and Groenwold, 1986). These devices (rather than refinements) did not, of course, allow for the effects of compaction and differing soil strengths (Feddes et al., 1988), but root extension was prevented below a moisture volume fraction of 0.15 (Sharp and Davies, 1985; Wraith and Baker, 1991).

35.3.4 Ideotyping

The term 'ideotype' is used to label the set of characteristics that would seem best to suit an organism to its environment. Simulation models are often seen by modellers as a good means to identify ideotypes for particular environments, although Ng and Loomis (1979) perceived models as being simply 'an aid in genotype evaluations'. It must be said that plant breeders seem to be less enthusiastic about the idea, perhaps because there is generally no direct link between the characteristics identified as parameters in a mechanistic model and particular genes that might be selected in a breeding programme. But the advocates of modelling argue that their systems offer a cheaper means of testing that changes, in which characteristics might be used to achieve a particular end than simply conducting a multiplicity of field trials.

So, for example, Spitters and Schapendonk (1990) addressed the question of the influences of both sustained water stress and a short-duration water stress imposed at differing times, and they did this for differing maturity classes of potato cultivar. Certainly, the amount of experimental work that would have been entailed actually to conduct these studies experimentally would have been enormous, and it would have been difficult, too, to impose such carefully defined water stresses. However, their results 'emphasised the complexity of selection for drought tolerance'. They found that the simulated reduction in yield from water stress could be lessened when growth was averaged over the whole season and that the optimal strategy to cope with drought stress varies with timing and severity of the stress. Their discussion concluded that crop simulations were necessary for the analysis of breeding strategies for drought tolerance.

On the contrary, Van der Zaag and Doornbos (1987) tried to explain differences in the ability of several potato cultivars to yield in terms of their ability to intercept light, to convert that energy into dry matter, and their harvest indices. Their work was experimental, done in both the Netherlands and Israel, and the authors found differences in light interception and in LCC that were statistically significant for both cultivar and place, and the interactions were non-significant, suggesting consistent characteristics of the cultivars. As the authors recognized in their introduction to the work, 'It is evident that late cultivars, with a longer duration of leaf area, intercept more light . . . '.

Neither 'physiological' approach seems to have identified characters that a breeder might want to be able to use in a screening programme.

Jefferies (1993), using the model of Jefferies and Heilbronn (1991), considered three strategies for achieving improved drought resistance: improving water supply, improving leaf growth and improving photosynthetic performance. He concluded that the greatest effect on tuber yield was to improve the relation between leaf expansion rate and soil moisture status. Rather than invoking potato breeding as a motivation for that study, Jefferies argued that physiological studies should concentrate on examining the basis underlying the ability to maintain leaf growth in the face of drought.

An idea of the value of modelling from the breeder's viewpoint was provided by Ellissèche and Hoogendoorn (1995) who discussed limitations on the use of models in breeding programmes but foresaw their usefulness in a number of categories of problem including durability of types of resistance to plant pests and diseases.

A wider application of modelling to defining ideotypes was made by Haverkort and Kooman (1997) when they applied their model to the estimation of yields of a standard potato genotype to a wide range of conditions of temperature and day length. They examined how yield defining, limiting and reducing factors influence the genotypically determined the length of the growth cycle and then estimated the optimum times of tuber initiation for potatoes being grown in such conditions. This application of modelling has been developed further (Haverkort and Grashoff, 2004) to incorporate a wider range of 'physiological opinions' to allow the more detailed characterization of the 'ideal' cultivar for a particular environment. The flexible nature of the model in this latest development poses some risks: values of parameters can be selected that are, potentially, mutually incompatible, and parameters may be included or required that are not relevant to the problem in hand. There is, therefore, a serious need for the model to be tested against a wider range of real crops.

35.3.5 Forecasting

One of the potential strengths of the semi-mechanistic model is the possibility of using it as a tool for prediction. There are many classes of forecast that can be addressed by such modelling. Some will operate over only a few days or hours, as for disease forecasting, or over a whole growing season as for estimates of fertilizer requirements. Other systems lend themselves to frequent updating and correction of the forecast, e.g. irrigation scheduling. Still others result in once-in-a-season decisions such as the choice of an appropriate seed rate. The position of a set of possible applications was outlined by MacKerron et al. (1990) when the topics of seed rates, yield, nitrogen requirement, pests and diseases were explored as potential applications of models in forecasts. Several of these topics have since been the subject of DSSs as mentioned briefly in this sub-section and more fully in Chapter 36.

The possibility of forecasting yields has, itself, three sets of application with very different possibilities for inputs and requirements for outputs. First, semi-mechanistic models can be used to forecast the likely yield from a single crop in the current year (MacKerron et al., 2004). Second, they can be used to provide estimates of regional yields in the current year (Heilbronn and MacKerron, 1989; MacKerron et al., 1990; MacKerron, 1992) to provide market information. Third, they can be used in an exploratory manner to provide estimates of likely yields in other regions and with several putative starts to the growing season (Haverkort, 1986; Haverkort et al., 2004). The inputs to each of these applications are necessarily different.

In forecasting the performance of a single specific crop, the inputs appropriate to that crop are used together with local estimates of forecast weather (or local long-term averages). Successive updates can be made as the season elapses and the forecast period shortens. This particular exercise has a feature that is nearly unique in modelling growth in that the answers from the model are given before the simulated crop is harvested and the user – a grower – believes that he/she then has the 'right' answer. There is little forgiveness for error. Only the forecasting of the risk of disease (Raatjes et al., 2004) has a greater pressure from the user.

Forecasting yields at a regional and national level (Heilbronn and MacKerron, 1989) represents a completely different scale of application from the other examples of forecasting being considered here and is worth examination. The method used by Heilbronn and MacKerron (1989) involved the selection of representative weather stations and making successive runs of the model during the growing season, progressively substituting actual weather as it occurred for values of the long-term-average weather (MacKerron, 1987). There were several difficulties in extending this method to regions of the UK (MacKerron, 1992). First, it was necessary that the regions examined were the same as those used administratively so that independent data on actual production would be available for comparison with the simulations. Second, such regions are large and contain heterogeneous weather conditions and husbandry practices giving difficulties in selecting representative meteorological stations. Third, the model is non-linear so that simulations made using 'average' values of, say, planting or harvesting and 'representative' weather conditions do not provide the same answers as averages of simulations made with the several sets of conditions. Accepting these limitations, over a period of 12 years, there was a quite good relation between estimates of regional yields and observed values (MacKerron et al., 1990; Fig. 35.1A). When averaged across the whole of the UK (MacKerron et al., 1990; Fig. 35.1B), many of the errors cancelled out (i.e. the differences between regions were not systematic), but there was an underlying systematic error in that the model overestimated yield. When that systematic error was removed, the system allowed forecasts of yield to be made (MacKerron et al., 1990; Fig. 35.2) that, in most years, correctly predicted the divergence from long-term average at the end of the growing season by the end of June. The resulting estimates of national yields can provide a basis for judging the market supply of potato and for making early assessments of differences from average.

Precise and up-to-date information on agricultural production is an essential component of a market economy. For a long time, the European Union had a common agricultural policy, the management of which relied on such market information. The European Community (as it was called earlier) supported the development of agrometeorological models for regional monitoring of crop state and for quantitative yield predictions on a

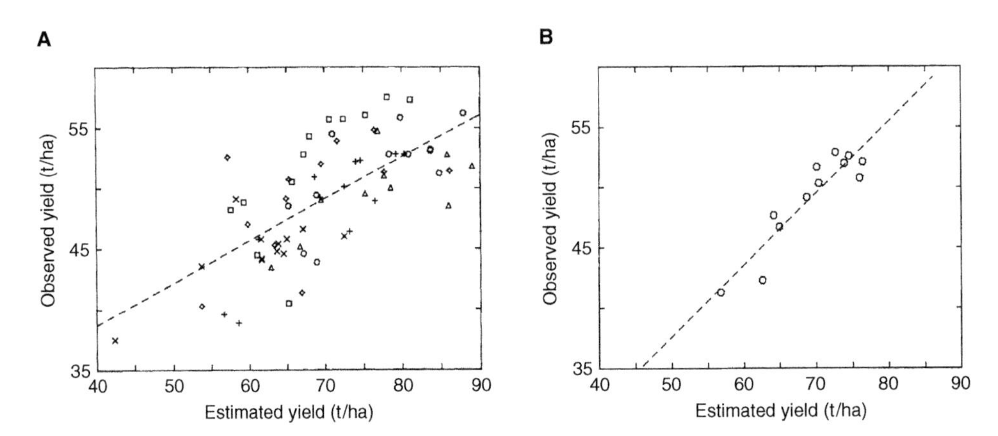

Fig. 35.1. (A) Observed and estimated values for the yield (t/ha) of the UK national potato crop, by region, 1977–88. × – Scotland, + – North, □ – West Midlands, ◇ – East Midlands, ○ – East Anglia and △ – South. (B) Observed and estimated values for the yield (t/ha) of the UK national potato crop, 1977–88.

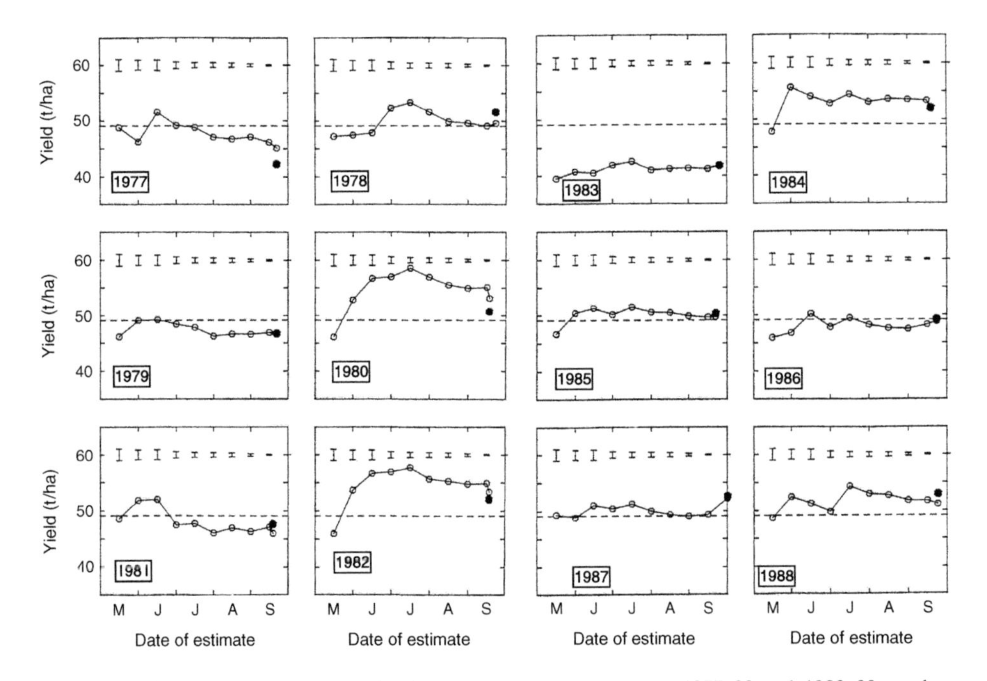

Fig. 35.2. Estimates of the final yield (t/ha) of the national potato crop, 1977–82 and 1983–88, made at fortnightly intervals (○), the observed yield (●) and the LTA yield (- - - - -). Error bars show the 95% confidence limit.

national scale. The potato crop was the topic of several such projects, e.g. MacKerron (1992), De Koning et al. (1993) and Boons-Prins et al. (1993).

There have been several noteworthy examples where two or more semi-mechanistic models have been combined into a DSS that is capable of being used in a commercial context. These include the PLANT-PLUS system for forecasting the incidence of disease, particularly late blight of the potato, and for providing a scale of changing risk of infection as time elapses since the previous protection, the crop grows and the weather changes (Raatjes et al., 2004). That DSS is used on a commercial scale. An educational tool was developed to allow users to explore the integrated control of potato cyst nematode programme (Elliott et al., 2004). Among the reasons why that package did not go further than the educational use is the difficulty in providing good, field-specific data on existing nematode populations. However, the package has been released to potato growers in the UK for general use. MacKerron et al. (2004) developed MAPP (Management Advisory Package for Potatoes), a package to help potato growers optimize their seed rates and harvest dates in the light of the cultivar being used, the size and cost of the seed tubers and the expected prices of the harvested crop in a range of sizes. That package combined simulation of water-constrained growth of the crop with cultivar specific characteristics of tuber multiplication and a simulation of tuber size distribution. It also indicates any need for irrigation.

These and other DSSs will not be discussed further here, but the references will provide a lead for further interest.

35.4 CALIBRATION/VALIDATION AND OTHER DIFFICULTIES

35.4.1 Sensitivity analysis

Sensitivity analysis is an important step in the assessment of any mathematical model that simulates crop growth. In many (most?) cases, the sensitivity of growth models cannot be tested from the first derivative of their driving functions as several of the variables appear in several discrete functions. In such cases, sensitivity has to be assessed by running the model first with 'typical' values of the variables and crop husbandry and a standard set of meteorological data. Then, the model is tested by running it repeatedly with differing values of the principal variables.

MacKerron and Waister (1985) and MacKerron (1985) recognized two forms of sensitivity that they termed 'internal' and 'external'. MacKerron and Waister (1985) examined the sensitivity of the output of the model to changes in functions within the model whether these should originate from a mistaken understanding of the processes or from differences in performance, e.g. between varieties. They called this 'internal sensitivity'. MacKerron (1985) then examined the sensitivity of the output of the model to changes in the input values, e.g. date of planting or harvest, or changes in insolation. He called this 'external sensitivity'. Similarly, Jefferies and Heilbronn (1991) tested the internal sensitivity of their model and found that it was least sensitive to the changes in the value of the LCC and most sensitive to values of water-use efficiency. Similarly, Van den Broek and Kabat (1995) reported that their model was particularly sensitive to water-use efficiency. Fishman et al. (1995) also found that sensitivity had to be assessed by running their model with controlled variation in values of parameters. Therefore, they tested only the four parameters that had the greatest effect on calculated yield.

Although all modellers recognize the importance of the sensitivity of their models to particular variables and/or parameters (e.g. Ritchie et al., 1995; Van den Broek and Kabat, 1995), it seems that few have set out the sensitivity of their models in as explicit a manner as did MacKerron and Waister (1985) and Jefferies and Heilbronn (1991). This relates to suggestions for further work.

35.4.2 Calibration and validation

The contents of this short section are very much a personal view and may be at variance with what is generally accepted practice in the modelling community. I will take this opportunity to express that opinion.

The construction of a mathematical model of crop growth involves the formal statement of hypotheses about the real system. Implementation of the model, preferably in a computer program, leads to a solution of the hypotheses for given inputs and should lead to the critical step of validation (Loomis et al., 1979). Validation involves comparisons of predictions with the results from independent experiments. It is a matter of some regret that so many of the models that are presented in the literature have never been validated by their authors and that still others have been given only the most cursory check. One could reasonably take the view that until a model has been validated it has limited worth.

Several models of crop growth contain variable 'constants' that can be adjusted to make the answers fit reality in particular circumstances. This is a dubious practice (variously called 'calibration' or 'tuning' by its practitioners) that might be tolerable under stringent, pre-set conditions. It is believable that a crop may be sensitive to particular variables, e.g. soil type, which the modeller has not been able to simulate as yet. Under these circumstances, it might be permissible to use an arbitrary constant or function to make the model reflect reality, on the firm condition that, once adjusted, these 'variables' become constant. Thus, a model might be run in comparison with one set of data that is associated with a particular area, and the results of the run might then be used to adjust variables to force agreement between model and reality. The model should then be run with unchanged 'tuning variables' and should be checked against several sets of independent data from the same area (e.g. from different fields or from different years). If further 'tuning' or 'calibration' is needed, then the model should be set aside until it has been redesigned.

35.4.2.1 An exercise in calibration

A workshop on potato modelling was held in Wageningen in May 1990 (Kabat et al., 1995b). The key feature of that workshop was that several models were run and tested against paired data sets. That is, there were four sets of weather, crop and soil data presented, two were from successive years of one experiment on a similar soil in each year and the other two sets came from a second experiment also with the same soil in the 2 years. The participants had each been given access to the first set of each pair and had been allowed to 'calibrate' their models. In general, model calibration involved adjusting parameter values until acceptable simulation results were produced for the calibration data set. The purpose of the exercise was to examine how well the models then simulated the results of the second data set in each pair. The developers of each model reported on the performance of their model in both the calibration and the validation stages. De Koning et al. (1995) showed yield as a curvilinear function of maximum rooting depth. They 'calibrated' their model for maximum rooting depth using the data from 1 year – that is, they found the value that worked and applied it to the data from the subsequent year. They found that their model appeared to overestimate drought stress even though the simulated soil moisture depletion proceeded more slowly than in the calibration year. In their own words, '. . . caution has to be taken when applying a calibrated model to other conditions, . . . '. Later, Kabat et al. (1995a) prepared a most thorough comparison of the simulations and evaluation of the parameterizations. A few of their observations can be quoted: with the exception of two models, the rate at which the crop proceeds from one developmental event to another is described by a function of air temperature (soil temperature before emergence) although the functions differ. Partitioning of dry matter is one of the most difficult to quantify satisfactorily, being influenced by a whole range of environmental influences. Four of the models being compared used a cumulative partitioning coefficient; the others used some form of instantaneous partitioning coefficient. All the models in the comparison were calibrated for dry matter partitioning and all but one for senescence processes. The number of other parameters that were adjusted in the calibration exercise differed between models. The variations between predicted and measured values of cumulative total dry weight and cumulative tuber dry weight were large as were the differences between models, with the exception of two of the models that gave values almost identical to the measured ones.

The model of Jefferies and Heilbronn (1995) was not included in the general set of comparisons because, it was thought, the model was partly developed on the data set used for verification of the other models. That was not strictly correct, as the model was completely independent of one of the pairs of data sets. Jefferies and Heilbronn's model was the only one not to be 'calibrated' (ever) in the sense that is used here and in the workshop.

35.4.2.2 Must a model give the 'right' answer?

Through all the plethora of efforts to make models 'work', it is important to remember that major purposes of a model are to explain and to confer understanding. Users should always be aware of the bounds of the models they use. They should think about the answers that models give, only in terms of the models themselves. Therefore, they have to know the basic principles of the model. That may be a challenge where the user is not the developer. For example, when the model predicts a lower yield than the user would have thought, he/she should be able to explain that by the way the model calculates this yield and to check whether it was his/her perception that was wrong or has the model been taken beyond its proper domain. All his/her other knowledge can be used to qualify the model's answer. So, for example, Rijneveld (1997) found differences in the amount of agreement between simulation and reality in two sites: one with a sandy clay loam soil and the other with a sandy soil. For the site with sandy soil, there was a significant difference between measurement and observation and he was able to indicate the parameters that led to the disparity. Yet, he concluded that it would be wrong simply to alter the parameters to make the output of the model fit the data from that site. To do so would not improve our understanding and would almost certainly make the model wrong in its simulation of crop growth and water use in soils that are less sandy. This presumes that the origins of the differences in operation of the model at the two sites are due to soil-related water supply, and it sets a target to improve the model so that it will accommodate differing soils without interference in the model.

35.4.3 Modellers tend to believe their models

Developers of models run the risk that they will 'believe' in their own models. When examining or interpreting their results, modellers tend to say 'is' or 'does' (e.g. Spitters and Schapendonk, 1990). 'The effects of water shortage *are*: . . . less leaf growth . . . (which) *is* caused by a lower rate of photosynthesis, (*is* caused by) a decrease in specific leaf area, and (*is* caused by) a lower proportion of assimilates allocated to haulm growth'. Well, are these things correct? They may well be, but there were no experimental observations made whatsoever in that particular study. These authors were in no way atypical, but what they should have written was that their model *simulated* these effects, or that it *appeared* that these would be the effects if the experiments were done. [Or some similar phrasing to remind the reader (and the authors) that what is being reported is not an experimental fact.]

One side-issue in the question of calibration is the difference between cultivars. Where experimental studies show differences between cultivars, e.g. in the base temperature for sprout growth (e.g. MacKerron, 1984), the modellers must consider whether to include such genotypic variation in their model. If they do so and the data are not generally available for all cultivars, then they must offer a default value and then admit that the output from the model is liable to be 'wrong'. Is it not better to select a 'typical' value for such a characteristic and accept that it is one source of uncertainty in the model? If there is general knowledge that a particular characteristic featured in the model does differ between cultivars, then they must test for the significance of any effect. Is the effect large or small? Is it linear or non-linear? If the effect is large enough to make a significant difference to the output, then they must give the option to set a known value but warn users of the sensitivity.

Such effects are among those that modellers seek to eliminate when they 'calibrate' their models for a particular location and cultivar. I argue that it is better to accept that the model is 'wrong'.

35.5 FUTURE WORK

Most models, and all that have had any general acceptance or application, have strong points where they simulate a part of the real world adequately. But, again, all have limits to their applicability. An associated weakness of modellers and those who use these models is that they come to believe in the models. When there is a mismatch between simulation and observation, there is a tendency to believe either that the observation is subject to sampling error (which it is) or that a small adjustment of the parameters in the model is all that is required to produce coherence between simulation and observation. What is not generally considered is that the model may, quite simply, be inadequate for the conditions obtaining. This leads to an idea for constructive collaborative work on simulation models of potato. The idea hinges on accepting that each model has some validity and then testing them exhaustively against field data such as those sets indicated in the GCTE Report No. 9 (1997) to establish for each model a domain of conditions within which the model is capable or relevant or pertinent.

REFERENCES

Abbaspour K.C., 1994, *Agric. Forest Meteorol.* 71, 297.

Abbaspour K.C., J.W. Hall and D.E. Moon, 1992, *Agric. Forest Meteorol.* 60, 33.

Allen E.J. and R.K. Scott, 1980, *J. Agric. Sci. Camb.* 94, 583.

Allmaras R.R., W.W. Nelson and W.B. Voorhees, 1975, *Soil Sci. Soc. Am. Proc.* 39, 771.

Belmans C., L.W. Dekker and J. Bouma, 1982, *Agric. Water Manage.* 5, 319.

Boons-Prins E.R., G.H.J. de Koning, C.A. van Diepen and F.W.T. Penning de Vries, 1993, Crop specific simulation parameters for yield forecasting across the European Community. Simulation Reports CABO-TT, no.32. CABO-DLO, Wageningen.

Campbell M.D., G.S. Campbell, R. Kunkel and R.I. Papendick, 1976, *Am. Potato J.* 53, 431.

De Koning G.H.J., M.J.W. Jansen, E.R. Boons-Prins, C.A. van Diepen and F.W.T. Penning de Vries, 1993, *Crop Growth Simulation and Statistical Validation for Regional Yield Forecasting Across the European Community.* Centre for Agrobiological Research, Wageningen.

De Koning G.H.J., C.A. van Diepen and G.J. Reinds, 1995, In: P. Kabat, B. Marshall, B.J. van den Broek, J. Vos and H. van Keulen (eds), *Modelling and Parameterisation of the Soil–Plant–Atmosphere System – A Comparison of Potato Growth Models.* Pudoc, Wageningen.

De Wit C.T., 1965, *Photosynthesis of Leaf Canopies.* Pudoc, Wageningen.

De Wit C.T., G. Brouwer and F.W.T. Penning de Vries, 1970, In: I. Svetlik (ed.), *Prediction and Measurements of Photosynthetic Productivity.* Pudoc, Wageningen.

Demetriades-Shah T.H., M. Fuchs, E.T. Kanemasu and I. Flitcroft, 1992, *Agric. Forest Meteorol.* 58, 193.

Dewar R.C., 1993, *Funct. Ecol.* 7, 356.

Durant M.J., B.J.G. Love, A.B. Messem and A.P. Draycott, 1973, *Ann. Appl. Biol.* 74, 387.

Elliott M.J., D.L. Trudgill, J.W. McNicol and M.S. Phillips, 2004, In: D.K.L. MacKerron and A.J. Haverkort (eds), *Decision Support Systems in Potato Production - Bringing Models to Practice.* Wageningen Academic Publishers, Wageningen.

Ellissèche D. and J. Hoogendoorn, 1995, In: A.J. Haverkort and D.K.L. MacKerron (eds), *Potato Ecology and Modelling of Crops Under Conditions Limiting Growth.* Kluwer Academic Publishers, Dordrecht.

Farquhar, G.D., S. von Caemmerer and J.A. Berry, 1980, *Planta* 149, 78.

Feddes R.A., M. der Graaf, J. Bouma and C.D. van Loon, 1988, *Potato Res.* 31(2), 225.

Feddes R.A., P.J. Kowalik and H. Zaradny, 1978, Simulation of field water use and crop yield. Simulation Monographs, Pudoc, Wageningen.

Feddes R.A. and P.E. Rijtema, 1972, *J. Hydrol.* 17, 33.

Fishman S., H. Talpaz, M. Dinar, M. Levy, Y. Arazi, Y. Roseman and S. Varshavski, 1984, *Agric. Syst.* 14, 159.

Fishman S., H. Talpaz, Y. Roseman, S. Varshavski and Y. Arazi, 1995, In: P. Kabat, B. Marshall, B.J. van den Broek, J. Vos and H. van Keulen (eds), *Modelling and Parameterisation of the Soil–Plant–Atmosphere system – A comparison of Potato Growth Models.* Pudoc, Wageningen.

Fishman S., H. Talpaz, R. Winograd, M. Dinar, Y. Arazi, Y. Roseman and S. Varshavski, 1985, *Agric. Syst.* 18, 115.

Global Change and Terrestrial Ecosystems (GCTE), 1997, Report No. 9 GCTE Focus 3 Potato Network: Model and Experimental Metadata (1996), D.K.L. MacKerron and J. Ingram (eds), Wallingford, UK.
Goudriaan J., 1986, *Agric. Forest Meteorol.* 38, 249.
Greenwood D.J., J.J. Neeteson and A. Draycott, 1985a, *Plant Soil* 85, 163.
Greenwood D.J., J.J. Neeteson and A. Draycott, 1985b, *Plant Soil* 85, 185.
Gregory P.J., M. McGowan and P.V. Biscoe, 1978a, *J. Agric. Sci. Camb.* 91, 103.
Gregory P.J., M. McGowan, P.V. Biscoe and B. Hunter, 1978b, *J. Agric. Sci. Camb.* 91, 91.
Gregory P.J. and G.R. Squire, 1979, *J. Agric. Sci. Camb.* 91, 91.
Gu Y.-Q., J.W. Crawford, D.R. Peiris and R.A. Jefferies, 1994, *Environ. Pollut.* 83, 87.
Gu Y.-Q., J.W. McNicol, D.R. Peiris, J.W. Crawford, B. Marshall and R.A. Jefferies, 1996, *AI Appl.* 10, 13.
Hackett C. and D.A. Rose, 1972, *Aust. J. Biol. Sci.* 25, 669.
Hamer P.J.C., M.K.V. Carr and E. Wright, 1994, *J. Agric. Sci. Camb.* 123, 299.
Hamza M.A. and L.A.G. Aylmore, 1992, *Plant Soil* 145, 187.
Haverkort A.J., 1986, *Potato Res.* 29(1), 119.
Haverkort A.J. and C. Grashoff, 2004, In: D.K.L. MacKerron and A.J. Haverkort (eds), *Decision Support Systems in Potato Production – Bringing Models to Practice.* Wageningen Academic Publishers, Wageningen.
Haverkort A.J. and P.L. Kooman, 1997, *Euphytica* 94, 191.
Haverkort A.J. and D.K.L. MacKerron (eds), 1995, *Potato Ecology and Modelling of Crops Under Conditions Limiting Growth.* Kluwer Academic Publishers, Dordrecht.
Haverkort A.J., A.M.W. Verhagen, C. Grashoff and P.W.J. Uithol, 2004, In: D.K.L. MacKerron and A.J. Haverkort (eds), *Decision Support Systems in Potato Production – Bringing Models to Practice.* Wageningen Academic Publishers, Wageningen.
Heilbronn T.D. and D.K.L. MacKerron, 1989, Forecasting yield of the national potato crop from weather and soil data and agronomic practice. Report to the Potato Marketing Board, London, UK.
Hertog M.L.A.T.M., B. Putz and L.M.M. Tijskens, 1997a, *Potato Res.* 40, 69.
Hertog M.L.A.T.M., L.M.M. Tijskens and P.S. Hak, 1997b, *Postharvest Biol. Technol.* 10, 67.
Hillel D. and H. Talpaz, 1976, *Soil Sci.* 121, 307.
Hsiao T.C., 1973, *Annu. Rev. Plant Physiol.* 24, 519.
Ingram K.T. and D.E. McCloud, 1984, *Crop Sci.* 24, 21.
Jefferies R.A., 1993, *New Phytol.* 123, 491.
Jefferies R.A., 1989, *J. Exp. Bot.* 40, 1375.
Jefferies R.A. and T.D. Heilbronn, 1991, *Agric. Forest Meteorol.* 53, 185.
Jefferies R.A. and T.D. Heilbronn, 1995, In: P. Kabat, B. Marshall, B.J. van den Broek, J. Vos and H. van Keulen (eds), *Modelling and Parameterisation of the Soil–Plant–Atmosphere System – A Comparison of Potato Growth Models.* Pudoc, Wageningen.
Jefferies R.A., T.D. Heilbronn and D.K.L. MacKerron, 1991, *Agric. Forest Meteorol.* 53, 197.
Jefferies R.A. and D.K.L. MacKerron, 1989, *Field Crops Res.* 22, 101.
Jensen K.H., 1983, Simulation of water flow in the unsaturated zone including the root zone. Series Paper No. 13. Institute of Hydrodynamics Engineering, Technical University of Denmark, Lungby.
Kabat P., B. Marshall and B.J. van den Broek, 1995a, In: P. Kabat, B. Marshall, B.J. van den Broek, J. Vos and H. van Keulen (eds), *Modelling and Parameterisation of the Soil–Plant–Atmosphere System – A Comparison of Potato Growth Models.* Pudoc, Wageningen.
Kabat P., B. Marshall, B.J. van den Broek, J. Vos and H. van Keulen (eds), 1995b, *Modelling and Parameterization of the Soil–Plant–Atmosphere System – A Comparison of Potato Growth Models.* Wageningen Pers, Wageningen.
Karvonen T. and J. Kleemola, 1995, In: P. Kabat, B. Marshall, B.J. van den Broek, J. Vos and H. van Keulen (eds), *Modelling and Parameterisation of the Soil–Plant–Atmosphere System – A Comparison of Potato Growth Models.* Pudoc, Wageningen.
Karvonen T., D.K.L. MacKerron and J. Kleemola, 2000, In: A.J. Haverkort and D.K.L. MacKerron (eds), *Management of Nitrogen and Water in Potato Production.* Wageningen Pers, Wageningen.
Kooman P.L., 1995, Yielding ability of potato crops as influenced by temperature and daylength. PhD Thesis, Wageningen Agricultural University, Wageningen.
Kooman P.L. and A.J. Haverkort, 1995, In: A.J. Haverkort and D.K.L. MacKerron (eds), *Potato Ecology and Modelling of Crops Under Conditions Limiting Growth.* Kluwer Scientific Publishers, Dordrecht.

Kooman P.L. and C.J.T. Spitters, 1995, In: P. Kabat, B. Marshall, B.J. van den Broek, J. Vos and H. van Keulen (eds), *Modelling and Parameterisation of the Soil–Plant–Atmosphere System – A Comparison of Potato Growth Models*. Pudoc, Wageningen.

Kunkel R. and G.S. Campbell, 1987, *Am. Potato J.* 64, 355.

Lesczynski D.B. and C.B. Tanner, 1976, *Am. Potato J.* 53, 69.

Loomis R.S., R. Rabbinge and E. Ng, 1979, *Annu. Rev. Plant Physiol.* 30, 339.

MacKerron D.K.L., 1984, Abstracts of the 9th Triennial Conference, EAPR, p. 364. EAPR, Interlaken.

MacKerron D.K.L., 1985, *Agric. Forest Meteorol.* 34, 285.

MacKerron D.K.L., 1987, *Acta Hortic.* 214, 85.

MacKerron D.K.L., 1992, *Agrometeorological Aspects of Forecasting Yield of Potato within the European Community*. Commission of the European Communities, Luxembourg.

MacKerron D.K.L., D.J. Greenwood, B. Marshall, R. Rabbinge and B. Schöber, 1990. Proceedings of the 11th Triennial Conference of the European Association for Potato Research. EAPR, Edinburgh.

MacKerron D.K.L. and G.J. Lewis, 1995, In: A.J. Haverkort and D.K.L. MacKerron (eds), *Potato Ecology and Modelling of Potato Crops Under Conditions Limiting Growth*. Kluwer, Dordrecht.

MacKerron D.K.L., B. Marshall and J.W. McNicol, 2004, In: D.K.L. MacKerron and A.J. Haverkort (eds), *Decision Support Systems in Potato Production – Bringing Models to Practice*. Wageningen Academic Publishers, Wageningen.

MacKerron D.K.L. and Z.Y. Peng, 1989, *Aspects of Applied Biology 22, Roots and the Soil Environment*. Association of Applied Biologists, Wellesbourne.

MacKerron D.K.L. and P.D. Waister, 1985, *Agric. Forest Meteorol.* 34, 241.

Marshall B., J.W. Crawford and J. McNicol, 1995, In: A.J. Haverkort and D.K.L. MacKerron (eds), *Potato Ecology and Modelling of Crops Under Conditions Limiting Growth*. Kluwer Academic Publishers, Dordrecht.

Monteith J.L., 1986, *Phil. Trans. Roy. Soc. Lond. B* 316, 245.

Monteith J.L., 1994, *Agric. Forest Meteorol.* 68, 213.

Neeteson J.J., D.J. Greenwood and A. Draycott, 1987, A dynamic model to predict yield and optimum nitrogen fertiliser application rate for potatoes. Proceedings No. 262, The Fertiliser Society of London, London.

Ng E. and R.S. Loomis, 1979. A simulation model of potato crop growth. Final report for co-operative agreement No. 12-14-5001-287 with the USDA Snake River Conservation Research Center, Kimberley, Idaho.

Ng E. and R.S. Loomis, 1984. *Simulation of Growth and Yield of the Potato Crop*. ISBN 90-220-0843-6. Pudoc, Wageningen.

Parker C.J., M.K.V. Carr, N.J. Jarvis, M.T.B. Evans and V.H. Lee, 1989, *Soil Till. Res.* 13, 267.

Parker C.J., M.K.V. Carr, N.J. Jarvis, B.O. Puplampu and V.H. Lee, 1991, *J. Agric. Sci. Camb.* 116, 341.

Phillips M.S., D.L. Trudgill, C. Hackett, M. Hancock, M. Holliday and A.M. Spaull, 1998, *J. Agric. Sci. Camb.* 130, 45.

Raatjes P., J. Hadders, D. Martin and H. Hinds, 2004, In: D.K.L. MacKerron and A.J. Haverkort (eds), *Decision Support Systems in Potato Production – Bringing Models to Practice*. Wageningen Academic Publishers, Wageningen.

Regel P.A. and P.J. Sands, 1983, *Field Crops Res.* 6, 1.

Rijneveld W., 1997, Potato-3: finishing, testing, analysing and validating a decision-supporting simulation model. Report to SCRI, Dundee, UK & Dept of Agronomy, Agricultural University, Wageningen

Rijtema P.E. and G. Endrödi, 1970, *Neth. J. Agric. Sci.* 18, 26.

Ritchie J.T., T.S. Griffin and B.S. Johnson, 1995, In: P. Kabat, B. Marshall, B.J. van den Broek, J. Vos and H. van Keulen (eds), *Modelling and Parameterisation of the Soil–Plant–Atmosphere System – A Comparison of Potato Growth Models*. Pudoc, Wageningen.

Roth O., J.O. Derron, A. Fischlin, T. Nemecek and M. Ulrich, 1995, In: P. Kabat, B. Marshall, B.J. van den Broek, J. Vos and H. van Keulen (eds), *Modelling and Parameterisation of the Soil–Plant–Atmosphere System – A Comparison of Potato Growth Models*. Pudoc, Wageningen.

Sale P.J.M., 1974, *Aust. J. Plant Physiol.* 1, 283.

Sands P.J., C. Hackett and H.A. Nix, 1979, *Field Crops Res.* 2, 309.

Schans J., 1991, *Agric. Syst.* 37, 387.

Schapendonk A.H.C.M., C.S. Pot and J. Goudriaan, 1995, In: A.J. Haverkort and D.K.L. MacKerron (eds), *Potato Ecology and Modelling of Potato Crops Under Conditions Limiting Growth*. Kluwer, Dordrecht.

Schapendonk A.H.C.M., M. van Oijen, P. Dijkstra, C.S. Pot, W.J.R.M. Jordi and G.M. Stoopen, 2000, *Aust. J. Plant Physiol.* 27, 1119.
Sharp R.E. and W.J. Davies, 1985, *J. Exp. Bot.* 36, 1441.
Sinclair T.R. and N. Seligman, 2000, *Field Crops Res.* 68, 165.
Spitters C.J.T., 1986, *Agric. Forest Meteorol.* 38, 231.
Spitters C.J.T., 1987, *Acta Hortic.* 214, 71.
Spitters C.J.T. and A.H.C.M. Schapendonk, 1990, *Plant Soil* 123, 193.
Spitters C.J.T., H.A.J.M. Toussaint and J. Goudriaan, 1986, *Agric. Forest Meteorol.* 38, 231.
Spitters C.J.T., H. van Keulen and D.W.G. van Kraalingen, 1989, In: R. Rabbinge, S.A. Ward and H.H. van Laar (eds), *Simulation and Systems Management in Crop Protection. Simulation Monographs.* Pudoc, Wageningen.
Spitters C.J.T. and S.A. Ward, 1988, *Euphytica* 37, 87.
Taylor H.M. and B. Klepper, 1978, *Adv. Agron.* 30, 99.
Tinker P.B., 1976, *Phil. Trans. Roy. Soc. Lond. B* 273, 445.
Van Bavel C.H.M. and J. Ahmed, 1976, *Ecol. Model.* 2, 189.
Van de Geijn S.C. and P. Dijkstra, 1995, In: A.J. Haverkort and D.K.L. MacKerron (eds), *Potato Ecology and Modelling of Potato Crops Under Conditions Limiting Growth.* Kluwer, Dordrecht.
Van Delden A., M.J. Kropff and A.J. Haverkort, 2001, *Field Crops Res.* 72, 119.
Van den Broek B.J. and P. Kabat, 1995. In: P. Kabat, B. Marshall, B.J. van den Broek, J. Vos and H. van Keulen (eds), *Modelling and Parameterisation of the Soil–Plant–Atmosphere System – A Comparison of Potato Growth Models.* Pudoc, Wageningen.
Van der Zaag D.E., 1984, *Potato Res.* 27, 51.
Van der Zaag D.E. and J.H. Doornbos, 1987, *Potato Res.* 30, 551.
Van Keulen H. and N.G. Seligman, 1987, *Simulation of Water Use, Nitrogen Nutrition and Growth of a Spring Wheat Crop.* Pudoc, Wageningen.
Van Keulen H. and J. Wolf (eds), 1986. *Modelling of Agricultural Production: Weather, Soils and Crops.* Pudoc, Wageningen.
Van Wijk A.L.M. and R.A. Feddes, 1986, Agricultural water management. Proceedings, Symposium on Agricultural Water Management, Arnhem, 18–21 June 1985: Technical Bulletin no. 40, ICW, Wageningen. A.A. Balkema, Rotterdam.
Vos J. and J. Groenwold, 1986, *Plant Soil* 94, 17.
Webb R.A., 1972, *J. Hort. Sci.* 47, 309.
Wraith J.M. and J.M. Baker, 1991, *Soil Sci. Soc. Am. J.* 55, 928.
Yandell B.S., A. Najar, R.M. Wheeler and T.W. Tibbitts, 1988, *Crop Sci.* 28(5), 811.

Chapter 36

Decision Support Systems in Potato Production

B. Marshall

Scottish Crop Research Institute, Invergowrie, Dundee DD2 5DA, United Kingdom

36.1 DEFINITION

Hayman (2004) reviewed the definition of decision support systems (DSSs) in some detail. Their purpose is self-evident from the term, namely to support the decision maker rather than make decisions in themselves. Whether this is what the end user always expects is debatable – we will come back to this later. DSSs come in many forms, from the spoken word, pamphlets and booklets, the neighbour, spreadsheets, telephone and Internet to advanced software. In its broadest definition, there is nothing to separate the term DSS from the agricultural extension service. For the purposes of this chapter, DSSs will be restricted to computer-based systems and their role in the wider context of decision support.

36.2 OPPORTUNITY

Since the introduction of the personal computer in the early 1980s, the sustained exponential growth in computing power has had a profound effect on the way we all work and not least in the way and speed with which we communicate. The email and mobile telephone have replaced the written letter and office-bound telephone for most communications. The quantity, detail and speed with which research data in crop physiology and agronomy can be captured, analysed and processed have increased many fold. A new discipline of numerical modelling has emerged and, among many diverse areas, has been applied to crops to predict effects of weather and husbandry (irrigation, nutrients and pests) on development, growth and yield and to meteorology, producing more accurate and longer-range forecasts. MacKerron (Chapter 35, MacKerron, this volume) reviews many of the developments in potato crop modelling. Weather is a key input in most models, and its variability from year to year and the uncertainty in forecasting means that no 2 years are alike and hence husbandry and crop performance will vary. The ability to forecast the consequences of changes in husbandry and of weather in a timely and accurate manner creates significant opportunities for computer-based DSSs to assist the grower in potato production and agriculture in general. Both the public-funded research and the commercial sector insist on demonstrable knowledge transfer to the industry

Potato Biology and Biotechnology: Advances and Perspectives
D. Vreugdenhil (Editor)

and look for public good and environmental benefit. Plant breeders have an output that, while challenging to produce, is easily quantified in the form of a new cultivar that they release. New cultivars are independently assessed for a range of traits including yield, tuber number, maturity date and resistance to a range of pests and diseases. The benefits of the breeders' research are clear to the end user (Woodruff, 1992). In contrast, the crop physiologists have a much more difficult task to demonstrate effective knowledge transfer (Hayman, 2004). DSSs offer a 'tempting' opportunity to facilitate knowledge transfer by integrating a large body of knowledge together in a 'user friendly package'. The danger is that the desire for knowledge transfer, although justifiable and to be encouraged, is driven more by the provider of the DSS than by the end user. The risk is that the product addresses the questions that can be answered rather than the priority needs of the customer.

Rapid developments in Information Communication Technology (ICT) have brought large volumes of information through the Internet to our fingertips at speeds that were not dreamt of a decade or so ago. And while the power of computing has risen, costs have fallen by orders of magnitude. In the 1980s in the UK, less than one-fifth of farm businesses had a personal computer that stood alone; now the vast majority have access to computers and the Internet. The technologies are available for the capture of site-specific data, for collation with external information sources and for processing it into an easily digestible form.

36.3 CURRENT AVAILABILITY

A search of the World Wide Web in 2005 for the words 'potato' and 'decision' and 'support' returned over 1.7 million hits. A refinement of the search to 'potato' and 'decision support' reduced this to 90 000. At the time of the search, Google also provided an optional means of filtering out similar pages, which brought this figure down to 8000. Visiting the first 200 ranked best matches and then the first 10 in each 100 thereafter up to 1001–1010 showed that initially half of all hits were relevant to the topic, and this fell to less than one-quarter by the 200th hit and to less than 1 in 10 by the 1000th hit (Fig. 36.1). Sorting the relevant hits by year from this sub-sample of 280 sites visited indicated that there was little activity in this area reported on the web before 1985, activity remained small in the next decade before taking off exponentially in the late 1990s and it was still gaining momentum in 2005. It can be argued that the distribution in time will be skewed towards current activity as webpages are maintained and updated. A similar search of CAB Abstracts revealed a much smaller number of scientific publications but nevertheless reflected a similar time course of relative activity (Fig. 36.2). More detailed analysis of the websites visited revealed that decision support was dominated by pest and disease issues and in particular late blight (Table 36.1). Over one-third of all relevant sites referred to late blight, and nearly three-quarters of all pest and disease issues were late blight. The next most frequent areas of crop-specific decision support were fertiliser, dominated by nitrogen, and irrigation. Another significant area of activity was in the use of modelling for determining the suitability of land, particularly in developing countries, for the cultivation of potato and alternative crops. Analysis of the CAB Abstracts revealed

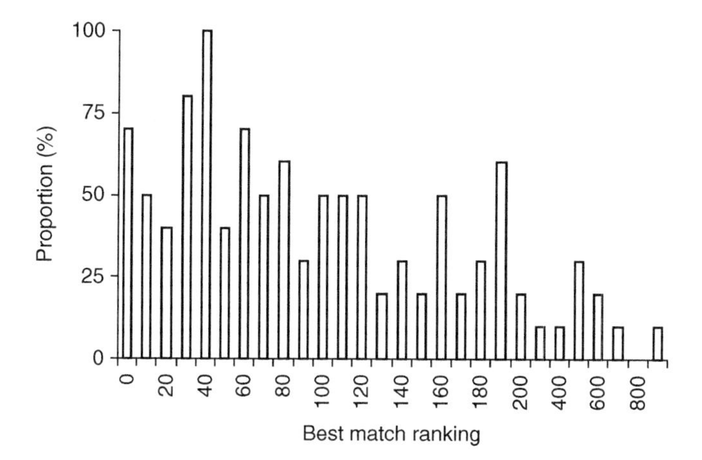

Fig. 36.1. Summary of the results of searching the World Wide Web using Google and the search words 'potato' and 'decision support' in April 2005. The figure shows the proportion of hits that correctly referred to decision support systems (either in use or under development) in potato. Samples of 10 consecutively ranked best matches were taken from the top 200 matches and then the first 10 matches in each 100 thereafter up to 1000–1010 (see Table 36.1 for further details of the type of decision support referred to).

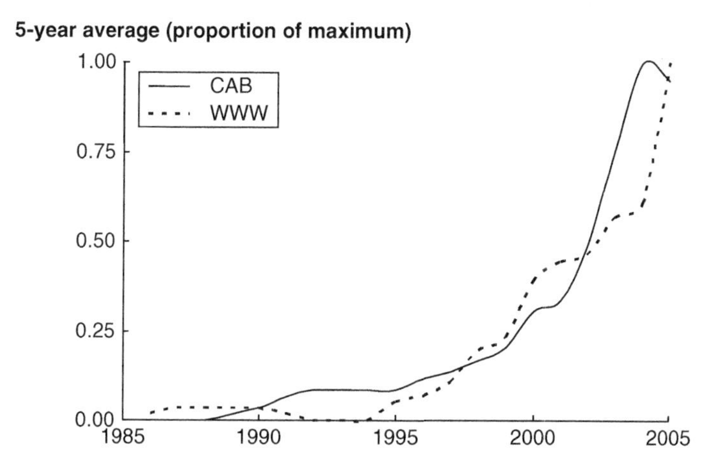

Fig. 36.2. The increasing number of references to decision support and potato both reported in the literature (solid line, source CAB Abstracts using search criteria 'potato AND decision AND support', 62 references returned) and the World Wide Web (broken line, over 8000 returned). Results are 5-year running average expressed as a proportion of the maximum number reported in any 5-year period.

Table 36.1 Summary of the results of searching the World Wide Web using Google and the search words 'potato' and 'decision support' in April 2005. A sub-sample was taken from the best 1000 matches (see Fig. 36.1 and text). Only those hits that correctly relate to the topic are included.

Crop specific	86%		
Pest and disease		58%	
Late blight			72%
Pest			8%
PCN			6%
Early blight			3%
Colorado beetle			2%
Common scab			2%
Verticillium spp.			2%
Weeds			2%
General IPM			5%
Fertilizer		17%	
Water		12%	
Irrigation			92%
Quality			8%
Other		14%	
Development			19%
Precision farming			19%
Seed rate			19%
Tuber size distribution			19%
Yield			13%
Cultivar			6%
Organic production			6%
Miscellaneous	14%		
Land use		63%	
DSS Book		32%	
Education		5%	

DSS, decision support system; IPM, integrated pest management, PCN, potato cyst nematode.

a remarkably similar breakdown of activity (Table 36.2). Most abstracts referred to just one decision area – of the 62 abstracts found, only 12 decision areas specific to potato were referred to. Overall, late blight was ranked top in both, followed by decisions on fertiliser and irrigation alongside land use and then potato cyst nematode (PCN) and tuber size distribution. All of these are common issues in both temperate and warmer climates. Next ranked came early blight and Colorado beetle, the former being restricted to the warmer regions. Decisions on seed rate are perhaps lower than one would expect. However, this decision is often part of the overall management of tuber size, and seed rate may well be subsumed within it. There is little to choose between the remaining

Table 36.2 Summary of the results of searching the CAB abstracts in May 2005 using search criteria 'potato AND decision AND support'.

Crop specific	88%		
Pest and disease		71%	
Late blight			71%
PCN			7%
Early blight			7%
Colorado beetle			7%
Weeds			4%
Virus			4%
Nitrogen		11%	
Water		8%	
Irrigation			100%
Other		10%	
Precision farming			17%
Seed rate			17%
Tuber size distribution			50%
Breeding			17%
Miscellaneous	13%		
Land use		78%	
DSS general		22%	

DSS, decision support system; PCN, potato cyst nematode.
references were returned referring to 72 decision areas detailed.

issues. There has clearly been a significant amount of effort applied in the development of decision support tools for the potato industry.

The four main areas of current decision support, late blight, fertiliser, irrigation and tuber size distribution, are presented by way of example.

36.3.1 Late blight

Late blight (caused by *Phytophthora infestans*) is the major disease of potato. It is ever present and must be managed in all years to avoid premature death of the crop and the disease entering the tubers with subsequent loss in storage (Chapter 7, Simko et al., and Chapter 29, Termorshuizen, this volume). As the consequences of failing to control the disease are so devastating, it would not be surprising for growers to adopt a prophylactic strategy, with the sole decision being when first to start the spraying programme. Nevertheless, there is much activity both in the public and in the commercial sectors in developing DSSs for the control of late blight. There has been and continues to be an extensive research effort into all aspects of the disease from epidemiology to genetics of breeding resistance. The research is highly organized with well-established international networks exemplified by the Global Initiative on Late Blight (GILB) (http://gilb.cip.cgiar.org/). Although GILB's primary aim is to improve management of late blight in developing countries, it draws together a worldwide partnership that is well experienced in all aspects of late blight including its management in both developed and developing countries. Schepers (2002) presented an overview of DSSs for the control of late blight in industrialized countries at the GILB 2002 conference in Hamburg. That paper emphasized that the control of late blight involves an integrated package of methods including tackling the sources of inoculum, the effective deployment of disease resistance in cultivars and the use of DSS tools. This contrasts with the complementary paper for developing countries where direct education of the grower in the management of late blight is the main priority, with no significant role for computer-based DSSs (Mizubuti and Forbes, 2002; Ortiz, 2002). Much more important was to understand the factors that influenced the farmers in their decision-making. Of the five factors including resources, perception and prioritization of problems, experience and knowledge, only the fifth related to the actual agro-ecology of the disease itself. It is the agro-ecology that is usually the main topic of the DSS in industrialized countries. In Europe, one can find examples of all forms of delivery of DSSs for late blight from a visit by the extension officer, telephone, fax, email and text message, to PC and the Internet. Web-Blight (www.web-blight.net), developed in Denmark and based on the NegFry model, has now been expanded to many of the countries surrounding the Baltic, including Estonia, Finland, Lithuania, Latvia, Norway, Poland and Sweden. It provides an online warning and prediction service. In the UK, there are at least three websites (DARD BlightNet, British Potato Council and Syngenta-UK) available to assist in the control of potato late blight. In Germany, a number of websites (ISIP Demo Daten, Bayerische LFL and Syngenta-DE) provide recommendations based on the DSS Simphyt. And in Switzerland, an Internet site uses a DSS based on PhytoPRE+2000. There are three basic ingredients: monitoring of the incidence of disease in the current season, weather data and a method of disease prediction. All include direct monitoring of the presence of the disease models, and weather data alone are not sufficient. A few use monitoring alone. Web-Blight in Denmark is delivered as Pl@nteInfo on the Internet (www.planteinfo.dk). The map that forecasts outbreaks is shown alongside a map of the monitoring data for ease of comparison. It describes itself as an information system and DSS for Danish farmers and agricultural advisers. A subscription system enables personalized information to be stored and used. For example, because the geographical location of the user's home is known after login, local weather data are used for model calculations. It can also store previously entered information on fields, crops and actions taken by the user for later use of the system. The Danish Agricultural Advisory Service (DAAS) organizes about 50 advisors in a formal survey network. The target is to monitor all conventional and organic fields, field trials and experimental trials. These 50 advisors cannot monitor all possible occurrences. Therefore, the farmers are also asked to act as additional scouts and call in an advisor when they suspect an infection. If the advisor recognizes symptoms as potentially late blight, a plant sample is sent to a specialist for verification. The advantage of this system is that all information on the Internet site is guaranteed to be about late blight. They found that even experienced potato advisors can be wrong in the interpretation of late blight symptoms. In 1998, there was an 8% error rate. Farmers in Denmark are confident of the verification method, actively support the survey system and trust the DSS sufficiently to delay fungicide applications until the warnings are generated, whereas previously, rumours about early attacks could get many farmers in a region on the tractor and applying their fungicide controls only to find later that it was a false alarm. The disease forecasts are used to identify the time when scouting should be intensified. This can differ by up to 3 weeks (mid-June to early July) for Danish conditions.

Companies such as Dacom, Opti-Crop and Pro-Plant have developed DSSs for control of late blight and offer these services – after registration and payment. Dacom's service is called Plant-Plus and has been used on-farm since 1994 for the management of *P. infestans*. It offers an integrated 5-day weather forecast that provides a predictive risk assessment for the coming days. The disease models require the availability of on-farm, automatic, weather data. Plant-Plus uses a biological model that is based on the life cycle of the fungus and incorporates infection events on an unprotected part of the crop. The model will recommend when to apply a new spray and what type of chemical to use, whether contact, translaminar or systemic. Dacom has always closely monitored its performance and the growers' confidence in it. The benefits are clearly demonstrated in field trials and commercial evaluations, from northern temperate Europe to the Mediterranean, up to Scandinavia and across the USA and Japan (Raatjes et al., 2004).

Access to local weather data in a timely fashion is often a problem for DSSs. Whereas the Danish appear to have an excellent publicly sponsored network of daily weather data, other countries do not, or users face high commercial charges for access to such weather information. Some commercial companies have addressed this head-on, providing a weather collection service alongside their DSS. In the Netherlands, Dacom maintains a network of about 100 meteorological stations, claiming to provide complete coverage of the potato-growing regions. Every 15 min, these weather stations measure a large amount of data, including the temperature, relative humidity, solar radiation and wind speed and direction, that is those factors that affect crop performance and disease progression. They have developed similar networks in other countries where they operate. Anywhere in the world, wherever a meteorological station is connected to a telephone line or the Internet, information can be downloaded to the Dacom databank and shared with its members. Thus, loss of data from one station due to a temporary failure can be replaced by interpolation from nearby stations.

Opti-Crop has similar services. Opti-Crop claims that farmers need help to assimilate the plethora of data thrown at them in the precision agricultural industry. Interestingly, in Australia at least, alongside their DSS, they also provide an agronomist to work closely with the grower throughout the entire season, allowing appropriate management practices and technologies to be included in the crop production system.

The European network for development of an integrated control strategy of potato late blight (EU.NET.ICP) was a 4-year concerted action that delivered in 2000 (for a project description, see http://ec.europa.eu/research/agro/fair/en/nl0206.html) with the major objective to review and report on the role of DSSs in integrating all the available information on the life cycle of *P. infestans*, the effect of weather both past and future on the disease progress, its interactions with plant development and growth, the use of fungicides, cultivar resistance and the sources of inoculum and disease pressure and improving the control of the disease. As part of the review, several DSSs were validated over a 3-year period (Hansen et al., 2002). They included Simphyt (Germany), Plant-Plus (Netherlands), NegFry (Denmark), ProPhy (Netherlands), Guntz-Divoux/Milsol (France) and PhytoPre+2000 (Switzerland). The use of a DSS reduced fungicide input by 8–62% compared with routine treatments. The level of disease at the end of the season was the same or lower using a DSS compared with a routine treatment in 26 of 29 validations. Similar results were observed by Dowley and Burke (2003) and Leonard et al. (2001) in separate studies of four European DSSs. Two were in the public sector and the other two were available through commercial outlets. Field experiments carried out over a 3-year period in Ireland showed that a prophylactic 7-day routine applied 13.7 applications of fungicide, whereas the DSS programmes varied between 5.7 and 12.3 applications. NegFry and SimPhyt produced the greatest reduction of 50%. ProPhy and Plant-Plus were more conservative at 10 and 25% reduction. All DSSs controlled the disease as well as the standard prophylactic, control treatment.

In conclusion, there are both economic justification and environmental benefits to be gained by using DSSs for late blight control. The variation between DSSs in the number of applications while all still maintaining control suggests that there is still room for improvement in some, but this must not be achieved at the cost of significantly increased risk. A DSS forms part of an integrated management plan. It requires accurate and extensive monitoring of disease pressure in the region, timely and accurate local daily weather records and tried and tested models of disease forecasting as well as good husbandry (cultivar resistance, hygiene, rotation and correct disposal of potential sources of future inoculum). It can be further improved by also taking into account weather forecasts looking several days ahead. This is used in some of the commercially available services to reduce the risk of an unnecessary fungicide application, should the forecast be confident that weather conditions will not be conducive to propagation of the disease.

36.3.2 Fertiliser

Nitrogen has been the main focus in the development of electronic DSSs for fertilisers. Of the three major nutrients, crop yield is most responsive to nitrogen and in some cases can respond to a split application. Environmental concerns related to nitrogen pollution of ground water, reinforced by legislation, are probably the main drivers for the development of DSSs for the application of nitrogen in its various forms, both inorganic and organic, including manures. In the past, cost had been less important, but this is also changing with the rising cost of fuels. Phosphorus has received less attention, perhaps because it is less prone to leaching. Once applied, it is quickly bound to the soil structure; typically, only 10% of that applied is taken up by the potato crop in that year. It can, therefore, build up in the soil and enter streams through surface run-off and soil erosion resulting in eutrophication. Algal growth is phosphorus-limited rather than nitrogen-limited. Phosphorus is already a concern in the USA, with states developing phosphorus indices to identify and determine phosphorus management measures (Sharpley et al., 2003). Phosphorus fertilisers will become a concern for all, not only as a source of pollution but as the world's supply of inorganic phosphorus runs out over the next few decades.

In more arid regions that rely on irrigation, for example centre pivot irrigation in Idaho, nutrients are applied throughout the growing season in the irrigation water (fertigation) in amounts that often follow a tried and tested 'blueprint' or prescription.

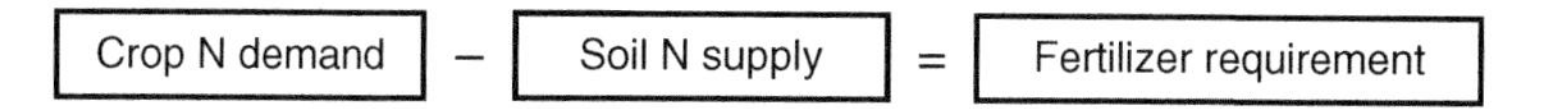

Fig. 36.3. Fertiliser requirement is essentially the nitrogen required to achieve the target yield less the nitrogen supplied from the soil. The greatest uncertainty is estimating the soil supply.

Traditionally, optimum nitrogen fertilisation rates were determined from the average response of field trials employing a range of nitrogen application rates over several sites and years. This ignored the variability in soil nitrogen supply (SNS) (Hofman and Salomez, 2000), and alternative approaches such as the budget method are replacing it. Whichever nutrient is being considered, the problem of estimating nutrient requirement is essentially the same; it is equal to the crop demand for that nutrient less the amount supplied by the soil during the season (Fig. 36.3). Crop nutrient demand is directly related to yield which in turn is determined by the weather, principally soil and air temperatures, solar radiation (sunshine), precipitation (rainfall) plus irrigation available and the length of growing season. Simulation models can be used to estimate water-limited yield; however, the grower will often have a target-harvested yield based on past experience. This harvested fresh weight yield of tubers needs to be converted to an estimate of total crop dry matter first. Adjustments are required for field leavings (small tubers left in the ground), dry matter concentration of the tubers and finally the proportion of biomass found in tubers compared to the whole plant. Given this figure, there is a robust relation determining the average nutrient concentrations and hence nutrient requirement of the crop. The same relation was shown to hold for nitrogen in 22 crop species (Greenwood, 1982; Greenwood et al., 1985; Lemaire, 1997). The origin of the relation was demonstrated by Marshall and Porter (1991). It was later extended to phosphorus and potassium, and at the same time, the optimum time courses of uptake for the three major nutrients were established (Greenwood and Draycott, 1995). This highlighted the need for high rates of nitrogen uptake to be maintained during canopy expansion, with uptake rates falling towards canopy closure. In contrast, there was a steadily increasing demand for phosphorus throughout canopy expansion.

Estimating soil supply is the major challenge in deriving a nitrogen requirement. There are two parts to the estimation: the amount of mineral nitrogen in the soil at planting and the amount of nitrogen that becomes available during the season through nitrogen mineralization. The latter will depend on soil type, especially the organic matter content. The simplest approach is a look-up table. In the UK, ADAS (agricultural advisory service for England and Wales) developed such a table in the 1970s and revised it several times, with the 7th edition published in 2000 (Anonymous, 2000). A three-level soil nitrogen index has been replaced by a new seven-level index of SNS. However, in practice, for potato, these are in fact reduced back to three levels! SNS takes into account soil type and organic matter content as well as previous cropping in an attempt to estimate current soil mineral nitrogen and future soil nitrogen mineralization. Combining SNS with annual rainfall, cultivar maturity type and planned length of growing season results in fertiliser recommendations ranging from 0 for early-lift, late maturing cultivars with high SNS to 250 kg N ha^{-1} for late-lift early maturing cultivars with low SNS. This 'field assessment' method of estimating soil mineral nitrogen at planting, N_{min}, does not take account of fluctuations that occur among fields with the same soil type and also from year to year. If these fluctuations are likely to be significant, then it is strongly recommended that N_{min} is measured directly (Neeteson, 1995). ADAS likewise supports this view when soil mineral nitrogen is likely to have a high and uncertain value. Sampling strategies for soil nitrogen, the need for DSSs and the incorporation of sampling into DSSs were reviewed by Johnson et al., (2000), Hofman and Salomez (2000), and Shepherd (2000) all in the one book.

An alternative strategy to measuring N_{min}, or in addition to it, is to apply less than the recommended rate of nitrogen fertiliser at planting, then measure the crop nitrogen status once the canopy is established and apply the remainder if it is deemed necessary. This has the added advantage that one is getting an indication of the combined effects of the soil mineral nitrogen present at planting and the additional supply of nitrogen through soil mineralization since planting less any losses by leaching.

Booij et al. (2000) reviewed options for measuring crop nitrogen status and concluded that invasive techniques that measure nitrogen content directly were not practicable. The only alternatives were non-destructive surrogates, for example petiole sap nitrogen measurements, chlorophyll meters (e.g. SPAD meter, Minolta, Osaka, Japan) looking at individual leaf tissues and crop reflectance methods (CROPSCAN Inc., Rochester, MN, USA) which can estimate leaf area index and total chlorophyll per unit ground area. Tractor-mounted crop reflectance methods are being used in the precision farming of cereals to adjust, in real time, the amounts of nitrogen fertiliser applied as the spreader passes over the field. So far, there has been no similar commercial development for potatoes. 'Blueprints' of the optimum nitrogen status of the crop over time have been drawn up, which are used to determine whether the crop is short of nitrogen or not (MacKerron, 2000). When a shortage is identified, the problem remains as to how much needs to be applied. With fertigation or foliar application, one can apply a little and often until the symptoms of deficiency are removed. The window of opportunity to detect, take remedial action and for that action to take effect is especially narrow for granular applications, and inevitably, one has to take a cruder action.

The following examples illustrate some of the methods referred to above which are currently in use in DSSs. Azobil, a DSS developed in France, uses what is referred to as the 'forecast balance' method (Chambenoit et al., 2004). It is essentially an expansion of Fig. 36.3. The nitrogen absorbed by the crop is equivalent to crop demand, and the soil supply is broken down into N_{min} that is measured and soil mineralization. Mineralization is a function of soil type and organic matter content, which is divided into active and longer-term pools depending on management policies for crop residues and the application of manures. They have one additional term, 'soil mineral nitrogen not available to the crop', which is determined empirically. It, apparently, reflects the roots inability to capture all the available nitrogen and is dependent on soil type. They have confirmed that the critical nitrogen curve, developed by Greenwood, Lemaire and colleagues, also applies to their cultivars. A model predicts potential growth determined by temperature and radiation and takes account of variation in planting date as a result of changes in soil conditions from year to year and planned changes to harvest date. Combining the critical curve and estimated potential growth determines the crop nitrogen demand.

In Belgium, they also use the same balance method as in Azobil (Goffart and Olivier, 2004) but only apply 70% of the full recommendation at planting. Then between 25 and 55 days after crop emergence, a chlorophyll meter is used to determine the crop nitrogen status. If it falls below a defined critical level, then the remaining 30% of the full recommendation is applied, avoiding the issue of how much to apply for remedial action. In New Zealand, a DSS called Potato Calculator (Jamieson et al., 2004) is currently being used to reduce the amount of nitrogen applied to early potato crops while still maintaining yield. The drivers for this, which are common throughout the world, are three-fold: the cost of nitrogen fertiliser is rising, local and national governments are increasingly concerned about non-point-source pollution of ground and surface water and growers need to demonstrate to local and international markets that they are growing their crops in an environmentally sustainable manner. Interestingly, Potato Calculator supersedes previous DSSs that were based on crop monitoring of nitrogen status (e.g. petiole sap testing). The crop-monitoring approach was seen as retrospective, dealing with the problem once nitrogen deficiency was manifest. The new approach seeks to anticipate nitrogen deficiency by using a detailed simulation model of the effects of temperature, light, nitrogen and water supply on potato development, growth and partitioning of carbon and nitrogen using recent weather and scenarios of future weather to estimate whether and when water and nitrogen will become limiting. The reliance on crop simulation alone is an unusual step, but the group has had previous successes in developing the Sirius Wheat Calculator in close partnership with growers (Jamieson et al., 2003). The experience gained has been incorporated into the Potato Calculator.

In conclusion, the means of determining a crop's demand for nitrogen based on target yield is sound and robust. Measurement of the soil mineral nitrogen present at planting removes the large uncertainties in estimates based on previous cropping and soil type alone. Although many models of soil mineralization have been developed by scientists to achieve the required accuracy, they tend to demand too much data input for practical purposes. Measurement of crop nitrogen status combined with split application should improve the accuracy with which supply meets demand and will work best when combined with small and frequent supplements of nitrogen applied through fertigation. For supplements applied in granular form, further problems remain, namely how much is required and how quickly will it take effect will depend on future rainfall or irrigation. The best response to supplemental nitrogen occurs if the initial supply is adequate to maintain rapid leaf expansion and canopy closure and if supplements keep the crop growing at maximum thereafter (Gianquinto and Bona, 2000).

36.3.3 Irrigation

It is perhaps surprising that DSSs for irrigation scheduling did not come out top in the searches on the Internet and in CAB Abstracts. In part, this may be due to the insistence that the term 'potato' was present. Water is so important that DSSs for irrigation are probably more frequently developed for a range of crops including potato; 'potato' may not be mentioned explicitly in all texts that were searched.

Irrigation accounts for 60–80% of the world's fresh water withdrawals applied to 15% of the agricultural land but accounting for half of total production. In contrast, late blight is associated specifically with the potato crop. Water is the single most limiting factor in potato production, even in cooler climates as in Finland with the short growing season, and in some countries, the crop receives the largest share of the irrigation (Johnson and Colauzzi, 2004). Within Europe, there is considerable variation in the amount of water applied, the proportion of total area under irrigation and method of application, even within similar climatic regions. For example, whereas England has nearly 80% of its potato crop under some form of irrigation, Belgium has little (Martins, 2000). Weatherhead and Danert (2002) reported that the amount of irrigation applied in England to a potato crop almost doubled in the 1980s as did the area irrigated. The average amount applied per unit area then stabilized, whereas the irrigated area increased by a further 60%. The increases have come entirely from the greater amounts applied after 31 July in each year in maincrop production. The irrigation of grass, cereals and sugar beet has declined significantly. Overall, the results indicate that growers are prioritizing irrigation on higher value crops and using water more efficiently. In over half of the irrigated potato crops in the UK, irrigation is now scheduled 'scientifically', reflecting the growing attention to water conservation. Slightly more growers use in-field soil moisture measurements, for example neutron probes or tensiometers, with the remainder using water balance calculations by either hand or computer. These are essentially the two methods for deciding when to irrigate: water balance and direct measurement of soil moisture. In the water balance method, the soil is often assumed to be at field capacity at planting, certainly in temperate regions where this is a reasonable assumption. Evaporation losses are then calculated on a daily basis by taking the potential evaporation and modifying it by the amount of ground cover of the crop. Potential evaporation is usually calculated from daily weather data, for example from dry- and wet-bulb temperatures, maximum and minimum air temperatures, soil temperature, solar radiation and average wind speed using the now-standard method of Penman–Monteith or similar method – itself a DSS. It provides an estimate of what a well-watered crop such as potato would transpire if it fully covered the ground. In the early stages of growth, much of the soil is bare. Evaporation from the soil is generally less than that from a crop, especially as the soil surface dries out, and this is taken into account. Allowances can also be made when water has become limiting and the crop reduces its rate of water loss. After the appropriate adjustments have been made, one is left with an estimate of the actual daily evaporation. There are many models available which have been validated in their own climates and have proved to be robust. In essence, with a few constraints such as not exceeding field capacity and capillary rise from water table, by summing the actual daily evaporation estimates and subtracting any precipitation (rainfall), one is left with an estimate of the soil moisture deficit, usually expressed in equivalent depth of net amount of water lost. The alternative is to measure some aspect of soil moisture directly, usually a potential indicating how much suction pressure has to be applied by the roots to withdraw the water from the soil. In either case, there are some simple thresholds that are used to determine when to apply water. The optimum thresholds, maximum soil moisture deficit or minimum water potential, vary over time. For example, during tuber initiation, too great a soil water deficit can allow common scab to develop and can reduce the number of daughter tubers produced. Too wet a soil can make the crop prone to powdery scab. Once the tubers have set, the crop can tolerate greater deficits before growth is affected.

The decision of when and how much water to apply is a tactical one that is made regularly throughout the growing season. Some examples of tactical DSSs in irrigation are Plant-Plus (Raatjes et al., 2004), Pl@nteInfo (www.planteinfo.dk), MAPP (MacKerron et al., 2004) and Potato Calculator (Jamieson et al., 2004). It is not surprising that Dacom having initially developed Plant-Plus and the associated network of weather stations for controlling late blight then extended it into irrigation management of potato. Plant-Plus offers both methods, the soil water balance based on calculations of PE and estimated actual evaporation and/or direct monitoring through the use of soil moisture sensors. Both the approaches will require some additional sensors to be attached to their data capture system, developed originally for late blight, but the basic infrastructure is in place. The ability to set the levels for field capacity and refill point, combined with the graphical outputs, allows the user to define when to irrigate and the most appropriate amount of water to apply.

While one may develop a DSS for irrigation on its own, when developing a DSS for nitrogen application and especially when considering the use of split applications, it is sensible to incorporate support for both irrigation and nitrogen because the two are inextricably linked. Nitrogen will not be taken up unless it is dissolved in the soil solution surrounding the roots. In fertigation, the two resources are physically packaged together. In the case of granular application of fertiliser, it is essential that rain and/or irrigation follow to ensure the nitrogen becomes available to the plant. Haverkort et al. (2003) describe the building of a nitrogen/irrigation DSS based on LINTUL-Potato which is intended to be available over the Internet. Another example is the use of SIMPOTATO in the Pacific Northwest of the USA, with the specific aims of improving the economic efficiency of production and minimizing the risk to the environment (Alva, 2006).

There are also strategic decisions as to how much and what type of irrigation equipment to purchase and where to deploy it. Where there is little rainfall, permanently installed irrigation systems such as centre pivot or drip-lines are preferred. In wetter climates, there is trade-off between capital investment and the logistics of moving equipment between fields to meet each crop's requirements. Indeed, there is a question of whether to irrigate throughout the growing season or just through the period of tuber initiation or not at all, set against the increasing risk of tuber disorders and yield loss. This leads to a different set of DSS tools that cover the range of crops that may be irrigated and is more likely to be used by an advisor or company selling the irrigation equipment, in discussion with the grower. Zaliwski and Górski (2001) reported on their evaluation of the cost-effectiveness of irrigation of potato in Poland using models, weather data and geographic information systems (GIS) to estimate the risk and degree of drought and impact on potato yields.

Concerns of over-abstraction, reduced flows in the rivers, falling levels of ground water and increased pollution risk mean that the cost of water is rising in many countries and new legislation is being imposed to restrict the amount of water that can be abstracted. This introduces yet another level of DSS, water management at the catchment scale and its consequences for land use. Although these are not likely to involve the growers directly, the decisions often have a major impact, for example should one grow potatoes at all. DHI Water and Environment (www.dhigroup.com), an independent research and consultancy organization developed from former institutes in Denmark, is a highly focused group looking at the availability and management of water on a wide range of scales, both marine and fresh water around the world.

The increasing concerns about climate change and the increasing demand for water for domestic, industrial and agricultural purposes mean that governments are introducing tight controls on its use, and costs are rising. The models of water use are relatively simple, based on sound physical principles, validated and robust. DSSs have an important role to play in optimizing the strategic deployment of irrigation and its day-to-day tactical use. As for control of late blight, local weather is required on a daily basis, and in this case, direct observations of crop canopy development and/or soil moisture may be required for some methods.

36.3.4 Tuber size distribution

An estimated 8% of the UK crop failed to meet size specifications, the largest cause of wastage, in 2000 with an estimated cost to the industry £24 m per annum. The size of the tuber is usually defined as either the minimum mesh size through which it will pass or its fresh weight – there is a simple logarithmic relation between the two. The economic value of a tuber can change dramatically as its size moves from one grade to another. Potatoes intended for the salad or canning market can halve in value if the size of the tuber increases by just 5 mm. Even in the potato starch industry, there are financial penalties if the sample delivered to the factory has too many small potatoes (van Haren, personal communication). Size specifications frequently change both for existing markets and when new markets develop. A tool that can predict how changes in husbandry affect the size distribution would be of help to the grower.

The distribution of tuber sizes in a crop is influenced by a complex of weather and agronomic variables and intrinsic variation within the plant, so that potato tubers are produced in a wide range of sizes in most if not all crops. However, the basic concept of managing tuber size is simple and exploits the fact that the spread of tuber sizes is proportional to the average tuber size, that is the ratio of the characteristic spread to the average tuber size is constant when expressed on a fresh weight basis (Sands and Regel, 1983; Marshall et al., 1998). Nemecek et al. (1996) found the same relation between spread and average tuber size using a slightly different function to forecast the proportions of seed potato produced. There is some evidence for variation between cultivars in this ratio, but the variation is small for practical purposes (Marshall and Thompson, 1986). There is little the grower can do to achieve a tighter more uniform size distribution. It has been claimed that bed systems or using smaller seed, for example mini-tubers, could achieve greater uniformity by better control of stem number and greater uniformity of spacing. At one time, the national list trials in the UK had a character for scoring absolute uniformity and then it was abandoned. It turns out that wherever a cultivar or treatment was thought to produce a more uniform size distribution, the effect had come about by increasing the number of daughter tubers and/or reducing yield. They were uniformly small! However, if the planting rates of the cultivars or treatments are adjusted, so that they produce similar numbers of daughter tubers per unit area and similar yields, then there is little or no difference in the resultant size distributions. The problem reduces to managing the average tuber size, and the spread will follow.

Average tuber size is the fresh weight yield of the potato crop divided by the number of daughter tubers present (see Fig. 3 of Marshall et al., 1998). When making a decision on what planting density to use, the grower is aiming to match the probable number of daughter tubers produced to the anticipated yield to achieve the economic optimum size distribution for the intended market. The number of daughter tubers produced is dependent on the stem density and on the cultivar but also on external environmental conditions, many of which are beyond the control of the grower. The one size seed, managed in store in the one way, may nevertheless produce significantly different numbers of daughter tubers from one year to the next. The reason is not necessarily that the seed is intrinsically different from one year to the next. The growing conditions during tuber initiation can significantly influence the number of daughter tubers produced, even when water is non-limiting. Experiments manipulating the physiological age of the seed tubers have failed to resolve the extent to which growing conditions in the previous crop from which the seed was derived, storage conditions and early growing conditions in the subsequent crop in which the seed is used contribute to this variation. Likewise, attempts to assess the condition of the seed at planting by counting the number of eyes that are open, number of stems per eye, etc. have been equally unsuccessful. Therefore, the aim of the decision on planting density is to get the crop into the right ballpark. It is then essential to make direct observations of the number of daughter tubers actually produced in the crop – one cannot rely on simulation alone.

Fortunately, the final number of daughter tubers to be produced is determined quite early, shortly after canopy closure and the end of tuber initiation. Once the final number has been fixed, it should then be a simple matter to take a series of samples to determine the actual number. Knowing the actual number, one can determine the yield one needs to achieve the desired size distribution and hence decide any shift in harvest date that might be needed. These samples can be spread over time: early samples shortly after canopy closure, then successive samples to confirm that the tuber number has stabilized and, by combining these stable samples, one can increase the precision of the estimate. It is common practice to dig such samples, but often, these are done only at end of the season and with only a visual inspection. By combining models of tuber bulking over time with models of tuber size distribution and sequentially sampling the observed tuber numbers, one can determine the optimum time for lifting with increasing precision. The DSS Management Advisory Package for Potatoes (MAPP) offers such a system for determining the time to stop the crop growing as well as support for seed-rate decisions based on target yield and size specification (MacKerron et al., 2004). MAPP physically prevents the grower from accessing support on when to harvest until at least one sample dig has been entered.

These calculations can also be done using spreadsheets, but that relies heavily on past experience, previous combinations of daughter tuber numbers and yields and typical bulking rates. For a specialist grower group working with a limited number of cultivars and markets, it is possible to develop suitable tables. Some commercial growers already use this approach (Anon, personal communication). It works for them because they have learnt by experience what the optimum combination of average tuber size and total yield has to be to meet their specific need. If the need changes, for example the target for total yield is changed or market size specification changes, then, in the absence of an underlying model of the size distribution, it is not possible to predict accurately how much average tuber size and hence plant density may need to change. However, such changes are readily accommodated by a DSS such as MAPP.

The models of tuber size distribution have been shown to be robust and applicable across a range of cultivars in Europe and Australia. Apart from the need to calibrate new cultivars, there is no reason to suspect that DSSs based on this type of model cannot be used anywhere in the world for supporting decisions on planting density and harvest dates. They are particularly useful when the system of production is being altered, for example different size of seed, different cultivar, market specification or yield level. But again, when it comes to deciding when to lift as the season progresses, data collection is an issue. In addition to the usual weather data required for predicting water-limited yield and hence irrigation scheduling, samples of the tubers have to be taken and analysed. This takes time. A means of semi-automating this process using a simple digital camera has been proposed and piloted by Marshall and Young (2004), and significant interest has been expressed by the industry.

36.4 TAKE-UP IN GENERAL

The use of ICT by German farmers increased by nearly 17-fold in the 16 years from 1982 to 1998, four-fold in the 1980s and another four-fold in the 1990s (Kuhlmann and Brodersen, 2001). All sectors increased as the number one driving force for uptake of ICT was the legal requirements for financial bookkeeping. Farmers hardly ever used formal DSSs, not even to exploit their financial accounting data as a means of business analysis and planning (Ohlmer et al., 1998). Of the 86 000 software installations reported in the German survey in 1998, 80% were mainly for retrospective calculations split approximately equally between financial bookkeeping, crop record keeping and livestock record keeping. Only 20% were installed for anticipative calculations, of which one-half was for livestock production and only one in seven for crop production, with the remainder used for marketing. There is some hope for DSS usage for operational planning which increased by 250-fold over the 16-year period, although the evidence suggested that the rates of uptake for crop and livestock production had fallen in the second half, with marketing being in ascendancy.

Clearly, legal requirements for finance and audit trails are inescapable drivers for computer usage on the farm. So, what are the constraints to the uptake of DSSs for strategic and tactical decision-making on the farm?

36.5 THE WAY FORWARD

36.5.1 Barriers to uptake

36.5.1.1 Investment

Employing a DSS on the farm requires an investment in both the direct costs of purchasing the software and the indirect costs of the time required to use the system, which can be significant. Unless the return on investment is clearly quantifiable and large enough, uptake of the product will be small.

36.5.1.2 Training

The initial costs of training in the use of computers and then of the software are potential obstacles to the uptake of DSSs. The need to use computers for financial bookkeeping, the increasing need to have accurate record keeping for audit trails and the early exposure of the children to computers both at home and at school mean that this first barrier, the use of computers per se on the farm, has all but disappeared in many cultures. Training costs to use the software remain.

36.5.1.3 Access

The person using the computer may not be the person making the operational decisions. On smaller farms, the spouse often maintains the computer records, whereas in larger production units, a secretary is often employed. The computer is generally fixed in the office, whereas the decision maker may well be out of the office for most of the time. The ideal would be a simple-to-use system that can be accessed from anywhere. Perhaps, this is one of the reasons why precision farming has had some success. For example, as referred to earlier, variable rate of application of nitrogen fertiliser can be achieved by placing crop detectors and computer processing on board the tractor, and the rate of application is automated. In fact, it is then no longer decision support: it is decision-making.

36.5.1.4 Credibility and understanding

To plan a set of actions, one needs a set of options, the outcomes of which are predictable. To make a prediction, one needs a model, but what is an appropriate model? Passioura (1996) warns against the rush of the enthusiastic scientist to be seen to be actively delivering knowledge transfer by taking a model developed for scientific purposes and implementing it directly into a DSS to advise farmers or policy makers – 'These quite different aspirations require quite different models . . . The best engineering models are based on robust empirical relations between plant behaviour and the main environmental variables. Because of their empirical nature, we should not expect them to apply outside the range of the environmental variables used in their calibration. Within their calibrated ranges, however, some have proved useful in providing sound management advice'.

Even given an appropriate model, it is crucial that it does not become a black-box. The farmer needs to understand how the models work *in the language of the farmer* – how a change of inputs interacts and produces a different outcome. Even with relatively few variables for the grower to manipulate, the range of possible outcomes can rapidly expand. There is a Catch 22 – if the system is overly simplified, then it is seen not to respond to local conditions, but if it is overly complex, then its behaviour becomes impenetrable to all but the developer (Cox, 1996; Kuhlmann and Brodersen, 2001). Both result in a lack of credibility, trustworthiness and reliability. Cox concluded that DSS technology, intended to support routine decision-making, is over-engineered. The need was and still is for models with a degree of resolution, appropriately validated, that cope adequately with the emergent properties of systems in the context of substantial background noise. This, Cox states, argues for simpler models, not more complex ones, if the interpretation of model output is to be done by people other than those who built the models. Enough complexity emerges through interaction of the problem owners with the problematic situation. Our own experiences in the development of MAPP confirm this view. The development of MAPP involved end users from the very first stage and was tested by them at various stages during its development. Nevertheless, on its release, some users can have the first seeds of doubt sown when the predictions of the underlying models apparently differ from their experience. In their eyes, the model is wrong! 'I would not use that seed-rate' or 'That's not the tuber size distribution I get from my crop' might be typical first-time responses. A few minutes spent reviewing the problem with the grower usually reveals that an input entered by the user was not the same as on the farm. For example, when there is a perceived discrepancy on the effect of seed rate (planting density) on the predicted tuber size distribution, it is not uncommon to find that the anticipated yield level entered by the grower is not a true reflection of past yield levels. Although advised of its importance in text that accompanies the entry of anticipated yield, the first-time user can easily miss the link between an input and its consequence, resulting in a loss of confidence. Confidence is quickly restored when the cause is recognized, and a change of input confirms that prediction and experience are after all consistent. The learning process is an important part of using a DSS. It is not easy to anticipate all the interactions between a user and a DSS, especially if it supports several decision areas that influence each other. Human interaction is therefore an important part of the learning process for all but the simplest DSS.

36.5.2 The customer

'Who are the users?', 'What do they want?', 'Why do they want it?', 'How do they use it?' and 'How do they vary?' are all important questions to consider before developing a DSS. Glenz (1994) reported that thousands of copies of DSSs for the selection of cultivars, crop protection and fertilisation of winter wheat were sold, but cautious questioning revealed that the clients rarely used them for direct decision-making. They actually used them as learning machines to explore the decision space they had to operate in (Kuhlmann and Brodersen, 2001). This was no bad thing as this is one area that was identified as deficient. For example, when the grower is trying to increase the proportion of the total yield in a specified market, for example smaller seed or more bakers, or is switching to a new market with a different set of size specifications, the required change in planting density is easily underestimated. A DSS allows the grower to explore the consequences and understand the factors that influence size distribution, what is possible and what is not. Indeed, some DSSs are built primarily for this purpose. For example, the primary purpose for a DSS for managing PCN developed by Elliot et al. (2004) is to educate the grower in how changing the number of years and the cultivars between successive crops of potato in the one field has dramatic effects on the populations of PCN over several rotations. Having used the system, the grower is then much more aware of the roles of resistance and tolerance to PCN, nematicides and length of rotation on managing the pest. It was released in the UK as a CD through the British Potato Council.

Hayman (2004) has written an excellent reflection on the experiences with DSSs in Australian dryland farming, which should be read by all those involved in their development. He cites Dreyfus and Dreyfus (1989) who drew important distinctions between novices and experts; the former are learning, whereas the latter are applying in earnest and feel deeply responsible for their decisions. 'Developers of DSSs often point to undergraduate education as a significant market for their software. The success of DSSs with 19-year-old undergraduate students may give a clue to the lack of success with 50-year-old farmers; DSSs are more appropriate for novices'. It raises the question of whether the same tool should be used by an undergraduate and an experienced farmer. There is also considerable variation within the experienced farming community. We have been involved recently in developing an online cultivar selection DSS with AveBe for the starch potato growers in the Netherlands and Germany (Optiras, http://optiras.agrobiokon.eu/). A primary objective of the DSS is to raise awareness of the management of PCN by appropriate use of rotation interval and cultivars. The industry has identified differing grower types, essentially classifying them by performance criteria, low and high quality of product and low and high yield. They particularly wish to target the lower yield, lower quality groups. The problem is that these are the least comfortable with the new technologies. Logging user transactions online, surveys and direct observational studies are revealing differences in how the four groups use the system. It is not just best practice that needs to be passed on but the psychology of how different grower groups think, ask questions and share knowledge. It was concluded that an additional section should be added which explains the underlying model and its influences. The user then has the option to learn first through a structured set of pages and then use the DSS. Without these pages, the user could still experiment and learn, but for some, this could take too much time and interest is lost. This particular example has a narrow focus, and it is feasible to provide a 'training service' online alongside the DSS.

Kerr (2004) investigating factors affecting the adoption of a DSS called DairyPro in the Australian dairy industry concluded that developers need to have a good working knowledge of the target industry and to understand the types of decisions that are made and the diversity of user types. It is important to have decision makers involved in the development of a DSS which also encourages a feeling of greater ownership of the eventual endproduct. Interestingly, they also advocate the use of domain experts to provide estimates of expected production levels rather than the traditional approach of using the results from simulation models to make these estimates.

Should DSSs be expected to be used in isolation? The answer is likely to be, 'not in the first instance at least'. Despite best efforts, each DSS is unlikely to capture all the expertise that could be applied nor all the specific circumstances of each farm business that may be relevant. It is, therefore, likely that DSSs will be used in conjunction with a consultant, at least in the early stages. Hayman (2004) quotes from Kilpatrick (1999) about a farmer who was willing to pay someone to travel to his remote farm in the Northern Territory of Australia. The farmer explained the advantage of one-to-one learning, 'If you can talk to them face to face, you can question them and that sort of thing, and . . . they'll tend to give you more information and a clearer picture on things, rather than just reading about things . . . When you're speaking to someone in person, you're getting also their ideas, and different ways of . . . putting together a breeding program, etc. It is very useful'. Farm consultants were also seen as people who had seen a range of businesses operating a particular system and have the ability to apply those experiences to a particular client's situation. Another perception by clients was that people who used consultants were 'thinkers', and thus, consultants were coming into contact with a whole range of innovative ideas. A similar conclusion was reached about growers in the study of the use of the online potato cultivar selection referred to earlier – growers belonging to the highest performing group were willing to share their knowledge and ask questions in contrast to the least performing group who did not share and did not ask penetrating questions.

The grower, after an initial training period, is likely to use DSSs designed for tactical (day-to-day) decisions on their own, unless the unexpected happens. As mentioned earlier, differences between model prediction and reality do occur. Interpretation of the reasons for the discrepancies often requires an expert. The consultant therefore has a key role to play – at least in the initial, learning phase.

And a final cautionary note: Robinson and Freebairn (2001) re-tell the story of an old farmer, tinkering with a rusty harrow who is approached by a young advisor offering a new manual on soil conservation and new farming techniques. After politely listening to the polished speech, the old farmer replied 'Son, I don't farm half as good as I know how already'.

36.6 CONCLUSION

DSSs are in use and being successfully deployed. Nevertheless, the lessons from Australia in the late 1980s and 1990s are still very relevant if one is to avoid the same mistakes. The agro-ecology, the scientific knowledge, is only a small part of the development of a successful DSS. Understanding the users, knowing who they are, their priorities, how they approach a problem and how they vary are all important considerations before starting to develop the DSS. The educational role of a DSS is very important, possibly even more so than aiding a specific decision. Sometimes education may turn out to be its sole use, especially with strategic decisions where more information than just the DSS is likely to be accessed in reaching a final conclusion. DSSs that support tactical decisions that respond to local conditions that are changing will be used for both purposes. A common feature of many tactical DSSs is the need for local daily weather data, and many require some form of monitoring either in the field, such as crop cover, tuber samples, soil moisture and soil mineral nitrogen, or in the surrounding region as in the case of outbreaks of disease such as late blight and pests such as Colorado beetle and aphids. How this information is to be gathered in a timely and efficient way can be a stumbling block in the uptake of a DSS and needs careful consideration. Finally, the work is not complete once the DSS has been constructed and distributed. It needs to be supported and evaluated to identify users' issues, additional training needs and possible improvements. Support for the product is probably the most difficult aspect of such projects, requiring significant resources. A DSS forms part of management strategy and works best alongside the other sources of information and advice. The farm advisor or local agronomist is an integral part of the wider decision support. Electronic DSSs are there to enhance, not replace, the local advisor.

It is still early days in the development and use of DSSs. Each group seeking to transfer their knowledge to the industry often specializes in a single crop and/or scientific discipline and often has to find its own expertise to develop a DSS. Consequently, DSSs still tend to be developed in relative isolation from each other and to focus on a limited number of decisions. Typically, a DSS will be built for a single crop and perhaps only one or two decision areas for that crop. The risks are that the individual DSSs do not integrate with each other to provide a package that efficiently maps on to the farm business. With separate DSSs for each crop type, unless developed by the same company, there is nothing to ensure that data entered into one system can be easily accessed by another. A farm comprises fields that are probably subject to a rotation, so that soil data entered in one year, say for growing a wheat crop, should be available the next year when it is put into potato. If the weather data or soil descriptions entered into one DSS are not accessible by another, then it can quickly become a significant frustration. This is not a new problem. In the UK, considerable effort has been put from the mid 1990s onwards to build an overarching shell that would capture such generic farm data, along with local and historical weather, and make it available to all compliant DSS modules. However, a recent search on the Internet did not reveal what the final outcome was. An alternative approach is for the industry, both commercial interests and agents of the state, to agree to a set of standards for communication of data, so that products from different sources can communicate. This would make it easier for smaller, specialist groups to contribute their knowledge to the industry and would be more efficient and less frustrating for the end user.

The great majority of DSSs are still based on a relatively simple approach of an underlying deterministic model with a graphical user interface to the end user. A window pops up asking the user for a set of inputs, the model runs and a set of outputs is created. A compiled help file or pop-up tips may assist the first-time user in what the various inputs and outputs are, possibly giving typical examples or 'tutorials' as to how it works. And with packages that deal with more than one topic, an underlying rule-base can ensure that the user is only asked for information that the DSS does not already know or cannot compute and yet requires to answer the specific question. This technique is employed in MAPP, using a rule-based language called CLIPS.

The problem with only using deterministic models is that the DSS is restricted to only using knowledge that has already been translated into such models. There is much more information available that is less certain and or less precise. There are methods to handle this more qualitative and uncertain information, for example rule-based and Bayesian modelling (Marshall et al., 1995). As yet, there are few examples of such approaches being used in DSSs. Frito-Lay Inc. has been working with Mindbox Inc. using ARTEnterprise™ to develop a Potato Expert System (PES) to share the best practices of their largest growers amongst all their growers (Ownby et al., 2002). Their system contains modules to assist with new site selection and evaluation, crop management and disease/disorder diagnosis. A fourth module was planned to provide advice on potato storage. It employs rule-based reasoning for questions such as site selection and fuzzy logic to aid the diagnosis of 80-plus potato diseases.

Future weather is a classic example of uncertainty, which has a strong influence on tactical decision-making. 'What is the probable yield of the crop and therefore what rate of fertiliser application and planting density should be used?', 'Should I prepare to irrigate in three days time or is there a likelihood of significant rain?', 'Should a blight spray be applied or will conditions remain favourable to avoid significant risk of blight in the next five days?' and 'When should I prepare to stop the crop growing in order to optimise tuber size distribution without significant risk of the tubers becoming infected?' are all examples of questions the answers to which are influenced by the weather and, therefore, subject to varying degrees by uncertainty. The problem is not so much how to calculate the consequences of this uncertainty but rather how to present it to the grower. Calculation involves some means of generating the probable future weather scenarios, using either historical weather records or weather generators for longer-term forecasts or meteorological models for shorter-term forecast, usually no more than 5 days in temperate climates. Whichever method is employed, the result is a probability distribution of possible outcomes. How best to convey this information to the user requires further study.

Since the late 1990s the use and content of the Internet has expanded rapidly. With the faster connections is increasingly becoming the first choice for dissemination of information, for advisors at least. It has the advantage that information and products can be updated at a single point and are immediately available to all. However, farm businesses, being in rural locations, often still find the speed of connection to the Internet a significant barrier. Portals, locations on the Internet, that link the user to a whole range of information sources and DSSs are becoming increasingly popular among advisers who will generally have faster connections. WebInfo, managed by the DAAS, provides an impressive example of such a portal. Analysis of accesses revealed that the local farm advisors are the principal users and are the primary if not sole means of keeping up to date professionally (Hansen, 2004). During a typical month, there are more than 15 000 users. It holds approximately 70 000 documents including documents concerning legislation and rules and regulations. Users are able to access more than 50 different DSS tools covering crop protection, feed planning, irrigation scheduling and more. During the 12 months from August 2003 to July 2004, there were 1.3 million external sessions viewing 6.8 million pages, with close to 50 000 different pages being viewed in total.

In an article in *Crop Protection Monthly* (Anonymous, 2005) discussing the slow uptake of DSSs in Europe, David Martin of Plantsystems Limited (www.plantsystems.co.uk) was cited as saying, DSSs 'can often increase the complexity of decision making and sometimes do not fit comfortably into pressurised production situations particularly if they require detailed attention'. The article continued, 'Many growers rely for crop protection advice on agronomists who often see the introduction of a DSS as a threat. Many DSSs also require the use of real-time weather data and forecasting, all of which involve additional costs. The full implementation of a DSS also requires training and ongoing support from the provider, without which usage can decline rapidly'.

The particular DSSs referred in this chapter derive from both commercially and state-funded work and have been included for illustrative purposes only. The selection is not exhaustive nor has any value judgement been placed on their suitability for any particular task. This work has been supported in part by the Scottish Executive Environment and Rural Affairs Department, Mylnefield Research Services and the British Potato Council. I also thank Donald MacKerron and Philip Smith for suggested improvements to the manuscript.

REFERENCES

Alva A.K., 2006, Verification monitoring of nitrogen and irrigation BMPs for horticultural crops and crop simulation models as aids to decision support system, Proceedings of the 5th International Symposium on the Agricultural Environment, Ghent, Belgium.

Anonymous, 2000, *Fertiliser Recommendations for Agricultural and Horticultural Crops*. RB 209, 7th edition, MAFF Publications, HMSO, London.

Anonymous, 2005, Decision Support Systems, *Crop Protection Monthly* 182, 9, available at www.crop-protection-monthly.co.uk

Booij R., J.L. Valenzuela and C. Aguilera, 2000, In: A.J. Haverkort and D.K.L. MacKerron (eds), *Management of Nitrogen and Water in Potato Production*, p. 72. Wageningen Pers, The Netherlands.

Chambenoit C., F. Laurent, J.M. Machet and H. Boizard, 2004, In: D.K.L. MacKerron and A.J. Haverkort (eds), *Decision Support Systems in Potato Production*, p. 54. Wageningen Academic Publishers, Wageningen, The Netherlands.

Cox P.G., 1996, *Agric. Syst.* 52, 355.

Dowley L.J. and J.J. Burke, 2003, Field validation of four decision support systems for the control of late blight in potatoes, Report 4922, Crops Research Centre, Oak Park, Carlow, Irish Agriculture and Food Development Authority.

Dreyfus H. and S. Dreyfus, 1989, In: T. Forester (ed.), *Computers in the Human Context: Information Theory, Productivity and People*, p. 125. Basil Blackwell, Oxford.

Elliot M.J., D.L. Trudgill, J.W. McNicol and M.S. Phillips, 2004, In: D.K.L. MacKerron and A.J. Haverkort (eds), *Decision Support Systems in Potato Production*, p. 142. Wageningen Academic Publishers, The Netherlands.

Gianquinto G. and S. Bona, 2000, In: A.J. Haverkort and D.K.L. MacKerron (eds), *Management of Nitrogen and Water in Potato Production*, p. 35. Wageningen Pers, The Netherlands.

Glenz H., 1994, Entscheidungsmodelle für den Anbau und die Bestandesführung von Winterweizen (Decision Models for Winter Wheat Management). Dissertation, Giessen, Germany.

Goffart J.P. and M. Olivier, 2004, In: D.K.L. MacKerron and A.J. Haverkort (eds), *Decision Support Systems in Potato Production*, p. 68. Wageningen Academic Publishers, The Netherlands.

Greenwood D.J., 1982, *Philos. Trans. R. Soc. Lond. B Biol. Sci.* 296, 351.

Greenwood D.J. and A. Draycott, 1995, In: P. Kabat, B. Marshall, B.J. van den Broek, J. Vos and H. van Keulen (eds), *Modelling and Parametrization of the Soil–Plant Atmosphere System. A Comparison of Potato Growth Models*, p. 155. Wageningen Pers, The Netherlands.

Greenwood D.J., J.J. Neeteson and A. Draycott, 1985, *Plant Soil* 85, 163.

Hansen J.P., 2004, Conference Proceedings, New information techniques in science, education and advisory for rural areas and agriculture, p. 51. Brwinów, Poland.

Hansen J.G., B. Kleinhenz, E. Jörg, J.G.N. Wander, H.G. Spits, L.J. Dowley, D. Michelante, L. Dubois and T. Steenblock, 2002, Proceedings of the 6th Workshop of a European Network for Development of an Integrated Control Strategy of Potato Late Blight, p. 231. Edinburgh, Scotland.

Haverkort A.J., J. Vos and R. Booij, 2003, Proceedings of XXVI International Horticultural Congress: Potatoes, Healthy Food for Humanity, ISHS, *Acta Horticulturae* 619, 213.

Hayman P.T., 2004, Decision support systems in Australian dryland farming: a promising past, a disappointing present and uncertain future. 4th International Crop Science Congress, Gosford NSW, Australia, available online at www.cropscience.org.au

Hofman G. and J. Salomez, 2000, In: A.J. Haverkort and D.K.L. MacKerron (eds), *Management of Nitrogen and Water in Potato Production*, p. 219. Wageningen Pers, The Netherlands.

Jamieson P.D., T. Armour and R.F. Zyskowski, 2003, Solutions for a better environment, Proceedings of the 11th Australian Agronomy Conference, Geelong, Victoria, Australian Society of Agronomy.

Jamieson P.D., P.J. Stone, R.F. Zyskowski, S. Sinton and R.J. Martin, 2004, In: D.K.L. MacKerron and A.J. Haverkort (eds), *Decision Support Systems in Potato Production*, p. 84. Wageningen Academic Publishers, The Netherlands.

Johnson P.A. and M. Colauzzi, 2004, In: A.J. Haverkort and D.K.L. MacKerron (eds), *Management of Nitrogen and Water in Potato Production*, p. 263. Wageningen Pers, Wageningen, The Netherlands.

Johnson P.A., R. Postma, J.P. Goffart and J. Salomez, 2000, In: A.J. Haverkort and D.K.L. MacKerron (eds), *Management of Nitrogen and Water in Potato Production*, p. 155. Wageningen Pers, Wageningen, The Netherlands.

Kerr D., 2004, *Artif. Int. Rev.* 22, 127.

Kilpatrick S., 1999, Learning on the job: How do farm business managers get the skills and knowledge to manage their businesses? CRLRA Discussion Paper D3/1999, University of Tasmania, Centre for Research and Learning in Regional Australia.

Kuhlmann F. and C. Brodersen, 2001, *Comput. Electronics Agric.* 30, 71.

Lemaire G., 1997, *Diagnosis of the Nitrogen Status of Crops*. Springer-Verlag, Berlin, Germany.

Leonard R., L.J. Dowley, B. Rice and S. Ward, 2001, *Potato Res.* 44, 327.

MacKerron D.K.L., 2000, In: A.J. Haverkort and D.K.L. MacKerron (eds), *Management of Nitrogen and Water in Potato Production*, p. 103. Wageningen Pers, The Netherlands.

MacKerron D.K.L., B. Marshall and J.W. McNicol, 2004, In: D.K.L. MacKerron and A.J. Haverkort (eds), *Decision Support Systems in Potato Production*, p. 118. Wageningen Academic Publishers, The Netherlands.

Marshall B., J.W. Crawford and J.W. McNicol, 1995, In: A.J. Haverkort and D.K.L. MacKerron (eds), *Potato Ecology and Modelling of Crops under Conditions Limiting Growth*, p. 323. Kluwer Academic Press, Dortrecht.

Marshall B., D.K.L. MacKerron, M.W. Young, J.W. McNicol and J. Wiltshire, 1998, *Aspects Appl. Biol.* 52, 11.

Marshall B. and J.R. Porter, 1991, In: J.R. Porter and D.W. Lawlor (eds), *Plant Growth: Interactions with Nutrition and Environment*, p. 99. Cambridge University Press, U.K.

Marshall B. and R. Thompson, 1986, *Potato Res.* 29, 261.

Marshall B. and M.W. Young, 2004, In: D.K.L. MacKerron and A.J. Haverkort (eds), *Decision Support Systems in Potato Production*, p. 100. Wageningen Academic Publishers, The Netherlands.

Martins F., 2000, In: A.J. Haverkort and D.K.L. MacKerron (eds), *Management of Nitrogen and Water in Potato Production*, p. 233. Wageningen Pers, The Netherlands.

Mizubuti G.S.G. and G.A. Forbes, 2002, Potato late blight IPM in the developing countries, Proceedings of Late Blight: Managing the Global Threat, GILB Conference, Hamburg, Germany, p. 93.

Neeteson J.J., 1995, In: P.E. Bacon (ed.), *Nitrogen Fertilization in the Environment*, p. 295. Marcel Dekker Inc., New York, U.S.A.

Nemecek T., J.O. Derron, O. Roth and A. Fischlin, 1996, *Agric. Syst.* 52, 419.

Ohlmer B., K. Olson and B. Brehmer, 1998, *Agric. Econ.* 18, 273.

Ortiz O., 2002, Teaching farmers IPM in developing countries, Proceedings of Late Blight: Managing the Global Threat, GILB Conference, Hamburg, Germany, p. 98.

Ownby J., D. Binney and P. Mazzoni, 2002, Artificial intelligence at work: using intelligent technology to automate agricultural knowledge and best practices, available at www. Mindbox.com

Passioura J.B., 1996, *Agron. J.* 88, 690.

Raatjes P., J. Hadders, D. Martin and H. Hinds, 2004, In: D.K.L. MacKerron and A.J. Haverkort (eds), *Decision Support Systems in Potato Production: Bringing Models to Practice*, p. 168. Wageningen Academic Publishers, The Netherlands.

Robinson J.B. and D.M. Freebairn, 2001, The decision maker has an important, but often neglected, role in model use, Proceedings of the International Conference on Modelling and Simulation, MODSIM 2001, Canberra, Australia.

Sands P.J. and P.A. Regel, 1983, *Field Crops Res.* 6, 25.

Schepers H.T.A.M., 2002, Potato late blight IPM in the industrialized countries, Proceedings of Late Blight: Managing the Global Threat, GILB Conference, Hamburg, Germany, p. 89.

Sharpley A.N., J.L. Weld, D.B. Beegle, P.J.A. Kleinman, W.J. Gburek, P.A. Moore and G. Mullins, 2003, *J. Soil. Water Conserv.* 58, 137.

Shepherd J.M., 2000, In: A.J. Haverkort and D.K.L. MacKerron (eds), *Management of Nitrogen and Water in Potato Production*, p. 165. Wageningen Pers, The Netherlands.

Weatherhead E.K. and K. Danert, 2002, Survey of irrigation of outdoor crops in 2001 – England, Report from Cranfield University, U.K.

Woodruff D.R., 1992. *Aust. J. Agric. Res.* 43, 1483.

Zaliwski A.S. and T. Górski, 2001, Determination of cost-effectiveness of irrigation of the potato in Poland with the aid of GIS, Proceedings of EFITA 2001, Montpellier, France, p. 181, available at www.efita.net

Index

Printed and bound by CPI Group (UK) Ltd, Croydon, CR0 4YY
08/05/2025
01864931-0001